GEOLOGY AT M.I.T. 1865-1965

GEOLOGY AT M.I.T. 1865–1965

A History of the First Hundred Years
of Geology at
Massachusetts Institute of Technology

I. The Faculty and Supporting Staff

ROBERT RAKES SHROCK
Professor of Geology, Emeritus

The MIT Press
Massachusetts Institute of Technology
Cambridge, Mass., and London, England

Copyright © 1977 by
The Massachusetts Institute of Technology

All rights reserved. No part of this book may be reproduced or utilized in any form or by any means, without permission in writing from the Publisher, except by a reviewer who wishes to quote a brief excerpt in connection with a review.

Composition by Charles V. Mahlmann

Printed and bound in the United States of America
by The Murray Printing Company

Library of Congress Cataloging in Publication Data

Shrock, Robert Rakes, 1904-
   Geology at M.I.T., 1865-1965.

   Contents: v. 1. The faculty and supporting staff.
   1. Geology--Study and teaching (Higher)--Massachusetts. 2. Geologists--Massachusetts. 3. Massachusetts Institute of Technology. I. Title.
QE47.M4S53   551'07'117444   77-71235
ISBN 0-262-19161-X

Dedicated

to
all the persons
Faculty and Supporting Staff alike
who made Geology what it was
at M.I.T.
during the Institute's first century
1865-1965

iv   GEOLOGY AT M.I.T. 1865-1965

A CENTURY OF GEOLOGY AT M.I.T. 1865-1965
(Sketch by Percy Lund)

PREFACE

The purpose of this work is to describe the role that geologists and geology played at the Massachusetts Institute of Technology (M.I.T.) during the Institute's first century, 1865-1965. My plan has been to prepare two closely related but independent volumes; one on the geology faculty and their supporting staff; the other on the departmental operations carried out by them as members of the Department of Geology (Course XII). The first, now completed as this volume, consists primarily of biographical sketches of the first fifty-three professors of geology. Supplementary discussions review the founding of the Institute, the development of the geology faculty and curriculum, and the nature and extent of the assistance rendered by the supporting staff. The second volume, now in preparation, will be devoted to the activities and accomplishments of the professors and their supporting personnel. The following are typical of the subjects to be discussed: 1) graduates and their theses; 2) books and other major publications by faculty members and alumni; 3) new ideas, methodologies, and instruments developed by Course XII personnel that have advanced the study of the earth; and 4) the overall impact of M.I.T.'s activities in geology on the broad fields of science and technology. An integral part of the second volume will be discussion of how M.I.T.'s Department of Geology actually operated in respect to such matters as curriculum, finances, space, and facilities.

The reportorial style used in this volume, as well as in the preparation of the second volume, was adopted so that particular sections of the text can be complete in themselves. Even though this format leads to some repetition, it seems to me justified in that it relieves the reader of having to refer to preceding or following pages for related material.

Published sources of information are used extensively as a basis for statements in the text. All such quoted material is fully referenced so that the reader can know who should be credited for the information, and where and by whom it was published. Copyrighted materials are reproduced with the kind permission of the respective publishers in cases where permission is required.

Most of the illustrations carry a credit line; exceptions are the fifty pen-and-ink sketches drawn by the late Henry B. Kane and based on photographs provided by the M.I.T. Graphic Arts Service; the seven photographs of professors kindly provided by Warren A. Seamans and his cooperative assistants from the M.I.T. Historical Collections; and a few miscellaneous uncredited photographs from my own files.

No <u>Index</u> of this volume was considered necessary because it would have been little more than a list of the names that appear repeatedly throughout the text. Instead, the detailed tables of Contents and list of Illustrations were prepared as guides for the interested reader who might wish to locate certain information.

In planning and preparing this volume I have had the aid and encouragement of many persons; far too many to list here individually. In general their assistance is acknowledged in a specific way where appropriate. There are a number of persons, however, who deserve recognition here, because of the nature and extent of their aid. My departmental colleagues without exception made my task much easier by providing information requested of them and in so doing helped to make their respective biographies more complete. I have found a constant source of stimulation and encouragement in a group of other colleagues who, like myself, have been engaged in recording some aspect of M.I.T.'s history. This group includes the late John E. Burchard, Dean of Humanities and Social Sciences, Emeritus; the late Carle R. Hayward, Professor of Metallurgy, Emeritus; C. Richard Soderberg, Dean of Engineering, Emeritus; Karl L. Wildes, Professor of Electrical Engineering, Emeritus; and Roy Lamson, Professor of Literature, Emeritus. And I have profited greatly and in many ways from repeated conversations with President Emeritus Julius A. Stratton and Miss Loretta Mannix, his indefatigable researcher, who have shared my deep interest in the people, activities, and events of the Institute's past history.

Three persons merit my special gratitude for contributions that have greatly enhanced the quality of this volume. The late Sylvia Bateman, as my Technical Assistant for five years, 1960-1965, searched widely and diligently for source materials and found many important items that I surely would never have discovered. The late Henry B. ["Chick"] Kane (IX-B S.B. 1924), a longtime friend and colleague, prepared the sketches of professors that add a special quality to the biographies; his untimely death prevented him from making sketches of most of the more recently appointed professors. With his typical generosity and modesty he insisted on doing them gratuitously and without signature. The handiwork of the third person, Charles V. Mahlmann (II S.M. 1953), is evident on every page. He prepared the final typescript from my handwritten copy, scaled the illustrations, arranged the page layouts, and proofread the final "camera-ready" copy. His enthusiastic interest in preparing the copy was a constant stimulus for me to keep on the job!

Finally, I hope that all who have served Geology at M.I.T. will be considered as included in the dedicatory statement on a preceding page.

Robert R. Shrock

Lexington, Massachusetts
1 April 1977

CONTENTS

Chapter	Page

Honors and Memberships 336
Bibliography of Thomas Augustus Jaggar, Jr. 336
  [Regular Bibliography] 336
  Departmental Reports for M.I.T.'s President's Reports,
    1905-1912 342

(8) CHARLES HYDE WARREN (1876-1950)   [Geochemist] 343

  MIT: 1900-1922 [Précis] 343
  [Biography] 343
  Bibliography of Charles Hyde Warren 348
  Sources of Biographical Information about Charles Hyde Warren . 349

(9) DOUGLAS WILSON JOHNSON (1878-1944)   [Geomorphologist] 351

  MIT: 1903-1907 [Précis] 351
  [Biography] 351
  Partial Bibliography of Douglas Wilson Johnson 353
  Biographies of Douglas Wilson Johnson 354

(10) HERVEY WOODBURN SHIMER (1872-1965)   [Paleontologist] 357

  MIT: 1903-1942-1965 [Précis] 357
  [Biography] 358
    Birth, Ancestry, and Education 358
    M.I.T. Years (1903-1942-1965) 359
      As Teacher 360
      As Author 362
    Professional and Civic Activities 364
    Personal Characteristics and Activities 364
    Retirement Years (1942-1965) 365
    Marriage, Family, and Family Life 366
    The Shimer Memorial Fund 367
    Summation 367
  Bibliography of Hervey Woodburn Shimer 368

(11) REGINALD ALDWORTH DALY (1871-1957) [Physical Geologist] 373

  MIT: 1907-1912-1915 [Précis] 373
  [Biography] 373
    Birth, Ancestry, and Early Education 374
    The Early Years at Harvard (1892-1901) 375
    Work along the 49th Parallel (1901-1907) and Marriage
      (1903) 375
    The M.I.T. Years (1907-1915) 376
    The Harvard Years (1892-1901; 1912-1942-1957) 379
    Summation 380
  Partial Bibliography of Reginald Aldworth Daly 380

(12) WALDEMAR LINDGREN (1860-1939)   [Economic Geologist] 385

  MIT: 1912-1939 [Précis] 385
  [Biography] 387
    Birth, Ancestry, and Early Education 387
    The First Great Step in a Distinguished Career (1878) 390
    Student Years at Freiberg's Bergakademie (1878-1883) . 393
    Lindgren emigrates from Sweden to America (1883) 398
    His First Jobs (1883-1884) 401
    The U.S. Geological Survey Years (1884-1912-1915) 402
    The M.I.T. Years (1908-1912-1933-1939) 411
      As Lecturer in Economic Geology (1908-1912) 411
      As William Barton Rogers Professor of Economic Geology
        and Head of Course XII, Department of Geology, (1912-
        1920; 1927-1933) 411

xii   GEOLOGY AT M.I.T. 1865-1965

Chapter	Page
Lindgren the Teacher	412
Lindgren as Invited Lecturer outside M.I.T.	413
Lindgren the Administrator	415
As Department Head	415
As Chairman of the Division of Geology and Geography of the National Research Council (1927-1928)	417
Founder and First Editor of the Annotated Bibliography of Economic Geology	417
National Research Council Committee on Processes of Ore Deposition	420
Lindgren as Geological Consultant	423
The Last Decade at M.I.T. (1928-1939)	424
Publications - Contributions	425
Lindgren's Part in Founding Economic Geology	426
Honors	427
Professional Societies - Memberships, Offices, and Committee Services	427
Dedicated Publications	429
Medals	430
Penrose Medal of the Society of Economic Geologists (1928)	430
Penrose Medal of the Geological Society of America (1933)	430
Gustave Trasenster Medal of the Université de Liége Belgium (1936)	432
Wollaston Medal of the Geological Society of London (1937)	432
Honorary Degrees	432
Special Personal Honors	433
William Barton Rogers Professor of Economic Geology (1912-1933)	433
Lindgren Library at M.I.T. (1932)	433
Lindgren Memorial Fund at M.I.T. (1959)	434
Waldemar Lindgren Citation Awards for Excellence in Research (1960)	434
Lindgrenite, a new Mineral, $Cu_2Mo_4 \cdot Cu(OH)_2$ (1935)	435
Summation	435
Lindgren the Scientist	436
Lindgren the Educator, Author, and Editor	436
Lindgren the Man	438
Bibliography of Waldemar Lindgren	439
Introduction	439
Regular Publications	440
Abstracts	450
Discussions of the work of other Geologists	450
Reviews	451
Departmental Reports included in the annual M.I.T. President's Reports for SYs 1911-12 to 1932-33, inclusive	452
Appendix A: Martinsson's "Waldemar Lindgren goes West"	453
Appendix B: Biographies and Biographical References to Waldemar Lindgren (1860-1939)	460
Appendix C: Lindgren's Editorial - "The Education of the Geologist" from Econ. Geol. 18/4: 405-409, (1923)	464
Appendix D: Annual Reports on Lindgren's work with the U.S. Geological Survey, 1884-1893-1912	466
(13) FREDERIC HENRY LAHEE (1884-1968) [Petroleum Geologist]	469
MIT: 1912-1918 [Précis]	469
[Biography]	471

Chapter	Page
Birth, Ancestry, and Youth	471
The Harvard Years (1906-1912)	472
The M.I.T. Years (1912-1918)	473
World War I and Employment in the Petroleum Industry	475
Services to and Awards from the American Association of Petroleum Geologists	479
On Statistics and the Oil Depletion Allowance	479
Services on the American Petroleum Institute Committee on Oil Reserves	481
Dancing and Skating	482
Summation	483
Bibliography of Frederic Henry Lahee	484

(14) LOUIS CARYL GRATON (1880-1970) [Economic Geologist] 491

- MIT: 1914-1918 [Précis] 491
- [Biography] 492
- Partial Bibliography of Louis Caryl Graton 497

(15) WILLIAM FRANCIS JONES (1885-1941) [Petroleum Geologist] 499

- MIT: 1918-1928 [Précis] 499
- [Biography] 499
- Bibliography of William Francis Jones 503

(16) JOSEPH LINCOLN GILLSON (1895-1964) [Economic Geologist] 507

- MIT: 1922-1930; 1961-1963 [Précis] 507
- [Biography] 508
- Bibliography of Joseph Lincoln Gillson 525

(17) WALTER HARRY NEWHOUSE (1897-1969) [Economic Geologist] 529

- MIT: 1923-1946 [Précis] 529
- [Biography] 530
- Bibliography of Walter Harry Newhouse 541

(18) MARTIN JULIAN BUERGER [Mineralogist; Crystallographer] 545

- MIT: 1920-1929; 1929-1973 [Précis] 545
- [Biography] 547
  - Birth, Ancestry, and Early Education 547
  - The Student Years at M.I.T. (1920-1929) 551
  - The Teaching Years at M.I.T. (1925-1973) 552
    - Student Teaching (1925-1929) 553
    - He becomes a Faculty Member (1929) 553
    - Early Field Work that almost produced an Economic Geologist 554
    - He finds a Research Interest in X-ray Crystallography 556
    - He founds a Laboratory and quickly learns 557
    - As a Teacher he gladly taught 560
    - As a Consultant on Mineralogy and Crystallography 562
    - As Chairman of the M.I.T. Faculty, Institute Professor, and A. P. Sloan Awardee 562
    - As Director of M.I.T.'s School for Advanced Study 563
  - As Visiting Professor or Guest Lecturer away from M.I.T. 564
  - Publications 566
    - Introduction 566
    - Books 567
      - [Twelve books so far] 567
      - Why and how he wrote books 568
  - Research Results: New Ideas and Theories; New Methods and Processes; New Instruments 571
    - Theoretical Contributions to Crystallography 571

Chapter	Page

        Theory useful in Diffraction Instrumentation 572
        Theory useful in Analysis of Crystal Structures 572
        Determining the Arrangement of Atoms in Crystals 573
        Theory related to Crystal Properties 574
        Patented and Unpatented Instruments and Processes
          [25 in all so far] 575
        How he came to invent Instruments 581
    Publishing and Editorial Activities 582
    As a Co-Founder and Member of Professional Organizations .. 583
    Honors 588
        Medals 588
        Honorary Degree 589
        Memberships and Activities in American Academies 589
        Memberships in Foreign Scientific Societies 590
        Buerger Bay in Frobisher Bay, Baffin Island, Canadian
          Arctic 591
        Buergerite, a ferric variety of Tourmaline 591
        Martin J. Buerger Festschrift 1968 593
        Martin J. Buerger Festband 1973 593
    Marriage and a Family of Six Daughters 594
    Summation 599
        As a Teacher 599
        As an innovative Investigator 600
        As a highly productive Author 600
        As an Inventor of Laboratory Instruments and Methods .. 601
        Religious Activities 602
    Bibliography of Martin Julian Buerger 607

(19) FREDERICK KUHNE MORRIS (1885-1962)   [General Geologist] 621

    MIT: 1927-1950 [Précis] 621
    [Biography] 621
        Birth, Ancestry, and Youth 621
        College Years and Early Teaching 623
        World War I and Civilian Service 624
        Teaching and Exploration in China 625
        Marriage to Florence Elisabeth Eddowes (1922) 627
        Work in Mongolia - The Third Asiatic Expedition of the
          American Museum of Natural History 628
        The M.I.T. Years (1927-1950) 629
          Teacher, Investigator, Consultant 635
          Public Lectures 640
          Early Travels and a Trip around the World 642
          World War II 644
          Organizational Memberships and outside Activities ... 644
        The Closing Years (1950-1962) 645
        Frederick K. Morris the Man 646
        The Frederick K. and Florence E. Morris Memorial Fund . 648
    Bibliography of Frederick Kuhne Morris 648

(20) WALTER LUCIUS WHITEHEAD (1891-1969)   [Economic Geologist] 653

    MIT: 1928-1962 [Précis] 653
    [Biography] 653
    Bibliography of Walter Lucius Whitehead 662

(21) LOUIS BYRNE SLICHTER   [Geophysicist] 665

    MIT: 1931-1945 [Précis] 665
    [Biography] 667
        Birth, Ancestry, and Youth 667
        The M.I.T. Years (1931-1941-1945) 668
        World War II (1940-1945) and the California Years (1946-) 672
        Retirement Years and Summation 673
    Bibliography of Louis Byrne Slichter 674

Chapter	Page
(22) WARREN JUDSON MEAD (1883-1960) [Engineering and Structural Geologist]	679
MIT: 1934-1949-1954 [Précis]	679
[Biography]	679
Birth, Ancestry, and Early Education	680
Student Years at the University of Wisconsin (1902-1908)	681
Teaching, Research, and Consulting Work during Wisconsin Years (1906-1934)	683
Teaching, Administration, Consulting, and Research during M.I.T. Years (1934-1954-1960)	688
Summation	691
Teacher of Geology and Administrator	691
Geological Consultant and Engineering Geologist	692
Research Scientist and Inventor	693
Marriage and Family	693
Honors and Memberships	693
The Mead Memorial Funds	694
Bibliography of Warren Judson Mead	694
[Regular Bibliography]	694
Departmental Reports included in the M.I.T. President's Reports for SYs 1934-35 to 1947-48	696
(23) HAROLD WILLIAMS FAIRBAIRN [Petrologist]	697
MIT: 1937-1972- [Précis]	697
[Biography]	699
Birth, Ancestry, and Early Education	699
College Years (1925-1932)	700
A Post-Doctoral Interim (1932-1934)	701
As Instructor at Queen's (1934-1937)	702
M.I.T. Years (1937-)	703
As Teacher	703
As Research Investigator	705
Other Departmental Activities	707
Field Work and Travel	708
Marriage and Family	709
Summation	710
Bibliography of Harold Williams Fairbairn	711
(24) ROBERT RAKES SHROCK [Paleontologist; Stratigrapher]	723
MIT: 1937-1970-1975 [Précis]	723
[Biography]	725
Birth, Ancestry, and Early Education	725
Early Training for Teaching	727
Wisconsin Years (1928-1937)	728
M.I.T. Years (1937-1970-1975)	729
Introduction	729
As Department Head (1949-1965)	732
As Teacher (1937-1970)	743
As Professional Geologist	743
As Author and Editor	744
As an Officer in Professional Societies	746
Government Service	747
Marriage, Family, and Social Activities	749
Memberships, Fellowships, Honors, and Awards	749
Retirement and Summation	751
Bibliography of Robert Rakes Shrock	753
Additional Miscellaneous Publications of Robert Rakes Shrock	760
1. Reports as an Officer of a Scientific Society (1938-1970)	760
2. REGISTERs of Course XII Alumni (1955-1970)	761

Chapter	Page
3. Departmental Reports in the annual M.I.T. President's Reports (1949-1964)	761
4. International Conference on the Earth Sciences on the occasion of the Dedication of the Cecil H. and Ida F. Green Building, Center for Earth Sciences, on 30 September - 2 October 1964	762

(25) ROLAND DANE PARKS (1900-1972) [Mineral Economist] 763

    MIT: 1940-1946 [Précis] 763
    [Biography] 764
    Bibliography of Roland Dane Parks 771

(26) JOHN NATHANIEL ADKINS [Geophysicist] 773

    MIT: 1946-1950 [Précis] 773
    [Biography] 773
    Bibliography of John Nathaniel Adkins 774

(27) PATRICK MASON HURLEY [Geochronologist] 775

    MIT: 1937-1942-1946- [Précis] 775
    [Biography] 776
        Birth, Ancestry, and Early Education 776
        Undergraduate Work in British Columbia (1927-1934) 777
        A Period of Practical Experience (1931-1937) 778
        Graduate Work at M.I.T. (1937-1940) 779
        The Beginning of an Academic Career (1940-1942) 779
        War Work and a Year at the University of Wisconsin (1942-1946) 780
        As a Member of the M.I.T. Faculty (1946-) 781
            As Professor of Geology (1946-) 781
            As a Research Scientist and Director of Research .. 783
            As Geological and Educational Consultant 787
            As a Public Lecturer 787
            As a Participant in National and International Conferences and Symposia on Geochronology and Plate Tectonics 788
        Marriage and Family 788
        Summation 788
    Bibliography of Patrick Mason Hurley 793

(28) WILLIAM HENRY DENNEN [Mineralogist; Physical Geologist] 805

    MIT: 1946-1967 [Précis] 805
    [Biography] 805
    Bibliography of William Henry Dennen 811

(29) LOUIS HERMAN AHRENS [Experimental Spectrochemist] 815

    MIT: 1948-1953; 1962-1963 [Précis] 815
    [Biography] 815
    Bibliography of Louis Herman Ahrens (Regular Bibliography) 819
        Publications as a Co-Editor 826

(30) THEODORE RICHARD MADDEN [Geophysicist] 827

    MIT: 1950- [Précis] 827
    [Biography] 827
    Bibliography of Theodore Richard Madden 832

(31) ELY MENCHER [Petroleum Geologist] 835

    MIT: 1952-1967 [Précis] 835
    [Biography] 835
    Bibliography of Ely Mencher 838

CONTENTS

Chapter	Page

(32) STEPHEN MILTON SIMPSON, JR.   [Geophysicist]   841

- MIT: 1952-1964 [Précis] 841
- [Biography] 842
- Bibliography of Stephen Milton Simpson, Jr. 849

(33) WILLIAM FRANCIS BRACE   [Structural Geologist]   851

- MIT: 1954- [Précis] 851
- [Biography] 851
  - Birth and Early Education 851
  - Student Years at M.I.T. (1943-1953) 852
  - As a Faculty Member at M.I.T. 853
  - Marriage and Family 854
  - Summation 854
- Bibliography of William Francis Brace 855

(34) GORDON JAMES FRASER MACDONALD   [Geophysicist]   859

- MIT: 1954-1958 [Précis] 859
- [Biography] 859
  - Birth and Early Education 859
  - The Harvard Years (1946-1954) 860
  - The M.I.T. Years (1954-1958) 861
  - The Years after M.I.T. (1958-) 862
  - Summation 863
- Bibliography of Gordon James Fraser MacDonald 863

(35) WILLIAM HAMET PINSON   [Geochemist - Meteoritics]   866

- MIT: 1956- [Précis] 866
- [Biography] 867
  - Birth, Ancestry, and Early Youth 867
  - Early Education 870
  - World War II Service 871
  - College Education (1946-1949) - (Emory University) 871
  - Graduate Student Years at M.I.T. and Harvard (1949-1952) .. 872
  - Teaching and Research at M.I.T. (1953-) 873
  - Marriages and Family 874
  - Summation 874
- Bibliography of William Hamet Pinson 877

(36) RICHARD RAYMAN DOELL   [Geophysicist - Paleomagnetism]   885

- MIT: 1956-1959 [Précis] 885
- [Biography] 885
- Bibliography of Richard Rayman Doell 888

(37) JOHN WIDMER WINCHESTER   [Nuclear Geochemist]   889

- MIT: 1956-1966 [Précis] 889
- [Biography] 889
- Bibliography of John Widmer Winchester 893

(38) WILLIAM STELLING VON ARX   [Physical Oceanographer]   899

- MIT: 1956-1970 [Précis] 899
- [Biography] 900
  - Birth, Ancestry, and Early Education 900
  - The College Years - Brown, Yale, and M.I.T. (1938-1945) ... 902
  - As Physical Oceanographer at W.H.O.I. (1945-1963; 1968-) . 903
  - As Professor of Oceanography at M.I.T. (1956-1970) 904
  - Marriage and Family 909
  - Summation 910
- Bibliography of William Stelling Von Arx 911

xviii   GEOLOGY AT M.I.T.   1865-1965

Chapter   Page

(39) ARTHUR JAMES BOUCOT   [Invertebrate Paleontologist]   915
    MIT: 1957-1961 [Précis]   915
    [Biography]   915
        Birth and Early Education   915
        Military Service in World War II   916
        The Influence of Honess, Swartz, Gordon, Howell, and Cloud   916
        Harvard, the U.S. Geological Survey, and M.I.T.   917
        Caltech, University of Pennsylvania, and Oregon State ...   919
        Summation   920
    Bibliography of Arthur James Boucot   921

(40) HARRY HUGHES   [Geophysicist]   925
    MIT: 1957-1964 [Précis]   925
    [Biography]   925
    Bibliography of Harry Hughes   927

(41) THOMAS CANTWELL, JR.   [Economic Geophysicist]   928
    MIT: 1960-1965 [Précis]   928
    [Biography]   928
    Bibliography of Thomas Cantwell, Jr.   931

(42) FRANCIS BITTER   [Experimental Physicist]   933
    MIT: 1934-1960-1967 [Précis]   933
    [Biography]   933
    Bibliography of Francis Bitter   938

(43) DAYTON ERNEST CARRITT   [Chemical Oceanographer]   939
    MIT: 1960-1968 [Précis]   939
    [Biography]   939
    Bibliography of Dayton Ernest Carritt   944

(44) JOHN HOWER, JR.   [Clay Mineralogist]   946
    MIT: 1960-1961 [Précis]   946
    [Biography]   946
    Bibliography of John Hower, Jr.   948

(45) RAYMOND HIDE   [Geophysicist - Fluid Mechanics]   949
    MIT: 1960; 1961-1967 [Précis]   949
    [Biography]   949
    Bibliography of Raymond Hide   952
        [Regular Bibliography]   952
        Miscellaneous writings, chiefly Research Reports, of
            limited circulation, not submitted for publication ..   955

(46) GIORGIO FIOCCO   [Geophysicist]   957
    MIT: 1963-1968 [Précis]   957
    [Biography]   957
    Bibliography of Giorgio Fiocco   960

(47) LEE WALLACE DEAN III   [Geophysicist]   964
    MIT: 1960-1967 [Précis]   964
    [Biography]   964
    Bibliography of Lee Wallace Dean III   966

(48) ANTHONY FRANK GANGI   [Geophysicist]   967
    MIT: 1964-1967 [Précis]   967
    [Biography]   967
    Bibliography of Anthony Frank Gangi   969

Chapter	Page
(49) WILLIAM CLAIR LUTH [Experimental Petrologist]	970
MIT: 1964-1968 [Précis]	970
[Biography]	970
Bibliography of William Clair Luth	977
(50) DAVID WILLIAM STRANGWAY [Geophysicist - Paleomagnetism]	979
MIT: 1965-1968 [Précis]	979
[Biography]	979
Bibliography of David William Strangway	982
(51) M. NAFI TOKSÖZ [Geophysicist - Seismology]	984
MIT: 1965- [Précis]	984
[Biography]	984
Bibliography of M. Nafi Toksöz	986
(52) GENE SIMMONS [Geologist - Geophysicist]	989
MIT: 1965- [Précis]	989
[Biography]	989
Bibliography of Gene Simmons	992

POSTSCRIPT - THE SECOND CENTURY BEGINS: 1965

This brief chapter summarizes the activities of the Department of Geology and Geophysics (Course XII) during the last fifteen years, 1950-1965, of M.I.T.'s first century, and provides an introduction to the accomplishments of the Department of Earth and Planetary Sciences (Course XII) during the first ten years, 1965-1975, of the Institute's second century, under the leadership of Prof. Frank Press, whose biography is added to the preceding fifty-two on page 997 994

----

(53) FRANK PRESS [Geophysicist]	997
MIT: 1965- [Précis]	997
[Biography]	998
Birth, Ancestry, and Early Education (1924-1944)	998
The Columbia Years (1944-1955)	999
The Caltech Years (1955-1965)	1002
The M.I.T. Years (1965-)	1003
Research and Publications (1946-1975)	1007
Public Service	1011
Public Lectures	1012
Membership in Professional Societies	1012
Editorial Services	1012
Services on State, National, and International Organizations of different kinds for a variety of purposes	1013
Consultant to Government Bureaus	1014
Honors and Awards	1014
The Frank Press Family	1015
Summation	1017
Bibliography of Frank Press	1020
Special Note	1032

ILLUSTRATIONS

	Page
A Century of Geology at M.I.T., 1865-1965 (Sketch)	iv
Chart showing by bars the rank and periods of service of Geology Faculty Members at M.I.T., 1865-1975	20
Chart showing order of appointment of Course XII Professors and number of M.I.T. Alumni serving on the Faculty during 1865-1972	22
Chairmen of the [Geology] Department [and earlier Heads of the Program in Geology before 1890] (Photos)	24
Mining and Metallurgy Faculty of Course III, 1922 (Photo)	30
Geology Faculty of Course III, 1922 (Photo)	30
Faculty of Course XII, Geology, 1952 (Photo)	32
Faculty of Course XII, Geology and Geophysics, 1957 (Photo)	32
Faculty of Course XII, Geology and Geophysics, 1960 (Photo)	36
Faculty of Course XII, Earth and Planetary Sciences, 1970 (Photo)	36
J. A. Cushman at Sharon, Massachusetts (Photo)	56
Profs. D. J. MacNeil and W. L. Whitehead at Crystal Cliffs Field Camp near Antigonish, Nova Scotia (Photo)	57
H. E. Hawkes, Jr. (Kane sketch)	58
N. A. Haskell (Kane sketch)	62
David Greenewalt (Photo)	63
J. B. Hersey (Kane sketch)	64
J. B. Hersey and colleagues, with rock specimens dredged from the Puerto Rico Trench (Photo)	65
C. O'D. Iselin (Kane sketch)	65
A Woods Hole Scene, by Joan T. Kanwisher (Sketch)	68
J. W. Kanwisher (Photo)	69
H. A. Morss, Jr. (Kane sketch)	69
Chart showing annual expenditures in Course XII as related to size of Faculty and number of Supporting Staff, of Graduate Students, and of Graduate Degrees awarded, 1950-1974	72
Chart showing order of appointment of Course XII Professors and number of M.I.T. Alumni serving on the Faculty during 1865-1972	88
Professors of Geology at M.I.T., 1865-1965 (Tabular List)	89
William Barton Rogers (Kane sketch)	99
View of the entrance to M.I.T. at 77 Massachusetts Avenue, Cambridge (Photo)	102
The Rogers Brothers' "Classification and Nomenclature of the Palaeozoic Formations" of Virginia and Pennsylvania (Table)	116
Principal structural features of the Valley and Ridge province in the middle and southern Appalachians (Map)	120
Idealized section across the Appalachians, showing the increase in intensity of folding toward the southeast (Cross section)	121
Plaque on the front of Kennedy's store at 16 Summer Street, Boston [where first M.I.T. classes met in early 1865] (Photo)	150
The former Mercantile Building on Summer Street, Boston (Lithograph)	150

ILLUSTRATIONS

Page

The second Rogers Building, facing westward, at 77 Massachusetts Avenue, Cambridge (Photo)	172
The first Rogers Building, facing southward, on Boylston Street, Boston (Photo)	172
Alpheus Hyatt (Kane sketch)	215
The Rogers Building and the Museum of the Boston Society of Natural History, looking northwest from the intersection of Berkeley and Boylston Streets, Boston (Photo)	228
William Harmon Niles (Kane sketch)	229
Thomas Sterry Hunt (Kane sketch)	237
An early classroom in "Boston Tech" in which Profs. Hunt, Hyatt, and Niles taught geology classes (Photo)	269
Chemical and Geological Essays, a book by T. Sterry Hunt (Photos)	270
William Otis Crosby (Kane sketch)	271
An example of the careful plans that Crosby prepared for his field excursions (Pamphlet)	280
George Hunt Barton (Kane sketch)	301
Prof. Barton and his students on a geology field trip to western Massachusetts (Photo)	312
Prof. Barton and colleagues waiting for the train after a field trip (Photo)	312
An Outline typical of the kind Prof. Barton prepared for field trips with students in the Teachers' School of Science (Pamphlet)	314
Barton typing his "Reminiscences" circa 1930 (Photo)	324
Thomas Augustus Jaggar, Jr. (Kane sketch)	325
My Experiments with Volcanoes, a book by T. A. Jaggar	326
Charles Hyde Warren (Kane sketch)	343
Douglas Wilson Johnson (Kane sketch)	351
View of a part of the classroom in "Boston Tech" used for instruction in geology and mining subjects (Photo)	355
View of another part of the classroom in "Boston Tech" used for instruction in geology and mining subjects (Photo)	356
Hervey Woodburn Shimer (Kane sketch)	357
Reginald Aldworth Daly (Kane sketch)	373
Harvard and M.I.T. Seals	384
Waldemar Lindgren (Kane sketch)	385
Lindgren's birthplace in Vassmolösa, Sweden (Photo)	388
Lindgren's sketch of King Bede's Burial Mound in Balestrand, Sogenfjord, Norway (Sketch)	392
Lindgren's sketch of Nystuen in Jotunheimen, Norway (Sketch)	392
Bergakademie buildings in Freiberg, Saxony (Photos)	394
Sketches and Notes from Lindgren's Bergakademie notebook in Lagerstättenlehre	397
Sketches and Notes from Lindgren's Bergakademie notebook in Maschinenlehre	397

	Page
Lindgren's Certificate of Appointment as Chief Geologist of the U.S. Geological Survey, 1911 (Photo)	409
Waldemar Lindgren: schoolboy in Kalmar, 1877; Bergakademie student, 1883; geologist with the U.S. Geological Survey, 1911; in the field, 1933 (Photos)	410
Two views of Lindgren with a graduate student (Photos)	414
A page of Lindgren's Bolivian field notes	424
The days of "Horse-and-Buckboard" field work at the turn of the century (Photo)	468
A jaunty Lindgren in the Hawaiian Islands, in 1903? (Photo)	468
Frederic Henry Lahee (Kane sketch)	469
Frederic H. Lahee and family, *circa* 1925 (Photo)	478
Lahee, at 83, skating with a "Pro" (Photo)	483
Louis Caryl Graton (Kane sketch)	491
William Francis Jones (Kane sketch)	499
Building 4 in the Cambridge M.I.T. where the Geology Department was assigned space in 1916 (Photo)	505
Joe Gillson in the field (Photo)	506
Joseph Lincoln Gillson (Kane sketch)	507
Gillson receives the Jackling Lecture Award (Photo)	518
Bruno Figallo's sketch of Gillson (Sketch)	518
Gillson, President of the American Institute of Mining and Metallurgical Engineers (Photo)	518
Gillson presents the Hardinge Award to R. B. Ladoo (Photo)	518
Walter Harry Newhouse (Kane sketch)	529
Newhouse in the field in Wyoming (Photo)	544
Martin Julian Buerger (Kane sketch)	545
Buerger with his favorite model (Mark II) of his precession camera (Photo)	558
Precession cameras (Photos)	576
Weissenberg cameras (Photos)	578
Cylindrical X-ray Powder-Diffraction cameras (Photos)	580
Chart showing Buerger Bay in the Canadian Arctic (Map)	592
Lila Buerger's Christmas cards (Sketches)	596
Prof. Buerger adjusts an instrument in his X-ray Crystallography Laboratory in Building 24 (Photo)	619
Crystal-model-making apparatus and crystal models constructed with it (Photos)	620
Frederick Kuhne Morris (Kane sketch)	621
Sketches by F. K. Morris (Blackboard sketch and Christmas cards) (Photo and sketches)	630
A Whimsical Map of M.I.T. (Sketch by F. K. Morris)	634
Geraldine Sullivan and Irving Breger in Prof. Whitehead's geochemical laboratory in Building 24 (Photos)	652
Walter Lucius Whitehead (Kane sketch)	653

ILLUSTRATIONS

xxiii

	Page
Prof. Whitehead and class in Oilfield Reservoirs (Photo)	661
Louis Byrne Slichter (Kane sketch)	665
Warren Judson Mead (Kane sketch)	679
Harold Williams Fairbairn (Kane sketch)	697
Prof. Fairbairn with students W. F. Brace and H. Eugster (Photo)	704
Cold cathode x-ray units built by departmental instrument maker in the Geology Shop (Photo)	710
Prof. Fairbairn's class in Petrology (Photo)	721
Prof. Fairbairn adjusting equipment in his X-ray Laboratory in Building 24 (Photo)	722
Robert Rakes Shrock (Kane sketch)	723
Prof. Shrock and Departmental Secretary, Pauline Richmond, at work in Building 24 (Photo)	731
Cecil H. Green, Robert C. Dunlap, Jr., and Robert R. Shrock at the 16th annual Earth Sciences Orientation Conference of the Geophysical Service Inc. Student Cooperative Plan, in Dallas, Texas, 1966 (Photo)	733
Percy Lund's sketch of the Green Building (Sketch)	734
Chart showing order of appointment of Course XII Professors and number of M.I.T. Alumni serving on the Faculty during 1865-1972	738
Chart showing annual expenditures in Course XII as related to size of Faculty and number of Supporting Staff, of Graduate Students, and of Graduate Degrees awarded, 1950-1974	739
Charts showing growth of Course XII Endowment Funds (Principal and Income), 1915-1975	740
Prof. Shrock with students in the Paleontological Laboratory in Building 24 (Photo)	742
Research Vessel R. R. SHROCK at Lewis' Wharf, Boston (Photo)	750
Roland Dane Parks (Kane sketch)	763
Prof. Parks and students (Photos)	768
John Nathaniel Adkins (Kane sketch)	773
Patrick Mason Hurley (Kane sketch)	775
Prof. Hurley and his Geochronology Laboratory (Photos)	782
First solid-source mass spectrometer built in Hurley's Geochronology Laboratory (Photos)	784
Prof. Hurley with significant rocks from Africa and Brazil (Photo)	786
Prof. Hurley adjusting equipment used for determining the ages of minerals and rocks (Photo)	792
William Henry Dennen (Kane sketch)	805
Prof. Dennen lecturing to a mineralogy class (Photo)	813
Prof. Dennen speaking at an Alumni Association Meeting (Photo)	814
Prof. Ahrens in the Godfrey L. Cabot Spectrographic Laboratory (Photo)	814
Louis Herman Ahrens (Kane sketch)	815
Theodore Richard Madden (Kane sketch)	827
Prof. Madden at the blackboard (Photo)	829

	Page
Madden with colleague Dr. T. Cantwell, Jr. (Photo)	830
Prof. Mencher prepares for a lecture (Photo)	834
Ely Mencher (Kane sketch)	835
Dr. Stephen M. Simpson, Jr. discussing a function of two variables (Photo)	840
Stephen Milton Simpson, Jr. (Kane sketch)	841
Ms. Kate Hadley (XII Ph.D. 1975) conducting an experiment in Prof. Brace's laboratory (Photo)	850
William Francis Brace (Kane sketch)	851
Gordon James Fraser MacDonald (Kane sketch)	859
William Hamet Pinson (Kane sketch)	866
Prof. Pinson lecturing to his astronomy class in front of the Green Building (Photo)	876
Richard Rayman Doell (Kane sketch)	885
John Widmer Winchester (Kane sketch)	889
William Stelling Von Arx (Kane sketch)	899
Von Arx adjusting equipment aboard ship at sea (Photo)	909
Arthur James Boucot (Kane sketch)	915
Harry Hughes (Kane sketch)	925
Thomas Cantwell, Jr. (Kane sketch)	928
Francis Bitter (Kane sketch)	933
Dayton Ernest Carritt (Kane sketch)	939
Eel Pond Bridge, Woods Hole, a sketch by Joan Kanwisher	945
John Hower, Jr. (Photo)	946
Raymond Hide (Kane sketch)	949
Giorgio Fiocco (Kane sketch)	957
Lee Wallace Dean III (Photo)	964
Anthony Frank Gangi (Photo)	967
William Clair Luth (Photo)	970
David William Strangway (Photo)	979
M. Nafi Toksöz (Photo)	984
Gene Simmons (Photo)	989
The Green Building in 1965 (Photo)	996
Frank Press (Photo)	997

I. INTRODUCTION

The purpose of this work is to record for future historians the kind of information they will need to understand and evaluate the activities and accomplishments of geologists and their colleagues at M.I.T. during the first century of the Institute's existence, 1865-1965.

How, by whom, and under what circumstances was M.I.T. founded, and who were the professors and supporting personnel that participated in its inception and subsequent development, with specific reference to Geology? After Course XII was established in 1890, how did the Department of Geology develop during the next 75 years, 1890-1965, and what of the accomplishments of its graduates and of its faculty members? These and related questions will be discussed in due course.

In the following pages I first review very briefly the origin of M.I.T., and how it happened that Geology and Mining were an essential part of the educational plan of the Institute from its very beginning. Then follows a brief discussion of how Geology fared during the Institute's first decade, 1865-1875, when a faculty was being gathered and an instructional program established. I next point out the lack of a definitive history of Geology at M.I.T., and of a similar history of the Institute itself, then cite the most important sources of historical information that do exist. Using these sources, I next sketch the activities and accomplishments of Geology at the Institute during the latter's first century of existence, 1865-1965, which divides rather naturally into four 25-year periods: 1865-1890; 1890-1915; 1915-1940; and 1940-1965. Following the foregoing review, which will be greatly expanded in a planned second volume, I describe the M.I.T. of 1965, as I viewed it as both professor and department head. In my analysis, I consider the faculty, together with its supporting staff, as the primary component, the raison d'être, of the Institute; I take this position because I hold that a "college" is by definition a society of scholars - a faculty - and that students, though of great importance, cannot by themselves alone constitute a college, university or institute. Holding such a view, however, I quickly accord our students their appropriate importance, for in today's "colleges" they are partly the reason for the institutions' existence. For the aforementioned reasons, the remainder of this volume is devoted to the development of the Course XII faculty and supporting personnel, and to individual biographies of each of the fifty-two professors who served as Course XII faculty members during M.I.T.'s first century, 1865-1965.

In a second volume I plan to discuss in detail how Course XII, established in 1890 and first designated the Department of Geology, developed; how its students and faculty members contributed to the advancement of science and technology; how they contributed to national and international organizations; how they brought recognition to the Institute; and how the Department in its totality made all this possible because of strong support that came both from within the Institute and from outside sources.

ORIGIN OF M.I.T.

M.I.T. was founded by a practical scientist, William Barton Rogers, who served as its first President and as its first Professor of Physics and Geology. Before coming to Boston, Rogers had been Professor of Natural Philosophy and Chemistry in the venerable College of William and Mary, from 1828 to 1835, and then Professor of Natural Philosophy in the University of Virginia, from 1835 to 1853. He had also taken the leading role in getting the Geological Survey of Virginia established in 1835, and had served as its Director and as State Geologist from 1835 to 1841. In the aforementioned activities, Rogers had lectured on all the sciences - physics, chemistry, geology, and biology - and had also taught mathematics. He had during the same period conducted field and laboratory investigations into problems in each of the sciences, and had repeatedly demonstrated the practical value of the geological and chemical work he and his assistants did on the Survey. In almost every aspect of his scientific work Rogers kept foremost in mind the possible use of his discoveries and ideas. It was to be expected, therefore, that when he and his associates drew up the curriculum for the proposed Massachusetts Institute of Technology in Boston, that curriculum would include an extensive foundation in mathematics and the physical and natural sciences, with emphasis on the practical application of all the subjects. A little farther on I will discuss the details of the aforementioned curriculum, so that the importance of geology and mining will be apparent, but I think it appropriate at this point to diverge and consider how Rogers first got and then developed the idea that ultimately became the basic plan of M.I.T.

In the following discussion it will quickly become evident that William's brother, Henry Darwin Rogers, was probably as responsible as he was in conceiving the idea of an independent polytechnic institute - a "School of the Arts" in which students would be trained to use the knowledge of the sciences - physics, chemistry, geology, and biology - in solving the problems that arose in the rapidly developing technology of the time. Evident, also, is the fact that Henry shared with William the strong conviction that Boston was the one place in the world where their kind of school was most likely to succeed. Finally, had Henry remained in Boston, instead of removing to Scotland, where he subsequently attained international distinction as a structural geologist, he almost certainly would

I. INTRODUCTION

have had a major role, with William, in organizing and founding M.I.T. As it happened, however, that role was not to be played; William had to carry on alone.

The concept of a polytechnic school in Boston - a school of practical science in which young men could be trained for professional careers in the industrial arts - originated in 1846 in the minds of William Barton Rogers, then Professor of Natural Philosophy at the University of Virginia, and his brother, Henry Darwin Rogers, who was residing and lecturing in Boston after having resigned as Professor of Geology at the University of Pennsylvania. As related farther on, in my biographical sketch of William, the two brothers first developed an interest in practical education in their college years, no doubt in part as the result of watching and helping their father make the equipment he needed for his classroom demonstrations. Patrick Kerr Rogers, the father, and Professor of Natural Philosophy and Chemistry in the College of William and Mary, taught his four sons, who later became the famous scientist "Brothers Rogers," not only the rudiments of science, in preparation for college, but also the use of tools for working wood and metal. Soon after leaving college, in 1826, William and Henry established a school of their own at Windsor, Maryland, and both, as well as James, gave science lectures at the Maryland Institute. The governors of the institution were so impressed by William's lectures that in early 1828 they asked William and Henry to draw up a plan and regulations for a proposed High School to be established in the Institute. In the light of later developments, it is significant to note that William included in his plan proficiency in mathematics and knowledge of the principles of astronomy, mechanics, natural philosophy, and chemistry.*
As implied in a preceding sentence, all four Rogers brothers took up scientific and educational careers, and among them they covered the whole field of pure and applied science. Inasmuch as they frequently exchanged views by correspondence and at family gatherings, and no doubt discussed their educational problems and ideas on such occasions, it may be presumptive to conclude that only William and Henry inclined to the practical use of scientific knowledge. Be that as it may, as early as 1837 William and Henry had jointly prepared "A Memorial to the Legislature of Pennsylvania" on behalf of the Franklin Institute for a School of Arts. Furthermore, during the next five years they conferred many times in connection with their work as directors of the geological surveys of Virginia and Pennsyl-

---

* Details of William's earlier educational activities are included in Volume 1 of his wife's Life and Letters of William Barton Rogers, cited in the list of references at the end of this chapter. Hereafter, citations of this invaluable work will be contracted to Life and Letters. William's plan mentioned above appears on page 49 of Volume 1.

vania, respectively, and repeatedly experienced the need for practically trained field and laboratory assistants. As a matter of fact, they persuaded their other brothers, James and Robert, to act as their assistants, from time to time, and to train others for field work.

In the spring of 1842, the two brothers attracted worldwide attention by the paper they presented at the Boston meeting of the Association of American Geologists and Naturalists entitled "On the Structure of the Appalachian chain, as exemplifying the laws which regulated the elevation of great mountain chains generally." By this time they were already well acquainted with educational, industrial and political leaders in both America and Europe, and were also aware of the growing need of business, industry, and science for technically trained personnel. As Prescott* summarized:

> "The career of the Rogers brothers coincided with the rapid development of science, technology, and industry in the United States in the first half of the nineteenth century. In their early manhood while teaching in Baltimore they had shared the popular excitement over the projected canal and railway. They had participated in the geological surveys that supplied the basic information for mines and railways throughout the East, and they had witnessed the beginnings of large-scale industry in New England. With their interest in science and in education, it was natural that they should sense the need for a new kind of education to serve as the basis for the great industrial and technological development that they foresaw. It was also natural that they should choose Boston as the most promising location for an institute of technology."

At the time of William's first visit to Boston in 1842, the city had a population of some 100,000, mostly of British ancestry. Together Boston and Cambridge constituted the leading literary center in North America, dominated by Harvard College. Bancroft, Emerson, Hawthorne, Longfellow, Prescott and Whittier were in their prime, and Holmes, Lowell and Thoreau were soon to become equally famous.

At the same time, New England was a thriving industrial and mercantile center with its great mill cities of Lawrence, Lowell and Manchester, and with Boston, its metropolis, as a center of industrial development, and a source of capital, business entrepreneurs, and skilled workmen. While manufacturing flourished on land, ships from the New England coast sailed the seven seas and brought wealth and foreign goods to Boston and its coastal neighbors.

Little wonder, then, that the two Rogers brothers concluded in the early 1840s that Boston was the most promising site for a school of in-

---

* "When M.I.T. was Boston Tech," p. 21; see complete reference in the bibliography at the end of this chapter.

I. INTRODUCTION

dustrial science of the kind they had outlined for the Franklin Institute in "A Memorial to the Legislature of Pennsylvania," presented about 1837.* In 1846, Henry resigned his professorship at the University of Pennsylvania, took up residence in Boston, and became a candidate for the Rumford Professorship at Harvard. When he expressed to John Amory Lowell his views regarding the value of a School of Arts as a branch to the Lowell Institute, he was asked by Lowell to set them down in writing. Lowell had obviously been impressed by what Henry had written to the Harvard corporation regarding the importance of teaching science in its applied form in the Boston community. Typically, Henry called on William for help in preparing his statement for Lowell, and typically William produced his now well-known plan for a Polytechnic School in Boston.** In writing to Henry about his request, William reveals his high opinion of Boston and Bostonians as follows (Life and Letters I, p. 259):

> "Ever since I have known something of the knowledge-seeking spirit, and the intellectual capabilities of the community in and around Boston, I have felt persuaded that of all places in the world it was the one most certain to derive the highest benefits from a Polytechnic Institution. The occupations and interests of the great mass of the people are immediately connected with the applications of physical science, and their quick intelligence has already impressed them with just ideas of the value of scientific teaching in their daily pursuits."

Henry failed to receive the Rumford Professorship at Harvard, and Lowell decided not to add a School of Arts to the Lowell Institute, so for the time being the two Rogers brothers turned their attention to other matters. William became so disconsolate with the situation at the University of Virginia that he tendered his resignation in 1848. This action so aroused his colleagues and other friends that he decided to withdraw it for the time being. The next year he married Emma Savage, the eldest daughter of a prominent Boston family, whom he had met in New Hampshire in 1845, and there was now another strong incentive to move to Boston. Things came to a head in 1853; he resigned his professorship for the second and last time, and he and Emma moved to Boston, where they spent the remainder of their lives.

Now William was free of all the disturbing incidents and situations in Charlottesville and could proceed with the planning of the polytechnic school he and Henry had been so much interested in seven years earlier.

---

* See Life and Letters I, p. 258 and 263.

** This appears as Appendix C with the title "A Plan for a Polytechnic School in Boston. 1846" in Volume I of Life and Letters, p. 420-427. See also p. 259 in the same volume.

Only now he would have to go it alone because Henry, probably disheartened by his failure at Harvard and with Lowell, as mentioned earlier on, became disenchanted with Boston, and in the summer of 1855 left for Scotland, to spend the remainder of his life there, after 1857 as Regius Professor of Natural History and Geology at the University of Glasgow. Thereafter, Henry seems not to have played any part in the founding or organizing of the polytechnic school that became M.I.T.

Now residing in Boston, and with Emma reunited with her family, William quickly made friends among the influential leaders of the community. He convinced them of the need for a polytechnic institution, took a leading role in petitioning the Massachusetts Legislature for a block of land in the recently filled Back Bay area, and chaired the committee that successfully petitioned the Legislature for a charter for the proposed Massachusetts Institute of Technology. The requested land was granted and the "Act to Incorporate the Massachusetts Institute of Technology" was approved by Governor John A. Andrew on 10 April 1861. However, the fall of Fort Sumter four days later and the outbreak of the Civil War delayed the opening of the Institute until the war ended. Then at long last, William Barton Rogers, geologist, and founder and organizer of the newly established M.I.T., could write in his diary for 20 February 1865 that oft-quoted entry:

> "Organized the School! Fifteen students entered. May not this prove a memorable day!"

From the foregoing discussion it should be clear that while the idea of a polytechnic institute for Boston was conceived jointly by both Rogers brothers, William Barton and Henry Darwin, the actual organizing and founding of M.I.T. was the work of William alone. While it seems reasonable to assume that the other two Rogers brothers, James Blythe and Robert Empie, as well as their teacher-father, Patrick Kerr Rogers, may also have contributed earlier to the idea of a school of applied science, whether they did, and to what extent, must be left uncertain, because of the lack of evidence for or against the assumption.

M.I.T.'s FIRST CURRICULUM, AND GEOLOGY FACULTY AND GEOLOGY, DURING THE FIRST DECADE, 1865-1875

Rogers' years of teaching at William and Mary (1828-1835) and at the University of Virginia (1835-1853), and of organizing and directing the Geological Survey of Virginia (1835-1841), had convinced him that if the foundling polytechnic institute that he and Henry envisioned for Boston were to be successful it would need a diversified curriculum, with subjects of instruction in the sciences and humanities that were attuned to

I. INTRODUCTION

the needs of industry and the mechanical arts, and a faculty competent to do the instructing. Geology and Mining were to be strongly represented in both curriculum and faculty. Rogers saw to it that they were.

The <u>First Annual Catalogue</u> of the "School of the Massachusetts Institute of Technology," dated 1865, contains a statement of the School's objectives, a list of the officers, a list of the students, a list of the officers of instruction, and a detailed discussion of the curriculum, etc. under the heading of "Regular Course."

The objects of the School were stated to be:

> "<u>First</u>, to provide a full course of scientific studies and practical exercises for students seeking to qualify themselves for the professions of the Mechanical Engineer, Civil Engineer, Practical Chemist, Engineer of Mines, and Builder and Architect.
>
> "<u>Second</u>, to furnish such a general education, founded upon the Mathematical, Physical, and Natural Sciences, English and other Modern Languages, and Mental and Political Science, as shall form a fitting preparation for any of the departments of active life; and, -
>
> "<u>Third</u>, to provide courses of Evening Instruction in the main branches of knowledge above referred to, for persons of either sex who are prevented, by occupation or other causes, from devoting themselves to scientific study during the day, but who desire to avail themselves of systematic evening lessons or lectures."

The requirements for a degree or diploma were as follows:

> "The studies and exercises of the first and second years, and the course of general studies in the third and fourth years, are required of all regular students. At the beginning of the third year, each regular student may select one of the following six courses, with a view of obtaining the corresponding degree or diploma: -
> 1. A Course in Mechanical Engineering.
> 2. A Course in Civil and Topographical Engineering.
> 3. A Course in Practical Chemistry.
> 4. A Course in Geology and Mining.
> 5. A Course in Building and Architecture.
> 6. A Course in General Science and Literature."

French was required in the first year, and both French and German in the second year. The mathematics of the first year included algebra, plane and spherical trigonometry, and solid geometry; of the second, analytic geometry and calculus. Professional subjects, including those in geology and mining, were taken in the third and fourth years, to meet the degree requirements of one of the six aforementioned courses. Tuition was $100 for first-year students, $125 for second-year students, and $150 for upperclassmen.

Thirteen diplomas were awarded to the first graduating class in June 1868; six of these were to mining students in Course IV. The Bachelor of Science degree was first specified on the 1871 diplomas; earlier diplomas merely certified that the recipients were graduates of M.I.T.

By 1873, eight years after its founding, M.I.T. had established ten regular courses and had renumbered (with Roman numerals) and renamed the original six as follows:

```
*     I.  Civil and Topographical Engineering
*    II.  Mechanical Engineering
    III.  Geology and Mining Engineering
*    IV.  Building and Architecture
*     V.  Chemistry
     VI.  Metallurgy
    VII.  Natural History
* VIII.   Physics
     IX.  Science and Literature
      X.  Philosophy
```

Those marked with an * still have the same numbers.

The first two graduates to do geological theses were William Otis Crosby (VII S.B. 1876) and George Hunt Barton (III S.B. 1880), both of whom later became professors of geology in Course XII.

It was not until 1890, a quarter century after M.I.T.'s founding, that Course XII was established as the Department of Geology and authorized to offer a Bachelor of Science Degree in Geology. Dixie Lee Bryant (XII S.B. 1891) was the first graduate to receive an S.B. degree in Course XII.

Although there was no Department of Geology as such, with its own degree program, until 1890, lectures on geological subjects were given from the very beginning of the Institute by William Barton Rogers. As Professor of Physics and Geology, from 1865 to 1868, he gave the first lectures in both these subjects, and it is worth repeating that 6 of the 13 members of the first graduating class, Class of 1868, received their diplomas in Course IV, Geology and Mining.

Soon, however, the combination of administrative and teaching duties proved too heavy for Rogers, particularly as he was not in good health, and on 29 July 1868 he resigned the Thayer Professorship in Physics, happily relinquishing it to Professor Edward C. Pickering, who had much to do with development of the Rogers Laboratory of Physics mentioned elsewhere in this history. On 24 October Rogers suffered a slight stroke, and when his condition worsened he was granted leave of absence in early December and taken to Philadelphia. Soon after, Professor John D. Runkle was appointed Acting President, to serve until Rogers recovered his health, and although he did get better he gave up teaching altogether and resigned

I. INTRODUCTION

from the presidency in May 1870. Thereafter, until his death on 30 May 1882, he limited his activities to administrative matters, becoming Professor Emeritus in 1881.

However, because of his and Henry's foresight in their earlier planning, a sound curriculum in geology and mining had been established, and we may be sure that he participated actively in recruiting the first geology professors, even though he was ill much of the time.

In 1870 Alpheus Hyatt was appointed Professor of Palaeontology, to be followed in 1871 by William H. Niles, as Professor of Physical Geology, and in 1872 by T. Sterry Hunt, as Professor of Geology. Three years later William O. Crosby, previously mentioned, was appointed Assistant in Paleontology and in 1883, the second M.I.T. alumnus mentioned earlier, George H. Barton, was appointed Assistant in Geology. Thus by the end of the Institute's first decade, Geology was represented by four faculty members - Hyatt, Niles, Hunt, and Crosby (as mentioned earlier, Rogers had given up regular lecturing in 1868) - and by a list of subjects that included physical and historical geology, mineralogy and petrology, paleontology, and economic and regional geology. Although instructional space, facilities, and materials were minimal in the Institute's first building, the Rogers Building on Boylston Street in Boston, students and faculty had access to the collections of the next-door museum of the Boston Society of Natural History. Furthermore, good surface transportation by trains and trolley cars, followed by short walks, made frequent field trips possible.

Thus by 1875, a decade after Rogers met that first class in Mercantile Hall in Boston on 20 February 1865, Geology was a firmly established discipline in the young Institute, with three professors and an assistant, a substantial list of geological and mining subjects, and a fair number of students interested in practical geology and mining. The Rogers' dream of a polytechnic institution was coming true, even though financial difficulties were besetting the young Institute, and the next seventy-five years would see it grow into one of the world's leading technological institutions. Geology, and the other closely related earth sciences, would participate actively in this growth, so that by the end of M.I.T.'s first century, the Institute could point with satisfaction to its impressive strength in the several disciplines to which the two Rogers brothers devoted their professional careers. Before discussing the actual events and persons involved in the history of Geology at M.I.T. during the Institute's first century, I think it appropriate to comment briefly on the sources of information that describe what actually happened during those eventful hundred years.

SOURCES OF HISTORICAL INFORMATION *

A definitive history of geology at M.I.T. has not been written; as a matter of fact there are only scattered references and a few unpublished manuscripts that deal in any detail with how geology fared as the Institute grew. The most informative sources are the reports on Course XII in the annual President's Reports, articles on geology in The Technology Review and Tech Engineering News, items of geological interest in The Tech, and Richards' autobiography, Robert Hallowell Richards: His Mark, published in 1936.

Nor is there a definitive history of the Institute itself. The ingredients of such a history, to the extent that they still exist, await the serious investigator in a great variety of source materials. By far the most informative of the sources is Life and Letters of William Barton Rogers, edited by his wife (with the assistance of William T. Sedgwick) and published in Boston in 1896. Through letters and documents written by Rogers, and descriptions of his plans and actions, the early history of the Institute is well portrayed. Most of the other accounts of the early years of M.I.T. are contained in the more comprehensive biographies of Rogers that appeared after his death. These are listed in his biography, which constitutes Chapter 1 farther on, and the interested reader's attention is directed to the following references in that list that are particularly informative: M.I.T. Society of Arts (1882); Rives (1883); Ruschenberger (1883); Walker (1887); The Tech (1904); Munroe (1888 and 1904); Anonymous (1905); and the Dedication Number of The Technology Review for 1916, the year the Institute moved across the Charles River from Boston to Cambridge. Important aspects of M.I.T.'s history are included in Munroe's A Life of Francis Amasa Walker (1923) and in Pearson's Richard Cockburn Maclaurin (1937). M.I.T.'s contributions to the two great world wars have been summarized in Ruckman's Technology's War Record (1920) and Burchard's Q.E.D., M.I.T. in World War II (1948). Departmental histories of special interest to geology are Davis' Chemistry at M.I.T. (Tech. Rev. 35/7, April 1933) and Goodwin's Physics at M.I.T. (Tech. Rev. 35/8, May 1933). The only history thus far written that covers a major part of M.I.T.'s first century is Prescott's brief but informative, though largely undocumented, When M.I.T. was Boston Tech published in 1954.

Perhaps the richest store of historical information lies awaiting discovery in the several publications of the Institute - the annual re-

---

* Full references to the most important items cited in the immediately following paragraphs are listed in the Special Bibliography at the end of this chapter.

I. INTRODUCTION

ports of the President, Treasurer, et al.; the annual Catalogues and Bulletins; The Technology Review; The Tech; Tech Engineering News; and more recently, Tech Talk.

Finally, only a few weeks ago, in early 1976, appeared two indispensable references on M.I.T.'s history - The Alumni Association's M.I.T. Alumni Register 1975 - Alumni Centennial Edition (1975) and Wylie's M.I.T. in Perspective (1976). Inasmuch as several individuals are now working on histories of different parts of M.I.T., the next decade should produce important contributions to that aspect of the Institute that has been neglected much too long.

GEOLOGY AT M.I.T., 1865-1965

What products came from the activities of M.I.T.'s geologists during the Institute's first century of existence? To answer this question it is first necessary to identify and define the activities, then describe the results of those activities.

As teachers, the geologists taught their students the facts, theories, hypotheses and speculations of geology and trained them to evaluate and use such knowledge to solve problems. From this activity came students trained to do practical geological work.

As a second part of their teaching they suggested and supervised investigations by students that led to original research which culminated in theses for degrees and articles for publication in official reports and current journals. In these several kinds of reports the students proposed new ideas and principles, described new minerals and fossils, suggested new and improved methodologies, and reported on novel devices and instruments.

As investigators themselves, the geologists also reported their research results in a variety of publications, and organized the scientific literature of broad fields of geology into textbooks for classroom use.

The most important products or results, therefore, were: 1) graduates trained to use geological knowledge in a practical way; 2) publications that organized and improved existing knowledge; 3) novel ideas and methodologies that extended the frontiers of research in the earth sciences, thereby adding new knowledge; and 4) devices and instruments that made possible more and better measurements and analyses of physical and chemical characteristics and phenomena of geological interest.

During the Institute's first quarter century, 1865 to 1890, there was no organized curriculum in geology and no degree in the discipline. There were, however, lecture-laboratory subjects offered covering the traditional aspects of geology - mineralogy and petrology, paleontology, and structural, economic, and regional geology - and these were required in several of the Courses that were organized in this period. Rogers, in planning his proposed polytechnic institute, had seen to it that M.I.T. graduates would take some geology subjects as a required part of their curriculum.

In 1890, at the beginning of M.I.T.'s second quarter century, 1890-1915, the numerous geological subjects were organized into a definite curriculum leading to a Bachelor of Science degree, and receiving the designation Course XII, Geology. Thus came into existence M.I.T.'s first Department of Geology* with Prof. William H. Niles as Head. Course XII's first bachelor's degree in geology was awarded Dixie Lee Bryant in 1891, as mentioned earlier on.

During Course XII's first quarter century, and M.I.T.'s second, 1890 to 1915, ending with the Institute's move across the Charles River from Boston to Cambridge in 1916,** the Department of Geology grew slowly but steadily, and continued to offer a broad range of subject matter to students in other courses while at the same time developing advanced subjects that would later constitute a foundation for graduate degrees. The first Master of Science Degree in Geology was awarded in 1909, and the first doctorate in 1910. Thus by 1915, the end of Course XII's first quarter century, the Department of Geology had a strong and diversified curriculum leading to both undergraduate and graduate degrees, although up to this time few actual degrees had been awarded - 14 S.B.s, 4 S.M.s, and 4 Ph.D.s. It was in fact fully prepared for the impressive growth that was to come in the next twenty-five years under the leadership of Waldemar Lindgren and Warren J. Mead.

The Department's second quarter century, 1915-1940, was roughly bounded by the two great world wars. World War I was underway when the Institute moved across the Charles in 1916, and World War II had begun in

---

* As mentioned elsewhere in this history, the First Annual Catalogue (p. 10) included "4. A Course in Geology and Mining" as one of the courses in which a student could obtain a degree or diploma. It has also been noted that 6 of the 13 members of the first graduating class, the Class of 1868, earned their degrees in "Geology and Mining." These graduates, however, were actually trained for professional careers in the mining industry. The only students who earned "geological" degrees before 1890 were William O. Crosby (VII S.B. 1876) and George B. Barton (III S.B. 1880); they had to meet requirements in the geology options of Courses VII (Natural History) and III (Geology and Mining Engineering), respectively.

** For a detailed account of this move see the Dedication Number (1916) of The Technology Review, Vol. 18, No. 7: iii + 463-800, il., (July 1916).

I. INTRODUCTION

Europe by the time Course XII was fifty years old. Waldemar Lindgren, Chief Geologist of the United States Geological Survey and one of the world's foremost economic geologists, had come to head the Department in 1912. In the same year, the preceding head, Thomas A. Jaggar, Jr., had resigned to become Director of M.I.T.'s newly established Hawaiian Volcano Observatory. Under Lindgren's leadership the Department produced an impressive group of graduates who would attain continental and worldwide fame in physical and economic geology. His own research, on which he based his classic textbook, <u>Mineral Deposits</u>, has had a fundamental and long-lasting influence on the study of ore deposits around the world; it still remains an important reference work more than sixty years since it first appeared in 1913.

Lindgren retired in 1933, although he remained active until his death in 1939, and was followed by Prof. Hervey W. Shimer who served as Acting Chairman during SY 1933-34. Then with the clouds of World War II beginning to gather, Dr. Warren J. Mead, a renowned structural and economic geologist at the University of Wisconsin, came to head Course XII in the fall of 1934. By Pearl Harbor Day, 7 December 1941, Mead had brought about a broadening and diversification of the curriculum and research program of the Department, obtained funds for construction of the Cabot Spectrographic Laboratory, and encouraged the program in geophysics earlier initiated by Prof. Louis B. Slichter. The hall corridors of the third and fourth floors of Building 4 became lined with an impressive group of glass cases holding exhibits of choice mineral and fossil specimens. And at the north end of the third-floor corridor, in rooms numbered 8--304-308, was the Lindgren Library, named in honor of Waldemar Lindgren and dedicated as he reached retirement in 1933. Once again Course XII was on the move and ready for the next decade, but World War II intervened and brought most of the Department's program to a halt for the next decade, 1940 to 1950.

The third quarter century of Course XII's existence was nearly ten years gone before post-war conditions were stabilized and the Institute back to normal. By this time, the beginning of SY 1949-50, the Department had been moved from Building 4, where it had occupied space since the move across the river in 1916, to Building 24, a new building hurriedly constructed for war research. Mead had retired as Chairman in June 1949, and I (Robert R. Shrock) had been appointed Acting Chairman for SY 1949-50; appointment as Chairman came the following year. During the next fifteen years, 1950-1965, the Department recovered the rapid growth in graduate students and research that was underway when World War II interrupted all academic programs in 1940 and for almost a decade thereafter. Among the more important happenings during this 15-year period were the following: 1) the Cabot Spectrographic Laboratory was enlarged with addition of a

Hilger emission spectrograph; 2) a new mass-spectrograph was procured and space made available for radiometric measurements of certain stable isotopes for age determinations of minerals and rocks; 3) a special laboratory was created to investigate induced polarization; 4) an industry-supported program, the Geophysical Analysis Group (GAG), was organized to investigate the signal-to-noise problem in geophysical exploration, using Norbert Wiener's recently developed time-series analysis; 5) two summer-training programs for undergraduates were initiated - one in Canada, in cooperation with the Nova Scotia Research Foundation and the Provincial Department of Mines, and the other in the Southwest, directed by Cecil H. Green (VIA S.B. 1923; S.M. 1923), President of Geophysical Service Inc. of Dallas, Texas; 6) a program of instruction and research in oceanography was started, with indispensable staff assistance from the Woods Hole Oceanographic Institution and critical financial support from the Office of Naval Research; and 7) the first subject in planetary science was offered by Dr. Irwin I. Shapiro, now Professor of Geophysics and Physics, in the spring of 1965. And in 1964, as the Institute's first century was drawing to a close, and Course XII's 75th year would soon end, a new 20-story building, funded by the Cecil H. Greens of Dallas, was occupied by Geology and Meteorology, with dedication ceremonies on 2 October 1964.

As the Institute ended its first century of existence and began its second, and as Course XII started its fourth quarter century, the Department of Geology and Geophysics, soon to be renamed the Department of Earth and Planetary Sciences, was ready for the decade of impressive change that followed. The new Green Building provided excellent space for the earth sciences - geology, geochemistry, geophysics, oceanography and meteorology; the numerous laboratories were equipped with the latest research facilities; McDermott Hall provided an excellent lecture room for almost 300 persons; the Lindgren Library, soon to have more than 20,000 volumes (and space for another 20,000) and 200 current earth science journals, and the Schwarz Memorial Map Room, were reestablished on one complete floor, with room left for 20 carrels and other study areas; and a lounge (later dedicated as the Ida Green Room in 1974) and adjoining conference room on the ninth floor provided a much-needed meeting place for students and faculty. Student enrollment, both undergraduate and graduate, was at an all-time high; a group of post-doctoral fellows supplemented the teaching and research staff; and the Department's total budget was more than a million dollars (in contrast to $150,000 for FY 1949-50) - it would quadruple to more than $4,000,000 by the end of the decade, in FY 1974-75.

The time was right for change of leadership and redirection of Course XII's overall program. After 16 years as Head, 1959-1965, I asked to be relieved of the chairmanship in order to devote my remaining five years

I. INTRODUCTION 15

to teaching and preparing for the writing of this history. The Administration, through the offices of then Dean of Science Jerome B. Wiesner, now M.I.T.'s President, induced Dr. Frank Press, renowned geophysicist and Director of Caltech's Seismological Laboratory, to come to the Institute as the new Head of Course XII.

Thus when Course XII began its fourth quarter century in the autumn of 1965 it had a new head; a new building; a growing enrollment, especially of graduate students; a new program in oceanography that would soon be officially conjoined with the Woods Hole Oceanographic Institution; and a varied and vigorous research program with growing financial support. So once again, as in the past, with change of leadership, the earth sciences at M.I.T., including geology, started on another period of change and growth which still continues in 1975 under Frank Press' able chairmanship. During the decade since Press came, a dozen professors have joined the Course XII faculty, to replace the same number who have retired, or resigned to go elsewhere. The name of the Department has been changed to Earth and Planetary Sciences to recognize the broadened program of Course XII and addition to the staff of physicists and chemists who are specialists in the space sciences. In the fall of 1966 a Joint Doctoral Program in Oceanography between M.I.T. and the Woods Hole Oceanographic Institution was officially approved by both institutions, and on 8 May 1968 the appropriate documents were signed by M.I.T. President Howard W. Johnson and W.H.O.I. President Paul M. Fye at a special ceremony at Woods Hole.

Throughout M.I.T.'s first century, geology and geologists have been an integral part of the Institute. Altogether 52 full-time and 5 part-time professors, as well as numerous instructors, lecturers, and assistants, served as geology teachers during the century. Together they published more than 50 major textbooks and at least 1,000 important articles that appeared in leading geological journals. Finally, they recommended and saw granted a total of 295 S.B. degrees, 155 S.M.s, 15 Sc.D.s and 177 Ph.D.s, for a total of 642 degrees in the three geosciences - geology, geochemistry, and geophysics. In addition to these degrees, 56 S.B.s were granted to students in the Geology option of Course III (III_3) during the years 1904 to 1922. Adding these 56 to the 642 total, and including the 2 degrees in Geology granted before 1890, gives a total of an even 700 geology degrees awarded through 1965.

THE M.I.T. OF 1965

Analysis of the program of a typical Department at M.I.T. in 1965, a hundred years after the first class was held in Boston in early 1865, reveals a dynamic system in which individuals interacted vigorously with one another and with the tools and facilities that were available throughout the Institute.

Such a departmental analysis also shows that the Institute as a whole was actually composed of four interacting groups: 1) The Corporation, composed largely of prominent business and industrial leaders; 2) the Administration, composed of President, Chancellor, Provost, Vice-Presidents and their staffs; 3) the several Schools with their respective Deans; and 4) the Departments with their respective faculties, one of whom served as Chairman or Department Head. Noteworthy is the fact that the M.I.T. faculty, in the broad definition of that word, was monolithic; i.e. every professor, instructor, lecturer and others who taught, as well as the deans, the provost, the chancellor and the president, were members of at least one department. While the Chairman and faculty of a particular department were responsible to a dean, they did not constitute a part of a school faculty - as a matter of fact the Graduate School, although it had a dean, had no departments under its control; it coupled with all the departments of the Institute through the Committee on Graduate School Policy, which consisted of a representative from each department, and had the dean as its chairman. In short there was no such thing as a Faculty of Science, a Faculty of Engineering, etc., per se; there was only an M.I.T. Faculty. It is obvious, therefore, that at M.I.T. the individual departments, acting through their chairman and individual faculty members, played a dominant role in determining the nature and quality of their teaching and research activities, and in the recruitment and selection of departmental personnel at every level. It is with this situation in mind that I should now like to examine in somewhat more detail what is actually involved in the operations and composition of M.I.T.'s present-day (1976) Department of Earth and Planetary Sciences.

Viewing the Department as the dynamic system of interacting components mentioned in a preceding paragraph, we see that the Corporation, President, Chancellor, Provost, administrative staff, and supervisors guide and control overall operations; faculty members teach, conduct research, and commonly carry on a variety of both intra- and extra-M.I.T. activities; students acquire knowledge, develop professional skills, and learn how to carry on research. Machine and electronic shops, computers, and other special facilities provide the support so vital for instruction and experimental investigation; libraries provide the indispensable reservoir of existing knowledge on which both teaching and research depend; and field camps, observatories, and other off-campus facilities provide opportunities for both student and staff training and research. Supporting personnel in departmental headquarters, teaching and research laboratories, shops, and special research facilities keep the entire program moving

I. INTRODUCTION

along in scheduled activities - curriculum, research projects, financial and personnel management, etc. - that together constitute the overall program. Space has to be provided for classrooms and laboratories; for collections, machines, and research facilities; and to house personnel. Finally, funds have to be found to finance the entire operation.

The quality and quantity of the products of a Department are determined in large measure by how effectively the many components of the loosely bound system of people, facilities, and machines interact. Let us now look specifically at the faculty members of Course XII, and their supporting staff, the personnel group that I consider of first importance in our Department of Geology.

PRIMARY IMPORTANCE OF THE FACULTY

In the unabridged second edition of Webster's New International Dictionary of the English Language, published in 1937, the third definition of college is given as follows on page 525:

> "3. A society of scholars, or friends of learning, incorporated for study or instruction, esp. in the higher branches of knowledge. ... In the United States the college is primarily an institution of higher learning receiving approved graduates of preparatory schools and offering instruction in arts, letters, and science, leading to the bachelor's degree; there is, however no clear line of demarcation, as institutions have retained the name college while extending their instruction to university scope."

The same Dictionary gives as one definition of faculty, the following on page 909:

> "10. The body of persons to whom are entrusted the government and instruction as of a university or college; the president, or principal, and teaching staff of a university, or of a college ..., or of a school."

In this history of geology at M.I.T., I consider the professors, or faculty, as the most important single group of individuals in the Institute, because without them there could be no college or university (or institute) according to the preceding definitions. I would place the students second in importance because in most North American schools of higher education today they are the raison d'être for the faculty, and by extension for the college or university as a whole. Accordingly, I devote the remainder of this volume to M.I.T.'s Geology college, as that term is defined in the first of the preceding definitions, or to the Institute's Geology faculty, as that term is defined in the second of the preceding definitions. I do this because the professors determine the breadth and quality of the curriculum, the rigor and relevance of the subject matter

taught and the research conducted, and the standards which they expect
their students to attain. In short, they play the dominant role in determining the nature and quality of the Department's products, hence I have
chosen to devote the major portion of this volume to biographies of the
fifty-two professors and five part-time professors who taught regularly
scheduled subjects in geology during M.I.T.'s first century. In a second
volume I intend to write about the accomplishments of our seven hundred
graduates, and how Course XII conducted the program that made it possible
for these graduates to get the knowledge and skills they took with them
when they left with their degrees.

As background for the biographies of the individual professors, which
follow as numbered chapters, it is appropriate to review how the Course XII
Faculty was actually developed; how supporting staff were recruited, incorporated into the teaching team, and funded; and how supplementary services such as those of the department's secretariat, our Lindgren Library,
and our machine shops and other special facilities were established.
This review constitutes Chapter IV.

In final analysis, it is people who make a Department go, and it is
to the people of Course XII that I devote the remainder of this volume,
first in brief review and then in individual biographies.

SPECIAL BIBLIOGRAPHY FOR CHAPTER I

Following are the more important works that contain information on the
organizing, founding, and subsequent growth and development of M.I.T.
This is by no means a complete list, but it is believed to include most
of the more important references.

1) First Annual Catalogue of the Officers and Students, and Programme of the Course of Instruction, of the School of the Massachusetts Institute of Technology, 1865-6. Boston: John Wilson and Sons, 39 p., (1865).

2) The Beginning of the Massachusetts Institute of Technology, by James P. Munroe. Tech. Quart. 1/4: 285-297, (1888).

3) Life and Letters of William Barton Rogers, edited by his wife [Emma Savage Rogers] with the assistance of William T. Sedgwick, in two volumes: Vol. I, p. 1-427, il.; Vol. II, p. 1-451, il. Boston: Houghton, Mifflin and Co., (1896).

4) John Daniel Runkle (1822-1902), by H. W. Tyler. Tech. Rev. 4/3: 276-306, port., (July 1902). Reprinted as John Daniel Runkle (1822-1902): A Memorial. Boston: George H. Ellis Co., 32 p., port., (1902).

5) Dedication Number (1916) of The Technology Review, published by the M.I.T. Alumni Association. Tech. Rev. 18/7: iii + 463-800, il., (July 1916).

6) Technology's War Record, John H. Ruckman, ed. Cambridge: Published by the War Records Committee of the Alumni Association of Massachusetts Institute of Technology, ix + 747 p., il., (1920).

I. INTRODUCTION   19

7) *A Life of Francis Amasa Walker*, by James Phinney Munroe. New York: Henry Holt & Co., vii + 449 p., il., (1923).

8) *Chemistry at M.I.T. - A History of the Department from 1865-1933*, by Tenney L. Davis. Tech. Rev. 35/7: 250-252, 264, 266, 268, 270, 272, il., (April 1933).

9) *Physics at M.I.T. - A History of the Department from 1865-1933*, by H. M. Goodwin. Tech. Rev. 35/8: 287-291, 312-313, (May 1933).

10) *Robert Hallowell Richards: His Mark* [an autobiography]. Boston: Little, Brown, and Co., 329 p., il., (1936).

11) *Richard Cockburn Maclaurin: President of the Massachusetts Institute of Technology 1909-1920*, by Henry Greenleaf Pearson. New York: The Macmillan Co., 302 p., il., (1937).

12) *Q.E.D.: M.I.T. in World War II*, by John Burchard. New York: John Wiley & Sons, Inc., xiv + 354 p., il., (1948).

13) *When M.I.T. was "Boston Tech" 1861-1916*, by Samuel C. Prescott. Cambridge: The Technology Press, xvii + 350 p., il., (1954).

14) *Register - Department of Earth and Planetary Sciences, 1865-1970*. Cambridge: M.I.T. Graphic Arts Service, iv + 74 p., il., (May 1970).

15) *MIT Alumni Register 1975 - Alumni Centennial Edition*. Published by the M.I.T. Alumni Association. Cambridge, Mass., 778 p., il., (1975).

16) *M.I.T. in Perspective: A Pictorial History of the Massachusetts Institute of Technology*, by Francis E. Wylie. Boston: Little, Brown, and Co., x + 220 p., il., (March 1976).

NOTE:

The reader who is interested in the part that the geological sciences played in the overall history of the Institute will find much on this subject in the biographies, memorials, and the like that are listed following the bibliographies in many of my own biographical sketches that follow Chapter IV. I also hope to discuss many aspects of geology at M.I.T., other than its faculty, in a later volume.

----

20 GEOLOGY AT M.I.T. 1865-1965

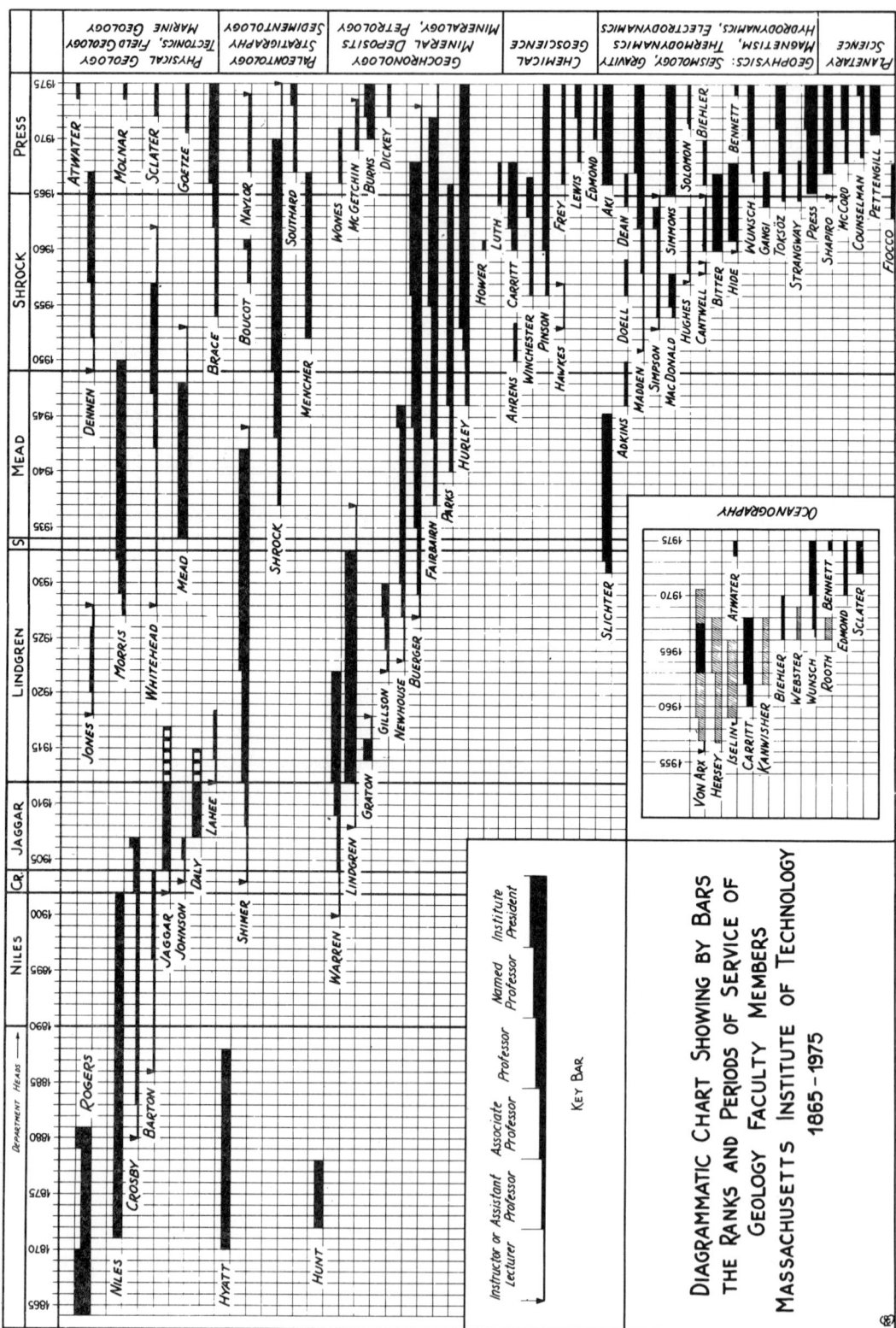

II. FACULTY OF COURSE XII, GEOLOGY, ETC. AT M.I.T.
1865-1970 (inclusive)

INTRODUCTION - DEVELOPMENT OF A FACULTY

A typical faculty of a major geology department anywhere in the United States will probably include members falling in most of the following five categories:

A. Those who remained at their alma mater as staff members after having done both their undergraduate and graduate work at the same school.

B. Those who received their undergraduate training in the institution where they now teach, but who attended another school for graduate work before returning to their alma mater.

C. Those who came to a particular school for graduate work, having done their undergraduate work elsewhere, and remained after receiving their doctor's degree to develop a professional career at their second "alma mater."

D. Those who did all of their academic work at one or more other schools before joining the department where they developed their professional career.

E. Part-time staff members of every rank, who fall in one of the preceding four categories, and visiting scientists, with a variety of titles, who join a departmental staff on a short-term basis (generally only a year, a term, or a few weeks) to deliver special lectures or conduct special seminars, colloquia, conferences, etc. These are commonly distinguished scientists who are appointed to named professorships or lectureships. They may also be exchange or temporary professors brought in to fill a position left vacant by a staff member on leave. Finally, some may come from federal bureaus or industrial organizations, or may be independent consultants.

For our present purpose, I shall consider as _faculty_ members only full-time teaching personnel who have attained at least the rank of assistant professor. In a somewhat different "faculty" category are Lecturers, Instructors, Visiting Lecturers and Visiting Professors, and certain other scientists, all of whom had short-term appointments and fall in category E in the preceding list. The individuals in these two categories constitute the major and senior component of a faculty. Inasmuch as an active and progressive departmental faculty is always in a state of flux, it follows that new persons come to replace those who have retired or have resigned to go elsewhere, and that the departmental roster changes from time to time. Although it is not an invariable rule, it is rather likely that a departmental staff that remains the same for long periods will not be as active and progressive as one that is constantly changing by filling vacancies due to retirements and by appointing new staff members to replace those hired away by competing colleges.

22 GEOLOGY AT M.I.T. 1865-1965

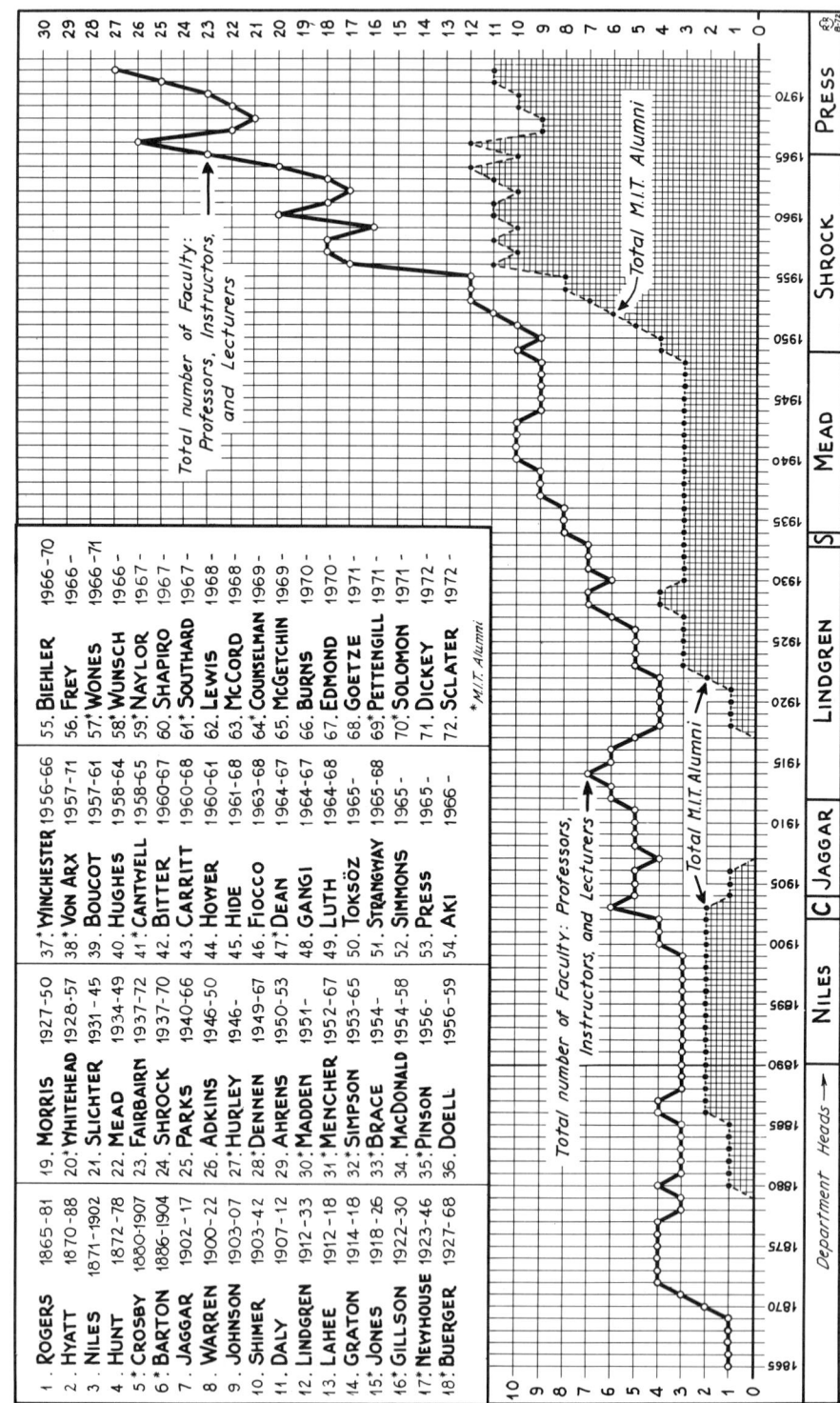

Chart Showing Order of Appointment of Course XII Professors and Number of M.I.T. Alumni Serving on the Faculty during 1865-1972

II. COURSE XII FACULTY

Following a brief discussion of the five categories included in the foregoing list is a section which gives pertinent appointment information about each of the first 53 persons who served as professors in Course XII during the first century of M.I.T.'s existence.

The minor and junior component of the typical departmental staff, and a very important one for the instructional and research parts of the academic program, are the <u>student assistants</u>. These are graduate students who perform a variety of services for the department, and who constitute the reservoir of talent from which faculty members of the future are drawn. They are discussed in the chapter on <u>Staff supporting Course XII Faculty and Students</u>.

A. Faculty from M.I.T. Baccalaureates, etc. only

A common and widespread policy among many, if not most, colleges and universities in North America is appointment of their own better graduates to the lower ranks of teaching positions in which they are given an opportunity to demonstrate their promise of becoming distinguished professors. (See chart on accompanying page.)

One aspect of this policy is appointment of their own recipients of graduate degrees who also did all their undergraduate work in the same department. While such a policy does permit an institution to develop its own "home-grown" product, it can well lead to undesirable inbreeding if not conducted with continuous care and concern for maintenance of highest quality. An almost unavoidable result of this policy, when carried to excess, is perpetuation of traditional curricula, areas of interest, methods of teaching and research, and a general euphoria which is likely to develop into a static situation in which needed changes are difficult to make. In contrast, if care is exercised, a department can build great strength from its own outstanding graduates and can be sure that its standards and policies, being known to the individual, are maintained and improved.

The M.I.T. Geology Department, through its first seventy-five years, <u>i.e.</u> from 1890 through 1965, hired many of its own undergraduate alumni (See accompanying chart). A considerable number of these soon left to follow their professional interests elsewhere, but a goodly number remained to become distinguished professors and to aid greatly in diversifying and strengthening the Department's academic and research programs. This group includes:

```
        W. O. Crosby        (VII S.B. 1876)
        G. H. Barton        (III S.B. 1880)
        W. F. Jones         (Class of 1909; no degree)
        W. L. Whitehead     (III S.B. 1913; XII Ph.D. 1918)
```

24   GEOLOGY AT M.I.T.   1865-1965

THE FIRST PROFESSOR OF GEOLOGY
William Barton Rogers
President of the Institute
1862-1870; 1879-1881
Professor of Physics and Geology
1865-1881

CHAIRMEN OF THE DEPARTMENT

Thomas Sterry Hunt William Harmon Niles William Otis Crosby
1872-1878   1878-1902   1902-1904

Thomas Augustus Jaggar, Jr. Waldemar Lindgren Hervey Woodburn Shimer
1904-1912   1912-1933   1933-1934

Warren Judson Mead   Robert Rakes Shrock   Frank Press
1934-1949   1949-1965   1965-

II. COURSE XII FACULTY

M. J. Buerger	(III S.B. 1925; XII S.M. 1927, Ph.D. 1929)
W. H. Dennen	(XII S.B. 1942, Ph.D. 1949)
T. R. Madden	(VIII S.B. 1949; XII Ph.D. 1961)
W. F. Brace	(XIII S.B. 1946; I S.B. 1949; XII Ph.D. 1953)
D. R. Wones	(XII S.B. 1954, Ph.D. 1960)
T. Cantwell, Jr.	(X S.B. 1948, S.M. 1949; XII Ph.D. 1960)
L. W. Dean, III	(VI-A S.B. 1956; S.M. (unspecified) 1957; VIII Ph.D. 1960).

B. Faculty from M.I.T. Baccalaureates who went elsewhere for Graduate Work

Some leading American educational institutions encourage their most promising seniors to continue graduate work at another school in order to broaden their professional background and diversify their academic training. Those students who follow this advice and do well then become prime prospects for departmental positions at their alma mater as they start on a professional career.

Two recently appointed professors, J. B. Southard (XII S.B. 1960; Ph.D. 1966, Harvard) and R. S. Naylor (XII S.B. 1961; Ph.D. 1967, Caltech.) belong in this category.

C. Faculty from M.I.T. Graduate Degree Recipients only

Some of the Department's most distinguished members have been former graduate students who came to work towards an advanced degree after completing undergraduate work elsewhere. These very important members added greatly to the Department's strength by bringing a wide variety of interests and then developing those interests into productive professional careers.

Included in this group are:

J. L. Gillson	(B.S. 1917 and M.A. 1920, Northwestern Univ.)
W. H. Newhouse	(B.S. 1921, Penn State Univ.)
P. M. Hurley	(B.A. and B.A.Sc. 1934, Univ. British Columbia)
E. Mencher	(B.S. 1934, City College of the City of N.Y.)
H. E. Hawkes, Jr.	(A.B. 1934, Dartmouth)
S. M. Simpson, Jr.	(S.B. 1950, Yale)
W. H. Pinson, Jr.	(B.S. 1948 and M.A. 1949, Emory Univ.)
W. S. Von Arx	(A.B. 1943, Brown; Sc.M. 1943, Yale).

D. Faculty from non-M.I.T. Sources

Perhaps the policy most commonly used to strengthen and expand a departmental faculty is to recruit both junior and senior staff members from other schools. Many of the more distinguished staff members of past years came to M.I.T. from other schools, and because they had already established a well-known reputation as geologists, they added immediate strength to the geology program and to the Geology Department.

During M.I.T.'s first quarter century, from 1865 to 1890, it was William Barton Rogers, Alpheus Hyatt, Thomas Sterry Hunt and William Harmon Niles, all renowned geologists when they came to M.I.T., who gave the geology program high quality and professional distinction.

In the second quarter-century, 1890-1915, it was Jaggar and the promising younger men, including Shimer, Johnson, Daly, Warren, and Lahee, who built an academic program strong in the physical and historical aspects of geology.

The third quarter-century, (1915-1940), was ushered in with the coming of the internationally renowned economic geologist, Waldemar Lindgren (1912-1933), whose reputation attracted many able students interested in mineral deposits. After Lindgren's retirement in 1933, the Department not only continued to be strong in physical and historical geology, as well as in mineral deposits, but also added to its strength and diversity by getting as its new head in 1934 the distinguished structural and engineering geologist, Warren J. Mead. The brilliant young geophysicist, Louis B. Slichter, had come from the University of Wisconsin only a few years before Mead's arrival from the same institution, and within the next few years, as the third quarter-century came to an end, Mead recruited three additional junior staff members, who had been at Wisconsin as graduate student or faculty member, Harold W. Fairbairn and Robert R. Shrock in 1937 and Roland D. Parks in 1940.

E. Development of post-World War II Faculty

World War II disrupted the beginning of the fourth quarter-century of geology at M.I.T., as student enrollment dropped, courses were suspended for the duration, and staff members left or went on leave to do war-related work.

By the time the country was again at peace, and the academic program of Course XII reestablished, Mead had reached retirement age in 1949, and Shrock was the new head. Recruitment of new staff was pursued vigorously, as the departmental staff agreed to build a more quantitative and experimentally-directed program, and during the 1950s and early 1960s a considerable number of exceedingly promising young geoscientists were added to the staff. Some were M.I.T. graduates; others had been trained at other schools. Some remained for only a few years before going elsewhere to develop distinguished careers--Ahrens in geochemistry, MacDonald in geophysics, Winchester in geochemistry, Doell and Strangway in paleomagnetism, Hide in fluid mechanics, and Luth in petrology. Others elected to remain in the Department and developed their careers in their respective fields--Brace in rock mechanics, Madden in geophysics, Toksöz in geophysics, Simmons in heat flow, and Wunsch in fluid mechanics.

The first century of geology at M.I.T. ended and the second began in 1965, when Shrock resigned as Chairman, and Frank Press succeeded him on 1 September 1965. Press, an internationally known seismologist, brought great distinction to the Department and during the next five years recruited a group of outstanding younger staff members who by 1970 made the Department, now renamed Earth and Planetary Sciences, one of the leading ones anywhere. There our history leaves Course XII with a future promise that was never better.

PROFESSORS OF GEOLOGY, etc. in COURSE XII

Introduction

During M.I.T.'s first century of existence, i.e. from 1865 to 1965, fifty-three full-time professors of "geology" (geochemistry, geology, geophysics, and subdisciplines of these) were appointed, as shown in the chart on a preceding page. The first was William Barton Rogers, Founder and first President of the Institute, who served as Professor of Physics and Geology from 1865 to 1881. The 53rd was Frank Press, Professor of Geophysics and Head of Course XII, from 1965 to the present (1975). When Press joined the Course XII faculty in 1965, the course was called the Department of Geology and Geophysics. In early 1969 the name was changed to Department of Earth and Planetary Sciences, and in 1971 Press became the first occupant of the distinguished professorial chair endowed by Cecil H. Green (VI-A S.B. 1923) to honor Robert R. Shrock (Honorary '26), who preceded Press as Course XII Chairman from 1949 to 1965 and was a longtime friend of Mr. Green.

In the following faculty roster, these 53 professors (as well as those added between 1965 and 1970) are listed in the chronological order of their first faculty appointment (i.e. some rank of professor). Under each is pertinent information as to pre-professorial appointments, length of service, promotions, resignations, retirement dates, etc. Some of this information is also displayed on the accompanying charts. Somewhat detailed biographies of the first 53 professors in the following list appear as numbered chapters following Chapter IV, Biographies of M.I.T.'s Geology Professors 1865-1965 (p. 87-99).

<u>Professors of Geology, etc. 1865-1970</u>
(in order of appointment)*

1. ROGERS, WILLIAM BARTON
1865-1870	President of the Institute
1879-1881	President of the Institute
1865-1881	Professor of Physics and Geology
1881-1882	Professor Emeritus Died 30 May 1882

2. HYATT, ALPHEUS
1870-1878	Professor of Palaeontology (also full-time at the Museum of the Boston Society of Natural History)
1878-1888	Professor of Zoology and Palaeontology
1888	Resigned, to continue full-time at the Museum of the Boston Society of Natural History
	Died 15 January 1902

3. NILES, WILLIAM HARMON
1871-1878	Professor of Physical Geology and Geography
1878-1902	Professor of Geology and Geography
1890-1902	Head, Course XII, Department of Geology
1902-1910	Professor Emeritus Died 13 September 1910

4. HUNT, THOMAS STERRY
1872-1878	Professor of Geology
1872-1878	Head, Course IV, Geology and Mining
1878	Resigned Died 12 February 1892

5. CROSBY, WILLIAM OTIS, (VII S.B. 1876)
1875-1876	Student Assistant in Paleontology
1876-1880	Assistant in Paleontology and Geology
1880-1881	Instructor in Geology
1881-1883	Instructor in Geology, Paleontology and Mineralogy
1883-1892	Assistant Professor of Mineralogy and Lithology
1892-1902	Assistant Professor of Structural and Economic Geology
1902-1906	Associate Professor of Geology
1902-1904	In charge of Course XII, Department of Geology
1906-1907	Professor of Geology
1907	Retired because of deafness
1907-1925	Professor Emeritus Died 31 December 1925

---

* The members of the list are numbered in the chronological order in which they were appointed to <u>professorial</u> rank. Teaching faculty members who served only as full-time or part-time Lecturers, or as part-time professors either on or off the campus, are not given a serial number nor are they counted as regular faculty on the accompanying charts; rather they are discussed in a following section.

Although only faculty members who held a professorial appointment are included in this roster, there were many other persons who participated in the teaching and research activities of Course XII in one way or another. These are discussed in another section of this chapter, and in the chapter on <u>Staff Supporting Course XII Faculty and Students</u>.

For M.I.T. alumni in the list, appropriate information is enclosed in parentheses following their name--<u>e.g</u>. BUERGER, (III S.B. 1925; XII S.M. 1927; Ph.D. 1929), and on an accompanying chart M.I.T. alumni have an asterisk before their name--<u>e.g</u>. 5. * CROSBY.

II. COURSE XII FACULTY

6. BARTON, GEORGE HARMON, (III S.B. 1880)
   - 1880-1881   Assistant in Drawing
   - 1883-1886   Assistant in Geology
   - 1886-1892   Instructor in Determinative Mineralogy and Geology
   - 1892-1896   Instructor in Geology
   - 1896-1904   Assistant Professor of Geology
   - 1904   Resigned   Died 25 November 1933

7. JAGGAR, THOMAS AUGUSTUS, JR.
   - 1902-1904   Lecturer in General Geology (coming from Harvard)
   - 1904-1917   Professor of Geology [last 5 years in Hawaii]
   - 1904-1912   Head, Course XII, Department of Geology
   - 1912-1917   Director, Hawaiian Volcano Observatory
   - 1917   Resigned (1 July 1917)   Died 17 January 1953

8. WARREN, CHARLES HYDE
   - 1900-1903   Instructor in Geology
   - 1903-1904   Instructor in Mineralogy
   - 1904-1909   Assistant Professor of Mineralogy
   - 1909-1912   Associate Professor of Mineralogy
   - 1912-1915   On staff of Hawaiian Volcano Observatory
   - 1912-1922   Professor of Mineralogy
   - 1920-1922   In charge of Course IX, General Science and General Engineering
   - 1922   Resigned, to become Dean of Sheffield School of Science at Yale University   Died 18 August 1950

9. JOHNSON, DOUGLAS WILSON
   - 1903-1905   Instructor in Geology
   - 1905-1907   Assistant Professor of Geology
   - 1907   Resigned, to become Assistant Professor of Physiography at Harvard University
                                               Died 24 February 1944

10. SHIMER, HERVEY WOODBURN
    - 1903 (Spring) Visiting Lecturer (from Columbia)
    - 1903-1908   Instructor in Geology
    - 1908-1912   Assistant Professor of Paleontology
    - 1912-1922   Associate Professor of Paleontology
    - 1922-1942   Professor of Paleontology
    - 1927-1928   Acting Head, Course XII, Department of Geology
    - 1931 (March-June) Acting Head, Course XII, Department of Geology
    - 1933-1934   Acting Head, Course XII, Department of Geology
    - 1942-1944   Honorary Lecturer
    - 1942-1965   Professor Emeritus   Died 13 December 1965

11. DALY, REGINALD ALDWORTH
    - 1907-1912   Professor of Physical Geology
    - 1912   Resigned, to become Sturgis Hooper Professor of Geology at Harvard University
    - 1912-1915   On Hawaiian Volcano Observatory Staff
    - 1915   Resigned from Hawaiian Volcano Observatory Staff
                                               Died 19 September 1957

12. LINDGREN, WALDEMAR
    - 1908-1912   Visiting Lecturer in Economic Geology
    - 1912-1933   William Barton Rogers Professor of Economic Geology
    - 1912-1920   Head, Course XII, Department of Geology
    - 1920-1926   Head, Course III, Department of Mining and Metallurgy, and Head, Course XII, Department of Geology
    - 1927-1933   Head, Course XII, Department of Geology
    - 1933-1937   Honorary Lecturer in Geology
    - 1933-1939   Professor Emeritus   Died 3 November 1939

MINING AND METALLURGY FACULTY OF COURSE III
Department of Mining, Metallurgy and Geology, 1922
Waldemar Lindgren, Chairman
Front row, left to right: W. Lindgren, H. O. Hoffman, and C. E. Locke
Back row, left to right: H. R. Aldrich, C. R. Hayward, R. C. Reed,
E. E. Bugbee, and B. B. Tremere, Jr.   (Photo by Notman)

GEOLOGY FACULTY OF COURSE III
Department of Mining, Metallurgy and Geology, 1922
Waldemar Lindgren, Chairman
Front row, left to right: H. W. Shimer, W. Lindgren, and C. H. Warren
Back row, left to right: H. E. McKinstry, W. F. Jones, and L. W. Currier
(Photo by Notman)

II. COURSE XII FACULTY

13. LAHEE, FREDERIC HENRY
    1912-1914   Instructor in Geology
    1914-1918   Assistant Professor of Geology
    1 December 1918 Resigned, to join staff of the Sun Oil Company
                                                    Died 3 December 1968

14. GRATON, LOUIS CARYL
    1914-1916   Professor of Mining Geology   Resigned 30 June 1916
    1916-1918   Lecturer in Mining Geology   Resigned 30 June 1918
    1918   Resigned, to resume professorship of Geology at
                       Harvard University   Died 22 July 1970

15. JONES, WILLIAM FRANCIS
    1918-1920   Instructor in Geology
    1920-1926   Assistant Professor of Structural Geology
    1926-1928   Lecturer on Petroleum Geology
    30 June 1928   Resigned, to enter the petroleum industry
                                                    Died 9 May 1941

16. GILLSON, JOSEPH LINCOLN, (XII S.M. 1921; Sc.D. 1923)
    1922-1924   Instructor in Mineralogy and Petrography
    1924-1927   Assistant Professor of Mineralogy and Petrography
    1927-1928   Associate Professor of Petrography
    1928-1930   Associate Professor of Mineralogy and Petrography
    1 August 1930 Resigned, to join staff of the E. I. DuPont de
                       Nemours & Company
    1961-1963   William Otis Crosby Lecturer   Died 4 August 1964

17. NEWHOUSE, WALTER HARRY, (XII S.M. 1923; Ph.D. 1926)
    1923-1927   Instructor in Mineralogy
    1927-1928   Assistant Professor of Mineralogy
    1928-1930   Assistant Professor of Economic Geology
    1930-1944   Associate Professor of Economic Geology
    1944-1946   Professor of Economic Geology
    30 June 1946   Resigned, to join staff of the U. S. Geological
                       Survey   Died 21 September 1969

18. BUERGER, MARTIN JULIAN, (III S.B. 1925; XII S.M. 1927; Ph.D. 1929)
    1925-1927   Assistant in Mineralogy
    1927-1929   Instructor in Mineralogy
    1929-1935   Assistant Professor of Mineralogy
    1935-1944   Associate Professor of Mineralogy
    1944-1956   Professor of Mineralogy and Crystallography
    1956-1963   Director, School for Advanced Study
    1956-1968   Institute Professor
    1 July 1968   Institute Professor Emeritus
    1968-1973   Senior Lecturer (part-time)
    1973-1975   Senior Research Associate

19. MORRIS, FREDERICK KUHNE
    1927-1928   Assistant Professor of Geology
    1928-1931   Associate Professor of Geology
    1931-1950   Professor of Geology
    1 July 1950   Professor Emeritus   Died 5 October 1962

20. WHITEHEAD, WALTER LUCIUS, (III S.B. 1913; XII Ph.D. 1917)
    1913-1914   Assistant in Geology
    1928-1942   Lecturer on Coal and Petroleum
    1942-1947   Assistant Professor of Geology
    1947-1957   Associate Professor of Geology
    1 July 1957   Associate Professor Emeritus
    1957-1962   Lecturer (part-time)   Died 2 December 1969

32 GEOLOGY AT M.I.T. 1865-1965

FACULTY OF COURSE XII, GEOLOGY, 1952
R. R. Shrock, Chairman
(from left to right)
Seated: R. D. Parks, W. L. Whitehead, P. M. Hurley, R. R. Shrock,
M. J. Buerger, H. W. Fairbairn, E. Mencher
Standing: E. A. Robinson, T. R. Madden, S. M. Simpson, Jr., L. H. Ahrens,
W. H. Dennen, N. A. Haskell, H. E. Hawkes, Jr.

(Photo by Jackman)

FACULTY OF COURSE XII, GEOLOGY AND GEOPHYSICS, 1957
R. R. Shrock, Chairman
(from left to right)
Seated: H. W. Fairbairn, W. L. Whitehead, P. M. Hurley, R. R. Shrock,
M. J. Buerger, R. D. Parks
Standing: T. R. Madden, S. M. Simpson, Jr., W. F. Brace, A. J. Boucot,
R. R. Doell, E. Mencher, H. Hughes, J. W. Winchester, W. H. Pinson, Jr.,
J. Savage, G. J. F. MacDonald.

(Photo by M.I.T. Graphic Arts)

II. COURSE XII FACULTY

21. SLICHTER, LOUIS BYRNE
    1931-1932   Associate Professor of Geophysics
    1932-1945   Professor of Geophysics
    1 October 1945 Resigned, to accept professorship in Department of
                    Geology, University of Wisconsin

22. MEAD, WARREN JUDSON
    1934-1949   Professor of Geology
    1934-1949   Head, Course XII, Department of Geology
    1 July 1949   Professor Emeritus
    1949-1954   Lecturer (part-time)   Died 16 January 1960

23. FAIRBAIRN, HAROLD WILLIAMS
    1937-1943   Assistant Professor of Geology
    1943-1955   Associate Professor of Geology
    1955-1972   Professor of Geology
    1 July 1972   Professor Emeritus
    1972-1977   Senior Lecturer (part-time)

24. SHROCK, ROBERT RAKES
    1937-1943   Assistant Professor of Geology
    1943-1949   Associate Professor of Geology
    1949-1970   Professor of Geology
    1946-1949   Executive Officer, Course XII, Department of Geology
    1949-1950   Acting Head, Course XII, Department of Geology
    1950-1952   Head, Course XII, Department of Geology
    1952-1965   Head, Course XII, Department of Geology and Geophysics
    31 August 1965 Resigned as Head of Course XII
    1 July 1970   Professor Emeritus
    1970-1975   Senior Lecturer (part-time)

25. PARKS, ROLAND DANE
    1940-1946   Assistant Professor of Geology
    1941-1946   On leave to War Production Board (Washington)
    1946-1966   Associate Professor of Mineral Industry
    1955-1956   On leave to Indian School of Mines, Dahnbad, India
    1961-1962   On leave to University of Assuit, Egypt
    1 July 1964   Joined staff of U.S. Treasury Department (on leave
                    1964-1966)
    1 July 1966   Associate Professor Emeritus
                                                   Died 18 December 1972

26. ADKINS, JOHN NATHANIEL
    1946-1950   Assistant Professor of Geophysics
    1948-1950   On leave with Office of Naval Research
    29 March 1950   Resigned, to take position in Washington with the
                    Office of Naval Research

27. HURLEY, PATRICK MASON, (XII Ph.D. 1940)
    1938-1939   Teaching Fellow
    1940-1941   Teaching Fellow
    1941-1942   Research Associate
    1946-1951   Assistant Professor of Geology
    1951-1953   Associate Professor of Geology
    1950-1963   Executive Officer, Course XII, Department of Geology
                    and Geophysics
    1953-1977   Professor of Geology
    1 July 1977   Professor Emeritus

28. DENNEN, WILLIAM HENRY, (XII S.B. 1942; Ph.D. 1949)
    1946-1948   Teaching Fellow
    1948-1949   Research Assistant
    1949-1952   Instructor in Geology
    1952-1957   Assistant Professor of Geology
    1957-1967   Associate Professor of Geology
    1963-1965   Executive Officer, Course XII, Department of Geology
                         and Geophysics
    31 August 1967   Resigned, to become Professor and Chairman, Department of Geology, University of Kentucky

29. AHRENS, LOUIS HERMAN
    1948-1950   Research Associate
    1950-1953   Assistant Professor of Geology
    31 December 1953 Resigned, to take Readership at Oxford University,
                         England
    1962-1963   Visiting Professor of Geochemistry

30. MADDEN, THEODORE RICHARD, (VIII S.B. 1949; XII Ph.D. 1961)
    1950-1951   Teaching Fellow in Geology
    1951-1952   Instructor in Geology
    1952-1953   Instructor in Geology and Geophysics
    1953   Research Assistant in Geology and Geophysics
    1953-1955   Instructor in Geology and Geophysics
    1955-1958   Lecturer in Geology and Geophysics
    1958-1962   Assistant Professor of Geophysics
    1962-1967   Associate Professor of Geophysics
    1967-   Professor of Geophysics

31. MENCHER, ELY, (XII Ph.D. 1938)
    1935-1938   Teaching Fellow
    1952-1967   Associate Professor of Geology
    31 August 1967   Resigned, to become Professor and Chairman, Department of Geology, City College of the City of New
                         York

32. SIMPSON, STEPHEN MILTON, JR., (XII Ph.D. 1953)
    1952-1953   Research Assistant
    1953-1954   Instructor in Geophysics
    1954-1962   Assistant Professor of Geophysics
    1962-1964   Associate Professor of Geophysics
    1964-1965   Lecturer in Geophysics
    30 June 1965   Resigned, to go into private consulting

33. BRACE, WILLIAM FRANCIS, (XIII S.B. 1946; I S.B. 1949; XII Ph.D. 1953)
    1950-1952   Teaching Fellow
    1954-1962   Assistant Professor of Geology
    1962-1966   Associate Professor of Geology
    1966-   Professor of Geology

34. MACDONALD, GORDON JAMES FRASER
    1954-1955   Assistant Professor of Geology
    1955-1958   Associate Professor of Geology
    30 June 1958   Resigned, to become Professor of Geophysics at the
                         University of California, Los Angeles

35. PINSON, WILLIAM HAMET, JR., (XII Ph.D. 1951)
    1950-1951   Research Assistant
    1953-1955   Research Associate
    1956-1960   Assistant Professor of Geology
    1960-   Associate Professor of Geology

II. COURSE XII FACULTY

36. DOELL, RICHARD RAYMAN
    1956-1959   Assistant Professor of Geophysics
    31 August 1959 Resigned, to join staff of U.S. Geological Survey
                    at Menlo Park, California

37. WINCHESTER, JOHN WIDMER, (V Ph.D. 1955)
    1956-1963   Assistant Professor of Geochemistry
    1963-1966   Associate Professor of Geochemistry
    31 December 1966 Resigned, to become Associate Professor in Department of Meteorology, University of Michigan

38. VON ARX, WILLIAM STELLING, (XIX Ph.D. 1955)
    1956-1957   Visiting Lecturer in Oceanography (9 months)
    1957-1959   Associate Professor of Oceanography (part-time)
    1959-1963   Professor of Oceanography (part-time)
    1963-1967   Professor of Physical Oceanography
    1968-1970   Professor of Physical Oceanography (part-time)
    1 January 1971 Resigned to resume full-time duties as a Senior Scientist at Woods Hole Oceanographic Institution

39. BOUCOT, ARTHUR JAMES
    1957-1960   Assistant Professor of Geology
    1960-1961   Associate Professor of Geology
    30 June 1961   Resigned, to become Associate Professor in Division of Geological Sciences, California Institute of Technology

40. HUGHES, HARRY
    15 September 1957 Visiting Lecturer in Geophysics (9 months)
    1958-1964   Assistant Professor of Geophysics
    30 June 1964   Resigned, to return to England to become a Scientist with the U.K. Atomic Energy Authority, Springfield Laboratory, Preston, Lancashire

41. CANTWELL, THOMAS, JR., (X S.B. 1948; S.M. 1949; XII Ph.D. 1960)
    1958-1959   Research Associate in Geophysics
    1960-1964   Assistant Professor of Geophysics
    1964-1965   Lecturer in Geophysics
    30 June 1965   Resigned, to spend full time as President of his company, Geosciences, Inc.

42. BITTER, FRANCIS
    1934-1945   Associate Professor of Metals (Metallurgy)
    1945-1951   Associate Professor of Physics
    1951-1960   Professor of Physics
    1956-1960   Associate Dean, School of Science
    1960-1967   Professor of Geophysics   Died 26 July 1967

43. CARRITT, DAYTON ERNEST
    1960-1962   Associate Professor of Chemical Oceanography
    1962-1968   Professor of Chemical Oceanography
    30 June 1968   Resigned, to become Professor in the Department of Oceanography, Nova University in Florida

44. HOWER, JOHN, JR.
    1960-1961   Assistant Professor of Geochemistry
    30 June 1961   Resigned, to return to Department of Geology, Montana State University

45. HIDE, RAYMOND
    1 September 1960 Visiting Lecturer (5 months)
    1961-1967   Professor of Physics and Geophysics
    30 June 1967   Resigned, to return to England to join staff of National Meteorological Office

FACULTY OF COURSE XII, GEOLOGY AND GEOPHYSICS, 1960
(From left to right)
Front Row: W. H. Dennen, R. D. Parks, P. M. Hurley, R. R. Shrock (Head), M. J. Buerger, H. W. Fairbairn, Ely Mencher, and J. W. Winchester.
2nd Row: W. F. Brace, T. R. Madden, S. M. Simpson, Jr., A. J. Boucot, Harry Hughes, Thomas Cantwell, H. A. Morss, Jr., and W. H. Pinson, Jr.
Insets: W. L. Whitehead (deceased), D. E. Carritt, and Francis Bitter (deceased).   (Photo by M.I.T. Graphic Arts)

FACULTY OF COURSE XII, EARTH AND PLANETARY SCIENCES, 1970
(From left to right)
Front Row: W.H.Pinson,Jr., T.R. Madden, M.J. Buerger, Frank Press (Head), R. R. Shrock, H. W. Fairbairn, K. Aki, and W. F. Brace.
2nd Row: C. C. Counselman, N. M. Toksöz, D. R. Wones, C. I. Wunsch, T.B. McCord, F. A. Frey, H. A. Morss, Jr., and T. R. McGetchin.
3rd Row: R. S. Naylor, C. S. Cox (Visiting Prof.), J. S. Lewis, Shawn Biehler, and J. B. Southard.
Insets: M. G. Simmons, I. I. Shapiro, P. M. Hurley, and W. S. von Arx.
(Photo by M.I.T. Graphic Arts)

II. COURSE XII FACULTY

46. FIOCCO, GIORGIO
    1963-1968   Assistant Professor of Geophysics
    1968   Research Associate
    15 April 1968   Resigned, to return to Italy

47. DEAN, LEE WALLACE, III, (VI-A S.B. 1956; VIII S.M. 1957; Ph.D. 1960)
    1960-1964   Instructor in Physics
    1964-1967   Assistant Professor of Geophysics
    30 June 1967   Resigned, to join staff of Naval Research Laboratory in Washington, D.C.

48. GANGI, ANTHONY FRANK
    1964-1967   Associate Professor of Geophysics
    31 August 1967   Resigned, to become Associate Professor of Geophysics at Texas Agricultural and Mechanical University

49. LUTH, WILLIAM CLAIR
    1964-1968   Assistant Professor of Geochemistry
    30 June 1968   Resigned, to become Associate Professor in Department of Geology, Stanford University

50. TOKSÖZ, M. NAFI
    16 January 1965 Assistant Professor of Geophysics
    1 July 1967   Associate Professor of Geophysics
    1 July 1971   Professor of Geophysics

51. STRANGWAY, DAVID WILLIAM
    1 February 1965 Assistant Professor of Geophysics
    31 August 1968   Resigned, to become Associate Professor in the Department of Physics (Earth and Atmospheric Physics), University of Toronto

52. SIMMONS, GENE
    1 July 1965   Professor of Geophysics

53. PRESS, FRANK
    1 September 1965 Professor of Geophysics
                     Head, Course XII, Department of Geology and Geophysics
    15 March 1969   Head, Course XII, Department of Earth and Planetary Sciences
    1 July 1971   Robert R. Shrock Professor of Earth and Planetary Sciences

54. AKI, KEIITI
    1 July 1966   Professor of Geophysics

55. BIEHLER, SHAWN
    1 July 1966   Assistant Professor of Geophysics
    30 June 1970   Resigned, to join faculty of the Department of Geological Sciences, University of California, Riverside

56. FREY, FREDERICK AUGUST
    1 July 1966   Assistant Professor of Geochemistry
    1 July 1972   Associate Professor of Geochemistry

57. WONES, DAVID ROBERT, (XII S.B. 1954; Ph.D. 1960)
    1954-1957   Teaching Assistant
    1 July 1966   Associate Professor of Geology
    30 June 1971   Resigned, to return to the United States Geological Survey as Chief of the Branch of Experimental Geochemistry and Mineralogy

58. WUNSCH, CARL ISAAC, (VIII S.B. 1962; XII Ph.D. 1967)
    1962-1963   Research Assistant
    1 October 1966   Lecturer in Oceanography
    1 July 1967   Assistant Professor of Oceanography
    1 July 1970   Associate Professor of Oceanography

59. NAYLOR, RICHARD STEVENS, (XII S.B. 1961)
    1 July 1967   Assistant Professor of Geology
    1 July 1973   Associate Professor of Geology
    31 August 1974   Resigned, to accept Chairmanship of Department of Geology at Northeastern University

60. SHAPIRO, IRWIN IRA
    1965 (Spring) Visiting Lecturer
    1 July 1967   Professor of Geophysics and Physics

61. SOUTHARD, JOHN BRELSFORD, (XII S.B. 1960)
    1 July 1967   Assistant Professor of Geology
    1 July 1973   Associate Professor of Geology

62. LEWIS, JOHN SIMPSON
    1 July 1968   Assistant Professor of Geochemistry and Chemistry
    1 July 1972   Associate Professor of Geochemistry and Chemistry

63. MCCORD, THOMAS BARD
    1 July 1968   Assistant Professor of Planetary Physics
    1 July 1971   Associate Professor of Planetary Physics
        Director, G. R. Wallace, Jr. Astrophysical Observatory

64. COUNSELMAN, CHARLES CLAUDE, III, (VI S.B. 1964; S.M. 1965; Ph.D. 1969)
    1965-1968   Teaching Assistant and Research Assistant (XVI)
    1 January 1969   Assistant Professor of Planetary Sciences
    1 July 1974   Associate Professor of Planetary Sciences

65. MCGETCHIN, THOMAS RICHARD
    1 July 1969   Assistant Professor of Geology
    31 December 1973 Resigned, to join the staff of the Los Alamos Laboratories, Los Alamos, New Mexico

66. BURNS, ROGER GEORGE
    1 July 1970   Associate Professor of Geochemistry
    1 July 1972   Professor of Geochemistry

67. EDMOND, JOHN MARMION
    1 July 1970   Assistant Professor of Oceanography

68. GOETZE, CHRISTOPHER
    1969-1971   Research Associate
    1 January 1971   Assistant Professor of Geology

69. PETTENGILL, GORDON HEMENWAY, (VIII S.B. 1948)
    1 January 1971   Professor of Planetary Physics

70. SOLOMON, SEAN CARL, (XII Ph.D. 1971)
    1971-1972   Instructor (part-time)
    1 January 1972   Assistant Professor of Geophysics

II. COURSE XII FACULTY 39

71. DICKEY, JOHN SLOAN, JR.
    1 July 1972   Assistant Professor of Geology

72. SCLATER, JOHN GEORGE
    1 July 1972   Associate Professor of Marine Geophysics

73. ATWATER, TANYA MARIA
    1 January 1973   Assistant Professor of Marine Geology

74. MOLNAR, PETER
    1 January 1973   Assistant Professor of Marine Geology

VISITING AND NAMED PROFESSORS AND LECTURERS

Introduction

As discussed briefly in the introduction to this chapter, it has long been the practice of M.I.T. to have visiting professors, special lecturers, and distinguished scientists and engineers as invited guests for limited periods of time. These distinguished individuals, who typically spent a few weeks to a few months at the Institute, came from colleges and universities near and far, from industrial organizations, from federal bureaus, and from private research institutions. In general they lectured on subjects not regularly offered in the M.I.T. curriculum, hence they greatly enriched the subject matter customarily offered in the standard curriculum.

Visiting Professors and Lecturers

During its seventy-five years of existence, from 1890 to 1965, Course XII had a fair number of visiting professors, special and named lecturers, and earth science colleagues as active participants in its instructional program. As will be seen from the following discussion, some of these persons came for only a few lectures, others spent weeks or months giving a series of lectures, and a few stayed on for years, some continuously and others intermittently as their services were needed. Some came as visiting or exchange professors; some as temporary appointees to fill a position left vacant by a staff member on leave or by one who had resigned; and many, from industry, government, and independent research institutions, as invited lecturers to present reports on subjects of special interest at the time.

For the record, and to emphasize the important contributions these distinguished persons made to the instructional program of Course XII, there follow in roughly chronological order brief comments about who these persons were, why they were invited to come, what subject matter they discussed, and how long they spent in residence, or, if elsewhere, how long they were connected with the Geology Department.

The first need for instructional assistance from the outside came in June 1902 when Niles retired as both professor and department head and Barton retired from active teaching. The Department was left with only three experienced instructors--Associate Professor Crosby, who was given the responsibility of directing the program of Course XII, but was not named Head, and two Instructors, C. H. Warren and F. G. Clapp--and a lone student assistant, H. L. Sherman.

Obviously, arrangements had to be made in a hurry so that the students who would be returning for the fall term would have a full list of subjects from which to choose. To meet this immediate need, the Institute made an arrangement with Harvard whereby interested students were permitted to take subjects offered by five geology professors, but they had to go from Boston to Cambridge to attend the classes. T.A.Jaggar, Jr. gave the M.I.T. students lectures in General Geology, Experimental Geology, and Field Geology; N. S. Shaler offered courses in General Geology and Paleontology; W. M. Davis lectured in Physiography; R. deC. Ward, in Climatology; and J. B. Woodworth, in Glaciology. During the second term, H. W. Shimer came from Columbia University two days each week to give the students lectures in Paleontology and Stratigraphy.

As might be expected, the students who went to Harvard for the lectures offered in the cooperative experiment lost so much time in transit and also encountered such frustrations in arranging their schedules that the arrangement was abandoned at the end of the school year in June 1903. The needs for the following school year, 1903-04, were met by bringing D. W. Johnson and H. W. Shimer into the Department as Instructors in Geology, and arranging to have Jaggar come from Harvard to Boston to give lectures in General Geology.

No attempt was ever again made to have M.I.T. geology undergraduates go to Harvard for special subjects, but the occasional undergraduate was permitted to attend classes there if he was prepared to pay the required fractional tuition and could arrange a satisfactory schedule. On the other hand, graduate students at both M.I.T. and Harvard were permitted and encouraged to attend colloquia, seminars, and regularly scheduled lectures at one another's institutions, this privilege being reciprocally extended without any tuition or other fee. By this cooperative agreement, which is still in force, many graduate students have been able to take courses with distinguished professors at both institutions while being registered at only one.

II. COURSE XII FACULTY

When increasing deafness forced Crosby to give up directing Course XII in 1904 and discontinue lecturing after 1907, there was no one in the Department who could take his place in economic geology, a subdiscipline in which he had gained a national reputation as a consulting geologist. In an attempt partially to overcome the lack of courses in this field, the Head of the Department, Prof. T. A. Jaggar, Jr., who had been induced to come from Harvard permanently in 1902, arranged to have specialists give a five-week series of lectures on economic geology each year.

During SY 1907-08, the special lecturer was Prof. James F. Kemp, the famous economic geologist from Columbia University, who lectured in the advanced subjects given earlier by Crosby; for the following three years, 1909-12, it was Waldemar Lindgren, at the time Geologist, in charge of the sections of Mining Geology and Metal Statistics, of the United States Geological Survey, and widely recognized as the world's leading student of mineral deposits. Lindgren came for five weeks each autumn during which he lectured and conducted conferences on ore deposits and economic geology. According to the M.I.T. President's Report for 1909-10, p. 110,

> "... a very liberal contribution from Mrs. William Barton Rogers ... made it possible to continue the lectures in economic geology by Mr. Waldemar Lindgren."

By 1912, Lindgren had been appointed Chief Geologist of the U. S. Geological Survey, and had so impressed the M.I.T. Administration that when Jaggar resigned from his position as Head of Course XII, Lindgren was asked to succeed him in that position and to accept appointment to the newly established William Barton Rogers Professorship in Geology. That is how Lindgren happened to come to M.I.T. on a permanent basis; an excellent example of how a special lecturer, appointed for a special task and a limited time, can prove to be just the right person at the right moment! And this was not the last time a visiting lecturer later became a permanent member of the geology faculty, as witness the cases of Jones, Whitehead, Madden, Hughes, Hide, and Shapiro.

Although Lindgren's coming on a permanent basis gave M.I.T.'s Department of Geology an eminent position in mineral deposits, there was no one to offer lectures or to supervise research in petroleum geology, which was becoming a rapidly expanding subdivision of economic geology in the early decades of the Twentieth Century. The temporary solution was again a special lecturer, this time Dr. Ralph Arnold, an authority on petroleum. Through a cooperative arrangement between the geology departments of M.I.T. and Harvard, Arnold delivered ten lectures on the "Geology of Petroleum," five at M.I.T. during the period 5-10 April 1915, and five at Harvard the same month. This arrangement seems to have been continued for two or three additional school years because the 1967 M.I.T. ALUMNI REGISTER shows Arnold to have been on the Geology staff from 1915 to 1918. (Also see M.I.T. President's Report for 1914-15, p. 102.)

Another cooperative arrangement between the two geology departments resulted in an exchange of personnel during the spring term of 1917 in which Lindgren gave a series of lectures at Harvard on "Gold-bearing Ore Deposits" and L. C. Graton, as a Special Lecturer, gave a short course at M.I.T. on "Ore Deposits." Earlier in the same school year, during October and November of 1916, Dr. E. S. Bastin of the U.S. Geological Survey had also given a series of lectures on economic geology to M.I.T. students in Courses XII and III.

It has been reported by two of his biographers* that Richard M. Field, late professor at Princeton, after receiving his doctorate in paleontology from Harvard in 1918, spent a short period of time on the M.I.T. staff, but his name is not included in the Faculty and Staff list of the 1967 M.I.T. ALUMNI REGISTER, nor is there any mention of him in the M.I.T. President's Reports or in the annual volumes of The Technology Review.

By 1917, the future importance of petroleum products as fuels, no doubt emphasized by the great demand for them in the World War that was going on, was widely recognized, and M.I.T.'s Department of Geology again went outside for help from an expert. William F. Jones '09, a former student in Course XII, who had gained considerable experience in exploration for petroleum, was appointed a Special Lecturer, and during the spring term of SY 1916-17 he gave an informal course of thirty lectures on "Oil and Coal Deposits." He continued as Special Lecturer for the next three years, presenting essentially the same material under the subject designation "Geology of Coal and Oil," and at the end of SY 1919-20 was appointed Assistant Professor to take the place made vacant by Lahee's resignation a year earlier. Upon assuming responsibility for the subjects formerly given by Lahee, Jones relinquished the petroleum course to another M.I.T. alumnus, Paul M. Paine (III S.B. 1904), a consulting petroleum engineer, who came from Tulsa, Oklahoma during the fall term of 1921 to deliver a

---

* "Richard Montgomery Field," an obituary notice by H. H. Hess in Amer. Geophys. Union, Tr. 43: 1-3, (1962):

"After serving on the geology faculty at Massachusetts Institute of Technology ..." (p. 1), and

"Richard Montgomery Field," an obituary notice by E. C. B. [Edward C. Bailey], in Geol. Soc. London, Pr. No. 1602, Session 1961-62: 154-155, (1962):

"... He then spent short periods on the staff at the Massachusetts Institute of Technology and at Brown University ..." (p. 154).

series of thirty lectures on "Oil and Gas Production" (Tech. Rev. 23, p. 14, 1921).

When in 1920 Courses III and XII were brought together under Lindgren's chairmanship as the Department of Mining, Metallurgy and Geology, the Department not only continued to depend on special lecturers for petroleum geology but also extended this practice to mining and to regional geology. Charles A. Mitke of Bisbee, Arizona, gave twenty lectures on "Mining Methods, Mine Valuation, and Prevention of Fires" during SY 1920-21; Prof. Emmanuel de Margerie came from Europe in SY 1921-22 to deliver twelve lectures on the "Geology of France and adjacent countries;" and T. A. Jaggar, Jr., formerly Head of Course XII, and now Director of the Hawaiian Volcano Observatory, returned for a series of lectures on "Volcanoes and Earthquakes" during February 1923.

Still depending on outside lecturers for help with petroleum geology, Jones having resigned in 1923 to return to the petroleum industry, Lindgren arranged for Prof. Roswell H. Johnson from the University of Pittsburgh to give a series of lectures on "Oil Production" during SY 1923-24, and for F. B. Tough, Chief Petroleum Technologist of the United States Bureau of Mines, to give a series of fifteen lectures on the same subject.

On the completion of his lectures, Tough generously returned $465 of his fees to the Institute to be used as a fund for aiding students in oil technology. A fuller discussion of this Tough Fund is included in the Chapter on The Financing of the Geological Sciences at M.I.T. (Volume 2).

As interest broadened beyond geology and mining, several special lecturers met those new interests in one way or another. Because of a recent shower of meteorites in New England in the autumn of 1925 Dr. G. P. Merrill, Curator of the United States National Museum, gave four lectures in December of that year on "Meteorites" (Tech. Rev. 28, p. 136, 1925-26) under the auspices of the Department of Mining, Metallurgy and Geology, and Dr. Robert B. Sosman (VIII S.B. 1904) gave a series of lectures on "The Earth's Composition." Sosman returned the following April (1926) to deliver ten lectures on "Elastic Waves and the Earth," seemingly the first lecturer from the outside who discussed one of the fundamental aspects of seismology, a subject that would not become of importance in Course XII for another decade when L. B. Slichter would join the faculty and start research on the seismicity of the earth. Only a year would pass, however, until another visiting lecturer would discuss the practical importance and use of seismic and other physical methods in oil exploration. This lecturer was Donald E. Barton, son of Course XII's Prof. George H. Barton (III S.B. 1880), a leading geologist and geophysicist from Houston, who gave twelve lectures on the "Use of geophysical methods in the location of oil" during SY 1927-28.

Another well-known oilman from the Southwest, millionaire geologist Everette L. DeGolyer,* was invited to give a series of lectures on "Minerals" (meaning petroleum, no doubt) in early 1929 (The Tech for 15 March 1929), and earlier in the same year, Dr. C. P. Berkey, the widely known engineering geologist from Columbia, had lectured on "Boulder Dam," a project on which he was one of the consultants at the time. (The Tech for 15 February 1929).

Immediately after the end of World War I, during which it became evident that mechanized warfare and its dependence on petroleum would have to be reckoned with in the future, to say nothing of the rapidly growing automobile industry and the thirst of its machines for the products of petroleum, it was not surprising that exploration for oil took on an accelerated pace. Every aspect of geology that promised to be helpful in the quest for new oil fields was eagerly seized upon and tested. One of these aspects was micropaleontology, the study of minute fossils so small that a microscope was needed to determine their diagnostic properties and refer them to useful categories or taxa.

One of the first North American paleontologists to show the importance of microfossils for correlation was Rufus M. Bagg who, as early as 1895, was describing Cretaceous and Tertiary foraminifers from the middle Atlantic slope, and was reporting them from well cuttings. Numerous American paleontologists took up the study of microfossils during the first two decades of the Twentieth Century, as pointed out by Galloway** and Croneis,** and by the early 1920s, study of microfossils for both scientific and utilitarian purposes was well underway in numerous places-- educational and research institutions, federal and state bureaus, and petroleum companies--and by a considerable number of individuals. One of these latter was Joseph A. Cushman, a 1903 biology graduate and a 1909 doctorate in the same subject from Harvard. After spending some twelve years of largely scientific work on the Foraminifera, followed by a short two-year period as a consulting geologist with several petroleum companies, he severed connections with the latter and henceforth devoted the remainder

---

* Mr. De - A biography of Everette Lee DeGolyer, by Lon Tinkle, with a Foreword by Norman Cousins. Boston: Little, Brown and Co., xix + 393 p., port., (1970).

** For discussions of the early work and workers in micropaleontology, the interested reader is referred to the following articles: Galloway, J. J., "Methods of correlation by means of Foraminifera," Am. Assoc. Petroleum Geol., B. 10: 562-567, (1926); and "The change in ideas about Foraminifera," Jour. Paleontology 2: 216-228, (1928); and Croneis, C., "Micropaleontology, past and future," Am. Assoc. Petroleum Geol., B. 25: 1208-1255, (1941); and "New frontiers in micropaleontology, with especial reference to petroleum exploration," Econ. Geol. 37: 16-38, (1942).

of his life to pure research in the Cushman Laboratory for Foraminiferal Research which he had established in 1923 in nearby Sharon, Massachusetts. Here in a sylvan environment he proceeded to become one of the world's leading authorities on the Foraminifera, both fossil and living.*

When Course XII students expressed an interest in getting practical first-hand training in micropaleontology, it was only natural that Shimer, a general paleontologist, should think immediately of soliciting the aid of his distinguished colleague in nearby Sharon, especially since the M.I.T. Department of Geology had neither the library nor the collections necessary for the study of microfossils, particularly the Foraminifera which was then the taxon under most intensive study. Action followed thought, and Cushman became a Special Lecturer in Micropaleontology for SY 1926-27. As Course XII students subsequently indicated a desire to study with Cushman, he would be reappointed as Special Lecturer so that they could receive academic credit for their work with him. Thus he served M.I.T. during three separate periods, 1926-1928, 1929-1933, and 1938-1939. In this happy and informal arrangement between Cushman and M.I.T., Course XII students travelled to the Laboratory at Sharon and immediately became accepted members of the friendly and hospitable "Laboratory Family." The present author experienced this same friendliness and hospitality in 1940 when he sought Cushman's assistance in revising the Foraminifera for INDEX FOSSILS OF NORTH AMERICA (Shimer and Shrock, 1944).

It is worth noting here that two of the seven women doctorates in Course XII, Katherine W. Carman (1933) and Louise Jordan (1939) did their thesis work under Cushman's supervision, as did one of the women masterates, Frances L. Parker (1930).

In lieu of a permanent appointment in the field of mineral fuels, but in recognition of the growing importance of the petroleum industry, as well as the long-established importance of the coal industry, the Department of Geology continued to rely on an outside Special Lecturer to conduct courses in the general field of fossil fuels. Accordingly, when Jones resigned from his temporary appointment in 1928, Dr. Walter L. Whitehead (III S.B. 1913; XII Ph.D. 1917), a protégé of Lindgren and an independent consulting geologist at the time, was appointed Special Lecturer on the Geology of Coal and Petroleum. For the next fourteen years,

---

* The outstanding achievements of Cushman, his students, and his Laboratory are described in a <u>Memorial Volume</u> published in April 1950 as a Contribution from the <u>Cushman Laboratory</u> for Foraminiferal Research, Sharon, Massachusetts. This 68-page booklet contains photographs of Cushman and his laboratory, and a complete list of his 554 publications. A brief biography of Cushman is also included in a later section of this Chapter titled "They Also Served."

until 1942, Whitehead continued as Special Lecturer, combining his responsibilities for instructional work on coal and petroleum, which required his residence during the term he lectured, with consulting assignments that often required absence from the Institute for a few weeks or months. After 1942, when he was appointed an Assistant Professor, he became a full-time member of the Department of Geology. An extended biography of Whitehead will be found in another part of this history.

The M.I.T. President's Report for 1928-29, p. 18, has the statement that a Mr. V. G. Gabriel, an advanced student, gave a course in "Prospecting by seismic methods," but there seems to be no other information about this significant statement--significant because it marked the first "course" in seismic prospecting offered to M.I.T. students up to that date (1928 or 1929), and pointed up the lack of instruction in geophysics that was alluded to in the M.I.T. President's Report for 1929-30, on p. 66:

> "The problem of adequate instruction in the science of Geophysics still awaits a satisfactory conclusion ..."

After the 1928 appointment of Whitehead as Lecturer, which would continue until World War II, a period of more than twenty years ensued before another Lecturer was appointed, this being Prof. D. J. MacNeil of Nova Scotia's St. Francis Xavier University. This twenty-year period, which represented the last five years of Lindgren's chairmanship of Course XII, and the 15-year incumbency of W. J. Mead in the same position, was a period of uncertainty about the future of geology at M.I.T., with the result that there was little incentive to invite special lecturers from the outside, particularly since several professors had been added to the departmental roster to cover fields previously ignored or partly covered by visiting scientists. The situation in Course XII definitely changed for the better, however, when in 1948 the Administration decided to support a strengthened and expanded geology program at both undergraduate and graduate levels. With all faculty members back full-time after leaves for war work, it was decided to strengthen and diversify the undergraduate program and to develop a series of subjects in geophysics.

One of the requirements of the revised undergraduate program was summer field work in the second or third year. To provide a way for Course XII students to meet this requirement, M.I.T., largely through the imagination and initiative of Prof. W. L. Whitehead, who was the chief undergraduate thesis supervisor in the Department, negotiated a summer school of geology with the Province of Nova Scotia. This school was started in the summer of 1948 under Whitehead's direction, and in 1950 M.I.T. invited

II. COURSE XII FACULTY

Dr. Donald J. MacNeil,* Professor of Geology and Head of the Department of Geology at St. Francis Xavier University at Antigonish, to accept an appointment as Lecturer at the summer school. From that year on until the school closed after the 1961 summer session, he was a valuable member of the resident M.I.T. faculty at the Nova Scotia Centre for Geological Sciences at Crystal Cliffs, some nine miles north of Antigonish, where the school was located. He lectured at the Centre, visited students in their field camps, accompanied them on field trips, helped them adapt to life in the Antigonish community, and contributed in many other ways to the success of the M.I.T. Summer School of Geology at Crystal Cliffs. This cooperative international enterprise is discussed more fully in another chapter of this history, and has also been described in two published articles--Rowlands, John J., "Acadia - Place of Plenty," Tech. Rev. 53: 345-350, (1951), and Shrock, R. R., "Ten Years in Nova Scotia: The Massachusetts Institute of Technology Summer School of Geology, 1948-1957," published by the M.I.T. Department of Geology and Geophysics, Cambridge, Mass., 96 p., (1951).

After Whitehead's retirement in 1957 as Associate Professor of Geology and Director of the M.I.T. Summer School of Geology, responsibility for directing the 1958 School fell to Boucot, who was aided by Parks and Nathaniel McL. Sage (XII S.B. 1941; S.M. 1951; Ph.D. 1953), who had assisted at the Centre during the previous summer and had the title of Lecturer during the 1958 session. During the 1959 session, Boucot was aided by Parks and Charles F. Hickox, Jr. from Colby College, who had earlier worked from the Centre as a Yale doctoral candidate. Like Sage, Hickox was given the title of Lecturer for the 1959 session, and again for the 1960 session, shortened to seven weeks, which he directed by himself, with advice and aid from Whitehead and MacNeil. Dennen directed the next and last of the 14 summer sessions in 1961 and like his predecessors had the aid of Whitehead and MacNeil. A more extended account of the personnel who served at Crystal Cliffs Centre is given in the chapter on Summer Field Camps and Programs (Volume 2), and a brief biography of MacNeil appears later on in this chapter under the heading "They Also Served."

A mounting interest in the use of geochemical methods for mineral exploration led to the appointment in 1953 of Dr. Herbert E. Hawkes, Jr. (XII Ph.D. 1940) as Lecturer in Geology. Hawkes was well-known for his expertise and field experience in the application of the latest geochemical

---

* Inasmuch as MacNeil's services were confined almost exclusively to the Department's activities in Nova Scotia, and were given only during the summer, he is not included as a regular member of the Course XII Faculty; rather, he is included here as a part-time non-resident member who served in an off-campus albeit important role.

methods being used in mineral exploration, having served on the United States Geological Survey for a number of years, and he brought to Course XII students the kind of practical knowledge they would need if they entered the broad field of mineral exploration and evaluation. After four years as Lecturer, Hawkes resigned to become Professor of Geology at the University of California, Berkeley. A brief biographical sketch of Hawkes appears later on in this chapter under the heading "They Also Served."

When, in 1956, it was decided to develop a program in oceanography in Course XII, it was of course necessary to find lecturers from the outside because at that time there were no members of the Course XII staff competent to offer instruction in the field. As related elsewhere in this history, Course XII and Course XIX were authorized to seek assistance from the Woods Hole Oceanographic Institution. Acting in consort, the two Departments were fortunate to interest several of the senior scientists at Woods Hole in an arrangement by which at least one of them would come to M.I.T. every term, as a visiting professor on a part-time basis, while the others stood ready to supervise the research of any students who wished to work at Woods Hole. During the nine years from 1956 to 1965, M.I.T.'s program in oceanography depended heavily on the scientists from Woods Hole who travelled to Cambridge to deliver their lectures. Their terms of service were as follows:

* *William S. Von Arx, General and Physical Oceanography
    Visiting Lecturer   1956-1957
    Associate Professor (part-time)   1957-1959
    Professor (part-time)   1959-1963
    Professor (full-time on M.I.T. faculty)   1963-1967
    Professor (part-time)   1968-1970

* *J. Brackett Hersey, Geophysical Oceanography
    Associate Professor (part-time)   1957-1963
    Professor (part-time)   1963-1968

* *Columbus O'D. Iselin, Introductory Oceanography
    Visiting Lecturer   1959
    Professor (part-time)   1959-1966

* Henry Stommel, Physical Oceanography
    Professor (part-time) in Course XIX   1959-1961
    Professor (full-time) in Course XIX   1963-197-

* Willem Malkus, Physical Oceanography
    Professor (part-time in Course XIX)   1959-1961

* George Veronis, Physical Oceanography
    Associate Professor (part-time in Course XIX)   1961-1964

* *John W. Kanwisher, Biological Oceanography
    Associate Professor (part-time)   1963-1968

---

*   Brief biographies of Hersey, Iselin, and Kanwisher appear farther on in this chapter under the heading "They Also Served." A somewhat more extended biography of Von Arx appears farther on (Biography 38) following the chapter on Biographies of the Course XII Faculty, because he served as a full-time professor for four years (1963-1967).

II. COURSE XII FACULTY

Ferris Webster, Physical Oceanography
    Assistant Professor (part-time) 1966-1969

Claës G. H. Rooth, Physical Oceanography
    Associate Professor (part-time) 1966-1968

By the end of SY 1965-66, Iselin had asked to be relieved of any further call for the lecture series he had given during four previous autumn terms; Hersey had left Woods Hole to accept a position with the Office of Naval Research in Washington; and both Von Arx and Stommel had become full-time professors at M.I.T. Although Kanwisher continued as a Course XII part-time associate professor, he did not come to Cambridge for any scheduled series of lectures. Furthermore, by 1966 several additional oceanographers had been added to the staffs of both Geology and Meteorology, and plans were underway to work out a joint graduate program with the Woods Hole Oceanographic Institution. When this joint program was officially announced in 1968, there was no further reason for having part-time appointments.

During SY 1956-57, Dr. Geoffrey Nicholls, Reader in Mineralogy at England's University of Manchester, was a visiting professor in the Department, coming especially to work with Dennen in the Cabot Spectrographic Laboratory. The next year, 1957-58, Dr. Takeshi Kiyono, Professor of Electronic Engineering in Japan's Kyoto University, came as a Fulbright Professor to spend the school year doing research work with Madden in the field of electromagnetic theory as applied to geophysics.

The geophysics program in Course XII that was just beginning to develop under Slichter before World War II disrupted it did not get started again until the early 1950s. Slichter left during the War and later resigned his appointment to go to the University of Wisconsin. Dr. John N. Adkins, who served as his successor from 1946-1950, left for the Office of Naval Research in Washington. As a consequence, the Department had to recruit Lecturers from outside in order to have competent persons teaching geophysics and at the same time to give us an opportunity to evaluate these visitors as candidates for permanent appointments.

At first, by a kind of bootstrap operation, several of our most promising graduate students were called upon to share in the teaching and research and were designated instructors or lecturers--_e.g._ Madden, Simpson, and Cantwell, all of whom were ultimately appointed assistant professors. In 1957, Dr. Harry Hughes came from Cambridge, England, as a Visiting Lecturer for nine months before being appointed an Assistant Professor, and Dr. Raymond Hide, from the same place, came in 1960 as a Visiting Lecturer for 5 months before being appointed Professor of Geophysics and Physics.

During this critical decade of the 1950s and early 1960s, when the Department was struggling to build a strong staff and program in geophysics, Dr. Norman Haskell* provided wise counsel and guidance as Research Associate (1950-58) and Lecturer (1958-64). Dr. James Savage spent one year (SY 1957-58) in the Department as Lecturer before leaving for a post at the University of British Columbia, and Dr. David Greenewalt* (XII Ph.D. 1960) aided greatly with introductory subjects, serving successively as Research Assistant (1956-58), Instructor (1961-64), and Lecturer (1964-66), before leaving to take a position with the Naval Research Laboratory in Washington, D.C.

Early in 1958 it was decided to try an exchange plan with Columbia University to see if a way could be found by which each of our Departments could enjoy lectures by distinguished professors from the other. To initiate the exchange Prof. M. J. Buerger spent the first week of January 1958 at Columbia, and Prof. Paul M. Kerr came to Cambridge for lectures on 9, 10, and 11 April. Kerr discussed ore deposits in general and gave an evening public lecture on "The Uranium Deposits of the Colorado Plateau." Although it was agreed by both Departments that the visits had been much worthwhile, no further exchanges were arranged for a variety of reasons.

In the mid-1950s, when Course XII was making strenuous efforts to upgrade its curriculum, expand its faculty, and build up its student body, it was decided to invite distinguished geoscientists to come for a few days to confer with faculty members and give lectures in their special fields. During this period the Department heard a wide range of subjects discussed by Dr. Carl Eckhart, physicist, and Dr. Walter Munk, geophysicist, from the University of California at San Diego; Dr. Walter Elsasser, physicist, from Princeton; and Dr. Philip Kuenen, marine geologist, from the University of Groningen.

When administrative duties in Departmental Headquarters increased so much that assistance was needed by the Chairman, Dr. Henry A. Morss, Jr. (VIII Ph.D. 1934) was appointed <u>Lecturer</u> in 1959 to render administrative help. He continued to hold the title until 1964 when it was changed to the more appropriate designation of <u>Administrative Assistant</u>, the title he held until resignation on 30 June 1972. A brief biography of Morss is included in a later section of this chapter under the heading "They Also Served," and his services are also discussed briefly in the chapter on <u>Supporting Staff for Course XII Faculty and Students</u>.

---

* Brief biographies of Haskell and Greenewalt appear farther on in this chapter under the heading "They Also Served."

II. COURSE XII FACULTY

Named Lectureships and Professorships

During the century from M.I.T.'s founding to 1970, one lectureship and three professorships were established in geology to honor individuals. These are designated <u>named</u> appointments and were the following: 1) the William Otis Crosby Lectureship, established in 1961; 2) the William H. Mason Professorship, established in 1868, but with a mysterious history thereafter; 3) the William Barton Rogers Professorship of Economic Geology, established by the M.I.T. Corporation in 1912 specifically for Waldemar Lindgren; and 4) the Robert R. Shrock Professorship in Earth and Planetary Sciences, established in 1970. Following are brief comments on each of these; more extended discussions are included in the chapter on <u>The Financing of the Geological Sciences at M.I.T.</u> in my Volume 2.

The William Otis Crosby Lectureship

The William Otis Crosby Lectureship Fund was established in early 1961 to receive a bequest from the late Irving B. Crosby (XII S.B. 1917) in conformity with an item in his will stipulating that the bequest be used to establish a lectureship in Course XII in memory of his father, William Otis Crosby (VII S.B. 1876), long-time member of Course XII's teaching staff and internationally known consulting geologist.

Irving stipulated in his will that:

> "... the income to be used to engage special scientists of note to give lectures in the field of geology, not including geophysics."

He carefully stipulated geology, and excluded geophysics, for the specific reason that both he and his father had been practical geologists and, as related to the present author, he wished to encourage geology in all of its aspects. This wish has been strictly adhered to in the selection of the geoscientists invited to come as Crosby Lecturers. Following are brief statements about the ten scientists who came during the first decade, 1961-1970.*

The first Crosby Lecturer was Dr. Joseph L. Gillson (XII S.M. 1921; Ph.D. 1923), retired Geologist with E. I. Du Pont de Nemours & Co. in Wilmington, Delaware. A former graduate student and faculty member of Course XII, Gillson spent two school years in the Department, 1961-62 and 1962-63. He divided his time about equally between a series of lectures on

---

* Further details about the Crosby Lecturers are given in the chapter on <u>The Financing of the Geological Sciences at M.I.T.</u> in my planned second volume.

industrial minerals, conferences and informal discussions with students interested in economic geology, and research on the Lindgren Collection of Ore Minerals.

The second Lecturer was Dr. Claire C. Patterson, Research Geochemist of the California Institute of Technology, who spent the spring term of 1964 lecturing and conducting research on lead pollution in water and air, a subject about which he was deeply concerned and on which he was regarded as one of the leading academic authorities.

Three years passed before the third Crosby Lecturer was appointed, and like Patterson he came from Caltech's Department of Geological Sciences. He was Dr. Barclay Kamb, Professor of Crystallography and a distinguished investigator of crystal structure. During the spring term of 1967 he presented ten lectures on crystal physics and the crystallography of ice.

A year later, during the spring term of 1968, the fourth Lecturer to come was Dr. W. Gary Ernst, Associate Professor of Geology on the Los Angeles campus of the University of California. He presented a series of ten lectures on the areal geology of the California Coast Ranges and the problems encountered in studying the metamorphism of the Franciscan rocks.

In the spring term of 1969, three petrologists came as Crosby Lecturers, each for about a week. Dr. David B. Stewart, Geologist with the U.S. Geological Survey in Washington, gave lectures on alkali feldspars and on the volcanic rocks of Maine's Penobscot Bay and talked to several regularly scheduled petrology seminars. Dr. D. H. Green, Petrologist from the Australian National University, spent the week of 4-12 May in the Department, during which he gave three lectures on metamorphism, fractionation of basaltic magmas, and composition of the upper mantle. The next week Dr. Wallace M. Cady, another Geologist from the U.S. Geological Survey, and an authority on the geology of New England, presented several special lectures in which he discussed some of the recent research in this region.

In October of 1969, the late John A. Gower (XII Ph.D. 1955), Associate Professor of Geology at the University of British Columbia, gave four lectures on modern mineral exploration with the following titles: 1) Modern mineral exploration in Western Canada; 2) Some applications of hydrothermal studies to mineral exploration; 3) Porphyries; and 4) Carbonatites.

A month later, in the first week of November, Paul L. Lyons, Chief Geophysicist of the Atlantic-Richfield Company's staff at Tulsa, Oklahoma, presented a series of six lectures on the following subjects:

1) Faulting in the Mid-Continent of the United States; 2) The oil potential of Alaska's North Slope; 3) Ancient continental spreading; 4) Total exploration for minerals and petroleum; 5) Geothermal exploration; and 6) Significance of the Gulf of Mexico in earth history.

A week later, 18-21 November, Dr. Paul K. Sims, geologist with the Minnesota Geological Survey, gave the following lectures on mineral deposits, a field of geology on which W. O. Crosby wrote many articles and in which he was professionally active as a consulting geologist for more than thirty years: 1) Taconite and iron ores of the Mesabi Range; 2) Colorado Mineral Belt; 3) Relation of massive sulfide deposits to Precambrian volcanism; and 4) Copper-nickel deposits in the Precambrian Duluth complex.

During the months of February and March 1972, the tenth Crosby Lecturer came in the person of Dr. E-An Zen, one of the U.S. Geological Survey's most distinguished investigators of the processes involved in rock metamorphism. During his stay he gave a series of lectures and led a workshop on methods of improving the available thermo-chemical data for rock-forming minerals as these data apply to reconstruction of phase diagrams.

In April and May of the same year, Dr. Donald L. Shreve, Professor of Geology and Geophysics, University of California at Los Angeles, gave the following lectures on theoretical geomorphology involving the evolution of landforms: 1) Channel networks; 2) Soil creep; and 3) River profiles.

The William Powell Mason Professorship of Geology

The W. P. Mason Professorship of Geology, first mentioned in the Report of the M.I.T. Treasurer for 1868 is a mystery! It is recorded that the income from the Mason "Legacy of $18,800" was to be used for a <u>Professorship of Geology</u>, and from 1868 to 1883 it seems to have been so used, although no professor was given the title of the professorship nor is there any record of whose salary was paid from the income.

In spite of the clearly stated opinion of the benefactor's son in 1883 that the income from the original bequest should be

> "... applied to the support of the Chair of the Professor of Geology,"

this stipulation somehow got lost in the next sixty years because by 1945, the amount of the original bequest, $18,800, is simply listed in the Treasurer's Report, under <u>Fund for Salaries</u>, as

> "Wm. P. Mason $18,800"

and in the most recent Treasurer's Report (1970), the following is all that remains as a description of the original fund:

> "2420   MASON, William P., 1868, $18,800. Bequest, the income for a professorship." (p. 295)

No mention is made of geology in this latest Report, and there seems to be no record that would indicate if any geology professor was ever designated the William Powell Mason Professor of Geology! Until a record can be found, the seeming disappearance of this first chair of geology at M.I.T. must remain a mystery. A more extended discussion of this first "Chair" and Professorship is given in the chapter on The Financing of the Geological Sciences at M.I.T. in my Volume 2.

The William Barton Rogers Professorship of Economic Geology

According to the President's Report for 1911-1912, in which the appointment of Waldemar Lindgren as Head of the Department of Geology was announced (p. 12):

> "Professor Lindgren [Waldemar Lindgren, Chief Geologist of the United States Geological Survey] being one of the most distinguished members of the National Academy of Sciences and one of the foremost economic geologists in the world, it seemed appropriate to associate his office with the name of William Barton Rogers, the founder of the Institute, who was a pioneer in economic geology in this country and a president of the National Academy of Sciences. For this purpose your Corporation in June last [i.e. June 1912] made provision for the establishment of the William Barton Rogers Professorship of Economic Geology, and Professor Lindgren was appointed as its first occupant."

Not only was Lindgren the first occupant of this professorship; he was also the last because after his retirement in 1933 all efforts to have the professorship reestablished by the M.I.T. Corporation met with failure, presumably because there were no candidates with distinction in economic geology equal to that enjoyed by Lindgren.

The Robert R. Shrock Professorship of Earth and Planetary Sciences

On January 29, 1970, the Chairman of the M.I.T. Corporation, James R. Killian, Jr., made the following announcement:

> "To Members of the Faculty:
>
>   I am pleased to announce that Mr. and Mrs. Cecil H. Green of Dallas, Texas, have made a gift of $1,200,000 to the Massachusetts Institute of Technology for the establishment of two endowed distinguished chairs.

One of the chairs will be called the Robert R. Shrock Professorship of Earth and Planetary Sciences in honor of Dr. Shrock, former department head who will retire as Professor of Geology on June 30. It is the first fully endowed professorship to be established in the department. Professor Shrock, who received A.B., A.M. and Ph.D. degrees from Indiana University, joined the M.I.T. faculty in 1937 and was head of the Department of Geology from 1950 to 1965. He is an authority on sedimentary rocks and invertebrate paleontology."

Early in December 1970 it was announced that Dr. Frank Press had been named the first Robert R. Shrock Professor of Earth and Planetary Sciences to occupy the distinguished chair endowed by the Greens earlier in the year. The appointment of Professor Press, who was then head of Course XII, Department of Earth and Planetary Sciences, commemorated a twenty-year friendship of Mr. Green and Professor Shrock. There is further discussion of this Professorship in the chapter on The Financing of the Geological Sciences at M.I.T. in my Volume 2.

THEY ALSO SERVED

In a preceding part of this chapter reference is made to the practice of appointing locally based geologists as staff members, either on a part-time professorial basis or on a full-time non-professorial basis (i.e. instructor, lecturer, or research associate), in order to have them present lectures and supervise student research in their special fields. This practice started early in the Institute's history, and continues to this day.

Because of the special services rendered M.I.T.'s Department of Geology by nine of these part-time or non-professorial "faculty" members, brief biographical sketches follow in which an attempt is made to point out the nature and value of those services. The individuals involved, in chronological order of first appointment, are Cushman in micropaleontology; MacNeil at the M.I.T. Summer School of Geology in Nova Scotia; Hawkes in Geochemistry; Haskell and Greenewalt in Geophysics; Hersey, Iselin, and Kanwisher in Oceanography; and Morss as Lecturer and Administrative Assistant.

JOSEPH AUGUSTINE CUSHMAN
(1881-1949)

CUSHMAN at Sharon

MIT: 1926-1939

Joseph Augustine Cushman, internationally known for his publications on Foraminifera, and founder of the Cushman Laboratory for Foraminiferal Research in Sharon, Massachusetts, was a Special Lecturer in micropaleontology at three different times, 1926-1928, 1929-1933, and 1938-1939. These were periods when one or more graduate students at M.I.T. wished to prepare themselves for professional careers as micropaleontologists in the petroleum industry. By special arrangement Course XII students were able to go to Sharon, where they heard special lectures on the Foraminifera by Cushman and became familiar with these commercially important microfossils by working in his famous Laboratory under his direct supervision.

Cushman, more than any other American paleontologist, brought the potential use of the Foraminifera to the attention of the petroleum industry and convinced the companies of their practical value in determining stratigraphy. By the time World War I was ended in 1918 he had demonstrated that the smaller Foraminifera could be used to delineate stratigraphic zones and to determine ecologic conditions. Within a decade, the use of the Foraminifera in the petroleum industry became widespread, Cushman established his private Laboratory at Sharon (1923), and students across the country began the serious study of Foraminifera in preparation for careers in micropaleontology.

At M.I.T., the first student to study with Cushman was Frances L. Parker, recently retired as Research Paleontologist at the Scripps Institution of Oceanography, where she established herself as one of the leading North American specialists on Tertiary and Recent Foraminifera. She entered M.I.T. in September 1928 and received a Master of Science degree in geology in 1930, with a thesis on the "Foraminifera of the east coast of South America." Next came the late Louise Jordan, in 1929, to work with Cushman, with whom she did a Master's thesis in 1930-1931 on "Foraminifera from the Pliocene of New Guinea" and a doctoral thesis in 1938-1939 on "A study of the Miocene Foraminifera from Jamaica, the Dominican Republic, and Republics of Panama, Costa Rica, and Haiti." Dr. Jordan had a distinguished career as a teacher and research micropaleontologist in the petroleum industry. She died in Norman, Oklahoma, on 22 November 1966, and a brief biography of her is included in the chapter on Women in Geology at M.I.T. The third and last of the Course XII students to study with Cushman, all women it should be noted, was Katherine Woodley Carman, recently retired after a long and successful career as a consulting micropaleontologist and petroleum geologist. She entered Course XII in 1927 and received a Ph.D. degree in paleontology in 1929, having done her doctoral thesis with Cushman on "The shallow-water Foraminifera in Bermuda." Other Course XII students studied micropaleontology with Cushman at his Sharon Laboratory, but only the three women mentioned in the preceding comments wrote theses under his supervision. Brief biographies of these are included in the chapter on Women in Geology at M.I.T. in my Volume 2.

The reader interested in Dr. Cushman is referred to the "Memorial Volume" of the Cushman Laboratory for Foraminiferal Research, published at

II. COURSE XII FACULTY 57

Sharon in April 1950. This excellent publication contains the following biographical sketches and a bibliography:

Joseph Augustine Cushman, by Ruth Todd (p. 5-16 and a full-page portrait).

History of the Cushman Laboratory, by Alice E. Cushman (p. 17-22).

Joseph A. Cushman - The Teacher, by James A. Waters (p. 23-24).

Dr. Joseph A. Cushman - As one to work with, by Ruth Todd (p. 25-26).

Joseph A. Cushman and the United States Geological Survey, by John B. Reeside, Jr. (p. 27-28).

Joseph A. Cushman and the National Museum, by Waldo L. Schmitt (p. 29-35).

The Cushman Laboratory and Harvard University, by Kirtley F. Mather (p. 36-37).

Joseph A. Cushman: In retrospect, by Percy E. Raymond (p. 38-39), and

Bibliography of Joseph Augustine Cushman, which lists 554 titles of publications and 3 additional articles categorized as "Partially completed and in preparation:" (p. 40-68).

The interested reader will also want to read J. A. Waters' "Joseph Augustine Cushman, 1881-1949," (Am. Assoc. Petroleum Geol., B. 33: 1457-1468, 1949), and L. G. Henbest's "Joseph Augustine Cushman and the contemporary epoch in Micropaleontology" (Geol. Soc. Am., Pr. 1951: 95-102, port., 1952).

It should be mentioned that Prof. Hervey W. Shimer gave the usual paleontological subjects at M.I.T., but not being a specialist in the Foraminifera, whose great practical value he recognized, he arranged to have his interested students get their special training direct from Cushman at his Sharon Laboratory.

DONALD JONATHAN MACNEIL
(1903-1968)

MIT: 1948-1961

Donald Jonathan MacNeil, longtime head of the Department of Geology at Saint Francis Xavier University in Antigonish, Nova Scotia, initiated with M.I.T.'s Prof. Walter L. Whitehead the cooperative venture known as the M.I.T. Summer School of Geology at Crystal Cliffs near Antigonish. From the day the school was conceived, in 1948, to its termination at the end of its 14th session in August 1961, MacNeil gave invaluable service to M.I.T. and our Geology Department in many ways. He led the students on

PROF. MACNEIL pointing out something of geological interest to PROF. WHITEHEAD, as the two of them sit on the guard log of the plank bridge at Crystal Cliffs.

field trips, lectured on the local geology, repeatedly resolved all sorts of neighborhood problems involving the students and local farmers and townspeople, and served on the School's Governing Board. For more than a decade, 1948-1961, he was an active member of the teaching staff at the Nova Scotia Centre for Geological Sciences, where we held our Summer School, serving as Lecturer, Consultant, Part-time Professor, and during some summers without any special title; but he was always available and ready to help us in any way he could.

Two short accounts of MacNeil's association with the M.I.T. Geology Department are given in my "Ten Years in Nova Scotia: The Massachusetts Institute of Technology Summer School of Geology, 1948-1957," (published by the Dept. Geology and Geophysics, M.I.T., Cambridge, Mass., 1957), and in M. Y. Williams' "Donald Jonathan MacNeil, 1903-1968," Roy. Soc. Canada, Pr. 4/7: 86-89, port., (1969). The interested reader will also find a fuller discussion of MacNeil's services in the chapter on <u>Summer Field Camps</u>. Prof. MacNeil died on 21 November 1968 after a long period of failing health.

HERBERT EDWIN HAWKES, JR.

MIT: 1954-1957

HERBERT EDWIN HAWKES, JR.

When in the early 1950s the Department of Geology petitioned the M.I.T. Administration for permission to expand its program in geochemistry, by adding instruction and research in geochemical prospecting for mineral deposits, permission was granted to seek a scientist experienced in the field. The Department turned to one of its own doctorates, Herbert Edwin Hawkes, Jr. (XII Ph.D. 1940) who had established a national reputation in the field while serving as Chief of the Geochemical Exploration Section of the U. S. Geological Survey at the Denver Federal Center. Hawkes joined the M.I.T. Faculty in January 1954 as Lecturer in Geochemistry, and for the next three and one-half years, until he resigned in June 1957, he offered lectures, laboratory work, and research involving the use of chemical techniques in exploring for various kinds of mineral deposits. He attracted a goodly number of Course III and Course XII students, and supervised the thesis investigations of six graduates. He left M.I.T. in 1957 to take a position in the Division of Mineral Technology at the University of California in Berkeley where he ultimately became Professor.

Herbert Edwin Hawkes, Jr. was born in New York City on 11 December 1912. He was the oldest son of Herbert Edwin Hawkes, Professor of Mathematics and longtime Dean of Columbia College (Columbia University), and Nettie M. (Coit) Hawkes. From Deerfield Academy, where he took his college preparatory training, he went to Dartmouth for a Bachelor of Arts degree in geology, awarded in 1934, followed by a fall term at Columbia in 1935.

After a year's experience as Field Party Chief with Hans Lundberg of Toronto, 1936-1937, Hawkes entered M.I.T. in September 1937. At the end of three years of graduate study, and having written a thesis under the supervision of W. J. Mead on the

> "Structural geology of the Plymouth-Rochester area, Vermont,"

he was awarded a Ph.D. degree in Geology in June 1940.

Immediately after receiving his doctorate he joined the staff of the United States Geological Survey and for the next thirteen years he was associated with the scientists responsible for the geophysical and geochemical programs of the Survey. He worked out of Washington in the earlier years (1940-1945), in Washington for the next period (1945-1950), and finally in Denver (1950-1954) until leaving in 1954 to come to M.I.T. At the time he left Denver he was Chief of the Geochemical Exploration Section of the Survey at the Denver Federal Center.

Immediately after his arrival at M.I.T. in January 1954 Hawkes was given the responsibility for organizing instruction and research in practical ore finding. He first offered Mineral Deposits Laboratory (12.41--0-4-0) in SY 1955-56, a subject formerly given by P. M. Hurley, then a year later combined several existing subjects into a single lecture-laboratory subject, 12.42 Ore guides (2-4-2), with one two-hour lecture and one four-hour laboratory per week. The lectures were to serve

> "... as a review of all techniques currently in use in exploration for mineral deposits (excluding fuels) with particular emphasis on the effective integration of these techniques in a commercial exploration program. Geological and mineralogical ore guides, and practical techniques of mineral identification, geophysics, geochemistry, and photogeology will be reviewed."

The laboratory work was designed to give students the opportunity to learn techniques useful in field and laboratory investigations and to interpret and apply their findings in the form of written reports.

Ore Guides was first offered in the second term of SY 1956-57 after Hawkes had conducted an experimental laboratory (12.41) the preceding year. The subject matter interested both undergraduate and graduate students, and during his brief period as Lecturer, 1 January 1954 to 30 June 1957, Hawkes directed the theses of six Course XII degree candidates--Paul W. Richardson (Ph.D. 1955), David H. Anderson (S.M. 1956), Herbert S. Jacobson (S.B. 1954), Charles T. Prewitt (S.B. 1955), Robert B. Clark (S.B. 1956), and Peter D. Engels (S.B. 1956).

Outside the classroom Hawkes conducted a vigorous program of research, and during his three and a half years in the Department wrote half a dozen papers, of which the most important was probably U.S.G.S. Bulletin 1000-F, "Principles of geochemical prospecting," published in 1957 and immediately adopted as a valuable handbook by those exploring for mineral deposits. (See the Bibliography for other publications.)

Although reappointed Lecturer in Geology and Geophysics for SY 1957-58, Hawkes resigned as of 30 June 1957 and accepted a professorial appointment in the Division (later Department) of Mineral Technology at the University of California in Berkeley. After some eight years there, meanwhile advancing to the rank of Professor, he resigned and for a year served as Editor for the Geological Society of America. But this was only interim employment, for within a year he returned to Washington, this time as an independent geological consultant. In 1970 he reported that he and his wife, Evelyn (Vorhees) Hawkes, were living in Washington, and in 1973 word has come that he now lives in Willoughby, Vermont.

Bibliography of Herbert Edwin Hawkes, Jr.

Symbols and abbreviations used in the following references are explained on p. 91-98; in general, abstracts are listed separately as well as with the references to the complete article. This bibliography begins with Hawkes' doctoral thesis (T--1940) and includes all known publications through 1959, a year and a half after he left M.I.T.

T--1940   Structural geology of the Plymouth-Rochester area, Vermont. 11 + 186 + 1 p., pls., maps, diagrs., (1940). (Ph.D. Thesis at M.I.T. in Course XII, June 1940.)

1--1940   (with Fairbairn, H. W.) Petrofabric analysis of dolomite [abst.]. Geol. Soc. Am., B. 51: 1926, (1940).

2--1941   Roots of the Taconic fault in west-central Vermont. Geol. Soc. Am., B. 52/5: 649-666, il., (1941).

3--1941a (and Wheeler, D. P., Jr.) Chromite deposits of the Del Puerto area, California [abst.]. Geol. Soc. Am., B. 52/12/2: 1950, (1941). (See also item 5--1941c, 1942.)

4--1941b (with Fairbairn, H. W.) Dolomite orientation in deformed rocks. Am. J. Sci. 239: 617-632, (1941).

5--1941c (and Wells, F. G., and Wheeler, D. P., Jr.) Chromite and
   1942   quicksilver deposits of the Del Puerto area, Stanislaus County, California. U. S. Geol. Surv., B. 936-D: iv, 70-110 (‡), il., (1942); [abst.], Geol. Soc. Am., B. 52/12/2: 1950, (1941) [See also item 3--1941a above].

6--1945   (with Balsley, J. R., Jr., and others) Aeromagnetic map showing total intensity 1000 feet above the surface of part of the Oswegatchie quadrangle, St. Lawrence County, New York. U. S. Geol. Surv., Geophys. Invs. Prelim. Map 1,... (magnetic data by Herbert Edwin Hawkes, Jr., et al.)..., (1945).

7--1946   (with Balsley, J. R., Jr., and others) Aeromagnetic survey at three levels over Benson Mines, St. Lawrence County, New York. U. S. Geol. Surv., Geophys. Invs. Prelim. Map 2,... (magnetic data and maps by H. E. Hawkes, Jr., et al.), (1946).

8--1946a Olivine from northern California showing perfect cleavages. Am. Mineral. 31/5-6: 276-283, il., (1946).

9--1947   Magnetic exploration for Adirondack iron ore, [N.Y.] [abst.]. Wash. Ac. Sci. J. 37/10: 373-374, (1947).

10--1947a The Geological Survey's geochemical prospecting unit [abst.]. Wash. Ac. Sci. J. 37/10: 375-376, (1947).

11--1947b Research on geochemical prospecting by the Geological Survey [abst.]. Econ. Geol. 42: 414, (1947).

12--1947c (and Hotz, P. E.) Drill-hole correlation as an aid in exploration of magnetic deposits of the Jersey Highlands, New York and New Jersey. U. S. Geol. Surv., B. 955-A: 1-17, il., (1947).

13--1948 Annotated bibliography of papers on geochemical prospecting for ores. U. S. Geol. Surv., Circ. 28: 6 p. (‡), (1948).

14--1949 (and Lakin, H. W.) Vestigial zinc in surface residuum associated with primary zinc ore in East Tennessee. Econ. Geol. 44: 286-295, (1949).

15--1949a Geochemical prospecting for ores: a progress report. Econ. Geol. 44/8: 706-712, (1949); [abst.], Ibid. 44/1: 80-81, (1949).

II. COURSE XII FACULTY

16--1950 Geochemical prospecting for ores [p. 537-555]: Chap. 30 in Trask, P. D., *Applied Sedimentation*, xi, 707 p., il., New York, John Wiley & Sons, (1950).

17--1951 Magnetic exploration for chromite. U. S. Geol. Surv., B. 973-A: 1-21, (1951).

18--1951a Geochemistry, a symposium on the prospector's newest tool-- Pt. 1: Min. Cong. 37/9: 55-61, 84, (1951); Pt. 2: 62-65, (1951). (Contains papers by Hawkes, H. E., Jr. and others, which are cited individually--see next item.)

19--1951b What geochemistry is and what it can do, in Pt. 1 of Hawkes, H. E., Jr., Geochemistry, a symposium on the prospector's newest tool: Min. Cong. J. 37/9: 55-58, il., (1951); reprinted, Precambrian 26/9: 8-9, 30, (1953), South African Min. Eng. J. 64/2, no. 3172: 485, 487, Johannesburg, Nov. 28, 1953.

20--1951c Geochemical prospecting--an aid to locating new mineral deposits. South African Min. Eng. J. 62/2, no. 3072: 765, 767, 769, il., Johannesburg, Dec. 29, 1951.

21--1952 Geochemical prospecting in the Blackbird cobalt district, Idaho [abst.]. Econ. Geol. 47/7: 771-772, (1952); Geol. Soc. Am., B. 63/12/2: 1260, (1952).

22--1953 (and Wedow, H., Jr., and Balsley, J. R., Jr.) Geologic investigation of the Boyertown magnetite deposits in Pennsylvania. U. S. Geol. Surv., B. 995-D, 135-149, il., (1953).

23--1954 (with Balsley, J. R., Jr., and others) Total intensity aeromagnetic and geologic map of Cranberry Lake quadrangle, New York. U. S. Geol. Surv., Geophys. Inv. Map GP 118, with text, (1954).

24--1955 (with Balsley, J. R., Jr., and others) Total intensity aeromagnetic and geologic map of Stark, Childwold, and part of Russell quadrangle, New York. U. S. Geol. Surv., Geophys. Inv. Map GP 117, (1955).

25--1955a (and Bloom, H.) Geologic application of a test for citrate-soluble metals in alluvium. *Science* 122: 77-78, (1955).

26--1956 (and Bloom, H.) Heavy metals in stream sediment used as exploration guides. Min. Eng. 8/11: 1121-1127, il., (1956); A.I.M.E., Tr. 1956, v. 205, (1957); reprinted in part, in Can. Inst. Min. Metal., Comm. Geophysicists, Methods and case histories in mining geophysics, p. 42-44, [1957].

27--1957 Principles of geochemical prospecting. U. S. Geol. Surv., B. 1000-F: 225-355, tabs., (1957).

28--1957a (and Bloom, H., and Riddell, J. E.) Stream sediment analysis discovers two mineral deposits [New Brunswick], in Can. Inst. Min. Metal., Comm. Geophysicists, Methods and case histories in mining geophysics, p. 259-268, il., [1957].

29--1957b Trends in geochemical exploration [p. 86-93], in Snelgrove, A. K., ed., Geological exploration--Institute for Lake Superior Geology, 109 p., with discussion, Houghton, Mich. Coll. Min. and Tech. Press, (1957).

30--1958 (with Anderson, D. H.) Relative mobility of the common elements in weathering of some schist and granite areas [New England]. Geochim. Cosmochim. Acta 14/3: 204-210, il., (1958).

31--1958a (and Richardson, P. W.) Adsorption of copper on quartz. Geochim. Cosmochim. Acta 15/1-2: 6-9, il., (1958).

32--1959   (with Salmon, M. L.) Fluorescent X-ray spectrographic analysis in geochemical prospecting [abst.]. Min. Eng. 11/1: 40, (1959).

33--1959a   (with Cantwell, T., and Rasmussen, N. C.) Nuclear detector for beryllium minerals. Min. Eng. 11/9: 938-940, il., (1959).

34--1959b   Geochemical prospecting [p. 62-78], in Abelson, P. H., ed., Researches in Geochemistry, 511 p., New York, John Wiley & Sons, (1959).

Translations

1--1946   [From Russian] The development and applications of airborne magnetometers in the U.S.S.R., by Logachev, A. A. Geophysics 11: 135-147, (1946).

2--1948   (with Sokoloff, V. P.) [Translations of six Russian papers on geochemical prospecting for ores.] U. S. Geol. Surv., Open File Rept., released 28 December 1948.

NOTE:

Publications after 1959 will be found in the annual Bibliographies published by the U. S. Geological Survey.

NORMAN ABRAHAM HASKELL
(1905-1970)

NORMAN ABRAHAM HASKELL

MIT: 1950-1964

Norman Abraham Haskell was for fifteen years a great help to the Department during a critical period when we were trying to develop a program of instruction and research in geophysics. First as Research Associate (1950-1958) and then as Lecturer (1958-1964), he gave the Department about a day a week while a full-time Physicist with the Air Force Cambridge Research Laboratory. Broadly trained in tectonophysics, seismology, geophysical exploration, and mechanics of deformation of granitic rocks, he brought to our developing program the fundamental knowledge and practical experience needed to organize and guide the geophysical program that was then evolving. He gave valuable advice on curriculum development and laboratory equipment, acted as thesis advisor when appropriate, participated actively in a weekly seminar for geophysics graduate students, and gave valuable counsel to the Geophysical Analysis Group in its early stages. Being a specialist on seismic wave propaga-

II. COURSE XII FACULTY

tion, he provided graduate students excellent counsel in this phase of geophysics.

Looking back on the decade from 1950 to 1960, when we had no regular professor of geophysics and only one or two young scientists starting academic careers in the field, it is surprising that the students and staff members of that period have done so well; that they have can be credited in part to the wise and helpful counsel given by Haskell and others like him.

Norman Abraham Haskell was born in Alton, Illinois on 30 June 1905. After completing preparatory education at the St. Louis (Missouri) Country Day School in 1923, he entered Harvard and was awarded a Bachelor of Science degree, magna cum laude, in Mining Engineering in 1927. He followed this with graduate work at Harvard and M.I.T., and received from Harvard two graduate degrees in geology, an M.A. in 1935 and a Ph.D. in 1936. Immediately thereafter he pursued a career in seismic exploration geophysics until 1941 when he left commercial work and joined the research group of the National Defense Research Committee (NDRC) to do war research on underwater ballistics. With war's end in 1946, he spent the next two years directing a special research laboratory for U.S. Smelting and Refining Company in Boston, then joined the Air Force Cambridge Research Laboratory in September and continued as a Geophysical Research Scientist in the Terrestrial Sciences Laboratory from then until retirement in 1968. It was during his long period of employment at AFCRL that he was willing to devote about a day a week to helping us develop the geophysical part of our overall geosciences program, as described in the foregoing paragraphs.

After retirement, Dr. Haskell moved from Cambridge to Cape Cod, where he died at Cummaquid on 11 April 1970.

DAVID GREENEWALT

DAVID GREENEWALT

MIT: 1961-1966

With a bachelor's degree in physics (1953) from Williams College, Greenewalt entered M.I.T. in the fall of 1953, wrote a doctoral thesis on "The Origin of Remanent Magnetism in Sedimentary Rocks," and received a Ph.D. degree in geophysics in 1960. A year later he was appointed an Instructor in Geophysics and from 1961 to 1964 he was responsible for instruction in the geophysical laboratory and assisted with lectures in the introductory course in geophysics. From 1964 to 1966 he served as Lecturer, continuing to assist with lecture and laboratory work in beginning geophysics, where his help was much appreciated by the students. In July 1966 he left M.I.T. to accept a position as research geophysicist with the U.S. Naval Research Laboratory in Washington, D.C. where he became involved in magnetic field measurements close to the ocean bottom.

JOHN BRACKETT HERSEY

JOHN BRACKETT HERSEY

MIT: 1957-1968

As mentioned in several other places in this history, we in Courses XII and XIX turned to the Woods Hole Oceanographic Institution for assistance when it was decided to initiate a program of instruction and research in oceanography at M.I.T. It was obvious that we should seek help from some of W.H.O.I.'s leading scientists - Hersey, Iselin, and Von Arx for Geology; Malkus and Stommel for Meteorology; and this is what was done.

In the spring term of 1957, with rising student interest in oceanography, Dr. J. B. Hersey, better known as "Brackett," was invited to be a Visiting Lecturer in Oceanography for the three months of April, May and June, with the understanding that he would come up to Cambridge twice a week to give lectures on the geophysical exploration of the ocean bottom.

Trained in physics at Princeton (A.B. 1934; A.M. 1935) and Lehigh (Ph.D. 1943), and experienced in geophysical field work with the U.S. Coast and Geodetic Survey (1935-1936), Phillips Petroleum Co. (1936-1939), Naval Ordnance Laboratory (1941-1944), and Woods Hole Oceanographic Institution (since 1946), Hersey brought to his M.I.T. students a wealth of professional experience backed by an impressive list of published articles, mostly dealing with seismic exploration of the suboceanic part of the earth's crust.

Hersey's lectures provided some of the stimulus needed to get our program in oceanography under way, and our two Departments were authorized to offer part-time appointments to Hersey, Iselin, and Von Arx in Geology, and to Malkus and Stommel in Meteorology. Accordingly, Hersey was appointed an Associate Professor (part-time) in July 1957, and advanced to Professor (part-time) in 1963, an appointment he held until 30 June 1968 when he resigned to resume full-time duties as a Senior Scientist at Woods Hole.

During his eleven years as a part-time professor, Hersey gave regularly scheduled lectures at M.I.T. twice a week during four different school terms, and supervised the thesis work of our M.I.T. graduate students on a more or less full-time basis once they were far enough advanced to start independent research. Altogether, he supervised the thesis research of 9 students--3 S.B.s, 4 S.M.s, and 2 Ph.D.s, and helped many others with their work at sea. Without his devoted assistance, it would have been impossible to develop the oceanographic program of Course XII that led ultimately to the joint M.I.T.-W.H.O.I. Graduate Degree Program established in 1968. In addition to his valuable participation in our instructional program, both at M.I.T. and at sea, Hersey provided financial aid to a number of M.I.T. students, in the form of research assistantships, help that made it possible for several sorely pressed students to meet the extra costs imposed on them by having to commute between Boston and Woods Hole.

II. COURSE XII FACULTY 65

Hersey can rightfully feel that he played a major role in starting and then developing the oceanographic part of the earth sciences' present program of instruction and research at M.I.T.

Dr. J. B. Hersey (right) holds a piece of rock dredged from one of the steep walls of the Puerto Rico Trench.
(Photo by Jan Hahn)

COLUMBUS O'DONNELL ISELIN
(1904-1971)

MIT: 1958-1966

More than any other person, Columbus O'D. Iselin convinced me that M.I.T. should initiate a program in oceanography after World War II ended. His counsel guided us in deciding which aspects of the broad field the Geology Department should attempt to cover, and his aid was invaluable when we sought instructors in physical, chemical and geophysical oceanography. Columbus first came as a Visiting Lecturer in the spring of 1959; thereafter, as Professor of Oceanography (part-time), from 1959 to 1966, he came every other spring, in 1961, 1963 and 1965 before resigning in 1966. Twice weekly during the term he came from his home on Martha's Vineyard to Cambridge, going first to Harvard for an early lecture, then to M.I.T. to present the same lecture before lunch. The weather had to be unimaginably bad to keep Columbus from making the trip. In his M.I.T. lectures he presented oceanography

COLUMBUS O'DONNELL ISELIN

in such an interesting and stimulating way that students were soon seeking his advice and assistance. Although he actually supervised only two theses at M.I.T., both for Master's degrees in oceanography, he was a kind and helpful counselor to many M.I.T. students who worked at Woods Hole. It was behind the scenes, however, where he quietly but firmly supported the M.I.T.-W.H.O.I. educational experiment, that he helped us most, because he strongly held that research institutions like W.H.O.I. should be closely linked with academic institutions like M.I.T. and Harvard.

Columbus O'Donnell Iselin* was one of America's giants in oceanography. Like his mentor, Henry Bryant Bigelow, longtime professor of biology at Harvard and Founder of the Woods Hole Oceanographic Institution, Iselin was educated at Harvard and was closely tied to the Woods Hole Oceanographic Institution throughout his entire scientific career from the Institution's opening in 1930 to his death in early 1971. His distinguished career and the recognition he received are fully described in the references cited in the accompanying footnote and need not be discussed here. Rather, it is more appropriate to emphasize the important role he played in helping us initiate the oceanographic part of Course XII's current program of instruction and research in the Earth Sciences. He was an outstanding teacher, and I can not describe him better than by repeating below what I wrote about him shortly after his death (Oceanus XVI/2: 34, June 1971).

I first became well acquainted with Columbus Iselin soon after the end of World War II when I became a member of the Panel on Oceanography of the Research and Development Board, of which Columbus was also a member. Many times at meetings during our period of service (1946-1952) we discussed how academic institutions could best go about training students

---

* As yet (1974) no extended biography of Iselin has been published, but much information about him appears in a number of articles and resolutions which are listed below:

1) [Oceanographer Columbus Iselin] Ocean Frontier [An essay on the oceans, and on Iselin's contributions to oceanography], Time LXXIV/1: 44-54, il., and Cover including picture of Columbus, July 6, 1959. (Also see Time, July 16, 1928.)

2) Hahn, Jan. Columbus O'Donnell Iselin [A brief biographical sketch printed for the occasion on which The Henry Bryant Bigelow Medal was awarded to Iselin, by the W.H.O.I. Board of Trustees on 22 June 1966. 9 p., port., 1966.]

3) Fye, P. M. Commemorative Resolution in Memory of Columbus O'Donnell Iselin (delivered by Fye at the Winter Meeting of the W.H.O.I. Corporation in Boston on 20 January 1971). Printed in Woods Hole Notes 3/1: 3, (Feb. 1971), and in W.H.O.I.'s 1970 Annual Report, p. 10-11, [1971].

4) Oceanus XVI/2: 1-50, June 1971. (C. O'D. I. Issue, containing brief comments by more than 50 friends and colleagues of Iselin.) Published by the Woods Hole Oceanographic Institution, Woods Hole, Massachusetts.

5) Carruthers, J. N. The late Dr. Columbus O'Donnell Iselin (1904-1971). Cahiers Oceanographique 23: 494-496, (1971).

6) Fye, P. M. Ocean Policy and Scientific Freedom (Columbus O'Donnell Iselin Memorial Lecture, delivered at the meeting of the Marine Technology Society at Woods Hole on 11 September 1972). A 19-page pamphlet containing biographical information about Iselin and printed for the meeting of the Society.

II. COURSE XII FACULTY

> *Go forth and be like Columbus*
>
> COLUMBUS O'DONNEL ISELIN was an ususual man in many respects, and some of his finest personal traits became evident when he assumed the role of a teacher. So naturally and effectively did he teach, that he was as likely to leave an impact on an informal group of coffee drinkers, or a crew aboard ship, as he was to stimulate the members of a formal class. He taught as much by example as by speech, and his stories of the sea convinced his listeners that he loved that realm of nature above all others.
>
> The oceanographic part of M.I.T.'s present program in the earth sciences could not have started and then developed the way it did without the help of that incomparable quintet from Woods Hole— Iselin, von Arx, Hersey, Stommel, and Malkus. And of these, Columbus stood first, because he came early to help us get instruction started, stayed on to help guide our academic program, and continued even after retirement to provide a subtle but definite influence that motivated more than one student to do a better job than he would have done otherwise.
>
> It was one of the strengths of Columbus that he never needed to raise the decibel level of his voice in order to be heard or understood; so gently, yet so firmly and logically, did he present an idea or express an opinion that it was rare to hear disagreement with him. Although courteous and considerate, at times almost to a fault, he never left his audience in doubt as to what he thought or where he stood on an issue, but this was done with a disarming gentleness and humility that belied the firm conviction that he had reached only after long and careful thought.
>
> His was not the rapid and facile rhetoric of the glib lecturer; rather, he seemed to carve his phrases out of a complex of thoughts, but they came out loud and clear. His ideas, his theories and hypotheses, his explanations, all could stand alone by their logic and clarity.
>
> Thus he taught an informative and convincing course, and conveyed to his students a strong sense of enthusiasm for the sea and devotion to the profession of oceanography. Perhaps the best indication of how deeply he felt his responsibility for meeting his classes was the fact that only the foulest of weather could keep him from making the long trip from Martha's Vineyard to Cambridge.
>
> Being deeply sensitive, and having great compassion and respect for people, it was natural that Columbus was so often sought out for assistance and counsel, both of which he always gave cheerfully and in good measure. And perhaps the greatest single impact he made on his students and peers was the desire generated in them "to go forth and be like Columbus."
>
> Cambridge, Mass.
> Robert R. Shrock

for professional careers in oceanography, and from these discussions between ourselves and with other interested panel members came ultimately a paper* that outlined what we considered a desirable curriculum. Although Columbus did not participate in the writing of the paper, his ideas were incorporated in the published version.

For one thing, Columbus strongly held that oceanographic research institutions like Woods Hole and Scripps could not attain their full potential unless they were closely coupled with one or more academic institutions that were strong in mathematics and the physical and biological sciences. He felt equally strongly that if an academic institution wished to develop a strong program in oceanography it should either have a substantial seaside laboratory and ships for research at sea or should be closely coupled with an independent research institution like W.H.O.I.

As a consequence of discussions with members of the R.D.B. Panel, and particularly with Columbus and some of his colleagues at W.H.O.I., when I was appointed Head of Course XII in 1950 I decided to initiate a program of instruction and research in oceanography at M.I.T. as soon as possible. (The history of this effort is discussed in the chapter on Oceanography at M.I.T. in my Volume 2.)

---

* Knudsen, V. O., Redfield, A. C., Revelle, R., and Shrock, R. R., Education and Training for Oceanographers, <u>Science</u> 111: 700-703, (1950).

Columbus had been coming up to Harvard for many years as a part-time professor, to give lectures in oceanography, and I knew from his extensive knowledge of the sea, and from his publications on oceanography,* that he would be an ideal candidate to offer an introductory course in oceanography to M.I.T. students. Accordingly, he was appointed a Visiting Lecturer for the spring term of 1959 and was asked to give biweekly lectures on general oceanography. Meanwhile the M.I.T. Administration authorized Courses XII and XIX to develop a cooperative program in oceanography and to offer professional appointments on a part-time basis to five W.H.O.I. scientists: Iselin, Von Arx and Hersey in Geology; Stommel and Malkus in Meteorology. From June 1959 to June 1966 Columbus was Professor of Oceanography (part-time) and delivered his biweekly lectures every other spring term, coming in 1961, 1963 and 1965. During his seven-year appointment he was always ready to help both students and staff members with their problems, and we could always count on his support and wise counsel. He was a tower of strength in those early years as our program began to develop. The interested reader will find further discussion of Columbus' M.I.T. activities in the chapter on <u>Oceanography at M.I.T.</u> (See Volume 2.)

To my knowledge no comprehensive bibliography of Columbus' writings has yet been compiled, but most of his more important publications (40) are listed in several of the articles cited in a preceding footnote. I understand that his scientific papers are being carefully preserved at the Woods Hole Oceanographic Institution.

A Woods Hole Scene, by Joan T. Kanwisher

---

* One of his most interesting and instructive articles is "The Development of Oceanography" which he presented at a meeting of The Newcomen Society in North America at Falmouth on 6 June 1957. This lecture was later published as: Matthew Fontaine Maury (1806-1873) "Pathfinder of the Seas," the development of oceanography. Newcomen Publ. in North America, Princeton Univ. Press, p. 8-26, (1957).

II. COURSE XII FACULTY

JOHN WILLIS KANWISHER

MIT: 1962-1968

JOHN WILLIS KANWISHER

As Course XII's program in oceanography got under way in the early 1960s, a need arose for at least some introductory lectures in biological oceanography. To meet this need, M.I.T. turned to the staff of the Woods Hole Oceanographic Institution, and appointed Dr. John W. Kanwisher Associate Professor of Oceanography (part-time) on 1 July 1962. For the next six years, 1962-1968, Kanwisher made himself available to M.I.T. students interested in learning about the biology of the oceans, and came to Cambridge during the fall term of SY 1963-64 to deliver biweekly lectures on "Life in the Sea." Although his appointment was discontinued shortly after formal inauguration of the Joint M.I.T.-W.H.O.I. Graduate Program in 1968, Kanwisher has continued his assistance and again came to Cambridge during January 1974 to participate in M.I.T.'s Independent Study Project.

HENRY ADAMS MORSS, JR.

MIT: 1959-1972

HENRY ADAMS MORSS, JR.

In the fall of 1959, when Course XII was just beginning to develop a program in oceanography, Dr. Henry ["Harry"] A. Morss, Jr. expressed a desire to participate and assist in developing the program and was appointed Lecturer with the thought that he might offer an introductory course in oceanography. He was a longtime lover of the sea, an experienced sailor, and both a scientist and a businessman. Course XII got itself an unexpectedly versatile and valuable member, who proved so knowledgeable and helpful in other ways that he never got an opportunity to deliver any lectures on oceanography. Rather, his talents were immediately put to use in a variety of ways and he quickly became an indispensable member of the Geology Department.

Harry was born in Boston on 1 February 1911, the son of Henry Adams and Edith (Sherman) Morss. He received his first college degree from Harvard in 1932, an A.B. in physics accompanied by honorary appointment to M.I.T. as a Traveling Fellow in Physics for the next academic year. Four years later, in December 1936, he received a Ph.D. degree in physics from M.I.T., and immediately joined the managerial staff of the Simplex Wire & Cable Company in Cambridge. He became factory manager, vice president, and board director in 1939 and continued in these positions until he resigned in 1958 to pursue other interests, which included a longtime interest in sailing boats and in oceanography.

Soon after he joined Course XII as Lecturer, Harry took charge of COMPASS (Committee on Planetary and Space Science), an interdepartmental group organized to develop a program of instruction and research in planetary and space science. Acting as secretary of the Committee, he organized symposia, scheduled lectures, and in other ways kept the group of interested scientists actively involved in what ultimately developed into the "planetary sciences" part of the current Course XII academic program.

When the Green gift of five million dollars for a new earth sciences building was announced in 1959, Harry was soon deeply involved in the planning of space and facilities, and then became our indispensable agent as construction proceeded. By the time the building was completed, and ready for occupancy and dedication, Harry's title had been changed to Administrative Assistant, a much more appropriate designation of his actual work, for by this time (July 1964) he had added to his duties as general trouble-shooter for the new building the further responsibility for preparing and monitoring the departmental budget, which was expanding exponentially. Add to these responsibilities an increasing involvement in Course XII's growing oceanographic program, not to mention the ticklish task of interviewing and recruiting new personnel for an ever-growing secretarial staff, and it is clear that he had his hands full. His scientific training in mathematics, physics and chemistry proved of great value, not only in following construction of the Green Building but also in monitoring the many proposals going to Washington for research funds, and the managerial experience he had gained at Simplex Cable was a valuable asset in trying to mesh secretaries, professors, job requirements and salaries into a workable combination, negotiate contracts and grants of research monies, administer the funds when received, and keep track of materials and personnel. He was even called on to assist in labor negotiations. When he decided to resign on 30 June 1972, he could look back on thirteen years of service with the assurance from his colleagues that he had played an important role* in developing Course XII into a better and bigger Department.

NOTE:

JAMES CRAMPTON SAVAGE (VIII S.B. 1952) was a Lecturer during SY 1957-58, but gave no regularly scheduled subject; rather, he assisted whenever and wherever needed in the overall program in geophysics. (See Departmental Photo on page 32.)

----

* Harry and his position as Administrative Assistant were featured in Mosaic 1/4: 24, (1970), an information quarterly published by the National Science Foundation, in a short piece titled "The Scientific Community--I. A gallery of some members. Henry A. Morss, Jr."

III. STAFF SUPPORTING COURSE XII FACULTY AND STUDENTS

INTRODUCTION

The faculty of any leading earth sciences department of the post-World War II period had to have a supporting staff of considerable size and diversity to conduct its instructional and research programs. This staff consisted of teaching and research personnel--<u>instructors</u>, both part-time and full-time; <u>teaching fellows</u>, and <u>teaching</u> and <u>research assistants</u>; <u>technical</u> and <u>administrative assistants</u>; departmental <u>librarians</u>; <u>machinists</u> and <u>instrument makers</u>; a <u>secretarial staff</u>; and more recently a special group of post-doctoral scientists called <u>research associates</u>.

The staff supporting the faculty and students of Course XII included people in all of the categories mentioned in the preceding paragraph, and it became larger and more diversified as the years passed, as the financial support from the outside increased, as the faculty and student body grew in size and range of interest, and as the Department's academic program broadened and became much more diversified (e.g. with the addition of instruction and research in oceanography and planetary sciences). Let us then consider the different kinds of supporting staff--what did they do? who were they? and when did they serve?

INSTRUCTORS AND TEACHING, RESEARCH, TECHNICAL, AND ADMINISTRATIVE ASSISTANTS IN COURSE XII, GEOLOGY, etc. 1865-1970 (inclusive)

Introduction

From the very beginning of instruction in geology and allied sciences at M.I.T., graduate students have constituted an important component of the teaching staff. Until World War II, most of the work that they did had a direct connection with teaching, and they were designated <u>Instructors</u>, <u>Teaching Fellows</u> or <u>Teaching Assistants</u>, or simply <u>Assistants</u>. Instructors usually were responsible only for lectures and for organization and supervision of laboratory work. Assistants, on the other hand, performed a wide variety of duties, often time-consuming and perhaps a bit boring, but nevertheless important, such as preparing materials for lectures and laboratory exercises; returning these materials to their proper storage cubicles after use; operating projectors during lectures; assisting actively in the laboratory while the students were studying specimens or preparing maps; and grading laboratory exercises, examination papers, and term reports. They, along with the instructors, also helped

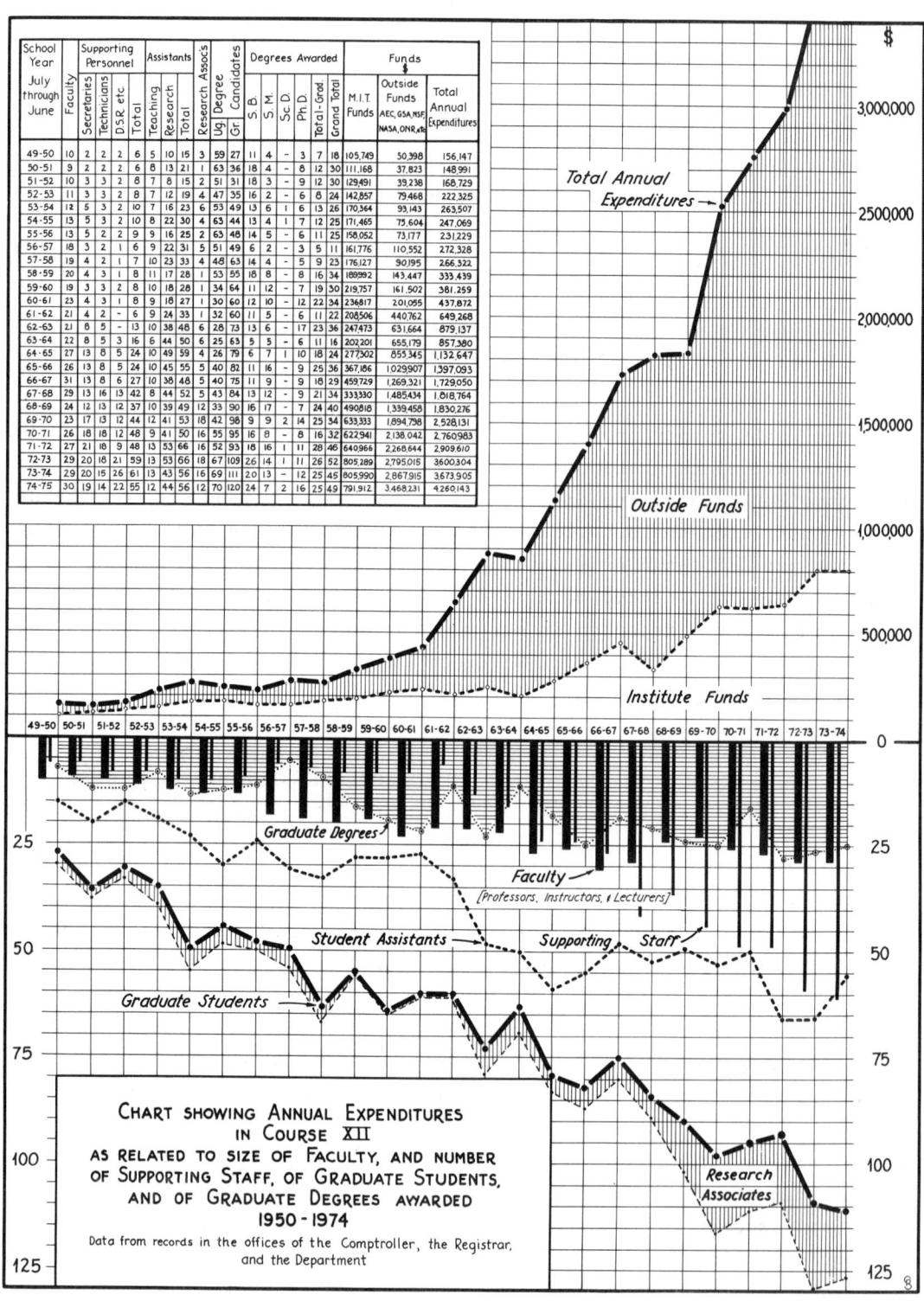

III. COURSE XII SUPPORTING STAFF

with field trips on regular school days as well as on weekends. From time to time they were also assigned curatorial tasks such as cleaning out drawers, replacing worn-out cardboard trays with new ones, preparing new labels for specimens, bringing specimen catalogues up to date, preparing special exhibits for hall exhibition cases, preparing materials for thin sectioning and chemical analysis, and many other similar tasks that were necessary to keep teaching and research collections of minerals, rocks, ores, fossils, and maps in good condition for student use.

Immediately after the end of World War II, a new kind of assistant came on the scene--the Research Fellow or Research Assistant. This was a graduate student who was paid to spend most of his time working on some research problem or program in a laboratory supervised by one of the senior professors in the Department. These assistants grew in number as Federal funds for support of research grew in amount, with the result that research assistants, particularly, soon outnumbered teaching assistants (See Chart on the opposite page). This imbalance was due to the fact that the research assistantship was a much more attractive appointment than the teaching assistantship, partly because of the difference in financial return (the former paid more), and partly because the research assistant could make progress on his thesis while earning his expenses, being permitted to use the results of his research work for his thesis in many cases, whereas the teaching assistant could only gain experience for a future career in teaching. While the latter could make about as rapid academic progress as the research assistant, so far as course work and general degree requirements went, he could not make progress on his thesis problem while teaching. As a consequence, students often requested an appointment as research assistant after they had served a year or two at teaching, and this request was generally granted if funds were available and the student was in good standing academically.

Staff List - Instructors and Assistants

The following list contains the names of former graduate students and certain other persons who served the Department of Geology (Course XII) as instructors, or as assistants of one kind or another (as well as both), during the years indicated. Included are the names of former Instructors (A single + indicates persons, and years served, who held a part-time or full-time instructorship for a short time before leaving the Institute for another position; a double ++ indicates persons who, after holding an instructorship for a period of time, advanced to an assistant professorship and thus became a faculty member), Teaching Fellows (designation used for only a few years), Teaching Assistants, Technical Assistants, Research Assistants, and Administrative Assistants. The data for this list were compiled from the 1967 M.I.T. ALUMNI REGISTER and from records in the Institute Archives and Course XII departmental files.

Abbot, Argyle C.	1928-1929	Chapman, Jamie C.		1961-19(
Adams, Arthur K.	1905-1906	Chintakrindi, Sankara R.		1964-19(_
Adey, Walter H.	1956-1959	Chipman, Ralph O.		1967-1969
Ahrenholz, H. William	1938-1939	Chung, Andrew C.		1966-1970
+Aldrich, Henry R.	+1920-1921	Cid-Dresdner, Hilda		1964
Aldridge, Keith D.	1963-1966	Cisowski, Stanley M.		1969-1970
Allan, John A.	1908-1911	Claerbout, Jon F.		1960-1965
Allan, John D.	1941-1942	+Clapp, Charles H.	1907-1908,+1908-1910	
Allsopp, Hugh L.	1964	+Clapp, Frederick G.	1900-1901,+1901-1904	
Anderson, David H.	1954-1955	Cloke, Paul E.	1951-1952,1954	
Anderson, Philip J.	1954-1959	+Cobb, Collier		+1890-1892
Arkani-Hamed, Jafargholi	1965-1969	Cohen, Elmer		1930-1931
Azároff, Leonid V.	1951-1953	Cole, James C.		1969-1970
		Combs, James B.		1968-1970
Backus, Milo M.	1952-1955	Copeland, Richard A.		1964-1970
+Barry, John G.	+1916-1920	Cormier, Randall F.		1952-1956
++Barton, George H.	1883-1886,++1886-1896	Coussa, Michael R.		1969-1971
Bateman, Sylvia (Technical)	1960-1965	Coyle, John P.		1948-1949
+Beall, George H.	1958-1960,+1960-1962	Crockett, James H.		1959-1961
Beamish, Peter C.	1962-1964	++Crosby, William O.	1875-1880,++1880-1883	
Beardsley, George F.	1964-1966	+Currier, Louis W.		+1920-1921
Beasley, Fannie M.	1950-1951			
Beaton, Neil S.	1932-1935	Dakin, Frances		1966-1968
Bediz, Peter	1942-1943	++Dean, Lee W., III		++1960-1964
Beger, Richard M.	1964-1968	Delnore, Victor E., Jr.		1967-1968
Beiser, Erna	1962-1964	++Dennen, William H.	1946-1948,++1949-1952	
Bell, Christopher	1948-1950	Dennen, William L.		1919-1920
Bence, Alfred E.	1964-1966	de Steiguer, William G.		1906-1907
+Benedict, Platt C.	+1921-1923	Devorkin, Donald B.		1957-1958
Benson, David G.	1954-1956	DeWolf, John B.		1964-1969
Benttinen, Theodore H., Jr.	1967-1970	Dick, Sheldon L.		1955-1956
Berrian, David B.	1967-1968	Dingler, John R.		1966-1968
Binkley, Christopher H.	1968-1969	Dodge, Francis B.		1912-1914
Binkley, Patricia (Technical)	1952-1953	Dollase, Wayne A.		1961-1966
Bishop, William M.	1960-1961	Dorris, James E.		1936-1937
Blackburn, William H.	1964-1967	+Doten, Robert K.	1929-1931,+1931-1932	
Bless, Stephan J.	1966-1970	Drapeau, Georges		1960-1961
Bloom, Barbara H. (Technical)	1956-1958	Dulaney, Ernest N.	1958-1959,1960-1961	
Bogert, Joseph C.	1938-1941			
Bohlen, Walter F.	1966-1969	Eckhardt, Donald H.		1958-1960
Bombolakis, Emanuel G.	1956-1957,1958-1960	England, Anthony W.	1963-1964,1967	
Bottino, Michael L.	1959-1963	Enzmann, Robert D.		1954-1955
Bowker, David E.	1953-1954,1956-1959	Erickson, Albert J.		1964-1970
Bowman, Robert	1953-1955	Estrada, J. E. M.		1969-1970
+Boydell, Harry C.	1923,1925-1926,+1926-1928			
++Brace, William F.	++1950-1952	Fahlquist, Davis A.		1956-1957
Brady, Brian T.	1961-1964	Fairborn, John W.		1964-1968
Bray, Joseph M.	1937-1940	Faramarzpour, Faramarz		1964-1966
Breeding, Roger	1962-1965,1968-1969	Faure, Gunter		1957-1959
Breslau, Lloyd R.	1957-1959,1961-1964	Fink, Donald G.		1933-1934
Bridge, William D.	1968-1969	Fink, Don R.		1954-1956
Briscoe, Howard W.	1953-1954	Finn, Ronald S.		1960-1961
Brookins, Douglas G.	1958-1963	Fitzgerald, W. F.		1966-1968
Brown, Levi	1932-1933	+Flaherty, Gerard F.	1929-1931,+1931-1932	
Brownlow, Arthur H.	1955-1959	Fowler, Frederick B.		1951-1952
++Buerger, Martin J.	1925-1927,++1927-1929	Franck, Mona L. (Technical)		1951-1953
Buerger, Newton W.	1934-1937	Franklin, Virginia (Ms.V.F.Ross)	1950-1951	
Buller, A. E.	1939-1940	Frasier, Clint W., Jr.		1965-1969
Buma, Grant	1969-1970	Frondel, Clifford		1936-1938
Burnham, Charles W.	1957-1961	Frueh, Alfred, Jr.		1946-1949
Burt, Donna G. (Technical)	1952-1954	Fryer, Brian J.		1968-1971
Burton, Virginia (Ms. V. Lang)	1947-1951	+Fuller, Myron L.	1897-1898,+1897-1900	
Butler, Robert D.	1932-1936	Furnival, George M.		1934-1935
Byerlee, James D.	1963-1966			
Byrd, William E., Jr.	1964-1967	Galbraith, James N.,Jr.	1958-1960,1961-1963	
		Galvin, Cyril J.	1957-1959,1961-1963	
Callahan, William H.	1926-1927	Gates, Michael T.		1968-1971
Calnan, Irene J. (Technical)	1953-1955	Gilbert,J.Freeman,Jr.	1953-1955,1956-1957	
Camfield, Paul A.	1964-1966	++Gillson, Joseph L.		++1922-1924
Canney, Frank C.	1948-1951	Goldstein, Myron A.		1967-1968
Carlisle, Donald B.	1964-1965	Goodspeed, George E.		1911-1912
Carter, Richard M.	1967-1969	Gordon, Bruce M.		1967-1970
Chang, Robert P. H.	1965-1966	Gore, Roger C.		1956-1958
Chapman, Clark R.	1970-1971	Gorfinkle, Lorraine G.		1947-1950

III. COURSE XII SUPPORTING STAFF   75

Goulet, Julien R.	1963-1965	MacIntyre, Ferren	1965-1966
Gowen, Walter K., Jr.	1957-1958	+MacKenzie, John D.	+1912-1916
Gower, Evelyn F. (Admin.)	1954-1965	Madariaga, Paul I.	1967-1971
Gower, John A.	1952-1954	++Madden, Theodore R.	1950-1951,++1951-1955
+Grabau, Amadeus W.	1892-1896,+1896-1897	Maehl, Richard H.	1955-1957
Gray, Bernard N.	1965-1967	Magnell, Bruce A.	1966-1970
Green, Edward J.	1958-1961,1963-1965	Maher, Francis X.	1962-1963
+Greenewalt, David	1956-1958,+1961-1964	Mao, Nai-Hsien	1962-1963
Greenfield, Roy J.	1962-1965	Marshall, Donald J.	1955-1957,1958
Grine, Donald R.	1954-1958	Martin, Randolph	1964-1968
+Gunning, Henry C.	+1926-1927	Matson, Wayne R.	1964-1965
		McIntire, William L.	1955-1956,1957-1958
Haddock, Elizabeth L.(Technical)	1954-1956	+McKinstry, Hugh E.	+1920-1921
Hagen, John C.	1951-1954	McMullen, Donald J.	1944-1945
Hall, Flemmon P.	1924-1925	McNutt, Robert H.	1961-1964
Hallof, Philip G.	1952-1954	Mead, Judson	1940-1941
Halverson, Ward D.	1961-1965	++Mencher, Ely	++1935-1938
Harper, Charles W., Jr.	1959-1961	Metes, Jon S.	1961-1962
Harris, Margaret R. (Technical)	1958-1960	Miller, Richard C.	1965-1968
Hart, Stanley R.	1957-1960	Mills, Joseph W.	1939-1942
Hauck, Anthony M., III	1959-1960	Moon, Warren D.	1965-1966
Heath, Edward W.	1958-1959	Moore, John M., Jr.	1957-1960
Heath, Marla Moody	1962-1965	Morrow, Dorothy E. (Technical)	1949-1951
Heath, Stanley A.	1963-1965	Morss, Henry A., Jr. (Admin.)	1964-1972
Heineken, Paul A.	1968-1969	Moss, John H.	1942-1943
Hendricks, Walter J.	1956-1957	+Muller, Charles J.	+1921-1923
Herzog, Leonard	1951-1952	Murray, Bruce C.	1954-1955
Hill, Claude P. T.	1952-1953		
Hingston, Townsend H.	1922-1924	Nalbandian, Mihran	1964-1966
Hogg, Nelson	1967-1971	Neece, Neal, Jr.	1950-1951
Holyk, Walter	1950-1952	Nelson, Philip H.	1962-1967
Hookstra, Carl R., Jr.	1967-1968	Ness, Norman F.	1955-1958
Horodyski, Robert J.	1965-1968	Netarwala, Minoo P.	1943-1944
Horowicz, Leon	1965-1966	Neves, Antonio S.	1954-1955,1956-1957
Horwood, Hereward C.	1933-1934	++Newhouse, Walter H.	++1923-1927
Howard, Michael L.	1965-1966	Nichols, Henry W.	+1893-1894
++Hurley, Patrick M.	1938-1939,++1940-1941	Niizeki, Nobukazu	1952-1957
+Hurst, M. Ewart	+1921-1922	Nikhanj, Y.	1968-1970
Hwang, Jae-Young	1963-1966	Nourbehecht, Bijan	1959-1961
		Nur, Amos M.	1966-1969
Ilsley, Ralph	1932-1934		
		Pan, Cheh	1961-1963
Jackson, David C.	1967-1969	Papworth, Elaine	1969-1971
Jackson, David S.	1961-1962	Parrish, William	1935-1938
Jacobson, Herbert S.	1954-1955	Paulding, Bartlett W., Jr.	1960-1965
Jarrell, Richard F.	1938-1942	Paulson, Richard W.	1964-1966
Jensen, Mead L.	1948-1950	Payson, Harold, Jr.	1962-1965
Johnson, Donald H.	1949-1950	Peacor, Donald R.	1958-1960,1960-1961
++Johnson, Douglas W.	++1903-1905	Pearsall, Cortland S.	1938-1939
Johnston, William G.	1948-1949	Pelletier, Yves J. A.	1965-1971
Jokela, Arthur W.	1963-1965	Perrett, Robert F.	1964-1966
		Perry, Eugene C., Jr.	1959-1963
Kaiser, Edward P.	1937-1940	Philpotts, John A.	1961-1965
Kearns, Margaret M.	1950-1951	Phinney, William C.	1955-1959
Kelley, Danford G.	1951-1955	Phipps, Donald	1962-1964
Kelly, Arthur M.	1959-1961	Pines, Ira R.	1969-1971
Ketchum, Carl B.	1965-1968	++Pinson, William H., Jr.	++1950-1951
King, Lewis H.	1950-1954	Podolsky, Terence	1950-1951,1952-1953
Kitrosser, David F.	1965-1967	Pollard, Melvin	1953-1954
Klubock, Morse H.	1951-1953	Pollock, James P.	1938-1940
Koch, Roger E.	1966-1967	Posadas, Veronica Gomez de	1964-1966
Kolitz, Byron L.	1967-1968	Posen, Harold	1955-1956
Koons, Harry C.	1963-1968	Powell, Richard M.	1956-1957
Kranck, Svante H.	1957-1958	Prewitt, Charles T.	1956-1962
Krogh, Thomas E.	1960-1964		
Krotser, Donald J.	1960-1966	Quigley, Robert M.	1957-1958
Krueger, Harold W.	1959-1960		
Kuenzler, Howard W.	1965,1967-1968	Racer, Charles W., Jr.	1959-1960
		Rahn, Kenneth A.	1962-1963
++Lahee, Frederic H.	++1912-1914	Raymond, Louis C.	1931-1932
Larner, Kenneth L.	1962-1969	Read, Motte A.	1902-1903
Latta, David B.	1969-1971	Redden, Martha	1967-1968
Lau, Ying Kai	1966-1967	+Reed, Rufus C.	+1922-1924
Lee, C. Y.	1967-1970	Reesman, Richard H.	1962-1964,1966-1968
Leith, Thomas H.	1950-1953	Regal, Donald	1970-1971
Lewis, Lloyd F.	1965-1966	Reid, John B., Jr.	1966-1970
Lightner, Leo F.	1957-1958	Richardson, Paul W.	1953-1954
Lindsey, Hildegarde	1964-1965	Ritter, Charles J.	1959-1962
Little, Roger G.	1962-1964	Robertson, Forbes S.	1941-1942
Loring, Douglas H.	1956-1957	Robin, Pierre-Yves F.	1967-1968
+Loughlin, Gerald F.	1903-1904,+1906-1912	Roe, Glenn D.	1962-1964

Root, Frank E.	1963	Trojer, Felix J.	1964-1969
Ross, William P.	1961-1962	Tsai, Yi-Ben	1965-1969
Rossby, Hans T.	1962-1966	Tupper, William M.	1955-1959
Rostoker, Mendel D.	1954-1957	Turyn, May S.	1953-1954
Rove, Olaf N.	1935-1938	Tuttle, O. Frank	1940-1942
Roy, David C.	1961-1967		
Russell, James D.	1960	Unger, John D.	1966-1967
		Utter, Robert R.	1961
Saint Amant, Marcel H. Y.	1967-1968		
Sammel, Edward A.	1956-1957	Van Leer, John C.	1965-1968
+Sanford, Thomas B.	1963-1966,+1966-1967	Vassar, Helen E.(Research Chem.)	1921-1923
Saull, Vincent A.	1948-1951	Volfovsky, Regina	1961-1964
Sawyer, Charles B.	1962	Vozoff, Keeva	1952-1955
Sax, Robert L.	1956-1959		
Schatz, John F.	1965-1971	Waldrop, Ann L.	1966-1969
Schilling, Jean-Guy E.	1964-1966	Walsh, William P.	1953-1954
Schneider, William A.	1958-1960	Walters, Lester J., Jr.	1963-1967
Schnetzler, Charles C.	1959-1963	Ward, Robert D.	1902-1904
Scholz, Christopher H.	1964-1967	Ward, Ronald W.	1966-1971
Schroeder, Gerald L.	1962-1965	++Warren, Charles H.	++1902-1904
Seele, Gordon D.	1954-1955	Watts, Raymond D.	1966-1967
Seraphim, Robert H.	1949-1951	Wayman, Cooper H.	1956-1957
Shaw, John W.	1904-1905	Weaver, Leonard W.	1940-1941
Sherman, Herbert L.	1902-1903	Webb, Marilyn M. (Technical)	1955-1956
Shields, Robert M., Jr.	1963-1964	Webber, G. Roger	1952-1955
++Shimer, Hervey W.	++1903-1908	Webster, Thomas F.	1957-1959
Short, Nicholas M.	1955	Welby, Charles W.	1949-1950
Sill, William	1961-1963,1964-1967	++Whitehead, Walter L.	++1913-1914
++Simpson,Stephen M.,Jr.	1952-1953,++1953-1954	Whiting, Francis	1949-1951
Slade, Martin	1967-1971	Whitney, Philip R.	1956-1958,1962
Smith, George F., Jr.	1955-1956	Wiggins, Ralph A.	1961-1965
Smith, Markwick K., Jr.	1953-1954	Wilder, Frederick M.	1902-1903
Smitheringale, William V.	1927-1928	Williams, Wilson G.	1965-1968
Smitheringale, William G.	1957-1959	Wing, Charles G.	1963-1966
Southwick, Peter F.	1948-1951	Wing, Lawrence A.	1951-1952
Southwick, Stanley H.	1948-1951	Wittels, Mark C.	1947-1949
Spilhaus,Athelstan F.,Jr.	1960-1962,1963-1964	Wolf, Eric C.	1969
Spooner, Charles M.	1965-1969	Wolfe, Jack C.	1967-1969
Stacy, Maurice C.	1950-1951	++Wones, David R.	++1954-1957
Stanley, Everett M.	1964	Wood, John A., Jr.	1954-1956
Stevenson, John S.	1931-1934	Woodside, John M.	1964-1967
Stewart, Keith J.	1954-1955	+Woodward, Amos E.	+1888-1890
Stoiber, Richard E.	1934-1935	Wright, Frances W.	1957-1958,1962-1963
Stoll, Walter C.	1940-1942	Wright, Hubert A., Jr.	1962-1963
Strickland, Lawrence	1952-1955	Wright, Livingston	1923-1924
Sullivan,Geraldine R.(Technical)	1947-1952	Wuensch, Bernhardt J.	1958-1963
Swift, Charles M., Jr.	1966-1967	++Wunsch, Carl I.	++1962-1963
		Wylie, Robert W.	1954-1958
Tang, Alice C. C.	1963-1965		
Taxer, Karlheinz J.	1965-1966	Yearsley, John R.	1962-1963,1964-1965
+Thompson, James B., Jr.	+1946-1949	Young, Edward J.	1948-1950
Thompson, Keith F.	1961-1966	Young, Theodore S. T.	1967-1971
Thompson, William B.	1959-1963	Yules, John A.	1963-1966
Tooley, Richard D.	1954-1956		
Towell, David G.	1963	Zeigler, Marilyn A. (Technical)	1951-1952
Towse, Donald F.	1948-1950	Zidle, Tobias	1963-1966
Treitel, Sven	1953-1958	Zoltai, Tibor	1955-1959

Instructors

The following individuals were Instructors for the school years indicated. Those who served for several years gave lectures, whereas those who held one-year appointments were actually much like the teaching assistants of more recent years. These individuals, whose names are also included in the preceding <u>Staff List</u>, have been separated from that list and given special mention here in order to recognize the valuable service they rendered geology as student instructors.

III. COURSE XII SUPPORTING STAFF

Woodward, Amos E.	1888-1890	Benedict, Platt C.	1921-1923
Cobb, Collier	1890-1892	Hurst, M. Ewart	1921-1922
Nichols, Henry W.	1893-1894	Muller, Charles J.	1921-1923
Grabau, Amadeus W.	1896-1897	Reed, Rufus C.	1922-1924
Fuller, Myron L.	1897-1900	Boydell, Harry C.	1926-1928
Clapp, Frederick G.	1901-1904	Gunning, Henry C.	1926-1927
Loughlin, Gerald F.	1906-1912	Doten, Robert K.	1931-1932
Clapp, Charles H.	1908-1910	Flaherty, Gerard F.	1931-1932
MacKenzie, John D.	1912-1916	Thompson, James B., Jr.	1946-1949
Barry, John G.	1916-1920	Beall, George H.	1960-1962
Aldrich, Henry R.	1920-1921	Greenewalt, David	1961-1964
Currier, Louis W.	1920-1921	Sanford, Thomas B.	1966-1967
McKinstry, Hugh E.	1920-1921		

Technical Assistants

In the 1950s and 1960s it was common practice to employ college graduates in geology, or college graduates who had had considerable course work in the subject, to do library research and prepare special reports on a variety of problems that arose in the conduct of the Department's academic program. Such work required considerable knowledge of geology and of geology departments around the country, and also some training in collecting and evaluating statistical data. The following women served Course XII as Technical Assistants, and performed tasks that contributed importantly to the Department's overall program. One of these, the late Sylvia Bateman, deserves special notice because as stated in the Preface she collected many of the items that are included in the present work.

Dorothy E. Morrow	1949-1951	Marilyn M. Webb	1955-1956
Marilyn A. Zeigler	1951-1952	Barbara H. Bloom	1956-1958
Patricia Binkley	1952-1953	Margaret R. Harris	1958-1960
Donna G. Burt	1952-1954	Sylvia Bateman	1960-1965
Elizabeth L. Haddock	1954-1956		

In addition to the nine women technical assistants listed above, all of whom did exclusively library, secretarial or editorial work, there was a second group of technical assistants, generally attached to and paid by a research project, who were employed for their special skills that could be used in the project research. Some of these were graduate students, some were not, but all worked on research projects where their special skills were needed. Their names are not listed here because they came and went on such irregular schedules, and in such variable numbers, that no systematic record was kept of them. Obviously, however, when as many as 10 were employed in a single school year, as was the case in SY 1969-70, these assistants gave much valuable service to the Department.

Administrative Assistants

As the fiscal management problems of the Department Head greatly increased in the later 1950s, in part because the Institute was undergoing major changes in its own fiscal policies and methods of control, aid had to be sought in meeting the new situation. During this period the amount of outside financial aid for research greatly increased, as did the number of graduate students, and by 1959 it was necessary to employ an academically trained assistant who had had experience in both financial management and personnel administration.

The Department was most fortunate in finding Dr. Henry A. Morss, Jr. (VIII Ph.D. 1934), a doctorate in physics, who had been with Simplex Wire and Cable Company of Cambridge for many years and who had the very kind of experience needed by the Department Head, who was then the present writer.

First appointed as Lecturer, in 1959, Dr. Morss was soon given the more appropriate title of Administrative Assistant (1964), which he continued to have until retirement in June 1972. He was first given responsibility for financial and personnel matters. Soon the problems of space were added to his worries, and later, when the new Green Building began to near completion in early 1964, he was assigned a major role in getting the Geology and Meteorology departments moved and settled down in the new space. Following dedication of the building on 2 October 1964, Morss became "Mr. Trouble-Shooter"--the one person that everyone went to with complaints. In this important capacity, he next took on the administrative responsibility for the new oceanography program, which had grown to a two-million dollar effort by 1965 and would eventuate in the Joint Doctoral Program with the Woods Hole Oceanographic Institution in 1968. When Press succeeded Shrock as Head of Course XII in 1965, Morss was reappointed Administrative Assistant, and continued to be that most important servant with the "passion for anonymity" that former President Franklin D. Roosevelt so assiduously sought in his assistants. All who had occasion to call on "Harry" Morss for assistance, particularly the occupants of the Green Building, know of his great devotion and efficient service to the earth sciences and to all those, teachers and students alike, engaged in their pursuit in one form or another. From January 1970 through June 1972 he was ably assisted by Lynn Hodges, also an Administrative Assistant. (See also the brief sketch on p. 69-70.)

D.S.R. Staff

A third group of persons with special skills, who spent full-time on project research, were categorized as D.S.R. Staff (= Division of Sponsored Research Staff) because their salaries were paid from project funds that came from outside the Institute. These individuals were generally employed because of their research skills and devoted all of their time to project work. Like machinists, instrument makers, and secretaries,

III. COURSE XII SUPPORTING STAFF

they kept the research program moving by doing their special assignments. In the later 1960s they numbered as many as a dozen persons, and constituted an important group in the departmental research team. Their names are not listed here, however, because they themselves did not participate directly or actively in the Department's instructional program.

RESEARCH ASSOCIATES--POST-DOCTORAL FELLOWS

The dramatic increase in financial support from foundations and federal agencies in the later 1960s led to greatly increased numbers of graduate degree candidates, and inevitably to larger and larger annual groups of doctorates (See Chart on a preceding page). In the earth sciences, this trend is obvious from the accompanying list which shows the total number of doctorates granted each year in the geosciences, from 1950 through 1970, and the number produced by M.I.T. (Geology, etc., Oceanography and Meteorology).

Partly because not all of the doctorates could be absorbed by industry, federal agencies, and academic faculties, and partly too because many doctorates found it highly satisfying to stay on in a department on a research project if financial support was available, still another category of appointment--the research associate or post-doctoral fellow--came into wide use in the 1960s. Although these associates, or "post-docs" as they came to be designated, seldom remained more than a year or two, chiefly to complete work started while they were doctoral candidates, some did stay on longer and a few tended to become almost fixtures in the Department even though they did not participate in the instructional part of the Department's overall program. By 1970 Course XII had a dozen such post-doctoral members, and they constituted an important group in the Department's whole research team. (See Chart on a preceding page.)

The following list contains the names of post-doctoral scientists who spent a few years in M.I.T.'s Department of Geology, etc. (Course XII), as paid Research Associates. The list does not include those post-doctorals who were financed from the outside while conducting research in the Department. With few exceptions, the persons listed below seldom stayed more than two or three years, regardless of the source of their financial support. They came to M.I.T. primarily to carry on research, but some of them also participated to a limited extent in the Department's instructional program, and a few held one-year appointments as part-time Instructors (e.g. Frasier, Sanford, et al.), or were promoted to faculty rank (e.g. Ahrens, Cantwell, Goetze, Hurley, and Pinson).

GEOLOGY AT M.I.T. 1865-1965

Adams, J. B.	1971-1972	Keevil, Norman B.	1938-1939
Ahrens, Louis H.	1948-1950	Kolbe, Peter	1965-1966
Andrews, Dudley J.	1971-1972		
		Larner, Kenneth	1969-1970
Baines, Peter G.	1970-1971	Larson, Edward E.	1965-1966
Beers, Roland F.	1943-1946		
Bloom, Mortimer C.	1935-1937	Mack, Harry	1969-1970
Bowden, Michael	1962-1964	MacMahon, Beverly E.	1966-1968
Breger, Irving A.	1946-1952	Maruyama, Takuo	1970
Byerlee, James D.	1966-1967	Minear, John	1968-1969
		Mogi, Kiyoo	1965-1966
Cantiez, Nezihi	1969-1971	Mohsen, Lotfi	1969-1970
Cantwell, Thomas, Jr.	1958-1960		
Chatterjee, Niranjan	1970-1971	Nur, Amos	1969-1970
Chinnery, Michael A.	1965-1966		
Chung, Dae-Hyun	1968-1972	Parrish, William	1939-1940
Combs, James B.	1970	Pekeris, Chaim L.	1933-1934, 1936-1947
Corless, James T.	1963	Pinson, William H., Jr.	1953-1955
Crockett, James H.	1964	Powell, Richard M.	1963
Davies, David	1969-1972	Rankin, John M.	1971-1972
Deer, John S.	1969-1970	Reasenberg, Robert D.	1969-1972
Duce, Robert A.	1964-1965	Roberts, Carol Lee	1968-1969
		Robinson, Enders A.	1952-1954
Eden, Henry F.	1962-1964		
Erickson, Albert J.	1970	Saito, Masanori	1966-1968
		Sanford, Thomas B.	1966-1967
Faul, Henry	1947-1949	Scarlet, Richard I.	1971-1972
Faure, Gunter	1963-1964	Schnetzler, Charles C.	1962-1963
Fiocco, Giorgio	1968-1969	Simmons, William F.	1968-1969
		Singh, Hausila P.	1969-1970
Gilmore, Robert S.	1971-1972	Slater, Gary L.	1971-1972
Goetze, Christopher	1969-1970	Stetson, Harlan T.	1936-1938
Grams, Gerald W.	1966-1968	Susse, Peter	1967-1970
Hahn, Theodor	1953-1956	Tacheuchi, Yoshio	1952-1954
Halverson, Ward D.	1965	Tsai, Yi-Ben	1969-1971
Hashimoto, Yoshikazu	1964-1965	Tossell, John A.	1971-1972
Haskell, Norman A.	1950-1958		
Helmberger, Donald V.	1967-1969	Urry, William D.	1937-1938
Hermance, John	1967-1968		
Herzog, Leonard F.	1952-1956	Vaughan, David J.	1971-1972
Hewitt, David	1969-1971		
Hirasawa, Tomowo	1968-1970	Walsh, Joseph B.	1964-1972
Horai, Ki-iti	1966-1972	Washken, Edward	1946-1949
Hurley, Patrick M.	1941-1942	Wasson, John T.	1962-1963
Husebye, Eystein	1967-1968	Watkins, Joel S.	1966-1967
		Wawersik, Wolfgang	1968-1969
Ida, Yoshiaki	1971-1972	Wiggins, Ralph A.	1966-1970
		Wing, Charles G.	1966-1971
Jaffe, Jack	1971-1972	Wood, Henry O.	1912-1915
Julian, Bruce R.	1969-1970		

DEPARTMENTAL LIBRARIANS

The history of the library facilities available to geology professors and students during M.I.T.'s first century is recounted in my Volume 2. Here brief reference is made only to the eighteen persons who served as geology librarians during this period.

During the first few years of M.I.T.'s existence, and long before a Department of Geology was established, students taking geology subjects found the books and maps they needed either in the private collections of their professors, or in the small collection of books brought together in the general Institute library. By 1870 the great collections of the Boston Public Library and of the Boston Society of Natural History were made available to the young Institute, and for the next twenty years these collections were much used by the Technology students and staff.

III. COURSE XII SUPPORTING STAFF

When Course XII was established in 1890 as the Department of Geology, and William H. Niles appointed as its Head, the Institute already had a collection of about 1,000 geological books and almost 900 books on mining, in addition to a sizeable collection of maps. Previous to 1890 these had been under the care of the Institute Librarian who looked after the books in a students' reading room on the main floor of the Rogers Building.

With establishment of the new Department, and appointment soon after (1900) of its first Secretary, Mabel Hodgkins, the geology library came under the direct supervision of the Department. Since Miss Hodgkins had supervision of this library she can be said to have been the first Departmental Librarian for Course XII. When she left to become Librarian of New Hampshire State College about 1910, she was soon succeeded by Hilda Williams, who became Departmental Secretary and the second Departmental Librarian in 1911 and continued in this dual role until 1932, three years before she retired as Secretary.

After the Institute had been moved across the Charles from Boston to Cambridge in 1916, and the Department of Geology settled down in Building 4, the Geology Library, now numbering more than 6,000 volumes and 4,000 pamphlets and maps, was split into two collections. The little used items were sent to the Central Library under the Great Dome; the remainder were shelved in a third-floor bay (Room 4-344) directly adjoining Chairman Lindgren's office, whence the Departmental Secretary and Librarian, the previously mentioned Miss Williams, could keep an eye on the books and maps.

In 1932 the separate libraries maintained by the Departments of Geology and of Mining and Metallurgy, respectively, were merged to form the Lindgren Library, so named to honor Dr. Waldemar Lindgren, who for a short time had headed both Courses III and XII, and were brought together in a three-bay room on the third floor of Building 8 (8--304-308). At this time the new branch library came under the direct supervision of the Central Library, (See President's Report for SY 1932-33, p. 65), and Carolyn Warren was appointed the first Lindgren Librarian.

When Miss Warren resigned to marry Professor Alexander Magoun, in the fall of 1936, she was succeeded by Grace Bogart, who came from the University of Wisconsin where she had been in charge of the Geology Department's library. Miss Bogart thus became the second Lindgren Librarian and served until 1945 when she resigned to take charge of a large industrial library. She was replaced by Sibyl E. Warren, like Carolyn before her, a sister of Physics Professor Bertram E. Warren (VIII S.B. 1923). This second Miss Warren served as the Lindgren Librarian until 1952, when the Lindgren Library, including the connecting Schwarz Memorial Map Room, was moved to and incorporated in the new Science Library in the Hayden Memorial Building.

Next followed a twelve-year period, 1952 to 1964, during which Course XII, housed in Building 24, was without its own separate library and librarian. This unhappy and unsatisfactory situation ended in the summer of 1964 when the Lindgren Library, and the associated Schwarz Memorial Map Room, were reestablished on the second floor of the new Green Building.

During this twelve-year period, seven persons followed Sibyl Warren, who left in 1935, and successively looked after the geology books in the Science Library. These seven, each designated Earth Science Librarian, were:

Anne-Marie Hartmere	1955-1958
Diana Jorjorian	1958-1959
Cynthia Leventhal	1959-1961
Diane Keenan	1961-1962
Paul R. Brayton, Jr.	1962-1963
Barton L. Wimble	1963-1964
Eleanor G. McGonagle	1964 (2 mos.)

Once the decision was made to reestablish the Lindgren Library and the Schwarz Memorial Map Room in a specially prepared floor in the Green Building, Eleanor L. Bartlett, Institute Archives Librarian, was assigned the complicated task of reconstituting what would be in fact a new and different Lindgren Library. The efficient way Miss Bartlett carried out her assignment, as an Acting Lindgren Librarian, as well as her full-time responsibilities in the Institute Archives, are related in my Volume 2.

After the dedication of the Green Building on 2 October 1964, Margaret Otto was appointed the first Lindgren Librarian in the Green Building, in the late fall of 1964, and continued in that position until July 1968. During these four years she had the help of two Library Assistants, Sara K. Landau (1964-1966) and Suanne Muehlner (1966-1968). The latter became Acting Lindgren Librarian in July 1968, and Lindgren Librarian in September 1971, after a year's leave to go on a special library mission to Germany. Arthuree McLaughlin was appointed Acting Lindgren Librarian for the year Mrs. Muehlner was on leave (i.e. Sept. 1970-Sept. 1971).

During the period from June 1968 through June 1972, three women served in the Lindgren Library as Library Assistants: Martha B. Zuska (January-June 1968), Anet Krzentz (July 1968-June 1969), and Elizabeth Butkus (July 1969-June 1972).

In summary, the books and maps belonging to the Geology Department have had the care and supervision of the following 18 persons referred to in the preceding paragraphs:

III. COURSE XII SUPPORTING STAFF

1.	Institute Librarian	1865-1900
2.	Mabel Hodgkins, Course XII Secretary & Librarian	1900-1910
3.	Hilda Williams, Course XII Secretary & Librarian	1911-1932
4.	Carolyn Warren, Lindgren Librarian	1932-1936
5.	Grace M. Bogart, Lindgren Librarian	1936-1945
6.	Sibyl E. Warren, Lindgren Librarian	1945-1955
7.	Anne-Marie Hartmere, Earth Science Librarian	1955-1958
8.	Diana Jorjorian, Earth Science Librarian	1958-1959
9.	Cynthia Leventhal, Earth Science Librarian	1959-1961
10.	Diane Keenan, Earth Science Librarian	1961-1962
11.	Paul R. Brayton, Jr., Earth Science Librarian	1962-1963
12.	Barton L. Wimble, Earth Science Librarian	1963-1964
13.	Eleanor G. McGonagle, Lindgren Librarian	1964 (2 mos.)
14.	Eleanor L. Bartlett, "Acting Lindgren Librarian"	1964 (9 mos.)
15.	Margaret A. Otto, Lindgren Librarian	1964-1968
16.	Suanne Muehlner, Acting Lindgren Librarian	1968-1970
17.	Arthuree McLaughlin, Acting Lindgren Librarian	1970-1971
18.	Suanne Muehlner, Lindgren Librarian	1971-1973
19.	Hedy Mattson, Lindgren Librarian	1973-

SHOP PERSONNEL

A walk through the corridors of the century-old M.I.T. as it was in 1965, and continued to be in 1970, quickly convinced the observer that there was a large and diversified activity, displayed in machine shops and special supply rooms, which was operating in support of the instructional and research program that engaged the attention of professors and students. The geology program has had two of these all-important facilities for many years: a machine shop and an electronics shop. Our research program could not have been mounted or carried on without these two shops and the skilled mechanics and technicians in them.

Previous to 1930 the program of instruction and research offered by Course XII was such that there was little need for a machinist or instrument maker, and none for an electronics engineer because this was years before the appearance of precise measuring devices such as are commonplace today.

For twenty years after the Institute was moved from Boston to Cambridge (i.e. from 1916 to 1936), the shop needs of the Department of Geology were few and far between. They usually were met by getting the work done outside the Institute (e.g. refurbishment of microscopes by a local optical company), or arranging to have it done in one of the departmental shops.

The first geology machine shop came into existence in 1934 as a result of efforts by M. J. Buerger, then Assistant Professor of Mineralogy. The Department had been asked to apply for a $10,000 grant-in-aid from the Rockefeller Foundation, and Buerger had submitted a three-fold request for 1) an X-ray tube and associated instrumentation, 2) funds to build a machine for making crystal models, and 3) funds for establishing a machine shop and employing a machinist. With funds in hand he initiated all three of his projects, and by advertising his need for a machinist he was fortunate to attract Otto von der Heyde, a skilled instrument maker who helped to set up the first shop in Room 4-344, in May 1934. For the next eight years, until mid-1942, when he resigned to establish his own shop, he served all members of the Department of Geology, to say nothing of his long-remembered piano concerts at the famous "bean suppers" given by the Meads each spring. For Otto was as skilled at the piano as at the lathe and workbench!

When he decided to leave M.I.T., von der Heyde recommended a fellow machinist, Charles E. Supper, who had been working for the Mining Department since 1934. Supper became the second machinist in the Geology Department and stayed until mid-1946 when, like von der Heyde, he left to devote full time to his private company. While employed by the Department, Supper built several of the instruments invented and designed by M. J. Buerger, who had foreseen his future need for the shop he helped to establish.

Supper's place was taken immediately by John Solo, another able instrument maker who, like von der Heyde, was a skilled musician, his instrument being the violin. Two years later, in 1948, as the Department's research program began to grow with new research projects and outside funds, a need arose for a second shopman, and John Annese was employed as a Machinist. For the next six years, from 1948 to the fall of 1954, these two shopmen, Solo and Annese, contributed valuable services to the Department. They gave indispensable aid to members of the Geology Department in establishing Buerger's crystallography laboratory, Hurley's geochronology laboratory, the Cabot Spectrographic Laboratory under the supervision of Ahrens and later Dennen, the geochemistry laboratory under the supervision of Pinson and Winchester, and the laboratory for geophysical research under the direction of Hurley and Madden.

When Solo left in the fall of 1954 to take charge of an industrial shop near Kendall Square, the Department was again fortunate to find a versatile and inventive instrument maker in the person of Kenneth Harper, with whom Annese continued to be an ideal team member in the Geology shop. Charles Crowley joined Harper and Annese in 1958, to do several specific jobs, and left in 1960 when these were completed. Shortly thereafter, in May 1962, Werner W. Burrows, a machinist, became Geology's third regular shopman.

When it came time for the Geology Department to move from Building 24 to its space in the new Green Building (54), Harper supervised the moving of the shop from the ground floor of Building 24 (24-031) to Building 54 in the spring of 1964. On the 6th floor, he and his two shop associates, machinists Annese and Burrows, reassembled the old machines, placed

III. COURSE XII SUPPORTING STAFF 85

several new machines purchased specifically for the new space, and by the summer of 1964 had all of them in full operation, ready to help the different professors as they moved their laboratories from Building 24 to their new space in the Green Building. It was under such circumstances as those attending the move of the Department from Building 24 to 54 that the quick and efficient help from the shopmen often saved much time and cooled rising tempers.

As different professors got their research laboratories established, the need for additional space and additional personnel arose even though it had seemed a few years earlier that the new space in the Green Building would be adequate for the foreseeable future. As a consequence, Hide found it necessary to create his own shop adjacent to his hydrodynamic laboratory in Building 20, and Brace soon had a special shop next door to his high-pressure laboratory on the 5th floor of the Green Building.

Then came the interdepartmental decision in early 1967 to combine the main geology and chemistry shops and to locate the facility in chemistry space in the basement of Building 6 (6-026).

At this juncture Harper resigned and left for California, where he was soon busy organizing a machine shop for a commercial company; Annese and Burrows moved with the Geology machines to the new Chemistry-Geology Shop in Building 6 (6-026); and a third machinist from Geology, Richard J. Letendre, who had come in April 1966 to work on a Simmons research project, transferred to the new shop. As a consequence, August 1967 found the three geology machinists--Annese, Burrows, and Letendre-- located in the new Chemistry-Geology Shop, along with their associates from the Chemistry Department, and a foreman who was soon thereafter succeeded by Paul J. Gabriel.

The history of our geology machine shops should include several small shops developed by professors for a special research project. One such was established by Prof. Hide in Building 20 to provide service for his hydrodynamics research group. John Burke and Jack Dubrin, mechanical technicians, worked in this shop, which was first located in 20--226-230 and later in 54--511. They joined Hide's research team in SY 1963-64 and left in 1968 when his laboratory and shop were disbanded after his resignation and return to England.

One additional shop in the Geology Department deserves to be mentioned. When Madden's geophysical research team had reached the stage where he and his colleagues and students needed an electronics technician, Clair Hendryx was found and hired in early 1961 and was assigned a one-bay space (54-617) for the electronic parts that he continues to provide.

DEPARTMENTAL SECRETARIES

It is assumed that the earliest geology teachers at M.I.T. wrote their own letters, prepared their own class notes and laboratory exercises, and individually did all those small but numerous jobs that today are given to a secretary without further thought.

The first person known to have been appointed as a Course XII Departmental Secretary was Miss Mabel Hodgkins who, according to Shimer (Unpublished typescript, History of the Department of Geology, M.I.T., in the Institute Archives, p. 37, date about 1944), served

> "... for some years before and after 1902 ..."

and ultimately left to become Librarian of the New Hampshire State College. Continuing from Shimer's history, p. 37:

> "... Then Miss Marjorie Fay, now of Cohasset, was here for a short time. Miss Hilda Williams came in 1911 remaining until 1935, a short time before her death."

There seems to be no record of how long Miss Fay worked, but Miss Williams served as both Secretary and Librarian until 1932, and as Secretary alone until 1935, when she resigned.

By this time a single secretary could no longer handle the work required by the enlarged faculty and expanded curriculum; consequently, in 1935, soon after he assumed the chairmanship of Course XII, Prof. Mead appointed two secretaries--one to be the Senior Secretary and the other, the Junior Secretary. This policy of two departmental secretaries would continue for more than thirty years until 1965, by which time departmental activities had so increased that additional persons had to be added to the secretarial force. Following is a list of the persons who served Course XII, in some secretarial capacity, from 1890 to 1970, in order of appointment:

Senior Secretaries		Junior Secretaries	
Mabel Hodgkins	1890+-1910?	Edna Howley	1935-1941
Marjorie Fay	(pre 1911)	Jeanne Trioulerye	1941-1944
Hilda Williams	1911-1935	Virginia King	1944
Dorothy Rove	1935-1937	Alma Koujian	1946-1949
Sylvia Bateman	1937-1946	Martha Ramsay	1949-1953
Irene Mugar	1946-1947	Joan Bartram	1953-1954
Pauline Richmond	1947-1965	Martha Ramsay Robes	1954-1956
Sylvia Chiesa	1965-1968	Eleanor Clebnik	1956-1957
Martha Africa	1968-1970	Eleanor Clebnik Levingston	1957-1960
Philadelphia Andrews	1970-1971	Gail Krupp	1960
Helen Fable	1971-	Eve Mayberry	1960-1961
		Deborah Jope	1961-1963
		Dianne Nissen	1963-1964
		Susan Renzi	1964-1965
		Mariann Pilch	1965-1969
		Paula Mere	1969
		Deena Pletsky	1969-1970
		Gail Meadors	1969-1970
		Martha Bacon	1970

87

IV. BIOGRAPHIES OF M.I.T.'s GEOLOGY PROFESSORS
1865-1965

INTRODUCTION

In the preceding chapter it was emphasized that the professors and their supporting staff in an M.I.T. department constitute the more important of the two most essential personnel groups of the Institute, their students being the other, so far as instruction and research are concerned. For this reason I have deemed it appropriate, in this history of the first hundred years of geology at M.I.T., to prepare a biography of each of the fifty-two geologists who held one or more full-time professorial appointments (i.e. assistant professor, associate professor, professor, Institute professor, or named professor) at the Institute during its first century, 1865-1965. The individual professors are discussed in the chronological order of their first appointment to a professorial rank, and their names are so ordered on the chart and accompanying typed list a little farther on. On both the chart and the list their period of service is indicated by the years they served, and M.I.T. alumni are marked with an asterisk (*). The list begins with William Barton Rogers, M.I.T.'s Founder, Organizer, first President, and first Professor of (Physics and) Geology, and ends with Frank Press, who came as Professor of Geophysics and Chairman (Head) of Course XII: Department of Geology and Geophysics, in September 1965, as M.I.T.'s second century began. I have also included in the chart the names of professors who joined our Department during SYs 1965-72, but I do not include biographies of them. In the preceding chapter I have also listed the successive appointments of each of the fifty-two professors (and of Prof. Press in addition) of whom a biography follows farther on.

NATURE OF THE BIOGRAPHIES

A general form has been followed in preparing the biographies, though this has been modified somewhat in some cases for special reasons that are generally obvious. An attempt has been made to present a brief biographical sketch of each professor and to cite sources, if such exist, where additional biographical information can be obtained by the interested reader.

Each biography consists typically of four or five items, each with its own special kind of information. These are: 1) a portrait sketched by the late Henry ("Chick") Kane (IX-B S.B. 1924) from a variety of

88 GEOLOGY AT M.I.T. 1865-1965

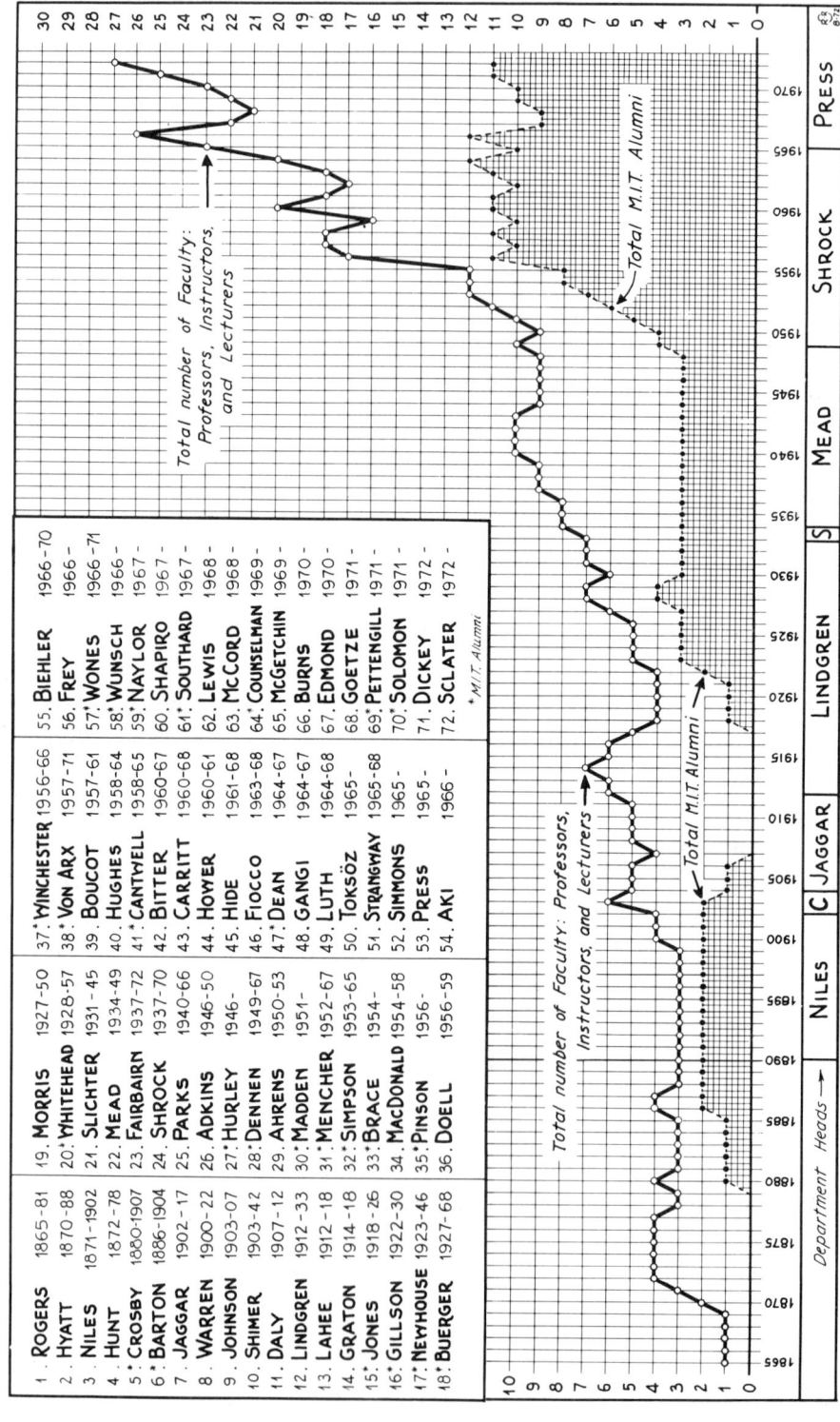

Chart Showing Order of Appointment of Course XII Professors and Number of M.I.T. Alumni Serving on the Faculty during 1865-1972

IV. COURSE XII BIOGRAPHIES

PROFESSORS OF GEOLOGY AT M.I.T., 1865-1965

	Name	Years of Service		Name	Years of Service
1	Rogers, W. B.	1865-1881	28 *	Dennen, W. H.	1949-1967
2	Hyatt, A.	1870-1888	29	Ahrens, L. H.	1950-1953
3	Niles, W. H.	1871-1902	30 *	Madden, T. R.	1951-
4	Hunt, T. S.	1872-1878	31 *	Mencher, E.	1952-1967
5 *	Crosby, W. O.	1880-1907	32 *	Simpson, S. M.	1953-1965
6 *	Barton, G. H.	1886-1904	33 *	Brace, W. F.	1954-
7	Jaggar, T. A., Jr.	1902-1917	34	MacDonald, G. J. F.	1954-1958
8	Warren, C. H.	1900-1922	35 *	Pinson, W. H.	1956-
9	Johnson, D. W.	1903-1907	36	Doell, R. R.	1956-1959
10	Shimer, H. W.	1903-1942	37 *	Winchester, J. W.	1956-1966
11	Daly, R. A.	1907-1912	38 *	Von Arx, W. S.	1957-1971
12	Lindgren, W.	1912-1933	39	Boucot, A. J.	1957-1961
13	Lahee, F. H.	1912-1918	40	Hughes, H.	1958-1964
14	Graton, L. C.	1914-1918	41 *	Cantwell, T., Jr.	1958-1965
15 *	Jones, W. F.	1918-1926	42	Bitter, F.	1960-1967
16 *	Gillson, J. L.	1922-1930	43	Carritt, D. E.	1960-1968
17 *	Newhouse, W. H.	1923-1946	44	Hower, J., Jr.	1960-1961
18 *	Buerger, M. J.	1927-1968	45	Hide, R.	1961-1968
19	Morris, F. K.	1927-1950	46	Fiocco, G.	1963-1968
20 *	Whitehead, W. L.	1928-1957	47 *	Dean, L. W., III	1964-1967
21	Slichter, L. B.	1931-1945	48	Gangi, A. F.	1964-1967
22	Mead, W. J.	1934-1949	49	Luth, W. C.	1964-1968
23	Fairbairn, H. W.	1937-1972	50	Toksöz, M. N.	1965-
24	Shrock, R. R.	1937-1970	51	Strangway, D. W.	1965-1968
25	Parks, R. D.	1940-1966	52	Simmons, M. G.	1965-
26	Adkins, J. N.	1946-1950	53	Press, F.	1965-
27 *	Hurley, P. M.	1946-			

* M.I.T. Alumni

photographs;* 2) a précis that describes the education and professional activities of each professor, what he contributed to the Geology Department and to M.I.T., and what impact he made on the earth sciences as teacher and investigator; 3) an extended and detailed discussion of the ancestry and early education of the individual, his more advanced training, his professional activities, accomplishments and honors, and comments on his personal attributes, these items generally constituting by far the largest part of the biography; 4) a bibliography as complete as possible, listing all the books, articles, reports and the like that he published while he was a member of M.I.T.'s Geology Department - the bibliography of those who are deceased includes all publications that I could find; for those who left the Institute after a period of service the list of publications is generally complete through the year of leaving or the year following; for professors emeriti, whether still active in teaching or research or retired, the list is complete only through the year of normal retirement, i.e. the year in which the emeritus rank was reached; finally, for those professors who are currently active (1975) and not yet at the age of retirement, I have tried to include all their publications through 1970; and 5) if appropriate, a list of necrologies, memorials, and biographies.

CITATIONS OF AUTHORS AND THEIR PUBLICATIONS

Several different methods of citing authors and their publications are used, depending on the place where the author or publication appears. In the regular text the author-date-page citation (e.g. Brown 1896, p. 30) is commonly used to cite a reference in the following bibliography. In the bibliographies of the biographies, the following practice is followed in citing references: number of article, date of publication in chronological order, (joint author or authors, if any), title, place of publication, volume, pages, and date in ( ). Examples of several such citations follow on the next page:

---

* Kane generously and enthusiastically entered into the spirit of my project by transforming ordinary photographs, largely supplied by M.I.T.'s Graphic Arts Service, into sketches carrying his own individualistic touches. His generosity is attested by his refusal to accept any fee, and he also declined my request that he initial his sketches. He passed off these actions in his characteristically modest way by saying, "Making the sketches has been fun and has been something I could do for M.I.T. and for the Geology Department!" The last sketch was completed only a few months before his death on 12 February 1971. I hardly need mention that Chick's sketches add something special to the biographies, and I owe a debt of gratitude to my longtime friend and colleague for his important contribution to this work.

IV. COURSE XII BIOGRAPHIES

8--1898 Title, place of publication, pagination, etc.
This indicates that the biographee wrote the paper alone, that it is the eighth paper he has written (either alone or with others), and that it is the first paper he published in 1898 (1898a would indicate the second, 1898b the third and so on).

8--1898 (and Jones, H. R.) Title, etc.
This would indicate that the biographee shared authorship jointly with Jones, and that he was the first author.

8--1898 (with Jones, H. R.) Title, etc.
This would indicate that the biographee shared authorship jointly with Jones, but that he was the second author.

8--1898 (and Jones, H. R. and Smith, A. H.) Title, etc.
This would indicate that the biographee, as first author, shared authorship jointly with Jones and Smith in that order. If there were more than two joint authors the authorship would read (and others).

8--1898 (with Jones, H. R. and Smith, A. H.) Title, etc.
This would indicate that the biographee was either the second or third author.

8--1898 (with Jones, H. R. et al.) Title, etc.
This would indicate that Jones was the first author, and the biographee was one of more than two other joint authors.

ABBREVIATIONS, ACRONYMS, AND SYMBOLS USED IN THE TEXT, FOOTNOTES, AND BIBLIOGRAPHIES

In order to condense the information in references to authors and their publications in this work - particularly in regard to the names of academies, agencies, associations, boards, bureaus, commissions, committees, institutes, institutions, surveys, societies, and the like; and to the names of magazines, journals, quarterlies, serials and similar publications - I have used the abbreviations, acronyms, symbols, and other devices listed following this paragraph. The reader is urged to consult this list if uncertain about the meaning of a particular item in a reference.

Abbreviations and Acronyms

-A-

AAAS; A.A.A.S.; or Am. Assoc. Adv. Sci.	American Association for the Advancement of Science
AAcAS; A.Ac.A.S.; or Am. Acad. Arts Sci.	American Academy of Arts and Sciences
A.A.G.N.; or Assoc. Am. Geol. Nat.	Association of American Geologists and Naturalists
AAPG; A.A.P.G.; or Am. Assoc. Petroleum Geol.	American Association of Petroleum Geologists
Abh.; or Abhandl.	Abhandlungen
Abst.; or [abst.]	Abstract; Abstracts

Abt.	Abteilung
Ac.; or Acad.	Academie; Academy
ACA; or A.C.A.	American Crystallographic Association
AEC; or A.E.C.	Atomic Energy Commission
AERE; or A.E.R.E.	Atomic Energy Research Establishment
AFCRL	Air Force Cambridge Research Laboratory
AGU; A.G.U.; or Am. Geophys. Union	American Geophysical Union
AIME; A.I.M.E.; A.I.M.M.E.; or Am. Inst. Min. (Metall.) Eng.	American Institute of Mining (and Metallurgical) Engineers
AIOP	Association Internationale de Oceanographie Physique
Am.; or Amer.	America; American
Am. J. Sci.	American Journal of Science
AMS; A.M.S.; or Am. Meteorol. Soc.	American Meteorological Society
AMSOC	American Miscellaneous Society
An.	Annal; Annals
Angew.	Angewissen
Ann.	Annual; Annuals
App.	Appendix
ARPA; or A.R.P.A.	Advanced Research Projects Agency
Assoc.	Association
ASXRED	American Society for X-Ray and Electron Diffraction

-B-

B.; or Bull.	Bulletin; Bulletins
Bd.	Band
Beitr.	Beiträge
Ber.	Berichte
BEW; or B.E.W.	Bureau of Economic Warfare
Biog.	Biographical; Biography
Biol.	Biological; Biology
Bol.	Boletin
Bos.	Boston
Br.	Branch
Bur.	Bureau

-C-

c	circa
Can.	Canada; Canadian
Chem.	Chemical; Chemistry
Chmn.	Chairman

Comm.	Commission; Committee
Contr.	Contribution; Contributions
C.R.; or c.r.	Comptes rendus
CSA; or C.S.A.	Crystallographic Society of America

-D-

DIC; or D.I.C.	Division of Industrial Cooperation
Div.	Division
Doc.	Document; Documents
DSR; or D.S.R.	Division of Sponsored Research

-E-

Econ.	Economic
Ed.; ed. (context)	Editor; Editors
ed.	edition
Edinb.	Edinburgh
Educ.	Education; Educational
e.g.	for example
Eng. Min. J.; or E. & M. J.	Engineering and Mining Journal
Eng.	Engineer; Engineers; Engineering
EOS	Former Transactions of the American Geophysical Union
EPA	Environmental Protection Agency
ESSA; or E.S.S.A.	Environmental Science Services Administration
et al.	et alii = and others
Ex.; or ex.	Example; Examples
Exhib.	Exhibit; Exhibits
Exp.; or Exper.	Experiment; Experiments; Experimental
Ext.; or Extr.	Extract; Extracts

-F-

ff.	following
FY	Fiscal Year
FY 1965-66	Fiscal Year 1965-1966

-G-

GAG; or G.A.G.	Geophysical Analysis Group
Geochem.	Geochemist; Geochemists; Geochemistry; Geochemical
Geochim.	Geochimica
Geod.	Geodesy; Geodetic
Geogr.	Geographer; Geographers; Geography; Geographical

Geol.	Geologist; Geologists; Geology; Geological
Geol. Soc. Am., B. 84: 175-186, (1973)	To be read: The Geological Society of America, Bulletin Volume 84, pages 175 to 186, 1973
Geophys.	Geophysicist; Geophysicists; Geophysics; Geophysical
Govt.	Government; Governmental
GSA; G.S.A.; or Geol. Soc. Am.	The Geological Society of America
GSC; or G.S.C.	Geological Survey of Canada

-H-

H.; or h.	Heft
Hist.	History; Historical

-I-

IAPSO	International Association of the Physical Sciences of the Ocean
ibid.	ibidem (Latin) = in the same place
idem	idem (Latin) = the same as that mentioned above
i.e.	id est (Latin) = that is
IEEE; or I.E.E.E.	Institute of Electrical and Electronic Engineers
il.	illustrated; illustration; illustrations
IMA	International Mineralogical Association
Inst.	Institute; Institution
Int.; or Inter.	International
Invest.	Investigation; Investigations
IUCr	International Union of Crystallography
IUGG; or I.U.G.G.	International Union of Geodesy and Geophysics

-J-

J.; or Jour.	Journal; Journals
Jb.	Jahrbuch
Jg.	Jahrgang
J.G.R.; or J. Geophys. Res.	Journal of Geophysical Research
J. Geol.; or Jour. Geol.	Journal of Geology
J. Geol. Educ.	Journal of Geological Education

-K-

Kryst.	Krystallographie

-L-

Lab.	Laboratory
Lett.	Letter; Letters

IV. COURSE XII BIOGRAPHIES

-M-

M.; or Min.	Miner; Miners; Mining
Mag.	Magazine
MCZ; or M.C.Z.	Museum of Comparative Zoology (Harvard)
Mem.	Memoir; Memoirs
Metal.; or Metall.	Metallurgist; Metallurgists; Metallurgy; Metallurgical
Miner.; or Mineral.	Mineralogist; Mineralogists; Mineralogy; Mineralogical
MIT; or M.I.T.	Massachusetts Institute of Technology
Mitt.	Mitteilungen
Mod.	Modern
Mon.	Monograph; Monographs
MSA; or M.S.A.	Mineralogical Society of America
Mtg.	Meeting; Meetings
Mus.	Museum

-N-

NAE; or N.A.E.	National Academy of Engineering
NAS; or N.A.S.	National Academy of Sciences
NASA; or N.A.S.A.	National Aeronautical and Space Administration
NAS-NRC	National Academy of Sciences - National Research Council
Nat.	National; Naturalist; Naturalists; Nature; Natural
N.E.L.	Naval Electronics Laboratory
No.; or no.	Number; Numbers
NOAA; or N.O.A.A.	National Oceanographic and Atmospheric Administration
Nov.	Novitates
NRC; or N.R.C.	National Research Council
ns; or n.s.	new series (e.g. Science n.s. 2)
NSF; or N.S.F.	National Science Foundation

-O-

Oc.; or Oc. Pap.	Occasional; Occasional Papers
ONR; or O.N.R.	Office of Naval Research
OSRD; or O.S.R.D.	Office of Scientific Research and Development

-P-

P.; or p.	Page; Pages
Pal.; or Paleont.	Paleontologist; Paleontologists; Paleontology; Paleontological
Pam.	Pamphlet

Pet.; Petr.; or Petrol.	Petrologist; Petrologists; Petrology; Petrological
Petrog.	Petrographer; Petrographers; Petrography; Petrographical
Petrol. (in context)	Petroleum
Phil.; or Philos.	Philosophy; Philosophical
Pl.; or pl.	Plate; Plates
Planet.	Planetary
Port.; or port.	Portrait; Portraits
PP; or P.P.	Professional Paper; Professional Papers
Pr.; or Proc.	Proceedings
Prog.	Progress
Pt.; or pt.	Part; Parts
Pub.	Publication; Publications

-Q-

Q.; or Quart.	Quarterly
Q.J.; or Q. Jour.	Quarterly Journal

-R-

RDB; or R.D.B.	Research and Development Board
Rec.	Record; Records
Res.	Research
Rev.	Review; Reviews; Revue; Revues
RLE; or R.L.E'.	Research Laboratory of Electronics (M.I.T.)
Rp.; Rep.; or Rept.	Report; Reports

-S-

S.	South
s.; or ser.	series
Sci.	Scientist; Scientists; Science; Sciences; Scientific
Sec.; or Sect.	Section; Sections
Senckenb. Leth.	Senckenbergiana Lethea
Seism.; or Seismol.	Seismologist; Seismologists; Seismology; Seismic; Seismological
SEPM; or S.E.P.M.	Society of Economic Paleontologists and Mineralogists
[sic]	Thus; [sic] inserted to note original spelling
SIO; or S.I.O.	Scripps Institution of Oceanography (La Jolla)
Soc.	Society; Societies
SpP; or Sp.P.	Special Paper; Special Papers
Sp. Pub.	Special Publication; Special Publications

Summ.   Summary; Summaries
Sup.; or Suppl.   Supplement; Supplements
SY   School Year; e.g. SY 1936-37
Symp.   Symposium

                                    -T-

T.; or t.   Tome; Tomes
T--1960   Thesis--1960
Tab.; or tab.   Table; Tables
Tech.   Technical
The Tech   The Tech (M.I.T.)
Tech. Rev.   Technology Review (M.I.T.)
TEN; or T.E.N.   Tech Engineering News (M.I.T.)
Tr.; or Trans.   Transaction; Transactions
T.S.S.   Teachers' School of Science (Boston)

                                    -U-

Un.   Union
Univ.   University
U.S.   United States (of America)
U.S.A.E.C.   United States Atomic Energy Commission
U.S.A.F.   United States Air Force
U.S.B.M.   United States Bureau of Mines
U.S.C.& G.S.   United States Coast and Geodetic Survey
U.S.G.S.   United States Geological Survey
U.S.N.   United States Navy
U.S.N.M.   United States National Museum

                                    -V-

Vol.; vol.; or v.   Volume; Volumes
Verh.   Verhandlungen

                                    -W-

W.P.B.   War Production Board
WHOI; or W.H.O.I.   Woods Hole Oceanographic Institution
W-SP; or W.-S.P.   Water-Supply Paper; Water-Supply Papers

                                    -Y-

Yearb.   Yearbook; Yearbooks

                                    -Z-

Z.; Zs.; or Zeitschr.   Zeitschrift

Symbols

...	Ellipsis, indicating material omitted
*	Asterisk, usually denoting a footnote
**	Double asterisk, denoting a second footnote on the same page
†,††	Daggers, used to denote a second or third footnote on the same page
(‡)	Mimeographed; multilithed; or Xeroxed
(3) 40	read: series 3, volume 40
40/3	read: volume 40, part or number 3
(1960)	at end of reference indicates the year of publication
( )	parentheses, generally used to enclose supplementary or parenthetical information in lieu of commas or quotation marks
[ ]	Brackets, generally used to enclose corrective or additional material
" "	Double quotation marks, used to enclose quoted matter
' '	Single quotation marks, used to enclose a quote within a quote, e.g.: "He cited Shakespeare, who wrote 'To be ...,' and then went on to say, 'That is the question.'"
8 vo	Octavo; eight leaves or sixteen pages, to a sheet

----

(1)
WILLIAM BARTON ROGERS
(1804-1882)

WILLIAM BARTON ROGERS

MIT: 1861-1865-1882

Founder, organizer, and the first president of M.I.T.; first professor of physics and geology; internationally known geologist renowned for his work in the Virginias; scientist, scholar, lecturer, and innovative educational philosopher; William Barton Rogers was preeminent as an applied scientist for half his lifetime, and since his death almost a century ago he has become a legend and the subject of many biographies, memorials, commentaries and such. Almost without exception his personal characteristics, technical skills, and professional accomplishments are described in superlatives. When at his best he was without a peer among the scientific men of his age in addressing a sophisticated audience. His public lectures, noted for their felicity of language, choice of phrase and clarity of expression, always attracted a large attendance. His classroom presentations and laboratory demonstrations were often so impressive and dramatic as to draw spontaneous applause from the students who customarily crowded into the lecture halls to hear him. His extemporaneous speeches and carefully prepared petitions and memorials were critically effective in getting the first Geological Survey of Virginia established, in preventing the loss of the University of Virginia annual annuity, in creating the American Association for the Advancement of Science from the preexisting Association of American Geologists and Naturalists, and in founding and organizing the Massachusetts Institute of Technology. In the last-named endeavor he introduced instructional innovations that laid the foundation for the great schools of science and engineering in America that followed M.I.T. as a model.

William Barton Rogers was by almost any measure an intellectual giant of his age, and most historians would consider the founding and organizing of M.I.T. as his crowning achievement. It is not to my present purpose, however, to construct yet another biographical sketch of Rogers as the Institute's founder - already more than a dozen exist, and his own postmortem Life and Letters ..., edited by his wife, is an incomparable source of information on his struggles to found his polytechnic school in Boston, as well as a mine of information on his many other activities. In this sketch I attempt primarily to discuss the scientific activities of William, and to some extent those of his three brothers - James, Henry and Robert - as these activities influenced the teaching and practical application of geology, first in Maryland, Pennsylvania, and Virginia; next on a broader geographic scale throughout North America and Europe; and finally in New England and at M.I.T.

William was born in Philadelphia, Pennsylvania, on 7 December 1804, the second of seven children of Irish-born Patrick Kerr Rogers and Hannah (Blythe) Rogers of Scottish descent. His mother, about whom very little is known except that she too emigrated to America from Ireland, died when he was only sixteen years old, leaving four sons between the ages of seven and eighteen - James Blythe (b. 1802), William Barton (b. 1804), Henry

Darwin (b. 1808) and Robert Empie (b. 1813); the other three children, two girls and a boy, had died in infancy. The four survivors just mentioned all became distinguished scientists - James and Robert as chemists, William and Henry as geologists. Because of their achievements they were commonly lauded and referred to collectively as the "Brothers Rogers." Henry and Robert figured prominently at times in William's life and will be mentioned frequently in discussion of the latter's scientific activities.

William received most of his pre-college education from his father, and continued to study independently under his guidance, at the College of William and Mary, while also pursuing regular studies at the College, where his father was Professor of Natural Philosophy and Chemistry. When the latter suddenly died of malaria in August 1828, young William, not yet twenty-four years of age, succeeded him and for the next seven years was Professor of Natural Philosophy and Chemistry, and of Mathematics, having already at such a young age established a reputation as an accomplished public lecturer. At first confining his lectures to physics and chemistry, he quickly turned to geology in 1833 when Henry, who had become interested in the field while in England, returned to Williamsburg and aroused his interest in the subject. Soon he and Henry were agitating for geological surveys in their respective states, Virginia and Pennsylvania. Shortly after the first Geological Survey of Virginia was established in 1835, William was appointed State Geologist. At about the same time Henry was appointed to the same office on the Geological Survey of New Jersey (1835) and a year later on the Geological Survey of Pennsylvania. In 1835 the two brothers were also appointed to prestigious chairs - William as Professor of Natural Philosophy at the University of Virginia; Henry as Professor of Geology and Mineralogy at the University of Pennsylvania.

Now started an intensive program of teaching and geological field studies on the part of both brothers, who worked in close concert and kept one another fully informed of the progress being made. By 1842 they had established the stratigraphy and structure of the great Appalachian Chain in their respective states, and in April of that year, at the third annual meeting of the Association of American Geologists and Naturalists held in Boston, they presented the results of their joint work, and a theory to explain the deformation of the Chain, in their classic paper titled "On the Physical Structure of the Appalachian Chain, as exemplifying the Laws which have regulated the Elevation of Great Mountain Chains generally." (See reference 27--1842f, 1843 in the Bibliography.) Although both brothers had published numerous reports and articles on their respective Survey programs, and were already highly regarded for the impressive volume and excellent quality of their work, the 1842 paper with its novel and imaginative theory of folding attracted immediate and enthusiastic response in America, and enjoyed a similar reaction when read later before the 1842 meeting of the British Association in Manchester.

The field work on which the brothers based their determination of the stratigraphy and structure of the Appalachian Chain was widely acclaimed for its high quality and accuracy, and today is regarded as classic for the region. Not so, however, with their imaginative but rather "bizarre" (as one writer, John Rodgers, 1949, puts it) theory to explain the deformation of the Appalachians - that folded mountains are formed by "wave-like motions of the crust" in response to similar motions in an underlying layer of lava. Although many geologists of the time enthusiastically accepted or supported the speculative explanation (e.g. J. W. Gregory, 1916), leading structural geologists both in America and abroad found aspects of the theory quite unacceptable, and present day tectonic specialists hold a similar view. Nevertheless, the overall result of the paper at the time it was presented and for following decades was to bring to both William Barton and Henry Darwin accolades that gave both highest rank among the world's geologists of the 19th Century.

In due course, after a period of both troubles and accomplishments, Henry resigned his professorship at the University of Pennsylvania in 1846 and settled in Boston, where he was soon giving scheduled lectures on geology at the Lowell Institute. Meanwhile, William turned to vigorous research in chemistry, when his brother Robert joined the faculty of the University of Virginia in 1842. During the next decade the two brothers published more than a dozen joint articles on chemical subjects before Robert left to succeed their lately deceased brother James as Professor at the University of Pennsylvania. This was 1852, one year before William left Charlottesville, after eighteen years at the University of Virginia, to take up permanent residence in Boston with his wife, Emma (Savage) Rogers, whom he had married in 1849.

By 1853, at forty-nine years of age, William had distinguished himself as a practical research scientist (chemist, physicist, and geologist) and geological field investigator, as public lecturer and college professor, and as a state survey and educational administrator. At M.I.T. he would add luster to all of these careers as research scientist, as Professor of Physics and Geology, and as Founder, Organizer, and the first President of the Institute.

Now began the last twenty-nine years of William's life (September 1853 to May 1882) during which he would continue to carry on research and publish numerous articles on physics and geology, give quite a few public lectures, and most of all attain his greatest success by first founding, then organizing, and finally guiding and directing the newly incorporated Massachusetts Institute of Technology through its formative years. In this latter endeavor he saw to it that the Institute's four-year curriculum had appropriate subjects in geology and mining and that a student, if it was desirable, could prepare himself for a professional career in mining geology. It is worth noting that six of the thirteen graduates of the first class, the Class of 1868, took their degree in Course IV, Geology and Mining!

In summary, scientists know William Barton Rogers for his accomplishments in geology (100 articles and reports and two textbooks), chemistry (21 articles, 17 jointly authored with his brother Robert), and physics (28 articles on physical phenomena). Academicians remember him as a professorial colleague at the College of William and Mary (1828-1835), University of Virginia (1835-1853), and M.I.T. (1865-1882). Administrators recall his record as State Geologist and Director of the Geological Survey of Virginia (1835-1842), Chairman of the Faculty at the University of Virginia during a year (1844) of student trouble, and President of M.I.T. (1865-1870; 1878-1881). Historians of science recognize his important services to scientific societies as founding member and one-time president of the Association of American Geologists and Naturalists (1847), founding fellow and 1876 president of the American Association for the Advancement of Science, and charter member and later the third president of the National Academy of Sciences (1879-1882). Three colleges awarded him the honorary LL.D. degree - Hampden-Sidney College (1848), his alma mater, the College of William and Mary (1859), and Harvard (1866). And in final tribute to his memory are a dozen memorials bearing his name - a mountain, a library, a physical laboratory, three professorships, several scholarship and research funds, and four buildings! In comparison with the great scientists who have followed him, he maintains his high standing among them after more than a century. He was truly an intellectual giant of his time.

Dramatic as was his life, even more so was his death as it came suddenly in mid-sentence while he was addressing the graduates of the Class of 1882 in Huntington Hall. His remains are interred in the James Savage burial lot in the Mount Auburn Cemetery in Cambridge. On the back of the massive granite monument of his father-in-law, JAMES SAVAGE, is his simple epitaph:

WILLIAM BARTON ROGERS
Born December 7, 1804
Died May 30, 1882

Close by in the same burial plot lie the remains of Emma (Savage) Rogers (March 4, 1824 - May 18, 1911), William's longtime wife and helpmate (1849-1882), who was of inestimable assistance to him while he labored to get the Institute established and its faculty and curriculum into action. After William's death, Emma continued to support the struggling Institute with funds, gifts of books, and otherwise as long as she lived. Much of the estate she left was willed to the Institute, and the <u>Emma Rogers Room</u> (10-340) perpetuates her name as the first of M.I.T.'s "<u>leading ladies</u>." Surely one of her most enduring monuments, for which every historian of M.I.T. has been or will continue to be deeply grateful, is the two-volume biography of her distinguished husband, <u>Life and Letters of William Barton Rogers</u> (edited by his wife, with the assistance of William T. Sedgwick, in two volumes, I and II). Cambridge: The Riverside Press; Houghton Mifflin, and Company; (1896).

In 1975, a century and a decade after its founding, the Massachusetts Institute of Technology stands as Rogers' enduring monument. This fact is declared to all the world by the four words cut boldly into the limestone architrave that surmounts the great Ionic columns at the 77 Massachusetts Avenue entrance (see photograph below):

WILLIAM BARTON ROGERS FOUNDER

View of the entrance to M.I.T. at 77 Massachusetts Avenue, looking east, with the Green Building rising conspicuously between but behind the two large domes. The "Rogers Building" (Building 7) includes the portion under the near dome with the columns and the wing to the left that houses the School of Architecture.

(MIT Photo by Campbell, 1975)

(1) WILLIAM BARTON ROGERS

BIRTH, ANCESTRY, FAMILY LIFE, AND EARLY EDUCATION*

William Barton Rogers, the Founder, Organizer, and first President of M.I.T., and the Institute's first Professor of Physics and Geology, was born in Philadelphia, Pennsylvania on 7 December 1804.

He was the second of seven children of Patrick Kerr Rogers (1776-1828) and Hannah (Blythe) Rogers (1775-1820), only four of whom survived to adulthood. The children, in order of birth, were James Blythe (1802), William Barton (1804), Henry Darwin (1808), two girls who died in infancy, Robert Empie (1813), and Alexander (1815) who died in infancy. The four sons who lived to adulthood - James, William, Henry and Robert - all achieved distinguished careers and were often referred to as "the brothers Rogers." More will be said about them as a group farther on.

The father of the Rogers children, Patrick Kerr Rogers, came from the north of Ireland, not far from Londonderry, and was the eldest of twelve children of Robert Rogers (1753-1803), the fourth of the name in lineal descent, and Sarah Kerr (1753?-1790). Robert has been described by Ruschenberger (1886, p. 106) as:

> "... a well-to-do Irish gentleman, liberal in his views, hospitable, convivial and duly appreciative of education and learning,"

and Sarah, according to tradition, writes the same biographer (1886, p. 106):

> "... was sprightly, conspicuous in conversation, and ever ready to discuss and advocate the New-Light doctrines of the Presbyterian Church, of which she was a member."

---

* The achievements of William Barton Rogers as research scientist; as professor of natural philosophy, and of geology and physics; as an educator and academic innovator; and especially as founder, organizer and first president of M.I.T. are well and widely known. Many biographical sketches, memorials, necrologies and laudatory statements have been published. Of special importance as information sources are "A Sketch of the Life of Robert E. Rogers, M.D., LL.D., with biographical notices of his father and brothers," by W. S. W. Ruschenberger (1885) and "Life and Letters of William Barton Rogers," by his wife [Emma Savage Rogers], with assistance of William T. Sedgwick (1896).

In preparing the following biographical sketch I have drawn heavily on the two references specifically cited in the preceding paragraph, but have also used a number of the other references listed under Biographies and Biographical References to William Barton Rogers (1804-1882) following Rogers' Bibliography at the end of this sketch.

In crediting information and quotations I use the author-date-page method of citation: e.g. Ruschenberger (1886, p. 104) in most cases, but for Mrs. Rogers' "Life and Letters of William Barton Rogers," I use the following citation: Title-volume-page; e.g. (Life and Letters I, p. 15).

When the time came for Patrick to decide on a profession, he eschewed a clerical career, because his opinions were not rigidly orthodox, and instead started on a commercial career in Dublin. This lasted a few years until about the time the Irish Rebellion broke out in the spring of 1798. Holding views hostile to the Government, he decided to flee for fear of arrest, and after almost three months at sea he arrived in Philadelphia in August, 1798. Shortly thereafter he was appointed a tutor in the University of Pennsylvania, and in the following three years he pursued studies that led to the degree of Doctor of Medicine from that institution.

While still a medical student, Patrick married Hannah Blythe on 2 January 1801. She has been described by Ruschenberger (1886, p. 107) as:

> "... an intelligent woman a year older than himself [i.e. Patrick], endowed with a cheerful and affectionate disposition."

She was the youngest daughter of James Blythe, a Glasgow-born publisher and stationer living in Londonderry, and Bessie (Bell) Blythe, a daughter of James Bell, a mathematical instrument-maker and an English citizen of Londonderry. After the death of their parents, Hannah, with her two older sisters, emigrated to Philadelphia, where she and Patrick Rogers first met.

Thus it can be seen from the foregoing genealogical information that the seven children of Patrick K. and Hannah B. Rogers carried in their veins a mixture of Irish, Scotch and English blood. All seven Rogers children were born in the United States, and the four brothers who survived infancy were first taught at home by their father, then at William and Mary, where he was professor. Later they all left Williamsburg to pursue their own respective careers, but William soon returned to occupy the chair left vacant by his father's death in 1828.

Times were hard for the growing Rogers family, and all the members were harassed at one time or another by ill health; as a matter of fact, all of the family seem to have had rather delicate constitutions, and this may well account for the fact that only four of the seven children survived to adulthood.

In spite of his training for a medical career, Patrick's practice of medicine was adversely affected by the necessity of a trip back to Ireland, and after his return, being unable to start anew successfully, he contracted debts as a consequence of which he was plagued by creditors for the remainder of his life.

Failing to be appointed to the chair of chemistry at the University of Pennsylvania, and to the Professorship of Natural Philosophy, Chemistry, and Mineralogy at the newly established University of Virginia, his tender of service being declined by no less a person than Thomas Jefferson himself, Patrick finally was elected Professor of Natural Philosophy and Chemistry in the venerable College of William and Mary (founded in 1693 and second only to Harvard in age) in 1819.

The next nine years, until his death on 1 August 1828, were difficult in many ways and were saddened by the death of Hannah Blythe Rogers* in July 1820, scarcely a year after he assumed his professorship. At the college, finances were so straitened that the salaries of the staff were often in arrears, in one instance as much as thirteen months, and Patrick had to make all of the apparatus for his lectures. In this latter task he regularly elicited the aid of his young sons, and they were thus able to develop unusual skill in operating tools for shaping wood and metals.

Being an enthusiastic teacher, and deeply devoted to his children, Patrick not only taught his four surviving sons the manual skills just mentioned but also inspired in them a love of knowledge and a desire to teach. In these intimate relationships, made even closer when the mother of the family died, father and sons were drawn together so closely that their mutual love and concern for one another persisted throughout their entire lives and was a constant characteristic of their prolonged correspondence, as is evident in <u>Life and Letters of William Barton Rogers</u> (see complete reference in Bibliography farther on).

As a result of this early home teaching, and of their later close personal relationships, the four Rogers brothers were so much alike in thought and action, and so closely knit in their personal affairs and concerns for one another, that one can hardly discuss the life of one of the four without some reference to the other three. For this reason there follows a brief discussion of the careers of William's three brothers.

THE FOUR BROTHERS ROGERS

As stated earlier, only four of the seven children of Patrick Kerr and Hannah (Blythe) Rogers survived to adulthood. These four - James, William, Henry, and Robert - all became eminent men of science, after their own early periods of frustration and adversity, and as such were often referred to as the "Brothers Rogers."** In the decades immediately

---

* Little seems to be known about the life and character of Hannah Blythe Rogers, partly perhaps because she died before her sons attained maturity - William was only 16 - and they seem not to have recorded much of their remembrance of her.

** See W. H. Ruffner's "The Brothers Rogers" (1898) cited in the list of Biographies, Biographical References and Sources of Information <u>re</u> William Barton Rogers (1804-1882) - Appendix B.

following the Civil War, when science and technology began the impressive development that would lead to the amazing Twentieth Century, there were probably nowhere in the United States four brothers who commanded more respect from their scientific peers, than did the four Virginia Rogers.

Strongly influenced and motivated by their teacher-father, and by a cheerful and affectionate mother who was taken from them too early to have left the full impact of her personality, the brothers shared an unusual devotion to one another and to their father. Again and again they sought the advice and assistance of one another, and the correspondence in Life and Letters gives ample proof of how much they depended upon one another throughout their entire lives.

While it is not appropriate to follow the career of each brother, interesting as that would be, I do feel it necessary to summarize the lives of James, Henry and Robert, because all of them at one time or another, and in one way or another, shared important relationships with their brother William, the subject of this sketch.

James Blythe,* the eldest son, earned a Doctor of Medicine degree from the University of Maryland in Baltimore in 1822, but after a few years of practice left medicine and worked as an industrial chemist for a short time. This occupation did not prove to his liking, however, and he next took up teaching, and in this he found his life work. After lecturing on chemistry in Baltimore for a short time, he went to Cincinnati where for four years (1835-1839) he was Professor of Chemistry in the Cincinnati College, meanwhile assisting William in the geological survey of Virginia during the summers. In 1840 he returned to the East Coast, and served successively as Lecturer on Chemistry in the Philadelphia Medical Institute (1841), Professor of General Chemistry in the Franklin Institute (1844-1847), and finally as Professor of Chemistry in the University of Pennsylvania until his death in 1852, in his fifty-first year.

The youngest of the four brothers, Robert Empie, also trained for a medical career and was graduated in medicine from the University of Pennsylvania in 1837, but he did not enter practice. Instead he assisted William in his geological survey work and worked as a medical chemist for five years until appointed Professor of Chemistry in the University of Virginia in 1842. Upon brother James' death in 1852, he succeeded him as Professor of Chemistry in the University of Pennsylvania. Within four years he also became Dean of the Faculty, and continued in this double office for the next twenty-five years. In May, 1877, he resigned to accept

---

* See Joseph Carson, A Memoir of the life and character of James B. Rogers, M.D. Philadelphia: T. K. and P. G. Collins, Printers, [pamphlet], 22 p., (1852).

appointment as Professor of Medical Chemistry and Toxicology in the Jefferson Medical College of Philadelphia, where he continued until his death in 1884, at age seventy-one.*

Henry Darwin, the third oldest of the four brothers, and the closest to William in professional activities, received his education in part from his father and then at the College of William and Mary. In 1831 he became Professor of Physical Sciences in Dickinson College (Carlisle, Pennsylvania), and afterward Professor of Geology in the University of Pennsylvania, an appointment he held for many years. He began his active geological career with a survey of New Jersey, publishing a report and geological map of the State in 1835. He then directed the Geological Survey of Pennsylvania from 1836 to 1856, publishing annual reports and a final report on the geology of the State consisting of two volumes and an atlas - a combined work that still stands as one of the major geological works of the age. In 1857 he was appointed Regius Professor of Geology and Natural History in Scotland's University of Glasgow, where he continued until his death in 1866 in his fifty-eighth year. His biographer, J. W. Gregory,** writes of his work as follows (p. 33):

> "His contributions to structural geology rank as of the highest importance. He laid the foundation of modern work on the origin and distribution of mountain chains and the major relief of the globe. He, more than any other man, was the founder of modern structural geology."

The close relationship between the four Rogers brothers is well summarized by J. P. Munroe in his article <u>William Barton Rogers - Founder of the Massachusetts Institute of Technology</u> (Tech. Rev. 6/4: 8-9, 1904) as follows:

> "As will appear, William and Henry were, of the brothers, those most closely associated in scientific work and in planning the Institute of Technology. The extraordinary intimacy and interdependence of these two, however, was only a little closer than that which bound together all four of the brothers Rogers, each to each, in a personal and scientific friendship most notable and beautiful. A warmth of phrase due to their Keltic ancestry and to their Southern residence gives to their familiar letters a tone which to New England ears savors of hyperbole. But there is no touch of

---

* The interested reader will find a detailed discussion of Robert's career, along with much information on his father and brothers, in <u>A Sketch of the Life of Robert E. Rogers, M.D., LL.D., with Biographical Notices of his Father and Brothers</u>, by W. S. W. Ruschenberger, Am. Phil. Soc., Pr. 23: 104-146, il., (1886).

** <u>Henry Darwin Rogers</u>. Glasgow (Scotland): James MacLehose and Sons, 38 p., (1916). See complete reference in Appendix B, (1916); also Ms. Gerstner's 1975 paper in the same Appendix.

insincerity, no suspicion that every word of solicitude in matters of health, or of rejoicing in scientific and professional progress is not as deeply felt as is glowingly expressed. The early years of all of them were filled with hardships and were often made discouraging by the unjust or unappreciative acts of others. Moreover, the fortunes of each brother greatly fluctuated, one meeting with success while another was encountering adversity, one advancing rapidly in public recognition while another, for the time, seemed destined to obscurity. Never, however, into their correspondence, never into their relationship, did there enter a word or thought of envy on the one hand or of exultation on the other. The success of one was the joy of all, the passing failure of another was the sorrow of all. Such harmony, such warmth of affection, such mutual helpfulness as this, could have but one result, - that together they should achieve results in science and reach positions of responsibility and honor which separately they would hardly have been likely to attain. Better than this, their rare concern in one another's affairs gave them a breadth of outlook and a diversity of interests unique in their generation. All of them followers of science, their paths were sufficiently diverse to cover practically the whole scientific field. And, while each perfected himself in his specialty, he followed also with such absorbed interest the work of the others as to attain in all branches of science a marked proficiency. This breadth of view gave them that authority and that power which made them such conspicuous leaders in the scientific development of the United States.

"More marked and most fruitful was this closeness of interest, this breadth of knowledge and this harmony of investigation in the case of William and Henry Rogers. Not only did it place them among the great leaders in geology, not only did it secure to them important State offices in scientific work and the friendship of learned men throughout the world, but it led them also to survey the whole field of pure and applied science with such clear and prophetic vision as to enable them to conceive, as early as 1846, a plan for a polytechnic school which, many years later, was to develop, along those very lines and under the guidance of William, into a school of technology such as we know today."

WILLIAM'S EARLY ACADEMIC CAREER (1819-1835)

There seems to be little known about the childhood or boyhood of William Barton Rogers and his brothers. As mentioned earlier on, they received most of their pre-college education at home, at the hands of their devoted father, but William recorded* that he and his older brother James did receive instruction in Latin at Baltimore College (probably Baltimore City College, the Public High School of Baltimore). When Patrick Rogers

---

* Life and Letters I, p. 15.

moved his family from Baltimore to Williamsburg, in October 1819, in order to assume his professorship in the College of William and Mary, William Barton and his older brother James were able to attend regular classes at the College during SY 1819-20. Thereafter, for several years, William seems to have experienced ill health much of the time, but did find time to read widely* and even to deliver an oration at the celebration of the third "Virginiad," at Jamestown, Virginia in May 1822. The youthful oration, delivered by 17-year old William, and printed in the Richmond [Va.] Enquirer of June 4, 1822, clearly indicates how early he was developing that "felicity of expression" so often mentioned by his later biographers:

> "... The first Virginian colonists bade a final adieu to the thronged land of their nativity. Having taken an affectionate farewell of their friends and dearest relations, they steered toward the ample shores of America ...
>
> "As they sailed into the Chesapeake, they viewed this spacious bay with admiration and delight, and found themselves enclosed in a vast amphitheatre formed by the distant forests which skirted its blue waters. The jutting points of land opened, as they advanced, into broad extended shores, or retired as if by enchantment. While the eye surveyed the rich exuberance of vegetation, and the diversified tints of the foliage which painted the varied landscape on every side, the heart dilated with the exulting anticipation of unequalled felicity, and the enraptured imagination dwelt only on dreams of delight. ..."

By the time he had reached twenty William had prepared four propositions to be included in his father's textbook, An Introduction to the Mathematical Principles of Natural Philosophy, and was reported by his father, in a letter to Thomas Jefferson, to have a "very extraordinary passion for physico-mathematical sciences." (Life and Letters I, p. 26.)

In 1826, when only 22 years old, he and his younger brother Henry opened a school at Windsor, near Baltimore. Here he gained valuable experience in both teaching and administration, while also delivering public lectures on the side. His regular lectures, on "Natural Philosophy," had to be delivered without a single specimen or piece of apparatus, yet they were well attended and enthusiastically received, clearly indicating that even at that early age he already possessed unusual powers of expression.

At this time, the autumn of 1826, Robert, now thirteen years old, left his father alone in Williamsburg and attended the school in Windsor

---

* Munroe (1904, p. 10) states that during the early 1820s William studied mathematics, physics, the Classics and modern languages, mainly under his father's guidance, but does not indicate whether these studies were carried on at William and Mary or at home. This independent study period seemingly ended in 1825 when William entered a mercantile firm in Baltimore.

operated by his older brothers Henry and William. Meanwhile, James had been graduated in medicine in 1822, from the University of Maryland in Baltimore, and had opened an office and formed a partnership for medical practice with a friend. His efforts, however, met with little success and he turned to chemical work later on, as previously mentioned.

Within a year of the time the two brothers had initiated their school at Windsor, William was giving lectures on astronomy and railroads at the Maryland Institute in Baltimore, and though his lectures were well attended and enthusiastically received, his hope for a permanent position in the Institute went for naught. Instead, he was given the responsibility for organizing and managing a High School to be opened by the Institute in May 1828. He and Henry had earlier visited the High School of the Franklin Institute, presumably to gain knowledge that would help them manage the Maryland School. The latter school was just getting underway, with favorable prospects, when Patrick Rogers suddenly died of malaria on 1 August 1828. Two months later William was chosen to succeed his father at William and Mary as Professor of Natural Philosophy and Chemistry.

Although he was not yet twenty-four years of age when he was elected to this prestigious chair (an appointment further enhanced two years later when the students petitioned the Board of Visitors to elect him to the then vacant chair of mathematics), William had had, in actuality, quite a little experience in both teaching (lecturing) and administration, and seemed to be confident that he could meet the responsibilities of the position. Nevertheless, he expressed in his "inaugural"* address the humility he felt in trying to fill the post that had been occupied so ably by his father during the preceding nine years (1819-1828). Once again William summoned his ever increasing felicity of language in expressing his feelings upon assuming the post so recently held by his much revered father and the responsibilities of his appointment. According to an account that appeared in the November 12, 1828 issue of the Phoenix Ploughboy (of Williamsburg, Va.), William included the following remarks in his Introductory Address at the official opening of the fall term of the College of William and Mary on 27 October 1828:

---

* It was the custom at William and Mary for a professor to initiate the formal opening of the new school year by an appropriate address. In 1827 Professor Patrick Rogers had delivered this address and had included in his remarks a eulogy of his lately deceased friend and colleague, Dr. Wilmer, President of the College. Now, in 1828, William, as the successor to his recently deceased father, had been selected to give the "Introductory Address," which could, in fact, be considered his "inaugural" address. (Life and Letters I, p. 58.)

"ADDRESS OF PROFESSOR ROGERS

"In entering upon the duties which have been devolved upon me by the governors of this institution, I am impressed with feelings which it is difficult to describe, - feelings that arise from the peculiar relationship in which I stand to the revered individual whom I have succeeded.

"To have returned to the scenes of my early youth - scenes hallowed in my bosom by every fond and pleasurable sentiment; to be enabled to renew the delightful associations which even the absence of several years has but slightly impaired; to tread again within these consecrated precincts, where at every step the remembrances of former years are awakened into animated existence, and where the very air I breathe seems almost to speak of companions dear to my affections, of social study and collegiate ambition - is, I confess, attended with emotions of the purest and liveliest satisfaction. And I may be permitted to add that these sentiments are heightened by reflecting on the circumstances in which I am about to renew my connection with these scenes, and to become again an inmate in the halls of my venerable Alma Mater.* But, alas! mournful considerations sadden these reflections, and, indulging in them, gratification is converted into grief."

Thus far in his career William had not yet developed an interest in geology. His teaching and public lectures dealt largely with the mathematics, astronomy, physics and chemistry of the time, all subjects being taught without the benefit of any laboratory work and almost no apparatus for lecture table demonstrations. For the next four years, 1829 to 1833,

---

* Although William referred to the College of William and Mary as his Alma Mater, I have been unable to find any evidence that he was actually graduated from that or any other college. In reply to my question about a baccalaureate degree, addressed to the College of William and Mary, Ms. Kay J. Domine, College Archivist, responded as follows in a letter dated September 8, 1975:

> "William Barton Rogers attended the College here from 1819 to 1824, but not always continuously and not always full-time. 1822 seems to be the last year of his regular studies as a student. As far as his actually graduating, there is indeed a large question. I have found no indication whatsoever that Rogers received any degree here (except for his later honorary degree). He was in the habit of referring to this as his Alma Mater and so are all of his biographers as you mentioned. William G. Guy gave the address at the 20th Annual Honors Convocation at William and Mary on Nov. 16, 1955. In his speech, entitled 'William Barton Rogers,' he said that Rogers graduated with great distinction in 1822. I have no idea where he got this information, but since I can find no verification of this, and since we do know so much about Rogers otherwise, it seems possible he never actually, officially graduated. I cannot say that for sure, of course."

his two professorships kept him fully occupied, and he engaged himself in teaching with enthusiasm and with all the vigor that his persistent ill health permitted. All this was to change, however, in a drastic way in 1833 when brother Henry returned from England.

WILLIAM BECOMES INTERESTED IN GEOLOGY (1833)

While William was rapidly gaining both teaching and administrative experience in his new position in Williamsburg, Henry was becoming interested in the socialism then being espoused by Robert Owen and his followers in England and supported by many prominent scientists of the time. In 1831 he went to England with Owen's son, Robert Dale Owen, to work with the former for "purely philanthropic objects" (Life and Letters I, p. 103). While in England, according to his biographer, J. W. Gregory (1916, p. 7), Henry studied chemistry under Edward Turner, then Secretary to the Geological Society of London, who seems to have stimulated anew the interest in geology that he had previously developed when making a survey for the projected railroad between Boston and Providence. No doubt attending lectures on geology by de la Beche, going on field trips, and attending meetings with other leading geologists of the day, further heightened his interest and soon led to his election as a Fellow of the Geological Society of London. By this time, 1833, he was associating with many of the intellectually élite of the time - Brewster, Darwin, de la Beche, Faraday, Lyell, Murchison, Phillips, Playfair, Sedgwick, Wheatstone and many other lesser figures. Soon William would join Henry in close friendship with these and many other leading European scientists, friendships that would continue throughout the lives of both brothers. Here I shall leave Henry's continuing career, which is presented in detail by J. W. Gregory in his excellent biographical sketch, Henry Darwin Rogers (see Bibliography and List of Biographies farther on), and return to William and his reaction to Henry's renewed enthusiasm for geology.

When Henry returned from Europe in the summer of 1833, he was so enthusiastic about geology that he quickly stimulated William to develop a similar interest, with the result that the latter was soon investigating the thermal spring waters and commercially useful mineral deposits of Virginia. These investigations, being of obvious practical nature, soon brought both William and Henry into close friendship and correspondence with leading politicians and scientists all over the United States, and marked the beginning of their respective lifelong careers in geology.

Within a year or so William had published four papers in the Farmer's Register, all dealing with the marls and greensands of Virginia; an article on artesian wells; and several papers on apparatus for chemical analysis of carbonates (see Bibliography farther on). In calling attention to the use and potential importance of marls and greensands for agricultural purposes, William aroused widespread interest throughout Virginia, an interest that no doubt soon stood him in good stead when he joined others* in agitating for the establishment of a geological survey.

ESTABLISHMENT OF THE FIRST GEOLOGICAL SURVEY OF VIRGINIA (1835)**

By late 1834 both William and Henry were convinced that the Commonwealth of Virginia should have a geological survey like those lately established in Massachusetts (1830), Tennessee (1831), and Maryland (1831). On 9 February 1835 William appeared before a special committee appointed by the Legislature to report upon a geological survey, and so impressed the members that he was asked to address the Legislature publicly the next day. According to his own letter to Henry, dated February 11, 1835 (Life and Letters I, p. 116-117), he made extemporaneously a convincing speech of about an hour's length, particularly emphasizing the benefits that a survey could bring to Virginia. Immediately thereafter he drew up the report for the special committee, and on the basis of this report the General Assembly, on 6 March 1835, passed an Act establishing the first Geological Survey of Virginia and providing for a State Geologist.

William was the obvious choice to organize and direct the newly established bureau, and within a month after passage of the founding Act he was appointed State Geologist, a position he held from 1835 through 1841.

1835, A CRUCIAL YEAR IN A PRODUCTIVE DECADE

The year 1835 was to be one of the most crucial in William's life for it brought three outstanding honors and a doubling of his responsibilities. Soon after being appointed State Geologist, and charged with the task of

---

* According to Bevan (1935, p. 65), several persons, including William Barton Rogers, were, as early as the fall of 1833, suggesting the establishment of a Geological Survey of Virginia. He quotes a letter from Peter A. Browne, of Philadelphia, the Corresponding Secretary of the Geological Society of Pennsylvania, to Governor Floyd, dated September 30, 1833, in which Browne discusses how the work of a geological survey could be of great benefit to the State and its citizenry.

** At the Richmond meeting of the Virginia Academy of Science on 3 May 1935, Arthur Bevan, then State Geologist, presented a paper before the Geological Section in commemoration of the centennial of the establishment of the first Geological Survey of Virginia, and of the appointment of W. B. Rogers as State Geologist. This paper, titled "William Barton Rogers, First State Geologist of Virginia (1835-1841)" was published in the Proceedings of the Virginia Academy of Science for 1934-1935, pages 63-67, and is the source of much of the information in this section.

making a reconnaissance of the State, he was elected to the chair of Natural Philosophy in the University of Virginia. Acceptance of this appointment meant moving from the malarial climate of Williamsburg to the much more healthful one of Charlottesville. This fortunate change of residence may well have saved him from an early death, for he had been plagued with frequent illness; even in the healthier climates of Charlottesville, and later on in Boston, his health remained precarious to the end of his life.

William was now in his thirty-first year, having been for the preceding seven years Professor of Natural Philosophy and Chemistry in the College of William and Mary. Before the year was out he was elected to membership in the Virginia Historical and Philosophical Society, and to the venerable and prestigious American Philosophical Society of Philadelphia. In short, by 1835, at the age of thirty-one, he had arrived as a distinguished scholar!

Now began one of the busiest, most challenging, and most productive decades of his life as a geologist. Not only did he have to face many frustrating and distasteful situations at the University, but simultaneously he had repeatedly to confront unfriendly politicians in Richmond in planning, organizing, staffing, and finally initiating the actual geological reconnaissance called for by the Act of 6 March 1835 that established the Survey.

At the University of Virginia student insubordination accompanied by rioting created a disquieting and disruptive atmosphere that adversely affected his teaching and research, further impaired his health, and soon caused him to yearn for a more congenial and better disciplined environment. His academic responsibilities were made even more distasteful in his later years because as Chairman of the Faculty he had to prepare the institution's report on the students' misbehaviors - a task that greatly distressed his gentle and courteous spirit.

His efforts to establish a geological survey and to organize and conduct a useful program of work were equally frustrating because of having to carry on a never-ending battle with unfriendly politicians in Richmond in order to persuade the legislators there to maintain and financially support the struggling young survey. He would almost certainly have failed, to say nothing of what might have happened to his health, had it not been for advice and active help from his two brothers, Henry and Robert. Henry, as State Geologist of both New Jersey and Pennsylvania, was conducting his program in closest cooperation with the work that William was directing in Virginia. During the same period, Robert temporarily suspended his training for a medical career and went into the field, not only as an assistant himself but also to train others how to conduct field work. Being without trained field assistants - there simply weren't any in those days - William had literally to start from scratch; this meant training young men to perform the actual work of exploration, observation, recording of pertinent geological information, and mapping. Without Robert's active help and Henry's advice, it is almost a certainty that the first Geological Survey of Virginia would not have lasted very long, particularly in the light of what was taking so much of William's thought and energy at the University.

As emphasized in preceding sections, a common characteristic of the brothers Rogers was their devotion to one another and their willingness to come to each other's assistance when the situation demanded. Again and again in later years, William and Henry would seek one another's advice and exchange ideas on geological matters, and both would come to Robert's assistance.

AS STATE GEOLOGIST AND DIRECTOR OF THE FIRST GEOLOGICAL SURVEY OF VIRGINIA (1835-1841)

Appointment as the first State Geologist of Virginia in August 1835 brought to William a host of problems that demanded much of his time and energy for the next six years, at the end of which time the Survey was prematurely terminated on 1 January 1842 for want of legislative appropriations.

The difficulties faced by Rogers and the work he accomplished in spite of them are best summarized by Bevan (1935, p. 67) as follows:

> "When we consider the vast area surveyed by Rogers - Virginia then contained almost 64,500 square miles of land - and the lack of base maps, and when we visualize the arduous conditions of travel and field work, we can but marvel at both the quantity and quality of the work accomplished. Even with our present understanding of the principles of geology and especially of the character of Appalachian structure and stratigraphy, coupled with all the advantages of modern transportation and cartography, it would be no easy task for one man with a few untrained assistants to accomplish a corresponding amount of work in the same length of time.
>
> . . . . .
>
> "William Barton Rogers, as the first State Geologist of Virginia, should thus have appropriate rank among those pioneer geologists in the United States who built with indefatigable labor and intuitive skill the broad foundations of our modern knowledge of the geology and mineral resources of the Virginias."

Suffice it to add that in the opinion of most present day geologists William Barton Rogers' most important contribution to geology was the work that he and his associates accomplished while he was State Geologist of Virginia.

APPENDIX D. (Page 196, Vol. I.)

ON THE NOMENCLATURE OF THE PALEOZOIC FORMATIONS INTRODUCED BY WILLIAM B. AND HENRY D. ROGERS.

SIXTY years ago, when the Geological Surveys of Virginia and Pennsylvania were first organized, the true correlation or equivalence of the Paleozoic terranes of the Eastern United States was only very imperfectly known, and a great diversity of nomenclature was the inevitable result. The young geologists of Virginia and Pennsylvania, the brothers William B. and Henry D. Rogers, in order to avoid the use of names which implied precise correspondence, not yet established, with European or other formations, adopted a system of their own, consisting of numbers. Later they proposed names to correspond with these numbers, the whole period being conceived of as a kind of geological day, and the terms applied varying accordingly, from the earliest formations upwards; viz.: No. 1, *Primal*; No. 2, *Auroral*; No. 3, *Matinal*; No. 4, *Levant*; No. 5, *Scalent*; No. 6, *Pre-meridian*; No. 7, *Meridian*; No. 8, *Cadent*; No. 9, *Ponent*; No. 10, *Vespertine*; No. 11, *Umbral*; No. 12, *Seral*.

Some of these names still survive in common use, and it is interesting to note that the Rogers system of numbers was retained in the Second Survey of Pennsylvania made by Professor Lesley, while the United States Geological Survey has recently adopted these numbers for the folios or final reports of the Appalachian region.[1] The equivalents of the Rogers numbers and names are shown in the following table, which is slightly modified, in the first and second columns only, from the tables in Macfarlane's Geological Railway Guide, prepared for Virginia by William B. Rogers:—

[1] See, also, Major Jed. Hotchkiss in *The Virginias*, vol. i. pp. 96, etc.

CLASSIFICATION AND NOMENCLATURE OF THE PALEOZOIC FORMATIONS.

General Groups.	Subdivisions in Virginia and Pennsylvania.	Rogers Numbers.	Rogers Names.
Carboniferous.	Upper Barren Group.	XVI, XVII.	Seral.
	Upper Coal Group.	XV.	Seral.
	Lower Barren Group.	XIV.	Seral.
	Lower Coal Group.	XIII.	Seral.
Millstone Grit.	Great or Pottsville Conglomerate.	XII.	Seral.
Sub-Carboniferous.	Greenbriar Shales. (Mauch Chunk Red Shale.)	XI.^{b}	Umbral.
	Greenbriar Limestone.	XI.^{a}	Umbral.
	Pocono Gray Sandstone.	X.	Vespertine.
Devonian.	Catskill.	IX.	Ponent.
	Chemung.	VIII.^{f}	Vergent.
	Portage.	VIIIe	Vergent.
	Genesee.	VIII.^{d}	Cadent.
	Hamilton.	VIII.^{c}	Cadent.
	Marcellus.	VIII.^{b}	Cadent.
	Corniferous.	VIII.^{a}	Cadent.
Silurian.	Oriskany.	VII.	Meridian.
	Lower Helderberg.	VI.	Pre-meridian.
	Salina.	V.^{c}	Scalent.
	Niagara.	V.^{b}	Scalent.
	Clinton.	V.^{a}	Surgent.
	Medina.	IV.^{b}	Levant.
	Oneida.	IV.^{a}	Levant.
Lower Silurian. (Ordovician.)	Hudson River.	III.^{c}	Matinal.
	Utica.	III.^{b}	Matinal.
	Trenton.	III.^{a}	Matinal.
	Chazy.	II.^{b}	Auroral.
	Calciferous.	II.^{a}	Auroral.
Cambrian.	Upper Cambrian. (Potsdam.)	I.	Primal.
	Middle Cambrian.		
	Lower Cambrian.		

The Rogers' Brothers' Classification and Nomenclature of the Paleozoic Formations of Virginia and Pennsylvania as given on pages 437 and 438 of *Life and Letters II*. (See footnote on the following page for further discussion of the nomenclature.)*

In his classic 1940-1941 monograph on the "Geology of the Appalachian Valley in Virginia," Butts (1940, pt. I, p. xxxi) states:

> "Use has been made of past contributions to the knowledge of the geology of the Valley. Particular mention should be made of the pioneer work of William B. Rogers while at the University of Virginia and first State Geologist of Virginia. He is still regarded as one of the foremost American geologists. Professor Rogers and his brother, H. D. Rogers, in Pennsylvania, were the first to interpret correctly the structural geology of the Valley, as much as 100 years ago. Professor Rogers' annual reports were illustrated by a series of excellent geological cross sections of the Valley that have been improved only by refinement through later detailed work. The main advance upon Rogers' work has been the subdivision of his large stratigraphic units into smaller units which have been distinguished by suitable names." See reference to Butts (1940) in Appendix B farther on.

A similar opinion was more recently expressed by Woodward (1961, p. 1626) in reappraising the geology of the Appalachians:

> "One could select 1838 as a turning point in Appalachian geology, as W. B. Rogers' report for the Virginia Geological Survey in that year included a section titled 'Geology of the Appalachian Region in Virginia.' If there are earlier uses of this title I do not know them for this article is the grandfather of all such that I have personally read and used. Indeed when I first started geologic field work in Virginia in 1925, for various parts of that State and West Virginia there were no more recent detailed studies than Rogers' classic reports. As with Eaton's geologic map of New York state (my personal copy is dated 1830), I am always startled at the keen perception of these early giants and their ability to grasp the gross elements of the Appalachian region."

---

* (See opposite page)
In their cooperative stratigraphic work in Virginia and Pennsylvania, William and Henry had a dual task: to determine the age and stratigraphic sequence of the marine beds in the Appalachians; and to name the different formations. The result of their joint efforts is shown in the table on the page opposite.

In the above matter, a letter from Henry to William, dated "Philadelphia, October 16, 1841" and reproduced on pages 196 and 197 in <u>Life and Letters I</u>, is of interest. In the letter, Henry suggested the <u>classification</u> and nomenclature shown on the page opposite, and suggested to William that he, with a Dr. Harrison's assistance, "compound three words equivalent to 'morning of day' before the coal, or simply 'ancient morning,' etc.?" This was done, and the brothers used the Latin-derived terms shown in their table, along with the Roman numerals I to XVII. The numerals were used for many years, before being replaced by the more familiar geographic names shown in the second column of the table, but the fanciful Latin-derived terms were essentially ignored by the geologists of the times and never gained wide acceptance.

William published seven annual Reports, and tens of separate articles, on the geology of Virginia, as listed in the Bibliography farther on. His and Henry's numbering of groups of formations west of the Blue Ridge, I to XVII (references 37--1844c and 156--1896a), continued to be used until modern subdivisions of Appalachian formations were made. He interpreted the formations as a great series of strata that had been deposited in an ancient ocean. He recognized the synclinal structure of Massanutten Mountain and described "the extraordinary phenomenon of inversion" of strata exhibited along North Mountain (see reference 14--1838 in the Bibliography farther on). He subdivided the Tertiary formations of Tidewater Virginia into Eocene and Miocene and made the first report on diatomite in the United States as a bed in the Tertiary (reference 31--1843d).

Soon after the Survey was terminated he and Henry combined their geological findings as the basis for a revolutionary and controversial interpretation of the origin of Appalachian structure in a paper "On the physical structure of the Appalachian chain, as exemplifying the laws which have regulated the elevation of great mountain chains generally," (about which more will be said farther on), and in the same year he correctly pointed out the connection of thermal springs in Virginia with anticlinal axes and faults (reference 29--1843b).

All the aforementioned papers, including the seven annual reports, and numerous other separately published articles were reprinted in 1884 in a single volume, Geology of the Virginias, thanks to efforts of William's wife, Emma Savage Rogers (see reference 153--1884).

AS PROFESSOR OF NATURAL PHILOSOPHY IN THE UNIVERSITY OF VIRGINIA (1835-1853)

The University of Virginia, founded by ex-President Thomas Jefferson, first opened its doors to students on 7 March 1825, hence it had been in existence only a decade when William joined its faculty in August 1835. For the ensuing eighteen years, until 1853 when he resigned to come to Boston, he faithfully devoted his energy to a variety of academic activities in addition to his regularly scheduled lectures. These are alluded to at intervals in Life and Letters (reference 155-1896) and need not be repeated here. Suffice it to note that there were times of trouble when, as Chairman of the Faculty, he had to report the facts about student misbehavior (Life and Letters I, Appendix B, p. 413-419), and to argue strongly against the Legislative proposal to withdraw the $15,000 annuity from the University (Life and Letters I, Appendix A, p. 399-412). In spite of these and other distressing events, however, William quickly established himself as one of the outstanding lecturers at the University, and found time to carry on a vigorous research program that yielded more than fifty published articles on a wide variety of scientific subjects.

In recognition of William's services as Professor and as Chairman of the Faculty, the University of Virginia established two different chairs in his honor: one in 1907 and one in 1965. These are discussed farther on in the section titled "Honors."

The interested reader will find an excellent account of William's activities while a professor at Charlottesville in "The Brothers Rogers," by W. H. Ruffner, who was a onetime Chaplain to the University of Virginia. His informative and sympathetic essay was printed in The Alumni Bulletin (published quarterly by the Faculty of the University of Virginia), Vol. V, No. 1, p. 1-13, (May 1898).

A TRIP TO BOSTON, AN IMPORTANT PAPER DELIVERED, AND A CONTROVERSIAL THEORY OF MOUNTAIN BUILDING PROPOSED (1842)

As William in Virginia and Henry in Pennsylvania conducted their respective geological surveys, they accumulated stratigraphic and structural evidence that led in due course to recognition of the great Paleozoic sequence of marine sediments in the Appalachian Chain, and of deformation of those strata in the form of northeasterly trending parallel anticlinal and synclinal folds dying out gently to the northwest but becoming more intensely deformed and broken by faults along the southeastern border of the folded belt. Their initial report on their findings, and their explanation of how the once flat-lying strata were buckled and folded, created a sensation in geological circles in both North America and Europe, and have continued to this day to be cited among the important earlier classic works of North American geology. It should be noted, however, that whereas their field work has stood the test of time, some 130 years, their theory of how the Appalachians were deformed has been the subject of continuous controversy, and remains so even today.

As stated in the preceding paragraph, the field work that led the Rogers brothers to formulate their theory of Appalachian deformation is as valid today as when first described more than 130 years ago. As Rodgers wrote in 1949 (see footnote a little farther on):

> "The first men to comprehend the structure of the Appalachian Mountains were the brothers Henry D. and William B. Rogers, who from 1835 to 1842 studied the mountains from northern New Jersey to southwest Virginia. They worked out the Paleozoic stratigraphy of the area and by means of it deciphered the folds of the middle Appalachians and the thrust faults of the southern Appalachians. [Underlining by R. R. Shrock.] They at-

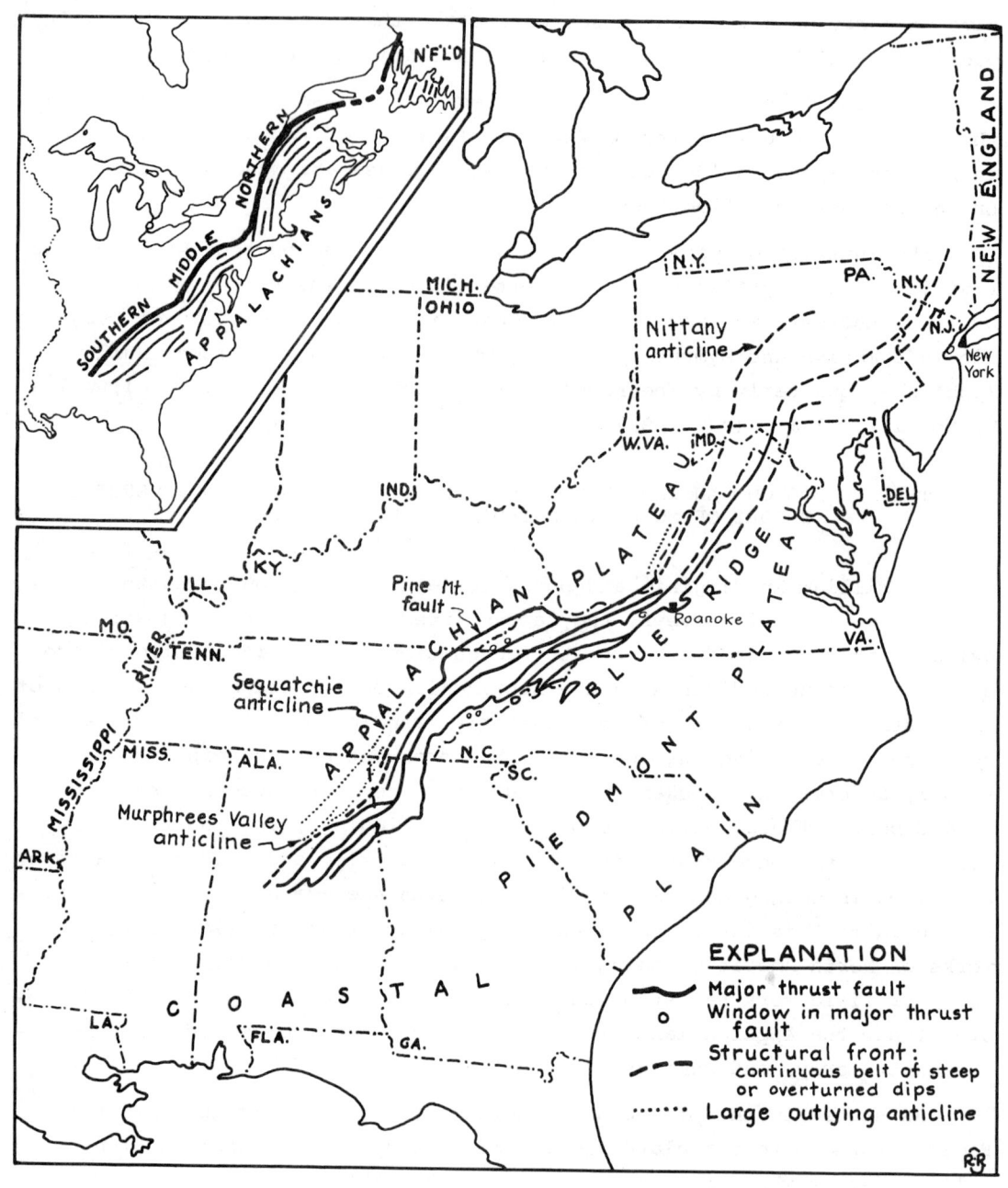

Principal structural features of the Valley and Ridge province in the middle and southern Appalachians. Map redrawn after Francis X. Bland, in John Rodgers' "Evolution of thought on structure of middle and southern Appalachians," A.A.P.G., B. 33/10: 1644, (1949).

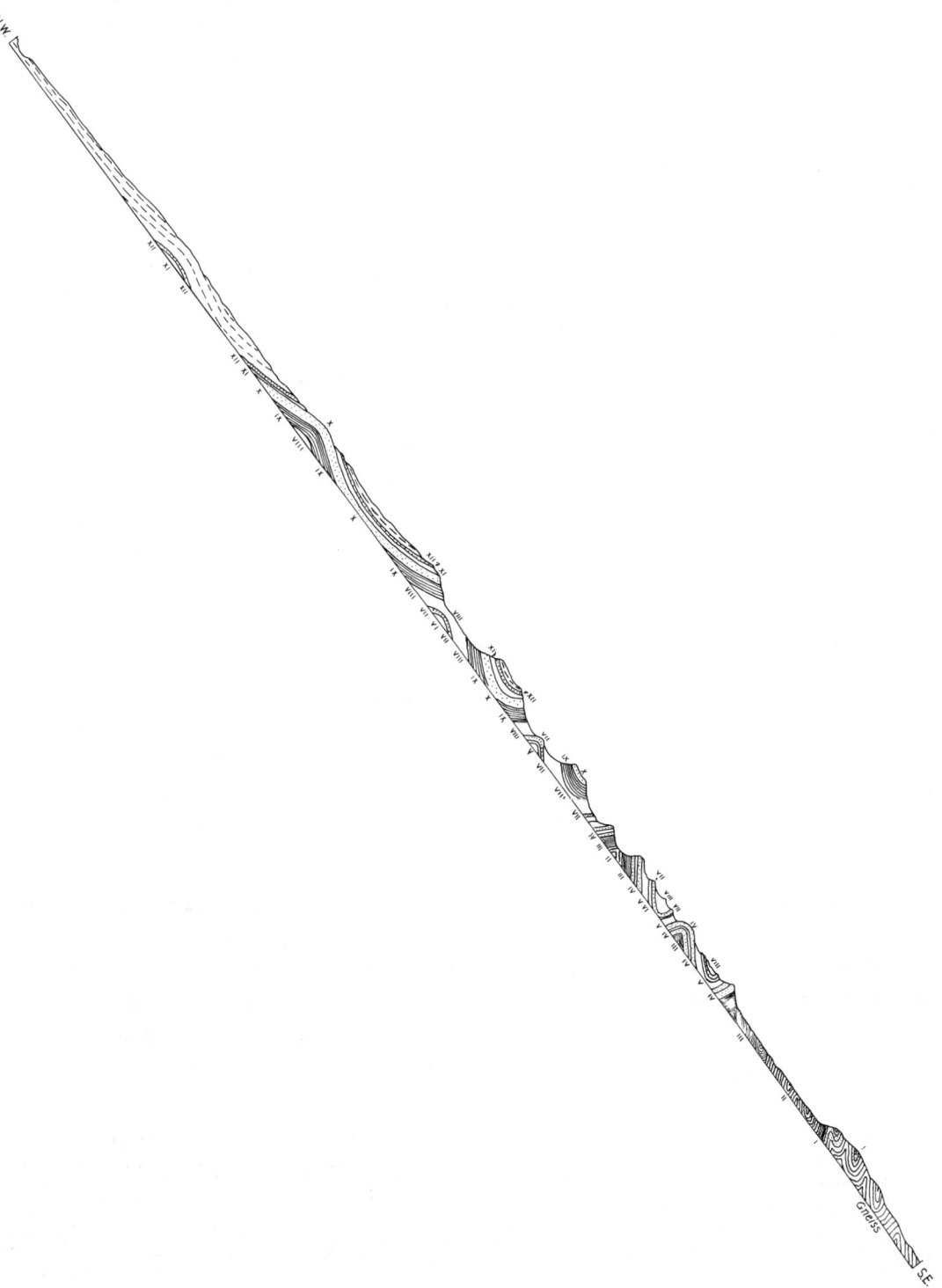

Idealized section across the Appalachians, showing the increase in intensity of folding toward the southeast. Redrawn after Plate XVI in Trans. Assoc. Am. Geol. and Nat. 1, 1840-1842; see Ref. 27--1842f, 1843, in Bibliography.

> tempted to explain the structure as the result of great explosions on the southeast; beginning with Dana, however, later geologists have generally ascribed it to lateral compressions." (p. 1643.)

Latest field investigators have repeatedly commented on the accuracy and high quality of William's field observations, confirming Merrill's 1906 (p. 342) evaluation:

> "When one considers the condition of the country at the time Rogers did his work - the lack of facilities for transportation, the entire lack of maps sufficiently accurate for purposes of plotting, the deep mantle of residuary material that nearly everywhere obscured the more solid rocks, and that there were no railroad cuts or other artificial exposures, such as exist today - one can but admire its accuracy."

Further comments on the quality of William's geological work are included in the preceding section on "As State Geologist, etc."

As stated in a preceding paragraph, the deformation of the Appalachians is as lively a subject of discussion and controversy today, in the 1970s, as it was in the 1840s, with at least three alternate hypotheses under serious consideration, hence the single all-embracing and fully satisfactory theory still remains to be formulated.*

Almost without exception, the biographers of both William B. and Henry D. Rogers unreservedly consider the 1842 paper as the most important single geological contribution of either brother, and one is led to a similar conclusion by the treatment given to the presentation and subsequent reactions in Life and Letters I, p. 208-218 ff.

It should be pointed out, however, in addition to my preceding comments, that whereas geologists around the world enthusiastically hailed and praised the outstanding contribution that the brothers had made in determining the stratigraphic sequence and deciphering the complex structure of the folded and faulted Appalachian Chain, many leading structural geologists of the day - e.g. de la Beche, Lyell, Murchison and Sedgwick - did not accept the theory proposed by the two brothers.

---

* The interested reader is referred to the following papers that discuss the subject: Evolution of thought on structure of Middle and Southern Appalachians, Am. Assoc. Petroleum Geol., B. 33: 1643-1654, (1949); Evolution of thought on structure of Middle and Southern Appalachians: Second paper, p. 1-15 in Appalachian Structures, Origin, Evolution and possible potential for new exploration frontiers: A Seminar, March 3-5, 1971, edited by P. Lessing, Ruth I. Hayhurst, J. A. Barlow and L. D. Woodfork. W. Virginia Univ. and W. Va. Geol. & Econ. Survey, xiii + 322 p., il., (1972); and The Tectonics of the Appalachians, New York: Wiley-Interscience, xii + 271 p., (1970), both by John Rodgers; and Structural patterns in the Southern Appalachians: Evidence for a gravity slide mechanism for Alleghanian deformation, Geol. Soc. Am., B. 86: 1316-1320, (1975) by R. C. Milici.

Because of the great importance given to the Rogers' theory by both geologists and biographers, and particularly because of the theory's relevance to the lively controversy that is still going on as to the mechanics of deformation of the Appalachians, I have chosen to include the somewhat formal essay that follows in order to give the interested reader a more complete history of how the two Rogers brothers produced what is generally regarded today as one of the classic contributions to the geology of North America. I am fully aware that the essay repeats much of what I have included in preceding paragraphs. Hopefully, however, the interested reader will not object, and the uninterested one will skip the whole discussion and go on to the next section.

In evaluating the theory, as discussed in the following section, I would strongly urge the serious reader to consult the following four references: 1) Benjamin (1899, p. 456), who wrote:

> "It has been well said that 'together they [i.e. William and Henry] unfolded the historical geology of the great Appalachian chain.'"

2) Gregory (1916), who was quite supportive of the brothers' theory; 3) Barrell (1918) (see Dana, E. S. et al. 1918), who was quite critical of some of their claims and of their theory; and 4) Willis (1943), who summarized and evaluated the Rogers brothers' contribution in what seems to me a quite fair-minded commentary, as follows (p. 39-40):

> "In America the belt of the Appalachian Mountains presents an unusually simple system of long narrow folds. The folded rocks are bedded sandstones, shales, and limestones, which were piled up to thicknesses exceeding ten thousand feet. They were spread in flat sheets. They are now bent into arches and troughs, in alternation, like wrinkled paper. The brothers, H. D. and W. B. Rogers ... during two decades before 1850, observed the folding. ... The effect is that which would be produced if the belt had been narrowed by compression, but the Rogers brothers thought it impossible that simple compression could have so acted. They conceived a compound action, a combination of vertical and horizontal forces, like rising and falling waves; and they found a not unreasonable cause in the agitation of a thin crust, floating upon a molten interior. Their inferences were consistent with the then ruling ideas of a molten globe and of catastrophic activity of terrestrial forces. Unfortunately for the theory the assumption of a thin crust floating on a fluid interior was mistaken, as was subsequently demonstrated by the geophysicists."

THE ROGERS BROTHERS' DESCRIPTION AND EXPLANATION OF THE STRUCTURE OF THE APPALACHIAN CHAIN

William Barton Rogers and his geologist brother, Henry Darwin Rogers, first attracted international attention in 1842 when they described the stratigraphy and structure of the Appalachian Mountains and proposed an imaginative and speculative theory as to how the originally horizontal strata, deposited in the sea, were uplifted and then folded and faulted into parallel flexures as they exist today.

Their field discoveries, and the theory they developed to explain the deformation of the originally horizontal strata, were first jointly presented in a paper, entitled "On the Structure of the Appalachian chain, as exemplifying the laws which regulated the elevation of great mountain chains generally," read at the Third Annual Meeting of the Association of American Geologists and Naturalists in Boston on Friday, 29 April, 1842. An outline of the paper was published in the American Journal of Science for April-May, 1842 (Vol. 43/1: 177-178), and the paper itself was published later as "On the Physical Structure of the Appalachian Chain, as exemplifying the Laws which have regulated the Elevation of great Mountain Chains, generally," by W. B. Rogers and H. D. Rogers. It appeared in full in the Reports of the First, Second, and Third Meetings of the Association of American Geologists and Naturalists at Philadelphia, in 1840 and 1841, and at Boston in 1842, embracing its Proceedings and Transactions. Boston: Gould, Kendall & Lincoln, viii + 9-544 p., pl., maps, sections, (1843). The paper has been reprinted in whole or in part in several publications as noted in reference 27--1842f, 1843 in the Bibliography farther on.

It is appropriate at this point to comment briefly on how the brothers came to join forces in their geological field work and to produce their classic 1842 paper, which I will quote from at some length a little farther on.

As stated in the preceding section, the brothers, as a result of prosecuting their respective duties as state geologists - Henry in New Jersey and Pennsylvania and William in the Virginias - during the 1830s and 1840s, found that the stratigraphy and structure of the folded and faulted Appalachians were common to their respective regions and beyond. In comparing their discoveries and discussing how the deformation of the strata might have taken place, they developed the theory referred to in the preceding section and quoted at length farther on in this section.

However, just how, when, and by whom the ideas making up the theory were first thought of is not altogether clear to me. We know for certain, from their extensive correspondence, some of which is reported in Life and Letters, that William and Henry were quite close, as were all four of the Rogers brothers for that matter, so it seems reasonable to assume that the two shared in developing the theory to its final form as set forth in the paper read at Boston in 1842. But which brother got the idea first is not altogether obvious; however, my guess is that Henry first originated the theory because, while in England in the early 1830s, he had become acquainted with de la Beche and other geological leaders, had become aware of the vigorous discussion of the dynamics of mountain building then going on, and had no doubt brought back some definite ideas of his own on the subject in 1833. Furthermore, I have concluded, from perusal of correspondence as well as of earlier publications by the two brothers, that Henry was the more speculative of the two, whereas William was the more accomplished expositor.*

We know that Henry developed a keen interest in geology during a visit to Britain in the early 1830s,** and that soon after he returned to the United States in 1833 he gave lectures in geology at the Franklin Institute. In 1835 he was appointed the first occupant of the Chair of Geology and Mineralogy in the University of Pennsylvania, and was made director of the Geological Survey of New Jersey, which was discontinued, however, in 1839. He also became State Geologist of Pennsylvania in 1836. Meanwhile, William had been appointed State Geologist of Virginia in 1835, after a vigorous and successful campaign to get a Geological Survey of Virginia established, and he and Henry wrote several joint papers on the geology of the State in 1837 and 1839 (see Bibliography farther on in this sketch). Also, while Henry was publishing reports on the geology of New Jersey (1836, 1840) and of Pennsylvania (1836-1842),† William was publishing similar reports on the geology of Virginia (1836-1842). Obviously,

---

* From many comments in the published record of the two brothers, it seems to me that Henry was the more imaginative and speculative, whereas William was the more conservative and the more dramatic and convincing in oral presentations. It should be noted further, that in the earliest announcements about the structure of the Appalachian Chain, it was Henry who was the first author, whereas in the detailed and documented papers that followed William's name came first.

Furthermore, Gerstner (1975, p. 26) has recently concluded that:

> "Although the theory discussed here is usually thought of as the joint work of Henry and his brother William, I believe, as will be shown below, that it arose primarily out of Henry's experiences in England in the early 1830s."

She goes on to point out that the dynamics of mountain elevation was being vigorously discussed in England in the early 1830s, when Henry was there, and that Henry's acquaintance and friendship with de la Beche and other structural geologists almost certainly made him aware of the several different theories of mountain elevation then being discussed.

I am inclined to agree with Ms. Gerstner that Henry probably came back from England in 1833 with definite ideas about the dynamics of mountain building and that he, not William, had the primary role in formulating the theoretical part of the 1842 paper.

** Gregory, J. W., "Henry Darwin Rogers - An Address to the Glasgow University Geological Society, 20th January, 1916." Glasgow: James MacLehose & Sons, 38 p., port., (1916).

† See footnote on next page.

both brothers were actively engaged in field work, and we can only conclude that out of their respective investigations and many discussions came the imaginative theory that brought them fame throughout the geological profession. Now let me return to the theory itself and how it was treated by the brothers' peers.

The theory, first published* in full as Part II, p. 507-531, in the Reports of the First, Second, and Third Meetings of the Association of American Geologists and Naturalists ... (see reference 23--1842f, 1843 in the Bibliography farther on), appears as follows (with omissions that I have indicated):

---

† (See preceding page).

The image of the State Geologist and his assistants and of their field activities, of at least one journalist, soon after the Pennsylvania Survey was established, is indicated by the following excerpts from the Philadelphia Public Ledger of March 30, 1839. The following letter, addressed to the newspaper, stimulated the amusing commentary of the Ledger reporter that follows the blacksmith's letter:

> "Friend Ledger -
> Will you have the kindness to inform a constant reader of your paper, where he can obtain a copy of the State Geologist Report that was lately ordered to be printed, by the Legislature? [Presumably the writer was asking about Henry Rogers' Third Annual Report on the geological exploration of the State of Pennsylvania of 1839.] By so doing, you will oblige many mechanics.
> A Blacksmith"

The Ledger reporter then commented as follows:

> "Professor Rodgers (sic) is now superintending the printing of the Report, which is in press and will shortly appear. We learn that it will be more valuable than its predecessors. The corps will shortly commence their summer campaign and renew their delightful wanderings amid the glorious scenes of our mountain districts. We cannot imagine a more pleasant office than this same - to be Locomotive Geologist of the State, well paid and well fed - to wander up and down the country, catching trout and shooting grouse, enjoying the unrivalled scenery of our interior, and nothing to do but now and then hammer a stone to pieces and look wisely at the fragments."

Might the arm-chair journalist, probably musing in his Philadelphia office, have written differently about Henry and his field assistants had he waded the streams, struggled through the underbrush, climbed the rocky slopes of the Pennsylvania mountains, and then carried his pounds of rock specimens back home at the end of a long day? I wonder!

* Originally published: Rogers, W. B. and Rogers, H. D., "On the Physical Structure of the Appalachian Chain, as exemplifying the Laws which have regulated the Elevation of great Mountain Chains, generally." Repts. First, Second and Third Meetings of the Assoc. Am. Geol. & Nat. at Philadelphia in 1840 and 1841, and at Boston in 1842, embracing its Proceedings and Transactions. Boston: Gould, Kendall & Lincoln, viii + 9-544, pl., map, sections, 1843. Rogers article - p. 474-553.

"THEORY OF THE FLEXURE AND ELEVATION OF THE STRATA, FOUNDED ON THE PRECEDING PHENOMENA, — COMBINED UNDULATORY AND TANGENTIAL CHARACTER OF THE MOVEMENT.

"That the movement which produced the permanent flexures was compounded of a wave-like oscillation, and a tangential or horizontal pressure, both propagated northwestward across the disturbed belt, is plainly indicated by the oblique character of nearly all the anticlinal and synclinal curves, both those which are closely folded, and those which are obtuse. This oblique inflection of the strata will, we confidently believe, be found to prevail as the regular form of all anticlinal axes, in every part of the world. It appears to imply a powerful tangential movement, always operating in the same direction for the same region, during the epoch of disturbance. A merely vertical force, exerted either simultaneously or successively, along a system of parallel lines, could only produce the same number of <u>symmetrical</u> anticlinal arches, while again, a horizontal or tangential pressure, uncombined with an alternate upward and downward motion, at regular intervals, could not possibly result in a system of parallel folds, or axes, or lead to any change in the position of the strata, beyond an imperceptible bulging of the whole tract, or else a confused rumpling and dislocation, dependent on local inequalities in the thickness or resistance of the crust, in different spots.

"That the <u>wave-like</u> flexures of our Appalachian strata are the result <u>of an actual onward, billowy movement</u>, proceeding from beneath, and <u>not</u> of a folding due simply to some <u>great horizontal or lateral compression</u>, will appear from the following considerations. In the first place, it is absolutely impossible to conceive, that <u>any</u> force, of an intensity however vast, exerted in the direction of a tangent to the earth's surface, could by itself shove a thick and imperfectly flexible crust into a system of close <u>alternate</u> folds. Beyond the imperceptible bulging of the whole tract laterally from the line of application of the force, no flexure could arise, other, perhaps, than some diminutive, but <u>irregular</u> plications, caused by inequalities in the strata or crust, and these, it is needless to remark, would be destitute of any law of parallelism and gradation, such as that which strikingly characterizes the Appalachian and other regions. No <u>system of narrow waves</u> of the strata, however flat, could originate from the most enormous lateral pressure, if unaccompanied by some vertical oscillation, producing parallel lines of easy flexure. Precisely such an alternate movement would ensue, if a succession of <u>actual waves</u> on the surface of the subterranean fluid rock rolled in a given direction beneath the bending crust.

"The inadequacy of the tangential or horizontal force, as a cause of the Appalachian axes, is still further obvious, when we consider, that no igneous rocks, of any sort, were thrust to the surface, except in the belt of country bordering this broad system of flexures on the southeast, and that, therefore, if the axes or foldings were produced solely by lateral pressure, the whole force must have been propagated from the lines, where the wedging in of the igneous matter occurred in this southeastern region, to the remotest of the axes, through all the intervening folds. But, consistently with mechanical analogies, such a transmitted force, instead of producing the gentle gradation of flexure, which we behold, would have expended itself in merely compressing or crushing the contiguous tracts across a narrow belt, a little widened by a succession of these tangential actions. The narrow disturbed belt would abound in irregular contortions, and beyond it we should suddenly come to the strata in their original horizontality.

"That such would really be the effect of the supposed horizontal action, is clearly proved by the singularly undisturbed condition, already stated, of the strata immediately, and for some distance, northwest of all our greatlines of dislocation. Along these lines, the uniform inversion, and the crushed and contorted state of the higher rocks, immediately northwest of the fracture, indicate plainly an enormous lateral thrust in that direction from the fault. Yet, even where the greatest energy of this force is manifested, the inversion or other disturbance extends only for a few hundred yards northwest of the fissure, while a little beyond, the horizontal posture of the rocks has been even less changed than in parts of the same region, where no fault exists.

"Even granting, that such a force, transmitted to a great distance across the chain, were capable of bending the strata of the remoter tracts into gentler undulations, the flexures on their northwestern sides ought to be relatively still steeper than they are, for in that quarter the curves are almost symmetrical. On the other hand, this near approach to a symmetry of curvature in the remoter axes, is an obvious consequence of the greatly reduced force and size of the nearly exhausted waves.

"The widening of the interval between the axes, as we go to the northwest, is another general fact, which, while it finds a ready explanation in the hypothesis of a violent undulation of the strata, would seem to be wholly at variance with the operation of a gradual and prolonged pressure, exerted northwestward. Conceiving the various degrees of inflection witnessed in different parts of the chain to have resulted from a long-continued pressure, we should be compelled to admit, that the southeastern side of the tract had had impressed upon it successively all the different gradations of flexure met with throughout the chain, and thus we should have to suppose, that the closely folded, crowded axes of the great valley were slowly developed by a force that, in its earlier stages, produced every where only wide and gentle arches. Yet, if such was the case, why do we not recognize a yet more uniform or gradual transition in the dimensions of the axes, than our Sections show. If the steepness of the flexures measures thus their age, why, it may be asked, are those of the same group so various in this respect, while their intimate relations to each other, in respect to parallelism, gradation of distance, and dip, plainly prove them to have had a contemporaneous origin? If a long period was consumed in their production, why did there not take place, by virtue of the simultaneous denudation and deposition which must have been in progress, a constantly unconformable superposition of the new deposits, as the axes slowly rose above the level of the water?

"But, while the observed variety in the magnitude and steepness of the flexures thus makes it incumbent on the advocates of such a theory of the gradual formation of axes, to admit, that the folded and closely crowded ones have arisen out of broader and normal curves, the general tenor of their doctrine of progressive and cumulative actions, implies, that the short and narrow flexures were produced first, and that some of them were enlarged into the vastly bolder and longer axes, which abound in many parts of the same region. This, however, seems an insuperable difficulty, since, if we suppose the breadth and length thus steadily to increase, a great number of intervening flexures and foldings would be necessarily obliterated or reversed.

"But, quitting the theory of a gradual horizontal pressure, another hypothesis suggests itself, as likely, in the present stage of geological speculation, to be offered in explanation of the structural laws we have described. It may be urged that a prolonged upward tension, or pressure exerted along a single line, might gradually create a broad and lofty anticlinal flexure, and might, by a mere shifting of the line, into positions always parallel to its first one, accomplish in time the elevation of all the axes of any of our Appalachian groups. Such a supposition would, doubtless, account for the simple features of a symmetrical flexure; but it would afford no clue to an explanation of those beautiful relations, which prevail between the form of the flexures and their position in the groups, to which they appertain, or to the fact of their assemblage into groups; and these are among the most interesting general facts, which a theory of flexures is called upon to explain. How could a merely vertical force, applied to the interior surface of the crust, either along a narrow line, or over an elongated elliptical, narrow zone, produce that oblique form of the anticlinal arch, which we find to be its normal configuration; or how could it give rise to the regular horizontal bending of the axis-line, as seen in the curving districts of the chain. Again, in what way can it explain the occurrence of the great lines of fault only on the northwestern side of the axes, or the close oblique foldings, in all the southeastern side of the belt. But, apart from all these objections, on what principle or analogy are we entitled to assume, that the supposed successive shifting of the upward force would be in parallel lines. Should the elevation theory be modified so as to suppose the upward force to have been exerted simultaneously along all the present anticlinal lines, but not in the manner of an undulation, the equally formidable difficulty arises of accounting for the production of any flexures; since, by the close contiguity of the parallel lines of upward pressure, the sole effect would be a nearly uniform diffused bulging of all that portion of the crust, upon which the tension was exercised.

"OF THE ORIGIN OF THE SUPPOSED SUBTERRANEAN UNDULA-
TIONS, AND OF THE MANNER IN WHICH THE STRATA BECAME
PERMANENTLY BENT AND DISLOCATED.

"The parallel flexures of the crust, so strikingly exhibited in the Appalachian chain, and recognizable, we believe, in nearly all disturbed mountainous districts, we conceive to have originated in the following manner. We assume, that in every region, where a system of flexures prevails, the crust previously rested on a widely extended surface of fluid lava. Let it be supposed, that subterranean causes competent to produce the result, such, for example, as the accumulation of a vast body of elastic vapors and gases, subjected the disturbed portion of the belt to an excessive upward tension, causing it to give way, at successive times, in a series of long parallel rents. By the sudden and explosive escape of the gaseous matter, the prodigious pressure, previously exerted on the surface of the fluid within, being instantly withdrawn, this would rise along the whole line of fissure in the manner of an enormous billow, and suddenly lift with it the overlying flexible crust. Gravity, now operating on the disturbed lava mass, would engender a violent undulation of its whole contiguous surface, so that wave would succeed wave in regular and parallel order, flattening and expanding as they advanced, and imparting a corresponding billowy motion to the overlying strata. Simultaneously with each epoch of oscillation, while the whole crust was thus thrown into parallel flexures, we suppose the undulating tract to have been shoved bodily forward, and secured in its new position by the permanent intrusion, into the rent and dislocated region behind, of the liquid matter injected by the same forces that gave origin to the waves. This forward thrust, operating upon the flexures formed by the waves, would steepen the advanced side of each wave, precisely as the wind, acting on the billows of the ocean, forces forward their crests, and imparts a steeper slope to their leeward sides. A repetition of these forces, by augmenting the inclination on the front of every wave, would result, finally, in the folded structure, with inversion, in all the parts of the belt adjacent to the region of principal disturbance. Here, an increased amount of plication would be caused, not only by the superior violence of the forward horizontal force, but by the production in this district of many lesser groups of waves, interposed between the larger ones, and not endowed with sufficient momentum to reach the remoter sides of the belt. To this interpolation we attribute, in part, the crowded condition of the axes on the side of the undulated district, which borders the region where the rents and dykes occur, and to it we trace the far greater variety which there occurs in the size of the flexures.

"In the progress of this bending and folding of the strata, throughout the undulated district, the continual introduction and consolidation in the fissured district, of fresh materials from the liquid mass beneath, rising in intrusive dykes, and filling the wide interstices of the broken strata, would permanently retain the inflected crust in the new attitudes into which it had been forced, and compensate for the reduction of horizontal breadth arising from the flexures. Permanent axes might even be produced without the fracturing of the crust being in all cases apparent at the surface, since innumerable fissures, of sufficient size to permit the sudden escape of an enormous quantity of elastic vapor, could temporarily form, and yet close again superficially, and still the strata be braced and retained in their flexured state by the dislodgement of fragments, and the intrusion and congelation of much lava matter in the lower parts of the rents.

"This theory agrees strikingly with the singularly undisturbed condition of the strata, northwest of our great lines of fault. When describing, under a preceding head, some of these enormous dislocations, especially those of southwestern Virginia, an account was given of the gradual transition of structure, from the normal to the folded or inverted form, and thence, to a successive ingulfing of certain groups of strata, into a line of fault, presenting sometimes, for the distance of seventy miles, an actual inversion of the lower Appalachian limestone or slate, upon either the carboniferous limestone or the next inferior group. The commencement in all cases of these faults, in the steeply folded synclinal part of the flexure, immediately on the northwest of the finally inverted anticlinal curve, would seem to prove conclusively, that the fracture has been due to a profound folding in and inversion of the rocks, carried to the extent of producing an actual snapping asunder of the beds where most incurved, followed by a squeezing downward of the opposite side of the trough, by the horizontal northwestward thrust of the anticlinal portion, causing the lower strata of the latter to lie directly upon geologically higher groups. The enormous mass of rocky material, thus forcibly pressed down and firmly held there, would, we conceive, constitute a vast <u>subterranean barrier or dam</u>, capable of arresting, in <u>some degree, the progress</u> of the succeeding waves, and of protecting the region for a moderate distance, towards the northwest, or the leeward side of the fault, from the undulations to which it would otherwise have been exposed. In confirmation of this view, it may be stated, that in tracing a line of dislocation toward either extremity, while the extent of strata thrust down, as indicated by the amount of the hiatus at the fault, is inferred to grow progressively less and less, or, what is the same thing, the supposed subterranean dam, presumed to diminish in depth, the region behind it, on the northwest, becomes more and more undulated, until, when we pass beyond the extremity of the fault, to where the normal form of the flexure is restored, we find the strata to the northwest reared into bold anticlinal and synclinal curves. Such is remarkably the fact with the fault at the northwest base of the Peters's and East river mountain, in Virginia, as well as with that which lies parallel to, and southeast of, the Cumberland mountain; and, in a word, with all the faults and crushed axes of great length throughout Virginia, Pennsylvania, and Tennessee. Even where two such lines of dislocation occur, parallel to each other, at an interval of not more than eight or ten miles, the central parts of the intervening tract exhibit unusually little disturbance, notwithstanding their proximity to the lines of violent disruption on each side.

"The assumed combination of the wave-like oscillation, and horizontal or tangential movement, will explain, we believe, all those general structural phenomena, which we have described as characterizing our Appalachian chain in all its length and breadth, and which obviously exist in many other mountain chains possessing numerous axes. It will account for all the varieties of flexure, normal, inverted, or dislocated, which are any where observable in the chain, since a mere difference in the ratio of the tangential to the undulatory movement, would produce every grade and form of inflection we have had to record.

"The theory explains, moreover, the remarkable law of diminishing steepness in the flexures, as we cross the whole belt northwestward from the region of intrusive veins and dykes, which has evidently been the quarter of extensive and violent actual disruptions of the crust. It moreover affords a reason for the striking parallelism which prevails between the axes in every division of the chain, and the veins and dykes in the corresponding tracts to the southeast. In this rent and dislocated zone of country, beginning with the chain of the Blue ridge, the incalculably numerous and greatly extended dykes and veins that every where penetrate and fill the altered and hypogene rocks, comprise, we believe, an ample quantity of in-wedged material, to balance the horizontal contraction of the whole plicated chain.

"The mere fact of a regular gradation in the amount of flexure, is of itself a proof, that the axes thus related had a common source, while the direction of this gradation, clearly establishes, that the southeast was the quarter from whence the movement proceeded.

"The views here entertained of the nature of the elevating action, afford a satisfactory cause for the arrangement of the axes in groups, since we have merely to imagine successive sets of pulsations of varying magnitude and momentum, to have followed each other in the same general period of disturbance, and we are supplied with a cause sufficient to produce all the diversity which we behold in the distances and directions of the flexures. The almost exact parallelism of these in each group, and the general parallelism of all that enter into the same division of the chain, are the necessary results of that wave-like movement in which we conceive the axes to have originated; and we confess ourselves at a loss to imagine how any other action, but an undulation of the crust, propagated in parallel lines, either straight or curving, could give rise to this extraordinary feature in these enormously extended anticlinal and synclinal lines.

. . . . .

"In conclusion, we would express our belief, founded on the phenomena referred to in this memoir, and on numerous similar geological facts, of recent as well as ancient date, which cannot be mentioned in this place, that all great paroxysmal actions, from the earliest epochs, to the present time, have been accompanied by a wave-like motion of the earth's crust."

This was an original, highly imaginative, and quite speculative theory, and as Sir Arthur Holmes has recently written:*

> "The possibility of an 'actual undulation of the supposed flexible crust of the earth ... propagated in the manner of a horizontal pulsation' was first suggested by the brothers W. B. and H. D. Rogers in 1842 after their pioneer investigation of the Appalachians. In Europe the image of an orogenic wave was introduced by Marcel Bertrand in 1887 and soon became familiar. These ideas have since served as the foundations for several working hypotheses devised to explain mountain building. One of the clearest examples of wave-like deformations of the crust is to be found in the Sunda region of Malaysia and Indonesia. It has been presented in admirable detail by R. W. van Bemmelen."

Regardless of which brother actually got the basic idea of their theory first - my guess is it was Henry - there seems little doubt that it was William, with his noted felicity and clarity of expression and overall oratorical skill, who impressively presented the basic data on which their theory rested, and who later on presented both data and theory to the scientific world. On the other hand, it should be noted that it was Henry who was listed as first author in the notice that appeared in the American Journal of Science for April-June 1842 (Vol. 43/1: 177) and of the abstract published in the Proceedings of the 12th Meeting of the British Association in 1842 (see reference 27--1842f, 1843 in the Bibliography farther on).

The brothers together made the first public presentation of their theory on 29 April 1842 at the third annual meeting of the Association of American Geologists and Naturalists convened in Boston during the week 25-30 April 1842. Their joint presentation came on Friday, the next to the last day of the week-long convention, and it is interesting and informative to read what a member of the audience, amateur geologist John L. Hayes of Cambridge, Massachusetts, later wrote about it (Life and Letters I, p. 210):

> "Notwithstanding the able address of Professor [Benjamin] Silliman, the elaborate paper of Professor [Edward] Hitchcock, and the frequent and interesting remarks of Mr. [later Sir Charles] Lyell [one of England's leading geologists], the marked feature of this meeting, which continued for a week, was the reading of the joint paper by the brothers Rogers ..."

Significantly, Hayes continues (p. 211-212):

> "... In making their joint exposition - for the 'paper,' as delivered was purely an oral statement -

---

* p. 1138 in Principles of Physical Geology, 2nd Ed. New York: The Ronald Press Co., (1965).

> William Rogers took upon himself the more modest but really more difficult part, of describing the phenomena, leaving to his brother the part of explaining the theory* of the phenomena. [The two underlinings are mine -- R.R.S.] ... Those who know the elegance of diction and manner which characterized the later address of the elder Rogers [i.e. William] can partially conceive of the effect he produced by the fluent and graceful oral statement of the complicated phenomena of this hitherto mysterious mountain chain, - a statement in which there was not one word of hesitancy, nor a word which was not the most fitting."

That William's oratorical skill added a special dimension and impact to his presentation is clearly evident from Hayes' further comment (p. 212):

> "This paper, or what purports to be the same, is published in the 'Transactions' of the Association. I have frequently read it since. To me it is now comparatively tame in expression. It lacks the inspiration of the scene and the men, the illustrative diagrams [which are included in the Coal Reports, reference 27--1842f of the Bibliography], the emphasis of voice and finger pointing out the distinguishing phenomena, and the fervour of the spontaneous utterance."

Such is the power of the spoken word when a commanding speaker embellishes his statements with deliberate oratory!

Shortly after the Boston presentation, Charles Lyell suggested that the brothers immediately submit an abstract of their paper to the British Association for presentation at the 1842 meeting in Manchester and for publication in that organization's Annual Report. Henry advised William by letter, dated May 30, 1842 (p. 215 in Life and Letters I), that he had prepared the suggested abstract and dispatched it to Prof. Phillips at Manchester. The abstract was presented and evoked vigorous discussion, most of which was laudatory, but some was critical of the proposed theory. A brief abstract was published shortly thereafter in Rept. British Assoc. 1842 [pt. 2], p. 40-42, (1843),** and a much longer commentary was later published under Section C, Geology and Geography, in the Proc. 12th Meeting British Assoc. (1843). The latter version, reprinted in the Am. Jour. Sci. 44: 359-368, (1843) is an excellent summary of and commentary on the Rogers' paper. It is highly recommended for reading by those seriously

---

* It seems that Henry was much more given to speculation than William, if one can judge from several of his articles, as pointed out by Joseph Barrell, "A Century of Geology - The growth of knowledge of earth structure," p. 157, 171, 174-175, in A Century of Science in America. New Haven, Conn.: Yale Univ. Press, ix + 458, il., (1918). (See reference [1918] Dana, E. S.)

** Rept. 12th Meeting, British Assoc. held at Manchester in June, 1842. (London: John Murray, p. 40-42, 1843.)

interested in getting a condensed but accurate description of the Rogers' complete paper, particularly as it includes comments by Murchison, Sedgwick, and de la Beche.

As a result of some adverse reaction to their theory as presented at the June 1842 meeting in Manchester, Henry suggested to William that they prepare a second paper – one concerning the bearing of certain earthquake phenomena on their theory and on the fundamental idea of an interior fluidity – and send it to the British Association to lend support to their theory. This paper was subsequently presented at the 4th Session of the Assoc. Am. Geol. & Nat., on 2 May 1843, and abstracted in the Am. Jour. Sci. 45: 341-347, (1843). It appears as reference 33--1843f in the Bibliography farther on.

The Rogers' theory stimulated much discussion and controversy at the time and firmly established their names among the leading geologists of the day. Most British geologists were high in their praise of the brothers' stratigraphic and structural work, including Murchison, Sedgwick, de la Beche and Lyell, but the latter four found fault with some aspects of the Rogers' theory, and Adams* mentions Lyell as having written:

> "According to the theory of the Professors Rogers, these wave-like flexures are to be explained by supposing the strata, when in a plastic state, to have rested on a widely extended surface of fluid lava and elastic vapours and gases. The billowy movement of this subterranean sea of melted matter imparted its undulations to the elastic overlying crust, which was enabled to retain the new shapes thus given to it by the consolidation of the liquid matter injected into fissures. For my own part [he goes on] I can not imagine any real connection between the great parallel undulations of the rocks and the waves of a subjacent ocean of liquid matter."

Two of the ablest and most penetrating evaluations of the Rogers' theory of origin of the folded Appalachians are those of J. W. Gregory** and Joseph Barrell,† as viewed in the light of physical and geological knowledge of the time (1916 and 1918, respectively). Gregory, who is Henry Rogers' biographer, writes approvingly of the Rogers brothers' theory whereas Barrell is highly critical and finds no basis for it.

---

* Adams, F. D., Sir Charles Lyell – His place in geological science and his contributions to the geology of North America. Science 78: 177-183, (1933).
** "Henry Darwin Rogers, An Address to the Glasgow University Geological Society, 20th January, 1916, ... with Bibliography by Colin M. Leitch." Glasgow: James MacLehose and Sons, 38 p., port., (1916).
† See Part IV. A Century of Geology – The growth of knowledge of earth structure, by Joseph Barrell, p. 153-192, in A Century of Science in America (E. S. Dana, ed.). New Haven, Conn.: Yale Univ. Press, vii + 458 p., il., (1918).

Today it seems to me that the fair-minded critic must evaluate the Rogers brothers' theory in terms of what was and was not known about the physical and chemical properties of the earth's crust in the 1830s.

With respect to the whole work, of which the theoretical explanation is only a part, I would evaluate the Rogers brothers' contribution to geology from two different aspects. Their pioneering field observations in Pennsylvania and Virginia established for the first time the regional stratigraphy and structure of the Appalachians - a succession of essentially parallel marine strata deformed into a belt of roughly parallel, northeast-southwest trending folds in which the strata of the deformed belt are openly and gently folded along the northwestern flank, but closely folded and faulted along the southeastern flank. Steep easterly dips and overturning characterize the latter flank as a result of overthrusting in a northwesterly direction. (See map and cross-section on preceding pages.)

The aforementioned stratigraphy and structure were determined by field work done under primitive conditions when there were few roads and trails, only the crudest of maps, and none of the mechanical vehicles and optical measurements so easily at hand today, 140 years later. It was field work of excellent quality and continues to be so regarded to the present time as pointed out by Woodward (1961, p. 1626) and Rodgers (1970). (See Appendix B.)

As to the second aspect of their contribution - the theory of how the folds and faults were produced - their imaginative and manifestly speculative idea of how the earth's outer crust behaved can hardly be supported by what is now known about the physical and chemical properties and behavior of the materials of the earth's outer crust. But again it must be remembered that in 1830 little was known about those basic physical sciences that are so critical in gaining an insight into geologic processes and the results they produce.

This is not to say that earth scientists should refrain from speculation; they should continue to speculate as vigorously as ever, because speculations have a way of stimulating the search for deeper insight and more complete understanding of perplexing problems. One can hardly desist from wondering how some of the most speculative aspects of the currently favored concept of sea-floor spreading and plate tectonics, as they bear on mountain building, may be regarded by earth scientists of the twenty-first century, armed with all of the new knowledge they presumably will have accumulated during the ensuing century and more!

THE 1840s, YEARS OF GREAT SIGNIFICANCE

For William the year 1842 was in many ways the most critical one of his life, because he would become acquainted with Boston and Bostonians, they would learn about him for the first time, and he would return to marry one of Boston's prominent young ladies, and ultimately to found M.I.T.

William first visited Boston in 1842 when he and Henry presented their classic paper on the Appalachian chain discussed in the preceding section. At the time of his visit, Boston was regarded as the "Athens of America," and William was deeply impressed by the bustling city and its heady intellectual atmosphere when compared with the rather depressing situation at the University of Virginia, where riotous students were causing him and his colleagues much distress.

As mentioned earlier on, William's work as State Geologist of Virginia (1835-1841) and Henry's as State Geologist of New Jersey (1835-1839) and Pennsylvania (1836-1856), conducted with the usual close cooperation of the Rogers brothers, had resulted in much new information on the geology of the three States, and in particular on the stratigraphy and structure of the folded Appalachians. It was this cooperative work that formed the basis for their joint paper at Boston in 1842, and their second important paper the next year titled "On the phenomena of the great earthquakes which occurred during the past winter ..." (see reference 33-1843f in Bibliography farther on). Numerous other articles based on work done while State Geologist of Virginia came from William's pen, and before the decade of the 1840s was over, William and Henry were numbered among the leading geologists in North America. And in Europe, after publication of their classic paper on the Appalachian Chain, they were widely known and respected in Great Britain as well as on the Continent, but their explanation of how the folded Appalachians were deformed did not meet with universal approval, as implied in William's letter to Henry on May 27, 1843. It seems that de la Beche and Sedgwick both took exception to some aspects of the hypothesis, and other leading geologists, though impressed by the originality and ingenuity of the "wave theory," were rather reluctant to accept it in its totality. Nevertheless, the brothers Rogers became well known in Europe immediately following presentation of their paper, and William was accorded great respect and enthusiastic receptions by many of the scientific "greats" when he and his bride toured Great Britain and Europe on their honeymoon in the summer of 1849 (Life and Letters I, p. 304-309).

Two years after presentation of their classic paper on mountain building, William and Henry returned to Boston for a visit, and in the fall of 1844 Henry gave a series of lectures on Geology at the Lowell Institute in Boston, while William returned to Charlottesville, where his colleagues had chosen him to serve as Chairman of the Faculty. School Year 1844-45 was to be a hectic and trying one for him.

Hardly had he taken up his duties as Chairman of the Faculty when there arose the possibility of repeal of the Act of the Assembly granting an annuity of $15,000 to the University. William had to prepare a memorial* to the Legislature of Virginia in defense of the University and its annual appropriation, and in this defense he clearly showed the educational breadth and insight that later on was so conspicuous in his successful efforts to found M.I.T.

Then in early 1845 came a renewal of the student disturbances of previous years. These became so serious by April that the civil authority had to be called upon to suppress the riotous students. And of course William, as Chairman of the Faculty, had to prepare a report** on the disturbances and subsequent actions by the University.

By the time the school year was about over, William was so worn out and depressed with the situation at the University that he wrote to Henry in late April as follows (Life and Letters I, p. 250):

> "... I feel so little sure of exemption from like humbling and disgraceful disorders in future that I intend earnestly to look about me for some other and more tranquil home ..."

Immediately following the close of the spring term, Henry and William made a trip through the White Mountains of New England, and it was on this journey that William first became acquainted with the family of James Savage, of Boston, whose eldest daughter, Emma, would later become his wife and play an important role in the founding and subsequent history of M.I.T.

By 1846, Henry was doing consulting work for some Boston clients and discussing with John Amory Lowell the possibility of a School of Arts as a

---

* Report of the Committee of Schools and Colleges [of the Legislature of Virginia] against the expediency of withdrawing the fifteen thousand dollars annuity from the University of Virginia, 1845. Document No. 41. (Prepared by W. B. Rogers, Chairman of the Faculty.) Reprinted in Life and Letters I as Appendix A, p. 399-412.

** In late April, 1845, William, as Chairman of the Faculty, prepared and issued a circular letter on "Student Riots in the University of Virginia," which appears as Appendix B (p. 413-419) in Life and Letters I. The contents of this interesting letter clearly indicate that student dissent was not unique to the 1960s!

branch to the Lowell Institute.* At Henry's request William prepared a plan for a Polytechnic School in Boston [This plan is included in <u>Life and Letters I</u>, Appendix C, p. 420-427], having previously expressed his own interest in such a school in Boston to Henry as follows (<u>Life and Letters I</u>, p. 259):

> "Were this or any other promotion of your views to lead hereafter to a closer union of our labours by placing me also in the congenial air of Boston, I would indeed rejoice. Under circumstances so auspicious for effort in teaching and research, we could, I am sure, both of us be more productive and far happier in our labours than can be now. Ever since I have known something of the knowledge-seeking spirit, and the intellectual capabilities of the community in and around Boston, I have felt persuaded that of all places in the world it was the one most certain to derive the highest benefits from a Polytechnic Institution."

There now followed another half a dozen years during which William became more and more interested in the idea of a polytechnic institute in Boston, and less happy with his situation in Charlottesville. William, Robert, and James were all now well established in professorships, and Henry was doing well with his geological consulting and public lecturing in Boston. However, by 1846 William was urging Henry to sound out leading practical business men in Boston to determine if they might be interested in a separate and independent polytechnic institute in their city (<u>Life and Letters I</u>, p. 277), and as implied from the several letters mentioned earlier, William was looking more and more hopefully to some sort of position in Boston, possibly in part because of his growing interest in Emma Savage.

---

* In a letter to William, dated March 8, 1846 (<u>Life and Letters I</u>, p. 258), Henry wrote:

> "How much I want you near me at this time to aid me in digesting and submitting my views on this important scheme to Mr. Lowell! If you and myself could be at the head of this Polytechnic School of the Useful Arts, it would be pleasanter for us than any college professorship, for there would be less discipline, indeed no more than with medical students. At no distant day, if not indeed soon, Mr. Lowell will, I hope, organize such a branch in his Institute; and if he does not, you and I can surely get one founded here by going about it in the right way. Let us give this matter our earnest and sober thoughts ... Can you send me a copy of our memorial on behalf of the Franklin Institute for a School of Arts?" [Reference is here made to "A Memorial to the Legislature of Pennsylvania," presented about 1837 (see <u>Life and Letters I</u>, p. 258 and 263).]

CHEMICAL RESEARCH WITH BROTHER ROBERT (1844-1852)

It is worthy of notice that once William had completed and published on most of the work done while he was State Geologist of Virginia, and after he and Henry had published their several papers on the Appalachians, he turned his research interest to chemical problems. This change in interest was no doubt due to the fact that Robert was appointed Professor of Chemistry at the University of Virginia in 1842, an action that brought the two brothers together on the same campus. Between 1844 and 1852, when Robert left to succeed James as Professor of Chemistry in the University of Pennsylvania, and a year before William resigned from his professorship to join Henry in Boston, the two Rogers brothers jointly wrote seventeen papers on chemical subjects (see references 34--1844 to 67--1854b). During this same interim, William published his first articles on New England geology, sharing authorship with Henry on some of them involving work that they had carried on together during William's frequent visits to New England.

ROMANCE, MARRIAGE, AND A EUROPEAN HONEYMOON (1849)

As mentioned earlier on, William first met the members of the James Savage family in the spring of 1845 while he and Henry were on a trip to the White Mountains. This acquaintance led to his becoming a frequent guest in the Savage home, particularly at their Lunenburg (Mass.) home during summer vacations, and ultimately to his asking for the hand of Emma, the eldest Savage daughter.

William and Emma were married in Boston on 20 June 1849 after he had been granted early leave from the University. On the same day they sailed from Boston for England on the Cunard steamer Europa. For the next three months the recently married couple travelled widely throughout Great Britain and on the Continent (see reference 58--1849e in Bibliography). Everywhere they went they were offered the most cordial hospitality, and William everywhere received the plaudits he deserved as one of America's leading geologists. He wrote back in his several letters of the summer that he had met many of the famed scientists of the day - Darwin, de la Beche, Faraday, Lyell, Mantell, Hugh Miller, Murchison, Oldham, Phillips, Ramsey, Sedgwick and Sowerby, especially at the Birmingham meeting of the British Association in mid-September.

At a dinner attended by more than seven hundred British scientists and their wives, William was seated conspicuously between Phillips and de la Beche and near the President (Dr. Robinson) of the Association. Near the end of the evening's festivities, William responded to a toast from

Murchison, who lauded the work of Henry and himself, in what he refers to as "quite a respectable speech, which was often and loudly applauded" [Life and Letters I, p. 305], and it was obviously a great success to judge from the comments of one of the diners, J. W. Mallet, who wrote [Life and Letters I, p. 306]:

> "It [the speech] came late in the evening, after much, perhaps most, of the matter appropriate to the occasion had been already utilized by others; yet it was clearly the success of the banquet."

Four days later, after delivering to the assembled geologists an hour-long communication on the geology of Virginia, he and Henry (who was not present) were lavishly praised for their geological work by Murchison, Lyell and de la Beche. As William put it, in a letter to Henry [Life and Letters I, p. 307]:

> "Indeed, they laid on the compliments so thick that I could hardly stand up under them. But it was a real triumph and joy to hear them successively declare that our development of the great law of flexures [see the discussion of this 'law' in the quoted article on a preceding page] was one of the grandest contributions to geology ever made, and to find that they gave us the entire and exclusive credit of having thus furnished a clue to the most difficult problems in European geology."

There can be no doubt that by this time in their professional careers, William and Henry were regarded both at home and abroad as among the most distinguished of geologists, and were being lauded for their pioneering field work in unravelling the stratigraphy and structure of the folded Appalachians.

THE LAST YEARS IN VIRGINIA (1849-1853)

Upon returning home in October from their exciting honeymoon, William and Emma settled down in Charlottesville, and William quickly resumed his various activities at the University.

For the next four years, until the spring of 1853, he attended to his regular responsibilities during the school year, devoted summer vacations to travel and to geological investigations, and worked away steadily on the many papers that would ultimately be brought together in his classic Geology of the Virginias in 1884.

Reference has already been made to the seventeen chemical papers that William published jointly with Robert, between 1844 and 1852, and to the first papers that he published alone and with Henry on New England geology.

During his eighteen years at the University (1835-1853), William came to be regarded as one of the ablest and most popular lecturers on the campus, in spite of the fact that he temporarily aroused the animosity of a small group of dissident students who created riotous situations in the 1840s. He lectured on astronomy, chemistry, geology and physics, and used both classroom demonstrations and laboratory experiments to supplement his formal lectures, an innovation in college teaching at the time.

In 1838 he published An Elementary Treatise on the Strength of Materials, based on his lectures on the subject, and in 1852 he published a textbook, Elements of Mechanical Philosophy, for third year students. He and Henry were also thinking of writing a general textbook on geology, as discussed in the section on textbooks farther on, but they reached only the point of preparing a preliminary outline.

Although he had resigned from the University of Virginia in 1848, a year before his marriage to Emma Savage, he was persuaded to reconsider and to withdraw his resignation, which he did. However, at the close of the school year 1852-53, he resigned again, this time definitely, and he and Emma moved to Boston, where both spent the rest of their lives. It was not a difficult decision for William, in the light of the events of the previous year. It seems likely also that Emma wished to return to Boston to be near her father, who was in ailing health.

On 12 June 1852 the close association of the four Rogers brothers was broken by the death of James, the eldest of the brothers. He had earlier on served both William and Henry, as a field assistant, when they were State Geologists of Virginia and of Pennsylvania, respectively, and then in 1840 became a permanent resident of Philadelphia where he served as Professor of Chemistry, first in the Philadelphia Medical Institute, then in the Franklin Institute, and finally in the University of Pennsylvania, where he was teaching at the time of his death.* As noted earlier, Robert was appointed to the chair left vacant by James' death, which led to his resignation from the University of Virginia and removal to Philadelphia. Henry was established in Boston, and this left William alone in Charlottesville. Little wonder then that William decided to resign his professorship in the University of Virginia and join Henry in Boston in 1853.

---

* According to Life and Letters I, p. 325, Joseph Carson published A Memoir of the Life and Character of James B. Rogers, M.D., in Philadelphia, in 1852, soon after his death, and an excellent account of his life is also included in W. S. W. Ruschenburger's The Brothers Rogers, published in Philadelphia in 1885 (see references following the Bibliography farther on).

THE BOSTON YEARS (1853-1882) - INTRODUCTION

With permanent residence established in Boston, making it possible for Emma to be near her father, William hoped to carry on three major projects that he had had to postpone because of heavy duties at the University of Virginia. He hoped to complete a number of papers on Virginia geology and to prepare a final report on the Geology of the Virginias, as mentioned in a preceding paragraph. He desired to pursue new attractive geological problems in New England. And thirdly, he hoped to found a polytechnic school in Boston. In due course, but posthumously, there was published his comprehensive Geology of the Virginias, now considered by many as his most important geological contribution. He did find time to carry on geological field work in New England and to publish a fair number of articles on his findings. It was on his third, and surely his most pressing project, the polytechnic school, however, that he would devote the major part of his time and energy, and M.I.T. would be the result.

While pursuing the aforementioned projects, William also was soon participating in the activities of the American Academy of Arts and Sciences and the Boston Society of Natural History, and in 1861 was a member of the Committee of Examinations of the Boston Public Library. In 1861 he was appointed Inspector of Gas and Gas Meters for the State of Massachusetts by Governor John A. Andrew, and during his several years in the office he developed an excellent system of inspection, which he discussed in two publications - references 135--1863 and 136--1864 in the Bibliography.

THE FOUNDING OF M.I.T. (1859-1865)*

The intent of this brief section is not to give an account of how William Barton Rogers went about founding and organizing M.I.T., for that is a long story that has already been recounted in numerous biographies. Rather, I propose to show how William and Henry, cooperating constantly in a variety of endeavors, came to realize the need of a polytechnic school in which young men and women** could be trained in the practical arts for careers in industry and commerce. I wish also to sketch William's activities up to the time he seriously took on the responsibility of preparing his several plans for such a polytechnic school in Boston, in order to

---

* Most of the source material for this section of Rogers' biographical sketch came from Life and Letters I and II, but I also found useful, in important ways, the numerous biographical sketches and related historical items listed following the Bibliography.

** See footnote on next page.

demonstrate how uniquely prepared he was for that task. Finally, I intend to point out the prominent place that William accorded geology in the program of his new school, the future M.I.T.

As early as 1826, when William was only 22 years old and Henry but 18, the two brothers had opened a school in Windsor, near Baltimore. Within a year thereafter William was lecturing on astronomy and railroads at the Maryland Institute, and was assigned the task of organizing and managing a high school to be opened by the Institute in 1828. Hardly had he organized the school, however, before his father died and he was chosen, at the age of 24, to succeed his father as Professor of Natural Philosophy and Chemistry at the College of William and Mary. For the next four years he gained valuable experience in academic work and probably would have continued indefinitely as a professor of the physical sciences had not Henry, in 1833, on his return from a visit to England, got him interested in geology.

Again, as in past years, the two brothers developed a common interest, this time in geology, and quickly sensed how important the discovery of natural resources could be to the general public. By 1835 William had been instrumental in getting the Legislature to establish the first Geological Survey of Virginia, followed by his appointment as State Geologist, and the same year Henry was made Director of the Geological Survey of New Jersey, and in 1836 was appointed State Geologist of the newly established Geological Survey of Pennsylvania.

Now, in order for the two brothers to conduct the geological field work called for in their appointments, they had to find assistants who were competent to identify rocks, minerals, and fossils, determine structure, and map the regional geology. No such competent young men were available, so the two brothers William and Henry, called on their other

---

** (See preceding page).
It is noteworthy, in these days of widespread concern about the liberation of women, that Rogers, almost a hundred thirty years ago, thought of <u>both</u> sexes as potential students in his proposed polytechnic school. In <u>his</u> "A Plan for a Polytechnic School in Boston, 1846," (p. 421-422 in <u>Life and Letters I</u>), he wrote:

> "According to my present notions of expediency and usefulness, the two professors in the scientific, or more properly the mixed department, should so frame their general courses of lectures as to make them acceptable and useful to the public at large, and thus furnish annual courses on general physics, chemistry and geology, which might draw <u>all the lovers of knowledge of both sexes</u> [My underlining - R.R.S.] to the halls of the Institute, whether they proposed or not, continuing their studies in the other and directly practical branches of the Institution."

brothers James and Robert for aid, and also set about training other assistants. Quickly they recognized the need for training young men to do practical work. Going back to their earlier experiences with the Windsor School and the Maryland Institute, they began to formulate the idea of a polytechnic school of applied science. Both had had experience at lecturing without the advantage of appropriate apparatus even to demonstrate experiments to a class in physics or chemistry, and now they saw that not only physical scientists, but also geologists, desiring to do practical work, would need to be trained in applied science if they were to pursue successfully professional careers in that field.

It is not clear from available information as to how and when the Rogers brothers got the idea of a polytechnic school of applied science, nor is it clear as to which brother thought of such a school first. We do know that in 1837 William and Henry presented a memorial to the Legislature of Pennsylvania on behalf of the Franklin Institute for a School of Arts (Life and Letters I, p. 258), and in 1846 they prepared "A Plan for a Polytechnic School in Boston" for consideration by J. A. Lowell as a new branch in his Institute.

One wonders, however, how much the Rogers brothers knew about the Gardiner Lyceum, which opened in 1823, and the Rensselaer School, which was established in 1824, and whether or not the announced purpose of these early schools initiated and influenced their thinking about a polytechnic school of applied science.

The Lyceum at Gardiner, Maine, which was established in 1823, was the first school of practical science in America, according to Ricketts (1933, p. 7),* and proposed to offer instruction "in the chemistry of agriculture and the arts" and the application of the laws of mechanics. It lasted only ten years, however, and seems to have had little effect on scientific and engineering education.

In striking contrast to the Gardiner Lyceum, the Rensselaer School, established in 1824 at Troy, New York, by Stephen van Rensselaer of Albany, was successful from the start, and has contributed importantly to polytechnic education in applied science ever since. In founding his school, van Rensselaer announced by letter, dated November 5, 1824 (Ricketts 1933, p. 8):

---

* Ricketts, P. C., "Rensselaer Polytechnic Institute - Amos Eaton, Author, Teacher, Investigator: The First Laboratories for the Systematic Individual Work of Students in Chemistry, Physics and Botany, to be created in any country, Established at Rensselaer School by Amos Eaton in 1824: B. Franklin Greene and the Reorganization in 1849-50. Troy, N.Y.: 32 p., port., (1933).

> "I have established a school in the north end of Troy, for the purpose of instructing persons, who may choose to apply themselves, in the application of science to the common purposes of life. My principal object is <u>to qualify teachers for instructing</u> [underlining by R. R. S.] the sons and daughters of farmers and mechanics, by lectures or otherwise, in the application of experimental chemistry, philosophy and natural history to agriculture, domestic economy, the arts and manufactures."

It is obvious that van Rensselaer's primary object was to establish a school of high quality for the training of teachers of science who would go forth and apply science to almost every branch of human activity.

Van Rensselaer chose geologist Amos Eaton, well known as a stimulating lecturer at Williams College and for his geological survey along a line from Boston to Lake Erie, to be Professor of Chemistry and Experimental Philosophy and lecturer on geology and land surveying. The Founder knew Eaton well; he had employed him as a geologist and as a lecturer on scientific subjects. Given the responsibility of organizing what would become the first American engineering school (Mann 1918),* Eaton promptly put into practice his original method of teaching, which was to have his students perform experiments and record observations, themselves, in the school's several laboratories, and to lecture, themselves, on the experiments they made. Furthermore, Eaton created laboratories with the best equipment and collections available for study in physics, chemistry, geology, and biology. As Ricketts (1933, p. 12) has emphasized:

> "If Eaton had no other claim to fame than the creation of <u>the first laboratories, established in any country, for the individual use of students themselves</u> [Ricketts' italics], his reputation would be assured."

Eaton died in 1842, eighteen years after he created America's first successful school of engineering, and the institution he left had become "the Mecca for teachers of applied science." (Mann 1918.)**

Inasmuch as so many of the ideas and so much of the educational philosophy of Amos Eaton, as carried out at the Rensselaer School in the 1820s and 1830s, became a part of the plan of a polytechnic school of applied science that the Rogers brothers developed in the middle of the nineteenth century, one can hardly resist assuming that William and Henry were influenced in their educational philosophy by their distinguished geological colleague, Amos Eaton, who has been called "The Father of American Geology." It should be added that B. Franklin Greene, the third

---

* Mann, C. R., "A Study of Engineering Education," The Carnegie Foundation for the Advancement of Teaching, B. 11, (1918).

** See reference in preceding footnote.

director of the Rensselaer School, completely reorganized the School's program in 1849-1850, which until then had placed chief emphasis on training teachers of applied science. He developed specific curricula in engineering and is given credit for establishing a pattern of instruction that Rensselaer graduates carried with them to departments of science and engineering in many of the great academic institutions of America and abroad. In 1855 Greene published an 87-page pamphlet "Rensselaer Polytechnic Institute. Its Reorganization in 1849-50; Its condition at the Present Time; Its Plans and Hopes for the Future," which Ricketts (1933, p. 29) characterized as follows:

> "This is a classic in the history of scientific education in this country." [Ricketts' italics.]

At the time of the reorganization there were only three engineering schools, other than Rensselaer, in existence in the United States - the Lawrence Scientific School at Harvard and the Sheffield Scientific School at Yale, both established in 1847, and the beginning of a school at the University of Michigan in 1848.

Inasmuch as Greene's "True Idea of a Polytechnic Institute" was so broad in concept and so perceptive of the future needs of engineering education, one assumes, with Ricketts (1933, p. 32) that

> "... the curriculum outlined in his pamphlet must have been known to every one interested in the establishment of the later schools." [Ricketts' italics.]

Recalling that William and Henry Rogers were actively promoting a polytechnic school of applied science for the Boston area - either as a division of the Lowell Institute or as an independent Institute of Technology - during the same and immediately following years that Greene was reorganizing the Rensselaer School into the very kind of institution envisioned by the Rogers brothers, one can not help but wonder how much the brothers were influenced by Greene's educational philosophy. I have not found the answer to this query in either Life and Letters or in any of the numerous biographies of William Barton Rogers.

Soon after establishing his home in Boston, in 1853, William was elected a member of the Boston Society of Natural History, of the American Academy of Arts and Sciences, and of the prestigious Thursday Evening Club. The meetings of these organizations gave him the opportunity to meet many of the academic and industrial leaders of Boston. Soon he had gathered about himself a small but enthusiastic group of influential Bostonians willing to aid him in establishing the polytechnic school that he and Henry had dreamed about since their youthful educational experiences at the Windsor School and the Maryland Institute.

However, within four years, in 1857, Henry accepted appointment as Regius Professor of Natural History and Geology in the University of Glasgow, and removed to a permanent home in Scotland where he lived until his death in 1866. Henceforth, from 1857, William worked alone with his small group of supporters and was ultimately successful in establishing M.I.T. as recorded in <u>Life and Letters I</u> and <u>II</u>, and discussed in numerous of his biographies.

During the previous fifteen years, 1844-1859, both Henry and William had been making their thoughts about the need of a polytechnic school known to the public through lectures delivered at the Lowell Institute and before other public audiences. In 1848-1849 Henry gave 24 lectures at the Lowell Institute on "Application of Science to the Useful Arts." During the next decade William also gave numerous public lectures. These included two on "Physical Forces," delivered in November 1855, before the Mercantile Library Association to an audience of twenty-five hundred persons; and at the Lowell Institute he delivered twelve lectures "On Water and Air in their mechanical, chemical and vital relations" (1858-1859), followed by a similar number on "Application of Science to Art" (1861-1862).*

On 18 February 1859 the leaders of the move to found the new school called a meeting at the Boston Society of Natural History, out of which came a special committee to memorialize the Legislature of Massachusetts for authority to establish a "Conservatory of Art, Science, and Historical Relics." William seems not to have helped much, if at all, in the preparation of the aforesaid Memorial (actually, he was away from Boston for part of the time during which it was being prepared). The Memorial failed to gain the approval of the Legislature because of its incompleteness and vagueness in certain respects. Undaunted, the Committee appointed a subcommittee to draw up a revised plan for an institution to advance the industrial arts and sciences and practical education in the Commonwealth. William did serve actively on this subcommittee as Chairman, and prepared his well-known <u>Objects and Plan of an Institute of Technology</u> in 1860. From this important pamphlet four years later came his definitive document entitled <u>Scope and Plan of the School of Industrial Science of the Massachusetts Institute of Technology</u>. (See Appendix A farther on.)

The Act of Incorporation of the Massachusetts Institute of Technology was signed by Governor John A. Andrew on 10 April 1861, only two days before the attack on Fort Sumter. The actual beginning of M.I.T., however,

---

* See <u>The History of the Lowell Institute</u>, by Harriette Knight Smith. Boston: Lamson, Wolffe and Co., 125 p., il., (1898).

was delayed by the four years of the Civil War, and William, as President and Professor of Physics and Geology, did not meet his first class until 20 February 1865, when he wrote in his diary that oft-quoted and prophetic statement (Life and Letters II, p. 224):

> "Organized the school! Fifteen students entered. May not this prove a memorable day!"*

Judging from the announcement that was made shortly before the opening of the School of Industrial Science, the classes that were to be offered during the spring of 1865 were:

Elementary mathematics, with practice in the use of the chain, level, etc.
Elementary physics;
Elementary chemistry, with manipulations;
Drawing;
The French Language (Life and Letters II, p. 224).

---

* Fifty years later, Eben S. Stevens, class of 1868, and one of President Rogers' original "Fifteen," recalled in a letter to the Editor of The Technology Review (16/7: 430-431, July 1914) what happened that memorable day "at 16 Summer Street [Boston] in the Mercantile Library Building [long since demolished] where the foundling Institute had rented three long narrow rooms":

> "Here assembled fifteen young men as the Class of '68. They were a 'picked-up lot' in that there was no preparatory school for such an institution of learning in those days and little or no examination as the writer recalls. The Faculty consisted of ten gentlemen with Rogers as professor of physics; a most remarkable man, who left his impress upon everyone with whom he came in contact, whether business men who furnished the sinews of war or students who revered him beyond words to express.

> "Rogers was genial, attractive, with a pleasant smile upon his strong face never to be forgotten, especially that prominent nose. A scientist of broad culture with such command of exceptional English that the students were forced to obtain Jenkins' Vest Pocket Lexicon, by the aid of which we were enabled to elaborate the professor's meanings. Of course, we watched such a mind for some flaw which was never discovered except that he always spelled balance with two l's.

> "We put the work of the first year into four months under Rogers, Runkle, Storer, and Carlton.

> "Returning in the fall, the advent of '69 crowded the Class of '68 over to an abandoned brick dwelling house on the west side of Chauncey Street where were added to the Faculty, Professors Watson and Bocher, the former just returned from Paris with Ph.D. added to his name, a very courteous gentleman of polished manners, but slightly deficient in executive qualities to lead a body of active men to fully appreciate his many excellencies. Professor Bocher taught us French, inviting us at times to his house where we recited and were received as guests of a gentleman of learning and character."

(The above appeared on page 49 of The Technology Review for February, 1965 [Vol. 67/4: 49, 1965]).

Plaque on the front of KENNEDY'S Store at 16 Summer Street, Boston (site of the former Mercantile Building, shown below).
(M.I.T. Historical Collections)

The former Mercantile Building on Summer Street in Boston.

The Massachusetts Institute of Technology began its work of instruction on the upper floor of this building, in the rooms of the Mercantile Library Association, at 16 Summer Street, Boston, on 20 February, 1865.
(J. H. Huffords, Lith.)

There is no mention of classes in geology or mining, but since these subjects were included in the first annual catalogue, it can probably be assumed that Rogers lectured on them during the first regular school year 1865-66.

Thus ends an exceedingly brief and incomplete account of how William Barton Rogers came to be interested in and actually founded a polytechnic school of applied science, the major achievement of a long and varied career of greatest distinction.*

In discussing the numerous meetings and communications that led up to the establishment of M.I.T., Tyler (1902, p. 276-306) writes:

> "Professor William Barton Rogers came to Boston in 1853, - in his forty-ninth year, bringing with him not indeed a matured plan for an Institute of Technology, but rather that enthusiasm, insight, breadth of scientific attainment, skill in popular exposition, and fitness for leadership which enabled him to organize success.
>
> "He occupied himself in writing and lecturing on scientific subjects, and became the natural leader of a group of enlightened citizens eager for the development of comprehensive plans for educational and scientific institutions in the land then being reclaimed from the tidal waters of the Back Bay. ...
>
> . . . . .
>
> "In April, 1862, Mr. Runkle, as first Secretary of the Institute, notified Professor Rogers of his election as President of the Massachusetts Institute of Technology, to serve until the first annual meeting, at which time the Government for the ensuing year would be elected. ..."

Subsequently, the Scope and Plan of the School of Industrial Science, which had been prepared by Rogers, was adopted in May, 1864, and to this day (1975) remains the "intellectual charter" of M.I.T.

The First Annual Catalogue of the Institute, for 1865-1866, shows William Barton Rogers LL.D. as President, and as Professor of Physics and Geology, under Officers of Instruction.

William Barton Rogers' accomplishments as Founder, Organizer, and first President of M.I.T. have been so aptly described by a recent M.I.T. President, also a Southerner, James R. Killian, Jr., that I take the liberty of quoting from one of his several commentaries (The M.I.T. Bulletin, The General Catalogue Issue for SY 1957-58, p. 1, 1957):

---

* Readers interested in a detailed account of Rogers' prolonged efforts in founding M.I.T. are referred to the numerous accounts and biographical sketches listed after the Bibliography, particularly Life and Letters of William Barton Rogers I and II (1896), Runkle (1882), Ruschenberger (1883), Walker (1887), The Tech (1904), Munroe (1904), Killian (1957), and Stratton (1965).

> "With a lively and prophetic vision of the part that science was to play in advancing human welfare, William Barton Rogers, geologist and natural philosopher of the University of Virginia, planned and worked for nearly a decade prior to the Civil War for the establishment of a great institute of technology. As its first president, he set the Massachusetts Institute of Technology on the course of intellectual leadership and public service that has given it continuing vitality and that has made it an institution of national and international influence.
>
> "Since its establishment, M.I.T. has constantly kept before it three objectives - the education of men [and women],* the advancement of knowledge, and service to industry and nation. In its early years the Institute pioneered in extending the laboratory method of instruction as an indispensable educational technique. ..."

Killian commented further on William's achievements as an educator in an address at the College of William and Mary in 1957,** and a later M.I.T. President, J. A. Stratton,† emphasized William's important role as an educational philosopher in helping to lay the foundation for engineering education in America.

GEOLOGY IN ROGERS' SCHOOL OF INDUSTRIAL SCIENCE (M.I.T.)

In all of their discussions of and plans for a polytechnic school in Boston, both William and Henry included instruction in geology (and mining) as an integral part of their proposed School of Industrial Science. The second edition of William's 1861 Objects and Plan of an Institute of Technology (Boston: John Wilson and Son, 1861) included five "schools" in the proposed School of Industrial Science and Art:

>   School of Design
>   School of Mathematics
>   School of Physics
>   School of Chemistry
>   School of Geology

---

* As evidence that M.I.T. looked favorably on women students early in its history, the following action is interesting. On 5 Feb. 1879 the Committee on School of Industrial Science voted:

> "Women who may have been or may hereafter be admitted to departments of instruction in the School may be, if they so desire, examined for a degree and if found qualified to pass under the same conditions that are applied in the examination of male students shall be entitled to graduation and to the usual diploma."

** "William Barton Rogers," Tech. Rev. 60: 105-108, ff., (1957).

† "The New Academy of Engineering," Tech. Rev. 67/9: 41-44, il., (July 1965); "Advice to a new Academy," Science 149: 1206-1208, (10 September 1965).

In this document William wrote as follows about the last-named school (p. 25):

"School of Geology

"Lastly, to complete the circle of instruction essential to the plan of industrial education here contemplated, provising should be made for systematic teachings in <u>Physical Geology and Mining</u>, with so much of the general science as is important to illustrate the more practical branches of the subject. With the aid of maps, sections, and specimens, it would be the aim of this Department to teach whatever is known of the laws of succession of rock-formations, and of the various faults, flexures, and other disturbances or modifications, by which they are affected; and to point out more especially the geological structure and mineral characteristics of each of the great divisions of our own and the neighboring territories, the position and extent of the coal-fields, the belts of iron-bearing and other metalliferous rocks, and the geological relations and ranges of the various granites, sandstones, slates, soapstones, limestones, marbles, marls, clays, and other mineral aggregates which have been found available for building or other purposes, or which may hereafter be brought into profitable use.

"In this connection, the method of conducting geological and mineral surveys would be brought under view. The student would be taught whatever related to the opening and extension of quarries, shafts, tunnels and drifts, and other details of mine-work, and to the winning, raising, and purification of the crude product; together with the drainage, ventilation, underground planning, and entire economy, of the mine; while, by the systematic study of specimens, he would be instructed in the characteristic appearances and properties of the more useful building-rocks, ores, coals, and other mineral materials; and would be prepared to test and estimate their value as applied to particular uses, and to make a proper selection, where necessary between the similar products of different quarries, deposits, or mines.

"Thus various in its practical instructions, both as regards principles and details, this Department of the School of Applied Science would not only offer facilities for professional training to those engaging in the pursuit of practical geology and mining, but, by innumerable facts and suggestions, would render most important service in agriculture, architecture, engineering, and most of the industrial arts."

At this juncture it might be appropriate to review briefly the state of geology in the 1840s and 1850s and to mention the different geological activities carried on by William that motivated him to include geology in his planning for the future M.I.T.

Until the end of the Civil War and the beginning of M.I.T., Geology was largely a matter of observation and description, and usually treated as an incidental subject in college curricula. It would come into its own, however, immediately after 1865, and by 1890 there would be numerous departments of geology in the United States, including the one at M.I.T. (Course XII).

By the time he was ready to draw up his plan for an Institute of Technology in Boston, William had measured stratigraphic sections and identified the minerals and fossils in many strata; mapped the folds and faults in the Appalachian Chain; discovered the first diatomites in the United States and pointed out the agricultural value of the Virginia glauconitic greensands; analyzed the water of hot springs, which he showed were closely related to anticlinal axes and faults; measured the temperatures of mine waters; and noticed how the quality of coal varied with the degree of metamorphism of the associated strata. He reported such observations in the journals of the day, discussed them with his students at the University of Virginia, incorporated them in his seven Progress Reports of the Geological Survey of Virginia, and ultimately brought many of them together in his so-called Coal Reports (1850). After his death in 1882, many of his earlier articles were brought together in his highly regarded Geology of the Virginias (1884).

With Geology such an important part of his scientific background and experience, and also because he always stressed the practical or useful aspects of all science, he predictably gave geology as he knew the science a prominent place in the curriculum of his proposed polytechnic school because he foresaw the need of trained field geologists for future field work throughout the United States.

It is especially noteworthy that he required aspiring geology and mining students to build a firm foundation in mathematics, physics and chemistry in their first two years before starting their specialized course work in the third year. In much the same way he foresaw the need for students in physics and chemistry to perform their own experiments in laboratories established for that purpose.

The First Annual Catalogue (1865-1866) of the Institute lists six Courses in which a student could obtain a degree or diploma, the fourth being GEOLOGY and MINING (p. 10). In the same Catalogue, on pages 22-23, appears the following curriculum to be taken in the third and fourth years by students wishing to major in Course IV:

> IV. - COURSE OF PRACTICAL GEOLOGY AND MINING
>
> Chemical Analysis, Quantitative;
> Descriptive and Determinative Mineralogy; Use of the Blowpipe.
> Lectures on Combustion and Fuel, and on Warming, Ventilation, and Lighting.
> Historical Geology, and Paleontology, -
>   Successive Systems, Groups and Formations, with their leading fossils.

Detailed Study of the Geology of North America.
Special Geology of Coal, Iron, Copper, Salt, Plaster, etc., with
  particular reference to North-American localities; and an account of important Mines, Quarries, etc.
Lectures on Mining [Specific subjects omitted here].
Drawing, - Topographical and Geological Sections and Maps; Conventional Representation of Rocks; Coloring of Maps and Sections; Plans and Sections of Mines, Quarries, and other open
  Workings; Mining, Machinery, and Implements.

Also included in a list of seven media of instruction are:

6. Practical Exercises in Surveying, Levelling, Geodesy, and
   Nautical Astronomy.
7. Excursions for the Inspection and Study of ... Geological
   Sections, Quarries, and Mines.

Once having achieved the legal authority, desired land, and financial support for his proposed Institute of Technology, and having developed a complete four-year program of study, Rogers had to recruit a faculty to conduct the instruction in the six Courses.

Although he himself was Professor of both Physics and Geology, as well as President of the Institute, and gave lectures in both sciences at the start, within six years he recruited three distinguished geologists to help with instruction in Course IV, Geology and Mining. These were Alpheus Hyatt (1870), William H. Niles (1871) and T. Sterry Hunt (1872). <u>Alpheus Hyatt</u>, renowned for his research on fossil cephalopods, was appointed Professor of Palaeontology in 1870. In 1871 <u>William H. Niles</u>, a widely known and popular public lecturer, was appointed Professor of Physical Geology and Geography, and became the Head of Course XII, Geology, when that Department was established in 1890. In 1872 <u>T. Sterry Hunt</u>, the foremost chemical geologist in North America, who could well be called the "Father of North American Geochemistry," was appointed Professor of Geology.

These three professors assumed responsibility for instruction in geology when Rogers became inactive in 1871 and continued to serve the Institute for varying lengths of time during the next two decades until the geological subjects were organized into a separate course, Course XII, Geology, in 1890, as discussed in another part of this history.

AS COLLEGE PROFESSOR AND PUBLIC LECTURER

William Barton Rogers was surely one of the outstanding class-room lecturers and public speakers of his time in America. Every one of his major biographies* is full of superlatives regarding his facility of

---

* The more important of William's biographies, together with numerous
  commentaries on his ability as a teacher, lecturer, and public speaker,
  are listed farther on in Appendix B following the Bibliography.

speech, his rich and impressive vocabulary, his imaginative phraseology, and his power and clarity of expression. It seems clear, not only from the statements of his one-time students and his numerous biographers but also from his published speeches, that he was one of the leading scientific orators of his time.

He left his audiences deeply impressed by the clarity and logic of his presentation as well as by the literary nature of his language. These unusual qualities appeared early. When he was only eighteen years old he delivered an impressive oration at the celebration of the third "Virginiad" at Jamestown in 1822, referred to on a preceding page. By 1827 he was lecturing at the Maryland Institute in Baltimore. A year later, when he succeeded his lately deceased father as Professor of Natural Philosophy and Chemistry at the College of William and Mary, in the fall of 1828, he delivered an "inaugural" oration which when read in its entirety leaves no doubt about the validity of the laudatory statements of his biographers.*

William's concept of the teacher-student relationship was clearly formed at the very beginning of his professorial career. He outlined it with impressive phraseology when, not yet twenty-four, he addressed the students of the College of William and Mary on 27 October 1828 upon succeeding his late father as Professor of Natural Philosophy and Chemistry (<u>Life and Letters</u> I, p. 64):

> "In assuming my functions in the college, it is natural that I should be desirous of conciliating your respect and kind regard. I would fondly hope that the mantle which has descended to me, though no longer graced by the paternal character with which age had invested my predecessor, may still, through a zealous devotion to your interests, be viewed with reverence and affectionate estimation. From my own experience as a student of this college, I am aware of the feelings with which, under certain circumstances, even the noblest and most ingenuous youths are accustomed to regard the collegiate authorities. I know they do not always advert to the community of interest by which the preceptor and pupil are naturally united to each other, but sometimes look with dissatisfaction, if not hostility, upon those who certainly should be among their best and most valued friends. Such feelings are much to be deprecated, and I sincerely desire never to become the object of them. It is, therefore, that I would here willingly begin that intercourse of kindness and mutual confidence which I shall ever labor to maintain, by giving you the assurance that I shall esteem it my duty, as it will be my delight, by every means within my power, to contribute to the success of your studious pursuits, and to your general happiness and welfare,

---

* Reference is made to this impressive address in a preceding section describing his early academic career and again in my comments about his concept of the teacher-student relationship which follow.

and by claiming from you in return a share of that cordial good-will which, with generous ardor, you dispense to your associates in letters, and your participants in study, emulation and honour."

How well the aforementioned concept flowered at the new Institute of Technology in Boston is clearly indicated by the remarks of James P. Tolman, a member of the earliest M.I.T. class (1868), of which each graduate was simply certified as a "Graduate of the School of the Massachusetts Institute of Technology." The familiar diploma would come later! Speaking as president of the M.I.T. Alumni Association, at the Memorial Meeting of the Society of Arts on 12 October 1882, he had the following to say about William Barton Rogers as a teacher (1882, p. 8):

"As an instructor Professor Rogers always came to his classes with his lectures thoroughly prepared. I cannot recall any instance of his appearing without his plan for the session being entirely matured, or any suggestion of his having felt his way in the arrangement of his course. He generally had the blackboards filled with copious notes, written by himself, thus greatly taxing his physical strength. His absorption in his subject made him almost impatient of the restraint imposed by models and apparatus, and at times interfered with the smoothness of the experiments which he had always carefully prepared, and which were so lucidly explained that ocular demonstration seemed superfluous. He almost always introduced illustrations referring to other branches of science than that in particular discussion; and thus greatly contributed not only to the profound respect entertained toward his own accomplishments, but also to the stock of general information of the student, which is so much to be desired in a technical education.

"This fact that he always visibly led his pupils, and never seemed in any manner to lean upon them, aided largely in the development of that discipleship of feeling which was so characteristic of the relation of his classes toward him.

"But it was not alone the earlier students who loved him. My meetings with later classes have been frequent enough to let me learn that even where there was none of that personal intercourse which was the great advantage of the beginning days, he was always known as the students' best friend."

Little wonder that his students and the general public crowded into the lecture hall when he was scheduled to speak, and that business leaders and legislators listened with respect and approval when he petitioned them for financial or legal assistance in support of some major project: *e.g.* establishment of the first Geological Survey of Virginia, or founding a polytechnic school in Boston's Back Bay.

That he was a superb lecturer when a professor at the University of Virginia is attested by the following statement of one of his students there. Said Wm. LeRoy Brown, President of the Agricultural and Mechanical

College of Alabama (as reported by Francis A. Walker before the National Academy of Sciences in April 1887 - see Appendix B):

> "I remember well the very great interest in and enthusiasm for science he excited among the students by his brilliant lectures. Often, especially when it was announced that he would begin his lectures on astronomy, have I seen his lecture-hall crowded with students from other departments, including those of law and medicine; indeed, so crowded with young men, eager to hear the eloquent presentation of the subject by the professor whom they so greatly admired, that not even standing room could be found in the hall. All the aisles would be filled, and even the windows crowded from the outside with eager listeners....
>
> . . . . .
>
> "His manner of presenting the commonest subject in science - clothing his thoughts, as he always did, with a marvelous fluency and clearness of expression and beauty of diction unsurpassed - caused the warmest admiration, and often aroused the excitable nature of Southern youth to the exhibition of enthusiastic demonstrations of approbation. Throughout Virginia - and, indeed, the entire South - his former students are scattered, who even now regard it as one of the highest privileges of their lives to have attended his lectures.
>
> . . . . .
>
> "... Tall in stature; with a figure of the type known to us through the pictures of Henry Clay; with a face that, destitute of all assumption or arrogance, was singularly commanding; with a voice whose compass and quality were capable of producing at once the largest and the finest effects of speech, Wm. Barton Rogers was, in the heighth of his powers, without a peer among the scientific men of his age in addressing an intelligent and cultivated audience."

That he lost none of his oratorical skills with age is demonstrated by the felicity of his language when, on 14 April 1855, he delivered an address at Williamstown (Mass.) on the twentieth anniversary of the foundation of the Lyceum of Natural History of Williams College and on the day of the dedication of Jackson Hall, a new museum of natural history. The address, which Mrs. Rogers states (Life and Letters I, p. 343) is <u>the only one her husband ever wrote out in full for publication</u> (underlined for emphasis - R.R.S.), is noteworthy not only for the quality of the language but also for the breadth of <u>biological</u> knowledge. His 'felicity of expression' is evident from the following brief excerpts of what must have been an hour-long lecture (Life and Letters I, p. 343-357):

> "In the midst of scenery whose picturesque beauties are but the varied repetition of the landscape which in another region for so many years spread its quickening charms around me, I have the privilege of renewing, though but for an hour, that living intercourse of speech which in the lecture-room every enthusiastic teacher so

much enjoys, and which for a large part of my life has been an almost daily recurring pleasure.

"The college bell that for nearly twenty years summoned me at this hour to my pleasant morning task seems even now with its inspiring music to fill the air around me. Let me then feel as if I were but obeying its customary call, and look upon you, young gentlemen of the Lyceum, as familiar lecture-room friends, that my heart unrepressed may take its share in whatever I may say to quicken your love of natural science, or to raise your thoughts to the contemplation of the grandeur and harmony of the universe. ...

"As the relationship and interdependence among the different departments of natural science, although recognized in principle, is often practically overlooked or disregarded. I have thought that I might not unprofitably employ the present occasion in illustrating its importance, and in urging upon the young votaries of science whom I address the enlarged and catholic spirit of study and research which, in the present advanced state of science, is as necessary to eminent success in any one department as it is essential to form the character of the philosophical naturalist. ...

"Honour to the memory of the illustrious Swede [Linnaeus], whose vast and accurate knowledge in each of the great realms of Nature afforded the materials for a systematic structure of the whole of natural science, - whose comprehensive genius planned, and whose unfaltering zeal built up and completed its sublime proportions. And honour, too, to the courageous, indefatigable men who, catching from his lips or his writings the inspiration of the true naturalist, left the calm retreats of study and the enjoyments of society, to brave the toils and perils of distant and inhospitable lands, in quest of new products of nature, or fresh materials for investigation. What isle so remote, what mountain so rugged or lofty, as not to have been the scene of their explorations? What sea so wide or continent so vast as to have been left untraversed by these enthusiastic adventurers in behalf of science and humanity?"

William's ability as a speaker was not something that just happened; letters of advice to his brothers about how to be a good lecturer show clearly that he not only greatly enjoyed public speaking but also gave much thought as to how best he could interest and impress his audience and make his points.

In a letter of May 1, 1841 addressed to his brothers, who were just getting started as professors, William gave the following good advice (<u>Life and Letters I</u>, p. 188-189):

"I am delighted to hear that James is so rapidly winning his way to fame as a lecturer, and I doubt not, my dear Robert, that you will find the task easy and pleasant. You ought to have an ample blackboard, or other equivalent space, for chemical formulae and drawings. I presume you will make use of brief heads of your lectures as reminders. Do not attempt to crowd

> too much in a single lecture, and avoid the common error of experimenting for the eye and not for understanding. Every experiment ought to be accompanied by a full and clear explanation, and this cannot be rendered too explicit and elementary. Cultivate a deliberate and distinct enunciation without sacrificing earnestness and animation of manner. Above all, do not attempt to be over choice in your phraseology, but use the language suggested at the moment. I believe that many an one has failed in making an interesting speaker from being thus fastidious at the beginning. There is nothing in which habits are sooner formed and more difficult to remove than public speaking. ... I think we have all of us erred in reading too little ... The highest eminence as a lecturer cannot be attained without a general culture of mind."

A year later, when writing to Henry (<u>Life and Letters I</u>, p. 207), he expressed his own delight when giving a lecture as follows:

> "I am glad that you are paying more attention to your lectures in the University. Nothing but practice is wanting with any of us to excel in this offhand kind of composition. In fact, I think that is the genius of the family, and depends upon a peculiarity in which we all share. <u>For my own part, I find that when I am strong, as I have been this winter, I absolutely revel in some of my better themes.</u> ..." [I have underlined the foregoing for special notice.]

At exercises in Huntington Hall, in the Rogers Building in Boston, on December 7, 1904, commemorating the one hundredth anniversary of William Barton Rogers' birth, one of his first M.I.T. students, Robert H. Richards (Class of 1868), spoke as follows (<u>The Tech</u> 24/31 [Wed. Dec. 7, 1904], p. 1, 2, 4; reprinted in <u>Tech. Rev.</u> 7: 39-42, 1905):

> "No student of the present day [1904] can feel the thrill of discovery in quite the same way as the first seven of us who were the nucleus of the embryo school [<u>i.e.</u> M.I.T.].
>
> "We were Professor Rogers's children, on whom he tried his experiments in education, - naughty children sometimes, teasing our professors like other students, but, I can truly say, without malice. That was not possible with that gracious presence, dignified, polished, courteous, albeit with a twinkling eye, ever before us.
>
> • • • • •
>
> "... We found ourselves attending Rogers's lectures in physics, illustrated on the blackboard by drawings and on the table by experiments.
>
> • • • • •
>
> "It is incredible [in 1904] that only forty or fifty years ago men were just beginning to believe that fossils were the record of past life preserved in rock books. Geology was a new and most fascinating subject, as presented in Rogers's lectures, although hardly recognized as a subject of collegiate rank. Mining and metallurgy

were arts, showing but dimly the scientific form they were to assume later.

. . . . .

"One of my classmates describes Rogers as the most wonderful example he ever knew of knowledge, of kindness, of wisdom, and of eloquence. So full of zeal he was that the student must work to the limit of ability to please him. Thus was set the pace we have kept. But with all this knowledge he was intensely practical; that is, he had the true scientific spirit which brings all truth to the service of all.

"As a teacher, he was unrivalled for clearness of statement and elegance of expression. Added to the verbal charm was a wonderful skill in blackboard illustration. Rapidity and accuracy were added to grace of line. His drawing of a perfect circle would always bring down the house in the Society of Arts. ..."

Ten years later, in an interview on When We Were Freshmen (Tech. Rev. 16: 575, 1914), Richards recalled:

"I entered a new world when I began to listen to the lectures in which President Rogers described wonderful phenomena in a most beautiful way. From him I also learned the geological history of the earth."

William Barton Rogers gave almost the last third of his life, 23 of 78 years, to M.I.T. as founder, organizer, professor, and president. The record of his activities is as follows:

1859-61	Leader in petitioning Legislature for land, funds, and charter for proposed School
1862-65	President of M.I.T. (Formally notified of election as President on April 19, 1862)
1865-67	Professor of Physics and Geology and President of M.I.T.
1867-68	Thayer Professor of Physics and Geology and President of M.I.T.
1868-70	Professor of Geology and President of M.I.T.
1870-71	Professor of Geology; resigned as President 3 May 1870
1871-78	Listed in Corporation, but not in Officers of Instruction or Faculty
1878-81	Listed in Corporation and Faculty, and President
1881-82	Professor Emeritus of Physics and Geology Died 30 May 1882.

SCIENTIFIC RESEARCH AND PUBLICATIONS

William's scientific research and publications covered a wide range of science - physics, chemistry, geology and botany. Nevertheless, his geological research and publications are most often emphasized over those in the other sciences, and for good reason. Perusal of his Bibliography shows that 93 (60%) of his 156 references are concerned with geological subjects. Twelve of the articles were co-authored with his brother Henry. Even though account is taken of the fact that there is duplication of some titles, it is still evident that more than half of William's publications involved geological subjects.

However, he wrote 28 papers on physics subjects and 21 on chemistry subjects, seventeen of the latter being co-authored with his brother Robert. Another 14 articles are on miscellaneous scientific subjects as widely different as the nature of dew and observations on the coiling of the tendrils of the winter squash vine.

His geological contributions fall into three general categories. The first, and in my judgment the most important, was the field work he carried on while getting the Geological Survey of Virginia established and then serving as State Geologist, a decade or more, from 1834 to 1844. The results of his field and laboratory investigations were described in his 1836 Reconnaissance Report; in his six annual Reports of Progress (1837-1842); in three important articles on the Tertiary formations of Virginia, prepared jointly with his brother Henry; and in a dozen articles on a variety of geological subjects, including greensands and infusorial earths and their uses, coals, thermal springs, and mine temperatures. I assign primary importance to the stratigraphic and structural field work that William conducted in Virginia, and correlated with similar work done by his brother Henry, State Geologist of Pennsylvania, particularly those investigations of the mountainous areas of their respective states, because the brothers demonstrated for the first time the folded and faulted structure of the Appalachian Chain, not only in their own respective states, but also for its full length of a thousand miles or more. This work led to development by the two Rogers brothers of an original theory of origin for the Appalachian Chain that was widely acclaimed in the 1840s and is discussed at some length in a preceding section.

In the conduct of his field work, Rogers made important discoveries of greensands that could be used by farmers to enrich their soil, of infusorial deposits that could be used for various commercial purposes, and of coals that had economic importance. In such work he was an early leader in demonstrating to the citizens of Virginia, and perhaps even more importantly, to the State's legislators, the great economic value and potential use of the various natural resources that were readily available.

In addition to the numerous items in his annual Progress Reports of the Geological Survey of Virginia, and the stratigraphic and structural work on the Appalachian Chain, William published many other articles on a surprisingly broad range of geological subjects: e.g. artesian wells (1843); descriptions of eight new species of Tertiary shells (1839); coalmine temperatures (1842, 1843); thermal and alkaline springs (1842, 1843); coal (1854); the New Red sandstone of the Connecticut Valley (1855); discovery of the Middle Cambrian trilobite Paradoxides at Braintree, Massachusetts (1856); rate of sedimentation in the South Joggins (1859); and deformed pebbles in the conglomerate at Newport, Rhode Island (1861, 1875).

The second contribution to geological literature, shared equally with his brother Henry, was development and exposition of the novel theory of origin of the folded and faulted Appalachian Chain mentioned in the preceding paragraph and discussed at some length in a preceding section. Although publication and subsequent defense of the theory itself brought both fame and controversy, and stimulated other geologists to extend and test their own theories, I would rate it of second importance only, because it lacked the fundamental scientific validity of the field observations on which it was based. Nevertheless, because many of William Barton Rogers' biographers have emphasized the importance of his and Henry's theory, I have discussed it at some length in a preceding section.

The third type of contribution to geology that William made was in applying the principles and laboratory methods of chemistry to a variety of geological problems. As stated earlier on he wrote 21 articles of a chemical nature, 17 of which he co-authored with his chemist brother Robert. Some of his earliest papers deal with the analysis of natural materials - shells (1834), marl and other carbonates (1834), and bi-malate of lime in sumac berries (1835). After his period of service as State Geologist, he published additional articles with Robert dealing with a variety of chemical subjects - new instruments and processes for carbonate analysis (1844); chemical equivalents and determination of iron in iron ores, cast iron, etc. (1844); new process for obtaining pure chlorine gas (1846) and formic acid (1846); absorption of carbonic acid by water, saline solutions, and other liquids (1848); new method of determining carbon in native and artificial graphite (1848); on decomposition of rocks by meteoric waters (1848); and the volatility of potassa and soda (1849).

William's lifelong interest in physics led to 28 articles on physical phenomena of which the following are typical: experiments on voltaic batteries (1835); binocular vision (1855, 1860); ozone in the atmosphere (1856, 1858); sonorous flames (1858); auroras (1859); a registering thermometer (1860); photometric measurement of Boston's electric illumination (1863); phosphorescence (1880); and an Address on the scientific work and the character of Joseph Henry, delivered in the Hall of Representatives, Washington, D.C. (1880).

In addition to the many papers on geology, chemistry, and physics, William also published a dozen or more brief articles on a surprisingly diverse group of subjects as mentioned on a preceding page.

During his years of active teaching Rogers published two books for use by his students - <u>An Elementary Treatise on the Strength of Materials</u>, being the substance of the lectures on the subject delivered in the School of Engineering of the University of Virginia. Charlottesville: Tomkins and Noel, 50 p., (1838); and Elements of Mechanical Philosophy, for the use of the Junior Students of the University of Virginia. Boston: Thurston, Torry & Emerson, Printers, 339 p., (1852).

Ten years before William founded M.I.T., he wrote Henry about the need of a manual or textbook of American Geology, pointing out that such a book would inform the public on their stratigraphic nomenclature, earth dynamics, and other views, and would be acceptable for college use and to men of science in the United States and abroad. William, presumably, went so far as to prepare a brief outline of 8 pages, titled "Division of the Subjects in the 'American Text Book of Geology,'" with four parts: I. Introduction; II. Geological Dynamics; III. Historical and Descriptive Geology; and IV. Practical Geology. This manuscript is preserved in Rogers' papers at M.I.T. So far as I have been able to determine, nothing ever came of this preliminary idea, though the brothers mentioned it several times in letters to one another between 1834 and 1855 (Life and Letters I: Henry to William, 28 November 1834, p. 111-112; William to Henry, 12 August 1855, p. 341 and 20 November 1855, p. 359).

It is obvious from the preceding lists that William Barton Rogers was interested from an early age in science (perhaps we should say "natural philosophy" for his time) in its broadest sense - physical, chemical, geological, biological - and more in the practical than in the theoretical aspects of the problems he investigated, though he did not neglect explanations and theories altogether. Because his scientific interests were so broad, he would probably be regarded today as a generalist rather than a specialist.

W. B. ROGERS' ACTIVITIES IN SCIENTIFIC ORGANIZATIONS:
A.A.G.N. - A.A.A.S. - B.S.N.H. - A.Ac.A.S. - N.A.S. - I.G.C. - A.A.P.S.S.

William Barton Rogers was clearly a scientific activist in that he participated in the founding and subsequent activities of a number of important associations, societies, academies and the like which had their beginnings during the 19th Century in which he lived his entire life.

The first geological organization in the United States was the American Geological Society (A.G.S.) which was founded at Yale on 6 September 1819.* Rogers was not yet fifteen and his education in science at that date was not sufficiently advanced to qualify him for participation in the founding of the organization, which lasted only until 1828.

---

* See p. 66-67 in G. B. Goode's "The Origin of the Natural Scientific and Educational Institutions of the United States." (Reprinted from the Papers of the Am. Historical Assoc. [p. 90-202]). New York: Putnams, 112 p., (1890).

The second geological organization to be formed was the Association of American Geologists and Naturalists (commonly shortened in references by dropping the last two words; hence, A.A.G.), a national society founded in 1840 to advance scientific interests on a national scale. William was one of the founders and served as president in 1847. At the Boston meeting in September 1847 he presided when the Association voted to change its name to American Association for the Advancement of Science at the next annual meeting, which was scheduled for Philadelphia in 1848.

At the Philadelphia meeting, on 20 September 1848, he was chairman throughout the meeting and at the end turned over the office to William C. Redfield, president-elect and the first president of the newly organized American Association for the Advancement of Science.* This memorable event is described in the first Proceedings of the Association (A.A.A.S.) published in 1849. William served on several committees of the Association following the Philadelphia meeting, was president in 1876, and was host chairman of the memorable Boston meeting in 1880, referred to farther on.

William and Henry were elected honorary members of the Boston Society of Natural History in June 1842. A little later, in May 1845, he and Robert were elected honorary fellows of the American Academy of Arts and Sciences, and subsequently he served on several of the Academy's committees. He also published a number of geological articles in the Academy's Proceedings (see Bibliography farther on).

---

* Although some biographers have stated that William was the first president of the A.A.A.S., because he chaired the Philadelphia meeting, the facts argue otherwise, as reported in J. R. Ritchie, Jr.'s "The American Association for the Advancement of Science," <u>The New England Magazine</u> 18/6: 638-661, port., il., (1898):

> "... William B. [Rogers] ... was ... of record as participating in the third meeting [of the American Association of Geologists, the immediate forerunner of the A.A.A.S.], but when he took up the work [participation in the A.A.G.'s affairs] he did it with a will, becoming president at the sixth meeting and coming again to the chair at the eighth. Under his leadership the vote was cast to form the new society [A.A.A.S.], under his eye the committee prepared the new constitution, one that was so broad that it remains today unchanged in any important particular, and he did not lose sight of the matter till it was finally settled and the new constitution was accepted. He was chairman at the assembling of the meeting in Philadelphia [in 1848], and was thus the <u>first actual president</u> [my underlining] of the new association, for he retained his place in the chair until the meeting had fully organized, until the association had become in fact the American Association for the Advancement of Science, - and then gave way to Dr. W. C. Redfield of New York, the president-elect." (p. 652.)

He was elected a founding member of the National Academy of Sciences, when that prestigious organization was incorporated by Congress in 1863, and was elected its third president in 1879, thereafter serving in that office until his death in 1882.

According to J. Brian Eby,* William was a member of the Founding Committee, under the presidency of James Hall, that established the International Geological Congress in 1875, and he helped to arrange the first meeting in Paris in 1878. Professor J. Capellini of the University of Bologna had first suggested such a Congress in Italy in 1874, but it was 1881 before a meeting took place in that country.

Soon after settling down in Boston William was invited to become a member of the exclusive "Thursday Evening Club" of Boston (originally called the "Warren Club" after its founder, Dr. John C. Warren), and later on became its president. (Life and Letters I, p. 310.)

HONORS

Recognition came to William Barton Rogers in many forms and for various achievements, both while he was living and posthumously. I have searched diligently in order to include every recognition of any consequence, hence any omission has been unintentional or an oversight.

MEMBERSHIPS IN PROFESSIONAL SOCIETIES

William's active participation in the founding and subsequent affairs of numerous professional organizations are discussed in the preceding section. The reader's indulgence is asked as I refer again to these activities in order to record as completely as possible both the honors accorded William and the services he gave to the organizations that honored him.

He was elected a correspondent of the Academy of Natural Sciences of Philadelphia in 1833, a member of the Virginia Historical and Philosophical Society in 1835, and a member of the American Philosophical Society in 1835. He was elected a member of the National Institution for the Promotion of Science in 1840 (Life and Letters I, p. 168-173), and in the same year he became a founding member of the Association of American Geologists and Naturalists. He served the latter as president in 1847 and presided at the 1848 Philadelphia meeting when the organization voted to

---

* "23rd International Geological Congress, Prague, August 1968." A.A.P.G., B. 53: 236-240, (1968). See also Rept. XVI Session, I.G.C., Vol. I and II, U.S.A., (1933).

change its name to American Association for the Advancement of Science (the well-known A.A.A.S.). He became a member of the newly named Association, served on several of its committees, was elected president in 1876, and was the hospitable host chairman in 1880, when the Association met in Boston.* [See also Science 166: 1553, (1969).] In 1842, soon after he and his brother Henry had presented their classic paper on Appalachian structure at the Boston meeting of the Association of American Geologists and Naturalists, both brothers were elected honorary members of the Boston Society of Natural History. On 1 June 1844 the brothers were informed that they had been elected foreign members of the Geological Society of London. Later the same year William was informed of his election to the Royal Society of Northern Antiquaries (Copenhagen). In May 1845 he was elected a fellow of the American Academy of Arts and Sciences. He was selected a charter member of the National Academy of Sciences in 1863 and elected the third president in 1879, following in office the illustrious Joseph Henry, and served in that office until his death in 1882.**

When the American Association for the Promotion of Social Science was organized in Boston, on 4 October 1865, William was elected its first president. Almost forty years later, in 1903 and 1904, the Association again held its meetings in Boston and took note of the services of Rogers as a founding member and the Association's first president.

COLLEGE DEGREES

Although it is known that William attended the College of William and Mary as a full-time student during SY 1819-20, and pursued studies there intermittently from 1820 to 1824, I have been unable to find any record of his having graduated or having received a baccalaureate degree despite statements to the contrary by several of his biographers.† Furthermore, as pointed out in a footnote earlier on in the section on William's Early Academic Career (1819-1838), William and Mary's College Archivist, Ms. Kay J. Domine, has found no indication thus far (1975)

---

* See "The American Association for the Advancement of Science," by John Ritchie, Jr. The New England Magazine n.s. 18/6: 638-661, (1898). This is an excellent account of the earlier history of the A.A.A.S. and particularly of the part that the Rogers brothers and M.I.T. played in the founding and subsequent development of the Association.

** See "Biographical Memoir of Wm. Barton Rogers [1804-1882]," by F. A. Walker, read before the National Academy [of Sciences], April, 1887, and printed by Judd and Detweiler, Printers, Washington, D.C., 13 p., (1895). See also Nat. Ac. Sci., Biog. Mem. 3: 3-13, (1895).

† See footnote on facing page.

that William actually received any earned degree from his so-called Alma Mater, though he did receive an honorary LL.D. degree in 1859.

William's first honorary degree came on 4 August 1848 when he was informed that the LL.D. degree had been conferred on him by Hampden-Sidney College in Virginia. The same honorary degree was awarded to him by his Alma Mater, the College of William and Mary, in 1859, and by Harvard in 1866.

MEMORIALS - BUILDINGS AND LABORATORIES;
SCHOLARSHIPS AND PROFESSORSHIPS; ETC.

Following William's death in 1882, numerous memorials were erected or established to honor him and to perpetuate his name in Massachusetts and Virginia. Nineteen such are described briefly in the following paragraphs; there could well be some others that I have missed in spite of my diligent search.

The Rogers Laboratory of Physics at M.I.T. (1869)

One of William Barton Rogers' most important contributions to the education of scientists and engineers was his idea of having the student learn by doing, as a supplement to his listening to lectures and watching the lecturer perform experiments before the class; of having him go into a laboratory and conduct his own experiment, make his own observations

---

† (Footnote from facing page)

Wrote a close friend and colleague, Jedediah Hotchkiss (1882, p. 28-29), soon after William's death:

> "After he had graduated at the College of William and Mary, at the age of twenty-five [i.e. it must have been in 1829], he began a course of lectures upon natural science at the Maryland Institute, at Baltimore."

A. H. H. Stuart, a student of William's father, Professor Patrick Rogers, and a life-long friend of William, recollected that

> "... William had graduated with great distinction a year or two before I entered college, and was looked up to with the respect and almost reverence with which college boys regard those who have won high college honours." (Life and Letters I, p. 56.)

Finally, on p. 59 of Life and Letters I appears the statement:

> "... William Barton Rogers, ... himself a graduate of distinction of William and Mary College ..."

In spite of the preceding statements, however, it is worth noting that in M.I.T.'s first annual catalog, for school year 1865-66, Rogers' name in the list of Officers of Instruction (p. 4) is followed by only the degree LL.D.!

and measurements, and write up the results. In so doing the student combined mind and hand, as implied in the motto on the M.I.T. seal – MENS ET MANUS.

A bronze plaque (see diagram below) on the corridor wall outside Rooms 4-313 and 4-317 records that William originated and developed the plan for an instructional laboratory of physics at M.I.T. as early as 1869, only four years after the founding of the Institute.* It would seem from some of William's own letters, however, that a decade or more earlier, while he was still at the University of Virginia, he had initiated the lecture-student laboratory method of instruction in physics, and probably also in geology, just as his brother Robert across the campus was following a similar procedure in teaching chemistry, a science in which laboratory work by students had long been an integral part of the instructional program in that science.

In actuality, it should be noted here that it was Prof. E. C. Pickering, to whom William transferred the teaching of physics in the fall of 1867, who put the laboratory into operation in 1869 as reported by Prescott (1954: 93-94):

THE ROGERS LABORATORY OF PHYSICS
ESTABLISHED AD-MDCCCLXIX

THE PLAN OF
A PHYSICAL LABORATORY
FOR PURPOSES OF INSTRUCTION
ORIGINATED WITH

WILLIAM BARTON ROGERS
FOUNDER OF

THE MASSACHUSETTS INSTITUTE OF TECHNOLOGY

AND WAS DEVELOPED
IN THIS INSTITUTION

Wording on the plaque that hangs on the corridor
wall outside rooms 4-313 and 4-317 at M.I.T.

---

* The interested reader is referred to H. M. Goodwin's informative article "Physics at M.I.T.: A History of the Department from 1865-1933," (Tech. Rev. 35/8: 287 ff., May, 1933), for a detailed discussion of how the Rogers Laboratory of Physics was developed.

> "The establishment of the laboratory of physics at M.I.T. at this time [1869] was an important contribution to American technical education. Before President Rogers' severe illness in 1868, he and Professor Pickering had given much thought to the establishment of a physical laboratory. It was impossible to discuss this matter with the president while he was so seriously ill, but .... Professor Pickering proceeded with the plans, which were approved by the Committee on Instruction on May 11, 1869, and the first physical laboratory in America was set up in time for the opening of the school year in September, 1869. On February 14, 1872, the Corporation passed the following resolution:
>
>> "Resolved that as a slight recognition of the eminent services which Professor William B. Rogers has rendered the Institute of Technology, the Physical Laboratory of the Institute shall be designated and hereafter known as
>
>> 'The Rogers Laboratory of Physics.'"

When Rogers was informed of this vote he accepted the honor modestly, told of his original reasons for planning such a laboratory, and gave Pickering the credit for organizing and directing it. (Goodwin 1933: 287 ff., and Prescott 1954: 94.)

In actuality, the historical record does not support the credit implied in the statement on the M.I.T. plaque. Rather, there is clear evidence that Amos Eaton, another great American geologist and educational innovator, had antedated Rogers by forty years in initiating laboratory work for students in physics, as well as in chemistry and geology. This he did at the Rensselaer School in 1824, as pointed out emphatically by P. C. Ricketts in 1933. In the foreword to his pamphlet on Eaton, Ricketts (1933) states:

> "I have written this little pamphlet for three reasons:
>
>> "To do justice to the memory of Amos Eaton, teacher, author, investigator: a man of genius whose great contributions to knowledge of the natural sciences, in the first half of the nineteenth century, should be better understood.
>
>> "To prove that the first chemical laboratory and the first physical laboratory established in any country for the use of students, themselves, in their daily work was created by Amos Eaton at Rensselaer School in Troy, N.Y. in 1824.
>
>> "For the purpose of helping to perpetuate the name of B. Franklin Greene who in 1849-50 developed curriculums in engineering which formed the pattern for nearly every school in this country established after that time."

Ricketts then proceeds to validate the preceding claims in rather convincing manner, and interested readers will find his fact-filled discussion well worth careful reading for the information it provides on the early development of science and engineering education in America.

(1) WILLIAM BARTON ROGERS

As mentioned in the preceding section on "The Founding of M.I.T. (1859-1865)," I am puzzled by the lack of any reference to the Rensselaer School by the Rogers brothers in their numerous letters and other writings, or for that matter by Goodwin (1933: 287 ff.) or Prescott (1954: 93-94).

Rogers Buildings in Boston (1883) and Cambridge (1938)

The first permanent building constructed for the newly incorporated M.I.T. was located in Boston, and was built alongside the recently completed Museum of the Boston Society of Natural History (now occupied by Bonwit Teller at 234 Berkeley Street), in the block bounded by Boylston, Clarendon, Newbury and Berkeley streets (see accompanying illustrations). Soon after William's death the Corporation named the new structure the ROGERS BUILDING,* and it was so called until razed in 1939 to make space for the massive building now occupied by the New England Mutual Life Insurance Company.

When, in 1916, the Institute was moved from Boston across the Charles River to its present site in Cambridge,** it found awaiting occupancy one great continuous complex of "buildings" directly east of Massachusetts Avenue, with a great court (now called Killian Court in honor of the Institute's tenth President, James Rhyne Killian, Jr.), open on the south side toward the Charles. As the years passed, additional sections ("buildings") were added to the original complex. One of these later additions, Building No. 7, was dedicated in 1938 and was named the ROGERS BUILDING. This is the part of the imposing facade along the east side of Massachusetts Avenue, with the street number 77, that has the great fluted Ionic columns of Indiana limestone flanking the main entrance and supporting the smaller of the Institute's two conspicuous domes (see accompanying photograph). Carved into the limestone of the architrave above the columns are the bold letters:
>   MASSACHUSETTS INSTITUTE OF TECHNOLOGY
>   WILLIAM BARTON ROGERS, FOUNDER

---

* At a meeting of the M.I.T. Corporation on 23 May 1883, the following resolution was adopted:
> "That the original building of the Institute of Technology be called hereafter the 'ROGERS BUILDING,' in recognition of the eminent services of Professor WILLIAM B. ROGERS, as the founder and organizer of the Institute; and that the announcement of this vote be made by the President at the graduating exercises of the year."

As called for by the resolution, President Runkle made the announcement at the close of the graduating exercises on 29 May 1883, just a year after William's stunning death on the same platform in Huntington Hall.

** The ceremonies attending the move are described in detail in <u>The Technology Review</u> 18/7: 462-800, (July 1916).

The second Rogers Building, facing westward at 77 Massachusetts Avenue, Cambridge, consists of the central section with the Ionic columns and the lefthand section (Architecture), and is designated Building 7 in the present M.I.T. complex.   (M.I.T. Photo by Campbell, 1975)

The first Rogers Building, facing southward on Boylston Street, Boston, with the Museum of the Boston Society of Natural History (Bonwit Teller in 1975) on the right to the east. View taken from the corner of Clarendon and Boylston streets.   (M.I.T. Historical Collections)

(1) WILLIAM BARTON ROGERS

William Barton Rogers Memorial Fund at M.I.T. (1883)

The Report of the M.I.T. Treasurer for the year ended June 30, 1974, lists the following on page 341:

> "560 A ROGERS (William Barton) MEMORIAL, 1883-85, $250,225. Contributions from 91 persons, the income for the support of the Institute."

Presumably the money in this fund was raised during the three years after his death in 1882, to perpetuate his memory at the Institute.

The William Barton Rogers Scholarship at The College of William and Mary (1905)

According to Ewing (1938, p. 25):

> "In 1905, the Massachusetts Institute of Technology established a scholarship, known as the William Barton Rogers Scholarship [presumably at M.I.T.], to be awarded annually to a student of William and Mary selected by its [i.e. W. & M.'s] faculty."

No such scholarship has been listed in the regular M.I.T. Catalogue during the past thirty years, and the reason is evident from a recent communication from (William and Mary's) College Archivist, Ms. Kay J. Domine (Letter to R. R. Shrock dated Sept. 8, 1975):

> "The William Barton Rogers scholarship was started by M.I.T. in either 1905 or 1904, most likely 1905. Again there is some confusion as to the exact date. First awarded in 1906, it was for a student who had 'taken sufficiently advanced work for entering the Institute of Technology with advantage.' The term 'with advantage' was dropped by 1929. At first the scholarship was for $300.00, but later was changed to $400.00. The scholarship was to be 'awarded to a student nominated by the William and Mary faculty.' It is in furtherance of the insistence of William Barton Rogers upon a liberal culture in connection with technical training that the Institute has set up a 'combined plan of study' with a number of liberal arts colleges, including the College of William and Mary. After 1945-46 it is no longer listed as being available."

William Barton Rogers Chair of Economic Geology at the University of Virginia (1907)

According to a brief item under "General Institute News," on pages 43 and 44 of The Technology Review for 1908 (Vol. 10), a chair honoring Rogers was established at the University of Virginia in 1907. The President of the University, Edwin A. Alderman, wrote to Mrs. Rogers as follows about the chair:

Oct. 24, 1907.

"My dear Mrs. Rogers, - On the 15th of October I had the honor to propose the following resolution to the rector and visitors of the University of Virginia:-

"Resolved, That the Chair of Economic Geology recently established at the University be designated the 'William Barton Rogers Chair of Economic Geology,' in recognition of the eminent and devoted services of William Barton Rogers to the University of Virginia.

"It gives me great pleasure to inform you that this resolution was adopted unanimously and with great enthusiasm by the rector and visitors of the University of Virginia. The Chair of Economic Geology was established in the University last June. The first incumbent is Thomas Leonard Watson, who came to us from the Virginia Polytechnic Institute. He is a man of distinction and eminence in geological work in this country. We are hoping to induce the State to enter upon the work of the geological survey. If we can succeed in this effort, the work which your distinguished husband inaugurated will be carried to full completion.

"I trust that the William Barton Rogers Chair of Economic Geology in this University will prove a source of unmeasured strength to the intellectual and economic development of the State.

Very sincerely,

(Signed) Edwin A. Alderman,
President."

William Barton Rogers Professorship of Economic Geology at M.I.T. (1912)

When Waldemar Lindgren was invited to come to M.I.T. as Professor and Head of the Department of Geology in 1912, the M.I.T. Corporation felt special recognition was due him, and this was recorded as follows in the President's Report for 1911-1912 (p.13):

"Professor Lindgren being one of the most distinguished members of the National Academy of Sciences and one of the foremost economic geologists in the world, it seemed appropriate to associate his office with the name of William Barton Rogers, the Founder of the Institute, who was a pioneer in economic geology in this country and a president of the National Academy of Sciences. For this purpose your Corporation in June last [June 1911] made provision for the establishment of the William Barton Rogers Professorship of Economic Geology, and Professor Lindgren was appointed as its first occupant. We have high hopes that he will establish a tradition of scientific achievement that will be a powerful stimulus for generations to come." [Underlining by R. R. Shrock.]

At the same time The Technology Review commented on the special professorship (Tech. Rev. 14: 365, 1912) as follows:

> "This [i.e. the professorship] will be a memorial peculiarly fitting to the great Virginian who founded the Institute of Technology. Himself a geologist, he was impressed with the value of economic geology. The more possible is it now to establish such a memorial through the great gift to the Institute of his wife and helpmeet, Mrs. Rogers, for the Institute can now afford to lay aside a portion of the bequest for the endowment of the new chair, which it is expected will be devoted largely to research."

Prof. Lindgren held this special professorship as long as he was active at M.I.T., from 1912 to 1933, but when he retired as Professor Emeritus in 1933, the professorship lapsed and the chair seems to have been discontinued altogether.

The William Barton and Emma Rogers Library at M.I.T. (1916)

The Minutes of the 362nd meeting of the M.I.T. Corporation held on 13 December 1916, only a few months after the Institute had been moved from Boston to Cambridge, record the following action:

> "It was also voted that the General Library in the dome of the Administration Building [Building 10] be named in honor of William Barton and Emma Rogers."

So far as I have been able to determine, the action voted was never executed! Existing records of the Institute's libraries show that during the past forty years the library in the dome has been called successively the Central Library, Engineering Library, and Vail Library of Engineering, and currently the James Madison Barker Library of Engineering.

In a recent history of "The Massachusetts Institute of Technology Libraries," prepared by Ms. Natalie N. Nicholson (Director of M.I.T. Libraries in 1974) for publication in the Encyclopedia of Library and Information Service (Allen Kent, editor), there appears the following statement (p. 4-5) in the 35-page mimeographed typescript dated June 20, 1974:

> "Rationally, a central library was planned and assigned to space at the very center of the building, on the first floor, where it could relate equally to all disciplines. Unfortunately, as construction of the building proceeded, budget strictures developed. An auditorium, which was to have been placed under the central dome, had to be eliminated and the library was installed in its place."

There is no mention, in the history, of the Corporation vote to name the library in honor of William Barton and Emma Rogers, but possibly that familiar phrase "budget strictures developed" may be the clue. Still, one wonders why the proposed names were not given to the library that was established beneath the great dome.

The William Barton Rogers Memorial Science Hall at The College of William and Mary (1927)

Among the major memorials created to honor Rogers is the William Barton Rogers Memorial Science Hall that stands on the campus of the College of William and Mary, in Williamsburg, Virginia.* It was dedicated in 1927, having been built at a cost of $300,000 raised publicly by an Advisory Committee of distinguished New Englanders and other Easterners. The solicitation brochure sent out by the Committee contains the following statement on page 3:

> "WILLIAM BARTON ROGERS, the founder of the Massachusetts Institute of Technology, was graduated from the College of William and Mary in Virginia a century ago.
>
> "'The spot where I caught the inspiration of science,' was the way he described his Alma Mater.
>
> "To honor this pioneer of scientific instruction and to commemorate this bond between the Massachusetts Institute of Technology and the College of William and Mary, it is proposed to erect on the campus of the latter the William Barton Rogers Memorial Science Hall."**

According to Ewing (1938, p. 26):

> "In the main hallway is mounted a bronze tablet whose appropriate sentiment was composed by the late President of Harvard University, Charles W. Eliot, who was professor of chemistry in Rogers' first faculty at M.I.T."

The tablet bears the statement shown on the following page according to Ewing (1938, p. 15).

William received his earliest college education at William and Mary, first attending as a full-time student during SY 1819-20, and then continuing intermittently, because of much poor health, for the next four years, 1820 to 1824. It has been assumed that he was graduated at the end of this period, inasmuch as his earliest biographers consistently state that he was a "graduate," but, as mentioned earlier on, I have found no specific reference to a baccalaureate degree (see preceding section on "College Degrees"). In 1828 he returned to the College, which he referred

---

* An anonymous article in The Technology Review for November, 1925 (Vol. 28/1, p. 9, 15-16), entitled "A Rogers Memorial," has a sketch of the proposed William Barton Rogers Memorial Science Hall (p. 9), and reports that the building, which is to be financed from funds raised by M.I.T. alumni and others, will provide space for botany, chemistry, geology, and physics (p. 15-16).

** See the 16-page pamphlet - William and Mary College, Williamsburg, Va. The William Barton Rogers Memorial Science Hall. [Williamsburg, Va., College of William and Mary, n.d.] in M.I.T. Archives - T-F/R731. See also "Early teaching of science at the College of William and Mary in Virginia," by Galen W. Ewing. Bull. Coll. William and Mary, vol. 32, No. 4, 29 p., (April, 1938).

(1) WILLIAM BARTON ROGERS

WILLIAM BARTON ROGERS

MEMORIAL HALL

IN THIS HALL

STUDENTS OF WILLIAM AND MARY

WILL HAVE THE OPPORTUNITY

AND THE MEANS OF

TRAINING THEIR SENSES

TO ACCURATE OBSERVATION

AND THEIR MINDS TO SOUND INFERENCE

FROM THEIR OBSERVATIONS.

HERE THEY WILL PUT IN PRACTICE

THE LOVE OF LIBERTY AND TRUTH

THE JOY IN DISCOVERY

AND THE DELIGHT IN WORK BELOVED.

HERE THEY WILL LEARN THE VALUE

OF INTIMACY WITH KINDRED SPIRITS

FRANK, COURAGEOUS, AND UNSELFISH.

HERE THEY WILL ACQUIRE

THE METHOD OF SCIENTIFIC RESEARCH

WITH ITS POWER, ITS ACHIEVEMENTS,

AND ITS INFINITE PROMISE.

Charles W. Eliot

A bronze tablet with the above statement originally hung in the main hallway of the Memorial Hall at the College of William and Mary.

to as his Alma Mater, to succeed his lately deceased father, as Professor of Natural Philosophy and Chemistry, and continued in that post until 1835 when he resigned to accept a professorship at the University of Virginia and simultaneously become the first State Geologist of Virginia.

An interesting piece of information about the Rogers Hall came recently to hand from the College Archivist, Ms. Kay F. Domine, in a letter to me dated Sept. 8, 1975:

> "The dedication of Rogers Hall has been another problem around this College [William and Mary] just recently. A lengthy search has been on to try to find out whether or not the Hall was actually dedicated, and again from a lack of any mention of it, it was generally decided that Rogers Hall was never formally dedicated. At that time the College built several buildings and we cannot locate any evidence of a dedication for any of them. This has recently become a problem because this fall the College is opening a new science building and is going to name that one William Barton Rogers Hall and thereby remove his name from the old building. This new Rogers Hall will be dedicated this fall during Homecoming. It has not been decided what the new name of the old Rogers Hall will be. There are no brochures on the old Rogers Hall. It seems to be just one of several buildings the College raised money for at that time."

William Barton Rogers School, Hyde Park, Boston (1928)

On 3 December 1928 the Boston School Committee honored William by the following action:

> "ORDERED, That the Hyde Park Intermediate School located in the former Hyde Park High School building is hereby named the William Barton Rogers School ..." (See 1928 reference in list following Bibliography farther on.)

The Hyde Park Intermediate School is still referred to as the William Barton Rogers School; it has been mentioned recently in newspaper stories about the turbulent busing problem in some of the Boston schools.

A Memorial Plaque at the University of Virginia (1930)

On 7 December 1930 the Technology Club of Virginia presented to the University of Virginia a metal tablet memorializing William for his services to that institution as Professor of Natural Philosophy and to M.I.T. as Founder and President. The tablet, which was hung in the Chemical Laboratory, has the statement shown on the accompanying illustration, which appears in L. B. Hitchcock's article "A Memorial to Rogers - The Founder of the Institute is honored at the University of Virginia," (Tech. Rev. 33/5: 245, 258, 260, Feb. 1931).

(1) WILLIAM BARTON ROGERS

> WILLIAM BARTON ROGERS
> DEC. 7, 1804 - MAY 30, 1882
>
> PROFESSOR OF NATURAL PHILOSOPHY AT THE
> UNIVERSITY OF VIRGINIA
> 1835 - 1853
>
> FOUNDER 1859 AND
> PRESIDENT 1865 - 1870, 1878 - 1881 OF THE
> MASSACHUSETTS INSTITUTE OF TECHNOLOGY
>
> IN WHOSE MEMORY
> THIS TABLET IS PLACED BY THE
> VIRGINIA ALUMNI OF THE INSTITUTE
> DEC. 7, 1930

Rogers Plaque in the Chemical Laboratory,
University of Virginia.
Tech. Rev. 33/5: 245 (1931).

In addition to the tablet there are other mementoes of William at the University of Virginia according to L. G. Hoxton (in Hitchcock 1931, p. 258), who spoke at the presentation:

> "... in those days the professor of Natural Philosophy had a wide perspective. He was free to roam at will in the extended fields of chemistry, geology, astronomy, and physics. Rogers left his footprints in each. Today we commemorate a tablet to him in a laboratory of chemistry; in the physical laboratory are to be found his bust, his apparatus cabinet, clock, and other personal instruments; his name is graven upon the Geological building; and a clock from his department is now installed and running in the Astronomical observatory."

William Barton Rogers Scholarships at M.I.T. (1890-1935-1947)

The Report of the M.I.T. Treasurer (M.I.T. Bull. 110/3: 340, Oct. 1974) for the year ended June 30, 1974, lists the following on page 340:

> "5400 A ROGERS, William Barton, $36,505. Established by subscriptions of members of the Alumni Association, through Professor Robert H. Richards '68 [1868] for loans to students. By vote of the Executive Committee, in March 1935, approved by the Alumni Council, the income, now not needed for loans, is made available for special scholarship aid."

Presumably this Fund was originally established in 1890 with contributions solicited from M.I.T. alumni through the efforts of Robert H. Richards (1844-1945) (Class of 1868) to memorialize President Rogers and to provide funds for tuition and other student expenses.

According to records in the M.I.T. Student Financial Aid Office, the balance in this Fund was transferred to Undergraduate Scholarships in 1948:

> "... so that the income may be used for the newly established William Barton Rogers Scholarships." (Memorandum to Mr. Killian from T. P. Pitre, Chairman, Undergraduate Scholarship Committee, dated April 5, 1948.)

The scholarships referred to in the preceding quoted statement were established in 1947. In a three-page memorandum to Mr. James R. Killian, Jr. from T. P. Pitre, Chairman, Faculty Committee on Undergraduate Scholarships, dated March 28, 1947 and preserved in the files of the M.I.T. Student Financial Aid Office, it is proposed:

> "1. To establish five scholarships to be known as the William Barton Rogers Scholarships.
>
> "2. Open to all entrance applicants with no geographical restriction."

The funds were to come in part from income of W. B. Rogers funds then set up as prizes* and from certain funds to be underwritten by the Alumni Fund.

At a meeting on 9 May 1947 it was agreed to implement the foregoing proposal and to announce a four-year scholarship program in the fall of 1947. (Data in the files of the M.I.T. Student Financial Aid Office.)

It was to this new four-year scholarship program that the $36,505 was transferred from the William Barton Rogars Loan and Scholarship Fund mentioned earlier in this section. The name of the latter fund was retained however, and is described as follows on page 76 of the current General Catalogue Issue, September 1974 (M.I.T. Bulletin 1974-75):

> "William Barton Rogers Scholarships (1947)
> A limited number of four-year grants established in honor of William Barton Rogers, founder of the Institute, enable young men of superior ability and future

---

* In 1940, such prizes were given to six seniors (as reported in Tech. Rev. 42/4: 160, Feb. 1940):

> "Awards of $300 each, given annually in memory of William Barton Rogers, the founder of the Institute, in recognition of high scholarship, character, and leadership in student affairs, were recently presented to six seniors. ... Presentation was made by President Compton in the presence of the Faculty Committee on Undergraduate Scholarships and the Heads of Academic Departments in which the students are studying."

promise but of limited financial means to pursue an education in engineering, science, or architecture. Stipends are awarded according to demonstrated need. These scholarships are open to all freshmen with no restriction of geographical location."

Mount Rogers, Rogers Memorial, and Mount Rogers Recreational Area (1883; 1938; 1966; respectively)

William Barton Rogers, geologist, is thrice memorialized in the Rogers Recreational Area in southwestern Virginia: by a mountain bearing his name; by a special memorial on that mountain; and by the designation of the area surrounding the mountain.

Mt. Rogers, the highest mountain in Virginia, elevation 5,729 feet* above sea level, located in the Mt. Rogers National Recreational Area near Damascus, was named after Geologist William Barton Rogers in 1883, a year after his death, in recognition of his original geological work on mountain formation.

The naming of the mountain was anticipated by William's long-time friend, Virginian Jedediah Hotchkiss, who was one of the speakers at the 288th meeting of the M.I.T. Society of Arts on October 12, 1882, which was designated as a Memorial Meeting to recognize the death of Rogers on the preceding May 30th. Reported Hotchkiss (1882, p. 34-35):

> "Today the United States government is engaged, for the first time, in a great geological survey that is to extend over all the States. The surveyors are now upon the mountains of Virginia. When they shall have measured all over that broad area, and determined its highest altitude, but adding nothing of moment to the great sections that William B. Rogers made, <u>that highest summit in the grand Appalachian ranges is to be called Mount Rogers</u>." [Underlining by R.R.S.]

In response to my request for information about Mt. Rogers, Ms. Kay J. Domine, Archivist of the College of William and Mary, wrote the following in a letter dated Sept. 8, 1975:

---

* At least three different elevations for Mt. Rogers have been published, as follows:
    5,729 ft. - Forest Service of the U.S. Dept. of Agriculture (1975 pamphlet) as cited above; and National Geographic Atlas of the World, 2nd ed. (1966), Virginia map on p. 30.
    5,720 ft. - Rand McNally Cosmopolitan World Atlas (1958), Virginia map on p. 111; LIFE Pictorial Atlas of the World (1961), Virginia map on p. 192; and The World Book Encyclopedia (1955), Virginia map following p. 8510 (vol. 17).
    5,719 ft. - The Encyclopaedia Brittanica (14th ed., 1929; vol. 23, p. 184).

As indicated above, I accept the figure of 5,729 feet.

> "In 1883 Mt. Rogers was named in honor of Rogers because
> of his theory on mountain formation. As a result of
> his work on the Appalachian Chain in the Virginias he
> was, along with his brother Henry, to present to a
> meeting of the A.A.G.N. [Association of American Geol-
> ogists and Naturalists] in Boston in 1842, a paper on
> the 'Laws of Structure of the More Disturbed Zones of
> the Earth's Crust.' This paper embodied a theory of
> mountain formation which ... was described at the time
> as 'the most original and brilliant generalization
> recorded in the annals of American Geology.' This work
> was recognized by Virginia when she named her very
> highest peak, Mt. Rogers."

Comment on Mt. Rogers also appeared in Tech Talk, 25 Feb. 1971, p. 3.

On 18 December 1938, at the meeting of the Virginia section of the American Chemical Society in Richmond, it was proposed that a memorial be placed on the top of Mt. Rogers, in honor of William Barton Rogers, Virginia's first State Geologist. The dedication of the memorial took place in late December during the home-coming Virginia convention of the American Association for the Advancement of Science. This story appeared in a newspaper report from Richmond, Virginia, under a date line of Dec. 18 [1938], under the heading "Mountain Memorial set for Geologist - Virginia Section of American Chemical Society Plans Honor for Dr. Rogers." A typescript of the story will be included in William's personal file in the M.I.T. Historical Collections, but at the moment I do not know where it was published.

According to a recent (1975) folder published by the Forest Service of the U.S. Department of Agriculture:

> "The Mount Rogers Recreational Area [underlining
> by R.R.S.] consists of 154,000 acres in southwestern
> Virginia. It was created by Congress in 1966. It has
> an alpine setting unique to the Eastern United States.
> Mount Rogers itself is the highest point in Virginia
> at 5,729 feet. It was named for William Barton Rogers,
> a native Virginian [sic!; actually born in Philadelphia,
> Pa.] who later became President of the Massachusetts
> Institute of Technology. He documented its elevation
> in 1840."

William Barton and Emma Savage Rogers Fund at M.I.T. (1937; 1971)

The M.I.T. Treasurer's Report for the year ended June 30, 1974 (M.I.T. Bull. 110/3: 341, October 1974) lists the following on page 341:

> "1740 A ROGERS, William Barton and Emma Savage, 1937 and
>   1970-71, $1,302,104. Bequest of Dr. Francis H.
>   Williams '73 [1873] including unvalued land at
>   Truro, Massachusetts. Income was added to prin-
>   cipal until 1957, after which 80 per cent of the
>   income is being used for research in pure sci-
>   ence with the balance added to the fund. ..."

(1) WILLIAM BARTON ROGERS

William Barton Rogers Hall at the University of Virginia (1950)

At the University of Virginia, where William Barton Rogers was Professor of Natural Philosophy from 1835 to 1853, one of the institution's numerous "halls" bears the name ROGERS HALL in his honor. The naming of the hall is discussed in the Foreword (by the President of the University of Virginia) of an article by Clemson (1961)* who also includes a brief biographical sketch of Rogers.

William Barton Rogers Chair in Physics at the University of Virginia (1965)

According to a recent letter from John Sullivan, Assistant Provost of the University of Virginia (Sullivan to Shrock dated August 18, 1975), a William Barton Rogers Chair in Physics was established at that institution on April 9, 1965, by the action of the Board of Visitors of the University. The first and present holder of the chair is Professor John W. Mitchell of the Department of Physics.

William Barton Rogers Book Collection at M.I.T. (1975)

When Rogers founded M.I.T. he made his library available to faculty members and students, and after his death the bulk of his books came to the Institute in one way or another. Until a few years ago William's books were scattered through several libraries (e.g. Science Library, Lindgren Library, etc.) and shelved like other similar books. Recently (SY 1974-75), however, an effort has been made to locate and sequester, in a special collection, as many of his books as could be found, so that they might be preserved.

Although no formal action seems to have been taken to give a special designation to these books, Ms. Natalie M. Nicholson, the Director of Libraries until the end of June 1975, was accustomed to refer to them as the William Barton Rogers Book Collection. In June 1975 some of the books were placed on exhibit in special cases in the corridor of Building 14W (Hayden Memorial Library), and were mentioned as belonging to the William Barton Rogers Book Collection, so hopefully that designation will come into general use and help to perpetuate William's longtime love of books.

---

* Clemson, H., "Notes on the Names at the University of Virginia: Notes on the Professors for whom the University of Virginia Halls and Residences are named." Charlottesville, Va.: Univ. Va. Press, 143 p., il., (1961).

William Barton Rogers Pooled Income Fund at M.I.T. (1975)

In the summer of 1975 the M.I.T. Administration created a special fund to honor the Institute Founder, Organizer and first President. The Fund is described as follows on page 3 of a booklet dated September 2, 1975 and titled William Barton Rogers Pooled Income Fund of the Massachusetts Institute of Technology:

> "The William Barton Rogers Pooled Income Fund (the 'Fund') is a trust through which donors can make irrevocable gifts to the Massachusetts Institute of Technology while retaining the income for themselves or for other living individuals for life. ... Upon termination of the income interests, the proportionate share of the principal of the Fund is withdrawn and transferred to the Institute to be used in carrying out its educational, scientific, research, and other purposes."

DEATH AND BURIAL (1882)

Death came suddenly to William Barton Rogers by heart failure on 30 May 1882, in mid-sentence, as he addressed the graduates of the Class of 1882 in Huntington Hall. Francis A. Walker, M.I.T.'s third President, described the dramatic event as follows (1895, p. 13):*

> "On the 30th of May, 1882, he rose to deliver the diplomas to the graduating class, most of whose course had been passed under his presidency. His voice was at first weak and faltering, but, as was his wont, he gathered inspiration from his theme, and for the moment his voice rang out in its full volume and in those well-remembered, most thrilling tones; then, of a sudden, there was silence in the midst of speech;** that stately figure suddenly drooped; the fire died out of that eye, ever so quick to kindle at noble thoughts, and, before one of his attentive listeners had time to suspect the cause, he fell to the platform - instantly dead. All his life he had borne himself faithfully and heroically, and he died, as so good a knight would surely have wished, in harness, at his post, and in the very part and act of public duty."

---

* F. A. Walker, "Biographical Memoir of Wm. Barton Rogers," Nat. Ac. Sci., Biog. Mem. 3: 3-13, port., (1895).

** Inasmuch as so many biographers refer to this dramatic event without mentioning the substance of Rogers' remarks, it seems appropriate to record here what William actually said in his last speech. According to Runkle (1882, p. 20-21), Rogers' actual remarks, as written down from memory, on the day after their delivery, by George W. Blodgett ('73), and slightly changed by H. A. Carson ('69), were as follows:

> "The manner in which I have been received, and the words you have uttered, even if I were in the vigor of early manhood, affect me so deeply as to make reply difficult. I confess to being an enthusiast on the subject of the Institute, but I am not ashamed of this enthusiasm when I see what it has come to be. It is
> (continued on next page)

On 2 June 1882, after funeral services in the same Huntington Hall where he died, the body of William Barton Rogers was carried to the Mount Auburn Cemetery in Cambridge. There it lies buried in the James Savage lot at 178 Walnut Avenue close beside the small white marble headstone, with the inscription, Emma Savage, wife of William Barton Rogers, that

---

** (continued from preceding page)
true that we commenced in a small way, with a few earnest students, in some rooms fitted up in Summer Street, while, as your president has said, the tides rose and fell twice daily where we now are. Our early labors with the legislature in behalf of the Institute were sometimes met not only with repulse but with ridicule, yet we were encouraged and sustained by the great interest manifested by many in the enterprise. Formerly a wide separation existed between theory and practice; now in every fabric that is made, in every structure that is reared, they are closely united into one interlocking system, - the practical is based upon the scientific, and the scientific is solidly built upon the practical. You have not been treated here today to anything in the nature of oratorical display; no decorations, no flowers, no music, but you have seen in what careful and painstaking manner these young men and women have been prepared for their future occupations in life. And although the extracts from the theses which have been presented have been unavoidably largely stated in the technical terms of science, yet they have shown a marvelous thoroughness and accuracy, and in some instances are valuable contributions to our knowledge of the subjects of which they treat. What you have seen has been no research under the direction of a tutor and by his assistance, or prepared for display on this occasion, but it has been the ordinary work of the students, built upon the principles they have acquired in the earlier years of their scientific course, and they show how thoroughly they are equipped for the practical industries, either in the laboratory or in the field.

"As I stand here today and see what the Institute is, what it has already accomplished, and what it is at present doing, I call to mind the beginnings of science. I remember that one hundred and fifty years ago Stephen Hales published a pamphlet on the subject of illuminating gas, in which he stated that his researches had demonstrated that 128 grains of bituminous coal - "

Runkle (1882, p. 21) then adds:

"'Bituminous coal,' these were his last words on earth. Here he bent forward, as if consulting some notes on the table before him, then slowly regaining an erect position, threw up his hands, and was translated from the scene of his earthly labors and triumphs to 'the tomorrow of death,' where the mysteries of this life are solved, and the disembodied spirit finds unending satisfaction in contemplating the new and still unfathomable mysteries of the infinite future."

marks the site of her ashes. The only indication that William is buried in the lot is the epitaph cut into the back of the massive gray granite stone that has the name JAMES SAVAGE, Emma's father, engraved on the front. The simple epitaph reads:

> WILLIAM BARTON ROGERS
> Born December 7, 1804
> Died May 30, 1882

WILLIAM BARTON ROGERS, SCIENTIST: A SUMMATION

To summarize adequately the complete range and nature of the accomplishments of William Barton Rogers as scientist and educator would require careful reading of a great volume of published and unpublished matter. Fortunately, his wife Emma, aided by William T. Sedgwick, edited many of his letters and published them, together with inter-letter commentary, in the well-known two-volume <u>Life and Letters of William Barton Rogers</u> (1896) listed in the Bibliography. A second reference, which is a valuable source of information on William, his siblings, and his parents is W. S. W. Ruschenberger's "A Sketch of the Life of Robert E. Rogers, M.D., LL.D., with Biographical Notices of his Father and Brothers" (1886). In addition to the two foregoing sources, more than twenty-five biographical sketches of variable length and comprehensiveness have been published during the ninety years since William's death in 1882. These are listed farther on in Appendix B, following the Bibliography, but I feel rather certain that I may well have overlooked a few publications of a similar sort. In spite of this rather substantial volume of writing about William, I think it fair to say that the <u>definitive</u> biography of William Barton Rogers still remains to be written, because no one of the references cited in my list alluded to above brings together, correlates, and evaluates <u>all</u> aspects of his distinguished career in a single comprehensive work. Fortunately for the history of science, several individuals are now involved in preparing such a definitive biography.

As stated in the introduction to this sketch, I have limited my remarks in large part to Rogers the scientist, and more particularly to him as geologist, and to the part he played in the development of geology in America and at M.I.T. I have deliberately limited my discussion of his greatest contribution, the founding of M.I.T., preferring to leave that formidable task to others much more competent to write about all aspects of the Institute, with which he concerned himself. Here I shall attempt to summarize as briefly as possible only the more important activities and events that brought him worldwide distinction and a position in the front rank of the leading geologists of his time.

(1) WILLIAM BARTON ROGERS

William Barton Rogers started his scientific career as public lecturer and college professor quite early. He had not yet reached eighteen when he attracted favorable newspaper comment on his oration at the 1822 "Virginiad" in Jamestown, and four years later he and his brother Henry were conducting a school at Windsor, near Baltimore. By 1825 he was delivering lectures on astronomy and railroads at the Maryland Institute, and he had barely organized a high school for that Institute when he was chosen to succeed his lately deceased father as Professor of Natural Philosophy and Chemistry at the College of William and Mary in the fall of 1828. This appointment launched him on a professorial career that only ended in 1881, more than fifty years later, when he finally retired as M.I.T.'s first emeritus professor with the title, Professor Emeritus of Physics and Geology.

William's first lectures at William and Mary dealt with physical and chemical phenomena, and five years passed before his brother Henry, freshly arrived home from England in 1833, aroused his interest in geology. Fascinated by the subject, he immediately turned his full attention to the study of minerals, rocks and fossils, and by 1835 had persuaded the General Assembly of the Virginia Legislature to establish a geological survey. Then came two great steps forward in his geological career.

He was appointed State Geologist of Virginia shortly after the Survey was established, and in the same year (1835) moved to Charlottesville to become Professor of Natural Philosophy at the University of Virginia.

Then followed a decade of troubles that tried him to the limit of body and spirit. He had to organize and conduct a program for the geological survey, train field assistants, and constantly fight the Legislature for financial support. Although the Survey existed only six years (1835-1841), being terminated because of lack of appropriations, William and his assistants were able to accomplish an impressive amount of geological field work, most of which was ultimately reported on in one form or another. His seven Progress Reports and the numerous articles published elsewhere, which were brought together posthumously in the Geology of the Virginias, won him national recognition. His joint work with Henry, who was State Geologist of Pennsylvania, established for the first time the stratigraphy and folded and faulted structure of the great Appalachian Chain. The joint paper by the two brothers, "On the [physical] structure of the Appalachian chain ...," delivered before the 1842 meeting of the American Association of Geologists in Boston, received enthusiastic acclaim and wide approval in both America and abroad, although the explanation of how the folds and faults were produced was not accepted by some of the leading geologists of the day, as mentioned in the preceding section on this matter. Nevertheless, the geological work done by William and Henry in their respective States brought them world renown and a position of highest standing among the stratigraphers and structural geologists of the world.

Few knew William Barton Rogers better than his lifelong friend and fellow field geologist and engineer, Major Jedediah Hotchkiss of Virginia, who reminisced nostalgically about their many days in the field together when the M.I.T. Society of Arts devoted its 288th meeting on 12 October 1882 to memorializing the life and scientific work of Rogers. Remarked Hotchkiss (1882, p. 33-34):

> "Professor Rogers ... had a way of inspiring all that came within the reach of his magic eye. The sound of his wonderful voice gave a charm to whatever he said. All over the State of Virginia even now you will continually meet people in the country, old men and old women, who recollect the days when Professor Rogers drove up with his gig, with Levi, his negro servant, behind him on horseback, accompanying him in his geological rambles, - recollect with pleasure that familiar lecture in the morning from the doorstep; for he never went away without leaving with each one that he visited a new vision of that which before they had seen with sealed eyes that it was his delight to unseal. One of the best of our living structural geologists, one of the same Scotch-Irish race, when a flaxen-haired boy, heard Professor Rogers describe to a group of listeners one of the grand arches of one of Virginia's mountain ranges, when, stooping down, like another great teacher, he wrote its structure in the sand, but wrote for all time. The boy never forgot the graphic rock story ...
>
> "It would furnish material for a singular study, - that primal geological circle. Levi, the negro servingman was in it. He became a geologist. He learned to think as his master thought; and when the great French geologist, Daubeny, came to visit Professor Rogers, the Frenchman was not up to riding horseback, so he took a seat in the gig, and Levi drove him, and as they rode through the grand sections of Appalachian structure there displayed, Levi gave him lessons in American geology. 'Dis sar,' said he, 'we call number one [I of the stratigraphic sequence Rogers worked out; rocks of Cambrian age]. Mighty fine crap [outcrop] ob it 'long here.' He had so well learned the lesson from the great master of American geology, he could teach it to the one of French."

One can see them now, William in the gig, and Levi following astride his horse, riding up and down over the successive parallel ridges of the Appalachians on the crudest of trails and roads; stopping now and then to identify a rock or determine the dip of a layer; traversing mile after mile the nearly four hundred miles across the entire width of the State of Virginia; and building formation on top of formation - limestone, sandstone, shale, coal - until they had determined seventeen different "formations," which he and brother Henry numbered I to XVII, in age Cambrian to Permian [See Appendix D in Life and Letters II, p. 437-438]. Little wonder, then, that many geologists today regard William's most important geological work as that done while State Geologist (1835-1841) in revealing the stratigraphy and structure of the folded and faulted Appalachians in the Virginias. (See also Table on page 116.)

As mentioned earlier, William's ninety publications on geological subjects, quite apart from his classic papers on the Appalachian Chain with Henry, were characteristically of a practical nature and called to the attention of the citizens of Virginia and elsewhere the potential value of their natural resources. Many of his discoveries were "firsts," and though they seem obvious and commonplace today, they were quite the opposite in the 1840s and 1850s.

As mentioned in a preceding section, there were disquieting student disturbances at the University of Virginia, from time to time, during his years as Professor of Natural Philosophy (1835-1853). Although he became one of the most popular professors on the campus, delivering dramatic and impressive lectures on astronomy, physics, chemistry and geology, he seems not to have stimulated any students to become enthusiastic disciples of geology. Nevertheless, those of his students who have recorded their recollections of his lectures, and others who attended his public lectures, invariably use superlatives in describing the organization, content, and clarity of subject matter and the felicity of his language. As many of his biographers have written, he was one of the outstanding scientific orators of his time, and as such he brought geology and related sciences to a wide public audience in an extraordinarily impressive manner.

While his lecturing was going well, he was greatly distressed by the student dissension of the 1840s, and longed for a more tranquil environment. Then, too, he had to fight the Legislature which was proposing to eliminate the annual annuity to the University. Finally, in 1848, he resigned his professorship, but was prevailed on by his close friends to withdraw his resignation. In 1853, however, he definitely decided to leave the University and Charlottesville and join Henry in Boston.

Trying as were the turbulent years in Virginia, William never abandoned his interest in physics, chemistry, and geology, and continued to publish articles in all three fields almost to the year of his death. Although for a time he devoted much of his effort to chemical research with his brother Robert, with whom he shared joint authorship of 17 articles, when Robert left the University in 1852 to join the faculty of the University of Pennsylvania, William turned back to geology and physics, and thereafter most of his publications involved some aspect of these two sciences.

Soon after resigning from the University of Virginia in 1853, and establishing his home in Boston, he was giving lectures at the Lowell Institute and gathering around him a small group of leading Bostonians interested in helping him establish the polytechnic school that he and Henry had envisioned as early as 1837. When Henry left Boston in 1858 to become Regius Professor at Scotland's Glasgow University, William put renewed efforts into agitation for the new school, and in 1861 the Massachusetts Legislature voted favorably on his Committee's petition to found the Massachusetts Institute of Technology. Planning went on vigorously in spite of the war years, 1861 to 1865, and William could finally write in his diary, for 20 February 1865, those prophetic words so often quoted by his biographers:

> "Organized the school! Fifteen students entered. May not this prove a memorable day!"

Today few would disagree with the statement that William Barton Rogers' crowning achievement was the founding, organizing, and directing of M.I.T., a story that has been related again and again by distinguished biographers, and a story that I have touched on only briefly.

I have tried, rather, to sketch his life and achievements as a scientist - geologist, physicist, and chemist - who while always thinking of how science could be put to practical use for the benefit of mankind, never abandoned the uncompromising view that training for practice of the arts must start with thorough training in mathematics, physics and chemistry, together with broad exposure to what today is called the humanities.

Great though his contributions were to the advancement of geology as a broadly applied earth science, many would regard as more important the educational philosophy and methodology that he and his first faculty members firmly established at M.I.T. with its significant motto - MENS ET MANUS. As a result, the fundamental training that our thousand "geology" graduates have received during the first century of the Institute's existence has been a controlling factor in their professional achievements and has demonstrated repeatedly the wisdom of Rogers' educational innovations that he included in his "Plan" for the polytechnic school that is now M.I.T.

BIBLIOGRAPHY OF WILLIAM BARTON ROGERS

This Bibliography, which contains only articles about geology and related sciences (chemistry and physics),* is more comprehensive and complete than any other known to exist. The references have been carefully checked with those listed in the following publications:

1) Geologic Literature on North America, 1785-1918. U.S. Geol. Surv., B. 746 - Pt. I. Bibliography (1922) and B. 747 - Pt. II. Index (1924).

2) The so-called Coal Reports [of the Virginias], (1850+). Reference 62--1850+.

3) The Virginias, Vols. 1-5, (1880-1884).

4) Geology of the Virginias, (1884). Reference 153--1884.

5) W. S. W. Ruschenberger's A sketch of the life of Robert E. Rogers, M.D., LL.D., with biographical notices of his father and brothers. Am. Phil. Soc., Pr. 23: 104-146, (1886).

6) Appendix C - A list [incomplete and not altogether accurate as to titles and dates] of publications by William Barton Rogers, in Vol. II, p. 430-436, of Life and Letters mentioned in the footnote below.

7) J. W. Gregory's Henry Darwin Rogers - An address to the Glasgow University Geological Society, 25th January, 1916. Glasgow: James MacLehose and Sons, 34 p., port., (1916).

References not included in the foregoing have come to my attention in one way or another as a result of my own searching or of help from others. I am especially grateful to Ms. Loretta Mannix of M.I.T. for bringing several of William's earliest articles to my attention, and to one of my former students and a longtime colleague, Dr. Wilson D. Michell, Geologist with the Reynolds Metals Company of Richmond, Virginia, for securing complete data on William's early articles in the Farmers' Register (References 2-6, 1834 and 1835).

Directly following the regular Bibliography is a short list of publications on which William made comments of some sort, generally in a discussion of the particular publication. This list is quite incomplete because William is reported to have discussed or commented on many papers, delivered at meetings of scientific societies, but no further information accompanies such reports to indicate what he said.

Appendix A, following the foregoing bibliographic items, consists of a brief discussion of The Founding Documents of M.I.T., in the preparation of which William had a prominent, in many cases even a critical, role.

Appendix B is a comprehensive list of biographies, biographical references, and other sources which contain important information on the life and accomplishments of William Barton Rogers, and added information on his three brothers - James, Henry and Robert - and the father, Patrick Kerr Rogers.

Symbols and abbreviations used in the following references, and in similar bibliographies in the fifty or more faculty biographies that follow, are

---

* The many other items written by Rogers, such as letters, reports, petitions, memorials, and the like, are listed or referred to in Life and Letters of William Barton Rogers, written by his wife [Emma Savage Rogers] with the assistance of William T. Sedgwick, in two volumes of 427 and 457 pages, respectively, and published in 1896 by Houghton Mifflin and Company of Boston. See reference 155--1896 farther on.

explained on preceding pages 91-98; abstracts may be listed separately as well as with the references to the complete article. For many of the earlier references, which had no formal title, an explanatory phrase in brackets [ ] indicates the nature of the subject matter. Every effort has been made to list the references in chronological order.

1--1830   Paper on Dew. The Messenger of Useful Knowledge 1/2: 17-20, (1830).

2--1834   Chemical analysis of shells. Farmers' Register 1/10: 589-591, (March 1834); Silliman's Jour. 26: 361-365, (1834).

3--1834a   On the discovery of green sand in the calcareous deposit of eastern Virginia, and on the probable existence of this substance in extensive beds, near the western limits of our ordinary marl. Farmers' Register 2/3: 129-131, (Aug. 1834); also in Geology of the Virginias, p. 1-9, (1884).

4--1834b   Apparatus for analyzing marl, and the carbonates in general. Farmers' Register 2/6: 364-365, (Nov. 1834); also in Geology of the Virginias, p. 9-11, (1884). See also reference 9--1835c.

5--1834c   Observations and queries respecting artesian wells. Farmers' Register 2/7: 451-455, (Dec. 1834).

6--1835   Further observations on the green sand and calcareous marl of Virginia. Farmers' Register 2/12: 747-751, (May 1835); also in Geology of the Virginias, p. 11-20, (1884).

7--1835a   (with Rogers, H. D.) Experimental enquiry into some of the laws of the elementary voltaic battery. Silliman's Jour. 27: 39-61, (1835).

8--1835b   On the existence of the bi-malate of lime in the berries of the sumach, and the mode of procuring it from them in the crystalline form. Silliman's Jour. 27: 294-299, (1835).

9--1835c   Apparatus for analyzing calcareous marl and other carbonates. Silliman's Jour. 27: 299-301, (1835). See also reference 4--1834b.

10--1835d   A self-filling syphon for chemical analysis. Silliman's Jour. 27: 302-303, (1835).

11--1836   Report of the geological reconnaissance of the State of Virginia. 143 p., Philadelphia, (1836); Another edition, Doc. No. 24, 52 p. [Richmond, Va., (1836)]. Reprinted in Geology of the Virginias: 21-122, (1884). Extr., ["The Great Western bituminous coal and salt region of the Virginias" by W. B. Rogers]--The Virginias 3: 135, 138-139, (1882); 4: 110-111, (1883).

12--1837   Report on the progress of the geological survey of the State of Virginia for the year 1836. (Doc. No. 34), 14 p. [Richmond, Va., (1837)]. Reprinted in Geology of the Virginias: 123-146, (1884); Another edition [with second report], 30 p., Philadelphia, (1838). Extr., ["The Vespertine or Formation No. X coals of the Virginias" by W. B. Rogers]--The Virginias 4: 110-112, (1883).

13--1837a   (with Rogers, H. D.) Contributions to the geology of the Tertiary formations of Virginia. Am. Phil. Soc., Tr. 5: 319-341, (1837); 6: 347-350, 371-377, il., (1839). Reprinted in Geology of the Virginias: 659-674, (1884). Abst., Am. J. Sci. 38: 182-184, (1840).

14--1838   Report of the progress of the geological survey of the State of Virginia for the year 1837 (Doc. No. 45), 24 p., [Richmond, Va. (1838)]. Reprinted in Geology of the Virginias: 147-188, (1884). Another edition [with first report], p. 31-87, Philadelphia, (1838). Extr., The Virginias 4: 112-113, (1883).

15--1838a   An Elementary Treatise on the Strength of Materials, being the substance of the lectures on that subject, delivered in the School of Engineering of the University of Virginia. Charlottesville [Va.]: Tomkins and Noel, 50 p., (1838).

16--1839   Report of the progress of the geological survey of the State of Virginia for the year 1838 (Doc. No. 56), 32 p., [Richmond, Va., (1839)]. Reprinted in Geology of the Virginias: 189-244, (1884). Extr. ["The Great Coal-field of the Virginias" by W. B. Rogers], The Virginias 3: 158-159, 164, (1882); 4: 113-115, (1883).

17--1839a   (and Rogers, H. D.) Contributions to the geology of the Tertiary formations of Virginia, second series. Am. Phil. Soc., Pr. 1: 88-90, (1839); Tr. 6: 347-350, (1839).

18--1839b   (and Rogers, H. D.) Contributions to the geology of the Tertiary formations of Virginia, second series continued, - Being a description of several species [8] of Miocene and Eocene shells, not before described. Am. Phil. Soc., Tr. 6: 371-377, il., (1839).

19--1840   Report of the progress of the geological survey of the State of Virginia for the year 1839, 161 p., [Richmond, Va., (1840)]. Reprinted in Geology of the Virginias: 245-410, (1884). Extr., ["The coal basins between the Alleghany Front Ridge and the Backbone Mountain in Mineral, Grant and Tucker counties, W. Va." by W. B. Rogers], The Virginias 3: 71-73, 77, 158-159, 164, (1882); 4: 115-116, (1883).

20--1841   Report of the progress of the geological survey of the State of Virginia for the year 1840, 132 p., [Richmond, Va., (1841)]. Reprinted in Geology of the Virginias: 411-536, (1884). Extr., The Virginias 2: 58-59, (1881).

21--1842   Report of the progress of the geological survey of the State of Virginia for the year 1841, 12 p., [Richmond, Va., (1842)]. Reprinted in Geology of the Virginias: 537-546, (1884).

22--1842a   (and Rogers, H. D.) Observations on the geology of the western peninsula of Upper Canada, and the western part of Ohio [1841]. Am. Phil. Soc., Pr. 2: 120-125, (1842); Tr. 8: 273-284, (1843).

23--1842b   On the porous anthracite or natural coke of eastern Virginia. Am. J. Sci. 43: 175-176, (1842)--Assoc. Am. Geol., Rp.: 68, (1843)--Abst., Geologist 1843: 39, (1843).

24--1842c   Observations on subterranean temperature made in the mines of eastern Virginia [abst.]. Am. J. Sci. 43: 176, (1842)--Assoc. Am. Geol., Rp.: 69, (1843).

25--1842d   [On erosion of strata underlying the Oriskany sandstone] [abst.]. Am. J. Sci. 43: 181-182, (1842)--Assoc. Am. Geol., Rp. 3rd An. Meeting: 36-37, (1842); Rp.: 73-74, (1843).

26--1842e   [On the age of the coal formation of Richmond, Va., and of the Fredericksburg sandstone]. Ac. Nat. Sci. Philadelphia, Pr. 1: 142, 250, (1842).

27--1842f (with/and Rogers, H. D.) On the [physical] structure of the
   1843   Appalachian Chain as exemplifying the laws which have regu-
            lated the elevation of great mountain chains, generally. Am.
            J. Sci. 43: 177-178 [abst.], (1842).--Geologist 1842: 235-240,
            (1842).--Rept. 1st, 2nd and 3d Meetings Assoc. Am. Geol. Nat.
            at Philadelphia in 1840 and 1841, and at Boston in 1842, em-
            bracing its Proc. and Trans. Boston: Gould, Kendall & Lincoln,
            viii + 9-544 p., pl., map, sections (1843) [Rogers' article on
            p. 474-553].--Abst., Rept. 12th Meeting British Assoc. held at
            Manchester in June 1842 (London: John Murray, p. 40-42, 1843);
            condensed version in Rept. 12th Meeting British Assoc. held at
            Manchester in June 1842, (1843) (from the Report in the London
            Athenaeum) [reprinted in Am. J. Sci. 44: 359-368, (1843)].--
            Reprinted in Coal Reports: 474-531, (1850+).--Reprinted in
            Geology of the Virginias, p. 599-642, (1884).--Excerpted (p.
            338-345) in Mather and Mason's Source Book of Geology. New
            York: McGraw-Hill Book Co., Inc., xxii + 702 p., (1939).

28--1843a   On the age of the coal rocks of eastern Virginia. Abst., Am.
            J. Sci. 43: 175, (1842); Geologist 1843, 38-39, (1843)--Assoc.
            Am. Geol., Rp.: 68 [abst.]; Tr.: 298-316, il., (1843)--Re-
            printed in Coal Reports: 298-316, (1850+)--Reprinted in Geology
            of the Virginias: 643-658, il., (1884).

29--1843b   On the connection of thermal springs in Virginia with anti-
            clinal axes and faults. Abst., Am. J. Sci. 43: 176, (1842);
            Geologist 1843: 39, (1843)--Assoc. Am. Geol. Rp.: 69, 323-347,
            (1843)--Reprinted in Coal Reports: 317-347, (1850+)--Reprinted
            in Geology of the Virginias: 575-598, (1884).

30--1843c   Observations of subterranean temperature in the coal mines of
            eastern Virginia. Abst., Am. J. Sci. 43: 176, (1842)--Assoc.
            Am. Geol. Rp.: 532-538, (1843)--Reprinted in Coal Reports:
            532-538, (1850+)--Reprinted in Geology of the Virginias: 567-
            574, (1884).

31--1843d   On the limits of the infusorial stratum in Virginia. Am. J.
            Sci. 45: 313-314, (1843).

32--1843e   (with Rogers, H. D.) Theory of earthquake action. Am. J. Sci.
            45: 341-347, (1843).

33--1843f   (with Rogers, H. D.) On the phenomena of the great earth-
            quakes which occurred during the past winter ... and on a
            general theory of earthquake motion, by which they propose to
            elucidate several points in geological dynamics. Am. Phil.
            Soc., Pr. 3: 64-67, (1843)--British Assoc. Rp. 1843 (pt. 2):
            57-59, (1843). See also the preceding reference, 32--1843e.

34--1844   (and Rogers, R. E.) An account of some new instruments and
            processes for the analysis of the carbonates. Am. J. Sci.
            46: 346-359, (1844).

35--1884a   (with Rogers, R. E.) [On chemical equivalents of certain sub-
            stances, included in Article VIII. - Abstract of the Proceed-
            ings of the 5th session of the Assoc. Am. Geol. and Nat.] Am.
            J. Sci. 47: 105-106, (1844).

36--1844b   (with Rogers, R. E.) [Determination of the amount of iron in
            iron ores, cast iron, etc.] Am. J. Sci. 47: 106, (1844).

37--1844c   (with Rogers, H. D.) On a system of classification and nomen-
            clature of the Palaeozoic rocks of the United States, with an
            account of their distribution more particularly in the Appa-
            lachian chain [abst.]. Am. J. Sci. 47: 111-112; also p. 137-
            160 and 247-278, (1844). See also Appendix D, p. 437-438, in
            Vol. II of Life and Letters, and Table on page 116.

38--1845   (with Rogers, H. D.) [On boulder trains in Berkshire Co., Mass.] Boston Soc. Nat. Hist., Pr. 2: 79-80, (1845). See also reference 44--1846d.

39--1845a Preface (p. iii-v) and possibly Introduction (p. 1-5) in Elements of Chemistry, including the history of the imponderables and the inorganic chemistry of the late Edward Turner, M.D., F.R.S.L.& E. Seventh Edition [the so-called "American Edition"], and the Outlines of Organic Chemistry by William Gregory, M.D., &c., Professor of Chemistry, University of Edinburgh, with notes and additions by James R. Rogers, M.D., Professor of General Chemistry, Franklin Institute, and Lecturer on Medical Chemistry, &c. and Robert E. Rogers, M.D., &c., Professor of Chemistry and Materia Medica, University of Virginia, &c. Philadelphia: Thomas, Cowperthwait & Co., xxi + 848 p., (1846).

The Preface and Introduction in this, the so-called "American Edition," are signed (R.), and I assume that the (R.) indicates William Barton Rogers, although there is no explanation to that effect in the book; however, on page 255 of Volume I of Life and Letters is the statement: "Amidst his regular duties, William found time to write the preface to an American edition of 'Turner's Chemistry,' which had been prepared by his brothers James and Robert and which was now [late 1845] issuing from the press." The Preface was written in Philadelphia in November 1845.

40--1846   (with Rogers, H. D.) On the geological age of the White Mountains. Am. J. Sci. (2) 1: 411-421, (1846).

41--1846a (with Rogers, R. E.) On a new process for obtaining pure chlorine gas. Am. J. Sci. (2) 1: 428, (1846).

42--1846b (with Rogers, R. E.) On a new process for obtaining formic acid, and on the preparation of aldehyde and acetic acid by the use of bichromate of potassa. Am. J. Sci. (2) 2: 18-24, (1846).

43--1846c The Tertiary infusorial formation of Maryland. Am. J. Sci. (2) 2: 141-142, (1846).

44--1846d (with Rogers, H. D.) An account of two remarkable trains of angular erratic blocks in Berkshire, Massachusetts, with an attempt at an explanation of the phenomena. Boston J. Nat. Hist. 5: 310-330, (1846)--Reprinted in Coal Deposits [Item 7: 1-20, pl.], (1846; 1850+).

45--1848   (with Rogers, R. E.) On the absorption of carbonic acid by water, saline solutions, and various other liquids. Abst., Am. J. Sci. (2) 5: 114-115, (1848)--A.A.A.S., Pr. 4: 298-308, (1850).

46--1848a On the transporting power of currents [abst.]. Am. J. Sci. (2) 5: 115-116, (1848).

47--1848b (with Rogers, R. E.) New method of determining the carbon in native and artificial graphites, etc. Am. J. Sci. (2) 5: 352-359, (1848)--British Assoc. Rept. 1848 (pt. 2): 59-60, (1848)--Edin. J. Prak. Chem. 1: 411-413, (1850)--J. de Pharm. 1851: 67-68, (1851).

48--1848c (and Rogers, R. E.) On the decomposition and partial solution of minerals, rocks, etc., by pure water, and water charged with carbonic acid. Am. J. Sci. (2) 5: 401-405, (1848)--Edinburgh New. Phil. Jour. 45: 163-164, (1848)--Froriep Notizen 9: 96-110, (1848)--Abst., British Assoc. Rept. 1849: 40-42, (1849).

49--1848d (and Rogers, R. E.) On the absorption of carbonic acid gas by liquids. Am. J. Sci. (2) 5: 96-110, (1848).

50--1848e (with Rogers, R. E.) Oxydation of the diamond in the liquid way. Am. J. Sci. (2) 6: 110-111, (1848)--Edinburgh New Phil. Jour. 45: 388-389, (1848)--British Assoc. Rept. 1848 (pt. 2): 60-61, (1848).

51--1848f (and Rogers, R. E.) On the decomposition of rocks, etc., by meteoric waters, and on the action of the mineral acids upon feldspars, etc. Am. J. Sci. (2) 5: 401; (2) 6: 396-397, (1848)--A.A.A.S., Pr. 1: 60-62, (1849).

52--1848g (and Rogers, R. E.) On the absorption of carbonic acid gas by sulphuric acid. Chem. Gazette 6: 477-480, (1848).

53--1849 (and Rogers, R. E.) On the volatility of potassa and soda, and their carbonates. A.A.A.S., Pr. 1: 36-38, (1849).

54--1849a (and Rogers, R. E.) On the absorption of carbonic acid by Leibig's dilute solution of phosphate of soda. A.A.A.S., Pr. 1: 62, (1849).

55--1849b Observations on the southern shore of Lake Superior. A.A.A.S., Pr. 1: 79-80, (1849).

56--1849c On acid and alkaline springs. A.A.A.S., Pr. 1: 94-95, (1849)--Am. J. Sci. 9: 123-126, (1850).

57--1849d (and Rogers, R. E.) On the comparative solubility of the carbonate of lime and the carbonate of magnesia. A.A.A.S., Pr. 1: 95-97, (1849).

58--1849e Fragments of notes of travel [A diary kept by W. B. Rogers during his honeymoon to Europe, June - October 1849]. Ms., 60 p., (1849). [M.I.T. Institute Archives.]

59--1850 [Discussion of measurement, origin, and classification of mechanical powers.] A.A.A.S., Pr. 4: 16-17, (1850).

60--1850a [On the gold belt in the United States.] A.A.A.S., Pr. 4: 21-22, (1850).

61--1850b (and Rogers, R. E.) On the absorption of carbonic acid by acids and saline solutions. A.A.A.S., Pr. 4: 298-308, (1850).

62--1850+ (and Rogers, H. D.) [Coal Reports--A bound volume of nine papers, dates of some of which are uncertain, but inasmuch as the last two items carry the date 1850, the bound volume was produced either that year or later, hence the date used here is indicated as 1850+. The following nine articles are generally collectively referred to as the Coal Reports.]

   1. (W.B.R.) On the age of the coal rocks of eastern Virginia; p. 298-316, (1843; 1850+). See also preceding reference 28--1843a.

   2. (W.B.R.) On the connection of thermal springs in Virginia with anticlinal axes and faults; p. 317-347, (1843; 1850+). See also reference 29--1843b.

   3. (H.D.R.) An inquiry into the origin of the Appalachian coal strata, bituminous and anthracitic; p. 433-474, (1850+).

   4. (W.B.R. and H.D.R.) On the physical structure of the Appalachian chain ...; p. 474-531, (1843; 1850+). See also reference 27--1842f, 1843.

5. (W.B.R.) Observations of subterranean temperature in the coal mines of eastern Virginia; p. 532-538, (1843; 1850+). See also reference 30--1843c. [Following p. 538, p. 539-544, is "Explanation of the Plates I-XVII," which plates refer to fossils and other geological items not included in the bound Coal Reports; these are followed on p. 544 by "Explanations for Plates XVIII-XXI, inclusive," showing Sections of the Appalachian Chain, which sections are bound between p. 544 and the next item, No. 6.]

6. (H.D.R.) Address delivered at the meeting of the Association of American Geologists and Naturalists, held in Washington [D.C.], May, 1844; 58 p., (1844). Printed by B. L. Hamlen, New Haven [Conn.], (1844; 1850+).

7. (H.D.R. and W.B.R.) An account of two remarkable trains of angular erratic blocks ...; 22 p., (1846). [Extracted from Boston Jour. Nat. Hist., June, 1846.] See also reference 44--1846d.

8. (H.D.R.) Report on the Coal Lands of The Zerbe's Run and Shamokin Improvement Company ... with the Charter; 20 p. Boston: Thurston, Torry, and Company, Printers, 1850, (1850+).

9. (H.D.R.) Report on the Coal Lands of The Mahanoy and Shamokin Improvement Company ... with the Charter; 20 p. Boston: Thurston, Torry, and Company, Printers, 1850, (1850+).

63--1852   Elements of Mechanical Philosophy, for the use of the Junior students of the University of Virginia. Boston: Thurston, Torry & Emerson, Printers, 339 p., (1852).

64--1853   [On vertical beams of light.]   A.A.A.S., Pr. 3: 81-82, (1853).

65--1854   [On coke.]   A.A.A.S., Pr. 3: 106-107, (1854).

66--1854a [Geological relations of the New Red Sandstone of the Middle States and Connecticut Valley to the coal-bearing rocks of eastern Virginia and North Carolina.] Boston Soc. Nat. Hist., Pr. 5: 14-18, (1854)--Reprinted in Geology of the Virginias, 765-768, (1884)--M. Mag. 5: 128-132, (1885).

67--1854b (and Rogers, R. E.) On the use of hydrogen gas and carbonic acid, to displace sulphuretted hydrogen in the analysis of mineral waters, etc. Am. J. Sci. (2) 18: 213-216, (1854).

68--1854c [On the natural coke in the vicinity of Richmond, Virginia.] Boston Soc. Nat. Hist., Pr. 5: 53-56, (1854)--A.Ac.A.S., Pr. 3: 106-107, (1857)--The Virginias 4: 158-159, (1883)--Reprinted in Geology of the Virginias: 675-678, (1884).

69--1854d Report on the Pridevale coal and iron ore of West Virginia. New York: 42 p., (1854). Reprinted in Geology of the Virginias: 679-706, (1884).

70--1854e The property of the Pridevale Iron Company [Preston Co., W.Va.]. M. Mag. 3: 355-370, 489-499, (1854)--Reprinted in Geology of the Virginias: 679-706, (1884).

71--1855   Results of calculations ... of the terminal velocity of raindrops of different diameters. Boston Soc. Nat. Hist., Pr. 5: 266-267, 282-283, (1855).

72--1855a On the relations of the New Red sandstone of the Connecticut Valley and the coal-bearing rocks of eastern Virginia and North Carolina. Am. J. Sci. (2) 19: 123-125, (1855). See also reference 66--1854a.

73--1855b   Observations on binocular vision. Am. J. Sci. (2) 20: 86-98, 204-220, 318-335; (2) 21: 80-95, 173-189, 439, (1855)--A.Ac.A.S., Pr. 3: 213-214, (1855).

74--1855c   Address before the Lyceum of Natural History of Williams College, August 14, 1855. Boston: 8 vo. 34p., (1855). [A copy is in the M.I.T. Institute Archives]. [An example of Rogers' reputed "felicity of language." Excerpted in Life and Letters I, p. 343-357.]

75--1855d   On the form of the curve resulting from the binocular union of a straight line with a circular arc, or of two equal circular arcs with one another. A.Ac.A.S., Pr. 3: 213-214, (1855)--Edinburgh. New Phil. Jour. 3: 210-217, (1856). See also reference 73--1855b.

76--1855e   [Note on the coal-bearing rocks near Richmond, Va., and the New Red sandstone of North Carolina (with discussion by C. T. Jackson).] Boston Soc. Nat. Hist., Pr. 5: 186, (1855)--Am. J. Sci. (2) 19: 123-125, (1855).

77--1855f   [On lignite from Lancaster, Pa., and eastern Virginia.] Boston Soc. Nat. Hist., Pr. 5: 189-190, (1855).

78--1855g   [On Mesozoic rocks in Virginia.] Boston Soc. Nat. Hist., Pr. 5: 201-202, (1855).

79--1855h   [On the metamorphic influence of trap rocks on adjacent sedimentary strata in Prince William Co., Va.] Boston Soc. Nat. Hist., Pr. 5: 202-204, (1855).

80--1855i   [On binocular combinations.] A.Ac.A.S., Pr. 3: 213-214, (Dec. 1855).

81--1856   [On the ozonometer.] A.Ac.A.S., Pr. 3: 220, (Jan. 1856).

82--1856a   On the origin and accumulation of the protocarbonate of iron in coal measures [and on the color of rocks]. Boston Soc. Nat. Hist., Pr. 5: 283-288, (1856)--Am. J. Sci. (2) 21: 339-343, (1856)--M. Mag. 6: 201-207, (1856).

83--1856b   [Proofs of the Protozoic age of some of the altered rocks in eastern Massachusetts from fossils recently discovered.] A.Ac.A.S., Pr. 3: 315-318, (Aug. 1856).

84--1856c   Observations on the variations of ozone in the atmosphere. Boston Soc. Nat. Hist., Pr. 5: 319-321, (1856)--Am. J. Sci. (2) 22: 141-142, (1856).

85--1856d   [Nitrates in cave earths.] Boston Soc. Nat. Hist., Pr. 5: 334-335, (1856)--Reprinted in Geology of the Virginias: 763-764, (1884).

86--1856e   [On the growth of stalactites.] Boston Soc. Nat. Hist., Pr. 5: 336-337, (1856)--Reprinted in Geology of the Virginias: 764-765, (1884).

87--1856f   [On trilobites from Braintree, Mass., and on the geologic relations of the district.] Boston Soc. Nat. Hist., Pr. 6: 27-29, 40-41, (1856)--M. Mag. 7: 371-373; 454, (1856).

88--1856g   Discovery of Paleozoic fossils in eastern Massachusetts. Am. J. Sci. (2) 22: 296-298, (1856).

89--1856h   On the discovery of Paradoxides in the altered rocks of eastern Massachusetts. Edinburgh New Phil. Jour. 4: 301-304, (1856); Abst., Ibid. 6: 314-315, (1856)--Excerpted in Mather and Mason's Source Book in Geology, p. 336-337. New York: McGraw-Hill Book Co., Inc., xxii + 702 p., il., (1939).

90--1857 Ruhmkorff's apparatus constructed by E. S. Ritchie of Boston. Am. J. Sci. (2) 23: 451-452, (1857)--A.Ac.A.S., Pr. 4: 376-382, (1860). See also reference 122--1860n.

91--1857a [Sketch of the life of Michael Tuomey.] Boston Soc. Nat. Hist., Pr. 6: 185-186, (1857).

92--1857b [On the scientific work of William C. Redfield.] Boston Soc. Nat. Hist., Pr. 6: 186-191, (1857).

93--1857c [On the faults and joints of the slate rocks of Governor's Island in Boston Harbor.] Boston Soc. Nat. Hist., Pr. 6: 217-218, (1857).

94--1857d Brief account of the construction and effects of a very powerful induction apparatus devised by Mr. E. S. Ritchie of Boston, United States. British Assoc. Rep. 1857 (pt. 2): 15-16, (1857).

95--1858 On ozone observations. Edinburgh New Phil. Jour. 6: 35-42, (1858).

96--1858a Some experiments on sonorous flames, with remarks on the primary source of their vibration. Am. J. Sci., (2) 26: 1-15, (1858)--Boston Soc. Nat. Hist., Pr. 6: 333-335, 339-340, 346-352, (1858).

97--1858b Experiments on some sonorous flames. Phil. Mag. 15: 261-263, (1858).

98--1858c On the origin of sonorous vibrations produced under certain conditions by flames from wicks or wire-gauze. Am. J. Sci. (2) 26: 240-241, (1858).

99--1858d On the formation of rotating rings by air and liquids under certain conditions of discharge. Am. J. Sci. (2) 26: 246-258, il., (1858).

100--1858e [On anticlinal flexures.] Boston Soc. Nat. Hist., Pr. 6: 332-333, (1858).

101--1858f [On the Clinton iron ores of the Appalachian belt.] Boston Soc. Nat. Hist., Pr. 6: 340-341, (1858).

102--1859 [On the thickness of the earth's crust.] A.Ac.A.S., Pr. 4: 218-219, (1859)--Boston Soc. Nat. Hist., Pr. 7: 47-48, (1859).

103--1859a Results of an examination of Japanese vegetable wax. Boston Soc. Nat. Hist., Pr. 7: 58-59, (1859).

104--1859b [On the infusorial earth from the Tertiary of Virginia and Maryland and the geological relations of the strata.] Boston Soc. Nat. Hist., Pr. 7: 59-64, (1859).

105--1859c [On the age of the rocks of Perry, Me. (with discussion by C. T. Jackson).] Boston Soc. Nat. Hist., Pr. 7: 86, (1859).

106--1859d [On the rate of accumulation of deposits in the South Joggins in Nova Scotia.] Boston Soc. Nat. Hist., Pr. 7: 168-170, (1859).

107--1859e [On the parallelism of the Lower Carboniferous of Pennsylvania and Virginia and of Nova Scotia and New Brunswick.] Boston Soc. Nat. Hist., Pr. 7: 170-173, (1859).

108--1860 [On the geology of the Eastport region, Maine.] Boston Soc. Nat. Hist., Pr. 7: 227-228, (1860).

109--1860a [On the geology of western Vermont.] Boston Soc. Nat. Hist., Pr. 7: 237-239, (1860)--Can. Nat. 6: 326-328, (1860).

110--1860b   [On the stratigraphical relations of deposits formed in an ocean under the conditions of stationary, subsiding, and rising position of the sea bottom (with discussion by L. Agassiz).] Boston Soc. Nat. Hist., Pr. 7: 246-249, (1860).

111--1860c   [On the coal vein at the Albert mine, New Brunswick (with discussion by C. T. Jackson).] Boston Soc. Nat. Hist., Pr. 7: 294-295, (1860).

112--1860d   A report on the registering thermometer of Dr. Lewis, of Mohawk, N.Y. Boston Soc. Nat. Hist., Pr. 7: 317-319, (1860).

113--1860e   [On passage beds.] Boston Soc. Nat. Hist., Pr. 7: 319-322, (1860).

114--1860f   On the Aurora of 1859 at Lunenburg, Massachusetts. Am. J. Sci. (2) 24: 255-256, (1860).

115--1860g   Notes on the Aurora of 28 August [1859] and several subsequent nights, as observed at Lunenburg, Massachusetts, Lat. $42°35'$ [North]. Edinburgh New Phil. Jour. 11: 90-99, (1860)--Am. J. Sci. (2) 29: 255-256, (1860).

116--1860h   Some experiments and inferences in regard to binocular vision. Am. J. Sci. (2) 30: 387-390, (1860)--A.A.A.S., Pr. 14: 187-192, (1860).

117--1860i   On our inability from the retinal impression alone to determine which retina is impressed. Am. J. Sci. (2) 30: 404-409, (1860)--A.A.A.S., Pr. 14: 192-198, (1860).

118--1860j   [On a new stereoscopic slide.] A.Ac.A.S., Pr. 4: 360-364, (Nov. 1860).

119--1860k   Experiments and conclusions on binocular vision. British Assoc. Rept. 1860 (pt. 2): 17-18, (1860).

120--1860ℓ   On the phenomena of electrical vacuum tubes, in a letter to Mr. Gassiot. British Assoc. Rept. 1860 (pt. 2): 30-31, (1860).

121--1860m   On the albertite of New Brunswick. Ac. Nat. Sci. Philadelphia, Pr. 12: 98, (1860).

122--1860n   Ruhmkorff's apparatus constructed by E. S. Ritchie of Boston. A.Ac.A.S., Pr. 4: 376-382, (1860). See also reference 90--1857.

123--1860o   [Camera lucida for obtaining twin drawings suitable for combination in the stereoscope.] A.Ac.A.S., Pr. 5: 73, (Nov. 1860).

124--1861   [On fossiliferous Potsdam pebbles in Carboniferous conglomerate in eastern Massachusetts.] Boston Soc. Nat. Hist., Pr. 7: 389-391, (1861).

125--1861a   [On elongated form and parallel arrangement of pebbles.] Boston Soc. Nat. Hist., Pr. 7: 391-394, (1861)--Abst., Am. J. Sci. (2) 31: 440-442, (1861).

126--1861b   On the group of rocks constituting the base of the Paleozoic series in the United States. Boston Soc. Nat. Hist., Pr. 7: 394-395, (1861).

127--1861c   [On the age of sandstones of St. Croix, N.B. and Perry, Me.] Boston Soc. Nat. Hist., Pr. 7: 398-399, (1861).

128--1861d   [Boulder with Devonian fossils from Saco River, Maine.] Boston Soc. Nat. Hist., Pr. 7: 409, (1861).

129--1861e   Observations on the coiling of the tendrils of the winter squash vines. Boston Soc. Nat. Hist., Pr. 7: 409-411, (1861).

130--1861f   [On the Paleozoic rocks of Dennis River in Maine.] Boston Soc. Nat. Hist., Pr. 7: 419, (1861).

131--1861g   [On the primordial fauna and the Taconic System.] Boston Soc. Nat. Hist., Pr. 7: 419-422, 427, (1861).

132--1861h   Notes on the geographical structure of western Vermont [abst.]. In Report on the geology of Vermont (Hitchcock) 1: 326-327, (1861).

133--1862   (with Jackson, C. T. and Blake, J. H.) [On] the frozen well of Brandon, Vermont. Boston Soc. Nat. Hist., Pr. 9: 72-81, (1862).

134--1862a   Working power of coal. Mechanics Mag. (London), 1861--Am. J. Pharm. 1862, p. 90, (1862).

135--1863   Electric illumination at Boston photometrically measured. Am. J. Sci. (2) 36: 307-308, (1863).

136--1864   An account of apparatus and processes for chemical and photometrical testing of illuminating gas. British Assoc. Rept. 34: 39-40, (1864).

NOTE: The lack of geological publications in the decade 1865-1874 is no doubt ascribable to the fact that during this period William was fully occupied with getting the proposed Massachusetts Institute of Technology organized and its educational program firmly established.

137--1875   On the Newport conglomerate [Rhode Island]. Boston Soc. Nat. Hist., Pr. 18: 97-101, (1875)--Abst., Am. J. Sci. (3) 10: 479, (1875).

138--1875a   On the gravel and cobblestone deposits of Virginia and the Middle [Atlantic] States. Boston Soc. Nat. Hist., Pr. 18: 101-106, (1875)--Abst., Am. J. Sci. (3) 11: 60-61, (1876)--The Virginias 3: 58-59, (1882)--Reprinted in Geology of the Virginias: 707-714, (1884).

139--1875   Hotchkiss' Geological Map of Virginia and West Virginia, the geology by Prof. W. B. Rogers ... Scale 24 miles to 1 inch. Richmond [Va.] 1875; 2nd ed. (1885).

NOTE: In The Virginias I/6: 85, (1880) Hotchkiss wrote as follows about his Map:

> "The Geological Map of Virginia and West Virginia, that accompanies this number of The Virginias, was constructed by Prof. Wm. B. Rogers, chiefly from data obtained from the Geological Society [Survey] of Virginia, which he, as State Geologist, so ably conducted from 1835 to 1841, and from his later observations ... [This Map] was gratuitously and with a noble generosity, furnished to the editor of this journal [i.e. Jed Hotchkiss], in 1873, to accompany a Geographical and Political Summary of Virginia, that he had prepared for the Board of Immigration, and illustrates a chapter on the Geology of the State. It is Rogers' Geological Map of the Virginias, as it came from his own hand ... the editor's only claim to it is that of ownership resulting from its copyright [1875] publication, and from the preparation of the topographical map on which it is based. This much by way of explanation, lest anyone should misunderstand the title to this map as it is printed."
> [Printed as Hotchkiss' Geological Map of Virginia and West Virginia; The Geology by Prof. William B. Rogers, chiefly from the Virginia State Survey, 1835-1841, "with later observations in some parts."]

140--1878? Catalogue of the note books of the Geological Survey of Virginia, 1835 to 1877, by Prof. Wm. B. Rogers and his assistants. A typescript of 3 pages, probably prepared in 1878.

141--1879   List of geological formations found in Virginia and West Virginia, in Macfarlane's Geologists' Travelling Hand-book, or American Geological Railway Guide. New York: D. Appleton & Co., (1879)--See The Virginias 1: 14-15, (1880).

142--1880   The Pigg River Mining Co. The Virginias 1/1 : 14, (1880).

143--1880a Table of the geological formations found in Virginia and West Virginia. The Virginias 1: 14-15, (1880); 3:61, (1882). See also 141--1879.

144--1880b Address on the scientific work and the character of Joseph Henry, delivered in the Hall of Representatives, Washington, D.C. Printed, as A Memorial to Joseph Henry, by order of Congress, 4 to., (1880).

145--1880c The iron ores of Virginia and West Virginia. The Virginias 1: 128-130, 138-140, 152-153, 160-161, 170-171, 174-175, 182-183, 186-188, (1880).

146--1880d Annual Report of the National Academy of Sciences [for 1879-1880]. Washington: Government Printing Office, 22 p., (1880). [46th Congress, 2d. session, House of Representatives Misc. Doc. No. 43.]

147--1880e Phosphorescence, as illustrated by Balmain's luminous paint, An address. M.I.T. Soc. Arts, Pr. 1880-1881: 77-81, (1880-1881).

148--1881   Infusorial stratum and associated Tertiary beds in the vicinity of Richmond, Virginia. The Virginias 2: 58-59, (1881).

149--1882   The infusorial deposit of Virginia in the Fort Munroe artesian well. The Virginias 3: 151-152, (1882)--Reprinted in Geology of the Virginias: 733-736, (1884).

150--1882a The fossils of formation No. III in Virginia. The Virginias 3: 175, (1882).

151--1882b Notes on the geology of the Virginias. The Virginias 3: 190, (1882); 4: 12-13, 38-39, 59-61, 71-72, 88-90, (1883).

152--1882c Report of Proceedings of the National Academy of Sciences from November 16, 1880 to the close of the year 1881 [W. B. Rogers, President]. Washington, D.C.:Judd and Detweiler, Printers, 24 p., (1882).

153--1884   [A Reprint of the annual reports and other papers on the] Geology of the Virginias, by the late William Barton Rogers [compiled and edited by his widow, Emma Savage Rogers]. New York: D. Appleton & Co., xv + 832 p., maps and sections, (1884). The contents of this Reprint are as follows:

   P. i-xv   Title page, other front matter, and Preface.

   P. 1-9   [Some observations on the Tertiary marl of lower Virginia.] (From the Farmers' Register 2/3: 129-131, (1834). See reference 3--1834a.

   P. 9-11   Apparatus for analyzing marl and the carbonates in general. (From the Farmers' Register 2/6: 364-365, (1834). See reference 4--1834b.

   P. 11-20   Further observations on the green sand and calcareous marl of Virginia. (From the Farmers' Register 2/12: 747-751, (May 1835). See reference 6--1835.

P. 21-122   Report of the geological Reconnaissance of the State of Virginia, 1835, (1836). See reference 11--183b.

P. 123-146   Report of the progress of the Geological Survey of the State of Virginia for the year 1836, (1837). See reference 12--1837.

P. 147-188   Report of the progress of the Geological Survey of the State of Virginia for the year 1837, (1838). See reference 14--1838.

P. 189-244   Report of the progress of the Geological Survey of the State of Virginia for the year 1838, (1839). See reference 16--1839.

P. 245-410   Report of the progress of the Geological Survey of the State of Virginia for the year 1839, (1840). See reference 19--1840.

P. 411-536   Report of the progress of the Geological Survey of the State of Virginia for the year 1840, (1841). See reference 20--1841.

P. 537-546   Report of the progress of the Geological Survey of the State of Virginia for the year 1841, (1842). See reference 21--1842.

P. 547-564   Analyses of waters of the principal mineral springs of Virginia.

P. 565-566   Temperatures of the warm, hot, and sweet springs, as observed by J. A. Chavallie, Esq., in 1806, and by myself and Dr. J. B. Rogers in 1834 and 1838.

P. 567-574   Observations of subterranean temperature in the coal mines of eastern Virginia. See references 24--1842c and 30--1843c.

P. 575-598   On the connection of thermal springs in Virginia with anticlinal axes and faults. See reference 29--1843b.

P. 599-642   (with H. D. Rogers) On the physical structure of the Appalachian chain, as exemplifying the laws which have regulated the elevation of great mountain chains generally. See reference 27--1842f, 1843.

P. 643-658   The age of the coal rocks of eastern Virginia. See reference 28--1843a.

P. 659-674   (and H. D. Rogers) Contributions to the geology of the Tertiary formations of Virginia. See references 13--1837a, 17--1839a, and 18--1839b.

P. 675-678   Observations on the natural coke and the associated igneous and altered rocks of the Oölite coal region in the vicinity of Richmond, Virginia. See reference 68--1854c.

P. 679-706   Report on the Pridevale coal and iron ore, West Virginia, (1854). See references 69--1854d and 70--1854e.

P. 707-714   On the gravel and cobble-stone deposits of Virginia and the Middle States. See reference 138--1875a.

P. 715-728   Notes from Macfarlane's Geological Railway Guide [corrected to 1883], (1879). See reference 141--1879.

P. 729-736 [Artesian borings at Fortress Monroe.] The Virginias 3: 151-152, (1882).

P. 737-746 Glossary of geological and other scientific terms. (From Lyell's Principles of Geology; appended to Virginia Geological Survey Report for 1840). See reference 20--1841.

P. 747-770 Appendix:
- A. History of the origin of the Survey (p. 749-763).
- B. Assistants employed on the Survey.
- C. Reference to a peculiar iron ore.
- D. Correction of the arrangement of certain paragraphs.
- E. Nitrates in cave earths (p. 763-764). See reference 85--1856d.
- F. The growth of stalactites in caves (p. 764-765). See reference 86--1856e.
- G. Secondary formations in Virginia and North Carolina. See reference 66--1854a.
- H. Letter to T. Sterry Hunt regarding unconformity of certain strata in Azoic rocks.
- I. Recent names of certain mineral springs of Virginia.

P. 771-832 Index.

154--1885 Geological sections on Coal River, W.Va. The Virginias 6: 153-154, (1885).

155--1896 Life and Letters of William Barton Rogers, edited by his wife [Emma Savage Rogers], with the assistance of William T. Sedgwick, in two volumes - I: x + 427 p., port., maps; II: 451 p., port. Boston: Houghton, Mifflin and Co., (1896).

156--1896a On the nomenclature of the Palaeozoic formations introduced by William B. and Henry D. Rogers. Appendix D, p. 437-438 in Life and Letters (preceding reference, vol. II). See also reference 37--1844c.

Henry D. Rogers described the nomenclature he and William proposed to use, in an address he delivered at the meeting of the Association of American Geologists and Naturalists held in Washington, D.C., in May, 1844. This address was published as Art. X in Am. J. Sci. 47/1: 137-160, and 47/2: 247-278, (1844). On pages 154-160 he discusses the different series names he and William proposed and which are included in the table on page 438 of Appendix D cited above. (See also my page 116.)

References to Comments Made by William Barton Rogers About Addresses and Publications by Other Geologists

(1855) Perrey, A.
(On the frequency of earthquakes and their cause [with discussion by Charles Stodder, C. T. Jackson, Charles Pickering, and W. B. Rogers].) Boston Soc. Nat. Hist., Pr. 5: 137-142, (1855).

(1855) Rogers, H. D.
(Footprints and other impressions on Carboniferous red shale of Pennsylvania [with discussion by J. C. Warren, W. B. Rogers, and C. T. Jackson].) Boston Soc. Nat. Hist., Pr. 5: 182-186, (1855).

(1860)   Agassiz, L.
　　　　(Consecutive faunae and their corresponding geologic formations [with discussion by W. B. Rogers].) Boston Soc. Nat. Hist., Pr. 7: 241-245, 250-252, (1860).

(1861)   Hitchcock, C. H.
　　　　(On the geology of Vermont, chiefly in connection with the Taconic system [with discussion by W. B. Rogers and Jules Marcou].) Boston Soc. Nat. Hist., Pr. 7: 426-427, (1861).

(1861)   Marcou, J.
　　　　(On the occurrence of silver and gold in the Rocky Mountains and California [with discussion by C. T. Jackson, W. B. Rogers, and A. A. Hayes].) Boston Soc. Nat. Hist., Pr. 8: 172, (1861).

(1880)   Hotchkiss, Jed
　　　　Geological Map of Virginia and West Virginia. The Geology by Prof. William B. Rogers, chiefly from the Virginia State Survey, 1835-1841. Scale - 1 : 1,520,640. In The Virginias 1/6: 85, (1880). See also 139--1875 in the preceding bibliographic references.

(1880- Hotchkiss, Jed
1884)　　The Virginias, a mining, industrial and scientific journal, devoted to the development of Virginia and West Virginia - Edited and published by Jed Hotchkiss, and printed at Staunton, Virginia. Vol. I (1880); II (1881); III (1882); IV (1883); and V (1884).
　　　　　Hotchkiss published numerous reprints, extracts, and quotations of articles published elsewhere by William Barton Rogers. A few of these are indicated as Extracts (Extr.) in the preceding Bibliography; many others, however, are not included because they are little more than references to the original articles.

NOTE: In addition to the preceding "discussions," there are numerous references to comments by William in the minutes or reports of meetings of several different societies - e.g. Association of American Geologists and Naturalists; American Association for the Advancement of Science; American Academy of Arts and Sciences; and Boston Society of Natural History.

APPENDIX A

DOCUMENTS INVOLVED IN THE FOUNDING OF MASSACHUSETTS INSTITUTE OF TECHNOLOGY

　　William Barton Rogers, alone, and in earlier years with brother Henry's suggestions and assistance, wrote numerous unofficial and official letters, comments, expositions, memorials, petitions and committee reports regarding a polytechnic school of applied science in the Boston community. The more important of these, indicated by an asterisk (*), are included in the following list of documents that had some bearing on the founding and organization of M.I.T. It should be noted that the first and second items were prepared long before Henry and William came to Boston and concerned the Franklin Institute in Philadelphia and the Lowell Institute in Boston, respectively. Nevertheless, they belong in my list of documents because they contain ideas later incorporated in plans prepared by William for the polytechnic school in Boston that ultimately became M.I.T.

*1. About 1837 William and Henry prepared a memorial for a School of Arts on behalf of the Franklin Institute of Philadelphia and submitted it to the Legislature of Pennsylvania. This memorial is mentioned on pages 258 and 263 of <u>Life and Letters I</u>.

*2. In 1846, at Henry's request, William prepared for J. A. Lowell a plan for a School of Arts as a branch to the Lowell Institute in Boston. This request is contained in a letter from Henry to William dated March 8, 1846 (<u>Life and Letters I</u>, p. 257-260), and the document titled "A Plan for a Polytechnic School in Boston, 1846" is reprinted as Appendix C on pages 420-427 of <u>Life and Letters I</u>.

*3. The earliest of the documents pertaining explicitly to the precursor of M.I.T. was a memorial petitioning the 1859 Massachusetts Legislature for "... a reservation of State land in the Back Bay for a Conservatory of Art and Science." This memorial was drawn up by Dr. Samuel Kneeland and presented to the Legislature by the "Associated Societies or Institutions ..." (also called "Massachusetts Conservatory of Art and Science") on 30 March 1859 as House Document No. 260 but was not approved. [<u>Life and Letters II</u>, p. 3.]

William's name appears in the list of petitioners but he seems to have taken little part in the effort, probably because he was away from Boston much of the time during which the document was being prepared. However, when the decision was made to submit another petition, this time to the 1860 Legislature, William was asked to prepare a new memorial. This he did, devoting the later part of the summer of 1859 to the task. The document that resulted (Appendix A in <u>Life and Letters II</u>, p. 403-418) bears the title:

*4. Memorial of the Associated Institutions of Art and Science to the Massachusetts Legislature 1860, Asking for a Reservation of Lands on the Back Bay. (Prepared by William B. Rogers.) House ... No. 13, January, 1860, Commonwealth of Massachusetts.

As with the previous petitions, this one was also not approved, but undaunted the Association of Institutions decided to make still another effort. A sub-committee was appointed for the purpose and was charged, writes Runkle (1882, p. 16):

"... with the duty of preparing and reporting a plan of an institution designed for the advancement of the industrial arts and sciences and practical education in the Commonwealth."

William was asked to prepare the requested plan, which he did during the summer of 1860. The resulting document was his well-known

*5. <u>Objects and Plan of an Institute of Technology</u>: including a Society of Arts, a Museum of Arts, and a School of Industrial Science, proposed to be established in Boston; prepared [by W. B. Rogers] by direction of the <u>Committee of Associated Institutions of Science and Arts</u>. Boston: 29 p., (1860), 2d ed., (1861), printed by John Wilson & Son.

The foregoing is the document commonly referred to, in discussions of the beginning of the Massachusetts Institute of Technology, as the founding document.

It was accepted by the Committee of Associated Institutions, read at a public meeting at the Boston Board of Trade on 5 October 1860, and submitted soon after to the Massachusetts Legislature.

Action came swiftly from the Legislature in the form of a joint standing committee, and again William was asked to prepare an appropriate report, which appeared in March 1861 as:

> *6. Report of the Joint Standing Committee of the Massachusetts Legislature of 1861 on the Memorial of the Associated Institutions of Science and Art. House Document 171, dated March 19, 1861. [A condensed form of this Report appears as Appendix B on pages 424-429 of Life and Letters II.]

On 10 April 1861 the Massachusetts Legislature passed and Governor John A. Andrew approved:

AN ACT

> 7. To incorporate the Massachusetts Institute of Technology, for the promotion of Arts, and Industrial Sciences, Agriculture, Manufactures and Commerce, and to grant aid to said Institute and to the Boston Society of Natural History.

The Act provided for a square of Back Bay land between Newbury and Boylston streets, of which the western two-thirds were assigned to M.I.T. and the eastern one-third to the Boston Society of Natural History. [M.I.T. Early Papers.] There on that block soon appeared the first building of Technology which became the familiar Rogers Building of Boston Tech.

The memorial of 1860, which led to the Act of 1861, was strongly supported by many influential Bostonians and by the Boston Society of Natural History, mentioned in the preceding paragraph, Boston Board of Trade, American Academy of Arts and Sciences, Massachusetts Charitable Mechanics Association, and the New England Society for the Promotion of Manufactures and Mechanic Arts. Statements of support from the foregoing organizations are reprinted on pages 419-423 of Life and Letters II as part of Appendix A, with the separate title "Petitions in Aid of Memorial."

The many activities centering around the memorials and reports to the Massachusetts Legislature regarding the Back Bay land and the chartering of M.I.T. are related by William in an 1861 account titled:

> *8. An account of Proceedings preliminary to the organization of the Massachusetts Institute of Technology. Boston, (1861). [Preserved in M.I.T. Early Papers.]

Runkle (1882, p. 17-19) mentions that the Society of Arts first acted as the "Corporation," but was later dissolved in favor of an official Corporation, limited to officers of the Society. The first meeting was held on 8 April 1862 at the Boston Board of Trade, at which the members organized themselves and then elected William Barton Rogers the President of the newly chartered Massachusetts Institute of Technology.

As noted earlier on in the section headed "The Founding of M.I.T. (1859-1865)," the Act of Incorporation of the Institute was signed by Governor John A. Andrew only two days before the attack on Fort Sumter (12 April 1861). Almost four years were to pass, because of the Civil War, before William could actually start his polytechnic school, but he used much of that time to lay the foundation for his polytechnic school and to prepare what Runkle (1882, p. 17) and others have called the "intellectual charter" of M.I.T. because of the Institute's close adherence to the essential points of the Plan. This final document carries the title:

> *9. Scope and Plan of the School of Industrial Science of the Massachusetts Institute of Technology, as reported by the Committee on Instruction of the Institute, and adopted by the Government May 30, 1864. [M.I.T. Early Papers.]

APPENDIX B

BIOGRAPHIES, BIOGRAPHICAL REFERENCES, AND OTHER SOURCES OF INFORMATION re WILLIAM BARTON ROGERS (1804-1882)

NOTE: Although a rather careful and prolonged search was made for biographies of and biographical references to William Barton Rogers, it is highly unlikely that the following list includes all such items. Inasmuch as he was a widely known scientist and educator, as well as the Founder of M.I.T., his death was recorded in many notices both in North America and abroad. It is hoped, however, that the more important and more informative biographies and source references are included in the following list.

(1852) Carson, J.
Memoir of the life and character of James B. Rogers. 22 p., (1852). [M.I.T. Library B/R72/Pam.]

(1870) Tribute to Prof. William Barton Rogers on his resignation as President of the Massachusetts Institute of Technology, copied from The Boston Daily Advertiser of June 2nd, 1870--Professor William B. Rogers. [Copy in the M.I.T. Institute Archives.]

(1870) [Tribute in the Boston] Daily Transcript for June 1, 1870.

(1870-1872) Other similar tributes, notices, and news items are mentioned in Life and Letters I and II (1896 reference listed earlier and farther on).

(1876) Walker, F. A.
Sketch of Prof. William B. Rogers. Pop. Sci. Month. 9: 606-611, (Sept. 1876). Philadelphia: Philadelphia Press, May 31, 1882.

(1882) [Obituary of] Prof. W. B. Rogers. Nature 26: 182-183, (June 22, 1882).

(1882) M.I.T. Society of Arts. In memory of William Barton Rogers, LL.D., Late President of the Society. Boston, 39 p., port., (1882). [Copy in the M.I.T. Institute Archives: T-F/R731.]

(1882) At the 288th meeting of the M.I.T. Society of Arts on 12 October 1882, held to memorialize William Barton Rogers, the following persons presented eulogies:
M.I.T. President Francis Amasa Walker,   p. 3- 4
Prof. W. P. Atkinson,   p. 5- 7
James P. Tolman, President of the
    M.I.T. Alumni Association,   p. 7- 9
Dr. John Daniel Runkle,   p. 10-21
Prof. C. R. Cross,   p. 22-26
Major Jedediah Hotchkiss,   p. 26-35
Letter from W. L. Brown,   p. 35-36
Letter from Prof. Francis H. Smith,   p. 36-39

(1883) Rives, W. C.
William Barton Rogers, LL.D.--An address delivered before the Society of the Alumni of the University of Virginia on Commencement Day, June 27, 1883. Cambridge, Mass.: John Wilson and Son, 32 p., (1883). [Copy in M.I.T. Institute Archives; see also Life and Letters II, p. 343, (1896), listed farther on.]

(1883) Ruschenberger, W. S. W.
A sketch of the life of Robert E. Rogers, M.D., LL.D., with biographical notices of his father and brothers [with cover titled "The Brothers Rogers"], 45 p., (1885). Am. Acad. Arts Sci., Pr. 18: 428-438, (1883)--Am. Phil. Soc., Pr. 23: 104-146, (1886).

(1) WILLIAM BARTON ROGERS

(1883) Cooke, J. P.
Notice of William Barton Rogers, Founder of the Massachusetts Institute of Technology. Am. Acad. Arts Sci., Pr.14: 426-438, (1883).

(1883) M.I.T.
The Massachusetts Institute of Technology in memory of Prof. William Barton Rogers, LL.D., late president. Boston, Mass.: M.I.T., 3 p., (1883).
This reprint, which is in the M.I.T. Institute Archives, contains the resolution of the Corporation and the announcement –
> "That the original building of the Institute of Technology be called hereafter the 'Rogers Building' in recognition of the eminent services of Professor William B. Rogers ..."

(1883) Wilder, M. P.
Address at the 1883 annual meeting in The New England Historic and Genealogical Register, Vol. 37, p. 130, (1883). [Reference is made to W. B. Rogers' death.]

(1885) Holland, J. W.
A Eulogy on the Life and Character of Prof. Robt. E. Rodgers [sic], M.D. Introductory to the course of 1885-86 at Jefferson Medical College, delivered September 30th, 1885. Philadelphia: Press of W. F. Fell & Co., 26 p., (1885). [M.I.T. Library: B/R72/Pam.]

(1887) Walker, F. A.
Biographical memoir of Wm. Barton Rogers. Read before the National Academy [of Sciences], April, 1887. Washington, D.C.: Judd and Detweiler, Printers, 13 p., (1887); Nat. Ac. Sci., Biog. Mem. 3: 1-13, (1895).

(1892) [Rogers, William Barton] in National Cyclopaedia of American Biography, Vol. 7, p. 410, (1892). (Original – New York: James T. White & Co., 1892; Reprint edition – Ann Arbor, Michigan: Univ. Microfilms, 1967.)

(1896) Life and Letters of William Barton Rogers, edited by his wife [Emma Savage Rogers], with the assistance of William T. Sedgwick, in two volumes--I: x + 427 p., port., maps; II: 451 p., port. Boston: Houghton, Mifflin and Co., (1896).

(1897) Mendenhall, T. C.
Life and Letters of William Barton Rogers [a review]. Science 6: 1-9, port., (1897).

(1898) Ritchie, John, Jr.
The American Association for the Advancement of Science. New England Magazine 18: 638-661, (1898). [Tells of Rogers' part in founding and guiding the A.A.A.S.]

(1898) Smith, Harriette Knight
The History of the Lowell Institute. Boston: Lamson, Wolffe and Co., 125 p., il., (1898). [Lists lectures by William B. and Henry D. Rogers.]

(1898) Ruffner, W. H.
The Brothers Rogers. The Alumni Bulletin (published quarterly by the Faculty of the University of Virginia), Vol. V, No. 1, p. 1-13, (May 1898). [Reprinted from the author's privately printed paper, with his permission.]

(1899) Benjamin, M.
Early presidents of the American Association [for the Advancement of Science] [Annual Address]. Easton, Pa.: Chem. Pub. Co., 62 p., port., (1899). Also A.A.A.S., Pr. 48: 379-459, port., (1899).

(1902) Tyler, H. W.
John Daniel Runkle (1822-1902). Tech. Rev. 4/3: 276-306, port., (July 1902). Reprinted as "John Daniel Runkle (1822-1902): A Memorial." Boston: George H. Ellis Co., 32 p., port., (1902).

(1904) The Tech
The issue of The Tech for December 7, 1904 (v. 24/31) carried seven short articles by early M.I.T. alumni, who recalled interesting incidents involving Rogers and the beginnings of the Institute.

    Eli Forbes '68 - A story of President Rogers, p. 1 and 3.
    J. P. Munroe '82 - Life of William Barton Rogers, p. 1 and 4.
    R. H. Richards '68 - Beginnings of the Institute of Technology, p. 1, 2, and 4.
    C. R. Cross '70 - Some reminiscences of President Rogers, p. 2.
    J. P. Tolman '68 - The human side of President Rogers, p. 3.
    Anonymous - An appreciation of the heroism of William Barton Rogers, p. 3.
    Anonymous - President Rogers and the Appalachian Club, p. 3.

(1904) Munroe, J. P.
William Barton Rogers, Founder of the Massachusetts Institute of Technology. Tech. Rev. 6/4: 501-550, port., il., (1904). [Reprinted as William Barton Rogers, 1804-1904. Boston: Geo. H. Ellis Co., Printers, 52 p., il., (1904).]
    This article contains excellent pictures of the buildings in Boston that were used by M.I.T. in its earliest decades.

(1904) Smith, F. H.
William Barton Rogers, as seen by an old pupil. Supplement to The Tech 24/35, 2 p., (December 16, 1904).

(1905) Anonymous
Centennial Commemoration of William Barton Rogers 1804-1904, held in Huntington Hall, Rogers Building [M.I.T.] on December 7, 1904, 25 p., including addresses by:

    Henry S. Pritchett, President, M.I.T., (p. 3-7) [26-30];
    Lyon G. Tyler, President, William and Mary College (p. 7-11) [30-34].
    Francis H. Smith, Professor of Natural History, University of Virginia, (p. 11-16) [34-39].
    Robert H. Richards '68 [1868], Professor of Mining Engineering and Metallurgy, M.I.T., (p. 16-20) [39-43]; and
    Norman Lombard '05, who read an extract from F. A. Walker's "Memoir of William Barton Rogers" presented to the National Academy [of Sciences] in April 1887, (p. 20-25) [43-48]. [See Walker, F. A. (1887) above.]

The preceding report was first published in the Technology Review, Vol. 7/1, p. 26-48, (1905), then reprinted from the Review by Geo. H. Ellis Co., Printers, of Boston, in 1905, with the changes in pagination indicated above. The Review pagination is included in brackets [ ].

(1906) Merrill, G. P.
Contributions to the history of American Geology. U.S. Nat. Mus., Rp. 1904: 189-734, il., (1906). [Rogers is mentioned on p. 341-344, 369-373, 402-405, and 710.]

(1913) True, F. W., Editor
The National Academy of Sciences, 1863-1913. (A history of the first half-century of the National Academy of Sciences, 1863-1913.) Washington, D.C.: Baltimore Press, xi + 399 p., (1913). [Rogers is mentioned on p. 21, 48, 61, 176-178, 272, 280, 286 and 293.] (Both William Barton and Robert Empie Rogers were corporators of the Academy when it was incorporated on 22 April 1863.)

(1914) Anonymous
When we were freshmen. Tech. Rev. 16: 571-581, (1914). (Includes the next article by Richards.)

(1914) Richards '68, R. N.
When we were freshmen. Tech. Rev. 16: 574-577, (1914).

(1915) Munroe, J. P.
William Barton Rogers, The Founder. Tech. Rev. 17: 1-6, (1915).

(1916) Gregory, J. W. [with Bibliography by Colin M. Leitch]
Henry Darwin Rogers--An Address to the Glasgow University Geological Society, 20th January, 1916, 38 p. Glasgow: James Mac Lehose and Sons, Publishers to the University, (1916).

(1918) Mann, C. R.
A study of Engineering Education. The Carnegie Foundation for the Advancement of Teaching, B. 11: xi + 139 p., (1918). New York.

(1918) Dana, E. S. and 14 other authors
A Century of Science in America. New Haven, Conn.: Yale University Press, vii + 458 p., il., (1918). [Of special interest regarding the contributions of Henry D. and William B. Rogers is Part IV. A Century of Geology - The growth of knowledge of earth structure by Joseph Barrell, p. 153-192.]

(1920) Merrill, G. P.
Contribution to a history of American state geological and natural history surveys. U.S. Nat. Mus., B. 109: xvii + 549 p., 37 pls., (1920). [See p. 507-512.]

(1924) Merrill, G. P.
The First One Hundred Years of American Geology. New Haven, Conn.: Yale Univ. Press, 773 p., 130 fig., 36 pl., (1924). [Rogers is mentioned on p. 183, 185, 218, 222, 249-250, 258, 377, 397 and 398.]

(1928) City of Boston (School Committee)
In the Rep. Boston School Committee for Dec. 3, 1928, p. 294, appears the following:

> "ORDERED, That the Hyde Park Intermediate School located in the former Hyde Park High School building is hereby named the William Barton Rogers School ..."

(1931) Hitchcock, L. B.
A Memorial to Rogers - The Founder of the Institute is honored at the University of Virginia. Tech. Rev. 33/5: 245, 258, 260, (Feb. 1931).

(1933) Eby, J. B.
23rd International Geological Congress, Prague, August 1968. Am. Assoc. Petroleum Geol., B. 53: 236-240, (1968). See also Rept. XVI Session, I.G.C., Vol. I and II, U.S.A., (1933).

(1933) Adams, F. D.
Sir Charles Lyell - His place in geological science and his contributions to the Geology of North America. Science 78: 179-183, (Sept. 1, 1933). [Adams, p. 182, quotes Lyell as doubting the Rogers' theory of origin for the folded Appalachians.]

(1933) Ricketts, P. C.
Rensselaer Polytechnic Institute - Amos Eaton, Author, Teacher, Investigator: The First Laboratories for the Systematic Individual Work of Students in Chemistry, Physics and Botany, to be created in any country, Established at Rensselaer School by Amos Eaton in 1824: B. Franklin Greene and the Reorganization in 1849-50. Troy, N.Y., 32 p., (1933).

(1935) [Authors]
Rogers, William Barton, in Dictionary of American Biography 16: 115-116, (1935).

(1935) Bevan, Arthur
William Barton Rogers, First State Geologist of Virginia (1835-1841). [In commemoration of the centennial of the appointment of W. B. Rogers as State Geologist, and the creation of the first Geological Survey of Virginia.] Presented to the Virginia Academy of Science, Geological Section, Richmond Meeting, May 3, 1935. Va.Ac.Sci., Pr. 1934-1935: 63-67, (1935).

(1935) Roberts, J. K.
William Barton Rogers and his contributions to the geology of Virginia, read before the National Academy of Sciences at the University of Virginia Meeting on Tuesday, November 19, 1935, (‡). Typescript of 7 pages in the archives of the National Academy of Sciences. See also the next reference.

(1935) Roberts, J. K.
William Barton Rogers (1804-1882) and his contribution to the geology of Virginia. Va. Geol. Surv., B. 46-C: 23-28, port., il., (1935)--Geol. Soc. Am., Pr. 1935: 305-310; Abst.: 99, (1936).

(1938) Bevan, Arthur
William Barton Rogers--pioneer State geologist. Presented before the American Association for the Advancement of Science, Section E, William Barton Rogers Memorial Program, Richmond [Va.] Meeting, Dec. 28, 1938. (‡) Typescript of 12 p. preserved in the Manuscript Department, University of Virginia Library, Charlottesville, Va.

(1938) Bryan, J. S.
William Barton Rogers--organizer and educator. Presented before the American Association for the Advancement of Science, Section E, William Barton Rogers Memorial Program, Richmond [Va.] Meeting, Dec. 28, 1938. (‡) Typescript of 13 p. preserved in the Manuscript Department, University of Virginia Library, Charlottesville, Va.

(1938) Roberts, J. K.
William Barton Rogers--student and teacher [of geology]. Read at the Richmond [Va.] Meeting of the American Association for the Advancement of Science, December 28, 1938. (‡) A typescript of 7 p. preserved in the Manuscript Department, University of Virginia, Charlottesville, Va.

(1938) Ewing, G. W.
　　　Early teaching of science at the College of William and Mary in Virginia. College William and Mary, B. 32/4: 29 p., il., (1938)--J. Chem. Ed. 15/1: 3-13, (1938).

(1939) Roberts, J. K. and Bloomer, R. O.
　　　Catalogue of topographic and geologic maps of Virginia. Richmond, 17 + 1 + 246 p., port., (1939). (Dedication to William Barton Rogers, p. 16-17. An account of the work of Prof. Rogers for the Geological Survey of Virginia, p. 10-11. Table of geological formations of Virginia and West Virginia by William Barton Rogers, p. 221-222.)

(1940) Bevan, S. C.
　　　William Barton Rogers, a pioneer American scientist. Sci. Monthly 50/2: 110-124, il., map, (1940).

(1940-1941) Butts, Charles
　　　Geology of the Appalachian valley in Virginia. Prepared in cooperation with the United States Geological Survey. University of Virginia, (1940-1941), 2 vols., port., il., (1941). Bibliographic footnotes. Issued in commemoration of William Barton Rogers. Va. Geol. Surv., B. 52: Pt. I - Geologic Text and Illustrations, xxxii + 568 p., il., (1940); Pt. II - Fossil Plates and Explanations, iv + 271 p., il., (1941).

(1942) Roberts, J. K.
　　　Biographical sketches of Virginia geologists, in J. K. Roberts' Annotated geological bibliography of Virginia, p. 28-63, (1942).

(1943) Willis, B.
　　　American Geology, 1850-1900. Am. Phil. Soc., Pr. 86: 34-44, (1943).

(1944) Jaffe, B.
　　　Men of Science in America. The role of science in the growth of our country. New York: Simon and Schuster, xl + 600 p., il., (1944).

(1948) Struik, D. J.
　　　Yankee Science in the Making. New York: Collier Books, 1st ed., (1948); rev. ed., 544 p., il., (1962). [Comments on the Rogers brothers, especially on p. 436-441.]

(1949) Rodgers, John
　　　Evolution of thought on structure of Middle and Southern Appalachians. Am. Assoc. Petroleum Geol., B. 33: 1643-1654, (1949). See also (1971) Rodgers.

(1954) Prescott, S. C.
　　　When M.I.T. was "Boston Tech." 1861-1916. Cambridge: The Technology Press, xviii + 350 p., port., il., (1954).

(1957) Killian, J. R., Jr.
　　　William Barton Rogers. Tech. Rev. 60: 105-108, 124, 126, 128, 130, (Dec. 1957).

(1961) Maslanka, J. S.
　　　William Barton Rogers' conception of an institute of Technology. M.I.T. ms 1961: 4 + 54 p. [S.B. Thesis in Course XXI-B, Humanities, (1961).]

(1961) Woodward, H. P.
　　　Reappraisal of Appalachian geology. A.A.P.G., B. 45: 1625-1655, (1961).

(1964) Anonymous
William Barton Rogers [A leaflet prepared for Exhibit shown in the Hayden Library during the International Conference on the Earth Sciences, held in conjunction with the Dedication of the Cecil and Ida Green Building on 2 October 1964. (‡) Typescript in M.I.T. Institute Archives.

(1964) Eaton, C.
The Mind of the Old South - Chap. 8, Rogers of Virginia and Le Conte of Georgia, the scientific mind, p. 137-157. Baton Rouge, La.: La. State Univ. Press.

(1965) Stratton, J. A.
The new Academy of Engineering - An address given this spring [1965] at a dinner in Washington [D.C.] of special interest to many M.I.T. alumni. Tech. Rev. 67/9: 41-44, il., (July 1965). [Contains comments on Rogers' and his ideas about the founding of the National Academy of Sciences.] (See also the following reference.)

(1965) Stratton, J. A.
Advice to a new Academy - The Engineering Academy, founded on the same principles as the NAS [National Academy of Sciences], faces difficult, important tasks. Science 149: 1206-1208, (10 Sept. 1965). [This article was originally an address delivered at the annual dinner of the National Academy of Sciences, of which he was then vice president, on 27 April 1965. In it Stratton refers to Rogers' and his ideas about the founding of the National Academy of Sciences.] (See also the preceding reference.)

(1969) Berl, W. G.
Out of the Ivory Tower. Science 166: 1553 (19 Dec. 1969). Reprinted in BOSTON magazine (December 1969).

(1970) Rodgers, John
The Tectonics of the Appalachians. New York: Wiley- Interscience, 271 p., (1970).

(1971) Rodgers, John
Evolution of thought on structure of Middle and Southern Appalachians: Second paper, p. 1-15 in Appalachian Structures, Origin, Evolution and possible potential for new exploration frontiers: A Seminar, March 3-5, 1971, edited by P. Lessing, Ruth I. Hayhurst, J. A. Barlow and L. D. Woodfork. W. Va. Univ. and W. Va. Geol. & Econ. Surv., xiii + 322 p., il., (1972). See also (1949) Rodgers.

(1974) Ernst, W.
William Barton Rogers, Ante Bellum Virginia Geologist. Virginia Cavalcade 24: 13-21, port., (1974). Richmond, Va.: Virginia State Library.

(1975) Gerstner, Patsy A.
A dynamic theory of mountain building: Henry Darwin Rogers, 1842. Isis 66/231: 26-37, (March 1975).

(1975) Milici, R. C.
Structural patterns in the Southern Appalachians: Evidence for a gravity slide mechanism for Alleghanian deformation. Geol. Soc. Am., B. 86: 1316-1320, il., (Sept. 1975).

(1975) Rogers, William Barton, by John Rodgers, in Dictionary of Scientific Biography, Charles C. Gillespie, Ed. p. 504-506 in Vol. 11, (1975). New York: Charles Scribner's Sons.

----

(2)
ALPHEUS HYATT
(1838-1902)

ALPHEUS HYATT

MIT: 1870-1888

Being a geologist with a broad interest in that discipline, William Barton Rogers, as Founder and First President of M.I.T., quickly saw to it that professors in this science were appointed to the faculty. The first of these was Alpheus Hyatt, at the time Custodian of the Boston Society of Natural History, who was appointed Professor of Palaeontology in 1870. Hyatt was trained in classical biology and paleontology at Harvard by famed zoologist Louis Agassiz, and brought to the fledgling Institute a broad knowledge of both zoology and paleozoology. A year after he came, the Institute established Course VII, Natural History, designed to prepare those "whose ulterior objective is the special pursuit of geology, mineralogy, botany, zoology, or to prepare for medicine, pharmacy, or rural economy." Hyatt occupied the chair of paleontology in this new Course for eighteen years, devoting his time almost exclusively to lecturing on paleontology and zoology, but it was as Custodian, and later as Curator, of the Boston Society of Natural History and Co-Founder of the Teachers' School of Science that he made his major impact on science and the teaching of science in the New England community. Furthermore, Hyatt did most of his research and published most of his more important works in connection with his duties at the Museum of the Boston Society of Natural History.

Alpheus Hyatt, the first professor in geology at M.I.T. after President Rogers, was primarily a biologist and paleontologist, and was one of that small group of fortunate students who studied under famed Louis Agassiz at Harvard, then went forth to distinguished careers, each in his own special discipline. Hyatt's fame rested on his innovative research on fossil cephalopods, but his accomplishments were far broader than that. His researches covered a wide range of animals, with critical work on sponges, bryozoans, pelecypods, gastropods, cephalopods and insects, and he published numerous purely philosophical articles mostly having to do with evolution. He taught both zoology and paleontology, served as a museum administrator, participated in organizing scientific societies and special schools, and maintained a seaside laboratory for instruction and research on marine life. More will be said about these different activities a little farther on, but first something about his early life and the experiences that prepared him for his later accomplishments.

He was born in Washington, D.C. on 6 April 1838, descendant of an old and honored Maryland family, and was almost sixty-four years old when he died suddenly of a heart attack, on 15 January 1902, while on his way to a meeting of the Boston Society of Natural History, with which he had been closely associated for more than thirty years.

After primary school he was sent to the Maryland Military Academy, then entered Yale as a member of the Class of 1860. After completing the freshman year, he left for a year's travel in Europe, and while in Rome his family tried to persuade him to enter the Church, but soon after, in 1858, he entered Harvard's Lawrence Scientific School. He intended to be an engineer, but he soon heard about famous Professor Louis Agassiz and his museum full of exciting biological treasures, and decided to abandon engineering for zoology. He joined an enthusiastic group of Agassiz's students and quickly found a place among them as a denizen of Zoological Hall, a small two-story wooden building which stood on the site of the present Peabody Museum of American Archeology. Hyatt, however, had quarters in nearby Divinity Hall. Although by this time deeply devoted to zoology and geology, he took full advantage of Harvard's educational opportunities by attending lectures in disciplines other than the natural sciences, and he read good literature. In later years this broad training proved valuable in his own educational and administrative activities.

As a student of Agassiz he graduated with high honors in 1862, then enlisted for service in the Union Army and soon rose to the rank of Captain in the 47th Massachusetts Regiment. In 1865, when he returned from the Civil War, he resumed his studies with Agassiz and was placed in charge of the fossil cephalopod collection of the Museum of Comparative Zoology. Thereafter, until his death thirty-seven years later, he was the unofficial curator of these fossils, and published articles on many of them.

In 1867 he married Ardella Beebe and joined three of his former schoolmates--E. S. Morse, A. S. Packard, and F. W. Putnam--in Salem (Massachusetts), where they worked together in the Essex Institute. Hyatt served as one of the curators in the Institute, and in 1869 helped to establish the Peabody Academy of Science, in which he also served as one of the curators. Here also in Salem the four former Agassiz students, who would become both lifelong friends and distinguished scientists, founded the American Naturalist, the first American journal devoted to biological sciences. Hyatt served as one of the editors for four years, 1867 to 1871. Then came the two events in 1870 that set his course for the remainder of his life. He was now thirty-two years old, had published more than twenty papers, and was well known in New England scientific circles for both geological and paleontological researches.

First came the call from the Boston Society of Natural History to be Custodian, a position he held from 1870 to 1881, at which time he was made Curator, and thereafter he continued as the scientific head of the Society until his death in 1902. Most of the research that he did on fossil cephalopods was done at the Society's Museum, and his more important papers were published in the Proceedings of the Society (See the Bibliography farther on). Here too he took the leading role in organizing the Society's Teachers' School of Science, and for more than thirty years (1870-1902) until his death he directed the School, wrestled with its many problems, shared lecturing with a dozen colleagues, and brought biology and geology to many public-school teachers of the Boston community and beyond. His biographers report that more than 1200 teachers attended lectures and participated in carefully organized field excursions to the seashore and to places of geological interest.

At his home in Annisquam (Massachusetts) he founded and organized a seaside laboratory, under the auspices of the Womens Education Association of Boston, and during the summer months, when the laboratory was open, he conducted dredging operations along the New England Coast. This pioneering summer laboratory started in 1880 and existed until 1886, when it was moved to Woods Hole. According to D. J. Zinn,* the present Marine Biological Laboratory was the immediate outgrowth of Hyatt's seaside laboratory at Annisquam, though it is true that Agassiz's earlier seaside program

> "... at the Anderson School of Natural History on Penekese Island was in essence the forerunner of all of the marine and fresh-water biology laboratories in this country, [but] it was not the direct ancestor of the Marine Biological Laboratory in Woods Hole."

In recognition of the important role he played in founding MBL, he was elected its first President and thereafter served as a leading member of its Board of Trustees.

The second event of 1870 came when Hyatt was appointed Professor of Palaeontology at M.I.T. on 10 November 1870. But there was much more to this event than his appointment, as is shown by the following excerpt from Vol. I (p. 69) of the Records of the Committee on Instruction:

> "The President read memoranda of an agreement, which had already received the sanction of the Boston Society of Natural History, between that Society and the Institute, by which on the payment of $400 per annum by the Institute, the Natural History Society agree to allow the Professors of the Institute to give all their courses in Natural Science in the Society's lecture room ... and to allow the free use of the Society's collections and library ... This will permit much needed lecture

---

* Zinn, D. J., "Early days at Woods Hole," Science 159: 484, (1968).

> room to be occupied by other departments of the School, by the removal of the Rogers' collection of geological specimens ... to the Society's building, there to be kept separate and distinctly marked, and to be returned to the Institute when the agreement between the institutions is terminated.
>
> It was also voted that Mr. Alpheus Hyatt be appointed Professor of Paleontology in the Institute, without a seat on the Faculty, ... salary of $500 per annum, to hold office at the pleasure of the Corporation ..."

It is evident from the foregoing excerpt that classroom space and collections in the Society's building were to be made available to M.I.T. staff and students, and also that only a part of Hyatt's time was to be allocated to the Institute (his salary of $500 per annum was only a half or less of the going salary for a professor at the time, and he was not given a seat on the faculty). Shortly after, the President's Report of 30 September 1874 states (p. xiv):

> "The Zoological and Palaeontological Laboratory ... has been fitted up in the building of the Boston Society of Natural History for the joint use of the Society and the Institute. Here Prof. Hyatt gives his instruction in Palaeontology, and also his course in Comparative Zoology in the third year, and the laboratory work of the fourth year to students in the course of Natural History."

The 1870s were a difficult time financially for the struggling young Institute, and at the meeting of the Committee on the School of Industrial Science on 7 November 1877, and again on 12 February 1878, the following action was taken:

> "Voted: that the Professorship of Zoology and Physiology and the Professorship of Palaeontology be united and named the Professorship of Palaeontology and Zoology." (Inst. Comm. Rec. I: 132, 1877.)

Shortly thereafter, at the meeting of the Committee on 9 October 1878, it was

> "Voted: that Prof. Hyatt receive for this year $1000, and have a seat on the Faculty." (Inst. Comm. Rec. I: 146, 1878.)

As a result of the foregoing actions, Hyatt's title was changed to Professor of Palaeontology and Zoology in 1878 and remained so until 1888 when the Executive Committee of the Institute concluded that

> "... it had become desirable that the classes in Zoology and Palaeontology should be taught within the lecture-rooms and laboratories of the Institute itself. In accordance with this view, so much of the arrangement between the Institute of Technology and the Boston Society of Natural History as provided for the instruction of our classes by the Custodian of the latter society, was ... terminated; ...

"In thus terminating an arrangement which had existed, with mutual advantage to both institutions, through a long term of years, the Faculty and the Corporation were not insensible of the extraordinary merits of Prof. Hyatt, Custodian of the Boston Society of Natural History, whether in scientific discovery or in the popular exposition of natural laws." (President's Report for 1887-1888, p. 22.)

Thus came to an end Hyatt's long association with and service to M.I.T., which lasted for eighteen years (1870-1888). These were critical years for the young Institute, and for the natural sciences in particular, inasmuch as the latter were an important and integral part of the curriculum. Hyatt's service, therefore, was of utmost importance to the development of both geology and biology at the Institute, and instruction in paleontology lapsed after his departure until Shimer joined the faculty in 1903.

Important as his services were to M.I.T. and they were important, as emphasized in the preceding paragraph, Hyatt's fame rested on a much broader base. It would prolong this discussion unduly to describe further his many activities and accomplishments outside M.I.T. Fortunately, however, these have been recorded by his numerous biographers, whose memorials are listed following his Bibliography later on. Suffice it here to list briefly the more important of them.

1. He was a stimulating teacher of paleontology and biology at M.I.T. (1870-1888), Boston University (1877-1902), and in the Teachers' School of Science (1870-1902) in Boston.

2. As Custodian first, and then Curator, of the Boston Society of Natural History, he was an outstanding museum administrator for more than thirty years (1870-1902).

3. He took an active part, often the leading role, in organizing and directing scientific societies, schools, and laboratories. He participated in the founding of the Peabody Academy of Sciences at Salem, The Teachers' School of Science in Boston, the seaside summer laboratory at Annisquam, and the Marine Biological Laboratory at Woods Hole, serving the last-named as first President.

4. He was a co-founder of the American Naturalist, serving as one of the editors for four years, and was one of the founders and the first president of the Society of Naturalists of Eastern United States.

5. Most important of all, in the judgment of his scientific peers, were his brilliant studies of fossil cephalopods from which he formulated new principles bearing on the mechanics of evolution. Probably the best known of these was his "law of embryonic acceleration," according to which "features appearing at or near the adult period are inherited at earlier and earlier stages in successive generations."

Perhaps less impressive, but certainly no less important, were his contributions to our knowledge of sponges, bryozoans, pelecypods, gastropods, and insects.

6. Recognition of his accomplishments came from scientists everywhere. He was elected a member of the American Academy of Arts and Sciences in 1869, of the National Academy of Sciences in 1875, and of the American Philosophical Society in 1895. Brown gave him the honorary degree of LL.D. in 1898. Abroad he was a correspondent of many learned societies.

It was to scientists like Alpheus Hyatt that President Rogers turned for help in getting his young Institute of Technology started. And it was to them in no small measure that the institution owed its survival through the early years of travail to become the famed Boston Tech of the 19th Century and the M.I.T. of the Twentieth.

On 7 January 1867 Hyatt married Miss Ardella Beebe, of Kinderhook, N.Y. In their years in Cambridge their large wooden house on Francis Avenue [which he named "Norton's Wood"] was a center of intellectual social life characteristic of the times. None of their three children followed science, but both their daughters became sculptresses, and it is of interest that the scientific accuracy of their work was remarkable quite apart from its artistic merit. Daughter Harriet (Mrs. Alfred G. Mayer) prepared a bas-relief and memorial tablet in bronze of her father, which was unveiled in the reading-room of the Marine Biological Laboratory on 4 September 1928 (See Conklin, E. G. in Bibliography). A photograph of Mrs. Mayer's bas-relief is included in Jackson's biography of 1913 (See items following the Biography).

BIBLIOGRAPHY OF ALPHEUS HYATT

The abbreviations and symbols used in the following references are explained on p. 91 - 98. In general, abstracts are listed separately as well as with the complete article. Every effort has been made to include all published items that concern geology and paleontology, starting with Hyatt's earliest known article about the Polyzoa, and all of the more important publications dealing with biological subjects. To my knowledge this is the most complete list of his published works yet brought together in a single bibliography.

Many of the earlier titles are merely descriptive references to verbal reports, and for this reason are generally enclosed in []. Also, because so many of his reports and articles appeared in publications of the Boston Society of Natural History, I have abbreviated the name of the Society to B.S.N.H.

The numerous reports that Hyatt wrote as Curator and Custodian of the Boston Society of Natural History during his long administration (1870-1902) are listed separately following the regular bibliography.

Finally, inasmuch as numerous memorials, necrologies, and other biographical sketches have been written about Hyatt, I have listed these in chronological order at the end of this sketch. No doubt some items have been missed, but I hope I have included all of the more informative ones. It is on the body of information included in them that I have drawn for the foregoing biography.

1--1864-1868   Observations on Polyzoa: Suborder Phylactolaemata. Essex Inst., Pr. 4: 197-228, il., (1864); B. 5: 97-112, il., (1866); Ibid: 145-160, il., (1867); Ibid: 193-232, il., (1868).

2--1865   Remarks on the Beatriceae, a new division of Mollusca. Am. J. Sci. (2)39: 261-266, (1865).

3--1865a   [On Beatricea and Pasceolus.] B.S.N.H., Pr. 10: 19, (1865).

4--1865b   [On the structure of the shells of Cephalopoda.] B.S.N.H., Pr. 10: 24, (1865).

5--1866   On the parallelism between the different stages of life in the individual and those in the entire group of the molluscous order Tetrabranchiata. B.S.N.H., Mem. 1: 193-209, (1886).

6--1866a   [Fresh-water sponges.] Essex Inst., Pr. 5: 59, (1866).

7--1866b   [On the agreement between the different periods in the life of the individual shell and the collective life of the tetrabranchiate cephalopods.] B.S.N.H., Pr. 10: 302-303, (1866).

8--1867   The fossil cephalopods of the Museum of Comparative Zoology. Harvard Coll., Mus. Comp. Zool., B. 1: 71-102, (1867). See also 43--1872.

9--1867a   [Temperatures in deep mines and caves.] Essex Inst., Pr. 5: 84, (1867).

10--1867b   [On Eozoön canadense.] Essex Inst., Pr. 5: 110, (1867).

11--1868   The Moss-animals or fresh-water Polyzoa. Am. Nat. 1: 57-64, 131-136, 180-186, il., (1868).

12--1868a   Rock ruins [Niagara Falls]. Am. Nat. 2: 77-85, (1868).

13--1868b   The animal nature of sponges. Am. Nat. 2: 101-102, (1868).

14--1868c   Sponges. Am. Nat. 2: 303-306, (1868).

15--1868d   The chasms of the Colorado. Am. Nat. 2: 359-365, (1868).

16--1868e   [Composition of our stone walls.] Essex Inst., Pr. 5: 154-155, (1868).

17--1868f   [Water.] Essex Inst., Pr. 5: 157, (1868).

18--1868g   [Affinities of Beatricea.] Essex Inst., Pr. 5: 187, (1868).

19--1868h   [Geological structure of the Adirondacks.] Essex Inst., Pr. 6: 5-6, (1868).

20--1868i   [Deltas and how formed.] Essex Inst., Pr. 6: 37, (1868).

21--1868j   [On a disintegrated rock at Salem.] Essex Inst., Pr. 6: 51-52, (1868).

22--1868k   [Meteoric shower of November 13, 1868.] Essex Inst., Pr. 6: 55, (1868).

23--1868*l*   [On the absence of distinct evidence of glaciation in the Yukon Valley, Alaska.] B.S.N.H., Pr. 12: 149-150, (1868).

24--1869   Fresh-water Polyzoa occurring in Maine. Portland Soc. Nat. Hist. 1: 157-162, (1869).

25--1869a   [Development of Nautili.] B.S.N.H., Pr. 12: 217, (1869).

26--1869b   [Eozoön canadense from Newburyport.] Essex Inst., B. 1: 142, (1869).

27--1869c   [Discussion of paper by E. Bicknell.] Essex Inst., B. 1: 141-142, (1869).

28--1870   [Eozoön in Essex County.] Essex Inst., B. 2: 93, (1870).

29--1870a   [On a raised beach at Marblehead, Mass.] Essex Inst., B. 2: 111, (1870).

30--1870b   [Report on the Cretaceous fossils from the Province of Sergipe, Brazil, in collection of Prof. C. F. Hartt, p. 385-393,] in Hartt, C. F., "Geology and Physical Geography of Brazil," in Thayer Expedition--Scientific Results of a Journey to Brazil, by Louis Agassiz and his travelling companions. Boston: Fields, Osgood & Co. xxiii + 620 p., il., (1870).

31--1870c   [On Morse's classification of the Brachiopoda.] B.S.N.H., Pr. 13: 417, (1870).

32--1871   [On reversions among the ammonites.] B.S.N.H., Pr. 14: 22-43, (1871); Pr. 17: 23-28, (1874).

33--1871a   [On the embryology of fossil nautiloids.] B.S.N.H., Pr. 14: 58, 396-399, (1871).

34--1871b   [Some geologic features in the vicinity of Salem, Mass.] B.S.N.H., Pr. 14: 91-92, (1871).

35--1871c   [Discussion of N. S. Shaler's "On the causes which have led to the production of Cape Hatteras."] B.S.N.H., Pr. 14: 110-121, (1871). See also next title.

36--1871d   [On Atlantic shore changes.] B.S.N.H., Pr. 14: 122-123, (1871); Am. Nat. 5: 182-183, (1871). See also preceding title.

37--1871e   On the geological survey of Essex County [Massachusetts]. Essex Inst., B. 3: 49-53, (1871).

38--1871f   [Affinities of the Brachiopoda and Polyzoa.] B.S.N.H., Pr. 14: 136-137, (1871).

39--1871g   [On natural selection.] B.S.N.H., Pr. 14: 146-148, (1871).

40--1871h   Catalogue of the ornithological collection of the Boston Society of Natural History. B.S.N.H., Pr. 14: 237-253, (1871).

41--1871i   [Review of J. W. Dawson's "Modern ideas of derivation."] Am. Nat. 4: 230-237, (1871).

42--1871j   Position of the brachiopods. Am. Nat. 5: 310-312, (1871).

43--1872   Fossil cephalopods of the Museum of Comparative Zoology: Embryology. Harvard Coll., Mus. Comp. Zool., B. 3: 59-111, il., (1872). See also 8--1867.

44--1872a   On the embryology and development of the shells of ammonoids and nautiloids. B.S.N.H., Pr. 14: 396-399, (1872).

45--1872b   The non-reversionary series of the Liperoceratidae, and remarks upon the series of the allied family Dactyloidae. B.S.N.H., Pr. 15: 4-21, (1872); Pr. 17: 29-33, (1874). See also 49--1874c.

46--1874   Evolution of the Arietidae. B.S.N.H., Pr. 16: 166-170, (1874).

47--1874a   Genetic relations of the Angulatide. B.S.N.H., Pr. 17: 15-23, (1874).

48--1874b   Appendix to communications on reversions among ammonites. B.S.N.H., Pr. 17: 23-28, (1874).

49--1874c   Appendix to communications on "The non-reversionary series of the Liperoceratidae." B.S.N.H., Pr. 17: 29-33, (1874). See also 45--1872b.

50--1874d Note on Aptenodytes patagonica Forst. B.S.N.H., Pr. 17: 94, (1874).

51--1874e Ascidian larvae. B.S.N.H., Pr. 17: 130-131, (1874).

52--1874f On the hollow-fibred horny sponges. B.S.N.H., Pr. 17: 204-205, (1874).

53--1875 Remarks on two new genera of ammonites, Agassiceras and Oxynoticeras. B.S.N.H., Pr. 17: 225-235, (1875).

54--1875a Abstract of a memoir on the "Biological relations of the Jurassic ammonites." B.S.N.H., Pr. 17: 236-241, (1875); Am. J. Sci. (3)10: 344-349, (1875).

55--1875b The Jurassic and Cretaceous ammonites collected in South America by Professor James Orton, with an appendix upon the Cretaceous ammonites of Professor Hartt's collection. B.S.N.H., Pr. 17: 365-372, (1875).

56--1876 Review of E. Mojsisovic's "Das Gebirge um Hallstatt." Am. J. Sci. (3)11: 412-413, (1876).

57--1876a [On the porphyries of Marblehead, Mass.] B.S.N.H., Pr. 18: 220-224, (1876).

58--1876b The genetic relations of Stephanoceras. B.S.N.H., Pr. 18: 360-400, (1876).

59--1876c Embryology of sponges. B.S.N.H., Pr. 19: 12-17, (1876).

60--1877 [Resolutions on F. B. Meek.] B.S.N.H., Pr. 19: 76-77, (1877).

61--1877a Revision of the North American Poriferae, with remarks upon foreign species: Pts. 1 and 2. B.S.N.H., Mem. 2: 399-408, 481-554, (1877).

62--1878 Ammonitoid forms in the Upper Trias of Nevada. U.S. Geol. Surv., Rept. 40th Parallel Survey 4/1: 105, (1878).

63--1878a Guides for Science Teaching No. 1. About pebbles. Boston: Boston Soc. Nat. Hist., 25 p., (1878). Boston: Ginn, Heath & Co., (1879) and (1883); Boston: D. C. Heath & Co., (1898).

64--1878b On a new species of sponge. Ac. Nat. Sci. Phila., Pr. 1878: 163-164, (1878).

65--1879 Guides for Science Teaching No. 3. Commercial and other sponges. Boston: Ginn, Heath & Co., 43 p., il., (1879).

66--1879a Guides for Science Teaching No. 5. Common hydroids, corals, and echinoderms. Boston: Ginn, Heath & Co., 32 p., il., (1879); Boston: D. C. Heath & Co., (1893).

67--1879b Cephalopoda (remarks on the genus Meekoceras and its species). U.S. Geol. Geogr. Surv. 5: 111-116, (1879); U.S. Geol. Geogr. Surv., 12th Ann. Rept.: 112-117, (1883).

68--1880 The genesis of the Tertiary species of Planorbis at Steinheim. B.S.N.H., Anniv. Mem. (1880); Am. Assoc., Pr. 29: 527-550, (1880). See also 71--1882.

69--1880a Moulting of the lobster Homarus americanus. B.S.N.H., Pr. 21: 83-90, (1880).

70--1880b Guides for Science Teaching No. 6. The oyster, clam, and other common mollusks. Boston: D. C. Heath & Co., 65 p., il., (1880), (1888), and (1894).

71--1882 Transformation of Planorbis at Steinheim, with remarks on the effects of gravity upon the forms of shells and animals [abst.]. Am. Nat. 16: 441-453, (1882). See also 68--1880.

72--1882a Guides for Science Teaching No. 7. Worms and Crustacea. Boston: Ginn, Heath & Co., 68 p., il., (1882); D. C. Heath & Co., (1889).

73--1883   Joachim Barrande. Science 2: 699-701, 727-729, (1883).
74--1883a   Genera of fossil cephalopods. B.S.N.H., Pr. 22: 253-338, (1883).
75--1884   The evolution of the Cephalopoda. Science 3: 122-127, 145-149, (1884).
76--1884a   Larval theory of the origin of cellular tissue. B.S.N.H., Pr. 23: 45-163 (1884); Am. Nat. 18: 460-464, (1884). See also 86--1885d and 88--1886.
77--1884b   The primitive concoryphean. Science 4: 351, (1884).
78--1884c   Fossil Cephalopoda in the Museum of Comparative Zoology (Preliminary Abstract). Am. Assoc., Pr. 32: 323-361, (1884).
79--1884d   The protoconch of Cephalopoda. Am. Nat. 18: 919-920, (1884).
80--1884e   On the nautiloid genus Enclimatoceras Hyatt, and a description of the type species. U.S. Geol. Surv., B. 4: 16-17, il., (1884); B.S.N.H., Pr. 23: 270, (1884).
81--1884f   Remarks on the statoblasts of Chalinula arbuscula. B.S.N.H., Pr. 23: 214, (1884).
82--1885   Tentaculites. Nat. Hist. Soc. N. Brunswick 1/4: 102, (1885).
83--1885a   [On Hyolithes from the St. John group.] Nat. Hist. Soc. N. Brunswick 1/4: 102, (1885); Can. Rec. Sci. 1: 141, (1885).
84--1885b   Cruise of the Arethusa [to Newfoundland]. Science 6: 384-386, (1885); B.S.N.H., Pr. 23: 316-319, (1886).
85--1885c   Structure of the siphon in the Endoceratidae [abst.]. Am. Assoc., Pr. 33: 490-491, (1885).
86--1885d   Larval theory of the origin of tissue [abst.]. Am. Assoc., Pr. 33: 491-492, (1885). See also 76--1884a.
87--1885e   Structure and affinities of Beatricea [abst.]. Am. Assoc., Pr. 33: 492, (1885).
88--1886   Abstract of larval theory of origin of tissue. Am. J. Sci. (3)3: 332-347, (1886). See also 76--1884a.
89--1886a   [On a supposed fossil from Lake Superior.] B.S.N.H., Pr. 23: 210-211, (1886).
90--1886b   [Expedition to Newfoundland.] B.S.N.H., Pr. 23: 316-319, (1886). See also 84--1885b.
91--1886c   Notes on C. vancouverensis Whiteaves. Geol. Surv. Canada, n.s. 2(3): 111, (1886).
92--1886d   [Remarks on the life of Miss Lucretia Crocker.] B.S.N.H., Pr. 23: 330-333, (1886).
93--1887   On primitive forms of cephalopods. Am. Nat. 21: 64-66, (1887).
94--1888   Values in classification of the stages in growth and decline with propositions for a new nomenclature [abst.]. B.S.N.H., Pr. 23: 396-407, (1888); Am. Nat. 22: 872-884, (1888).
95--1888a   Sketch of the life and services to science of Prof. Spencer F. Baird. B.S.N.H., Pr. 23: 558-565, (1888).
96--1888b   Evolution of the faunas of the Lower Lias. B.S.N.H., Pr. 24: 17-30, (1888).
97--1888c   [The Taconic question.] Am. Geol. 2: 137, (1888).
98--1889   [Notes on Popanoceras, Acrochordiceras, Trachyceras, Arniotites, and Badiotites] in Whiteaves, J. F., On some fossils from the Triassic rocks of British Columbia. Contr. Can. Pal. I/2: 127-149, (1889).

99--1889a   Genesis of the Aretidae. Harvard Coll., Mus. Comp. Zool. Mem. 16/3: xi + 238 p., il., (1889); Smiths. Contr. Knowledge 26/673: xi + 238 p., il., (1889); Reviewed by J. Marcou, Am. Geol. 6: 128-133, (1890).

100--1889b   Modes of evolution in fossil shells. N.Y. Ac. Sci., Tr. 8: 114-115, (1889).

101--1890   (and Arms, Jennie M.) Guides for Science Teaching No. 8. Insecta. Boston: Boston Soc. Nat. Hist., 23 p., 300 p., il., (1890); Boston: D. C. Heath & Co., 300 p., il., (1904).

102--1891   Carboniferous cephalopods. Texas Geol. Surv., Ann. Rept. 2: 327-356, il., (1891). See also 109--1893a.

103--1891a   Report of Mr. Alpheus Hyatt (Division of Lower Mesozoic Paleontology), p. 97-100, in U.S. Geol. Surv., 11th Ann. Rept. 1889-1890, Pt. I. Geology, 757 p., il., (1891).

104--1891b   Report of Prof. Alpheus Hyatt (Division of Lower Mesozoic Paleontology), p. 111-112, in U.S. Geol. Surv., 12th Ann. Rept. 1890-1891, Pt. I. Geology, 675 p., il., (1891).

105--1892   Report of Mr. Alpheus Hyatt (Division of Lower Mesozoic Paleontology), p. 142-143, in U.S. Geol. Surv., 13th Ann. Rept. 1891-1892, Pt. I. Rept. Director, 240 p., (1892).

106--1892a   Remarks on the Pinnidae. B.S.N.H., Pr. 25: 335-346, (1892).

107--1892b   Jura and Trias at Taylorville, California. Geol. Soc. Am., B. 3: 395-412, (1892). Abst., Am. Nat. 27: 470-471, (1892); Am. Geol. 10: 183, (1892); Am. J. Sci. (3)44: 330, (1893).

108--1893   The fauna of Tucumcari. Am. Geol. 11: 281, (1893).

109--1893a   Carboniferous cephalopods. Texas Geol. Surv., Ann. Rept. 4/2: 377-474, il., (1893). See also 102--1891.

110--1893b   The terms of bioplasty. Zool. Anzeiger 16: 317-323, 325-331, (1893).

111--1893c   Bioplastology and the related branches of biologic research. B.S.N.H., Pr. 26: 59-125, (1893).

112--1893d   Phylogeny of an acquired characteristic. Am. Phil. Soc., Pr. 32: 349-647, il., (1893). See also next title.

113--1893e   Phylogeny of an acquired characteristic. Am. Nat. 27: 865-877, il., (1893). See also preceding title.

114--1893f   Bemerkungen zu Schulze's "System einer deskriptiven Terminologie." Biol. Centralblatt. Bd. 13: 504-511, (1893). See also next title.

115--1894   Remarks on Schulze's "System of Descriptive Terms." Am. Nat. 28: 369-379, (1894). See also preceding title.

116--1894a   Trias and Jura in the Western States. Geol. Soc. Am., B. 5: 395-434, (1894); Abst., Am. Geol. 13: 148, (1894); Am. J. Sci. (3)47: 142-143, (1894).

117--1895   Remarks on the genus Nanno, Clarke. Am. Geol. 16: 1-12, il., (1895).

118--1896   (and Arms, Jennie M.) The meaning of metamorphosis. Nat. Sci. 8: 395-403, (1896).

119--1896a   Terminology proposed for description of the shell in Pelecypoda [abst.]. Am. Geol. 16: 252-254, (1895); Am. Assoc., Pr. 44: 145-148, (1896).

120--1896b   Lost characteristics. Am. Nat. 30: 9-17, (1896).

121--1896c   Report on the Mesozoic fossils [from Alaska]. U.S. Geol. Surv., Ann. Rept. 17/1: 907-908, (1896).

122--1897   Cycle in the life of the individual (ontogeny) and in the evolution of its own group (phylogeny). Am. Ac. Arts Sci., Pr. 32: 209-224, (1897); Science n.s. 5: 161-171, (1897).

123--1898   Evolution and navigation of Hawaiian land shells [abst.]. A.A.A.S., Pr. 47: 357-358, (1898).

124--1898a A new classification of fossil cephalopods [abst.]. A.A.A.S., Pr. 47: 363-365, (1898); Science n.s. 8: 398, (1898).

125--1898b General Guide to the Museum of the Boston Society of Natural History. Boston: Boston Society of Natural History, 47 p., il., (1898).

126--1899   Some governing factors usually neglected in biological investigations. Marine Biol. Lab. [Woods Hole, Massachusetts] Lectures, p. 127-156, 1899.

127--1899a Jules Marcou. Am. Ac. Arts Sci., Pr. 34: 651-656, (1899).

128--1900   [Chapter on Cephalopoda, p. 583-689, in] Text-book of Palaeontology, by Karl von Zittel. London: Macmillan & Co., Ltd., xi + 839, il., (1900).

129--1903   Pseudoceratites of the Cretaceous, edited by T. W. Stanton. U.S. Geol. Surv., Mon. 44: 351 p., il., (1903).

130--1905   (and Smith, J. P.) The Triassic cephalopod genera of America. U.S. Geol. Surv., P.P. 40: 394 p., il., (1905).

Reports of Hyatt as Curator and Custodian at the Boston Society of Natural History

Alpheus Hyatt was Custodian of the Boston Society of Natural History from 1870 to 1881, and Curator from 1881 to 1902, the year he died. Following are the reports he made during these years (B.S.N.H., Pr. = Boston Society of Natural History, Proceedings):

[Report of the Custodian]	B.S.N.H., Pr. 14: 207-214,	(April 1871)
[Report of the Custodian]	B.S.N.H., Pr. 17: 1- 11,	(Sept. 1874)
[Report of the Custodian]	B.S.N.H., Pr. 18: 1- 14,	(Aug. 1875)
[Report of the Custodian]	B.S.N.H., Pr. 18: 332-344,	(April 1876)
[Report of the Custodian]	B.S.N.H., Pr. 19: 186-192,	(May 1877)
[Report of the Custodian]	B.S.N.H., Pr. 20: 1- 9,	(Dec. 1878)
[Report of the Custodian]	B.S.N.H., Pr. 20: 244-257,	(May 1879)
[Report of the Custodian]	B.S.N.H., Pr. 21: 1- 14,	(May 1880)
[Report]	B.S.N.H., Pr. 21: 175-186,	(May 1881)
[Report of the Curator]	B.S.N.H., Pr. 22: 1- 13,	(May 1882)
[Report of the Curator]	B.S.N.H., Pr. 22: 339-353,	(May 1883)
Report on the Museum	B.S.N.H., Pr. 24: 1- 11,	(July 1888)
Report on the Museum	B.S.N.H., Pr. 24: 243-254,	(May 1889)
Report on the Museum	B.S.N.H., Pr. 25: 1- 19,	(May 1890)
Report of the Curator	B.S.N.H., Pr. 25: 269-283,	(May 1891)
Report of the Curator	B.S.N.H., Pr. 25: 425-445,	(April 1892)
Report of the Curator	B.S.N.H., Pr. 26: 127-137,	(April 1893)
Report of the Curator	B.S.N.H., Pr. 26: 275-290,	(May 1894)
Report of the Curator	B.S.N.H., Pr. 26: 505-522,	(April 1895)
Report of the Curator	B.S.N.H., Pr. 27: 107-117,	(May 1896)
Report of the Curator	B.S.N.H., Pr. 28: 45- 52,	(May 1897)
Report of the Curator	B.S.N.H., Pr. 28: 275-289,	(May 1898)
Report of the Curator	B.S.N.H., Pr. 29: 1- 14,	(May 1899)
Report of the Curator	B.S.N.H., Pr. 29: 223-233,	(May 1900)
Report of the Curator	B.S.N.H., Pr. 29: 347-358,	(May 1901)

MEMORIALS AND BIOGRAPHICAL SKETCHES OF ALPHEUS HYATT (1838-1902)

NOTE:

Unless specifically indicated, no bibliography is included.

(1885) Tarr, R. S.--Sketch of Professor Alpheus Hyatt. Pop. Sci. Month. 28: 261-267, (1885).
    A brief but excellent biography.

(1899) Zirngiebel, G. F.--Teachers' School of Science. Pop. Sci. Month. 55: 451-465, 640-652, (1899).
    An excellent account of the famous "School," which was founded as a result of a suggestion of Alpheus Hyatt to John Cummings in 1870.

(1902) Boston Transcript, Jan. 16, 1902.

(1902) Beecher, C. E.--Obituary. Am. J. Sci. (4)13: 164, (1902).

(1902) Henshaw, S.--Alpheus Hyatt. Science n.s. 15: 300-302, (1902).

(1902) Dall, W. H.--Alpheus Hyatt. Pop. Sci. Month. 60: 439-441, port., (1902).

(1902) Minot, C. S. et al.--Memorial of Alpheus Hyatt. Boston Soc. Nat. Hist., Pr. 30/4: 413-433, (1902); printed in part in Tech. Rev. 4: 323-327, port., (1902).
    Contains comments by C. S. Minot, E. S. Morse, A. S. Packard, W. M. Warren, A. C. Boyden, and F. W. Putnam.

(1902) Crosby, W. O.--Memoir of Alpheus Hyatt. Geol. Soc. Am., B. 14: 504-512, port., (1902).
    This biography includes an almost complete list of Hyatt's publications.

(1903) Packard, A. S.--Alpheus Hyatt. Am. Ac. Arts Sci., Pr. 38: 715-727, (1902-1903).
    An excellent discussion of Hyatt's fundamental contributions to biology and paleontology.

(1903) Stanton, D. W.--Alpheus Hyatt, 1838-1902. Wash. Ac. Sci., Pr. 5: 389-391, (1903).

(1909) Brooks, W. K.--Biographical memoir of Alpheus Hyatt, 1838-1902. Nat. Ac. Sci., Biog. Mem. 6: 311-325, port., (1909).
    An excellent biography and bibliography, especially strong on Hyatt's contributions to biology.

(1911) Mayer, A. G.--Alpheus Hyatt, 1838-1902. Pop. Sci. Month. 78: 128-146, port. (1911).
    An excellent and sympathetic memorial by a close friend; one of the best of published accounts of Hyatt's life and character.

(1913) Jackson, R. T.--Alpheus Hyatt and his principles of research. Am. Nat. 47: 195-205, photograph of a bas-relief of Hyatt made by his daughter, Harriett (Mrs. Alfred G. Mayer), (1913); Abst., Geol. Soc. Am., B. 24: 105, (1913).
    An excellent description and evaluation of Hyatt's biological and paleontological researches.

(1928) Conklin, E. G.--Memorial of Alpheus Hyatt. Science n.s. 68: 291-292, (1928).

(1928) Welsh, L. W.--Ancestral Colonial Families: Genealogy of the Welsh and Hyatt Families of Maryland and Their Kin. Independence, Mo.: Lambert Moon Printing Co., (1928).

(1930) Creed, Capt. P. R. (ed.)--The Boston Society of Natural History, 1830-1930, 117 p., il., (1930). Boston: Updike, The Merrymount Press.
    Pt. I. Milestones (p. 1-64) includes an excellent history of the Society with brief references to numerous of its former officials, including Hyatt, p. 44-54.

(1932) HYATT, Alpheus, Dictionary of American Biography 9: 446-447, (1932).

(1944) Lillie, F. R.--The Woods Hole Marine Biological Laboratory. Chicago: Univ. Chicago Press, ix + 284 p., port., (1944).

[195?] Anonymous--Alpheus Hyatt. M.I.T. Biographies of Members of the Faculty and Officers of Administration, Vol. I, p. 12, port., [195?].

(1968) Zinn, D. J.--Early days at Woods Hole. Science 159: 484, (1968).
    Sets the record straight on the relation of Hyatt's seaside laboratory at Annisquam to the Marine Biological Laboratory at Woods Hole.

The Rogers Building of the Massachusetts Institute of Technology (background) and the building of the Boston Society of Natural History (foreground), the latter now occupied by Bonwit Teller, looking northwest from the intersection of Berkeley and Boylston Streets, Boston.
(M.I.T. Historical Collections)

Classes were held in the Institute's first building - it was not named the Rogers Building until 1883 - as early as 1866, while it was still being completed, and the last class was held in 1938 shortly before the building was demolished to make room for the massive structure subsequently built for the New England Mutual Life Insurance Company. In the later 1860s some classes were also held in the rooms of the adjacent Boston Society of Natural History, and Prof. Hyatt, who was also on the staff of the Society, arranged for Institute students to study the collections of minerals, rocks, and fossils in the Museum of the Society.

(3)
WILLIAM HARMON NILES*
(1838-1910)

WILLIAM HARMON NILES

MIT: 1871-1902

William H. Niles was appointed Professor of Physical Geology and Geography in 1871, soon after completing an undergraduate program that included study with Louis Agassiz, Josiah P. Cook and Asa Grey at Harvard, and with James D. Dana, O. C. Marsh and G. J. Brush at Yale. He was awarded a B.S. degree by Harvard in 1866, a Ph.B. degree by Yale in 1867, an A.M. degree by Wesleyan in 1870, and an honorary LL.D. degree by Temple in 1903.

Known widely as an excellent public lecturer, he divided his time in his early years at M.I.T. (1871-1878) between teaching at the Institute and public lecturing throughout Massachusetts. In some years he delivered as many as a hundred public lectures on physical geography and glaciation. In 1878, when T. Sterry Hunt resigned and relinquished direction of the instructional program in geology, Niles was appointed Professor of Geology and Geography and given the responsibility Hunt had had. Twelve years later, in 1890, when Course XII, Geology, was established and a Bachelor of Science degree in Geology authorized, he became Head of the Course, and continued in this capacity until retirement in 1902. Aided by his capable student assistant and later faculty colleague, William Otis Crosby (VII S.B. 1876), he accumulated a collection of minerals, rocks, ores, fossils and maps, totalling more than 30,000 specimens, that became the basic laboratory materials for the numerous geology subjects developed during his thirty years of service to the Institute.

Outside M.I.T. Niles was a leading geologist in the Boston area. He was Professor of Geology at Boston University from 1872 to 1902, and taught at Wellesley College from 1882 to 1908, occupying the chair of geology, and heading the Department of Geology, from 1891 to retirement in 1908. In those days it was quite possible for one professor to serve more than one institution, because most of the work to be done was lecturing and overseeing laboratory study and field excursions.

---

* For the biographical part of this sketch I have drawn principally on the following sources: "William Harmon Niles," by George H. Barton (III S.B. 1880), Tech. Rev. 4: 417-424, port., (1902), and "Memoir of William Harmon Niles," by George H. Barton, Geol. Soc. Am., B. 22: 8-14, (1911). For the bibliographical part I have compiled the list of publications from two sources: "Bibliography of W. H. Niles," by George H. Barton, Geol. Soc. Am., B. 23: 34-36, port., (1912), and "Geologic Literature on North America, 1785-1918," by John M. Nickles, U. S. Geol. Surv., B. 746 (Pt. I. Bibliography), 1167 p., (1922) and B. 747 (Pt. II. Index), 658 p., (1924). Additional sources are cited in the text, and there is a brief biographical sketch on page 26 of Volume I of "Massachusetts Institute of Technology. Biographies of Members of the Faculty and Officers of Administration," [195?], which is to be found in the M.I.T. Institute Archives.

Even with his responsibilities at M.I.T., Boston University and Wellesley College, Niles still found time to participate actively in many of the scientific societies of the day and to deliver as many as a hundred public lectures in a single year. For five years he was president of the Boston Society of Natural History, three times he was president of the Appalachian Mountain Club, and he served the New England Meteorological Society in a similar capacity.

In the last third of the 19th Century, when the young M.I.T. was growing into a great technological institution, when geology departments came into existence at Boston University and Wellesley College, as well as at M.I.T., and when the general public had to depend on the public lecture to learn what was going on in the academic world, Niles was a towering figure in the Boston area. His fame rested not on publications, though he did publish several important articles; rather, it rested on his outstanding ability as a lecturer on a public platform and a teacher in the class room. For more than thirty years he served well the Institute, the Boston academic community, and the sciences of the earth.

William Harmon Niles, one of the three earliest professors of geology at M.I.T., was born in Northampton, Massachusetts, on 18 May 1838, and died at the Copley Square Hotel in Boston on 12 September 1910, eight years after retiring from M.I.T. as an emeritus professor.

He inherited a remarkably retentive memory from his father, Rev. Asa Niles; and his early interest in nature and later felicity in speaking reflected the intellectual gifts of his mother, Mary A. (Marcy) Niles. When he was less than four years old, his parents moved from Northampton to Worthington, Massachusetts, and here he received his early education in a district school. During his boyhood, his interest in minerals and plants led him into the countryside, and by the time he was sixteen he had a good collection of minerals from Worthington, as well as from four neighboring towns. These specimens he arranged and labelled, thus foreshadowing the lifelong habit of organization that was so evident in his public lectures and formal teaching later on.

Being a member of the family of a New England clergyman meant living in a household with limited means; as a consequence, young William had to work his way through secondary school and college. At the age of seventeen he started teaching in Worthington, and after four successive winters, he moved to North Blanford for two terms, and then to North Becket for one term. During the summers between terms he worked regularly on his father's farm. Walking five miles across the hills of western Massachusetts to his first school, and keeping the attention of students older and bigger than he was, left a vivid impression that he carried through life and no doubt helped him to mature rather rapidly.

His first formal instruction in science did not come until he was twenty and had entered the Wesleyan Academy at Wilbraham, Massachusetts.

Lack of means, however, prevented him from attending consecutive terms, but he did get some instruction and, what was much more important, he got encouragement from his mother's brother, Dr. Oliver Marcy, who later went to Northwestern University. On Marcy's advice he travelled to Cambridge to study with famed Louis Agassiz, then at Harvard's Museum of Comparative Zoology. Although he met the strictest requirements of that famous teacher, whose rather unorthodox methods quickly eliminated all but the best and most determined students, he did not follow zoology as a career. After devoting considerable time to the study of modern corals and fossil mollusks and crinoids, the last named group becoming the subject of his thesis later on, he turned to geology, and especially physical geography, and these became his favorite sciences.

While at the Museum of Comparative Zoology Niles was one of a group of students, attracted to Harvard by the fame of Agassiz, who were later to become leading teachers and investigators--J. A. Allan, A. S. Bickmore, C. F. Hartt, Alpheus Hyatt, Horace Mann, A. S. Packard, F. W. Putnam, O. H. St.John, S. H. Scudder and A. E. Verrill.

After receiving his B.S. degree in zoology from Harvard in 1866, he spent a year and a half in Yale's Sheffield Scientific School, at the end of which he was awarded a Ph.B. degree, in 1867. Three years later he was given an A.M. degree by Wesleyan University, and many years later, following retirement from M.I.T., he was given an LL.D. degree by Temple University.

While a student at Yale, he studied mineralogy with G. F. Brush, geology with James D. Dana and O. C. Marsh, and physical geography with Daniel C. Gilman. These teachers obviously aroused the love of nature and interest in mineralogy that he had had from boyhood, and it was easy for him to turn from zoology to geology.

Even before leaving Cambridge for New Haven he had been appointed instructor and lecturer in natural science at the State Teachers' Institutes of Massachusetts. Over a period of ten years he lectured in every part of the Commonwealth, and became widely and popularly known for his presentations of geological and geographical lectures, often illustrated by excellent lantern slides.

When President William Barton Rogers was seeking a professor of geology for the young M.I.T., Niles was one of the best known lecturers on the subject in the Boston area and was an obvious candidate. He had had sound training in biology and geology, had taught in several private schools, and was an experienced and popular lecturer. Accordingly, upon Rogers' recommendation, the M.I.T. Corporation appointed him

> "Professor of Physical Geology and Geography, at a salary of $500, without a seat on the Faculty, on June 14, 1871." (Minutes of the Corporation, Vol. II.)

For eight years, 1871-1878, Niles confined his teaching to the second half of each school year in order that he could continue giving public lectures during the fall term. This part-time arrangement no doubt explains the wording of the minute quoted above--$1000 a year was the customary salary for a beginning professor, and only full-time professors held a seat on the faculty.

Niles travelled extensively in Europe, reporting on some of his observations in short articles in the Proceedings of the Boston Society of Natural History (See his Bibliography farther on), and used the knowledge gained from firsthand observation in his teaching at the Institute and in his many public lectures, which are reported to have

> "... yielded an important part of his resources for travel and extended geographical study." (Barton 1911, p. 11.)

Niles became such a well-known public lecturer and so much in demand during his earlier years at the Institute, i.e. 1871-1878, that he spoke as many as fifty to a hundred times in a single season. In addition to many special public lectures, he gave lecture courses under the sponsorship of a number of Boston scientific organizations. For example, he gave three 12-lecture courses at the Lowell Institute in Boston, on "Geological history, ancient and modern," "The atmosphere and its phenomena," and "Physical geography of the land," respectively. Other courses in geological and geographical subjects were given for the Boston Society of Natural History, the Teachers' School of Science and the Appalachian Mountain Club.

In 1878, when T. Sterry Hunt retired from the chair of geology at the Institute, Niles was appointed Professor of Geology and Geography and given the responsibility of directing the instructional program in geology. It should be noted that there was neither a degree program nor a degree in geology at this time; these would not come until 1890. When Course XII, Geology, was established in 1890, however, he became the Head of the Department of Geology, and continued in that position until he retired in 1902, after 31 years of service to the Institute.

When Niles assumed responsibility for instruction in geology in 1878, William Otis Crosby (VII S.B. 1876) was appointed assistant in geology to help him in laboratory and field work. At this early date there were no teaching collections of rocks, minerals, ores, fossils, or maps, and very little in the way of instruments or equipment at the Institute. Professor and assistant, acting in concert, with Niles looking after paleontology and Crosby after geology, started to gather specimens wherever they could find them. By 1902 they had assembled a collection of more than 30,000 specimens that could be used in teaching mineralogy, petrology, economic geology, structural geology, and paleontology. These specimens had been carefully segregated into natural groups for teaching purposes, and most of the specimens had been labelled by the time Niles retired as Department Head. This joint effort is the more impressive when it is recalled that the two collectors did their work before the day of the automobile and Ward's Catalog of Minerals, Rocks and Fossils!

Niles is credited by his biographer, George H. Barton (III S.B. 1880), as having been the first person to suggest that General Francis A. Walker be considered as the successor to President John D. Runkle. He is reported to have gone, on his own accord, but with the permission of the authorities, to New Haven for a personal interview with Walker that opened the way for the negotiations that resulted in the latter's becoming M.I.T.'s third President on 1 November 1881.

In addition to being professor, and later on department head, at M.I.T., Niles simultaneously held other important positions in the Boston community. He was Professor of Geology at Boston University from its first graduating class in 1872 to his year of retirement from M.I.T. in 1902. He taught at Wellesley College for more than twenty-five years, 1882-1908, and his services to that institution were acknowledged in the [Wellesley] College News 10/6, (1908) at the time of his retirement from that institution, in June 1908, as follows:

> "William Harmon Niles, B.S., Ph.B., M.A., LL.D., joined the faculty of Wellesley College in 1882 as lecturer in geology. Classes at once responded to his skilled touch. Interest so increased and work so strengthened that in 1888 the one course broadened into a department of which Doctor Niles was made the head. In 1891 Doctor Niles accepted the chair of geology, which was then established, and he has remained in full charge of the work, now expanded into four courses ...
>
> "The services of this esteemed officer have not been confined to class-room duties merely. Professor Niles came to Wellesley in a day of beginnings. His standing among scientists, the weight of his judgment, the intimacy of his connection with a great technological school, all lent themselves effectively to the work of framing suitable laws of growth for the young college. In all its succeeding history the college has enjoyed from Professor Niles sympathy, support, and counsel, which have been highly appreciated."

Outside the academic halls he was president of the Boston Society of Natural History for five years (1892-1897); three times the president of the Appalachian Mountain Club; president of the New England Meteorological Society; and president of the Lawrence Scientific School Alumni

Association. He was also a fellow of the American Academy of Arts and Sciences, of the Geological Society of America, and of the American Association for the Advancement of Science.

Niles was primarily a lecturer and teacher, and in both he excelled as attested by both published and unpublished statements of those who heard his public lectures and attended his regularly scheduled classes at M.I.T. or elsewhere. A few typical examples will serve to bring out his personal appearance, kind and generous personality, concern for students, and devotion to nature and the out-of-doors.

Howard L. Coburn (II S.B. [1887] 1888), in reminiscing about "When we were freshmen," (Tech. Rev. 16: 217-218, 1914), wrote as follows regarding Prof. Niles. [Sept. 1883. September examinations for entrance to M.I.T.]:

> "Well, somehow or other I got through them; have always credited the examiners with much kindness of heart for I certainly do not think my papers were really passable. God bless them, I say, especially dear old Prof. Niles who let me through in geography, a subject I never knew anything about. I remember that years afterward he told me that in all probability he did not look at my paper."

The late Carle R. Hayward (III S.B. 1904), longtime Professor of Metallurgy in Course III, jotted down for me the following recollection of Niles as a teacher (personal letter dated November 27, 1957):

> "He was a jolly rotund individual with a white beard. He used numerous excellent slides in his lectures on general geology. He was especially interesting in his discussion of glaciers with accounts of personal experiences. His lectures stimulated my interest in physical geography and surface changes brought about by glaciers, streams and internal forces. My recollection is that practically all members of my section felt the same."

Another of his students, reminiscing in his adult years, wrote as follows [F. H. Newell (III S.B. 1885), William Harmon Niles. Passing of the first Professor Emeritus of the Institute--An appreciation by a former pupil. Tech. Rev. 12/4: 425-427, (Oct. 1910)]:

> "Every institution of learning, as well as every large enterprise, needs a man of the type of Professor Niles, with broad sympathies, genial bearing, quick to perceive the work to be done and willing to do the little things as well as the big, and notably who possesses as Professor Niles did, that large common sense which solves difficulties and smooths out the paths for others.
>
> "In looking back over a period of many years of personal acquaintance, ripening into friendship, we may not recall the title of a single article or scientific paper by Professor Niles or any particularly

striking or original work of his, yet the memory retains the impression of a long series of acts of kindness, of sound advice, of cheering yet direct criticism founded upon a full knowledge of many subjects, all of which in the aggregate has been of indescribable value to the students who came in contact with him."

As mentioned in a preceding paragraph, Niles was awarded an A.M. degree by Wesleyan University (Connecticut) in 1870 and an honorary LL.D. by Temple University in 1903. A 4000' peak in the Aleutian Islands* was named Mount Niles by T. A. Jaggar in 1907, as reported in his "Journal of the Technology Expedition to the Aleutian Islands, 1907" (Tech. Rev. 10: 1-37, 1908). The entry in the "Journal" reads (p. 22):

"July 27.--We have named the white dome Mount Niles in honor of the distinguished Professor Emeritus of the Institute, ..."

and there is a picture of the peak opposite page 22.

On 31 December 1868 Niles married Helen M. Plympton, the daughter of a prominent Cambridge physician, and for forty years she was his constant and devoted companion. Her death left him deeply saddened and alone, as they had no children, and his own end came soon after, on 12 September 1910.

BIBLIOGRAPHY OF WILLIAM HARMON NILES

Symbols and abbreviations used in the following references are explained on pages 91 - 98; in general, abstracts are listed separately as well as with the references to the complete article. This bibliography is believed to include all of the more important publications of William Harmon Niles (1838-1910).

1--1865		[On the systematic position of Pasceolus.] Boston Soc. Nat. Hist., Pr. 10: 19-20, (1865).
2--1865a		[Remarks on the relations between the vegetation and geology in the hills of western Massachusetts.] Boston Soc. Nat. Hist., Pr. 10: 49-50, (1865).
3--1866		(and Wachsmuth, C.) Evidence of two distinct geological formations in the Burlington limestone. Am. J. Sci. (2)42: 95-99, (1866).
4--1866a		[On the subdivision of the Burlington limestone of Iowa.] Boston Soc. Nat. Hist., Pr. 11: 6-7, (1866).
5--1869		[On the occurrence of shells of existing species in a boring at Fort Warren, Boston Harbor.] Boston Soc. Nat. Hist., Pr. 12: 244, 364, (1869).

---

* The 4000-foot peak to which Jaggar gave the name Mount Niles is located at the east end of Korovinski Bay, on Atka Island, $52°+$ N. Lat. and $170°+$ W. Long.

6--1870   [Discussion of] Note on the glacial moraines of the Charles River Valley near Watertown [Mass.] by N. S. Shaler. Boston Soc. Nat. Hist., Pr. 13: 277-279, (1870).

7--1871   On the physical features of Massachusetts. Boston Soc. Nat. Hist., Pr. 13: 414-415, (1871).

8--1871a   Some interesting phenomena observed in quarrying. Boston Soc. Nat. Hist., Pr. 14: 80-87, (1871); Pr. 16: 41-43, (1873).

9--1871b   [Discussion of] On the causes which have led to the production of Cape Hatteras by N. S. Shaler. Boston Soc. Nat. Hist., Pr. 14: 110-121, (1871). (Also see Am. Nat. 5: 178-183, 1871.)

10--1871c   [On the conglomerates of Montague and Brighton, Mass. with discussion on conglomerates by Charles Pickering, W. T. Brigham, and C. T. Jackson.] Boston Soc. Nat. Hist., Pr. 14: 128-129, (1871).

11--1872   Peculiar phenomena observed in quarrying. Am. J. Sci. (3)3: 222-223, (1872).

12--1872a   [Metamorphism of pebbles in conglomerate rocks at Chestnut Hill Reservoir, Boston, Mass. with discussion by N. S. Shaler and C. T. Jackson.] Boston Soc. Nat. Hist., Pr. 15: 1-2, (1872).

13--1873   Some remarks upon the agency of glaciers in the excavation of valleys and lake basins. Boston Soc. Nat. Hist., Pr. 15: 378-381, (1873).

14--1874   On some expansions, movements, and fractures of rocks, observed at Monson, Mass. Am. Assoc., Pr. 22/2: 156-163, (1874).

15--1875   The physical features of the State of Massachusetts (with discussion by T. S. Hunt). Boston Soc. Nat. Hist., Pr. 17: 507-508, (1875).

16--1876   The geological agency of lateral pressure exhibited by certain movements of rocks. Boston Soc. Nat. Hist., Pr. 18: 272-284, (1876).

17--1878   Upon the occurrence of zones of different physical features upon the slopes of mountains. Boston Soc. Nat. Hist., Pr. 19: 324-330, (1878).

18--1878a   Upon the relative agency of glaciers and subglacial streams in the erosion of valleys. Boston Soc. Nat. Hist., Pr. 19: 330-336, (1878); Am. J. Sci. (3)16: 366-370, (1878).

19--1884   On the cause of turns in lava streams. Boston Soc. Nat. Hist., Pr. 22: 490, (1884).

20--1894   A geological study of Lake Mohonk and Lake Minnewaska, N.Y. [abst.]. Am. Geol. 13: 211, (1894).

21--1903   Review of "Handbook of Climatology: Part I. General Climatology" translated by R. DeC. Ward from J. Hann's "Handbuch der Klimatology." New York: The Macmillan Co., (1903), in Tech. Rev. 5: 414-415, (1903).

22--1907   Address at the celebration of the One Hundredth Anniversary of the birth of Louis Agassiz, Sanders Theatre [Harvard University], May 27, 1907. Cambridge Hist. Soc., Pr. II: 92-98, (1907).

----

(4)
THOMAS STERRY HUNT
(1826-1892)

MIT: 1872-1878

THOMAS STERRY HUNT

Thomas Sterry Hunt, whom many would consider the "Father of North American Geochemistry," came to M.I.T. in 1872 following a brilliant career of 25 years with the Geological Survey of Canada. Known widely in both chemical and geological circles by virtue of his speaking and writing activities, and for his numerous original ideas involving chemical aspects of earth materials, he was invited by the M.I.T. Corporation to join Alpheus Hyatt and Wm. H. Niles in the Institute's program of geological instruction. Hyatt offered the work in paleontology, Niles lectured in physical and historical geology, and Hunt would assume responsibility for mineralogy and petrography. Although appointed Professor of Geology in January 1872, he did not enter upon his duties until October, at which time he not only commenced his lectures but also assumed responsibility for directing the instructional program in geology, there being no Department or Course of Geology at the time, and carried on this administrative chore until he left on a year's leave of absence starting on 1 October 1878. Early in 1879 he resigned from his M.I.T. appointment. He did not again take on either an academic position or employment in industry or government, but he did continue to do research and to publish as vigorously as ever until illness brought an end to his activities shortly before his death in 1892. His original ideas and concepts, in many cases provocative and controversial, were set forth boldly and vigorously in five books and more than 360 articles, and gained for him an international renown that was acknowledged by his peers in the form of elected positions and honorary memberships in professional organizations, honorary degrees, and other recognitions. Those of his contemporary colleagues who knew him and his work well, and were competent to judge, considered him one of the leading American scientists of his time.

BIRTH, ANCESTRY, AND EARLY YEARS (1826-1845)

Thomas Sterry Hunt,* who preferred to be known as T. Sterry Hunt, was born in Norwich, Connecticut on 5 September 1826 and died of a heart attack in New York City on 12 February 1892, following almost a decade

---

* I have drawn freely on several of the biographies of Hunt that are listed farther on at the end of his Bibliography, particularly those of Douglas, Dawson, and Crosby. These latter are especially informative about the ancestry and early life of our subject.

of failing health. He was the oldest son among the six children of Peleg and Jane Elizabeth (Sterry) Hunt, both of whom were descendants of famous Puritan ancestors. On his father's side, his forebears and their descendants have left an impressive record in art and letters, and one of the earliest of his American ancestors, William Hunt, was one of the founders of Concord, Massachusetts in 1635. The Sterry family, on his mother's side, was a famous one which traced its ancestry back to Puritan England in the time of the Commonwealth, when several members with the family name were renowned preachers. One such, Peter Sterry, was Chaplain to Oliver Cromwell; another, Thomas Sterry, wrote the well-known <u>A Riot Among the Bishops; or, A Terrible Tempest in the Sea of Canterbury</u>. About 1753, one branch of the Sterry Family, consisting of three brothers--Roger, Robert, and Cyprian--and a sister, came to America and settled in Providence, Rhode Island. Two of Roger's sons, John and Consider, edited and published <u>The True Republican</u>, a leading organ of the old Jeffersonian party, and wrote textbooks of arithmetic and algebra that were well-known in New England in the 1790s. Consider's daughter, Jane Elizabeth Sterry, was married to Peleg Hunt in 1823, as mentioned earlier, and our subject was their eldest son. With such famous maternal ancestors it is easy to understand why Hunt preferred to be known as T. <u>Sterry</u>.

    The Hunt family moved from Norwich to Poughkeepsie on the Hudson when Sterry was still a child. There the father died a few years later (1838), and the mother, left in rather straitened circumstances, decided to move her family of six young children back to their former home in Norwich, Connecticut. Sterry, the oldest son, was only twelve at the time, and after attending the local grammar school for a brief period had to find employment to help support his widowed mother and her family. He first worked in a printing office, then successively in an apothecary's shop, a bookstore, and later in a country grocery store in the village of Greenville near Norwich. These jobs didn't last long, however, six months at the most in each, but the last one was different. As clerk in the grocery store that did not have much business and whose owner was not too exacting, young Sterry had his first opportunity to indulge his interest in science. Intending to study medicine, a career then popularly followed by young men who had an interest in things scientific, he kept a skeleton under the counter and assembled home-made apparatus for chemical experimentation. Encouraged by the local physicians, who loaned him books, Sterry literally educated himself, but the knowledge he gained from them was not enough. Although a brilliant learner, he must also strike out on his own. This he did by investigating the properties of hyriodic acid (hydrogen iodide), anticipating to some extent the work of Deville and, according to popular tradition in Norwich, by improving the dyestuffs and methods of dyeing then used in the local mills.

THE YALE YEARS (1845-1847)

At eighteen, and as yet without formal contact with science, Hunt went to New Haven and obtained work as a reporter on a New York paper. This was 1845, and the Sixth Annual Meeting of the Association of American Geologists and Naturalists, the progenitor of the present American Association for the Advancement of Science, was being held in New Haven. In reporting on these meetings to his newspaper he got his first formal introduction to the world of science and was unanimously elected a member of the Association.

While in New Haven, Sterry visited the elder Benjamin Silliman, whom he had met earlier in Norwich after one of the professor's lectures, and so impressed the famous scientist by his proficiency in chemistry and mineralogy that Silliman secured his admission to the Scientific School of Yale University. Soon thereafter he became the paid assistant of the younger Silliman, Prof. Benjamin Silliman, Jr., who at the time was making an extended series of water-analyses. For an hour or two each day Hunt assisted Silliman in connection with his lectures; the rest of the time he devoted to analytical work in the laboratory. Now he had found work to his liking, had been accepted into the Silliman household and laboratory, and no longer had to struggle for a livelihood. His career as a geochemist was launched.

Adams (1932, p. 208) in his biography of Hunt, quotes the following about these Yale days from Douglas (1898, p. 7):

> "Writing to a friend in 1845 he [Hunt] says, 'I have seated myself in the laboratory with the flasks by my side so as to work and write at the same time ... I have free access to Professor Silliman's cabinet and a key to unlock all the cases ... I am boarding in a club of students at $1.25 a week. We have little or no meat. I do not like this very well, but it is cheaper, though I think I will board myself after a while. The room I expected to have has been occupied, as I was uncertain whether I was coming and so I have taken up my lodgings in the loft of the laboratory building and am so quite at home with chemical apparatus and preparations all around, [but] 'they are congenial spirits,' as Mr. Silliman remarked when he showed me the room.'"

During his Yale years, 1845-1847, Hunt quickly demonstrated his abilities as a chemical analyst, imaginative investigator, indefatigable worker, and accomplished author. During this period he contributed more than twenty short articles and notes (See the Bibliography later on) to Silliman's Journal (now the American Journal of Science), founded by the elder Silliman in 1818, and one of his very early articles, "A Description and Analysis of a New Mineral Species Containing Titanium with Some

Remarks on the Constitution of Tellurium Minerals," foreshowed the fields of science, the geochemistry of earth materials, to which he would devote a long lifetime of assiduous thought and effort. His considerable knowledge of chemistry, even at this early stage when he was not yet twenty and essentially without formal college training in the subject, is attested by the fact that he wrote the part dealing with organic chemistry in Silliman's First Principles of Chemistry (1846). In the preface to the first edition of his book, Silliman wrote prophetically of his young assistant as follows:

> "The author takes pleasure in acknowledging the important aid derived in this portion of his work from his friend and professional assistant, Mr. Thomas Sterry Hunt, whose familiarity with the philosophy and details of chemistry will not fail to make him one of its ablest followers. The labour of compiling the organic chemistry has fallen almost solely upon him."

Sometime in 1846, when Dennison Olmstead, Jr. resigned his post as Chemist on Prof. C. B. Adams' Geological Survey of Vermont, to fill a similar post on the recently organized Geological Survey of Canada, Hunt was appointed to fill the vacancy. This employment proved to be short-lived, however, for early in 1847 Olmstead returned to Vermont from Montreal to die, and soon thereafter Hunt once again filled a vacancy left by Olmstead, this time on the Canadian Survey as discussed in the following section.

TWENTY-FIVE YEARS (1847-1872) WITH THE GEOLOGICAL SURVEY OF CANADA

The Geological Survey of Canada was established in 1842, and the next year William E. Logan (later Sir William Logan) was appointed its first Director. Logan quickly sensed the need for an able chemist and mineralogist on his staff and after Olmstead's resignation mentioned above he wrote to the younger Silliman, asking him if he knew of anyone suitable for the position. Silliman, supported by Profs. C. Upham Smith and James D. Dana, strongly recommended Hunt, whose services were engaged forthwith. Now the die was cast and Hunt would enter on the career that would lead to almost limitless opportunities in a virgin field of science--the geochemistry of earth materials--and eventually to international renown as a leading scientist of his time.

Accordingly, in February 1846, and not yet twenty years of age, Hunt went to Montreal and took up his duties as Chemist and Mineralogist to the Geological Survey of Canada. Thus began that intimate association with Sir William Logan that would last for twenty-five years, from 1847 to 1872, and would result in Hunt's name thereafter being associated in a special way with the history of geology in the Dominion of Canada.

Living as a bachelor in a small two-room apartment that opened off the laboratory in St. Gabriel Street, Montreal, and without any laboratory assistant, Hunt not only did the routine chemical analytical work of the Survey, but also conducted a broad program of geochemical research that reached into many of the perplexing and controversial geological problems of the day. Soon after commencing his studies, he joined with other officers of the Survey in bringing together and correlating the immense amount of information on the geology and mineral resources of Canada that had accumulated before 1850. This information was published in the monumental 1000-page Geology of Canada in 1863, of which those parts dealing with petrography, mineralogy, the mineral waters, and much of the chapters on economic geology were written by Hunt.

Hunt's strictly official work as Chemist and Mineralogist to the Survey would have been more than enough for most men; but being an indefatigable worker accustomed to long hours, and ambitious to make known his ideas and research results to the world, he produced an impressive number of communications. His Bibliography, which appears later on in this sketch, includes more than 200 titles distributed among Silliman's Journal, the Canadian Naturalist, the Canadian Record of Science, the Philosophical Magazine, the Transactions of the Royal Society, the French Academy of Sciences, the Proceedings of the American Association for the Advancement of Science, and the Reports of Progress of the Geological Survey of Canada.

It is not the purpose of this sketch to describe and evaluate the contributions that Hunt made to geology during his years in Canada, though these years were manifestly his most productive in number and diversity of publications. Suffice it to note here the broad range of subject matter and of problems on which Hunt communicated ideas and comments, leaving it to the interested reader to seek more details in the several biographical sketches cited later on. His communications included, among others, articles on the origin, composition and places of origin of petroleum, sulphate and phosphate deposits, metallic ores (gold, copper, iron, nickel), and igneous and metamorphic rocks; chemistry of the primeval earth and early atmosphere; composition of natural gases and mineral waters; lithology of stratiform rocks along the Ottawa and upper lakes; Precambrian stratigraphy; the Taconic Problem; and his controversial "Crenitic Hypothesis," to be mentioned later on.

Hunt was long ago given credit for pointing out, as early as 1861, the accumulation of petroleum in anticlinal structures.* He was one of the

---

* The Oil and Gas Journal Petroleum Panorama Number, Jan. 1959.

first to investigate the chemical composition of mineral waters and many of his articles contain analytical data.

He made the first systematic attempt to subdivide and classify geologically the stratiform crystalline rocks of Canada and we owe to him the names Laurentian, Norian, Huronian, Montalban, Taconian and Keweenian. In attempting to determine how the early Precambrian rocks were formed, he developed his "Crenitic Hypothesis," which he vigorously defended on chemical grounds. Briefly stated the hypothesis proposes that the condensation of an atmosphere still containing all the volatile elements and the reaction of its solvents upon undifferentiated basic rock was the key to the genesis of crystalline rocks, an hypothesis that can hardly be accepted in the face of modern chemistry.

To the end Hunt hoped to:

> "... expand, in a connected treatise, his splendid generalizations on stellar and telluric chemistry, and especially to trace the influence of water under heat and pressure in decomposing the primitive basic crust of the earth, and in creating out of the primary elements the older crystalline rocks, and again in re-creating from their sterile ashes, through decay and death, the newer life-supporting and life-entombing strata which contain, written in generation after generation, the newer history of the earth." (Crosby 1892:370.)

But his proposed The History of the Earth, in which he hoped to accomplish the above stated goal, was frustrated by his death.

It should be noted that his most important contributions of these years were later incorporated in his five volumes--Chemical and Geological Essays (1874 and 1878); "Azoic Rocks" (Special Report on the Trap Dikes and Azoic Rocks of Southeastern Pennsylvania) (1878); Mineral Physiology and Physiography (1886); A New Basis for Chemistry (1887); and Mineralogy According to a Natural System (1891). To the reader interested in learning more about the nature and importance of Hunt's prodigious literary activity during his Canadian years, it is suggested he consult the biographical sketches by Frazer (1893), Pumpelly (1893), Douglas (1898), and Adams (1932) listed after the Bibliography later on.

In addition to his work as Survey Chemist and Mineralogist, as administrative and literary assistant to Logan, Director of the Survey, and as investigator and science reporter extraordinary, Hunt managed to spend several months of each year in the field.

Finally, every year from 1856 to 1862 he spent the spring months in Quebec where he lectured on chemistry to the students at Laval University in French that was said to be as fluent and elegant as his own native

English. Following this he occupied the chair of applied chemistry and mineralogy at McGill University in Montreal from 1862 to 1868.

M.I.T. YEARS (1872-1878)

In 1872 Hunt severed his connections with the Geological Survey of Canada to accept an appointment at M.I.T. as Professor of Geology. During his six-year tenure he also directed the instructional program in Geology, there being no Department or Course of Geology at the time; they would not come until eighteen years later in 1890.

When Hunt came to M.I.T. in October 1872,* he was known throughout the scientific world as one of the leading scientists of his time, being particularly noted for his more than 200 publications, which represented original and imaginative contributions in the fields of chemistry and of mineralogy and chemical geology (geochemistry of today), and for 25 years of distinguished service as Chemist and Mineralogist to the Geological Survey of Canada and as a close colleague of its great Director, Sir William Logan.

As third Professor of Geology, after Founder William Barton Rogers, Hunt joined Alpheus Hyatt and William H. Niles to form the three-man staff that would henceforth offer the program of instruction in Geology. The catalog for SY 1872-73 shows Hyatt teaching paleontology, Niles responsible for physical and historical geology, and Hunt scheduled for mineralogy and petrography. This assignment of subjects prevailed during the six years of Hunt's tenure, SY 1872-73 through SY 1877-78, at the end of which he resigned to take up private life as a geological consultant. During this period, as mentioned earlier, he also directed the program of instruction in Geology.

As an M.I.T. Professor, Hunt continued his geochemical activities as vigorously as ever. He wrote 60 publications, chiefly on geological and geochemical subjects, included in which were two important separately published volumes--Chemical and Geological Essays (1875) and Special Report

---

* According to the Records of the Government [Corporation] of M.I.T. for June 14, 1871, it was voted:

> "... that Professor T. Sterry Hunt be offered the position of Professor of Geology at a salary of $1000 for the present year."

But seemingly this offer was never actually made to Hunt because the next item that concerns Hunt appears in the minutes for Jan. 3, 1872 to the effect that:

> "... Thomas Sterry Hunt was appointed Professor of Geology at a salary of $2500, to commence his duties in October, 1872."

on the Trap Dikes and Azoic Rocks of Southeastern Pennsylvania (1878). In the Essays he collected some 545 of his previously published articles, unpublished lectures, and miscellaneous comments and notes in an effort to establish his priority for a large number of ideas involving chemical and geological phenomena.* Hunt was openly jealous of the credit he felt due him for his original ideas, and the Essays were an effort to document that credit. As one of his biographers has written (Adams 1932, p. 219):

> "A characteristic which impressed many people unfavorably was his rather too frequent and definite insistence on priority for his many views and opinions. It is to be feared that, like some other great men, he possessed, together with other undesirable qualities, a certain amount of vanity in his disposition."

Both Frazer (1893) and Adams (1932) report Hunt to have been an excellent lecturer, and the latter lists some of these at the end of his biography as "Extra Lectures:"

1866;-Lectures on chemical and physical geography,
1867 delivered before the Lowell Institute, Boston.

1872--Twenty lectures on chemistry, delivered before the Ladies' Educational Institute, Montreal.

1874--Six lectures on chemistry of the waters, delivered before the Boston Society of Natural History, Boston.

1875--One lecture on the constitution of water as related to modern chemistry and physics, before the Examiner Club.

1875--One lecture on the glacial period, delivered before the Literary and Historical Society of Quebec.

1875--Eighteen lectures on the practical Geology of the United States, Boston.

1875--The relations of Chemistry to Pharmacy and Therapeutics, an Address before the Massachusetts College of Pharmacy, Boston, 1875.

1876--Eighteen lectures on elementary geology. [Boston?]

1876--A course on the older rocks, before the Boston Society of Natural History.

1876--The building of the earth, delivered in Salt Lake City.

1877--Chemical history of the earth, before Chestnut Street Club.

1881--Coal. [Boston?]

1883--Twelve lectures on mineral physiology, before the Lowell Institute.

---

* For a complete statement of Hunt's purpose in collecting in one volume both published and unpublished material, the interested reader is referred to the Preface of the Essays.

1884--The manufacture of iron, before the Finance
      Club, Cambridge, Mass.

1885--On Arbor Day, Montreal.

1886--On the Alps, Liverpool, England.

1888--Theory of volcanoes, Montreal.

1889--Goethe and modern science, the Concord School
      of Philosophy.

THE POST-M.I.T. PERIOD (1878-1892)

Adams (1932, p. 217) states that Hunt did not find teaching a congenial exercise because he took no particular interest in the dry details of historical geology nor did he have the patience and sympathetic understanding so often required of the good teacher. Because his talents lay elsewhere, as evidenced by his long list of publications, his application later for the chair of Geology at Columbia was rejected in favor of Newberry in spite of the fact that he was supported by some of the giants of geology of his time--Lyell, Murchison, Dana, Silliman and the Rogers brothers. As a result of this rejection, he retired to private life and spent the last decade of his life doing consulting work and continuing his writing. In the latter he attempted to bring together many of his ideas in two more separately published volumes--A New Basis for Chemistry: A Chemical Philosophy (1887), subsequently translated into French and published in Paris and Liége as Un Système Chimique Nouveau (1889), and Systematic Mineralogy Based on a Natural Classification with a General Introduction (1891). Raymond (1891, p. 527) also reported that much of Hunt's A New Basis for Chemistry was translated into Russian.

Although failing health forced him to spend his last two years in his room at the Park Avenue Hotel in New York, or at St. Luke's Hospital, the discipline and indomitable courage that had driven him onward through a lifetime sustained him to the end. Until the day before he died he worked as he could on yet another book, to be left unfinished, in which he hoped to present an even more convincing defense of his long-held and often controversial views.

Hunt remained a bachelor until late in life, but in 1877 before leaving M.I.T. he married Miss Anna Rebecca Gale, a member of a well-known family in Montreal.

SUMMARY AND THE RECORD OF A LIFETIME

Thomas Sterry Hunt was endowed with an imaginative mind, an unusually retentive memory, and an elegance of speech and diction in both English and French. He was a proud man with a driving ambition to discover and make public a deeper understanding of the earth and its materials. He possessed almost unbounded energy and a stubborn belief in his own carefully formulated original ideas. Because he was the first North American scientist to apply the principles of chemistry to the perplexing problems of geology, he can well be called the "Father of Chemical Geology, or Geochemistry, in North America."

For his time he was a prodigious contributor to the journals of his day. His interests ranged widely among problems that involved the chemistry of earth materials--gases, waters, petroleum, minerals, ore deposits, and rocks of every kind--as can be readily determined by even a cursory examination of his impressive bibliography of some 360 titles. Taking his publications as a whole, they were primarily chemical, geochemical, or petrological, for Hunt was both chemist and geologist. His life ambition was to bring chemistry and geology into such a close relationship that the principles of the first science could be used to gain deeper insight into the problems of the second.

While the Bibliography that follows this sketch is believed to include all of his publications that concern geological subjects, it is almost certain that I have missed articles of a strictly chemical nature that appeared in chemical journals and periodicals unfamiliar to me. It is hoped, however that these latter publications, admittedly incomplete in number, nevertheless support the contention that Hunt made important contributions to Chemistry as well as to Geology.

In writing of Hunt as a chemist, Davis (<u>Tech. Rev.</u> 28/7: 264,266, [1933]) had the following to say:

> "Thomas Sterry Hunt, one of the most distinguished chemists of his time, was Professor of Geology from 1871-1878. Although he was not a member of the [M.I.T.] Chemistry Department, it seems impossible that he should have been without influence upon chemistry at the Institute. Let it be recalled that he was probably the first to define organic chemistry as the chemistry of carbon and its compounds. (p. 264.) ... He stood on the advancing frontier of chemistry. He originated the theory of simple water types, and the germs of the ideas usually attributed to Gerhardt may be found in his earlier papers. His researches upon the equivalent volumes of liquids and solids were a remarkable anticipation of Dumas. He had definite and significant ideas of the real molecular complexity of mineral substances. Among his more practical achievements may be mentioned his invention and patenting in 1859, of the permanent green ink which has found wide use in the printing of greenback currency. He was a charter member of the American Chemical Society and President of that organization in 1879 and 1888."

As might be expected of a person of his nature, Hunt took an active, sometimes even aggressive, role in many of the controversial geological discussions of the day, and is reported by all of his biographers to have been a careful investigator and a convincing speaker.

In his later years he turned toward more theoretical, speculative, and philosophical approaches in his writing, and near the end tried to marshal some of his early innovative work in support of several hypotheses that he hoped would ultimately be accepted as fundamental laws of chemistry and geology. That this hope did not eventuate has been ascribed by even his most sympathetic biographers to his stubborn insistence on the validity of his ideas too long after they had been superseded by newer discoveries. His fierce belief and pride in his own work made it exceedingly difficult for him to abandon it and accept the ideas of a colleague. Merrill commented on this characteristic as follows:

> "His early papers, as summarized in his volume, Chemical and Geological Essays, were inspiring and full of suggestive matter, and had he but held himself in check he might have passed into history as an honored leader.
>
> "Unfortunately he early developed an erratic tendency, and a disregard for facts that in any manner conflicted or failed to substantiate his views."
> (p. 447 in The First One Hundred Years of American Geology, 1924.)

Inasmuch as most of Hunt's original and highly speculative ideas were formulated before he came to M.I.T., and also because he brought many of his chemical and mineralogical ideas together into volumes published separately after he left the Institute, I shall leave it to the interested reader to seek full description and evaluation of these ideas in the biographies and memorials cited in the list following the Bibliography.

Perhaps as good an evaluation as any of his life's work as a geochemist is that of Pumpelly, who wrote of Hunt's record as follows (1892, p. 384-385):

> "A review of his recorded work shows that he was a brilliant and original thinker, and that his speculations in chemical geology were based on a large amount of original laboratory research and on a skillful use of that of others. Such a review brings out to light also a lack of that experience in detailed field-work, both original and critical, especially in structural geology, which is essential in building hypotheses and in testing them step by step. One cannot but feel that he was seriously limited by this deficiency, and that this limitation caused him to continue through the world's half century of progress in geology to construct a history of the early globe on a plan circumscribed by conceptions formed early in his career. Throughout his time he was the leading representative of chemical geology in America,

and his works contain, both on the side of original research and of speculation, very much of the material necessary to construct the same history on lines more in accord with the present requirements. On its suggestive side, Dr. Hunt's work in chemical geology has ranked high in both hemispheres and its influence will long continue to be felt, and in a growing science this is perhaps the rarest and most important side."

In summary, it seems fair to state that T. Sterry Hunt, in his own time, and employing the known science of his time, made a significant impact on both chemistry and chemical geology. The ideas and concepts that he formulated in his earlier years received respectful consideration whether approved or criticized. As would be expected at this date, more than a century later, some of his ideas now seem naive and others have had to be drastically revised, yet much of his analytical work has stood the test of time and has become an important part of the early literature of North American geochemistry.

In view of the preceding statement, it should be recorded here that Hunt can be credited with the following accomplishments:

1) Described several new minerals--enceladite, algerite, loganite.
2) Analyzed many samples of ores and mineral waters.
3) Reported on the rocks and minerals of the Ottawa region.
4) Developed new techniques for chemical analysis.
5) Contributed importantly to the preparation of Logan's <u>Geology of Canada</u>.
6) Reported extensively on the composition and formation of saline deposits (chlorides, sulphates, nitrates).
7) Is credited with first defining organic chemistry as the chemistry of carbon and its compounds.
8) Was one of the first to suggest that petroleum is indigenous to the sedimentary rocks in which it is found.
9) First proposed the theory that petroleum was associated with anticlinal structures.
10) Reported extensively on the mineralogy of crystalline rocks--igneous, sedimentary and metamorphic.
11) Developed with J. Douglas a new process for extraction of copper from its ores.
12) Prepared an important report on the Hocking Valley coal field of Ohio.
13) Reported on the "Azoic" rocks of southeastern Pennsylvania.
14) One of the first to attempt to work out the Precambrian sequences in Canada, which gave us the names Laurentian, Huronian, Norian, etc.
15) Discovery of the green ink used in U.S. "green-back" paper currency.
16) Numerous novel techniques for determining elements by chemical analysis.

HONORS, DEGREES, AND DECORATIONS

Hunt's international renown brought him numerous honors, degrees, and decorations. The more important of these may be mentioned briefly. He was one of the international jurors at the great industrial exhibitions in Paris in 1855 and 1867, and at our own Centennial Exhibition in Philadelphia in 1876. He was elected a Fellow of the Royal Society in 1859 at age 33 and was at the time of his election the youngest of the members; and in 1873 he was elected a member of the National Academy of Sciences. The French Government appointed him a Chevalier of the Legion of Honor and later advanced him to Officer, and after the meeting of the International Geological Congress in Bologna in 1881, King Humbert appointed him an Officer of the Italian Order of St. Mauritius and St. Lazarus. He was a Vice President of the American Association for the Advancement of Science in 1871, the President of the American Institute of Mining and Metallurgical Engineers in 1871, and twice President of the American Chemical Society, in 1879 and again in 1888. He was one of the Founders of the Royal Society of Canada in 1882 and served as its first elected President (1884-1885). He suggested the idea of an International Geological Congress in 1875 and served as the first Secretary (1875), then as Vice President at meetings in Paris (1878), Bologna (1881) and London (1888). He helped to organize Laval University in Quebec and was awarded an honorary LL.D. degree at its first Convocation. Harvard awarded him an M.A. degree early in his career, McGill awarded him an honorary LL.B. later, and still later, in 1881, Cambridge honored him with its LL.D. degree. It is interesting to note, in passing, that Hunt held no earned college degree, in fact not even a high school diploma; his formal education was limited to a few short months in the public schools of Norwich, Connecticut, and less than three years at Yale during which time he could only snatch a few hours now and then from his otherwise full-time duties as a laboratory assistant to attend lectures. He was indeed a self-educated man! And at the height of his career he stood tall among the best of the formally educated scientists of his time.

BIBLIOGRAPHY OF THOMAS STERRY HUNT

Abbreviations and symbols used in the following references are explained on p. 91-98 ; in general, abstracts are listed separately as well as with the complete article. This bibliography is believed to include all of Hunt's important geological writings published in North American journals, periodicals, etc. but probably lacks some published abroad. Furthermore, while numerous of his articles on chemical subjects are included, no special effort was made to compile a complete list of these articles. Finally, numerous of the articles listed below were later included in his Chemical and Geological Essays (1875) with either the same or a somewhat different title; these are designated by adding [C. & G.E.] to the reference. A few were also reprinted in his Mineral Physiology and Physiography (1866); these are designated by adding [M.P. & P.]. During his productive years, 1847-1892, Hunt was at M.I.T. during the period 1872-1878. It has seemed appropriate, however, because of his eminence as North America's first geochemist, to record here as complete a bibliography as possible of his geological and geochemical publications.

1--1846   Description and analysis of a new mineral species, containing titanium [enceladite]; with some remarks on the constitution of titaniferous minerals. Am. J. Sci. (2)2: 30-36, (1846).

2--1846a   Ozone. Am. J. Sci. (2)2: 103-110, (1846).

3--1864b   (with Silliman, B., Jr.) On the meteoric iron of Texas and Lockport. Am. J. Sci. (2)2: 370-376, il., (1846).

4--1846c   On the artificial formation of specular iron. Am. J. Sci. (2) 2: 411-412, (1846).

5--1847   Report [analyses of minerals], in Adams, C. B., Third Ann. Rept. on the Geology of the State of Vermont, App. A: 23-27, (1847).

6--1847a   Review of the Organic Chemistry of M. Charles Gerhardt. Am. J. Sci. (2)4: 93-100, (1847).

7--1847b   On the relations of glycocoll and alcargene. Am. J. Sci. (2) 4: 108, 266-267, (1847).

8--1847c   On the action of sulphuretted hydrogen upon nitric acetone. Am. J. Sci. (2)4: 350-353, (1847).

9--1847d   [Analyses of ores and mineral waters.] Can. Geol. Surv., Rept. Prog. 1845-46: 122-125, (1847).

10--1848   On the chemical constitution of gelatine and its transformations. Am. J. Sci. (2)5: 74-78, (1848).

11--1848a   Scientific Intelligence. I. Chemistry and Physics. (Under this heading Hunt published numerous short items--reviews, comments, notes, etc. as listed below.) Am. J. Sci. (2)5: 116 ff., (1848).

         1. "On some new compounds of phosphorus," by A. Wurtz, p. 116-117.

         2. Sulphoxyarsenic acid, p. 117.

         3. "On some combinations of nitric and hyponitric acids," by C. Gerhardt, p. 117-118.

         4. "On the products of the decomposition of gelatine by chromic acid," by A. Schlieper, p. 118; Valeracetonile, p. 119.

         5. On the products of the oxydation of caseine, p. 119.

         6. "On the decomposition of cyanid of ethyle," by C. Frankland and Dr. H. Kolbe, p. 119.

         7. "On the presence of sugar of milk in the milk of the Carnivora," by A. Bensch, p. 119-120.

         8. Silicates in the blood of fowls, p. 120.

         9. On the existence of copper, arsenic, antimony, and tin in mineral waters, p. 120.

         10. "On the transformation of tannic into gallic acid," by C. Wetherell, p. 121-122.

         11. "On two new alkaloids obtained from aldehyde," by Wöhler and Liebig, p. 122-123.

12. "On the action of anhydrous acid upon the amides and ammoniacal salts," by Dumas et al., p. 263-265.

13. "On xyloidine, pyroxyline, and some analogous products," by F. Dumonte and Mendrel, p. 265-266.

14. "On the conversion of neutral nitrogenous bodies, as fibrine and caseine, into fatty substances," by M. Blondeau, p. 266.

-----

2. "General formulas for the silicates and borates," by A. Laurent, p. 405-407.

3. "On phosphamid," by M. Gerhardt, p. 407-408.

4. "On the nature of hydrofluoric acid," by M. Louyet, p. 408.

5. "On the quantitative determination of sulphur in organic substances," by W. Heintz, p. 408.

6. "On isomorphism," by Laurent and Berzelius, p. 409.

7. "On the products of the action of cyanic acid upon alcohol and aldehyde," by Liebig and Woehler, p. 409-410.

8. "On the preparation of crystallized bile," by M. Plattner, p. 410-411.

9. Test for strychnine, p. 411.

12--1848b On the analysis of chromic iron. Am. J. Sci. (2)5: 418-419, (1848).

13--1848c On the anomalies presented in the atomic volume of sulphur and nitrogen; with remarks on chemical classification, and a notice of M. Laurent's theory of binary molecules. Am. J. Sci. (2)6: 170-178, (1848). [C. & G.E.]

14--1848d Note to a paper on the chemical nature of gelatine, published in the American Journal [of Science] of Jan., 1848, p. 74. Am. J. Sci. (2)6: 259-260, (1848). [See item 10--1848.]

15--1849 Report on the rocks and minerals of the Ottawa, especially apatite, and analyses of mineral waters. Can. Geol. Surv., Rept. Prog. 1847-48: 125-165, (1849).

16--1849a On the acid springs and gypsum deposits of the Onondaga salt group. Am. J. Sci. (2)7: 175-178, (1849); Edinb. New Philos. J. 47: 50-53, (1849).

17--1849b On some principles to be considered in chemical classifications. Am. J. Sci. (2)7: 399-405; (2)8: 89-95, (1849). [C. & G.E.]

18--1849c Chemical examination of algerite, a new mineral species ... including a description of the mineral by F. Alger. Boston J. Nat. Hist. 6: 118-123, (1849); Am. J. Sci. (2)8: 103-106, (1849).

19--1849d Chemical examination of the water of the Tuscarora sour spring, and some other mineral waters of western Canada. Am. J. Sci. (2)8: 364-372, (1849).

20--1849e On the decomposition of aniline by nitrous acid. Am. J. Sci. (2)8: 372-375, (1849).

21--1849f On the geology of Canada. A.A.A.S., Pr. 2: 325-334, (1849).

22--1849g On a new mineral algerite. Boston Soc. Nat. Hist., Pr. 3: 259-260, (1849).

23--1850   Report [on examination of mineral waters and ores, with analyses]. Can. Geol. Surv., Rept. Prog. 1848-49: 47-65, (1850).

24--1850a   Report [on the examination of soils, mineral waters, and ores, with analyses]. Can. Geol. Surv., Rept. Prog. 1849-50: 73-106, (1850).

25--1850b   On the Geology of Canada. A.A.A.S., Pr. 2: 325-334, (1850); Am. J. Sci. (2)9: 12-19, (1850).

26--1850c   Remarks on the constitution of Leucine, with critical observations upon the late researches of M. Wurtz. Am. J. Sci. (2) 9: 63-67, (1850). [C. & G.E.]

27--1850d   Chemical examinations of the waters of some of the mineral springs of Canada. Am. J. Sci. (2)9: 266-275, (1850).

28--1850e   [On algerite from Franklin, N.J.]. Boston Soc. Nat. Hist., Pr. 3: 150-151, (1850).

29--1850f   Researches upon some derivatives of the benzoic series; by G. Chancel. Am. J. Sci. (2)9: 275-276, (1850).

30--1850g   On some saline springs containing baryta and strontia. A.A.A.S., Pr. 4: 153-154, (1850).

31--1850h   On the determination of phosphoric acid. A.A.A.S., Pr. 4: 338, (1950).

32--1850i   On the presence of malate of lime in the sap of the Acer saccharinum. A.A.A.S., Pr. 4: 389, (1850).

33--1850j   Report on soils of Canada East and mineral springs. Can. Geol. Surv., Rept. Prog. 1849-50: 73-106, (1850).

34--1851   On the Taconic System. A.A.A.S., Pr. 4: 202-204, (1851).

35--1851a   On the mineral springs of Canada. Am. J. Sci. (2)11: 174-181, (1851).

36--1851b   On the chemical composition of the mineral warwickite. Am. J. Sci. (2)11: 352-356, (1851).

37--1851c   Examinations of some Canadian minerals. Philos. Mag. (4)1: 322-328, (1851).

38--1851d   Description and analysis of loganite, a new mineral species. Philos. Mag. (4)2: 65-67, (1851).

39--1851e   On the homologies of the alcohols and their derivatives. A.A.A.S., Pr. 6: 216-217, (1851).

40--1851f   Columbite of Haddam. A.A.A.S., Pr. 6: 243, (1851).

41--1852   Report [on the examination of minerals and mineral waters]. Can. Geol. Surv., Rept. Prog. 1850-51: 35-54, (1852).

42--1852a   Report [of examination of minerals, soils, etc.]. Can. Geol. Surv., Rept. Prog. 1851-52: 93-121, (1852).

43--1852b   On the compound ammonias, and the bodies of the cacodyle series. Am. J. Sci. (2)13: 206-211, (1852).

44--1852c   On octahedral oligist iron. Am. J. Sci. (2)13: 370-373, (1852).

45--1852d   Examination of some American minerals. Am. J. Sci. (2)14: 340-346, (1852).

46--1852e   Remarks on the lithological and palaentological characters of the Potsdam sandstone. A.A.A.S., Pr. 6: 271-273, (1852).

47--1852f   On the economical uses of the skin of the white porpoise. A.A.A.S., Pr. 6: 386-387, (1852).

48--1852g	Examinations of phosphatic matters, supposed bones, and coprolites, occurring in the Lower Silurian rocks of Canada. Geol. Soc. London, Q.J. 8: 209-210, (1852).

49--1853	Correspondence of T. S. Hunt of Montreal, on atomic volume. Am. J. Sci. (2)15: 116, (1853).

50--1853a	Considerations on the theory of chemical changes, and on equivalent volumes. Am. J. Sci. (2)15: 226-234, (1853). [C. & G.E.]

51--1853b	On the constitution and equivalent volume of some mineral species. Am. J. Sci. (2)16: 203-218, (1853). [C. & G.E.]

52--1853c	The theory of chemical changes and equivalent volume. Am. J. Sci. 25: 226-234, (1853); Philos. Mag. 5: 526-535, (1853); Ueber chemische Zersetzung, über die Constitution und das Atomvolum einiger Minerale. Chem.-Pharm. Central Blatt, 3 Dec. 1853, No. 54: 849-858, (1853).

53--1854	Report [on the examination of minerals, mineral waters, etc.] Can. Geol. Surv., Rept. Prog. 1852-53: 153-179, (1854).

54--1854a	Parophite. Am. J. Sci. (2)17: 127, (1854).

55--1854b	On the theoretical relations of water and hydrogen. Am. J. Sci. (2)17: 194-199, (1854). [C. & G.E.]

56--1854c	(with Logan, W. E.) On the chemical composition of recent and fossil Lingulae and some other shells. Am. J. Sci. (2) 17: 235-239, (1854); Can. J. 2: 264-265, (1854).

57--1854d	Remarks on the mineral species algerite. Am. J. Sci. (2)17: 351-352, (1854).

58--1854e	On some of the crystalline limestones of North America. Am. J. Sci. (2)18: 193-200, (1854); Can. J. 3: 36-38, (1854).

59--1854f	Illustrations of chemical homology. A.A.A.S., Pr. 8: 237-247, (1854); Am. J. Sci. (2)18: 269-271, (1854).

60--1854g	On the composition and metamorphism of some sedimentary rocks. Philos. Mag. (4)7: 233-238, (1854); Erdman J. Prakt. Chem. 52: 174, (1854).

61--1855	Thoughts on solution and the chemical process. Am. J. Sci. (2)19: 100-103, (1855). [C. & G.E.]

62--1855a	[Extract from a letter from T. S. Hunt to J. D. Dana, dated Montreal, Canada, March 12, 1855.] On the equivalent of some species. Am. J. Sci. (2)19: 416-417, (1855).

63--1855b	On the so-called talcose slates of the Green Mountains. Am. J. Sci. (2)19: 417, (1855).

64--1855c	On a newly discovered meteoric iron. Am. J. Sci. (2)19: 417, (1855).

65--1855d	On some ores of nickel from Lake Superior. Am. J. Sci. (2)19: 417-418, (1855).

66--1855e	Mineralogical notes. Am. J. Sci. (2)19: 428-429, (1855).

67--1855f	Examinations of some feldspathic rocks. Philos. Mag. (4)9: 354-363, (1855); Erdman J. Prakt. Chem. 56: 149-154, (1855).

68--1855g	Observations sur les roches magnésiennes du groupe de la rivière Hudson ... Soc. Geol. France, B.(2)12: 1029-1032, (1855).

69--1855h	Note sur les sources acides et les gypses du Haut-Canada. Acad. Sci. Paris, C.R. 40: 1348-1351, (1855).

70--1855i	Sur les volumes atomiques. Acad. Sci. Paris, C.R. 41: 77-81, (1855).

71--1855j   Recherches sur les eaux minérales du Canada. Acad. Sci. Paris, C.R. 41: 300-304, (1855).

72--1855k   Sur les rapports entre quelques composés différant par H_2 et par O_2. Acad. Sci. Paris, C.R. 41: 1167-1169, (1855).

73--1855*l* (with Logan, W. E.) Esquisse géologique du Canada ... à l'exposition universelle de Paris, 1855. [Can. Geol. Surv.]: 100 p., Paris, 1855. Map in Soc. Geol. France, B.(2)12: opposite p. 1316, (1855).

74--1857   Report for the year 1853 [on mineral waters, etc.] Can. Geol. Surv., Rept. Prog. 1853-56: 347-371, (1857).

75--1857a   Report for the year 1854 [on metamorphic rocks]. Can. Geol. Surv., Rept. Prog. 1853-56: 373-390, (1857).

76--1857b   Report for the year 1855. Can. Geol. Surv., Rept. Prog. 1853-56: 391-429, (1857).

77--1857c   Report for the year 1856 [on the mineralogy of metamorphic rocks]. Can. Geol. Surv., Rept. Prog. 1853-56: 431-494, (1857); Extract, with title, Contributions to the history of ophiolites. Am. J. Sci. (2)25: 217-226, (1857); (2)26: 234-240, (1858); Erdman J. Prakt. Chem. 75: 457-458, (1858).

78--1857d   On the serpentines of Canada and their associated rocks. Roy. Soc. London, Pr. 8: 423-425, (1857).

79--1857e   On the part which the silicates of the alkalis may play in the metamorphism of rocks. Roy. Soc. London, Pr. 8: 458-461, (1857); Philos. Mag. (4)15: 68-70, (1858); Am. J. Sci. (2) 25: 287-289, (1858).

80--1857f   On the chemical composition of the waters of the St. Lawrence and Ottawa rivers. Philos. Mag. (4)13: 239-245, (1857).

81--1857g   On the reactions of the alkaline silicates. Am. J. Sci. (2) 23: 437-438, (1857).

82--1857h   On the probable origin of some magnesian rocks. Am. J. Sci. (2)24: 272-273, (1857); Roy. Soc. London, Pr. 9: 159-164, (1858).

83--1857i   Note on the cherokine of C. U. Shepard. Am. J. Sci. (2)24: 275, (1857).

84--1857j   On the origin and metamorphosis of some sedimentary rocks. Can. J. n.s. 2: 355-357, (1857); Abst., Can. J. 2: 261-262, (1857).

85--1857k   On serpentine and some of its uses. Can. Nat. 2: 28-34, (1857).

86--1857*l* In the Can. Geol. Surv., Rept. Prog. 1853-56, published in 1857, numerous short articles were included, and these also appeared as abstracts in Can. Nat. 3: 91-97 and in Am. J. Sci. (2)25: 217-226 and (2)26: 234-240, under the title, "Contributions to the history of ophiolites," etc.

87--1857m   Adams (1932, p. 224--see List following Bibliography) reports that articles on the following subjects appeared in the Geol. Surv. Can., Rept. Prog. 1853-56, (1857), which is referred to in the preceding reference, 86--1857 : [Triclinic feldspars of the Laurentian series; Silurian rocks; nickel ores; metallurgy of iron ores; extraction of salt from sea-water; magnesian waters; plumbago; and peat.]

88--1857n   Origin of magnesian rocks [abst.]. Can. Nat. 2: 258, (1857).

89--1857o   On some euphotides and other feldspathic rocks [abst.]. Edinb. Nat. Philos. J., n.s. 5: 366-367, (1857).

90--1857p   On the serpentines of the Green Mountains and some of their associates [abst.]. Edin. Nat. Philos. J.,n.s. 5: 367, (1857).

91--1857q   General considerations on the metamorphism of the sedimentary rocks [abst.]. Edinb. Philos. J., n.s. 6: 350, (1857).

92--1858   Report on the year 1857 [dolomites, limestones, fish manures, etc.]. Can. Geol. Surv., Rept. Prog. 1857: 193-229, (1858).

93--1858a   On the theory of igneous rocks and volcanoes. Can. Nat. 3: 194-201, (1858); Can. J.,n.s. 3: 201-208, (1858). [C. & G.E.]

94--1858b   On the chemistry of the primeval earth. Am. J. Sci. (2)25: 102-103, (1858).

95--1858c   On the extraction of salts from sea water. Am. J. Sci. (2) 25: 361-371, (1858); Can. Nat. 3: 97-110, (1858).

96--1858d   On the origin of the feldspars and on some points of chemical lithology. Am. J. Sci. (2)25: 435-437, (1858).

97--1858e   On euphodite and saussurite. Am. J. Sci. (2)25: 437, (1858).

98--1859   Report for the year 1858 [intrusive rocks, magnesian limestones, etc.]. Can. Geol. Surv., Rept. Prog. 1858: 171-218, (1859). In part, Can. J., n.s. 5: 426-442, (1860). Abst., Am. J. Sci. (2)31: 124, (1859).

99--1859a   Fish manure. Can. Nat. 4: 13-23, (1859).

100--1859b Contributions to the history of euphodite and saussurite. Am. J. Sci. (2)27: 336-349, (1859); Erdman J. Prakt. Chem. 80: 333-336, (1860).

101--1859c On some reactions of the salts of lime and magnesia and on the formation of gypsums and magnesian rocks. Am. J. Sci. (2)28: 170-187, 365-383, (1959).

102--1859d On the formation of magnesian limestones [abst.]. Can J., n.s. 4: 184-186, (1859).

103--1859e Contributions to the history of gypsums and magnesian rocks. Abst., Can. Nat. 4: 294-295, (1859); A.A.A.S., Pr. 13: 227-247, (1860). [C. & G.E.]

104--1859f Formation of siliceous rocks [abst.]. Can. Nat. 4: 295-296, (1859).

105--1859g On some points in chemical geology. Geol. Soc. London, Q.J. 15: 488-496, (1859); Can. Nat. 4: 414-425, (1859); Min. Mag. (2)2: 14-24, (1860). Abst., British Assoc., Rp. 30: sec. 83-84, (1861); Am. J. Sci. (2)30: 133-137, (1860); Philos. Mag. (4)17: 148-149, (1859). [C. & G.E.]

106--1860   Review.--On some points in the geology of the Alps [Mémoire sur les terrains liassique et keuperian de la Savoie par Alphonse Favre]. Am. J. Sci. (2)29: 118-124, (1860).

107--1860a On some of the igneous rocks of Canada. Am. J. Sci. (2)29: 282-284, (1860).

108--1860b Notes on the dolomites of the Paris Basin, etc. Am. J. Sci. (2)29: 284-285, (1860).

109--1860c New Palaeozoic fossils, by J. H. McChesney. Am. J. Sci. (2) 29: 285-286, (1860).

110--1860d On the formation of gypsums and dolomites. Geol. Soc. London, Q.J. 16: 152-154, (1860).

111--1860e Analysis of Canadian wolfram [Lake Couchiching, Ont.]. Can. J., n.s. 5: 303, (1860).

112--1860f   On the intrusive rocks of the district of Montreal. Can. J., n.s. 5: 426-442, (1860).

113--1860g   On the titaniferous iron ores of Canada. Can. Nat. 2: 41-42, (1860).

114--1860h   Sur les relations les matières et albumenoides. Acad. Sci. Paris, C.R. 50: 1186-1187, (1860); J. Pharm. 38: 122-123, (1860).

115--1861   Notes on the history of petroleum or rock oil. Can. Nat. 6: 241-255, (1861); Smith. Inst., Ann. Rept. 1861: 319-329, (1862); Can. News 6: 5, 6, 16-19, 35-36, (1862). [C. & G.E.]

116--1861a   Geological Survey of Canada.--Report of Progress for 1858. (A review). Am. J. Sci. (2)31: 122-124, (1861).

117--1861b   On the theory of types in chemistry. Can. Jour. 6: 120-129; Acad. Sci. Paris, C.R. 52: 247-250, (1861); Philos. Mag. 22: 15-23, (1861); Am. J. Sci. (2)31: 256-264, (1861). [C. & G.E.]

118--1861c   On some points in American geology. Am. J. Sci. (2)31: 392-414, (1861); Can. Nat. 6: 81-105, (1861). [Reprinted in part as "The origin of mountains" in C. & G.E.]

119--1861d   Note on chloritoid from Canada. Am. J. Sci. (2)31: 442-443, (1861).

120--1861e   On ozone, nitrous acid and nitrogen. Am. J. Sci. (2)32: 109-110, (1861).

121--1861f   On the origin of some magnesian and aluminous rocks. Am. J. Sci. (2)32: 286-288, (1861); Can. Nat. 6: 180-184, (1861); Can. News 6: 158-160, (1861).

122--1861g   On the unity of geological phenomena in the Solar System, by L. Saemann [translated by T. S. Hunt]. Can. Nat. 6: 444-451, (1861).

123--1861h   Mr. Barrande on the Primordial zone in North America and on the Taconic system of Emmons. Can. Nat. 6: 374-383, (1861); In part (Taconic System), Am. J. Sci. (2)32: 427-430, (1861).

124--1862   [On the names of certain rocks.] Can. Nat. 7: 17-19, (1862).

125--1862a   Note on the Taconic system of Emmons. Can. Nat. 7: 78-80, (1862); Am. J. Sci. (2)33: 135-136, (1862).

126--1862b   Considérations sur la chimie du globe. Acad. Sci. Paris, C.R. 54: 1190-1194, (1862); Can. Nat. 7: 201-205, (1862).

127--1862c   On the various theoretical views regarding the origin of the primitive formations: note. Can. Nat. 7: 262-263, (1862).

128--1862d   Note on the occurrence of glauconite in the Lower Silurian rocks. Am. J. Sci. (2)33: 277-278, (1862).

129--1862e   (with Logan, W. E.) Descriptive catalogue of a collection of the economic minerals of Canada [by W. E. Logan] and of its crystalline rocks [by T. S. Hunt]; London International Exhibition, 1862. Can. Geol. Surv.: 88p., Montreal, (1862).

130--1863   Contributions to the chemical and geological history of bitumens and of pyroschists or bituminous shales. Am. J. Sci. (2)35: 157-171, (1863).

131--1863a   On the gold mines of Canada and the manner of working them. Can. Nat. 8: 13-19, (1863).

132--1863b   On the chemical and mineralogical relations of metamorphic rocks. Geol. Soc. Dublin, J. 10: 85-95,(1864); Dublin Quart. J. Sci. 3: 220-230, (1863); Am. J. Sci. (2)36: 214-226, (1863); Can. Nat. 8: 195-208, (1863). [C. & G.E.]

133--1863c	On the earth's climate in Paleozoic times. Am. J. Sci. (2) 36: 396-398, (1863); Can. Nat. 8: 323-325, (1863); Philos. Mag. (4)27: 236-237, (1864).	
134--1864	Note sur la nature de l'azote et la théorie de la nitrification. Acad. Sci. Paris, C.R. 55: 460-462, (1864); Philos. Mag. 25: 27-29, (1864).	
135--1864a	On the nature of jade, and on a new mineral species described by Mr. Damour. [Sur la nature du jade. Acad. Sci. Paris, C.R. 56: 1255-1257, (1864);] Am. J. Sci. (2)36: 424-428, (1863).	
136--1864b	Contributions to lithology. Am. J. Sci. (2)37: 248-266; (2) 38: 91-104, 174-185, (1864); Can. Nat., n.s. 1: 16-36, 161-189, (1864).	
137--1864c	Laurentian rhizopods of Canada. (Extract of a letter from T. Sterry Hunt, F.R.S., to J. D. Dana, April 2, 1864.) Am. J. Sci. (2)37: 431, (1864).	
138--1864d	Notes on the silicification of fossils. Can. Nat., n.s. 1: 46-50, (1864).	
139--1864e	(and Hall, J. and Logan, W. E.) On the geology of eastern New York. Can. Nat., n.s. 1: 368-369, (1864); Am. J. Sci. (2)39: 96-97, (1864).	
140--1864f	On peat and its uses. Can. Nat., n.s. 1: 426-441, (1864).	
141--1864g	On organic remains in the Laurentian rocks of Canada. British Assoc. Repts. 34: 58 [title only], (1864).	
142--1865	Canada: a geographical, agricultural, and mineralogical sketch. [Canada], Bur. Agriculture, 33 p., Quebec, (1865).	
143--1865a	Petroleum: its geological relations considered with especial reference to its occurrence in Gaspé. 19 p., map, Quebec, (1865).	
144--1865b	Vorkommen des Apatits in Canada. Neues Jahrb. Mineral. p. 845; Halle, Zeitschr. Gesammt, Naturwiss. 25: 297, (1865).	
145--1865c	Contributions to the chemistry of natural waters. Am. J. Sci. (2)39: 176-193; (2)40: 43-60, 193-213, (1865); Can. Nat., n.s. 2: 1-21, 161-183, 276-299, (1865). [C. & G.E.]	
146--1865d	On the mineralogy of <u>Eozoon canadense</u>. Can. Nat., n.s. 2: 120-127, (1865).	
147--1865e	A geographical sketch of Canada. Can. Nat., n.s. 2: 356-363, (1865).	
148--1865f	On the mineralogy of certain organic remains from the Laurentian rocks of Canada. Geol. Soc. London, Q.J. 21: 67-71, (1865); Philos. Mag. 29: 76-77, (1865).	
149--1866	Report [on the gold of Lower Canada]. Can. Geol. Surv., Rept. Prog. 1863-66: 79-90, (1866).	
150--1866a	Report [Laurentian limestones, minerals, petroleum, salt, porosity of rocks, peat, etc.]. Can. Geol. Surv., Rept. Prog. 1863-66: 181-291, (1866). Reprinted in part, with additions, under the title, "On the mineralogy of the Laurentian limestones of North America," N.Y. St. Cab., Ann. Rept. 21: 47-98, (1871). [C. & G.E.]	
151--1866b	Further contributions to the history of lime and magnesia salts. Am. J. Sci. (2)45: 49-67, (1866).	
152--1866c	On the primeval atmosphere. Can. Nat., n.s. 3: 117-120, (1866); A.A.A.S., Pr. 15: 34-37, (1867).	

153--1867 On the metallurgical system of Messrs. Whelpley and Storer. A.A.A.S., Pr. 15: 30-34, (1867); Am. J. Sci. (3)43: 305-309, (1867).

154--1867a On petroleum [abst.]. A.A.A.S., Pr. 15: 29-30, (1867); Can. Nat., n.s. 3: 121-123, (1866).

155--1867b On the Laurentian limestones and their mineralogy [abst.]. A.A.A.S., Pr. 15: 54-57, (1867); Can. Nat., n.s. 3: 123-125, (1866) [1867].

156--1867c Report on the gold region of Hastings. Can. Geol. Surv., Reports on the gold region of the County of Hastings: 3-6, (1867).

157--1867d On the objects and method of mineralogy. Am. J. Sci. (2)43: 203-206, (1867); Can. Nat., n.s. 3: 110-114, (1866) [1867]; A.A.A.S., Pr. 7: 238-242, (1868). [C. & G.E.]

158--1867e On the chemistry of the primeval earth. Geol. Mag. 4: 357-369, 432, 477-478, (1867); Can. Nat., n.s. 3: 225-234, (1867); Arch. Sci. Phys. Nat., n.p. 31: 5-14, (1868); R. Inst., Pr. 5: 178-185, (1869). [C. & G.E.]

159--1867f Sur les pétroles de l'Amerique du Nord. Soc. Géol. France, B. (2)24: 570-573, (1867).

160--1867g Terrains anciens de l'Amérique du Nord (with discussions by J. Marcou). Soc. Géol. France, B. (2)24: 664-669, (1867).

161--1867h Sur la théorie de l'origine des montagnes. Soc. Géol. France, B. (2)24: 687-689, (1867).

162--1867i Sur la formation des gypses et des dolomies. Acad. Sci. Paris, C.R. 64: 815-817, (1867).

163--1867j Sur quelques réactions de sels magnésiens et sur les roches magnésifères. Acad. Sci. Paris, C.R. 64: 846-849, (1867).

164--1867k (with others) Description géologique du Canada. Esquisse Géologique du Canada, p. 3-35, (1867). [À l'exposition universelle de 1867.]

165--1867ℓ On the mineralogy of crystalline limestones. Geol. Mag. 4: 175-176, [abst.], (1867); 357-369, 432-477, 478, (1867); Les Mondes, 15: 17-24, (1867); Arch. Sci. Phys. Nat. 31: 5-14, (1868); Can. Nat., n.s. 3: 225-234, (1868); Roy. Inst., Pr. 5: 178-185, (1869); Smith. Rept. 1869: 182-207, (1869); N.Y. St. Cab. Nat. Hist., Rept. 1867, App. E, (1867).

166--1867m On the Laurentian limestones and their mineralogy [abst.]. A.A.A.S., Pr. 15: 54-57, (1867); Can. Nat., n.s. 3: 123-125, (1866) [1867].

167--1868 Report on the gold region of Nova Scotia. Can. Geol. Surv.: 48 p., (1868).

168--1868a On some points in the geology of Vermont. Am. J. Sci. (2)46: 222-229, (1868).

169--1868b Notes on the geology of southwestern Ontario. Am. J. Sci. (2) 46: 355-362, (1868); Can. Nat., n.s. 4: 11-20, (1869).

170--1868c A notice of the chemical geology of Mr. D. Forbes. Geol. Mag. 5: 49-59, (1868).

171--1869 On the probable seat of volcanic action. Geol. Mag. 6: 245-251, (1869); Can. Nat., n.s. 4: 166-173, (1869); Am. J. Sci. (2)50: 21-28, (1869). [C. & G.E.]

172--1869a [Beloeil Mountain, Quebec.] Can. Nat., n.s. 4: 220-222, (1869).

173--1869b   Volcanoes and earthquakes. Can. Nat., n.s. 4: 387-397, (1869). [See 1870a.]

174--1869c   The magnetic iron sands of Canada. Can. Nat., n.s. 4: 467-469, (1869).

175--1869d   [Description of the New England granite formation.] Essex Inst., B. 1: 106-107, (1869).

176--1869e   Borings for oil in southwestern Ontario region [abst.]. Am. Nat. 2: 388, (1869).

177--1869f   On the geology of northeastern America [abst.]. Am. Nat. 3: 442, (1869).

178--1870   Report [on the Goderich salt region: on iron and iron ores]. Can. Geol. Surv., Rept. Prog. 1866-69: 211-304, (1870). In part, Can. Nat., n.s. 6: 70-89, (1871).

179--1870a   Volcanoes and earthquakes [abst.]. Am. Geogr. Stat. Soc., J. (2)2: 89-98, (1870); Can. Nat., n.s. 4: 387-397, (1869 [1870]). [See 1869b.]

180--1870b   On American iron sands [abst.]. A.A.A.S., Pr. 19: 131-132, (1870).

181--1870c   On Laurentian rocks in eastern Massachusetts. Am. J. Sci. (2)49: 75-78, (1870); Can. Nat., n.s. 5: 7-10, (1870).

182--1870d   Contributions to the chemistry of copper. Am. J. Sci. (2)49: 153-157, (1870); Acad. Sci. Paris, C.R. 69: 1357-1360, (1869); Osterreische Zs. Berg. Hüttenw. 18: 157-159, (1870); Abst., Can. Nat., n.s. 4: 324, (1869).

183--1870e   On norite or labradorite rock. Am. J. Sci. (2)49: 180-186, 398, (1870); Can. Nat., n.s. 5: 31-38, (1870).

184--1870f   On the geology of eastern New England. Am. J. Sci. (2)50: 83-90, (1870); Can. Nat., n.s. 5: 198-205, (1870).

185--1870g   On Laurentian rocks in Nova Scotia. Am. J. Sci. (2)50: 132-134, (1870).

186--1870h   Notes on granite rocks [abst.]. A.A.A.S., Pr. 19: 159-161, (1870). [See 1871b.]

187--1870i   On granite and granitic vein stones. Can. Nat., n.s. 5: 388-406, (1870); Am. J. Sci. (2)50: 82-89, 182-191, (1871); (3) 1: 115-125, (1872).

188--1870j   (On the black iron sand of sea beaches [abst.]). Am. Nat. 4: 569-570, (1870).

189--1870k   The liquefaction of rocks. Geol. Mag. 7: 60-61, (1870).

190--1870l   Review of Hart's Geology of Brazil. The Nation, (Dec. 1, 1870).

191--1871   On the chemistry of the earth. Smith. Inst., Ann. Rept. 1869: 182-207, (1871).

192--1871a   [Hunt's Presidential Address to A.A.A.S.] I. The Geognosy of the Appalachian System, and II. The Origin of Crystalline Rocks. A.A.A.S., Pr. 20: 1-59, (1871); Am. Nat. 5: 450-509, (1871). Abridged, Nature 5: 15-17, (1871); Abst., Geol. Mag. 9: 76-78, (1872). [C. & G.E.]

193--1871b   Notes on granitic rocks. Am. J. Sci. (3)1: 82-89, 182-191, (1871); (3)3: 115-125, (1872); Can. Nat., n.s. 5: 388-406, (1870 [1871]). Abst., A.A.A.S., Pr. 19: 159-161, (1871). [C. & G.E.]

194--1871c   On a mineral silicate injecting Paleozoic crinoids. Am. J. Sci. (3)1: 379-380, (1871); Can. Nat., n.s. 5: 449-451, (1870 [1871]).

195--1871d   On astronomy and geology [abst.]. Can. Nat., n.s. 5: 460-462, (1870 [1871]).

196--1871e   On the oil-bearing limestone of Chicago. Am. J. Sci. (3)1: 420-424, (1871); Can. Nat., n.s. 6: 54-59, (1871); A.A.A.S., Pr. 19: 157-159, (1871); Am. Chem. 2: 27-29, (1872). [C. & G.E.]

197--1871f   Mineral silicates in fossils. Am. J. Sci. (3)2: 57-58, (1871); Am. Nat. 5: 445-447, (1871).

198--1871g   On the oil wells of Terre Haute, Indiana. Am. J. Sci. (3)2: 369-371, (1871); Am. Nat. 5: 576-577, (1871); Ind. Geol. Surv., Ann. Rept. 2: 135-136, (1871).

199--1871h   Messrs. King and Rowney on Eozoon canadense. Roy. Irish Acad., Pr. (2)1: 123-127, (1871).

200--1871i   [On the porphyries of the coast of Massachusetts.] Essex Inst., B. 3: 53-54, (1871).

201--1871j   The mountain of Montarville and its geological history [abst.]. Can. Nat., n.s. 6: 224-226, (1871).

202--1871k   Notes on the Hunt and Douglas new process for extraction of copper from its ores. Am. Chem. 1: 199-200, (1871).

203--1872   Report [on silver ores from Eureka mine, near Port Hope, B.C., and on coal and lignites]. Can. Geol. Surv., Rept. Prog. 1871-72: 66-67, (1872).

204--1872a   History of the names Cambrian and Silurian in geology. Can. Nat., n.s. 6: 281-312, 417-448, (1872); Geol. Mag. 10: 385-395, 453-461, 504-510, 561-566, (1873). [C. & G.E.]

205--1872b   On Alpine geology. Am. J. Sci. (3)3: 1-15, (1872). [C. & G.E.]

206--1872c   Remarks on the late criticisms of Prof. Dana. Am. J. Sci. (3)4: 41-52, (1872).

207--1872d   The origin of crystalline rocks. Am. Chem. 2: 291-292, (1872).

208--1872e   On labradorite rocks in New Hampshire and Colorado. Geol. Surv. N.H., Rept. Prog. 1871: 13-14, (1872).

209--1872f   On the geology of the vicinity of Boston. Boston Soc. Nat. Hist., Pr. 14: 45-49, (1872).

210--1872g   Osservazioni intorno alla geologia del gruppo del Monte Bianco. Transl., Firenze, B. Com. Geol. Ita. 3: 131-140, (1872).

211--1872h   Remarks on the extraction of bismuth from certain ores. A.I.M.E., Tr. 1: 260-261, (1872).

212--1873   The geognostical history of the metals. A.I.M.E., Tr. 1: 331-342, (1873).

213--1873a   [Iron ores of the ancient crystalline rocks of northern New York.] A.I.M.E., Tr. 1: 370-371, (1873).

214--1873b   Remarks on an occurrence of tin ore at Winslow, Maine. A.I.M.E., Tr. 1: 373-374, (1873).

215--1873c   The origin of metalliferous deposits. A.I.M.E., Tr. 1: 413-426, (1873); Van Nostrand's Eng. Mag. 11: 326-334, (1874); Ky. Geol. Surv., Rept. Prog. 2 n.s.: 301-317, (1877); also in Half-hour recreations in popular science (Dana Estes, ed.) No. 10: 375-391, Boston [1873].

216--1873d   On some points in dynamical geology. Am. J. Sci. (3)5: 264-270, (1873). [C. & G.E.]

217--1873e   On the copper deposits of the Blue Ridge. Eng. Min. J. 16: 25-26, 89-90, 106-107, (1873); in part, Am. J. Sci. (3)6: 305-308, (1873).

218--1873f   On the various theories to account for the phenomena of volcanism. Boston Soc. Nat. Hist., Pr. 15: 250-252, (1873).

219--1873g   [On concentric lamination in rocks.] Boston Soc. Nat. Hist., Pr. 15: 261-262, (1873).

220--1873h   [Discussion of C. H. Hitchcock's "Classification of the rocks of New Hampshire."] Boston Soc. Nat. Hist., Pr. 15: 304-309, (1873).

221--1873i   On the crystalline schists of the Green and White Mountain series. Boston Soc. Nat. Hist., Pr. 15: 309-310, (1873).

222--1873j   [Eulogy of John Torrey.] Boston Soc. Nat. Hist., Pr. 15: 312-315, (1873).

223--1873k   On the Eozoon canadense. Irish Acad., Pr. 1: 123-127, (1873-1874).

224--1873l   [Progress in] Geology. In Annual Record of Science and Industry for 1872: 32-39, (1873); ... 1873: 44-54, (1874); ... 1874: 67-76, (1875); ... 1875: 99-114, (1876); ... 1876: 89-104, (1877); ... 1877: 165-182, (1878); ... 1878: 287-312, (1879).

225--1873m   Commemorative notice of Adam Sedgwick. Ann. Rept. Council, A.Ac.A.S., 1873, 4 p., (1873).

226--1874   Century's progress in theoretical chemistry. Philadelphia, Penna., 1874, 15 p., (From Am. Chem., Aug. & Sept. 1874).

227--1874a   The coal and iron of southern Ohio considered with relation to the Hocking Valley coal field and its iron ores ... 78 p., Salem, Mass., (1874).

228--1874b   The paleogeography of the North American continent. Am. Geogr. Soc., J. 4: 416-431, (1874).

229--1874c   Decomposition of crystalline rocks. Boston Soc. Nat. Hist., Pr. 16: 115-117, (1874); Abst., Am. J. Sci. (3)7: 60-61, (1874).

230--1874d   On the stratification of rock masses. Boston Soc. Nat. Hist., Pr. 16: 237-239, (1874).

231--1874e   [On the geologic occurrence of glauconite and fossil resins.] Boston Soc. Nat. Hist., Pr. 16: 301-302, (1874).

232--1874f   The deposition of clays. Boston Soc. Nat. Hist., Pr. 16: 302-304, (1874).

233--1874g   On Dr. Genth's researches on corundum and its associated minerals. Boston Soc. Nat. Hist., Pr. 16: 332-335, (1874).

234--1874h   Supplementary note on the geology of the north shore of Lake Superior. A.I.M.E., Tr. 2: 58-59, (1874).

235--1874i   The Ore Knob copper mine and some related deposits [Ashe Co., N.C.] (with discussion by R. W. Raymond). A.I.M.E., Tr. 2: 123-129, (1874).

236--1874j   The coals of the Hocking Valley, Ohio. A.I.M.E., Tr. 2: 273-278, (1874); Eng. Min. J. 17: 182-183, (1874).

237--1874k   [Discussion of J. C. Smock's "The magnetic ores of New Jersey ..."] A.I.M.E., Tr. 2: 314-333, (1874); Eng. Min. J. 17: 293-294, 306-307, 326-327, (1874).

238--1874l   The disintegration of rocks and its geological significance [abst.]. A.A.A.S., Pr. 22: B39-B41, (1874 [1875]).

239--1874m   On wet processes of copper extraction [abst.]. A.A.A.S., Pr. 22: B78-B79, (1874 [1875]).

240--1874n   Notes on the geology and economic mineralogy of the southeastern Appalachians [abst.]. A.A.A.S., Pr. 22 (2): 113-115, (1874).

241--1874o   The metamorphism of rocks [abst.]. A.A.A.S., Pr. 22 (2): 115-116, (1874); Can. Nat., n.s. 7: 162, (1874).

242--1874p   Geology of southern New Brunswick [abst.]. A.A.A.S., Pr. 22 (2): 116-117, (1874).

243--1874q   Breaks in the American Paleozoic series [abst.]. A.A.A.S., Pr. 22 (2): 117-119, (1874); Can. Nat., n.s. 7: 160-161, (1874).

244--1874r   Remarks on Prof. Newberry's paper on "Circles of deposition, etc.," A.A.A.S., Pr. 22 (2): 196-198, (1874).

245--1875   Chemical and Geological Essays. xxii + 489 p., Boston, Mass.: James R. Osgood & Co., (1875). Notice by J. D. Dana, Am. J. Sci. (3)9: 102-109, (1875). 2nd ed., xlvi + 489 p., Salem, Mass., (1878). 3rd ed., xlvi + 489 p., New York, N.Y., (1891). 4th ed. (with new preface), xlvi + 489 p., New York, N.Y., (1891).

246--1875a   The development of our mineral resources. Harper's Magazine, p. 82-94, (1875).

247--1875b   Report [on Hoosac Tunnel]. In Boston, Hoosac Tunnel, and Western Railroad Company, Report of the Corporators (Mass. House Document No. 9), Appendix, Boston, (1875).

248--1875c   On the Boston artesian well and its waters. Boston Soc. Nat. Hist., Pr. 17: 486-488, (1875).

249--1875d   [Discussion of W. H. Niles', "The physical features of Massachusetts."] Boston Soc. Nat. Hist., Pr. 17: 507-508, (1875).

250--1875e   Remarks on the relations of primordial and crystalline works in New England and elsewhere. Boston Soc. Nat. Hist., Pr. 17: 508-510, (1875).

251--1875f   The decayed gneiss of Hoosac Mountain [Mass.]. Boston Soc. Nat. Hist., Pr. 18: 106-108, (1875); A.I.M.E., Tr. 3: 187-188, (1875).

252--1875g   Prof. J. D. Dana on the alteration of rocks. Boston Soc. Nat. Hist., Pr. 18: 108-112, (1875).

253--1875h   The disintegration of rocks and its geological significance [abst.]. A.A.A.S., Pr. 22: B39-B41, (1875); Am. Nat. 9: 471-473, (1875). [See 1874ℓ.]

254--1875i   On the cement of some natural and artificial stones [abst.]. A.A.A.S., Pr. 23: 106-107, (1875).

255--1875j   The sewage question chemically considered [abst.]. A.A.A.S., Pr. 23: 107-109, (1875).

256--1875k   [Discussion of E. C. Pechin's "The minerals of southwestern Pennsylvania."] A.I.M.E., Tr. 3: 399-408, (1875); Eng. Min. J. 19: 146-147, 226, (1875).

257--1875ℓ   [Discussion of F. Prime, Jr.'s "On the occurrence of the brown hematite deposits of the Great Valley."] A.I.M.E., Tr. 3: 410-417, (1875); Am. J. Sci. (3)9: 433-440, (1875); Eng. Min. J. 20: 285-287, (1875).

258--1875m   The geological survey of Missouri. Am. Nat. 9: 240-245, (1875).

259--1875n   Remarks on hematite iron ores, eastern United States. A.I.M.E., Tr. 3: 417-422, (1875).

260--1875o   Deposition of sediment. Am. J. Sci. (3)9: 61-62, (1875).

261--1875p   Celestial chemistry. Pop. Sci. Monthly, 6: 420-422, (1875).

262--1876   The Cornwall iron mine and some related deposits in Pennsylvania. A.I.M.E., Tr. 4: 319-325, (1876).

263--1876a   A new ore of copper and its metallurgy. A.I.M.E., Tr. 4: 325-328, (1876).

264--1876b   A century's progress in theoretical chemistry. Am. Chem. 5: 46-51, (1876); Pop. Sci. Month., 7: 420, (1876).

265--1877   The Goderich salt region [Ont.]. A.I.M.E., Tr. 5: 538-560, (1877); Eng. Min. J. 23: 167-168, 185-186, 204, 215-217, (1877); Can. Geol. Surv., Rept. Prog. 1876-77: 221-243, (1878).

266--1877a   The Quebec group in geology. Boston Soc. Nat. Hist., Pr. 19: 2-4, (1877).

267--1877b   On the history of the crystalline stratified rocks [abst.]. A.A.A.S., Pr. 25: 205-208, (1877).

268--1877c   Geology of eastern Pennsylvania. A.A.A.S., Pr. 25: 208-212, (1877).

269--1877d   The geology of the older rocks of western America [abst.] Geol. Mag. (2) 4: 574, (1877); A.A.A.S., Pr. 26: 265-266, (1878).

270--1877e   Progress of Geology. Harper's Annual of Science, p. 164-182, (1877).

271--1878   Special Report on the Trap Dikes and Azoic Rocks of Southeastern Pennsylvania. 2nd ed., xxi + 253 p., il. Harrisburg, Pa.: Pa. Geol. Surv., (1878).

272--1878a   [Introductory Remarks (as President) at Philadelphia Meeting.] A.I.M.E., Tr. 6: 18-20, (1878).

273--1878b   [Discussion of J. C. Smock's "The fire-clays and associated plastic clays, kaolins, feldspars, and fire-sands of New Jersey, ...] A.I.M.E., Tr. 6: 177-192, (1878); Eng. Min. J. 25: 185, 200, (1878).

274--1878c   [Discussion of W. E. C. Eustis' "The nickel ores of Oxford, Quebec, Can."] A.I.M.E., Tr. 6: 209-213, (1879); Eng. Min. J. 25: 187, (1878).

275--1878d   On the geology of the Eozoic rocks of North America. Boston Soc. Nat. Hist., Pr. 19: 275-279, (1878).

276--1878e   The origin and succession of the crystalline rocks. Geol. Mag. (2) 5: 466-473, (1878); Nature 18: 443-445, (1878); British Assoc., Rept. 48: 536, (1878).

277--1878f   The geological relations of the atmosphere. Acad. Sci. Paris, C.R. 87: 452-454, (1878); Abst. in Nature 18: 475, (1878); British Assoc., Rept. 48: 544, (1878).

278--1878g   Progress of Geology. Harper's Annual of Science, p. 287-312, (1878).

279--1879   The history of some pre-Cambrian rocks in America and Europe. Am. J. Sci. (3)19: 268-283, (1879); Can. Nat., n.s. 9: 257-275, (1880); A.A.A.S., Pr. 28: 279-296, (1880); Abst., Boston Soc. Nat. Hist., Pr. 20: 140-141, (1879); also Am. J. Sci. (3)19: 268-283, (1880).

280--1879a   The coal and iron of the Hocking Valley, Ohio. A.I.M.E., Tr. 7: 313-315, (1879); Abst., Eng. Min. J. 27: 200-201, (1879).

281--1879b   [Discussion of J. F. Blandy's "The Lake Superior copper rocks in Pennsylvania."] A.I.M.E., Tr. 7: 331-339, (1879).

282--1879c   Table of Geological Formations (p. 51); Eozoic (p. 10) and Dominion of Canada (p. 52-55), <u>in</u> Macfarlane's American Geological Railway Guide, (1879).

283--1880   On the iron-bearing and associated rocks of the Marquette region, and comparisons with the Archean of Canada and of the eastern United States. [Wis. Geol. Surv.]. Geol. Wis. 3: 657-660, (1880).

284--1880a   The Taconic system in geology [abst.]. Am. Nat. 15: 494-496, (1881); Can. Nat., n.s. 9: 429-431, (1880).

285--1880b   Sur les limites du terrain cambrien. Int. Geol. Cong., Paris 1878, C.R. 99-100, (1880).

286--1880c   Des terrains pré-Cambriens dans l'Amérique du Nord. Int. Geol. Cong., Paris 1878, C.R. 229-233, (1880).

287--1880d   The chemical and geological relations of the atmosphere. Am. J. Sci. (3)19: 349-363, (1880).

288--1880e   On the recent formation of quartz and on silicification in California. Am. J. Sci. (3)19: 371-372, (1880); Can. Nat., n.s. 9: 435-437, (1880); Eng. Min. J. 29: 369, (1880).

289--1880f   The genesis of certain iron ores [abst.]. <u>Science</u> 1: 209, (1880); Can. Nat., n.s. 9: 431-433, (1880).

290--1880g   On the origin of anthracite. <u>Science</u> 1: 303, (1880); Can. Nat., n.s. 9: 434-435, (1880).

291--1881   Coal and iron in southern Ohio; the mineral resources of Hocking Valley ... xii + 152 p., map, Boston, Mass.: S. E. Cassino, (1881).

292--1881a   Pre-Cambrian rocks. Can. Nat., n.s. 10: 126-127, (1881).

293--1881b   Historic notes on cosmic physiology [abst.]. A.A.A.S., Pr. 30: 48-50, (1881).

294--1881c   Remarks on Pre-Cambrian rocks of Great Britain. Boston Soc. Nat. Hist., Pr. 30: 104-141, (1881).

295--1881d   The hydrometallurgy of copper and its separation from the precious metals. A.I.M.E., Tr. 10: 11-25, (1881).

296--1881e   The domain of physiology, or Nature in thought and language. London, Edinburgh and Dublin Philos. Mag. 12: 233-253, (1881). [M.P. & P.]

297--1882   Mineral physiology; an address delivered before Vassar Brother's Institute, Poughkeepsie, N.Y., Nov. 28, 1882, 21 p. [privately published?, 1882?].

298--1882a   Sur les terrains éozoiques ou précambriens. Soc. Géol. France, B. 3 (10): 26-28, (1882).

299--1882b   [On the pre-Cambrian or Eozoic rocks of Europe as compared with those of North America][abst.]. Geol. Soc. London, Q.J.: Pr. 4-5, (1882); Geol. Mag. (2)9: 38-39, (1882).

300--1882c   Celestial chemistry from the time of Newton. Philos. Soc. Cambridge, Pr. 4: 3, (1882); Am. J. Sci. (3)23: 123-133, (1882). [M.P. & P.]

301--1882d   The relations of the natural sciences--Inaugural Address before the Royal Society of Canada. Can. Nat., n.s. 10/5: 257-264, (1882); with additions, Pop. Sci. Monthly, 22: 165-172, (1882).

302--1882e   [A.A.A.S. Meeting at Montreal, Quebec, in August 1883.] [Address of welcome], Pr. 31: 613-616; [Remarks and Notices], Pr. 31: 618-639, (1882).

303--1882f   The Eozoic rocks of central and southern Europe--The serpentines of Italy. A.A.A.S., Pr. 31: 419, (1882). [Title only.]

304--1883   The geological history of serpentines, including notes on pre-Cambrian rocks. Roy. Soc. Can., Pr. Tr. 1 (iv): 165-215, (1883). [M.P. & P.]

305--1883a   A historical account of the Taconic question in geology, with a discussion of the relations of the Taconian series to the older crystalline and to the Cambrian rocks. Roy. Soc. Can., Pr. Tr. 1 (iv): 217-270, (1883); Pr. Tr. 2 (iv): 125-157, (1885); Abst., Science 3: 675-676, (1884). [M.P. & P.]

306--1883b   Coal and iron in Alabama. A.I.M.E., Tr. 11: 236-248, (1883); Eng. Min. J. 35: 113-115, (1883); Abst., Science 1: 101-102, (1883).

307--1883c   The decay of rocks geologically considered. Am. J. Sci. (3) 26: 190-213, (1883); Abst., Science 1: 324-325, (1883); Am. Nat. 17: 645-646, (1883).

308--1883d   The geology of Port Henry, N.Y. Can. Nat., n.s. 10: 420-422, (1883); Sci. Am. Suppl. 8: 3096, (1883).

309--1883e   The geology of Lake Superior. Science 2: 218-219, (1883).

310--1883f   The Pre-Cambrian rocks of Wales. Science 2: 403, (1883).

311--1883g   Notes on Prof. James Hall's address [Contributions to the geological history of the American continent]. A.A.A.S., Pr. 31: 69-71, (1883).

312--1883h   A classification of the natural sciences. A.A.A.S., Pr. 32: 29-31, (1883). [M.P. & P.]

313--1883i   The pre-Cambrian rocks of the Alps. A.A.A.S., Pr. 32: 239-242, (1883); Am. Nat. 17: 1099-1102, (1883); Abst., Science 202: 322-333, (1883).

314--1883j   Note on age of rocks on border of Trias near the iron mines of Pennsylvania. Am. Philos. Soc., Pr. 21: 458, (1883). [Discussed by P. Frazer.]

315--1883k   The serpentine of Staten Island, N.Y. [abst.]. Science 2: 242-243, (1883); Am. Nat. 17: 1037-1039, (1883); A.A.A.S., Pr. 32: 242-243, (1884); Science 202: 323, (1883).

316--1884   The apatite deposits of Canada. A.I.M.E., Tr. 12: 459-468, (1884); Eng. Min. J. 37: 138-140, (1884); Can. Rec. Sci. 1: 65-75, (1885).

317--1884a   The genesis of crystalline rocks. Am. Nat. 18: 605-607, (1884).

318--1884b   On Cambrian rocks of North America [abst.]. Am. Nat. 18: 409-411, (1884); Can. Rec. Sci. 1: 77-81, (1884).

319--1884c   On the chemistry of the natural silicates [abst.]. British Assoc., Pr. 48: 679, (1884).

320--1884d   The Eozoic rocks of North America [abst.]. Can. Rec. Sci. 1: 82-88, (1884); Geol. Mag. (3)1: 506-510, (1884); British Assoc., Rept. 54: 727-728, (1885).

321--1884e   [Record of recent scientific progress in] Geology. Smith. Inst.,Ann. Rept. 1882: 126, 325-345, (1884); Extr. in The Virginias 5: 141,161, (1884).

322--1884f   The origin of crystalline rocks. Roy. Soc. Can., Pr. Tr. 2 (iii): 1-67, (1885); Abst., Can. Rec. Sci. 1: 75-77, (1884); Science 3: 674-675, (1884).

323--1885   Les divisions du système éozoique de l'Amérique du Nord. Soc. Géol. Belgique, An. 12: Mem 3-10, (1885).

324--1885a   The classification of natural silicates. Am. Nat. 19: 795-798, (1885). [Also see 1884c.]

325--1885b   The geognosy of crystalline rocks [abst.]. Can. Rec. Sci. 1: 147-148, (1885).

326--1885c   Remarks on natural gas in Canada. A.I.M.E., Tr. 13: 782, (1885).

327--1885d   Biographical Notice of Benjamin Silliman [Jr.]. A.I.M.E., Tr. 13: 782-785, (1885).

328--1885e   [Record of scientific progress, 1883] Geology. Smith. Inst., Ann. Rept. 1883: 443-464, (1885).

329--1885f   The geology of the Scottish Highlands. Science 5: 87-89, (1885).

330--1885g   Observations sur les roches magnésiennes du groupe de la rivière Hudson, que M. Logan a décrites dans la séance du 7 mai 1884, p. 104: Soc. Géol. France, B. (2) 12/2: 1029-1032, (1885).

331--1886   Mineral Physiology and Physiography: a second series of chemical and geological essays ... xvii + 710 p., Boston, Mass.: S. E. Cassino, (1886) [Reviewed in Am. Geol. 8: 110, (1886)]; 2nd ed. with new preface, (1890).

332--1886a   A natural system of mineralogy, with a classification of native silicates, p. 279-401 in [M.P. & P.]; see preceding reference. See also: Roy. Soc. Can., Pr. Tr. 3 (iii): 25-93, (1886); Abst., Can. Rec. Sci. 1: 129-135, 244-247, (1885), Ibid. 2: 116-119, (1886); Am. J. Sci. 132: 410, (1886).

333--1886b   An electrical furnace for reducing refractory ores. A.I.M.E. Tr. 14: 492-495, (1886); Can. Rec. Sci. 2: 52-55, (1887).

334--1886c   Note on the apatite region of Canada. A.I.M.E., Tr. 14: 495-496, (1886); Can. Rec. Sci. 1: 65, (1886).

335--1886d   Apatite deposits in Laurentian rocks [abst.]. A.A.A.S., Pr. 34: 199, (1886).

336--1887   The genetic history of crystalline rocks. Roy. Soc. Can., Pr. Tr. 4 (iii): 7-37, (1887); Abst., Can. Rec. Sci. 1: 147, (1887).

337--1887a   The Law of Volumes in Chemistry. Can. Rec. Sci. 2: 261-264, (1887).

338--1887b   Supplement to "A natural system in mineralogy, etc. ..." Roy. Soc. Can., Pr. Tr. 4 (iii): 63-80, (1887).

339--1887c   The Taconic question re-stated. Am. Nat. 21: 114-125, 238-250, 312-320, (1887).

340--1887d   Elements of primary geology. Geol. Mag. (3)4: 403-500, (1887); Abst., British Assoc., Rept. 57: 704-705, (1888); Eng. Min. J. 34: 219, (1887); Nature 36: 574, (1887).

341--1887e (and Douglas, J.) The Sonora [Mexico] earthquake of May 3, 1887. Am. Nat. 21: 1104-1106, (1887); British Assoc., Rept. 57: 712-713, (1888); Seism. Soc. Japan, Tr. 12: 29-31, (1888).

342--1887f Remarks on Hocking Valley District. A.I.M.E., Tr. 15: 754, (1887).

343--1887g Further notes on the hydrometallurgy of copper. A.I.M.E., Tr. 16: 80-82, (1887).

344--1887h Gastaldi on Italian Geology and the crystalline rocks. Geol. Mag. (3)4: 531, (1887); Abst., British Assoc., Rept. 57: 703-704, (1887).

345--1887i Chemical integration. Am. J. Sci. (3)34: 116, (1887).

346--1887j A New Basis for Chemistry. A Chemical Philosophy, xiv + 165 p. Boston, Mass.: S. E. Cassino, (1887). Translated by Prof. W. Spring into French and published in Paris and Liége as Un Système Chèmique Nouveau in 1889.

347--1888 On subdivisions, unconformities, characteristics, etc. of the Lower Paleozoic formations. Int. Cong. Geol., Am. Comm. Repts., 1888A, p. 68-69, (1888).

348--1888a On the study of mineralogy. British Assoc., Rept. 58: 627-630, (1889); Can. Rec. Sci. 3: 236-242, (1888).

349--1888b The theory of solution [abst.]. British Assoc., Rept. 58: 636-637, (1888); published in full in Chem. News, Sept. 28, 1888.

350--1888c Mineralogical evolution [abst.]. British Assoc., Rept. 58: 682-684, (1888); Can. Rec. Sci. 3: 242-245, (1888).

351--1888d (with Frazer, P., Jr. et al.) Report of the Subcommittee on the Archean, in International Congress of Geologists, American Committee, Repts. ... A: 74 p., Philadelphia, Pa., (1888); Am. Geol. 2: 143-192, (1888); Int. Geol. Cong., IV, London, 1888, C.R. App. A: 13-86, (1891).

352--1888e (with Winchell, N. H. et al.) Report of the Subcommittee on the Lower Paleozoic, in International Congress of Geologists, American Committee, Repts. ... B: 37 p., Philadelphia, Pa., (1888); Am. Geol. 2: 193-224, (1888); Int. Geol. Cong., IV, London, 1888, C.R. App. A: 87-120, (1891).

353--1889 The classification and nomenclature of metalline minerals. Am. Philos. Soc., Pr. 25: 110-180, (1889); Abst., Roy. Soc. Can., Pr. Tr. 6 (iii): 61-63, (1889).

354--1889a On garnet-veins of the Laurentian formation. A.I.M.E., Tr. 17: 594, (1889).

355--1890 The iron ores of the United States. A.I.M.E., Tr. 19: 3-17, (1891); Eng. Min. J. 50: 601-602, 622-624, (1890); Iron Steel Inst. J. 1890, II: 628-644, [1891].

356--1890a [Letter concerning Peter von Turner.] A.I.M.E., Tr. 19: xxi, (1890).

357--1890b The geological history of the Quebec Group. Am. Geol. 5: 212-225, (1890).

358--1890c Notes on Geology of Eastern New York, p. 137, in Macfarlane's American Geological Railway Guide, 2nd ed., (1890). (See 1879c.)

359--1890d Co-efficient of mineral condensations in chemistry. Am. Chem. J. 12: 565-585, (1890).

360--1891   [On the use of the term Ordovician.] Int. Geol. Cong., IV, London 1888, C.R.: 225-226, (1891).

361--1891a   Systematic Mineralogy Based on a Natural Classification, xvii + 391 p. New York, N.Y.: Scientific Pub. Co., (1891).

BIOGRAPHICAL SKETCHES AND MEMORIALS OF
THOMAS STERRY HUNT (1826-1892)

(1891) HUNT, Thomas Sterry, in The National Cyclopaedia of American Biography 3: 254; New York, N.Y.: James T. White & Co., (1891).

(1891) [Anonymous]--Prominent men in the mining industry. Eng. Min. J., p. 527, (Nov. 7, 1891).

(1892) Raymond, R. W.--The late Thomas Sterry Hunt. Eng. Min. J., p. 224-225, (Feb. 1892). [No bibliography.]

(1892) Crosby, W. O.--Thomas Sterry Hunt. Am. Acad. Arts Sci. (Biographical Notices), p. 367-372, (May 1892). [No bibliography.]

(1892) Dawson, J. W.--Thomas Sterry Hunt. Can. Rec. Sci. 5: 145-149, port., (1892). [No bibliography.]

(1893) Douglas, J.--Biographical notice of Thomas Sterry Hunt. A.I.M.E., Tr. 21: 400-410, (1893). [No bibliography.]

(1893) Frazer, P.--Thomas Sterry Hunt, M.A., D.Sc., L.L.D., F.R.S. Am. Geol. 11: 1-13, port., (1893). [No bibliography.]

(1893) La Flamme, J. C. K.--Le Docteur Thomas Sterry Hunt. 16 p. (Extract de l'Annuaire de l'Université Laval 1892-1893), Quebec [P.Q.], (1892).

(1893) Pumpelly, R.--Memorial of Thomas Sterry Hunt. Geol. Soc. Am., B. 4: 379-393, (1893).
    This memorial has an extensive bibliography, but some details of titles are lacking, and many of the journal references are incomplete.

(1898) Douglas, J.--A memoir of Thomas Sterry Hunt, M.D., L.L.D. (Cantab.). (Obituary notice of Thomas Sterry Hunt.) Read before the American Philosophical Society, April 1, 1898. Reprinted from Pr. Am. Philos. Soc. Mem. Vol., (1898), as a separate booklet; 61 p., port., Philadelphia, Pa.: MacCalla & Co., Inc., (1898).
    This is by far the most complete and comprehensive biography of T. Sterry Hunt known to have been published. It contains a sympathetic but candid description of Hunt as a person and as a scientist and includes an excellent bibliography, although some of the titles and dates are incorrect.

(1906) Merrill, G. P.--Contributions to the history of American Geology. No. 135--From the Report of the U.S. National Museum for 1904, p. 189-734, with 37 plates, (1906). Washington: Govt. Printing Office.
    Hunt is referred to briefly on half a dozen different pages.

(1924) Merrill, G. P.--The First Hundred Years of American Geology. 773 p., il.; New Haven, Conn.: Yale Univ. Press, (1924).
    Hunt is referred to or his work discussed on half a dozen different pages, especially p. 446-448.

(1926) The Technology Review has several interesting items about Hunt. The number for December 1925 (Vol. 28, No. 2) has on page 74: "A picture of Thomas Sterry Hunt: Professor of Geology from 1871 to 1878; portrait painted by Horace R. Burdick, and presented to the Institute by William E. Nickerson, '75."

A notice of his death appeared on p. 194-195 of the number for February 1926 (Vol. 28, No. 4), and there is reference to him in T. L. Davis' History of M.I.T.'s Department of Chemistry in Vol. 35, p. 264, (1933) [See Davis (1933) a little farther on in this list.].

(1932) Adams, F. D.--Thomas Sterry Hunt, 1826-1892. Nat. Ac. Sci., Biog. Mem. 15: 205-238, port., (1932).
    This memorial contains an extensive bibliography which, however, is somewhat difficult to use because of incomplete data on many references.
    The biographical part draws heavily on Douglas' 1898 memorial cited earlier on.

(1933) Davis, T. L.--Chemistry at M.I.T.--A history of the Department from 1865-1933. Tech. Rev. 35: 250-252, 264, 266, 268, 270, 272, (1933).
    Hunt's influence at M.I.T. as a chemist is discussed on page 264 of this article and is quoted on a preceding page in this present biography.

[195?] Massachusetts Institute of Technology. Biographies of Members of the Faculty and Officers of Administration. Vol. 1, [195?].
    There is a brief biological sketch of Hunt on page 5.

View of part of an early classroom in "Boston Tech" in which Profs. Hunt, Hyatt, and Niles gave lectures and conducted laboratory instruction in geological subjects.

(M.I.T. Historical Collections)

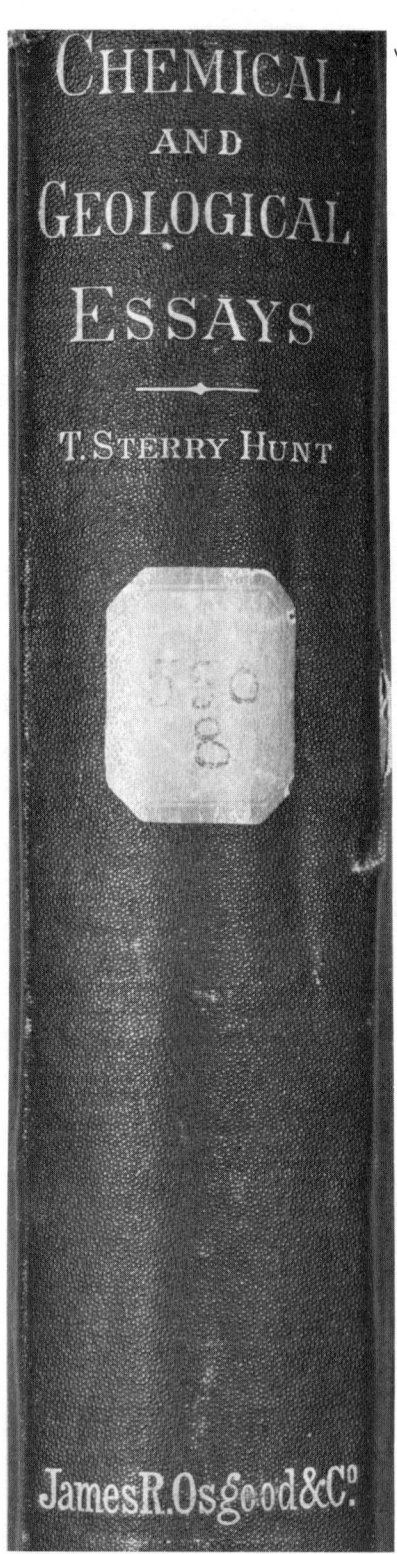

CHEMICAL AND GEOLOGICAL

ESSAYS

BY

THOMAS STERRY HUNT, LL.D.,

Fellow of the Royal Society of London; Member of the National Academy of Sciences of the
United States, the Imperial Leopoldo-Carolinian Academy, the American
Philosophical Society, the American Academy of Sciences,
the Geological Societies of France and Belgium
and of Ireland; Officer of the Order
of the Legion of Honor,
etc., etc., etc.

BOSTON:
JAMES R. OSGOOD AND COMPANY,
Late Ticknor & Fields, and Fields, Osgood, & Co.
1875.

Hunt's Geochemistry book.

(5)
WILLIAM OTIS CROSBY
(1850-1925)

WILLIAM OTIS CROSBY

MIT: 1880-1907

William Otis Crosby, Course VII--Natural History--1876, the fifth professor of geology at M.I.T., was more or less intimately connected with the Institute for more than half a century. He first entered as a student in 1871 and died an emeritus professor fifty-four years later on the last day of 1925. In this long span of years as teacher, author, department head, and consulting geologist, he ably served his students, the Department of Geology, the Institute, and the public clientele that sought his professional advice. His students found him an enthusiastic and stimulating teacher who drew heavily on his diverse geological experiences and rich accumulation of specimens carefully collected for teaching purposes. The Department and Institute received his conscientious direction of the geology program during the interval, 1902-1904, between department heads Niles (1878-1902) and Jaggar (1904-1912). While serving the Institute he also carried on curatorial and research work at the Museum of the Boston Society of Natural History, from 1875 to 1902, and participated actively in the Society's Teachers' School of Science by lecturing and conducting field excursions. For half of his life he carried on an extensive consulting practice as an economic and engineering geologist, gaining a highly respected reputation in both fields. He was one of the very first geologists to work with engineers on excavation and construction projects and was a pioneer in engineering geology. He left a record of hundreds of well-trained students; thousands of valuable specimens at M.I.T. and the Museum of the Boston Society of Natural History; some 150 publications; and more than 100 confidential reports prepared for his clients. The high esteem in which he was held by all who knew him is evident in the tributes paid him both before and after death in laudatory articles, biographical sketches, resolutions, necrologies, and memorials. In these are extended accounts of his ancestry, early family life, travel experiences, professional assignments, etc. The interested reader will find these sources well worth consulting. They are listed at the end of this biography, following the Bibliography.

INTRODUCTION

This sketch concerns itself with six closely related aspects of the life of William Otis Crosby: 1) his birth, ancestry, and early training; 2) his services to M.I.T. as teacher, department head, and collector-curator of specimens for teaching and research (1875-1907-1925); 3) his role as collector and curator, and as author of guidebooks, at the Museum of the Boston Society of Natural History (1875-1902); 4) his authorship of some 150 published contributions to the literature of geology; 5) his activities as a geological consultant on mineral deposits, construction problems, and engineering projects involving tunnels, dam sites and other water supply problems; and 6) his influence, closely coupled with that of his engineer classmate, John R. Freeman (I S.B. 1876), in bringing the importance of geology to the attention of civil engineers.

BIRTH, ANCESTRY, PARENTAGE, AND EARLY TRAINING

William Otis Crosby was born in a log cabin on the northern bank of the Ohio River in Decatur, Ohio, on 14 January 1850 and died of pneumonia in Jamaica Plain (Boston) in the early dawn of 31 December 1925.

He was the first of six children, two boys and four girls, of Francis William Crosby (1823-1909) and Hannah Everett (Ballard) Crosby (1824-1909). His parents, distantly related in a complicated genealogy going back to Edward Everett (1794-1865), traced their ancestry through numerous Crosbys, Ballards and Everetts to English forebears who emigrated to America in the 1630s. Details of this genealogy may be found in Prindle's unpublished Ancestors and Descendants of Timothy Crosby, Jr., Soldier of the American Revolution, constituting item 7 in the Biographical References listed at the end of this biography.

Francis William Crosby, Otis' father, was teaching school in Decatur, Ohio, when he married one of his students, Hannah Everett Ballard, on 3 April 1849 near the end of the school year. The young married couple decided to settle down in Decatur for the time being, and William Otis was born there the following January.

Otis' father was an independent and strong-willed man. He held deep anti-slavery convictions and helped fugitive slaves along toward freedom by maintaining a station in the "Underground Railroad." He also had similar feelings about the sacredness of life and the wrong of needless killing, and invariably refused to eat flesh food. He loved nature, liked nothing better than travel and climbing, and was an able and enthusiastic collector of minerals and rocks.

It is quite obvious from genealogical data and family history that Otis' mother was also an unusually independent person with strong social and intellectual attributes, and endowed with great patience, tolerance and understanding. She lost her fourth and fifth children when they were only a few years old; managed to hold her family together through the hard four years of the Civil War when her husband served in the Union Army;

and many years later efficiently managed Crosby House, a well-known and highly regarded hotel, in Washington, D.C., while her husband, now Capt. Francis William Crosby, was in Europe most of ten years travelling about, collecting geological specimens for several museums, and writing many letters to the folks back home.

During this decade, 1891-1901, when he was 70 years old (having been born in 1823), he wrote that he had climbed Vesuvius 19 times and experienced 21 earthquakes! Even at 83 years he was strong enough to accompany Otis on a 1906 trip to Alaska.

It is easy, therefore, to understand how William Otis derived his own deep convictions about killing animals and eating their flesh. Like his father he refused to eat flesh food, choosing instead to consume "copious" helpings of cheese and bread to assuage his hunger. He had the same love of nature and of adventurous travel as his father, and their climbs together in the Colorado Rockies are interestingly recorded in a journal he kept in 1871, when but 21 years old. From both parents he inherited a strong intellectual curiosity, a deep desire to understand what he saw, and an insatiable thirst for knowledge. He learned poetry by the thousands of lines, read books at every opportunity, carefully prepared himself for every class, field trip, or consulting assignment by reading the pertinent literature, and through his research added substantially to the literature of geology. And last but not least, he was as inveterate a collector of minerals, rocks, and fossils as his father. In short, the finer attributes of both father and mother were passed on in full measure to their first-born, and he used them wisely and well.

It may be medically significant that when death came to Capt. Crosby at age 86 in 1909, to his wife Hannah at age 85 only eight days later, to their son, William Otis at age 75 in 1925, and to Otis' son, Irving Ballard, the last of the line and childless, at age 68 in 1959, it was pneumonia that brought the end in every case!

Young Crosby is reported to have been an excellent pupil in secondary school, to have been "a wonder in mathematics," and to have beaten his teacher regularly in chess. (Lane 1930, p. 520.) Having to serve as man of the house in his very early teens, 1861-1865, while his father was in the Union Army, young Otis matured early and learned how to accept responsibility and manage his own affairs.

His first serious interest in geology and the out-of-doors was aroused when at eighteen the Crosby family moved to Concord, North Carolina. Here Capt. Crosby managed one of the gold mines he had heard about on his march to the sea with General Sherman, and young Otis had his first opportunities to become acquainted with mineral deposits, mining and milling operations, and the great outdoors. He began to read books on geology at this time and shortly afterward, when he worked as a clerk at the Pension Bureau in Washington, he spent so much of the night reading that after two years poor health forced him to return to outdoor life in North Carolina in order to regain his strength. Then came a turning point in his life and the events that ultimately shaped his future as a geologist.

Having acquired a silver mill in Georgetown, Colorado, Capt. Crosby decided to spend the summer of 1871 operating the mill and testing out a special furnace that he planned to design and build for smelting the gold and silver ores of the district. So it came about that father and son set out for Colorado by rail from New York on 18 March 1871.

What our geology neophyte, only recently turned 21, saw, did and thought during the trip West and thereafter until mid-October, was recorded in a clearly written "Journal," titled Geological and other sketches by W.O.C. This interesting and informative record, now preserved in the M.I.T. Institute Archives, was reproduced verbatim in my recent work, The Geologists Crosby of Boston (Shrock 1972). It has been drawn on extensively for the following narrative.

As soon as they got settled down in the Georgetown area, both Crosbys went immediately to work on the new furnace, called the "Ballard" or "stick-in-the-mud" furnace, and ultimately produced a design which proved to be the best in the district. But Otis also found time to visit many of the mines and mills in the Georgetown district and to climb some of the higher peaks. Being a keen observer and learning rapidly from his travels, Otis quickly gained the reputation of being one of the most knowledgeable geologists in the Georgetown area.

It was not surprising, therefore, that he was recommended as a guide when visitors needed such service. So when an M.I.T. party of twenty-one students, four professors, and President John D. Runkle appeared in Georgetown on 22 July 1871, and inquired about a guide, young Crosby was recommended. Soon he was conducting the group from one mill to another, showing them the newly developed Ballard furnaces, guiding them through tunnels and underground workings, and explaining the local geology.

Runkle was greatly impressed with Crosby's practical knowledge of mining methods, smelting practices, and regional geology. One afternoon, as the two of them walked down a road to Georgetown, Runkle asked Crosby if he would be interested in coming to M.I.T. to build two of the Ballard furnaces for the new metallurgy laboratory. While doing this, Runkle pointed out, he could also register as a student and work towards a degree, meanwhile receiving a modest salary for his efforts. Crosby wrote down the discussion with Runkle in the journal he kept that summer, and the following excerpt is an example of how a chance encounter set the course of one young man's life for the rest of his years--more than half a century.

> "Mr. Runkle seems to have taken quite a fancy to me from the start & now as we were walking down to Georgetown, he made me the following offer. That if I would come to their Institute in Boston, my tuition should cost me nothing the first year & the second year he would give me an assistant professorship with a comfortable salary. A few days afterward he promised me a free pass from here to Boston. Now, considering that he had known me but four days and had not had more than fifteen minutes conversation with me, that was certainly a remarkable proposition & I could not but consider it quite complimentary to myself. If the proposition was made in good faith & I believe it was, I doubt if every young man could get as good. Still the old adage holds true; that the prime mover in all human actions is self interest, & so it is in this case. The President contemplates fitting up their laboratory with all the necessary appliances for treating gold and silver ores. Among other things he proposes to build two of our furnaces & a stamp mill, & he undoubtedly sees that with my practical experience in those things I could be of considerable use to them. I am glad that it is so, in fact I could not consent to go if it were otherwise, for it would seem too much as if I was a charity scholar. But as it is, I am determined to go if it is a possible thing, for it is probably the best offer I will ever get from a similar institution."
> (p. 95 of the Journal)

AN M.I.T. UNDERGRADUATE

As a result of his encounter with the M.I.T. party, and the offer from Runkle, Crosby returned to Boston in the fall of 1871 and registered as an M.I.T. freshman in the geology and mineralogy option of Course VII, Natural History, as there was no separate course or degree in geology at that early date.

He found his first year's work interesting and exciting, no doubt in part because of the field experience he had already had, and by the end of the year had definitely decided that he wanted to be a geologist. Being without funds at term's end, however, he had to drop out during the school year of 1872-1873 in order to earn enough money to continue his studies.

Returning to Georgetown at the close of the spring term, he again took up work with his father during the remainder of 1872 and the spring of 1873. By mid-summer, however, he was back in the East, having been attracted by the opportunity to study field science at a new summer school for science teachers being organized by Harvard's famed Louis Agassiz.

So it was that in July of 1873 Otis found himself duly registered in the new school on the Island of Penikese off the Massachusetts coast. This first summer school of natural history in America, later to be known as the Anderson School of Natural History, offered rigorous training for aspiring teachers of science. The student group of 28 men and 16 women would, according to Agassiz' advanced announcement, be offered a

> "Programme of a Course of Instruction in Natural History, to be delivered by the Seaside, in Nantucket,* during the Summer months, chiefly designed for Teachers who propose to introduce the Study into their Schools and for Students preparing to become Teachers."**

Agassiz wished especially to put emphasis on field observation, for he admonished his future students as follows about his plans:

> "... Penikese Island is not to be regarded as a place of summer resort for relaxation. I do not propose to give much instruction in matters which may be learned from books. I want, on the contrary, to prepare those who shall attend to <u>observe for themselves</u>. I would therefore advise all those who wish only to be taught natural history, in the way in which it is generally taught, by recitations, to give up their intention of joining the school."†

The summer's work at Penikese profoundly influenced young Crosby, because he was a keen observer and was just the kind of enthusiastic and serious student that Agassiz wanted for his new school. The habits of collecting, observing, recording, and classifying that he developed at Penikese served him well for the next fifty years as he carried on geological work around Boston, in Massachusetts and other states of the United States, and in many foreign countries.

After Penikese, Crosby reentered M.I.T. in the fall of 1873 and proceeded to graduation in June 1876 without further interruption. By the time he reached his senior year he had become deeply interested in the

---

* As the result of a munificent offer of an utter stranger, a Mr. John Anderson of New York, of the island of Penikese in Buzzard's Bay, along with a dwelling house and barn and an endowment of fifty thousand dollars, Agassiz changed the location of his proposed school, named it the Anderson School of Natural History at Penikese, and expanded certain of his original plans. (See Wilder, p. 190-191, 1898.)

** Quoted by Burt G. Wilder <u>in</u> "Agassiz at Penikese," in the Agassiz Memorial Number of the <u>American Naturalist</u> for March 1898, p. 189-196. The quoted statement (p. 189) was contained in a circular issued by Agassiz on 14 December 1872.

† Quoted by Burt G. Wilder <u>in</u> "Agassiz at Penikese," in the Agassiz Memorial Number of the <u>American Naturalist</u> for March 1898, p. 190.

geology of the Boston area, in part because of work he carried on at the Museum of the Boston Society of Natural History during non-school hours. As a consequence of this interest, he chose to write his bachelor's thesis on the "Geology of Eastern Massachusetts," a subject on which he would write more than thirty articles and reports during the next thirty years. His was the first geology thesis at M.I.T., but because there was no course or degree in geology until 1890, his degree was in the "Geology and Mineralogy" option of Course VII, then called Natural History.

In the early years of the Institute, it was common practice to offer a teaching position to the more promising graduates, partly because they were well trained to carry on and maintain the rigorous academic standards insisted on by Rogers and his colleagues, and partly because scientists of "professorial" stature and reputation were few in number. Like other new academic institutions struggling with the problem of building a faculty with limited funds, M.I.T. had to grow many of its own young professors by recruiting prospective teachers from its annual crop of graduates, after testing them as undergraduate student assistants.

While yet an undergraduate student, Crosby's industry and marked ability at teaching and research attracted the attention of Alpheus Hyatt, then Professor of Paleontology at M.I.T. and Curator of the Museum of the Boston Society of Natural History. Hyatt offered him an assistantship at the Museum, where he could work after hours, and later saw to it that he was appointed a Student Assistant in Paleontology at M.I.T. for his senior year, 1875-1876, and an Assistant in Paleontology for the two years immediately following his graduation, 1876-1878.

MARRIAGE TO ALICE ALZINA BALLARD

On 4 September 1876, soon after receiving his Bachelor of Science degree from M.I.T., William Otis Crosby of Boston and Randolph, New Hampshire, married Alice Alzina Ballard of Lansing, Michigan. The couple were first cousins, their respective genealogies going back to common Ballard and Everett ancestors. Alice was the daughter of Appleton Ballard and Epiphene (Ellenwood) Ballard, and the granddaughter of Frederick Ballard (1780-1868) and Achsah (Everett) Ballard (1784-1857). William Otis was the son of Hannah (Everett) Ballard (1824-1909), who was the daughter of Frederick Ballard (1780-1868) and Achsah (Everett) Ballard (1784-1857) and hence the grandson, as Alice was the granddaughter, of the two last-named ancestors. William Otis and Alice Alzina had two children, Alpheus Ballard Crosby, who died in infancy, and Irving Ballard Crosby, born 4 January 1891. Alice died in Randolph, New Hampshire, on 31 July 1949, and Irving, in the same place on 18 September 1959.

Little is known about the family life of the W. O. Crosbys, except that they lived in Jamaica Plain during the school year and in a family house near Randolph, New Hampshire during the summer months. William Otis was away from home and family quite frequently, because of his consulting assignments, but when at home, he worked long hours, and the Crosbys often entertained their friends in the evening and on Sundays.

WORK AT THE MUSEUM OF THE BOSTON SOCIETY OF NATURAL HISTORY

Crosby's appointment as an assistant at the Museum, starting in 1874, when he was only a sophomore at M.I.T., gave him at an early age the unusual opportunity of working under two renowned scientists-- Alpheus Hyatt, paleontologist, who served the Museum as Curator, as well as being Professor of Paleontology at M.I.T.; and Thomas Tracy Bouvé,* an early Boston naturalist widely known for his important role in building up the Museum's collections and expanding its public services before and during his incumbency as President, i.e. during the 1860s and 1870s.

In typical fashion, young Crosby went promptly to work on his first assignment, which was to bring some order to the Museum's geological collections then in a rather chaotic state. The Eser Paleontological Collection awaited thorough revision and adequate display; the collection of Henry D. Rogers was in great disorder because of careless packing in Glasgow after his death there; and the collection of brother William Barton Rogers lay untouched because of his heavy administrative duties and recurrent illnesses. Altogether, Crosby and his assistant, a Miss Carver, had a formidable task; that they performed satisfactorily is attested by Hyatt's reports to that effect in his Annual Reports on the Museum, starting with the one for 1874-1875, and his appointment of Crosby, upon graduation from M.I.T. in 1876, as Assistant in Geology and Mineralogy, in charge of the Museum's geological and mineralogical collections.

Under the supervision and guidance of Hyatt and Bouvé, Crosby and his assistants worked long and diligently on the Museum's collections, learning how to care for specimens properly, how to identify and classify them, how to exhibit them most effectively for public view, and how best to organize them for laboratory study. So well did they carry out their assignments that teachers of natural science in the Boston schools began to bring their classes to the Museum, and hundreds of students, accompanied by their teachers, visited the Museum annually. One of the teachers is reported to have said:

---

* Crosby's deep appreciation of Bouvé's advice and guidance are well expressed in the brief necrology he published--"Thomas Tracy Bouvé," Amer. Acad. Arts Sci., Pr. July 1897, p. 340-344, 1897.

> "These collections Professor Crosby made into a demonstration of the evolutionary processes of inorganic nature. To the teacher they were an illuminating guide, to the child a story that goes right on."
> (Shimer and Lindgren 1927, p. 38.)

While working at the Museum, chiefly in the later 1880s, Crosby prepared and published "Guides" to a number of the collections. These are listed in the Bibliography (1881, 1886, 1886a, 1889b, and 1892).

One additional task that Crosby performed while working for the Museum was preparation of the first three volumes of a proposed 8-volume work on the "Geology of the Boston Basin." He had first become interested in the geology of the Boston area while an M.I.T. undergraduate, and chose to write his Bachelor's thesis on the "Geology of Eastern Massachusetts." Later, as he gained familiarity with the rocks and structure of the Boston area, in part from field trips with Bouvé, he was encouraged to make an intensive study of the geology of the Boston Basin, and when he took up this study seriously, he was actually paid a salary to carry on the field work. This is supposed to have been the first case of an American scientific society--in this instance the Boston Society of Natural History--actually paying a salary for a specific investigation. This record is included in Hyatt's 1895 Report on the Museum, in which he states:

> "Last year (1894), special attention was called to the scientific investigations of Professor Crosby in the Department of Geology, and to the need of keeping up and providing for such a class of work, if we desired to do our part in the history of Science in New England. We are essentially a local society and ought to strive to do a large share of the local work of investigation. For the past year this Society has paid one of its salaries to this investigator and permitted him to count his purely scientific work in the field and in this building as a return for the money paid him. The amount expended has been very small and the returns as seen by this report, large. So far as I know, however, this Society is the first institution of its class in this country to grant a salary, even though it be a small one, for investigation ..."

During the years he was associated with the Museum, he also participated in the Society's Teachers' School of Science, delivering series of lectures, offering field-lessons on the geology of the Boston Basin, and leading other field excursions to local areas of geological interest. (See the announcement of the 1895 Field-Lessons on the following page.)

Crosby continued to work at the Museum until 1902, when he resigned to accept the added responsibilities of directing the program of the M.I.T. Department of Geology on the retirement of Prof. Wm. H. Niles, Head, late in 1901. The Museum of the Boston Society of Natural History,

Teachers' School of Science.

GEOLOGY OF HINGHAM.

By W. O. Crosby.

Program of Field-Lessons.

May, 1895.

The party will meet each time in the Hingham Railway Station at 2.10 P.M., or on the arrival of the train leaving Boston at 1.15 and Braintree 1.44 P.M.

For every lesson, except the second, a barge will be used, to economize time and minimize the walking; and none of the lessons, probably, will require more than two miles of walking. The expense of the barge will be twenty-five cents per lesson for each member of the party, payable to the driver at the close of each lesson.

Each lesson will be finished in time for the train leaving Hingham at 5.52 P.M., due at Braintree 6.13, and Boston 6.35 P.M.

The districts to be studied in the different lessons are, in a general way, as follows:—

May 15.—Rocky Neck and Planter's Hill. See Part I. (Nantasket and Cohasset.)

May 18.—The Village area, between Main Street and the Hockley district.

May 22.—Melville Garden, Planter's Fields, and Huit's Cove, including also Otis Hill, and the red felsite district.

May 25.—The Beal's Cove area, including also the Stoddard's Neck eskers, and Baker's Hill.

May 29 and June 1.—These two lessons will be devoted to the glacial geology of Hingham, and especially to tracing the successive shores, outlets, and deltas of Lake Bouvé, the great glacial lake of the South Shore.

An example of the careful plans that Crosby prepared for his field excursions.

a close neighbor of M.I.T.'s Rogers Building on Boylston Street, no longer exists, and its exhibitions and collections have long since been dismantled and fragmented. During its existence, however, the imaginative and creative handiwork of William Otis Crosby was much in evidence, and many a school child of the decades at the turn of the century gained its first glimpse of the natural world from his exhibitions and "Guides."

AN M.I.T. FACULTY MEMBER

As previously mentioned, Crosby was carrying a full academic program at M.I.T. during the period 1874-1876 while working at the Museum. Furthermore, Hyatt had seen to it that he was appointed a Student Assistant in Paleontology during his last year, 1875-1876, and when he graduated that he was appointed an Assistant in Paleontology for the two years immediately following his graduation, $i.e.$ 1876-1878.

Thus Crosby passed without interruption from the status of an undergraduate student assistant to that of a regular full-time assistant on the teaching staff. Thereafter, he climbed steadily upward on the academic ladder until 1907 when he was forced to retire because of deafness. His successive appointments were as follows:

```
Student Assistant in Paleontology                          1875-1876
Assistant in Paleontology                                  1876-1878
Assistant in Geology and Paleontology                      1878-1880
Instructor in Geology and Paleontology                     1880-1881
Instructor in Geology, Paleontology, and Mineralogy        1881-1883
Assistant Professor of Mineralogy and Lithology            1883-1892
Assistant Professor of Structural and Economic Geology     1892-1902
Associate Professor of Geology                             1902-1906
Director* of the Department of Geology                     1902-1904
Professor of Geology                                       1906-1907
Emeritus--retired because of deafness in 1907              1907-1925
                                        Died 31 December 1925
```

It is obvious from the preceding list of titles that Crosby taught about every aspect of geology during his career--paleontology, general geology, mineralogy, petrology, structural geology, and economic geology.

---

* When Prof. Niles tendered his resignation as Professor of Geology, and Head of the Department, late in 1901, no successor was appointed immediately; rather, the work of the Department was continued under Crosby's general direction. He wrote and signed the Report of the Geology Department for the school year 1902-03. He relinquished his responsibilities when Dr. Thomas A. Jaggar, Jr. was appointed Head on 12 April 1904. Jaggar also wrote and signed the Report of the Geology Department for the school year 1903-04.

From these data, gleaned from the President's Reports for the three school years, 1901-02, 1902-03 and 1903-04, it seems that Crosby, while in general charge of the Department from late in 1901 to the spring of 1904, was never appointed Head.

He was, in fact, a broadly trained geological generalist, who subsequently developed special expertise as a consultant on mineral deposits, dam and reservoir sites, and foundation problems. When almost total deafness forced him to resign from teaching in 1907, he turned to a full-time career as a consulting geologist in the fields just mentioned and continued an active and highly successful practice until death at the end of 1925.

Crosby's students and memorialists are unanimous in emphasizing three outstanding traits that he exhibited as a teacher. In his lecturing, he clearly and logically presented the subject matter in language befitting a student of literature and now and then spiced with pithy remarks or whimsical poetry which he could recite by the yard. He was also able to make the subject matter more interesting and stimulating because his wide travels permitted him to speak of so many places, features, rocks, minerals, and fossils from firsthand knowledge--a situation that tends to make students have more respect for what their teacher says!*

In the laboratory he was an equally effective and inspiring instructor, emphasizing the importance of observation--the lesson he learned so well with Agassiz at Penikese--and of drawing correct conclusions from the real world, i.e. from the fossils, minerals, rocks, and ores that he so often brought back from his excursions, field trips, and travels to distant places. Already an avid collector when he came under Bouvé's influence at the Museum of the Boston Society of Natural History, he was further stimulated by that enthusiastic collector to make every specimen he collected a thing of very special personal interest, and he happily shared that special interest with his laboratory students.**

Thirdly, Crosby was an enthusiastic leader of excursions and field trips, whether for "proper Bostonian" ladies and gentlemen, M.I.T. students, school children, or his own professional peers. He carefully prepared special leaflets for these trips, instructing participants how to reach the day's "field laboratory," which was generally by train or street car and then afoot, and informing them of what they were supposed to see. Many of these "field lessons," as he called them, were conducted under the

---

* Amadeus W. Grabau (XII S.B. 1896), one of Crosby's students who attained worldwide recognition in stratigraphy and paleontology, wrote of his former professor in the Preface to his Principles of Stratigraphy (1913) as follows:

> "My own interest in ... the principles of Orogenesis and Geodynamics, goes back to the days when, as a student, I listened to the illuminating expositions of that versatile and accomplished exponent of rational geology, Professor William Otis Crosby, at the Massachusetts Institute of Technology."

** See footnote on following page.

auspices of the Teachers' School of Science which was conducted by the Boston Society of Natural History. A typical leaflet is reproduced on a preceding page.

PUBLICATIONS - TEACHING AIDS, REPORTS, AND JOURNAL ARTICLES

In Crosby's day, 1880-1905, there were few textbooks of mineralogy and geology and almost no laboratory manuals of the sort so common today. As a consequence, he wrote his own books and manuals to stimulate his students to develop their own observational powers and to aid them in identifying, organizing and correlating the materials they were studying. Some of these were published by M.I.T.: e.g. A Syllabus ... in Structural Geology (1903); Laboratory Guide to Elementary Lithology (undated); and Notes on Chemical Geology (1897). Others were published outside: e.g.

** Crosby's enthusiasm for minerals and rocks and love of collecting them became manifest when he spent the summer of 1871 working with his father in the famous Georgetown Mining District in the Colorado Rockies. In the "journal" he kept that summer (See Appendix A in Robert R. Shrock's The Geologists Crosby of Boston, 1972), he recorded with delight and enthusiasm the occasional successes they had in discovering good mineral occurrences. One of his senior colleagues, W. H. Niles, wrote as follows about his collecting tendencies as he ended his teaching career in 1907 because of increasing deafness:

> "In 1878, as an Assistant in geology and mineralogy he began to teach classes in these subjects. Each day when he returned from his field studies he brought with him specimens which gave character and practical value to his work as instructor. At first the Institute possessed no geological collections of value, nor could it afford to make appropriations for their purchase. The district about Boston is a rich field for one in quest of the different kinds of rock specimens. These he gathered until he has been enabled to place before each member of his classes selected specimens of each of the leading kinds of rocks of the globe. These have been studied in classes under his personal direction, and the students have thus acquired a practical knowledge of these characteristics. Thus his instructions have been conducted in the fullest spirit of the educational work at our Institute.

> "Professor Crosby has been called a born collector, but the writer thinks of him as an experienced, enthusiastic, and scientific collector. His gatherings of the numerous specimens of the characteristic rocks of so many species in Eastern Massachusetts ... has had another bearing upon his work at the Institute. Exchanges were freely made, and in that way he acquired for the Institute a considerable amount of material for the collections in mineralogy and structural geology."

[Tech. Rev. 9: 174-175, (1907).]

Common Minerals and Rocks (1881; 1893); and Tables for the determination of Common Minerals (1887-1895).

In his original research Crosby's interests ranged far and wide, and his publications exemplify this broad range of interest. His S.B. thesis on the "Geology of Eastern Massachusetts" was the first of more than 30 articles on the subject, and his 3 volumes on the Geology of the Boston Basin, the first three of a projected 8-volume work on which he was working when he died, are still invaluable references. A short camping trip to Trinidad in 1878 and a trip to Cuba in 1881 resulted in eight papers dealing with geological features in those islands. His early interest in ore deposits, no doubt stimulated and encouraged by his father, led to investigations of mineral properties in many parts of the world. From these travels and investigations he gained new knowledge that he set forth in a variety of publications, and brought back large suites of ores and host rock to enrich the Institute's teaching and research collections. Because of these investigations Crosby came to be regarded as one of the leading economic geologists of his day in the United States.

CONSULTING ENGINEERING GEOLOGIST

Among engineers, Crosby's advice on practical geologic matters was highly valued. He was an expert on clay deposits and mineral fuels; he was often called on in court cases to help solve legal controversies involving geological principles; and in his later years he gained an international reputation as an expert on water supply and construction of dams for water storage. His first work in this field started in 1895 when he was appointed as consulting geologist to the Metropolitan Water Commission of Boston. This assignment led to investigation of the foundations for the Wachusett Dam, the North Dike, and conditions along the route of the proposed aqueduct tunnel. Later he conducted the geological work in preparation for construction of the dam across the tidal part of the Charles River, for excavation of the Boston subways and sewage tunnels, the South Boston Dry Dock, and for the new M.I.T. buildings in Cambridge. Farther afield, Crosby was called on for advice on dam foundations and construction by many persons and organizations, and his class-mate, a well-known civil engineer in his day, John R. Freeman, commented *

> "The Director of the U. S. Geological Survey told the writer [i.e. Freeman] a few years ago that Crosby had become probably the best adviser to be found in the United States upon geological conditions at a dam site presenting difficult conditions."

---

* Quoted in "Memorial of William Otis Crosby" by H. W. Shimer and Waldemar Lindgren; Geol. Soc. Am., B. 38: 39, (1927).

Crosby served as consulting geologist for the United States Reclamation Service in connection with several dam sites, including that for the Arrow Dam in Idaho, then the highest in the world, and for the U. S. War Department in regard to the foundations for the Muscle Shoals Dam across the Tennessee River. He was consulting geologist to the Board of Water Supply of New York City for six years, 1906-1912, making geological investigations for the Ashokan and Kensico dams, the deep Hudson River siphon, the Catskill aqueduct, and the underground water of Long Island. Certainly W. O. Crosby deserves to be called one of the earliest engineering geologists of North America, if not, indeed, the father of them all!

CONSULTING WORK WITH JOHN R. FREEMAN

In the waning decades of the 19th century, when far-sighted planners saw the need for greatly increased water and power supplies for the cities of the 20th century, engineers were called upon, especially in eastern North America, to design the dams, tunnels, and other works that would be needed for the storage and distribution of future water supplies.

Among the engineers who would have a prominent part in this planning was John Ripley Freeman, an 1876 graduate from M.I.T.'s Department of Civil Engineering. Freeman served his post-academic apprenticeship under three of the foremost hydraulic engineers of his time--Hiram F. Mills, James B. Francis and Charles S. Storrow--studying the water supply problems of factories in Massachusetts, New Hampshire and Vermont. From this decade of study, 1876-1886, followed by another decade of employment with the Associated Factory Mutual Fire Insurance Companies of Boston, came a series of fundamental papers on the hydraulics of fire streams, the nozzle as an accurate water meter, and friction losses in water-pipe systems, that gained him international recognition.

In 1895 he became a member of the Metropolitan Water Board of Boston, and during the next year he made studies for bringing water to the city by

---

* Crosby's early consulting assignments are discussed at length in my The Geologists Crosby of Boston (1972). Suffice it to remark here that most (75%) of his consulting assignments (a few more than 100) before 1907, when he resigned and retired from teaching at M.I.T. because of deafness, involved mineral deposits; thereafter, he devoted much more time to water-supply problems. It should be noted, however, that he was already involved with problems of the Boston water supply as early as 1895 (e.g. Wachusett dam, Wachusett aqueduct tunnel, etc.), with Strickler tunnel in Colorado in 1900; with underground water of Long Island by 1903; and with the problems of the New York Board of Water Supply by 1906. Most of his major assignments, however, were to come later.

** John Ripley Freeman in The Tech Engineering News for February, 1923, v. 3, p. 241, 248, and 250.

gravity flow in such manner as to do away largely with the necessity for pumping. The development of the Wachusett dam and the great Quabbin Reservoir would utilize these ideas half a century later.

Freeman was the first consulting engineer of the New York Board of Water Supply, as well as its <u>first</u> employee, and his 600-page report on the New York Water Supply, made to Bird S. Coler, Comptroller of the City in 1900, was generally acknowledged to be the most elaborate water works report ever prepared by one engineer (Eng. News-Record, 19 January 1922).

As Chief Engineer of the Committee appointed in 1901 to consider the advisability of constructing a dam across the tidal estuary of the Charles River between Boston and Cambridge, to form a larger water park, Freeman submitted another 600-page report that still commands the admiration and respect of his engineering and scientific colleagues after 70 years.

Among Freeman's numerous achievements, the one of special interest in this biography of W. O. Crosby was his early recognition of the importance of geology to engineering structures and his employment of geologists as consultants. This recognition is well stated in <u>The Technology Review</u> for 1932 (v. 35, p. 64) as follows:

> "He was one of the first engineers to realize the importance of geology in connection with engineering structures and he was probably the greatest factor in America for bringing engineering geology to the service of the engineering profession."

It is in this connection that Freeman enters the story of W. O. Crosby. As classmates, Freeman and Crosby walked together to classes at the old Rogers Building from Trinity Hall on Trinity Avenue to save car fare. Years later, when he wished advice on geological matters, Freeman naturally turned to Crosby for help, and from this developed a long and successful relationship between engineer and geologist that ended only with Crosby's death in 1925.

Freeman, in a much quoted but as yet undiscovered document, commented as follows about this cooperative relationship:

> "The writer many years ago became impressed with the possibilities of use of a trained geologist in helping a consulting engineer to more complete and accurate knowledge of conditions underground; and when he became a member of the original Metropolitan Water Commission in Boston in 1895, he brought about the appointment of Professor Crosby as consulting geologist to the engineers of the Board. Crosby was set at work studying foundations for the Wachusett Dam, the North Dike and conditions along the route of the aqueduct tunnel. There were some particularly puzzling problems as to the formations at the dam sites and the possibilities of deep leakage through pre-glacial gorges filled with sand and gravel. Nothing had been

> said about salary, and when his first account for services was rendered, it came in at the surprisingly small per diem of $10. per day. Upon remonstrating that this was hardly worth the dignity of a Tech professor, his characteristic reply was that he found so many interesting geological problems scattered all along the line that he wanted to spend so much time studying them, that he thought it best to make his per diem small."

And Crosby is reported by Lane (1930, p. 524) to have said:

> "... that it was Freeman who had dragged him from the professor's chair and workshop into his pleasant contacts with construction engineering." [As stated elsewhere, this 'dragging' actually started when Freeman had Crosby appointed to the Metropolitan Water Board of Massachusetts in 1895, and to the Board of Water Supply of New York City in 1906.]

Lane, who knew both Crosby and Freeman quite well, commented as follows about Crosby's work (1930, p. 524):

> "He was one of the first geologists, if not the first, to be regularly employed in helping the civil engineer to a more complete and accurate knowledge of underground conditions, such as were encountered in large waterwork propositions, building the largest dry dock in America, and various large dams."

In the light of the preceding comments, and other published information, it is surprising and perplexing that William Otis Crosby and his many important consulting assignments are essentially ignored in three fairly recent reviews of the early history of engineering geology in America -

1. Mead, W. J., Engineering Geology, p. 571-578 in Geology, 1888-1928, 50th Anniversary Volume of The Geological Society of America, 578 p., 1941.
2. Paige, Sydney, Chmn., Application of Geology to Engineering Practice: Berkey Volume, 327 p., 1950. (Published by The Geological Society of America.)
3. Grout, F. F. and Aldrich, H. R., Memorial to Charles Peter Berkey (1887-1955). Geol. Soc. Am., Bull. 76, p. 45-51, following p. 474, 1965.

It seems to the present author that Freeman the civil engineer and Crosby the consulting engineering geologist, both widely known at the turn of the century, deserve a definite and important place among the very earliest pioneers who started engineering geology in the United States by joining their respective talents.

It is worth a supplementary note here to mention that Crosby's only living son, Irving Ballard Crosby (XII S.B. 1917), was not only a geologist; he also followed rather closely in his father's footsteps by gaining an international reputation as an expert on water-supply and dam sites.

For further details on Irving, the interested reader is referred to the following: "Memorial to Irving Ballard Crosby (1891-1959)" by Robert R. Shrock, Geol. Soc. Amer, Pr. 1959, p. 117-120, portrait, (1960); and The Geologists Crosby of Boston by Robert R. Shrock (1972).

MEMBERSHIPS, AWARDS, AND OTHER RECOGNITIONS

W. O. Crosby was a member of numerous scientific organizations: American Academy of Arts and Sciences, American Association for the Advancement of Science, American Institute of Mining and Metallurgical Engineers, Boston Society of Natural History, Geological Society of America, Seismological Society of America, Society of Economic Geologists, and Appalachian Mountain Club.

He was twice awarded the Walker Prize of the Boston Society of Natural History, given annually for the best memoir on a scientific subject. He and M. L. Fuller (XII S.B. 1896) were awarded a silver medal at the 1900 Paris Exposition for an unusually fine collection of minerals and building stones from the United States. (See The Technology Review, v. 3, p. 91, 1901.)

On his retirement in 1907, students and colleagues in Course III, Mining and Metallurgy, gave him a gold charm consisting of crossed picks with a tiny ruby inset at the crossing point. On his 60th birthday, eleven of his geological colleagues tendered him a dinner and presented him with a two-handled silver loving cup. (See The Technology Review, v. 12, p. 150, 1910.)

A 4,000-foot serrated peak at the east end of Korovinski Bay, on Atka Island in the Aleutians, was named Mount Crosby by M.I.T. Prof. Thomas A. Jaggar, Jr., who led the Technology Expedition to the Aleutian Islands in 1907. (See The Technology Review, v. 10, p. 1-37, 1908.)

Shortly after Crosby's forced retirement because of deafness, Prof. H. W. Shimer led his colleagues in collecting contributions for the Crosby Honorary Fund, which provides income to be used to improve the Department's collections. Crosby's name is further perpetuated at M.I.T. by the William Otis Crosby Lectureship Fund which was established in 1961 to receive a bequest from Otis' son, Irving B. Crosby, who died on 18 September 1959. The fund provides income that is used to bring distinguished scientists to the Department to lecture on geological subjects. Further details on these two Funds can be found in the Chapter on "The Financing of the Geological Sciences at M.I.T." in my Volume 2.

SUMMARY

William Otis Crosby, the fifth professor of geology at M.I.T., served the Institute for 54 years, starting as a student assistant in 1875, retiring as Professor in 1907, and dying as Emeritus Professor in 1925. During this long period of service he did the following -

1. Organized and taught courses in paleontology, mineralogy, lithology, physical geology, structural geology, economic geology and engineering or practical geology.
2. Developed helpful laboratory exercises and prepared useful manuals and determinative tables as teaching aids.
3. Collected specimens of fossils, minerals, rocks, ores and structures for demonstration use in lectures and for study in laboratory.
4. Organized field trips and excursions to geologically important places, where students could investigate the real world, and prepared helpful leaflets for these trips.
5. Acted as a geological consultant on soils, mineral deposits, rock structures, water supply, tunnel excavation and dam sites, thus rendering valuable service to a public clientele and bringing back his experience and important collections to enhance his teaching.
6. Wrote more than 150 articles, books, reports, reviews and other publications, making his discoveries, ideas and opinions known to the scientific world.
7. Delivered evening lectures in the Lowell Free Course of Instruction for the general public, thereby bringing knowledge to persons who could not attend classes during the day.
8. Served as an assistant in mineralogy and geology at the Museum of the Boston Society of Natural History for more than twenty-five years, 1875 to 1902, during which he brought about improvement and increase in size of the collections of both the Museum and M.I.T., wrote useful "Guides" to the geological exhibits, gave public lectures, and led field excursions.

In all of these activities Crosby brought credit and prestige to the Institute, to the Museum, to himself, and to his chosen science, Geology.

BIBLIOGRAPHY OF WILLIAM OTIS CROSBY

Symbols and abbreviations used in the following references are explained on pages 91-98; in general, abstracts are listed separately as well as with the references to the complete article. The list has been checked against the following:

1. Geologic Literature on North America, 1785-1918 (Nickles), U. S. Geol. Surv., Bull. 746 and 747, 1924.
2. Bibliography of North American Geology, 1919-1928 (Nickles), U. S. Geol. Surv., Bull. 823, 1931.

3. Memorial of William Otis Crosby (Shimer and Lindgren), Geol. Soc. Am. Bull., v. 38, p. 34-45, 1927.

4. A Register of Publications of the Institute [M.I.T.] and its Officers, Students and Alumni, 1862-1893, on pages 39-42, 1893. [M.I.T. Publication.]

5. A hand-written list kept by W. O. Crosby himself on pages 1-14 and 21-23 in a "Record" ledger now in the M.I.T. Institute Archives.

6. Discussions with the late Irving B. Crosby, only living child of Prof. and Mrs. W. O. Crosby.

7. The Geologists Crosby of Boston (Shrock). M.I.T. Graphic Arts Service, Cambridge, p. 53-63, 1972.

The following Bibliography is believed to contain all of W. O. Crosby's important publications, as well as a few unpublished manuscripts of historical interest. Reprints of many of the articles listed are bound into several volumes that are included in the Crosby papers in the M.I.T. Institute Archives.

1--1871 Subterranean explorations [Discussion of cave at Hannibal, Missouri]: Sci. Am. 25: 405, (1871). [The initials of the author should be W.O.C., not R.O.C.]

2--1872 Middle Park, Colorado. Sci. Am. 26: 149, (1872). [The initials of the author should be W.O.C., not R.O.C.]

3--1872 a-e Society of Arts of the Massachusetts Institute of Technology [Reports of Meetings]. Sci. Am. 26, as follows:
a) Meeting of Dec. 14, 1871, p. 21, (Jan. 6),
b) Meeting of Jan. 11, 1872, p. 116, (Feb. 17),
c) Meeting of Feb. 8, 1872, p. 164, (Mar. 9),
d) Meeting of Mar. 14, 1872, p. 229, (Apr. 6),
e) Meeting of Apr. 11, 1872, p. 292, (May 4).

4--1873 Affection and sagacity of a dog. Am. Nat. 7: 237-238, (1873).

5--1874 Life and work at Penikese. New York Tribune, (1874). [Issue not determined.]

6--1875 Light of the sky. Am. Acad. Arts Sci., Pr. 9: 425-428, (1875).

7--1875a Review: Chemical and Geological Essays [by T. S. Hunt]. Am. Nat. 9: 416-417, (1875).

8T--1876 Geology of Eastern Massachusetts, 166 p., il., (1876). (S.B. Thesis at M.I.T. in Course VII, Natural History, June 1876.) [Thesis in M.I.T. Institute Archives.] Published in enlarged form as Contributions to the Geology of Eastern Massachusetts, Boston Soc. Nat. Hist., Oc. Papers III, iv + 286 p., il., (1880).

9--1876a Report on the geological map of Massachusetts (Massachusetts Commission to the Centennial Exposition). Pamphlet, 52 p., Boston, (1886).

10--1877 Notes on the surface geology of eastern Massachusetts. Am. Nat. 11: 577-587, (1877).

11--1878 Physical Geography and Geology of the Island of Trinidad. Boston Soc. Nat. Hist., Pr. 20: 44-55, (1878).

12--1879 On the occurrence of fossiliferous boulders in the drift of Truro, Cape Cod, Mass. Boston Soc. Nat. Hist., Pr. 20: 136-140, (1879).

13--1879a On a possible origin of petrosiliceous rocks. Boston Soc. Nat. Hist., Pr. 20: 160-169, (1879).

14--1879b   Native bitumens and the Pitch Lake of Trinidad. Am. Nat. 13: 229-246, (1879); Sci. Am. Sup. 7: 2771-2772, 2785-2786, (1879).

15--1879c   Bitumen. Abstract of a lecture before the Boston Scientific Society. Science Observer 2: 54-55, (1879).

16--1879d   Asphaltum and the pitch lake of Trinidad. Swiss Cross 1: 172, (1879).

17--1879e   How the appearance of a fault may be produced without fracture. Geol. Mag. n.s. 6: 269-298, (1879).

18--1879f   Review: MacFarlane's Geologists Travelling Hand-book. Am. Nat. 13: 251-252, (1879).

19--1880   Contributions to the geology of eastern Massachusetts. Boston Soc. Nat. Hist., Oc. Papers III, 286 p., map, (1880). (Also see 8T--1876.)

20--1880a   On the evidence of compression in the rocks of the Boston basin [with discussion by M. E. Wadsworth, p. 313-318]. Boston Soc. Nat. Hist., Pr. 20: 308-313, (1880).

21--1880b   Distorted pebbles in conglomerates. Boston Soc. Nat. Hist., Pr. 20: 368-378, (1880).

22--1880c   On the age and succession of the crystalline formations of Guiana and Brazil. Boston Soc. Nat. Hist., Pr. 20: 480-497, (1880).

23--1880d   Biographical sketch of Levi S. Burbank. Boston Soc. Nat. Hist., Pr. 21: 49-51, (1880).

24--1880e   Pinite in eastern Massachusetts, its origin and geological relations. Am. J. Sci. (3) 19: 116-122, (1880).

25--1880f   (and Barton, G. H.). Extension of the Carboniferous formation in Massachusetts. Am. J. Sci. (3) 20: 416-420, (1880).

26--1880g   Geological Report upon the Narraguagus Mine, Washington County, Maine. Pamphlet of 8 pages, (1880). [Privately printed.]

27--1880h   Geological Report on the Pembroke Mine. Pamphlet of 7 pages, (1880). [Privately printed.]

28--1880i   Geology of Frenchman's Bay, Maine. Boston Soc. Nat. Hist., Pr. 21: 109-120, (1880).

29--1881   Common Minerals and Rocks: Guides for Science Teaching, no. 12. Boston, Boston Soc. Nat. Hist., 130 p., (1881); [4th ed.], 205 p., (1886). (Also see 44--1886a and 62--1890b.)

30--1881a   On the absence of joint structure at great depths and its relations to the forms of coarsely crystalline eruptive masses. Geol. Mag. (2) 8: 416-420, (1881).

31--1881b   On the classification of the textures and structures of rocks. Boston Soc. Nat. Hist., Pr. 21: 280-288, (1881).

32--1883   On the classification and origin of joint structures. Boston Soc. Nat. Hist., Pr. 22: 72-85, (1883); Abst., Am. Assoc. Adv. Sci., Pr. 21: 280-288, (1882).

33--1883a   On the elevated coral reefs of Cuba. Boston Soc. Nat. Hist., Pr. 22: 124-130, (1883).

34--1883b   Origin of continents. Geol. Mag. (2) 10: 241-252, (1883); (3) 1: 46-47, (1884).

35--1883c   On the mountains of eastern Cuba. Appalachia 3: 129-142, (1883).

36--1883d   Probable occurrence of the Taconian system in Cuba. Science 2: 740, (1883).

37--1884   On the chasm called "Purgatory" in Sutton, Mass. Boston Soc. Nat. Hist., Pr. 22: 434-436, (1884).

38--1884a   Origin and relations of continents and ocean basins. Boston Soc. Nat. Hist., Pr. 22: 443-485, (1884).

39--1884b   On the relations of the conglomerate and slate in the Boston basin. Boston Soc. Nat. Hist., Pr. 23: 7-27, (1884).

40--1884c   Chemical geology [under Letters to the Editor]. Science 3: 59, (1884).

41--1884d   The colors of natural waters. Science 3: 445-446, (1884).

42--1885   Colors of soils. Boston Soc. Nat. Hist., Pr. 23: 219-222, (1885).

43--1886   Geological Collections: Mineralogy. Guides to the Museum of the Boston Society of Natural History. Boston, published by the Society, 184 p., (1886). (See 59--1889b.)

44--1886a   Common Minerals and Rocks (4th ed. with 75 new pages). Boston, Boston Soc. Nat. Hist., 205 p., (1886). (See 29--1881 and 62--1890b.)

45--1886b   Notes on joint structure. Boston Soc. Nat. Hist., Pr. 23: 243-248, (1886).

46--1886c   (and Barton, G. H.). On the great dikes at Paradise, near Newport [Rhode Island]. Boston Soc. Nat. Hist., Pr. 23: 325-330, (1886).

47--1887   Tables for the determination of common minerals. Boston. 1st ed., 74 p., (1887); 2nd ed., 84 p., (1888); 3rd ed., 106 p., (1895).

48--1887a   The elevated pot-holes near Shelburne Falls, Massachusetts. Tech. Quart. 1: 36-38, (1887).

49--1887b   Asphaltum and the Pitch Lake of Trinidad. Swiss Cross 1: 72, (1887).

50--1888   Geology of the outer islands of Boston Harbor. Boston Soc. Nat. Hist., Pr. 23: 450-457, (1888).

51--1888a   Geology of the Black Hills of Dakota. Boston Soc. Nat. Hist., Pr. 23: 488-517; Pr. 24: 11, (1888).

52--1888b   Methods of instruction in mineralogy and structural geology, in the Massachusetts Institute of Technology [Read before the American Society of Naturalists, at New Haven, Dec. 29, 1887]. Tech. Quart. 1: 187-194, (1888).

53--1888c   On the joint structure of rocks. Tech. Quart. 1: 245-250, (1888).

54--1888d   Quartzites and siliceous concretions. Tech. Quart. 1: 397-407, (1888); Sci. Am. Sup. 2: 10466-10468, (1888).

55--1888e   (and Greeley, J. T.). Vesuvianite from Newbury, Massachusetts. Tech. Quart. 1: 407-408, (1888).

56--1888f   (and Brown, C. L.). Gahnite or zinc spinel from Rowe, Massachusetts. Tech. Quart. 1: 408, (1888).

57--1889   Physical history of the Boston Basin. Boston Soc. Nat. Hist., Teachers' School of Science, Lowell Free Courses 1889-1890 [Ten lectures]. Boston, J. Allen Crosby, 22 p., (1889, 1890).

58--1889a   Relations of the pinite of the Boston Basin to the felsite and conglomerate. Tech. Quart. 2: 248-252, (1889).

59--1889b   Geological Collections: Mineralogy. Guides to the Museum of the Boston Society of Natural History, 2nd ed., Boston, Boston Soc. of Nat. Hist., 184 p., (1889). (First edition was published in 1886; see 43--1886.)

60--1890   Geological history of the Boston Basin [Massachusetts]. Boston Soc. Nat. Hist., Pr. 25: 10-17, (1890).

61--1890a   [Discussion of] Note on glacial climate (with discussion by Warren Upham and W. O. Crosby). Boston Soc. Nat. Hist., Pr. 24: 460-467, (1890).

62--1890b   Common Minerals and Rocks. Guides for Science-Teaching. No. XII. Boston, D. C. Heath & Co., 205 p., (1890). (Also see 44-1886a and 29--1881.)

63--1890c   The kaolin in Blandford, Massachusetts. Tech. Quart. 3: 228-237, (1890).

64--1890d   The Madison bowlder [New Hampshire]. Appalachia 6: 61-70, 105, (1890).

65--1891   On the contrast in color of the soils of high and low latitudes. Am. Geol. 8: 72-82, (1891); Tech. Quart. 4: 36-45, (1891).

66--1891a   Composition of the till or bowlder-clay. Boston Soc. Nat. Hist., Pr. 25: 115-140, (1891).

67--1892   Geological Collections: dynamical geology and petrography. Guides to the Museum of the Boston Society of Natural History. Boston, Boston Soc. Nat. Hist., 302 p. [including 86 plates containing 186 figures], (1892).

68--1892a   Geology of Hingham, Massachusetts [abst.]. Boston Soc. Nat. Hist., Pr. 25: 499-512, (1892).

69--1892b   Biographical sketch of Thomas Sterry Hunt. Am. Acad. Arts Sci., Pr. 19: 367-372, (1892).

70--1893   Geology of the Boston Basin: Nantasket and Cohasset. Boston Soc. Nat. Hist., Oc. Papers IV, 1/1: 1-177, (1893).

71--1893a   The origin of parallel and intersecting joints. Tech. Quart. 6: 230-236; Am. Geol. 12: 368-375, (1893).

72--1894   Geology of the Boston Basin: Hingham. Boston Soc. Nat. Hist., Oc. Papers IV, 1/2: 179-288, (1894).

73--1894a   (and Ballard, Hetty O.). Distribution and probable age of the fossil shells in the drumlins of the Boston Basin. Am. J. Sci. (3) 48: 486-496, (1894).

74--1894b   A classification of economic geological deposits, based on origin and original structures. Tech. Quart. 7: 27-48; Am. Geol. 13: 249-268, (1894).

75--1894c   Origin of the coarsely crystalline vein granites or pegmatites [abst.]. Am. Geol. 13: 215-216, (1894).

76--1894d   (and Grabau, A. W.). Record of Geological Observations made in Nova Scotia by Students of the Massachusetts Institute of Technology, July 1894; 52 p. and Supplement, 24 p., with numerous photographs, etc. [Typewritten copy in the M.I.T. Institute Archives.]

77--1895   Notes on the geology of the reservations: Report of the Metropolitan Park Commission [Boston], January 1895: 55-68, (1895).

78--1895a   Sandstone dikes accompanying the great fault of Ute Pass, Colorado. Essex Inst., B. 27: 113-147, (1895).

79--1895b   A classification of economical deposits. Eng. Min. J. 59: 28-29, (1895).

80--1895-c-e, 1898   [Short reviews and abstracts of publications containing chemical and mineralogical data.] Geological and Mineralogical Chemistry (included as a section of Review of American Chemical Research [= RACR]), Tech. Quart.:

	v. 8, RACR 1, p. 114-118, (1895) [13 publications reviewed] v. 8, RACR 1, p. 206-210, (1895) [11 publications reviewed] v. 8, RACR 1, p. 411-421, (1895) [35 publications reviewed] v. 9, RACR 2, p. 28- 31, (1896) [10 publications reviewed] v. 9, RACR 2, p. 56- 59, (1896) [13 publications reviewed] v. 9, RACR 2, p. 103-106, (1896) [15 publications reviewed] v. 10, RACR 3, p. 115-120, (1897) [21 publications reviewed] v. 10, RACR 3, p. 161-163, (1897) [5 publications reviewed] v. 11, RACR 4, p. 14- 24, (1898) [20 publications reviewed] [A total of 143 publications reviewed; also see item 86--1897.]
81--	[Undated] Laboratory Guide to Elementary Lithology [at M.I.T.], 33 p. (Privately printed, and undated, but probably used in class in the late 1890s and early 1900s.)
82--1896	Glacial lakes of the Boston Basin. Am. Geol. 17: 128-130; Science, n.s. 3: 212-213, (1896).
83--1896a	(with Crosby, F. W.). The sea mills of Cephalonia. Tech. Quart. 9: 6-23, (1896).
84--1896b	Englacial drift. Am. Geol. 17: 203-234; Tech. Quart. 9: 116-144, (1896).
85--1896c	See references in preceding item 80--1895-c-e, 1898.
86--1897	Notes on chemical geology--for Massachusetts Institute of Technology, 121 p., Boston, (1897). [Also see item 80--1895-c-e, 1898.]
87--1897a	Geological and mineralogical work of Thomas Tracy Bouvé. Boston Soc. Nat. Hist., Pr. 27: 236-239, (1897).
88--1897b	[Biographical sketch of] Thomas Tracy Bouvé. Am. Acad. Arts Sci., Pr. 32: 340-344, (1897).
89--1897c	Contribution to the geology of Newport Neck and Conanicut Island. Am. J. Sci. (4) 4: 230-236, (1897).
90--1897d	(and Fuller, M. L.). Origin of pegmatite. Am. Geol. 19: 140-180; Tech. Quart. 9: 326-536, (1897).
91--1897e	The great fault and accompanying sandstone dikes of Ute Pass, Colorado. Science, n.s. 5: 604-607; Geol. Soc. Am., B. 10: 141-164, (1897). (Also see item 78--1895a.)
92--1898	History of the Blue Hills complex [abst.]. Am. Assoc. Adv. Sci., Pr. 47: 130-135; Am. Geol. 22: 263-264, (1898).
93--1898a	Geology: South Shore, in Grabau, A. W. and others--Guide to localities illustrating the geology ... of the vicinity of Boston. Am. Assoc. Adv. Sci., 50th Anniversary Meeting, Boston, August 1898. Cambridge, F. W. Putnam, 100 p., (1898). [Crosby's article is on p. 21-31.]
94--1898b	See last reference in preceding item 80--1895-c-e, 1898.
95--1899	Geology of the Wachusett dam and Wachusett aqueduct tunnel of the Metropolitan Water Works in the vicinity of Clinton, Mass. [Massachusetts]. Tech. Quart. 12: 68-96, (1899).
96--1899a	Geological history of the Nashua Valley during the Tertiary and Quaternary periods. Tech. Quart. 12: 288-324, (1899).
97--1899b	Archean-Cambrian contact near Manitou, Colorado. Geol. Soc. Am., B. 10: 141-164; Abst., Am. Geol. 23: 92; Science, n.s. 9: 101; Ottawa Nat. 12: 198, (1899).
98--1899c	The glacial lake of the Nashua Valley [abst.]. Am. Geol. 23: 102-103; Science, n.s. 9: 106, (1899).
99--1900	Geology of the Boston Basin: the Blue Hills complex. Boston Soc. Nat. Hist., Oc. Papers IV, 1/3: 289-694, (1900).

100--1900a   Outline of the geology of Long Island in its relations to the public water supply. Tech. Quart. 13: 110-119, (1900). [In J. R. Freeman's "Report upon New York's Water Supply," p. 553-572.]

101--1900b   Outline of the geology of Staten Island in relation to the Public Water Supply, in J. R. Freeman's "Report upon New York's Water Supply," p. 573-581.

102--1900c   Notes on the geology of the proposed dams in the valleys of the Housatonic and Ten-Mile rivers. Tech. Quart. 13: 120-127. [In J. R. Freeman's "Report upon New York's Water Supply," p. 582-587.]

103--1900d   On the origin of phenocrysts, and the development of the porphyritic texture in igneous rocks. Am. Geol. 25: 299-310, (1900).

104--1901   Are the amygdaloidal melaphyres of the Boston Basin intrusive or contemporaneous? Am. Geol. 27: 324-327, (1901).

105--1901a   The tripolite deposit of Fitzgerald Lake, near Saint John, New Brunswick. Tech. Quart. 14: 124-127, (1901).

106--1901b-1902   Geological history of the hematite iron ores of the Antwerp and Fowler belt in New York. Tech. Quart. 14: 162-170, (1901); Am. Geol. 29: 233-242, (1902).

107--1902a   Memoir of Alpheus Hyatt. Geol. Soc. Am., B. 14: 504-512, port., (1902).

108--1902b   Origin of eskers. Boston Soc. Nat. Hist., Pr. 30: 375-411; Am. Geol. 30: 1-38, (1902).

109--1902c   Origin and relations of the auriferous veins of Algoma (western Ontario). Tech. Quart. 15: 161-180, (1902).

110--1902d   A study of hard-packed sand and gravel. Tech. Quart. 15: 260-264, (1902).

111--1903   A syllabus of the course of instruction in Structural Geology given to the students in Mining Engineering [at M.I.T.]. Boston, 32 p., (1903). [Privately printed.]

112--1903a   A syllabus of the course of instruction in Structural Geology given to the students in Civil Engineering, General Studies, Sanitary Engineering, Landscape Architecture, and Biology [at M.I.T.]. Boston, 17 p., (1903). [Privately printed.]

113--1903b   A syllabus of the course of instruction in Building Stones given to the students in Architecture [at M.I.T.]. Boston, 23 p., (1903); 24 p., (1908). [Privately printed.]

114--1903c   A study of the geology of the Charles River estuary and the formation of Boston Harbor: Massachusetts Report of Committee on Charles River dam, Boston; Appendix No. 7, p. 345-369, (1903).

115--1903d   A study of the geology of the Charles River estuary and Boston Harbor, with special reference to the building of the proposed dam across the tidal portion of the river. Tech. Quart. 16: 64-92, (1903).

116--1903e   The hanging valleys of Georgetown, Colorado. Am. Geol: 42-48; Tech. Quart. 16: 41-50; Abst., Science, n.s. 17: 227; Jour. Geol. 11: 117; Sci. Am. Sup. 55: 22666, (1903).

117--1903f-1904   Structure and composition of the delta plains formed during the Clinton stage of the glacial lake of the Nashua Valley. Tech. Quart. 16: 240-254; 17: 37-75, (1903-1904).

118--1904a   (and La Forge, L.). [Notes on water resources of] Massachusetts. U. S. Geol. Surv., W-S.P. 102: 94-117, (1904).

119--1904b   [Notes on water resources of] Rhode Island. U. S. Geol. Surv., W-S.P. 102: 119-125, (1904).

120--1904c   Geology of the Weston Aqueduct of the Metropolitan Waterworks in Southboro, Framingham, Wayland, and Weston, Massachusetts. Tech. Quart. 17: 101-116, (1904).

121--1904d   (and Loughlin, G. F.). A descriptive catalogue of the building stones of Boston and vicinity. Tech. Quart. 17: 165-185, (1904).

122--1905   [Underground waters of] Massachusetts and Rhode Island. U. S. Geol. Surv., W-S.P. 114: 68-75, (1905).

123--1905a   Water supply from the delta type of sand plain. U. S. Geol. Surv., W-S.P. 145: 161-178, (1905).

124--1905b   The limestone granite contact deposits of Washington Camp, Arizona. Tech. Quart. 18: 171-190, (1905); Am. Inst. Min. Eng., Bi-monthly Bull. 6: 1217-1238, (1905); Tr. 36: 626-646, (1906).

125--1905c   Genetic and structural relations of the igneous rocks of the lower Neponset Valley, Massachusetts. Am. Geol. 36: 34-47, 69-83; Tech. Quart. 18: 386-409, (1905).

126--1906   (with Veatch, A. C. et al.). Chap. 5. Results of sizing and filtration tests, in Underground Water Resources of Long Island, New York. U. S. Geol. Surv., P.P. 44: 338-360, (1906).

127--1907   Ore deposits of the eastern gold-belt of North Carolina. Tech. Quart. 20: 280-286, (1907); Am. Inst. Min. Eng., Bi-monthly Bull. 19: 171-178, (1908); Tr. 38: 849-856, (1908).

128--1907a   Volcanic activity in Alaska. Science, n.s. 26: 78, (1907).

129--1908   Outline of the geology of Long Island, N.Y. N.Y. Acad. Sci., An. 18: 425-429; Abst., Science, n.s. 28: 936, (1908).

130--1908a   See preceding item 113-1903b.

131--1912   Dynamic relations and terminology of stratigraphic conformity and unconformity. J. Geol. 20: 289-299, (1912).

132--1913   Report on the new site of the Massachusetts Institute of Technology, 1913. 26 typewritten pages included as an Appendix in E. S. Sheiry's "Methods for determining certain physical properties of subsoils," Civil Engineering S.B. Thesis at M.I.T., 1926. [Thesis in Science Library at M.I.T.]

133--1914   Physiographic relations of serpentine, with special reference to the serpentine stock in Staten Island, N.Y. J. Geol. 22: 582-593; Abst. with discussion, Geol. Soc. Am., B. 25: 87-88, (1914).

134--1914a   Buried gorge of the Hudson River and geologic relations of Hudson syphon of the Catskill aqueduct [abst.]. Geol. Soc. Am., B. 25: 89-90, (1914).

135--1919   Certain aspects of glaciation in Alaska [abst.]. Geol. Soc. Am., B. 30: 115, (1919).

136--1920   Certain aspects of glaciation in Alaska [abst.]. Geol. Soc. Am., B. 31: 132, (1920).

137--1924   Soils of Coos County [New Hampshire]. Boston Soc. Nat. Hist. Pr. 37: 34-68, (1924).

138--1924a   (and Crosby, I. B.). Keystone faults. Abst., Geol. Soc. Am. B. 35: 94, (1924); 36: 623-640, (1925); Pan-Am. Geol. 41: 147, (1924).

139--1928   Certain aspects of glacial erosion. Geol. Soc. Am., B. 39: 1171-1181, (1928). [Published posthumously.]

Addenda to the
Bibliography of William Otis Crosby

Not included in the foregoing Bibliography of William Otis Crosby (1850-1925) are numerous papers that he prepared but either published unsigned or without a date, or left unpublished in manuscript form. The most important of these are listed in the following paragraphs. They include: 1) some 20 unsigned reviews, notes, and reports published in Science in the years 1884 to 1888;* 2) an undated, privately printed 33-page booklet on lithology--Laboratory Guide to Elementary Lithology [at M.I.T.], 33 p.; privately printed, probably at M.I.T., and used in the 1890s and early 1900s; 3) a diary titled Geological and other sketches by W.O.C. covering a trip he made from Washington, D.C. to Georgetown, Colorado, during the spring of 1871; 4) a set of 66 handwritten Notes on the Geology of the Rocky Mountains; 5) a set of handwritten class notes taken in Professor T. Sterry Hunt's course in geology in school-year 1873-1874; and 6) a 38-page typewritten manuscript titled Water Resources of the Boston Basin.

These six items are considered worth brief discussion because of their historical interest, and also because several of them provide further evidence of William Otis' methods of work and attitudes toward man and nature.

1. Unsigned Reviews, Notes, and Reports published in Science during the years 1884-1888 inclusive. Volume number of Science, page numbers, and date of publication follow the title.

    1884   [Review of the] Geology of the Susquehanna River Region, by I. C. White, Second Survey of Pennsylvania; v. 4, p. 120-121, 1884.

    1884a   [Review of the] Geology of South-eastern Pennsylvania by P. Frazer; v. 4, p. 447-448, 1884.

    1885   The Tenth Volume of the Census Report, [a review of Production, technology, and uses of petroleum and its products, by S. F. Peckham; The manufacture of coke, by J. D. Weeks; Building-stones of the United States, and statistics of the quarry industry for 1880; Census Report X, 1884]; v. 5, p. 239-240, 1885.

    1885a   Some State Geological Reports, [a review of The Geology of Minnesota, vol. 1, by N. H. Winchell and Warren Upham, 1884; Indiana Department of Geology and Natural History, 12th and 13th Annual Reports, by John Collett, 1883, 1884]; v. 5, p. 529-530, 1885.

---

* It is known that W. O. Crosby wrote these pieces because he listed them on pages 21 to 23 of a Record "ledger" in which he also listed chronologically the titles of his books, articles and other publications. Furthermore, his son, the late Irving Ballard Crosby, confirmed the listing to the present author. This Record ledger is now in the M.I.T. Institute Archives.

1885b   Geology of the Virginias, [A review of "A reprint of geological reports and other papers on the Geology of the Virginias," by W. B. Rogers, 1884]; v. 6, p. 17-18, 1885.

1885c   Geological Society of Canada, [A review of the "Geological Survey of Canada, Report of Progress for 1882-1884"]; v. 6, p. 521-523, 1885.

1886   The U. S. Geological Survey, [A review of the "Fourth Annual Report of the U. S. Geological Survey ... 1882-1883," by J. W. Powell]; v. 7, p. 158-160, 1886.

1887   Mining industries and mineral resources of the United States, [A review of the "Tenth Census of the United States," vol. XV: Report on the mining industries of the United States (exclusive of the precious metals), by Raphael Pumpelly]; v. 9, p. 347-348, 1887.

1887a   [Review of] "Mineral resources of the United States" in the Tenth Census of the United States, by D. T. Day, 1885; v. 9, p. 348, 1887.

1887b   Geology of Minnesota, [Review of 13th and 14th Annual Reports of the Geological and Natural History Survey of Minnesota, for the years 1884 and 1885, by N. H. Winchell]; v. 9, p. 401-402, 1887.

1887c   Explosions in coal-mines, [Review of an article by W. N. and J. B. Atkinson, inspectors of coal mines for the north of England]; v. 9, p. 429-431, 1887.

1887d   [Review of Orton's "Preliminary Report on oil and gas in Ohio"]; v. 9, p. 504-505, 1887.

1887e   Geology of New Jersey, [Review of the Annual Report of the Geological Survey of New Jersey for 1886]; v. 9, p. 595-596, 1887.

1887f   [Review of "Geological History of Lake Lahontan, a Quaternary Lake of North-western Nevada," by I. C. Russell]; v. 10, p. 79-80, 1887.

1887g   [Review of] "On the relation of the Laramide molluscan fauna to that of the succeeding fresh-water Eocene and other groups," by C. A. White, (U. S. Geol. Surv., Bull. 34); v. 10, p. 126-127, 1887.

1887h   [Review of] "Synopsis of the flora of the Laramie Group," (an extract from the 6th Annual Report of the U. S. Geol. Surv.), by L. F. Ward; v. 10, p. 150-151, 1887.

1887i   [Review of] "Types of the Laramie flora," (U. S. Geol. Surv., Bull. 37), by L. F. Ward; v. 10, p. 151-152, 1887.

1887j   [Review of] Sixth Annual Report of the United States Geological Survey (1884-85), by J. W. Powell; v. 10, p. 235, 1887.

1887k   [Review of] "The Driftless Area of the Upper Mississippi," by T. C. Chamberlin and R. D. Salisbury; v. 10, p. 306-307, 1887.

1887l   [Review of] "Mount Taylor and the Zuni Plateau," by C. E. Dutton; v. 10, p. 317-319, 1887.

1888   [Review of] Annual Report of the Geological Survey of Pennsylvania for 1886, parts I and II; v. 11, p. 45, 1888.

2. Laboratory Guide to Elementary Lithology

This 33-page, privately printed booklet was no doubt prepared for class use at M.I.T. in the late 1890s and early 1900s when W. O. Crosby was teaching courses in rocks and minerals. The booklet lacks a date and any indication of where or when published, but it seems reasonable to suppose that it was printed at M.I.T. sometime in the 1890s. A copy is included in the M.I.T. Institute Archives.

3. Geological and Other Sketches by W.O.C.

The most interesting and important of Crosby's unpublished works is a handwritten diary, or journal as it would have been called at the time, titled Geological and other sketches by W.O.C. In this journal he describes a trip he made during the spring of 1871 from Washington, D.C. to Georgetown, Colorado, to start work in a silver mill located in Georgetown and owned by his father. Farther on he records his thoughts and activities as he goes about learning how to be a mining engineer.

The narrative begins as he leaves Washington by train on 3 March 1871 and ends with an entry dated 30 December (1871) stating that he is moving his "bedding, cooking utensils & grub" to the North Empire Mill to take charge of its operations.

The narrative appears on pages 50 to 135 of a 9" × 11" hard-back ledger and is written in easily readable script, though punctuation and capitalization are a bit informal. The original is preserved in the M.I.T. Institute Archives; an exact typescript of the original is included in my recent The Geologists Crosby of Boston (1972).

It is this narrative that contains Crosby's account of how he was asked to guide M.I.T. President Runkle's party of professors and students through the mills and mines around Georgetown, and how, as a result of his impressive service, he was invited by Runkle to enter M.I.T., which he did soon after.

4. Notes on the Geology of the Rocky Mountains

Included in William Otis Crosby's papers is a set of 66 pages of handwritten Notes on the Geology of the Rocky Mountains, unsigned and undated, but easily recognizable from the characteristic script and positively identified by his son, Irving Ballard Crosby, who preserved the handwritten sheets in one of his own envelopes labelled "Notes on the Geology of the Rocky Mountains--W.O.C." It is not known whether William Otis prepared these notes before or after he went to Georgetown, Colorado, in the spring of 1871. The "Notes" are now among the Crosby Papers in the M.I.T. Institute Archives.

5. Class Notes taken in Prof. T. S. Hunt's Geology Course

A Crosby item of special academic interest is a set of 224 pages of handwritten and illustrated class notes taken while a student in a series of geology lectures delivered by Professor T. Sterry Hunt, starting on 17 November 1873 and ending on 28 April 1874.

These notes, of special interest in the 1970s a hundred years later, constitute 224 pages of a $6\frac{1}{2}$" × 8" hardback notebook and give an excellent outline of geology as it was taught at M.I.T. a century ago. The notebook is preserved in the Crosby Papers in the M.I.T. Institute Archives.

6. Water Resources of the Boston Basin

The final unpublished Crosby item considered worth special mention here is a short typewritten report titled Water Resources of the Boston Basin, by W. O. Crosby. It contains 38 pages of text and 12 additional pages of tables of data. Although undated, it was obviously written sometime after 1904 because the data in some of the tables are totalled through that year. It is worth noting that William Otis was actively engaged in the study of a number of water-supply problems in the early 1900s (see bibliographic items 115--1903d, 116--1903e, 118--1904a, 120--1904c, 122--1905) and may well have prepared this report in connection with some of his studies during these years. The typescript is included in the Crosby Papers in the M.I.T. Institute Archives.

BIOGRAPHICAL REFERENCES

For detailed discussions of W. O. Crosby's life, the reader is referred to the following items, which have been used extensively in preparing this sketch:

1. Niles, W. H., A Sketch of Professor Crosby's work, thirty-five years associated with the Massachusetts Institute of Technology. Tech. Rev. 9: 194-195, (1907).
2. [Resolution]. [M.I.T.] Faculty Records 19: 430-432, (20 October 1926). [In M.I.T. Registrar's vault.]
3. [Historical Volumes of former M.I.T. Professors], V. 1, p. 59, no date, (in M.I.T. Institute Archives).
4. Johnson, D. W., William Otis Crosby. Science, n.s. 63: 609-610, (1926).
5. Shimer, H. W. and Lindgren, W., Memorial of William Otis Crosby. Geol. Soc. Am., B. 38: 34-35, port., (1927).
6. Lane, A. C., William Otis Crosby (1850-1925). Am. Acad. Arts Sci., Pr. 64: 518-526, (1930).
7. Creed, P. R., I. Milestones (p. 3-64) in The Boston Society of Natural History, 1830-1930, Boston. [Crosby is referred to on p. 45 ff.]
8. Prindle, P. W., Ancestors and Descendants of Timothy Crosby, Jr., Soldier of the American Revolution, a mimeographed compilation, unpublished, by P. W. Prindle, 55 Noroton Ave., Darien, Connecticut, (1970).
9. Shrock, R. R., The Geologists Crosby of Boston: William Otis Crosby (1850-1925) and Irving Ballard Crosby (1891-1959). Cambridge, Mass., Massachusetts Institute of Technology, xii + 96 + A1-A79, (1972).

(6)
GEORGE HUNT BARTON
(1852-1933)

GEORGE HUNT BARTON

MIT: 1883-1904

In the last three decades of the 19th century and the first three of the 20th there was great interest in developing instruction in natural science for school children and schoolteachers of the Greater Boston community. Several M.I.T. geology professors participated actively in the movement and played important teaching or leadership roles. Among these was George H. Barton (III S.B. 1880), who became an important participant in the movement while a faculty member in Course XII (1883-1904), and then left the Institute to devote full time to leadership of the Teachers' School of Science (1887-1933) and of the Children's Museum of Boston (1913-1933).

After graduation from M.I.T. in 1880, and following a two-year stint of geodetic mapping with the Hawaiian Government Survey, Barton returned to the Institute as Assistant in Geology to help Prof. W. O. Crosby and Prof. Alpheus Hyatt with laboratory and field instruction. In 1886 he was appointed Instructor in Determinative Mineralogy; in 1892, Instructor in Geology; and in 1896, Assistant Professor of Geology, an appointment he continued to hold until he resigned in 1904. During his 21-year period as an M.I.T. faculty member he taught numerous existing subjects--mineralogy, blowpipe analysis, and physical and historical geology; developed new courses and teaching aids in structural geology, lithology, and field geology; and assisted other departments in summer field courses. In addition, he also served on the faculty of several other educational institutions in the Greater Boston area--Boston University (1893-1904; 1915); Wellesley College (1921-1922); and Tufts College, from time to time.

In 1887, at the suggestion of his colleague, Prof. W. O. Crosby, he became involved in what later developed into his life's work--the popularizing of elementary geology and natural history among school children and secondary school science teachers in New England. Crosby suggested that he organize a private course of field lessons in the spring to follow his own lectures delivered during the preceding winter in the Teachers' School of Science. Such a course was offered, and it proved so popular that it was repeated in the fall. Barton was also asked to organize a night course in mineralogy, and this too proved popular with schoolteachers of science. Soon Barton was working closely with Prof. Alpheus Hyatt, who was directing the Teachers' School of Science on the side as part of his responsibilities as Curator of the Boston Society of Natural History which sponsored the School. In 1891 Hyatt asked Barton to take charge of the School and develop it into a normal school with a four-year program and a certificate at the end. By the autumn of 1895 he had the four-year program organized and ready to offer for the first time. The program consisted of Saturday lectures on mineralogy the first year, lithology the second, dynamical and structural geology the third, and historical geology the fourth. The lectures given during the winter months were supplemented by field lessons in the spring and fall months. Thousands of New England teachers of science were introduced to practical elementary geology by Barton during the forty-some years that he taught in the Teachers' School of Science, and death came suddenly as he started to lecture to his T.S.S. class in historical geology. He had taken another class on a field lesson only a few weeks earlier, and he was past 81 years old!

In April 1913 he became Chairman of the Board of Directors of the Children's Museum, an outgrowth of the activities of the Teachers' School of Science. Because of his understanding and patience as a mediator, the museum was made an independent organization, officially named "Children's Museum of Boston," and moved to a new home in Olmsted Park, Jamaica Plain. He served as President for two different periods, 1914-1917 and 1927-1929, and in between as Vice President for seven years, 1920-1927. After 1929 he was elected Honorary President for life. Although he cannot be said to have founded the Children's Museum, he surely can be given the credit for establishing it on a firm and enduring basis, recognition attested by the bas-relief of him that hangs on a wall in the present Museum, and by Adelaide B. Sayles' dedication to him of her The Story of The Children's Museum of Boston, Boston: G. H. Ellis Co., Inc., xii + 87 p., port., (1937).

From early manhood, Barton was a popular and stimulating public lecturer--he gave his first public lecture on 19 April 1880, while a senior at M.I.T., and his last more than 50 years later, in his 82nd year. Between these two dates he gave hundreds of public lectures, generally illustrated by slides from his large collection, on glacial geology, on minerals and rocks, and on the geologic processes and features he had seen firsthand throughout New England, in Nova Scotia, in Newfoundland and Greenland, and in Hawaii. These were generally given to groups of school children, business men, church organizations, and social groups, and more often than not without a fee.

Finally, Barton kept a diary throughout his adult life, from 1868 to 1933, and the last entry was typed only two days before his death. He condensed these diaries into a typescript of some 1167 pages, which he titled Reminiscences. He also prepared numerous other typescripts of an autobiographical nature. All of these important and informative documents are now preserved in the George Hunt Barton Memorial Room of the Goodnow Library in Sudbury, Massachusetts, his birthplace. More is written on these memorabilia in the footnote a few pages farther on and in the Note following the Bibliography at the end of this sketch.

In summary, while Barton carried a full load of academic work as an M.I.T. faculty member and contributed importantly to the instructional program in geology from 1883 to 1904, his true life work and his major contribution to education in science lay in his 46 years of service to the Teachers' School of Science, from 1887 to 1933, his 20 years of devoted leadership to the Children's Museum of Boston, from 1913 to 1933, and his innumerable public lectures delivered to a remarkably varied group of audiences, more often than not consisting of children and teachers.

Barton found in his devoted wife, Eva May (Beede) Barton, an enthusiastic teacher of children and a willing companion on field excursions. Of their two children who lived to adulthood--their first child died in infancy--Donald Clinton Barton gained international renown as an exploration geophysicist, and daughter Helen Mary (Mrs. Harold French Eastman) passed on to her own two daughters the love and wonder of nature so faithfully held by "Gramp," as Barton is affectionately called by his grandchildren.

BIRTH AND ANCESTRY*

George Hunt Barton, the fifth professor of geology at M.I.T. after Rogers, was like W. O. Crosby before him a native son of Massachusetts and a graduate of M.I.T. (III S.B. 1880). He was born in Sudbury on 8 July 1852 and died in Cambridge on 25 November 1933 as he started to lecture to a class in historical geology in the Teachers' School of Science. In his 82nd year when death came, he was starting his 52nd year of teaching geology, as mentally alert as ever but beginning to fail physically. Even then he had led one of his classes on a field excursion less than two months earlier.

He was the first of three children of George Washington Barton, who was born in Concord, Massachusetts, but moved to Sudbury in 1851, and Mary Susan Hunt, the youngest daughter of Israel and Ruth (Wheeler) Hunt of Sudbury. The Bartons had three children: George H. (our subject), born in 1852; Frank P., 1857, and Alice M., 1859.

Barton's ancestors were of old New England stock and were largely English, but one line was supposed to have been French. On his father's side were numerous ancestors with names of various forms, among them

---

* In preparing the biographical part of this sketch I have drawn heavily on Barton's own Reminiscences, an unpublished typescript of 1167 pages, condensed from diaries he kept for 65 years, 1868 to 1933, and on several other of his typescripts, which are listed after the Bibliography farther on. These valuable documents, together with his diaries and other memorabilia, are now preserved in the George Hunt Barton Memorial Room of the Goodnow Library in Sudbury, Massachusetts, his birthplace. A Xerox copy of Reminiscences is preserved in the M.I.T. Historical Collections.

I have also found quite helpful A. C. Lane's "Memorial of George H. Barton (1852-1933)," Geol. Soc. Am., Pr. 1934: 161-172, port., (1935). Lane was a close personal friend of Barton's and added his own comments to the voluminous information in the Reminiscences. He tramped with Barton on the latter's last field lesson only a few weeks before he (Barton) died.

There are brief references to Barton's activities in the M.I.T. President's Reports for SYs 1883-84 to 1903-04; in Frances Zirngiebel's "Teachers' School of Science [Boston]," Pop. Sci. Month. 55: 451-465, 640-652, (1899); in A. Hyatt's "Report of the Curator," Boston Soc. Nat. Hist., Pr. 29: 357, (May 1, 1901); in the 1905 TECHNIQUE, p. 27 with photograph; in a brief death notice on p. ii under Adversaria following p. 160, Tech. Rev. 36: 4, (Jan. 1934); and in Adelaide B. Sayles' "The Story of The Children's Museum of Boston, from its beginnings to November 18, 1936." Boston: George H. Ellis Co., Inc., xii + 87 p., port., (April 1937).

There were also numerous obituarial notices in both local and national newspapers. Typically, these were brief and contained information that is much more fully discussed in the references cited above. Some of the more important of these obituaries are cited at the end of the biographical part of this sketch.

Barton, Barden, Bardens, and Bardeen. These forebears traced their ancestry back to the Plymouth Colony in the mid-1600s. His mother, Mary Susan Hunt, was descended in the direct paternal line from William Hunt, one of the first settlers of Concord, Massachusetts, who came from Yorkshire, England, in 1635.

With sixty years of his life behind him, Barton became much interested in genealogy, as many an older person has been wont to do. He joined the New England Historic Genealogical Society, the Colonial Families Society, the Edward Rice Descendants, the Stone Family Association, and the Descendants of Elijah Wheeler. When Barton developed an interest in something, he was likely to follow that interest in a big way! His own ancestry is recorded at some length in Volume 88 (p. 211-214, July 1934) of the Register of the New England Historical and Genealogical Society, and in several other publications cited in the following footnote.* In addition, Barton himself left several genealogical typescripts, including one on his father, Ancestral Families of George Washington Barton, and one on his mother, Genealogy of Mary Susan Hunt. These are listed at the end of the Bibliography farther on.

YOUTH, EARLY EDUCATION, AND PROFESSIONAL TRAINING

Barton's youthful years seem to have been happy and exciting ones. He kept a diary from a very early age, 16, to the week of his death, and when past 70 years of age he condensed these into a typescript of some 1167 pages, which he titled Reminiscences (See Note following the Bibliography farther on). In these daily accounts he recorded what he did, where he went, what he saw, how he felt, and what the weather was. He gives a vivid account of life and work on a New England farm (his father's) in the 1860s and 1870s—how at age 5 he drove a one horse wagon behind his father's, which was drawn by a yoke of oxen; drove the cows to pasture at daybreak and back home at nightfall; and dropped kernels of corn in hills for planting. Later on, at nine years, he drove a one-horse wagon alone to the grist mill and to nearby Maynard, and when scarcely

---

* Detailed genealogical information about George Hunt Barton and his ancestors can be obtained from the following sources—"Genealogy of the Hunt Family," by T. W. Wyman, Jr. Boston: John Wilson & Son, 414 p., (1862).—"The History of Sudbury, Massachusetts, 1638-1889," by A. S. Hudson. Published by the Town of Sudbury, xxii + 660 p., il., (1889). —"Acton in History," compiled for "The Middlesex County History," by J. Fletcher. Philadelphia and Boston: J. W. Lewis & Co., p. 238-301, il., (1890).—"Annals of Sudbury, Wayland and Maynard," by Alfred Hudson, 40 p., il., (1891) [Publisher unknown].—"Walter Haynes of Sutton Mandeville, Wiltshire, England, and his descendants (1583-1928)," by Frances Haynes. Haverhill, Mass.: Record Pubs., 219 p., il., (1929).

twelve he took an ox-drawn wagon full of cordwood, apples, or other farm produce to market in neighboring towns, often starting well before dawn. In his teens he worked on his father's farm from March to December, and on most Saturdays during the winter months when he attended school. Little wonder that he decided to forego a career on the farm and train to be a teacher instead!

He began his education at the Pantry District School in the northeastern part of Sudbury, about half a mile from his home, before he was six, and he records that he was an eager and interested pupil. When he was $10\frac{1}{2}$ years old his mother died (8 January 1863), and the family was broken up temporarily, but his father soon married "a woman of high ideals" with two sons older than he was who became true brothers. At 14 he was sent to A. J. Lathrop's private school in Waltham for a term, and then to Chester Academy, Chester, Vermont, for three terms when he was 17 and 18. He next attended the Rev. L. P. Frost's private school in Maynard (Mass.), for three terms, and finally at 20 spent one term at the Maynard High School (SY 1872-73). During these years he always kept books close at hand, on the wagon seat, beside his chair at the dining table, or at his side when watching the cattle, so that he could read in spare moments. He records that he learned to read "easy" French and German in this way and also by study in the evenings. He also read much poetry during these youthful years, as indicated by his collection of favorite poems (See Note following the Bibliography farther on). And he also wrote a few poems of his own when the spirit moved him. Again and again throughout his Reminiscences he recalls the happy days of his youth, along with the hard work on the farm, and it was evident that he enjoyed every passing day whether spent at labor, visiting relatives, or with friends at play. His diaries record the reactions of a gentle, affectionate, and compassionate young man of high moral and ethical character who was rapidly being transformed from the "green country boy," as he described himself, to the much more knowledgeable city youth of the kind he encountered when he left home.

As he approached 21, in the spring of 1873, already having made clear to his father that he wanted to strike out on his own, he left home to work at carpentering and painting in Worcester. This didn't last long, however, and in October he decided to visit his Uncle Gorham Barton, who was building a lumber mill in Eau Claire, Wisconsin. His stop at Niagara Falls on the way gave him his first chance to see some of the geology of the hinterland and to become acquainted with frontier people who differed from the New Englanders he knew in their religions and political views. Again he took up carpentry work and learned blacksmithing as well. This kind of work, however, had no more appeal than it had had back in Sudbury, so he decided he would like to be a teacher.

Accordingly, in the spring of 1874, he returned to Massachusetts, entered the Warren Scientific Academy in Woburn, and prepared for entrance to M.I.T. Although he passed the entrance examinations in full in June 1875, he decided to spend an additional year at Warren and then take the second year examinations at M.I.T. in June 1876. With these examinations behind him, he entered M.I.T. in September 1876 as a member of the Class of 1879. Failing health forced him to drop out for a year with the result that he graduated a year late, in June 1880. His S.B. degree was in Course III, Geology and Mining Engineering, and his thesis was on "The Geology of the Norfolk County Basin." His was the second M.I.T. thesis on a geological subject, W. O. Crosby's having been the first, and he was the second "geology" graduate, again Crosby having been the first.

After graduation he stayed on for SY 1880-81 as an Assistant in Drawing. He then obtained employment with the U. S. Geological Survey on the Hawaiian Government Survey, and for the next two years, 1881-1883, he gained valuable experience in geophysical field work, doing civil engineering, and in geological observation. Not only did he participate in the geodetic work being done, which showed that

> "the deflection of the vertical was caused by the great mass of the mountains being much nearer the sea on one side than on the other," (Lane 1935: 163),

but he also made many geological observations and took numerous photographs, both of which he used frequently in his later lectures.* After two years, however, he decided to leave Hawaii (to which he would return years later as leader of a tourist excursion) and return to Massachusetts to join the geological staff of the Institute, which he would serve for the next 21 years, 1883-1904.

M.I.T. YEARS (1875-1881; 1883-1904)

As recorded elsewhere in this history, both in narrative and on charts, Barton was first an undergraduate student at M.I.T. (1876-1880); next the second to write a senior thesis on a geological subject and to graduate as a "geologist" (III S.B. 1880) [Course III was called Department of Geology and Mining Engineering in 1880; Course XII, Geology, would not be established until 1890!]; then Assistant in Drawing for a

---

* A lengthy diary of his activities during the two years in Hawaii is preserved in the Barton Room of the Goodnow Library in Sudbury, Massachusetts. See Note following the Bibliography farther on.

year (1880-1881). After a two-year absence (1881-1883) to work for the U. S. Geological Survey on the Hawaiian Government Survey, he returned to M.I.T. and held successively the following appointments during his 21-year period of service: Assistant in Geology (1883-1886); Instructor in Determinative Mineralogy and Geology (1886-1896); and finally Assistant Professor of Geology (1896-1904), the appointment from which he resigned to devote full time to lecturing and conducting field lessons, for the Teachers' School of Science, and to public lecturing. These latter activities are discussed at some length farther on.

Barton started his geological teaching career at M.I.T. as an assistant to Prof. W. O. Crosby, and thereafter until he resigned in 1904 he devoted essentially all of his time and efforts to teaching! This was long before the "publish or perish" dictum became the imperative for promotion. He was paid to teach, and that is what he did. As a consequence, his bibliography shows few major publications during the twenty-one years (1883-1904) he served on the Course XII faculty. In contrast, he taught a large number of different subjects and carried a teaching load that would be considered inordinately heavy today. In addition, he often accompanied Crosby on his field trips, acting as a field assistant, and devoted considerable time to week-end field excursions with his own students. One of these students, the late Prof. Carle R. Hayward (III S.B. 1904), wrote me as follows some years ago:

> "Like Niles and Crosby, Barton wore a beard. I took only one short course with him in lithology which was quite interesting and well presented. I did not take his course in field work which I understand was where he shone best. He gave field courses summers for a modest fee. These were taken mainly by public school teachers who were enthusiastic about them."

He did, however, find some time for one research project--mapping drumlins. He had become interested in the eskers, kames and drumlins that he saw on his numerous excursions, and when N. S. Shaler offered him the opportunity to map the drumlins of Massachusetts, he gladly accepted the offer. Shaler was head of the New England Division of the United States Geological Survey and was interested in having the glacial features of Massachusetts more fully investigated. Barton spent most of five summers, 1890-1895, mapping the drumlins of Massachusetts, and though he published little on what he found, his map has been used extensively by writers on glacial geology ever since.

TEACHERS' SCHOOL OF SCIENCE (1887-1933)

Important as were Barton's services at M.I.T. in Course XII, however, of far greater importance to education in geology were his activities in the Teachers' School of Science and in the Children's Museum of Boston,* both sponsored by the Boston Society of Natural History, and his many illustrated lectures given to public audiences.

Soon after returning to M.I.T. in 1883, and as a result of being Crosby's assistant, Barton became involved in what would develop into his true life work--teaching elementary geology and natural history to public school teachers of science and school children, under the sponsorship of the Boston Society of Natural History, and popularizing geology by public lectures to a wide variety of audiences throughout the Greater Boston area.

During the winter of 1886-1887, W. O. Crosby had been giving a course in geology in the Teachers' School of Science, and when it ended in the spring, many of his students asked if they could not be given some field lessons after the indoor instruction ended. Crosby suggested to Barton that he try giving such lessons, and even wrote a letter that he could use to inform prospective students. The substance of the letter was that he, Barton, would conduct a series of field lessons to geologically interesting localities for which there would be a charge of one dollar per person plus car-fare. The first lesson was at Parker Hill in Roxbury, on Saturday afternoon, 4 June 1887, and 17 students participated, yielding Barton $17, which Barton wrote in his Reminiscences (p. 264) made him "extremely pleased." Thus began his first connection with the Teachers' School of Science, a connection that would continue for 46 years, until his death in 1933, during which period he not only conducted field lessons but also gave evening lectures and ultimately became Curator and later Director (1902-1933).

The first field lessons proved so successful that they were continued in the fall, as well as in future years and were followed by a private

---

* The Teachers' School of Science, and its offspring, The Children's Museum, have played a major role in the education of the general public of Greater Boston for a full century, beginning in 1870. Furthermore, M.I.T. geologists have had a critical role in organizing, directing, and participating in the educational activities of these two institutions, which from their beginning have been sponsored by the Boston Society of Natural History. Suffice it here to mention that Alpheus Hyatt founded the Teachers' School of Science in 1870, and that George H. Barton not only gave courses in the School for more than 45 years and directed it for 31, but he was also largely responsible for getting the Children's Museum established as an independent institution in 1914, after which he provided much of its leadership until his death in 1933. There is a more detailed and extended discussion of these two rather unique educational institutions in the chapter titled "M.I.T. Geology Professors--Boston Society of Natural History--Teachers' School of Science--Children's Museum--Marine Biological Laboratory." (See Volume 2.) Interested readers are also referred to the publications cited in the preceding footnote.

course of twelve lessons on Mineralogy given on winter evenings in 1888 and early 1889.* These latter lessons, as well as the field lessons, were later incorporated in the program of the Teachers' School of Science.

It should be remembered that at this time, in the late 1880s and early 1890s, there were few opportunities for persons interested in nature and the out-of-doors to learn about minerals, rocks, and fossils by direct contact with the real thing. Barton's field lessons and winter lectures met this need inexpensively for many individuals, particularly for public school teachers of science.

In 1870 the Boston Society of Natural History had organized the Teachers' School of Science, to be financed by John C. Cummings and directed by Alpheus Hyatt, for the purpose of introducing natural science into the public schools of the Boston area. This was to be accomplished over the long term by first providing free lectures on science to public school teachers. So when Barton's field lessons and winter evening lectures became so popular, Hyatt asked Barton if he would take general charge of the Teachers' School of Science and develop it into a four-year normal school with some sort of certificate or diploma as credit for work done. Actually, one of Barton's students, Miss Mary F. Thompson had earlier suggested that regular examinations be held in his courses and that the teachers be given credit for their work, something that had not been done previously.

By the fall of 1891, Hyatt, Barton and others had decided to offer in the School four-year courses in Botany, Geology, and Zoology, with regular examinations and a certificate at the end. Hyatt would offer the zoology lectures; Barton, those in geology; and Robert W. Greenleaf, those in botany. The first lecture under this new arrangement was given at the Boston Society of Natural History on 7 November 1891. This was only the start, however, because Barton had to develop the program suggested by Hyatt step by step.

By the autumn of 1895 he had the four-year course in geology fully organized and ready to be offered for the first time. It would consist of

---

* Barton wrote in his <u>Reminiscences</u> (p. 274):

> "On July 1 [1889] I received notice from Prof. Hyatt that Mr. Augustus Lowell, Trustee of the Lowell Institute Fund, who had for a few years paid the expenses of the Lectures in the Teachers' School of Science of which Prof. Hyatt was Curator, had told the latter to pay me $500.00 with which to carry on my field lessons for the coming season and so allow the members of the class free tuition. This was my first salary received in the School ... a very welcome addition to my Institute Salary."

Mineralogy the first year; Lithology the second; Dynamical and Structural Geology the third; followed by Historical Geology in the fourth year. When the first class (in Mineralogy) was called to order at 2 P.M., on Saturday, 5 January 1895, almost 200 applicants crowded into the Lecture Hall of the Boston Society of Natural History. Inasmuch as such a large class could not be properly handled, it was divided into two equal sections, one coming from 1 to 3 P.M. and the second from 3 to 5 P.M. Each student was asked to pay three dollars for the course! Thus began the four-year geology course of the Teachers' School of Science, and it continued until 1910 when it became a part of the University Extension at Harvard and elsewhere. It should be added that the regular course in the T.S.S. program was supplemented by Barton's field lessons given in the fall and in the spring.

As mentioned previously, Barton gave the lectures during the winter months and conducted the field lessons in the fall and spring. The nature of the lecture courses is clearly set forth in Hyatt's 1901 Annual Report of the Boston Society of Natural History (Pr. 29: 257, May 1, 1901) as follows [by this time they were called "Lowell Free Lectures" because of the Lowells' financial support]:

Lowell Free Courses

A course on structural geology of fifteen lessons of two hours each was given by Prof. Geo. H. Barton in the lecture room of the Society during the past winter. This was the third series in a four years' course. The instruction was given by means of lectures, illustrated by diagrams, charts, and a large number of specimens both for class and table use. The stereopticon was also used whenever good illustrations of the subject in hand were to be obtained. Examinations as usual were an essential part of the work, one of twenty minutes' length being held at each exercise; and a final examination of three hours was held at the close of the term. The average rank in these examinations was high and shows an extreme interest in the work when it is recognized that this is carried on in the spare time of the teachers who are also heavily burdened with the necessary work of the week in school. The average attendance for the term was 99.1.

CHILDREN'S MUSEUM (1913-1933)

The second public enterprise in which Barton participated for many years (1913-1933), after leaving M.I.T., was the Children's Museum, which was a direct outgrowth of the Teachers' School of Science. In April 1913 he was made Chairman of the Board of Directors of the Museum, then a part of the Science Teachers' Bureau, which itself was also an outgrowth of the T.S.S. Acting as mediator, Barton succeeded in getting the Science

Teachers' Bureau to vote that the Children's Museum be made an independent organization, and on 19 May 1914 it was so incorporated as the "Children's Museum of Boston," having been moved meanwhile from the Refectory Building in Franklin Park to "Pine Bank," the Edward Perkins home, in Olmsted Park on the shore of Jamaica Pond, in Jamaica Plain.

Barton was elected President of the new Museum and served for three years, August 1914 to January 1917, refusing re-election for a fourth time. He then served as Vice President from November 1920 to November 1927, at which time he again accepted the responsibilities of President for another two years. At the end of this period, feeling that he could no longer be sufficiently active, he was made Honorary President, which office he held until his death four years later in 1933. Although Barton did not actually start the Children's Museum, it is surely fair to state that without his patience and persistence the institution would never have been established on a firm and enduring basis. Recognition of his contribution is attested by the bas-relief that hangs on the wall in the present Museum, and by the dedication to his memory of Adelaide B. Sayles' The Story of The Children's Museum (of Boston from its beginnings to November 18, 1936). Boston: G. H. Ellis Co., Inc., xii + 87 p., il., (1937). Some additional information on the history of the Museum and on the part Barton played in that history is included in his own Report of the President [1915], (Children's Museum of Boston, Bull. 2/3: 19-21, 1916).

When the headquarters of the Teachers' School of Science was moved to Cambridge from Boston, Barton's influence was responsible for the founding of the Children's Museum in Cambridge, which also became the headquarters for the Teachers' School of Science.

Thus it is seen that for 46 years, 1887-1933, Barton was deeply involved in bringing geology to the general public, and particularly to the children and science teachers of the Greater Boston Community.

In his brief unpublished "History of the Department of Geology, M.I.T.," written in the late 1930s or early 1940s, Shimer comments as follows about Barton as Director of the Teachers' School of Science and President of the Children's Museum (p. 15):

> "Professor Barton retired to succeed Hyatt as Director of the Teachers' School of Science. There for a generation his genial and contagious enthusiasm made the School of Science the center of activity for bringing geology to the teachers and youth of Boston and vicinity. During this time [1902-1933] Professor Barton had perhaps the most influence of any man of his time in spreading geological knowledge and interest among the people of Boston. [Underlining by R.R.S.] He took

Prof. George H. Barton (S.B. III 1880), fourth from the left with a pick on his right shoulder, led a geology field trip to western Massachusetts in the late Nineteenth Century. Note that the ladies outnumbered the gentlemen!

A party of geologists waiting for the train after several days in the field. Prof. W. O. Crosby (S.B. VII 1876) is second from the left and Prof. G. H. Barton (S.B. III 1880) in the middle. The group obviously had success collecting large geodes!

(Both photographs from negatives in the Barton Collection of the Goodnow Library in Sudbury, Massachusetts)

them on innumerable trips, gave innumerable public lectures, and made the geology of Boston a subject of human concern among its citizens." (Page 15 of Shimer's typescript, which is preserved in the M.I.T. Institute Archives.)

TRAVEL AND FIELD EXCURSIONS

Barton was deeply interested in both field work and travel regardless of whether the trip was to a nearby outcrop or a distant glacier. He used his own observations and photographs to illustrate his lectures, and he was one of the first to conduct scientifically-oriented and commercially-conducted excursions for tourists, as well as regularly scheduled field lessons which he offered through the Teachers' School of Science.

His first trip away from home, when only twenty-one, took him to Wisconsin by way of Niagara Falls, and fired his ambition to train for a teaching career. Two years of geodetic surveying in Hawaii gave him unusual opportunities to observe active volcanoes and experience earth tremors first hand and to study the dynamic action of sea upon land. In 1885 he accompanied Alpheus Hyatt to Newfoundland and Labrador, where they collected fossil cephalopods as much as ten feet long. A decade later, in 1896, he went with Peary on his sixth Arctic expedition to Greenland (thirteen years before his eventful dash to the North Pole), and there gained firsthand knowledge of continental glaciation and the Arctic.*

Later on he revisited Hawaii several times, and made trips to Nova Scotia, to numerous National Parks, and to Alaska, all for the purpose of acquainting his tourist groups with the geological wonders of nature. Meanwhile, at home, he was continually leading field excursions to local places of special geological interest. These latter trips, or "field lessons" as he liked to call them, took his students to most of the specially significant geological areas in the Greater Boston region, and brought him tributes from many quarters.

Again and again in his Reminiscences, Barton mentions how he took "tramps," as he called his walks and field trips, with interested friends. It is obvious from these comments that he enjoyed nothing more than a tramp through the fields and woods and the opportunity to explain whatever

---

* See "The Sixth Peary Expedition," by Albert Operti, in The Home Magazine 8/4: 329-337, (April 1897). The scientists are pictured on page 331. See also references 14--1897a and 16--1898 in the Bibliography farther on.

Teachers' School of Science.

OUTLINE FOR COURSE
—OF—

Field Lessons in Geology,
—BY—

PROF. GEORGE H. BARTON.

Appendix A.

Origin of the Topographic Features of Massachusetts.

Massachusetts = simply a part of the Atlantic Slope.
Present Topography = all produced since close of Palaeozoic Time.
Denudation during Jurassic and Cretaceous = destroyed all previous topography.
Whole surface = worn down to lowland of faint relief = a peneplain.
Only large elevations remaining above this level =
 White Mountains of New Hampshire.
 Black Mountains of North Carolina.
 Blue Ridge of Virginia.

Elevation took place at beginning of Tertiary.
The Cretaceous peneplain = became the Tertiary plateau = the upland surface of our highlands = Massachusetts plateau.
Tertiary rivers eroded in this plateau the present valleys and open lowlands.
The broad lowland of the Triassic belt, or Connecticut Valley = general level of 200 feet or less above sea-level.
The upland valley of the Berkshire limestones = 600 to 1200 feet above sea.
The narrow valleys of the crystalline rocks = Deerfield.
Westfield, Miller's River Valley = types.
A new but moderate elevation = took place at close of Tertiary.
Rivers cut deeper into the Tertiary lowlands.
A pre-Triassic peneplain underlies the Triassic sandstone of the Connecticut Valley.
Exposed on west side of the Valley by erosion of the sandstone = the oldest topographic form on the Atlantic Slope.
Form of the peneplain or plateau = well seen from many places.
At Chester, on B. & A. R. R., ascend from Westfield Valley southward = a fine view of it is obtained from Blandford.
At Shelburne Falls, on F. R. R., ascend from Deerfield Valley northward = an equally good view from Hawley.
On eastern slope of Hoosac Mountains = an especially good view.

From carriage road, just below the eastern crest = the same general level stretches as far as the eye reaches.
A few prominent peaks rise above it.
The valleys, like the Deerfield = cut into it.
Appearance from the train in the valley = very mountainous.
Appearance from the side of Hoosac = wide plateau cut by valleys.
Wachusett, Greylock, and the crests of the Hoosac Range = rise above general plateau level = relics of the older higher land.
Monadnock = a characteristic mountain of this type, hence Monadnocks = term to these older remnants.
Literature of the Massachusetts Plateau.
W. M. Davis, Bull. Geol. Soc. Am. Vol. II, 1891, pp. 545-81.
 Am. Jour. Sci. 3d Series, Vol. 37, 1889, pp. 423-434.
Literature of Greylock and Hoosac Mountains.
J. E. Wolff, et al., Monograph XXIII, U. S. Geol. Survey.

Appendix B.

Points for Observation on Hoosac Tunnel Excursion.

Boston to Belmont = Slates of Boston Basin underlying drift.
Claypits in Champlain clays at West Cambridge.
Crystalline rock hills = northern boundary of Basin seen on right.

A typical Outline of the kind Barton prepared for the field trips he conducted for his students in the Teachers' School of Science.

geology there was to see. And he seemed willing to go whether accompanied by only a single individual or a group.

The Outlines that Barton prepared for his T.S.S. lectures and his "Field Lessons in Geology" show clearly the pedagogical methods of the times and also how carefully he prepared for the lectures and excursions.

PUBLIC AND OTHER LECTURES

Throughout his professional life, Barton gave public lectures on minerals and rocks, origin of the scenery, and the regional geology of the places he had visited. These lectures, the first given on 19 April 1880 when he was a college senior, were given mainly to school children and church groups and more often than not were without a fee. In addition, he gave many public lectures to a wide variety of audiences, but these generally dealt with his travel experiences in Hawaii and Greenland, rather than with geology alone. His main purpose in all his lectures was to popularize geology by sharing his photographs, knowledge, and enthusiasm for the out-of-doors with his listeners. It was in these lectures particularly that he drew on the thousands of lantern slides he had accumulated from his travels and excursions mentioned earlier.

In addition to these public lectures, his regularly scheduled courses at M.I.T., his courses at the Teachers' School of Science, and his many activities at the Children's Museum, he still found time to serve three other educational institutions in the Greater Boston area. He served Boston University as instructor in geology from 1893 to 1899, as assistant professor from 1900 to 1904, and as lecturer on geology in 1915. During SY 1921-22 he was lecturer on geology at Wellesley College, and at times he delivered lectures at Tufts College.

Barton not only gave innumerable lectures; he also attended an impressive number by others, judging from the numerous entries in his Reminiscences.* If one listed the names of the dozens of public lecturers

---

* In his Reminiscences Barton records the lectures he heard and in many cases adds interesting comments of one sort or another, e.g. (p. 260):

> One evening in December [1886] ... I heard Henry M. Stanley [of Livingstone fame] give an account of his travels in Africa in a most graphic manner that held the audience spell-bound for nearly two hours. He was introduced by Mark Twain in the latter's own peculiar manner, beginning with, as nearly as I can give it,
>
> > "I am asked to introduce Henry M. Stanley, but why I should be asked to introduce a man like Stanley I do not know unless that doing a totally unnecessary thing in a totally inadequate manner I am just the one!"

he heard, the list would include many distinguished individuals of the times--Theodore Roosevelt, Henry M. Stanley, Mark Twain, Admiral Peary, Louis Agassiz, Charles Lyell, Archibald Geikie, Booker T. Washington, and Lew Wallace, to name only a few. Clearly, Boston in the late 1800s and early 1900s attracted many of the foremost figures of the times, and Barton heard as many of them as he could.

MARRIAGE AND FAMILY

On 18 September 1884 in Stow, Massachusetts, George H. Barton married Eva May Beede, who was born at Royalton, Vermont, on 29 August 1855 and died a year before her husband on 8 January 1932. Eva May was the daughter of George Sloan and Helen Mary (Sanborn) Beede, and was a kindergarten teacher (an early graduate of Miss Elizabeth Peabody's school of kindergarten training and child psychology) at the time of her marriage. Like her husband, she was deeply interested in education, especially of children, and often accompanied him and his students on field excursions.

The Bartons, George and Eva, had three children--Harold Beede, who died in infancy; Donald Clinton, who became an outstanding exploration geophysicist* in the petroleum industry; and Helen Mary, who married Harold French Eastman. Both Barton children, no doubt influenced strongly by their parents, went far with their education; Donald won A.B. (1911), A.M. (1912), and Ph.D. (1914) degrees from Harvard, and Helen Mary received an A.B. (1914) from Radcliffe and an S.B. (1917) from Simmons. The older of the two Eastman daughters, Eleanor Beede Eastman (Mrs. Ernest R. Spinney) of Concord, Massachusetts has been most helpful in providing information about her grandfather, the subject of this sketch, who idolized his grandchildren, and is affectionately referred to by them as "Gramp." The tremendous impact that a great teacher can have on a receptive young mind is well exemplified by the influence Barton had on his aforementioned grand-daughter, who wrote the following for this biography on 10 July 1974:

---

* See "Dr. Donald C. Barton." (The Humble Sales Lubricator [The Humble Oil Company Magazine] 9/15: 8, July 20, 1939); "Donald Clinton Barton. An appreciation," by W. E. Pratt. (Min. & Metall., p. 529, Nov. 1939; "Obituary, Donald C. Barton [1889-1939]," by L. T. B_____ (Inst. Petroleum J. 26/195: 40, 1940); "Donald C. Barton [1889-1939]," by F. H. Lahee (Am. Assoc. Petroleum Geol., B. 24: 1521, 1940); "Memorial to Donald Clinton Barton [1889-1939]," by W. E. Pratt (Geol. Soc. Am., Pr. 1939: 153-166, port., 1940); and "Donald C[linton] Barton [1889-1939]," by R. von Zwerger (Beitr. Angew. Geophysik, B. 8: 258, 1940).

"Musings of a Grand-daughter"

"Re: George Hunt Barton
By: Eleanor Beede Spinney

"I knew George Barton only until I was thirteen years old. Much of the insight I now have, at fifty-four, has developed over the years as I have matured. At the time, he was simply my beloved Gramp. But we had more than a simple grandfather-child relationship, we knew one another intellect to intellect.

"How did this happen? Gramp was a teacher and a teacher is interested in each other person. Most particularly, he was interested in me as a person. I can't remember his ever 'talking-down' to me; he always spoke with me as a peer and one he liked as well as loved. Our conversations and our relationship stretched me; somehow, he always brought out the best in me. For instance, he would use the correct word, not a 'cute' or minimizing word - yet he always put that correct word in context of other words I could understand. So I learned. And I learned by having him help me to grow not by telling me. He was a true teacher.

"He had a child-like fascination with Creation. The Universe was his wondrous discovery to share with others. There was a night when he - 80? 81? - and I - 11? 12? - stood under the dome of heaven on a hillside in Stow and he shared his knowledge of the Universe with me. That night, he gave me God. It was one of the great experiences of my life and it was the beginning of God's presence in it. Although as a young man he had pulled away from the Church, he believed firmly in the Creator-Father-God. Between us there was never any talk of Christ but I now realize from my readings of what he wrote in Journals and to others that he had a tremendous curiosity about Christ as divine.

"His concern for all peoples was an important part of his personality to the extent that I would almost call him militant in his lack of prejudice. There was no patience in him for racial or other prejudicial remarks or actions. He was open in his approach to all matters, there being no false-modesty in his conversation or action. For instance, for him bodily differences were simply a part of God's creation and no reason for simpering. It was through him, I believe, that I began to develop my own interest in biology and medicine.

"He and I walked a lot and, as we "tramped" (his preferred phrase), I became aware of his angina as fact - not as a reason for anxiety. It was just a part of life with Gramp that occasionally he would sit, take a pill, and need to be left alone for a bit. He never transmitted to me any concern for his own life or for pain.

"My memories of him: the distinctive whistle he used to call to people; waking up early on a hot summer morning to see him swinging a scythe with beauty and grace; lecturing on top of Mt. Tom; collecting train tickets on a field trip; typing interminably on the porch of our cottage in Stow; his ever-present chocolate drops; hands-on-hips listening or watching; his beautiful white hair as I smoothed his brow. I remember the feeling I had when I heard he was dead - overwhelming thankfulness that he had died as he wished - in the middle of a word on the lecture platform; a sense of loss but tempered with a knowledge of the orderliness of his going. It was in the fulness of time. I loved to visit the apartment on Trowbridge St. in Cambridge and each time, Gramp and Gee asked me to leave a picture I had done or a writing or something else of myself until I should come again.

"But best of all, I remember being in museums with him ... the old Natural History, Agassiz, the little building on Jarvis Street where the Children's Museum started (and where Gramp died), the brick Cildren's Museum on the isthmus jutting out into Jamaica Pond. The wonder of being back-stage in a museum with him still overwhelms me. The special musty smell still comes back; the childish sense of a 'secret-place;' the joy of seeing exhibits prepared. He was sharing an important part of his life with me and I knew it.

"One cannot write in this way about Gramp without referring further to his wife, Eva, my beloved grandmother always called Gee (hard 'G'). Hers was a personality that beautifully complemented his; she was a serene person who took his more ebullient personality in stride. I have never known anyone with as much quiet innerdignity and outward calm coupled with real interest. For instance, she was terrified of thunderstorms; so she undertook to bring me to love them - we would stand on the porch of the cottage while she pointed out the gathering clouds, named them, informed me about the process taking place as the lightning and the thunder came, thrilled to the turbulent beauty. It wasn't until long after she was dead that I learned from my mother what Gee really did for me - her terror was completely concealed. She shared her interest in ferns, small animals, weeds; she loved to play the piano and sing to me - I can still hear 'Go Tell Aunt Rhodie' in her true, clear voice. She and Gramp loved one another and together they loved me, without question, and also knew me as a person.

"It interests me that the greatest influence I feel from George Barton is the sense of wonder. He didn't just 'teach me about geology;' despite his influence on me, his subject has held little interest for me except in a general way. He did something much more important: he gave me a giant step toward becoming my own person - a sense of inner worth and personhood, a sense of the delight in using my own brain in my own way.

"In this way, George Barton was a true teacher."

BARTON, THE MAN - A SUMMATION

Barton is remembered by those who knew him as an affectionate and compassionate person who tended to elicit similar responses from those around him. He was always willing to lecture, to guide an individual or a group on an excursion or a collecting trip, to help prepare an exhibition or label specimens, or to lend a helping hand to a child in difficulty with some project. He had the patience and understanding to be an ideal scientific guide for tourists, and these same attributes made him the successful mediator in resolving the recurring problems of the Children's Museum. He loved people and was never happier than when in the midst of a social group or the leader of a party of men and women on a field excursion. He was active in behalf of woman's suffrage and welcomed women in his classes and on his field trips. He was a Free Thinker, yet he had respect for the "devout and sincere of all denominations."

Perhaps the best summary of his life is the one delivered by Harvard President Lowell at a reunion dinner in 1927 celebrating Barton's 25th anniversary as Director of the Teachers' School of Science:

> "Since the death of Professor Alpheus Hyatt [1902] the care of the school and its welfare have depended wholly on Professor Barton. It is he who has directed the spirit of the enterprise; it is he whose devotion has made it what it is. Through his work he has given to a generation of teachers a familiarity with the fundamental principles of science and through them he has influenced the teaching of countless children in Boston and vicinity."

A more specific tribute appears on the bronze plaque or bas-relief by Cyrus E. Dallin that hangs on the wall at the head of the stairs leading to the second floor of the Resource Center (former Mitton residence) of the Children's Museum. The tribute reads:

GEORGE HUNT BARTON
1852-1933

First President of Children's Museum
Beloved and Inspiring Teacher
A Pioneer in Geological Field Education

In 1889 seven women who had been fellow students in his T.S.S. course in mineralogy formed the "Barton Chapter" of The Agassiz Association,* and later on both women and men who had taken his field lessons were invited to join. The chapter flourished for many years.

As would be expected of a person who loved people and enjoyed working for and with them, Barton held membership in many organizations, and was often called upon to serve in some leadership role. When only 24 he helped to organize "The Boston Amateur Philosophical Association," which soon became "The Boston Amateur Scientific Society" and finally simply "The Boston Scientific Society." Started on April 17, 1876, with Barton

---

* The Agassiz Association was organized by Harland H. Ballard of Pittsfield, Massachusetts, in 1889, to bring together young people interested in studying natural history. Chapters were organized throughout the United States, and the Boston Chapter was named after Barton. (Barton's Reminiscences, p. 275.)

as its first President, it lasted until 1922. He was active in organizing the Teachers' School of Science Association in 1901 and served as its president in 1904. He was President (1903) and Natural History Councilor of the Appalachian Mountain Club and served on the Council of the Boston Society of Natural History. He was elected a fellow of the American Academy of Arts and Sciences in January 1916 and was the first fellow elected to the Geological Society of America after it was organized in 1890. He was a member of numerous organizations that were concerned with the out-of-doors: the Alpine Club of America, the American Forestry Association, the Arctic Club, the Harvard Travellers' Club, and the Massachusetts Forestry Association. As mentioned earlier, an interest in genealogy led to membership in a number of genealogical organizations, including the New England Historic Genealogical Society, and he served the latter as a committee member and as Councilor for a number of years.

Finally, few scientists have revealed their inner selves and thoughts as completely as Barton did in his 1167-page typewritten Reminiscences representing a condensation of the diaries he kept throughout 65 years of his life. This amazing document, now preserved along with the diaries themselves in the Goodnow Library in Sudbury, Massachusetts, describes the life of an unusually perceptive and sensitive individual whose intellectual powers and fine personal traits emerge steadily with the passing of the years. Simultaneously it portrays in vivid and charming detail what life was like in New England in both country and city during the half century and more from the Civil War to the decade following World War I (1860-1930).

How better can an individual's strength of character and philosophy of life be discerned than by reading his own thoughts as he knows that his days are numbered? Upon reaching his 81st birthday, on 8 July 1933, Barton wrote in his Reminiscences for that day:

> "My 81st birthday. It is very difficult for me to appreciate that I am an old man, 11 years older than my Father or any of his family reached and within two years of the age at which my Grandfather Hunt passed away. I am still in very good condition and carrying on all my regular teaching and supervising of the Teachers' School of Science but leaving most of the manual labor to Miss Denton [Carrie D. Denton, his Assistant] and my janitor. I can still walk many miles but have to climb hills slowly. [He long suffered from angina pectoris]. Most of my generation have passed away ... Well so it goes and I am not troubled at the thought of my own passing on but [look] forward to it with pleasant anticipation except for the leaving behind my dear ones but as the dearest of all [his beloved wife Eva who had died almost two years before on 8 January 1932] has gone I am more than ready to go, to join her if there is a continued existence, if not I shall not be conscious of anything,

either past, pres[e]nt or future. Still it seems to me that the existence in the present condition is not comprehensible as to its meaning and use if all existence terminates with the termination of our lives here."

Less than five weeks later, on 25 November 1933, as he started to lecture to his T.S.S. class in historical geology, he died as his once strong heart stopped beating in mid-word.*

BIBLIOGRAPHY OF GEORGE HUNT BARTON

Symbols and abbreviations in the following references are explained on pages 91-98. Abstracts are listed as separate references, and may also be cited with the complete article. Every effort has been put forth to make this bibliography as complete as possible.

T--1880 The Geology of the Norfolk County Basin, 63 p., il., (1880). (S.B. Thesis at M.I.T. in Course III, Geology and Mining Engineering, June 1880.)

1--1880a (with Crosby, W. O.) Extension of the Carboniferous formation in Massachusetts. Am. J. Sci. (3) 20: 416-420, (1880).

2--1881 Geology of the Norfolk County basin [Mass.]. Science Observer 3: 41-42, (1881).

3--1884 Notes on the lava flow of 1880-81 from Mauna Loa. Science 3: 410-413, (1884).

4--1886 (with Crosby, W. O.) On the great dikes at Paradise, near Newport [Rhode Island]. Boston Soc. Nat. Hist., Pr. 23: 325-330, (1886).

5--1889 A preliminary paper on the drift in portions of Middlesex County [Mass.]. Tech.Quart. 2: 316-321, (1889).

6--1889a Geology of Sudbury, p. 644-652 in The History of Sudbury, Massachusetts, 1638-1889, by A. S. Hudson. Published by The Town of Sudbury, xxii + 660 p., il., (1889).

---

* The obituarial notice that appeared in the Boston Sunday Globe of 26 November 1933 stated:

> "A heart attack occurring as he lectured before a class of 20 men and women at the Cambridge Museum for Children on Jarvis St., Cambridge, yesterday resulted in the sudden death of Prof. George H. Barton."

Actually, as related to me by his grand-daughter, Eleanor Beede (Eastman) Spinney, he had started to lecture to his class when his voice failed in mid-word and he fell forward, but a physician (a Dr. Crockett) sitting on the front row, and alertly noticing his behavior, caught him before he fell all the way to the floor. He had died in the instant marked by the failure of his voice.

Obituarial notices also appeared in numerous local and national newspapers, as follows:

> Boston Globe, 25, 26 and 29 November 1933
> Boston Traveler, 25 November 1933
> Boston Post, 25 or 26 November 1933
> Cambridge Chronicle, 1 December 1933
> Maynard Enterprise (Sudbury edition), 29 November 1933
> New York Times, 26 November 1933

7--1890   Geologic Sketch of Acton, p. 280, in Acton in History (compiled for the Middlesex County History, published by J. W. Lewis & Co., of Philadelphia, with maps and illustrations additional, by Rev. James Flecher). Philadelphia and Boston: J. W. Lewis & Co., p. 238-301, (1890).

8--1892   Boulders formed in situ. Tech. Quart. 5: 401-405, (1892).

9--1893   [Discussion of "The origin of drumlins," by W. Upham.] Boston Soc. Nat. Hist., Pr. 26: 2-25, (1893).

10--1894   Glacial origin of channels on drumlins [abst.]. Am. Geol. 13: 224, (1894); Am. J. Sci. (3) 48: 349-350, (1894). (See also next item.)

11--1894a Glacial origin of channels on drumlins. Geol. Soc. Am., B. 6: 8-13, (1894). (See also preceding item.)

12--1896   Evidence of the former extension of glacial action on the west coast of Greenland in Labrador and Baffin Land. Am. Geol. 18: 379-384, (1896).

13--1897   Glacial observations in the Umanak district, Greenland [abst.]. J. Geol. 5: 89-92, (1897); Science n.s. 5: 89, (1897). (See also next item.)

14--1897a Scientific work of the Boston party on the Sixth Peary Expedition to Greenland. Report B. Glacial observations in the Umanak district, Greenland. Tech. Quart. 10: 213-244, il., map, (1897). (See also preceding item.)

15--1897b Lieutenant Peary's Expedition. Science n.s. 5: 308-310, (1897).

16--1898   Lieutenant Peary's last Greenland Expedition. The National Magazine 8: 313-323, (1898).

17--189?   Lessons in Structural Geology (Inside title: Outline for Course of Lessons in Structural Geology). Boston, 78p., (189?). Outline of Dynamical and Structural Geology (Inside title: Outline for Course of Lessons in Structural Geology). Boston, 160 p., (1901).
    1901

18--1900   Outline of Elementary Lithology (Inside title: Outline for Course of Lessons in Lithology). Boston, 54 p., (1900); Boston, 112 p., (1901).
    1901a

19--1901b Questions on Dynamical and Structural Geology. Boston, 29 p., (1901).

20--1902   Outline of Historical Geology (Inside title: Outline for Course of Lessons in Historical Geology). Boston, 1 + 78 + 3 p., (1902).

21--1902a A biographical sketch of William Harmon Niles. Tech. Rev. 4: 417-424, port., (1902).

22--1902? Outline for Course of Lessons in Mineralogy. 67 p., [? Boston, 1902?]

23--1903   The general geographical features of Boston and vicinity. J. Geogr. 2/6: 277-285, (1903).

24--1911   Memoir of William Harmon Niles, 1838-1910. Geol. Soc. Am., B. 22: 8-14, (1911).

25--1912   Bibliography of W. H. Niles. Geol. Soc. Am., B. 23: 34-35, port., (1912).

26--1912a A nature museum for children. Boston: Broadside. 1912 (Not seen, hence pagination unknown; item catalogued in the library of the Museum of Science in Boston).

Annual Reports as Curator or Director of the Teachers' School of Science

As Curator and later Director, of the Teachers' School of Science, Barton reported annually on the School's activities. These reports were published for a decade in the Proceedings or the Museum and Library Bulletin of the Boston Society of Natural History, as follows:

1) Proc. Ann. Meeting, May 7,1902. B.S.N.H.,Pr. 30/5: 435-439, July 1902
2) Proc. Ann. Meeting, May 6,1903. B.S.N.H.,Pr. 31/2: 25- 32, June 1903
3) Proc. Ann. Meeting, May 4,1904. B.S.N.H.,Pr. 32/1:   6- 11, Oct. 1904
4) Proc. Ann. Meeting, May 3,1905. B.S.N.H.,Pr. 32/5: 126-127, June 1905
5) Proc. Ann. Meeting, May 2,1906. B.S.N.H.,Pr. 33/1:   7- 8, July 1906
6) Proc. Ann. Meeting, May 1,1907. M. & L.,   B. 4:   6- 7, June 1907
7) Proc. Ann. Meeting, May 6,1908. M. & L.,   B. 8:   6- 7, July 1908
8) Proc. Ann. Meeting, May 5,1909. M. & L.,   B. 10:   6- 7, May 1909
9) Proc. Ann. Meeting, May 4,1910. M. & L.,   B. 12:   5- 7, June 1910
10) Proc. Ann. Meeting, May 3,1911. M. & L., B. 17:   5- 7, June 1911

Thereafter, information about the yearly activities of the Teachers' School of Science was included in the Annual Reports of the University Extension. (See chapter on "M.I.T. Geology Professors--Boston Society of Natural History--Teachers' School of Science--Children's Museum--Marine Biological Laboratory " in my Volume 2.

NOTE:

As mentioned in the foregoing biographical sketch, Barton left an impressive number of typescripts on a variety of subjects. These are more than passing reflections; they represent the thoughts of an individual of unusual sensitivity and compassion, who truly loved people and demonstrated that affection by his attitudes and actions. Included in his private papers, now happily preserved in the George Hunt Barton Memorial Room of the Goodnow Library in Sudbury, Massachusetts, the town where he was born, are the following typescripts:

1) Reminiscences, a narrative of 1167 pages compiled from diaries and other records that he kept throughout his life starting in his earliest years. Lane, in his 1935 Memorial of Barton (p. 161), characterizes them as follows: "These will be an invaluable primary source of information regarding New England life, the geology and geologists of the period from Agassiz and Hyatt to his death, and growth of the movement for adult and popular science education (children's museums) in which he was a leader." The last entry was made only two days before his death, which occurred on 25 November 1933.

2) The Teachers' School of Science. An Account of its Conception and of its Work and its Results. 22 p., (undated).

3) My association with The Teachers' School of Science. 36 p., (undated).

4) Autobiography of George Hunt Barton. 115 p., (undated).

5) The Religion of a Scientist. An essay of 21 pages, typed about 1930, in which he philosophizes about religion and sets forth his own views.

6) Hawaii 1881-1883. A 67-page typescript describing his two years in Hawaii, from departure on 9 August 1881 to arrival back in Sudbury on 21 September 1883.

7) Poems. Collected in Youth by George Hunt Barton. A compilation of Barton's favorite poems, including a dozen or more by himself, and a few by his future wife, Eva M. Beede. Approximately 240 pages, (undated).

8) Ancestral families of George Washington Barton [his father], (with Laurilla [Hunt] Sanders as the second author).

9) Genealogy of Mary Susan Hunt (later Mrs. George W. Barton) [his mother]. 84 p., (undated).

10) Introduction to Genealogy of Mary Susan Hunt (Mrs. George W. Barton) [his mother]. 112 p., (undated).

11) Miscellaneous Notes on Families. Approximately 25 pages; parts separately paged, (undated).

12) Notes by the Way 1873-1874. 10 p., (1873-1874). A daily account of his trip to Wisconsin.

13) [Miscellaneous Items including]
    Letter to "my darling daughter," dated Feb. 9, 1921.
    "To my children, Donald and Helen--My early memories of their Mother." (undated).
    "Tales of a Grandfather." 2 p., (undated).

Barton typing his "Reminiscences": Photograph by some member of his family circa 1930.

(7)
THOMAS AUGUSTUS JAGGAR, JR.
(1871-1953)

MIT: 1902-1917

THOMAS AUGUSTUS JAGGAR, JR.

Inspired at Harvard by Shaler, Wolff, Cooke and Daly, and in the field by Emmons, Hague and Palache, to investigate the dynamic processes of the earth, Thomas Augustus Jaggar, Jr. chose to study volcanoes and earthquakes by directly and constantly recording volcanic and seismic activity at a special observatory on the lip of one of the world's great active volcanoes, Kilauea in Hawaii.

Coming to M.I.T. in 1904, seven years after completing his doctorate at Harvard, and with almost ten years of teaching and field experience in geology, Jaggar had definitely determined to devote his life to the study of earthquakes and volcanoes. He headed Course XII, Geology, from 1904 to 1912, then resigned as head but continued as professor of geology until 1917. During his first eight years he was an outstanding lecturer and a productive investigator, and he published more than thirty articles during this period. His major interest, however, centered not on teaching but on observing and measuring the dynamic processes at work in the earth. Accordingly, when M.I.T. received from the estates of Edward and Caroline Whitney a bequest of $25,000 for research and teaching in geophysics, with a view to the protection of human life and property, Jaggar proposed that the bequest be used to establish and operate a year-round observatory on the active volcano Kilauea in Hawaii. With the Institute's approval, he resigned as head of Course XII, took a leave of absence in early 1912, and went to Hawaii to found the Hawaiian Volcano Observatory. Acting as Director, and soliciting further funds through the Hawaiian Volcano Research Association, of which he was also the Director, Jaggar established and then directed the activities of the Observatory from 1912 to 1917, at which time he resigned his professorship at M.I.T. However, he continued as Director of the Observatory until 1919, at which time it was taken over by the U.S. Weather Bureau and he became an employee of the U.S. Government. When operation of the Observatory was passed on to the U.S. Geological Survey in 1924, and a Section on Volcanology established in 1926, Jaggar was made Chief of the Section. Nine years later, in 1935, when the Section was disbanded and the volcano observatories in Hawaii (Kilauea) and California (Lassen) taken over by the National Park Service, Jaggar became Chief Volcanologist with the Park Service and remained in that position until retirement in 1940. Thereafter, until death in 1953, he continued to write and lecture on volcanoes and earthquakes. At the end, the Hawaiian Volcano Observatory became his monument, and his more than 300 publications, including his charmingly written autobiography, <u>Experiments with Volcanoes</u>, gave reason for calling him the Dean of American Volcanologists.

Throughout his professional life of half a century he was deeply concerned about the devastation wrought by earthquakes and volcanic eruptions. Although he travelled widely to acquaint himself with the nature and magnitude of this devastation, he felt strongly that little could be done about predicting earthquakes and volcanic eruptions until systematic and

326   GEOLOGY AT M.I.T.   1865-1965

My Experiments With
Volcanoes
T. A. JAGGAR

See the footnote on the page opposite for additional information on the above book.

sustained records were accumulated in areas of seismic and volcanic activity. As a result, the systematic publications of the Hawaiian Volcano Observatory are the first ever maintained. He gave much thought to the development of methods that could be employed in protecting people and property during earthquakes and from ash falls, glowing clouds, and lava flows. He suggested bombing and the construction of barriers to divert lava flows, and gave much attention to how structures could be made more resistant to earthquakes. In summary, he sought a better understanding of earthquakes and volcanoes so that this understanding could be used to ameliorate the disastrous effects of these awesome natural phenomena.

BIRTH, ANCESTRY, AND EARLY EDUCATION

Thomas Augustus Jaggar, Jr.,* dean of American volcanologists, was born in Philadelphia, Pennsylvania, on 24 January 1871 and died in Honolulu, Hawaii, on 17 January 1953, a single week short of his 82nd birthday. He was the only child of Reverend Thomas Augustus Jaggar, Episcopal bishop of Southern Ohio and 113th in succession to the American episcopate, and Anna Louisa (Lawrence) Jaggar, the daughter of John W. Lawrence of Flushing, New York. He traced his ancestry back to Jeremy Jaggar who came from England, probably with the second Winthrop colony, settling at Watertown, Massachusetts, previous to 1634.

He received his early education in public and private schools in Philadelphia (Delancey School) and Cincinnati, and in Montreux, Switzerland.

YEARS AT HARVARD (1889-1906) AND M.I.T. (1904-1917)

He entered Harvard in 1889, received an A.B. degree in geology in 1893 and an A.M. in 1894. For most of the next two years he continued his

---

* Most of the information in this biography was drawn from four sources: 1) Jaggar's informative and charmingly written autobiography--My Experiments With Volcanoes (Honolulu: Hawaiian Volcano Research Association, xii + 198 p., port., 1956); 2) Gordon A. Macdonald's brief memorial statement in The Volcano Letter No. 519, p. 1-4, (January-March 1953) entitled "Thomas Augustus Jaggar;" 3) his longer necrology, with a partial bibliography, entitled "Thomas Augustus Jaggar (1871-1953†), Nécrologie," which appeared in Bull. Volcanologique Sér. II, Tome XIV, p. 198-209, port., (1953), Naples, Italy; and 4) his obituarial note, "Dr. T. A. Jaggar," in Nature 171/4357: 774, (2 May 1953). In addition to the preceding references, briefer statements about Jaggar will be found in the following: The Technique [M.I.T.] for 1905, p. 19 ("Thomas Augustus Jaggar, Jr."); "Dr. T[homas] A[ugustus] Jaggar [1871-1953]" in Nature 171: 774, (1953) [Anonymous]; "Thomas A[ugustus] Jaggar [1871-1953]" in Earthquake Notes 24: 7, (1953) [Anonymous]; and Tech. Rev. 30/2: 88-89, (Dec. 1927).

education in Europe where he studied mineralogy with Prof. Paul Groth at Munich and petrography with Prof. Harry Rosenbusch at Heidelberg. He then resumed graduate work at Harvard, under Prof. J. E. Wolff, and received a Ph.D. degree in 1897. While still a doctoral candidate, he was appointed Instructor in Geology at Harvard in 1895 and continued to hold that appointment until 1903, when he was advanced to Assistant Professor. A year later, in 1904, he accepted M.I.T.'s offer to become Professor of Geology and Head of its Geology Department, but also continued as Assistant Professor at Harvard until 1906. What sort of scientist was this new professor?

Jaggar's love of the out-of-doors, and of plants and animals, was stimulated by his perceptive father, who made many a hike with him, and acquainted him with the wonders of nature. His scientific interests were greatly heightened at Harvard where he studied with Nathaniel Shaler, Josiah Cooke, Oliver Huntington, John E. Wolff, and Robert Jackson, and was a fellow graduate student with Reginald A. Daly. Field work (on the U.S. Geological Survey from 1898 to 1904) with S. F. Emmons in the Black Hills, with Arnold Hague in the Yellowstone region of Wyoming, and with Charles Palache in the Bradshaw Mountains in Arizona, greatly excited his natural interest in geological phenomena. It was not until he met Frank Alvord Perret, however, that he set the course for what would become his life's work--experimenting with volcanoes. He first met Perret in 1906 on the slope of Vesuvius, and quickly sensed that his new acquaintance, as he wrote in My Experiments With Volcanoes (p. xi),

> "... was the world's greatest volcanologist. His skill was taking pictures. Mine was making experiments. We agreed that these two skills in action would accomplish what theories never could approach."

From then on Jaggar decided to devote his life to understanding the earth by investigating the phenomena and processes he himself could observe, experiment with, and measure. As he wrote on page xii of his autobiography, referred to in the previous sentence:

> "I have been called geologist and seismologist, volcanologist and geophysicist. I am none of these. I am interested in the evolution of what Hoyle calls 'This quite incredible universe.' I am just as interested in Bergson's 'Creative evolution' as in Hoyle and Lyttleton's 'New cosmology.' And more interested in life than either. The elements for fruition are a thick earth's crust, a comparable pattern for earth and moon, and a mechanism for earth core."

This then was the relatively young and highly ambitious professor who came to head Course XII, Geology, in 1904. Actually, Jaggar served both Harvard and M.I.T. from 1902 to 1906: at Harvard he was Instructor (1902-1903) and Assistant Professor (1903-1906); at M.I.T. Lecturer in General

Geology (1902-1904), followed by Professor and Head of the Department of Geology, from 1904 to 1912, after which he became Director of the Hawaiian Volcano Observatory until he resigned from M.I.T. on 1 July 1917.

Being a dynamic and enthusiastic lecturer, and a strong advocate of field work and travel, Jaggar's impact on M.I.T.'s temporarily weakened Department of Geology was soon felt. As SY 1902-03 started, Course XII was without a Head, Niles having retired as Professor Emeritus on 30 June 1902 and Barton having resigned to succeed Hyatt as Director of the Teachers' School of Science. Crosby, the only full-time professor left, was given charge of the geology program until a new Head could be found, and professors at Harvard were temporarily engaged to give the Course XII students lectures in a number of geology subjects. In this arrangement Jaggar was Lecturer in General Geology, Experimental Geology, and Field Geology. Although the cooperative arrangement with Harvard lasted only one year (SY 1902-03), because loss of students' time in transit and scheduling conflicts made its continuance impossible, Jaggar was willing to travel to Boston to give lectures in General Geology during SY 1903-04. At the end of the school year he was engaged as Professor and Head of Course XII and thereafter, until 1912, when he resigned as Department Head to become Director of the Hawaiian Volcano Observatory, he quickly brought life into geology, as reported to me by one of my former colleagues, Prof. Carle R. Hayward (III S.B. 1904):

> "Jaggar, coming from a distinguished reputation at Harvard, gave the department [Geology] a shot in the arm. New courses were introduced and the interest already existing in many mining students was capitalized by the formation of a new option, Mining Geology, in the mining department. This course started the year [1905] before I came on the teaching staff and was an important subject of discussion among the students. Jaggar had a striking personality and those who came to know him best felt that impact strongly. A group of students which accompanied Jaggar on an Alaskan trip one summer [1907] came back full of enthusiasm for the experience.
>
> "My contacts with Jaggar were confined to occasional meetings in the old M.I.T. lunch room on Trinity Place [Boston] where he always added to the interest of the general conversation.
>
> "There were many expressions of regret when he left to take up his work in Hawaii. His services there in studying volcanic action need no comment."

Jaggar's term of service as chairman of Course XII, though short (1904-1912), was most effective in reestablishing Geology as an exciting and promising field of study. He himself gave the lectures in physical geology--"dynamic geology" as he liked to call it--and Warren gave the work in mineralogy and petrography, while H. W. Shimer and D. W. Johnson came on from Columbia to aid Crosby in paleontology and physiography, respectively. Jaggar's annual reports on the Department, as included in the annual M.I.T. President's Reports (see Bibliography farther on), indicate that instruction in geology was lively and attracted a large number of undergraduate students, particularly those interested in practical geology and mining.

During SY 1904-05, Course XII was reorganized and its name changed from Geology to Geology and Geodesy. Two options were provided: the first in <u>Geology</u>, the second in <u>Geodetic Surveying</u>. The latter permitted a student to take the S.B. degree in the astronomical and topographic branches of Civil Engineering. Students in mining engineering were attracted to the first option, with three getting bachelor's degrees in 1905, and ten enrolled in the option in the Class of 1906.

Jaggar's approach to geology was a dynamic one; he was deeply interested in geological processes, especially the more visible and spectacular ones. As would be expected, he was irresistibly attracted to earthquakes and volcanoes, and because of this interest he repeatedly went to see these phenomena in action or to view the devastation produced by them. In 1902 he went on the <u>S.S. Dixie</u> expedition to the Antilles to view the results of the catastrophic eruptions of Mont Pelée and La Soufrière and came away with the decision that he would devote his scientific career to experimental geology--to applying the experimental methods of physics and chemistry to geology--for the benefit of mankind.

In 1906, as mentioned earlier, he visited Italy to study Vesuvius in action (See references 20 and 21 in the bibliography farther on) and there met Perrett who would later become known, like Jaggar, as one of the great volcanologists of the twentieth century. In 1907 he led an M.I.T. expedition to the Aleutian Islands to study Makushin and Bogoslof volcanoes (See references 30 and 31 in the bibliography farther on). In 1909 he and R. A. Daly visited the Hawaiian volcanoes, after which he went on to Japan to visit the facilities of the Imperial Earthquake Investigation Committee. It was on this visit that he first met and became acquainted with Prof. F. Omori, one of Japan's greatest seismologists, who became one of his most admired scientific friends. In 1910, accompanied by Prof. C. M. Spofford of M.I.T.'s Civil Engineering Department, he visited Costa Rica, to study its volcanoes and view the devastation wrought by the 1910 Cartago earthquake, and then Jamaica to examine what was being done to enforce earthquake-proof construction following the disastrous Kingston earthquake of 1907.

As a result of these expeditions and visits, and of the observations he made on them, he became convinced of the inadequacy of the after-the-fact expeditionary method of investigating volcanoes and earthquakes. He saw the necessity for permanent observatories to keep a constant daily record, year after year, of volcanic and seismic activity. Impressed by the paucity of knowledge possessed by scientists regarding the prediction of earthquakes and volcanic eruptions and the alleviation of the destruction produced by them, he decided to do something about the matter.

The opportunity came on 1 July 1909 when M.I.T. received from the estates of Edward and Caroline Whitney of Boston a bequest of $25,000 to be used for research and teaching in geophysics, with a view to the protection of human life and property.*

For the record it should be noted that as early as 1898 there was interest in observational geophysics among a number of M.I.T. faculty members. In that year a small geodetic observatory** was built in Boston, and research followed in geodesy, astronomy, magnetism and geology. The research was conducted in part under the supervision of staff members of the Civil Engineering Department and in part under certain professors of geology. The result of this interest in geophysics on the part of M.I.T. is known to have influenced the trustees of the Whitney estates to make the bequest quoted in the preceding footnote.

Furthermore, as noted earlier, Jaggar had become much interested in volcanoes and earthquakes, had visited the sites of a number of recent

---

* The legal instrument (deed) by which the bequest was made is reproduced as follows, on p. 1 of T. A. Jaggar's "Report of the Hawaiian Volcano Observatory of the Massachusetts Institute of Technology and the Hawaiian Volcano Research Association," Boston: Soc. Arts of M.I.T., vii + 74 p., (1912):

> "We the undersigned, as trustees of the Estates of Edward and Caroline Whitney, hereby give to the Massachusetts Institute of Technology the sum of twenty-five thousand dollars ($25,000) for the establishment of a fund to serve as a memorial of Edward and Caroline Rogers Whitney of Boston and to be known as the Whitney Fund. The principal and interest of the said Whitney Fund is to be expended at the discretion of the Corporation of the Massachusetts Institute of Technology for the conduct of research or teaching in geophysics. We desire that the work carried on in any laboratories which may be wholly or partly supported by contributions from the said Whitney Fund shall include investigations in seismology, conducted with a view to the protection of human life and property. Without intending to control the discretion of the Corporation of the Massachusetts Institute of Technology in respect to the application of the said Whitney Fund, we state our present preference that some investigations in geophysics be undertaken in Hawaii."

** See the anonymous article, "The Geodetic Observatory," in Tech. Rev. 1: 170-171, sketch, (1899).

earthquakes and volcanic eruptions, and in April 1909 had visited Kilauea in Hawaii and witnessed the Japanese volcanoes Tarumai and Asama in eruption. On the same trip he had studied the recording facilities of the Imperial Earthquake Investigation Committee of Japan, so he returned to Cambridge with impressive observational data and with much knowledge of the excellent facilities that were available for accumulating critical data on both seismic and volcanic activity.

THE HAWAIIAN VOLCANO OBSERVATORY

On the basis of the knowledge he had acquired by travel and otherwise, Jaggar concluded that Kilauea was the best place to locate the kind of observatory he had in mind. Accordingly, he presented his reasons to the M.I.T. Administration, and with their approval proceeded to interest his colleagues in the program envisioned for the proposed observatory and to raise supporting funds from interested individuals and organizations, particularly in Hawaii and in the Greater Boston Community.

The interested reader is referred to Jaggar's 1912 "Report of the Hawaiian Volcano Observatory" for a detailed account of how the Observatory got established and of the nature of the earliest experiments made at Kilauea. Suffice it here to note that up to 1912 the work of establishing and instrumenting the Hawaiian Volcano Observatory was initiated and carried forward by M.I.T., with financial support that came in part from the Whitney bequest and in part from a dozen or more Boston contributors. Additional support was obtained in Hawaii, through the activities of the Hawaiian Volcano Research Association, which Jaggar organized and directed. In due course the Association and M.I.T. agreed that Jaggar be sent to Hawaii in 1912 to aid in preparing a plan of operations for the proposed Observatory that had been established the preceding year. Accordingly, Jaggar resigned as Head of Course XII and was granted leave of absence in December, 1911, and instructed to go immediately to Hawaii to continue at Kilauea the recording of volcanic activity started earlier by his esteemed colleague, F. A. Perrett. Immediately on Jaggar's arrival in Honolulu in early January, 1912, a fund-raising campaign was started and in less than a month sufficient funds were pledged to let a contract for a Volcano House for the use of the M.I.T. representative engaged in volcanologic research. Although Jaggar returned to Boston in March for a visit of a few months, for all practical purposes he would henceforth devote his full time and energies to volcanologic research in Hawaii as Director of the Hawaiian Volcano Observatory, even though he continued to hold the title Professor of Geology at M.I.T. until he resigned on 1 July 1917. Thus ended Jaggar's official connection with M.I.T., but he continued as Director of the Observatory until 1919.

In 1919, the Hawaiian Volcano Observatory was taken over by the U.S. Weather Bureau, which at the time was charged with investigating earthquakes in the United States, and Jaggar, as a volcanologist, became an employee of the U.S. Government. Five years later, in 1924, operation of the Volcano Observatory passed from the Weather Bureau to the U.S. Geological Survey, and in 1926 the Survey established a Section of Volcanology, with Jaggar as its Chief. Under his supervision the Lassen Volcano Observatory was established in California and a station was developed at Dutch Harbor, Alaska. When the Survey's Section of Volcanology was disbanded in 1935, and the Hawaiian and Lassen Volcano Observatories were taken over by the National Park Service, Jaggar became Principal Volcanologist with the Park Service and continued in that capacity until retirement in 1940.

RETIREMENT YEARS AND SUMMATION

After retirement Jaggar became a Research Associate in Geophysics at the University of Hawaii, an appointment he continued to hold until his death in 1953. During this period he spent much time compiling the results of his volcano studies at Kilauea, but he also thought more about the cause of explosive eruptions like that at Mont Pelée in 1902, and even returned to one of his earliest interests--that of precisely measuring the hardness of minerals and other substances (See reference 7--1898b in the Bibliography farther on).

Jaggar's most important contribution to geology is considered to have been the establishment in 1912 of the Hawaiian Volcano Observatory and the continuous record it kept from then until he retired thirty years later in 1941. This record includes routine seismic data, measurements of shifting and tilting of the ground, and descriptions of various eruptions and flows. He and his associates collected samples of lava and gas for analysis, measured the temperature of molten lava, and recorded the rise and fall of the lava in Hilemaumau. To make these measurements Jaggar himself invented and constructed many of the recording instruments.

Jaggar had a long-time interest in the sea bottom and though not the first to suggest drilling a hole into the suboceanic crust, as early as 1939 he did propose the idea, suggesting that several old warships be tied together and serve as a drilling platform. He thought petroleum engineers could design and build a special drill to mount on the ships, but did not suggest just how the clump of ships could be kept on station. (He could hardly have suggested the automatic devices used today by the GLOMAR CHALLENGER as they were not yet developed in 1939.)

In 1943, at 72 years of age and with half a century of geological field research behind him, he suggested a staggering program for exploring the sea bottom--drill 1,000 holes, each 1,000 feet deep, into the floor of the deeper parts of the world ocean on a worldwide pattern. He opined that the samples from the holes could yield "a century of scientists with priceless physical, chemical and biological information."* How pleased and excited he would be if he could stand on the deck of the GLOMAR CHALLENGER today and see how those "oil men," aided by electrical engineers, had solved the problems of drilling from a floating ship!

Jaggar was an excellent photographer. He was Arnold Hague's field assistant while Hague was doing the field work for the Absaroka Folio, and an example of his skill is a photograph of "The two great talus cones in south Stinking Water Canyon, looking West," taken on 9 September 1893 and featured on the cover of the April 1969 number of the Journal of Geological Education (Vol. 17, No. 2, April 1969). Years later, with the superb photographs of volcanologist Frank Alvord Perret as models, Jaggar himself made many a spectacular photograph in the course of his work on volcanoes, as exemplified in his Volcanoes Declare War.

Thomas Augustus Jaggar, Jr. was a combination of scholar, field investigator, inventor** and experimenter, convincing lecturer, accomplished author, and originator of innovative ideas. And perhaps above all he was a humanist, for he never forgot that science serves best when it serves the needs of mankind.

He lectured to train students, arouse public interest in volcanoes and earthquakes, and raise funds to build a field observatory and conduct crater-side observations. He wrote to acquaint the people of the world, both young and old, with the awesome and destructive power of volcanoes and earthquakes. He invented numerous ingenious devices to aid in his experiments with volcanoes. And near the end he summed up his life's work in an informative and charmingly written autobiography, My Experiments with Volcanoes (1956). He was indeed the Dean of American Volcanologists, and his double monument is the Hawaiian Volcano Observatory, founded in 1912, and his numerous communications about volcanoes and earthquakes-- more than 300 publications, a great number of formal lectures, and innumerable informal lectures and comments ephemerally reported or noticed in newspapers and other media around the world.

---

* See p. 109 in W. J. Cromie, Why the Mohole: Adventures in Inner Space. Boston: Little, Brown and Co., Inc. ix + 230 p., (1964).

** Jaggar developed an amphibian car in 1928 (The Tech 48/3: 1, Feb. 13, 1928) for field work, and mentions other devices he designed in his autobiography, My Experiments with Volcanoes (1956).

It is worth noting that the monthly bulletins of the Hawaiian Volcano Observatory and the Volcano Letter were the first systematic publications of a volcano observatory ever maintained. It is worth further mention that Jaggar was always concerned about the loss of human lives and destruction of property from earthquakes and volcanic eruptions.* He gave much thought to the possibility of predicting earthquakes and volcanic eruptions, to designing suitably resistant structures for lands subject to frequent earthquakes, and to developing methods of protecting communities from ash falls, glowing clouds, and lava flows (e.g. to divert flows by bombing).

Coming last to an evaluation of the scientific contributions he made, I can do no better than to quote the comments of one of his most competent and distinguished colleagues, volcanologist Gordon A. Macdonald (Nature 171: 774, May 2, 1933):

> "Dr. Jaggar's contributions to the science of volcanology have been very great. Studies of the Kilauea lava lake demonstrated its complex nature, with a shallow pool of hot fluid lava lying on a semi-solid but mobile plug. Measurements ascertained the distribution of temperature in the lake, and the composition of the liberated gases was determined. Tumescence and detumescence of the volcanic structure were demonstrated, probably resulting from changes in magmatic pressure beneath. The entire picture of volcanic activity was augmented and clarified. Perhaps most important of all, however, was the development by Jaggar and his associates of methods of aerial bombing to deflect lava flows, and the concept of huge diversion barriers to protect cities or harbours from inundation by lava. Jaggar never lost sight of the fact that the real purpose of science is to serve mankind."

MARRIAGES AND FAMILY LIFE

Jaggar married Helen Klein on 15 April 1903, and two children were born to their union--a son, Kline, and a daughter, Eliza Bowne. On 21 September 1917 he married Isabel Peyran Maydwell, who shared with him the physical work involved in his study of volcanoes and both the disappointments and joys of his numerous administrative activities.

---

* Jaggar's concern for his fellow humans and their property is exemplified by the motto of the Hawaiian Volcano Research Association, which he served as Director from its founding in 1912 to his death in 1953, "Ne plus haustae aut obrutae urbes,"--"No more shall the cities be destroyed." (Macdonald 1953: 206.)

HONORS AND MEMBERSHIPS

In recognition of his distinguished research Dartmouth awarded him an honorary Sc.D. degree in 1938, and the University of Hawaii awarded him an LL.D. degree in 1945. He received the Burr Prize of the National Geographic Society in 1945.

He was elected to membership in the American Academy of Arts and Sciences, Washington Academy of Sciences, Hawaiian Academy of Sciences, American Geophysical Union, and Seismological Society of America. He held membership in the Harvard Travellers Club and in the University Club of Honolulu.

BIBLIOGRAPHY OF THOMAS AUGUSTUS JAGGAR, JR.

Symbols and abbreviations used in the following references are explained on pages 91-98. Abstracts are listed separately and may also be included with the reference to the complete article. Every effort has been put forth to make this the most complete list of Jaggar's publications yet assembled. I am in special debt to Donald W. Peterson, Scientist-in-Charge of the Hawaiian Volcano Observatory, Hawaii National Park, Hawaii, for sending me a copy of the list of 291 titles of articles by Jaggar that are included in Chester Wentworth's (unpublished) 1958 Bibliography covering all publications of the Hawaiian Volcano Observatory to that year.

1--1894   Some conditions of ripple mark. Am. G. 13: 199-201, (1894).

2--1895   (with Jackson, R. T.) Studies of Melonites multiporus [abst.]. Am. G. 16: 239-240, (1895). (See also the following reference.)

3--1896   (with Jackson, R. T.) Studies of Melonites multiporus. Geol. Soc. Am., B. 7: 135-170, il., (1896). (See also the preceding reference.)

4--1896a   On the geological work of vortices and eddies [abst.]. Science n.s. 3: 375, (1896).

5--1898   An occurrence of acid pegmatite in diabase. Am. G. 21: 203-213, (1898).

6--1898a   Some conditions affecting geyser eruption. Am. J. Sci. (4) 5: 323-333, (1898).

7--1898b   Ein Mikrosklerometer zur Härtebestimmung. Zeit. Krystal. 29: 262-275, (1898).

8--1899   Experimental investigation of the formation of minerals in an igneous magma [a review]. J. Geol. 7: 300-313, (1899).

9--1901   The laccoliths of the Black Hills. U.S. Geol. Surv., Ann. Rept. 21/3: 163-290, il., (1901).

10--1902   The next eruption of Pelé. Science n.s. 16: 871-872, (1902).

11--1902a Field notes of a geologist in Martinique and St. Vincent. Pop. Sci. Month. 61: 352-368, (1902).

12--1902b The crater of the Soufrière volcano, St. Vincent. Harper's Weekly 46: 1281, (1902).

13--1903   Professor Heilprin on Mont Pelé. Science n.s. 17: 423-425, (1903).

14--1904   The eruption of Mont Pelé, 1851 (translated from the French of Le Prieur, Peyraud, and Rufz). Am. Nat. 38: 51-73, (1904).

15--1904a   The initial stages of the spine on Pelé [Martinique, W.I.]. Am. J. Sci. (4) 17: 34-40, (1904).

16--1904b   The eruption of Pelé, July 9, 1902. Pop. Sci. Month. 64: 219-231, (1904).

17--1904c   Economic resources of the Northern Black Hills; Pt. I, General Geology. U.S. Geol. Surv., P.P. 26: 7-41, il., (1904).

18--1905   (and Palache, C.) Description of Bradshaw Mountains quadrangle [Ariz.]. U.S. Geol. Surv., Geol. Atlas Bradshaw Mountains Folio (No. 126): 11 p., il., (1905).

19--1906   A comparison of Mont Pelée and Vesuvius. New York Evening Post, (April 14, 1906).

20--1906a   The volcano Vesuvius in 1906. Tech. Quart. 19: 104-115, (1906).

21--1906b   The eruption of Mount Vesuvius, April 7-8, 1906. Nat. Geogr. Mag. 17: 318-325, il., (1906).

22--1907   The Earth as a Living Organism. Abstracts of four lectures, published by the Twentieth Century Club. [Not seen, but reported on p. 165 of the President's Report, January 1907, M.I.T. Bull. 45/2, (1907).]

23--1907a   Reginald Aldworth Daly. Tech. Rev. 9: 178-181, (April 1907).

24--1907b   The Technology Expedition to the Aleutian Islands. Tech. Rev. 9: 182-183, (April 1907). (See also 30--1908b.)

25--1907c   How should faults be named and classified? Econ. Geol. 2: 58-62, (1907).

26--1907d   Current methods of observing volcanic eruptions [abst.]. Science n.s. 25: 764-765, (1907).

27--1907e   Experiments illustrating erosion and sedimentation [abst.].
28--1908   Science n.s. 25: 767, (1907). Also Harvard Coll., M.C.Z., B. 49 (g.s. 8): 285-306, (1908).

29--1908a   A theory of ore deposition. Discussion of a review by F. L. Ransome of paper by J. E. Spurr. Econ. Geol. 3: 529-532, (1908).

30--1908b   Journal of the Technology [M.I.T.] expedition to the Aleutian Islands, 1907. Tech. Rev. 10: 1-37, il., (1908). (See also 24--1907b.)

31--1908c   The evolution of Bogoslof Volcano. Am. Geogr. Soc., B. 40: 385-400, (1908). Also Abst., Science n.s. 28: 575, (1908).

32--1909   The Messina earthquake: prediction and protection. The Nation 88/2271: 22-23, (Jan. 7, 1909).

33--1909a   Physiography of North America. Chapter in Baedeker's United States 1909; revised from work of N. S. Shaler.

34--1909b   Report of the Committee on Earthquake and Volcano Observations [abst.]. Science n.s. 29: 630-631, (1909). Also Geol. Soc. Am., B. 20: 659-660, (1910).

35--1910   Course in Geology. The Tech (Special Mining and Geology Issue) 29: 23, 26, (Jan. 15, 1910).

36--1910a   Japanese volcanoes. Bull. Soc. Arts [M.I.T.], 8 pages, unnumbered, il., (Feb. 1910). (See also Ibid., Jan. 1910.)

37--1910b   Studying earthquakes: the remarkable work of the Japanese Earthquake Committee. Century Magazine 80: 589-596, il., (Aug. 1910).

38--1910c   The duty of New England at the present time, with reference to the Endowed Colleges and the Secondary Schools. New England Assoc. Colleges and Secondary Schools, Pr., (October, 1910).

39--1910d   Special problems and their study in economic geology. Econ. Geol. 5: 776-780, (1910).

40--1910e   Genetic classification of active volcanoes [abst.]. Science n.s. 32: 188-189, (1910); Geol. Soc. Am., B. 21: 768, (1910).

41--1911   The earthquakes in Costa Rica. Sci. Conspectus 1: 33-40, (1911).

42--1911a   The Costa Rica volcanoes and the earthquakes of April 13 and May 4, 1910. Assoc. Eng. Soc., J. 46: 49-62, il., (1911).

43--1911b   The Technology Station in Hawaii. Tech. Rev. 13: 509-515, (1911).

44--1911c   The observation of earthquakes. Seismol. Soc. Am., B. 1: 48-82, (1911).

45--1912   Structure of esker fans experimentally studied [abst.]. Geol. Soc. Am., B. 23: 746, (1912).

46--1912a   Succession in age of the volcanoes of Hawaii [abst.]. Geol. Soc. Am., B. 23: 747, (1912).

47--1912b   On the region of origin of the central California earthquakes of July, August and September, 1911. Seismol. Soc. Am., B. 2: 31-39, (1912).

48--1912c   Report of the Hawaiian Volcano Observatory of the Massachu-
    [1914]   setts Institute of Technology and the Hawaiian Volcano Research Association, T. A. Jaggar, Director, January-March, 1912. Boston: Soc. Arts (M.I.T.), 74 p., [1914].

NOTE:

Jaggar resigned as Head of Course XII at the end of 1911 and was given leave of absence to go to Hawaii and aid in the operation of the Hawaiian Volcano Observatory which he had helped to establish several years earlier. For the remaining 42 years of his life, 1912-1953, he gave full time and effort to the affairs of the Observatory, publishing some 280 papers, most of which appeared in the several publications of the Observatory, and are included in the next two items of this Bibliography. Articles appearing elsewhere than in the different publications of the Observatory make up the remainder of this Bibliography.

49--1912   [Accounts of volcanic activity and earthquakes on the island
  to 1927   of Hawaii from 1912 to 1927]: Hawaiian Volcano Observatory, Bulletins 1-15, (1912-1927).

50--[1913?] The Cross of Hawaii. Honolulu Chamber of Commerce Ann. Rept. 1912, 12 p. (pamphlet), [1913?]

51--1914   Education and the Philosophy of Change. (Commencement address at the University of Hawaii, June 1914. Honolulu: Star-Bulletin Co., 6 p., 1914.)

52--1915   The outbreak of Mauna Loa, Hawaii, 1914. Am. J. Sci. (4) 39: 167-172, (1915).

53--1915a   The diary of Kilauea. Sci. Amer. Supp. 79: 36-37, (1915).

54--1915b   Notes from a volcano laboratory [Hawaii]. Sci. Conspectus 5: 85-103, (1915).

55--1915c   Science. Volcanism: [Review of] "The problem of volcanism" by J. P. Iddings. New Haven: Yale University Press, xi + 273 p., il., (1914). The Nation 101/2613: 155-157, (1915).

56--1915d   Science. The Silliman Lectures: [Review of] "Problems of American Geology." New Haven: Yale University Press, (1915). The Nation 101/2618: 296-298, (1915).

57--1915e   Activity of Mauna Loa, Hawaii, Dec.-Jan., 1914-15. Am. J. Sci. (4) 40: 621-639, il., (1915).

58--1916   The proposed Hawaiian volcano museum. Hawaiian Volcano Observatory, Weekly Bull. 3/4: 23-52, (1915) [1916].

59--1917   Lava flow from Mauna Loa, 1916. Am. J. Sci. (4) 43: 255-288, (1917).

60--1917a   Volcanologic investigations at Kilauea. Am. J. Sci. (4) 44: 161-220, (1917).

61--1917b   Live aa lava at Kilauea. Wash. Ac. Sci., J. 7: 241-243, (1917).

62--1917c   On the terms aphrolith and dermolith. Wash. Ac. Sci., J. 7: 277-281, (1917).

63--1917d   Thermal gradient of Kilauea lava lake. Wash. Ac. Sci., J. 7: 397-405, (1917).

64--1918   The index of danger from volcanoes. Hawaiian Volcano Observatory, Weekly Bull. 6: 15-20, (1918).

65--1918a   Results of volcano study in Hawaii. Nature 101: 54-57, (1918).

66--1918b   (and Romberg, A.) An experiment in teleseismic registration. Seismol. Soc. Am., B. 8: 88-89, (1918).

67--1920   A New Zealand department of volcanic research. New Zealand J. Sci. Tech. 3: 162-167, (1920).

68--1920a   Seismometric investigation of the Hawaiian lava column. Seismol. Soc. Am., B. 10: 155-275, (1920).

69--1921   Experiences in a volcano observatory [Hawaii]. Nat. Hist. 21: 337-342, (1921).

70--1921a   The program of experimental volcanology. Pan-Pacific Sci. Conf. Pr., B. P. Bishop Mus. Spec. Pub. 7: 309-324, (1921).

71--1921b   When Kilauea was dangerous. Paradise of the Pacific 34: 77-79, (1921).

72--1922   Ten years of work, volcano research, Hawaii. Hawaiian Volcano Observatory, Bull. 10/4: 27-32, (1922).

73--1922a   A plea for geophysical and geochemical laboratories. Wash. Ac. Sci., J. 12: 343-353, (1922).

74--1922b   The fire pit of Kilauea. Mid-Pacific Mag. 23: 353-359, (1922).

75--1923   The Yokohama-Tokyo earthquake of September 1, 1923. Seismol. Soc. Am., B. 13: 124-146, (1923).

76--1924   (and Thurston, L. A.) The Hawaiian Volcano Research Association. Pan-Pacific Sci. Cong., Australia, 1923, Pr. 1: 847-850, [1924].

77--1924a   Sakurajima, Japan's greatest volcanic eruption. Nat. Geogr. Mag. 45: 441-470, il., (1924).

78--1924b   (with Finch, R. H. and Emerson, O. H.) The lava tide, seasonal tilt, and the volcanic cycle. Pan-Pacific Sci. Cong., Australia, Pr. 2: 1369-1375, (1924); also Month. Weather Rev. 52: 142-145, (1924).

79--1924c   The borings at Kilauea volcano. Month. Weather Rev. 52: 146-147, (1924).

80--1924d   Activity of Kilauea volcano. Science n.s. 60 (supplement): x, xii, (1924).

81--1924e   Predicting earthquakes. Scribner's Mag. 74: 370-383, (1924).

82--1924f   (and Finch, R. H.) The explosive eruption of Kilauea in Hawaii, 1924. Am. J. Sci. (5) 8: 353-374, il., (1924).

83--1925   Adventures at volcanoes. Pop. Sci. Month. 106: 53, (1925).

84--1925a   Sur les rives du lac de deu. L'Illustration 4287: 248, (May 2, 1925).

85--1925b   The Hawaiian volcanoes [abst.]. Wash. Ac. Sci., J. 15: 304, (1925).

86--1925c   Earthquake insurance, in Soc. Geogr. Genève, Matériaux pour l'étude des calamités, Ann. 2/7: 191-217, (1925).

87--1925d   Earthquake insurance. National Underwriter, p. 10, (Aug. 13, 1925).

88--1925e   Plus and minus volcanicity [abst.]. Wash. Ac. Sci., J. 15: 416-417, (1925); also Bull. volcanologique, Ann. 2: 327-328, (1925).

89--1925f   Progress of volcanology during 1924. Wash. Ac. Sci., J. 15: 424-425, (1925); also Bull. volcanologique, Ann. 2: 336-337, (1925).

90--1925 to 1940   [Accounts of volcano activity and earthquakes on the island of Hawaii from 1925 to 1940]: Volcano Letter, no. 1-469, (1925-1940).

91--1926   So-called volcanic earthquakes. Science n.s. 63: 414-415, (1926).

92--1926a   The section of volcanology of the United States Geological Survey. Science n.s. 64: 242-243, (1926).

93--1927   Engulfment in volcanism [abst.]. Wash. Ac. Sci., J. 17: 23-24, (1927).

94--1928   Volcano research of the United States Geological Survey. Wash. Ac. Sci., J. 18: 512-515, (1928).

95--1928a   Volcanic relations of great Pacific earthquakes [abst.]. Third Pan-Pacific Sci. Cong., Tokyo, 1926, Pr., p. 370, (1928).

96--1928b   (and Finch, R. H.) Tilting and level changes at Pacific volcanoes. Third Pan-Pacific Sci. Cong., Tokyo, 1926, Pr., p. 672-686, il., (1928).

97--1928c   (and Finch, R. H.) Temperatures of volcano borings [abst.]. Third Pan-Pacific Sci. Cong., Tokyo, 1926, Pr., p. 687, (1928).

98--1928d   Engulfment during Pacific explosive eruptions [abst.]. Third Pan-Pacific Sci. Cong., Tokyo, 1926, Pr., p. 687-688, (1928).

99--1928e   Earthquakes and volcanoes, p. 30-40 and 12 figs. in vol. 2 of New Human Interest Library. Chicago: Midland Press, (c1928).

100--1929   Mapping the home of the Great Brown Bear. Nat. Geogr. Mag. 55: 109-134, il., (1929).

101--1929a (and Finch, R. H.) Tilt records for thirteen years at the Hawaiian Volcano Observatory. Seismol. Soc. Am., B. 19: 38-51, il., (1929).

102--1929b Amateur seismology. Sci. Am. 141: 411-413, (Nov. 1929).

103--1929c   Graded swelling and shrinking of volcanoes. Hawaiian Ac. Sci., Pr. 1929, B. P. Bishop Mus. Spec. Pub. 15: 10-11, (1929); also Volcano Letter 264: 1-3, il., (1930).

104--1931   The mechanism of volcanoes: Physics of the Earth, 1. Volcanology. N.R.C., B. 77: 49-71, (1931).

105--1931a   The eruption cycles in Hawaii. Hawaiian Annual 1932: 83-93, (1931).

106--1931b   Volcanology. Sci. Am. 143: 176-179, (Sept. 1930).

107--1932   Volcanologic developments in 1931-32. Am. Geophys. Union, Tr. 13th Ann. Meeting, p. 271-273, (1932).

108--1932a   Twenty years of volcano study in Hawaii. Volcano Letter 381-384, (1932).

109--1933   Elevation changes, horizontal shift, and tilt at Kilauea volcano [abst.]. Wash. Ac. Sci., J. 23: 113-114, (1933).

110--1935   Shipboard plane-table and azimuth camera: An experiment in navigation. Hawaiian Ac. Sci. Pr., B. P. Bishop Mus. Spec. Pub. 26: 13-14, (1935).

111--1935a   Living on a volcano ("Tin Can Island"). Nat. Geogr. Mag. 68: 91-106, il., (1935).

112--1936   Memorial of Bundjiro Koto. Geol. Soc. Am., Pr. 1935: 263-272, port., (1936).

113--1936a   The coming lava flow, our most serious responsibility. Honolulu Daily Advertiser, (March 27, 1934); reprinted in Volcano Letter 440: 1-6, (1936).

114--1936b   The bombing of Mauna Loa, 1935. Military Engineer 28: 241-245, (1936); reprinted in Volcano Letter 442: 1-7, (1936).

115--1938   Our Hawaiian Volcanoes. Honolulu: Castle & Cooke, Ltd., 12 p., (1938).

116--1938a   Structural development of volcanic cones. Am. Geophys. Union, Tr. 19th Meeting, Pt. 1: 23-32 (‡), (1938).

117--1938b   A simple seismoscope. Am. Geophys. Union, Tr. 19th Meeting, Pt. 1: 125, (1938).

118--1940   Magmatic gases. Am. J. Sci. 238: 313-353, (1940).

119--1945   Protection of harbors from lava flow. Am. J. Sci. 243-A: 333-351, (1945).

120--1945a   Volcanoes Declare War: logistics and strategy of Pacific volcano science. Honolulu: Paradise of the Pacific, Ltd., iii + 166 p., il., (1945).

121--1946   The great tidal wave of 1946 [from Aleutian Deep to Hawaiian Islands]. Nat. History 55: 263-268, 293, il., (1946).

122--1946a   Union through the ages. [Collection of addresses delivered by T. A. Jaggar between August 1934 and June 1946.] Honolulu: Paradise of the Pacific, Ltd., 32 + 84 p., port., (1946).

123--1947   Origin and development of craters [Hawaiian volcanoes]. Geol. Soc. Am., Mem. 21, xvii + 508 p., il., (1947).

124--1949   Steam blast volcanic eruptions, a study of Mount Pelée in Martinique as type volcano. Hawaiian Volcano Observatory, Spec. Rept. 4, 137 p., il., (1949).

125--1950   Abrasion hardness. Hawaiian Volcano Observatory, 5th Spec. Rept., 43 p., il., (1950); also Industrial Diamond Rev. 12: 1-6, (1952).

126--1953   Synopsis of volcanism. Pacific Sci. Cong., 7th, New Zealand, 1949, Pr. 2: 333-346, il., (1953). [Wellington.]

127--1956   My Experiments With Volcanoes [with a brief comment on Jaggar's relation to the Hawaiian Volcano Research Association by staff members of that organization]. Honolulu: Hawaiian Volcano Research Association, xii + 198 p., port., il., (1956).

NOTE:

In Wentworth's list of Jaggar's publications, referred to in the note introducing this bibliography, are 291 titles, but duplications reduce this number to 288. Of this latter number, 78 are included in the foregoing list, leaving 210 unlisted. Most of these 210 titles refer to short items appearing in the several publications of the Hawaiian Volcano Observatory; and are collectively cited in my items 49--1912 to 1927, 72--1922, and 90--1925 to 1940. Some, but not all, are included in the Bibliography of North American Geology (1929-1939) [U.S. Geol. Surv., B. 937, 1546 p., (1944)].

Altogether Jaggar published a total of at least 320 papers, almost all of which concern volcanoes or earthquakes.

Departmental Reports for M.I.T. President's Reports, 1905-1912

As head of Course XII, Department of Geology, Jaggar prepared a brief report of the Department's activities following the end of each calendar year. These reports, included in the President's Report for each particular year, were published in the following January, and were as follows:

1) Dept. Geol. M.I.T. Pres. Rept. 1904, [M.I.T., B. 40/2: 75- 77].
2) Dept. Geol. M.I.T. Pres. Rept. 1905, [M.I.T., B. 41/2: 59- 62].
3) Dept. Geol. M.I.T. Pres. Rept. 1906, [M.I.T., B. 42/2: 63- 72].
4) Dept. Geol. M.I.T. Pres. Rept. 1907, [M.I.T., B. 43/2: 115-119].
5) Dept. Geol. M.I.T. Pres. Rept. 1908, [M.I.T., B. 44/2: 109-111].
6) Dept. Geol. M.I.T. Pres. Rept. 1909, [M.I.T., B. 45/2: 106-112].
7) Dept. Geol. M.I.T. Pres. Rept. 1910, [M.I.T., B. 46/2: 111-116].
8) Dept. Geol. M.I.T. Pres. Rept. 1911, [M.I.T., B. 47/2: 104-109].

----

(8)
CHARLES HYDE WARREN*
(1876-1950)

CHARLES HYDE WARREN

MIT: 1900-1922

Trained at Yale in chemistry (Ph.B. 1896) and mineralogy (Ph.D. 1899), with some teaching experience in both disciplines, C. H. Warren brought to M.I.T. at the turn of the century a unique combination of knowledge and skills eminently suitable for instruction and research in mineralogy and petrology. Starting as an Instructor in Geology in 1900, he advanced to Assistant Professor in 1904 and to Associate Professor in 1912, after which as Professor of Mineralogy he taught mineralogy, crystallography, and petrology until 1922 when he resigned to become Dean of Yale's Sheffield Scientific School and Professor of Geology--posts he held until he reached retirement age in 1945. During his M.I.T. years, 1900-1922, he published more than 20 articles on mineralogical and petrological subjects, wrote a widely used elementary text for laboratory work, A Manual of Determinative Mineralogy, supervised 15 theses, played a major role in shaping the M.I.T. educational program following World War I, and then served his last two years as Head of the Division of General Studies which included the Courses in General Science and in General Engineering. After this distinguished career in science at M.I.T. he returned to his alma mater, to Yale's Sheffield Scientific School, and there during another 22-year period he achieved a second distinguished career, this time in academic administration, before retirement in 1945. Born in Connecticut in 1876, he died from a heart attack in a hospital near his birthplace in 1950 as he neared the end of his 74th year.

Charles Hyde Warren was born in Watertown, Connecticut on 27 September 1876 and died from coronary insufficiency in a hospital in Torrington, Connecticut on 16 August 1950, five years after retirement from Yale. His father was Charles Alanson Warren, the son of Alanson and Sarah M. (Hickox) Warren. His mother was Frances Maria (Hyde) Warren, the daughter of Ira and Grace (Lum) Hyde. Charles, the second of two children, came of a long line of New England forebears, the first of whom was Richard Warren, who came over on the MAYFLOWER in 1620 as one of the forty "Strangers," members of the Church of England who came to the New World in search of economic opportunity. Richard, in turn, traced his ancestry back to the time

---

* The sources of information on which I drew in preparing this biographical sketch are listed at the end, directly following the Bibliography.

of the Norman conquest of England to Earl William de Varrene or Warrene, whose castle was situated on the river Varrene in Normandy, whence the family name. Some of Richard's descendants moved from Massachusetts to Connecticut in the 18th century and ultimately settled in Watertown, and here among the Connecticut hills Charles Hyde Warren was born and to them he returned when he retired from Yale in 1945.

Upon graduation from the Waterbury (Conn.) High School, Warren entered Yale's Sheffield Scientific School in 1893 and after a three-year course of study was awarded a Bachelor of Philosophy degree in Chemistry. He was an excellent student, being elected to Sigma Xi in his senior year, and graduated with honors with the class of 1896. Having studied chemistry under H. L. Wells and mineralogy under S. L. Penfield, it is not surprising that he pursued graduate work in mineralogy for this was a popular field of scientific study at Yale even in those days. He received the Ph.D. degree in 1899 after submitting a doctoral thesis titled "Investigations in mineralogy and crystallography, including a description of four new minerals from Franklin, New Jersey."

He started his university career while working for the doctorate, serving as a laboratory assistant in analytical chemistry during SY 1896-97 and as instructor in mineralogy under Penfield from 1897 to 1900. Then followed a brief period of industrial experience as a chemist with the Scovill Manufacturing Company in Waterbury, but when he received an offer of an appointment as Instructor in Mineralogy and Geology from M.I.T. within the year (1900), he returned to the academic world and never again left it.

With excellent training in chemistry and mineralogy, and field experience in New Jersey and in the Yellowstone National Park region, Warren was well prepared to assume the teaching responsibilities in mineralogy and petrology at M.I.T. He lost no time getting teaching materials together.* He also soon became involved in a number of research projects and started publishing his findings as short "Notes." He became interested in the basic rocks at Cumberland, Rhode Island, and in the Quincy granite south of Boston, and these investigations generated a series of important ideas which were set forth in ten articles between 1908 and 1915.

---

* In the Ann. Rept. Pres. & Treas. of Dec. 11, 1901, appears the following statement (p. 38):

> "A new mineralogical collection has been arranged by Dr. Warren, to be used particularly as a reference collection by the students in second-year mineralogy. He has also brought together a collection of crystals, and is mounting them upon small stands of wood for the purpose of teaching crystallography to a greater extent from the natural crystals."

The implications of these ideas and their importance to geology, especially to mineralogy and petrology, are pointed out by Knopf in his biographical sketch cited later on. Suffice it to state here that Warren's years at M.I.T., from 1900 to 1922, were notably productive, whereas when he left to become Dean of Yale's Sheffield Scientific School his scientific research essentially ended.

Starting as Instructor, in the fall of 1900, Warren advanced to Assistant Professor of Mineralogy in 1904, Associate Professor in 1909, and Professor from 1912 to 1922, when he left for Yale. From 1912 to 1915 he also served on the staff of Jaggar's Hawaiian Volcano Observatory, and during one school year, SY 1918-19, taught some classes at Wellesley College. During his M.I.T. years Warren engaged in a variety of activities which he described as follows in the Vicennial Record of his Yale Class of 1896 (See reference in list following Bibliography):

> "Have been much occupied with expert work for various mining and manufacturing chemical concerns in addition to duties connected with teaching, also have carried out and published considerable research of purely scientific character.
>
> "Have become actively interested in the Single Tax Question, being a member of the Single Tax League of Massachusetts.
>
> "Have been especially interested in applying the polarizing microscope to investigation of commercial products, particularly high-temperature furnace products, such as ceramics and refractories, also to problems in chemical work generally. Have developed courses of instruction in this subject for chemical engineers."

Warren and his activities were described to me as follows in a personal letter from one of his students, the late Carle R. Hayward (S.B. III 1904):

> "Warren came as a young beardless Ph.D. from Yale and at once became very popular with the students. He gave an excellent course in mineralogy and kept the interest and respect of all his classes. In later years, before M.I.T. established the office of executive officer he in fact carried on these duties for Lindgren leaving the latter free for more important duties and professional consulting." (Personal correspondence, 1957.)

Unable to find a book on mineralogy suitable for the kind of practical work he wanted to teach his M.I.T. students, Warren prepared and had privately printed in 1910 a brief but highly useful Manual, which was later published by McGraw-Hill Book Co., Inc. as <u>A Manual of Determinative Mineralogy</u> (163 p., 1921 and 1922). This little book helped many an M.I.T. student in geology and mining engineering identify the minerals given him for study in the laboratory.

Warren's interest in education extended well beyond his department and scientific discipline. When he showed obvious administrative ability, he was made Chairman, at the end of World War I, of the Institute's important Committee on Courses of Instruction. In this position he played a major role in shaping the curriculum for the post-war years. In recognition of the liberalizing effect that the new curriculum had on the formerly rather rigid undergraduate program of study, Warren was placed in charge of the Division of General Studies and of the Courses in General Science and Engineering, and served in this capacity during his last two years at M.I.T., 1920-1922.

With his excellent training in both chemistry and mineralogy, Warren brought to M.I.T. an approach to teaching and research that immediately attracted those students primarily interested in practical geology. His interest in the chemical aspects of minerals, however, led to the desire to gain deeper understanding of how minerals form and react under different conditions of pressure and temperature, and this interest led to study of ceramics and refractories. It was only natural, therefore, that when N. L. Bowen came from Canada (Queens University) to M.I.T. to work on a doctoral thesis, and turned for advice to Daly, Jaggar, Warren and Lindgren, he was encouraged to follow his interest in the physical chemistry of minerals. At the appropriate time (1910) it was arranged that he do his thesis work in Washington, at the Carnegie Institution's recently founded Geophysical Laboratory, where the desired laboratory facilities were available.* Thus Warren helped to start on the way a brilliant young student who would revolutionize the study of igneous rocks in the next three decades.

All together Warren supervised a total of 15 geology theses during his 22 years at M.I.T.--7 S.B.s, 3 S.M.s, and 5 doctorates, an impressive

---

* In his "Memorial to Norman Levi Bowen (1887-1956)" [Geol. Soc. Am., Pr. 1956: 117-121, port., (1957)],J. F. Schairer states (p. 117):

> "In the year 1906 the Geophysical Laboratory of the Carnegie Institution was established in Washington, D.C., for the purpose of studying the problems of rock and mineral composition and origin through systematic and sustained experimentation in the laboratory. The Director, Dr. Arthur L. Day, had brought together a staff of physicists, chemists, and geologists for a concerted attack upon the problems of rock and mineral genesis. In the year 1910 Norman L. Bowen came to this Laboratory as a young student from M.I.T. to use the facilities in making a phase-equilibrium study of the silicate system nepheline-anorthite. He was permitted to use the results of this study for a thesis for the degree of Doctor of Philosophy at the Massachusetts Institute of Technology in 1912. After graduation he joined the staff of the Geophysical Laboratory as Assistant Petrologist, ..."

number when there were few graduate students in Course XII. In addition he shared supervisory duties with Lindgren and advised many students on the mineralogical and chemical aspects of their problems.

Warren not only worked closely with students but also developed mutual research interests with his Cambridge colleagues at Harvard. He and Palache co-authored four mineralogical papers, three of which concern the Quincy granite, and during SY 1920-21 jointly supervised H. E. McKinstry's M.I.T. Master's thesis on the "Petrology of the granites and associated pegmatites of Rockport, Massachusetts." The joint publication by Warren and McKinstry, that was based in part on the latter's work and titled The granites and pegmatites of Cape Ann, Massachusetts, did not appear until 1924, two years after Warren had gone to Yale, and was the last scientific paper he published, which led Knopf to write in his biography (p. 161):

> "American science lost an able worker in a field too little cultivated."

In 1922, when R. M. Chittenden, Director of the Sheffield Scientific School of Yale University, retired, Warren was called by his alma mater to become Dean of the School, and in accepting he asked also to be appointed to a professorship so as to have tenure. Thus he became Dean of the famous Sheffield Scientific School and Sterling Professor of Geology, posts that he would hold until retirement--from the Deanship in 1945, from the Sterling professorship in 1938 (he was also Chairman of the Department of Geology from 1923 to 1938, when he resigned and simultaneously relinquished the Sterling professorship), and from his professorship in Mineralogy in 1945. Here we leave it to others to tell the story of Warren's distinguished career at Yale, but it is worth recording some of the retired dean's thoughts as he set them down for the 50-year Class Book of the Yale Class of 1896:

> "As an emeritus dean, I shall retire to my farm ... However, I shall probably do some writing by way of finishing up several pieces of petrographic and mineralogic research that I started years ago in Boston and worked at intermittently as my duties permitted, which was to my great regret very little. <u>The miscellaneous alleged intellectual activities of a dean are not exactly conducive to the pursuit of rigorous scientific work</u>." [Underlining is mine.]

Warren was married to Charlotte Wardner Lamson at Watertown, Connecticut, on 17 June 1903, and four children were born to their union: Richard (1905), Allen Johnson (1907), Rachel (1910) who died in 1912, and William Lamson (1912). He was survived by his wife, three sons, and four grandchildren.

BIBLIOGRAPHY OF CHARLES HYDE WARREN

Symbols and abbreviations used in the following references are explained on pages 91 - 98; in general, abstracts are listed separately as well as with the references to the complete article. This bibliography is believed to include all geological publications written by Warren, and it is worth noting that all but 10 of the 35 titles were prepared while he was at M.I.T. during the years 1900 to 1922.

1--1898 Mineralogical notes. Am. J. Sci. (4)6: 116-124, (1898); Zeitschr. für Kryst. 30: 595-604, (1899).

2--1899 (with Penfield, S. L.) On the chemical composition of parisite and a new occurrence of it in Ravalli Co., Mont. Am. J. Sci. (4)8: 21-22, (1899).

3--1899a Some new minerals from the zinc mines at Franklin, N.J., and note concerning the composition of ganomalite. Am. J. Sci. (4)8: 339-353, (1899); Yale Bicentenary Pubs. Contr. Mineral. Petrogr.: 325-342, (1901).

4--1901 Mineralogical notes. Am. J. Sci. (4)11: 369-373, (1901).

5--1903 Mineralogical notes. Am. J. Sci. (4)16: 337-344, (1903).

6--1904 Petrographical notes on the rocks of the Weston aqueduct [Mass.]. Tech. Quart. 17: 117-123, (1904).

7--1906 The mineralogical examination of sands. Tech. Quart. 19: 317-338, (1906).

8--1906a (with Hidden, W. E.) On yttrocrasite, a new yttrium-thorium-uranium titanate. Am. J. Sci. (4)22: 515-519, (1906).

9--1906b Note on the estimation of niobium and tantalum in the presence of titanium. Am. J. Sci. (4)22: 520-522, (1906).

10--1908 Contributions to the geology of Rhode Island. II. The petrography and mineralogy of Iron Mine Hill, Cumberland. Am. J. Sci. (4)25: 12-38, (1908).

11--1908a (with Palache, C.) Kröhnkite, natrochalcite (a new mineral) and other sulphates from Chile. Am. J. Sci. (4)26: 342-348, (1908).

12--1908b Note on the alteration of augite-ilmenite groups in the Cumberland, Rhode Island, gabbro (hessose). Am. J. Sci. (4)26: 469-477, (1908).

13--1908c Ueber das Vorkommen von Hortonolith bei Cumberland, Rhode Island, U.S.A. Zeitschr. für Kryst. 44: 209-211, (1908).

14--1909 Note on the occurrence of an interesting pegmatite in the granite of Quincy, Mass. Am. J. Sci. (4)28: 499-452, (1909).

15--1910 A Manual of Determinative Mineralogy. Printed privately for student use at M.I.T. (See item 29--1921a.)

16--1910a (and Palache, C.) Pegmatite in the granite of Quincy, Mass. [abst.]. Science n.s. 32: 220, (1910); with discussion, Geol. Soc. Am., B. 21: 784, (1910).

17--1911 The barite deposits near Five Islands, Nova Scotia. Econ. Geol. 6: 799-807, (1911). Abst., Geol. Soc. Am., B. 21: 786-787, (1911).

18--1911a (and Palache, C.) The pegmatites of the riebeckite-aegirite granite of Quincy, Mass., U.S.A., their structure, minerals, and origin. A.A.A.S., Pr. 47: 125-168, pl., (1911).

19--1911b (with Palache, C.) The chemical composition and crystallization of parisite and a new occurrence of it in the granite pegmatites at Quincy, Mass., U.S.A. With notes, on microcline, riebeckite, aegirite, ilmenite, octahedrite, fluorite and wulfenite from the same locality. Am. J. Sci. (4)31: 533-557, (1911).

20--1912 The ilmenite rocks near St. Urbain, Quebec; a new occurrence of rutile and sapphirine. Am. J. Sci. (4)33: 263-277, (1912).

21--1913 Petrology of the alkali granites and porphyries of Quincy and the Blue Hills, Mass., U.S.A. A.A.A.S., Pr. 49: 203-331, (1913).

22--1914 (and Powers, S.) Geology of the Diamond Hill-Cumberland district in Rhode Island-Massachusetts. Geol. Soc. Am., B. 25: 435-476, (1914).

23--1915 A quantitative study of certain perthitic feldspars. A.A.A.S., Pr. 51: 127-154, (1915).

24--1916 A graduated sphere for the solution of problems in crystal optics. Am. J. Sci. (4)42: 493-495, (1916).

25--1916a George Jarvis Bush (1831-1912). A.A.A.S., Pr. 51: 853-857, (1916).

26--1917 (and Allan, J. A.) A titaniferous augite from Ice River, British Columbia, with a chemical analysis by M. F. Conner. Am. J. Sci. (4)43: 75-78, (1917).

27--1918 On the microstructure of certain titanic iron ores. Econ. Geol. 13: 419-446, (1918).

28--1921 The crystalline characters of calcium carbide. Am. J. Sci. (5)2: 120-128, (1921).

29--1921a
1922 A Manual of Determinative Mineralogy. New York, McGraw-Hill Book Co., Inc., 1st ed. (1921), 2nd ed. (1922), ix + 163 p. (See item 15-1910.)

30--1924 (and McKinstry, H. E.) The granites and pegmatites of Cape Ann, Massachusetts. A.A.A.S., Pr. 59: 315-357, (1924).

31--1932 (with Morse, H. W. and Donnay, J. H. D.) Artificial spherulites and related aggregates. Am. J. Sci. (5)23: 420-439, (1932).

32--1933 Ernest Howe (1875-1932). Am. J. Sci. (5)25: 97-100, (1933).

33--1933a Louis Valentine Pirsson (1860-1919). A.A.A.S., Pr. 68: 659-662, (1933).

34--1940 William Ebenezer Ford (1878-1939). Am. J. Sci. 238: 63-66, (1940).

35--1940a William Ebenezer Ford (1878-1939). A.A.A.S., Pr. 74: 121-123, (1940).

36--1950 The Sheffield Scientific School from 1847-1947. The Centennial of the Sheffield Scientific School, Yale University Press, p. 156-157, (1950).

SOURCES OF BIOGRAPHICAL INFORMATION ABOUT CHARLES HYDE WARREN

The reader interested in C. H. Warren will find much more information about him in the following references, on which I have drawn freely, than I could include in my brief biography here:

1) Technique, p. 31, April 1905.

2) [M.I.T.] President's Report for SY 1921-22, p. 10.
    Includes comments on Warren's career at M.I.T., 1900-1922, before being appointed Dean of Yale's Sheffield Scientific School.

3) Chittenden, R. H.: History of the Sheffield Scientific School of Yale University, 1846-1922. New Haven: Yale Univ. Press, 2 vols.: x + 298 p. and x + 302 p., (1928).
    Information on Warren appears in volume 2 on page 400.

4) Longwell, C. R.: Charles Hyde Warren (1876-1950). Am. Philos. Soc. Yearb. 1950: 328-333, (1951).

5) Knopf, A.: Memorial to Charles Hyde Warren (1876-1950). Geol. Soc. Am., Pr. 1951: 159-164, port., (1952).

6) Records at Yale University include the following items, copies of which are in Warren's file in the M.I.T. Institute Archives:
    a) 1896S. Vicennial Record, Class of 1896 [Sheffield Scientific School, Yale.]
        Warren briefly reports on his activities at M.I.T.
    b) 1896S. Thirty-Year Record [Class of 1896, Sheffield Scientific School, Yale.], Hartford, 1928.
        Pictures and a brief statement by Warren appear on p. 151 and 152.
    c) 1896S. 50 Year Class Book, Nov. 1947.
        A three-page autobiographical sketch, with pictures of his children, in which Warren describes his trials and tribulations as Dean. Pages are unnumbered.

7) Obituary notices appeared in:
    The New York Times, p. 27, 17 Aug. 1950.
    School and Society 72: 143, 26 Aug. 1950.
    Science 112: 348, 22 Sept. 1950.
    Yale Univ. Obituary Records, Sheffield Scientific School, (1950-1951), p. 107-108, (1952).

(9)
DOUGLAS WILSON JOHNSON
(1878-1944)

DOUGLAS WILSON JOHNSON

MIT: 1903-1907

D. W. Johnson,* one of America's foremost geomorphologists, began his university career at M.I.T. in 1903 immediately after receiving his doctorate from Columbia. During his first year, as Instructor in Geology, he took post-doctoral work at Harvard with famed physiographer William Morris Davis, whose teachings and writings profoundly influenced his later work in geomorphology. During his short period of time on the M.I.T. faculty, as Instructor (1903-1905) and Assistant Professor (1905-1907), he taught courses in Physical Geology, Physiography, and Topographic Geology, and published a dozen or more articles on a variety of physical geological subjects. He left for Harvard in 1906, but continued to lecture at the Institute during SY 1906-07, and in 1912 moved on to Columbia where he achieved an illustrious career during the next thirty years. Although he remained at M.I.T. only four years on a full-time basis, and an additional year on a part-time basis, the dozen or more publications he produced in that short period presaged the impressive productivity that later brought him high rank among the world's greatest geomorphologists.

Douglas Wilson Johnson was born in Parkersburg, West Virginia on 30 November 1878 and died of a heart attack at Sebring, Florida on 24 February 1944. He was one of a family of six children, of Isaac and Jane Amanda (Wilson) Johnson, who were pioneers in the movement for national prohibition. Isaac edited "The Freeman" of Parkersburg, West Virginia, and lectured extensively in support of the temperance movement. Jane, although frail in health, was elected State President of the Women's Christian Temperance Union at five successive conventions.

Two of the four boys in the Johnson family died in childhood, while two girls younger than Douglas and his older brother Sam lived to reach maturity. The father died when Douglas was 11, and the mother followed 10 years later, with the result that Sam had to assume financial responsi-

---

* I have drawn heavily on several of the biographies of Johnson that are listed at the end of this biography, and on conversations with the late Prof. Hervey W. Shimer, a longtime friend of the Johnsons.

bility for the family. When Douglas was quite young, Miss Cynthia Martin became an adopted member of the Johnson household, and during a critical illness nursed him back to health when his mother was unable physically to take care of him. Miss Martin became a second mother to the Johnson children and elicited from Douglas a loyal and understanding friendship. It seems probable that this friendship was partially responsible for the solicitous attitude that Douglas had toward his wife, Alice Adkins (whom he married on 11 August 1903), when she became partially and then totally blind--a relationship so beautifully related by the Wrights in their biography cited later on.

Douglas was early recognized as an unusually alert and able boy, and his family encouraged and helped him to prepare for an academic career. His forebears were represented by men and women of intellectual interests and of high public spirit, so he came by his talents quite naturally. His distinguished career as a scholar started with his winning all the prizes for which he was eligible in the Parkersburg Public Schools and ending with an impressive list of honorary degrees, medals and prizes, and honorary memberships in learned societies. How often it has happened that an illustrious career is presaged by the winning of prizes and awards in the public schools and in college; so it was with Douglas Johnson!

From the Public Schools of Parkersburg, Douglas entered Denison University as a freshman and took his first course in geology with Prof. W. G. Tight. The next year he transferred to the University of New Mexico, for the sake of his health, and there came under the influence of C. L. Herrick, a former geology professor at Denison, who would have a profound effect on his future academic career.

His first college degree came in 1901, a Bachelor of Science in Geology from New Mexico. Then followed two years of graduate work at Columbia and a Ph.D. degree in 1903. Now he was ready to enter on the academic career that would lead to greatness, and it would start at M.I.T.

Johnson started teaching rather early in life. In the late 1890s, before entering the University of New Mexico, he taught in the country school at Long Reach, West Virginia, seat of the Johnson family. Later on, in New Mexico, he taught in the Albuquerque High School while attending the University and completing requirements for his bachelor's degree. Then two years later with a doctorate in hand from Columbia, he was appointed to his first college position.

Like a number of distinguished geologists, both before and after him, Johnson began his university career at M.I.T. He came to the Institute in the fall of 1903 as an Instructor in Geology and was accompanied by a fellow doctorate from Columbia, Hervey W. Shimer, who had come to the

Institute twice a week during the preceding term (spring 1903) to deliver lectures in Paleontology and Stratigraphy. Thus commenced the academic careers of two young and ambitious geologists, both of whom would achieve international distinction in their special fields--Johnson in Geomorphology, Shimer in Paleontology.

Johnson was given responsibility for much of General Geology and for what was then called Physiography. Shimer records in his unpublished "History of the Department of Geology, M.I.T." [1944?], preserved as a typescript in the M.I.T. Institute Archives, that

> "During this year [i.e. SY 1904-05] Johnson with the aid of Shimer and their wives, mounted on linen the hundreds of topographic maps which have been so useful ever since." (p. 16 of mss.)

During his first year at M.I.T., Johnson found time to take post-doctoral work at Harvard with the great physiographer, William Morris Davis, and was profoundly influenced by his teachings and ideas. Mutual respect and admiration quickly developed between the two, and lasted throughout their lives. It was not surprising, therefore, that a few years later (1907) Johnson was invited to join the Harvard Geology Department as Assistant Professor of Physiography. There he remained until 1912 when he returned to Columbia as Associate Professor of Physiography and entered on what became an illustrious career that ended only with his death in 1944. Let us now leave his later career and return to his short period at M.I.T.

Johnson served first as Instructor in Geology, from 1903 to 1905, then as Assistant Professor of Geology from 1905 to 1907. Although he joined the Harvard faculty in 1906, he continued to give Topographic Geology at M.I.T. during SY 1906-07. Even this early in his academic career he is reputed to have been an excellent teacher, and his dozen or more publications during his four years at M.I.T. presage his future productivity.

Inasmuch as Johnson achieved his distinguished reputation at Columbia, the reader interested in learning about that career is referred to the biographies cited later on following the partial Bibliography.

PARTIAL BIBLIOGRAPHY OF DOUGLAS WILSON JOHNSON

Symbols and abbreviations used in the following references are explained on pages 91 - 98 ; in general, abstracts are listed separately as well as with the references to the complete article. This bibliography, which includes only a very few of Johnson's 160 publications, is limited to the articles he wrote before and during the time he taught at M.I.T. His complete bibliography is included in several of the biographies listed on the following page.

1--1900   The geology of the Albuquerque sheet, New Mexico. Denison Univ. Sci. Lab., B. 11: 175-239, (1900).

2--1902   Notes of a geological reconnaissance in eastern Valencia County, New Mexico. Am. Geol. 29: 80-87, (1902).

3--1903   Block mountains in New Mexico. Am. Geol. 31: 135-139, (1903).

4--1903a
   1904   Geology of the Cerrillos Hills, New Mexico, Part I. General Geology, Columbia Univ. School of Mines Quart. 24: 303-350, 456-500; Part II. Paleontology, p. 173-246. Part III. Petrography, v. 25: 69-98, (1903-1904).

5--1905   The distribution of fresh water faunas as an evidence of drainage modifications. Science n.s. 21: 588-592, (1905).

6--1905a   The Tertiary history of the Tennessee River. J. Geol. 13: 194-231, (1905).

7--1905b   The scope of Applied Geology and its place in the Technical School. Econ. Geol. 1: 243-256, (1905).

8--1905c   The biological evidence of river capture. Am. Geogr. Soc., B. 37: 154-156, (1905).

9--1905d   Youth, maturity and old age of topographic forms. Am. Geogr. Soc., B. 37: 648-653, (1905).

10--1906   Technology-Harvard geological expedition. Science n.s. 24: 471-472, (1906).

11--1906a Relation of the law to underground waters. U.S. Geol. Surv., W-S.I&P. 122, 55 p., (1906).

12--1906b Report on the geological excursion through New Mexico, Arizona and Utah, summer of 1906. Tech. Quart. 19: 408-415, (1906).

13--1906c Geology of the Nantasket area. Science n.s. 23: 155-156, (1906).

14--1907   A recent volcano in the San Francisco Mt. region, Arizona. Geogr. Soc. Philadelphia, B., 7 p., (1907).

15--1907a Drainage modifications in the Tallulah district. Boston Soc. Nat. Hist., Pr. 33: 211-248, (1907).

16--1907b Volcanic necks of the Mount Taylor region, New Mexico. Geol. Soc. Am., B. 18: 303-324, (1907).

17--1907c River capture in the Tallulah district, Georgia. Science n.s. 25: 428-432, (1907).

NOTE:

The remaining 140 some publications are listed in the Bibliographies of North American Geology and in several of the biographies listed below.

BIOGRAPHIES OF DOUGLAS WILSON JOHNSON

Lobeck, A. K.: Douglas (Wilson) Johnson (1878-1944). Assoc. Am. Geogr. An. 34/4: 216-222, port., (1944).

Wright, J. K.: Douglas [Wilson] Johnson, 1878-1944. Geogr. Rev. 34/2: 317-318, (1944).

Sandford, K. S.: Douglas Wilson Johnson (1878-1944), obituary notice. Geol. Soc. London, Quart. Jour. 100/3-4: lx-lxi, (1945).

Berkey, C. P.: Douglas Wilson Johnson (1878-1944). Am. Philos. Soc. Yearbook 1944: 374-379, (1945).

Wright, F. J. and Anna Z.: Memorial to Douglas [Wilson] Johnson. Geol. Soc. Am., Pr. 1944: 223-239, port., (1945).

Martonne, E. de: Douglas (Wilson) Johnson. An. de Géographie, no. 297, 55 Année: 49-62, (1946).

Hsu, Giun-Tze: A biographical sketch of Douglas Wilson Johnson. Geogr. Soc. China Jour. 14/1: 23-41, port., (1947).

Bucher, W. H.: Biographical Memoir of Douglas Wilson Johnson, 1878-1944. Nat. Acad. Sci., Biog. Mem. 24/5: 197-230, port., (1947).

View of part of an early classroom at "Boston Tech" used for lectures, laboratory instruction, and library work in geology and mining subjects. See other views of this room on pages 269 and 356.

(M.I.T. Historical Collections)

Another view of the early classroom in "Boston Tech" in which Professors Crosby and Shimer gave lectures and conducted laboratory instruction in structural geology, stratigraphy, and paleontology. Compare with views on pages 269 and 355.

(M.I.T. Historical Collections)

In the earliest years of the Institute, before collections and other teaching aids had been obtained, Prof. Hyatt arranged with the Boston Society of Natural History next door to have his students in paleontology use the extensive collections in the Society's Museum. This convenient arrangement was possible because Hyatt was on the Society's staff as well as being Professor of Palaeontology in the young Institute.

(10)
HERVEY WOODBURN SHIMER
(1872-1965)

MIT: 1903-1942-1965

HERVEY WOODBURN SHIMER

When death came to Hervey Woodburn Shimer in his 94th year, on 13 December 1965, it brought to the close a long and productive life of teaching and writing in paleontology, historical geology and natural history, and ended a 62-year association with M.I.T.'s Geology Department. After four decades of active teaching (1903-1942), he became Professor Emeritus and held this title for the next 23 years until death. He was known around the world as the author of books on fossils and historical geology that played an important role for half a century and more in the academic training of students studying paleontology and historical geology.

Shimer first came to M.I.T. in the spring of 1903 as a non-resident lecturer to give courses in paleontology and stratigraphy. Later in the same year he was appointed an instructor in geology and remained in that rank for five years (1903-1908); thereafter he advanced in rank to assistant professor (1908-1912), associate professor (1912-1922), and professor (1922-1942), the rank held until retirement on 30 June 1942. He then became Professor Emeritus of Paleontology and held this title for the remainder of his life as stated above.

Born of Pennsylvania-German parents, brought up under the strict Lutheran creed, and educated broadly through the masterate at first Gettysburg College and then at Lafayette College, he entered Columbia University in 1901 with a deep interest in paleontology. He gained much from his association with Amadeus W. Grabau (XII S.B. 1896), and while completing his doctorate, which he received in 1904, he was offered and accepted the opportunity to join the Geology Department at M.I.T. as mentioned in the preceding paragraph.

During his four decades of service at M.I.T., from 1903 through 1944, the last eighteen months (after retirement) of which he taught geography to U.S. Army recruits, Shimer lectured, supervised and conducted research and wrote mainly in the fields of paleontology, historical geology, and stratigraphy. However, he initiated and taught at one time or another courses in evolution, paleoclimatology, sedimentology, and sedimentary petrography. In student research he supervised 16 theses--10 bachelor's, 2 master's, and 4 doctor's. From his own research he wrote more than 50 important articles on geological subjects and five major books which played an important role for more than half a century in the academic training of students studying paleontology, earth history, and evolution-- North American Index Fossils, with A. W. Grabau, (1909, 1910); An Introduction to the Study of Fossils (1914; 1933); An Introduction to Earth History (1925; 1936); Evolution and Man (1929); and Index Fossils of North America, with R. R. Shrock, (1944). He served for many years as Registration Officer for Course XII and as a member of the important Graduate Committee that recommended degrees and awarded scholarships and fellowships.

He was active in civic and intellectual affairs in his home town of Hingham, Massachusetts, and both before and after retirement was the mainstay of the South Shore Nature Club. He and his wife, Florence (Henry) Shimer (B.A. Cornell; M.A. Columbia), participated actively in the affairs of their home community, entertained faculty and students of the Geology Department at a fall picnic for many years, and reared two children who have followed them in the academic tradition--John Asa Shimer (S.B. Harvard 1955; S.M. and Ph.D. M.I.T. 1939 and 1942), until 1974 Professor of Geology at Brooklyn College of the City University of New York; and Mary Henry (Shimer) Mangat-Rai (B.A. Radcliffe 1936; M.A. and Ph.D. Bryn Mawr 1938 and 1944), teacher in the Art Department of Wheaton College (Norton, Mass.). Gettysburg College awarded Hervey an honorary Doctor of Science degree in 1916, in recognition of his distinguished work as a paleontologist, and in 1965, the Hervey W. and Florence H. Shimer Fund was established at M.I.T. by Course XII alumni as a perpetual memorial to the Shimers.

BIRTH, ANCESTRY, AND EDUCATION

Hervey Woodburn Shimer* was born on a farm near the village of Martin's Creek, Pennsylvania, on 17 April 1872, the second son of John Calvin and Maria Rebecca (Engler) Shimer. With his two brothers and two sisters he spent his early years on the farm under the strict Lutheran training of his Pennsylvania—German parents. Throughout his long life he followed much of the teaching of his parents--hard work, honesty, high moral and ethical standards--but he soon turned from their stricter creed to a more liberal attitude toward both man and beast. He was proud of his German ancestry and included genealogy among his serious hobbies. This interest led him in 1908 to publish a paper on "The Pennsylvania-German as Geologist and Paleontologist," in which he cited a number of Pennsylvania Germans who had attained distinction in geology. Today his own name would stand high on that list. He also wrote two articles on Shimer genealogy--

---

* Much of the biographical part of this sketch has been taken either verbatim, or with but slight alteration, from my "Memorial to Hervey Woodburn Shimer (1872-1965)" published in Geol. Soc. Am., B. 77: P73-P80, port., (1966). I have not, therefore, enclosed in quotation marks most of the material excerpted from this earlier biography because I myself wrote it in the first place, and to set aside such excerpts in quotation marks would, it seemed to me, disrupt unduly the appearance of the pages.

On the other hand, I have doubled the number of references in the Bibliography, thus including the complete list of publications promised in the statement at the end of the "Memorial" referred to at the beginning of this note.

Additional biographical details are to be found in Tech. Rev. 35: 300, (1933); American Men of Science (9th ed., 1955, etc.); Who's Who in America (1950-51); and in an obituarial notice, "Dr. Hervey W. Shimer, 93, of Hingham, Paleontologist," in The Boston Herald, Tuesday, Dec. 14, 1965.

"Joseph Shimer, a biographical sketch," in History and Genealogy of the Shimer family in America 4: 408-410, (1927); and "John Calvin Shimer and his family," in Ibid. 6: 564-568, (1946). [References 60--1927a and 90--1946b in the Bibliography farther on.]

He received his early education at Lorch's Preparatory School in nearby Easton, Pennsylvania, and then completed two years of study at Gettysburg College (1891-1893) before accepting a teaching position in the public schools of his home town. After four years of what would become a teaching career stretching over four decades, he resumed his own education, entering Lafayette College in 1897. He earned an A.B. degree in 1899 and an A.M. degree in 1901.

While pursuing his graduate studies at Lafayette (1899-1901) he gained further teaching experience as a tutor of modern languages, and this particular training no doubt aroused the interest that he developed and maintained throughout life in words and their meaning. This interest was of great use in his studies of plants and animals, both living and fossil, and led late in life to the preparation of two handy booklets on the subject--Origin and Significance of Plant Names (1943) and Plant Names, their origin and meaning (1950).

Having developed a liking for plants and animals as a boy on the farm, he entered Columbia University in 1901 with a first interest in paleontology and received a Ph.D. degree in that field in 1904. While working toward his degree, he once again did teaching, as an assistant in paleontology for two years (1901-1903), and during this time came under the influence of Amadeus W. Grabau (XII S.B. 1896), with whom he would later do some of his most important paleontological work. After receiving his doctor's degree at Columbia, he did post-doctoral work in paleontology at Harvard. In 1916, his first alma mater, Gettysburg College, awarded him an honorary Doctor of Science in recognition of his achievements in paleontology.

M.I.T. YEARS (1903-1942-1965)

Shimer's association with M.I.T. began in the spring of 1903 when he came from Columbia University two days each week to give Course XII students instruction in paleontology and stratigraphy. Later on in 1903 he joined the M.I.T. faculty as instructor, and immediately took charge of the instruction in historical geology, paleontology, and physiography (including climatology), at the same time assisting in the laboratory and with field work in lithology. He was also assigned responsibility for the Department's paleontological collections and its library. Quite a load for a young graduate student, not yet with the Columbia doctorate that would come in another year, as he started his career as a college teacher!

After serving as Instructor in Geology for five years, 1903-1908, Shimer advanced through the several professorial ranks in paleontology--assistant professor (1908-1912), associate professor (1912-1922), and professor (1922-1942)--to retirement in July 1942 as Professor Emeritus of Paleontology, a title he held for the next 23 years to the time of his death in his 94th year on 13 December 1965.

During his active years as a faculty member he twice served one-year appointments as Acting Chairman of the Geology Department: the first during SY 1927-28, while Lindgren was on leave in Washington (as Chairman of the Division of Geology of the National Research Council); the second during SY 1933-34, following Lindgren's retirement and before W. J. Mead came as department head in the summer of 1934. On several other occasions, for brief periods of a few months, he had responsibility for the Department while the chairman was on leave (e.g. from March 1st to end of spring term in 1931 while Lindgren was on leave).

Twice also he stepped out of his role as geology teacher to instruct recruits in training for military service. In 1918 he conducted the class work of a section in the Department of English and History as a part of the Student Army Training Corps (SATC), and during the early years of World War II (1942-1944), after retirement from active teaching, he returned as Lecturer to assist in several geography subjects given for the Meteorology B group of the U. S. Army Air Corps (M.I.T. Pres. Rept. 1943: 119).

He served for many years as Course XII's Registration Officer and a member of the Institute's Undergraduate and Graduate Committees. As a member of the latter's Subcommittee on Financial Aid he was especially effective in obtaining tuition scholarships (particularly for Canadian students) and other kinds of financial assistance for students who were short on funds.

As Teacher

Shimer was pre-eminently a teacher. As mentioned earlier he gained his first teaching experience in the public schools of his home town, Martin's Creek, Pennsylvania (1893-1897), with only two years of college study to his credit. As a graduate student at Lafayette College he gained another two years (1899-1901) of experience as a tutor of modern languages Still later, as a doctoral candidate at Columbia, he served as an assistant in paleontology from 1901 to 1903, and in the spring term of SY 1902-03 he also travelled to M.I.T. twice a week as a non-resident lecturer in paleontology and stratigraphy.

During the four decades (1903-1944) that he taught paleontology and historical geology at M.I.T., he also introduced new courses in sedimentology, paleoclimatology, and organic evolution, and from time to time offered special short courses in the geology curriculum. Inasmuch as classes were never large, seldom more than ten students at the most, he conducted most of his discussions in an informal manner, with his favorite pipe always close at hand. Thus he quickly established rapport with his students and created an atmosphere that encouraged free discussion. He also liked to give frequent one- or two-question quizzes to test the students and to keep them up to date in their reading assignments. He had almost infinite patience and always a sympathetic word for a student who needed encouragement.

He was one of the favorites among the students as a supervisor of thesis research, and during his 39 years of active teaching he supervised 16 theses--10 bachelor's, 2 master's, and 4 doctor's, and in addition aided many other thesis students who had fossils to identify, or sedimentary rocks to interpret.

As stated earlier, Shimer was at his best when teaching, whether he was giving a formal lecture, leading a scheduled field trip, or participating with friends in an informal walk or an impromptu picnic excursion. His former M.I.T. students, now scattered widely around the world, will remember him as a gentle, quietly spoken, pipe-smoking professor who gave carefully organized, up-to-date lectures based on extensive reading and wide-ranging field observations and travel. He taught during a time of descriptive generalism rather than quantitative specialism, so his lectures covered a broad range of subject matter in biology, paleontology, and earth history, disciplines now subdivided into many separate specialisms.

Like many of the naturalists of his time, he knew well the plants of the marshes, bogs and swamps, the trees of the forested plains and mountains, and the flowers of the meadows and gardens. Equally well he knew the animal life of the soil and the rock crevices, of the lakes, streams, and oceans, and of the air. He would happily identify trees, shrubs, and more lowly plants while conducting friends on "nature walks," and he taught his children to love nature by having them list all the plants they could identify on a hike. Being mindful of the world of plants on every hand, as well as of the animals hidden in their midst, he would point out to his hiking companions how organisms adapted to environments, how they brought about decomposition and disintegration of rocks, how they fouled the swamps and pond waters at certain times of the year, and how they could be eaten if one knew the difference between the edible and poisonous ones. One quickly sensed that he saw the world around him as a unity of living things, all trying to get along with each other as best they could under the existing circumstances, and as a constant struggle for survival.

Imbued with a strong work ethic that went back to his youth, possessed of a disciplined and orderly mind, and skilled as an observer of the natural environment, Shimer could hardly help being an excellent teacher and writer. Yet he was always concerned about his teaching in particular. In 1906, as a young and relatively inexperienced but conscientious assistant professor, he sent a questionnaire to several dozens of his colleagues in colleges and on the U.S. Geological Survey. He wanted to know what importance they attached to knowledge of stratigraphy and paleontology as part of the basic training for academic work and for work on the Survey. His replies included thoughtful and helpful comments and suggestions from some of the most distinguished geologists of the day--M. R. Campbell, N. H. Darton, G. H. Girty, C. W. Hayes, Andrew Lawson, Charles Schuchert, F. W. Stanton, T. Wayland Vaughan, A. C. Veatch, C. D. Walcott, and Bailey Willis. As might be expected, the suggestions showed the same range that one would probably get today--teach more (or less) of the subject matter; use more (or fewer) examples and specimens; devote more (or less) time to the particular subject; etc. Little seems to have come from his effort, interesting and instructive though it was, but it does show Shimer's genuine desire to seek the advice of his experienced colleagues in the hope of gaining suggestions as to how he might improve his teaching.

As Author

As a result of having the personal characteristics mentioned in the preceding paragraph, of gaining experience in teaching and field work through the years, and of having at hand the advice and comments of some of the most distinguished paleontologists and stratigraphers in the United States, Shimer was able to bring much knowledge and experience to his teaching and writing. He organized this knowledge primarily for teaching purposes, and from his class notes ultimately came a series of textbooks on paleontology and historical geology that were important and widely used in the training of geology students during the first half of this century.

The first of these was the two-volume North American Index Fossils, co-authored with A. W. Grabau, his former teacher at Columbia, and first published in 1909 and 1910 (See Bibliography farther on). This work quickly became an indispensable reference for both field geologists and students studying invertebrate and stratigraphic paleontology, and was in constant demand to the very day in 1944 when its successor, Index Fossils of North America (co-authored with R. R. Shrock assisted by a dozen specialists), made its appearance. Today, 1974, thirty years later, this latter book remains an important reference work, and the royalty from its surprisingly steady sale goes to support the publication costs of the Journal of Paleontology.*

An Introduction to the Study of Fossils was first published in 1914, then revised in 1933, and only recently (1965) declared out of print and discontinued after more than fifty years! It was one of the earliest American textbooks to give a reasonably balanced treatment of fossil plants, invertebrates, and vertebrates. Most geology students during the thirty years between World Wars I and II (1914-1944) learned about fossils from this widely used textbook.

The two books on Index Fossils and his "Introduction" have made Shimer's name familiar to several generations of geologists.

In An Introduction to Earth History (1925; rev. ed. 1936) Shimer brought together scattered information and focused it on the broad principles of evolution, cause and effect, and uniformitarianism. In Evolution and Man (1929) he discussed further man's origin, evolution, and place in nature. And as retiring President of the Boston Geological Society in 1939, he addressed the members on one of his favorite philosophical subjects, "Expanding consciousness and democracy" (Science 89: 325-329, 1939).

Although his major works, the five books mentioned in the preceding paragraphs, as well as numerous of his 100 articles, deal largely with fossils, earth history, and biological evolution, his many shorter articles and pamphlets show that his intellectual interests were much broader and more varied. These range in subject from igneous dikes of Vermont to Triassic coral reefs of British Columbia; from the geology of the Rhine Valley to that of his backyard South Shore near Hingham (Mass.); from kitchen middens, to cave houses in Arizona and man's ancestral home; and from letters to the editors of local newspapers to pamphlets for the South Shore Nature Club. And as late as his 87th year he was writing and lecturing with enthusiasm!

He also acted as scientific aid and critic for Vols. 1 and 11 (1919 and 1921) of The Wonder World (George L. Shuman & Co.), and for Vol. 1 of The New Wonder World (1932). He also contributed to A Short History of Science (rev. ed.) by W. T. Sedgwick, H. W. Tyler and R. P. Bigelow (1939).

---

* See "A 27-year report on sales of Index Fossils of North America," by R. R. Shrock, J. Pal. 46/3: 453-455, il., (1972).

PROFESSIONAL AND CIVIC ACTIVITIES

Shimer was active in many scientific, cultural, and civil organizations. He was elected a fellow of the American Academy of Arts and Sciences and served as Librarian for some years. In 1922 he was elected Vice-President of Section E, Geology and Geography, of the American Association for the Advancement of Science, and was a fellow of the Geological Society of America. He was a member of The Paleontological Society, Boston Society of Natural History, Boston Geological Society (President, 1939), Malacological Society of Boston, Washington Academy of Sciences, American Anthropological Association, and the American Forestry Association. He held memberships in The Association of Ph.D.'s of Columbia University, the American Association of University Professors, The Society of Sigma Xi, and the Twentieth Century Club.

In historic Hingham, south of Boston, where Shimer lived for nearly half a century, after moving his family there from Cambridge in 1920, he took an active part in the intellectual and civic activities of the town. He served Derby Academy and the Hingham Public Library as a trustee; was a leading member of The Manuscript Club; and one of the most active members of the Massachusetts Civic League, for whose annual reports he wrote numerous introductory or supplementary statements.

PERSONAL CHARACTERISTICS AND ACTIVITIES

In his own personal life he aspired always to be humane, gentle, and scholarly, and to seek the truth, wherever the search led and whatever the results. He tried constantly to improve himself as an individual and as a teacher. Tolerance, courtesy, kindness, humility, understanding--these were all qualities that Shimer had in full measure. It was his habit to keep in view on his desk printed cards and scribbled suggestions to remind him to keep personal improvement always in mind: e.g. "Maintain an attitude of reverence towards everything;" "Be truthful, kind and helpful;" "Correct your own weaknesses, mental and moral;" "Never speak ill of your rivals." It was because of his inherent kindness that Shimer was not a particularly good critic; he just found it too distasteful to find fault with the work of a colleague. He much preferred to say something complimentary about an individual. In my twenty-eight years of association with Hervey Shimer, I do not recall ever having heard him speak ill of any one.

When a youth he played the piano, organ, flute, guitar and clarinet, and liked to join his companions in singing. And he enjoyed listening to military bands. He liked nothing better than to fix things around the house and yard, "tinkering" he called it, and his specialty was constructing shelves to hold the rock specimens and books that were to be seen on every floor in the Shimer home. He was an insatiable reader of Western novels (Zane Grey and other authors), historical romances, nature books (Gene Stratton Porter was a favorite author), and detective stories, and would puff away contentedly on his pipe while devouring the pages at a rapid clip. And he was especially fond of poetry and classical literature. Characteristically he kept paper and pencil ready at hand to jot down some idea that came to mind--a thought to be used in a lecture, mentioned in a letter, or developed into a pamphlet or a scientific article. His last letters to me, only a year or so before his death at 93, showed no diminution in his mental alertness and little change in his quite legible handwriting.

RETIREMENT YEARS (1942-1965)

After retirement on 30 June 1942, and the additional 18-month stint of special teaching during the war years, Shimer turned to a long-standing interest in natural history and conservation, and literally ran the South Shore Nature Club in Hingham for many years. He participated actively in scheduled meetings, gave lectures, led nature-study hikes, and produced a series of informative pamphlets on natural history subjects for the membership. Included in some twenty such pamphlets that he was instrumental in getting written, and which were printed under the aegis of the Club, were the following that he himself wrote:

- Trees prepare for winter, No. 1, (1944)
- Reflections on the size of living things, No. 6, (1946)
- Coal, No. 7, (1946)
- The borrowed-time club, No. 10, (1948)
- Carnivorous plants, No. 12, (1950)
- Plants and crystals, living and non-living forms, No. 13, (1950)
- Turtles, No. 16, (1953)
- Hingham's River, No. 19, (1954)
- Spring miracle, No. 20, (1955)

His active mind and facile pen continued to produce articles on natural history subjects until he had passed his 87th year, and in 1960 he sent me a typescript of a lecture he gave to the Hingham Historical Society on the "Geology and Colonial History of Hingham." Thus he shared his knowledge, experience, and thoughts as readily and enthusiastically with his neighbors and townfolk as with his academic and professional colleagues.

MARRIAGE, FAMILY, AND FAMILY LIFE

No biographical sketch of Hervey W. Shimer would be complete without writing a little about his wife, Florence French (Henry) Shimer, to whom he was married on 1 June 1904. She was born in Sacramento, California, but her parents moved to Cortland, New York, when she was quite small, and she grew up in that city. Her only sibling, a sister, died at an early age. Florence graduated from the Cortland Normal School in 1897, and entered Cornell University that same autumn. She earned a bachelor's degree in the Classics in 1901, and the Cornell Class Book notes that "she received many honors because of her love for the Muses." Even this early she was writing poetry, an interest that culminated half a century later in a little book of verses, Twelve Poems (by Florence Shimer, Hingham, Mass.: Püterschein, 12 p., 1950). While Hervey worked on his doctorate at Columbia, Florence pursued her own intellectual interests and earned a master's degree from the same institution.

Florence was Hervey's intellectual companion and constant helpmate at home, enthusiastic companion on a picnic or hike, and skilled editor until her death at 83 on 16 January 1962. Their home was a tranquil and intellectual retreat from the noisy and mundane world, yet the Shimers opened it generously to a host of friends. Every autumn for many years they held a picnic that became a tradition in the M.I.T. Department of Geology. Many students of those days will remember the valuable role those picnics, and the short geological hikes commonly following the lunch, played in providing the opportunity for them to meet one another and to come to know faculty members and their wives under informal conditions.

It was in such a quiet and scholarly home environment that Hervey wrote his paleontological textbooks, scientific papers, and natural history pamphlets, and Florence her thoughtful poetry. Here, also, their two children learned to love science, art, literature and the world of nature and to appreciate the finer things in life.

Their elder child, John Asa Shimer, with a B.S. from Harvard (1935) and a masterate and doctorate from M.I.T. (XII S.M. 1939; Ph.D. 1942), has been following his father's footsteps as a geological teacher and author. He recently retired (June 1974) as Professor of Geology at Brooklyn College of the City University of New York and has already published four geological textbooks: Graphic Methods in Structural Geology (with W. L. Donn, 1943), This Sculptured Earth: The Landscape of America (1959), This Changing Earth; an Introduction to Geology (1968), and Field Guide to Landforms in the United States (1972).

John's wife, Genevieve, formerly a commercial illustrator, is now fully occupied as National Director of the Country Dance and Song Society of America, founded in 1915, with headquarters in New York.

Their younger child, Mary Henry Shimer, first earned a B.A. degree in fine arts at Radcliffe in 1936, and then an M.A. (1938) and a Ph.D. (1944) in art from Bryn Mawr. While doing research in India, she met and married Brigadier Charles R. Mangat-Rai, of the Indian Army, now retired.

Mary, Polly as she is known to her closer friends, has long been a teacher in the Art Department of Wheaton College (Norton, Mass.), and her husband has until recently been a teacher of mathematics at Thayer Academy (Braintree, Mass.). As a retired Brigadier of the Corps of Engineers, Indian Army, he has recently completed a section of the Regimental history of the Corps and is now teaching yoga at their home in Hingham. The Mangat-Rais' only child, John Nirmal, is now Assistant Pastor at the Bethel Revival Center (Pentecostal Church and Bible School) in Boston.

THE SHIMER MEMORIAL FUND

All whose lives were touched by the acts and thoughts of Hervey and Florence Shimer were the better for them. In 1965, as discussed in the chapter on "The Financing of the Geological Sciences at M.I.T." a fund was established to honor the memory of the Shimers. The deep respect and love held for them by former students, neighbors, fellow townsmen and other friends has become manifest from the hundred contributions and accompanying letters that have been received. Today the income from the Hervey W. and Florence H. Shimer Fund is being used to help students with field expenses. This Fund is discussed further in my planned Volume 2.

SUMMATION

The long and productive life of Dr. Shimer was aptly summed up at his memorial service by Rev. Donald F. Robinson, Second Parish Church, Hingham, as follows:

> "Hervey Shimer's view of the world was on a vast scale, a scale so vast that even the rocks, which to most of us are symbols of permanence, were manifestations of an eternal flux and change. But it was change with a purpose, a demonstrable goal. It was his lifework to read in the rocks the record of developing life, moving toward greater complexity of body and fuller awareness of mind, a development that progressed steadily through all obstacles and set-backs, indeflectable. He saw human life and present human circumstances in that context. ... Widely known and loved by all who knew him, he will be missed; for he was one of those gently independent individuals who can never really be replaced."

BIBLIOGRAPHY OF HERVEY WOODBURN SHIMER

Symbols and abbreviations used in the following references are explained on pages 91 - 98. If only an abstract was published, it is cited as a separate item; if a complete article with essentially the same title was also published, the abstract is generally cited along with the complete article. This bibliography is the most complete list of Shimer's scientific publications yet compiled.

1--1902 (and Grabau, A. W.) Stratigraphic and faunal succession in the Hamilton Group of Thedford, Ontario [abst.]. Science n.s. 15: 82-83, (1902); Geol. Soc. Am., B. 13: 149-186, il., (1902).

2--1902a Petrographic description of the dikes of Grand Isle, Vermont. Vt. St. Geol. Rept. 3: 174-183, (1902).

3--1902b Columbia University summer school. Am. Geol. 30: 69-71, (1902).

4--1903 Fall excursions of the geological department, Columbia University. Am. Geol. 31: 62-64, (1903).

5--1903a Columbia University geological department [excursion]. Am. Geol. 32: 130-131, (1903).

6--1903b Columbia University geological department excursion. Am. Geol. 32: 259-260, (1903).

7--1903c The survival of the strongest. The Tech 23: 98-99, (1903).

8--1903d Adaptations to aquatic, arboreal, fossorial and cursorial habits in mammals: III. Fossorial adaptations. Am. Nat. 37: 819-825, (1903).

9--1905 A peculiar variation of Terebratalia transversa Sowerby. Am. Nat. 39: 691-693, (1905).

10--1905a Upper Siluric and Lower Devonic faunas of Trilobite Mountain, Orange County, New York. N.Y. St. Mus., B. 80: 173-269, il., (1905).

11--1906 Old age in Brachiopoda, a preliminary study. Am. Nat. 40: 95-121, (1906). Abst., Science n.s. 23: 290, (1906); A.A.A.S., Pr. 55: 379, (1906).

12--1906a (with Grabau, A. W.) North American index fossils. Sch. Min. Quart. [Columbia] 27: 138-243; 38: 20-100, (1906).

13--1907 An almost complete specimen of Strenuella strenua (Billings). Am. J. Sci. (4) 23: 199-201, 319, il., (1907).

14--1907a A Lower-Middle Cambrian transition fauna from Braintree, Massachusetts. Am. J. Sci. (4) 24: 176-178, il., (1907).

15--1907b The broader features of the geologic history of North America in diagram. Tech. Quart. 20: 287-291, (1907).

16--1907c [Review of] Geology of Connecticut [Conn. St. Geol. Nat. Hist., B. 6 and B. 7, (1907)]. Tech. Quart. 20: 375-378, (1907).

17--1908 (and Blodgett, Mildred E.) The stratigraphy of the Mt. Taylor region, New Mexico. Am. J. Sci. (4) 25: 53-67, il., (1908).

18--1908a [Description, on p. 107-108, of invertebrates found by C. D. Walcott in Kanab Valley, Utah, 1879] in The Triassic portion of the Shinarump Group, Powell by W. Cross. J. Geol. 16: 97-123, il., (1908).

19--1908b Dwarf faunas. Am. Nat. 42: 472-490, (1908).

20--1908c The Pennsylvania-German as geologist and paleontologist. The Pennsylvania-German Society 9: 411-415, (Sept. 1908).

21--1909 Stratigraphic Geology - Laboratory and Lecture Guide. Boston: M.I.T., 139 p., il., (1909).

22--1909a (with Grabau, A. W.) North American Index Fossils, [Invertebrates]. New York: A. G. Seiler and Co., Vol. 1, viii + 853 p., il., (1909).

23--1910 (and Shimer, Florence H.) The lithological section of Walnut Canyon, Arizona, with relation to the cliff dwellings of this and other regions of northwestern Arizona. Am. Anthropologist 12: 237-249, (1910).

24--1910a (with Grabau, A. W.) North American Index Fossils, Invertebrates. New York: A. G. Seiler and Co., Vol. 2, xv + 909 p., il., (1910).

25--1910b Paleontology. The Tech (Special Mining and Geology Issue) XXIX (29)/85: 25, (Jan. 15, 1910).

26--1911 Lake Minnewanka section [Alberta]. Can. Geol. Surv., Sum. Rep. 1910, p. 145-149, (1911).

27--1911a (with Clapp, C. H.) The Sutton Jurassic of the Vancouver Group, Vancouver Island, B.C. Boston Soc. Nat. Hist., Pr. 34: 425-438, (1911).

28--1911b The small cave houses of Arizona. Sci. Conspec. 2: 16-18, il., (1911).

29--1912 Geology of the Rhine Valley. Sci. Conspec. 2: 108-114, il., (1912).

30--1912a Kitchen middens as ethnological records. Sci. Conspec. 3: 27-28, il., (1912). Abst., Sci. Am. Supp., p. 96, (8 Feb. 1913).

31--1913 Bergson's view of organic evolution. Pop. Sci. Month., p. 163-167, (Jan. 1913). Abst., Literary Digest, p. 454-455, (1 March 1913).

32--1913a Early man. Sci. Conspec. 3: 97-113, il., (1913).

33--1913b Spiriferoids of the Lake Minnewanka Section, Alberta. Geol. Soc. Am., B. 24: 112-113 [abst.], 233-240, (1913).

34--1913c (and Powers, S.) A new sponge from the New Jersey Cretaceous. U.S. Nat. Mus., Pr. 46: 155-156, il., (1913).

35--1914 An Introduction to the Study of Fossils. New York: The Macmillan Co., 450 p., il., (1914). (See also revision, xviii + 496 p., il., (1933), 69--1933.)

36--1914 a-c [Articles in Sci. Conspec. 4, (1914)]
  a) Our European population, p. 9
  b) Restoration of extinct reptiles, p. 108-109
  c) The tripod of evolution, p. 110-111

37--1914d (with Powers, S.) Notes on the geology of the Sun River district, Montana. J. Geol. 22: 556-559, (1914).

38--1915 Postglacial history of Boston. Am. J. Sci. (4) 40: 437-442, (1915).

39--1915 a-d [Miscellaneous articles in Sci. Conspec. 5, (1915)]
  a) Bateson's theory of evolution, p. 24-25
  b) When reptiles ruled the earth, p. 44-47
  c) The unsolved mystery of why the stomach does not digest itself, p. 52
  d) Plants and animals distinguished, p. 82-84

40--1916   Fossiliferous Miocene boulders from Block Island, Rhode Island. Am. J. Sci. (4) 41: 255-256, (1916).

41--1916 a-k   [Miscellaneous articles in Sci. Conspec. 6, (1916)]
a) Twin trees, p. 52
b) Nature's unity of physiological plan, p. 53-54
c) Flight in animals, p. 66-70
d) Self-direction in evolution, p. 70
e) Bird migration, p. 100-105
f) The beginnings of flight in birds, p. 106-110
g) Animals have outstripped plants, p. 111-112
h) The iconoclasm of fact, p. 131
i) Evolution of service, p. 132-136
j) Be a sport, p. 136
k) Evolution through contrasts, p. 137-138

42--1916*l*   The rôle of service in evolution. Sci. Month. 3: 191-195, (1916).

43--1916m   (and Lahee, F. H.) [Review of] A Textbook of Geology, by L. V. Pirsson and C. Schuchert. Science n.s. 43: 497-501, (1916).

44--1918   Postglacial history of Boston. A.Ac.A.Sci., Pr. 53: 441-463, (1918).

45--1919   Samuel Wendell Williston (1852-1918). A.Ac.A.Sci., Pr. 54: 421-423, (1919).

46--1919a   Permo-Triassic of northwestern Arizona. Geol. Soc. Am., B. 30: 155 [abst.], 471-497, (1919).

47--1920   Knowledge for service. Tech. Eng. News 1: 9, (1920).

48--1921   Man's upward climb. Tech. Eng. News 2: 3, 12, (1921).

49--1922   Early earth history. Tech. Eng. News 2: 239, 243, (1922).

50--1923   Why study history? The Boston Traveler, Sat. Dec. 29, p. 6 (in editorials by the people), (1923).

51--1924   [Review of] Climatic Changes by E. Huntington and S. S. Visher (1922). J. Geol. 32: 543-544, (1924).

52--1924a   Some forces in man's social evolution. Science n.s. 59: 199-203, (1924). (Address of the Vice-President and Chairman of Section E--Geology and Geography. A.A.A.S., delivered before the joint meeting of Section E at Cincinnati, December, 1923.)

53--1924b   [Review of] "The Devonian crinoids of New York State," by Winifred Goldring (1923). Wellesley Alumnae Mag. 9: 70-71, (1924).

54--1924c   Be original and spurn the crowd. The Boston Traveler, (Dec. 27, 1924).

55--1925   An Introduction to Earth History. Boston: Ginn and Co., viii + 411 p., il., (1925); revised edition, 1936, 411 p. (See also 72--1936.)

56--1926   Memorial of Frederick Burritt Peck. Geol. Soc. Am., B. 37: 111-114, port., (1926).

57--1926a   Upper Paleozoic faunas of Lake Minnewanka section, near Banff, Alberta. Can. Geol. Surv., B. 42: 1-84, (1926).

58--1926b   A Triassic coral reef in British Columbia. Can. Geol. Surv., B. 42: 85-89, (1926).

59--1927   (and Lindgren, W.) Memorial of William Otis Crosby. Geol. Soc. Am., B. 38: 34-45, port., (1927).

60--1927a   Joseph Shimer, a biographical sketch, in History and Genealogy of the Shimer family in America 4: 408-410, (1927).

61--1929   Evolution and Man. Boston: Ginn & Co., 273 p., il., (1929).

62--1929a Evolution everywhere. Tech. Rev. 31/3: 145-146, il., (Jan. 1929).

63--1929b Nature vs. Nurture [a review of Franz Boas's] Anthropology and Modern Life (1929). Tech. Rev. 31: 436, 438, (May 1929).

64--1929c Builders of the earth, a chapter in the series, Man and His World 2: 41-67. New York: D. Van Nostrand Co., (1929).

65--1930   Fossils of eastern Massachusetts. Tech. Eng. News 11: 222-223, il., (1930).

66--1931   The Niagara cave-in. Christian Science Monitor, p. 3, (Jan. 20, 1931).

67--1931a "Introductions" to Current Social Research in Massachusetts for Massachusetts Civic League. (Published each year by Town Room Research Committee, of which Shimer was Chairman for several years, both before and after 1931.)

68--1932   "Animals before man" (p. 274-288); "Early man" (p. 289-299); and "Peoples of long ago" (p. 300-319), in Vol. 1 of George L. Shuman and Co.'s The New Wonder World, (1932).

69--1933   An Introduction to the Study of Fossils [plants and animals], rev. ed. New York: The Macmillan Co., xviii + 496, il., (1933). (See also 35--1914.)

70--1934   Correlation Chart of Geologic Formations of North America. Geol. Soc. Am., B. 45: 909-936, pl. 118-122, (1934); Abst., Ibid., Pr. 1933, p. 108, (1934).

71--1934a Sir Archibald Geikie (1835-1924). A.Ac.A.Sci., Pr. 69: 507-508, (1934).

72--1936   Introduction to Earth History. Boston: Ginn & Co., 411 p., rev. ed. without changing pagination. (See also 55--1925.)

73--1936a (with Howell, B. F. and Lord, G. S.) New Cambrian Paradoxides fauna from eastern Massachusetts [abst.]. Geol. Soc. Am., Pr. 1935: 385, (1936).

74--1936b David White (1862-1935). A.Ac.A.Sci., Pr. 70: 600-602, (1936).

75--1938   Man's ancestral home. Sci. Month. 46: 249-254, (1938).

76--1938a Henry Fairfield Osborn (1857-1935). A.Ac.A.Sci., Pr. 72: 377-379, (1938).

77--1939   Expanding consciousness and democracy. Science 89: 325-329, (1939). (Address of the retiring President of the Boston Geological Society.)

78--1939a XII, Geology [M.I.T.--For what lines are our graduates preparing?]. Tech. Eng. News 20: 98, (1939).

79--1940   Waldemar Lindgren (1860-1939). A.Ac.A.Sci., Pr. 74: 141-142, (1940).

80--1940a [Review of] Prehistoric Life, by P. E. Raymond, (1939). Science 91: 143-144, (1940).

81--1940b (and Shrock, R. R.) Proposed revision of North American Index Fossils. J. Pal. 14: 286, (1940).

82--1942   Geology of the South Shore of Massachusetts Bay (including a geological sketch map of Hingham, Mass.). Hingham, Mass.: South Shore Nature Club, 28 p., il., (1942).

83--1943   Origin and significance of plant names. Hingham, Mass.: South Shore Nature Club, 64 p., (1943). (See also 99--1950a.)

84--1943a Dr. Grabau in China. Science 97: 555-556, (1943).

85--1944   (and Shrock, R. R.) Index Fossils of North America. (A publication of The Technology Press, M.I.T.). New York: John Wiley & Sons, Inc., ix + 837 p., 303 pl., (1944).

86--1944a Trees prepare for winter. Leaflet 1 of the South Shore Nature Club of Hingham, Mass., (1944).

87--1945   Note on Grabau in China. Science 102: 298, (1945).

88--1946   Reflections on the size of living things. Leaflet 6 of the South Shore Nature Club of Hingham, Mass., (1946).

89--1946a Coal. Leaflet 7 of the South Shore Nature Club of Hingham, Mass., (1946).

90--1946b John Calvin Shimer and his family, in History and Genealogy of the Shimer family in America 6: 564-568, (1946).

91--1946c Amadeus William Grabau [1870-1946], an appreciation. Am. J. Sci. 244: 735-736, (1946).

92--1947   Paleoecology, in symposium on "What in ecology is most significant to the biology teacher?" The Am. Biol. Teachers 9: 178-180, (1947).

93--1947a Geology of Rocky Woods Reservation [in Medfield, Mass.]. Published by the Trustees of Public Reservations, Medfield, Mass., 16 p., map, (1947).

94--1947b Memorial to Amadeus William Grabau [1870-1946]. Geol. Soc. Am., Pr. 1946: 155-166, port., (1947).

95--1948   Eternal changes, eternal truths [The Introduction to Rocky Woods Geology]. The Cornell Plantations 4: 46-47, (1948).

96--1948a Glacial boulder dedicated to Dr. Oliver Howe. The Hingham Journal, p. 5, (June 24, 1948).

97--1948b The borrowed-time club (Living relics, plants and animals). Leaflet 10 of the South Shore Nature Club of Hingham, Mass., (1948).

98--1950   Carnivorous plants. Leaflet 12 of the South Shore Nature Club of Hingham, Mass., 2 p., (1950).

99--1950a Plant names, their origin and meaning (enlarged "Plant Names" of 1943). South Duxbury, Mass.: Faulkner and Field, pubs., 40 p., (1950). (See also 83--1943.)

100--1950b Plants and Crystals, living and non-living forms. Leaflet 13 of the South Shore Nature Club of Hingham, Mass., (1950).

101--1953 Turtles. Leaflet 16 of the South Shore Nature Club of Hingham, Mass., (1953).

102--1954 Hingham's River. Leaflet 19 of the South Shore Nature Club of Hingham, Mass., (1953).

103--1955 Spring miracle. Leaflet 20 of the South Shore Nature Club of Hingham, Mass., (1955).

104--1958 The Agassiz Rock Reservation [Manchester, Mass.]. Published by the Trustees of Public Reservations, Medfield, Mass., 15 p., il., (1958).

105--1959 Looking at the New England Landscape. . The Massachusetts Audubon, p. 53-58, il., (Nov.-Dec. 1959).

106--1960 Geological and Colonial history of Hingham [Mass.]. Unpublished typescript, 10 p. (‡), (1960). [Copy in the M.I.T. Historical Collections.]

(11)
REGINALD ALDWORTH DALY
(1871-1957)

MIT: 1907-1912-1915

REGINALD ALDWORTH DALY

Reginald Aldworth Daly, one of North America's most distinguished geologists, was already well known, with more than 30 articles and several original theories to his credit, when he was appointed Professor of Physical Geology at M.I.T. in 1907. Born and educated through the first four years of college in Ontario, Canada, he entered on graduate work at Harvard in 1892, and for the next 65 years was continuously associated with that institution with the exception of three brief interludes: study and travel in Europe for two years (1896-1898); geologist with the Canadian International Boundary Commission for six years (1901-1907); and professor of geology at M.I.T. for five years (1907-1912), followed by service on the staff of the Hawaiian Volcano Observatory for three more years (1912-1915).

When Daly joined the M.I.T. faculty as Professor of Physical Geology in October 1907, he had already achieved continental renown as a tireless field geologist, a synthesizer of scattered geological data, and an imaginative innovator of new hypotheses to explain what he had seen in the field. He had just completed his tremendous investigation of the geology along the 49th parallel, to which he had devoted six field seasons, and he would prepare and publish his impressive 3-volume report on that work - Geology of the North American Cordillera at the Forty-ninth Parallel - while at the Institute. Earlier on he had made a field study of Mount Ascutney in Vermont, and from this work he developed the theory of magmatic stoping. Observations during a trip to Newfoundland and Labrador in 1900 aroused his interest in past changes in sea level, an interest that continued throughout his entire life and led to later theories on post-glacial uplift, and his glacial-control theory to explain coral reefs. Hence, by the time he joined the M.I.T. faculty, he had already demonstrated his dedication to field work, his unusual ability to correlate and synthesize his and the data of other investigators, and his remarkable imagination and ingenuity in developing original theories to explain the geological relations he observed in the field. All these attributes, and many others, became even more obvious after he left M.I.T. and as he came to be recognized as one of the most distinguished and controversial geologists of the first half of the Twentieth Century. Inasmuch as it is not the purpose of this sketch to discuss the life work of Reginald A. Daly, but rather to discuss only his activities and influence as a professor of geology at M.I.T. during a rather brief period, the reader who is interested in full accounts of his remarkable life as a geologist is referred to biographies by Birch, Billings, and others.*

At M.I.T., from 1907-1912, he taught courses in physical geology, petrography, and petrogeny; joined Warren in supervising four advanced theses; and devoted much time to developing a research laboratory in

---

* See footnote on following page.

physical geology. He joined Jaggar and Warren in emphasizing the need for investigating the physical and chemical aspects of minerals and rocks and of natural processes, and continued to publish articles in support of theories he had annunciated earlier--e.g. magmatic stoping, mechanics of intrusion, differentiation of secondary magma through gravitative adjustment, origin and classification of igneous rocks, limeless oceans of Precambrian times, sea-level change associated with Pleistocene glaciation, and the glacial-control theory of coral reef development. He also became much interested in volcanic action, and after resigning his M.I.T. professorship in 1912 to accept the prestigious Sturgis Hooper Professorship at Harvard he joined the staff of the newly founded Hawaiian Volcano Observatory under the direction of his colleague, Thomas Augustus Jaggar, Jr.

Probably the two most important tasks he accomplished at M.I.T. were preparation and publication of the 3-volume report on his 6-year investigation of the geology along the international boundary between Canada and the United States - Geology of the North American Cordillera at the Forty-ninth Parallel - and preparation of a set of notes which he later expanded into the first of his half dozen books - Igneous Rocks and Their Origin (1914).

Although he was an internationally known geologist when he left M.I.T. for Harvard in 1912, Daly continued to add to his professional stature at a steady pace for the next thirty years, and when he reached retirement in 1942 he stood among the most distinguished of the world's geologists. This distinction was made evident by the medals, honorary degrees, and numerous other recognitions accorded him. As stated earlier, the brief biographical sketch that follows is purposely limited to Daly's short appointment at M.I.T. Full accounts of his long and extraordinarily productive career are happily available in excellent biographical sketches by two of his distinguished Harvard colleagues, Francis Birch and Marland P. Billings, who are cited in the footnote below.

BIRTH, ANCESTRY, AND EARLY EDUCATION

Reginald Aldworth Daly was born on a farm near Napanee, Ontario, on 19 May 1871. He was the youngest of the family of four sons and five daughters of Edward Daly, an Irish tea merchant born in Dublin and educated at Trinity College, and Jane Maria (Jeffers) Daly, daughter of an Ontario preacher, William Jeffers. In 1876 the Daly family moved into

---

* The two most complete and informative biographies I have found are Francis Birch's Reginald Aldworth Daly, May 19, 1871-September 19, 1957 (Nat. Ac. Sci., Biog. Mem. XXXIV, p. 31-64, port., 1960) and M. P. Billings' Memorial to Reginald Aldworth Daly (1871-1957) (Geol. Soc. Am., Pr. 1958, p. 115-122, port., 1959); both contain essentially complete bibliographies of Daly's writings, and both were drawn on extensively in preparing my own biographical sketch that follows.

No effort was made to prepare a complete list of biographical sketches, memorials, necrologies, obituaries and the like, but the following may be cited - T. A. Jaggar, Jr.: Reginald Aldworth Daly (Tech. Rev. 9: 178-181, 1907); M. P. Billings: Reginald A. Daly, Geologist (Science 127(3288): 19-20, 3 January 1958) and Reginald Daly 1871-1957, Tribute to a stimulating and truly great geologist (Geotimes 2/11: 10, May 1958); Daly, Reginald A., in American Men of Science, 9th ed., (1955); and Reginald A. Daly, 1871-1957, Am. J. Sci. 255: 731, (1957) [Anonymous].

the town of Napanee, and there Reginald received his pre-college education, graduating from the Napanee High School in 1887. He then attended Victoria College at Cobourg, Ontario, graduating with an A.B. degree in 1891 and an S.B. degree in 1892, meanwhile instructing in mathematics during SY 1891-92. According to his own account his interest in geology was first aroused when Prof. A. P. Coleman held up a piece of granite and remarked, "This is made of crystals."

THE EARLY YEARS AT HARVARD (1892-1901)

Soon after graduation Daly visited Harvard, and when encouraged by interviews with those two great teachers of geology, J. D. Whitney and N. S. Shaler, he decided to undertake graduate work with them and their colleagues. At this time Harvard had one of the outstanding geology faculties in North America, with Whitney and Shaler, W. M. Davis, J. E. Wolff, J. B. Woodworth, T. A. Jaggar, Jr., Charles Palache, and R. DeC. Ward. Daly won an M.A. degree in 1893 and the Ph.D. degree in 1896, and got his first taste of teaching at Harvard as an assistant to Shaler during SY 1894-95. After two years of study and travel abroad, on a Parker Traveling Fellowship, he returned to Harvard as an instructor in physiography from 1898 to 1901.

WORK ALONG THE 49th PARALLEL (1901-1907) AND MARRIAGE (1903)

At this time two great changes took place in Daly's life; one professional, the other domestic. Although he had obtained an excellent foundation in the fundamentals of geology at Harvard, he felt the need to leave the lecture room and go into the field where he could study geology firsthand and have an opportunity to develop and test his own ideas. With this purpose in mind he returned to Canada to accept an appointment as geologist with the Canadian International Boundary Commission. For the next 6 years, from 1901 to 1907, he accomplished the tremendous task of investigating the geology along the 49th parallel from the Strait of Georgia, in British Columbia, to the Great Plains--a belt 400 miles long, 5 to 10 miles wide, and about 2,500 square miles in area. It was a truly stupendous task, and according to his own notes, "No geologically trained assistant was employed in any part of the field!" A quick sampling of his classic 3-volume report on the 6-year project, Geology of the North American Cordillera at the Forty-ninth Parallel (Geol. Surv. Can., Mem. 38, 3 vols., xxviii + 857 p., (1912), will suffice to indicate the high quality of Daly's field work and his ability as an author.

Soon after returning to Canada to commence work on the 49th parallel project, Daly married Louise Porter Haskell, a talented and gracious daughter of a distinguished South Carolina family and an 1897 graduate of Radcliffe College with a summa cum laude degree in history. Soon after their marriage in 1903 they established their home in Ottawa and lived there until 1907 when they returned to Cambridge, thereafter their home until Mrs. Daly's death in 1947 and Daly's in 1957. Daly became a citizen of the United States in 1920. The Dalys had only one child, a son named Reginald Aldworth Daly, Jr., who died at the age of three.

Louise Haskell Daly was a strong influence in her brilliant husband's life; she edited and typed his manuscripts, accompanied him in his travels, and stimulated in him an interest in art and music. Her death left him bereft of a companion of more than forty years, and he never recovered his characteristic buoyant bearing during the decade before his own death.

THE M.I.T. YEARS (1907-1915)

Daly joined the M.I.T. faculty as Professor of Geology in October 1907 at the end of a critical period in the history of Course XII, then called the Department of Geology. Crosby had been forced to give up teaching in 1902 because of increasing deafness; Barton had been asked to devote full time to the Teachers' School of Science after 1902; and Johnson had resigned in 1907 to accept an appointment at Harvard. Niles had retired in 1902 after serving as the Head of Course XII since its inception; Jaggar and Warren had been appointed Lecturers in 1902, and the former was appointed Professor and Department Head in 1904. By 1907 the departmental staff had been reduced to Jaggar in general geology, Warren in mineralogy and petrology, and Shimer in paleontology and historical geology. During the first two years of this critical interim, 1902-1907, the Institute had called on the Harvard Department of Geology for assistance, and M.I.T. students were permitted to attend certain geology classes at Harvard. It was to be expected that Jaggar, who was induced to accept a full-time appointment as professor and department chairman at M.I.T. after serving as a part-time lecturer from 1902 to 1904, would look to his former colleagues at Harvard for additional staff members for his M.I.T. Department. And that is how Daly came to be invited to join the M.I.T. Geology Department in 1907 as he completed his 6-year field project along the 49th parallel.

When he came to M.I.T. in October 1907, Daly added a tower of strength in igneous petrology and physical geology (formerly called dynamical geology), by virtue of his brilliant research on Ascutney Mountain and his impressive work along the 49th parallel. In his brief commentary on Daly, published in The Technology Review of April 1907 shortly after his acceptance of an appointment as Professor of Physical Geology, Jaggar wrote -

> "The new chair has a twofold significance,- it marks the importance of earth physics to engineers and inaugurates the establishment of a research laboratory of physical geology at the Institute. The policy of the Department of Geology is to serve with as great efficiency as possible the Courses in Mining and in Civil Engineering. The main work of both these professions deals with physical geology in all its phases.
>
> "The man called to occupy this post is a combined scholar, field worker, and thinker of new principles... Everything which Professor Daly has published has been based on extended field investigation. It is the kind of investigation, moreover, which attacks problems, not the sort which merely maps areas.
>
> .....
>
> "While he is thoroughly trained in the microscopical and chemical methods of the petrographer, his reasoning is based primarily on what the field shows as to the physical relations of one rock body to another. In this he has held fast to the broad principles taught by Dr. Shaler, and has not allowed himself to be warped into merely narrow laboratory methods, which by themselves are fatal to a strong grasp of the meaning of the earth's crust."

In the preceding statement can be read the educational philosophy of the new Course XII Head, Thomas A. Jaggar, Jr., who intends to change Course XII from a mere "service department," as it was from its inception in 1890 to 1904, to a department in which a research laboratory in physical geology will be coupled closely with problems investigated in the field.

By 1907 Jaggar was already planning for the research on volcanoes and earthquakes that he would initiate with the foundation of the Hawaiian Volcano Observatory in 1912, and Warren had already published half a dozen articles on the chemical characteristics of minerals and was soon to start one of his doctoral students, Norman L. Bowen (XII Ph.D. 1912), on laboratory research that would revolutionize the study of igneous rocks. With Daly's coming, the M.I.T. Department would have three of the leading research petrologists in America, and would initiate fundamental laboratory research to go with the traditional lecture-laboratory-field sequence.

Within three years, Course XII had developed an excellent program of instruction and research, and was coordinating that program with the work in Course III, Mining Engineering and Metallurgy, so effectively that The Tech for January 15, 1910 produced a "Special Mining and Geology Issue" which contained individual articles by professors in both departments. It is interesting and informative to read what Daly wrote on "Geological Research" (The Tech XXIX/65: 23,30, Jan. 15, 1910):

> "Dynamical geology represents the application of physics and chemistry to the problems of the earth's evolution. It is, therefore, highly appropriate that geological research should be and is, among our instructors and graduate students, an integral part of the Institute's work. Every year the principles of chemistry and physics are being enlarged or restated. Either of these sciences is in a state of flux. A great number of geological problems are being attacked with the new methods provided by physical chemistry. So rapid is the advance in all three of the basal sciences that revision of the principles of physical geology is a constant necessity. The geologist has an obvious advantage who has among his colleagues physicists and chemists, who will draw his attention to recent discoveries, or to improved statement of fundamental principles, and who will advise him where only the expert is a safe guide. Such is the opportunity of a geologist at a well equipped technical institution, or at an equally well equipped university. Research on the principles of physical geology is there more fittingly prosecuted than even in the government surveys.
>
> "Every advance made in general or dynamical geology is a direct or indirect gain to economic geology and, therefore, to the thoroughly trained, practical mining expert. A successful mining geologist has the research spirit. Routine and slavish adherence to text book, handbook, or lecture instruction are not for him. As millions of dollars may be interested in his report, he must go deeper into interpretation of local facts than anyone has ever done before. No two mining camps are alike, no two problems in finding or following an ore-body are alike; each case requires a new and special application of geological principles, and these are tested with each application."

The record shows that during the five years Daly was at M.I.T., from October 1907 to June 1912, he offered the following subjects:

> General Geology (856);
>
> Topographic Geology (874); and
>
> Geology of the Igneous Rocks (877);

and shared four other subjects with various colleagues:

> Structure and Field Geology (857, 870), with Jaggar;
>
> Geological Surveying (871), with Jaggar and Laughlin;
>
> Field Geology, Advanced (873), with Jaggar, Warren, and Laughlin; and
>
> Geology of North America (890), with Jaggar and Shimer.

He cosupervised 4 theses--1 S.M. (J. D. Trueman 1909) and 3 doctorates (C. H. Clapp Ph.D. 1910; J. A. Allan Ph.D. 1912 and S. J. Schofield Ph.D. 1912), and published more than a dozen important articles which added greatly to his rapidly growing renown in both North America and abroad. He prepared and published several short papers and his classic 3-volume report on the geology along the North American 49th parallel; speculated about the limeless oceans of earliest geologic time and the lack of calcareous fossils in the earliest Pre-Cambrian limestones; called attention to the evidences of sea-level changes along the Labrador coast and postulated that the Pleistocene glaciation was related to the growth of coral reefs; commented further on his several theories involving petrogenesis; and coauthored a report on the devastating landslide at Frank, Alberta. These dozen papers are a good sample of the 150 articles he wrote during his lifetime and indicate how broad-ranging his geological interests were. The major impact he made as an M.I.T. professor was clearly the leading role he took in formulating bold, provocative, and often controversial theories, and putting these before the geological profession with compelling clarity and logic. He continued along the same line for three decades as a professor at Harvard from 1912 to retirement in 1942.

As he developed his lectures on petrography and petrogeny he accumulated a large set of notes which he later organized and expanded into the first of his half dozen books--<u>Igneous Rocks and their Origin</u> (McGraw-Hill, 1914).

Although he resigned his professorship at M.I.T. in June 1912, he joined the staff of the newly created Hawaiian Volcano Observatory directed by M.I.T.'s Dr. Thomas A. Jaggar, Jr., and maintained this connection with the Institute until 1915, meanwhile visiting numerous of the Pacific islands and publishing conclusions drawn from his observations there.

THE HARVARD YEARS (1892-1901; 1912-1942-1957)

Daly's years at Harvard can be divided into three periods: years of preparation and early research; 30 years of astounding productivity that resulted in some 70 papers and 6 widely acclaimed books and countless lectures that stimulated hundreds of students to develop "imaginative muscle," as he expressed it; and 15 closing years, beyond three score and ten, during which he remained alert to and interested in every new advance in geology almost to the end. I remember driving him home one evening after a meeting of the Boston Geological Society in the old lecture room on the first floor of the Geology Building at Harvard; it was only two years before his death, he was then almost 85, yet he discussed the lecture of the evening with clarity and vigor. What an experience it would be for him to come back for just one day and hear the leading proponents discuss sea-floor spreading and plate tectonics! We can be certain his analytical and imaginative mind would quickly sense the kinds of research needed to gain deeper insight into these currently attractive phenomena; and what would he deduce from the lunar data, which have produced the same "drowning in facts" that he deplored in 1914 in his first book, Igneous Rocks and Their Origin?

SUMMATION

Billings, in his 1958 G.S.A. Memorial, put it well when he wrote (p. 118):

> "Daly was truly one of the great men of American geology. Probably no other American geologist was better known or more widely read about. Geology was his life. His accomplishments were the result of an exceptional personality and single-mindedness of purpose."

It can also be said that by his bold and imaginative theories he challenged and stimulated his most eminent colleagues to join him in the never ending quest to understand better the origin and behavior through time of our planet earth. Birch, in his 1958 Biographical Memoir cited on a preceding page, is more specific about Daly's influence (p. 52):

> "Perhaps the most enduring elements of his work will be the many contributions toward the quantification of the geological sciences, a transformation now conspicuously in process, in which large infusions of physics and chemistry have given geology new methods of observation and new powers of interpretation."

Although the time Daly spent at M.I.T. was but a brief interlude in a long career that is part of the geological greatness of Harvard, we count him, nevertheless, as one of the giants of geology, who passed our way and stopped long enough to leave an indelible mark in the record of geology at M.I.T.

PARTIAL BIBLIOGRAPHY OF REGINALD ALDWORTH DALY

Symbols and abbreviations used in the following references are explained on p. 91 - 98; abstracts are generally listed separately as well as with the complete article.

Inasmuch as Daly was a member of the M.I.T. Faculty for only eight of his more than sixty years of teaching and research as a geologist, I have listed below only those articles published through 1916, one year after he resigned from M.I.T. After serving as Professor of Geology from 1907 to 1912, he resigned his professorial appointment to return to Harvard as Sturgis Hooper Professor of Geology, a position he held until retirement in 1942. However, he was a staff member of the Hawaiian Volcano Observatory from 1912 to 1915, and inasmuch as this research facility was directed by M.I.T.'s Prof. T. A. Jaggar, Jr., it seemed appropriate to include in Daly's bibliography the articles he published on research done during these later years.

Daly's many publications after 1916 are included in the bibliographies in several of his biographies cited at the beginning of this sketch, and also in the U.S. Geological Survey's Bibliographies of North American Geology.

1--1896   The quartz porphyry and associated rocks of Pequawket Mountain (the eastern "Keasarge" of New Hampshire) [abst.]. Science 3: 752, (1896).

2--1897   Studies of the so-called porphyritic gneiss of New Hampshire. J. Geol. 5: 694-722, 776-794, (1897).

3--1898   Review of Le Granite des Pyrenées et ses Phénomènes de Contact, by A. Lacroix. J. Geol. 6: 759-762, (1898).

4--1899   The peneplain--a review. Am. Nat. 33: 127-138, (1899).

5--1899a   On the optical characters of the vertical zone of amphiboles and pyroxenes; and on a new method of determining the extinction angles of these minerals by means of cleavage pieces. A. Ac. A. S., Pr. 34: 309-323, (1899); Abst., Science 8: 919-920, (1899).

6--1899b   A comparative study of etch-figures: The amphiboles and pyroxenes. A. Ac. A. S., Pr. 34: 373-429, (1899). (Translated in Soc. Min. France, B. 23, 1900.)

7--1899c   On a new variety of hornblende. A. Ac. A. S., Pr. 34: 431-437, (1899).

8--1899d   Three days in the Caucasus. Acta Victoriana. Victoria Univ., Toronto, May, p. 3-15, (1899).

9--1899e, 1900   The Russo-Siberian Plain. Acta Victoriana. Victoria Univ., Toronto, December, 10 p., (1899); J. School Geog. 4: 81-90, (1900).

10--1899f, 1900a   Palestine as illustrating geological and geographical controls. Am. Geog. Soc., B. 31: 444-458, (1899); B. 32: 22-31, (1900).

11--1900b   The calcareous concretions of Kettle Point, Lambton Co., Ontario, J. Geol. 8: 135-150, (1900).

12--1900c   The deepest fiord on the Labrador coast. Science 12: 688, (1900).

13--1900d, 1901   Notes on oceanography. Science 12: 114-116, 148-150, 688-689, (1900); 13: 951-954, (1901).

14--1901a   The physiography of Acadia. Harvard Coll., M.C.Z., B. 38 (g.s. 5): 73-104, il., (1901).

15--1901b   Notes on Oceanography: An oceanographic museum; Marine currents and river deflection. Science 13: 951-954, (1901).

16--1902   The geology of the region adjoining the western part of the international boundary. Can. Geol. Surv., Sum. Rep. 1901 (Ann. Rep. 14): A39-51, (1902).

17--1902a   The geology of the northeast coast of Labrador. Harvard Coll., M.C.Z., B. 38 (g.s. 5): 205-270, (1902).

18--1902b   Report on geology [Brown-Harvard expedition to Nachvak, Labrador, in the year 1900]. Geog. Soc. Phila., B. 3: 206-212, (1902).

19--1903   The geology of Ascutney Mountain, Vermont. U.S. Geol. Surv., B. 209: 122 p., il., (1903).

20--1903a   The mechanics of igneous intrusion. Am. J. Sci. (4) 15: 269-298; (4) 16: 107-126, (1903); (4) 26: 17-50, (1908); Abst., J. Geol. 11: 101-102, (1903).

21--1903b   Geology of the western part of the international boundary (49th parallel). Can. Geol. Surv., Sum. Rep. 1902 (Ann. Rep. 15): A138-149, (1903).

22--1903c   Variolitic pillow-lava from Newfoundland. Am. Geol. 32: 65-78, (1903).

23--1903d   The high places of Labrador. Acta Victoriana. Victoria Univ., Toronto, Christmas Number, p. 175-185, (1903).

24--1904   Geology of the international boundary [British Columbia]. Can. Geol. Surv., Sum. Rep. 1903 (Ann. Rep. 15): A91-100, (1904).

25--1905   Geology of the western part of the international boundary (49th parallel). Can. Geol. Surv., Sum. Rep. 1904 (Ann. Rep. 16): A91-100, (1905).

26--1905a   The accordance of summit levels among alpine mountains; the fact and its significance. J. Geol. 13: 105-125, (1905).

27--1905b   The classification of igneous intrusive bodies. J. Geol. 13: 485-508, (1905).

28--1905c   Machine-made line drawings for the illustration of scientific papers. Science 22: 91-93, (1905); Am. J. Sci. (4) 19: 227-229, (1905).

29--1905d   The secondary origin of certain granites. Am. J. Sci. (4) 20: 185-216, il., (1905).

30--1906   The differentiation of a secondary magma through gravitative adjustment [Moyle sill in the Purcell Mountain Range, Idaho-Montana]. Festschrift zum siebzigsten Gebürtstage von Harry Rosenbusch, E. Schweizerbartsche Verlagsbuchhandlung, Stuttgart, p. 203-233, (1906).

31--1906a   The Okanagan batholith of the Cascade Mountain system. Geol. Soc. Am., B. 17: 329-376, (1906).

32--1906b   Abyssal igneous injection as a causal condition and as an effect of mountain building. Am. J. Sci. (4) 22: 195-216, (1906); Abst., Science 24: 367-368, (1906); A.A.A.S., Pr. 56-57: 267-268, (1907).

33--1906c   The nomenclature of the North American Cordillera between the 47th and 53rd parallels of latitude. Geog. J. 27: 586-606, (1906).

34--1906d   Report on field operations in the geology of the mountains crossed by the international boundary (49th parallel). Can. Dep. Interior, Rep. Chief Astronomer (pt. IX of Ann. Dep. Rep. for 1905): 278-283, (1906).

35--1907   The limeless ocean of pre-Cambrian time. Am. J. Sci. (4) 23: 93-115, (1907).

36--1907a   Report on field operations in the geology of the mountains crossed by the international boundary (49th parallel). Can. Dep. Interior, Rep. Chief Astronomer (pt. V of Ann. Dep. Rep. for 1906): 133-135, (1907).

37--1908   The mechanics of igneous intrusion. Am. J. Sci. (4) 26: 17-50, (1908). (Also see 20--1903a.)

38--1908a   The origin of augite andesite and of related ultra-basic rocks. J. Geol. 16: 401-420, (1908).

39--1908b   Review of Traité de Géologie, by E. Haug. Science 28: 886, (1908).

40--1909   First calcareous fossils and the evolution of the limestones. Geol. Soc. Am., B. 20: 153-170, (1909).

41--1909a	The geology and scenery of the northeast coast [of Labrador], in *Labrador, the country and the people*, by W. T. Grenfell and others, p. 81-139. New York, (1909), 2nd ed., (1913).	
42--1910	Geological research. *The Tech* (Special Mining and Geology Issue) XXIX (29), No. 85: 23, 30, (Jan. 15, 1910).	
43--1910a	Average chemical compositions of igneous-rock types. A. Ac. A. S., Pr. 45: 211-240, (1910).	
44--1910b	Origin of the alkaline rocks. Geol. Soc. Am., B. 21: 87-118, (1910); discussion, B. 21: 785, (1910). Abst., *Science* 32: 220, (1910).	
45--1910c	Pleistocene glaciation and the coral reef problem. Am. J. Sci. (4) 30: 297-308, (1910).	
46--1910d	Hawaiian volcanoes [abst.]. *Science* 32: 188, (1910); Geol. Soc. Am., B. 211: 767, (1910).	
47--1911	Origin of the coral reefs; a suggestion bearing on the question of the former mobility of the earth's crust under the deep oceans. *Sci. Conspectus* 1: 120-123, (1911).	
48--1911a	The nature of volcanic action. A. Ac. A. S., Pr. 47: 47-122, (1911).	
49--1911b	Magmatic differentiation in Hawaii. J. Geol. 19: 289-316, (1911).	
50--1911c	Relative erosive efficiency of ice caps and valley glaciers [abst.]. Assoc. Am. Geog., An. 1: 121, (1911).	
51--1912	Reconnaissance of the Shuswap lakes and vicinity (south central B.C.) [abst.]. Can. Geol. Surv., Sum. Rep. 1911: 165-174, (1912).	
52--1912a	Pre-Cambrian formations in south central British Columbia [abst.]. *Science* 35: 311, (1912); Geol. Soc. Am., B. 23: 721, (1912).	
53--1912b	Some chemical conditions in the pre-Cambrian ocean. Int. Geol. Cong., XI, Stockholm, 1910, C.R.: 503-509, (1912).	
54--1912c	(and Miller, W. G. and Rice, G. S.). Report of the commission appointed to investigate Turtle Mountain, Frank, Alberta. Can. Geol. Surv., Mem. 27: 34 p., il., (1912).	
55--1913	Geology of the North American Cordillera at the Forty-ninth Parallel. Can. Dep. Interior, Rep. Chief Astronomer, 1910, App. 6, 3 vols., xxvii + 857 p., (1912); Geol. Surv. Can., Mem. 38, 3 vols., xxvii + 857 p., (1912).	
56--1913a	Introduction to the geology of the Cordillera. Geol. Surv. Can., Guidebook 8, Toronto to Victoria (p. 105-167), Int. Geol. Cong., XII, Canada, (1913).	
57--1913b	Annotated guide, Golden to Savona. Geol. Surv. Can., Guidebook 8, Toronto to Victoria (p. 202-234), Int. Geol. Cong., XII, Canada, (1913).	
58--1913c	Sills and laccoliths illustrating petrogenesis. Int. Geol. Cong., XII, Canada, 1913, C.R. 189-204, (1914). [Advance copy, 1913.]	
59--1913d	[Discussion of A. L. Day and E. S. Shepherd's] Water and volcanic activity. Geol. Soc. Am., B. 24: 573-606, 707, (1913).	
60--1914	Geology of the Selkirk and Purcell Mountains at the Canadian Pacific Railway (main line). Geol. Surv. Can., Sum. Rep. 1912: 156-164, (1914).	
61--1914a	*Igneous Rocks and Their Origin*. New York: McGraw-Hill Book Co., Inc., xxii + 563 p., il., (1914).	

62--1915   Origin of the iron ores at Kiruna. Geology of the Kiruna district, No. 5. Vetenskapliga och praktiska Undersökningar i Lappland. Anordnade af Loussavaara-Kiiravaara Aktiebolag [Geology No. 5], Stockholm, p. 1-31, (1915).

63--1915a   A geological reconnaissance between Golden and Kanloops, B.C., along the Canadian Pacific Railway. Can. Geol. Surv., Mem. 68, 260 p., il., (1915).

64--1915b   The glacial-control theory of coral reefs. A. Ac. A. S., Pr. 51: 157-251, (1915).

65--1915c   Ores, magmatic emanations, and modes of igneous intrusion; discussion of paper by B. S. Butler. Econ. Geol. 10: 471-472, (1915).

66--1916   Homocline and monocline (with discussion by G. W. Stose and W. H. Hobbs). Geol. Soc. Am., B. 27: 89-92, (1916).

67--1916a   Problems of the Pacific Islands. Am. J. Sci. (4) 41: 153-186, (1916).

68--1916b   A new test of the subsidence theory of coral reefs. Nat. Ac. Sci., Pr. 2: 664-670, (1916); Abst., Geol. Soc. Am., B. 28: 151, (1917).

69--1916c   Petrography of the Pacific Islands. Geol. Soc. Am., B. 27: 325-344, (1916).

NOTE:

From 1917 to 1957, the date of the last article, Daly published some 85 additional articles and 6 books, which are listed in Birch's Biographical Memoir (Nat. Ac. Sci., Biog. Mem. XXXIV (34): 31-64, 1960).

(12)
WALDEMAR LINDGREN
(1860-1939)

MIT: 1912-1939

WALDEMAR LINDGREN

When Waldemar Lindgren came to Boston in 1912 to head M.I.T.'s Department of Geology he was one of the leading economic geologists in the world. He was internationally known for his work on ore deposits; was Chief Geologist of the United States Geological Survey, the world's largest such organization; and was a member of the nation's most prestigious honorary scientific society, the National Academy of Sciences. The Institute's Corporation recognized his eminence in science by creating for him a special chair - the William Barton Rogers Professorship of Economic Geology - in June 1912. During the next twenty-one years, until retirement in 1933, he added greatly to his professional stature as professor, research geologist, author, editor, administrator, and educational statesman. In the latter two categories he headed the Institute's Department of Geology from 1912 to 1920 and from 1927 to 1933, and the combined Department of Mining, Metallurgy, and Geology from 1920 through 1926.

Lindgren's life of almost eighty years (1860-1939) divides rather naturally into three periods of about the same duration. The first, 1860-1883, saw him born and reared in a cultured home in southeastern Sweden and educated in the excellent gymnasium at Kalmar, followed by five years of technical training at the world-famous Royal Mining School (Bergakademie) in Freiberg, Saxony, from which he was graduated with highest honors as "Bergingenieur" (Mining Engineer) and "Markscheider" (Mine Surveyor) in 1882.

The second period, 1883-1912, begins with his emigration to the United States; continues with his first three jobs, all within a year; and includes the next twenty-eight years as an employee of the United States Geological Survey during which he advanced from Assistant Geologist (1884), to Geologist (1895), and finally to Chief Geologist (1911). During the first twenty years of this period he was almost continuously in the field investigating the ore deposits and associated geology of western mining districts. As the end of the century approached he turned more of his attention to the fundamental problems of ore deposition and began synthesizing the vast amount of observational data he had accumulated during his earlier years of field work. From this synthesis came epochal papers on "Metasomatic processes in fissure veins" (1901), "The relation of ore deposition to physical conditions" (1907), and "Metallogenetic epochs" (1909), and his classic reference work and textbook, Mineral Deposits (1913). By the time he was appointed Chief Geologist of the Survey in 1911, he had become the dean of American mining geologists; the next year he would carry this distinction with him to M.I.T.

The third and last period of his life, 1912-1939, was spent in the academic world as the William Barton Rogers Professor of Economic Geology and Head of M.I.T.'s Department of Geology (Course XII). Actually, Lindgren first came to M.I.T. as a visiting Lecturer during the fall terms of 1908 to 1911, and continued beyond normal retirement, from 1933 to 1936, as Honorary Lecturer, and as Professor Emeritus from 1933 to his death in 1939. During this third period of his life, free of the restrictions and responsibilities of those employed in federal bureaus, Lindgren turned his attention to the broader aspects of his lifelong interest - mineralogy and mineral deposits - and in doing so added to his professional stature that of educational statesman. True, he had helped to found the journal Economic Geology in 1905, while still on the Survey; now he turned to other equally important tasks. His first, after coming permanently to M.I.T., was completion and publication of his monumental reference work Mineral Deposits, which was immediately adopted in most quarters as the authoritative work on the subject, and which remains today, some 60 years later, as an important, although now much out-of-date, reference work of major proportion. In spite of his teaching and administrative duties, he found the time and energy to continue a program of research that produced a steady flow of important articles, now broadened to include the vital need for basic as well as applied research and for better programs of education for aspiring geology students. While on leave from the Institute during SY 1928-29, to serve as Chairman of the Division of Geology and Geography of the National Research Council, in Washington, he initiated and almost single-handedly raised the money for the Annotated Bibliography of Economic Geology, thereafter serving until death as an Associate Editor. He submitted some 2,545 abstracts during the decade 1928-1938, which involved reading an astounding volume of geological literature in half a dozen languages not his native and adopted tongues. In a busy teaching schedule he took off short periods to carry out consulting assignments in Canada, Mexico, Chile, Bolivia, and at home in several western States. He supervised the thesis work of more than sixty M.I.T. graduates, many of whom went on to distinguished careers in economic geology, particularly in Canada and at home in the United States. Throughout his long active career, he had the loyal and sympathetic companionship of Ottolina Allstrin from Sweden, whom he married in 1886 and lost by death in 1929; there were no children. Lindgren himself died in his home in Brookline, Massachusetts, on 3 November 1939.

As with all men of greatness, recognition in full measure came to Lindgren in many forms and from many sources. He was an elected member of most of the geological and scientific societies of his time in North America, and foreign member of the leading such societies in Europe. He was a member of the National Academy of Sciences, the American Academy of Arts and Sciences, and the American Philosophical Society. He was elected Vice-President of the American Institute of Mining and Metallurgical Engineers for the year 1912-1913, President for 1920-1921, and Honorary Life Member in 1931. He was president of the Mining and Metallurgical Society in 1920. As one of the founders of the Society of Economic Geologists, he served as the Society's first president, in 1922, and received its third Penrose Medal in 1928 for his distinguished achievements in the earth sciences. He was elected president of the Geological Society of America in 1924 and was awarded its highest honor, the Penrose Medal, in 1933. In 1936 he was given La Médaille Gustave Trasenster by the Association of Alumni Engineers of the Université de Liége, Belgium, and the next year the Geological Society of London chose him to receive its prestigious Wollaston Medal. He was unanimously elected Honorary Chairman of the Sixteenth International Geological Congress held in the United States in 1933. The Honorary Degree of Doctor of Science was conferred upon him by Princeton in 1916 and by Harvard in 1935.

Lindgren himself published more than 250 reports, topical articles, reviews and the like, including four editions of Mineral Deposits, and an amazing 2,545 abstracts for the Annotated Bibliography of Economic Geology. In 1933 the American Institute of Mining and Metallurgical Engineers dedicated their 797-page book Ore Deposits of the Western States to Lindgren "as an expression of their appreciation and admiration," and designated it the "Lindgren Volume." On the centenary of his birthday, in early 1960, the journal he helped to found in 1905, Economic Geology, designated the first part of the January-February number of volume 55 (1960) the "Waldemar Lindgren Centennial Commemorative Number," in his memory. In the preface Jensen (1960, p. vii) presents the evidence for calling him the World's Foremost Economic Geologist, a designation few would dispute. Lindgren has also been memorialized in still other ways, both in America and abroad, and both while he lived and after death: e.g. the Lindgren Library (1932) and the Lindgren Memorial Fund (1959) perpetuate his name at M.I.T., and the Lindgren Citation Awards for excellence in research were established by the Society of Economic Geologists in 1960. No doubt Lindgren has also been honored in his native Sweden, but I have not found notice of any such honors in my limited search of the Swedish literature.

Waldemar Lindgren clarified much of the uncertainty that existed concerning the processes of mineralization and ore deposition, and demonstrated beyond argument the impressive power of the trained intellect when that intellect is properly guided in its attack on a challenging problem. Ninety years after he started his professional career, and thirty-five years after death brought it to a close, his place among the world's greatest geologists remains secure.

BIRTH, ANCESTRY, AND EARLY EDUCATION*

Waldemar Lindgren, a descendant of the storied Goths of history, was born in Sweden on 14 February 1860. His birthplace was the village of Vassmolösa, located about eighteen kilometers southwest of Kalmar, in the

---

* I joined the M.I.T. geology faculty in September 1937, two years before Lindgren's death on 3 November 1939, so had little opportunity to become acquainted with him, particularly as he was essentially an invalid during the last year of his life. I did, however, hear the last lecture he gave, which was titled "The black gold [=petroleum] of California," and greeted him a few times in the third floor corridor of the Institute's Building 4. Occasionally, too, I saw him sitting at his desk, coatless but with his hat still on his head, pince-nez in place, and cigarette in hand, reading some lately delivered journal. In such a situation his concentration seemed complete.

Accordingly, in preparing the preceding précis and the following biographical sketch, I have had to draw heavily on the body of information included in the twenty-five or so biographies, biographical sketches, necrologies, memorials, and obituarial notices that are listed chronologically at the end of this sketch following the Bibliography as Appendix B. Also listed are a few of many brief notices, regarding Lindgren's activities and recognitions, that appeared in The Technology Review and Tech Engineering News; a list of such items could easily be doubled or tripled by a thorough search of the two M.I.T. publications just mentioned as well as of The Tech and the annual M.I.T. Presidents' Reports.

I should point out that of the numerous sources of biographical information mentioned in the preceding paragraph, the articles of eight authors were of special assistance because of the variety and extent of the detailed information contained in them. Åhman (1948, 1953), Backman (1964), and Martinsson (1973) tell of Lindgren's youthful years in his homeland, Sweden; Schiffner (1935) writes of his college years as a student at the Bergakademie in Freiberg, Saxony; Graton (1933, 1939, and 1960) and Butler (1950) write at length about his long period of service with the U. S. Geological Survey; and Newhouse (1939) and Buerger (1940), two of his numerous distinguished students, discuss him as teacher, senior colleague, and department head at M.I.T.

extreme southeastern part of Sweden, on the mainland directly across the channel from the Island of Öland. The house in which he was born is pictured below.

Waldemar was the youngest of the five children of Johan M. and Emma B. Lindgren. The children were Gerda (1849-1904), Tora (1851-1928), Einar (1852-1900), Sigrid (1856-1934), and Waldemar (1860-1939).

His father, Johan Magnus Lindgren (1817-1890), was a descendant from a line of landholders in middle Sweden, and was a well-educated man. He served the County of Kalmar as a District Judge, and the library that he had accumulated for his own reading pleasure was much used by his youngest son.

(Häradshövdingebostället i Vassmolösa, där Waldemar Lindgren föddes år 1860 - Photograph by Erik Åhman, in his article "Professor Waldemar Lindgren, en nestor inom malmgeologien," published in Kalmar nations skriftserie xxix, p. 15-21, 1952.)

Waldemar Lindgren was born in the house pictured above on 14 February 1860. It was the residence of his father, Johan Magnus Lindgren (1817-1890), who was the District Judge (häradshövding) of Södra Möre in Vassmolösa, county of Kalmar, for some 45 years. The photograph was taken by E. Åhman in August 1947. (See Åhman's [1948, p. 350] brief note "The birth-place of Waldemar Lindgren," in Appendix B - A list of biographies and biographical sketches, following the Bibliography.)

Waldemar's mother, born Emma Bergman (1830-1906), descended from a line of celebrated ancestors who had played important roles in the history of Sweden back to the fifteenth century. She was the daughter of Rector and Pastor Carl Abraham Bergman, one of the important figures in Swedish culture, who worked for the advancement of secondary education and was also one of the founders of the Evangelical Alliance. Through his mother, Waldemar could trace his ancestry to the family of Vult von Steijern, whose members in many generations held high offices and made great contributions to Swedish culture.

Waldemar spent his youth with his parents in Vassmolösa and showed at an early age that he had an extraordinarily quick mind and keen intellect. In the gymnasium at Kalmar he excelled in scientific subjects, and in his father's extensive library he found great pleasure in reading books on exploration and similar subjects. There was little of geological interest in the flat country around Kalmar, but when his parents took him on a summer vacation trip to Lysekil, on the west coast of Sweden, when he was twelve years old, he was fascinated by the granite formations that he saw there.* Immediately after this visit he began to collect minerals and rocks and to read what he could find written about them.

At about this same time Waldemar's uncle, Aron Lindgren, who had been much interested in natural science before becoming a District Judge (like his brother Johan M.) began to encourage him in his scientific studies. Among other things he made available to Waldemar a small library of books on chemistry and geology that he had accumulated while studying law at Uppsala.**

Along with his regular studies of language and literature, Waldemar read the scientific books that his uncle had given him and became increas-

---

* In responding to his citation for the Wollaston Medal of the Geological Society of London, Lindgren reminisced as follows:
> "I think my interest in all these things [i.e. determining the life history of mineral deposits] was first kindled when as a lad of twelve I visited the west coast of Sweden, where the mighty breakers of Skagerak had polished off the granite knobs of islands and skerries and I could see for myself the large crystals of the pegmatite dykes that broke across the monotonous fine-grained granite." (Geol. Soc. London, Q.J. 93/4: cxxix, Feb. 1937.)

** Åhman (1952, p. 16) writes that in the early 1930s he himself found in the attic of a building that had once been an annex to the Kalmar gymnasium, where Waldemar had studied, an early 1840 edition of Graham & Otto's Lehrbuch der Chemie. This 4-volume work had belonged to Aron Lindgren, during his student years at the University of Uppsala, but was later turned over to Waldemar, whose name was written in the first volume.

ingly interested in mineralogy and the geology of the region around his home. When only thirteen, he visited small iron and copper-mining operations in neighboring regions. During the summer holidays of 1874 and 1875, with money provided by his father, he visited the mining regions of central Sweden, where he collected minerals and saw a variety of ore deposits.* By now, even though only sixteen, he had definitely decided to pursue geology as a profession. He spent the summer of 1876 visiting Kongsberg and other famous mines in Norway, and the next summer (1877) he went alone to Germany to visit the world-famous Royal Mining Academy at Freiberg, and while there visited mines in Saxony and elsewhere in Germany. Returning to Sweden, he resumed his studies at the gymnasium** in Kalmar and passed the maturity examination in 1878. Now he was ready for university. But before going on to further study he worked for two months in the mines of the great Swedish zinc deposit at Åmmeberg.

THE FIRST GREAT STEP IN A DISTINGUISHED CAREER (1878)

The fall of 1878 found Lindgren graduated from the gymnasium at Kalmar as a <u>student</u> (which in Sweden implies competence to attend a university); he would go on to the renowned Bergakademie (Royal Mining Academy)

---

* Dr. Anders Martinsson, paleontologist at Uppsala University, recently wrote me (letter dated 17 May 1976) that Pierre Backman's grandson, Per Backman, had inherited a collection of postcards that Waldemar sent his parents from time to time during his youthful travels, and also later on during his world travels when he visited numerous foreign countries as a consulting economic geologist. Even in his teens he used English, French, and German as easily as his native Swedish. His devotion to his parents and siblings that characterized his early messages persisted into adulthood as manifested by the following message on one of the postcards dated 1903:

> Steamship "Sonoma" July 29/03
>
> Tutuila, Samoa, South Pacific
> Lat. $14°$ S.
> Long. $168°$ W.
>
> Dear Mother: Just arrived here after a pleasant voyage. Stopped one day at Honolulu. Next stop at Auckland, New Zealand. Expect to arrive in Sydney, N.S.W. Aug. 7. Everything satisfactory and enjoying the trip very much. Love to my sisters!
> Your affectionate son,
> Waldemar

** According to Martinsson (1973, p. 280):

> "The type of school (högre allmänt läroverk) which Lindgren attended in Kalmar took him through what corresponds to the junior college in the present educational system in the United States."

at Freiberg, Saxony, and from there to a distinguished career in the United States. Before discussing the preceding, however, I think it worthwhile to interrupt my narrative and ask - what sort of person was this young Swede who would, at the end of his long professional career, be called the <u>Foremost Economic Geologist in the World</u>? (Jensen 1960, p. vii.) How did he feel, at age 18, about his future?

He had been born of parents whose ancestors included many distinguished persons who had contributed importantly to the political and cultural life of Sweden. He had all the advantages of growing up in a cultured family circle in which proper conduct, courteous behavior, broad education, and polylingual ability* were insisted upon. He had passed through a rigorous educational system which prepared him well for university study - in science, in languages, and in the subjects now called the humanities. Finally, he had travelled extensively to acquaint himself with activities in his chosen field of interest - mining geology in its broadest sense. In short, he was <u>ready</u> to leave home and start his career.

At this critical time in his life, and he surely thought of it as such, he provides us with an impressive image of himself and of his thoughts and ambitions, in a statement that he wrote on 2 June 1878, some four months after his 18th birthday and immediately following graduation from the gymnasium at Kalmar.**

> "At an important point in my life, the second of June 1878, the big step has been taken; the object for which I have striven for many years [<u>i.e.</u> graduation from the gymnasium] has been reached. ...I am not yet wholly happy. I am now sitting down to order my papers, look through my letters, pack everything and leave Kalmar. It is not without a feeling of sadness that I look back over these long years on all the happiness, all the small sorrows that I have had, the happy times with my friends. I must now part from them; perhaps I will not see most of them again. I think of my beloved parents, of all the love they

---

* It is known that Lindgren's parents were polylingual, and that they and Waldemar corresponded in several different languages. Dr. Anders Martinsson of Uppsala University has written me (17 May 1976) that Waldemar started to write to his parents in English even before he emigrated to the United States; that the parents used English between themselves at the time both German and French were preferred to English as a foreign language in Sweden; and that many years later Waldemar wrote to them from the United States in French! See also the footnote on a preceding page.

** We are told by Åhman (1953, p. 17) and Martinsson (1973, p. 280) that the statement was written soon after graduation. Dean Emeritus C. Richard Soderberg, my esteemed colleague at M.I.T., and another of Sweden's distinguished scientific sons (mechanical engineering), kindly prepared for me the translation quoted here.

King Bede's Burial Mound, in Balestrand, Sogenfjord, Norway. (Signed "Waldemar")

Nystuen in Jotunheimen, Norway, after a sketch taken from Jemmason 1876. (Signed "Waldemar Lindgren")

have shown me and all they have done for me. I hope
that it will be possible for me to show them that I
am worthy of them and one day let them see that their
concern has not been lost. I feel it. - I am standing at one of the landmarks of life. I have completed
one step; I stop for a moment and see many roads lying
open in front of me. But I do not hesitate; my future
is sharply defined for me. Forwards on it! New courage; new vigour! I am now going out into the world to
struggle alone. ... May the idea of the True, the Good,
and the Beautiful always be the light and goal towards
which I strive, for ever the sworn soldier of Science.
<u>Nec aspera terrent</u>!

Waldemar"

These reflections are clearly those of a sensitive, perceptive, and reflective youth who is grateful to family members and friends for all that they have done for him but realizes that the time has come when he must strike out on his own. He confidently chooses the road of science, though he could not have known what great honors would come to him along that road, and that they would be won in a country far across the ocean from his homeland.

In his reflections Waldemar must have thought of the many trips he had made throughout Scandinavia; of the mountains he had sketched (see accompanying illustrations); and of the verses he wrote that were stimulated by the mountains, glaciers, fiords, and lakes that he had seen during his travels in Norway and Sweden.*

STUDENT YEARS AT FREIBERG'S BERGAKADEMIE (1878-1883)

After some uncertainty as to whether he would continue his education in Sweden or abroad, Lindgren decided in favor of the Bergakademie in Freiberg, Saxony, which he had visited in the summer of 1877. Thus in the autumn of 1878, not yet nineteen, Lindgren entered the renowned institution (as Student No. 3017) and started on a four-year course of study that led to graduation as "Bergingenieur" (Mining Engineer) and "Markscheider" (Mine Surveyor) in 1882. He continued graduate study in metallurgy and chemistry until May 1883, at which time he left for America.

---

* While a student at the gymnasium in Kalmar, Lindgren belonged to a literary club called "Förbundet N3," founded in 1864, whose members competed for prizes in poetry and essays. He won the Club's First Prize in 1876 with an anonymous poem, "Zindari and Verouna," an Italian love story, and read a number of travel accounts to the Club in 1876 and 1877. Some of the latter are included in Backman's (1964, p. 99-108) <u>Skola för Skalder: Förbundet N3, 1864-1895</u>," together with an eleven-verse poem titled "En färd över en fjällsjö i Norge," ["Traversing a mountain lake in Norway"] written in 1877 (p. 108-110).

Institute on Prüferstrasse.

In the reddish brick building (middle) is still housed the Surveying Institute; in the gray building to the left, the facade of which was newly shaped in 1900, Lindgren had classes in machine design, mining law, and perhaps also electrical engineering, insofar as it was offered at that time.

Nonnegasse 14.

Lindgren lived in the middle building, which was repainted reddish-brown during fairly recent refurbishment.

RECENT (1976) PHOTOGRAPHS OF THE FAMOUS BERGAKADEMIE IN FREIBERG, SAXONY

The Old Elizabeth Mine.

In Lindgren's day this mine was used for instructional purposes by the Academy. Lindgren did all of his practical work here, including surveying, required of all mining engineers at that time.

All photographs kindly provided by Dr. Günter Freyer of Freiberg.

In Lindgren's student days at Freiberg, the Academy, already more than a century old, having been founded in 1765, was the oldest advanced technical institution in the world. Situated in the midst of the celebrated silver-lead-copper mines of the Erzgebirge and at the center of the mining and metallurgical operations in Saxony, yet within a celebrated medieval city of rich Teutonic culture and social tradition, the Academy was the ideal place for a talented and diligent student like Waldemar Lindgren. Here he would mingle with outstanding students from every part of the world and study with four of the Academy's most eminent professors - Theodor Richter, in metallurgy and blow-pipe analysis; Clemens Winkler in chemistry; Albin J. Weisbach (the younger) in mineralogy and crystallography; and Alfred W. Stelzner in geology and ore deposits. Of the four, Stelzner, who was far ahead of his time in the understanding of ore genesis, probably had the greatest influence in shaping and enlarging Lindgren's particular field of geology. Here too the eager young Swede, coming in close contact with Europeans and students from both Americas, developed further that unusual facility with languages that later on opened to him most of the principal geologic literature of the world as he prepared abstracts of articles and books, written in half a dozen foreign languages, for the Annotated Bibliography of Economic Geology.

The class notes that Lindgren prepared while attending the Bergakademie show that he was an attentive student who organized his notes carefully and supplemented them with formulae, equations, and sketches (see accompanying illustrations). His notebooks, which are preserved in the M.I.T. Institute Archives, are of much interest in that they show the nature of the subject matter taught a century ago at the leading mining school in the entire world. The training that he received was broad, rigorous, fundamental, and highly practical; it fitted him superbly for the unparalleled opportunities that would come to him as he investigated the newly opened mines throughout the Americas. Of special value was the knowledge of chemistry and mineralogy first acquired in his youth and steadily added to throughout his long professional career.* It was the

---

* Lindgren's keen interest in minerals was first aroused when he was only ten years old, and he never lost this interest as he grew older and gained knowledge of the whole field of mineralogy. In Weisbach and Stelzner at Freiberg, he found teachers who inspired him to extend his knowledge of minerals and ore deposition, and of geology in a broader sense; and as evidence of this stimulation, "... he often took his mineralogy and petrography text books to bed with him, and read from them till he fell asleep." (Buerger 1940, p. 184.) Many years later, when responding to the citation for the Wollaston Medal, he again revealed his lifelong love affair with minerals in stating:

> "And I shall ask for no greater joy than I experienced when I could prove that a zeolite (analcite) had crystallized from the basic magma of the Highwood Mountains,..."
> (Geol. Soc. London, Q.J. 93/4: cxxix, Feb. 1937.)

Facing pages in Lindgren's notebook in Lagerstättenlehre, — a subject which he studied while a student at Freiberg.

Notebook in M.I.T. Institute Archives

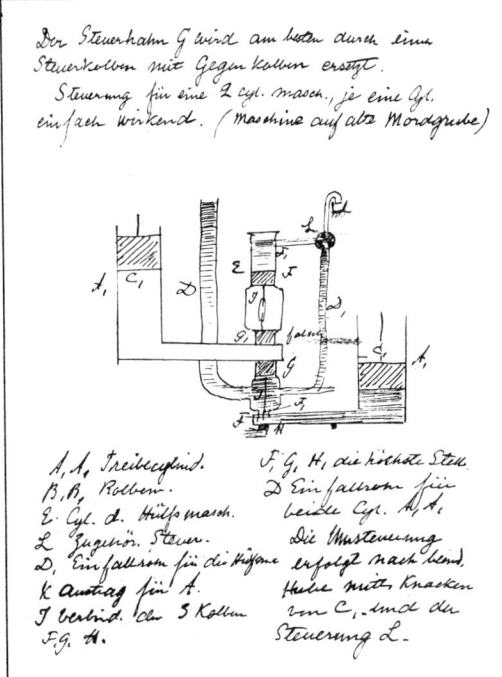

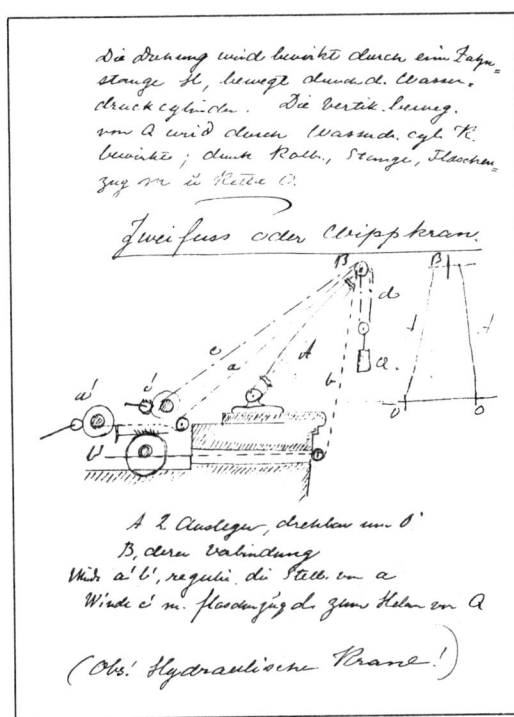

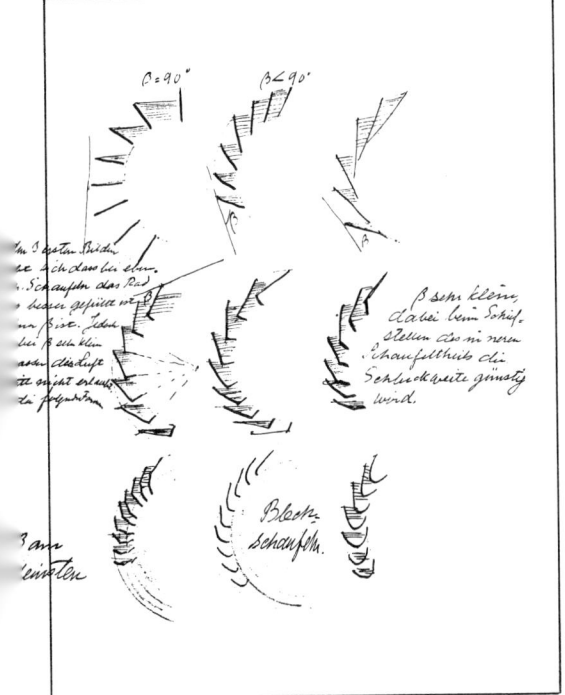

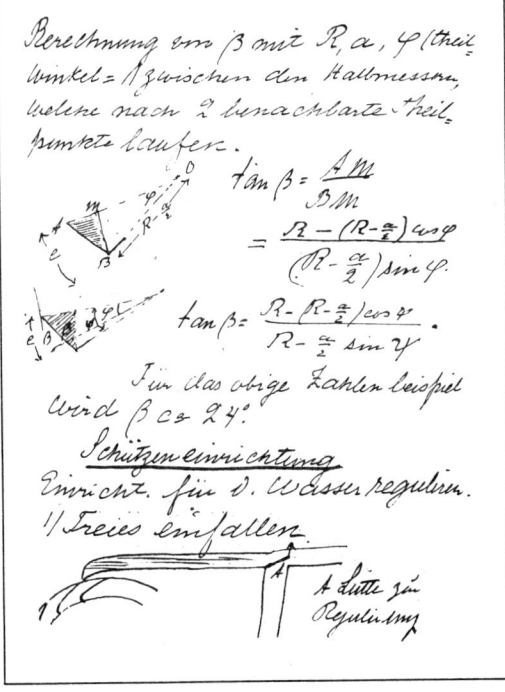

Sketches and Notes from Waldemar Lindgren's 1880 Notebook in Maschinenlehre at Freiberg's Bergakademie, Saxony.
Notebook in M.I.T. Institute Archives

application of this knowledge that in large part enabled him to recognize the fundamental importance of metasomatism and ultimately to construct a classification of mineral deposits that gained and continues to hold acceptance by most of the world's economic geologists.

After passing the examination for Diplomat of Mining Engineering in 1882, Lindgren stayed on at the Bergakademie for an additional year in order to get advanced training in chemistry and metallurgy. And as might be expected, he used his summer vacations to visit as many famous mines as possible in Germany, Austria, Bohemia and Italy. He also found time to write his first three geological papers - items 1, 2 and 3 in his Bibliography farther on - which deal with the mineralogy of Långban, one of the most interesting mineral deposits of Sweden.

Having gained knowledge of most of the better known mines of Europe, either by first hand examinations or as a result of reading the existing literature, Lindgren longed to widen his experience and extend his travels beyond his homeland and the part of Europe he knew. No doubt he read of the exciting geological and mining activities that were going on in the western part of the United States, and we can assume that his American schoolmates at the Bergakademie fired his imagination and stimulated his desire to participate in those activities. In spite of an offer of work in Chile, which he declined, he decided to come to the United States, hoping he could find employment with an opportunity to use the practical training he had obtained at the Bergakademie. By this decision Sweden lost one of its most promising young scientists, and the United States gained a superbly trained geologist who would later become internationally known for his knowledge of mineral deposits of the whole world.

LINDGREN EMIGRATES FROM SWEDEN TO AMERICA (1883)

The decision to leave his homeland and seek a career as mining geologist in America having been made, Lindgren, with a friend and travel companion, presumably V. Rapp, departed for America in early June 1883, with prospects for employment solely dependent on a few letters of recommendation that he carried; among these was one addressed to Prof. Raphael Pumpelly. I shall come back to Pumpelly a little farther on, but first I would like to mention one aspect of Lindgren's voyage across the Atlantic and of his experiences during the month following his arrival.

In a recent article titled "Waldemar Lindgren goes west," Anders Martinsson (1973) published four letters (in an English translation) that young Waldemar wrote to his parents during his voyage in June and his subsequent experiences during July. By the end of the latter month he had found employment and was camped among the bears and the pine trees in the Big Belt Mountains near Diamond City, in Montana Territory.

The first letter, written aboard H.M.S. Britannic, under way from Liverpool to New York, and dated 9th June, 1883, describes young Waldemar's departure from Freiberg, his impressions of Holland, and his crossing of the English Channel to Dover, where, he writes:

> "At Dover it was I who took the command, since my friend and travel companion did not know many words of English and it was I who had induced him to travel via England."

He goes on to comment on his train ride to London, where he bought some clothes,

> "... which in England are inexpensive and of excellent quality."

Thence to Liverpool and passage on H.M.S. Britannic. Then follows a delightful description of the ship, the food, the activities aboard, and the passengers - among them:

> "... about 800 emigrants, mostly Swedes and Irishmen,"

as deck passengers, and other passengers:

> "... about half of them Americans, the rest French or English,"

including some remarkable persons:

> "... there is Mr. Morgan, one of the wealthiest bankers in New York, and Mr. Vanderbilt, almost the wealthiest man in the world. The former is big and fat, and has a red nose, the latter the same, minus the red nose, looking very modest and good-natured."

The second letter, posted from Pittsfield, Massachusetts, and dated 5th July, 1883, informs his parents that he has been employed as Assistant Geologist on the Northern Transcontinental Geological Survey, and that he is on his way by a transcontinental train to Bozeman, Montana Territory, where he will work as a surveyor and geologist. He comments favorably on the people he has met, clearly revealing in his comments why he came to be so widely admired and respected in his later years:

> "I do not understand why it is that Einar [Waldemar's brother, the construction engineer] finds the people here disagreeable. My experience is that people are about the same in all countries and usually gentle and compliant if one takes them the right way."

The third letter, posted from Livingstone, Montana Territory, and dated July 19, 1883, describes his train trip from Massachusetts, via Albany, Chicago and St. Paul to Livingstone. He was deeply impressed by the rich land and bountiful crops he saw en route, and was moved to comment:

> "If I had been a farmer, I would have settled in Wisconsin. Now I do not wonder why so many Swedes have gone there."

His description of Bozeman in 1883 reads not unlike the Hollywood portrayal of a typical "frontier town" in a present-day "Western":

> "Bozeman is a little town with 5000 inhabitants, some brick houses, here regarded as the acme of elegance - lots of saloons and gambling houses - a population consisting four-fifths of men. There is a peculiar life in the streets, or rather the street. Half of the population is on horse-back, and every minute wild characters with broad-brimmed hats, long beards, and blue woolen shirts come dashing along the street on the spur."

At Bozeman he joins W. M. Davis, his immediate superior, then an instructor in geology at Harvard, and describes the camp life of his party as he and Davis do geological work along the canyon of the Yellowstone River.

In his fourth letter, written in camp on Boulder Creek (Big Belt Mountains) near Diamond City, Montana Territory, on July 29th, 1883, he describes further the impressive scenery in the Yellowstone region and paints a vivid picture of Diamond City, which sprang up when placer gold was discovered about 1865, and then declined to a sleepy town with

> "... a row of 30 small shacks inclined in different directions with a general store and a saloon."

He was impressed by the many bears that he saw - grey, brown and black - and by the ubiquitous rattlesnakes. Most of all, though, his comments clearly indicate that he is thoroughly enjoying the outdoor life in spite of the long hours and rather primitive camp life. He sums up this early phase of his professional life in words that will make every former field camper a little nostalgic:

> "Now the evening is approaching, and Hamner (Kamner?) is beginning to prepare the supper which seems intended to consist of oat porridge, corn, ham, bread, tea, and fruit preserves as dessert. Everything is boiled and fried over an open fire, and the bread is baked with baking-powder in a pot embedded in coal. Those days when we are in the neighbourhood of some ranch in the valleys are great feasts, since then there are milk, potatoes, and vegetables around. We are all doing well, however. The mountain air arouses a magnificent appetite, and the outdoor life is healthy. We have not made use of our tents yet and prefer to sleep under the stars, well wrapped in our canvas, rubber sheets, and blankets. Once - it was near Livingstone - there was sharp frost in the morning. The best part of the day is the evening around the immense camp fire, since there is plenty of dry wood, and we do not need to save it."

Gone for most of us, and from field work in most of the United States, are the kinds of experiences that Lindgren described for his parents almost a century ago! I have found Martinsson's translations of the four letters so interesting and so informative about the early 1880s that I have requested, and have been graciously given, the author's permission to include his article "Waldemar Lindgren goes west" as Appendix A farther on. I commend it to readers who would enjoy a sentimental journey through times and experiences now only a part of the history of our country.

HIS FIRST JOBS (1883-1884)

Now to return to Pumpelly and that letter of recommendation to him that Lindgren carried on his arrival in New York in early June 1883. At the time of Lindgren's arrival, Pumpelly was in general charge of the Northern Transcontinental Survey that had been established by the Northern Pacific Railway, which was then being built across Montana Territory. According to the second letter that Lindgren wrote to his parents, dated 5th July 1883, Pumpelly had already organized his staff and was then in Montana. It was then and there that a letter was sent to Pumpelly asking him about employment for Lindgren. Soon afterward came a telegram in reply saying that he, Lindgren, had been employed as an assistant geologist and assigned to work under W. M. Davis, from Harvard, himself Pumpelly's assistant.

Although Lindgren's biographers have almost invariably implied that he submitted his letter directly to Pumpelly, his own comments in his second letter to his parents suggest that several of his friends may have aided him in getting Pumpelly's favorable action. He wrote:

> "By a letter of recommendation from one of my friends, who is now in Mexico, to Mr. A. Thomas, President of the Northern Pacific Railroad, I obtained employment."

And a little farther on:

> "Partly Thomas, partly also the general manager in Drexel & Morgan's great banking firm, a Dane by the name of Christensen, were very kind and helped me."

Whatever the nature and extent of the aid he received, if any, Lindgren became associated with two great figures in American geology at the very beginning of his career in the United States. The first job did not last long, however, because the Northern Transcontinental Survey was discontinued early in 1884. Fortunately he was soon able to get a position as an assayer in the Gregory Smelting Works near Helena, Montana, and later in the same year a better position as a draftsman and designer of smelting furnaces for the Marcus Daly interests at Anaconda. (Graton 1933, p. xv.)

These latter short-term assignments were not to Lindgren's liking, however, for he did not want to remain in mining and metallurgy; his primary interest was in mining geology, the combination of his longtime interest in minerals and ores and the geology he had learned from his professors at Freiberg. So before the year was ended he applied to Dr. George F. Becker, then in charge of the Pacific Division of the U.S. Geological Survey, for employment with the Survey, and was immediately appointed a member of Becker's staff, partly because of his excellent training and partly because of Pumpelly's high opinion of his work in Montana. Thus it happened that in November 1884, three jobs and seventeen months after landing in New York, he was a member of the world's greatest geological organization, the United States Geological Survey, which he would serve with growing distinction for the next thirty years, reaching the highest scientific rank of Chief Geologist in 1911, the year before he would resign his Survey appointment and join the M.I.T. Faculty.

Many years later, in responding to his citation for the Geological Society of America's Penrose Medal, Lindgren rather amusingly refers to this early period of his career as follows (Geol. Soc. Am., Pr. 1933, p. 46, June 1934):

> "It seems to me, also, that this honor should have been conferred on one who had struggled with adversity, with illness, with non-appreciation, and come out triumphantly, rather than on one who has not passed through these tests of fire. I am quite sure that I have not had enough of hard knocks for my own good.
>
> "Can you conceive of anything more fortunate? A young chap comes to New York with a letter from a Philadelphia Quaker friend to a friend in New York who introduced the said young chap to a vice-president of the Northern Pacific, who wrote to a friend of his, named Pumpelly; and Pumpelly wrote back and said, 'Yes, I'll pay the boy twenty-five dollars a month if he will come out to Montana.' And the boy said: 'Sure thing,' and he was dumped right into the field party of William Morris Davis and into one of the most wonderful provinces of alkaline rocks in the world. How is that for a start?
>
> "Later, Pumpelly wrote a letter to G. F. Becker of the United States Geological Survey and still later Becker communicated with C. D. Walcott, and thus the chap, who was myself, landed in Washington, D.C. And to Walcott my everlasting gratitude! He trusted me enough to give me an important piece of independent work, and this policy was continued by my good friend, George Otis Smith.
>
> "So you see, I have proved my thesis. My friends just passed me along, and the rest was easy."

THE U.S. GEOLOGICAL SURVEY YEARS (1884-1912-1915)

In appointing to its staff the brilliant young Swede, less than two years away from his graduation from the world's leading mining school in

Freiberg, the U.S. Geological Survey acquired in Lindgren a scientist who was uniquely prepared to carry on the field investigations and laboratory analysis that would be required as successive mines were opened throughout the Western States in the closing decades of the Nineteenth Century. His was surely the opportunity of a lifetime - being there at the beginning and participating in the dramatic mining developments west of the Mississippi. He put it well in his response to the citation for the Penrose Medal mentioned in a preceding paragraph:

> "And another thing for which I am deeply grateful is that I have been allowed to witness and take part in that most wonderful of all epics - the winning of the West - an epic only comparable to the migration of my ancestors, the Goths - and that I have been intimately associated with that breed of self-reliant men - men who did not write to their native town for help, nor apply for a guide to the government, but who went right along with their ox teams and mule wagons, astride their horses, with the Winchester under the knee, fearing neither wilderness nor privations. That is a memory worth having."

Lindgren's first assignment as an assistant geologist was field work in western mining districts, the very kind of life he had looked forward to, once he decided to attend the Bergakademie and prepare for a career as a mining geologist. Now he would combine his boyhood interest in minerals and mines with later acquired interest in geology.

Soon after starting his Survey career, his life was made doubly enjoyable by his marriage to Miss Ottolina Allstrin of Gothenburg, Sweden, on 8 March 1886; she was his gracious homekeeper* - they had no children - and his constant and loyal companion until her death forty-three years later in 1929.

---

* The graciousness of the Lindgren home is clearly shown in a letter I received from the late L. C. Graton dated May 31, 1963:

> "As I intimated to you in my letter of April 27, my life was for many years definitely threaded through with the influence of Lindgren. Becoming his assistant in 1903 on the Survey's Cripple Creek re-study, I had a room immediately adjoining those in which Mr. and Mrs. Lindgren 'kept house.' To an extent that would have been embarrassing except that they made it seem so natural, they treated me almost as a son: countless suppers with them and still more evenings before their fireplace where he would at times sound me out and at others, in his modest but effective way, give me pointers from his own rich experience."

In his biography of Lindgren, Graton (1933, p. xxi) describes Mrs. Lindgren as

> "... a woman of rare charm and sympathetic understanding who, until her death in 1929, was inseparable companion and inspiring associate."

Year after year, for his first dodecade of work, 1884-1896, Lindgren spent almost all of his time investigating one western mining district after another, mainly in California.* In those days advancement on the Survey was slow, and Lindgren's kind of work kept him in relative obscurity, but he himself was frequently observing new features and new relationships in mineral deposits and their associated rocks, and, as Graton (1933, p. xv) points out, was systematically laying the foundation for later contributions of the greatest importance to the understanding of the nature and origin of mineral deposits. Even so, Lindgren's bibliography for this period lists more than 25 articles, including 15 on the "gold belt" of California; in the latter are included the Survey's Annual Report 17 on "The gold-quartz veins of Nevada City and Grass Valley districts, California," (reference 27--1896b in Lindgren's Bibliography farther on) and the descriptions of four folios: Placerville (#3, 1894), Sacramento (#5, 1894), Marysville (#17, 1895), and Smartsville (#18, 1895).

In 1895 Lindgren was advanced to the Survey rank of Geologist and soon thereafter was given more freedom and more independence in his choice of geological work. As an example, he was granted leave to serve as Associate Professor of Geology at Leland Stanford Junior University during SY 1897-98. Among the students at that time was Herbert C. Hoover, a mining engineering student, who had earlier served as Lindgren's first official assistant;** who himself would later on also become a famous engineer and in turn employ Lindgren as a consultant; and who would become the 31st President of the United States of America (1929-1933).

---

* The reader interested in the year by year work conducted by Lindgren while an employee of the Geological Survey, from 1884 to 1912, is referred to Appendix D, following the Bibliography, in which seasonal activities are reported in the Annual Reports of the Survey.

** After Herbert Hoover was elected the 31st President of the United States in 1929, Lindgren wrote an informal and interesting article, "A President in the making" for Tech Engineering News (Vol. 10/2: 53, 82, 1929) in which he recounted how he hired and trained young Hoover, a mining engineering senior at Stanford, and began a lifelong friendship and association with the young engineer who would later become our Chief Executive.

Years later, Hoover himself (as quoted in Butler's 1960 Memorial of Lindgren) recalled an amusing incident with a mule when he was Lindgren's young assistant (Geol. Soc. Am., Pr. 1949: 187, 1950):

> "I started professional life as a summer assistant to Waldemar Lindgren for the United States Geological Survey in 1893 and continued for three summers. He did me the then great honor of putting my name on the Pyramid Peak Quadrangle Geologic sheet.
>
> "Years afterwards I had the pleasure of employing him as a geologist to study the lava-buried stream systems in Victoria, Australia. We remained affectionate friends as long as he lived. (footnote continued on facing page)

As Geologist, with his new freedom and independence, Lindgren now turned to his primary interest, the origin of mineral deposits; and while seeing through publication a dozen or more descriptive articles left over from his earlier field work, he began to spend more time at the petrographic microscope and at the analysis and synthesis of the many observations he had made in the preceding decade.

In the brief period from 1895 to 1900, his 24 publications included 5 folio descriptions in addition to the 4 cited in the preceding paragraph: Nevada City (#29, 1896), Pyramid Peak (#31, 1896), Truckee (#39, 1897), Boise (#45, 1898), and Colfax (#66, 1900), and several mineralogical and petrographical papers. Of much greater importance and significance, however, were six articles on fissure veins: references 24--1895b, 27--1896b, 29--1896d, 33--1897c, 37--1898b, and 42--1900 (see the Bibliography farther on), and two lengthy Survey Annual Reports (27--1896b, and 36--1898a).

(footnote continued from facing page)

> "An incident will indicate his character. Each summer we camped with pack trains, cooks and teamsters. I was made the 'disbursing officer' - not an honor. One morning we discovered that one of our mules was dead. An autopsy by the teamsters and myself showed that the cork of a hind shoe was caught in the halter rope by which he was tied to a tree and as his neck was broken, we concluded that he was scratching his head with his hind foot to have caught the shoe in the rope.
>
> "Under the printed regulations, I was required to make an affidavit of the cause of death so as to account for the loss. This was signed by myself and the teamsters the first time we reached a Notary Public. The statement was disallowed by Washington auditors on the ground that mules did not scratch their heads with their hind feet - and my salary was docked $60.00 (a month's pay). Lindgren promptly gave me the $60.00 and assured me he would collect it when he returned to Washington. He never did get the money and always refused its repayment from me with the remark that it was his most valuable proof of the dumbness of bureaucracy, and he would hate to have his story destroyed."

Ten years later, on 11 February 1960, when Joseph L. Gillson (XII S.M. 1921; Sc.D. 1923) and I visited Mr. Hoover in his Waldorf-Astoria Towers apartment in New York City, to request a supporting letter that we could use in soliciting contributions to the Lindgren Memorial Fund we were establishing - he graciously sent us such a letter a little later - he told us the mule story with a twinkle in his eye, then added approximately as follows:

> "I often wondered if our mule, or any mule for that matter, actually did scratch his head with a hind foot. Then one day while in a curio shop I found a small statuette of a mule doing exactly as our teamsters and I had concluded - scratching his head with a hind foot. I felt fully vindicated thereafter, even if the Washington auditors had ruled otherwise those many years ago!"

The year 1900 is particularly significant in Lindgren's professional career because it was then that he first publicly discussed in detail his carefully formulated conclusions on the origin of fissure veins in his classic paper "Metasomatic processes in fissure veins" (reference 46--1901a). In the sixteen years preceding 1900, Lindgren had regularly published the descriptive details of his numerous and varied field studies, and in the last five years he had published several topical discussions in the current technical journals. Then in 1900 he synthesized all his pertinent observations and ideas in the now classic paper mentioned in the preceding sentence. The importance of this paper is clearly set forth in Graton's 1933 evaluation of Lindgren's scientific career as follows (see item 1933, Graton, L. C., p. xvi in Appendix B):

> "The carefully arranged and detailed results of his successive field studies were presented as official reports. To these were added even more significant topical discussions published in the current technical literature.
>
> "Gradually the soundness, penetration, and inventiveness of these writings began to receive recognition. Perhaps the general realization of Lindgren's outstanding ability came first as a result of his paper entitled 'Metasomatic processes in fissure veins.' This and other important papers presented at the Washington meeting of the American Institute of Mining Engineers in February, 1900, and published in the famous Volume XXX of the TRANSACTIONS, established a veritable milestone in the study of ore deposits. Particularly to be mentioned in this connection are: 'Some principles controlling the deposition of ores' by Prof. C. R. Van Hise, wherein were effectively arrayed the views of that forceful protagonist of meteoric waters as the dominant agent of ore formation and whereby that hypothesis may be said to have reached the pinnacle of its ascendancy; and 'The secondary enrichment of ore deposits,' in which S. F. Emmons, the scholarly dean of American students of ore deposits, systematically enunciated and clarified the fundamentals of that process of superficial reworking which is now recognized as of such great economic importance. It is not without significance that these two papers by older investigators embodied conceptions and proposals based on field studies of the more generalized and qualitative kind, whereas the contribution by the younger man, Lindgren, though reflecting no less thorough a grasp of the broader aspects of field occurrence, rested primarily on the detailed evidence yielded by that instrument of precision, the petrographic microscope. Significant, too, is the fact that Emmons and Van Hise, in these papers which stand among their major contributions to the science, dealt with the effect of descending waters. Lindgren, on the other hand, in this, his first contribution of really general import, though with characteristic caution and reservation admitting the possible rôle of 'circulating surface waters alone,' emphasized 'for many ... perhaps for the majority of fissure veins' close genetic connection with bodies of intrusive rocks, in accordance with the theory that had been originally proposed by de Beaumont, but had in

later years been relatively neglected. In the content and implications of this paper may be discovered advance glimpses of many of those ideas which, in Lindgren's later writings, have become established as important principles of the philosophy of ore deposition."

The impact of this paper on the mining and geology professions was profound; probably as great as that of his first edition of <u>Mineral Deposits</u> (1913), a comprehensive reference work in which he described and classified mineral deposits of all kinds from all parts of the world and discussed their origin. More is said about this now classic book farther on.

Henceforth, for the next dodecade, 1900-1912, Lindgren's professional life noticeably quickened as he added to his Survey's duties the responsibilities of leadership he was called upon to assume as a result of the dominant role he was playing in economic geology. From 1900 onward, Lindgren's rise in the Geological Survey was rapid, culminating in his appointment as Chief Geologist* in 1911 (see accompanying illustrations), and his publication record for the dodecade is most impressive: an amazing 90 publications including his classic paper on metasomatism, a first and second

---

* The impressive qualifications of Lindgren for Chief Geologist, the highest rank in the roster of the U.S. Geological Survey, and one of the highest recognitions that can be accorded a geologist of the U.S.A., were set forth by the Director of the Survey, George Otis Smith, in the following letter of recommendation:

November 2, 1911.

The Honorable, [Walter L. Fisher]
   The Secretary of the Interior.

Sir:

   I have the honor to recommend the appointment of Waldemar Lindgren, Geologist at $5000, as Chief Geologist at same.

   Mr. Lindgren is 51 years of age and received his education in Sweden and at the Freiberg School of Mines where he received the degree of mining engineer in 1883. On the following year he joined the United States Geological Survey as Assistant Geologist, since which time he has been continuously connected with the organization. ... During his connection with the Geological Survey he has also rendered services as the Associate Professor on Mining and Metallurgy at Leland Stanford University, and as lecturer on Economic Geology at the Massachusetts Institute of Technology for several years. Since 1907 he has been in charge of the investigations of the Survey in metalliferous deposits and during the same period has reorganized the metal statistics under the Division of Mineral Resources.

   Mr. Lindgren's success as a working geologist is best indicated by the accompanying list of some fifty reports included in the publications of this Survey. Prominent among these are his reports on the Gold Quartz District of California, Gold and Silver Mines in Idaho, copper deposits in Arizona, and especially his report resulting from the resurvey of the Cripple Creek District, together with Dr. Ransome, and his last publication, a monographic treatment of the tertiary gravels of the Sierra Nevada.   (footnote continued on next page)

edition of <u>Mineral Deposits</u>, lengthy reports on mining districts and mineral deposits in the Survey's several publications, as follows,

Annual Report 20 (reference 42--1900a in Bibliography)
              22 (reference 45--1901 in Bibliography)

Bulletin 202 (reference 49--1902 in Bibliography)
        254 (reference 67--1904d in Bibliography)

Professional 27 (reference 65--1904b in Bibliography)
Paper   43 (reference 71--1905a in Bibliography)
             54 (reference 90--1906d in Bibliography)
             68 (reference 116--1910 in Bibliography)
             73 (reference 121--1911 in Bibliography)

Contributions to 1907 (reference 109--1908g in Bibliography)
Economic Geology 1908 (reference 115--1909e in Bibliography)
                 1909 (reference 120--1910d in Bibliography)
                 1910 (reference 127--1911f in Bibliography)
                 1911 (reference 132--1913a in Bibliography)

Folio 103--Nampa   (reference 63--1904 in Bibliography)
      104--Silver City (reference 64--1904a in Bibliography)
      129--Clifton   (reference 70--1905 in Bibliography)

(footnote continued from previous page)
    In my opinion, in which I believe the majority of his associates would concur, Mr. Lindgren is easily the leading geologist of the Survey. As a stratigraphic geologist he has few superiors; as a mineralogist and petrographer he is probably surpassed by no member of the Survey, and as a mining geologist he has no equal in America, either within or without the Survey force. Added to this he is a trained mining engineer. Mr. Lindgren also has a world-wide reputation as an authority on ore deposits, and he is also a member of the National Academy of Sciences as well as of other scientific bodies. In addition to his Survey publications he has contributed not less than three score articles to the various scientific and technical journals.

    As an administrative officer Mr. Lindgren has shown exceptional ability in handling such problems as the development of the metal statistics, the supervision of the mineral examination of claims within the National Forests, and during the past two years the classification of mineral land within the Northern Pacific land grant. Mr. Lindgren has also acted for the Survey in connection with securing and presenting evidence in mining fraud cases, at the request of the Department of Justice and the Post Office Department. It is also worthy of note that Mr. Lindgren represented the Survey in the first case of cooperation between the Survey and the General Land Office in land classification in 1895, in which examination I served as his assistant.

    Mr. Lindgren's standing as an original investigator is such that I would regard it as a distinct loss to the public service for him to be assigned to administrative duties that would involve any reduction in his output as a geologist. To meet this contingency it is my purpose to so reorganize the Geologic Branch as to relieve the Chief Geologist of all administrative functions except the most important - those relating to the determination of scientific policy, the formulation of field plans, and general supervision of geologic work.

    The accompanying sheets give lists of Mr. Lindgren's contributions to Survey publications and also his personal record.

                                                  Very respectfully,
                                                    [George Otis Smith]
                                                        Director

Lindgren's Certificate of Appointment as Chief Geologist
of the United States Geological Survey

and many topical articles in current technical journals, particularly the Transactions of the A.I.M.E. and Economic Geology. (See Bibliography farther on.)

When the Survey's Division of Mineral Resources was reorganized in 1905, Lindgren was chosen to head the Section devoted to precious and semiprecious metals (gold, silver, etc.). Two years later, on 1 March 1907, Lindgren succeeded S. F. Emmons as Chief of the Division of Metalliferous Geology. Finally, in 1911 came appointment as Chief Geologist of the Survey, with still greater responsibilities, especially time-consuming administrative duties which he did not relish, and with less time and freedom for the kind of scientific work he had done in previous years. It was not surprising, therefore, that despite the prestige of his appointment he resigned from it in 1912,* to accept a professorship at M.I.T.

---

* Although Lindgren resigned his appointment as Chief Geologist in 1912, he continued his connection with the Survey for the next three years in order to complete the studies that were in progress when he left Washington for Cambridge. Finally, when he resigned from the Survey in 1915, and thus ended a 31-year period of service, his contact with it did not cease for he remained an interested observer and adviser, and on occasion even a temporary consultant, until his death in 1939. During his long period of service he had brought a new prominence to the Survey by virtue of his outstanding work as an economic geologist.

WALDEMAR LINDGREN

As a schoolboy in Kalmar, about 1877.
(Photo by Rosalie Sjoman, Kalmar)

As a Bergakademie student, about 1883.
(Schulz & Suck, Karlsruhe)

About 1911 when he was appointed Chief Geologist of the U.S. Geological Survey.
(U.S.G.S. Photo)

In the field, Santa Rita, New Mexico, 6 September 1933
(Photographer unknown)

It is likely that he was strongly influenced in his decision to accept by the enthusiasm with which his series of special lectures on ore deposits delivered during the three previous academic years, 1908-09 to 1910-11, had been received by the Institute's students and faculty members in geology and mining.

THE M.I.T. YEARS: 1908-1912-1933-1939

AS LECTURER IN ECONOMIC GEOLOGY (1908-1912)

Lindgren's association with M.I.T. began in the fall of 1908 when he was appointed Lecturer in Economic Geology for SY 1908-09. He secured a leave of absence from the U.S. Geological Survey in Washington, D.C., where he was Geologist in charge of the sections of Mining Geology and of Metal Statistics, so that he could reside in Boston, where the Institute was then located, and devote full time to his lectures. During his five-week stay in the late autumn he gave a series of lectures and conferences on ore deposits and economic geology that were open not only to students at M.I.T. but also to interested students and faculty members from Harvard and other colleges in the metropolitan Boston area. (Pres. Rept. 1907-1908, p. 109.)

This arrangement with Lindgren was so enthusiastically received that it was continued during SYs 1909-10, 1910-11, and 1911-12, whereupon it was terminated with his appointment as William Barton Rogers Professor of Economic Geology in 1912. It is of interest to note that the expenses of Lindgren's lectureship for several years were met by contributions from Emma Rogers (wife of M.I.T.'s founder, geologist William Barton Rogers), a strong supporter of the Department of Geology as long as she lived. (Pres. Rept. 1909-1910, p. 112.)

AS WILLIAM BARTON ROGERS PROFESSOR OF ECONOMIC GEOLOGY AND HEAD OF COURSE XII: DEPARTMENT OF GEOLOGY (1912-1920; 1927-1933)

When in 1912 Prof. R. A. Daly left M.I.T. to accept the prestigious Sturgis-Hooper Professorship at Harvard, and Prof. T. A. Jaggar, Jr. resigned as Head of Course XII to become Head of the Technology Volcano Observatory in Hawaii, the vacancies left by their resignations had to be filled with distinguished geologists to maintain the Department's strength.

President Maclaurin, as part of his address to the students at the June 1912 Commencement, emphasized the fundamental importance of geology to mining, agriculture, engineering and business, and discussed how Course XII would be changed in the reorganization of the Institute. He announced the establishment of a new chair, the <u>William Barton Rogers Professorship of Economic Geology</u>, which would be occupied by Waldemar Lindgren, M.E., recently appointed Chief Geologist of the U.S. Geological Survey and a renowned authority on mineral deposits. Maclaurin pointed out that William Barton Rogers, the Institute's founder, had been an outstanding economic geologist, hence it was altogether fitting that the chair established in his name be devoted to that aspect of geology. (<u>Tech. Rev.</u> 14/6: 365-367, June 1912; Pres. Rept. 1911-1912, p. 13, 25.)

When Lindgren first came to M.I.T. in the fall of 1908, to deliver a series of lectures on ore deposits and economic geology, he was already a renowned economic geologist and immediately attracted the best students to his lectures. His reappointment as Lecturer for the next three school years brought him into closer contact with both faculty members and students at M.I.T. as well as at Harvard and at other schools in the Boston area, with the result that his permanent appointment as William Barton Rogers Professor of Economic Geology and Head of Course XII, then called the Department of Geology, was enthusiastically acclaimed throughout the Boston area.

Lindgren resigned his position as Chief Geologist of the U.S. Geological Survey in 1912 (however retaining his connection with the Survey until 1915), and that fall took up full-time duties as professor and department head at the Institute. For the next thirty-one years, until he became Professor Emeritus in 1933, he gave regularly scheduled lectures and laboratory work on mineral deposits, conducted graduate seminars, carried on his own research and writing, supervised the thesis work of a host of graduate students, and at the same time acted as consultant to numerous industrial clients and government bureaus. Already renowned as one of the foremost economic geologists of the world, after he became a full-time professor of geology at M.I.T. in 1912, his fame continued to grow through the years, and by the time he reached retirement he was universally regarded as the <u>Nestor of Economic Geology</u> (Åhman 1953).

Lindgren the Teacher

Lindgren's students considered him a stimulating teacher because he constantly urged them to make their own observations, acquire their own facts first-hand, and develop self-reliance.* They held him in highest

---

* The following tribute to Lindgren appears on page 1 of Palmer C. Putnam's 1934 Course XII (Geology) M.S. thesis entitled "Reconnaissance among some volcanoes of Central America":

> "It is difficult for a pupil to express his admiration and his affection for his Master. Those who have
> (footnote continued on facing page)

respect for his vast fund of knowledge and for the modesty and earnestness with which he presented his own original ideas. His emphasis on the great importance of an understanding of the principles of chemistry and physics impressed them repeatedly because he demonstrated how valuable they had been to him in his own research. Finally, they admired him because of his sincere interest in them and his concern that they develop their own talents to the maximum. Lindgren himself supervised the thesis work of some 60 different students during his active teaching period from 1912 to 1933. Altogether he signed 64 theses - 25 for the S.B., 22 for the S.M., and 17 for the doctorate - impressive evidence of the heavy supervisory load he carried in addition to his other departmental duties. Little wonder, then, that so many of his students became outstanding economic geologists as teachers or as practitioners; he set the example for them.

Lindgren as Invited Lecturer outside M.I.T.

Once Lindgren had severed his formal connection with the U.S. Geological Survey, which was in 1915 - he remained on the Survey staff from 1913 to 1915 in order to complete work started before he resigned as Chief Geologist to join M.I.T. - and had entered the academic world, his renown soon brought him numerous invitations to give one or a series of special lectures outside metropolitan Boston.

(footnote continued from facing page)
        not also felt the guiding hand will not understand, and the Master will be sure to minimize. Yet I can not allow this occasion to pass without making acknowledgement of the deep debt in which I feel myself placed toward Dr. Waldemar Lindgren. More patience, more understanding, more devotion toward the men under his care can hardly be imagined. Perhaps my deepest sense of gratitude lies in the fact, that although Dr. Lindgren is always right, he never makes one uncomfortable by exposing the enormity of his stupidity on the spot. He arranges it so that full realization comes in the introspective hours of night, and not in the embarrassment of his presence. I can do little else than thank him heartily for his endless patience, and to register a sincere hope that in my applications of his teachings nothing will transpire unworthy of his tradition - a high hope admittedly; but at least a hope towards which one may be permitted to strive."

* Typical of Lindgren in this last respect was his remark that one of the most valuable lessons he himself learned as a student at Freiberg came from a reproof by one of his teachers, Weisbach, that the student should rely on himself and not on the professor. (Newhouse 1939, p. 584.)

(Upper)
Prof. Lindgren studies a thin section while Venezuelan graduate student, Victor M. Lopez (XII S.M. 1936, Ph.D. 1937) awaits the results of Lindgren's observation.

Note Lindgren's "pince-nez" eyeglasses, and the several items on the laboratory table.

(Lower)
Having looked at the thin section, Lindgren ponders what he will say to student Lopez.

Note the items in Lindgren's laboratory: cardboard file box; leather brief case; reference books; petrographic microscope; microscope light; thin sections and boxes; and slide rule.

These two photographs were probably taken in 1937, shortly before Lindgren retired permanently. They are preserved in the MIT Historical Collections.

In the spring of 1916 he delivered ten special lectures on ore deposits at Columbia, at the invitation of that Institution's Department of Geology (Pres. Rept. 1915-1916, p. 100, 1916); in March 1927 he gave four lectures at McGill, and he also gave several lectures at Queens University, also in Canada. In early 1923 he lectured on "The geology of Bolivia" in Hopkins Hall at Yale (Tech. Rev. 25: 245, Mar. 1923), and several times lectured to the students at Harvard.

Lindgren the Administrator

As Department Head

During his first eight years (1912-1920) as a full-time member of the M.I.T. faculty, Lindgren was head of Course XII, then called Geology and Geological Engineering. In his departmental comments included in the President's Report for SY 1919-20, he observed that:

> "Considering the number of our students it must naturally be borne in mind that the students of Course III, Option 3 [i.e. Department of Mining and Metallurgy, Geology Option], are very closely connected with the Geological Department and really may be considered to belong to it as much as they do to the Mining Department."*

This sentiment led to the consolidation of Course III and Course XII at the beginning of SY 1920-21, with the consolidated program numbered III and called the "Department of Mining, Metallurgy, and Geology," with Lindgren as Head. In reporting the next year on how the consolidation worked out, Lindgren commented as follows (Pres. Rept. 1921-22, p. 51):

> "As to administration this plan worked very well, but there is doubt in my mind whether it should be continued indefinitely."

As Lindgren thought possible, the consolidated plan did not prove successful, with the result that he included in his comments in the President's Report for SY 1926-27 (p. 18) the following:

> "The most important change during the year was the reestablishment of Mining and Metallurgy as a separate Department." [But with its same number III.]

With this action, Course XII, Geology, was also re-established, with Lindgren continuing as Head and at the same time relinquishing the chairman-

---

* In the several more recent Registers of Course XII Alumni, we have followed Lindgren's suggestion and have included the names of the more than 50 geologically inclined students who took their S.B. degrees in Option 3 of Course III, and wrote theses on geological subjects.

ship of Course III to Prof. W. Spencer Hutchinson.* From 1 January 1927, when the separation of Courses III and XII took place, until 30 June 1933, when he became Professor Emeritus, Lindgren continued to serve as Head of the Department of Geology. Subsequently, as the result of a request from the geology faculty, Lindgren was appointed Honorary Lecturer, a rank he held until 1936, after which he continued as Professor Emeritus until his death in 1939.

During his 21 years as a department head, Lindgren's renown and academic leadership attracted a continuous stream of outstanding graduate students from both the United States and Canada, as well as from abroad, many of whom went forth later to achieve distinguished careers in geology: e.g. Victor Dolmage, George Hanson, Henry C. Gunning, and others from Canada (See p. 24-25 of Course XII's REGISTER 1865-1970); Guillermo Zuloaga and Victor Lopez from Venezuela; J. L. Gillson, W. H. Newhouse, M. J. Buerger, and others, from the United States.

---

* Some years ago I asked one of my M.I.T. colleagues, the late Carle Hayward, to give me his remembrance of Lindgren, because he had been a professor in the Department of Mining and Metallurgy (Course III) when Lindgren served as chairman of both Course III and Course XII, from 1920 through 1926. Hayward wrote me as follows in a letter dated 27 November 1957:

> "Professor Lindgren
>
> "I can speak of Lindgren only as a colleague. At first he seemed cold and difficult to approach but as acquaintance ripened into friendship I learned that here was not only one of the world's greatest geologists but one of the choice characters on our staff.
>
> "I didn't come to know him well until after the retirement of Professor Hofman, the department of Mining and Metallurgy was combined with the Department of Geology for administrative purposes. Lindgren told me at that time that he was instructed to bring two men into Course III from practice and within a short time G. B. Waterhouse an employee of Bethlehem Steel Company was appointed Professor of Metallurgy and W. S. Hutchinson a consulting mining engineer, Professor of Mining.
>
> "There were a few 'growing pains' and 'headaches' connected with this re-organization and it needed a man of reputation and unimpeachable integrity to iron them out. His short term as head of the combined departments worked out very well during this transition period.
>
> "Lindgren was in great demand as a consultant particularly as expert in law suits. His work in connection with the famous Butte suits was epochal. The student chapter of A.I.M.E. was quite active in Lindgren's day and members of the staff usually attended the more important meetings. Lindgren was at least once a year one of the speakers and his wide practical experience which he called on in his talks made him exceptionally interesting. I don't recall a single instance when I did not listen to him with pleasure and profit."

As Chairman of the Division of Geology and Geography of the National Research Council (1927-1928)

In 1927 Lindgren took a year's leave of absence from M.I.T. to serve as Chairman of the Division of Geology and Geography in the National Research Council in Washington, D.C. Freed of his academic responsibilities for the time being, he immediately initiated two projects that would have great future impact on economic geology - 1), the founding of the Annotated Bibliography of Economic Geology; and 2), the review of the whole problem of ore genesis.

Founder and First Editor of Annotated Bibliography of Economic Geology

Soon after his arrival in Washington, Lindgren had his Division of Geology and Geography appoint a special committee

"to consider the advisability of compiling an annotated bibliography of economic geology."*

As Chairman of the Committee he took a leading role in raising the funds necessary to cover the expenses of abstracting and editing, and then of printing the proposed Bibliography.**

With the requisite funds in hand or in sight to publish the proposed Bibliography for five or six years, The National Research Council assumed responsibility for supervising preparation of the Bibliography and disbursing the funds collected.

A new Committee, again with Lindgren as Chairman, was established for actual preparation of the first volume. The responsibility for printing and distributing the Bibliography was assumed by the Economic Geology Publishing Company, which subsequently published all thirty-seven volumes during the years 1928 to 1966, the last year being the one in which the Bibliography ceased publication.

---

* An account of how the bibliography came into existence, and of its subsequent history, is included in a short explanatory note on p. ii of Volume 37, Number 2 (the last one published) of Annotated Bibliography of Economic Geology for July-December, 1964. Urbana, Ill.: The Economic Geology Publishing Co., (1966). See also Lindgren's article "Annotated Bibliography of Economic Geology," p. iii-viii of the first volume of the publication of the same title, 1928.

** Lindgren made a rather hard-hitting appeal to economic geologists and mining companies by letter (reproduced on an accompanying page), and in an editorial titled "Bibliographies, Annotated Bibliographies and Geological Abstracts" in Econ. Geol. 23/4: 564-568, (1928).

NEW YORK OFFICE
THE ENGINEERING FOUNDATION
29 WEST THIRTY-NINTH STREET

CABLE ADDRESS
NARECO
WASHINGTON, D. C.

NATIONAL RESEARCH COUNCIL

Established in 1916 by the National Academy of Sciences
under its Congressional Charter and organized with the cooperation of the
National Scientific and Technical Societies of the United States

B & 21ST STREETS, WASHINGTON, D. C.

March 26, 1928

TO ALL THOSE INTERESTED IN ECONOMIC GEOLOGY

      This is an earnest appeal for help towards establishing an Annotated Bibliography of the International Literature of Economic Geology.

      No country is sufficient unto itself and least of all in science. All geologists probably realize this; and those who employ geologists should likewise realize that the efficiency of their staff is largely dependent upon familiarity with the progress of science throughout the world. We can not afford isolation or imperfect knowledge. The enterprise represents a new step forward in Economic Geology and it is fitting that this continent with its almost unlimited resources should lead in it.

      A plan worked out by the Division of Geology and Geography of the National Research Council for such an Annotated International Bibliography has been approved by the Council itself, the U. S. Geological Survey, the U. S. Bureau of Mines, the American Institute of Mining and Metallurgical Engineers, and the Society of Economic Geologists. The plan is before several other technical societies and their approval is confidently expected at their next meetings.

      The plan includes a subscription fund of from $18,000 to $24,000 which will provide means for the editorial work for from five to seven years. The National Research Council has agreed to serve as treasurer and custodian of these funds to be expended in accordance with the usual customs of the Council. Tentative arrangements have been made with an experienced bibliographer to take charge of the work. The Economic Geology Publishing Company has agreed to publish the bibliography, which is to be issued in two volumes per year, aggregating about 400 pages, and to sell it for an equitable price which will probably be about $2.50 per volume, which should pay for paper, printing and distribution.

      Several years ago the Division of Geology and Geography of the Research Council proposed a plan to collect a permanent fund to establish an Abstract Journal of Geology. This project had to be abandoned because it proved impracticable to obtain the necessary funds estimated at $200,000. The pledges received for this enterprise from many individuals and companies are herewith released and cancelled by the National Research Council.

-2-

It will be observed that the present plan is restricted to economic geology and calls for funds to carry the work for a limited time only, but it is felt that the work will prove so useful, indeed, so indispensable, that little difficulty will be experienced in providing for its indefinite continuation. The subjects to be covered include articles on all metallic and non-metallic deposits (including petroleum and gas), hydrology, engineering geology and soils (in so far as related to geology). Publications in other branches of geology will be included if they have any bearing on economic geology.

There will be no failure if all of us who are interested work together earnestly and energetically. Your assistance is solicited. It will benefit yourself and all those connected with the mineral industry. Please use the attached blank in pledging contributions either in one sum or spread over three equal annual payments.

The payments will not be called for until the amount deemed necessary has been subscribed.

Very sincerely yours,

W. Lindgren

W. Lindgren
E. De Golyer   W. C. Mendenhall
G. F. Loughlin   R. A. F. Penrose, Jr.

..................1928

I we subscribe the sum of..Dollars

toward a fund for the establishment and maintenance of an Annotated

Bibliography of Economic Geology, and promise to pay the same to the

National Research Council, Washington, D. C., as Trustee, either in

one sum or
three equal annual payments, when the sum of $18,000 has been pledged.

$........................   (Address)   ....

Inasmuch as Lindgren is known to have initiated the idea of the Bibliography, no doubt in part because of his good knowledge of at least five languages other than Swedish, German and English, and also because he played a leading role in getting it started, he has been recognized as the founder of the publication. Surely of equal importance is the fact that from the beginning, he headed the Committee in charge of the Annotated Bibliography, continuing in this editorial capacity for 18 years until 1938, when he had to retire because of ill health (he died late in the following year, on 3 November 1939). During his decade as Chairman of the Committee, followed by his year of retirement, he not only prepared the almost unbelievable number of 2,545 abstracts by himself, many of them of articles and books in one of the six other languages besides Swedish and English that he read; he also saw to it that his M.I.T. professorial colleagues and graduate students lent a hand now and then, thereby demonstrating to them that the ability to read the scientific literature of geology in a foreign language had a practical value other than satisfying the foreign-language requirement for a doctoral degree! Lindgren's foresight and wisdom in initiating the Annotated Bibliography is attested by the fact that 27 more volumes appeared after his death before it ceased publication in 1966.

National Research Council Committee
on Processes of Ore Deposition

Lindgren early developed a deep interest in the origin of ore deposits, and in the first number of Economic Geology (vol. 1/1: 34-46, Oct.-Nov. 1905) he raised some fundamental questions in an article titled "Ore Deposition and Deep Mining," as follows (p. 34):

> "As men began to delve deep into the earth's crust instead of confining themselves to the shallow diggings and mines of earlier times many deposits were found to diminish in size or value, or both, and the question of maintaining the world's supply of useful metals came to be discussed with interest and sometimes with anxiety.
>
> "The newly established science of ore deposits was called upon to elucidate the mode of deposition of ores and the probabilities of finding workable ores in depths below the ore bodies of the surface. The new science proceeded with faltering steps tentatively groping its way and leaning in turn on many hypotheses and theories, and if even now the answers which it gives are often hesitating and filled with doubt, we may perhaps be justified in saying that we do now, better than before, understand the mode of deposition of ores. How lacking our knowledge is, nevertheless, none appreciate more than those whose vocation it is to attempt the solution of these problems.

"A forecast of the probable distance of continuity of any deposit beyond the limit of empirical rules obtained by the observation of many similar ore bodies must of necessity be based on some hypothesis concerning its origin; this hypothesis we are able, in most cases, to advance to the standing of a theory and in fewer cases to the position of assured facts.

"Vital and preliminary questions are: Has the ore been deposited together with the surrounding rock? Has it been introduced later from outside sources? Or has it been concentrated to a workable deposit from minute quantities originally contained in the rock?"

During the next twenty years, as he investigated ore deposits around the world and developed the classification first introduced in his famed textbook Mineral Deposits in 1913, he realized that there were many unanswered questions about how ore deposits are formed. When, in April 1926, at the annual meeting of N.R.C.'s Division of Geology and Geography, it was recommended that a committee be organized to consider the processes of ore deposition, Lindgren was the obvious person to head such a committee.

The Division's recommendation was acted upon in due course, and in the spring of 1927 the Committee on Processes of Ore Deposition was organized, with Lindgren as Chairman. Other members of the Committee were L. C. Graton, D. F. Hewett, Adolph Knopf and F. L. Ransome. In the fall of 1927, when Lindgren became Chairman of the Division of Geology and Geography, the Committee was increased to ten members by the addition of E. T. Allen, E. S. Bastin, W. H. Emmons, G. F. Loughlin, and R. C. Wells.

Lindgren, considering that his first duty as committee chairman was to review the current state of research on the subject of ore deposition, made a statistical study of this aspect and early in 1928 published an article titled "Research in Processes of Ore Deposition," as Tech. Pub. 78 (14 p.) issued with Mining and Metallurgy, Feb. 1928. Meanwhile, members of the Committee were meeting and corresponding with one another, and at the meeting of the Division on 28 April 1928, Lindgren presented the "Report of the Committee on Processes of Ore Deposition," subsequently published in the September 1928 number of Economic Geology (Vol. 23/6: 591-611, 1928). The Report cited a long list of problems that needed investigation; called for closer cooperation between geologists, and between them and physicists and chemists; and urged a more enlightened attitude and more financial support for research on the part of official bodies of investigation and by the metal producing industries.

The preceding two activities of Lindgren while at the National Research Council are in actuality only part of his total efforts; the latter are so impressive that I include on the next page the "Memorial" written by E. S. Bastin.

Excerpt from Minutes of the Annual Meeting of the Division of Geology and Geography, National Research Council, held April 27, 1940.

* MEMORIAL TO PROFESSOR WALDEMAR LINDGREN

Nearly six months have now elapsed since Waldemar Lindgren's passing on November 3, 1939. In that interval many of the devoted friends who knew him best have written at length concerning his achievements and paid glowing tribute to his qualities as a man. Some of these memorials have already been published and others will appear later. It would seem inappropriate here to attempt to do again what these friends and associates have done so supremely well. Yet this occasion cannot be allowed to pass without some expression of our abiding affection for Waldemar Lindgren and some grateful acknowledgment of the work that he accomplished in his connection through many years with the National Research Council.

Because of his membership in the National Academy of Sciences, Lindgren was doubtless familiar, at least in a general way, with the National Research Council and its objectives from its very inception in 1916. His active participation in the work of the National Research Council seems to have begun in 1925 as a member of the Committee on the Measurement of Geologic Time and he continued to be a member of that committee until his death. He was a member of the Committee on State Geological Surveys from its organization in 1928 to its reorganization in 1932.

His year as Chairman of the Division of Geology and Geography in 1927-28 was a most active one. It was marked by notable activity on the part of the Committee on the Processes of Ore Deposition to whose chairmanship Dr. Lindgren had been appointed by Dr. David White. Dr. Lindgren continued as chairman until 1931 and as a member until his death. Many of the projects undertaken by this committee were begun at his suggestion and actively shared in by him.

During his term of office the Committee on the Investigation of Clay Minerals was organized and a Committee on Potash-Soda Feldspars, under the chairmanship of others but with his interest and support. Both performed most useful work.

The most notable achievement, however, of his chairmanship and the one to which he devoted the greatest amount of personal effort and enthusiasm was the initiation of the Annotated Bibliography of Economic Geology. Largely through his personal efforts he obtained during the year of his chairmanship pledges of $13,000 for the project and this amount was later increased to more than $21,000. Over 99% of these pledges were paid in full and they sufficed to carry on the project for eight years. He arranged for the publication of the bibliography by the Economic Geology Publishing Company and continued to father the project until it was taken over by the Society of Economic Geologists in 1936. His labors were not confined to the financing of this project, but included the major responsibility for enlisting other geologists in the preparation of abstracts as well as an amazing amount of personal reviewing of publications.

These activities which touch us as members of the National Research Council particularly closely are of course not the most important

---

* Written by Professor Edson S. Bastin and, in his absence, presented by Dr. D. F. Hewett. Adopted by the Division by a rising vote.

achievements in a remarkably productive life that included nearly thirty years of association with the U. S. Geological Survey here in Washington - years during which he was acquiring and bringing into orderly arrangement in his mind those concepts of ore deposition that he was to pass on to generations of younger geologists and mining engineers during the next twenty-seven years through his epochal book on Mineral Deposits and through his personal teachings at the Massachusetts Institute of Technology. His association with the National Research Council clearly revealed, however, three outstanding features of his character. First, his devotion to his special field of ore deposits and his ability to utilize every new opportunity to further its progress. Second, the breadth of his understanding of and interest in other branches of the earth sciences, some of them far afield from his own speciality. And, third, his unselfish willingness to contribute no small fraction of his energies to cooperative enterprises for the benefit of his fellows. In short, he knew supremely well how to strike a proper balance between distinguished personal gallantry in scientific accomplishment and shoulder to shoulder cooperative effort. And like a crowning glory to his greatness was the unassuming kindliness that endeared him to all who were privileged to know him well.

The Division of Geology and Geography of the National Research Council does honor to his genius as a scientist and to his nobility as a man.

Lindgren as Geological Consultant

No sooner had Lindgren resigned his Survey appointment as Chief Geologist than he was called upon by various mining interests for consulting services. While true that he had, as early as 1904-1905, investigated the gold-quartz veins and buried gold-bearing gravels of Victoria, Australia (see references 84--1905n and 85--1905o in the Bibliography), for one of the great London mining concerns, of which his first regular field assistant, Herbert Hoover, was directing engineer, it was not until 1913 that he took on substantial consulting assignments. In that year he examined mines in Mexico (Hidalgo) for Boston interests. In 1917 he took leave during the spring term to spend from February to June making extensive studies of the great Braden and Chuquicamata mines in Chile for their American owners (see reference 158--1922 in the Bibliography); and in 1921 he returned to South America to investigate the tin and silver deposits in Bolivia (see references 165--1924a and 169--1924e in the Bibliography); sheets from his field notebook are reproduced on the next page. Nearer home, in Cuba and Canada, and at numerous sites in the United States, he investigated a wide variety of mineral deposits for numerous mining companies. One of his more challenging assignments was as an expert witness in the celebrated Utah-Apex legal case.

Page 93 of Lindgren's Notebook #1, Bolivia, 1921
Notebook in M.I.T. Institute Archives

THE LAST DECADE AT M.I.T. (1928-1939)

Upon returning to M.I.T. in the fall of 1928, after his year in Washington at the National Research Council, Lindgren immediately resumed his lectures and laboratory instruction, and started the fourth revision of Mineral Deposits. Upon reaching retirement age at 65 in 1930, he was persuaded to continue until 1933, when he became Professor Emeritus and was made an Honorary Lecturer, an appointment he held until 30 June 1936, when he retired permanently. Even after retirement he continued to give an occasional lecture and to publish. His broad interest in the field of economic geology, and his ability to recognize and accept new developments in the field, are no better shown than by the fact that his last lecture, which I was fortunate to hear, was on "The Black Gold [=Petroleum] of California - Oil" and his last article carries the title "Gold and petroleum in California."

PUBLICATIONS - CONTRIBUTIONS

Lindgren's publications are categorized by Butler (1960, p. 189-196) as follows, the number in ( ) indicating the number of articles, reports and the like in the category:

Local Descriptions (54)
Regional Descriptions (16)
Topical Descriptions - Types, Processes, and Relations (29)
Hot Spring Phenomena and Deposits (5)
Economic Considerations - Production and Resources (17)
State of Science and Profession (11)
General Geology (5)
Physiography (5)
Mineralogy (15)
Petrology (7)
Discussion (23)
Abstracts (1068) [Bateman (1960, p. ii) puts the number at 2,545!]
Textbooks (7, including 4 editions of Mineral Deposits)
Biographical Notices (11)

Excepting the Abstracts, which appeared in the earlier volumes of Annotated Bibliography of Economic Geology, and the Textbook items, Lindgren's reports and articles are to be found in 1) the several publications of the U.S. Geological Survey - Annual Reports, Bulletins, Professional Papers (6), Folios (12), Contributions to Economic Geology, and Mineral Resources; 2) the several publications of the American Institute of Mining and Metallurgical Engineers (Transactions, Bulletins, Technical Publications, etc.); and 3) in current technical journals, especially Economic Geology.

For an evaluation of Lindgren's most important contributions to geology, I have gone to the statements of Åhman (1953), Bastin (1934; 1940), Bateman (1960), Buerger (1939), Butler (1960), Graton (1933; 1938; 1960), Jensen (1960), McLaughlin (1939), and Newhouse (1939). From these I conclude: 1) that his greatest contributions lay in "... disclosing the broadest bearings of mineralogy [his greatest scientific love] upon the processes of ore deposition" (Graton 1933, p. xxii), and 2) in the construction of a logical and defensible genetic classification of mineral deposits, based on sound physico-chemical principles, as set forth in his classic textbook, Mineral Deposits. More specifically, his chief contributions to the science of ore deposition would include the following, not necessarily in the order of importance as listed (Graton 1933; Buerger 1939; Newhouse 1939; et al.):

1. Defense and added proof of the dominating role of igneous processes in ore formation.
2. Recognition and classification of the alterations or changes in rocks along fissure veins and igneous intrusions.
3. Nature and influence of physical conditions in ore deposition.

4. Ideas of replacement, particularly constancy of volume.
5. The importance of colloids in certain ore-forming processes.
6. Discovery of analcite as an original constituent of igneous rocks.
7. The classification of ore deposits into temperature groups and the use of geological thermometers.
8. Recognition of mineral zoning both radially and in depth.
9. One of the first to apply effectively the petrographic microscope to the study of ores and related alteration products, by using polished sections and reflected light.
10. First recognition of contact-metamorphic ore deposits in North America and further elucidation of the contact-metamorphic process.
11. Insistence on the great importance to the geologist of a thorough knowledge of chemistry, mineralogy, and physics.
12. Inauguration of unified study of the geology and ore deposits of extensive regions or provinces.
13. Amplification and clarification of the conception of metallogenetic epochs and provinces.
14. The idea of "persistent" and "non-persistent" minerals.
15. Geological analysis of gold production and silver production.
16. Sequence of mineral deposition in sulphide ores.
17. Effective linking of statistical and geological studies.
18. Service to his professional colleagues in the abstracting of the more important literature on economic geology in half a dozen foreign languages.
19. Leadership in founding Economic Geology and the Annotated Bibliography of Economic Geology, and in organizing and directing the National Research Council Committee on Processes of Ore Deposition.
20. Finally - and perhaps in many ways the most important contribution of all - the development and advocacy of a comprehensive philosophy of mineral genesis, set forth in his classic reference work Mineral Deposits, which for half a century has had a profound and far-reaching effect on a host of field workers and laboratory investigators.

LINDGREN'S PART IN FOUNDING ECONOMIC GEOLOGY

Lindgren was one of the chief instigators, if not indeed the principal one, in the founding of Economic Geology, which was started in 1905 as "A semi-quarterly journal devoted to geology as applied to mining and allied industries," and which long ago became one of the world's leading geological periodicals. In his touching memorial to John Duer Irving, whom he induced to serve as the journal's first Editor, Lindgren (1918, p. 416-417) wrote as follows about the birth of the journal:

"One day in 1905 a small group of geologists met at the Cosmos Club in Washington to discuss the possibility of establishing a medium of exchange of opinion and a place where short papers on subjects relating to the genesis of ore deposits could be discussed. Spurr, I believe, was one of the first to advance the suggestion. It was received with great approval but also with some doubt as to the possibility of success. Finally a stock company was formed and the founders received handsome certificates which indicated a financial sacrifice upon the altar of science. The next question was, who was to be the editor? It must be a hard worker and an unselfish man, a man willing to do much work for nothing, for the conditions of the company allowed then no recompense for such work, and has not done so for thirteen years. The writer of this proposed Irving as editor and he was partly instrumental in persuading him to accept. He admits this with strangely mixed feelings, for ever since some of the men of that group have felt that by that action they perhaps laid too heavy a burden on a fellow worker, a burden which once assumed he would not give up in the face of difficulties. The first years were hard. Papers were scarce and financial conditions doubtful. Conditions gradually became better and the journal became self-sustaining, but only because of the unselfishness of the editor. No one but those who were associated with Irving could appreciate the hard work, the drudgery and the worry connected with the enterprise. Perhaps no one knows it better than Professor W. S. Bayley, who as the business editor of the journal for these thirteen years has given equally freely of his scant spare time. But there was also the note of joy for success crowned the efforts. The journal became an acknowledged leader of applied geology and the papers published in its files constitute a record of the world's best work in this line. One need not look far in the pages of the journal to see the marks of Irving's personality. Many are the useful discussions on field methods, reports, ore shoots, etc., that Irving originated. Without wanting to do so, he has reared an enduring monument to his memory in these thirteen volumes of our journal."

Lindgren's name was the first in the list of Associate Editors, and he served in this capacity for the next thirty-five years until his death in 1939. Furthermore, the first number of the journal carried an article by him titled "Ore-deposition and Deep Mining" (Vol. 1/1: 34-46, Oct.-Nov. 1905). In the ensuing three decades he published 42 topical articles, 6 discussion articles, and 22 reviews in the journal he helped to found.

HONORS

PROFESSIONAL SOCIETIES - MEMBERSHIPS, OFFICES, AND COMMITTEE SERVICES

The achievements of Waldemar Lindgren, the Man, the Mining Geologist, the Teacher, the Scientist, and the devoted Society Member, were exceptionally impressive and they brought him highest recognition in both the

United States and abroad: election to leading organizations in this country and to highest office in them; election to honorary membership in numerous leading foreign societies; award of prestigious medals and of honorary academic degrees; and other special honors involving perpetuation of his name on a library and on two special funds.

He was a member of the following organizations and received the several different recognitions listed:

National Academy of Sciences

American Philosophical Society

American Academy of Arts and Sciences

American Association for the Advancement of Science

American Institute of Mining and Metallurgical Engineers (Vice-President 1912-1913; President 1920-1921; Honorary Member* 1931)

The Mining and Metallurgical Society of America (President 1920; Honorary Member)

The Geological Society of America (President 1924; Penrose Medalist 1933)

The Society of Economic Geologists (President 1922; Penrose Medalist 1928)

The Washington Academy of Sciences

The Geological Society of Washington

The Geological Society of Boston

Canadian Institute of Mining and Metallurgy (Corresponding Member)

Royal Academy of Science of Sweden (Foreign Member)

---

* Honorary membership in the A.I.M.E., the highest honor given by that Institute, was awarded Lindgren in 1931. The Technology Review for April 1931 (p. 358) reported on the award as follows:

"WALDEMAR LINDGREN

"A Tribute to Waldemar Lindgren

"At the Annual Dinner of the American Institute of Mining and Metallurgical Engineers on February 18, the following tribute was paid Waldemar Lindgren, Professor of Geology at the Institute, by Dr. George Otis Smith: 'Mr. President, it is a privilege and a pleasure to come to New York when speaking the truth in praise of a man is in order - and personally I keenly appreciate having the truth told about a geologist. The truth about Waldemar Lindgren might be concentrated into these few words - a great and good man. Why say more? Simply because we love to pay due tribute to one so eminent and so modest.

'... In honoring Waldemar Lindgren we are honoring not only American science but also the science of his native land, for he well represents that large body of Swedish geologists and engineers who have contributed so much to the world's store of knowledge and metals.

(footnote continued on facing page)

The Geological Society of Sweden (Honorary Corresponding Member)

Academy of Engineering of Sweden (Foreign Member)

The Geological Society of London (Foreign Fellow; Wollaston Medalist 1937)

The Geological Society of Belgium (Honorary Member)

Membre d'honneur étranger de la Association des Ingènieurs sortis de l'École de Liége

The Geological Society of Leningrad (Foreign Member)

Honorary Chairman, Sixteenth (XVI) International Geological Congress held in the United States in 1933.

DEDICATED PUBLICATIONS

Two specially designated publications were dedicated to Waldemar Lindgren: 1) the A.I.M.E.'s Ore Deposits of the Western States (see first 1933 item in Appendix B), printed while he was still living, and dedicated as follows:

To
WALDEMAR LINDGREN
Leader of American Economic Geologists
this volume is dedicated
by his professional peers,
as an expression of
their appreciation and admiration

2) Part 1 of Number 1 of Volume 55 of Economic Geology was designated the Waldemar Lindgren Centennial Commemorative Number. Three items in this latter publication make special reference to Lindgren and his work:

(footnote continued from facing page)

> ... His career has been notable in its volume and quality of output; as a Government geologist for 28 years when he did so much in raising the scientific standards of the United States Geological Survey; as a mining geologist in private employ, when he widened his knowledge of ore deposits by adding Australia and South America, Mexico and Canada to his roll of scientific conquests; and as a teacher at a great institution passing along the torch to those inspired by his words and example.
>
> 'While a field geologist of the Survey, Mr. Lindgren was the chief and friend of the young geologist whom now, after 35 years so many present here tonight delight to call 'The Big Chief', Herbert Hoover. Mr. President, it is my happy duty to present you, as the latest addition to a most distinguished list of engineers of worldwide fame, Professor Waldemar Lindgren, for the highest honor that our Board of Directors can bestow, Honorary Membership in the American Institute of Mining and Metallurgical Engineers.'"

A. M. Bateman's "Foreword - Waldemar Lindgren and Economic Geology"; M. L. Jensen's "Preface - Centenary of Waldemar Lindgren"; and L. C. Graton's "Editorial - If Lindgren were here."

MEDALS

Four prestigious medals were awarded to Lindgren: two American, the Penrose Medals of the Society of Economic Geologists and of the Geological Society of America, respectively; and two foreign, the Trasenster Medal from Liége and the Wollaston Medal of the Geological Society of London.

Penrose Medal of the Society of Economic Geologists (1928)*

Lindgren was the third recipient of the Penrose Medal of the Society of Economic Geologists; the recipients before him were T. C. Chamberlin and J. H. L. Vogt. The medal was presented to Lindgren at a dinner of the Society held in New York on 27 December 1928, in recognition of his

"... distinguished achievement in the geological sciences."

President S. F. Emmons called attention to his many achievements, citing as the outstanding ones his contributions to Contact Metamorphism, and to Metasomatic Processes in Fissure Veins, and his comprehensive textbook on Mineral Deposits. George Otis Smith mentioned Lindgren's years of fruitful work on the United States Geological Survey, and of his important role in training many of the leading geologists of the United States. H. Foster Bain discussed Lindgren's directing influence in developing the knowledge in the field of economic geology, in which he was the acknowledged leader.

Penrose Medal of the Geological Society of America (1933)

In 1933 the Geological Society of America awarded Lindgren its highest honor, the prestigious Penrose Medal, which recognizes "eminent research in pure geology" and "outstanding original contributions or achievements which mark a decided advance in the science of geology." As the sixth Penrose Medalist, Lindgren followed T. C. Chamberlin (1927), J. J. Sederholm (1928), F. A. A. La Croix (1930), W. M. Davis (1931), and E. O. Ulrich (1932).

As recorded in the Proceedings of the Chicago Meeting (Geol. Soc. Am., Pr. 1933: 42-47, 1934), the medal was presented to Lindgren at the

---

* See "Award of the Penrose Medal" in Econ. Geol. 24/1: 109, (Jan. 1929), and the news item on p. 214 of Tech. Rev. 31/4, (Feb. 1929).

Annual Dinner on 29 December 1933, on the recommendation of the Award Committee, whose Report constitutes a brief but comprehensive statement of Lindgren's accomplishments (p. 43):

"REPORT OF THE COMMITTEE ON PENROSE MEDAL AWARD

"Dr. Waldemar Lindgren has been an important and active contributor to American geology and economic geology since 1886. Few men have contributed so much to the advancement of our knowledge of North American geology. His contributions include among them many papers of outstanding merit that mark notable advances in the science of ore deposits. Among these are his papers on 'Metasomatic Processes in Fissure Veins,' 'The Character and Genesis of Certain Contact Deposits,' 'The Relation of Ore Deposition to Physical Conditions,' 'The Nature of Replacement,' 'Mineral Deposits,' and 'The Colloid Chemistry of Minerals and Ore Deposits.' These are pioneer papers in the field of contact metamorphism, replacement, colloid chemistry applied to ore deposition, and the role of physical-chemical conditions in ore deposition, the lines along which our greatest progress has been made in the science of ore deposits during the past 35 years. He was also one of the first to recognize the importance of the study of polished sections of ores in problems of ore genesis. But his influence is by no means limited to his own important contributions. For many years on the United States Geological Survey he guided his associates and especially the younger men along lines of fruitful research and in recent years as a professor at the Massachusetts Institute of Technology he has played the same role. Many of the highest grade investigators in this field at the present time received their start under his guidance and stimulation. There can be little question but that he outranks by far any living student of ore deposits. It is manifestly a difficult task to rate the work and influence of men who have been active in entirely different fields of geologic investigation, but in this case it seems that Professor Lindgren stands at the head of the list."

After a brief introduction by President C. K. Leith, E. S. Bastin first commented on his personal relationships with Lindgren, then presented the Medal with the following closing remark (p. 45):

"It is my privilege, Professor Lindgren, on behalf of the Geological Society of America to present to you the Penrose Medal in recognition of your position of world leadership as investigator and teacher in the field of ore deposits, your accomplishments and services in other branches of geology, your distinguished record in the public service, and in token of our respect and deep affection."

Dr. Lindgren then responded in a way that showed his sense of humility and delightful humor (p. 46); part of this response is quoted in a preceding section of this biography. (See quoted response on page 402.)

Gustave Trasenster Medal of the University of Liége, Belgium (1936)

Lindgren was the 1936 Gustave Trasenster Medalist, a recognition accorded him by the Association of Alumni Engineers of the Université de Liége, Belgium, for his research in applied geology. In discussing the award, the school's Revue Universelle des Mines (8me Série, Tome XII, N°. 3, 1936) reported as follows:

> "Cette année, le Conseil d'Administration de l'A. I.Lg. a voulu récompenser un géologue dont la carrière a été spécialement orientée vers les recherches de la geologie appliquée; il a fixé son choix sur Waldemar Lindgren, professeur émérite au Massachusetts Institute of Technology de Cambridge (Mass. U.S.A.)."

The award was also reported in the May 1936 issue of The Technology Review (Vol. 38/8: 350, 1936).

Wollaston Medal of the Geological Society of London (1937)

In 1937 the Geological Society of London awarded Lindgren the prestigious Wollaston Medal, its most distinguished international honor in the field of mineralogy. Among the other great men to receive the Medal was Charles Darwin. The award was announced as follows in the Quarterly Journal of the Society (Geol. Soc. London, Q.J. XCIII (93)/1: p. xv, Feb. 1937):

> "The Wollaston Medal to Professor Waldemar Lindgren of the Massachusetts Institute of Technology, for his researches 'concerning the mineral structure of the earth,' and especially concerning the problems of metasomatism, contact ore-deposits, and the application of physical chemistry to ore-deposition."

The citation and Lindgren's response appear on pages cxxvii-cxxviii of the preceding reference. The award was also reported in the March 1937 number of The Technology Review (Vol. 39/5: i, following p. 216, 1937).

HONORARY DEGREES

The Honorary Degree of Doctor of Science was conferred upon Lindgren by Princeton University at its Commencement on 13 June 1916, and was in recognition of his outstanding achievements as an economic geologist.

The Honorary Degree of Doctor of Science was conferred upon Lindgren by Harvard University at its Commencement on 20 June 1935 in one of Harvard's greatest days of recognizing outstanding individuals; included in the eleven honorary doctoral recipients of that memorable day, along with

Lindgren, were Albert Einstein, Charles Schuchert and Albert Sauveur in Science; Thomas Mann in Letters; and Norman H. Davis, John C. Merriam, William A. Neilson, George Sarton, Henry A. Wallace, and William Allen White in Laws. The Boston Evening Transcript for Thursday, 20 June 1935, featured the honorary degree recipients on its front page, and the Commencement exercises on pages seven and eight. Lindgren's citation read:

> "WALDEMAR LINDGREN: A geologist to whom all men turn for knowledge of the metallic secrets hidden in the rocks."

SPECIAL PERSONAL HONORS

In addition to having a special chair established for him - the William Barton Rogers Professorship of Economic Geology, four unusual honors accorded Lindgren deserve special mention because they were established to perpetuate his name and to provide a constant reminder to geologists and mining engineers that he was one of their truly great practitioners. These are: 1) the Lindgren Library and 2) the Lindgren Memorial Fund, both at M.I.T.; 3) the Lindgren Citation Awards for excellence in research, established by the Society of Economic Geologists; and 4) Lindgrenite, a new mineral.

William Barton Rogers Professor of Economic Geology (1912-1933)

Lindgren's accomplishments prior to his offer of a professorship from M.I.T. led the Corporation of the Institute to establish for his occupancy the William Barton Rogers Professorship of Economic Geology, thereby honoring both the Founder of M.I.T., himself a distinguished economic geologist, and Lindgren, who the year before had been promoted to the prestigious rank of Chief Geologist of the U.S. Geological Survey, as discussed earlier on in this biography.

Lindgren Library at M.I.T. (1932)

The Lindgren Library was established in 1932, by consolidating the holdings of the Department of Mining and Metallurgy and the Department of Geology, an action earlier recommended by Lindgren himself as a result of his service as Head of the combined departments during 1920 to 1926. When the library was moved from the Hayden Building to its present site on the second floor of the new Green Building (54-200), the Mining, Metallurgy, and Crystallography holdings were left behind in the Science Library, whereas the holdings in Meteorology and Oceanography were added. Today (1975) this outstanding library contains more than 40,000 books and reports, including complete sets of U.S. Geological Survey and U.S. Bureau of Mines publications; 400 major reference works; 450 current journals and serials; 450 atlases; and 15,000 maps and charts. It can truly be called an "earth sciences" library.

Lindgren Memorial Fund at M.I.T. (1959)

In 1959, William L. Dennen (XII S.B. 1917; S.M. 1920), Joseph L. Gillson (XII S.M. 1921; Sc.D. 1923), Walter L. Whitehead (III_3 S.B. 1913; XII Ph.D. 1918), and I established the Waldemar Lindgren Memorial Fund to perpetuate his name at M.I.T. Solicitation of funds, with a letter of support from former President Herbert Hoover, brought contributions from more than a hundred M.I.T. alumni and industrial friends, and in 1975 the Principal of $30,805 yielded an income of $2,685, which is being used to meet special needs of the geological part of Course XII.

Waldemar Lindgren Citation Awards for Excellence in Research (1960)

In 1960, on the 100th Anniversary of Lindgren's birthday, there appeared the following announcement by H. M. Bannerman, Secretary of the Society of Economic Geologists (Econ. Geol. 56/7: 1583-1584, Nov. 1960):

> "Waldemar Lindgren Citation Awards and Distinguished Research Lectureship
>
> "On advice of the Committee on Research, the Council voted to establish 'Waldemar Lindgren Citation Awards for excellence in research.' These Awards, named in honor of the late Waldemar Lindgren, will each consist of an appropriately inscribed scroll and a cash award of $100, and will be presented by the President at the Annual Meeting of the Society, 'to geologists whose researches as university students result in papers of high merit on a subject judged important to economic geology.' Not more than three Awards may be made in any one year. (This is, in fact, an extension of a plan outlined by the Committee and reported in the Proceedings for 1958 (Economic Geology, Vol. 54, No. 6 [p. 1144], Sept.-Oct. 1958) which provided for a single award.)"

A later Secretary of the Society of Economic Geologists (E. N. Cameron) reported on the Awards as follows (Econ. Geol. 57/2: 290, 1962):

> "The ... Awards were to be presented first in 1962, but the number of papers received in 1961 was insufficient to serve as a basis for granting the Award in 1962. Therefore, the first of the ... Awards will be made in Society year 1963."

Lindgrenite, A New Mineral

In 1935 Charles Palache* of Harvard University gave the name Lindgrenite to a molybdate of copper, $Cu_2MoO_4 \cdot Cu(OH)_2$, a green monoclinic mineral occurring in veinlets in limonitic quartz in the copper mine at Chuquicamata, Peru. The crystals are tabular and show a perfect cleavage parallel to the clinopinacoid.

SUMMATION

Waldemar Lindgren, born in Sweden in 1860, and educated at the gymnasium in Kalmar, left his homeland in 1878 to attend the world renowned Bergakademie (Royal Mining Academy) in Freiberg, Saxony, whence he was graduated as a Bergingenieur (Mining Engineer) and Markscheider (Mine Surveyor) in 1882. He emigrated to the United States in 1883, and the next year joined the United States Geological Survey as Assistant Geologist. During the next 28 years, 1884-1912, he advanced to Geologist in 1895 and then to Chief Geologist in 1911, from which he resigned in 1912 to accept the William Barton Rogers Professorship of Economic Geology at M.I.T. For the next 21 years, until retirement as Professor Emeritus in 1933, he was Head of the Institute's Department of Geology, from 1912-1920 and from 1927-1933, and of the Department of Mining, Metallurgy, and Geology from 1920 to 1926. He served as Honorary Lecturer after regular retirement, from 1933 to 1936, and died in Brookline, Massachusetts on 3 November 1939. He was married to Miss Ottolina Allstrin of Gothenberg, Sweden, on 8 March 1886, and was left widowed and childless upon her death in 1929.

At the height of his career, Lindgren was one of the world's leading economic geologists, and today is ranked high among the greatest of North American geologists. His greatness derives from outstanding field work and laboratory investigations in the mining districts of the Western States; from masterly use of the observations made during these activities in formulating fundamental and original concepts of ore deposition; and in publication of many major reports, which laid the foundation for his comprehensive reference work Mineral Deposits, first published in 1913 and still a valuable book more than half a century later. After joining the M.I.T. Faculty as William Barton Rogers Professor, he headed the Institute's Department of Geology from 1912 to 1933, and developed it into one of the world's leading departments for training economic geologists.

---

* Palache, C. - Lindgrenite, a new mineral. Am. Mineral. 20: 187, (1935).

In the following paragraphs, I comment briefly on Lindgren as scientist; as educator, author, and editor; and as a man - in each category of which he achieved highest distinction.

LINDGREN THE SCIENTIST

Lindgren was a scientist par excellence; a geologist who brought into focus on the problem he was investigating a remarkable set of attributes that any young scientist might well ponder and develop. He learned quite early to acquire and organize a broad range of knowledge - by reading extensively in his father's library, observing and questioning through travel, listening intently and understandingly to his great teachers at Freiberg, and mastering half a dozen languages not his own. Later he became the ardent and expert fact-gatherer, storing in his memory a mountain of factual detail which he would later retrieve and organize in support of some specific idea. His unusual knowledge of languages gave him easy access to most of the important geologic literature of the world, and he used this knowledge to supplement his own observations. Having gathered all the information possible, by whatever means and from whatever source, he then analyzed it to determine its significance, and followed his analysis with publication of his conclusions as soon as he felt it appropriate. Finally, possessed of almost unlimited mental and physical energy, and imbued with the desire to gain the deepest insight into how mineral deposits were formed, he left an impressive legacy of fact and theory for his successors to use, revise and refine. Butler (1950, p. 180) summarized his attributes well in stating:

> "Covering the fields to the extent that he did required a brain and an energy that few possess and still fewer have the will to apply. Lindgren's accomplishment was the result of a talent for sustained well-directed effort by a superior intellect working in a fertile field at a most opportune period."

Perhaps Lindgren the scientist was most aptly characterized in the citation for his honorary doctorate from Harvard:

> "A geologist to whom all men turn for knowledge of the metallic secrets hidden within the rocks."

LINDGREN THE EDUCATOR, AUTHOR, AND EDITOR

Almost from the beginning years of his professional career, Lindgren was an educator, in the broadest sense of that word. His first students were those young men who were his field assistants and who learned much by working with him in the field or laboratory, or listening to him around the campfire or at the fireplace in his home; men like Herbert Hoover, L. C. Graton, John M. Boutwell, W. L. Whitehead, and many others. Then when he assumed larger responsibilities as he moved up the ladder on the Survey, he set an example of industry, incisiveness, and personal integrity for those who worked with him: education by example is a powerful teaching technique. His participation in the activities of the American Institute of Mining and Metallurgical Engineers, the Mining and Metallurgical Society of America, the Society of Economic Geologists, the Geological Society of America, and the National Academy of Sciences' National Research Council, produced an impressive number of actions of an educational nature. Among those cited earlier on were the founding of the journals Economic Geology and Annotated Bibliography of Economic Geology, both powerful and far-reaching journals of condensed information that kept the economic geologist up to date as to what was being done in his field around the world. Lindgren's lifelong interest in mineral genesis and ore deposition led him to initiate the National Research Council's "Committee on Processes of Ore Deposition" and to help prepare the searching 1928 report of the Committee (references 183--1928a and 186--1928d), which includes a masterly-drawn research program as relevant to today's needs as for those of fifty years ago. As an academician Lindgren gave much thought to how best to motivate and train students for careers in geology, and took the trouble to set down his mature thoughts in the articles "Economic geology as a profession (Editorial)" and "The education of the geologist" (references 152--1919b and 162--1903a in the Bibliography), which like many of his other topical articles are fully as relevant in the 1970s as they were in the 1920s! Finally, Lindgren the lecturer and author, by both spoken and written words of unusual terseness and eloquence, and in a language not originally his own, not only made listening a learning and rewarding experience, but also left with his audience the impression that they had heard a master educator; and when they read his publications, they would again sense that he meant to inform and educate as well as merely to describe and explain or theorize. Those who knew Lindgren as a colleague and those who have knowledge of his achievements would probably agree with his one-time student and later colleague at M.I.T., Walter H. Newhouse, who, after stating briefly Lindgren's major contributions to the knowledge of ore deposition, concluded (biographical reference 1939):

> "Transcending all these in importance, however, was his comprehensive philosophy of mineral genesis and the far-reaching effect this has had on a host of other workers."

It was this comprehensive philosophy, generously shared through speaking and writing, that added a special luster to Lindgren's greatness.

LINDGREN THE MAN

In his prime, Waldemar Lindgren stood out in any crowd by virtue of his distinguished appearance, military bearing, courtly manner, linguistic facility, and social poise. His keen intellect, so able at acquisition of knowledge, at analysis and synthesis and at evaluation of the work of others, was never arrogant or even sharply severe, because he was at heart a sensitive and gentle man. He had a fine sense of humor, as evidenced by his refusal to accept Herbert Hoover's repayment for that dead mule (as recounted in an earlier section), relishing the Government's refusal of accepting Hoover's explanation of the mule's death as an example of bureaucratic stupidity; and as evidenced by his response on being cited for the Penrose Medal of the Geological Society of America - he modestly ascribed to good luck and kind friends what he actually accomplished by sound judgment, outstanding analytical ability, and indefatigable labor.

His deep sensitivity toward nature and humans appeared early when, still a gymnasium student in Kalmar, he sketched the mountains of Scandinavia and composed poems about lakes and Indian lovers for his N3 literary club. This subtle sensitivity stayed with him into maturity and was never more evident than in his poignant farewell to his close friend and colleague, Captain John Duir Irving, who had served so long and effectively as the first Editor of <u>Economic Geology</u> and who died in the line of duty in the A.E.F. in France in World War I (Econ. Geol. 13/6: 418, Sept. 1918):

> "And so, John Irving, among darkening shadows you had to take the last trail and cross the Great Divide! We bid you farewell - loving brother, and friend, distinguished investigator, faithful editor and teacher, gallant soldier - Chevalier sans peur et sans reproche.
> WALDEMAR LINDGREN"

Those biographers and colleagues who knew him well invariably mention his searching mind, his brilliant but considerate performance in heated debate, his personal dignity and kindly manner, his amazing energy, and his modesty with regard to the impressive record of accomplishments which brought him repeated honors of highest distinction not only from his adopted country but also from across the Atlantic where he was born and educated.

Inasmuch as I can not claim to have known Dr. Lindgren at all well, I have had to depend on his colleagues, both those now deceased and still

---

* See Lindgren's Editorial "Some remarks on Reviews and Criticisms" (Econ. Geol. 24: 650-653, 1929) for an example of his incisive yet kindly remarks about reviewing and criticizing the work of one's colleagues.

living, and on his many biographers, whose sketches are listed farther on, for comments regarding him as a man. For those who would like a more complete description of him and his work, I would recommend especially the several perceptive and inclusive biographical sketches by Graton (1933; 1939; 1960); the numerous excerpts in Butler's 1950 Memorial, as well as his own informative comments; and the warmly personal sketches by two of his students, W. H. Newhouse (1939) and M. J. Buerger (1939). All the aforementioned references are listed chronologically in Appendix B a little farther on following Lindgren's Bibliography and Appendix A.

It seems evident that Waldemar Lindgren, native son of Sweden and naturalized son of the United States, attained a level of accomplishment that fully justified such laudatory designations as: "en nestor inom malmgeologien" (Åhman 1953); "Lindgren's greatest contributions have lain in disclosing the broadest bearings on mineralogy upon the processes of ore deposition" (Graton 1933, p. xxii); "Mr. Lindgren is easily the leading geologist of the Survey" (George Otis Smith - letter of 1911); "World's foremost Economic Geologist" (Jensen 1960 p. vii); and "A geologist to whom all men turn for knowledge of the metallic secrets hidden in the rocks" (Harvard's Honorary Sc.D. citation, 1935).

BIBLIOGRAPHY OF WALDEMAR LINDGREN

INTRODUCTION

Symbols and abbreviations used in the following references are explained on pages 91 to 98. In general, abstracts are listed separately, as well as with the complete article, if the article itself was written by Lindgren; however, inasmuch as Lindgren also prepared more than two thousand (2,545) abstracts of articles and books for the Annotated Bibliography of Economic Geology, which he helped to found and then edited, in cooperation with the Society of Economic Geologists, I have included a special section on these particular abstracts a little farther on, and have discussed this special editorial activity of Lindgren at some length in the preceding biographical sketch. Following the section on these particular abstracts is a listing of discussions of the work of other geologists; discussions in which Lindgren took an active part and published his comments. Then follows a list of reviews of 23 books and major reports, all but one of which were published in Economic Geology, another journal he helped to found and edit. Finally, there is a list of the 22 Departmental Reports that Lindgren prepared for the annual M.I.T. President's Reports during his period of service, 1912-1933, as Chairman of the Department of Geology. Appendix A is a copy of Martinsson's article "Waldemar Lindgren goes west," which includes four letters that Lindgren wrote to his parents in 1883. They are reproduced by the kind permission of Dr. Martinsson and the publisher. Appendix B is a list of all the biographies, biographical sketches, necrologies, memorials, and the like that I have been able to find; I hope that the list is reasonably complete, though I am sure that I have missed such biographical items in some of the less widely distributed scientific journals both in the Americas and abroad. Appendix C is a copy of Lindgren's editorial in which he sets down his philosophy of education with special reference to the training of geologists. Finally, Appendix D is a list of references containing brief statements of Lindgren's field and office activities while an employee of the United States Geological Survey from 1884 to 1912.

REGULAR PUBLICATIONS

Not included in the following list are several poems that Lindgren wrote while a student in the gymnasium at Kalmar - these were published years later by Backman (1964) - and four letters that he wrote to his parents, the first during his voyage to America early in 1883 and the other three at intervals later in the same year as he worked as an assistant to Raphael Pumpelly on the Northern Transcontinental Survey in connection with the construction of the Northern Pacific Railroad. These four letters were recently translated into English from the original Swedish and were published by Anders Martinsson, with the title "Waldemar Lindgren goes west," in the Geol. Fören. Stockholm Förhandl. 95: 280-287, (June 1973); they are reproduced in toto in Appendix A.

1--1880 Mimetesit från Långban. Geol. Fören. Stockholm Förhandl. Bd. 5: 272, (1880).

2--1881 Om arsenaterna från Långban. Geol. Fören. Stockholm Förhandl. Bd. 5: 552, (1881).

3--1884 Ännu några ord om Berzeliiten. Geol. Fören. Stockholm Förhandl. Bd. 7: 291, (1884).

4--1886 Eruptive rocks of Montana. U.S. 10th Census 15: 719-737, (1886).

5--1887 The silver mines of Calico, California. A.I.M.E., Tr. 15: 717-734, il., (1887).

6--1888 Contributions to the mineralogy of the Pacific Coast. Cal. Ac. Sci., Pr. (2) 1: 1-6, (1888).

7--1889 Notes on the geology of Baja California, Mexico. Cal. Ac. Sci., Pr. (2) 1: 173-196, (1889).

8--1890 Petrographical notes from Baja California, Mexico. Cal. Ac. Sci., Pr. (2) 2: 1-17, (1890).

9--1890a (with Melville, W. H.) Contributions to the mineralogy of the Pacific Coast. U.S. Geol. Surv., B. 61: 40 p., (1890).

10--1891 Notes on the geology and petrography of Baja California, Mexico. Cal. Ac. Sci., Pr. (2) 3: 25-33, (1891).

11--1891a Eruptive rocks from Montana. Cal. Ac. Sci., Pr. (2) 3: 39-57, (1891).

12--1892 The gold deposit at Pine Hill, California. Am. J. Sci. (3) 44: 92-96, (1892).

13--1892a The glacial period: a discussion of Mr. Manson's theory. Min. Sci. Press 64: 94, (1892).

14--1893 Two Neocene rivers of California. Geol. Soc. Am., B. 4: 257-298, il., (1893).

15--1893a A sodalite syenite and other rocks from Montana. Am. J. Sci. (3) 45: 286-297, (1893).

16--1893b The auriferous veins of Meadow Lake, California. Am. J. Sci. (3) 46: 201-206, (1893); Min. Sci. Press 68: 118, (1894).

17--1893c The relation between ore deposits and their enclosing walls. Eng. Min. J. 55: 340-341, (1893).

18--1894   (and Turner, H. W.) Description of the gold belt [California]; description of the Placerville sheet. U.S. Geol. Surv., Geol. Atlas Placerville Folio 3: 3 p., il., (1894; reprinted 1914). Abst., J. Geol. 4: 248-250, (1896).

19--1894a   Description of the gold belt [California]; description of the Sacramento sheet. U.S. Geol. Surv., Geol. Atlas Sacramento Folio 5: 3 p., il., (1894; reprinted 1914; preliminary edition 1892). Abst., J. Geol. 4: 250-251, (1896).

20--1894b   The gold-silver veins of Ophir, California. U.S. Geol. Surv., An. Rep. 14/2: 243-284, map, (1894). Abst., Min. Sci. Press 71: 216, 233, (1895); J. Geol. 4: 373-374, (1896).

21--1894c   An auriferous conglomerate of Jurassic age from the Sierra Nevada. Am. J. Sci. (3) 48: 275-280, (1894).

22--1895   (and Turner, H. W.) Description of the Marysville sheet [California]. U.S. Geol. Surv., Geol. Atlas Marysville Folio 17: 2 p., maps, (1895). Abst., J. Geol. 3: 976-977, (1895).

23--1895a   (and Turner, H. W.) Description of the gold belt [California]; description of the Smartsville sheet. U.S. Geol. Surv., Geol. Atlas Smartsville Folio 18: 6 p., maps, (1895).

24--1895b   Characteristic features of California gold quartz veins. Geol. Soc. Am., B. 6: 221-240, map, (1895); Min. Sci. Press 70: 181-182, 213-214, 244, (1895). Abst., Science 1: 68, (1895).

25--1896   Description of the [Nevada City, California] special maps. U.S. Geol. Surv., Geol. Atlas Nevada City Folio 29: 7 p., maps, (1896). Abst., J. Geol. 5: 409-411, (1897).

26--1896a   Description of the gold belt [California]; description of the Pyramid Peak quadrangle. U.S. Geol. Surv., Geol. Atlas Pyramid Peak Folio 31: 8 p., maps, (1896).

27--1896b   The gold-quartz veins of Nevada City and Grass Valley districts, California. U.S. Geol. Surv., An. Rep. 17/2: 1-262, (1896).

28--1896c   The age of the auriferous gravels of the Sierra Nevada: with a report on the flora of Independence Hill, by F. H. Knowlton. J. Geol. 4: 881-906, (1896). Abst., Zeitschr. Prak. Geol. 1897: 226-227, (1897).

29--1896d   The gold quartz veins of California. Am. Geol. 17: 338-339, (1896).

30--1897   Description of the gold belt [California]; description of the Truckee quadrangle. U.S. Geol. Surv., Geol. Atlas Truckee Folio 39: 8 p., maps, (1897).

31--1897a   The granitic rocks of the Pyramid Peak district, Sierra Nevada, California. Am. J. Sci. (4) 3: 301-314, map, (1897).

32--1897b   Monazite from Idaho. Am. J. Sci. (4) 4: 63-64, (1897); Min. Sci. Press 75: 168, (1897).

33--1897c   Filling and replacement in gold bearing fissure veins. Eng. Min. J. 63: 573, (1897).

34--1897d   The granitic rocks of the Sierra Nevada [abst.]. Science 5: 361, (1897).

35--1898   Description of the Boise quadrangle [Idaho]. U.S. Geol. Surv., Geol. Atlas Boise Folio 45: 7 p., maps, (1898).

36--1898a   The mining districts of the Idaho Basin and the Boise Ridge, Idaho. U.S. Geol. Surv., An. Rep. 18/3: 617-719, maps, (1898).

37--1898b Orthoclase as a gangue mineral in a fissure vein. Am. J. Sci. (4) 5: 418-420, (1898); Min. Sci. Press 77: 32, (1898).

38--1898c The primary gold deposits of the Sierra Nevada. Min. Sci. Press 76: 258-259, (1898).

39--1898d The canyons of the Salmon and Snake rivers, Idaho [abst.]. Science 7: 71-72, (1898); Eng. Min. J. 65: 158, (1898).

40--1899 The copper deposits of the Seven Devils, Idaho. Min. Sci. Press 78: 125, (1899).

41--1900 Description of the Colfax quadrangle [California]. U.S. Geol. Surv., Geol. Atlas Colfax Folio 66: 10 p., maps, (1900).

42--1900a The gold and silver veins of Silver City, De Lamar, and other mining districts in Idaho. U.S. Geol. Surv., An. Rep. 20/3: 65-256, maps, (1900).

43--1900b Granodiorite and other intermediate rocks. Am. J. Sci. (4) 9: 269-282, (1900).

44--1900c Wood River mining district, Idaho [abst.]. Science 11: 348-349, (1900).

45--1901 The gold belt of the Blue Mountains of Oregon. U.S. Geol. Surv., An. Rep. 22/2: 551-776, maps, (1901).

46--1901a Metasomatic processes in fissure veins. A.I.M.E., Tr. 30: 578-692, (1901). See also 53--1902d.

47--1901b Rare minerals in gold quartz veins of eastern Oregon. Min. Sci. Press 82: 252, (1901).

48--1901c Trias in northeastern Oregon [abst.]. Science 13: 270-271, (1901).

49--1902 Tests for gold and silver in shales from western Kansas. U.S. Geol. Surv., B. 202: 21 p., (1902). Abst., Eng. Min. J. 74: 111-112, (1902).

50--1902a The character and genesis of certain contact deposits. A.I.M.E., Tr. 31: 226-244, (1902). See also 54--1902e.

51--1902b The gold production of North America, its geological derivation and probable future. Min. Sci. Press 85: 177, 193, 206, (1902).

52--1902c A deposit of titanic iron ore from Wyoming [abst.]. Science 16: 984-985, (1902).

53--1902d Metasomatic processes in fissure-veins, p. 498-612, il., in The Genesis of Ore-Deposits (2nd ed.), by Franz Pošepný. New York: A.I.M.E., xxi + 806 p., il., (1902). See also 46--1901a.

54--1902e The character and genesis of certain contact deposits, p. 716-733 in The Genesis of Ore-Deposits (2nd ed.), by Franz Pošepný. New York: A.I.M.E., xxi + 806 p., il., (1902). See also 50--1902a.

55--1903 The water resources of Molokai, Hawaiian Islands. U.S. Geol. Surv., W-S.P. 77: 62 p., (1903).

56--1903a Neocene rivers of the Sierra Nevada. U.S. Geol. Surv., B. 213: 64-65, (1903).

57--1903b Mineral deposits of the Bitterroot Range and Clearwater Mountains, Montana. U.S. Geol. Surv., B. 213: 66-70, (1903).

58--1903c Copper deposits at Clifton, Arizona. U.S. Geol. Surv., B. 213: 133-140, (1903); Eng. Min. J. 75: 705-707, (1903).

59--1903d   The geological features of the gold production of North America (with discussion by W. G. Miller, W. L. Austin, J. E. Spurr, and H. W. Turner). A.I.M.E., Tr. 33: 790-845, 1077-1083, (1903); Ibid. 34: 921, (1904). Reprinted in Emmons, S. F., Ore Deposits (published by A.I.M.E.): 424-449, New York, (1913).

60--1903e   The gold production of North America, its geological derivatives and probable future. Int. Min. Cong., 5th Pr.: 29-36, (1903).

61--1903f   Notes on the geology of Molokai, Hawaiian Islands [abst.]. Science 17: 309, (1903).

62--1903g   Metallic sulphides from Steamboat Springs, Nevada [abst.]. Science 17: 792, (1903).

63--1904   (and Drake, N. F.) Description of the Nampa quadrangle [Idaho-Oregon]. U.S. Geol. Surv., Geol. Atlas Nampa Folio 103: 5 p., maps, (1904).

64--1904a   (and Drake, N. F.) Description of the Silver City quadrangle [Idaho]. U.S. Geol. Surv., Geol. Atlas Silver City Folio 104: 6 p., maps, (1904).

65--1904b   A geological reconnaissance across the Bitterroot Range and Clearwater Mountains in Montana and Idaho. U.S. Geol. Surv., P.P. 27: 123 p., maps, (1904).

66--1904c   Gypsum deposits in Oregon. U.S. Geol. Surv., B. 223: 111, (1904).

67--1904d   (and Ransome, F. L.) Report of progress in the geological resurvey of the Cripple Creek district, Colorado. U.S. Geol. Surv., B. 254: 36 p., (1904).

68--1904e   (and Hillebrand, W. F.) Minerals from the Clifton-Morenci district, Arizona. Am. J. Sci. (4) 18: 448-460, (1904); U.S. Geol. Surv., B. 262: 42-54, (1905).

69--1904f   The genesis of copper deposits. Eng. Min. J. 78: 987-988, (Dec. 1904). An abstract of 75--1905e.

70--1905   Description of the Clifton quadrangle [Arizona]. U.S. Geol. Surv., Geol. Atlas Clifton Folio 129: 13 p., maps, (1905).

71--1905a   The copper deposits of the Clifton-Morenci district, Arizona. U.S. Geol. Surv., P.P. 43: 375 p., maps, (1905).

72--1905b   The production of gold in the United States in 1904. U.S. Geol. Surv., B. 260: 32-38, (1905).

73--1905c   The production of silver in the United States in 1904. U.S. Geol. Surv., B. 260: 39-44, (1905).

74--1905d   (and Ransome, F. L.) The geological resurvey of the Cripple Creek district, Colorado. U.S. Geol. Surv., B. 260: 85-98, (1905).

75--1905e   The genesis of the copper deposits of Clifton-Morenci, Arizona. A.I.M.E., Tr. 35: 511-550, (1905). Reprinted in Emmons, S. F., Ore Deposits (published by A.I.M.E.): 517-556, N.Y., (1913). See also 69--1904f.

76--1905f   The Hauraki goldfields, New Zealand. Eng. Min. J. 79: 218-221, (1905).

77--1905g   The occurrence of stibnite at Steamboat Springs, Nevada. A.I.M.E., Bi-Month. B. 2: 275-278, (1905); Tr. 36: 27-31, (1906). Reprinted in Emmons, S. F., Ore Deposits (published by A.I.M.E.): 629-632, N.Y., (1913).

78--1905h   Ore deposition and deep mining. Econ. Geol. 1: 34-46, (1905).
79--1905i   Occurrence of albite in the Bendigo veins. Econ. Geol. 1: 163-166, (1905).
80--1905j   Chemistry of copper deposits. Eng. Min. J. 79: 189, (1905).
81--1905k   (and others) Gold and silver. U.S. Geol. Surv., Mineral Res. 1904: 141-220; 1905: 113-341; 1906: 111-371, (1905-1907).
82--1905l   The great fault of the Bitterroot Mountains [abst.]. Science 21: 224, (1905).
83--1905m   The subterranean gases of Cripple Creek [Colorado] [abst.]. Science 21: 662, (1905).
84--1905n   The deep leads of Victoria. Eng. Min. J. 79: 314-316, (1905).
85--1905o   Characteristics of gold-quartz veins in Victoria. Eng. Min. J. 79: 458-460, (1905).
86--1906   Metasomatic processes in the gold deposits of Western Australia. Econ. Geol. 1: 530-544, (1906).
87--1906a   (and Graton, L. C.) A reconnaissance of the mineral deposits of New Mexico. U.S. Geol. Surv., B. 285: 74-86, (1906).
88--1906b   The Annie Laurie mine, Piute County, Utah. U.S. Geol. Surv., B. 285: 87-90, (1906).
89--1906c   The gold deposits of Dahlonega, Georgia. U.S. Geol. Surv., B. 293: 119-128, (1906).
90--1906d   (and Ransome, F. L.) Geology and gold deposits of the Cripple Creek district, Colorado. U.S. Geol. Surv., P.P. 54: 516 p., maps, (1906).
91--1906e   Ore deposition and deep mining. Min. Sci. Press 92: 41, (1906).
92--1906f   The Hamilton mine, New Mexico [abst.]. Science 23: 697-698, (1906).
93--1906g   Gold and pyrite. Min. Sci. Press 93: 226, (1906).
94--1906h   Discussion of paper by John A. Reid, Sketch of the geology and ore deposits of the Cherry Creek district, Arizona. Econ. Geol. 1: 698-699, (1906).
95--1906i   Gold and silver: Colorado; New Mexico; South Dakota; Southern Appalachian States; Texas; Wyoming. U.S. Geol. Surv., Mineral Res. 1905: 185-214, 275-284, 293-305, 337-341, (1906).
96--1907   The relation of ore deposition to physical conditions. Int. Geol. Cong. X, Mexico, 1906. C.R.: 701-724, (1907). Econ. Geol. 2: 105-127, (1907).
97--1907a   Review of the copper deposits of the Robinson mining district, Nevada, by A. C. Lawson (Cal. Univ., Dept. Geol., B. 4: 287-357); Econ. Geol. 2: 195-304, (1907).
98--1907b   Some gold and tungsten deposits of Boulder County, Colorado. Econ. Geol. 2: 453-463, (1907).
99--1907c   Present tendencies in the study of ore deposits. Econ. Geol. 2: 743-762, (1907); Min. Sci. Press 96: 567-571, (1908). Abst., Science 27: 349-350, (1908).
100--1907d The development of the metal mining industries in the Western States. Am. Min. Cong., 9th An. Sess., Rep. Pr.: 156-165, (1907).
101--1907e Methods of igneous intrusion [abst.]. Science 25: 623, (1907).

102--1908   Will the production of gold in the world keep pace with the increasing demands of commerce and trade? Am. Min. Cong., 10th An. Sess., Rep. Pr.: 265-271, (1908).

103--1908a Investigations relating to deposits of metalliferous ores. U.S. Geol. Surv., B. 340: 18-22, (1908).

104--1908b A geological analysis of the silver production of the United States in 1906. U.S. Geol. Surv., B. 340: 23-35, (1908).

105--1908c Notes on copper deposits in Chaffee, Freemont, and Jefferson counties, Colorado. U.S. Geol. Surv., B. 340: 157-174, (1908).

106--1908d A recent vein at Ojo Caliente, New Mexico [abst.]. Science 27: 348-349, (1908).

107--1908e New occurrence of willemite [New Mexico] and anhydrite [Newhouse, Utah] [abst.]. Science 28: 933-934, (1908).

108--1908f (and McCaskey, H. D.) Gold and silver. U.S. Geol. Surv., Mineral Res. 1907/1: 111-135; 1908/1: 157-183, (1908-1909).

109--1908g (with Hayes, C. W.) Contributions to economic geology, 1907: Pt. I. - Metals and nonmetals, except fuels. U.S. Geol. Surv., B. 340: 482 p., (1908).

110--1909   The localization of values in ore bodies and the occurrence of shoots in metalliferous deposits. Econ. Geol. 4: 56-61, (1909).

111--1909a The Tres Hermanas mining district, New Mexico. U.S. Geol. Surv., B. 380: 123-128, (1909).

112--1909b Resources of the United States in gold, silver, copper, lead, and zinc. U.S. Geol. Surv., B. 394: 114-156, (1909); Nat. Conserv. Comm. (60th Cong., 2d sess., Sen. Doc. 676). Rp. 3: 521-557, (1909).

113--1909c Metallogenetic epochs. Econ. Geol. 4: 409-420, (1909); Can. Min. Inst., J. 12: 102-113, (1910); Can. Min. J. 30: 430-434, (1909); Min. World 31: 1111-1113, (1909).

114--1909d [The discovery of a selenium mineral in the gold-quartz ores of the Republic district, Washington] [abst.]. Science 30: 972, (1909).

115--1909e (with Hayes, C. W.) Contributions to economic geology, 1908: Pt. I. - Metals and nonmetals, except fuels. U.S. Geol. Surv., B. 380: 482 p., (1909).

116--1910   (and Graton, L. C. and Gordon, C. H.) The ore deposits of New Mexico. U.S. Geol. Surv., P.P. 68: 361 p., maps, (1910).

117--1910a The hot springs at Ojo Caliente and their deposits. Econ. Geol. 5: 22-27, (1910).

118--1910b Anhydrite as a gangue mineral. Econ. Geol. 5: 522-527, (1910).

119--1910c Special problems and their study in economic geology. Econ. Geol. 5: 772-776, (1910).

120--1910d (with Hayes, C. W.) Contributions to economic geology, 1909: Pt. I. - Metals and nonmetals, except fuels. U.S. Geol. Surv., B. 430: 653 p., (1910).

121--1911   The Tertiary gravels of the Sierra Nevada of California. U.S. Geol. Surv., P.P. 73: 226 p., maps, (1911). Abst., Wash. Ac. Sci., J. 2: 191-193, (1912).

122--1911a (and Irving, J. D.) The origin of the Rammelsberg ore deposit. Econ. Geol. 6: 303-313, (1911).

123--1911b Copper, silver, lead, vanadium, and uranium ores in sandstone and shale. Econ. Geol. 6: 568-581, (1911).

124--1911c Some modes of deposition of copper ores in basic rocks. Econ. Geol. 6: 687-700, (1911).

125--1911d Geology of the National mining district, Nevada. Min. World 35: 1175-1176, map, (1911).

126--1911e Platinum and allied metals. U.S. Geol. Surv., Mineral Res. 1909/1: 595-601; 1910/1: 773-780; 1911/1: 987-1003, (1911-1912).

127--1911f (with Hayes, C. W.) Contributions to economic geology, 1910: Pt. I. - Metals and nonmetals, except fuels. U.S. Geol. Surv., B. 470: 558 p., (1911).

128--1912 The nature of replacement. Econ. Geol. 7: 521-535, (1912).

129--1912a The bonanza of National, Nevada [abst.]. Wash. Ac. Sci., J. 2: 107-108, (1912).

130--1912b Successive phases of mineralization in veins of volcanic regions. Can. Min. Inst., Tr. 15: 187-191, (1912).

130a-1912c See Hill, J. M. in section on "Discussions of the Work of Other Geologists" following this Bibliography. Lindgren wrote Pt. I. Geologic Introduction, p. 5-43, in Hill's The Mining Districts of the Western United States. U.S. Geol. Surv., B. 507: 309 p., (1912).

131--1913 Mineral Deposits. New York: McGraw-Hill Book Co., Inc., xv + 883 p., (1913); 2d ed., xviii + 957 p., (1919); 3d ed., xx + 1049 p., il., (1928); 4th ed., xvii + 930 p., il., (1933). See also 150--1919, 182--1928, and 199--1933.

132--1913a Contributions to economic geology (short papers and preliminary reports), 1911: Part I, Metals and nonmetals, except fuels. U.S. Geol. Surv., B. 530: 400 p., (1913).

133--1914 (and Turner, H. W.) Reprints from Placerville, Sacramento, and Jackson folios [California] Nos. 3, 5, and 11, respectively: 9 p., maps, U.S. Geol. Surv., (1914).

134--1914a (and Bancroft, H.) The Republic mining district, Washington. U.S. Geol. Surv., B. 550: 133-166, map, (1914).

135--1914b The origin of the "garnet zones" and associated ore deposits. Econ. Geol. 9: 283-292, (1914); A.I.M.E., B. 90: 949-956, (1914); Ibid., Tr. 48: 201-208, (1915).

136--1914c (and Whitehead, W. L.) A deposit of jamesonite near Zimapan, Mexico. Econ. Geol. 9: 435-462, (1914).

137--1915 Geology and mineral deposits of the National mining district, Nevada. U.S. Geol. Surv., B. 601: 58 p., maps, (1915). Abst., Wash. Ac. Sci., J. 5: 580-581, (1915).

138--1915a The igneous geology of the Cordilleras and its problems, p. 234-286 in Problems of American Geology. New Haven: Silliman Foundation [Yale], (1915).

139--1915b (and Ross, C. P.) The iron deposits of Daiquiri, Cuba (with discussion by Max Roesler, B. B. Lawrence, L. C. Graton, Harrison Souder, C. P. Berkey, A. C. Lane, and J. D. Irving). A.I.M.E., B. 106: 2171-2190, (1915); Tr. 53: 40-66, (1916).

140--1915c The origin of kaolin. Econ. Geol. 10: 89-93, (1915).

141--1915d Processes of mineralization and enrichment in the Tintic mining district [Utah]. Econ. Geol. 10: 225-240, (1915).

142--1916 Gold and silver deposits in North and South America. Pan Am. Sci. Cong. 2nd Washington, Pr. sec. 7, vol. 8: 560-577 (1917). A.I.M.E., B. 112: 721-746, maps, (1916); Tr. 55: 883-909, maps, (1917). Smiths. Inst., An. Rep. 1917: 147-173, (1919).

143--1917 [On the deposition of the various forms of silica.] A.I.M.E., B. 126: xvi, (1917).

144--1918 The occurrence of the halogen salts of silver. Econ. Geol. 13: 225-226, (1918).

145--1918a The Idaho peneplain (discussion). Econ. Geol. 13: 486-488, (1918).

146--1918b Volume changes in metamorphism. J. Geol. 26: 542-554, (1918).

147--1918c Genesis of the Sudbury nickel-copper ores (discussion). A.I.M.E., B. 136: 857, (1918).

148--1918d John Duer Irving. Econ. Geol. 13: 413-418, port., (1918).

149--1918e John Duer Irving: In Memoriam. Eng. Min. J. 106/6: 263-264, (Aug. 1918).

150--1919 Mineral Deposits, 2nd ed. New York: McGraw-Hill Book Co., xviii + 957 p., 284 fig., (1919). See also 131--1913, 182-- 1928, and 199--1933.

151--1919a (and Loughlin, G. F.) Geology and ore deposits of the Tintic mining district, Utah. U.S. Geol. Surv., P.P. 107: 282 p., il., (1919). Abst., Wash. Ac. Sci., J. 9: 316-317, (1919).

152--1919b Economic geology as a profession (Editorial). Econ. Geol. 14: 79-86, (1919).

153--1919c Certain iron resources of the world - Scandinavia. A.I.M.E., Tr. 61: 120-130, (1919).

154--1920 Gold production of the world, its future prospects and its relation to price. Min. Metall. Soc. Am., B. 13: 67-69, (1920).

155--1920a Vein filling at Bendigo, Victoria. Econ. Geol. 15: 312-314, (1920).

156--1920b Regarding magmatic nickel deposits. Econ. Geol. 15: 535-538, (1920).

157--1921 Present tendencies in the study of mineral deposits. Min. Metall. Soc. Am., B. 145: 42-49, (1921).

158--1922 (and Bastin, E. S.) The geology of the Braden mine, Rancagua, Chile. Econ. Geol. 17: 75-99, (1922); Copper Resources of the World 2: 459-472; 16th Int. Geol. Cong., Washington, D.C., 1933. Menasha, Wis.: George Banta Pub. Co., 2 vols., 855 p., (1935).

159--1922a (and Hamilton, L. F. and Palache, C.) Melanovanadite, a new mineral from Mina Ragra, Pasco, Peru. Am. J. Sci. (5) 3: 195-203, il., (1922).

160--1922b A suggestion for the terminology of certain mineral deposits. Econ. Geol. 17: 292-294, (1922).

161--1923? (and Ball, S. H.) Summary of Proceedings of the Society of Economic Geologists, 1921-1922, 24 p., [no date; 1923?].

162--1923a The education of the geologist (Editorial). Econ. Geol. 18: 405-409, (1923).

163--1923b Concentration and circulation of the elements from the standpoint of economic geology. Econ. Geol. 18: 419-442, (1923).

164--1924   The colloid chemistry of minerals and ore deposits, p. 445-465 (chap. 18, vol. 2), in The Theory and Application of Colloidal Behavior, Robert H. Bogue, Ed. New York: McGraw-Hill Book Co., (1924).

165--1924a The tin deposits of Chacaltaya, Bolivia. Econ. Geol. 19: 223-228, (1924). See also discussion, Ibid., p. 765-766, and 169--1924e.

166--1924b (and Davy, W. M.) Nickel ores from Key West mine, Nevada. Econ. Geol. 19: 309-319, (1924).

167--1924c Contact metamorphism at Bingham, Utah. Geol. Soc. Am., B. 35: 507-534, (1924).

168--1924d Microchemical reactions (Editorial). Econ. Geol. 19: 762-764, (1924).

169--1924e Fluorite in Bolivian tin mines. Econ. Geol. 19: 765-766, (1924). See also 165--1924a.

170--1925   The search for covered ore bodies. T.E.N. 5/6: 214, 230, (Jan. 1925).

171--1925a The cordierite-anthophyllite mineralization at Blue Hill, Maine, and its relation to similar occurrences. Nat. Ac. Sci., Pr. 11: 1-4, (1925).

172--1925b Gel replacement, a new aspect of metasomatism. Nat. Ac. Sci., Pr. 11: 5-11, (1925).

173--1925c Metasomatism [Presidential Address]. Geol. Soc. Am., B. 36: 247-261, (1925).

174--1926   Ore deposits of the Jerome and Bradshaw Mountains quadrangles, Arizona. U.S. Geol. Surv., B. 782: 192 p., il., (1926).

175--1926a Magmas, dikes, and veins (with discussion by J. T. Singewald, W. H. Emmons, J. C. Anderson, J. F. Kemp, R. J. Colony, A. M. Bateman, J. E. Spurr, C. A. Porter, and B. Stevens). A.I.M.E.; Tr. 74: 71-126, il., (1926); Pamp. 1575: 47 p., (1926); Eng. Min. J. 122: 125-133, (1926). Abst., Min. Metal. 7: 305-306, (1926). Extr., Can. Min. J. 47: 821-828, (1926).

176--1926b Replacement in the tin-bearing veins of Caracoles, Bolivia. Econ. Geol. 21: 135-144, (1926).

177--1926c World lead deposits. Min. Metal. 7: 244-245, (1926); Can. Min. J. 47: 908-909, (1926).

178--1927   James Furman Kemp: to his memory. Econ. Geol. 22: 84-90, (1927).

179--1927a (with Shimer, H. W.) Memorial of William Otis Crosby. Geol. Soc. Am., B. 38: 34-45, (1927).

180--1927b Hot springs and magmatic emanations. Econ. Geol. 22: 189-192, (1927).

181--1927c Paragenesis of minerals in the Butte veins (discussion). Econ. Geol. 22: 304-307, (1927).

182--1928   Mineral Deposits, 3d ed. New York: McGraw-Hill Book Co., Inc., xx + 1049 p., 317 fig., (1928). See also 131--1913, 150--1919, and 199--1933.

183--1928a Research in processes of ore deposition. A.I.M.E., Tech. Pub. 78: 14 p., (1928); Tr. 76: 290-307, (1928). Abst., Min. Metal. 9: 79, (1928).

184--1928b (and Creveling, J. G.) The ores of Potosi, Bolivia. Econ. Geol. 23: 233-262, (1928). See also discussion, Ibid., p. 459, (1928).

185--1928c Bibliographies, annotated bibliographies, and geological abstracts. Econ. Geol. 23: 564-568, (1928).

186--1928d Report of the Committee on Processes of Ore Deposition. Econ. Geol. 23: 591-611, (1928).

187--1928e Historical review of the study of polished sections of opaque minerals, p. 1-6, in The Laboratory Investigation of Ores: a Symposium, by E. E. Fairbanks, New York, (1928).

188--1929 A President in the making. T.E.N. 10/2: 53, 82, (March 1929).

189--1929a Some remarks on reviews and criticisms (Editorial). Econ. Geol. 24: 650-653, (1929).

190--1930 Pseudo-eutectic textures. Econ. Geol. 25: 1-13, il., (1930).

191--1930a The New Sweden: The Vikings have become able scientists and engineers. Tech. Rev. 32/7: 345-347, 384, 386, 392, 394, (May 1930).

192--1930b Departmental Notes - Department of Geology. T.E.N. XI: 144, 168, il., (1930).

193--1930c (and Lausen, C.) The Pre-Cambrian greenstone complex of the Jerome quadrangle, by Carl Lausen, a discussion. J. Geol. 38: 460-465, (1930).

194--1930d Discussion of the review of Annotated Bibliography of Economic Geology. J. Geol. 38: 566-567, (1930).

195--1931 Memorial of Claude Ellsworth Siebenthal. Geol. Soc. Am., B. 42: 138-146, port., (1931).

196--1931a (and Abbott, A. C.) The silver-tin deposits of Oruro, Bolivia. Econ. Geol. 26: 453-479, (1931).

197--1931b (with Bastin, E. S. et al.) Criteria of age relations of minerals. Econ. Geol. 26: 561-610, (1931).

198--1932 Memorial tribute to Pierre Termier [1859-1930]. Geol. Soc. Am., B. 43: 116-117, port., (1932).

199--1933 Mineral Deposits, 4th ed. New York: McGraw-Hill Book Co., Inc., xvii + 930 p., 332 fig., (1933). See also 131--1913, 150--1919, and 182--1928.

200--1933a Differentiation and ore deposition, Cordilleran region of the United States: Ore deposits of the Western States (Lindgren Volume), p. 152-180, A.I.M.E., (1933).

201--1933b Memorial of Richard Alexander Fullerton Penrose, Jr., July 17, 1863 - July 31, 1931; a tribute to his life and achievements. Am. Phil. Soc., Pr. 72: 101-114, (1933).

202--1933c Coronadite "redivivus". Am. Mineral. 18: 548-550, (1933).

203--1934 Response of Dr. Lindgren [on being cited for the Penrose Medal]. Geol. Soc. Am., Pr. 1933: 45-47, (June 1934).

204--1935 Biographical memoir of George Perkins Merrill 1854-1929. Nat. Ac. Sci., Biog. Mem. 17: 31-53, port., (1937); preprint, (1935).

205--1935a The silver mine of Colquijirca, Peru. Econ. Geol. 30: 331-346, (1935).

206--1935b Waters, magmatic and meteoric. Econ. Geol. 30: 463-477, (1935).

207--1935c Harry Cyril Boydell [1879-1935], in memoriam. Eng. Min. J. 136: 583, (1935).

208--1935d Frederic Leslie Ransome, 1868-1935, a memorial. Econ. Geol. 30: 841-842, (1935).

209--1936   Succession of minerals and temperatures of formation in ore deposits of magmatic affiliations. A.I.M.E., Tech. Pub. 713: 23 p., (1936); Tr. 126: 356-376, (1937). Abst., Min. Metal. 17: 270, (1936); Year Book 1936: 72, (1937).

210--1937   Memorial of Frederick Leslie Ransome [1868-1935]. Geol. Soc. Am., Pr. 1936: 249-258, port., (1937).

211--1937a   [Frederick Leslie Ransome, 1868-1935.] Geol. Soc. London, Q.J. 371, vol. 93/3: xcv-xcvi, (1937).

212--1938   Gold and petroleum in California. Calif. J. Min. Geol. 34: 27-32, (1938).

ABSTRACTS

In addition to the abstracts cited in the preceding references and the reviews listed a little farther on, Lindgren wrote 2,545 abstracts for the Annotated Bibliography of Economic Geology. Bateman (1960, p. ii) cited the above figure, which the alert reader will realize is more than twice the number mentioned by Graton (1933, p. xxxii) in his "Life and scientific work of Lindgren" in A.I.M.E.'s Lindgren Volume, Ore Deposits of the Western States. The discrepancy, of course, derives from the fact that Graton's count was made in 1933, while Lindgren was still busily turning out his monthly quota of abstracts, whereas Bateman's count was made long after Lindgren's death with his complete bibliography at hand.

DISCUSSIONS OF THE WORK OF OTHER GEOLOGISTS

Lindgren's great knowledge of and broad interest in mineral deposits of every kind led him to discuss publications by other geologists in many instances. He commented briefly on the following 23 references:

Report to Hon. J. W. Powell, Director, U.S. Geol. Surv., 15th An. Rept.: 174-175, (1895).

Rickard, T. A. et al., a discussion [by S. F. Emmons, W. H. Weed, J. E. Spurr, W. Lindgren, J. F. Kemp, F. L. Ransome, T. A. Rickard, C. R. Van Hise, C. W. Purington] republished from Eng. Min. J., 90 p., (1903).

Weed, W. H. et al., The genetic classification of ore bodies; a proposal and a discussion (by S. F. Emmons, J. E. Spurr, Waldemar Lindgren, F. L. Ransome) [see Rickard, T. A. above]. Eng. Min. J. 75: 553-554, (1903).

Weed, W. H., A genetic classification of ore deposits (abst., with discussion by J. E. Spurr and Waldemar Lindgren). Science 17: 273-274, (1903).

Emmons, S. F. et al., A further discussion on ore deposits (by S. F. Emmons, J. F. Kemp, F. L. Ransome, T. A. Rickard, J. F. Kemp, F. L. Ransome, T. A. Rickard, C. R. Van Hise, Waldemar Lindgren, W. H. Weed). Eng. Min. J. 75: 476-479, 594-595, (1903).

"Alchemist", Chemistry of copper deposits. Eng. Min. J. 78: 189, (1905).

Reid, J. A., Sketch of the geology and ore deposits of the Cherry Creek district, Arizona. Econ. Geol. 1: 698-699, (1906).

Irving, J. D. et al., The localization of values or occurrence of shoots in metalliferous deposits. Econ. Geol. 4: 59-61, (1909).

Irving, J. D., Special problems and their study in economic geology. Econ. Geol. 5: 772-776, (1910).

Hill, J. M., The Mining districts of the western United States with a geologic introduction [Pt. I, p. 5-43] by Waldemar Lindgren. U.S. Geol. Surv., B. 507: 5-57, (1912).

Graton, L. C. et al., To what extent is chalcocite a primary and to what extent a secondary mineral in ore deposits? A.I.M.E., Tr. 48: 194-200, (1914).

Leith, C. K., Recrystallization of limestone at igneous contacts. A.I.M.E., Tr. 48: 214, (1914).

Phalen, W. C., Salt making by solar evaporation. A.I.M.E., Tr. 50: 952, (1914).

Dresser, J. A., Asbestos in southern Quebec. A.I.M.E., Tr. 50: 962, (1914).

Billingsley, P., The boulder batholith of Montana. A.I.M.E., Tr. 51: 52, (1915).

Burgess, J. A., Halogen salts of silver. Econ. Geol. 13: 225-226, (1918).

Rogers, A. H. and Van Wagenen, H. R., The Chilean nitrate industry. A.I.M.E., Tr. 59: 26, (1918).

Rich, J. L., An old erosion surface in Idaho: is it Eocene? Econ. Geol. 13: 486-488, (1918).

Roberts, H. M. and Longyear, R. D., Genesis of the Sudbury nickel-copper ores as indicated by recent explorations. A.I.M.E., Tr. 59: 65-66, (1918).

Wheeler, H. A., Rapid formation of lead ore. A.I.M.E., Tr. 63: 318, (1920).

Winchell, A. N., Petrographic studies of limestone alteration at Bingham. A.I.M.E., Tr. 70: 900-901, (1924).

Howe, E., The gold ores of Grass Valley, California. Econ. Geol. 19: 620-621, (1924).

Bruce, E. L., Red Lake area of Patricia, Ontario. Can. Inst. Min. Metal. 29: 214-215, (1926).

REVIEWS

In addition to the impressive productivity indicated by the preceding lists, Lindgren found time to review the following 23 books and major reports:

Stelzner, A. W. and Bergeat, A. - Die Erzlagerstätten, pt. I. Econ. Geol. 1: 83-87, (1905-1906).

Spurr, J. E. - Geology of the Tonopah mining district. Econ. Geol. 1: 711-715, (1905-1906).

Knopf, A. - Notes on the Foothill copper belt of the Sierra Nevada. Econ. Geol. 2: 86-87, (1907).

Knopf, A. - An alteration of the Coast Range serpentine. Econ. Geol. 2: 87, (1907).

Beck, R. - Ueber die Beziehungen zwischen Erzgangen und Pegmatiten. Econ. Geol. 2: 87-88, (1907).

Stutzer, O. - Die Eisenerzlagerstätten bei Kiruna. Econ. Geol. 2: 88-90, (1907).

Stutzer, O. - Die Eisenerzlagerstätten Gellivare in Nordschweden. Econ. Geol. 2: 90-91, (1907).

452   GEOLOGY AT M.I.T. 1865-1965

Stutzer, O. - Turmalinführende Kobalterzgänge. Econ. Geol. 2: 194, (1907).

Ministerio de Agricultura (Argentina): Padron Minero de los Territorios Nacionales, 1890-1905. Econ. Geol. 2: 194-195, (1907).

Lawson, A. C. - The copper deposits of the Robinson mining district, Nevada. Econ. Geol. 2: 195-204, (1907).

Stelzner, A. W. and Bergeat, A. - Die Erzlagerstätten, pt. II. Econ. Geol. 2: 607-611, (1907).

Barrell, J. - Geology of the Marysville mining district, Montana. Econ. Geol. 2: 611-617, (1907).

McConnell, R. G. - Report on gold values in the Klondike high level gravels. Econ. Geol. 3: 650-652, (1908).

Ministerio de Agricultura (Argentina); Los Yacimientos de boratos y otros productos minerales explotables del Territorio de los Andes. Econ. Geol. 3: 652-653, (1908).

Mines Department (Transvaal): Report of the Geological Survey for the year 1906. Econ. Geol. 3: 653-656, (1908).

Krusch, P. - Untersuchung und Bewertung von Erzlagerstätten. Econ. Geol. 4: 65-67, (1909).

Sjögren, H. - Origin of the iron ores in older pre-Cambrian series of Sweden. Econ. Geol. 5: 494-498, (1910).

Beck, R. - Lehre von den Erzlagerstätten. Econ. Geol. 6: 79-82, (1911).

DuToit, A. L. - Copper-nickel deposits of the Insizwa, Mount Ayliff, East Griqualand. Econ. Geol. 8: 191-192, (1913).

Daly, R. S. - Igneous rocks and their origin. Science 41: 166, (1914).

Butler, Loughlin, Heikes, et al. - The ore deposits of Utah. Econ. Geol. 15: 683-685, (1920).

Wagner, P. A. - The iron deposits of the Union of South Africa. Econ. Geol. 24: 776-780, (1929).

Stansfield, J. - Assimilation and petrogenesis: separation of ores from magmas. Econ. Geol. 24: 782-783, (1929).

NOTE:

Hereafter, reviews similar to the preceding appeared in the Annotated Bibliography of Economic Geology.

DEPARTMENTAL REPORTS INCLUDED IN THE M.I.T. PRESIDENT'S
REPORTS FOR SYs 1911-12 TO 1932-33 INCLUSIVE

As head of Course XII (1912-1920) and (1927-1933), and of Course III (1920-1926), Lindgren prepared a brief report on Departmental activities following the end of each school year in June. These reports, included in the President's Report for each particular school year, were published in January of the following year or October of the same year, and were as follows:

1) M.I.T. Pres. Rept. SY 1911-12 (Jan. 1913), [M.I.T., B. 48/2: 111-115]
2) M.I.T. Pres. Rept. SY 1912-13 (Jan. 1914), [M.I.T., B. 49/2: 94- 97]
3) M.I.T. Pres. Rept. SY 1913-14 (Jan. 1915), [M.I.T., B. 50/2: 102-106]
4) M.I.T. Pres. Rept. SY 1914-15 (Jan. 1916), [M.I.T., B. 51/2: 102-106]
5) M.I.T. Pres. Rept. SY 1915-16 (Jan. 1917), [M.I.T., B. 52/2: 98-100]
6) M.I.T. Pres. Rept. SY 1916-17 (Jan. 1918), [M.I.T., B. 53/2: 89- 91]
7) M.I.T. Pres. Rept. SY 1917-18 (Jan. 1919), [M.I.T., B. 54/2: 69- 70]

```
 8) M.I.T. Pres. Rept. SY 1918-19 (Jan. 1920),  [M.I.T., B. 55/6:  63- 64]
 9) M.I.T. Pres. Rept. SY 1919-20 (Jan. 1921),  [M.I.T., B. 56/3:  72- 73]
10) M.I.T. Pres. Rept. SY 1920-21 (Jan. 1922),  [M.I.T., B. 57/3:  57- 59]
11) M.I.T. Pres. Rept. SY 1921-22 (Oct. 1922),  [M.I.T., B. 58/3:  51- 54]
12) M.I.T. Pres. Rept. SY 1922-23 (Oct. 1923),  [M.I.T., B. 59/3:  59- 62]
13) M.I.T. Pres. Rept. SY 1923-24 (Oct. 1924),  [M.I.T., B. 60/3:  57- 60]
14) M.I.T. Pres. Rept. SY 1924-25 (Oct. 1925),  [M.I.T., B. 61/3:  13- 14]
15) M.I.T. Pres. Rept. SY 1925-26 (Oct. 1926),  [M.I.T., B. 62/3:  15- 16]
16) M.I.T. Pres. Rept. SY 1926-27 (Oct. 1927),  [M.I.T., B. 63/3:  49- 51]
17) M.I.T. Pres. Rept. SY 1927-28 (Oct. 1928),  [M.I.T., B. 64/3:  47- 48]
18) M.I.T. Pres. Rept. SY 1928-29 (Oct. 1929),  [M.I.T., B. 65/3:  61- 63]
19) M.I.T. Pres. Rept. SY 1929-30 (Oct. 1930),  [M.I.T., B. 66/3:  65- 66]
20) M.I.T. Pres. Rept. SY 1930-31 (Oct. 1931),  [M.I.T., B. 67/3: 113-115]
21) M.I.T. Pres. Rept. SY 1931-32 (Oct. 1932),  [M.I.T., B. 68/3: 106-107]
22) M.I.T. Pres. Rept. SY 1932-33 (Oct. 1933),  [M.I.T., B. 69/3:  94- 96]
```

APPENDIX A

As stated in the preceding biographical sketch, the following article by Dr. Martinsson, titled "Waldemar Lindgren goes west," is so interesting, and so informative about Lindgren, as a recent college graduate engineer, and about the 1880s, when he emigrated from his homeland in Sweden to the United States, that I have appended it to my biographical sketch, with the gracious permission of the author and publisher.

Martinsson, A., 1973: Waldemar Lindgren goes west. *Geologiska Föreningens i Stockholm Förhandlingar*, Vol. 95, pp. 280—287. Stockholm, June 15, 1973.

Waldemar Lindgren goes west

ANDERS MARTINSSON

Four letters from Waldemar Lindgren (1860—1939) to his parents in Sweden give an account of his voyage across the Atlantic as an emigrant in 1883 and the initial phase of his first work as a geologist, in the Big Belt Mountains of Montana.

Anders Martinsson, Department of Palaeobiology, Box 564, S-751 22 Uppsala, 24th November, 1972.

Documents illustrating how the young Waldemar Lindgren became a geologist have had a remarkable tendency to fall into the hands of students of later generations at the same school in Kalmar and with the same scientific inclinations. Forty years ago, one of them, Erik Åhman, now a Senior State Geologist with the Geological Survey of Sweden, came across among the rubbish in a Kalmar attic Waldemar Lindgren's copy of Graham & Otto's *Lehrbuch der Chemie*, given to him by his uncle Aron Lindgren, a County Judge like his father. This find was the first step towards one of the Lindgren biographies (Åhman 1953). Waldemar Lindgren's several biographers have shown that his interest in geology dates back to a trip as a ten year old boy in 1870 to Lysekil on the western coast of Sweden and that the two judges supported his mineral collecting trips and his visits to mines in the following years. The type of school (*högre allmänt läroverk*) which Lindgren attended in Kalmar took him through what corresponds to the junior college in the present educational system in the United States. Immediately upon his graduation as a *student* (which in Sweden is a very definite concept, implying competence to attend a university proper), he wrote, on 2nd June, 1878, a remarkable statement with acknowledgements to his family and friends (Åhman 1953: 17) and a bold programme for the future:

"I feel it — I am standing at one of the landmarks of life. One step has been taken; I stop for a moment and see many roads lying open in front of me. But I do not hesitate; my course is clearly defined to me. Forwards on it. New courage, new vigour! I am now going out into the world to struggle alone... May the idea of the True, the Good, and the Beautiful always be the light and goal towards which I strive, for ever the sworn soldier of Science. *Nec aspera terrent!*"

The True and the Beautiful (*Sanna, sköna*) was the motto of the literary society of the college, founded in 1864 and called N3. While digging through its chaotic archives in 1949 to find material for a radio programme, I came across the writings of the originally anonymous winner of the society's First Prize in 1876 with the poem "Zindari and Verouna", an Indian love story, whose author particularly appealed to me as being the only predecessor in the society who had documented a tangible interest in the natural sciences. A number of travel accounts by the same author, who turned out to be Waldemar Lindgren, from the mountains, glaciers, and fiords of Norway, were read to the society in 1876 and 1877. Parts have later been published by Charles Pierre Backman (1964: 99—108) — who, by coincidence, had started his sixty-three year career in Swedish journalism in 1906 by reviewing a book given to him by Waldemar Lindgren's mother. He also published Waldemar Lindgren's eleven-verse poem entitled "Traversing a mountain lake in Norway", which was written in 1877.

These writings hardly contain geology enough to be reproduced in a geological publication, but four typed copies of letters found in Pierre Backman's remaining collections do. They form an account of Waldemar Lindgren's journey from Freiberg in Saxony to his first area of field work in Montana. After his graduation in 1882 from the *Bergakademie* as a mining engineer he stayed in Freiberg until the end of May, 1883, which is when his account starts. It is also largely concerned with the contents of a brief message from Liverpool which has not been traced. The copies of the letters follow very accurately Swedish usage before the spelling reform of 1906, although the typewriter is of a later date; this provides some guarantee that the copies are literal. Here they follow in an English translation which at least has much of its local and Swedish-German background in common with Lindgren's in the early eighties. The letters hardly need annotation. Several biographers, among them Graton (1933), Butler (1950), and Jensen (1960) have given detailed accounts of Waldemar Lindgren's subsequent career. Jensen even subdivided this career into three periods of about a quarter of a century, placing one of Lindgren's main "landmarks of life" exactly within the period of the letters presented here, and the other one on his appointment as William Barton Rogers Professor of Economic Geology at the Massachusetts Institute of Technology, which ensued shortly after his formal appointment by the Secretary of the Interior as Chief of the United States Geological Survey, an institution which he had then served for 26 years.

Acknowledgements. — Sincere thanks are due to Dr. Ellis L. Yochelson, U.S. Geological Survey, who prompted the publication of these letters, to Dr. Robert F. Lundin, Arizona State University, who congenially examined them in the English draft, and to Dr. Robert R. Shrock, Emeritus Professor at the M.I.T., who most generously supplied all the biographical information available at his Department and in the Lindgren Library.

From Freiberg to the Banks of Newfoundland

R.M.S. *Britannic,* Friday, 9th June, 1883, 400 miles from New York.

Dearest Parents: I hope that my letter from Liverpool arrived in order. I had intended to write more in detail and more extensively, but since contrary to my presumption we did not call at Queenstown, I had time to write only a few lines. Until today my journey has been excellent in all respects. Lightheartedly, without debts and followed by the cheers of my comrades I steamed off from Freiberg. The journey from Aix-la-Chapelle to Brussels and Ostend was as agreeable as could possibly be expected: it was entertaining to see north Belgium or Flanders, if even from a railway window. All the country is like one single large city, and one travels all the time between villas, factories, and gardens. Around Bruges and Ghent gardening is much advanced — even more than in Holland, they say, and it is true that one often travels right through flowerbeds. In Ostend we went right on board a little homely steamer which was to take us across to Dover. We had the most wonderful weather to be expected, and the Channel was literally like a mirror. At Dover it was I who took the command, since my friend and travel companion did not know many words of English and it was I who had induced him to travel *via* England. At Dover it was quite beautiful: high limestone cliffs, white like chalk [!], and sharply contrasting with the verdure, and the sea was green and smooth like a mirror.

In half an hour the steam-ferry arrived from Calais, a curious thing consisting of two vessels, chained together and united under one deck. The express train for London departed at once and went off at a desperate speed considerably faster than the fast trains in Germany. One sees very little of agriculture in England; almost everything seems to be green pasture with high hedges and large herds of sheep and cows. We arrived in London — Charing Cross — in the afternoon after three hours of railway journey. I had a direct ticket from Dresden, and my belongings were equally checked through from Dresden directly to London. After some difficulty — since it is the London season now and all the hotels are booked up — we found rooms at Craven's Hotel near the Strand — the main business street — and I arranged my business during Thursday and Friday — collected letters, equipped myself with clothes, etc., which in England are inexpensive and of excellent quality. I am now also enclosing some picture post-cards of London. On Friday we went with all our things up to Euston Station and boarded the Flying Express with Liverpool as our destination. After five hours we were in Liverpool. Already in London I had in the "American Exchange" ordered two tickets for the White Star liner *Britannic,* which was to depart from Liverpool at 3 p.m. We spent the night in Liverpool, strolled about in the streets and made some arrangements while waiting. We picked up our tickets and found to our great satisfaction that we had obtained excellent places. At this time the steamers from Europe to America are not particularly crowded, and through the influence of the "American Exchange" — in brackets a superb institution — we obtained places which should have cost 18 guineas, whilst we paid just 15 guineas or 310 Marks. These liners have only deck passengers and first class. The difference in price depends only on the situation of the berth and the number of persons in the cabin. All liners, except one or two, carry deck passengers, and consequently we have about 800 emigrants, mostly Swedes and Irishmen. At 3 p.m. last Saturday we went on board a little tender which carried the passengers out to the *Britannic* which was anchored a little farther out on the Clyde [!]. Three liners were bound to depart on the same day: the *Britannic,* the *Scythia* (Cunard Line), and the *Alaska* (Guion Line). The latter is the fastest ship on the Atlantic and will probably arrive 10—20 hours before us. Next to the *Alaska* comes the *Britannic,* which makes the passage between New York and Queenstown in about 7 days and 10 hours. Tomorrow afternoon we expect to enter the harbour of New York, I cannot for the moment state the dimensions of the ship, but it is a real giant, not very broad, but immensely long. V. Rapp and I have a spacious cabin together, comfortably equipped with a sofa and a chair and it is overall as comfortable and elegant as one could possibly desire. The prices include all meals, only wine and beer excluded, and are from 15 to 22 guineas. At the top is the "hurricane deck" where one walks or takes a lazy rest in the sunshine. Then comes the lower — main — deck with the smoking-room and the lounge, then the second deck with a magnificent dining-room and the cabins. Last comes the third deck with the barber's shop, bathrooms, and other installations. From 8.30 till 10 there is breakfast, which is announced on a big Chinese tam-tam [!]. One is entitled to eat anything one wants and as much as one wants, and everything served is excellent. Usually I have a breakfast consisting of coffee, fried fish, eggs, and fresh bread. At 1 o'clock comes the luncheon with soup, beef-steak or whatever one wants. At 6 o'clock is the dinner, announced by tam-tam — a terrible noise heard all over the ship. There is an abundant selection of everything one may want — fish, fowl, beef, and pudding as well as dessert, and the limit for eating is set by the appetite only. The waiters are numerous and attentive, are called "stewards" and wear blue uniforms with gold. After dinner one takes a coffee with taste in the smoking-room, and if one has appetite enough for it, there is a light supper of tea and sandwiches or whatever one wants, at about 10 o'clock, after which one goes to bed conscious of having fulfilled the duties of the day. I enclose the list of passengers — there are no particularly remarkable or agreeable passengers — about half of them Americans, the rest French or English. Yes, very true — there are remarkable persons on board, indeed: there is Mr. Morgan, one of the wealthiest bankers in New York, and Mr. Vanderbilt, almost the wealthiest man in the world. The former is big and fat, and has a red nose, the latter the same, minus the red nose, looking very modest and good-natured.

Last Saturday when we left Liverpool the weather was as great as could be. All three liners steamed off at one time. It looked really magnificent. Slowly we went down the muddy Clyde [!], and after a few hours we were out in the Irish Channel, smooth as a mirror and with the blue montains of Wales to the left [!]. We headed south, and next morning we had the steep coasts of Ireland to the right [!]. Here the *Alaska* caught up with us and passed us — cheers and wavings of hands without an end. The lighthouse of Fastnet, a tall tower on a high rock out in the sea, was the last we saw of Europe, and in a few hours we were out in the Atlantic with the sea forming the horizon all around us. It was warm and sunny and only a slight rocking movement from the swell, which never ceases here in the open sea. One walks up and down, lies on the sofas, and leans against the railing and enjoys life altogether.

After a couple of days, however, it was different. We approached the cold and foggy region which surrounds Newfoundland. Here, off Newfoundland, comes the Arctic current and crosses the Gulf stream, and the results are mists, storms, and cold weather. The wind freshened, and the ship was rocked by the long, mighty waves of the Atlantic in a very ungentle manner so that one had much to do to remain on foot during the day and in one's berth during the night. Rapp became sea-sick and stayed in his berth for 48 hours. I remained in excellent health and had a ravenous appetite. On the Newfoundland Bank we came into the mist and proceeded slowly, ejecting terrible howls with the siren. Here it is necessary to watch for ice-bergs which at this time come drifting down from Baffin's Bay and are dangerous to come together with. Every half our the temperature of the water is measured. Within a few hours it decreased from 12 to 3 degrees. The winter coats were taken out and everything was damp, cold, and unpleasant. Tonight we left the disagreeable Newfoundland Banks behind, the mist and the cold weather equally, and again we are steaming ahead in sunshine and high temperature both in the air and in the sea. Tomorrow at 4 o'clock we will probably enter the harbour. I enclose in this letter several things of interest, including maps of the U.S.

Kindest regards to all of you; do not be concerned about me, because I feel excellent and will manage everywhere without difficulties. More from New York.

Yours affectionately,
Waldemar

[Added with an asterisk, the position of which is not indicated in the letter:] We usually make about 380—400 miles in 24 hours or 16 miles an hour — 16 knots. The number of miles to be covered next day is the subject of much betting.

The assistant geologist

Pittsfield, Massachusetts, 5th July, 1883.

Dearest Parents: Now my course for the near future is laid and ready. I am on my way to the West, to Montana Territory, employed as Assistant Geologist at the Northern Transcontinental Geological Survey. By a letter of recommendation from one of my friends, who is now in Mexico, to Mr. A. Thomas, President of the Northern Pacific Railroad, I obtained employment. The Northern Pacific Railroad which is now being built and is to be ready this autumn, connects St. Paul in Minnesota with Portland on the Pacific coast. This railway has been granted a strip of land on each side, 200 English miles broad and 2000 miles long. The railway passes through the territories along the border of British North America, which are remarkable with respect to geology and minerals. These regions are little known, and therefore the Pacific Railroad has entrusted the geological survey to Professor Pumpelly, a geologist famous even in Europe. This survey (*undersökning*) will take many years. Professor Pumpelly had already organized his staff and is now in Montana. It was asked by letter whether he wanted and could take me on. Partly Thomas, partly also the general manager in Drexel & Morgan's great banking firm, a Dane by the name of Christensen, were very kind and helped me. The day before yesterday there came a telegram from Pumpelly in Montana that I had been engaged as a geologist — everything free — travel, accommodation, meals, and 150 crowns a month to begin with. This is not much, but it is just a beginning, and Pumpelly does not know whether I am any good; if I prove that I am, a rise will not be delayed. I am very satisfied — this is just the work I prefer: a new region, almost unknown to science; impressive views and tracts out among the Rocky Mountains; many opportunities to gain distinction; a healthy life, even if it is a little strenuous. I have had much to do in New York, since there has been much business to transact and many implements to procure that are necessary for the sojourn out in the wilderness of the West. A terrible heat of 30—33 degrees in the shade all the time in New York; it is one of the hottest cities in the world during the summer. My itinerary is as follows (at the cost of the Northern Pacific Railroad): New York—Chicago 30 hours, Chicago—St. Paul 24 hours, St. Paul—Bozeman (Montana) 61 hours, a grand total a five days and five nights by express. Tomorrow night I leave Albany and travel directly without breaks. However, owing to the sleeping-cars, drawingroom-cars, and dining-cars, the comfort on the railway is great, and it should not be too tiresome. Soon, then, I will be leaving civilization behind, to become one of its pioneers — a "frontier man" as they say here. Among the regions which we are to survey is the very famous, enigmatic Yellowstone River area with its hot springs and geysers. The papers enclosed show to you how the Pacific Railroad goes and how the geysers of Yellowstone appear.

I suppose you are wondering where I am now: Pittsfield, which you can easily find on the map, is a little beautiful town, situated fairly high and therefore used by many as a summer residence when it becomes so desperately hot in the large cities. The Mac Cays [!] are here and have rented a "cottage" for the summer, a little house with a garden, excellently beautiful and agreeable. I arrived here yesterday, was heartily welcomed and am staying here till tomorrow. It is most agreeable to rest a couple of days after the heat and noise in New York. They are as friendly to me as they have always been, and I was forced to settle down in their house at once. I would prefer to stay here longer, but Professor Pumpelly has cabled that I should come "at once". Here are Mr. and Mrs. Mac Cay and their three daughters, the eldest one is married to a lawyer who is at present in St. Paul, Minnesota, and there are two unmarried ones, whom I know from Dresden; one is 22 years and the other is 18. Young Mac Cay, who has just obtained a doctor's degree in chemistry, could not come because of important business in Princeton.

I find America delightful and feel superb. One eats very well and has a lot of comfort which is unknown in Europe. The hotels are excellent; one pays a total of 2 to 5 dollars (ususally 2.50 or 3 $), everything included. One may eat at any time, anything, and as much as one wants. Tipping is almost unknown. With regard to meals — there is breakfast at 8—9, coffee, sandwiches, beef-steak or fish, and fruits. Dinner is at 1 o'clock according to a long menu, and supper at 6 o'clock. There are lots of fruits here: bananas in large red and yellow bunches, particularly tasty when fried, pine-apples in big heaps and, in addition, all the common fruits. Most of it comes from Florida or the Bermuda Islands.

The journey from New York to Albany is very beautiful, near New York with houses in infinity and higher up impressive scenery where the river traverses the Catskill Mountains, an isolated, blue mountain region with sharp contours and covered by conifer forests. I recalled Irving's *Rip van Winkle* which Gerda used to read with me in the old days. On my way westward I shall be passing the Niagara Falls the day after tomorrow. The time difference between us will gradually increase — upon arrival at my destination it will be 10—11 hours, i.e., if you want to point to me, point right down towards the ground and just a little to the west.

I have not yet received any letter from you, although I had expected one in New York before I left. Address all letters c/o American Exchange 162, Broadway, N.Y. City, because I shall probably change my residence often, and they handle my mail.

The countryside here is rough, forested on the hills, and with corn, rye, and wheat fields in the valleys, fertile and beautiful. I have not seen a single poor house; most people seem to live in good circumstances. It makes one delighted to see the order and cleanliness which prevails everywhere. All the houses are wooden and the barns painted red as at home. All the small towns are as embedded in a garden, rows of trees along the streets, and every square is a park. Most houses have a *piazza* or a veranda. Now I cannot write more. Next time you hear from me I suppose I will be in Montana.

There has just been a little thunderstorm with rain, and now it is so fresh and cool and agreeable. I do not understand why it is that Einar [Waldemar's brother, the construction engineer] finds the people here disagreeable. My experience is that people are about the same in all countries and usually gentle and compliant if one takes them the right way. Einar is still in New York and is doing well. Most cordial greeting to all of you, Sigrid, Gerda, Tora, from

your affectionate Waldemar

On horseback to Gallatin

Livingstone, Montana Territory, July 19th, 1883.

Dearest Parents: Today I have a day off and can use my time for my correspondence. Many thanks for the letter of June 28th which, forwarded from Bozeman, arrived today. I was just beginning to wonder why it was that I had not received any letters. My long journey ended happily and was very interesting. Across the Niagara and the suspension bridge there, it went along the shore of Lake Erie through southern Canada and Indiana to Chicago in Illinois. I have not much to say about this part of the country, since I was fast asleep. At least I have one good characteristic: on a steamer or on the railway, travelling or resting, I always sleep and eat well. In Chicago I stopped just for a few hours, changed cars and proceeded through Wisconsin and Minnesota — through the most magnificent country I have seen in my life, a real paradise for the farmer; green meadows, lush pasture, with clover an ell deep, long corn and wheat fields and neat, beautiful houses. I hardly saw one poorly dressed person or a poor house. If I had been a farmer, I would have settled in Wisconsin. Now I do not wonder why so many Swedes have gone there. In St. Paul, the last big city one comes to, I changed cars and placed myself for three days in the comfortable Palace Car "Pyramid Park" which was to carry me to Bozeman. They have the usual first class cars here, very comfortable and practically arranged, and Pullman Palace Cars for long journeys, equipped with all possible comfort and luxury.

And so we proceeded towards the West. Northern and western Minnesota does not have soils as good as in the eastern part, but it is not really bad. So we arrived in Dakota Territory and traversed it for 24 hours, across the endless prairies, sometimes quite smooth, sometimes with undulating hills, "rolling prairie", with grass that was already beginning to turn yellow. Here and there lay a whitened buffalo skull; sometimes we caught sight of antelopes, which stood still for a while, looking at the train with curiosity at a safe distance. Sometimes passed by a settlement, with a row of yellow plank houses and men with broad-brimmed hats and woollen shirts.

"Bad lands" is the name for the western part of Dakota. It is a remarkably sterile, dissected region with the most absurdly shaped mountains of yellow sandstone, fire-red clay, and black, partly burning seams of lignite. One part of it, particularly bizarre and fantastic, is called Pyramid Park.

In Montana we followed the broad valley of the Yellowstone for one whole day: barren sandy hills with a green strip in the middle along the river. The railway is here less safely built than should be permitted — much more with regard to getting the business quickly started than to safety: sand banks which are in a disagreeably mobile state and wooden bridges which squeak in a suspicious manner upon crossing. However, I suppose things will soon be better; the railway was completed only a few months ago. Across the Missouri there is a magnificent, immensely long iron bridge, constructed by Einar's firm, Morrison, in New York.

At last the mountains are coming into sight — in the distance they lock like blue clouds with white spots. One approaches them ever more, and at Livingstone, "the Gate to the Mountains", one gets into them. Now the railway climbs rapidly; at Muir the pass is at 5000 feet; through a magnificent alpine landscape one proceeds to Bozeman. (I had it wrong in my previous letter regarding the difference in time — the difference between Montana and Sweden is just 8½ hours.) Bozeman is a little town with 5000 inhabitants, some brick houses, here regarded as the acme of elegance — lots of saloons and gambling houses — a population consisting four-fifths of men. There is a peculiar life in streets, or rather the street. Half of the population is on horseback, and every minute wild characters with broad-brimmed hats, long beards, and blue woolen shirts come dashing along the street on the spur. Bozeman is situated in a broad, level valley with sky-high mountains on all sides.

In Bozeman I was instructed to join my party, consisting of one geologist, Mr. Davis, a student from Harvard [W. M. Davis], a "packer", three mules and four horses; they were working then in Yellowstone lower *cañon,* a few miles south of Livingstone. I completed my equipment in Bozeman and proceeded from there to Livingstone and from there to our camp. We camped three days in Yellowstone *cañon* and investigated the rocks and their mode of occurrence some miles down the *cañon*. There are several coal seams, whose positions should be determined in particular. The work was hard, with long rides and difficult climbs — in addition there was a real scourge in the form of nasty mosquitoes which in the evenings became unbearable along the Yellowstone. We had to sit in smoke or have a net around our heads to be left in peace — and it was difficult anyway. We are camping in the open air — but I have become used to it now and like it very much. I first spread a rubber sheet on the ground, then a couple of blankets as a substratum, and then I wrap myself in a couple of long blankets with my saddle under my head and sleeps as calmly and safely as in the most magnificent bed — and do not awake until Kamner, our "packer", *factotum,* and cook, with the voice of a Stentor cries "Breakfast, boys!" We have complete field equipment, with a tent for rainy days and tinned provisions. Our dinner consist of tomatoes, coffee, bread baked by ourselves, fried ham, beans, and some fruit preserve. Cream is left to imagination — by this I also avoid "leather in the stomach". A couple of weeks of camping out has been very beneficial to me. You would hardly recognize me: broadbrimmed helmet-hat, blue woollen shirt, high leather gaiters with spurs, a broad leather belt with an immensely long revolver, and my instruments over my shoulder. The country is peaceful — no Indians or ruffians, but it is the habit of the country to be armed, and it is always safe to be convinced of one's superiority. Some days ago I procured fresh meat — very delicions — Tora is lucky not to be a geologist out here — by shooting a big prairie hen, and if I get close enough, the antelopes will have to take care of themselves, too. The country is an elevated plain, very rough and dissected by deep valleys or canyons (*cañons*) with small brooks or rivulets, along which there is a lush verdure; on the elevations everything is dry, dusty and yellow with sun-flowers, "sagebush" [!], and a low cactus species with large, yellowish red flowers. To the left and to the right rise the great mountain ranges, ultramarine against the deep blue, dry atmosphere. Here and there along the major rivers there is a ranch or "a cattle-farmers settlement" — the greater part of the country is still barren. A week ago we spent Sunday calmly and agreeably. We camped in a high valley, 6000 feet above the level of the sea. It is difficult for me to imagine that I really am so far from you and out in "the far, far West", in the land of the wonderful Indian books.

Since Sunday we have been roaming the country north of the railway and are now heading towards Diamond City and Fort Logan. The day before yesterday I parted from my company and rode alone 40 miles in a day down to Livingstone to clarify some geology; it was a strenuous day. Yesterday I rested in a proper hotel — according to western standards. And now I have joined my comrades in Gallatin. More soon.

Most cordial greetings,

Yours,
Waldemar

Amongst the bears of the Big Belt Mountains

Camping on Boulder Creek (Big Belt Mountains) near Diamond City, July 29th, 1883, Sunday.

Dearest Parents: Since I last wrote to you we have advanced considerably during our trip northwards. We are now heading for Fort Benton on the Missouri, in order to investigate on the way the Little Belt Mountains and the Highwood Mountains, whose geology is as yet completely unknown. From there we intend to go across the prairie to the Main Divide of the Rocky Mountains at Cadotte's Pass and return to Helena — the capital of Montana with 10 000 inhabitants, many mines, comparatively civilized — and then finally in September make a trip through the *cañon* of the Missouri in a boat. I shall probably spend the winter here at the recently discovered coal fields near Bozeman — however, this is as yet uncertain. They will have to raise my salary considerably — otherwise I shall abandon the N. T. Survey and try to find employment at the mines around Helena. I think, however, that they will prefer to keep me.

Today is Sunday, and we are camping high up on a slope under immense pine-trees — east of us are the Black Mountains, a forested part of the Big Belt, famous for its bears — grey, brown, and black. 200 feet below us rushes the Boulder Creek in a narrow, rocky gorge, and to the west is Confederate Gulch, a deep valley with numerous gold-washing enterprises and Diamond City. The latter is very modest and does not live up to its brilliant name: a row of 30 small shacks inclined in different directions with a general store and a saloon. The period of splendour is over for Diamond — *urbs fuit*. About 1865 the gold-bearing beds were discovered here and an *estampede* took place, similar to the one of 1849 in California. Millions of dollars were washed out of the gravel, and Diamond City witnessed many kinds of remarkable scenes. Now this is past, and only a few gold-washing enterprises on a large scale are still profitable. Here and there, however, one meets some old fellow who has kept going faithfully since 1865 and washes his ground for a few months each year when there is a supply of water, and then they often make 16—30 $ a day when they have good luck. In a little valley called Montana Gulch which was once very profitable, I met an old character, a Bavarian from *das bayerische Hochgebirge,* who had emigrated in 1849 and after many vicissitudes had ended up here. He had a claim, where he used to wash a little for gold, a little shanty, a little garden, and a number of poultry. A real hermit. However, it does not seem to have been very bad, because I saw myself how he washed out a nugget worth half a dollar in the late afternoon. All the valley is much dissected and looks very remarkable. From Diamond City to Ft. Logan we travel along the main road — a comparatively great comfort. Often, however, we must penetrate across mountains and valleys and find our own way as well as we can. Up on the plateaus there are many cattle and sheep. They are healthy and live well, and they are very wild and stay out all the winter, in spite of the fact that it is often very severe. Like the reindeer, they scratch away the snow and are as fat in the winter as in the summer. To raise cattle is a very profitable business here.

Last Sunday we had a very nice camp near Sixteen Mile Creek. We caught lots of trout there, extremely delicious, and we felt very fine. At the same place there came a big brown bear and paid us a visit. One of us stood up the slope and took care of the horses when he saw that they were disturbed and were looking around. There was Bruin, calmly watching the situation for a while; he thought it over a little and took a walk down to the creek, where one of us was fishing. As this man was only armed with a small pocket revolver, he retreated in a hurry. The bears are very nice and well-behaved here — if they are not attacked, they never touch a man and move around and pick berries peacefully. Since I was out on an expedition among the mountains, I did not see Bruin, but I met something else: two rattlesnakes on a hill in the sunshine — they are everywhere here — however, one does not come across them too often.

Today, being the astronomer of the expedition, I made observations to determine the time, the meridian, and the deviation of the compass. Since we mostly work with a compass we have to be familiar with it. Now the evening is approaching, and Hamner [Kamner?] is beginning to prepare the supper which seems intended to consist of oat porridge, corn, ham, bread, tea, and fruit preserves as dessert. Everything is boiled and fried over an open fire, and the bread is baked with baking-powder in a pot embedded in coal. Those days when we are in the neighbourhood of some ranch in the valleys are great feasts, since then there are milk, potatoes, and vegetables around. We are all doing well, however. The mountain air arouses a magnificent appetite, and the outdoor life is healthy. We have not made use of our tents yet and prefer to sleep under the stars, well wrapped in our canvas, rubber sheets, and blankets. Once — it was near Livingstone — there was sharp frost in the morning. The best part of the day is the evening around the immense camp fire, since there is plenty of dry wood, and we do not need to save it.

More from Fort Benton. Cordial greetings to everybody.

Yours affectionately,
Waldemar

REFERENCES

Åhman, E., 1953: Professor Waldemar Lindgren, en nestor inom malmgeologien. *Kalmar nations skriftserie 29 (1952)*, 15—21. Uppsala.

Backman, C. P., 1964: *Skola för skalder*. Förbundet N 3 1864—1895. 387 pp. Stockholm.

Butler, B. S., 1950: Memorial to Waldemar Lindgren. *Proc. Geol. Soc. Am. Annu. Rep. 1969*, 177—196. Washington, D. C.

Graton, L. C., 1933: Life and scientific work of Waldemar Lindgren. *Ore Deposits of the Western States (Lindgren Volume)*.

Jensen, M. L., 1960: Centenary of Waldemar Lindgren. *Econ. Geol. 55*, III—VIII.

APPENDIX B

BIOGRAPHIES AND BIOGRAPHICAL REFERENCES TO WALDEMAR LINDGREN (1860-1939)

It is hoped that the following list includes the most important and comprehensive biographies and biographical sketches of Waldemar Lindgren, particularly those in English. I feel sure, however, that the list fails in including such publications in countries other than the United States, and in languages other than English. For this lack I ask the reader's indulgence. Fortunately, several of Lindgren's countrymen (Erik Åhman, Charles Pierre Backman, Anders Martinsson, and C. Richard Soderberg) have provided considerable information about his youth and about the region of southeastern Sweden where he was born; for this indispensable assistance I am most grateful.

The following references are listed in chronological order so as to separate those written while he was living from those published after his death. Such order also shows the continued interest in Lindgren since his death in 1939, particularly in Sweden, his homeland.

(1912-1939) News Items

Tech. Rev. 14/1: 72 (Jan. 1912)	Lindgren appointed W. B. Rogers Professor of Economic Geology and Head of Course XII.
Tech. Rev. 14/6: 365-367 (June 1912)	Lindgren called to the Institute to head Department of Geology, succeeding Dr. Jaggar, who goes to Hawaii.
Tech. Rev. 26/5: 244 (March 1924)	Lindgren elected President of the Geological Society of America.
Tech. Rev. 31/4: 214 (Feb. 1929)	Receives Penrose Gold Medal of the Society of Economic Geologists. Portrait.

Tech. Rev. 33/7: i, f.p. 358 (April 1931)	Elected an Honorary Member of the A.I.M.E.
Tech. Rev. 35/8: 300-301	Retires and becomes Professor Emeritus and Honorary Lecturer.
Tech. Rev. 36/7: 389 (April 1934)	Awarded Penrose Medal of the Geological Society of America.
Tech. Rev. 38/1: i, f.p. 36 (Oct. 1935)	Receives honorary degree of Sc.D. from Harvard - "... a geologist to whom all men turn for knowledge of the metallic secrets hidden in the rocks."
Tech. Rev. 38/8: 350 (May 1936)	Receives silver Médaille Gustave Trasenster (Belgium).
Tech. Rev. 39/5: i, f.p. 216 (March 1937)	Receives Wollaston Medal of the Geological Society of London.
Tech. Rev. 42/2: 77 (Dec. 1939)	Obituarial notice and portrait.
T.E.N. 10/2: 53, 82 (March 1929)	Lindgren comments on Herbert Hoover, as one of his early field assistants.
T.E.N. 10/8: 409 (Jan. 1930)	Brief biography of Lindgren, with a portrait.
T.E.N. 11/4: 144, 168 (May 1930)	An informative report on the Department of Geology, by Lindgren.
The Tech (Nov. 7, 1939)	"Dr. Lindgren, Retired Prof., Dies Nov. 3rd. Distinguished Geologist at Technology Thirty Years. Held Many Degrees."

(1912) Anonymous
  Professor Lindgren called to the Institute. Tech. Rev. 14/6: 365-367, (June 1912).

(1931) Anonymous
  Waldemar Lindgren. Min. Metal. 12/291: 125, port., (March 1931).

(1933) The Committee on the Lindgren Volume (J. W. Finch, Chairman)
  Preface, p. xi-xii in Ore Deposits of the Western States (Lindgren Volume). New York: A.I.M.E., xxxiv + 797 p., port., il., (Feb. 1933).

(1933) Graton, L. C.
  Life and scientific work of Waldemar Lindgren, p. xiii-xxxii in Ore Deposits of the Western States (Lindgren Volume). New York: A.I.M.E., xxxiv + 797 p., port., il., (Feb. 1933).

(1934) Bastin, E. S.
  Presentation Address [Presentation of the Penrose Medal of the Geological Society of America to Waldemar Lindgren]. Geol. Soc. Am., Pr. 1933: 43-45, port., (June 1934).

(1935) Schiffner, C.
  Aus dem Leben alter Freiberger Studenten. Freiberg Sa.: Verlagsanstalt Ernst Mauckish, (1935). Lindgren's biography is on p. 300-304.

(1936) Anonymous
  M. Waldemar Lindgren: Titulaire pour 1936 de la Médaille Gustave TRASENSTER. Revue Universelle des Mines, 8me Série, Tome XII/3: 93-94, port., (March 1936).

(1937) Kitson, Sir Albert
[Citation of Waldemar Lindgren for the Wollaston Medal of the Geological Society of London, 19 February 1937.] Geol. Soc. London, Q.J. 93/4: 127-128, (1937).

(1939) Anonymous
Dr. Lindgren, Retired Prof., ... (for complete title of sketch see the list of M.I.T. publications on a preceding page). The Tech. (Nov. 7, 1939).

(1939) Ramdohr, P.
Waldemar Lindgren (1860-1939)
Zeitschr. prakt. Geol., Jahrg. 47, Heft 11: 187, (Nov. 1939).

(1939) Geijer, P.
Waldemar Lindgren, Feb. 14, 1860 - Nov. 3, 1939. Geol. Fören. Stockholm Förhandl. Bd. 61/4: 509-512, port., (Nov.-Dec. 1939).

(1939) Graton, L. C.
Waldemar Lindgren, 1860-1939. Econ. Geol. 34/8: 850a-850f, port., (1939).

(1939) Holland, T. M.
Prof. W. Lindgren (1860-1939). Nature 144/3661: 1083-1084, (Dec. 1939).

(1939) McLaughlin, D. H.
Waldemar Lindgren (1860-1939). Min. Metal. 20/396: 571-572, (Dec. 1939).

(1939) Newhouse, W. H.
Waldemar Lindgren, 1860-1939. Science 90/2347: 584-585, (Dec. 1939).

(1939) Anonymous
Waldemar Lindgren, 1860-1939. Tech. Rev. 42/2: 77, port., (Dec. 1939).

(1940) C_____, J. M.
Waldemar Lindgren (1860-1939). Soc. nac. minería (Chile). Bol. minero, Ano 46, Nr. 477: 5-8, port., (Jan. 1940).

(1940) Anonymous
(Waldemar Lindgren, 1860-1939.) Wash. Ac. Sci. Jour. 30/2: 92, (Feb. 1940).

(1940) Buerger, M. J.
Memorial to Waldemar Lindgren (1860-1939). Am. Mineral. 25/3: 184-188, port., (Mar. 1940).

(1940) Bastin, E. S.
Memorial to Professor Waldemar Lindgren: Excerpt from Minutes of the Annual Meeting of the Division of Geology and Geography, National Research Council, held April 27, 1940. (‡) 2 p., (Apr. 1940).

(1940) Loughlin, G. F.
(Waldemar Lindgren, 1860-1939). Wash. Ac. Sci. Jour. 30/11: 497-499, (Nov. 1940).

(1940) Shimer, H. W.
Waldemar Lindgren (1860-1939). A.Ac.A.S., Pr. Vol. 74/6: 141-142, (Nov. 1940).

(1940) Schneiderhöhn, W.
Waldemar Lindgren (1860-1939). Zentralblatt Mineralogie 1940, Abt. A, Nr. 3: 65-69, port., (1940).

(1942) Berkey, C. P.
Waldemar Lindgren (1860-1939). Am. Phil. Soc. Yearbook 1941: 386-389, (1942).

(1942) C.P.B. [= Berkey, C. P.]
[Obituarial Notice of Waldemar Lindgren.] Geol. Soc. London, Q.J. + Pr. 97/2-4: 78-79, (April 10, 1942).

(1948) Åhman, Erik
The birth-place of Waldemar Lindgren. Notiser in Geol. Fören. Stockholm Förhandl. Bd. 70/2: 350-351, il., (Mar.-Apr. 1948).

(1950) Butler, B. S.
Memorial to Waldemar Lindgren. Geol. Soc. Am., Pr. 1949: 177-196, port., (June 1950). (This important biography contains personal tributes from the following associates of Lindgren: John M. Boutwell, Eldred D. Wilson, W. H. Newhouse, Alan M. Bateman, Walter C. Mendenhall, Herbert C. Hoover, and George Otis Smith.)

(195?) R.P.B. (Robert P. Bigelow)
Waldemar Lindgren. M.I.T. Biographies of Members of the Faculty and Officers of Administration, Vol. 2: 63, (195?). (#T-F5/F-14B in M.I.T. Institute Archives.)

(1953) Åhman, Erik
Professor Waldemar Lindgren, en nestor inom malmgeologien. Kalmar nations skriftserie 29 (1952): 15-21, (Uppsala 1953). (Rough translation by Dean Emeritus Richard Soderberg in tape and typescript form in M.I.T. Institute Archives.)

(1957) Hayward, Carle
Personal Correspondence: Letter from Hayward to Shrock dated November 27, 1957. Letter in M.I.T. Institute Archives.

(1960) The Editor [Alan M. Bateman]
Foreword: Waldemar Lindgren - and Economic Geology. Econ. Geol. 55/1 (Waldemar Lindgren Centennial Commemorative Number): i-ii, (Jan.-Feb. 1960).

(1960) Jensen, M. L.
Preface: Centenary of Waldemar Lindgren. Econ. Geol. 55/1 (Waldemar Lindgren Centennial Commemorative Number): iii-viii, (Jan.-Feb. 1960).

(1960) Graton, L. C.
Editorial: If Lindgren were here. Econ. Geol. 55/1 (Waldemar Lindgren Centennial Commemorative Number): 192-200, (Jan.-Feb. 1960).

(1964) Backman, C. P.
Skola för Skalder: Forbundet N3 1864-1895. Stockholm: Kalmarklubben N3, 387 p., il., (1964). (Waldemar Lindgren is mentioned a number of times in the chapter on "N3-skald som blev världsberömd geolog," p. 94-111.) (This book, which is a history of an important literary club at the Kalmar gymnasium, has a number of travel accounts and an eleven-verse poem titled "En färd över en fjällsjö i Norge," written by Lindgren in the late 1870s.)

(1973) Martinsson, Anders
Waldemar Lindgren goes west. Notiser in Geol. Fören. Stockholm Förhandl. Bd. 95/2: 280-287, (Stockholm, June 15, 1973). This interesting article contains four letters from Waldemar Lindgren to his parents in Sweden (translated from Swedish to English by Martinsson) in which he gives an account of his voyage across the Atlantic as an emigrant to America in 1883 and the initial phase of his first work as a geologist in the Big Belt Mountains of Montana.

(1976) Mather, Kirtley
Lindgren, Waldemar, p. 370-371 in Dictionary of Scientific Biography, Charles C. Gillespie, Ed. New York: Charles Scribner's Sons, Vol. 8, (1976).

APPENDIX C

In an editorial titled "The Education of the Geologist," (Econ. Geol. 18/4: 405-409, June-July 1923), Lindgren clearly describes his philosophy of education and the training an aspiring young geologist should have. It is considered timely enough today, fifty years later, to include in this departmental history.

"THE EDUCATION OF THE GEOLOGIST."

To criticize the younger generation has ever been the favorite sport of those who are more advanced in years. At the outset I wish to disclaim any such intention, fully realizing that new conditions and new problems confront those that are entering upon their life's work. It can not be denied, however, that the tendency at the present time is strongly towards applied geology, and that the pure science is more or less neglected. There was a time (and it is still so in some European countries) when economic geology was considered something infra. dig. of geologists. Here, at least, this feeling has passed and the pendulum has begun to swing towards the other extreme. My experience has been mainly in a technical school, where the utilitarian view finds strong advocates but even in the larger universities the geological sciences as such appear to be neglected. The temptations are strong, it is true. Everywhere the young men see how substantial awards fall to the economic geologist and how little in comparison pure science has to offer. A suggestion of the worth of the scientific career of the teacher and investigator falls in the great majority of cases on deaf ears, or is dismissed with a smile. Practical life, action, money - these are the lights that seem to call the young men. So that even those who are deeply and genuinely interested in geology soon turn more or less definitely towards the applied science.

The recent great demand for petroleum geologists has resulted in vast numbers of young men following this branch, and too often they are turned out simply provided with the minimum of knowledge necessary for an "outcrop hound" or for the chasing of anticlinal structures, but quite useless for the advancement of geology.

I am certainly not belittling this tendency towards the practical. For many engineering professors a modicum of geology may suffice but, when the majority of geologists in a country concentrate on applications, pure science is sure to suffer.

No one can deny that we are approaching this dangerous stage in this country. Any one who has tried to obtain high grade teachers in mineralogy, petrography, or general geology will know how few and far between are the real first class men who have it in them to do pioneering work in their sciences.

We in this country have often enough been accused of hurrying, rushing, and trying to "get there" by the quickest road, and to some extent the change [charge?] is well founded. The student who has passed through a four year course in geology and allied sciences think [thinks?] that he has little more to learn, and is first of all looking for rapid advancement, not realizing that only years of painstaking work and hard study lead to the top of the ladder. Particularly this applies to the economic geologist in the mining field. Too many regard applied geology simply as an engineering study which once mastered during a few years of study, will serve for a life time. Many of these men never publish anything, never try to advance the science or benefit their colleagues by writing their experiences. They are really petrified geologists, and of no use for the advancement of science. It is a healthy sign that high grade scientific societies begin to object to admit such men as members.

What is there to be done to remedy such an unfortunate state of affairs? Increased remuneration for teachers and investigators will do something, but nothing in the world will produce a great geologist but an intense devotion to this science coupled with talent or genius. I have little faith in research made to order, nor in elaborate equipment as a necessary requirement. The only really effective remedy is on the part of teachers to try to instil love of science in the minds of their ablest scholars. I believe that we should select the fittest and spend our time on them instead of trying to pound knowledge into the heads of those who in spite of intellectual inferiority try to make their way through our courses. I confess to a hearty dislike for examinations, recitations, and quizzes, which are the approved means for accomplishing the object last stated.

I am not aware that there is any agreement as to the studies necessary for the education of the geologist. It will be found that the older men in geology have attained their position through many different paths of education. Some are purely self-taught. Others have passed through academic courses, and have been obliged to pick up their practical knowledge afterwards. Others have passed through engineering schools and there selected geology as their life work. Most of these men sooner or later find that they are deficient in some allied science, necessary for their work. Many find it necessary to study surveying, for instance. In former days little exact geological mapping was taught. Mapping was simply a case of hand compass, pacing, and a pre-existing topographic map. If the latter was wrong, so much the worse for the resulting geology.

Even now there is little uniformity in the requirements for a geological degree. Some universities will award such a degree with little reference to the fundamental sciences, deeming them only of secondary importance. I have known of cases in universities where a student has been allowed to take nearly every geological course offered, when he was totally innocent of any knowledge of analytical geometry, descriptive geometry, and trigonometry.

In the following lines I shall set forth what seems to be the necessary requirements for a first degree of B.S. in Geology. In the first place some knowledge in English, history, logic, and languages should be required. Those who first take their B.A. degree will, of course, be at a considerable advantage. One cultural language besides English should be required. For candidates for the Doctor's degree good reading knowledge of both French and German should be demanded.

Regarding mathematics, there is much diversion of opinion. It seems to me that calculus, including differential equations, is indispensable. Even if the student scarcely ever has an opportunity to use these subjects in research, he must be familiar with the processes by which so many of the problems of physics and chemistry are solved. The same applies to descriptive geometry, an instrument absolutely needful for the solutions of problems of structural geology. Many students I find are appallingly unfamiliar with these simple constructions. Physics and chemistry must be insisted upon, including some work in quantitative analysis.

Modern requirements include, at least, the principles of physical chemistry, with knowledge of the phase rule and ability to understand diagrams illustrating chemical equilibria. These latter subjects are becoming increasingly useful in the study of the relations of minerals. Courses in geology, paleontology, mineralogy, and petrography are, of course, essential. Strong emphasis should be placed upon the determination of minerals and on the conditions of mineral formation. Seminars in geology, comprising reading on various subjects are very effective in promoting interest and self-reliance. Stuffing and hand feeding should be discouraged and instead the student should be taught how to do the work himself. Practice in geological mapping needs not to be emphasized, but what must be insisted upon is a course in topographic surveying and familiarity with transit, level, and plane table.

Courses in economic geology, including ore deposits, non-metallic deposits and hydrology are necessary. These courses should be accompanied by laboratory work in which the student is taught the construction of sections, and maps, and the examination and description of specimens of rocks and ores. The latter I have found by experience to be a most beneficial practice. The student may be able to determine the various minerals in sections and specimens, and yet be wholly unable to describe what he has seen. He dearly loves to launch into more or less hypothetical accounts of the origin of the specimen, but to make him describe in good English the actual relations observed is most difficult.

The courses in the various subjects should be carefully balanced to avoid any specialization. That will come soon enough. His undergraduate years should be devoted to a preliminary study of all important phases of the science.

It is well, I believe, to allow a certain number of optional subjects according to individual tendencies in subjects of civil engineering, mining engineering, metallurgy, astronomy, meteorology and the like.

A course such as outlined above would fit the young geologist to take up a subordinate position in any branch of geology. If [he] has the time and the means, post-graduate studies are most desirable. They are not essential for if he has been taught self-reliance in thinking and studying he will be able to acquire the advanced subjects by himself.

I know well that there is considerable difference of opinions among teachers in regard to some of these questions. It is hoped that others, who have longer experience and who are better qualified than myself may contribute to a discussion of the subjects here briefly referred to.

WALDEMAR LINDGREN.

APPENDIX D

Lindgren's summer field work and winter office and laboratory work from 1 December 1884 to the end of fiscal year 1893 were reported on regularly by G. F. Becker in the Annual Reports of the Survey as follows:

6th An. Rp. 1884-85, p. 70, (1886) - Field work on quicksilver deposits and microscopic work in the laboratory.

7th An. Rp. 1885-86, p. 93-97, (1887) - Worked with Becker in the California Gold Belt.

8th An. Rp. 1886-87, p. 154, (1889) - Turner and Lindgren worked on the survey of the Gold Belt, and the field work for his (i.e. Lindgren's) 1887 A.I.M.E. paper (see Bibliography) "... was done by Mr. Lindgren at his own expense during a leave of absence."

9th An. Rp. 1887-88, p. 102, (1889) - Worked on the quicksilver deposits and Gold Belt.

10th An. Rp. 1888-89, p. 141 in Pt. I. Geology, (1890) - Becker mentions Lindgren's assistance in his own work that resulted in Monograph XIII on The Geology of the Quicksilver Deposits of the Pacific Coast (Becker 1888).

11th An. Rp. 1889-90, p. 95-96, (1891) - Spent season mapping Placerville and adjoining sheets, as well as work in the Gold Belt.

12th An. Rp. 1890-91, p. 104-106, (1891) - Worked in the Gold Belt, and published report on the eruptive rocks of Montana (see reference 11--1891a in the Bibliography).

13th An. Rp. 1891-92, Pt. I, p. 133-135, (1893) - Becker reports that the work of Turner and Lindgren in mapping Gold Belt structure "... has been performed with great intelligence and with the utmost fidelity."

14th An. Rp. 1892-93, Pt. I, p. 192, (1893) - Becker's last report states that Lindgren took packtrain to Truckee sheet and also worked on several Gold Belt sheets.

During FY 1893-94, Lindgren had charge of his first field party and reported on its activities in the 15th Annual Report. Thereafter, for the next twelve seasons, and after having been advanced to the rank of Geologist, he and members of his party conducted field work in a number of the Western States, as reported in the 16th to 26th Annual Reports.

15th An. Rp. 1893-94, p. 174-175, (1895) - Lindgren makes his first individual report as Assistant Geologist, stating that he worked on the Truckee sheet and in mining districts of Nevada City and Grass Valley. This was Powell's last year as Director of the U.S. Geological Survey.

16th An. Rp. 1894-95, Pt. I, p. 35-36, (1895) - Lindgren's party worked in the Gold Belt and spent the winter on maps and laboratory work. This was C. D. Walcott's first year as Director of the Survey.

17th An. Rp. 1895-96, Pt. I, p. 47-48, (1896) - Lindgren's party worked in the Pacific region; Lindgren was assisted by H. C. Hoover (later President Herbert Hoover); during the off season laboratory work was conducted.

18th An. Rp. 1896-97, Pt. I, p. 44-45, (1897) - Lindgren's party worked in Idaho and Wyoming, with office work on the California folios.

19th An. Rp. 1897-98, Pt. I, p. 47-49, (1898) - Lindgren's party mapped in Idaho. By special request of the Secretary of the Interior, Lindgren went to Arizona to examine the Chiricahua Range.

20th An. Rp. 1898-99, Pt. I, p. 48, (1899) - Party mapped in Idaho. The Justice Department asked Lindgren to examine a property near Oracle, Arizona.

21st An. Rp. 1899-1900, Pt. I, p. 80-81, (1900) - Mapping carried on in Montana, Idaho, and Washington. The Secretary of the Interior asked Lindgren to testify in Spokane, Washington, regarding Government lands.

22nd An. Rp. 1900-01, Pt. I, p. 90-91, (1901) - Party worked in Oregon, and Lindgren spent offseason writing reports in Washington, D.C.

23rd An. Rp. 1901-02, p. 51-53, (1902) - Party members worked on Neocene gravels in Nevada; mapped in Arizona; and investigated gold in Kansas Cretaceous shales.

24th An. Rp. 1902-03, p. 61-62, (1903) - Mainly office work, with three weeks in Placer County, California.

25th An. Rp. 1903-04, p. 50, (1904) - Lindgren's party including L. C. Graton as an assistant, worked in Cripple Creek district of Colorado; Lindgren was on leave for six months to do consulting work in Australia. (See footnote on page 390 earlier on.)

26th An. Rp. 1904-05, p. 44-45, (1905) - Lindgren spent most of his time on office work on Cripple Creek data, assisted by L. C. Graton. He also made some examinations in Georgia and the Carolinas.

Reports on parties ceased after the 26th Annual Report; little is reported about Lindgren's field activities after 1905. On 1 March 1907 he succeeded S. F. Emmons as Chief of the Section of Economic Geology of Metalliferous Deposits, and thereafter, until leaving the Survey in late 1912 to join the M.I.T. faculty, he had to assume a substantial load of administrative work, which inevitably reduced the amount of time available for his own research interests. Nevertheless, he was able to give a series of regularly scheduled lectures at M.I.T. during the fall term of 1908, 1909, 1910 and 1911, and to continue to publish the results of his earlier field work. Even as late as 1912, however, Lindgren spent some time doing field work in Utah, Nevada, California, Colorado, and South Dakota. (33rd An. Rp. 1911-12, p. 57, 1912.) In November 1912, when he left the Survey to join the M.I.T. faculty, H. D. McCaskey succeeded him as Chief of the Section of Metallic Resources. Lindgren, however, did not sever his connections with the Survey altogether in 1912; rather, he continued on a special part-time basis until 1915 in order to complete projects started before he resigned as Chief Geologist, the rank he had attained on 3 November 1911. (See copy of certificate on page 409.)

The Days of "Horse and Buckboard" Field Work at the Turn of the Century

Although the identification of the driver in the photograph is not altogether certain, he is believed to have been Waldemar Lindgren, inasmuch as Lindgren worked in the southwestern part of the United States when the photograph was taken. The view shows yuccas in bloom south of Deming, Luna County, New Mexico, circa 1905.

(U. S. Geological Survey Lindgren, W. 367)

A Jaunty Lindgren in the Hawaiian Islands, 1903?

The photograph was taken on the south coast of Molokai, Kalawao County, Hawaii. Note the pince-nez on the black cord, a distinguishing Lindgren accoutrement. The date when the photograph was taken is unknown, but a good guess would be in the early summer of 1903 when Lindgren was on his way to Australia for a six-months consulting assignment. (See page 467.)

(U. S. Geological Survey Lindgren, W. 242)

(13)
FREDERIC HENRY LAHEE
(1884-1968)

FREDERIC HENRY LAHEE

MIT: 1912-1918

Frederic Henry Lahee was a distinguished son of New England, born and brought up in Massachusetts, trained at Harvard, and launched on a geological career by teaching, first at Harvard (1906-1912) and then at M.I.T. (1912-1918). World War I brought an end to his short academic career, however, and in 1918 he joined the Sun Oil Company as a field geologist. Advanced to Chief Geologist in 1920, he held that position until 1947; then became Geological and Research Counselor, and continued with Sun in that capacity until he retired permanently in 1955 at age 71. Under his able guidance the exploration department of the Sun Oil Company expanded from a small staff occupying but two rooms to a large and successful organization. As a practicing petroleum geologist he quickly attained a position of prominence and respect among his professional peers, and after retirement from Sun he continued to contribute his time and services to a number of important committees of the American Association of Petroleum Geologists, the one professional organization he had served long and efficiently in many capacities during his thirty-five years of active employment with Sun. The Association recognized his services and professional achievements by appointing him editor of the Bulletin from 1929 to 1932, electing him president for the term 1932-1933, awarding him honorary membership in 1947, and presenting him its highest award, the Sidney Powers Memorial Medal, in 1953. Shortly after his death the Association made its last gesture of recognition by establishing the Frederic Lahee Memorial Fund in his honor.

The 84-year-old Dr. Lahee was one of the senior statesmen of petroleum geology and is credited with having made numerous important contributions to geology while serving as an educator, scientist, author, oil finder, oil-company executive, and recorder of the petroleum industry's successes and failures in drilling. His two major contributions are his world-famous textbook, Field Geology (McGraw-Hill Book Co., Inc.), first published in 1916 and still in wide use today, 60 years later, having gone through six editions and translation into Spanish and Russian; and the "Lahee wildcat-classification system," first developed when he served as chairman of AAPG's old Committee on Statistics of Exploratory Drilling, and still in use today as the basic method by which AAPG and other organizations classify exploratory wells.

Inasmuch as Lahee spent only 6 of his more than 60 years of geological activities at M.I.T., I would normally limit the following biographical sketch to that short period; however, I have chosen to diverge from such a course and to discuss his entire career for the following reasons. His potential as a teacher and author became manifest during his brief appointment at M.I.T., and one of his major contributions to geology, the preparation and publication of his widely used field manual, Field Geology, started while he was at the Institute, but this ability at collecting, organizing and publishing useful data persisted throughout his entire career of more than 60 years. Secondly, he is an excellent example of a talented individual who was born, brought up, and educated in the New England environment, and who might normally have been expected to remain in the academic world of Cambridge and Boston, yet circumstances intervened and he left to gain renown as a petroleum geologist in the competitive and aggressive atmosphere of the petroleum industry. Yet throughout his professional career, as a practical geologist in an intensively competitive business, he never lost the love of and respect for the life style that he knew as a youth. He continued to develop his teaching and writing skills throughout his life; he became an unusually graceful ballroom dancer; he achieved national recognition for his ability as an ice skater, almost qualifying for the Silver Skates award at age 83; and he maintained a living style of courtesy, shy aloofness, dignity, and personal behavior that marked him as an outstanding figure in any gathering. These characteristics led many people to address him as <u>Doctor</u> Lahee; only his closest friends were invited to call him Fred. Finally, when at my request he provided me with what I found to be a fascinating account of his life, with permission to quote from it as I wished, and because the story of his kind of industrial executive is so seldom recorded, I have taken advantage of his kindness and cooperation and have quoted extensively from his comments in order to describe how one sensitive and talented youth, educated in the benign academic atmosphere of a college city, achieved greatness in the hurly-burly and turbulent atmosphere of one of America's most aggressive and imaginative industries, without abandoning the standards, teachings, and interests of his youth.

Frederic Lahee's impact on geology at M.I.T., though limited to a brief period of but six years, 1912-1918, was important in three regards. He initiated and developed an excellent program of instruction in methods of geological field work for students in geology, mining, and civil engineering, and from this program as a base prepared his well-known textbook, <u>Field Geology</u>, which after 60 years is still used around the world as an indispensable field manual. He was an excellent teacher of physical geology as well as of field methods; students who took these courses have told me they still remember his crisp and terse language in class, his well-organized lectures, and overall program, and his insistence on neat and accurate drawing, precise diction, and clear and concise reports. He was, in fact, a perfectionist, whether in the field, on the dance floor, or on the ice. Finally, he was a careful observer and recorder, and during his six years at the Institute he published ten papers on physical geology that indicate he would no doubt have continued to be a productive author in the broad field of geology had he remained in academic work. As it turned out, when he entered the petroleum industry in 1918, he immediately directed his attention to the geology of oil and gas, and the 90 papers and 40 committee reports that he published thereafter deal almost exclusively with some aspect of the petroleum industry.

In preparing the preceding précis and the following more detailed biographical sketch I have drawn freely on a number of memorials, notices, news reports, and other sources, in addition to numerous discussions and considerable correspondence with Dr. Lahee. These are cited in the footnote below.*

----

* Lahee, F. H., You're never too old .... 111, From 1899 to 1957 [about Lahee's skating history]. <u>Skating</u>, p. 10-11, (February 1958). -- Private correspondence between F. H. Lahee and R. R. Shrock [1959-1968].-- Rettger, R. E., Frederic Henry Lahee, Honorary Member, A.A.P.G., B.31/5: 840-842, port., (1947). -- Levorsen, A. I., Frederic Henry Lahee, Powers Memorial Medalist, A.A.P.G., B. 37: 1822-1825, port., (1953). -- Lahee,
(continued on following page)

BIRTH, ANCESTRY, AND YOUTH

Frederic Henry Lahee was born in Hingham, Massachusetts, on 27 July 1884. His father, Henry Charles Lahee, was one of a family of nine children, all born in or near London, England, and his grandfather Lahee was a professor of music at the Royal Academy and was organist of one of the cathedrals in London. Of the nine children, 4 boys and 5 girls, all the girls became musicians and teachers of music (piano, organ, singing, violin, cello, flute, etc.)! The boys were given a choice between a two-year course on a training ship and then going to sea, on the one hand, or a college education, on the other. Fred's father and one brother chose the sea, and both spent several years traveling about over the globe in the old-fashioned square-rigged sailing ships of those days.

On one of his trips, Fred's father landed in San Francisco and then crossed the United States to Boston whence he called on a family of cousins, the Frederick Longs, who lived in nearby Hingham. There he met Selina Ida May Long, whom he married after a brief visit to London and a return to the United States for life. He was naturalized soon after as a citizen of the United States.

Selina Long had studied to become a concert pianist, but was dissuaded from such a career, and in the course of a few years presented her husband with three children - Frederic Henry, Florence Ellen, and Arnold Warburton.

Fred wrote to me as follows about his early years (letter dated May 4, 1966):

> "During my childhood and teenage years, we moved eleven times. We lived in Hingham, then Forest Hills, Newton Highlands, Brookline, and finally Cambridge.
>
> "My father was for several years Secretary of the New England Conservatory of Music. Then he organized a bureau for the placement of musicians and teachers of music, which kept him busy until his retirement. Occasionally he went on lecture tours, and he wrote many articles and six books, mostly on music and related subjects. He was a member of the Boston Authors' Club.

(continued from preceding page)

F. H., Response [to Levorsen's citation for Powers Medal]. A.A.P.G. B.37: 1825-1826, (1953). -- Frederic Henry Lahee, retired geologist dies. The Dallas Morning News, p. 3D, (December 4, 1968). -- Memorial Service for Frederic Henry Lahee, First Unitarian Church of Dallas, (December 7, 1968), 7 p. [in R.R.S. papers]. -- AAPG establishes Frederic Lahee memorial fund. The Oil and Gas Journal, p. 41, (Jan. 6, 1969). -- Rettger, R. E., Frederic Henry Lahee (1884-1968). Newsletter of Dallas Geophys. Soc. and Dallas Geol. Soc., p. 8, (Feb. 1969). -- Rettger, R. E., Frederic Henry Lahee (1884-1968). A.A.P.G., B. 53/9: 2014, port., (1969). -- Heroy, W. B., Jr., Memorial to Frederic Henry Lahee, 1884-1968. Geol. Soc. Am., Mem. Vol. 1: 38-41, port., (1973).

"My mother played the piano beautifully. In our childhood days we three children often played with our toys under our concert grand piano. My father played the cello, and on Sundays we children used to enjoy hearing them play together. Later, we three learned to play the piano, and also my sister and I took lessons on the cello, and my brother, on the violin. In our teens we frequently played trios together. By the time I had completed my High School course, I knew 'by heart' and could play pretty well, many of the compositions of Beethoven, Mozart, Chopin, etc. But I gave up practicing after entering college, for I had to earn my way through by winning scholarships, doing odd jobs, etc.

"I may say that at the Brookline High School I won all three available prizes--in English composition, in German, and in Mathematics. But I was secretly approached by a teacher on the committee which awarded the prizes, and she suggested that possibly I would be willing to let the Math. prize go to the student who actually came out second. I agreed, and this was done, and no one ever knew of the change except those on the committee and myself.

THE HARVARD YEARS (1906-1912)

"That I was to go to Harvard was unquestioned from my early childhood. The idea of my going anywhere else never entered any of our heads. And my sister was destined for Radcliffe, and my brother also for Harvard. I am proud to say that all three of us won scholarships for our freshman year and for each of the three subsequent years of our college careers.

"At the close of my freshman year at Harvard I won the second Bowdoin Prize for a paper entitled "The Theory of Evolution." Also I received a 'detur,' a book bound in crimson, and presented to each of the highest ranking scholars in the freshman class. In entering Harvard I had anticipated two courses--I believe German and English composition--so that, to receive my A.B. degree I had to pass 15.5 courses, instead of the full required 17.5 courses. These I passed with 6 B's and all the rest A's, so that I was elected a member of Phi Beta Kappa.

"My first course in Geology, and I may say my first introduction to the subject, was a half course given by Prof. Shaler. For me it was just a 'fill-in.' I knew that I wanted to make some branch of natural science my life's work, but I did not know which. So, through my first two years, I did some heavy thinking. I decided against astronomy (for which I had constructed a telescope that would magnify 30 diameters when I was 11 years old), because I didn't want to stay up late at night. I decided against botany because I didn't want to ruin my eyes by looking through a microscope for several hours a day. I decided against zoology after we students, in a second course, had to watch the dissection of a living frog. So what was left but geology! And of course geology offered a splendid chance for travel.

> "I therefore concentrated on this science, became assistant in it, under Prof. J. B. Woodworth, then instructor, and finally took an A.M. in it as my major (in 1908), and in 1911 I received my Ph.D. in Geology. Thus I taught at Harvard, first as assistant and then as instructor, from 1906 to 1912. In 1912, through the kind interest and recommendations of Prof. Reginald Daly, I was offered an instructorship at Mass Inst. Tech., which I accepted."

During his Harvard years Lahee, conscious of his extreme bashfulness and desirous of overcoming it, secretly joined a class in dancing and two years later became the Secretary and Manager of a Cambridge group known as the "Cheap and Hungries," which was composed largely of college widows and young professors. Fred reports that he livened up the group by bringing in a number of young men and Radcliffe girls. Included in the latter was a talented young lady, Lucasta (Louie) Karr Hodge, whom he married on 23 December 1912, shortly after taking up his new duties across the Charles River at M.I.T. in Boston. To this union were born four children: Genevieve, Henry, Ruth Holden, and John Aspinwall.

THE M.I.T. YEARS (1912-1918)

Upon assuming his duties as an instructor in geology at M.I.T. in 1912, Lahee immediately had responsibility for half a dozen different subjects. During his entire six years, SYs 1912-13 through 1917-18, he had sole responsibility for Dynamical Geology (12.31), Structural and Field Geology (12.34, 12.64), Field Geology (12.62), and Geological Surveying (12.65), and during his last three years he offered Physiography (12.35), Economic Geology (12.45), and Hydrography (12.68). He also shared the lectures in General Geology (12.36) with Shimer; those in Geology of North America (12.91) with Lindgren and Shimer; and Advanced Field Geology (12.67) with Lindgren and Warren.

He was an instructor for two years, 1912-1914, and an assistant professor from 1914 to December 1918, when he resigned to enter the petroleum industry. He left M.I.T. too soon to leave much of an impact on the Institute program or on its students; his brilliant career was to be developed outside academic halls, in the great petroleum industry, yet when he reached the end of his active professional career he commanded as much respect from his former academic colleagues as from his industrial associates, for he had remained a true academician while developing into a skillful and astute industrial executive. It is because this son of New England so successfully blended the academic with the industrial that I have attempted to portray the nature and activities of his whole life rather than limit his biography to the six short years he served the Institute.

Even though Lahee was kept fully occupied at Harvard and M.I.T. during his years in Cambridge, he still managed to teach occasional subjects at Radcliffe and was an instructor in geology at Wellesley from 1908 to 1918.

Let me now go back to his first months at M.I.T. and quote what he wrote me about them (letter dated 4 May 1966):

> "In my teaching at M.I.T. I had one large class--I believe over 100--which I think met twice a week. It included civil engineers, sanitary engineers, mining engineers, and geologists, and possibly some others. All were required to take this beginning course. Not all were much interested in geology. I soon noticed that, when my subject became boresome, a general sound of rustling developed but I soon learned how to stop it. For every lecture I armed myself with several examples of the practical application of each of the topics to be treated. When the rustling was noticeable, I would throw in a description of a practical application, and at once the room became as quiet as a tomb. I remember one particular example, that of meandering rivers. The meanders had no special interest for these boys, but when I explained how the fill on the inner curve and the erosion on the outer bank of each meander affected property rights and property boundaries, everyone listened intently."

In addition to the course described above, Lahee offered one in which he trained the students how to use the various methods and techniques then employed in geological field work. The notes he worked up for this course were later expanded into his well-known <u>Field Geology</u>, a book often dubbed "the geologists' bible," that was first published in 1916, has gone through six revisions, has been translated into Spanish* and Russian, and is still

---

* Typical of the high regard in which geologists have held <u>Field Geology</u> is the sentiment expressed by one of Lahee's former South American students who wrote from Bogotá in September 1958:

> "Dear Doctor: Since the beginning of my career back in 1935, I have been deeply inspired and encouraged by your excellent 'Field Geology.' Today, as a professor of Structural Geology in our National University, in Bogotá, I have the greatest pleasure and satisfaction of using the Spanish translation of your magnificent book (just out of the 'Omega' press, Barcelona), as reference and text-book for my Colombian students.

> "In the last few years I have been actively working and fighting for the improvement of the scientific and technical education in the fields of Earth Sciences in my country. My cooperation with Father Ramirez toward this objective has culminated in this book, perhaps the only one on the subject in Spanish, to date.

> "Please accept this copy, Dr. Lahee, as grateful homage of one of your students who, faithful to the tradition contained in your book and writings, strives for the development of the national geologic thought in this Andéan country."

used widely around the world today, almost 60 years later!

From the same field course that provided the inspiration for Field Geology came some interesting serendipity in 1965 in the form of a generous contribution in support of student field work in geology at M.I.T. It happened this way. An M.I.T. mining undergraduate, Richard T. Lyons (III S.B. 1917), had taken with Lahee certain geology subjects that were required in his Mining Course and as a consequence had become interested in geology by the time he entered service in World War I immediately after graduation. When he was discharged from service and could find no employment in mining, he turned to Lahee for advice, through him obtained a job with an oil company, and thus commenced a successful and lifetime career in the petroleum industry. Recalling how useful to him had been the practical instruction received in Lahee's courses, and wishing to encourage Course XII undergraduates to get training in field mapping, he and Mrs. Lyons established the Richard and Sammie Lyons Fund in early 1965 to yield income that could help students meet their field training expenses.

WORLD WAR I AND EMPLOYMENT IN THE PETROLEUM INDUSTRY

By 1917 the Lahees had three children and Fred was wondering how he might support his growing family and at the same time use his geological training to aid the war effort.

> "When the first World War involved our country, I had a wife and three children to support, and no money except my immediate earnings. So I decided that with my knowledge of geology I might be of some use in the war effort by exploring for oil. Already Donald Barton [son of George H. Barton (III S.B. 1880)], Sidney Powers [XII S.M. 1913], and others had gone west for this purpose. In the summer of 1917 I had worked for Myron Fuller [XII S.B. 1896], of the firm of Fuller and Clapp [F.G. Clapp XII S.B. 1901], Consulting Geologists, and had been stationed in West Virginia where Fuller had undertaken a survey for the Sun Co. of Philadelphia. In 1918, before the war ended, Fuller had been appointed by Sun Oil Co. to organize a geological department of small size, himself to be Chief Geologist. I worked for him during that summer, but returned in late Sept. to continue with my teaching. However, I soon received from Fuller an offer to join his staff and a request as to what remuneration I would want. Well, I was receiving $2300 from M.I.T. [as an Assistant Professor], but in addition I received enough for taking Wellesley and Radcliffe girls on their required geological field trips, for two short consulting jobs, and for summer employment, to bring my total for the year up to

---

* The M.I.T. Registrar's records show that Lyons took four different subjects with Lahee: Dynamic Geology (12.31), Structural and Field Geology (12.64), Field Geology (12.62), and Geological Surveying (12.65).

> $5700.00; so I answered that I would be glad to join his staff for $6000.00. This, I understand, was quite a shock to Mr. J. Edgar Pew and for Mr. Myron Fuller; but I got it! Two years later, in 1920, Fuller resigned; Robert W. Pack was made Chief Geologist; and after a few months Pack was made Manager of the Beaumont office, and I was promoted to the office of Chief Geologist, a position which I held until 1947, when I was made Geological Counsellor. I held this latter position until retirement, at the age of 71, in 1955."

When I last visited Fred in his home in Dallas in 1966, he proudly showed me the beautiful office furniture in his own study that the Sun Oil Company had generously allowed him to take from his Company office when he retired.

The detailed geological work that Fred carried on for the Sun Oil Company during his 37 years of service is presumably preserved in the Company's confidential files, or destroyed, but Fred describes his responsibilities and activities in a general way as follows:

> "During my tenure as Chief Geologist [1920-1947] I had many a problem to face in my recommendations for expansion of my department (which for several years included field exploration, subsurface mapping, core drilling, paleontology, chemistry, and geophysics). Gradually the department grew, and as it grew the several classes of work were put under the immediate supervision of specialists, all reporting to me. At a fairly early stage in this growth geophysics was separated from geology, and was made a separate department under Dr. Charles Bazzoni.

> "During my period of active duties with Sun Oil Company, I was sent on three foreign trips. The first of these was to Argentina. My assignment was to examine two large ranches, each of more than a million acres in area, and lying just east of the Andes Mountains. The nearest station was then San Rafael, 100 miles east of the ranch headquarters. The 100 miles had to be driven in an open truck of French manufacture. I arrived at this ranch at the end of July, 1923, and had head-quartered there for at least six months. A month or so after my arrival two other geologists were sent down by Sun Oil Co. to assist me ... Between us, we examined in some detail, and reported upon, the general geology, prospects for oil, prospects for shale oil, sulphur deposits, lead ores, available water for irrigation and for power, prospects for cattle raising and for farming, and so on. I need not go into the later history of this project, except to say that Sun Oil Co., as such, did not go into it. In recent years I am told that many of my recommendations have been followed in a broad development program by outside interests.

> "My second trip, in 1926, was to Europe. I took my oldest daughter, then 13 years old, with me, and we visited many of our relatives in England and in Switzerland. On a side trip by myself, I paid a visit to the Schlumberger people in Paris, and also

to the Nautik Company in Hanover, Germany. My object, as requested by Mr. Joseph Pew (now deceased), was to find out all I could about equipment and procedures used by these people in their determining the inclination of holes drilled off the vertical. I was treated courteously and allowed to see the apparatus in action. I don't now remember the nature of my report.

"My third trip, in the summer of 1927, was with J. Edgar Pew to Lake Maracaibo in Venezuela. Following this trip, Mr. Pew brought Sam Williston and Charles Nichols to Dallas, assigning to them a major problem (1) of determining methods of surveying drilled holes for their inclination, and direction of inclination, from the vertical, and (2) development of a method of electrical field surveying for oil. In the progress of this work they had already discovered interesting facts as to the conductivity (or resistivity) of different rock materials when observed in a drill hole. They were actually about to learn the basic principles used in the Schlumberger method, but they soon checked and found out that Schlumberger had already applied for patent rights in the U.S.

"Through these several approaches, and also through talking to operators and reading articles on 'crooked holes,' I became much interested in the problem. Within my department, W. E. Winn invented the forerunner of Sperry Sun Company's 'sypho;' and Williston and Nichols worked on their gyroscopic instrument for well surveying. I prepared a long and comprehensive article on 'Crooked Holes,' which I delivered at the Fort Worth convention of the Am. Association of Petroleum Geologists in the spring of 1929. This was the longest paper ever delivered at a meeting of the A.A.P.G. I talked for one hour and 10 minutes. And apparently I held my audience, filling the assembly hall and reaching out into the corridors, in wrapt attention all that time.* [See item 61--1929c, The problem of crooked holes. A.A.P.G., B. 13: 1095-1162, (1929).]

"I believe that I was one of the earliest advocates of the value of sharing non-confidential data among the

---

* In his "Response," when given the A.A.P.G.'s prestigious Sidney Powers Memorial Medal, Lahee recalled the following interesting incident (A.A.P.G., B. 37: 1825, 1953):

"In the late twenties, at our annual convention in Fort Worth, I presented a paper on crooked holes, a subject at that time of paramount interest. I had been talking for some time, illustrating my points by lantern slides, when I received a note on a small piece of paper that had travelled from one to another until it finally reached me. It read,

'You have been talking $1\frac{3}{4}$ hours. Better stop.
                                                    Sidney.'

So I soon brought my speech to a conclusion. I fear that this was the longest made at any of our conventions. And I was grateful to Sidney Powers for calling me down. I did not feel angry or hurt." [In 1912 Lahee was a young instructor at M.I.T. while Powers was a graduate student (XII S.M. 1913).]

Frederic H. Lahee and family, circa 1925, ready to start from home (Mockingbird Lane, Dallas, Texas) to Colorado. This car served both as company and family car. (Also published in A.A.P.G., B. 50: 1102, 1966.)

> companies. And I am still an advocate of that principle. For instance, I can see no reason whatever why a central reputable bureau should not gather all the basic data used in calculating oil and gas reserves, as long as these data are now available to the public, but only through separate state bureaus, etc. As long as such data <u>are now available</u>, their release to such a central bureau would in no way involve the distribution of <u>confidential</u> information. Yet, there are plenty of near-sighted individuals who militate against this plan."

Even though he served Sun Oil Company as their Chief Geologist for some 27 years, he planned his time and activities so well that he was able to indulge his innate desire to do scholarly work in geology. As a result, during those 27 years he published more than 40 articles, many of them devoted to specific oil fields or to problems unique to the petroleum industry, such as: crooked holes; origin, occurrence, and migration of gas and oil; oil-field waters; classification of holes drilled for oil and gas; and sharing of drilling information. He never lost interest in the education of young geologists, and he wrote several articles on this subject. Finally, he kept abreast of the latest advances in field exploration methods and techniques and incorporated them in successive revisions of his <u>Field Geology</u>, the last and sixth such revision appearing in 1961, 45 years after the book was first published, and only a few years before his death in 1968. The fifth edition was translated into Spanish in 1958 as Geologia Practica (Barcelona: Omega), and a Russian translation has been used widely in the Soviet Union. In addition to all of the preceding scholarly publications, he still found time to prepare an impressive number of Reports, 41 in all, for half a dozen committees of the American Association of Petroleum Geologists. These are listed farther on following his Bibliography. Only a person with unusual discipline and persistence, systematic planning, and deep dedication to his work could accomplish so much in addition to the work required daily in his position as Sun's Chief Geologist.

SERVICES TO AND AWARDS FROM THE AMERICAN ASSOCIATION OF PETROLEUM GEOLOGISTS

Lahee became a member of the American Association of Petroleum Geologists in 1919, only three years after it was founded, and one year after it changed its original name from The Southwestern Association of Petroleum Geologists at its 1918 annual meeting, and served it in many ways during his lifetime. He was editor of the Association's Bulletin from 1929 to 1932, and served as president in 1932-1933. He served as chairman of half a dozen important committees, for which he wrote annual reports: Public Relations (later called Applications of Geology), 1932-1934; Publications, 1938-1939; Wildcat Drilling, 1935-1944; Statistics of Exploratory Drilling, 1945-1955; College Curricula in Petroleum Geology, 1940-1943; and Radioactive Mineral Exploration, 1952-1955. The Reports he prepared on the activities of the foregoing committees are cited in lists following his regular Bibliography. In 1947 the Association made him an Honorary Member and in 1953 they gave him their highest award, the Sidney Powers Memorial Medal.

ON STATISTICS AND THE OIL DEPLETION ALLOWANCE

Starting in 1935 Lahee began to collect data on wildcat drilling of oil wells, and for ten years, 1935-1944, published annual reports on his data in the Bulletin of the A.A.P.G. (See list following his Bibliography farther on). When the A.A.P.G. organized its Committee on Statistics of Exploratory Drilling in 1945, Lahee was designated chairman, and for the next decade he prepared eleven annual reports, for years 1945-1955 inclusive, which, like his earlier reports on wildcat drilling, were also published in the Bulletin (See list following his Bibliography farther on). These statistics and the precise definitions of types of holes involved, for which Lahee constructed a systematic classification, designated by some as the "Lahee System," have long been used by the oil industry and by government bureaus as a guide to development trends. In citing Lahee for the Sidney Powers Memorial Medal in 1953, Levorsen wrote this of Lahee's statistics (A.A.P.G., B. 37: 1822-1825, 1953):

> "There is more to these statistical reports than merely figures, however. They are based on a solid understanding of the intricate problems involved in the exploratory effort. In the hands of a person with less understanding and less interest in accurate definitions and precise language, they could easily become the cause of confusion. As it now stands 'Lahee's figures' are a by-word in the oil industry and are accepted everywhere as by far the most careful and intelligent analysis of the results of our exploratory and discovery effort. ...No geologist has done more than Lahee in the direction of providing both the industry and the nation with these needed, accurate statistics of annual discovery, relative efficiencies of different methods of exploration and the relation of effort to discovery."

When the weekly magazine _Time_, for 24 May 1954, p. 91, made certain statements about the depletion allowance which Lahee considered "absurd," he wrote a letter of protest to the Editor of _Time_, but was informed that the letter could not be published because there was not enough space to accommodate the pertinent facts and figures he cited. Lahee's letter was then discussed in the August 1954 issue of _Sun News_ (Dallas) in a story, "Sun Geologist Assails Magazine Statement on Depletion Allowance," and was printed in _Sun Marketing News_ for September 1954 (p. 8) under the heading, "Production Sunman Refutes 'Time' Depletion Statements." The letter is quoted below with the kind permission of the Sun Oil Company:

> "Your implied conclusions with reference to the $27\frac{1}{2}$ per cent depletion allowance, as stated on page 91 of your issue of Time for May 24, 1954, are so absurd that they need correction by some actual facts, some actual statistics.
>
> "Statistics begin with enumerations of individual cases. They conclude with totals and averages. The only fair way to judge for a large number is to use averages. Thus, if among 10,000 wells, 9,998 are drilled to depths between 1,000 and 10,000 feet and two are carried down to 20,000 feet, it would be absurd to conclude that _all were_ carried to 20,000 feet, or that the cost of drilling all must be based solely on a 20,000-foot depth.
>
> "The fact that a _very few_ men have had great success in their drilling for oil does not by any means imply that all those who drill for oil are going to be successful. In the last ten years, the industry has drilled 46,586 exploratory wells in territory where oil has not been found previously. What have the results been? Of these 46,586 wells, 41,389 were dry. The remaining 5,197 discovered oil, or gas, or oil plus gas. Many of them yielded so little that they had to be abandoned. Many of the oil wells opened fields so small that their total reserves are estimated at less than one million barrels. One million barrels of oil may sound to the uninitiated like a huge volume of oil, but actually, on the average, a field with any less oil does not pay the costs of its development.

> "Where careful figures are available for fields discovered at least six years ago--i.e., with six years of development history on which to base estimates of their reserves--statistics show that only 2.4 per cent of the total exploratory wells drilled in territory where oil had not previously been found, discovered oil fields that are paying out. In other words: only one out of every 41.6 such wells discovered an oil field that paid out. More than 94 per cent of the total number drilled are economic failures, with 3.6 per cent being dry gas wells.
>
> "This is why the 27½ per cent depletion allowance is necessary. It offers a fair and reasonable opportunity for those who drilled the many thousands of failures to try again. And just write it down in your book of 'musts', that this country must keep searching for oil. Reduce this depletion allowance and you will remove the inducement to keep searching. The real losers will be the American people in terms of their national defense and standard of living."

And Lahee's protest is just as relevant today, in 1974, as it was 20 years ago.

SERVICES ON THE AMERICAN PETROLEUM INSTITUTE COMMITTEE ON OIL RESERVES

Lahee joined the American Petroleum Institute in 1936, and from 1946 to 1955 he served the Institute as chairman of its important Committee on Oil Reserves. This committee publishes an annual report on the changes in oil reserves in the United States, and the detailed studies that are reported provide the data that the petroleum industry and the government must have in order to determine whether the nation is discovering more or less oil than it is consuming - a determination that is of crucial importance in today's (1974-1975) energy crisis. In 1952 Lahee was awarded a "Certificate of Appreciation" by the API, with the following inscription that eloquently summarizes his life work and life style:

> "Petroleum geologist; educator; author; respected by all in his chosen profession; unassuming; kind; faithful; diligent; outstanding in long and effective service in the scientific search for oil and gas reserves; true guardian of basic concepts; untiring in his efforts in developing cooperative procedures for the estimation of national petroleum reserves; member since 1936, and chairman since 1946 of the American Petroleum Institute Committee on Petroleum Reserves.
> November 12, 1952."

Eight years after his retirement from Sun Oil Company, and three years after the preceding award was made, Lahee was given a second accolade, a second "Certificate of Appreciation," which summarizes his long and important concern with the problem of the occurrence and recovery of petroleum:

> "For outstanding service as a member (1931 to present) of the Advisory Committee on Fundamental Research on Occurrence and Recovery of Petroleum; as a member (1937 through 1941) of Research Project 4 Advisory Committee; as a member (1942 through 1952) of Research Project 43 Advisory Committee; and as a member and chairman of a Special Committee to develop the format, contents, and general procedure for initiating and handling the biennial publication on Fundamental Research on Occurrence and Recovery of Petroleum.
>
> "Dr. Lahee has been a faithful participant in the fundamental research activities, and has given freely of his time and knowledge to the forwarding of this important Institute program.
> November 15, 1955."

DANCING AND SKATING

In spite of the impressive amount of work that Lahee turned out each year as a geologist, he still managed to find time to enjoy his family and to indulge in his favorite hobbies, which included music and photography, and particularly dancing and skating. Of the last two, in which his skill was equal to that of the professionals, he wrote me as follows:

> "In the late 1930's my back was seriously injured in an automobile accident. Several of the incipient ribs, low on the back, were broken off and the connecting sinews were badly torn. Due to this injury I had to give up tennis, which I had greatly enjoyed, because the frequent sudden lunges for the ball did damage. The surgeon told me that the only exercises I could take were swimming, dancing, skating, and walking. I am not much of a swimmer, and certainly walking in Dallas is most monotonous and unenjoyable. So, dancing and skating were left."

He was a skilled and graceful ballroom dancer, and at several A.A.P.G. conventions took the floor with his partner to perform request numbers. He remembered these occasions in a 1966 letter to me as follows:

> "At the Los Angeles Convention of A.A.P.G. in March, 1947, Mrs. Raymond Stehr (Sally Stehr) and I were asked to dance before the banquet gathering in the Biltmore Bowl (about 1200 people). With a large name band playing, and with nobody else on the floor, and in the bright spot lights, we danced a Viennese waltz and a samba, and, as my notes read, 'we got a tremendous hand.'
>
> "Five years later, on March 27, 1952 [Fred was then almost 68 years old], Sally and I were again asked to dance in the Biltmore Bowl, again before a capacity audience and at the annual banquet of the A.A.P.G. This time we danced a fast foxtrot, a samba, and a Viennese waltz. A picture [was] taken of us in action ..."

Fred learned to skate on frozen ponds in Brookline (Massachusetts) when only six or seven years old and passed the bronze, silver, and gold

Fred, at 83, skating a "Swing Dance" with the Pro.

medal tests of the Cambridge Skating Club by 1907 when a senior at Harvard. Then followed a long period of inactivity until some 30 years later, when he again resumed one of his favorite pastimes, recalling that,

> "I did relatively little skating from 1914 until about 1940. I certainly lost at least 28 years with no skating during this period. In 1940 or 1941 I joined the Dallas Figure Skating Club, and since that time I have been in many shows and have also passed the United States Figure Skating Association dance tests--preliminaries, bronzes, pre-silvers, and all but one of the silvers. My last such test, the American Waltz, I took and passed on April 18, 1958, at the age of 74."

SUMMATION

Frederic Henry Lahee's life spectrum was one with many bands of high achievement. He was a dedicated teacher and editor with highest standards; he was an imaginative and successful petroleum geologist and industrial executive; he was a persistent and indefatigable collector of data on well drilling which he then systematized into a logical classification of exploratory wells; he wrote clearly and concisely, as well as voluminously, on a wide range of geological subjects and problems; and always he strove for perfection whether as teacher, practical geologist, committee chairman, consultant, family head, ballroom dancer or figure skater. In the judgment of his many friends he came as near as any person could to achieving that perfection. He was proud, but never arrogant - proud of his ancestry;* proud of his own achievements, which were due in no small measure to determination, resolution, stern discipline, constant persistence and that driving desire for perfection; and he was equally proud of the accomplishments of others - his family, his professional colleagues, and his partners at play. So well did he show what man can be!

BIBLIOGRAPHY OF FREDERIC HENRY LAHEE

Abbreviations and symbols used in the following references are explained on p. 91 - 98. In general, abstracts are listed separately as well as with the complete article. Lahee's first publication, 1--1902, was written when he was a Junior in Brookline (Mass.) High School, and the next 12 were written before he published his first geological article. In addition to the numbered references, 41 Committee Reports are included at the very end of this Bibliography.

1--1902		The Curse of Dido. Translated by F.H.L. from the Latin into English verse, from Book IV (lines 607-629) of Vergil's Aeneid, and published in Sagamore of Brookline [Mass.] High School, (April 1902).
2--1903		Signs of Spring - The Birds. Boston Evening Transcript, (27 Feb. 1903).
3--1903a		Signs of Spring - The Insects. Ibid., (3 March 1903).
4--1903b		Signs of Spring - The Plants. Ibid., (12 March 1903).
5--1903c		Wanton destruction of life in the Parks. Brookline [Mass.] Chronicle, (23 May 1903).
6--1903d		Prima ab origine. Sagamore of Brookline [Mass.] High School, (May 1903).
7--1903e		Should the naturalist employ truth or fiction? [pseudonym Sympycna Fusca]. Gold Medal Essay at Brookline [Mass.] High School, (June 1903).
8--1904		The calls of spiders. Psyche 11: 74, (Aug. 1904).
9--1904a		A glimpse of autumn. Gardner Journal, (Oct. 1904).
10--1905		Thunderstorms. Boston Evening Transcript, (July 1905).
11--1905a		The Dog-Day. Ibid., (Aug. 1905).
12--1905b		Dews and Frosts. Ibid., (Sept. 1905).
13--1906		Life, a poem by F. H. Lahee. Ibid., (3 Aug. 1906).
14--1908		The filling of Emerald Lake [B.C.] by an alluvial fan. Science 27: 752-753, (1908).
15--1908a		An alluvial fan near Field, in British Columbia. Am. Geog. Soc., B. 40: 340-344, (1908).
16--1908b		A fault in an esker. Science 28: 654-655, (1908).
17--1909		Theory and hypothesis in geology. Science 30: 562-563, (1909).

---

* Fred mentioned in one letter to me that his father could trace his ancestry back to the time when the Spanish Armada was destroyed off the coast of Ireland in 1588; and that his mother could trace her ancestry back to the Crusaders in 1000 A.D. He also added that his wife, Louie Long, had an ancestor on the Mayflower.

18--1910   Dodecahedral jointing due to strain of cooling. Am. J. Sci. (4) 29: 169-170, (1910).

19--1912   Crescentic fractures of glacial origin. Am. J. Sci. (4) 33: 41-44, (1912).

20--1912a Relations of the degree of metamorphism to geological structure and to acid igneous intrusion in the Narragansett Basin, Rhode Island. Am. J. Sci. (4) 33: 249-262, 354-372, 447-469, (1912).

21--1912b A new fossiliferous horizon on Blueberry Mountain, in Littleton, New Hampshire. Science 36: 275-276, (1912).

22--1913   Geology of the new fossiliferous horizon and the underlying rocks, in Littleton, New Hampshire. Am. J. Sci. (4) 36: 231-250, il., (1913).

23--1914   Late Paleozoic glaciation in the Boston Basin, Mass. Am. J. Sci. (4) 37: 316-318, (1914).

24--1914a Crystalloblastic order and mineral development in metamorphism. J. Geol. 22: 500-515, (1914).

25--1914b Contemporaneous deformation: a criterion for aqueoglacial sedimentation. J. Geol. 22: 786-790, (1914).

26--1914c Misuse of the term "eruptive." Econ. Geol. 9: 72-73, (1914).

27--1916   Review of L. V. Pirsson and C. Schuchert's A Textbook of Geology. Science 43: 497-501, (1916).

28--1916a Field Geology. New York: McGraw-Hill Book Co., Inc., 1st ed., 508 p., (1916); 2nd impression, with appendix on geologic mapping, 528 p., [1917].
2nd ed., 649 p., (1923); See also 39--1923.
3rd ed., 780 p., (1931); See also 68--1931.
4th ed., 853 p., (1941); See also 101--1941.
5th ed., 883 p., (1952); See also 113--1952.
   Geologia Practica [Spanish edition of 5th edition]. Barcelona, Spain: Ediciones Omega, S.A., 874 p., (1958). See also 122--1958a.
6th ed., 926 p., (1961); See also 123--1961.

29--1916b Origin of the Lyman schists of New Hampshire. J. Geol. 24: 366-381, (1916).

30--1919   Graphic determination of dip components where dips are measured in feet per mile. Econ. Geol. 14: 176-178, 262-263, (1919).

31--1919a Geologic factors in oil prospecting. Econ. Geol. 14: 480-490, (1919).

32--1920   The barometric method of geologic surveying for petroleum mapping. Econ. Geol. 15: 150-169, il., (1920).

33--1920a Relation of oil pools to ancient shore lines (discussion). Econ. Geol. 15: 350-354, (1920).

34--1921   Use of the terms "erosion," "denudation," "corrasion," and "corrosion." Science 54: 13, (1921).

35--1921a Discussion of C. T. Lupton and L. Wallace's Geology of the Cat Creek oil fields, Fergus and Garfield counties, Montana. A.A.P.G., B. 5: 327-328, (1921).

36--1921b Discussion of D. M. Collingwood's Graphic method for determining the surface projection of the axis and crest traces at any depth of an asymmetrical anticline. A.A.P.G., B. 5: 328-329, (1921).

37--1922   Field methods in petroleum geology, in D. T. Day's A Handbook of the Petroleum Industry, vol. 1: 167-201, il., (1922).

38--1922a   Temperature of fluids in wells. A.A.P.G., B. 6: 547-548, (1922).

39--1923   Field Geology, 2nd ed. New York: McGraw-Hill Book Co., Inc., 649 p., 457 fig., (1923). See also 28--1916a.

40--1923a   The Currie Field, Navarro County, Texas. A.A.P.G., B. 7: 25-36, il., (1923).

41--1923b   Discussion of R. V. A. Mills' Natural gas as a factor in oil migration and accumulation in the vicinity of faults. A.A.P.G., B. 7: 14-24, (1923).

42--1923c   (with Pratt, W. E.) Faulting and petroleum accumulation at Mexia, Texas. A.A.P.G., B. 7: 226-236, (1923); Oil Eng. and Finance 4: 119-122, il., (1923).

43--1923d   Sericitization and dolomitization compared with the fixed carbon ratio of coal as indices of metamorphism in oil-bearing formations. A.A.P.G., B. 7: 291-293, (1923).

44--1924   The New Richland [oil] field, Navarro County, Texas. Min. Met. 5: 379-380, (1924).

45--1924a   Note on the origin of petroleum. A.A.P.G., B. 8: 669-671, (1924).

46--1924b   Structural and stratigraphic data of northeast Texas. Econ. Geol. 19: 563-565, (1924).

47--1924c   The education of the geologist. Econ. Geol. 19: 684-686, (1924).

48--1924d   The permeability of rocks. Econ. Geol. 19: 768-769, (1924).

49--1924e   Sharing of drilling information by oil companies. Econ. Geol. 20: 199, (1924).

50--1924f   Review of operations in Texas outside of the Gulf Coast district (for 1924). [This paper was apparently read at an A.I.M.E. meeting but never published.]

51--1925   The Wortham and Lake Richland faults [Texas]. A.A.P.G., B. 9: 172-175, il., (1925).

52--1925a   Comparative study of well logs on the Mexia type of structure. A.I.M.E., Tr. [preprint] no. 1396, 21 p., il., (1925); [with discussion], Tr. 71: 1329-1350, (1925).

53--1925b   The rate of solution of gypsum. J. Geol. 33: 548-549, (1925).

54--1926   Further notes on the origin and nature of the Currie structure, Navarro County, Texas. A.A.P.G., B. 10: 61-71, il., (1926).

55--1926a   Discussion of L. L. Foley's The origin of the faults in Creek and Osage counties, Oklahoma. A.A.P.G., B. 10: 293-303, (1926).

56--1926b   Some features of red-bed bleaching. A.A.P.G., B. 10: 636-637, (1926).

57--1926c   Operations in Texas outside of the Gulf Coast district (in 1925). A.I.M.E., Tr. 73: xxiv (1926). [Read at 133d meeting in New York but apparently never published.]

58--1927   The petroliferous belt of central-western Mendoza Province, Argentina. A.A.P.G., B. 11: 261-278, (1927).

59--1928   Clay Creek dome, Washington County, Texas. A.A.P.G., B. 12: 1166-1167, (1928).

60--1929   Discussion of E. H. Griswold's Acid bottle method of subsurface well survey and its applications. A.I.M.E., Tr. 82 (Petroleum Dev. Tech.) 1928-29, p. 41-49, (1929).

61--1929a Oil and gas fields of the Mexia and Tehuacana fault zones, Texas, in A.A.P.G.'s Structure of Typical American Oil Fields 1: 304-388, il., (1929).

62--1929b Discussion of tension faulting. A.A.P.G., B. 13: 638-639, (1929).

63--1929c Discovery of new pools. A.A.P.G., B. 13: 849, (1929).

64--1929d Let us have more "Geological Notes." A.A.P.G., B. 13: 849-850, (1929).

65--1929e Outline for the study of the crooked hole problem. A.A.P.G., B. 13: 854-859, (1929).

66--1929f The problem of crooked holes. A.A.P.G., B. 13: 1095-1162, (1929).

67--1930   Unit operation and unitization in Arkansas, Louisiana, Texas, and New Mexico. A.I.M.E., Tr. 86: 34-42, (1930); Abst. Tr. 91: 382, (1930).

68--1931   Field Geology, 3rd ed. New York: McGraw-Hill Book Co., Inc., 780 p., 538 fig., 1931. See also 28--1916a.

69--1931a Clay Creek dome, Washington County, Texas. A.A.P.G., B. 15: 279-283, (1931).

70--1931b Discussion of cap-rock petrography. A.A.P.G., B. 15: 528, (1931).

71--1931c Clay Creek dome, Washington County, Texas. A.A.P.G., B. 15: 1113-1116, (1931).

72--1932   Frio clay, south Texas. A.A.P.G., B. 16: 101-102, (1932).

73--1932a The East Texas oil field. A.I.M.E., Tr. 98 (Petroleum Div.): 279-294, (1932).

74--1932b Bioherm and biostrome. A.A.P.G., B. 16: 484, (1932).

75--1932c Geology and the new conception in pool development [with discussion]. A.P.I. Prod. B. 209: 4-8, (1932).

76--1932d Oil seepages and oil production associated with volcanic plugs in Mendoza Province, Argentina. A.A.P.G., B. 16: 819-824, (1932).

77--1932e Introduction to symposium on reservoir conditions in oil and gas pools. A.A.P.G., B. 16: 861-863, (1932).

78--1932f Discussion of W. P. Z. German's Compulsory unit operation of oil pools. A.I.M.E., Tr. 92 (Petroleum Div.): 11-37, (1932).

79--1932g Discussion of J. E. Pogue's Economics of the crude oil potential in the United States. A.I.M.E., Tr. 92 (Petroleum Div.): 633-643, (1932).

80--1932h Discussion of E. Oliver's Stabilizing influences for the petroleum industry. A.I.M.E., Tr. 98 (Petroleum Div.): 22-37, (1932).

81--1932i Contributions of petroleum geology to pure geology in the southern Mid-Continent area. Geol. Soc. Am., B. 43: 953-964, il., (1932), [Abst., March 1932]; Pan-Am. Geol. 57: 75, (1932).

82--1933   Some facts concerning the occurrence of oil and natural gas. National Better Business Bureau, Inc., B. 89: 5 p., (1933).

83--1933a The East Texas oil field, p. 67-77, il.; The Keechi and Palestine salt domes, Texas, p. 77-82, il., in 16th Int. Geol. Cong., U.S. 1933, Guidebook 6, Excursion A-6, (1933).

84--1933b Petroleum Geology [A.A.P.G. Presidential Address]. A.A.P.G., B. 17: 548-557, (1933); in part, Oil and Gas J. 31: 33-34, (1933).

85--1933c (with Wilde, H. D., Jr.) Simple principles of efficient oil-field development. A.A.P.G., B. 17: 981-1002, (1933).

86--1934 (with Wrather, W. E. and others) Problems of Petroleum Geology (Sidney Powers Memorial Volume), a sequel to Structure of Typical American Oil Fields. A.A.P.G., 1073 p., il., (1934).

87--1934a (with Wrather, W. E.) Preface to Sidney Powers Memorial Volume: Problems of petroleum geology, A.A.P.G., p. ix-x, (1934).

88--1934b Foreword to Carbon ratios: Problems of petroleum geology (Sidney Powers memorial volume), A.A.P.G., p. 67, (1934).

89--1934c Foreword to Migration and accumulation of petroleum: Problems of petroleum geology (Sidney Powers memorial volume), A.A.P.G., p. 247-251, (1934).

90--1934d A study of the evidences for lateral and vertical migration of oil: Problems of petroleum geology (Sidney Powers memorial volume), A.A.P.G., p. 399-427, (1934).

91--1934e (with Washburne, C. W.) Foreword to Oil-field waters: Problems of petroleum geology (Sidney Powers memorial volume), A.A.P.G., p. 833-840, (1934).

92--1934f The occurrence of oil and natural gas. Sci. Monthly 38: 467-470, il., (1934).

93--1935 Ground waters in or associated with oil fields. Am. Geophys. Union, Tr. 16th Ann. Mtg. Pt. 2: 438-441 (‡), Nat. Res. Council, (1935).

94--1936 Lateral migration of oil at Van, Texas. A.A.P.G., B. 20: 615, (1936).

95--1938 Wildcat drilling [Gulf Coastal Plain] in 1935 and 1936. A.A.P.G., B. 21: 1079-1082, (1937); correction, B. 22: 1236, (1938). For later annual reports on wildcat or exploratory drilling see the list farther on following this Bibliography.

96--1938a Maps [in] Science of Petroleum, vol. 1: 275-283. Oxford Univ. Press, 1938.

97--1938b Chapel Hill pool, Smith County, Texas. A.A.P.G., B. 22: 1107, (1938).

98--1938c Ground-water problems related to production of oil. Am. Geophys. Union, Tr. 19th Ann. Mtg., Pt. 1: 347-348 (‡), Nat. Res. Council, (1938).

99--1940 Where will young graduates in petroleum geology acquire field experience? in Levorsen, A. I., Chmn., Symposium on new ideas in petroleum exploration. A.A.P.G., B. 24: 1386-1388, (1940).

100--1940a [Memorial of] Donald C. Barton [1889-1939]. A.A.P.G., B. 24: 1521, (1940).

101--1941 Field Geology, 4th ed. New York: McGraw-Hill Book Co., Inc., xxxii + 853 p., il., (1941). See also 28--1916a.

102--1941a This matter of estimating oil reserves. A.A.P.G., B. 25: 164-166, (1941).

103--1941b Robert Thomas Hill, 1858-1941. Science 94: 249-250, (1941).

104--1941c (and others) Discussion on influence of geophysics upon geology curricula. A.I.M.E., Tech. Pub. 1382, 5 p., (1941).

105--1944 Standardization in compiling and reporting data on oil reserves. A.A.P.G., B. 28: 1217-1219, (1944).

106--1945   Review of exploratory drilling statistics, 1938 to 1944. A.A.P.G., B. 29: 1581-1592, (1945). See also the list of annual reports on wildcat and exploratory drilling between 1935 and 1955 farther on following this Bibliography.

107--1947   The geologist in industry. School and College Placement 8/2: 4 p., (Dec. 1947).

108--1949   Depth classification of new-field wildcats. A.A.P.G., B. 33: 1283-1284, (1949).

109--1949a Overlap and non-conformity. A.A.P.G., B. 33: 1901, (1949).

110--1950   Our oil and gas reserves - their meaning and limitations. A.A.P.G., B. 34: 1283-1287, (1950); Oil and Gas J. 49: 69, 71-72, (1950).

111--1951   Degrees of success in wildcat drilling. A.A.P.G., B. 35: 138-140, (1951).

112--1951a Use of "air" and "aerial." A.A.P.G., B. 35: 2608, (1951).

113--1952   Field Geology, 5th ed. New York: McGraw-Hill Book Co., Inc., xxx + 883 p., (1952). See also 28--1916a.

114--1953   Radioactive mineral exploration [abst.]. A.A.P.G., B. 37: 1127, (1953).

115--1953a Response [to receiving the Sidney Powers Memorial Medal]. A.A.P.G., B. 37: 1825-1826, (1953).

116--1955   The terminology of petroleum reserves. World Petroleum Cong., 4th, Rome, 1955, Pr., sec. 2, p. 561-565, il., with discussion, (1955).

117--1955a Program of radioactive surveys in petroleum exploration, published by the Am. Inst. Chem. Eng., "on behalf of the Nuclear Eng. and Sci. Cong., Dec. 1955, in Cleveland, Ohio." [Included in Lahee's typescript list of his publications], (1955).

118--1956   The last twelve years of exploratory drilling in the United States. The Mines Mag., p. 36-40, (Nov. 1956).

119--1957   Statistics play vital role in exploration. World Oil 144: 129-130, 132, 138, il., (1957).

120--1957a Trends in exploration. (Presented at Shreveport, Louisiana on 21 March 1957; published by the American Petroleum Institute.)

121--1958   Statistics on natural-gas discoveries. A.A.P.G., B. 42: 2037-2047, (1958).

122--1958a Geologia Practica [Spanish translation of 5th edition of Field Geology [see 113--1952]. Barcelona, Spain: Ediciones Omega, S.A., 874 p., (1958). See also 28--1916a.

123--1961   Field Geology, 6th ed. New York: McGraw-Hill Book Co., Inc., 926 p., (1961). See also 28--1916a.

124--1961a Probability of success in exploration. Chap. 4 in Petroleum exploration handbook. New York: McGraw-Hill Book Co., Inc., p. 4-1 to 4-8, il., (1961).

125--1962   Statistics of exploratory drilling in the United States, 1945-1960. Tulsa, Okla.: A.A.P.G., 135 p., tab., (1962). See also Reports listed on a following page.

126--1968   Comments on J. A. Taylor's A geologist looks at the mineral depletion provision. The Professional Geologist, p. 4, 7, (March 1968); Lahee's comments are on p. 12 of the July number, (July 1968).

The following are Reports of Frederic H. Lahee in special capacities, including editor, chairman of committees, and representative on the National Research Council.

Reports of the Editor of the Am. Assoc. Petroleum Geol.:
    A.A.P.G., B. 14: 666-667, (1929)
    A.A.P.G., B. 15: 581-583, (1930)
    A.A.P.G., B. 16: 516-519, (1931)

Reports of the Chairman (F.H.L.) of the Committee on Public Relations, later called the Committee on Applications of Geology:
    A.A.P.G., B. 17: 605, (1932)
    A.A.P.G., B. 18: 708, (1933)
    A.A.P.G., B. 19: 747, (1934)

Reports of the Representative (F.H.L.) of the A.A.P.G. on the National Research Council:
    A.A.P.G., B. 21: 678, (1936)
    A.A.P.G., B. 22: 615, (1937)
    A.A.P.G., B. 23: 754, (1938)
    A.A.P.G., B. 24: 938, 1148, (1939)

Reports of the Chairman (F.H.L.) of the Committee on Publication:
    A.A.P.G., B. 22: 614-615, (1938)
    A.A.P.G., B. 23: 753-754, (1939)

Reports of Frederic H. Lahee on wildcat drilling:
    Wildcat drilling in 1935 and 1936. A.A.P.G., B. 21: 1079-1082 (1937)
    Wildcat drilling in 1937. A.A.P.G., B. 22: 645-648, (1938)
    Further data on wildcat drilling in 1937. Ibid., B. 22: 1231-1235, (1938)
    Wildcat drilling in 1938. Ibid., B. 23: 789-794, (1939)
    Wildcat drilling in 1939. Ibid., B. 24: 953-958, (1940)
    Wildcat drilling in 1940. Ibid., B. 25: 997-1003, (1941)
    Wildcat drilling in 1941. Ibid., B. 26: 969-982, (1942)
    Wildcat drilling in 1942. Ibid., B. 27: 715-729, (1943)
    Classification of Exploratory Drilling and Statistics for 1943. Ibid., B. 28: 701-721, (1944)
    Exploratory Drilling in 1944. Ibid., B. 29: 629-645, (1945)

Reports of the Chairman (F.H.L.) of the Committee on Statistics of Exploratory Drilling:
    For the year 1945. A.A.P.G., B. 30: 813-828, (1946)
    For the year 1946. Ibid., B. 31: 917-930, (1947)
    For the year 1947. Ibid., B. 32: 851-868, (1948)
    For the year 1948. Ibid., B. 33: 783-804, (1949)
    For the year 1949. Ibid., B. 34: 995-1013, (1950)
    For the year 1950. Ibid., B. 35: 1123-1141, (1951)
    For the year 1951. Ibid., B. 36: 977-995, (1952)
    For the year 1952. Ibid., B. 37: 1193-1210, (1953)
    For the year 1953. Ibid., B. 38: 971-987, (1954)
    For the year 1954. Ibid., B. 39: 787-804, (1955)
    For the year 1955. Ibid., B. 40: 1057-1095, (1956)

Reports of the Chairman (F.H.L.) of the Special Committee on Curricula, later called the Special Committee on College Curricula in Petroleum Geology:
    For the year 1940. A.A.P.G., B. 25: 969-972, (1941)
    For the year 1941. Ibid., B. 26: 942-946, (1942)
    For the year 1942. Ibid., B. 27: 694-697, (1943)
    For the year 1943. Ibid., B. 28: 670-675, (1944)

Reports of the Chairman (F.H.L.) of the Advisory Committee on Radioactive Mineral Exploration:
    A.A.P.G., B. 36: 1689-1690, (1952)
    Ibid., B. 37: 1177-1178, (1953)
    Ibid., B. 38: 1357-1359, (1954)
    Ibid., B. 39: 1185-1186, (1955)

(14)
LOUIS CARYL GRATON
(1880-1970)

LOUIS CARYL GRATON

MIT: 1914-1918

Louis Caryl Graton, one of North America's best known mining geologists, and a distinguished professor at Harvard for almost forty years (1910-1949), served M.I.T. briefly, from 1914 to 1918, under rather unusual circumstances. When Harvard and M.I.T. agreed to engage in a cooperative program of instruction and research in certain fields of engineering, including mining, starting in September 1914, Graton was to be Harvard's representative on the joint faculty. He was appointed Professor of Mining Geology at M.I.T. in 1914, meanwhile maintaining his similar position at Harvard, and it was assumed that he would be a regular participant in the joint program. However, when the joint arrangement was declared illegal by the Massachusetts Supreme Judicial Court in November 1917, and was terminated at the end of SY 1917-18, Graton's appointment, which was then Special Lecturer in Mining Geology, was also terminated in June 1918. Although his actual participation in the joint program was minimal, by his own statement, the story of how he became involved is considered worth inclusion in this history because of the part played by Waldemar Lindgren and the light it throws on the cooperative efforts being developed at the time by the two great educational institutions along the Charles River.

Fortunately for the present purpose, Dr. Graton kindly wrote me a brief autobiographical sketch in 1963, at age 83, in which he describes how he first became acquainted with Waldemar Lindgren, when he was hired as an assistant by the U.S. Geological Survey, and how, later on, when Lindgren came to M.I.T., he was asked to participate in the Harvard-M.I.T. joint program. Knowing my intention in requesting his comments, he graciously gave me permission to use them in any way I wished, and I am grateful to him because his recollections about this interesting period in the history of geology at M.I.T. are not likely to be recorded elsewhere.

Because Graton's formal association with M.I.T., from 1914 to 1918, was but a brief interlude in his long and distinguished career as a mining geologist on the U.S. Geological Survey and at Harvard, I confine the following sketch to a brief discussion of his activities before he joined the M.I.T. faculty and of those he carried on while a participant in the Harvard-M.I.T. joint program. In addition to quoting extensively from the autobiographical comments that Graton kindly sent me, I have also drawn freely on D. H. McLaughlin's excellent and comprehensive Memorial published in 1973 (Memorial to Louis Caryl Graton, 1880-1970. Geol. Soc. Am., Memorials I: 33-45, port., 1973). Readers interested in Graton's long and active life as a mining geologist are urged to read the Memorial by his former colleague at Harvard for it portrays in detail the characteristics and achievements of one of the more colorful American geologists of this century.

Louis Caryl Graton, one of Harvard's most distinguished mining geologists, and for four years a part-time professor at M.I.T., was born on 10 June 1880 in Parma, Monroe County, New York, and died on 22 July 1970, in his 91st year, at a nursing home in New Haven, Connecticut.

He was the son of Louis Graton, a native of Quebec, and A. Ella Gould, a young school teacher in the small town of Parma. His parents not only gave him the strong physique that led to his longevity, but also imbued him with the work ethic and a strict and clear sense of values that governed his life, and that led him at an early age to appreciate that he would have to work hard and long if he wished to attain his goals.

After early schooling at Hornell, New York, and being Valedictorian at his high school at age 16, he entered Cornell University in the fall of 1896 with a scholarship won by successful completion of the [New York] State Regents examination. By the time he was awarded the Bachelor of Science degree in 1900 he had become much interested in geology as well as in chemistry. After several years of work as a mining geologist in the Sudbury district, and a winter of graduate work at McGill University, during which as a demonstrator in chemistry he gained his first teaching experience, he won the Schuyler Scholarship at Cornell, which enabled him to resume graduate work during SY 1902-03. At the end of this year of study came a turning point in his life, and more than a quarter of a century would pass before he would return to Cornell to receive a Ph.D. degree in 1930.

In the spring of 1903 he had the good fortune to be engaged by Waldemar Lindgren to be a field assistant on the U.S. Geological Survey. This appointment gave him the opportunity to become acquainted with and work under the acknowledged world leader in research on mineral deposits, and led to a lifelong friendship that ended only with Lindgren's death in 1939.

The events of the earlier years, 1903-1912, of this association, and those that led both Lindgren and Graton to become colleagues at M.I.T. and Harvard, respectively, are best told by Graton himself (Letter from Graton to Shrock dated May 31, 1963):

TEL: 795-9383
(AREA CODE 203)

L. C. GRATON
274 PLEASANT HILL ROAD
ORANGE, CONNECTICUT

May 31, 1963

"Prof. Robert R. Shrock, Chairman
Department of Geology and Geophysics,
Massachusetts Institute of Technology,
Cambridge 39, Mass.

Dear Doctor Shrock:-

"At last I have accumulated all information likely to be secured -- though probably not all that exists -- regarding my relations during the Cooperation in Engineering between Technology and Harvard in the 'Teens.

"As contrasted with that of most from the Harvard group, my participation was definitely tenuous. Indeed, the surrounding circumstances were so unusual, yet so determinative, that I feel they should here be set down lest there might arise questions as to my concurrence with the idea of Cooperation, and my personal satisfaction and pride in becoming what I supposed would be a long-continuing member of Technology's Faculty as well as of Harvard's. So I impose on you the many parts of the lengthy story in order that you may decide what, if any, may merit reference ...

"As I intimated to you in my letter of April 27, my life was for many years definitely threaded through with the influence of Lindgren. Becoming his assistant in 1903 on the Survey's Cripple Creek re-study, I had a room immediately adjoining those in which Mr. and Mrs. Lindgren "kept house." To an extent that would have been embarrassing except that they made it seem so natural, they treated me almost as a son: countless suppers with them and still more evenings before their fireplace where he would at times sound me out and at others, in his modest but effective way, give me pointers from his own rich experience. In the Southern Appalachians in 1904 I was put largely on my own, but Lindgren came to expound and check. In 1905 we were together for much of the long field season in the New Mexico study -- a type of regional inquiry he had long been hoping for; it was there in particular that he found basis for integrating his experience and conclusions into what took shape as a systematic and comprehensive theory of ore deposits. On horseback or in camp he would (deliberately) think out loud, asking in effect, "are there holes in that idea?" When, in 1906, there appeared his paper "The Influence of Physical Conditions on Ore Deposition," I realized that I had been present at the later stages of gestation of what developed into the outstanding effort of his life and a contribution that will long stand unsurpassed.

"In 1906 the Division of Mineral Resources was reorganized, with Lindgren in charge of all products except fuels; copper was assigned to me, still sharing his Washington office and receiving his encouragement and guidance. Some time after starting on his first (1908) group of lectures for Technology, he told me about the arrangement, and from time to time, with an air that seemed to apologize for interrupting my work, he would give me an oral peep at this or that bit. Of course, I was thrilled at the arrangement; and glowed, as did he, when he returned from Boston -- my in absentia introduction to Technology.

"My connection with Harvard came as a surprise. Early in 1909 I had left the Survey and moved to New York to direct the newly-established Copper Producers' Association [C.P.A.]. It was understood that I should have time to complete the large chore pending when I left the Survey. In mid-summer of 1909, H. L. Smyth, Harvard Geologist and Head of the Mining School, asked me to lunch with him in New York. At the end of the conference I agreed to study his suggestion of a tryout as his assistant and to ascertain whether C.P.A. would allow me to spend part time each month at Cambridge. A C.P.A. Director and Vice President of Calumet & Hecla also a devoted alumnus of Harvard smoothed the grant of time. So, early in October I started bi-weekly two-day lectures and lab. work in mining geology, mutually on trial as Instructor, commuting from New York. [That was the first year of Lowell's tenure as President.] The trial being satisfactory to Smyth, C.P.A. and me, I was appointed Asst. Prof. as of 1910, still on part-time and commuting basis.

"Some time in that period I learned of my next tie-in with what was eventually to affect Technology, i.e. Gordon McKay money. The very first small allocations to Harvard from the McKay income had begun about 1905, and by 1909 mining geology was to be included for a modest slice and with some of it Smyth had engaged me. Later, the allocation somewhat increased; and early in 1912 I was given full rank as of July 1, with agreement by Harvard, C.P.A. and me that I would move to Cambridge early in 1913 and serve on "full time" except for correspondence and two days' absence in New York each month for C.P.A.

"Moving to Cambridge in April 1913 I gave much attention to organizing a study of Supergene Copper Enrichment as a combined field-laboratory investigation by Harvard Mining School, Geophysical Laboratory and principal western-hemisphere Copper Companies.* This involved my absence in the field for long periods. And although the investigation continued with such staff as did not enlist, my touch with it was materially diluted by being granted leave from Harvard to direct in New York the Copper Producers' Committee of the War Industries Board from September of '17 until April of '19. The point pertinent to the present subject is that within that 1913-19 span of my Harvard tenure fell the 1914-18 Cooperation with Technology.

"The agreement between the Governing Boards of the two institutions was to embrace the standard engineering subjects and mining and metallurgy; geology was not mentioned. Of the staff in the Harvard Mining School, only Smyth and I also gave courses (in mining geology) under the Faculty of Arts and Sciences, while Profs. Palache and Wolff of the latter gave courses (in mineralogy and petrography) to students in the Graduate Mining School. By the year '13-'14, I believe my salary, though received in a single check, was borne by both Faculties. Thus, I neither offered nor was equipped to give any non-geological or engineering course. This plainly made rather anomalous my technical position

---

* When material from the field began to pour in for intensive investigation, an effective member of the laboratory staff during 1914 was Walter Whitehead, holder of an Austin Fellowship, who later returned to Technology to become a stalwart in its Department of Geology -- my warm regards to him now.

with respect to the Merger. Moreover, because relatively new in residence and quartered in the Geology building, I had heard little of the facts and resulting Merger chit-chat current in the Engineering and the Mining buildings. Nevertheless, along with other professors listed in the Mining School, my name was included on the Technology rolls as of 1914 and so continued until 1918.

"Meanwhile, Lindgren, with whom I had retained contact on Survey and other matters, resigned from the Survey in time to take up residence as Head of Geology at Technology in the autumn of 1912. I was delighted thereby; and as occasion offered, I visited at his Brookline home or his office in the old Tech Buildings in Boston; the visits were more frequent after I moved to Cambridge and he to the new buildings on the Charles. At the outset of these chats, neither of us had definite knowledge about the proposed Merger, and each assumed that neither would be involved; for Tech geology was not embraced in the Merger. Later, when the plan became more evident, with the probability of my being included, I expressed chagrin and regret, telling him I would refuse to be put in the position of bringing my students or competing for his within his own institution. The remark about 'competing' with one of Lindgren's stature sounded absurd and silly as soon as I had uttered it. He immediately relieved the situation by assuring me that all would work out satisfactorily.

"Inasmuch as the merged group embraced no representative for geology from the Technology faculty, I regarded Smyth as my superior concerning the plans for my service under the new arrangement. He knew that I was about to be moved -- because my top-floor wing in the Museum had become loaded beyond the safety factor -- into space in Pierce Building being vacated by members of the Engineering staff moving to Tech. And I told him of my situation vis-a-vis Lindgren. We agreed that the only course was for me to seek and follow Lindgren's advice.

"In general and simplified terms, this was the program. Based in Pierce and continuing with my own courses and my personal research, my general work calendar was to be made known to Lindgren. I was to give lectures at Technology singly or in groups on subjects of my choosing, also laboratory demonstrations of techniques being developed at Harvard, at times mutually convenient to Lindgren and me, and specifically at intervals, sometimes lengthy, when he needed to be absent. Students from Lindgren's courses were free to attend such of my Harvard lectures or seminars as they might wish. Conversely, some of my advanced boys eagerly availed themselves of sitting in on certain lectures by Lindgren.

"From time to time Lindgren and I discussed how the arrangement was going. He seemed satisfied, and so was I. Insofar as I recall, except for the official listing in the Technology Directory, my name did not otherwise appear. My salary checks continued to come from the Harvard Bursar. All connection with the Technology arrangement ceased for me when I left for New York in September, 1917, as earlier mentioned. Two months later, the Massachusetts Supreme Court declared the Agreement invalid, not because of the concept but on grounds that it failed to conform with terms of the Gordon McKay Trust."

Most sincerely yours,

L. C. Graton

In the M.I.T. catalogues for SY 1914-15 and SY 1915-16, Graton is listed under the "Officers of Instruction" (p. 18) as <u>Professor of Mining Geology</u> and is included in the list of members of the Department of Geology (Course XII) for SY 1914-15 (p. 333). However, his name is not included in the similar Course XII staff list for subsequent years. In the catalogue for SY 1916-17, the last year he was an active participant in the joint program, he is listed under "Special Teachers and Lecturers" on page 35.

The abortive effort of Harvard and M.I.T. to forge a cooperative program of instruction and research in Civil Engineering (I), Mechanical Engineering (II), Mining Engineering and Metallurgy (III), Electrical Engineering (VI), and Sanitary Engineering (XI) is discussed briefly (p. 285-293) in Dean Samuel C. Prescott's <u>When M.I.T. was "Boston Tech," 1861-1916</u> (Cambridge, Mass.: The Technology Press, xi + 350 p., 1954), and at considerable length in a long chapter (Chap. 10) in H. G. Pearson's biography, <u>Richard Cockburn Maclaurin</u>, President of the Massachusetts Institute of Technology, 1909-1920. New York: The Macmillan Co., v + 302 p., port., (1937).

There is little record of Graton's activities at M.I.T., other than the autobiographical remarks quoted on preceding pages. From all I have been able to discover, he seems actually to have given only one set of lectures, and these during the spring of 1917. The following statement appears in the President's Report for 1917 (M.I.T., B. 52/2: 98, 1917):

> "By agreement with the Geological Department of Harvard
> University, Professor Louis C. Graton, as special lec-
> turer, will give a short course on 'Ore Deposits' during
> the coming spring term [<u>i.e</u>. 1917]. By the same agree-
> ment Professor Lindgren will give a series of lectures
> at Harvard College on 'Gold-Bearing Ore Deposits.'"

That these two series were actually delivered is confirmed by a statement to that effect on page 89 of the President's Report for 1918 (M.I.T., B. 53/2: 89, 1918).

L. C. Graton's achievements were aptly summarized when he was awarded an honorary LL.D. degree at the Charter Day ceremonies at the University of California, Riverside, on 11 February 1964, being presented to President Clark Kerr by his old colleague from Harvard, Regent Donald H. McLaughlin [See also McLaughlin's Memorial, mentioned in a preceding paragraph]. Graton's citation read as follows:

> "Distinguished earth scientist; Professor Emeritus of
> Mining Geology at Harvard University, who during a long
> career has contributed signally to both the academic
> and the practical aspects of his chosen profession.
> For fifty years a leader in the study of ore deposits
> and the processes by which they originate, and noted

also for his original work in mineralography and volcanology. An inspiring teacher, he has, through the accomplishments of his many outstanding students, added greatly to the impact of his own personal achievements."

PARTIAL BIBLIOGRAPHY OF LOUIS CARYL GRATON

Symbols and abbreviations used in the following references are explained on pages 91 - 98. Inasmuch as Prof. Graton was on the M.I.T. Faculty for only a short time, 1914-1918, and then on only a part-time basis and under unusual circumstances, which are discussed in the preceding biographical sketch, I have listed below only those articles published through 1918, at the end of which year he resigned from his M.I.T. appointment. His many publications after 1918 can be found in his several memorials, cited at the beginning of this sketch, and in the standard Bibliographies of North American Geology.

1--1898   (with Dodge, N.) Alcohol, water and potassium nitrate. J. Phys. Chem. 2: 498-501, (1898).

2--1903   Up and down the Mississiaga. Ont. Bur. Min., 12th Ann. Rept.: 157-172, (1903).

3--1903a   On the petrographical relations of the Laurentian limestones and the granite in the township of Glamorgan, Haliburton County, Ontario. Can. Rec. Sci. 9: 1-38, (1903).

4--1905   (with Hess, F. L.) The occurrence and distribution of tin. U.S. Geol. Surv., B. 260: 161-187, (1905).

5--1905a   The Carolina tin belt. U.S. Geol. Surv., B. 260: 188-195, (1905).

6--1905b   (with Schaller, W. T.) Purpurite, a new mineral. Am. J. Sci. (4) 20: 146-151, (1905); Zeitschr. Kryst. 41: 433-438, (1905).

7--1905c   Consanguinity in the eruptive rocks of Cripple Creek (Colorado) [abst.]. Science 21: 391, (1905).

8--1906   Description and petrology of the metamorphic and igneous rocks [Cripple Creek district]. U.S. Geol. Surv., P.P. 54: 41-113, (1906).

9--1906a   (with Lindgren, W.) A reconnaissance of the mineral deposits of New Mexico. U.S. Geol. Surv., B. 285: 74-86, (1906).

10--1906b (with Gordon, C. H.) Lower Paleozoic formations in New Mexico. Am. J. Sci. (4) 21: 390-395, (1906); Science 23: 590-591, (1906).

11--1906c Reconnaissance of some gold and tin deposits of the Southern Appalachians. U.S. Geol. Surv., B. 293: 9-118, (1906).

12--1907   Copper. U.S. Geol. Surv., Min. Res. 1906: 373-438; 1907 (1): 571-644, (1907-1908).

13--1907a (and Siebenthal, C. E.) Silver, copper, lead, and zinc in Central States. U.S. Geol. Surv., Min. Res. 1907 (1): 483-549, (1908).

14--1910   The occurrence of copper in Shasta County, California. U.S. Geol. Surv., B. 430: 71-111, (1910).

15--1910a (with Lindgren, W.) The ore deposits of New Mexico. U.S. Geol. Surv., P.P. 68: 361 p., (1910).

16--1913   Investigation of copper enrichment. Eng. M. J. 96: 885-887, (1913).

17--1913a   Ore deposits at Butte, Montana [discussion]. A.I.M.E., B. 83: 2735-2736, (1913).

18--1913b   (and Murdoch, J.) The sulphide ores of copper; some results of microscopic study (with discussion by J. F. Kemp, H. V. Winchell, and L. C. Graton). A.I.M.E., B. 77: 741-811, (1913); Tr. 45: 26-93, 529-530, (1914).

19--1913c   Notes on rocks from the Coppermine River region, Canada. Can. M. Inst., Tr. 16: 102-114, (1913).

20--1915   (and others) To what extent is chalcocite a primary and to what extent a secondary mineral in ore deposits? [discussion.] A.I.M.E., Tr. 48: 194-200, (1915).

21--1915a   Discussion of P. Billingsley's The Boulder batholith of Montana. A.I.M.E., B. 97: 31-47, (1915); B. 101: 1128-1137, (1915); Tr. 51: 31-56, (1916).

22--1915b   Discussion of A. P. Thompson's The occurrence of covellite at Butte, Montana. A.I.M.E., B. 100: 645-677, (1915); B. 108: 2464-2471, (1915); Tr. 52: 563-603, (1916).

23--1915c   Discussion of R. E. Somers' Geology of the Burro Mountains copper district, New Mexico. A.I.M.E., B. 101: 957-996, (1915); Tr. 52: 604-644, (1916).

24--1915d   Discussion of W. Lindgren and C. P. Ross' The iron deposits of Daiquiri, Cuba. A.I.M.E., B. 106: 2171-2190, (1915); Tr. 53: 40-66, (1916).

25--1916   Discussion of C. F. Tolman, Jr.'s Observations on certain types of chalcocite and their characteristic etch patterns. A.I.M.E., B. 110: 410-433, (1916); Tr. 54: 402-441, (1917).

26--1916a   Discussion of Y. S. Bonillas et al. Geology of the Warren mining district [Arizona]. A.I.M.E., B. 117: 1397-1465, (1916); Tr. 55: 284-355, (1916).

27--1916b   Discussion of M. Roesler's Geology of the iron-ore deposits of the Firmeza district, Oriente Province, Cuba. A.I.M.E., B. 118: 1789-1839, (1916); B. 123-125: 375-376, 439-448, 856-859, (1917); Tr. 56: 77-141, (1917).

28--1917   (and McLaughlin, D. H.) Ore deposition and enrichment at Engels, California. Econ. Geol. 12: 1-38, (1917). (See also 30--1918.)

29--1917a   Discussion of L. P. Teas' The relation of sphalerite to other sulphides in ores. A.I.M.E., B. 131: 1917-1931, (1917); Tr. 59: 68-87, (1918); A.I.M.E., B. 136: 844-845, (1918).

30--1918   (and McLaughlin, D. H.) Further remarks on the ores of Engels, California. Econ. Geol. 13: 81-99, (1918). (See also 28--1917.)

31--1918a   Discussion of H. M. Roberts and R. D. Longyear's Genesis of the Sudbury nickel-copper ores as indicated by recent explorations. A.I.M.E., Tr. 59: 57-67, (1918); B. 134: 555-584, (1918); B. 136: 848-858, (1918); Can. M. Inst., Tr. 21: 80-117, (1919).

32--1918b   The relation of sphalerite to other sulphides in ores (discussion). A.I.M.E., B. 136: 844-845, (1918).

NOTE:

Almost a decade passed before Graton published another article, but once he resumed publishing (in 1927), his name again became a familiar one in Economic Geology, in which appeared many of his more important shorter articles. He also shared authorship with a dozen or more of his colleagues in joint reports. The 40 or more articles and reports he had a hand in publishing after 1918 are listed in McLaughlin's Memorial (Geol. Soc. Am., Memorials II: 42-45, 1973) and in the several Bibliographies of North American Geology.

(15)
WILLIAM FRANCIS JONES
(1885-1941)

WILLIAM FRANCIS JONES

MIT: 1918-1928

Immediately after the end of World War I there was need for an instructor who had had practical field experience in the petroleum industry and could lecture on practical uses of geology. W. F. Jones '09, was brought into the Department in early 1918 and during the next ten years, until he resigned in June 1928, he instructed in general physical geology and petroleum geology. He brought to his instruction almost a decade of practical field work as a consultant to mining and petroleum companies, and during his years at M.I.T. took several short leaves of absence to accept consulting assignments in Mexico, the West Indies, and Venezuela. He left M.I.T. in 1928 to take charge of a small oil company in Tulsa, Oklahoma, but soon returned to Massachusetts where he spent the last decade of his life investigating the shoreline features of the islands off the coast of Cape Cod.

William Francis Jones was born in New York City on 13 December 1885 and died on Nantucket Island, Massachusetts, on 9 May 1941. He was the third of three boys in a distinguished family, and at an early age showed a strong interest in out-of-doors science--an interest that will be discussed a little farther on as it probably influenced him considerably in choosing to follow geology as a profession.

William's father, Bassett Jones, was a prominent architect in New York City, and his mother, Sarah Catherine (Oakey) Jones, was a member of the old Oakey family of New York. Both his brothers, listed in Who's Who in America, preceded him as students at M.I.T. Bassett Jr. attended the Institute from 1895 to 1898, and ultimately became a well-known consulting electrical engineer in New York. His second brother, Sullivan W. Jones, attended the Institute through the sophomore year, as a member of the Class of 1900, and later served as State Architect of New York from 1923 to 1928.

As mentioned in a preceding paragraph, William displayed an unusual interest in natural science at an early age, and when but fourteen he organized a group of New York City boys into the "Society of Young American Scientists." He and his youthful associates held scientific exhibitions, and spent much time at the American Museum of Natural History where they assisted in curating the conchological collection. The interest in modern shells developed as a boy remained with Jones throughout his adult years, and he later accumulated an excellent collection which he ultimately presented to the Institute. The choicest specimens in this collection were later given to Harvard's Museum of Comparative Zoology, because they were rare and it was concluded that they would be of greater scientific use at Harvard, there being at the time no neontologist or conchologist at M.I.T. interested in modern shells; the bulk of his collection, however, remains at the Institute as the nucleus of the study collection of mollusks used in historical geology and paleontology.

While still in high school Jones became much interested in Shakespeare, and once organized and staged a performance of Julius Caesar in which he acted the part of Cassius. His brother Bassett wrote to Miss Comstock (letter dated July 10, 1948, and preserved in the M.I.T. Institute Archives) that this performance received much creditable newspaper criticism. He further stated that William could recite Hamlet in toto, all parts, and often entertained his friends with recitations.

Jones registered at M.I.T. as a freshman in the fall of 1905, having prepared for college at the Stevens Preparatory School (now Stevens Academy) in Hoboken, and for the next three years was a special student in the Department of Mining Engineering and Metallurgy. He did not continue for the fourth year, however, choosing instead to travel across the continent to the West Coast for his fourth college year, SY 1908-09, which he spent at the University of California at Berkeley. He did not complete requirements for a Bachelor's degree, however; instead he joined the staff of J. E. Spurr, consulting geologist, and thus started on a career in consulting geology that took him on many domestic and foreign assignments in the next ten years before he returned to M.I.T. as an Instructor in early 1918.

He spent parts of 1909 and 1910 in Mexico and in northwestern United States. His work in Mexico led to deep interest in both the country and its people, and he was later described as "a student of Latin American History," probably because of his perceptive article in the Technology Review (vol. 29, 1927) with the admonitory title, "Let Mexico Alone!"

In 1910 and 1911 he acted as consulting geologist to coal mining companies in the State of Washington, and 1912 and 1913 found him doing field work in petroleum geology for the Southern Pacific Company. The next four years, 1914-1918, took him on a variety of assignments, largely exploration for oil, in the United States, the West Indies, Mexico, and Central America. He would return to the three last-named regions again in later years while teaching at M.I.T.

As World War I drew to a close in 1918, there was a great surge of interest across the country in exploration for minerals and mineral fuels, especially oil and gas, and Jones was invited to join the staff of Course XII early in the year as an Instructor.* From January to June of 1918 he was an Instructor in general geology; and during the first term of SY 1919-20 he instructed in geological surveying. When Courses III and XII were combined in 1920 under Lindgren's chairmanship, as a new department of Mining, Metallurgy, and Geology, he was appointed an Assistant Professor of Structural Geology. He served in this capacity for the next six years, with several short leaves of absence to resume consulting assignments with petroleum companies, then resigned from his professorship in 1926 to accept a less demanding appointment as Lecturer on Petroleum Geology. He continued to give lectures on this subject until January 1928 and in June resigned to accept a managerial position with a small oil company in Tulsa, Oklahoma.

In 1926, he and W. L. Whitehead '13 (III), who would succeed him later as Lecturer on Coal and Petroleum (1928-1942), organized an exploration party in Venezuela, which included two other M.I.T. geologists, W. B. Millar '26 and Guillermo Zuloaga '27. The party left from Caracas and for ten months did geological work in the almost trackless southwestern part of the country, penetrating as far as the Colombian border at the foothills of the Colombian Andes. This memorable expedition is described and illustrated in the five-page article, "Touring Venezuela in a Model T," mentioned in the Bibliography.

During the same period, 1926-1928, while serving as Lecturer, Jones made frequent trips to Mexico as an engineering and geological consultant, and became much interested in the economic and political changes that were then going on in the country, some of which he described in his Technology Review article, "Let Mexico Alone!" In the spring of 1927, following a visit to the country, he delivered a dozen lectures on the Mexican situation to various local organizations, civic and otherwise, including the Institute's Faculty Club.

---

* On Jones' record in the M.I.T. Registrar's Office (Record Book for 1918), there is a 1918 notation to the effect that he would have to take a number of specific subjects before he could meet the requirements for the S.B. degree in Course III. This notation seems to indicate that when he returned to M.I.T. early in 1918, as Instructor in Geology, he looked into the question of what subjects he would still have to pass to get an S.B. degree, and a Committee set the conditions as noted in his record. Inasmuch as there is no further information recorded, Jones must have decided against any more course work, with the result that he never received an S.B. degree, though he is always indicated as an alumnus of the class of 1909.

The new position in Tulsa did not work out satisfactorily for Jones, and when his health began to fail he moved his family back to Nantucket, where he had spent some of his earlier years, and was soon teaching science in the local high school.

As early as 1908 Jones had become much interested in the submarine and terrestrial topographical features along the Nantucket shorelines, and his first known geological report, which was never published, was a study of Haulover Beach at the head of Nantucket Harbor. According to Shimer (see note following the Bibliography), Jones became deeply interested in both coastal features and coastal processes around Nantucket Island, and his comments were frequently quoted in the local newspapers. When he moved his family from Tulsa to Nantucket in 1930, having been forced to give up active field work because of failing health, he turned his attention once again to the shoreline features of Nantucket Island.

For most of the eleven years that he had left, before death in May of 1941,

> "... he lived in a beautiful little house beside the lagoon north of Nantucket Harbor. Back of this little house was a grove of pines said to be the only pines on the Island and planted by his father whose ashes were scattered beneath these trees." (Letter from W. L. Whitehead to Miss Julia W. Comstock dated January 22, 1948, and now preserved in Jones' file in the M.I.T. Institute Archives.)

During these last years he devoted much of his time and energy to investigating the coastal features and processes on Nantucket, Martha's Vineyard, and along the shoreline of Cape Cod. According to a letter from his brother Bassett to Miss Comstock (dated July 10, 1948, and preserved in the M.I.T. Institute Archives), he had almost completed a paper on the most recent (*i.e.* post-glacial) geological history of this part of southeastern Massachusetts at the time of his death.

This report presents evidence to support the conclusion that during the past four thousand years there has been a 20-foot change [drop] in sea level relative to the land along the coast of southeastern Massachusetts, and that formerly offshore bars once draped the coast and are now represented by Coatue Beach on Nantucket, Muskeget Island, Cape Pogue on Chappaquiddick Island, and Long Beach off Barnstable on the Cape. Unfortunately, this extensive and intensive study of post-glacial shoreline history has not been published thus far; we have to be content with the 1937 and 1951 abstracts listed in the Bibliography!

It has been difficult to learn much about Jones' contributions to the academic program of Course XII during the decade that he served on the staff. Certain it is that he lectured on coal and petroleum geology, taught structural geology, and instructed students in geological surveying. But he was frequently away on leaves of absence of varying lengths of time and supervised very little student research, his record showing but six theses, all S.B.'s, as being supervised by him. It must be remembered, however, that in the decade following World War I, geology staff members did much consulting during the summer, as well as during the regular school year, for there were no outside project funds in those days and no summer programs in geology at the Institute.

Jones' professional career as a consulting geologist provides an excellent example of the kinds of experiences field geologists had in the days of the Model T and before the DC-3, when they walked, rode horses and mules, poled shallow-draft boats along the main drainage channels, ate food of uncertain quality wherever they could get food at all, and slept where night found them regardless of the quality of their quarters. Although exploration of strange and far places still holds its magic spell for those who like adventure, part of the glamour has gone with the coming of the helicopter, the jeep, tinned foods, iced beer, and the great array of insect repellents.

Jones died on 9 May 1941 from spinal cancer that his brother Bassett believed could be traced to some tropical disease that he contracted a decade earlier while doing geological field work in Yucatan and other countries of Central America. His death was noticed as follows in the Class Notes '09 in The Technology Review (44) for November 1941 (p. xxvii):

> "William F. Jones passed away on May 9 after a long illness. Jones had been living at Wauwinet, Nantucket, Mass., where he made his home for the past seven years after he was forced to give up active practice as a geologist."

BIBLIOGRAPHY OF WILLIAM FRANCIS JONES

(Symbols and abbreviations used in the following references are explained on p. 91-98; in general, abstracts are listed separately as well as with references to the complete articles. This bibliography begins with Jones' first publication in 1911 and includes one posthumous abstract published in 1951.)

1--1911   Coal-bearing Eocene of western Washington: Pierce Co. [abst.] Geol. Soc. Am., B. 25: 121-122, (1911).

2--1916   El Petroleo [in Spanish], published by Sauter and Co., San José, Costa Rica, 44 p., il., (1916). [Not seen, but reported by H. W. Shimer in a letter in the M.I.T. Institute Archives dated May 2, 1945.]

3--1918   Discussion of paper by A. W. Lauer, "The petrology of reservoir rocks and its influence on the accumulation of petroleum." Econ. Geol. 13: 147-149, (1918).

4--1918a   A geological reconnaissance in Haiti; a contribution to Antillean geology. J. Geol. 26: 728-752, (1918).

5--1918b   Intrusive origin of the Gulf Coast salt domes; its bearing on the accumulation of oil [discussion]. Econ. Geol. 13: 621-622, (1918).

6--1920   The relation of oil pools to ancient shore lines. Econ. Geol. 15: 81-87, (1920).

7--1922   A critical review of Chamberlin's "Groundwork for the study of Megadiastrophism." Am. J. Sci. (5) 3: 393-413, (1922).

8--1923   What is the origin of the earth? Tech. Eng. News 3/8: 233, 252, 254, (1923).

9--1924   Aspectos geológicos generales de la región de Tabasca en su relación con la existencia de petróleo. Bol. Petróleo Mexico 17: 349-352, (1924).

10--1925   Report by Professor William F. Jones, Assistant Professor of Structural Geology, Massachusetts Institute of Technology, in regard to the possibilities of the development and utilization of the coal deposits in Massachusetts, in Special Report of the Special Commission on the Necessaries of Life, relative to the Anthracite Deposits of Southeastern Massachusetts and of Rhode Island. House [Document] No. 1025 [The Commonwealth of Massachusetts], p. 24-28, (1925).

11--1926   The future of the petroleum industry in the United States. Tech. Eng. News 6: 267, 282, 290, (1926).

12--1926a   Replacement or displacement by dikes [Medford area, eastern Massachusetts]. Eng. Min. J.-Press 121: 250, (1926).

13--1927   Let Mexico alone! Tech. Rev. 29: 419-423, il., (1927).

14--1928   (and Whitehead, W. L.) Touring Venezuela in a Model T--An account of a geological expedition just north of the Equator. Tech. Rev. 30: 221-225, il., (1928).

15--1937   Late post-glacial history of the southeastern New England Coast [abst.]. Geol. Soc. Am., Pr. 1936: 81-82, (1937).

16--1951   (and Lucke, J. B.) Evolution of Nantucket shore lines [abst.]. Geol. Soc. Am., B. 62: 1453, (1951).

NOTE:

Jones' brother Bassett wrote to Miss Julia W. Comstock (in a letter dated July 10, 1948, and now preserved in the M.I.T. Institute Archives) that William became much interested in coastal forms and processes and by 1928 had completed a "Report on Nantucket Harbor and its improvement." This report is supposed to have been published in 1938; actually, however, no published version is known, but a typescript copy (of 20 pages and 5 plates) is included in the Nantucket files of the Division of Waterways, Massachusetts Department of Public Works, 100 Nashua Street, Boston 02114. It has the title "Report on Nantucket Harbor and its improvement, March 15, 1938" [20 p., 5 pls.].

It is this report that seems to have been referred to by W. H. Bradley and J. I. Tracey, Jr., in a 1949 U. S. Geol. Surv. open-file report on "Nantucket Harbor and the proposed cut at Chatham Bend," designated (200: R290), who state on page 1:

> "The location, physical features, geology, and history of Nantucket Harbor have been effectively treated in two earlier reports - 'Report on Nantucket Harbor and its Improvement' by William F. Jones, consulting engineer of Nantucket, March 15, 1938; ..."

This open-file report also contains a bathymetric chart of Nantucket Harbor that was compiled by William F. Jones, June 1938 and modified locally by W. H. Bradley, August 1948.*

In a letter from H. W. Shimer to Miss Julia W. Comstock, dated May 2, 1945 and now preserved in Jones' file in the M.I.T. Institute Archives, Shimer wrote:

> "In 1908 he [Jones] prepared 'A Study of Haulover Beach,' at the head of Nantucket Harbor, and reported it to Nantucket. Throughout this year there were many articles by him in regard to this beach in the Inquirer and the Mirror."

View looking northeast across the Great Court at Building 4 of the Cambridge M.I.T., in which Chemistry, Physics, and Geology shared space, after the move from Boston in 1916.
(M.I.T. Historical Collections)

---

* I am indebted to Dr. Clifford A. Kaye, Geologist, of the Boston Office of the U. S. Geological Survey for much appreciated assistance in locating and then having sent to me the open-file report by Bradley and Tracey.

Joseph L. Gillson (XII S.M. 1923; Sc.D. 1923), Chief Geologist of the E. I. DuPont de Nemours & Company of Wilmington, Delaware, in the field in the 1950s.

    Gillson served the DuPont Company as geologist for 32 years, 1928-1960, and was primarily responsible for the location and procurement of numerous non-metallic commodities collectively designated "industrial minerals and rocks:" ex. barytes, celestite, fluorspar, ilmenite, and sulphur. This work involved much travel and field work, both in the United States and abroad, and ultimately resulted in his being recognized as one of the leading authorities on industrial minerals and rocks, as pointed out in the following biographical sketch.

(Du Pont Company Photograph)

(16)
JOSEPH LINCOLN GILLSON
(1895-1964)

JOSEPH LINCOLN GILLSON

MIT: 1922-1930; 1961-1963

Joseph L. Gillson, a distinguished Course XII alumnus and an economic geologist known internationally for his publications on industrial minerals, spent a total of 13 years in M.I.T.'s Department of Geology during three separate periods of time: first, as a graduate student, 1920-1923; second, as a regular member of the geology faculty, 1922-1930; and third, as the William Otis Crosby Lecturer during SYs 1961-62 and 1962-63. As a graduate student he earned a masterate in 1921 and a doctorate in 1923, both under Waldemar Lindgren's supervision. With this training under Lindgren, in addition to earlier work with U. S. Grant at Northwestern University and with Charles Palache at Harvard, Gillson was well prepared to teach mineralogy, petrography, and mineral deposits. Consequently, when, in 1922, H. S. Warren left M.I.T. for Yale, Gillson, who had gained some teaching experience at Harvard during SYs 1920-21 and 1922-23, while carrying on to give Warren's courses. He was appointed an Instructor in Mineralogy and Petrography in 1922 and was advanced in rank to Associate Professor by 1928, at which time he accepted a position as Geologist with the E. I. du Pont de Nemours & Co. of Wilmington, Delaware.

During his six years as an active faculty member at M.I.T., Gillson carried a heavy teaching load, yet managed to supervise a total of 12 student theses--9 for the S.B. degree and 3 for the S.M. degree, and to publish 20 papers, chiefly on minerals and rocks. During the summers of this period, 1921-1927, he worked as a Junior Geologist with the U.S. Geological Survey, doing field work in half a dozen different States. Several of his publications were based on some of this field work.

During his 32 years at du Pont, first as Geologist and then as Chief Geologist, Gillson became an internationally recognized expert and authority on industrial minerals by virtue of a dozen or more important publications on the subject. By far the most important of these were his nine annual reviews devoted to Industrial Minerals (1951-1959 inclusive) and work as Editor-in-Chief of the completely revised third edition of the AIME's classic <u>Industrial Minerals and Rocks</u>, published in 1960. During these same years he played major leadership roles in the American Institute of Mining, Metallurgical, and Petroleum Engineers (AIME), the Society of Economic Geologists (SEG), the American Geological Institute (AGI), and the Mineralogical Society of America (MSA). In 1957 the AIME honored him with the Daniel C. Jackling Award and in 1963 with the Hal Williams Hardinge Award.

Always a scholar, even while devoting most of his energies to industrial problems, when the time came to retire from du Pont, he happily accepted a two-year appointment at M.I.T. as the first William Otis Crosby Lecturer, and then completed his long and active career as Professor of Geology at Arizona State University at Tempe, a post he held at the time of his death on 4 August 1964.

On the basis of his professional activities and accomplishments, it seems a fair evaluation to state that Gillson's greatest single contribution to geology was demonstrating to academic and industrial scientists and engineers the great diversity, commodity value, economic importance, and international significance of non-metallic industrial minerals. Second only was his energetic and effective leadership as an officer in his professional organizations.

Joseph Lincoln Gillson* was born in Evanston, Illinois on 12 February 1895. His father, Louis Kossuth Gillson (1852-1942), was a patent attorney for the Pullman Car Company and also had other clients in the railroad business. His mother, Ida (Bartholomew) Gillson (1854-1929), is remembered as a righteous and dignified lady who had more than her share of the brains and ability in the family. She was an active member of the First Baptist Church of Evanston, and it was at one of the meetings of that church's Young Peoples Group that Joe first met Grace Brown, who would become his wife in 1918. Louis and Ida, both of English ancestry, were married in 1875. Five children were born to their union: a daughter, Cornelia (1875-1884), and four sons, Charles B. (1877-1922), Milo (1879-1895), Louis K. (1893-1906), and Joseph L. (1895-1964). Only Joseph and his oldest brother Bert (Charles B.) lived to adulthood. The Gillsons also adopted a young girl, Florence Gillson Ranney, who was about the same age as Joe and very much his youthful playmate, being treated always as if she were actually a natural daughter in the family.

---

* I am deeply grateful to the three Gillson children--Jane (Gillson) Langton of Lincoln, Massachusetts; Patricia (Gillson) Baker of Modesto, California; and Joseph L. Gillson, Jr. of Wilmington, Delaware--for making available to me the papers of their father and mother, and for constructively criticizing this sketch.

   I have also drawn information from the following brief biographical sketches and notes:

   1957. The Presentation [of the 1957 Jackling Lecturer, J. L. Gillson], by Ian Campbell. Min. Mag., May 1957, p. 549.

   1960. Joseph L. Gillson--AIME President, 1960, by R. A. B. [R. A. Beals], in Jour. Metals 12: 118, (Feb. 1960); and Mining Eng., Feb. 1960, p. 116-117 [photograph on p. 116].

   1961. Editorial: A Tribute [to J. L. Gillson as he retired as President of AIME], A Welcome [to incoming President, R. R. McNaughton]. Jour. Metals, March 1961, p. 181.

   1964. J. L. Gillson dies in New Mexico; former AIME President was 69. Jour. Metals, Sept. 1964, p. 740, 744.

   1964. Joseph L. Gillson: An Appreciation by Robert M. Grogan, Min. Metal. Soc. Am., B. 321, Vol. LVII/2; 66-67, December 1964.

When Joe was still in grade school his parents moved from Evanston to Wilmette, where he completed his primary education and later graduated from the New Trier High School in neighboring Kenilworth (Illinois) in 1913. Although he did not distinguish himself in any way in his high school work, he did decide to go to college, probably because of his mother's influence. Northwestern was nearby and handy, so that the fall of 1913 found young Joe a freshman taking chemistry for his science, and commuting between Wilmette and Evanston on a motorcycle, even as students are still doing today, sixty years later. In spite of some typical collegiate high-jinks, such as driving the car that absconded with the President of the Sophomore class on the night of the Sophomore Hop, he was included in the six top students of the Freshman class and was given an award of a pair of books for his efforts. As an outside activity he played the flute well enough to be in a large symphony orchestra connected with the School of Music.

When Joe found he had to take a year of science other than chemistry, which he had not cared for, he followed the suggestion of his advisor and registered for a course in geology given by U. S. Grant. This was to be one of the crucial decisions of his life for two reasons: he took to geology like a duck to water, and found in Professor Grant a kind and helpful person who would become a close friend and counselor and one of the strongest influences in his professional life.

At the end of his junior year he again made a decision that would greatly influence his future career. He registered for the University of Missouri's 1916 summer field course, and 25 June found him at Camp Branson (named after E. B. Branson who directed the course.), located at Bull Lake, at the foot of the Wind River Mountains, in Wyoming. He greatly enjoyed the five weeks of field work, done from a tent camp as a base; found in Branson another geologist who would become a lifelong friend; and definitely decided that geology, especially field geology, was the career he wanted to follow. His "Report of Field Work in Course 105S, U. of Mo.," illustrated with many photographs and sections, and vivid descriptions of what he saw, illustrates the attention to detail and the penchant for accuracy that would so strongly characterize his later work on industrial minerals. A further result of the Wyoming experience was that it brought Joe and his earlier girl friend and later wife, Grace Brown, together with a common interest, as she had spent the summer of 1916 in Lander, from which the Missouri group had taken off for Camp Branson. Many are the romances that have bloomed in summer field courses in geology!

Returning to Northwestern for his senior year, Joe resumed his activities in a social group called "The Wranglers," which later became a chapter of Alpha Delta Phi, and in Alpha Chi Sigma, the chemical fraternity, and completed his undergraduate requirements by February of 1917. Election to Phi Beta Kappa and Sigma Xi followed, and his Bachelor of Science degree, "With highest distinction," was awarded in June 1917.

Meanwhile, World War I had been raging in Europe for several years, and when the United States declared war on Germany on 6 April 1917, Joe enlisted in the Illinois Naval Militia the next day. He was sent to the Philadelphia Navy Yard for preliminary training. He did manage to get leave to return for his Commencement, however, and appeared wearing his newly issued "middy suit."

Soon after arrival in Philadelphia, Joe had a chance to take an examination for Assistant Paymaster. Although he had never had any training in accounting, and came away from the examination feeling he had done poorly, he was shortly thereafter (4 August 1917) notified that he had been commissioned an Ensign. His first assignment was to the U.S.S. Indiana, on which he served from April to August; then soon after being commissioned he found himself on the old German liner George Washington, now the U.S.S. Washington, which was being equipped for transport duty, and of all things he was to be Commissary Officer! It was a far cry from ore bodies and granites and geological reports to ordering 20,000 dozens of eggs, 20,000 pounds of beef, hundreds of crates of oranges and dozens of other food items for the 8,000 troops and 500 officers that the transport would carry from New York to Brest, France every three weeks. Fourteen times he made the round trip from America to France, experiencing both comic and tragic events and almost rubbing elbows with many a dignitary.

One night, during a severe storm, the ten-gallon cream cans got loose in one of the big refrigerators, and Joe found the cold room knee-deep in cream! Another time the transport collided with a small freighter in mid-ocean, creating a lot of excitement, but no one got hurt. Once he watched one of the ships of his convoy go down from a torpedo, and on return trips he saw to it that the wounded being brought back from Europe had special meals.

Then there were all sorts of "top brass" who had to be looked after-- a group of senators, Assistant Secretary of the Navy Franklin D. Roosevelt, Secretary of State Stimson, and President Wilson, himself, with his staff. How did an officer aboard a transport prepare for such a group of notables? Joe hired a special crew from the Biltmore Hotel in New York, and thought everything was under control, until the night before the President was to come on board, when the Senior Paymaster boarded the ship and became frantic when he found that no special china had been ordered for the notables. Of course Joe's responsibility was only the food, not the china, but good navy man that he was, he somehow routed out the owner of a fancy china store on Fifth Avenue at two o'clock in the morning, and by six o'clock they had appropriately fancy tableware on board.

With eight round trips behind him, Joe and Grace Brown, that little girl from the church group back in Evanston, decided to get married. The ceremony was performed on 13 September 1918 in New York, while Joe was in port between voyages. But the newlyweds would have to wait out the end of the conflict until permanent reunion would come with Joe's discharge in May 1919. In the interim, Joe made half a dozen more round trips on the U.S.S. Washington, surviving the influenza epidemic that on a single trip took the lives of 80 men and two of the ship's officers, and came out of the service a Lieutenant, Junior Grade. Glad to end his military service, and eager to get back to geology, he decided to return to Northwestern for graduate work. Before following him back home, however, it seems appropriate to insert a short discussion of his wartime marriage.

As mentioned elsewhere in this biography, Joe first met his future wife, Grace Irene Brown, when they were members of a Young Peoples Group in the First Baptist Church of Evanston. Later on they found a common interest in Wyoming, Grace spending the summer of 1916 in Lander, whence Joe took off with the University of Missouri field course group for Camp Branson in the Wind River Mountains. It was 1918, however, and in the middle of Joe's period of service in the Navy, before they could get married. There was no time for a honeymoon then, however, for Joe had only three days before his ninth round trip to Brest. Meanwhile, Grace arranged to share a small apartment in New York with another woman, so that she could see Joe during the few days that his ship was in port between trips, which was about every three weeks. This was the time when the flu had reached epidemic proportions along the Atlantic Seaboard. When the malady took her apartment-mate, Grace was left alone to await the end of the war and reunion with Joe, which came the following May.

Grace Irene Brown was born in Eau Claire, Wisconsin, on 3 March 1898, one of three daughters of Harry F. and Carolyn (Pope) Brown. She was a graduate of Northwestern University and also had training as a nurse. Her two sisters were librarians, and Joe's daughter, Jane, remembers how her two aunts gave her children's books to read, and how she dreamed of writing such books herself, a dream that would actually come true many years later.

Three children were born to the Gillsons soon after they moved to Boston: Joseph L. Gillson, Jr. (b. 1921) now a research physicist in du Pont's Central Research Department in Wilmington, and unmarried; Jane, born in 1922, married to William G. Langton of Lincoln, Massachusetts, with three sons, and an authoress of children's books; and Patricia, born in 1926, with two daughters, and now organizational Development Specialist, Charmin Paper Products in Modesto, California.

With the war over, Joe and his bride of seven months returned to Evanston in time for him to start graduate work during the summer term of 1919. This gave him an opportunity to join U. S. Grant and his students on the annual field trip to the Lake Superior region, and he was delighted to resume work with his former professor. During the ensuing school year, he acted as a laboratory assistant in General Geology during the first semester, and as Instructor in one section of General Geology the second. Meanwhile, he completed all requirements for the Masterate during the year and received a Master of Arts degree in geology in June 1920. Now came the decision that would lead to his brief academic career, and then to his long industrial career as an economic geologist.

Gillson applied for admission to M.I.T.'s Graduate School to work toward a doctorate, and received a $300 scholarship to cover his tuition for SY 1920-21 (quite a contrast to the $2500 tuition charge in 1970 and $3200 in 1972). He was able to get an assistantship at Harvard, where he would assist Charles Palache in mineralogy. During the same period he took work with Palache in mineralogy at Harvard, and at M.I.T. he studied with Charles Warren and Waldemar Lindgren. At the end of the busy year he presented a thesis, supervised by Lindgren, on

> "Some notes on the geology of the Shoshone Canyon, Park County, Wyoming,"

and received a Master of Science degree in Geology in June 1921.

During the summer of 1921 Gillson and Palache went rock collecting in New Hampshire, and their experiences would add quite a few pages to this narrative. Suffice it to relate just one--a trip in Palache's old Dodge, which he didn't drive, to Red Hill, New Hampshire. With the old car heavily laden with rocks, they started for home, Joe driving, but didn't go very far before they had the first of their ten flat tires. These were the old clincher type that had to be pried off the wheel or rim with several discarded spring leaves. Once off, the inner tube had to be removed, patched, tested for leaks, and then put back into the casing, and the casing pried back onto the wheel. Finally came the job of hand pumping the tire to the desired pressure. Mrs. Gillson has written that Joe and Palache did not get home until dawn of the next day, and that Joe was just about worn out. Little wonder since he had both to drive and to fix the tires! For "oldtimers" like myself, who are thoroughly familiar from experience with those clincher tires, spring leaves, and the odor of patching cement, ten flats seem just a few too many to make any rock collecting trip a pleasant memory!

The fall of 1921 found Gillson again going back and forth between M.I.T. and Harvard. At M.I.T. he started work on a doctorate under Lindgren, and at Harvard he continued teaching the mineralogy course, but with the title of Instructor, inasmuch as he was in sole charge of the course substituting for Palache who was on a sabbatical leave.

By 1922 Joe was making good progress on his thesis and was getting excellent teaching experience at the same time. Having worked closely with Charles H. Warren at M.I.T., he was an obvious candidate to take over Warren's courses when he resigned to accept a Sterling professorship at Yale in 1922. Thus began the 8-year period in Gillson's life when he devoted his energies to teaching M.I.T. students in geology and mining. Starting as Instructor in Mineralogy and Petrography in the fall of 1922, he advanced to Assistant Professor in 1924 and to Associate Professor in 1927, the rank he held when he resigned on 1 August 1930. He gave lectures, conducted laboratory exercises and led field trips for students in mineralogy and petrography. At the same time he carried on his doctoral thesis work on contact metamorphism, supposedly under the supervision of Lindgren, but he saw little of the latter, who was then preparing the third edition of his classic Mineral Deposits. So, as with graduate students from time immemorial, Gillson turned to his fellow students for assistance, rather than disturb the busy professor, and by June 1924 had completed his thesis on

> "Certain phases of the geology and ore deposits of the Pend Oreille silver mining region, Bonner County, Idaho."

He received a Doctor of Science degree in 1924, along with promotion to Assistant Professor. Gillson had been working long and hard as both doctoral candidate and staff member, and Mrs. Gillson was to write some thirty years later -

> "The trouble with teaching in those days was that to get ahead you had to do research, all of the money to finance that research had to come out of your own pocket, and we were making $3600 then." [Letter to Ian Campbell dated Jan. 19, 1957--copy preserved in the M.I.T. Institute Archives.]

Quite the same, yet also quite different, in the early 1970s!

During his teaching years at M.I.T., Gillson spent the summers working as a Junior Geologist for the U. S. Geological Survey. The first two summers (1921 and 1922) were with Edward Sampson in Idaho; the next two (1923 and 1924) were with L. G. Westgate around Pioche, Nevada; and the next three (1925, 1926 and 1927) were on various assignments (emery in New York; talc in Vermont; etc.). A 1923 field notebook (U.S.G.S. 9-918) has lists of Pioche specimens collected and some interesting items in the daily expenses, e.g.

```
            5 gas @ 32           $ 1.60
            Movies                  .35
            Haircut                 .50
            1 doz oranges           .70
            Bd. & Lodging:
                16 meals  @ 50    8.00
                 6 nights @ 50    3.00
            Supper                  .85
```

Several articles resulted from this field work. (See Bibliography.)

During his academic years at M.I.T., Gillson gave courses in mineralogy (1923-1930), petrography (1923-1930), optical crystallography (1923-1927), geological surveying (1926-1929), and economic geology of non-metallics (1923-1930). When he returned after retirement, to serve two years as the first William Otis Crosby Lecturer (1961-1963), he gave lectures and informal seminars on industrial minerals, and spent many happy hours nostalgically re-examining drawer after drawer of mineral specimens in the great Lindgren collection, many of which he had studied forty years earlier as a fledgling instructor in the Department. He seemed to feel delight and happiness, as well as to experience a measure of sadness, in becoming re-acquainted with his old friends, especially those from Pioche, Nevada.

One of his duties as a faculty member was to help students find thesis problems and then supervise the work as the research progressed. Altogether, Gillson supervised twelve theses--9 for the S.B. degree and 3 for the Masterate.

Gillson's interest in industrial minerals began to develop about 1926. As he wrote me in one of his letters, he noted that Lindgren was giving little attention to the "non-metallics" in his general course on mineral deposits. So he asked Lindgren if he might offer a course in that field. Receiving a favorable reply, Gillson organized a series of lectures under the subject title "Economic Geology of Non-Metallics," and gave them for several years. One day an alumnus of Course III, Mining Engineering and Metallurgy, wrote to Lindgren that he was seeking a man to look for barytes for the du Pont Company. Lindgren gave the letter to Gillson, Gillson took the job, and thus began his 32-year association with the E. I. du Pont de Nemours & Co. of Wilmington, Delaware.

This 1926 summer job looking for barytes in the South led to more work in the next two summers and finally to full-time employment in the fall of 1928, when he took a year's leave of absence from M.I.T. This leave was extended for the SY 1929-30, and Gillson resigned from his M.I.T. associate professorship on 1 August 1930.* Thus ended his 8-year

---

* During SY 1929-30 Gillson served as Associate Professor of Economic Geology at his alma mater, Northwestern University, offering several lecture courses and laboratories dealing with mineral commodities and ore deposits.

period on the M.I.T. faculty though as noted in the preceding sentence he was on leave for the last two years, SYs 1928-29 and 1929-30. He would return again, however, some thirty-one years later, but more about that appointment later on in this narrative.

Following his first summer job with du Pont, in 1926, which involved a search for barytes, Gillson next turned his attention to titanium desired for pigments. This work led him to three trips to India, three or four to Brazil, one to West Africa, and many within the United States. By the time he finished this assignment he was probably one of the best informed geologists on titanium anywhere to be found, and his publications became authoritative references on the subject. However, he wrote me in late 1960 that

> "I am known as the lone wolf crying for a pneumatolytic origin of ilmenite and titaniferous magnetite deposits in the anorthosite."

Next came a need for fluorspar (fluorine for refrigerators and "bug bombs"), and this exploration sent Gillson to Newfoundland, Spain, and Southwest Africa. As a result of this work on fluorspar, when World War II came on, Joe was asked to serve the War Production Board as a member of the Industry Advisory Committee on Fluorspar Mining. After the War, in the late 1950s, he made numerous trips to Mexico in search of fluorspar and developed some deposits in the northern part of the country. These were not exploited by du Pont, however, because of certain political changes in the Mexican laws and because of the marginal size of the deposits.

Other field work as a du Pont geologist took him in search of sulphur, ilmenite, barytes, celestite, and fluorspar, and involved investigations of ground-water supplies, foundation problems, and general site studies.

By 1950 he had become particularly interested in the production and use of a large group of chiefly non-metallic commodities collectively designated "industrial minerals." These included aggregates, abrasives, andalusite, asbestos, natural asphalt, barytes, bentonite, beryl, borax, boron, calcium chloride, ceramics, clays, columbite, diamonds, diatomaceous earth, dolomite, feldspar and nepheline, fluorspar, gem stones, glass, gilsonite, graphite, gypsum, kyanite, light weight aggregates, lime, limestone, lithium, magnesite, mica, monazite, nitrates, perlite, pumice, refractories, rock wool, roofing granules, salt, glass sands and silica products, sillimanite, slag, sodium carbonate (trona), sodium sulphate, soapstone, pyrophyllite, sulphur, talc, titanium, zircon, and zirconium compounds.

Gillson's work at du Pont involved him in the search for a long list of raw materials referable to that catch-all category of "industrial minerals," and this search made him a world traveler. On his earlier trips, when the children were still small, he was sometimes able to take the whole family along (e.g. to Brazil and India); later on, when the children were older and could fend for themselves, Mrs. Gillson alone was his companion, and an enthusiastic one at that! The experiences they had on these trips would make an interesting story in themselves. They often visited places where accommodations left much to be desired, even for men, let alone a woman! In 1940 Joe spent most of the year as special geological advisor to the Government of Travancore, India. These travels provided him with unusual opportunities to see industrial minerals on a worldwide scale and to accumulate an ever-growing knowledge which he called on when preparing his numerous annual reports and other publications on them.

Starting in 1950, he published nine successive annual reports on the aforementioned industrial minerals, which led to his appointment as Editor-in-Chief of the completely revised third edition of AIME's comprehensive Industrial Minerals and Rocks published in 1960. And Gillson was no figurehead in this enterprise; as was typical of his pattern of work, he assiduously followed the progress of the volume from manuscript through page proof. When he accepted an assignment, he saw to it that the job got done, and he expected others to do the same.

Owing to these annual reviews, his expertise in the broad field of non-metallic industrial minerals was widely recognized in the mining industry, and he was frequently invited to speak on the subject to interested professional groups. He kept no systematic record of these speeches, but they certainly numbered several a year for the last ten years of his life, particularly while he was President-elect and then President of the AIME (1959 and 1960, respectively). Typical examples are the following, of which he preserved notes or rough drafts:

> "The Industrial Mineral Industry" delivered to a meeting of the American Mining Congress in Salt Lake City on 11 September 1957.
>
> "Supplies of mineral raw materials used by the chemical industry" delivered to a meeting of the American Institute of Chemical Engineers at San Francisco on 8 December 1959.
>
> "Industrial minerals in the Southwest" delivered to the Southwest Minerals Conference in Los Angeles on 21 April 1960.
>
> "Opportunities in geology other than petroleum" delivered to the University of Texas Geology Alumni on 9 October 1960.
>
> "Industrial Minerals," one of his last, delivered to the Maricopa Subsection of AIME in Phoenix, Arizona, on 6 February 1964.

In 1958 the following statement appeared below his picture accompanying one of his numerous articles on Industrial Minerals in the Mining Congress Journal (Vol. 44: 93 [1958]):

> "Joseph L. Gillson has long been an authority on industrial minerals. Through his efforts in part, the mining industry's interest in this field has been intensified in recent years. Consequently Joe Gillson's vast knowledge of the geology, economics and exploitation of non-metallic minerals and his ability to analyze and present this information have earned for him ever-increasing acclaim and stature by his colleagues in the mining industry."

Gillson's international reputation in the field of industrial minerals was recognized when he was asked to serve as technical adviser to the American delegation, and to present a paper on the subject, at the United Nations Conference on the Application of Science and Technology for the Benefit of the Less Developed Areas, held in Geneva in 1962.* From his several working papers came his last major contribution to the literature on industrial minerals--"Development of non-metallic mineral resources in a dominately [sic] agrarian economy" first reproduced as a working paper (23/A/697) for the Conference and later published in the U. S. Department of State Papers, Vol. 2, p. 106-117 in 1963. (See item 58--1963a in the Bibliography.)

Gillson joined the American Institute of Mining and Metallurgical Engineers in 1923, soon after receiving his doctorate and starting his teaching career at M.I.T. For more than forty years he was a loyal and active member, performing many special jobs assigned to him. Although he belonged to and served many other professional organizations, he gave the most service to AIME and considered it his favorite. According to Mrs. Gillson, Joe's interest in AIME and industrial minerals was stimulated by Benjamin Miller of Lehigh for whom he had a high regard. Miller was Chairman of the Institute's Industrial Minerals Division and asked Joe to head one of his subcommittees. Joe hit it off well with the AIME folks from the very beginning, and, as the years passed, and he became more widely known among the membership, he accepted one committee chairmanship after another, until his long and loyal service was honored by his election as Vice President (1951-1953 and 1956-1959), and then President (1960), and by selection for the Daniel C. Jackling Lecture Award in 1957, and The Hal Williams Hardinge Award in 1963. There are numerous published

---

* This Conference was discussed in Business Week for March 23, 1963, p. 89-92, and Gillson's picture appeared on page 89. Gillson's comments were also summarized under the title "Application of technology for the benefit of less-developed countries" on pages 495-498 in the Journal of Metals for July 1963.

518   GEOLOGY AT M.I.T.   1865-1965

J. L. Gillson (right), Vice President of AIME, receives the DANIEL COWAN JACKLING LECTURE AWARD of the Institute from Ian Campbell in 1957, for "... his significant contributions to the advancement of economic geology, his leadership, and his keen sense of professional responsibility."

Sketch of J. L. Gillson, drawn by Bruno Figallo for the cover of the March 1957 issue of GEOTIMES, published by the American Geological Institute, of which Gillson was President at the time.

(Photos from Gillson's papers)

(ABK Photo Service)

J. L. Gillson (right), President of AIME, hands the gavel of office to R. R. McNaughton, President Elect, in 1961.

J. L. Gillson holds plaque of AIME's THE HAL WILLIAMS HARDINGE AWARD, which he, as President of AIME, presented to Raymond Bardeen Ladoo (seated) in 1961. Gillson himself received the same Award in 1963 "... for pioneer work in industrial mineral resources, for wide dissemination of useful data thereon, and for generating systematic effort within this field."

notices of his services to the Institute, but it will suffice here to cite only the following two:

1. The Daniel C. Jackling Lecture--The Lecturer, J. L. Gillson: The Presentation, by Ian Campbell. Min. Eng., p. 549, May 1957.
2. J. L. Gillson dies in New Mexico; former AIME President was 69. J. Metals 16/9: 740, 744, September 1964.

Busy as he always was with his regular work at du Pont, Gillson nevertheless found time to serve a number of professional societies in one way or another. In his earlier years, when he was publishing articles in the American Mineralogist, he served the Mineralogical Sciety of America (MSA) as Vice President in 1931-1932. Twice he was elected Vice President of AIME, 1951-1953 and 1956-1959, and later President in 1960 as detailed in the preceding section. He was Vice President of the Society of Economic Geologists (SEG) in 1955 and President in 1959. He served as both Vice President (1956) and President (1957) of the American Geological Institute (AGI). He was also a member of the American Association of Petroleum Geologists (AAPG), the American Geophysical Union (AGU), the Geochemical Society, the Geological Society of America (GSA; Fellow), and the National Society of Professional Engineers (NSPE). In addition, he served on numerous professional committees, generally as Chairman; some of these are mentioned elsewhere in this biography.

As stated previously, when Gillson reached retirement age at du Pont in 1960, we were fortunate to have him accept appointment as the first William Otis Crosby Lecturer to give such lectures as he desired in the general field of practical geology. The fall term of 1961 found the Gillsons settled in an apartment in Boston and Joe busy with a group of undergraduates interested in industrial minerals. For the next three terms he not only continued lectures in this field but during the second year, SY 1962-63, he added instruction in polished section work and geological engineering, and even took on supervision of the thesis work of two Harvard students. There was to be no such thing as retirement for Joe Gillson!

From Boston the Gillsons moved to Tempe, Arizona, in 1963, where Joe was appointed a Professor of Geology at the Arizona State University. In probably his last photograph, he is pictured gazing at a chunk of rock on which his right hand rests, and captioned, "Geology's Dr. Joseph Gillson," on page 7 of the university's Arizona Statesman, Winter of 1964. This publicity, however, came after Joe's death on 4 August in a hospital in Carlsbad, New Mexico, following an operation. So ended a long, active, and happy life of geologizing, which he shared with Grace Brown Gillson who accompanied him on many of his travels abroad and to the annual conventions of his favorite professional organizations, and who followed him in death less than three years later in February 1967.

Over and above the professional achievements of a successful scientist and engineer is the impact that the individual makes on others and the impression he leaves for them to remember. First, let us consider Gillson's achievements as a geologist; then try to summarize the impact he made and the impression he left.

His rank as Associate Professor at M.I.T. after eight years of teaching shows that he was rapidly ascending the academic ladder; his 20 publications during the same period (1922-1930) are proof of intense industry and research ability; and the 12 student theses he supervised show he did not neglect his students. In short, his academic career had great promise when he left it to enter industry.*

As an employee of du Pont's for more than 30 years, Gillson proved to be an astute and highly successful engineering and economic geologist. He found the Company substantial sources of barytes, fluorspar, titanium and a host of other so-called "industrial minerals." These practical successes did not dampen his scholarly interests, however, for he published some 35 articles during his 32 years at du Pont, including nine successive annual reviews of Industrial Minerals. As a result of these activities he came to be regarded as the American authority on such mineral commodities.

In 1957 the AIME honored him with its coveted Daniel C. Jackling Award** for

> "his significant contribution to the advancement of economic geology, his leadership, and his keen sense of professional responsibility,"

---

* By the end of 1928 Gillson was so highly regarded by Lindgren that in a letter dated January 15, 1929, Lindgren wrote:

"My dear Professor Gillson:

    I expect to leave today, January 15th, and do not expect to return until Monday, February 4th. Will you be kind enough to act in my absence as Head of the Department.

                          Very sincerely yours,

                          Waldemar Lindgren, In-charge
                          Department of Geology"

WLW:W

[That Gillson did in fact so act is indicated by numerous memoranda prepared by him for Lindgren during this brief period.]

** Geotimes for March 1971 (Vol. 1, No. 9) had a Figallo sketch of Joe, as the Jackling awardee, on its front cover and mentioned reception of the award on pages 5 and 10. This sketch is reproduced on another page of this biography.

and in 1963 the same Institute selected him for its Hal Williams Hardinge Award

> "for pioneer work in industrial mineral resources, for wide dissemination of useful data thereon, and for generating systematic effort within this field."

Finally, the Alumni Association of his alma mater, Northwestern University, recognized his outstanding achievements in both the academic and the industrial worlds by selecting him for an Award of Merit on 17 June 1961.

But honors and citations and awards fell lightly on Joe Gillson's shoulders for he was fundamentally a humble person. Regardless of where he was, he was so informal and unpretentious that he was likely to be overlooked by those who like to bask in the reflected glory around a celebrity. A salutation as Professor Gillson or Doctor Gillson would often elicit no response, whereas simple "Joe" would bring a warm smile to his face and the invitation to start a discussion on some important matter. He seldom indulged in trivial discourse; there was too much of importance to discuss. Therein might well lie part of the reason why he was so successful in raising money* among the members of AIME for the new United Engineering Center building in New York, and why these same members chose him again and again for their highest offices and honors.

In reply to a letter, in which I asked him for some background information about himself, Joe included the following typically modest paragraph (in letter to Robert Shrock dated Dec. 31, 1960):

> "If I am known for any particular reason as president of AIME it has to do with the money-raising among the members for the new building [i.e. United Engineering Center in New York City]. Gus Kinzel who was president in 1958 gave me the job of being chairman of the fund-raising committee, and most of the time since I have been hounding the members for money. That part I will be glad to be through with. If the petroleum engineers had done as well as the miners and metallurgists have done, we would be just about over the top, but they have their society office in Dallas, and don't have the loyalty to the National organiza-

---

* In 1960, on the 100th Anniversary of Waldemar Lindgren's birth, I conceived the idea of establishing a memorial in the form of a scholarship and/or library fund to perpetuate his memory at M.I.T. When I raised the question of such a fund with Gillson, and also asked what he thought about asking Mr. Herbert Hoover to be Honorary Sponsor, his reaction was favorable and enthusiastic. Before January was over, he wrote to Hoover, later got a favorable reply, and on 11 February 1960 he and I called on the former President in his suite in the Waldorf-Astoria Towers in New York. The result was a letter from Hoover which he authorized us to use in soliciting contributions from Lindgren's former students and other professional friends. There is more on this particular incident in the chapter on Departmental Funds in my Volume 2.

> zation and to headquarters in New York. They are still
> $50,000 short of their share which was only $150,000.
> The miners and metallurgists have raised about $330,000
> of their $350,000 and I believe they will find the
> other $20,000 before the end of the campaign in February."

Joe never hesitated to express his feelings about such matters as loyalty, support, and willingness to do an assigned job, and this is clearly evident in a letter he addressed to the AIME membership in 1960 regarding the solicitation of funds for the United Engineering Center Building:

> "You have heard (almost ad nauseum) about our commitment
> to give our $500,000 share toward building our new Engineering Building. Our sister societies have met, or
> almost met, their quotas - but we still need $80,000.
>
> "Some of you haven't given a thin dime. Your fellow
> members who have given - including the younger men -
> have given generously. If you have already given,
> please give a little more. If you haven't given,
> please make up for it now."

(Quoted from an editorial on p. 181, Jour. Metals, March 1961.)

Joe Gillson liked people, particularly those who shared his own views and professional interests, and who were ready to lend a hand when help was needed. And people liked Joe, as indicated by a statement that

> "He was one of the most popular men ever to serve the
> Institute [AIME]....he was on a first-name basis with
> a tremendous number of members, not only of AIME, but
> also of other groups in the Mineral Industry." (Jour.
> Metals, Sept. 1964, p. 74.)

He was a member of numerous social and honorary organizations, among which were Alpha Delta Phi (social), Alpha Chi Sigma (chemistry), Cosmos Club (Washington, social), Phi Beta Kappa (honorary scholastic), Sigma Gamma Epsilon (national geological, etc.; associate member), and Sigma Xi (honorary scientific). As mentioned on preceding pages he also belonged to the American Association of Petroleum Geologists, American Geological Institute, American Geophysical Union, American Institute of Mining, Metallurgical and Petroleum Engineers, Engineers Club (New York), Geochemical Society, Geological Society of America, Mineralogical Society of America, Mining and Metallurgical Society of America, Mining Club of New York, National Society of Professional Engineers, Philadelphia Mineralogical Society, Society of Economic Geologists, and Walker Mineralogical Club (Toronto).

Taking into account his relatively short teaching career, his three decades of employment as industrial geologist, and his even longer period of membership in and services to the leading professional organizations in his field of interest, how and in what ways did he make his major impact as a professional geologist?

In my opinion, Gillson's greatest contribution to geology was making the scientists and engineers in academia and industry aware of the great variety and importance of that large group of chiefly non-metallic commodities collectively designated "industrial minerals." He spoke and he wrote throughout his entire professional career, anytime an opportunity arose, about the increasing number of mineral substances that were being sought by industry (his last article listed more than 45 commodities); he emphasized that, while no substance except cement produced a large number of dollars, the annual value of all industrial minerals now approaches three billion dollars, and that in tonnage and value of total production, many industrial minerals now outrank many of the metals. He repeatedly pointed out the great economic importance of industrial minerals as raw natural resources, and how these resources, or the lack of them, divided the nations into the "haves" and "have-nots," respectively, and made necessary some future sharing of them on an international basis. He delivered dozens of lectures on the subject, wrote numerous articles on individual minerals, thoroughly reviewed the world situation regarding them for nine successive years (1951 to 1959 inclusive), and wound up as Editor-in-Chief of the third and latest edition of AIME's classic Industrial Minerals and Rocks published in 1960, the year he retired from du Pont.

Of course Joe served his employer well or he would not have held his job at du Pont. He found for them the industrial minerals they needed, for that was what they were paying him for, hence he was a most successful industrial geologist. But his interests went far beyond the mere challenge of finding and acquiring deposits of desired raw materials for his employer. Joe was compelled by his strong academic attitude to speak and write about the materials, and by his deep interest in people to join with them in associations devoted to the discovery, recovery, and ultimate use of our mineral resources. It is no overstatement to call Gillson "Mr. Industrial Minerals," and his name will always be coupled with this ever-growing category of the world's natural riches, as recorded in the citation read when he was given The Hal Williams Hardinge Award in 1963, and in the A.I.M.E. Resolution on his death in 1964 (see next page).

I have known few geologists with as much "common sense" and practicality as Joe Gillson had, and well worth pondering is the advice he left for future geology students as he concluded his Jackling Lecture on "A Geologist Looks at Industrial Minerals" (Min. Eng., May 1957, p. 555 and AIME, Tr. 1957, Vol. 208):

> "Now, in conclusion, may I emphasize that the real purpose of this analysis is to arouse in students just entering the field of economic geology an interest in the industrial mineral field, and to give them a small glimpse of the challenge that this large field offers.

MEMORIAL RESOLUTION

345 EAST FORTY-SEVENTH STREET
NEW YORK

WHEREAS, with the death of JOSEPH LINCOLN GILLSON on August 4, 1964, the American Institute of Mining, Metallurgical, and Petroleum Engineers lost one of its most inspiring and beloved members who brought honor to it from all parts of the World; and

WHEREAS, he brought maturity to the application of economics to the geology of industrial minerals, and new industrial technology to their processing; and

WHEREAS, during his forty-one years of loyal service to the Institute as a teacher, scholar, and human example which touched both the embryo engineer, and the old timer, he challenged their minds and spirits toward excellence; and

WHEREAS, he served the Institute faithfully as a Director for two terms, as a Vice-President for two terms, as its honored President in 1960, and as the 1958-1959 Chairman of the Fund Raising Committee through which the Institute contributed most substantially to the construction of the United Engineering Center; and

WHEREAS, he was honored with the Daniel C. Jackling Award in 1957 and with the Hal Williams Hardinge Award in 1963; and

WHEREAS, he personally contributed to the heritage of mineral literature, and most recently served as Editor-in-Chief of "Industrial Minerals and Rocks," Third Edition; and

WHEREAS, since 1962 until his unexpected passing he served the Institute as Chairman of its Inter-Engineering Society Cooperation Committee, encouraging cooperation with other engineering disciplines and pointing the way to growth in professional stature;

BE IT THEREFORE RESOLVED that the American Institute of Mining, Metallurgical, and Petroleum Engineers records with deep sorrow the passing of one of America's most distinguished engineers; and

BE IT FURTHER RESOLVED that this Resolution be spread upon the Minutes of this Meeting and that copies be sent to Mrs. Gillson, to Joseph L. Gillson, Jr., and to his daughters Mrs. Jane Langton and Mrs. Patricia Baker.

Respectfully submitted,

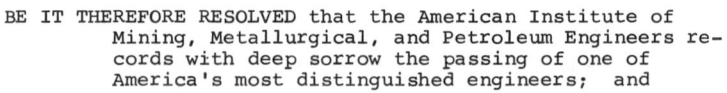

R. William Taylor, General Secretary

September 25, 1964

(From the official minutes of the Board of Directors of the American Institute of Mining, Metallurgical, and Petroleum Engineers)

> May I urge also that our major colleges and universities re-examine their curricula to find more room to give some of their students the wide and diverse training required for work with industrial minerals. These students may have to be relieved from the abstruse and erudite aspects of theoretical geological studies, so dear to the hearts of many of the more academic-minded of the staffs, and be given time to take courses in the mining and economic departments of the schools they are attending."

One would have to search diligently through the past comments of many wise and successful geologists to find equally relevant advice for the geology student of the 1970s!

Death came suddenly and unexpectedly on 4 August 1964, in Carlsbad, New Mexico, while he was convalescing after an operation. His constant companion of 46 years, Grace Brown Gillson, followed him in death three years later, and today the two lie buried in Carlsbad.

BIBLIOGRAPHY OF JOSEPH LINCOLN GILLSON

Symbols and abbreviations used in the following references are explained on p. 91-98; in general, abstracts are listed separately, as well as with references to the complete article. This bibliography begins with the titles of Gillson's three graduate degree theses (T), then continues with his first published paper (1925) and includes all known titles through 1964, the year of his death.

T--1920   A recomposed granite and associated rocks of Saganaga Lake, northeastern Minnesota. [A.M. thesis at Northwestern University.] xiii + 192 p., (1920).

T--1921   Some notes on the geology of the Shoshone Canyon, Park County, Wyoming. [S.M. thesis at M.I.T.] 46 p., (1921).

T--1923   Certain phases of the geology and ore deposits of the Pend Oreille silver mining region, Bonner County, Idaho, 2 vols. (text and plates). [Sc.D. thesis at M.I.T.] 5 + 230 p., 8 il., pl., maps, (1923).

1--1925   (and Shannon, E. V.) Szaibelyite from Lincoln County, Nevada. Am. Mineral. 10: 137-139, (1925).

2--1925a Zircon, a contact-metamorphic mineral in the Pend Oreille district, Idaho. Am. Mineral. 10: 187-194, (1925).

3--1926   Conichalcite from the Bristol mine, Lincoln County, Nevada. Am. Mineral. 11: 109-114, (1926).

4--1926a Optical notes on some minerals from the Mahopac iron mine, Brewster, New York. Am. Mineral. 11: 281-286, (1926).

5--1926b Pigeonite from the Triassic traps of the Connecticut Valley. Am. Mineral. 11: 317-319, (1926).

6--1926c (and Warren, E. C.) A preliminary petrographic study of Portland Cement. Am. Ceramic J. 9/12: 783-786, (1926).

7--1927   Granodiorites in the Pend Oreille district of northern Idaho. J. Geol. 35: 1-31, (1927).

8--1927a Origin of the Vermont talc deposits, with a discussion on the formation of talc in general. Econ. Geol. 22: 246-287, (1927).

9--1927b   The granite of Conway, New Hampshire, and its druse minerals. Am. Mineral. 12: 307-319, (1927).

10--1927c   Biaxial calcite. Am. Mineral. 12: 357-360, (1927).

11--1928   (and Callahan, W. H. and Millar, W. B.) Adirondack studies: the age of certain Adirondack gabbros, and the origin of the reaction rims and peculiar border phases found in them. J. Geol. 36: 149-163, (1928).

12--1928a   Horizontal versus vertical forces in crustal movements of the earth [report of address by Bailey Willis]. Science 67: 608-610, (1928).

13--1928b   On the origin of the alkaline rocks. J. Geol. 36: 471-474, (1928).

14--1928c   [Review of] Geology of Petroleum and Natural Gas, by E. R. Lilley. Tech. Eng. News 9: 212, (1928).

15--1929   Petrography of the Pioche district, Lincoln County, Nevada. U. S. Geol. Surv., P.P. 158-D: 77-86, (1929). Reviewed by F. F. Grout in Econ. Geol. 25: 667-670, (1930).

16--1929a   Contact metamorphism of the rocks in the Pend Oreille district, northern Idaho. U. S. Geol. Surv., P.P. 158: 111-121, (1929).

17--1929b   On use of the term deuteric. Econ. Geol. 24: 100-102, (1929).

18--1929c   Bathygenetic and orogenetic movements. Science 69: 194-195, (1929).

19--1929d   (and Williams, R. M.) Contact metamorphism of the Ellsworth schist near Blue Hill, Maine. Econ. Geol. 24: 182-194, (1929).

20--1930   (and Kania, J. E. A.) Genesis of the emery deposits near Peekskill, N.Y. Econ. Geol. 25: 506-527, (1930).

Note: Gillson did his first work for the E. I. du Pont de Nemours & Company in the summer of 1926, and became their Geologist in 1928; he took a year's leave from M.I.T. during SYs 1928-29 and 1929-30, and resigned his appointment as associate professor on 1 August 1930.

21--1932   Genesis of the ilmenite deposits of St. Urbain, County Charlevoix, Quebec. Econ. Geol. 27: 554-577, (1932).

22--1932a   (with Westgate, L. G. and Knopf, A.) Geology and ore deposits of the Pioche district, Nevada. U. S. Geol. Surv., P.P. 170, 79 p., (1932).

23--1933   A day in a Ceylon gem field. Am. Mineral. 18/7: 300-308, (1933).

24--1937   Talc, soapstone and pyrophyllite (Chap. 46, p. 873-892) in A.I.M.E.'s Industrial Minerals and Rocks, New York: Am. Inst. Min. Met. Eng., (1937).

25--1937a   (with Hess, F. L.) Titanium (Chap. 47, p. 893-910) in A.I.M.E.'s Industrial Minerals and Rocks, New York: Am. Inst. Min. Met. Eng., (1937).

26--1945   Fluorspar deposits in the western States. A.I.M.E. Tech. Pub. 1783, Min. Tech. 9: 28 p., (1945); Tr. 173: 19-46, (1947).

27--1947   (and Jahns, R. H.) Report of Committee on Basic Research in the Field of Industrial Minerals and Rocks. Econ. Geol. 42: 737-746, (1947).

28--1949   Titanium (Chap. 49, p. 1043-1073) in A.I.M.E.'s Industrial Minerals and Rocks, 2nd ed., New York: Am. Inst. Min. Met. Eng., (1949).

29--1950   Deposits of heavy minerals on the Brazilian Coast. Min. Eng. 187: 685-693, (1950).

30--1950a   Electrostatic methods of concentration (p. 1093-1095) in J. H. Perry et al., "Chemical Engineers' Handbook," 3rd ed., New York: McGraw-Hill Book Co., Inc., (1950). [Also see item 60--1963c.]

31--1951   Industrial Minerals: Domestic production satisfies most requirements as output maintains high level. Min. Cong. J. 37: 64-66, (1951).

32--1952   Industrial Minerals: Intense activity and development characterized the year in most of the many branches of the industrial mineral field. Min. Cong. J. 38: 134-137, 146, (1952).

33--1953   Industrial Minerals: Expansion in all branches of non-metallic industry was worldwide. Min. Cong. J. 39: 90-96, (1953).

34--1953a   (with Carpenter, J. H. et al.) Mining and concentration of ilmenite and associated minerals at Trail Ridge, Florida. Min. Eng. 5: 789-795, (1953); A.I.M.E., Tr. 1953, v. 196, (1954).

35--1954   Industrial Minerals: Many producers enjoyed boom year and several non-metallics remained on critical list. Min. Cong. J. 40: 67-74, 77, 79-81, (1954).

36--1955   Industrial Minerals in 1954: Complete report on fast growing field shows banner year for some; bright future for all. Min. Cong. J. 41: 100-110, (1955).

37--1956   Teaching industrial minerals to geologists. J. Geol. Education 3: 7-10, (1956).

38--1956a   Industrial Minerals: A busy year in a field where small businesses make big industries. Min. Cong. J. 42: 126-136, (1956).

39--1956b   Genesis of titaniferous magnetites and associated rocks of the Lake Sanford district, New York. Min. Eng. 8: 296-301, (1956); A.I.M.E., Tr. 1956, v. 205, (1957); discussion by A. Hubaux and reply by author, Min. Eng. 10: 379-380, (1958); A.I.M.E., Tr. 1958, v. 211, (1959).

40--1956c   [Memorial to] George Harold Anderson (1893-1956). Am. Assoc. Petroleum Geol., B. 40: 2035, (1956).

41--1957   Industrial Minerals: Following the fortunes of the construction industry to which it is closely tied, industrial minerals experienced an upsurge in activity during 1956. Min. Cong. J. 43: 141-149, 160, (1957).

42--1957a   Memorial of George Harold Anderson [1893-1956]. Am. Mineral. 42: 240-241, (1957); Geol. Soc. Am., Pr. 1956: 103, (1957).

43--1957b   [The 1957 Daniel C. Jackling Lecture] A geologist looks at industrial minerals. Min. Eng. 9: 550-555, (1957); A.I.M.E., Tr. 1957, v. 208, (1958).

44--1957c   The Untermanns of Vernal: a dedicated husband-wife geologist team. Geotimes 2/10: 10, 19, (1957).

45--1958   Industrial Minerals--where small businesses make a big industry--Its diversified activities continued through the year with amazing vigor. Min. Cong. J. 44: 93-103, 136, (1958).

46--1959   Industrial Minerals: Most industrial minerals, tied to the fortunes of the growing construction industry, had another successful year. Min. Cong. J. 45: 122-130, 145, (1959).

47--1959a   Sand deposits of titanium minerals. Min. Eng. 11: 421-429, (1959); A.I.M.E., Tr. 214: 421-429, (1959).

48--1960   (and others, as editors) Industrial Minerals and Rocks, 3rd
              ed., completely revised. New York: Am. Inst. Min. Met. and
              Petroleum Engs., (1960). [Includes papers by numerous authors
              which are cited individually.] [Reviewed in Econ. Geol. 55:
              853, (1960).]

49--1960a   Bentonite, in preceding reference 48--1960: 87-91, (1960).

50--1960b   (and others) The carbonate rocks, in preceding reference 48--
              1960: 123-201, (1960).

51--1960c   (with Castle, J. E.) Feldspar, nepheline syenite, and aplite,
              in preceding reference 48--1960: 339-362, (1960).

52--1960d   Sand deposits of titanium minerals. Am. Inst. Min. Met. and
              Petroleum Engs., Tr. 214: 421-429, (1960).

53--1960e   Intriguing examples of geology applied to industrial minerals.
              [President's Address before the Society of Economic Geolo-
              gists.] Econ. Geol. 55: 629-644, (1960).

Note: Gillson retired in 1960 from his position as Chief Geologist with
        E. I. du Pont de Nemours & Co. in Wilmington, Delaware. He had
        been in du Pont's employ for more than thirty years, (1928-1960).

54--1960f   President's report to the Society of Economic Geologists.
              Econ. Geol. 55: 1580-1581, (1960).

55--1961   The Presidential Address: State of the Institute [AIME].
              Min. Eng., p. 361-364, April (1961).

56--1961a   The Presidential Address, State of the Institute (A Summary).
              J. Metals 13/6: 439-441, (1961).

57--1963   The Northern Rhodesian copperbelt: is it a classic example of
              syngenetic deposition. Econ. Geol. 58: 375-390, (1963).

58--1963a   Development of non-metallic mineral resources for fertilizers
              in a dominantly agrarian economy, in Natural Resources--Miner-
              als and Mining, mapping....--United Nations Conference on Ap-
              plication of Science and Technology for the Benefit of the
              Less Developed Areas, Geneva, 1963, U. S. Papers, Vol. 2,
              Washington, D. C. U. S. Dept. State, p. 106-117, (1963).

59--1963b   Application of technology for the benefit of less-developed
              countries: Comments on the UN Conference on Application of
              Science and Technology for the Benefit of the Less Developed
              Nations. Also included are summaries of papers on resource
              development and the establishment of a steel industry in a
              developing economy. (Based on comments by Joseph L. Gillson,
              and summaries of conference papers.) Jour. Metals, p. 495-
              498, July (1963).

60--1963c   Electrostatic separation (p. 21--67-70), in J. H. Perry's
              "Chemical Engineers' Handbook," 4th ed. (prepared by a staff
              of specialists under the editorial direction of Robert H.
              Perry, Cecil H. Chilton, and Sidney D. Kirkpatrick). New
              York: McGraw-Hill Book Co., (1963). [Also see item 30--1950a.]

61--1964   Facilitating registration of AIME members as professional en-
              gineers. Min. Eng. 16/6: 46, 49, (1964).

NOTE:

In addition to the preceding references, Gillson wrote many of the reviews, especially those about non-metallic industrial minerals, for the
first three volumes of the Annotated Bibliography of Economic Geology,
(1928-1930).

(17)
WALTER HARRY NEWHOUSE*
(1897-1969)

MIT: 1923-1946

WALTER HARRY NEWHOUSE

Trained at Penn State in mineralogy and mining geology under the rigorous tutelage of A. P. Honess and his colleagues, Newhouse was well suited to take up the study of mineral deposits under the great master, Waldemar Lindgren, when he came to M.I.T. for graduate work in October 1922. For the next 25 years he would first complete his graduate training under Lindgren, then work with him as a colleague, and finally take his place as Professor of Economic Geology in 1933 when he retired. Still another decade would pass before he would leave M.I.T. during World War II. From 1923 to 1945, almost every year saw one or more important publications come from his pen, many of which were so original as to open up whole new fields of investigation of mineral deposits. During the same period, 33 students did their thesis work under his supervision; 9 Bachelors, 6 Masters, and 18 Doctors. Many of these went on to become leading academic and industrial economic geologists in the United States and Canada as well as abroad. While conducting his academic duties and pursuing his research interests in the Department he participated actively in reviewing articles on metalliferous deposits written in English, French, German and Spanish for the <u>Annotated Bibliography of Economic Geology</u>. During the first decade of that abstract journal, from 1929 to 1938 inclusive, Newhouse prepared hundreds of abstracts on a wide variety of subjects.

Newhouse's earliest research involved studies of ore minerals and their paragenesis, and his first two dozen papers dealt largely with these subjects. Next he showed how to determine the direction of flow of mineralizing solutions. He was the first in America to develop and use a high dispersion-resolution spectrograph as an analytical tool to determine minor elements in minerals. Turning to the study of how openings form because of movement along curved or irregular fault planes, he initiated a classic research project which ultimately resulted in the publication of a many-authored work dealing with the relation of ore deposits to structural features. All the foregoing research projects he conducted while at M.I.T. In 1943-1944 before leaving M.I.T., he also made a highly successful mission to Liberia, which resulted in the development of the Bomi Hill iron deposit, and sparked the present iron ore industry of that country.

---

* I am grateful to Mrs. Walter H. (Grace) Newhouse, Prof. A. F. Hagner, Drs. G. W. DeVore and T. P. Thayer, and Prof. M. J. Buerger for much personal information; and to Prof. H. W. Fairbairn for the opportunity to see his "Memorial" in advance of publication. [Memorial to Walter Harry Newhouse (1897-1969), by H. W. Fairbairn, Geol. Soc. Am., Mem. Vol. 1: 69-72, port., (1973); also see Memorial of Walter Harry Newhouse, December 13, 1897-September 21, 1969, by G. W. DeVore, Am. Mineral. 58: 380-382, il., (1973).]

He left M.I.T. in 1946 to accept a professorship in geology at the University of Chicago and for the next 17 years, until retirement in 1963, he continued to lecture on and investigate mineral deposits. From 1946 to 1957 he also served as Head of Chicago's Department of Geology. During his Chicago years, and for a short time after retirement, he devoted his research efforts almost exclusively to intensive study of the great anorthosite body in the Laramie Range, Wyoming. Partly because of the highly complex terrain, and perhaps more because of his "uncompromising perfectionism" as Fairbairn (1971) describes it, he was unable to bring his greatest research effort (he referred to it as his "Magnum Opus") to a successful synthesis. Indicative of what a masterpiece this might have been are the last four brief statements he published (see his Bibliography). In summary, Newhouse brought distinction to the Department of Geology by his original and imaginative publications on the broad problem of ore genesis and contributed importantly to the training of a generation of economic geologists.

Walter Harry Newhouse was born on 13 December 1897 at Fisher, Pennsylvania and died suddenly of pneumonia and a weak heart on 21 September 1969 in Tucson, Arizona in his seventy-second year. He was the second of nine children and the eldest son of Edward W. and Hattie (Elder) Newhouse, sturdy Pennsylvania-German farmers who brought up their large family in the disciplined household characteristic of the region and the times. Both the Newhouses and the Elders traced their ancestry back to European forebears. The ancestral Newhouses included Swiss and Welsh members; the Elders traced back to Prussian and Scotch ancestors. Both families had been in America for several generations and had established themselves in Pennsylvania as prosperous farmers.

Walter's siblings were an older and five younger sisters and two younger brothers. Being the oldest son in a large family, and subject to the strict discipline of the German household, he grew up a hard-working and serious-minded individual with the highest standards of integrity and personal behavior. While yet in high school he helped out on his parents' farm, worked for a dollar a day for a neighboring farmer, and when somewhat older took a turn as a coal miner. With such a background of experience, and with a firm belief in self dependence, he naturally sought out for his closest friends and most respected students those who were devoted to their work and of serious mind. His students quickly sensed his personal characteristics and lost no time getting to work in his classes, in the laboratory, or in the field. He in turn helped them in every way he could, and those who completed their work with him, or worked with him later in the field, became his lifelong friends. One of his field associates, Dr. A. F. Hagner, has the following to say (letter to Prof. H. W. Fairbairn dated June 17, 1970) -

"As you know, Walt was dedicated to geology. In fact, as far as I know he had no other interests and no hobbies except perhaps national politics and world affairs. Even these he pursued only in the current news - radio, TV, and newspapers. He had a good sense of humor, and was generally a very cheerful person to be with. He took delight in the achievements of his best students and frequently talked about them. ... His geologic standards were very high, and his inspiration brought out the best in his associates and students. I always found him to be very fair and considerate, although he was somewhat impatient with those who were unduly dogmatic, authoritative, or who held narrow views."

Newhouse started his education in the elementary grades in his birthplace, Fisher, a small town in Clarion County in the bituminous region of western Pennsylvania, and then entered Fisher High School, where in three school years of eight months each he completed the program that most students of the time took four years of nine months each. Only his great determination and intense devotion to getting an education drove him to put forth the effort needed to be certified for graduation in the spring of 1915. He then spent some time at the Clarion Normal School after which, in the fall of 1917, he entered the Pennsylvania State College. Here he registered as a degree candidate in Agriculture, but the next year he changed to Mining Geology, and in June 1921 received a Bachelor of Science degree in that course. Inspection of his academic record at Penn State reveals that he took some 20 different geology and mining courses, along with basic science and engineering subjects, and that he did well above average in most of the subjects. Here also he had the only military training of his life--about a month in October and November, 1918--in the Student Army Training Corps (SATC).

Choosing not to continue his college training for the moment, possibly in part because he wanted to get away from his neighborhood and see some of the outside world, but mainly because he needed to earn some money to meet the costs of further education, he accepted a position teaching science and mathematics in the high school in Artesia, New Mexico for the school year 1921-1922. Although he spent only the one year in New Mexico, he left with a love of the Southwest, and when he retired from active field work some 40 years later, he decided to spend his last years in neighboring Arizona with the hope that the dry climate would make life more comfortable.

Newhouse entered M.I.T. in October 1922 and completed requirements for the S.M. degree the next spring, being able to make such rapid progress because of his thorough undergraduate training at Penn State. The new Institute had been in Cambridge only six years, having moved from the Boston side of the Charles River in 1916. The Geology Department headed by Waldemar Lindgren was comfortably settled in Building 4, extensive study and research collections were available, a small but excellent library had been brought together next to Lindgren's office, and the whole departmental atmosphere was charged with an expectant and exciting air. Lindgren was a world leader in the study of mineral deposits, his <u>Mineral Deposits</u> first published in 1913 having become a classic authority; Warren, soon to go to the deanship of Yale's Sheffield Scientific School, taught mineralogy and petrology; and Shimer covered the fields of stratigraphy and paleontology. And there was always Harvard's great Department up the Charles River with Daly, Graton, Palache, Raymond and others.

Newhouse soon found a problem of interest for his Master's thesis, "Ore deposits of Kokomo, Colorado," and earned the S.M. degree in June 1923. After a summer as rodman on a U.S.G.S. party around Craig, Colorado, he resumed his work under Lindgren's supervision, next entering on the study of marcasite, and in January 1926 he was granted his Ph.D. degree with a thesis on "Paragenesis of certain occurrences of marcasite." Meanwhile, he had also been gaining experience in teaching, having been appointed Instructor in Mineralogy in 1923 after receiving his S.M. degree, and had been granted a $750 scholarship for SY 1925-26 to help him complete his thesis work.

With his graduate work behind him, and an assistant professorship only a year away, he was now launched on a career of teaching and research in economic geology that would occupy him fully for the remainder of his life and bring him satisfaction and renown for the students he trained, the new ideas he generated, and the successful field work he conducted on three continents.

To complete the academic record, Newhouse served successively as Instructor in Mineralogy (1923-1927); Assistant Professor of Mineralogy (1927-1928), Assistant Professor (1928-1930), Associate Professor (1930-1944), and finally Professor (1944-1946) of Economic Geology until he resigned on 30 June 1946 to join the staff of the U. S. Geological Survey. Almost immediately, however, he was called to the University of Chicago where he taught mineral deposits until retirement in 1963. During his 23 years at M.I.T. as a staff member of the Department of Geology, from 1923 to 1946, he took two short leaves of absence: the first, from 15 December 1928 to 15 March 1929, to head an exploration party in Venezuela; the second, from 1 November 1943 to 12 December 1944, to head a U. S. Geological Survey team to examine and evaluate the mineral resources of Liberia for the Liberian Government.

At the University of Chicago, Newhouse succeeded E. S. Bastin in economic geology, and from 1946 to retirement in 1962 he taught courses on mineral deposits. At the time he joined the staff, as Professor of Economic Geology, another M.I.T. alumnus, Norman L. Bowen (XII Ph.D. 1912) was Head of the Department of Geology, and this no doubt strongly influenced him to accept the Chicago position, for he could look forward to many years of work with the kind of research-minded scientist he so greatly admired and respected. Certainly he could not have known that Bowen would resign within a year and that he, himself, would be urged to assume the chairmanship of the Department of Geology.

As Fisher* relates it:

> "When he came to the University from M.I.T., Walt had no idea that he would shortly become Chairman. Bowen's resignation took all of us by surprise, and only after much urging and soul-searching did Newhouse accept this onerous task."

Having decided to accept the chairmanship, Newhouse immediately went about bringing to Chicago a group of distinguished geologists to join those already there and to reconstruct the Department and its curriculum along more quantitative lines. This was not accomplished, however, without difficulties and disappointments, as told elsewhere,** and in 1957 he resigned as Chairman. He continued, however, as Professor of Geology for the next six years, until retirement as Professor Emeritus in 1963, at which time he and Mrs. Newhouse moved to Tucson, Arizona permanently.

During the two decades from 1946 to 1967 he devoted most of his research time to the study of the anorthosite body near Laramie, Wyoming, returning summer after summer to the field after classes were over, generally accompanied by Mrs. Newhouse. He first had financial aid from the U. S. Geological Survey Mineral Deposits Branch, then from the University of Chicago, but mainly he paid his expenses out of his own pocket. Being scrupulously honest, he would not drive a Survey vehicle when Mrs. Newhouse accompanied him in the field because, in his words, "it would not look good."

Having briefly outlined Newhouse's long academic and professional career, from 1923 to 1967, it is now appropriate to review and evaluate his labors as a longtime member of M.I.T.'s Geology Department.

Being exceedingly original and imaginative in his geological thinking, keenly perceptive in his field and laboratory observations, and endowed with ability to sense the critical relationship or measurement that

---

* Fisher, D. J. The Seventy Years of the Department of Geology, University of Chicago, 1892-1961: Chicago, The University of Chicago (copyright by D. J. Fisher), xii + 147 p., il., (1963).

** Fisher, D. J., op. cit., 1963.

gave deeper insight into the fundamental nature of a problem, Newhouse quickly attracted many of the better graduate students to his areas of interest. As a consequence, he was one of the most sought after thesis supervisors in the Geology Department. He assumed this responsibility with a deep sense of obligation to the student, because he always felt that the professor himself should set the highest standards by his own personal conduct and professional work before asking the same from his students and peers.

It is not surprising, therefore, to learn from the record that as a teacher, Newhouse attracted many of the best-prepared students in the Department who came to M.I.T. to learn about mineral deposits. During Lindgren's last decade of teaching, 1924-1933, Newhouse shared with him the responsibility of suggesting and supervising thesis investigations for these students, many of whom came from neighboring Canada and from other countries farther away. As a consequence, Newhouse's name appears as Supervisor on 33 theses presented for degrees in geology between 1923 and 1946: 9 Bachelors, 6 Masters, and 18 Doctors. Many of the recipients of these degrees have become leading academic and industrial economic geologists in Canada, (indicated by an asterisk in the following list) and the United States, as well as abroad--e.g. *P. E. Auger, *N. S. Beaton, M. L. Brashears, Jr., R. D. Butler, *Neil Campbell, *G. M. Furnival, *W. C. Güssow, Nithipatana Jalichandra (Thailand), V. M. Lopez and Guillermo Zuloaga (Venezuela), *C. S. Lord, J. W. Mills, Minoo Netarwala (India), William Parrish, J. P. Pollock, *T. E. H. Sargent, J. A. Shimer, *J. S. Stevenson, R. E. Stoiber, and W. C. Stoll. These geologists are living evidence of Newhouse's effectiveness as a teacher.

In his own research, Newhouse hewed a rather narrow path through the forest of ore deposit problems, but where he worked he cleared away the brush and left standing the major trees that represented the fundamental truths. Only with the complex and extensive Laramie Range did he meet more than his match, and even then he might have left a woodland clean and clear if he had been able to overcome his lifelong propensity for perfectionism. One of his field assistants during the Wyoming work was George W. DeVore, who wrote H. W. Fairbairn as follows about Newhouse (letter dated June 1, 1970):

> "Dr. Newhouse is not an easy person to describe. He was fired by an absolute idealism, a perfectionism and an uncompromising devotion to his science. New ideas, originality, and fundamental insight were the only criteria he recognized. He was thoroughly endowed with these attributes himself. He was the most original man I ever knew, with the ability to make the most astute observations and arrive at the perceptive relationship that would restructure the whole

problem. He was a natural field man who insisted on looking at science from the 'Natural History Viewpoint' as he would put it. He had a basic instinct to identify the critical feature, measurement or correlation. His dedication and enthusiasm were contagious so that all who were even near him were inspired to put forth extreme efforts to achieve similar results. Unfortunately he was extremely possessive about his ideas and observations and refused to publish his work until he had the opportunity to adequately explore, exploit and develop them. His genius to see additional consequences in everything he did resulted in his never finding an end to the work and, so, he never published his results. He spent enormous efforts and time in protecting and defending his ideas and observations from being taken by others. Many of his ideas were stolen so that he had adequate justification to react as he did."

His personal research, based almost always on some combination of field work and subsequent laboratory study, characteristically started with an observational approach--he liked to call it a "natural history approach"--then proceeded to laboratory investigation and finally to an experiment or a synthesis that would test his conclusions.

His more important field studies include work on magnetite at Cornwall, Pennsylvania (1929), Gold mineralization in the Guyana Highlands of Venezuela (1928), Cu-Pb-Zn deposits of Buchans in Newfoundland (1927-1929), mineral zoning in the Triassic Highlands of New Jersey-Pennsylvania-Virginia (1933), zonal gold mineralization in Nova Scotia (1936), cordierite deposits of the Laramie Range, Wyoming (1948), and a geological map of the anorthosite areas in the southern part of the Laramie Range (1957).

Left unpublished were possibly the two most important of all his field projects: 1) the purposely unpublished 110-page "Report of the Geological Mission to Liberia, December 1943-May 1944," prepared as a confidential document for the U. S. Department of State; and 2) the great mass of field data on the Laramie Range that was intended to form the basis for a definitive monograph on that perplexing and complex igneous terrain. The first report resulted in the spectacular economic development of a major iron ore deposit. The second effort unhappily left an unfulfilled promise to himself and a long and tedious task for those of his colleagues who must try to retrieve something from the notes and maps and analyses left behind.

Newhouse's laboratory investigations include: 1) important contributions on criteria of replacement in the opaque minerals (1928; 1931); 2) a study of fluid inclusions as indicators of geothermometry (1932); 3) a comprehensive paper on opaque oxides and sulphides in common rocks (1936); and 4) a pioneering report on analysis of minor elements in minerals by means of high dispersion-resolution spectrography (1941). Other papers from his laboratory of the same period contain experimental data: 1) a 1927 attempt at a pyrrhotite-pentlandite phase diagram; 2) a 1940 study of shear textures in sulphides; and 3) a 1941 paper confirming direction of solution flow from observations of asymmetric crystal growth (Newhouse received a project grant--(#264-38)--of $1,160 from The Geological Society of America in 1938 to help finance the field studies for this research. This was one of the earliest of the G.S.A. grants for mineralogical research.) These laboratory investigations were imaginative pioneering efforts at a time when quantitative work on minerals and rocks was rare and at a much lower level of sophistication than today.

Dr. T. P. Thayer has kindly granted me permission to quote several paragraphs relevant to the Liberia mission which were included in a letter he wrote Prof. H. W. Fairbairn dated February 19, 1970.

> "Dear Professor Fairbairn:
>
> The experiences Walt Newhouse, Art Butler, and I had in Liberia are some of my favorite memories, and I am glad to share them with you.
>
> Dr. A. P. Butler, Jr., now with the USGS in Denver, and I were Walt Newhouse's companions on the Geological Mission to Liberia from December, 1943 to May, 1944. The purpose of the mission was to evaluate the Bomi Hill deposit as a potential source of iron ore and revenue for a port at Monrovia. These had been explored by a Dutch syndicate in 1936-37, which had been forced to relinquish the concession under American diplomatic pressure. After mapping the Bomi Hill deposits, the party made a 350 mile trek (in 25 days) into the interior to reconnoiter other deposits in the Wologisi Mountains. President Roosevelt had promised Liberia the port at Monrovia on his way back from the Casablanca Conference with Churchill and the Russians.
>
> No report was printed, but one entitled 'Report of the Geological Mission to Liberia, December 1943-May 1944' was placed in open file early in 1945. On the basis of this report, L. K. Christie negotiated a concession covering Bomi Hills, and in 1945 the Liberia Mining Company Ltd., began developing the deposits. Republic Steel Corporation acquired the majority interest in the Liberia Mining Company in 1949, and the first ore was shipped in August 1951.
>
> The spectacular financial success at Bomi Hill (45 miles of railroad and all plant costs were paid off in two years!) stimulated intense interest in other known deposits in the country, and in 1970 4 major mines together will ship more than 20 million tons. Total shipments since 1951 will approach or pass 100 million tons this year. The Newhouse report, in short, sparked the present iron ore industry in Liberia."*

---

* J. J. Davis described the nature and development of this unusually high-grade iron-ore deposit in "Iron boom in Liberia," in Steelways 7/5: 21-24, (Sept. 1951).

As the clouds of World War II were gathering, Newhouse turned away from the inviting field of spectrographic analysis that he had pioneered, and which six of his doctoral candidates exploited for their thesis work (see chapter on The Cabot Spectrographic Laboratory), and took up the study of the structural control of economic mineral deposits. Intrigued by the relations of ore deposits to openings caused by movement along a curved or irregular fault plane (1940), he accepted the chairmanship of a National Research Council Committee concerned with this and other aspects of ore deposition, and organized a symposium to study the problem. From this effort came a classic report, <u>Ore deposits as Related to Structural Features</u> (Princeton University Press, 280 p., il., 1942), consisting of contributions from sixty-five collaborators and including a masterful summary and discussion by Newhouse himself who served as Editor. Many who are competent to judge his contributions to the overall study of mineral deposits consider this work to be one of his most important.

Newhouse was an enthusiastic and dedicated field man who believed strongly that the geologist who would know mineral deposits should go to the field first, then study his specimens in the laboratory. As a consequence, he spent almost every summer in the field for 36 years, 1926-1962, on consulting assignments, on projects financed by federal or state bureaus, or on his own funds. The following list gives an excellent survey of this practical side of his professional career, and shows that his assignments took him to most states in the United States, to half of the provinces of Canada, and to South America, Europe and Africa. The more important of these were the following:

1. Assistant to Waldemar Lindgren in geological examination of the United Verde Extension Mine, Jerome, Arizona, (June-August 1926).
2. Assistant to Waldemar Lindgren doing calculations of ore reserves for the Federal Mining and Smelting Mines in the Tri-State Lead-Zinc District and in Idaho, (Sept.-Oct. 1926).
3. Geologist in charge of exploration for the Buchans Mining Company of Buchans, Newfoundland; (June-Sept. 1927, June-Sept. 1928, June-Aug. 1929). Results reported to American Smelting and Refining Company of New York.
4. Geologist in charge of exploration in the gold fields of the Guyana Highlands of Venezuela for Venezuelan interests, (Dec. 1928-March 1929).
5. Examination and Report on the Paymaster Mine near Porcupine, Ontario for the United Mineral Lands Corporation of Boston, (Aug.-Sept. 1929).
6. Examined structural features and mineralization in a number of metal mines in the western United States and Canada under a research grant from M.I.T., (June-Oct. 1930).

7. Geological work on metalliferous deposits of the Triassic Belt in New Jersey, Pennsylvania and Virginia, supported by a research grant from M.I.T., (June-Aug. 1931).

8. Geological work on the metalliferous deposits of the Triassic Belt in New Jersey, Pennsylvania, Virginia and Nova Scotia, supported by a Rockefeller Research grant, (July-Sept. 1932). For results see appended bibliography.

9. Geological work on the gold deposits of Nova Scotia. Also examination of tungsten deposits in the same Province. This work was carried on under a Rockefeller Research grant, (June-Sept. 1933). For results see appended bibliography.

10. Geological work on zinc, lead, antimony and barite deposits in Virginia, Tennessee, Arkansas, Kentucky and Illinois. Research supported by a Rockefeller Research grant, (June-Sept. 1934).

11. Geological work on granite pegmatites in the eastern United States, in the region extending from Maine to North Carolina. Research supported by a Rockefeller Research grant, (June-Sept. 1935).

12. Geological work in the Grenville province and other parts of southern Ontario, Canada, on iron, gold, and fluorspar deposits, (Aug.-Sept. 1936).

13. Visited mines in Norway and Sweden with Dr. Per Geijer, present Director, Geological Survey of Sweden, Nels Magnusson, Geologist, Geological Survey of Sweden, and Olaf Ödman, Geologist, Boliden Mine, Sweden, and also mines in the Ural Mountains in Russia with the International Geological Congress, (June-Sept. 1937).

14. Geological work was done on chromite deposits in Quebec and titaniferous magnetite deposits in Quebec and the Adirondack region, New York, (July-Sept. 1938).

15. Geological work on magnetite deposits in highlands region of New Jersey and New York states, under a grant from M.I.T., (July-Sept. 1939).

16. Geological work on zinc mineralizations in New Hampshire, New York, Ontario and Quebec, under a research grant from M.I.T., (Aug.-Sept. 1940).

17. Research on spectrographic studies of minerals and rocks, under a three-year research grant from the Carnegie Institution of Washington ($7,500), (1940-1942 inclusive).

18. Consulting work with the U. S. Smelting, Refining and Mining Company for eight months, on wall rock alteration at some of their mines, and in search for a new commercial source of beryllium (May-Dec. 1941).

In the laboratory, Newhouse carried on numerous consulting assignments that involved microscopic and other examinations of ores from widely scattered areas--Sudbury, Ontario; rutile deposits in Virginia; London Mine, Colorado; Rio Tinto, Nevada; Utah Apex Mine, Utah; Morenci and Jerome-Bradshaw Quadrangle, Arizona; Oiseau River nickel deposit in Manitoba; Butte, Montana; Bralorne Mine, British Columbia; Temescal tin deposit, California; tin ores from Bolivia; gold deposit near Parcoy, Peru; bauxite deposits in China; and Fantoche chromite mine in New Caledonia. These assignments show the versatility of Newhouse's laboratory research.

It is worth noting that by 1946 Newhouse had completely given up consulting work of any kind, and for the next twenty years, until his death in 1969, he devoted almost all of his research time and energy to the study of the anorthosite body near Laramie, Wyoming, as discussed on a preceding page.

Newhouse held membership in numerous scientific societies including the American Academy of Arts and Sciences, American Geophysical Union, The Geological Society of America, Mineralogical Society of America, Society of Economic Geologists, American Institute of Mining and Metallurgical Engineers, Canadian Institute of Mining Engineers, Norwegian Geological Society, Swedish Geological Society, and the Société Géologique de Belgique (Honorary Member).

Although he was never called to serve as an officer in any of the organizations mentioned in the foregoing paragraph, he did serve his profession effectively on the National Research Council's Committee on Processes of Ore Deposition, from 1932 to 1940 (as Chairman); as Editor of Ore Deposits as Related to Structural Features from 1938-1942; Society of Economic Geologists' Program Committee (Chairman, 1933); National Research Council's Committee on Geophysics (Chairman), Advisory to the Office of Naval Research (1947 ff.); Associate editor of Journal of Geology (1957-1962); and member of the Board of Natural Resources and Conservation of Illinois, by the Governor's appointment, (1946-1963).

Newhouse was a charming person with a good sense of humor; he told a story well and relished a lively and vigorous table conversation. Trivial matters and idle chatter bored him quickly, but a controversial subject, whether scientific, economic, political, or philosophical, caught his immediate interest and attention. For a number of years it was common to see him lunching with Buerger, Fairbairn and Shrock in the small faculty dining room at the northwest corner of Walker on the second floor. This was in the late 1930s and early 1940s before there was a Faculty Club or a regular Faculty Dining Room. These lunches usually followed an hour of handball on one of the courts on the top floor of Walker. Newhouse was a worthy opponent in handball; having been a runner in his college days, he possessed the stamina and coordination of the trained athlete. This ability was well shown one day when he ran down the long corridor of M.I.T.

and overtook a much younger man who was making off with his typewriter. Many years later, at age forty-six, he made a 350-mile packing trip in the Liberian rain-forest in 25 days. Those of us who knew him intimately came to accept his "low boiling point," which he often deprecated but could not overcome. In the field he always carried his fair share and was universally admired and respected by his associates from fellow geologists to cooks and wranglers.

In his college days at Penn State Walter became acquainted with a young schoolteacher, Grace Edna Brown, who was teaching grammar school in a neighboring county. Like Walter, she was of Pennsylvania-German ancestry, with forebears who came from Germany and Holland, and again like him she was the member of a family that engaged itself in farming, carpentry, and merchandising their farm produce. Grace was born in Pleasant Unity, Pennsylvania on 25 January 1894. She graduated from the high school at Mount Pleasant, Pennsylvania, and then took teacher training courses at normal school, and during summer school sessions at Penn State. From this training she obtained a teaching certificate that permitted her to teach grammar school in Pennsylvania, and later in Nebraska, Washington, and Massachusetts.

Although they had first become acquainted at Penn State in 1919, Walter and Grace did not get married until 30 June 1923. During this waiting period Grace taught school and Walter went off to M.I.T. to work on his S.M. degree in geology, which he was awarded in June 1923. Later in the month he and Grace were married in Steamboat Springs, Colorado, where Walter would do his first work for the U. S. Geological Survey as a geological assistant. For the next 44 years Grace was a constant inspiration, trusted helpmate, and intellectual partner. It was not unusual for her to don field clothes and boots and go into the field with him, especially in his post-M.I.T. days when he was working in the Laramie Range.

The Newhouses' only child, son W. Jan, attended the primary and secondary schools of Melrose, a northern suburb of Boston, then earned a Bachelor's degree in Botany from Dartmouth in 1948, a Master's degree in marine botany from the University of New Hampshire in 1952 and a Ph.D. degree in marine botany from the University of Hawaii in 1967. He carries on the academic tradition of the Newhouse family as Professor of Marine Biology and Chairman of the Department of General Science at the University of Hawaii.

BIBLIOGRAPHY OF WALTER HARRY NEWHOUSE

Symbols and abbreviations used in the following references are explained on pages 91 - 98. In general, abstracts are listed separately as well as with the references to the complete article. This bibliography begins with the title of Newhouse's S.M. thesis (T--1923) and includes all known publications through 1955, a decade after he left M.I.T.

T--1923   Ore deposits of Kokomo, Colorado, 65 p., (1923). (S.M. Thesis at M.I.T. in Course XII, June 1923.)

1--1925   Paragenesis of marcasite. Econ. Geol. 20: 54-66, il., (1925).

2--1926   An examination as to the intergrowth of certain minerals. Econ. Geol. 21: 68-69, (1926).

T--1926a Paragenesis of certain occurrences of marcasite, 159 p., il., (1926). (Ph.D. Thesis at M.I.T. in Course XII, January 1926.)

3--1927   Some forms of iron sulphide occurring in coal and other sedimentary rocks. J. Geol. 35: 73-83, (1927).

4--1927a The equilibrium diagram of pyrrhotite and pentlandite and their relations in natural occurrences. Econ. Geol. 22: 288-299, il., (1927).

5--1927b Intimate intergrowths and mutual boundaries as proof of contemporaneous deposition. Econ. Geol. 22: 403-407, (1927).

6--1927c (and Callahan, W. H.) Two kinds of magnetite? Econ. Geol. 22: 629-632, (1927).

7--1928   (and Buerger, M. J.) Observations on wood tin nodules. Econ. Geol. 23: 185-192, (1928).

8--1928a The time sequence of hypogene ore mineral deposition. Econ. Geol. 23: 647-659, (1928).

9--1928b The microscopic criteria of replacement in the opaque ore minerals (p. 147-161), in H. W. Fairbanks, Ed., The Laboratory Investigation of Ores: New York, McGraw-Hill Book Co., Inc., 262 p., il., (1928).

10--1929 The identity and genesis of lodestone magnetite. Econ. Geol. 24: 62-67, (1929).

11--1929a (and Callahan, W. H.) A study of the magnetite ore body at Cornwall, Pennsylvania. Econ. Geol. 24: 403-410, (1929).

12--1929b (and Zuloaga, G.) Gold deposits of the Guayana Highlands, Venezuela. Econ. Geol. 24: 797-810, (1929).

13--1930 (and Flaherty, G. F.) The texture and origin of some banded or schistose sulphide ores. Econ. Geol. 25: 600-620, il., (1930).

14--1931 Some relations of ore deposits to folded rocks. A.I.M.E. Tech. Publ. 422, 25 p., il., (1931); with discussion, Tr. 1931: 224-251, il., (1931).

15--1931a The geology and ore deposits of Buchans, Newfoundland. Econ. Geol. 26: 399-414, (1931).

16--1931b A pyrrhotite-cubanite-chalcopyrite intergrowth from the Frood mine, Sudbury, Ontario. Am. Mineral. 16: 334-337, il., (1931).

17--1931c (with Bastin, E. S., Graton, L. C., Lindgren, W., Schwartz, G. M., and Short, M. N.) Criteria of age relations of minerals with especial reference to polished sections of ores. Econ. Geol. 26: 562-610, (1931).

18--1932   The composition of vein solutions as shown by liquid inclusions in minerals. Econ. Geol. 27: 419-436, (1932).

19--1933   Gold, its relations to civilization. Tech. Eng. News 14/1: 4-5, 20, (Feb. 1933).

20--1933a   Mercury in native silver. Am. Mineral. 18: 295-299, (1933).

21--1933b   Mineral zoning in the New Jersey-Pennsylvania-Virginia Triassic area. Econ. Geol. 28: 613-633, il., (1933); Abst., Pan-Am. Geol. 60: 159-160, (1933); with discussion in 16th Int. Geol. Cong. 1933 Rept. 1: 460, (1936).

22--1933c   The temperature of formation of the Mississippi Valley lead-zinc deposits. Econ. Geol. 28: 744-750, (1933).

23--1934   The source of vanadium, molybdenum, tungsten, and chromium in oxidized lead deposits. Am. Mineral. 19: 209-220, (1934).

24--1935   [Review of] The geology and ore deposits of the Horne Mine, Noranda, Quebec, by Peter Price, 1934. Econ. Geol. 30: 326-327, (1935).

25--1936   [Review of] The minerals of Franklin and Sterling Hill, Sussex County, New Jersey by Charles Palache, 1935. Econ. Geol. 31: 531-532, (1936).

26--1936a   (and Glass, J. P.) Some physical properties of certain iron oxides. Econ. Geol. 31: 699-711, (1936).

27--1936b   A zonal gold mineralization in Nova Scotia. Econ. Geol. 31: 805-831, il., (1936).

28--1936c   Opaque oxides and sulphides in common igneous rocks. Geol. Soc. Am., B. 47: 1-52, il., (1936).

29--1938   Direction of solution flow and the formation of minerals. Science 88: 109, (1938).

30--1939   Waldemar Lindgren, 1860-1939. Science 90: 584-585, (1939).

31--1940   Openings due to movement along a curved or irregular fault plane. Econ. Geol. 35: 445-464, il., (1940).

32--1941   The direction of flow of mineralizing solutions. Econ. Geol. 36: 612-629, il., (1941).

33--1941a   Spectrographic studies of minor elements in minerals. Carnegie Inst. Washington Year Book No. 40, for 1940-41, p. 142-144, (1941).

34--1942   [Editor] Ore Deposits as Related to Structural Features. Princeton, N.J., Princeton Univ. Press, xi + 280 p., il., (1942). (Contains papers by numerous authors which are cited individually.)

35--1942a   Structural features associated with the ore deposits described in this volume, p. 9-53, in Newhouse, W. H. (Editor), Ore Deposits as Related to Structural Features. Princeton, N.J., Princeton Univ. Press, xi + 280 p., il., (1942).

36--1942b   (with Frondel, C. and Jarrell, R. F.) Spatial distribution of minor elements in single-crystals. Am. Mineral. 27: 726-745, il., (1942).

37--1945   Report to U. S. State Department on iron ore possibilities in Liberia, 110 p., (1945). (Unpublished.)

38--1945a   (and Hagner, A. F.) Structure of the Laramie Range anorthosite, Wyoming [abst.]. Geol. Soc. Am., B. 56: 1184-1185, (1945).

39--1948   (and Hagner, A. F.) Zoned metasomatic gneisses related to structure and temperature, Laramie Range, Wyoming [abst.]. Geol. Soc. Am., B. 58: 1212-1213, (1947); Am. Mineral. 33: 203, (1948).

40--1949   (and Hagner, A. F. and DeVore, G. W.) Structural control in the formation of gneisses and metamorphic rocks [Wyo.]. Science 109: 168-169, (1949).

41--1949a   (and Hagner, A. F.) Cordierite deposits of the Laramie Range, Albany County, Wyoming. Wyo. Geol. Surv., B. 41: 18 p., il., (1949).

42--1955   Compositional lineation and its relation to complex folding. Science 122: 284, (1955).

43--1957   (and Hagner, A. F.) Geologic map of anorthositic areas, southern part of Laramie Range, Wyoming. U. S. Geol. Surv., Mineral. Invest. Field Studies Map MF 119, (1957).

As stated earlier, Newhouse prepared hundreds of abstracts for the Annotated Bibliography of Economic Geology during its first decade from 1928-1938 inclusive.

NOTE:

After so many (25) years of concentrated field and laboratory study of the Laramie anorthosite body, Newhouse's perfectionism prevented him from publishing anything more than a few provocative abstracts on his work. How much he could have given to geology had he only set down his ideas as they evolved! Now his colleagues must try to sort out his findings from the voluminous notes he left. As G. W. DeVore recently wrote to Prof. Fairbairn (letter dated June 1, 1970):

> "The insight, the method of analysis, the dogged determination to explain every exception, the realization of the significance and generality of the interpretation that were his special contributions to research are not part of the public record; this is the tragic loss. This could have been part of the scientific heritage had he published as he went along developing the ideas and discovering the consequences of his interpretations rather than waiting until each task was 'completed.'"

The interested reader will find additional candid but sympathetic comments by DeVore in his excellent "Memorial on Walter Harry Newhouse," cited in an earlier footnote.

Newhouse in the Field in Wyoming

Newhouse spent most of his later summers of field work investigating the large body of anorthosite in the Laramie Range in Wyoming, and planned to describe it in what he hoped would be his "Magnum Opus." Being the perfectionist that he was, and always seeking that last important detail, he never got around to finishing his projected work. Knowing him as I did, I am sure that the report he envisioned would, indeed, have been one of his major contributions to geology had he lived to complete it.

Here we see him in typical field clothes with his ever-present cigarette and his characteristic smile. He enjoyed nothing as much as the search for new information that could only be obtained by studying the rocks themselves in the field. For many summers his wife Grace accompanied him into the field and was a valuable helpmate.

(Photograph by Grace Newhouse, 1951)

(18)
MARTIN JULIAN BUERGER

MARTIN JULIAN BUERGER

MIT: 1920-1929; 1929-1973

Martin Julian Buerger, one of M.I.T.'s most distinguished geology alumni and professors, is a uniquely Institute product. Born in Detroit, Michigan, and reared and educated in the State of New York, he came from the Morris High School in New York City to M.I.T. in 1920. He started his undergraduate work in chemistry, transferred to chemical engineering as a sophomore, dropped out for a year, transferred to mining engineering when he returned as a junior and completed the requirements for a bachelor's degree in the last-named discipline (III S.B. 1925). He transferred to geology upon taking up graduate study in 1925, completed requirements for a master's degree in two years (XII S.M. 1927) and won his doctorate in mineralogy in 1929 (XII Ph.D. 1929). He started his long teaching career while still a graduate student; first as a Teaching Assistant for two years, 1925-1927, then as an Instructor for a similar period of time, 1927-1929. Advanced to Assistant Professor in
1929 upon receiving his Ph.D. degree, he became a full-time faculty member and started up the professional ladder. He was appointed Associate Professor in 1935, Professor in 1944, and Institute Professor in 1956. Becoming emeritus in 1968, he continued thereafter as Institute Professor Emeritus for the following five years, 1968-1973, when at the final retirement age of 70 years he was appointed Senior Research Associate for an additional two years, 1973-1975. When he finally closed his long career at M.I.T. on 30 June 1975 he had been a member of the Institute family for 55 years, having started as a freshman in 1920; a member of the teaching staff for 50 years, having started as a teaching assistant in 1925; and a member of the faculty for 46 years, having been appointed an assistant professor in 1929. Few have served the Institute so long and with such distinction.

Martin Julian Buerger was born in Detroit, Michigan on 8 April 1903, the older of two sons of Martin J. G. and Julie E. R. (Weber) Buerger. Both parents were born in the United States and were descendants of devout German Lutherans who had emigrated from Germany early in the 19th century. Martin Julian, the subject of this biographical sketch, traces his forebears back to Rev. Ernst Moritz Buerger who had led his congregation from Saxony to Perry County, Missouri, in 1838 and helped to found there the Lutheran Church - Missouri Synod. A generation later Julie Weber's father came from Germany to the Concordia Seminary near St. Louis to study for the Lutheran ministry.

Although born in Detroit, young Martin had barely started his primary education when his father, a photoengraver, moved his family to Long Island, where Martin continued through several grades before the family again moved - first to Garden City, then in succession to Buffalo; Toronto, Canada; Jamaica, New York; back to Buffalo; and finally to New York City, where he completed his secondary school work at the Morris High School, whence he came to M.I.T.

Entering the Institute as a freshman in 1920, Buerger fully intended to be a chemist, but after leaving chemistry and trying chemical engineering for a term or so, he dropped out for a year, then came back to complete his bachelor's degree in mining engineering in 1925, having transferred to that discipline because he could use his chemistry to understand minerals. Taking mineralogy with Prof. W. H. Newhouse, as a budding mining engineer, convinced him he wanted to be a mineralogist. However, when he heard lectures on the nature and possible uses of x-rays by W. L. Bragg, who came to M.I.T. as a Visiting Lecturer in 1927, he was intensely interested in trying to use this newly discovered kind of ray to determine the crystal structure of a mineral. With x-ray equipment borrowed from his graduate schoolmate in physics, B. E. Warren, he succeeded in solving the structure of marcasite. So impressed and excited was he by this success that he turned away from ore deposits and economic geology and set about establishing an x-ray diffraction laboratory. In doing so he set the course of his teaching and research for the remainder of his academic career. Fifty years later, in 1975, when he finally closed out his career at M.I.T., he had long since become "Mr. Crystallography" to his students and peers.

On retirement in 1975 he could look back on a highly successful and satisfying career in crystallography and mineralogy that had brought him honors from around the world - medals; an honorary degree; membership in leading academies and professional societies in both America and abroad; a Festschrift on his 65th birthday (1968); a Festband on his 70th (1973); a bay in Arctic Canada bearing his name; and fittingly a variety of tourmaline named buergerite in his honor.

Why all the foregoing honors?; surely an appropriate question. Here is the answer. During almost half a century, as a member of the teaching staff at M.I.T., 1925-1973 (he stayed on for two more years as a Senior Research Associate), Buerger first created and then directed an x-ray diffraction laboratory devoted to crystal-structure analysis that became internationally renowned for its instruments, analytical methods, research publications, and the young scientists trained under his direction.

He invented two dozen new instruments and accessories - the Equi-inclination Weissenberg camera; Goniometer ball driller for crystal models; Powder cameras; Precession camera; Two wave-length microscope; Single-crystal diffractometer (Mark I, Mark II, and an automated design); etc. - and a half dozen or more new analytical methods and processes. With these, alone and with students, he solved the crystal structure of 21 minerals: e.g. marcasite, arsenopyrite, cubanite, berthierite, tourmaline, chalcocite, nepheline, pectolite, coesite, and rhodizite, and of the nonminerals terramycin, diglycine hydrobromide, and diglycine hydrochloride.

He developed theories to explain numerous crystal properties and behavior - lineage structure, polymorphism, genesis of twinning, disorder, crystal growth, etc., and pointed out the important role of temperature in mineralogy. His 10 major textbooks and more than 230 journal articles, in which his inventions, new methods and new ideas are discussed, have had a dominant role in changing Mineralogy and Crystallography from art to a science.

Finally, he directed the thesis work of 39 graduates (7 bachelor's, 12 master's, and 20 doctor's) and guided the research of a dozen or more post-doctoral guests; many of both categories, pre-doctorates and post-doctorates, came from abroad (Brazil, Chile, Norway, Germany, Switzerland, Spain, Greece, and Japan). Fourteen of his 20 doctorates and at least 8 of his numerous post-doctoral guests have chosen to follow him in his academic role and now hold professorships in some of the world's leading universities - Harvard, M.I.T., Michigan, U.C.L.A., McGill (Canada), University of Rio de Janeiro, Universidad de Chile, Universität Göttingen, Universität Aachen, Universität des Saarlandes (Saarbrücken), and University of Thessaloniki. Thus Buerger can be assured that his ideas, his analytical methods and procedures, his instruments and, most of all, his philosophy and standards of research will continue long into the future, thus giving him the immortality that outstanding students create for their great teachers.

When men achieve true greatness, they seldom do it alone; usually there is a woman who plays a major albeit silent role. Such has been the case with Martin Buerger. Lila MacAskill Buerger, his devoted wife of 37 years and mother of their 6 daughters, also deserves an accolade. She has managed their home with patience, skill and resourcefulness, and watched carefully over the health and activities of the children, so that Martin could concentrate on his scientific work; she has travelled with him to meetings around the world and has been the gracious partner he needed at social affairs. She has sat quietly and smiling at the banquet table while he received the medal or other honor; and on countless occasions she has been the cordial hostess at home with doigts de Fée when they have entertained faculty members and graduate students in living room and dining room made strikingly attractive by her own artistic handiwork. Lila well deserves this much too brief but very sincere recognition.

BIRTH, ANCESTRY, AND EARLY EDUCATION

Martin Julian Buerger,* the older of two sons of Martin J. G. and Julie E. R. (Weber) Buerger, was born in Detroit, Michigan, on 8 April 1903. His only sibling, Newton Weber, like himself followed an academic career in mineralogy and crystallography and will be mentioned again

---

* In preparing this biographical sketch I have had the invaluable assistance and cooperation of Prof. and Mrs. Buerger, both of whom patiently and willingly answered my many questions, even if in some cases the queries seemed a bit irrelevant to my primary purpose. I am deeply grateful to them for helping to make this sketch much more complete and accurate than I could have made it alone.

In addition, I have drawn to a limited extent on the following publications which include information of biographical nature: Anonymous 240, Mineraçao e Metalurgia 13/74: 130-132, port., (July-Aug. 1948). - Presentation of Day Medal to Martin J. Buerger, Citation by H. H. Hess, Geol. Soc. Am., Ann. Rept. 1951: 55, port., 55-56 (July 1952). - School for Advanced Study, Tech. Rev. 58/5: 245, (1955-1956). - Crystals made clear, The MIT Observer 5/3: 4, (Dec. 1958). - Presentation of the Roebling Medal to Martin J. Buerger, [Citation by] Clifford Frondel, Am. Mineralogist 44: 391-392, port., 393-395, (March-April 1959). - M. J. Buerger, Personal Reminiscences, p. 550-555 in Fifty Years of X-ray Diffraction, by P. P. Ewald, published for the International Union of Crystallography by N.V.A. - Oosthoek's Vitgeversmaatschappij, Utrecht, i-ix + 717 p., (July 1962). - [Class of '24 Notes] p. 96 in Tech. Rev. for July 1966. - Tech Talk, p. 3, port., (May 29, 1968). - [Prof. Buerger] under "11 Remarkable Colleagues," in Tech. Rev. 70/9: 80, (July/Aug. 1968). - "Martin J. Buerger, '24, Institute Professor of Mineralogy and Crystallography," by Robert R. Shrock, in Tech. Rev. 70/9: 81, port., (July/Aug. 1968). - "Mineralogy, Crystallography and M. J. Buerger," by L. V. Azároff, p. 3-4 and port., in Martin J. Buerger Festschrift 1968, Zeitschr. für Krystal., B. 127/1-4: 1-326, il., (1968). - "Buerger, Martin Julian; American crystallographer and mineralogist," p. 60-61 and port., in McGraw-Hill Modern Men of Science, Vol. 2, (1968). New York: McGraw-Hill Book Co., Inc. - WHO'S WHO IN AMERICA, with world notables, p. 302. Chicago: Marquis Who's Who Inc., Vol. 36, (1970-1971). - American Men of Science, 11th ed., The Physical and Biological Sciences: A-C, (p. 660). New York: R. R. Bowker Co., (1965).

farther on. Both parents were born in the United States and descended from German forefathers who immigrated to America early in the Nineteenth Century.

Martin John Gottfried Buerger, the father, was born in Buffalo, New York, in 1876. He was the grandson of Rev. Ernst Moritz Buerger, one of the clergymen who led a group of Lutheran immigrants from Saxony to the United States in 1838, and founded the Lutheran Church - Missouri Synod in Perry County, Missouri. Ernst's forefathers, as far back as 1679, had lived in Saxony, near Dresden, and had been pastors in the Lutheran Church.*

Martin John was the youngest of a family of 13 Buerger children, 12 boys and 1 girl, and became a skilled craftsman in the typical German tradition - meticulous, imaginative and perfectionist. He was a photoengraver all of his life, working during his earlier years for Doubleday Page, the publishers, located on Long Island.

When invited to join the aforementioned publishing company, in 1910 when Martin Julian was only seven years old, he moved his family from Detroit to Hempstead on Long Island so as to be near his new place of employment. After renting a house for a few years, he decided to design and build one of his own, and in subsequent years he built houses in both Hempstead and nearby Garden City, not as a full-time job, for his photoengraving kept him busy during the week, but more as a hobby and as a challenge to his abilities as a designer and builder. With such a father, it is perhaps not surprising that both his sons have exhibited similar abilities as imaginative builders whether the structures were houses or crystal models.

The mother, Julie Emma Rebecca (Weber) Buerger, described to me by her son Martin as "eine gute Hausfrau," was born in Wausau, Wisconsin, on 30 August 1878, the second oldest of 7 children, 6 girls and one boy (in addition several other children died in infancy). Her father, like Martin John's, was a descendant of German Lutherans, and as a young man had come

---

* Rev. Ernst wrote a fascinating account of his early years in Saxony and his later years in the United States, and from it one gets a clear impression of the religious zeal of a dedicated pastor who had the courage to leave the land of his birth for an unknown country across the Atlantic Ocean, and who to the last of his days continued to insist that the treasures of Lutheranism would be lost to the Lutheran Church if that Church gave up the German Language in favor of English! His account has been translated into English by one of his descendants, Rev. Edgar J. Buerger, and was privately published by the subject of this biographical sketch as Memoirs of Ernst Moritz Buerger, translated by Edgar Joachim Buerger and published by Martin Julian Buerger, Lincoln, Massachusetts, September, 1953, 94 p., and a portrait.

to America to study in the Missouri synod Concordia Seminary, which had been founded by the immigrant Rev. Ernst Moritz Buerger and his colleagues, as mentioned in a preceding paragraph, and was himself a pastor in the Lutheran Church in northern Wisconsin at Crandon.

Thus the Buerger and Weber families actually came together in America in the generation before Martin John Gottfried Buerger and Julie Emma Rebecca Weber were born.

Mother Julie had graduated from High School No. 16 in Buffalo and had taught in several parochial schools there before marriage to Martin John. Later, when she and her husband lived in New York City, she organized and carried on an informal manufacture at home, which involved making small doll-like objects for gifts and decorations. She herself created the designs for her products, and from the proceeds of their sale managed to maintain a home and provide financial support for her two sons, whose custody she was granted when divorced from her husband, Martin J. G. Buerger, in the early 1920s. Soon after son Martin entered M.I.T., she moved from New York City to Cambridge and found employment near the Institute with the General Radio Company, meanwhile living in a one-room flat nearby under conditions that today would be described as penurious. Yet she helped her two sons to go through M.I.T., from freshmen to doctorates, and made their cramped living quarters a place of "gemütlichkeit" and good food for the sons' Institute schoolmates. Her sturdy Teutonic heritage gave her a long and active life, and when death came on 10 March 1972, she was in her 94th year.

During his youthful years, Martin Julian attended a number of different primary and secondary schools because of the frequent moving about of his family. He began his primary schooling in the Detroit schools, continued for short periods in Hempstead, Long Island, and Toronto, Ontario, and finally completed the 8th grade in Jamaica, New York. Then followed two years at his mother's old school in Buffalo, High School No. 16, with completion of his secondary education at Morris High School in the Bronx, from which he graduated in 1920.

As Martin tells it, he came across one of his father's chemistry textbooks one day,* and though only seven years old at the time, he remembers that he became intensely absorbed in the subject matter, soon

---

* As Martin relates it in his "Personal Reminiscences" (see p. 550 in reference 203-1962j in the Bibliography farther on):

"My interest in science was started by a volume entitled The Complete Chemistry, a Text Book for High Schools and Academies by Elroy M. Avery (Sheldon and Company,
(footnote continued on next page)

mastered much of it, and astounded his elders when he had an opportunity to demonstrate his knowledge of chemistry. Yet, interestingly, he did not follow a career in chemistry for long, though his dominant interest in the order and symmetry manifested in the physical world would continue unabated throughout a lifetime of research in mineralogy and crystallography.

He was fortunate to have excellent teachers of physics* and mathematics, especially geometry, in high school (Morris H.S. in the Bronx), and he remembers that his deep interest in these two subjects, which continues to this day, was undoubtedly the result of the superb training he received from his high school teachers.

When the time came to think of going to college, Martin's early interest in that chemistry book led his father to recommend M.I.T., largely because he had heard that the Institute was the best college in the country to get training in analytical chemistry. And so an early fascination with a chemistry text, excellent high school instruction in mathematics and physical science, and a discerning father's recommendation all combined to bring Martin to the Institute in the fall of 1920. He had done well on the Regent's examination required of all graduating seniors of the State of New York, and had been offered a scholarship to any college in the State. He was expected to choose Cornell, which was the traditional college for the Buergers to attend, but he declined the scholarship when the opportunity came to enter M.I.T.

(footnote continued from preceding page)
    New York and Chicago, 1883). This was my father's high school chemistry text in 1893, and I discovered it in our family library about 1910, when I was seven years old. This explained a good deal about matter, physical and chemical changes, and of course, chemistry. This seemed to be just what my mind needed, for it explained the complicated things about me in terms of simpler units, so that it became obvious that the world was a matter of chemistry. I was soon well acquainted with the contents of the book, and astonished my parents and their friends with this curious knowledge.

    "In due time I studied chemistry in high school and found it, along with geometry, my most interesting study. My father, noting this interest, felt that M.I.T. was the place for me, and when I finished with high school, I continued my studies there."

* Martin recalls that a Mr. Pyle, who wrote the physics textbook he studied, let him do some independent experiments on Hertzian [radio] waves, which at the time heightened his interest in physical phenomena-- a perceptive action by a teacher that could well be followed by others when an inquiring student makes his interest known!

THE STUDENT YEARS AT M.I.T. (1920-1929)

As would be expected, in light of the preceding discussion, Martin looked forward to the introductory subjects in mathematics, physics and chemistry that he would be required to take as an entering freshman.

Entering the Institute in the fall of 1920, he decided to major in chemistry. Soon, however, he found that the excellent preparation he had received at the Morris High School made his classes in the physical sciences uninspiring and boring. Furthermore, he disliked the poor way mathematics was taught, and it was years later as a graduate student before he became interested enough to pursue mathematics beyond the simplest calculus. The result of this disenchantment with M.I.T. was that after transferring to chemical engineering for a term or so, he dropped out of the Institute at the end of the second year and spent a year as a technician in industry.

Having to work to support his mother and brother, he found employment with Western Electric in the Wall Street district where the first machine-switching telephone systems were being installed. At this late date about all that he remembers of his 13-hour work days is that he did learn how to join wires with a good solder joint!

A year as a technician in industry was enough to convince Martin that if he was ever going to get anywhere he needed to complete his college education, so the fall of 1923 found him back at the Institute to start his third year. Remembering that he had had an unpleasant experience in his quantitative analysis course in chemistry, when his professor lost his (i.e. Martin's) notes, and being no longer interested in regular chemistry, he sought and received advice as to what courses used chemistry. His attention was called to the importance of applied chemistry in mining and metallurgy, and this led him to transfer from Course X (Chemical Engineering) to Course III (Mining Engineering) - the first important turning point in his academic career. As he tells it in his "Personal Reminiscences"*:

> "Because of the uninspiring teaching of chemistry, I looked around for another related field. I found that mining engineering made use of chemistry through mineralogy, so I entered the study of mining. This was a fortunate choice, because I immediately came in contact with Mr. Walter H. Newhouse, at that time a graduate student studying for his doctor's degree under Professor Waldemar Lindgren. Every student needs a wise and inspiring teacher, and Newhouse became this to me. What I had missed in chemistry, Newhouse made good in mineralogy,

---

* See p. 550-551 in P. P. Ewald's <u>Fifty Years of X-ray Diffraction</u>; see also this reference, 203--1962j in the Bibliography.

for that was what he taught. But more than that, Newhouse had a feeling for relevance, and always stripped a matter of its extraneous wrappings and went directly to the core. I recognized this characteristic and tried to emulate him. Although I graduated with a degree of B.S. in mining engineering, I took my bachelor's thesis under him. At just about that time he was given funds for a research assistant and he offered this post to me. My job was to relieve him of teaching elementary mineralogy while he devoted the corresponding time to research.

"This was a turning point in my career. During the last two years, I had had tuition scholarships, but my mother had supported me. It had never occurred to me to go on with graduate study, since there was no money available for it, so this assistantship changed my career from that of a practising mining engineer to that of a teacher. In two years I took my master's degree in geology, and in two more my doctor's degree in mineralogy, all under Newhouse, who had been advanced to an assistant professor. Newhouse and I became close personal friends, and shared an office. We both basked in the research atmosphere created by Professor Waldemar Lindgren, dean of geologists and head of our Department of Geology."

Martin wrote his bachelor's thesis on the "Tin Ores of Oploca, Bolivia," with J. L. Maury (see references T--1925 and 1--1927 in his Bibliography) and received an S.B. degree in Mining Engineering in June 1935. Transferring from mining to geology for graduate work, and writing his master's thesis on "The deformation of ore minerals: a preliminary investigation," (see T_1--1927, and references 4--1928 and 5--1928a in his Bibliography), he received an S.M. degree in Geology in June 1927. Continuing on with graduate work, he wrote his doctoral thesis on "Translation-gliding in crystals," (see references T_2--1929 and 9--1930 in his Bibliography), and received a Ph.D. degree in Mineralogy in June 1929.

THE TEACHING YEARS AT M.I.T. (1925-1973)

Buerger's teaching career at M.I.T. actually began when he started graduate work in the fall of 1925. His complete 50-year record as a teaching assistant, faculty member, administrator, and senior research associate has been as follows:

Assistant in Geology	1925-1927
Instructor in Geology	1927-1929
Assistant Professor of Mineralogy and Petrography	1929-1935
Associate Professor of Mineralogy and Petrography	1935-1937
Mineralogy and Crystallography	1937-1944
Professor of Mineralogy and Crystallography	1944-1956
Chairman of the Faculty	1954-1956
Director of the School for Advanced Study	1956-1963

Institute Professor (and Professor of Mineralogy and Crystallography)	1956-1968
Institute Professor (and Professor of Mineralogy and Crystallography) Emeritus	1968-1973
Institute Professor Emeritus	1973-
Senior Research Associate	1973-1975

STUDENT TEACHING (1925-1929)

Buerger's first teaching experience was gained while he was a Master's candidate during SYs 1925-26 and 1926-27. He first assisted Newhouse, who wanted more time from teaching for research, by conducting classes in beginning mineralogy. Later on, at Lindgren's request, and as an Instructor, he taught a subject on the petrography of ceramics in which the students were taught to use the petrographic microscope for identifying ceramic materials such as mullite in spark plugs. He continued as an Instructor until 1929, when he was awarded his doctorate and was promoted to Assistant Professor.

HE BECOMES A FACULTY MEMBER (1929)

When J. L. Gillson took a year's leave of absence in the fall of 1928, to do consulting work for the Du Pont Company, there was no faculty member available to teach optical crystallography and petrography. Sensing this situation, Buerger asked Lindgren, then head of both Course III and Course XII, if he might add Gillson's two subjects to his own course in introductory mineralogy. Lindgren assented, and in this way Buerger assumed a full-time schedule of teaching and became a member of the Faculty when promoted to Assistant Professor in 1929. Eight years later, when H. W. Fairbairn joined the geology faculty, Buerger relinquished optical crystallography and petrography to Fairbairn and expanded his own instruction in x-ray crystallography and related subjects.

He was promoted to Associate Professor in 1935, to Professor of Mineralogy and Crystallography in 1944, and to the distinguished rank of Institute Professor in 1956, at the time he completed a two-year period as Chairman of the M.I.T. Faculty and began a seven-year period as Director of the Institute's newly organized School for Advanced Study, which is discussed at some length farther on. He resigned as Director of the School in 1963 and became an emeritus professor in 1968* but continued

---

* At the time he reached the emeritus stage I prepared a brief statement on Buerger for The Technology Review which appeared on page 81 of the Review for July/Aug. 1968 as: "Martin J. Buerger, '24; Institute Professor; Professor of Mineralogy and Crystallography." A shorter version appeared on page 80 of the same number of the Review under the heading: "11 Remarkable Colleagues."

for the next five years on a half-time basis as Institute Professor Emeritus. Normally he would have retired permanently at this time, having reached age seventy, but for the next two years, 1973-1975, he was appointed a Senior Research Associate. Finally, at the end of SY 1974-75, he did retire permanently after fifty-five years of association with M.I.T. as student, professor, administrator, Institute Professor, and lastly Senior Research Associate, with the final title of Institute Professor Emeritus.

From 1928 to 1968, when he reached normal retirement age, he carried on a vigorous and highly productive program of instruction and research, and this activity has continued essentially unabated during the ensuing seven years to the present time (December 1975). During this 47-year period, Buerger has published 244 articles and 12 books (16 articles and 3 books since retirement in 1968!); supervised a total of 39 theses for 7 bachelor's, 12 master's and 20 doctor's degrees; and developed a laboratory for x-ray diffraction work and crystal-structure analysis that has become known around the world and has attracted graduate and post-doctoral students from every continent except Australia. His scientific achievements have brought fame in good measure and with it the varied honors and awards reserved for only those who attain greatness in their profession.

In the following sections I will discuss the different activities engaged in by Buerger during his half century as an academician at M.I.T., as a visiting professor at other educational institutions around the world, as a leader in his professional societies, and as a renowned scientist in his chosen special fields of mineralogy and crystallography.

EARLY FIELD WORK THAT ALMOST PRODUCED AN ECONOMIC GEOLOGIST

Buerger might well have followed a career as a practical mineralogist and field geologist had he not become so deeply interested in and devoted to the whole broad field of crystallography.

His first experience in practical geology came during the summer of 1925 following his graduation in mining engineering. He and two of his classmates, Ariel F. Horle (S.B. III 1926) and J. L. Maury (S.B. III 1926), decided to seek employment in the mines at Butte, Montana, Maury's home. Buerger found his job at the Belmont Mine where he was taken on as a mucker. A week of this hard, lonely work on the hot 3000-foot level, with "copper water" dripping down his back from the roof, convinced him that he didn't want to do this kind of work all summer, so he applied for and got a job as a sampler and spent the rest of the summer working in this capacity.

A year or two later, having meanwhile transferred from mining to geology and feeling that he needed some field work, he took the Civil Service Examination in geology for employment with the U.S. Geological Survey. He did well enough on the examination to warrant a strong recommendation from Dr. Lindgren, who was consulted about his general qualifications, and was assigned to work with James Gilluly in the Oquirrh Mountains of Utah. He recalls that Jim was rather discouraged with him because he had not had any actual geological field work nor any experience in surface surveying; his only previous experience along these lines had been a summer-school course in underground surveying and that summer's work at Butte. In discussing with me the summer's work with Gilluly, he summed it up by saying:

> "I think Jim was pretty discouraged with me by the end of the summer, and I was not invited to come back!"

The next stint of field work came in 1928 when he was invited by Newhouse to join a summer field party that was to explore the region around the Buchans Mine in Newfoundland. Newhouse had spent the previous summer (1927) at Buchans and needed several field assistants to continue the program a second summer. Buerger recalls that he thoroughly enjoyed his assignment - working alone in the bush searching for outcrops, mapping the unknown interior of Newfoundland, and tracing rusty ore boulders "up glacier."

He spent the next summer prospecting for the Swedish-American Prospecting Company in the same general region. This was to be the last of his regular field projects, however, for he next turned to consulting on mineral deposits - assignments typically of only a few days duration - and he remembers that these frequently involved looking at the "feldspar mines" people thought they had in their backyards. One job that he particularly remembers was the tracing of a 100-foot wide dike of trap rock across Massachusetts for a client who was selling crushed rock for road metal. He recalls that he traced boulders of the traprock "up glacier" just as he had traced the rusty ore boulders in Newfoundland, and that he mapped the dike all the way to the Connecticut border. His mapping was refined by magnetometer surveys at intervals across the till-covered dike. The work, done in the mid-1930s, resulted in locating the trap-rock quarry at Holden, Massachusetts. Buerger had all but forgotten about it when, some forty years later, in 1970, several of his colleagues at the University of Connecticut, where he was serving as a Visiting Professor after retirement from M.I.T., wondered if a certain large dike in Connecticut continued across the state boundary into Massachusetts. He was able to assure them that it did, and that he had mapped it as far as the central part of Massachusetts. Today it is regarded as marking a major fracture in the earth's crust in Massachusetts and Connecticut, whence it disappears into the sea. Clearly a well-done job of mapping never goes out of date and can often serve in unexpected ways!

As late as 1937 Buerger still had the explorationist's urge in his blood, even though he was by this time more than fifty papers into his career as an x-ray crystallographer. In the summer of that year he was a member of Commander* Donald Mac Millan's expedition to Baffin Island in the Arctic Ocean. On land he was as interested as ever in the geology, but on shipboard his thoughts returned to crystallography, and he started the first of his major textbooks, X-ray Crystallography (1942), during the trip up and back. (See discussion farther on under "Books".)

Buerger's fertile mind works best when freed from the niggling and distracting pressures of his surroundings, and he has told me that some of his best ideas came when he was relaxing on a Florida beach or swimming in Lake Taconic, his yearly retreat since his boyhood days. Aboard MacMillan's ship he organized the subject matter for his first major book, as mentioned in the preceding paragraph. While on a four-month leave in Brazil, in 1948, as a Visiting Professor in the Faculty of Philosophy in Rio, he was able to generalize Dorothy Wrinch's primitive result of solving the vector set based upon a triangle and to tell his class how to solve a general vector set, before he left in January 1949 to return to Cambridge. As he so rightly comments, in his "Personal Reminiscences" (p. 553; see reference 203--1962j in his Bibliography):

> "I believe most scientists, including myself, do not spend enough time in an attempt to gain perspective. Usually we are too busy finishing the many projects we have started."

HE FINDS A RESEARCH INTEREST IN X-RAY CRYSTALLOGRAPHY

The first turning point in Martin's scientific career came, as mentioned earlier on, when he transferred from chemistry and chemical engineering to mining engineering and then to geology, at which time he developed an interest in mineralogy.

The second turning point, probably the most critical and important in his career, came while he was working on his doctoral degree, and led to a lifetime career in crystallography. Again let us quote from his "Personal Reminiscences," (see p. 551 of the reference cited in the two preceding footnotes and in his Bibliography):

---

* Later Rear Admiral.

"In 1927, before I obtained my doctor's degree, Bragg spent a term at M.I.T., and I attended his lectures. Bertram E. Warren, a fellow graduate student in physics, became a student of Bragg's at that time, and with him worked out the structure of diopside. After receiving his doctor's degree, Warren continued in crystal-structure analysis. I had no laboratory facilities, but also wished to try out crystal-structure analysis, so, with Warren's kind permission to use his equipment, I took a set of rotating-crystal and oscillating-crystal photographs of marcasite. I solved this simple two-parameter structure alone in 1930-31 by applying the techniques I read about in the papers appearing in the Zeitschrift für Kristallographie.

"This marked the second, and perhaps the most important, turning point in my career. I had been teaching mineralogy, optical crystallography, and petrology, so that up to this point I had been a mineralogist. But the structure of marcasite represented my first excursion into pure crystallography. I was delighted with the certainty of the conclusion reached by crystal-structure analysis, as compared with the arguable results published by the geologists and mineralogists of that period. Just following my delightful experience with marcasite, the new president of M.I.T., Karl Taylor Compton, obtained a large grant for research from the Rockefeller Foundation, and eventually I was asked by Professor Henry [Hervey] Shimer, acting head of our department at the time, what I needed for research. I immediately dreamed up $10,000 worth of X-ray equipment, had it approved, and I was shortly in the business of investigating the crystal structures of minerals."

Now began what would become in the following decades one of the leading x-ray laboratories for crystal-structure analysis in the entire world. The development of this laboratory is discussed in the following section.

HE FOUNDS A LABORATORY AND QUICKLY LEARNS

As a student, Buerger learned quickly from his graduate college mates, W. H. Newhouse and B. E. Warren (who later became distinguished professors at M.I.T.), from the great professors who fired his imagination, Waldemar Lindgren and Lawrence Bragg (later known as "Sir Lawrence" after being knighted), and from the geological literature of the day.

Possessed of remarkable ability to sense how the arrangement of atoms in crystals could have relevance to the determination and understanding of crystal-structure in natural and synthetic minerals, and having also the imaginative experimentalist's skill to devise and use the most effective research instruments, Buerger developed at M.I.T. a world-renowned laboratory in which x-ray diffraction was used for crystal-structure analysis in a most impressive way.

Prof. Martin J. Buerger with his favorite model of the precession camera: the commercial version of Mark II.
(Photo from M.I.T. Office of Public Relations)

Here in this laboratory were invented or redesigned and improved more than two dozen instruments for crystal-structure research; seven of the prototype instruments are now in the Smithsonian Institution in Washington, and improved models of them are now in use around the world. Inasmuch as I regard Buerger's invention and design of instruments as one of his major contributions to the transformation of mineralogy and crystallography from art to science, I devote a later section to this aspect of his professional career at M.I.T. The photograph on the facing page shows him with one of his favorite inventions, the precession camera.

From Buerger's laboratory came research results that formed the substance of 11 books and more than 240 articles by him and his students - he liked to share authorship with his students and more advanced associates, with the result that an even third of his articles are jointly authored.

Here, too, came students from foreign countries around the world to learn Buerger's research methods, to learn how to use the instruments and facilities of the laboratory, and possibly most important of all to enjoy the "gemütlichkeit" research and social atmosphere of the laboratory at M.I.T. and of the Buerger home in suburban Lincoln.

Buerger first initiated x-ray diffraction research on single crystals in 1931,* when he established his laboratory, and he constantly revised his methods and instrumentation for the next thirty or more years until his laboratory in the new Green Building was dismantled in 1968, when he retired, and the facilities moved to the laboratory of Prof. Bernhardt J. Wuensch (Ph.D. XII 1963), one of his doctorates, in the M.I.T. Department of Metallurgy and Materials Science.

Some idea of how Buerger conducted his research is clearly revealed in his own words:

> "The idea of a minimum function occurred to me as I was laboriously multiplying pairs of Patterson values to form the product function. The first minimum functions were therefore made by comparing pairs of values of the Patterson function as written down as if ready for contouring a Patterson projection of the crystal. I compared the two values at the ends of a line image as I allowed the image to range over the cell, selecting, at each location of the image, the minimum, and reading it aloud to my daughter Marla who wrote down this value in its proper place on another map, the M_2 map. This tedious procedure, after being practiced a few times, gave way to the graphical method. All this was stimulated by a visit to Rio, and I could mention other ideas developed by appropriate loafing in Florida, etc." (From p. 553-554 in "Personal Reminiscences"; see reference 203--1962j in his Bibliography.)

---

* (see footnote on next page)

AS A TEACHER HE GLADLY TAUGHT

Buerger was, and still is, a superb teacher, and is so regarded by the host of students who have heard him lecture. One of his students attributed his teaching success to

> "just avoiding complex ways of expressing simple things."
> (The M.I.T. Observer 5/3: 4, Dec. 1958).

Buerger himself has the following to say about teaching, comments that reveal far better than I can the attitudes and philosophy of a great teacher (quoted from p. 552 of his "Personal Reminiscences"; see reference 203--1962j in his Bibliography):

> "I have found teaching a worthwhile career. In the first place, it has been my experience that students react to me as I reacted to my teachers. They need the teacher not only to guide them in technical matters, but to transmit a philosophy to them, partly by precept, partly by providing an appropriate atmosphere. Students are susceptible, and the teacher has a great responsibility. Many times I have had the experience of having a former student unconsciously quoting back to me my own philosophy. On the other hand, I have also been taught much by my students. The close rapport between student and teacher makes it possible for the teacher to absorb from his students knowledge which has developed since the teacher was involved in formal study, or which the student, with his youthful viewpoint, has seen fit to cultivate. More generally, I have found that a group of students and their teacher are members of a small select society, and that they teach and inspire one another, especially if the group has reached the critical size of, say, five. Within this group it becomes fashionable to advance in knowledge and to publish newly acquired knowledge, so that the members of the group emulate and stimulate one another."

Reference is made elsewhere in this sketch to the number of students who did their thesis work under Buerger's supervision: 7 bachelor's, 12 master's, and 20 doctor's. A count of his doctorates in 1975 showed 15 in former or present academic positions, 3 in important research positions, and 2 deceased - a truly impressive record! In addition, 7 post-doctoral fellows and 7 seminar attendants spent time in Buerger's laboratory or participated in his seminars, and all of these now hold academic positions.

In the following list, seminar attendants are indicated by **; post-doctoral fellows by *; and the 20 M.I.T. doctorates (including Mortimer C. Bloom, who did not do his doctoral work with Buerger), are undesignated.

---

* (footnote from preceding page)
The President's Report for 1930-1931, p. 113, states that "A new x-ray laboratory, under the direction of Prof. Buerger, will be installed ... by means of an appropriation of $2,980 from the Rockefeller Research Fund."

Former Graduate Students (now doctorates) and Post-doctoral Fellows*
or Seminar Attendants,** with an indication
of their current professional activity

Date of Attendance or Degree	Name	Place of Employment
1934	H. C. Horwood	(Deceased; formerly business)
1938	Mortimer C. Bloom	Industrial Research (England)
1939	Clifford Frondel	Harvard University
1939	Newton W. Buerger (now retired)	Naval Post-Graduate School, Monterey, California
1946	Edward Washken (Sc.D.)	(Deceased; formerly private business)
* 1947	Elysiario Tavora	University of Rio de Janeiro
1948	Gabrielle (Hamburger) Donnay	McGill University
1949	Alfred J. Frueh	University of Connecticut
* 1951	José Luis Amorós	Universidad Complutence, Madrid, Spain
1951	Mead L. Jensen	University of Utah
* 1951	Luisa Bradhe	University of Oslo (Norway)
* 1952	Joseph Zemann	University of Vienna (Austria)
1953	Leonid V. Azároff	University of Connecticut
1953	Virginia F. Ross	M.I.T. (Senior Research Associate)
** 1954	Severino Garcia-Blanco	Consejo Superior de Investigaciones Cientificas (Spain)
* 1956	Theo Hahn	Universität Aachen
1957	Nobukazu Niizeki	Industrial Research (Japan)
** 1958	Erwin Parthé	University of Basle (Switzerland)
1959	Tibor Z. Zoltai	University of Minnesota
** 1959	Mohamed Gheith	Boston University
** 1959	Roberto Poljak	Johns Hopkins University
1961	Charles W. Burnham	Harvard University
* 1961	Karl Fischer	Universität des Saarlandes, Saarbrücken
1962	Donald R. Peacor	University of Michigan
1962	Charles T. Prewitt	SUNY, Stony Brook, N.Y.
* 1963	Panos Rentzeperis	University of Thessaloniki (Greece)
1963	Bernhardt J. Wuensch	M.I.T.
1964	Hilda Cid-Dresdner	Universidad de Valdivia (Chile)
1966	Wayne A. Dollase	University of California (Los Angeles)
** 1968	George Felsche	E.T. Hochschule (Switzerland)
** 1968	Hajo Onken	Saarbrücken, Aachen
** 1968	Peter Süsse	Göttingen
1969	Felix J. Trojer	Battelle Memorial Institute (Switzerland)
1970	A. Lyneve C. Waldrop	Siena College (New York)

During SYs 1942-43 and 1943-44, when World War II greatly disrupted regular academic work in the United States, Buerger joined Fairbairn in teaching elementary physics to officer candidates in training for later command responsibilities. In spite of his heavy teaching load during this period, however, Buerger managed to carry on an active research program and publish two books and a dozen articles (see his Bibliography for titles).

AS A CONSULTANT ON MINERALOGY AND CRYSTALLOGRAPHY

Reference is made earlier to numerous short-term consulting assignments that Buerger accepted in the 1930s, many of which involved small occurrences of feldspar in people's backyards - occurrences that the owners characteristically considered more valuable than they actually were.

In the 1940s he consulted on soap for Lever Brothers, from which work came four articles - references 81--1942e; 90--1945c; 92--1945e; 93--1945f; and 104--1947b. He also did limited consulting for A. D. Little and Owens-Illinois Glass Co. (glass) in the 1950s. The only consulting arrangement he has maintained for any length of time, and still does, has been with the Du Pont Company in Delaware, with which he has been associated for more than twenty years. Because of his intense devotion to the purely scientific aspects of an industrial problem, the monetary attraction of consulting has always been subordinated in favor of the opportunity to solve an especially challenging problem. Furthermore, if permissible, Buerger has characteristically published the results of his consulting activities. The story of how one of his books, Elementary Crystallography, grew out of his consulting experiences at Du Pont, is told in the section on "How he wrote books."

AS CHAIRMAN OF THE M.I.T. FACULTY, INSTITUTE PROFESSOR, AND A. P. SLOAN AWARDEE

Buerger served as Chairman of the M.I.T. Faculty for two years, SYs 1954-55 and 1955-56, and at the end of his appointment President Killian paid him a special tribute "for exceptional service rendered during his term of office," at the Faculty Meeting on 5 June 1956. At the same time the Corporation announced his appointment as an Institute Professor,

> "an honor bestowed by the faculty and administration of M.I.T. on a faculty colleague who has demonstrated distinction by a combination of leadership, accomplishment, and service in the scholarly, educational and general intellectual life of the Institute or wider academic community." (Quoted from a pamphlet dated 18 May 1973 distributed to the faculty.)

As a fitting climax to SY 1955-56 Buerger was granted an "Alfred P. Sloan Award for Outstanding Performance" in the amount of $1,000.00, and a similar award was repeated at the end of SY 1956-57.

AS DIRECTOR OF M.I.T.'s SCHOOL FOR ADVANCED STUDY

On Wednesday evening, 4 January 1956, at the Walforf-Astoria Hotel in New York, the M.I.T. Corporation gave a memorable dinner for some 1550 special guests as a tribute to Karl Taylor Compton, M.I.T.'s distinguished President from 1930 to 1949. The theme of the dinner, with President James Rhyne Killian, Jr. presiding, was "Science the Mighty Multiplier." One of the purposes of the dinner, and the one I am interested in here, was to announce the founding of a School for Advanced Study. In making the announcement, Killian stated that the Corporation's action had followed an idea suggested by Professor M. J. Buerger, then Chairman of the Faculty, and the next morning's New York Times carried a story reporting that Buerger would be Director of the School.

The idea of the School* was to provide for bringing together a small group, 7 to 10 per year, of outstanding young doctorates who would interact

---

* Buerger's idea of the proposed school was discussed briefly in Tech. Rev. 58/5: 245, (March 1956) as follows:

"In its initial embodiment the School will be simply an organizational entity, but it is hoped ultimately to provide a center and adequate housing for fellows and guests. Scholars who are invited to M.I.T. for advanced study will have the status of 'fellows' in the School. This year there have been approximately 100 such people from 15 countries, studying at the Institute and they have been registered as 'guests' or 'visiting fellows.'

"Dr. Buerger said that ordinarily visiting scholars do not enroll in courses or seek degrees, and M.I.T. has not asked them to pay tuition. The majority are supported by fellowships or grants, and, he added:

"'We welcome such scholars and believe we can be of greater help to them by establishing the new School. Much can be gained in science and engineering through the interchange of ideas. Close association and intimate discussion between men in the same field of research, or in different fields, can be productive of new insights.

"'By establishing a school, we will be able to bring the scholars closer together and closer to members of our own Faculty. Special programs can be arranged for them and arrangements can be made for them to meet in informal conferences. Plans for a special on-campus housing unit for visiting scholars are being considered.'

"The school will be similar in its objectives to the Institute for Advanced Studies at Princeton, but the Princeton center has a permanent staff of some size. Unlike the Princeton School, the M.I.T. School for Advanced Study will be an integral part of the Institute, and constitute an extension of the level of the programs of the Undergraduate and Graduate Schools."

with the faculty on the one hand, and with advanced students on the other. The School was to be financed for the first few years from funds to be provided by Alfred P. Sloan, Jr.

Buerger assumed the directorship of the School at its inception, in the spring of 1956, and guided it through its first few years from his office in Building 24 and with the help of a lone secretary. Difficulties soon arose as to who should be selected as fellows, and it became evident that the original plan was not going to succeed. In the midst of uncertainty about the School's future, particularly when there was no provision in the plan of the Second Century Fund for the separate building that Buerger thought the School should have, he resigned as Director in 1963. Thereafter enthusiasm for continuing the School subsided, as the objectives of the Second Century Fund engaged the attention of the Corporation, Administration and Faculty, and the educational experiment came to a quiet end. The idea was a good one, but it turned out to be unsuited for M.I.T.*

AS VISITING PROFESSOR OR GUEST LECTURER AWAY FROM M.I.T.

Being widely known internationally, in the fields of crystallography and mineralogy, for his articles and books, his leading roles in the activities of national and international professional societies and their journals, and for his numerous doctorates in crystallography, it is not surprising that Buerger has been invited to be a visiting professor or guest lecturer by universities around the world. The following list illustrates how extensively he has responded to such invitations.

<u>Brazil, 1948-1949</u>. He spent the fall term of SY 1948-49, from September 1948 through January 1949, as a Visiting Professor in the Faculdade

---

* Buerger's after-thoughts are worth including here, as given to me in December 1975, because they add interesting information on the demise of the School for Advanced Study at M.I.T.

> "A major reason for my resignation was that the original funding by Sloan eventually ran out, and I was looking for other sources. Although I have forgotten the details of what happened 12 years ago ... I do remember that I found a source of funding, but the Institute refused me permission to tap it, and also failed to fund the residence which I had taken the trouble to get an estimate on. ... I still maintain that the School for Advanced Study WAS right for M.I.T. Our lead inspired in other universities half-a-dozen Institutes for Advanced Study on our general design, and they still exist. The Institute should again return to the start we once had. The natural senior membership of the School should be the dozen Institute Professors and the holders of named professorships. Of course, they now need a much larger residence." (Personal communication, 1 December 1975.)

de Filosofía, Universidade de Rio de Janeiro. This appointment was arranged through Dr. Elysiario Tavora who had been a post-doctoral guest in Buerger's M.I.T. Crystallography Laboratory in 1947.

Spain, 1952. As Visiting Lecturer in the Consejo Superior de Investigaciones Cientificas, Buerger not only participated in the instructional program but also helped to get established an x-ray diffraction laboratory for crystal-structure analysis. One of his former post-doctoral guests, Dr. José Amorós, who wished him to visit Barcelona, which he in fact did do later, had first to get him appointed at the Madrid institution in order to make the week-long visit to Barcelona possible.

Chile, 1962. During our spring term of 1962, he was Guest Lecturer in the Instituto de Fisica y Matemáticas de Chile, in Santiago. The same year and later he had two graduate students from Chile who received graduate degrees in crystallography at M.I.T. under his direction: Hilda Cid-Dresdner (S.M. XII 1962; Ph.D. XII 1964) and Isabel Garaycochea-Wittke (S.M. XII 1966).

India, 1963. Invited by some of his Indian colleagues to give some lectures in the Winter School at Madras in December 1962, he not only complied with their request but also, at the request of the U.S. State Department, delivered lectures at several other Indian schools before returning home in January 1963.

Greece, 1964. In August and September 1964 he gave a series of lectures in the NATO School on the island of Zakinthos. This visit was arranged through the efforts of Prof. Panos Rentzeperis who had been a post-doctoral guest in Buerger's laboratory at M.I.T. in 1963.

Minnesota, 1970. He was a Visiting Professor in the Department of Geology and Geophysics at the University of Minnesota, Minneapolis, for the spring term of 1970. One of his doctorates, Tibor Zoltai (Ph.D. XII 1959), was Chairman of the Department at the time.

Kentucky, 1971. He was a Visiting Professor in the University of Kentucky's Department of Geology for two weeks in April 1971, where one of his former students, William H. Dennen (S.B. XII 1942; Ph.D. XII 1949), was Chairman of the Department. During his stay in Lexington he gave a course to advanced students in crystallography on the subject "Image Sets and Image Functions" and conducted two seminars on "The History of Crystallography."

Virginia, 1974. He spent April to June 1974 as Visiting Professor in the Department of Geological Sciences at Virginia Polytechnic Institute and State University at Blacksburg.

Connecticut, 1968-1975. In 1968, when he became Institute Professor Emeritus, he was invited to join the faculty of the University of Connecticut as a University Professor.* The invitation came from one of his most distinguished doctorates, Prof. Leonid V. Azároff (Ph.D. XII 1954), who was then and still is the Director of the University's Institute of Materials Science. From 1968 to 1973, when Martin reached mandatory retirement age, he participated actively in the instructional program of the Department of Geology. In addition to giving regularly scheduled lectures in crystallography, he acted as Graduate Advisor in the Institute of Materials Science, supervised the thesis investigations of several graduate students, and carried on his own research and writing as vigorously as ever.

Even though he had to retire in 1973, when he reached age 70, the University asked him to continue his connection, as a Visiting Professor, to come and go as he pleased, and that he continues to do at the present time (1975). To show their appreciation of his many contributions to the activities of the University, the Board of Trustees and Staff presented him with a Certificate of Recognition, a testimonial of their gratitude, when he retired on 1 October 1973.

Harvard, 1973. When he reached age 70, on 8 April 1973, and had to retire from both M.I.T., where he had been an Institute Professor Emeritus from 1968 to 1973, and from the University of Connecticut, as mentioned in the preceding paragraph, another of his doctorates stepped into the breach. Prof. Clifford Frondel (Ph.D. XII 1939) arranged for Buerger to be an Honorary Research Fellow in Harvard's Division of Geological Sciences, and there Buerger continues his research and writing as if his retirement date was ahead instead of behind him!

PUBLICATIONS

INTRODUCTION

During his professional career of half a century (1926-1975 inclusive) Martin J. Buerger has produced more than 250 publications; these are listed in his Bibliography farther on. He is the lone author of 173 of

---

* He was invited to serve on a full-time basis, but, because of his long association with M.I.T. he could not persuade himself to leave the Institute completely. Accordingly, he continued to teach at M.I.T. on a half-time basis, and accepted only a half-time appointment at the University of Connecticut. To carry out the program, Lila and Martin had an apartment in Storrs and spent half-a-week there and half-a-week at their home in Lincoln.

the publications listed and joint author of 83. The 256 publications fall into the following categories, which clearly indicate the aspects of crystallography and mineralogy in which he has been most productive:

Books on Crystallography	12
Articles on Crystallography and Mineralogy	188
1) Crystal structure of minerals — 89	
2) Crystallography - more general — 30	
3) Techniques and Methods — 26	
4) Ores and minerals — 15	
5) Instruments and Apparatus — 11	
6) Polymorphism and Metamorphism — 9	
7) Optics — 4	
8) Soaps — 4	
Abstracts	31
Reviews	19
Citations, Memorials, and Miscellaneous	6
Total:	256

This impressive record places Buerger high among graduates and faculty members of M.I.T.'s Department of Geology with respect to their publication record as indicated in the following list:

	Name	Status	Books	Total Publications
1)	Lindgren	Professor	1 (4 ed.)	247+1068 abst.
2)	Hunt	Professor	1	361
3)	Grabau	(S.B. XII'96)	13	287
4)	Buerger	(S.M. XII'25; Ph.D.'29); Professor	12	254
5)	Hurley	(Ph.D. XII'40); Professor	1 (12 translations)	200
6)	Rogers	President; Professor	2	160
7)	Shimer	Professor	8	100+
8)	Shrock	Professor	6	100+

BOOKS

Thus far (1975) Buerger has written twelve books (ten alone and two with other authors) concerning crystallography - surely one of the most impressive achievements of any crystallographer to date. These are included in his Bibliography, and are listed here in chronological order. Together they have played a major role in changing mineralogy and crystallography from descriptive to quantitative disciplines, and in providing the means by which students could be led step by step into a deeper and deeper insight into the nature and significance of the structure of crystalline substances.

1) The Optical Identification of Crystalline Substances (‡).
    M.I.T. Letter Shop, 100 p., (1939).
2) Numerical Structure Factor Tables.
    Geol. Soc. Am. Sp. P. 33, 119 p., (1941); reprinted, (1953).

3) X-ray Crystallography.
   Wiley, 531 p., (1942).
4) The Photography of the Reciprocal Lattice.
   ASXRED Mon. 1, 37 p., (1944). (Reprinted later.)
5) Elementary Crystallography.
   Wiley, 528 p., (1956).
6) (with Azároff, L. V.) The Powder Method in X-ray Crystallography.
   McGraw-Hill, 342 p., (1958).
7) Vector Space.
   Wiley, 347 p., (1959).
8) Crystal-Structure Analysis.
   Wiley, 668 p., (1960).
9) The Precession Method in X-ray Crystallography.
   Wiley, 276 p., (1964).
10) Contemporary Crystallography.
    McGraw-Hill, 364 p., (1970).
11) Introduction to Crystal Geometry.
    McGraw-Hill, 204 p., (1971).
12) (with Amorós, J. L. and Amorós, Marisa C. de) The Laue Method.
    Academic Press, 376 p., (1975).

Why and How He Wrote Books

I have often been asked: "Why and how do scientists go about writing their books?" The answer of one author to this interesting question can be found in Buerger's comments on why and how he prepared several of his ten major textbooks. Of special interest in this connection is the statement he made to me recently:

> "Each book was written under interesting circumstances, often because the time became available for writing."
> [my underlining.]

Available time for relaxation and contemplation, coupled with unusual circumstances that stimulate an imaginative and productive mind, seem to be one answer to the question posed at the beginning of this paragraph. Time and circumstance were powerful forces in motivating Buerger's efforts, and I regard it worthwhile to record here a few of his comments, particularly for the benefit of aspiring young authors.

In his "Personal Reminiscences" (see p. 551-552 in reference 203--1962j in the Bibliography), Buerger wrote:

> "In 1937 I was a member of Donald MacMillan's Expedition to Baffin Island in the Arctic. This trip up and back gave me the leisure to start my book X-ray Crystallography. The book had two objectives. The crystallographic preliminaries of crystal-structure analysis had never been treated in any detail, but had always been compressed into a small portion of a book covering all of crystal-structure analysis and the results the art had achieved to that date. X-ray Crystallography was

intended to pay attention to the important preliminaries
of unit-cell and space-group determination. At the same
time it was written to the mineralogical crystallographers
and as a protest against their current practices. At
that time, the mineralogists made optical goniometric
studies of the face development of crystals, and from
these they attempted to deduce, on the basis of various
theoretical ideas, mostly ill-founded, what the lattice
of the crystal was. They were so confused that they
tried to distinguish between a 'morphological lattice'
and a 'structural lattice.' X-ray Crystallography was
intended to present them with a fool-proof way of finding
the lattice. I wrote a preface in which I stated my
opinion of their groping methods. The preface was too
frank to publish, so I had to write another moderate one,
which now appears with the book, but I had the pleasure
of writing what I thought of the mineralogical crystallography of the day. Writing two prefaces (the second
publishable) was a practice I continued in later books."

Elementary Crystallography (1956) was the result of consulting experiences with the du Pont Company, and the boredom of the long train ride
to get to Wilmington, Delaware, for the periodic consultations. Buerger
wrote me as follows (personal communication, 1 December 1975):

"I was invited by du Pont to consult on investigating the
tiny crystals which occurred in certain plastics. This
role expanded into another department, known as the Central Research Department, which was interested in fundamental physical and chemical properties of many things.
However, the Company at the time had no one who knew
much about crystallography. Eventually I gave a course
of lectures from time to time as I visited them. I
gained lots of converts, fussed about establishing an
x-ray laboratory ... and now, almost all of a certain
large group of chemists are using powder photographs as
a routine part of their investigations, most of them
are familiar with space groups, and not a few determine
structures which are not too complicated.

. . . . .

"As I first traveled from Boston to Wilmington by train,
a trip which took about 7 hours, I found free time on
my hands; after a few rides I tired of reading detective
stories to pass the time and so began to outline Elementary Crystallography as I rode. This began as an
outline of lectures to du Pont chemists. It resulted
in a book which was published in 1956.

"The Powder Method [with L. V. Azároff, 1958] was
written because, having assigned my teaching assistant
L. V. Azároff the job of writing a laboratory manual
for our new x-ray laboratory course for mineralogists,
I observed that he did a superb job of writing, so I
suggested that, when he was free, we should put together
a book on the powder method. We even, in our enthusiasm,
outlined a set of chapters and assigned half of them to
each of us. After Azároff left to do research at Armour
Institute in 1953, he remembered but I forgot, so I was
astounded when he wrote that he had finished a certain

part of his assignment, and how was I coming with mine? With this jolt I got busy. The post office lost our completed manuscript in the Christmas rush of 1956, and we had to get the second copy entirely retyped and checked, but we did get the book out by 1958.

"I began writing Vector Space [1959] on a long 12-day boat trip from Rio [Rio de Janeiro, Brazil] to the U.S.A. in 1949, after lecturing on this subject. I worked on it again on the 16-day boat trips to Amsterdam and return in 1957 and 1958. It was published in 1959. But by the time I landed in the U.S.A. the last time, I had already realized that another book would be required for the generalization of the subject matter. I had nearly completed this new manuscript in 1962, but had to set it aside for a sabbatical at Santiago, Chile in 1963 (This old manuscript now needs revision and I hope to get at it shortly.)

"In going to Chile [for the 1963 sabbatical], I realized that I could not take any extensive set of reference material with me so, knowing that long trips by boat to Chile and return were in store for me (the return trip from Tocopilla to Pensacola, Florida was made on a saltpeter freighter and took a full month), I would have to get the material right out of my head. Since the original ASXRED Monograph on the precession method [see book list on a preceding page] was now much out of date, I decided to write an entire new book on this now popular method of recording x-ray data. This was easy since I had devised the method and there was very little literature on it but my own. In the time occupied by freighter travel and the time when I was not actually lecturing in Chile, I prepared the first draft of the manuscript on The Precession Method in X-ray Crystallography. It was published in 1964. It is now out of print [1975].

"As my retirement approached, I suddenly realized that all my books had been written for the benefit of my own students who were studying to become professional crystallographers or mineralogists. Meanwhile ... biologists, chemists, ceramists, metallurgists and physicists who didn't need or want to know crystallography in depth, nevertheless ought to have a sufficiently authoritative account of this subject instead of having to absorb ancient and often incorrect teachings usually found in their own textbooks. To this end I tried to write for them by working on Contemporary Crystallography. I gave the manuscript to McGraw-Hill and they received good reviews of it, but then asked me to reduce its length. This I did by abridging the early chapters. After production of the text began, I resubmitted the replaced part, now entitled Introduction to Crystal Geometry and also received good reviews of this part, so it was published as a separate volume. I had intended to round out this pair of volumes to a trilogy by writing a third volume to be entitled 'Structural Crystallography,' but McGraw-Hill did a curious thing which stopped my plans: after Contemporary Crystallography had been out three years, they back-charged me ... for what they called excessive changes in correcting the proof three years earlier. I could never get a detailed explanation of this most unusual treatment ... so, with so many other calls upon my time, I did not complete the third part of the trilogy."

RESEARCH RESULTS: NEW IDEAS AND THEORIES; NEW METHODS AND PROCESSES; NEW INSTRUMENTS

Buerger likes best to think of himself as a "pure" crystallographer, meaning by the adjective "pure" that his primary interest has always been in the broader and deeper theoretical aspects of the crystalline state of matter.

It is a fact, however, that because of W. L. Bragg's influence on him while yet a student, as mentioned earlier, most of his research and of his publications, both articles and books, deal with x-ray crystallography and crystal-structure determination and analysis. Furthermore, he has devoted considerable attention to the development of new methods and new instruments, but he has always regarded these as only means to an end - the end being to gain deeper insight into and better understanding of crystal relations and behavior.

Important as the foregoing contributions have been, and they have brought him national and international renown, he has always looked and thought beyond them to the grander view of crystallography, and some of his most important contributions have implications far beyond or little related to x-ray diffraction and crystal structure.

One needs to take an overview of Buerger's complete research record in order to appreciate the variety and extent of his contributions to the broad field of crystallography. This I will now do by discussing the effective way in which he has consistently used theory as a first step in his research technique, before going on to attack the problem itself.

THEORETICAL CONTRIBUTIONS TO CRYSTALLOGRAPHY

A point to emphasize about Buerger's approach to the solution of a problem is that he first creates the theory, then proceeds to test it by invention of an instrument, development of a method or process of analysis, or application of it or them to a body of existing data. This particular approach may produce a new theory, a new or improved method or process of analysis, or a new or modified instrument for a certain kind of measurement. Buerger's impressive Bibliography, which constitutes the second part of this biographical sketch, includes many examples of each of these analytical steps - new ideas, new methods or processes, and new instruments - and the particular theory behind them. Let us now consider how he has applied theory to the three aspects of crystallography to which he has made extraordinary contributions: 1) diffraction instrumentation; 2) crystal structure analysis; and 3) crystal properties.

THEORY USEFUL IN DIFFRACTION INSTRUMENTATION

Buerger has applied his theoretical approach to instrumentation in two ways: 1) modification and improvement of existing instruments; and 2) invention of new instruments. He developed the following important theories, among others:

1) Equi-inclination theory in Weissenberg cameras.
2) Theory of precision cell-geometry determination in Weissenberg (single-crystal) cameras.
3) Theory of precession motions for a single-crystal instrument.
4) Generalized de Jong-Bouman principle, for precession camera.
5) Theory of x-ray surface-reflection fields in Weissenberg photographs, (which was extended to precession photos by Takéuchi).

The instruments and processes that he invented as a result of applying the preceding theories are discussed in an accompanying section - "Patented and Unpatented Instruments and Processes."

THEORY USEFUL IN ANALYSIS OF CRYSTAL STRUCTURES

The following list includes the more important and better known examples of theories that Buerger has developed for and used in the analysis of crystal structures - end numbers in parentheses, e.g. (99--1946b) refer to references in his Bibliography:

1) Symmetry in Patterson space: (126--1950; 240--1969b).
2) Symmetry in Fourier space (extended by Veinstein in Russia): (121--1949; 122--1949a; 126--1950).
3) Theory of image space generally - image-seeking functions, including minimum functions: (132--1950g; 136--1950k; 139--1951a; 141--1952a; 143--1953; 144--1953a; 145--1953b; 164--1956e; 172--1958c; 174--1959a; 206--1962m; 212--1963b; 224--1964e; 230--1966b).
4) Theory of reduced cells: (166--1957a; 171--1958b; 178--1960a).
5) Theory of Patterson functions of substructures (extended by Takéuchi and Morimoto): (150--1954a; 173--1959).
6) Difference-Patterson Fourier syntheses: (80--1942d).
7) Partial Fourier syntheses: (164--1956e; 177--1960).
8) Diffraction symmetries of twins: (149--1954; 177--1960).

9) Fourier summations for symmetrical crystals: (122--1949a; 177--1960).
10) Interpretation of Harker syntheses: Implication theory: (99--1946b; 116--1948f; 119--1948i; 173--1959).
11) Lorentz and polarization correction factors: (73--1940a; 88--1945a; 98--1946a).
12) Geometry of crystal projections: (227--1965b).
13) Diffraction symbols: (232--1967a; 241--1969c).
14) Algebraic representation of sets of points and their images and convolutions: (191--1961f; 193--1962).

DETERMINING THE ARRANGEMENTS OF ATOMS IN CRYSTALS

After W. H. and W. L. Bragg demonstrated that the arrangements of atoms in crystals could be determined by x-ray diffraction data, interest in crystallography increased greatly, and Buerger was one of the earliest to become interested. As mentioned in a preceding section, Buerger attended the lectures by W. L. Bragg, when the latter came to M.I.T. as a Visiting Lecturer in 1927, and became convinced that an understanding of minerals and their inter-relations had to be based upon the arrangements of their atoms.

Up to this time Buerger had been studying certain physical properties of minerals, especially translation-gliding, but once he heard Bragg he became fascinated by the possibility of determining the crystal structure of minerals by x-ray diffraction. Lacking any facilities of his own, he got permission from his fellow graduate student in physics, B. E. Warren, to use his equipment. He quickly solved the structure of marcasite, and the great delight he experienced at the certainty that he had conclusively solved the structure gave him the kind of exhilaration that a boy might have upon entering a candy shop full of goodies for the taking.

Soon funds were procured for his own x-ray diffraction laboratory, and with his doctorate behind him, and an appointment as an assistant professor, he was launched on a career in crystal-structure analysis that would continue for the next 35 years. During this period he and his graduate students from around the world solved the crystal structure of the following 21 minerals and 5 non-mineral crystals (the number or numbers in parentheses following the name of the mineral refer to references in the Bibliography farther on in which that mineral or substance is discussed):

    marcasite (13, 51, 60)
    lollingite (15, 60)
    arsenopyrite (38, 44, 46, 60, 69)
    cubanite (39, 62, 94, 106, 155)
    berthierite (42, 156)
    manganite (47)
    valentinite (61)

tourmaline (63, 115, 133, 201, 202)
gudmundite (68, 69)
chalcocite (85, 199, 213, 215)
nepheline (100, 152, 154, 158)
livingstonite (146, 147)
pectolite (163, 190, 197, 216)
jamesonite (168)
coesite (172, 174)
cahnite (184, 188)
narsarsukite (200)
bustamite (205)
rhodizite (228, 234, 235)
serendibite (253)
rhodonite

Co_2S_3 (157)
diglycine hydrobromide (162)
diglycine hydrochloride (165)
potassium hexatitanite (207)
terramycin hydrochloride (182)

In addition to the preceding, Buerger and his students have published substantial structural information on the following 12 minerals:

realgar (28, 43)
claudetite (77)
kaliophilite (78)
orpiment (79)
tridymite (83, 237)
pyrrhotite (96, 105, 198)
wollastonite (160, 190, 197, 216)
pentlandite (161)
andalusite (183, 186)
pollucite (231)
pharmacosiderite (233, 236)
warwickite (249)

Furthermore, as might be expected, Buerger's doctorates have continued on their own to determine and describe the heretofore unknown crystal structures of numerous minerals.

THEORY RELATED TO CRYSTAL PROPERTIES

Many of his peers regard as the most important of Buerger's contributions the theories that he has advanced that are related to the chemical and physical properties of crystals. These are examples of Buerger's lifelong primary interest in what he likes to call "pure crystallography." They tend to come after the instruments have been constructed, the measurements made, and the accumulated data correlated. They are, in fact, the foundation blocks on which future advances are based.

Among the more important of his numerous theories in this category, he himself would list the following, not necessarily in order of decreasing importance - end numbers in parentheses, *e.g.* (25--1934e) refer to references in his Bibliography:

1) Theory of translation gliding and twin gliding: (4--1928; 5--1928a; 9--1930; 10--1930a).

2) Lineage structure in crystals: (16--1932a; 18--1932c; 25--1934e).

3) Genesis of twin crystals: (89--1954b; 179--1960b; 180--1960c).

4) Theory of phase transformations - displacive and reconstructive transformations: (112--1948b; 138--1951; 185--1961; 245--1971; 247--1971b).

5) Crystal polymorphism: (49--1936n; 50--1936o; 54--1937c; 55--1937d; 58--1937g; 185--1961).

6) Stuffed derivative structures: (29--1935a; 30--1935b; 102--1947; 151--1954b).

7) Importance of the several faces of a crystal: (107--1947e).

8) Symmetries of superstructure crystals: (44--1936i; 46--1936k).

9) Theory of exsolution textures: (22--1934b).

10) Role of temperature in crystal growth: (109--1947g; 112--1948b).

11) Significance of disorder in crystals: 109--1974g; 112--1948b).

PATENTED AND UNPATENTED INSTRUMENTS AND PROCESSES

Buerger has invented a number of ingenious devices and processes of particular use in his crystallographic research. Several of these are illustrated on accompanying pages, and the following have been patented:

2,362,430 - Nov. 7, 1944 - Martin J. Buerger, Arlington, Mass.
"Articles of nonmetallic mineral compounds and methods of producing same"

2,460,334 - Feb. 1, 1949 - Martin J. Buerger, Lincoln, and Edward Washken, Cambridge, Mass.
"Process of making bonded structures"

2,548,344 - Apr. 10, 1951 - Martin J. Buerger, Lincoln, and Edward Washken, Cambridge, Mass.
"Process of cementing plastically deformable bodies and products thereof"

3,108,185 - Oct. 22, 1963 - Martin J. Buerger, Weston Road, Lincoln, Mass.
"Precession instrument for use in the photography of the reciprocal lattice of a crystal"

Among the unpatented items may be listed the following which were developed in the chronological order indicated. The last figure or figures enclosed in ( ) refer to references in Buerger's Bibliography. The prototype of the items marked with an asterisk (*) has been deposited in the Smithsonian Institution in Washington, D.C. The prototype or dupli-

576 GEOLOGY AT M.I.T. 1865-1965

1

2

3

4

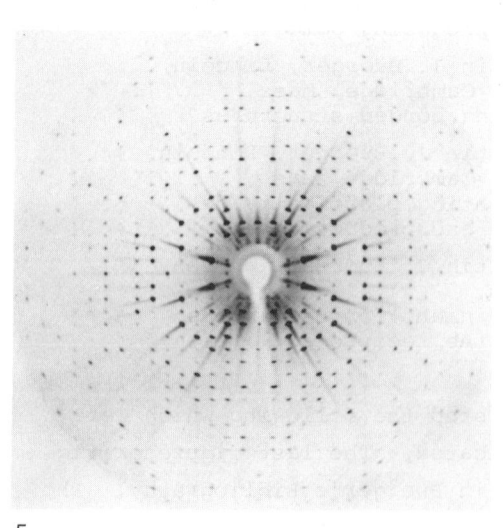

5

6

PRECESSION CAMERAS

cate of items marked with a dagger (†) has been deposited in the M.I.T. Historical Collections, and items now in the possession of Prof. Bernhardt J. Wuensch of M.I.T. (Room 13-4037) will ultimately be deposited in the same Collections.

1) "Very early American" oscillating-crystal camera (unpublished, but built about 1934; whereabouts unknown).

2)† American version of oscillating-crystal camera (built about 1936) - Described on p. 188-193 of X-ray Crystallography (76--1942).

3)* Equi-inclination Weissenberg camera (24--1934d; see also Zeit. für Kristallogr. 88: 195-220, 1934; 41--1936f).

4)† Device for interpreting Weissenberg photographs (31--1935c; 32--1935d).

5)* Gas x-ray tube (33--1935e).

6)* "Early American" powder camera (built about 1933) (35--1936).

7)* Goniometer ball driller for crystal models, Mark I (34--1935f; 37--1936b; 66--1938a).

8)† Goniometer ball driller for crystal models, Mark II (unpublished; 1936).

9)* Precision back-reflection Weissenberg (59--1937).

10)* Earliest (Mark I) precession camera (constructed about 1938 or 1939) - Described on p. 206-211 of X-ray Crystallography (76--142).

PRECESSION CAMERAS

1. Mark I, prototype. (Ref. 76--1942; X-ray Crystallography, p. 206-211, 1942. Ref. 220--1954a; Z. Kristal. 120: 11-12, 1964. Ref. 221--1964b; The Precession Method in X-ray Crystallography, p. 4-7, 1964.)

2. Mark II, first version (on which settings are being made by Dr. Luisa Brahde, for taking a photograph). (Ref. 86--1944a; The Photography of the Reciprocal Lattice, ASXRED Mon. 1: 30-34, 1944.)

3. Mark II, second version. (Ref. 86--1944a; Ibid., ASXRED Mon. 1: 34-41. Ref. 243--1970; Contemporary Crystallography, p. 149-185, 1970.)

4. Device for measuring the linear and angular parameters of the reciprocal lattice as they are available on precession photographs. (Ref. 221--1964b; The Precession Method in X-ray Crystallography, p. 184-185, 1970.)

5. Precession photograph of the h̄k0 net of the reciprocal lattice of berthierite, $FeSb_2S_4$.

6. Dr. Milan Rieder (of the Universita Karlava, Prague, Czechoslovakia) and Dr. Martin Buerger discussing Dr. Rieder's back-reflection precession camera which is based on the geometry given in The Precession Method in X-ray Crystallography, p. 24, 1970. The photograph was taken by Dr. Rieder on August 12, 1975, in Amsterdam, on the occasion of the 10th Congress of the International Union of Crystallography.

(M.I.T. Photos, other than 6)

1

2

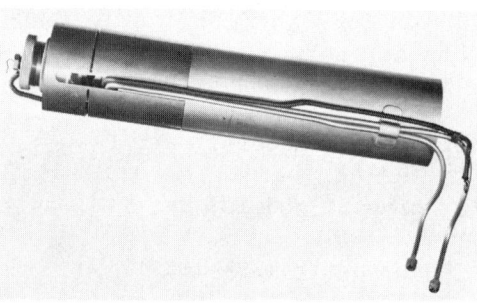

3

4

5

WEISSENBERG CAMERAS

1. Mark I equi-inclination Weissenberg camera. (Ref. 24--1934d; Z. Kristal. 88: 356-380, 1934, Ref. 41--1936g; Z. Kristal. 94: 87-99, 1936.)
2. Mark II equi-inclination Weissenberg camera. (Ref. 76--1942; X-ray Crystallography, p. 252-295, 1942.)
3,4. Accessories which attach to the layer-line screen which permit taking Weissenberg photographs at controlled, elevated temperatures. (Details unpublished, but noted as part of the research in Ref. 85--1944; Am. Mineral. 29: 55-65, 1944.)
5. Device for transforming locations of spots on an equi-inclination photograph to positions of points of the reciprocal lattice of the crystal. (Ref. 24--1934d; Z. Kristal. 88: 356-380, 1934.)

(M.I.T. Photos)

11)* American version of the de Jong-Bouman retigraph (constructed in 1938-1939) - Described on p. 331-346 of X-ray Crystallography (76--142).

12)*† Mark II Precession camera (86--1944a).

13) Two-wavelength microscope (built about 1940; now at the University of Southern Illinois in the Materials Science Division) (74--1941; 132--1950g; 134--1950i).

14)† High temperature powder camera (84--1943).

15)* Single-crystal diffractometer, Mark I (built in 1952, unpublished).

16) Single-crystal diffractometer, Mark II (built in 1953, unpublished). Item now in Wuensch's possession.

17) Accessory for changing crystal orientation on the diffractometer (built in 1953, unpublished). Item now in Wuensch's possession.

18) Automated single-crystal diffractometer (220--1964a). Manufactured by Charles Supper Company of Wellesley, Massachusetts.

19)† Dawton photometer: for determining the integrated intensities of x-ray reflection from x-ray photographs (built about 1940) (p. 95-102, fig. 15-16, in 175--1960).

20)† Ellipsograph: device for drawing ellipses, two models (built in 1950, unpublished).

21) Various devices for examining, orienting, and adjusting crystals for x-ray examinations, including a device for transferring a crystal from one goniometer head to another (unpublished). Last device now in Wuensch's possession.

22)† Device for cutting 5" × 7" photographic film into perfect circular form (designed about 1942).

23)† Device for measuring positions of lines on powder photographs (designed in the early 1940s).

24)† Devices for measuring positions of diffraction spots on Weissenberg photographs and then plotting them in reciprocal space, including special paper for plotting.

25)† Device for measuring spacings and angles between lines of spots in photographs of the reciprocal-lattice photographs, especially precession photographs. Now manufactured by Charles Supper Company of Wellesley, and copied by many other makers of x-ray diffraction apparatus.

The ball driller for making crystal models (#8 in preceding list) is now made by several manufacturers. The equi-inclination Weissenberg powder and precession camera is now manufactured in seven or more countries.

1

2

3

4

5

CYLINDRICAL X-RAY POWDER-DIFFRACTION CAMERAS

The improved version of the x-ray camera was copied by North American Philips and is now the American version of the powder camera. All the devices for measuring x-ray photographs have been copied by most manufacturers of x-ray diffraction equipment.

HOW HE CAME TO INVENT INSTRUMENTS

An intriguing question that commonly arises regarding new instruments and their inventors is: "How does an inventor get the idea for a new instrument, or for an improvement in an existing one?" Some insight into how invention actually takes place can be gained from what Buerger records about his own precession camera, a case in which the old proverb, "Necessity is the mother of invention," is apropos. In his "Personal Reminiscences" (see p. 554 in reference 203--1962j in the Bibliography) he writes:

> "The precession camera was developed as I was writing the part of Chapter 10 of X-ray Crystallography concerned with the limited symmetry information obtainable from oscillation photographs. It was evident that the symmetry of the oscillation motion limited the symmetry of the record. Why not increase the symmetry of the motion? Precession ideally provided a radially symmmetrical motion. The oscillation and precession motions are compared in a figure on p. 207 of X-ray Crystallography. It was easy to make the first camera, shown on p. 208, and the first photographs, shown on p. 209, justified

---

CYLINDRICAL X-RAY POWDER-DIFFRACTION CAMERAS

1. Set of four Mark I powder cameras on their mounts, arranged to receive x-ray beams from the four windows of a Hadding-type cold-cathode gas x-ray tube. The tube is partly described in Rev. Sci. Instruments 6: 385-386, (1935); the cameras are described in Amer. Mineralogist 21: 11-17, (1936).
2. Early form of Mark II powder cameras designed to photograph only the powder lines in the regions 0 to $90°$, since there are rarely lines beyond this for organic crystalline substances like soaps.
3. Internal construction of the camera illustrated in 2, showing an early form of the device for spreading the film against the inside of the camera cylinder, and an old form of the specimen-centering device. (See J. Appl. Phys. 10: 417-428, 1945.)
4. Set of two Mark II powder cameras and their mounts arranged to receive the x-ray beams from the line-focus ports of a Machlett hot-cathode x-ray tube. (See J. Appl. Phys. 10: 417-428, 1945.)
5. Internal construction of a Mark II powder camera, showing the device for maintaining the film in firm contact with the internal circumference of the cylinder, the powder-specimen holder, the device for centering the specimen, and the cones for reducing x-ray scattering by air to the photographic film near the entrance and exit ports (left and right, respectively).

(M.I.T. Photos)

the whole idea. All this was carried out in 1937. In 1938 de Jong and Bouman showed how to avoid radial distortion as well, and this improvement was quickly added to the primitive precession camera to give the first precession camera as we now know it. This just missed appearing in X-ray Crystallography, the manuscript of which was finished in 1940. It was described in Monograph No. 1, 1949, of the American Society for X-ray and Electron Diffraction."

PUBLISHING AND EDITORIAL ACTIVITIES

Reference is made in an earlier section to the fact that Martin personally published the Memoirs of Ernst Moritz Buerger, his great grandfather, who had led his congregation of Lutherans from Saxony to Perry County, Missouri, in 1838.

Buerger also took an active interest in getting several scientific publications established, as pointed out in the following paragraphs, and this interest led not only to the establishment of new publications but also to the founding of several crystallographic societies. As he recalls in his "Personal Reminiscences" (see p. 554-555 in reference 203--1962j in the Bibliography):

"Before World War II, most crystallographers published their serious papers in the Zeitschrift für Kristallographie. Editors of journals in the better recognized sciences did not always encourage crystallographic contributions. About 1939 I began to agitate among my colleagues for the establishment in the United States of a Journal of Crystallography. There was some support for the idea, but most were afraid to go ahead for fear of financial troubles, and certain societies, notably the Mineralogical Society of America, began to be alarmed at possible defection of some of its membership. Maurice L. Huggins and I corresponded about the possibility of a journal, and eventually I found myself a chairman of a subcommittee of the National Research Council Division of Chemistry charged with investigating the possibility of establishing a journal. In the end, the Americans did not want to proceed alone, but the episode drew together a group of X-ray crystallographers and this led to their forming a society, to be called the American Society for X-ray and Electron Diffraction (ASXRED), in July, 1941, at Gibson Island. I believe that Huggins and I pushed the rest into this."

When the American Society for X-ray and Electron Diffraction (ASXRED) was founded, the members had no place to publish their research, so Buerger headed a committee to seek funds for and establish a monograph series. His own The Photography of the Reciprocal Lattice (see reference 86--1944a in the Bibliography), ASXRED Monograph No. 1 in the series, was published by the Society in 1944. Subsequently, the Society published one other monograph; thereafter the American Crystallographic Association took over publication responsibility, published three other epoch-making monographs, and is currently seeking other appropriate manuscripts.

As mentioned in a preceding section, Buerger has served for almost thirty years as a Co-Editor of the IUCr's International Tables for X-ray Crystallography, the third edition of which is expected to be published in 1976. Among other things he designed and closely supervised preparation of many of the illustrations for this latest edition, and helped find funds to finance the preparation.

He has long been a Co-Editor of the Zeitschrift für Kristallographie (1953 to the present, 1975).

Some years ago he was induced to initiate and act as Editor of a series of Monographs in Crystallography for John Wiley & Sons, Inc. Thus far four monographs have been published under his editorship, one is in press (Dec. 1975), and two manuscripts are in preparation. The monographs that have reached publication are:

1) Polymorphism and Polytypism in Crystals, by A. R. Verma and P. Krishnan, (1966), and reviewed in Science 154: 1316-1317, (9 December 1966) by J. A. Kohn.

2) Molecular Crystals: Their Transforms and Diffuse Scattering, by José Luis Amorós and Marisa Amorós, (1968).

3) Fourier Methods in Crystallography, by G. N. Ramachandran and R. Scrinivason, (1970).

4) Color and Symmetry, by A. L. Loeb, (1971).

5) Crystallographic Groups of Four-Dimensional Space, by H. Brown, R. Bülow, J. Neübuser, H. Wondrotschek, and H. Zassenhaus, (in press, Dec. 1975).

AS A CO-FOUNDER AND MEMBER OF PROFESSIONAL ORGANIZATIONS

Membership in a professional society or association should imply a willingness to participate actively in the affairs of the organization. Buerger has amply met this obligation in those academies, societies, and associations in which he has held membership. What is even more important, in my opinion, is that he was among the first to agitate for, and assist as a co-founder in establishing, several of the earliest organizations designed to bring x-ray crystallographers together - e.g. the Crystallographic Society of America (CSA), The American Society for X-ray and Electron Diffraction (ASXRED), the American Crystallographic Association (ACA), the International Mineralogical Association (IMA),* and the International

---

* Buerger single handedly initiated the idea of the IMA and only took the idea to the Mineralogical Society of America to get wider support for it, which he did get.

Union of Crystallography (IUCr). Fortunately for the historical record, Buerger recently (19 August 1975) reviewed, at a meeting of the ACA at Pennsylvania State University, his own activities, as well as those of his colleagues, in founding the organizations mentioned above. He has kindly granted me permission to reproduce the lecture he gave on that occasion, and I am deeply grateful to him because his discussion not only involves history in the making - he was there!; it also records how important his own leadership was, and includes some interesting thoughts about the future potential of x-ray crystallography. His lecture follows:

Background and Early History of the American Crystallographic Association
by Martin Buerger
University of Connecticut, Storrs, Connecticut 06268

The American Crystallographic Association began life on January 1, 1950. It held its first meeting on April 10-12 of that year at Pennsylvania State College in State College, Pennsylvania. Today, a little over a quarter of a century later the society holds its 25th meeting in the same locality, to find that the place is called University Park, the home of Pennsylvania State University.

Six years after its first meeting the origin of the ACA was outlined by William Parrish and Betty Wood in volume IV of the Norelco Reporter. There it is recorded that the ACA is the successor of the American Society of X-Ray and Electron Diffraction and the Crystallographic Society of America. These two societies, in turn, had begun their lives as the second world war was beginning to disturb the world. Two separate groups had begun to raise the question of the usefulness of establishing a journal for crystallographic research and this eventually stimulated the formation of the International Union of Crystallography which then not only took over the International Tables for X-Ray Crystallography, but a short time later established Acta Crystallographica. Few of you here recall the birth of the ACA, and even fewer remember the origins and acts of the two societies which eventually merged to become the ACA. It seemed useful, then, while there are a few left who took part in the acts which led to our beginning, to briefly outline the history on this occasion.

One pre-existing branch of the ACA, the Crystallographic Society of America, had its beginning as a local organization of crystallographers in Cambridge, Massachusetts and vicinity. As I recall it, at first most of its members were a few mineralogists from the neighborhood who felt that the future of mineralogy lay in understanding the roles of crystal structures in the properties, genesis and relations between minerals. The membership included Harry Berman, Newton Buerger, Martin Buerger, William Dennen, Clifford Frondel, Cornelius Hurlbut, Joseph Lukesh as well as the chemists Cutler West, I. Fankuchen and others. According to a remark in a letter from M. J. Buerger to Paul Kerr dated May 1, 1941, this group began to meet in the winter of 1939 for the purpose of discussing their research results with each other. The meager records which remain show that I. Fankuchen addressed the second meeting, held May 1, 1941 at Harvard University, on the subject "Preparation and Handling of Small Crystals" in which, among other things, the prototype of all twiddlers was described. Volume 95 of Science, which appeared on January 2, 1942, contains a note from the secretary of the society, Clifford Frondel, to the effect that after the business meeting of November 17, 1941, held at M.I.T., Joseph Lukesh addressed the assembled crystallographers on "The Tridymite Problem." On the occasion of the fifth meeting on April 24, 1942, at M.I.T., Percy Bridgman gave an invited lecture on a crystallographic aspect of his high-pressure work, which he entitled "Polymorphism at High Pressures." After this there appears to have been a hiatus in the activities of the society due to preoccupation with war work. During this period the society lost its vice-president, Harry Berman, as a casualty of the war. When the war finally ended the society prepared for a large meeting, held at Smith College on March 21-23, 1946. On this occasion the small local society of crystallographers assumed the status of a national society. At the Smith College meeting, papers were presented by Richard S. Bear, J. D. H. Donnay, Howard Evans, I. Fankuchen, Samuel Gordon, Joseph Lukesh, Dan McLachlan, Benjamin Schaub, Newman Thibault, Edward Washken, Cutler West, and Dorothy Wrinch. The abstracts of their papers were published in the American Mineralogist, as were the abstracts of papers given at later meetings.

From the time of the earliest meetings of the Cambridge crystallographers, the possibility of launching a Journal of Crystallography was considered; indeed an estimate of possible subscribers was made and the financing of the undertaking discussed. Unfortunately it was found that a subsidy would be necessary to begin publication, and the society had no success in arranging for one.

The crystallographers of the American Crystallographic Society, which included biologists, ceramists, chemists, mineralogists and physicists, had banded together to present the results of their crystallographic research to one another because their own professional societies offered them little encouragement and limited discussion. In such meetings their results were overwhelmed by the routine and classical results, and their own results were little appreciated.

Meanwhile certain chemists had found some relief from this situation. Many interested in the new results coming from the study of the atomic arrangement in solid matter were members of the National Research Council's Division of Chemistry, Committee on X-Ray and Electron Diffraction, then chaired by Maurice L. Huggins. Stimulated by the efforts of the ACS to establish a journal devoted to crystallography, Huggins called a conference under the auspices of the N.R.C. Committee on X-Ray and Electron Diffraction to be held at the American Museum of Natural History in New York, June 10-11, 1941. Although no action was taken to go ahead with the journal, the attendants did agree to form a society of structure researchers. The NRC Committee on X-ray and Electron Diffraction (to which I had been appointed in 1940) took the initiative in organizing the society, and its constitution was adopted at the Gibson Island meeting of July 30, 1941. Huggins was elected its first president and Warren its first vice-president, to succeed to the presidency the next year. The second meeting was held in Cambridge, Massachusetts, December 31, 1941. Meanwhile, many names had been considered for the new society; the one finally chosen by written ballot from the many submitted was "The American Society for X-Ray and Electron Diffraction," which was abbreviated ASXRED. Thus the lineage of the ACA took its name from the name of the committee which brought it into existence.

There were now two societies concerned with different aspects of crystallography. Some believed that the societies were in competition and felt it would be to the advantage of both to join forces. Others pointed out that the ASXRED was named after a tool which, though important to crystallographic testing and research, did not begin to represent the whole science of crystallography, and could, indeed, pass out of existence, as did the optical goniometer. In 1948 a joint committee of the ASXRED and the CSA was formed to consider the consolidation of the two societies. Their report was sent to both societies. The membership of the ASXRED discussed the proposal for joining on December 19, 1948 and the CSA on April 8, 1949. Both memberships voted for consolidation. The two societies merged into the American Crystallographic Association on January 1, 1950. The new society held its first meeting on April 10-12 at Pennsylvania State College, as a continuation of the Conference on Computing Methods and the Phase Problem, which had been organized by Ray Pepinsky.

From all this it is apparent that, while the ACA is formally 25 years old, its roots range farther back. The ASXRED and the CSA are not the parents of the ACA, but rather they are earlier divisions of the ACA, so that our beginnings extend back at least to 1938 and 1941.

From this broader point of view, what has been accomplished until now? Perhaps most important is that the interests of the root societies in a journal devoted to crystallography was the stimulus which eventually launched the International Union of Crystallography and Acta Crystallographica. This came about in the following way: As president of the ASXRED in 1943, I realized the need of a vehicle to publish articles too long for the ordinary journal. Accordingly, I appointed a committee consisting of J. D. H. Donnay, George Tunell, and myself to see what could be done. We circulated a memorandum to the membership of the ASXRED sounding them out on their interest in a possible monograph series. They authorized us to go ahead. To finance the undertaking, I solicited contributions from companies who were in the business of making x-ray diffraction equipment and received support from The General Electric X-Ray Corp., Machlett Laboratories, North American Philips, and Picker X-Ray Corp. These contributions were placed in a revolving fund. The monographs were distributed free to all members of the ASXRED but sold to all others. Two monographs were published by the ASXRED before the advent of the ACA. The ACS published one monograph during the same period.

On instruction from the membership of the ASXRED, and as chairman of the Monograph Committee, I wrote on October 16, 1944 to H. Lipson, Secretary of the X-Ray Analysis Group of Cambridge, England, telling them of our Monograph Series and informing them of our interest in establishing a journal. I expressed the hope that the X-Ray Analysis Group would join with us in this project. Lipson called a meeting of that group on November 18, 1944 to consider this and other business. Eventually Sir Lawrence Bragg, chairman of the X-Ray Analysis Group, suggested that the Americans send a delegation to England to discuss the possibility of publishing a journal and to consider the formation of an International Union of Crystallography. The meeting was eventually set for July 12 and 13, 1946. The American delegation included Fankuchen, Germer, Harker, McLachlan, Zachariasen and myself. From this meeting arose the International Union of Crystallography, Acta Crystallographica and plans for a new edition of the International Tables.

Among the several noteworthy accomplishments of the ACA are the solution of numerous crystal structures whose reports have so filled the pages of Acta Crystallographica that the journal had to be split into two parts per volume and other material separated into another journal. It would be interesting to see an analysis of these contributions, but I make no attempt to do that here. I would, however, like to note that in one field of crystallography, the main channel has been outlined by members of the ASXRED, although contributions have come from many countries, namely, the direct methods of crystal-structure analysis.

I believe I had the honor of presenting the break which led to the hope of developing direct methods. At the Lake George meeting of the ASXRED in 1946 I presented the <u>implication diagram</u>. This results from a simple characteristic rotation and shrinking of the Harker section of the Patterson function. It has the property, in favorable space groups, of mapping the locations of atoms in a projection of the crystal structure. In less favorable space groups this desired result is accompanied by certain ambiguities and satellites. I demonstrated the use of the theory by solving the structure of the mineral nepheline, space group $\underline{P}6_3$. That a structure could be solved by use of reflection magnitudes alone was a surprising result and it was greeted with no little incredulity. But Fankuchen immediately pointed out that since this could be done it implied that there must exist phase information hidden among the collection of reflection magnitudes. This was a stunning conclusion, for up to this time it had been believed that since the phases were experimentally unobservable, they were hopelessly missing and there was not any direct route to the solution of a crystal structure. By the time of the next ASXRED meeting at St. Margarite in 1947, the Harker- Kasper inequalities were presented. These provided phase information under certain conditions, so there could no longer be any doubt that phase information was contained in the set of reflection magnitudes.

In subsequent meetings of the ASXRED many kinds of inequalities were reported. It is interesting that at the Conference on Computing Methods and the Phase Problem just preceding the first meeting of the ACA here in 1950, David Sayre showed that the comparison of the electron density with its square revealed that there existed some quite simple relations between the signs of certain structure factors. Sayre's conclusions inspired a spate of theoretical investigations of sign relations between structure factors. These culminated in the statistical work by Herbert Hauptman and Jerome Karle entitled "The Solution of the Phase Problem," published as Monograph No. 3 of the ACA. Later this was followed by a long series of papers, mostly in Acta Crystallographica by Isabella Karle and J. L. Karle, which taught crystallographers a routine called the "Symbolic-addition method" which led to a direct determination of the crystal structure.

Although the "direct" route from diffraction intensities to the crystal structure is commonly thought to lead only through Fourier space, it may also be followed through Patterson space. In the same Conference on Computing Methods and the Phase Problem 25 years ago, I showed that a Patterson function can be decomposed into images of the crystal structure. In another paper in Acta Crystallographica entitled "A New Approach to Crystal-Structure Analysis" I presented image-seeking functions by which a Patterson function can be transformed into an approximation to the electron-density function. Although many crystallographers have used this "direct" route through Patterson space, the route is less popular than the symbolic-addition method, probably because it has not been reduced to a computing routine. It is, however, a powerful "direct" method which has had no difficulty in solving an inorganic structure which has 15 non-heavy atoms and 20 oxygen atoms per asymmetric unit.

I have presented an outline of how the ACA arose from its two root societies, and some of the achievements of that period. Having solved some of the problems of the past so thoroughly that the solutions are available to other sciences as computer routines, let us not make the mistake of believing that crystallography has reached its zenith and is coasting downhill. Like all other sciences crystallography is open-ended, and the solution of crystal structures by diffraction is not its sole objective. I need only mention another phase problem as an example. It should be possible to theoretically predict the structure of a phase in any phase field, and not have to discover it experimentally. At present we can not even do this for one-component systems. Understanding phase fields and their structures is one of the many problems which should engage our attention during the next quarter century. Ladies and gentlemen of the American Crystallographic Association, please present your report on this phase problem before the end of nineteen hundred ninety nine.

From the preceding account, it is to be noted that Buerger was president of the CSA from its founding, in 1939, to 1946, and served the ASXRED as its third president in 1948. When the CSA and ASXRED merged to form the ACA on 1 January 1950, Buerger became an active member of the new Association and in 1971 received its Isador Fankuchen Award.

It should be noted also that he was invited by Sir. Lawrence Bragg to help found an International Union of Crystallography and was one of the six members of the delegation that went to England in the summer of 1946 to attend the preliminary organizational meeting. The results of that historic meeting were the founding of the IUCr, the establishment of its journal, Acta Crystallographica, and the planning for taking responsibility for a new edition, the second, of the International Tables for X-ray Crystallography.

Starting with the aforementioned meeting in 1946, Buerger served the Union as a Council Member from 1946 to 1951, and has been a Co-Editor of the International Tables for X-ray Crystallography since 1946. During the summer of 1975 he supervised preparation of a substantial part of the third edition of the International Tables and personally carried certain of the illustrations and accompanying typescript to the publisher in Europe. He served the U.S.A. National Committee on Crystallography in 1951, 1952 and 1956-1962.

He joined the Mineralogical Society of America in 1926, was elected a fellow in 1929, served as Vice-President in 1942, and was elected President in 1947. He was awarded the Society's prestigious Roebling Medal in 1958, as mentioned elsewhere in this sketch. He was the M.S.A. Representative of the United States to the International Mineralogical Association from 1968 to 1970.

He was elected a fellow of the Geological Society of America in 1946, served as a Vice-President in 1948, and was awarded the Society's Day Medal in 1951, as discussed elsewhere in this biographical sketch. His Numerical Structural Factor Tables was published by the Society as Special Paper No. 33 (see reference 75--1941a in the Bibliography) in 1941 and reprinted in 1953.

HONORS

Greatness in scientific as well as in other achievements is honored in the United States, as in other countries, by a variety of awards and recognitions - medals; honorary academic degrees; honorary memberships in academies, societies, and other similar organizations; election to office in professional societies; and having something named after one's self (in Martin's case, Buerger Bay in Baffin Island and Buergerite, a variety of the mineral tourmaline). Buerger has earned a fair share of such honors, of which the following are the more important ones.

MEDALS

Buerger was designated the recipient of the Arthur L. Day Medal of the Geological Society of America for 1951. This medal is awarded:

"in recognition of outstanding distinction in contributing to geologic knowledge through the application of physics and chemistry to the solution of geologic problems."

The medal was awarded at the annual meeting of the Society held in Detroit, Michigan, Buerger's birthplace, in November 1951. The Citation by H. H. Hess and the Response by Buerger were later published in the Proc. Vol. Geol. Soc. Am. An. Rept. 1951, p. 55-56, port., (July 1952).

He was chosen the 1958 recipient of the Roebling Medal of the Mineralogical Society of America, the highest honor that Society confers, and was presented the medal at the annual meeting of the Society in St. Louis in November 1958. The Presentation by Clifford Frondel (Ph.D. XII 1939), Buerger's first doctorate in mineralogy, to whom Martin himself presented the same medal in 1965 (see reference 224--1965a in the Bibliography), and the Acceptance by Buerger, were later published in Am. Mineralogist 44: 391-394, port., (March-April 1959).

HONORARY DEGREE

In appreciation of his outstanding work in crystallography, the Faculty of Sciences of the University of Berne, Switzerland, elected him a Doctor honoris causa (Dr.h.c.) in November 1958. The Dean of the Faculty at the time of the election was Prof. Werner Nowacki, who had been a longtime foreign colleague of Martin's. The citation* was published in Der Bund (Bern) in the issue for Monday, 24 November 1958, and short notices appeared in The Tech for Tuesday, December 9, 1958, and in The M.I.T. Observer 5/3, (December 1958).

MEMBERSHIP AND ACTIVITIES IN AMERICAN ACADEMIES

Buerger was elected a fellow of the American Academy of Arts and Sciences in 1945, and served as a Councilor from 1950 to 1954.

He was elected a member of the National Academy of Sciences in April 1953. The National Research Council of the Academy appointed him a Voting Delegate to represent the U.S.A. at the First Congress of the International Union of Crystallography held at Harvard University in July and August 1948. Subsequently, he served as the Academy's Delegate to the following General Assemblies: Second (Stockholm, 1951), Third (Paris, 1954), Fifth (Cambridge, England, 1960), and Seventh (Moscow, 1966). He was also a

---

* "In Anerkennung seiner grossen Verdienste um die Entwicklung der neueren Kristallographie und Kristallstrukturlehre in theoretischer und experimenteller Hinsicht, als Erforscher zalreicher Kristallstrukturen und als Verfasser von Lehrbüchern neuartiger Richtung."

Representative of the Academy at the XXIst Session of the International Geological Congress, held in Copenhagen in August 1960, and served as a member of the National Research Council's Committee on Solids for two different periods, 1949-1953 and 1956-1962.

MEMBERSHIPS IN FOREIGN SCIENTIFIC SOCIETIES

Buerger has been elected to honorary membership in eight foreign scientific academies, which indicates the recognition his contributions to crystallography have received abroad.

In 1952 he was elected a Foreign Member of the Academia Brazileira de Ciencias (Brazilian Academy of Sciences) in recognition of his contributions to science and most helpful collaboration with Brazilian research workers.

In March 1954 he was notified that he had been elected a Foreign Member of the Academia delle Scienze di Torino (Academy of Sciences of Torino).

In September 1957, his longtime Swiss mineralogical colleague, Prof. Fritz Laves, informed him that he had been elected to

> "Ehrenmitglied der Deutschen Mineralogischen Gesellschaft"

in Zurich. This election as an Honorary Member of the Society was:

> "in Anerkennung Ihrer bedeutenden wissenschaftlichen Leistungen auf Gebieten, die von der Mineralogie bis zur theoretischen Kristallstrukturforschung reichen; und in Anerkennung Ihres persönalichen und erfolgreichen Einsatzes für die internationale Zusammenarbeit aller Mineralogen."

In June 1960 he was elected a "Socio straniero" (Foreign Member) of the "Academia Nazionale dei Lincei" (Rome) in the "Classe di Scienze fisiche, matematiche e naturali."

In February 1961 Buerger was notified that he had been elected a "Korrespondierenden Mitglied" (Honorary Corresponding Member) in the Mathematical and Natural Sciences class of the Bayerischen Akademie der Wissenschaften in Munich.

May 1962 brought the notice that he had been elected an Honorary Corresponding Member of the Österreichische Akademie der Wissenschaften" (Vienna), again in "der mathematisch-naturwissenschaftichen Klasse der Akademie."

The latest notice came from Madrid in September 1966 informing him that he had been elected a "Miembro Honorio" of the Real Sociedad Española de Historia Natural."

BUERGER BAY IN FROBISHER BAY, BAFFIN ISLAND, CANADIAN ARCTIC

Buerger was a geologist member of the Donald MacMillan Expedition to Frobisher Bay, Baffin Island, in the Canadian Arctic in the summer of 1937. MacMillan attached Buerger's name to a small bay at the base of Terra Nivea Glacier along the southwest coast of Frobisher Bay at about Lat. 62°20' N. and Long. 66°30' W. The bay is shown on Chart No. 5854 (Canada: South Coast of Baffin Island; Frobisher Bay) of the Hydrographic Office, U.S. Navy, issued in May 1943. (See chart on next page.) MacMillan informed Buerger by a letter dated May 28, 1943 that he had

> "... taken the liberty of attaching your [i.e. Buerger's] name to a bay on the south side [of Frobisher Bay], which I hope some day to sound out and explore and map correctly."

After his return from the expedition Buerger prepared an illustrated article, "Spectacular Frobisher Bay," which was published in The Technology Review for 1938 and later reprinted in the Canadian Geographic Journal for the same year. (See reference 67--1938b in the Bibliography.)

BUERGERITE, A FERRIC VARIETY OF TOURMALINE

In early 1965 the Commission on New Minerals and Mineral Names, of the International Mineralogical Association (IMA), approved the name buergerite, in honor of Martin Julian Buerger, for a new sodium ferric variety of the mineral tourmaline. The idealized composition is:

$$NaFe^{+3}_3Al_6Si_6B_3O_{30}[OH,F]$$

The new variety was first reported in 1964 by Mason, Donnay and Hardie in Science* and in an abstract in Special Paper 76 of the Geological Society of America,* but no name was proposed at the time. Two years later, after the name buergerite had been approved, as stated above, the mineral was described in the American Mineralogist.*

One of Buerger's doctorates, Dr. Gabrielle (Hamburger) Donnay (Ph.D. XII 1948), was the prime mover in getting Buerger's name applied to the new variety which, incidentally, occurs as dark brown to black crystals of gem quality. Dr. Donnay, as Gabrielle Eva Hamburger (her maiden name)

---

* The three articles cited above are as follows: Mason, Brian; Donnay, Gabrielle; and Hardie, L. A. - Ferric tourmaline from Mexico, Science 144/3614: 71-73, il., tab., (1964). - Unusual iron tourmaline from Mexico [abst.], Geol. Soc. Am., Sp. P. 76: 109, (1964).

Donnay, Gabrielle; Ingamells, C. O.; and Mason, Brian. - Buergerite, a new species of tourmaline, Am. Mineralogist 50/1-2: 198-199, (1966).

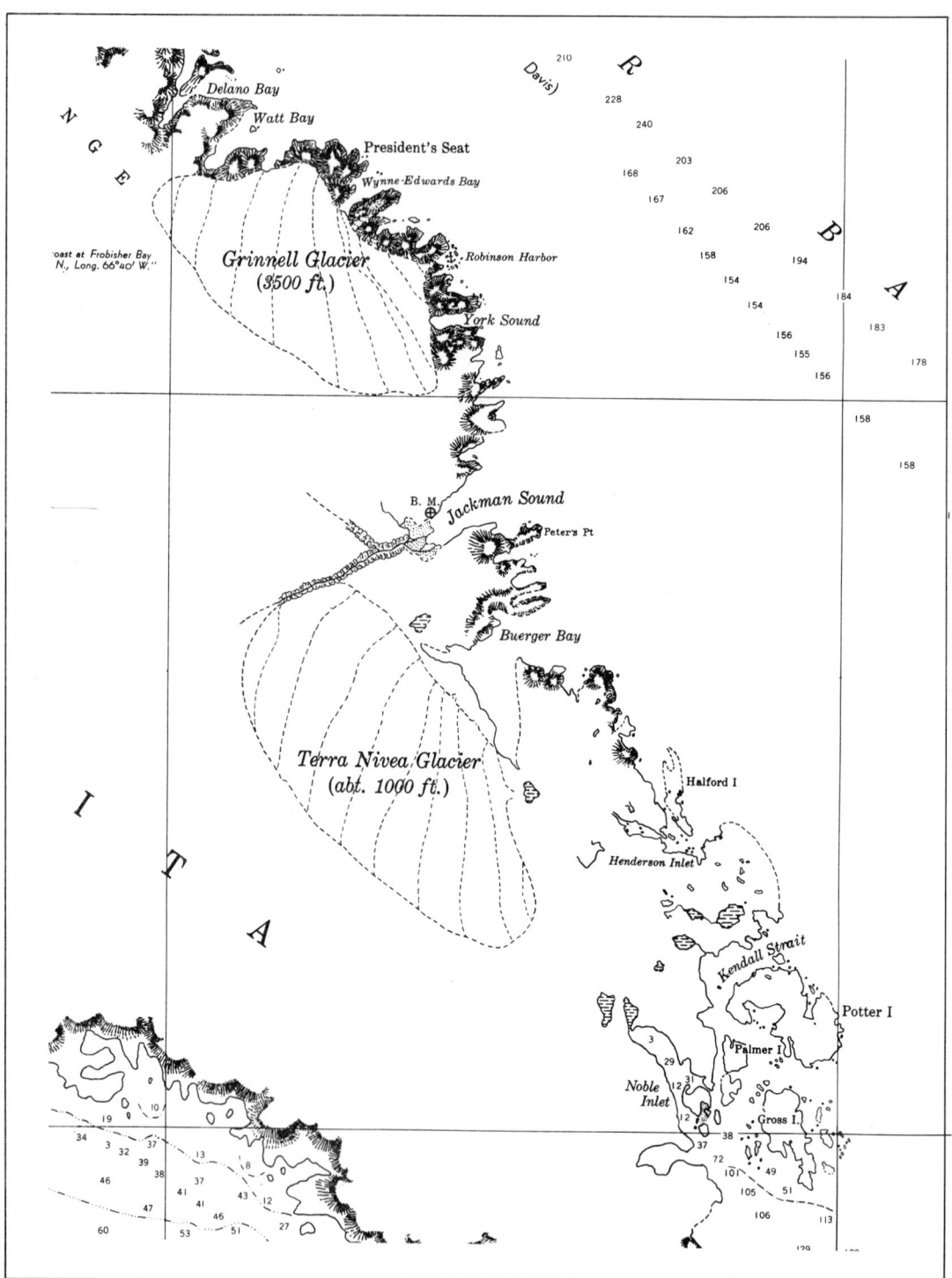

A small section of U.S. Navy's Hydrographic Office Chart No. 5854, of May 1943, showing Buerger Bay in the approximate center of the section, indenting the southwest coast of Frobisher Bay and lying directly below Terra Nivea Glacier. The location of the crossing coordinates directly southwest of Grinnell Glacier is: Lat. 62°30' N., Long. 67°00' W.

had done her doctoral thesis on "The Crystal Structure of Tourmaline" in Buerger's laboratory in 1945 to 1948. Later on, when more sophisticated methods of refining crystal structures became available, Buerger and two other of his graduate students, Charles W. Burnham (XII Ph.D. 1961) and Donald R. Peacor (XII Ph.D. 1962), continued research on tourmaline and ultimately were able not only to position very precisely the location of the atoms, but also to find their thermal ellipsoids. This process called "refining the structure" had become highly developed in the 1960s. (See references 63, 115, 133, 201 and 202 in the Bibliography.)

MARTIN J. BUERGER FESTSCHRIFT 1968

Following an Old World custom of honoring a distinguished teacher, 35 of Buerger's former students and associates contributed reports on their recent research to a single issue of the Zeitschrift für Kristallographie, Band 127, Heft 1-4, Seiten 1-326, (1968), a Festschrift dedicated to Professor Buerger on his 65th birthday (1968).

In the prefatory statement of the volume, one of his former students, L. V. Azaroff,* now Professor and Director of the Institute of Materials Science at the University of Connecticut, cites three important areas in the development of which Buerger has played an outstanding role by his contributions: 1) the advancement of crystal structure analysis; 2) the transformation of mineralogy from a purely descriptive science to one that seeks to relate the properties of minerals to their structure; and 3) the education of various scientists in the principles and practice of x-ray crystallography and of some thirty graduate students and an equal number of postgraduates in its applications to mineralogy. Furthermore, by 1968 he had published five books [by 1975, twelve books carried his name as author], most of which are recognized as the standard in its field, in which he systematized and expanded his contributions for more effective use by future students.

MARTIN J. BUERGER FESTBAND 1973

Martin's 70th birthday, 8 April 1973, was celebrated by an open house secretly arranged by Lila in conjunction with a number of his former and current graduate students. On this festive occasion, we guests were shown an unbound copy of what was soon published as the Martin J. Buerger Festband in the Zeitschrift für Kristallographie, B. 138: 1-459, 126 fig.,

---

* Mineralogy, Crystallography and M. J. Buerger, Zeitschr. Krystal. 127: 3-4, (1968).

(August, 1973). Although there is no prefatory biographical statement like the one in the 1968 Festschrift, there is the following informative dedicatory statement:

> ON THE OCCASION OF THE 70th BIRTHDAY OF
> PROFESSOR MARTIN J. BUERGER
> BORN 8th APRIL 1903
> AN EDITOR OF THIS JOURNAL FROM ITS REFOUNDATION
> IN 1954
> DEDICATED BY HIS FELLOW-EDITORS
> AND THE PUBLISHERS

This is a rather unusual dedication in that the Festband was seemingly initiated not by Buerger's former students but by his fellow editors and with the blessing of the publishers. The contributors to the volume are from all over the field of crystallography, and are not confined to Buerger's students, though many of them made contributions to the papers.

MARRIAGE AND A FAMILY OF SIX DAUGHTERS

In 1938 Martin married Lila Mae MacAskill, the youngest of three children of Angus Duncan and Margaret (Curran) MacAskill. Angus was born in Nova Scotia of Scotch parentage* and Margaret Curran was born in Louisa, Quebec, of Irish ancestry. Both had come separately from Canada to the Boston area, as young immigrants, and had married and settled in Newton Centre before the turn of the century. Lila was born there in 1910.

Lila's brother, the first of the three children, was born 14 years before her and died from a children's disease when only 3 or 4 years old. Her sister, 12 years older than herself, died of tuberculosis in 1914 only two weeks after the death of her mother, from the same malady, with the result that Lila was left motherless and without a sister or brother when only four years old. For the next three years she was cared for by relatives or neighbors, but when this arrangement proved unsatisfactory her father sent her to his home Province, Nova Scotia, to live with his two unmarried siblings, Christine and John MacAskill, who were living at home

---

* Like Martin's forebears, who came to Missouri by sailing ship from Saxony and thence up the Mississippi River to St. Louis, Lila's came also by sailing ship from Scotland to Cape Breton Island. Because of religious persecution in Scotland, a Scotch Presbyterian minister, one Norman MacLeod, persuaded some 500 of his parishioners on the Isle of Lewis of the Outer Hebrides off the northwest coast of Scotland to take to a dozen sailing ships and immigrate to Nova Scotia. Among the parishioners was Lila's great grandfather MacAskill. After a miserable year in Cape Breton, the minister and half of his flock sailed away to land ultimately in New Zealand, but great grandfather MacAskill stayed behind and lived out his life in Nova Scotia. One of his descendants, Lila Mae MacAskill, would meet and marry one of Ernst Moritz Buerger's descendants, Martin Julian Buerger, three generations later!

in Englishtown, Cape Breton Island, where they were taking care of their father, Lila's grandfather, Duncan MacAskill.

Lila was only seven when sent to her new home in Englishtown, and there she attended primary and secondary school and grew up among playmates and schoolmates of Scotch ancestry, with her aunt Christine taking the place of her deceased mother.

Immediately following graduation from Englishtown High School, after having taken special teacher's training during her senior year, she was certified to teach the primary grades in the local schools because of the scarcity of teachers, but for only one year. After completing the year, she decided to take her meager savings and return to Boston where she could train for secretarial work. So at 19 she entered the Hickox Secretarial School in Boston. Living with her father, now remarried, she completed in less than nine months the course of training that normally took a year or more, and did so well that she was asked to teach after completing the course. As she remarked to me, she simply had to concentrate her efforts to the maximum in order to complete the training program before her savings ran out.

After a year of teaching in Hickox she decided such work was not to her liking, and since jobs were scarce, it being in the depths of the Great Depression, she decided to return to Nova Scotia and resume training for teaching. After a year's study at the Nova Scotia Normal School in Truro, and special summer work at Dalhousie University, she was certified to teach kindergarten and grades 1 through 11. For the next three years she taught the primary grades in several schools in the Englishtown neighborhood, but impressed by the lack of future opportunities in teaching there, she decided once again to return to her home in Boston, particularly as economic conditions were improving in the United States.

We need to interrupt Lila's schedule of activities here in order to bring Martin into her life. While she was attending Hickox Secretarial School she met Martin through her cousin Ruth, who was dating Martin's brother Newton. Their acquaintance led to correspondence, when Lila returned to Nova Scotia to resume training for teaching. Late in 1935 Martin visited Lila and her relatives in Englishtown, saw the kind of thrifty and self-sufficient farm life they led, fell in love with Lila, and came home with some long-range plans on his mind. Soon after Lila returned to Newton Centre in 1937 they became engaged and would have been married there except that Martin, meanwhile, had contracted to go with MacMillan on his 1938 summer expedition to the Canadian Arctic. So marriage had to be postponed for the time being. Meanwhile Lila found employment in the local Treasurer's Office as a secretary.

Lila's 1972 Christmas Card. Turn 90° clockwise to read.

Lila's 1971 card - The Buergers were dividing their time about equally between M.I.T. and the University of Connecticut.

Soon after Martin returned from the Arctic he and Lila were married on 5 July 1938. At long last the years of loneliness, homelessness, and insecurity were over for Lila. After three years in Arlington, the Buergers built a new home in pastoral Lincoln in 1941-1942 as World War II was under way, and there they live today (1975) after rearing six daughters - Marla Christine, Julie Margaret (deceased), Laura Pauline, Janet Elizabeth, Dorothy Ruth, and Patricia Anne - six of the "seven jewels" (Lila being the seventh) to whom Martin dedicated his fifth major book, Vector Space (1959).

I have described Lila's youth in some detail because the experiences of those early years no doubt account for the gracious hospitality that she has offered for so many years to guests in her home.

The Buerger home has long been a pleasant meeting place where Lila and Martin have entertained visiting crystallographers and mineralogists, M.I.T. and Harvard faculty members, and particularly Martin's graduate students at the Institute.

The last-named group repaid some of the Buergers' hospitality when they hosted a surprise dinner party in the Stratton Student Center at M.I.T. on 6 April 1968 to celebrate Martin's 65th birthday that would come two days later on the 8th of April. Martin's remembrance of the memorable occasion, written as a thank-you letter to those who planned and executed the affair, merits inclusion here because it demonstrates the respect and admiration in which he and Lila are held by Martin's former students, and the list of the latter shows the worldwide impact of his influence as a teacher. He wrote as follows:

"April 6, 1968 was a day I will never forget, for about eight o'clock I found myself suddenly facing half-a-hundred old friends from near and far. The plan had been that the Frondels and the Buergers were going to meet the Parrishes, who were in town doing some house hunting, for dinner. And that turned out to be the truth, but not the whole truth. As we entered the Mezzanine Lounge at the Student Center, I got the impression that we must be in the wrong room because there were too many people there, and they were all standing watching us come in the door; but then I recognized somebody, and then another, and another ... But this pleasant surprise was almost immediately replaced by incredulity when I began to realize, one by one, that some of these old friends were a long ways from home. I instantly recognized that José and Marisa Amorós were from the Mississippi Valley, so the thought crossed my mind that maybe they were having a surprise party to celebrate the publication of their new book 'Molecular Transforms' which we had been expecting from day to day - But there was Theo Hahn from Aachen - Karl Fischer from Saarbrücken - Hajo Onken from the same place - Oh, no, that can't be Jose Zemann from Vienna, but it is - and what's Yoshio Takeuchi from Tokyo doing here - maybe there's some sort of crystallographic symposium which I missed. While I covered my ignorance and surprise by confused greetings, I rapidly formed a number of tentative hypotheses which I couldn't entertain long because they didn't fit the facts very well. Of course, those who lived nearby, like Charlie Burnham, Ed Gheith, Bernie Wuensch, Dick Beger, Felix Trojer, Lyneve Waldrop, Peter Süsse, Herbert Thurn, and Martha Redden, could be accounted for on the basis of trying to pull a fast one with an unscheduled seminar. But then, what were Charlie Supper and Otto von der Heyde doing? They never showed any interest in seminars before. And how about Tibor Zoltai from Minneapolis; Don Peacor from Ann Arbor; Mike Frueh from Montreal; Howard Evans, Gai Donnay, and even Farouk El Baz from Washington; Bill Statton, Paul Arthur, and Charlie Prewitt from Wilmington; Lee Azároff from Storrs, and wow! Le Roy Jensen from Salt Lake City! Besides, how about all the ladies - some of whom I didn't know and some I didn't recognize because they had changed their hair color -

"Well it wasn't a seminar, and it gradually dawned on me that it was a birthday party. The only trouble with this reasonable hypothesis was that it was obviously impossible for these friends from distant lands to attend. But it was still difficult to brush off facts, and it wasn't until the next day that some reasonable explanation for them was forthcoming. Meanwhile, I recognized the vulnerability of my position. I had had a certain amount of fun from time to time in the past by presenting some of the attendants with diplomas which were not universally accepted by the academic community, so I had a suspicion that this might be an occasion on which the tables were to be turned. Fortunately this fear was groundless, for the Marcasite Society gave no diplomas that night, although it did circulate a few choice Late Abstracts, which proved to me that they were still on their toes. And there was a lot of good-natured joshing about some generation-old gripes.

"After enjoying the best meal I can recall M.I.T.'s ever having served, the Marcasite Society got down to more serious matters. How can I ever thank you for Festschrift in my name; for two pieces of jewelry each with a beautiful crystal of the variety of tourmaline one of you named after me; for a book dedicated to me in words I consider most acceptable; for a bound volume of the bibliographies of my students, fellows, and coworkers along with some personal reminiscences. Your accomplishments as represented by the bibliography are a source of great satisfaction to me. On the evening of April 6, 1968, my cup was running over.

"Lila and I had the good fortune to see many of you at our home the following day, and a few of you a little longer. Only then did we learn something of the ramifications of the grand plan which explained the incredible features of the previous evening. We congratulate you on a superb job of organization and execution. We are both most grateful for the tremendous effort so many of you made in order that April 6 might be such a success. I noted that some fifteen of you who attended the party, and ten who couldn't make it, are now professors. Our wish is that, when each of you reaches 65, your own students will arrange for you an occasion such as you arranged for me."

SUMMATION

The achievements of the long and distinguished scientific career of Martin Julian Buerger have impelled his peers to refer to him as "Mr. Crystallography" and have brought him both national and international recognition of several different kinds. In America he has been awarded the prestigious Day and Roebling medals of the Geological Society of America and of the Mineralogical Society of America, respectively; elected to membership in the American Academy of Arts and Sciences and the National Academy of Sciences; and elected to the presidencies of three of his professional societies, several of which he helped to found. He has played a prominent role in the organization and administration of the leading American and international organizations for crystallographers and mineralogists. Abroad he has been honored by election to membership in academies and societies in Brazil, Italy (Torino and Rome), Switzerland (Zurich), Bavaria, Austria and Spain (Madrid). The University of Berne (Switzerland) awarded him an honorary degree in 1958. A variety of tourmaline, buergerite, and an inlet in Baffin Island (Canadian Arctic), designated Buerger Bay, will preserve his name in perpetuity. A 1968 Festschrift and a 1973 Festband record the scientific achievements of his doctorates and guest post-doctorates who spent time in his laboratory.

In evaluating and summarizing his career, I would divide it into four closely related but quite distinct activities or subcareers: 1) as an outstanding teacher; 2) as an innovative investigator; 3) as a highly productive author; and 4) as an imaginative inventor. And I rank their importance in the order given.

AS A TEACHER

Buerger's former students unanimously vote him one of their most outstanding teachers. Noted for his mastery of the subject, his logical organization of the subject matter, his clear and simple but rigorous presentations, and his obvious pleasure in lecturing and leading seminar discussions, he has been, and continues to be, a superb teacher and an excellent model for his students. Little wonder, then, that 14 of his 20 doctorates and 7 of his dozen or more post-doctoral guests now hold professorships in leading universities around the world.

Professor Leonid V. Azároff, one of Buerger's most distinguished doctorates, recorded an important quality of Buerger's teaching when he dedicated one of his recent books, Elements of X-ray Crystallography (McGraw-Hill, 1968) as follows:

> "This book is dedicated to
> MARTIN J. BUERGER,
> Who
> has found crystallography most enjoyable
> and,
> believing that man should enjoy his work,
> has spent most of his life
> making crystallography
> enjoyable
> for others"

I have chosen to give the first rank to Buerger's activities as a teacher for the reason that through his teaching he has not only given his students a superb model for their own future careers, but has stimulated them to go forth and exceed their master in assaying and extending knowledge, developing new ideas and theories, inventing new methods and instruments, and seeking always the deeper understanding of the physical world. In the long run I like to think that the great teacher has his immortality not in his books and articles and inventions but in the achievements of his students, who follow and improve on his philosophy and techniques of teaching, and who in their turn accept and pass on what they have in turn inherited from their professorial predecessors.

AS AN INNOVATIVE INVESTIGATOR

For half a century Buerger's eager and inquiring mind, aided by a quality of inventive ingenuity possessed by only a fortunate few, has probed the physical world for deeper understanding of how crystals grow and behave in their natural environment.

Using a laboratory that he initiated as he started his academic career in 1929, and inventing an imposing array of instruments and methods of analysis as he gained experience and sophistication, he carried on a program of crystal-structure analysis and related research that produced some 240 articles, 12 books, a dozen or more theories, an equal number of new instruments, determination of the crystal structure of 21 minerals, and 39 student theses (7 S.B.'s, 12 S.M.'s, and 20 Doctor's).

AS A HIGHLY PRODUCTIVE AUTHOR

In the laboratory Buerger not only conducted his own research with vigor and relish, consistently publishing results as rapidly as justified and preparing textbooks the better to organize and present material to his students; he also stimulated and strongly encouraged, perhaps I should say <u>insisted</u> that, his graduate students follow his example; which many of them have done! As a result, <u>his</u> students, unlike so many who publish little or nothing after completing graduate work, have followed his example and are today among the younger research leaders in reporting on their discoveries and producing helpful textbooks.

As mentioned in a preceding section Buerger ranks high among former Course XII professors and alumni in the number of books (12) and articles (244) published, and his doctorates outrank all others in their productivity.

AS AN INVENTOR OF LABORATORY INSTRUMENTS AND METHODS

As pointed out in a preceding section - <u>Patented and Unpatented Instruments and Processes</u> - Buerger invented, designed and got made more than 25 instruments and accessories that he needed to pursue his laboratory research on crystal-structure analysis. Several of these instruments are now in common use in x-ray diffraction laboratories throughout the world.

AS A PERSON

Martin Buerger has been an intellectually restless, aggressive and inquiring scientist almost from the day he first began to sense the world around him - recall his delving into his father's chemistry textbook when only seven years old and soon astounding his elders by how much of it he knew in a short time! He has been a scientist in a hurry! eager to understand the physical environment in which he lived. He has always had more ideas than he could possibly follow up, and even at age 72 he is working and producing at a rate that can justifiably be envied by persons several decades younger! He told me a few days ago that he has always been <u>lazy</u>, that repetitive activities bore him, that after doing something once he doesn't want to do the same thing, make the same instrument adjustment or read the same dial, again and again! So what does he do? He devises a new method or invents a new instrument and then makes it automatic; all this so that he can rush on to the next enticing problem. Such are some of the characteristics of the man who has done so much for his chosen field of crystallography.

But Buerger has always found time in his busy schedule for his students. He has even enticed some of them to help plant trees on his property in suburban Lincoln, but Mrs. Buerger made up for it by providing a fine meal afterwards. The Buergers have opened their home again and again to Martin's colleagues and students, and their living room has frequently been transformed into a domestic seminar room where the professor could exhibit his latest invention or expound on his latest idea. We should remember, though, that his helpmate Lila artistically arranged the room and expertly prepared the tasty repasts that have long since become a Buerger tradition. And if one is invited into Martin's study he sees there the same order and symmetry of things that the x-ray reveals among the atoms in crystals.

How his students have reacted to all the preceding was clearly shown when they enthusiastically cooperated to produce his Festschrift for his 65th birthday and his Festband for his 70th. And the mineral that they named Buergerite shines black from the jewelry of the Buerger ladies to remind them of the affection in which they and Martin are held by his former students.

RELIGIOUS ACTIVITIES

Considering the fact that Martin comes from a long line of ministers of the gospel and less active but equally devout Christians on both paternal and maternal sides, as pointed out at the beginning of this biography, it is not surprising that he also has an abiding faith in the Protestant creed and its concept of God. When we discussed the matter of religion recently, during a taped interview, he emphasized again his deep personal faith, his belief that he has been blessed by a divine power, and his conviction that prayer can aid the believing person. He called my attention to a commencement address that he delivered in 1964 at The Stony Brook School, Stony Brook, Long Island, New York, saying that it expressed fully his own personal convictions about "Scientists and God." I regard his address as a fitting conclusion to my discussion of his brilliant scientific career, and am grateful to him and Christianity Today (p. 6-8 of the issue dated August 28, 1968) for permission to quote it in full, as follows:

Scientists and God

MARTIN J. BUERGER

Many today tell us that the rapid advance of modern knowledge is giving man mastery of the world and that mastery of the universe is just around the corner. They imply, therefore, that man is self-sufficient. This viewpoint had begun to pervade our thinking well before the Russians launched Sputnik. And now that we have emulated them and have even orbited men, the feeling is becoming general that we have begun to master the universe itself. As a result, some think that we no longer need God and that he should therefore be dropped from our lives. Not only is this the Communist view; it is also that of many sophisticated Western thinkers.

But what would the world be like if we should succeed in eliminating God from consideration? A comparison of the West, which was deeply influenced by Christ, with that part of the world which was not influenced by him answers the question. The two worlds are not the same. The non-Christian world thinks differently about right and wrong, about the sanctity of life, and about the place of women in society. Without detailing such differences, let us simply note that the Christian idea of right and wrong derives from the concept of sin and is absolute; that the Christian view of life comes from the concept of the equality of all persons before God, as does also the Christian view of women. The non-Christian views on these subjects are based essentially on the idea of the tyranny of the stronger. Those in Western civilization who advocate the elimination of God overlook the fact that our very freedoms are part of our Christian heritage.

A position often implied and sometimes openly expressed is that every scholar knows man to be self-sufficient and that no real scientist believes in God any more. This is simply not true. Consider some evidence, beginning at my own institution of learning.

A few years ago a conference of M.I.T. faculty and religious leaders was called to discuss the best ways to meet the spiritual needs of the Protestant and Orthodox students on the campus. A number of professors attended. But what was particularly significant was which ones attended. At M.I.T. there is a category of distinguished faculty members known as Institute Professors. Three such professors had been named to this top honor at the time of the conference. Of the three, two took part in the conference. A year or so later, one of them gave the baccalaureate sermon, in which he took his stand as a professing Christian before the graduating class and attending faculty. Obviously it cannot be said that these Institute Professors are able to believe in God because they do not meet the standards of real scientists. On the contrary, they are leaders in their fields; all of them have pulled out of their fertile brains ideas that have created new fields for other scientists to follow.

SOME WHO BELIEVED

The generalization that scientists do not believe in God will not bear scrutiny, for some of the greatest scientists have believed. A seventeenth-century example was Sir Isaac Newton, who was such a pious man that he always doffed his hat when God was mentioned, and who wrote extensively on the Scriptures. Newton was clearly no run-of-the-mill scholar; on his work all physics rests, and even Einstein needed it as a start for his own great work. In the same century as Newton lived the French scientist and mathematician, Blaise Pascal. One of the greatest mathematical minds of all time, he was a profound Christian; his *Pensées*, one of the world's great books, is a landmark in Christian philosophy.

A modern example of a world-renowned scientist who was a believer was John von Neumann of the Princeton Institute for Advanced Study. It was he who led in the development of the high-speed digital computers that have so changed the course of present-day science. All mathematicians know him for his great contributions to pure mathematics, and in 1956 he received the $50,000 Enrico Fermi award for his basic scientific contributions. A Nobel Laureate is also among the believers—Professor Victor F. Hess, who in 1936 received the Nobel prize in physics for his discovery of cosmic rays.

These scientists happen to be among those who, in addition to being first-rank scholars, have made their religious convictions publicly known. But there are many more who have not made them known. After all, only occasionally in science does an opportunity arise for a man's convictions to stand revealed. In my own field of crystallography I know most of the several hundred internationally prominent scholars; yet save for a few instances when I was present for a revealing conversation, I have not learned their individual religious or anti-religious feelings. Nevertheless I can name some Christians among them who would not object to being counted. One of Switzerland's greatest crystallographers, Werner Nowacki, is a Christian, as is Spain's greatest authority in this field, José Luis Amorós, and also one of America's leaders, José Donnay. And J. H. Robertson, a well-known Scottish research scientist, is a Christian. I once heard Robertson, a Presbyterian, and Amorós, a Roman Catholic, argue a religious point; each spoke as a dedicated Christian and each was proud of his faith. England's leading woman crystallographer, Dame Kathleen Lonsdale, is a Quaker. She is a professor at the University of London, a fellow of the Royal Society, and vice-president of the International Union of Crystallography, and she was decorated by the Queen for her outstanding scientific work.

A curious fact is that Russia's most distinguished crystallographer, Academician N. V. Belov, is a believer who can no longer practice his Orthodox faith because Russia has banished God. Belov knows more about symmetry and, with his students, has determined the arrangements of atoms in more silicate minerals than any other man on earth. And is it not significant that, even in Russia, there is a believer among the greatest scientists? One is reminded of the biblical statement, "I have reserved to myself seven thousand men, who have not bowed the knee to Baal" (I Kings 19:18; Rom. 11:4).

Are there advantages for a scholar in being a Christian? As a scientist, I believe that there are immeasurable advantages. Consider a small boy who wants an answer to a question. Ordinarily he will go to his parents. Later, when he is in high school, he may go to his teacher for an answer. Still later, in college, he may approach his professor. If one day, however, he finds himself a professor and a question arises that he cannot resolve, then what is he to do? Perhaps he can obtain an answer from an authority greater than himself. But suppose that he himself becomes an authority in a specialized field, and there arises something in that field which puzzles him. What can he do to get an answer?

Fortunately the Bible gives us a method for such a problem: "If any of you lack wisdom, let him ask of God, that giveth to all men liberally, and upbraideth not; and it shall be given him" (Jas. 1:5). This marvelous invitation encourages all of us, scientists included, to approach not just a higher authority but *the Highest Authority*, the One who designed the universe, the interrelations of which are a part of what the scientist seeks to know. There is only one price for this service: one must believe the invitation. Anyone who believes is invited to ask and is assured an answer.

Asking implies prayer. Prayer is a way of getting in

personal touch with the Creator of the **unive**rse, by which is meant not only the physical universe but also the logic with which it is put together. The judgment with which a man handles the knowledge of the universe and its logic is called wisdom, and James invites us to improve this judgment by going to the Ultimate Authority. But this invitation must not be construed as a blanket promise that any prayer will be answered according to the petitioner's desire. If two persons have a contest—say, a wrestling match—and both pray for victory, it is difficult to see how both prayers can be answered as the petitioners wish. There may at times be two answers, one of which may be "No."

THE MAZE IN PERSPECTIVE
But Christians have another great promise that will help them even when the answer is "No." This one is stated in Romans 8:28: "And we know that all things work together for good to them that love God, to them who are the called according to his purpose." Here we learn that no matter how bad things appear to be, they happen for our eventual benefit if we love God. Only the Creator of the universe could fulfill this promise; yet a simple analogy may help our finite minds to understand it. If you are required to make your way through a maze, the path must sometimes be what seems a retrogression instead of an advance toward the goal. Only someone who sees the whole maze in perspective can direct you with certainty through it. God is in exactly this position, because he sees the whole, because all things are present before him, and because he sees the end from the beginning.

The Bible records many examples of this principle. A classical case is that of Joseph, who said of his brothers' selling him to the Ishmaelites: "But as for you, ye thought evil against me; but God meant it unto good, to bring to pass, as it is this day, to save much people alive" (Gen. 50:20).

But to return to the subject of prayer, although the answer to a particular prayer may be "No," nevertheless we are promised that a request for wisdom *is* generally answered. But is this really true? Can one depend on it? By experience I have found that one can, and I am sure that my experience is not unique.

On occasion I have invented theories. Now a theory never occurs to its author in complete and final form; it commonly arrives as a sudden flash of basic ideas that must be explored and developed. These nearly always present knotty problems. A number of these situations that have presented themselves to me have appeared to be quite beyond my powers to resolve; yet since the theories came out of my own imagination, I could hardly expect to consult someone else about them. But all such problems exist in the logic of the universe God created. Thus I knew for a certainty that I could get help from God, and whenever I asked for this help I received it.

One case was remarkable. I had struggled for days to resolve a curious dilemma in which I obtained different answers by two different routes. Finally I remembered to pray for wisdom, and while I was in the very act of framing my request to God, the solution came to me in wordless form. It seemed a fulfillment of Isaiah 65:24, "... before they call, I will answer; and while they are yet speaking, I will hear." To be sure, this verse is taken out of context, but it is very much like what Christ tells us so directly in Matthew 6:8: "... your Father knoweth what things ye have need of, before ye ask him."

Receiving wisdom is, of course, only one of the benefits God gives the believer. But one does not serve God primarily to obtain benefits. No one can be persuaded to serve God if he does not believe in God. If he does believe, then serving God is a natural consequence of being part of God's family, and the earthly advantages are wholly subsidiary to the family relationship.

Does the scientist need God? Can the scientist afford to ignore this shortcut to knowledge of the things that his curiosity drives him to study? Many do ignore it, but many believing scientists have found God's promise of wisdom to be true. ☐

Martin has been a member of the Board of Directors of the Institute for Advanced Christian Studies since its inception in 1967. This organization, among other things, funds scholarly writings of an evangelical Christian nature. And as mentioned early in this biography, in 1953 Martin himself privately published the Memoirs of Ernst Moritz Buerger, his great grandfather, who was one of the clergymen who led a group of Lutheran immigrants from Saxony to the United States in 1838 and founded the Lutheran Church - Missouri Synod in Perry County, Missouri. Martin has also been a faithful member of Boston's Park Street Church (Congregational) since his student days at M.I.T. and has served as Clerk of its Board of Deacons since February 1941.

As a kind of postscript to my own summation of Buerger's life and scientific achievements, I am pleased to include here a special summation and evaluation of his contributions to crystallography by Martin himself. He kindly prepared this in late 1975 at my special request.

Martin Buerger's Contributions to Crystallography
[prepared by Buerger himself]

"When I began to be interested in crystallography, there were available the older theories, particularly those of symmetry, and the older experimental apparatus and methods of study, especially the reflecting goniometer as used in the study of crystal forms, and also the polarizing microscope for use in studying crystal optics. After World War I, Bragg's lead in crystal-structure analysis had just opened that field, and I was attracted into it by a series of lectures by Bragg at M.I.T.; this inspired me to try my own hand at a simple structure. I chose the mineral sulfide, marcasite, whose structure at that time was unknown. My colleague, Bert Warren, permitted me to use his oscillating-crystal apparatus in the x-ray laboratory he and Bragg had set up. After a long grind of interpreting the large collection of oscillating-crystal photographs, I soon solved the structure, using methods I had found in the literature, and learning as I proceeded. My solution acted as a turning point for me, because no one could argue me out of the result, a characteristic I had not found in results of geological research generally.

"As I entered the field of crystal-structure analysis, I found myself not only solving structures, but improving the designs of accepted instrumentation and even inventing some new instruments. For example, the standard powder camera was the one originally designed by the inventors of the technique, Debye and Scherrer. Apparently it had never been improved; it was full of features which demanded adjustments for each new photograph, and the method of bringing the x-ray beam into and out of the camera, as well as the devices for holding and adjusting the crystal in the beam were quite crude. To me these things were intolerable, and I was not happy until I had built my own version. It wasn't long before I discovered that the North American Philips Company, whose men had seen my camera when they visited my laboratory, was manufacturing a duplicate of it, so I decided to publish my design. New designs of other apparatus, based upon new theoretical ideas, included the equi-inclination Weissenberg camera and the precession camera, which are now found in all laboratories doing crystal-structure investigation throughout the world, and are manufactured in half-a-dozen countries. The single-crystal diffractometer with Weissenberg geometry is also my design (first made in 1952-53) as well as its automated version. This instrument has a competitor, known as the 'four-circle' diffractometer, made by G.E. and by Picker.

"Incidental to my study of crystals by x-ray diffraction, I wrote both journal articles and books concerned with theories of crystal geometry, some of which are still in use. These include the symmetries of Fourier space and Patterson space, as applied to crystals, and the use of what I call 'image theory' to achieve an approximation of the electron density of a crystal through a manipulation of its Patterson function. The devices which do this are called 'image-seeking functions', whose best-behaved member is the 'minimum function'. This is in use by most crystallographers who use the Patterson function to solve structures.

"My entry into the field of the Patterson function resulted, I believe, in inspiring the development of the so-called 'direct methods' of crystal-structure analysis which are now routinely used by chemists to find the arrangements of atoms in their molecules. In the process of solving the structure of the mineral nepheline I was lead to the 'implication diagram' as a straightforward step in the solution of the Harker function. Curiously enough, this showed that the phase information necessary to the solution of a structure by Fourier synthesis, yet not directly available in the x-ray diffraction effects, is nevertheless contained, somehow, in the whole set of measureable intensities. This followed because for certain space groups the implication diagram is a projection of the crystal structure along a symmetry axis. It was this clue that inspired Harker and Kasper to search among the mathematical inequalities to find one which could relate $|\underline{F}|^2$'s to $\underline{F}$'s and thus permitted them to establish the Harker-Kasper inequalities. This was the beginning of 'direct methods' as carried out in Fourier space. Eventually these inequalities lead to methods, especially what is called the 'symbolic addition' method, which now permitted finding the arrangements of atoms in crystals which have up to about 100 atoms per cell.

"I am still interested in developments which are extensions or generalizations of the Patterson function. Indeed, I have finished a first draft (now in need of much revision) of a book entitled 'Image Sets and Image Functions' which brings together a lot of material and which will integrate the whole matter of 'image space' with the use of an image algebra.

"Yet, these developments do not represent my entire interest in crystallography. The realm of crystal science is extensive because it treats of the mathematics, physics, and chemistry of the crystalline state, all of which are just a generalization of the science we call mineralogy. Some of my contributions which are outside the x-ray diffraction field are gliding, crystal growth and habit, twinning, and the general theory of polymorphism and phase transformations. In the latter field I had the opportunity of giving my early views first in my retiring presidential address 'The Role of Temperature in Mineralogy' before the Mineralogical Society of America in 1947. In this paper I provided names for the two major types of phase transformation (displacive and reconstructive) and introduced most mineralogists of that era to the idea of disorder by bringing together the early ideas of Barth on one hand and Bragg and Williams on the other. Neither of these knew of the works of the other. I applied these ideas, for the first time, to the phase diagrams of the feldspars, and this appears to have inspired a lot of further research on the relations between the various members of this important group of minerals.

> "I should mention that my early work on lineage structure appears to me to have some importance. At the present it is unknown except to a few, mostly my students, and among them it is eclipsed by the fashionable dislocation theory and its experimental results. Lineage structure is present in nearly all crystals and cannot be ignored indefinitely, so I believe crystallographers will eventually return to this visible imperfection."

BIBLIOGRAPHY OF MARTIN JULIAN BUERGER

Symbols and abbreviations used in the following references are explained on p. 91-98. Abstracts may be listed separately as well as with the complete article. All publications are listed through 1975, two years after Buerger completed the two-year period as Honorary Research Associate, following his permanent retirement on 30 June 1973 as Institute Professor Emeritus.

T--1925   (and Maury, J. L.) Tin ores of Oploca, Bolivia, 4 + 38 p., il., (1925). (S.B. Thesis at M.I.T. in Course III, Department of Mining Engineering and Metallurgy, June 1925.) (See also 1--1927.)

T_1--1927   The deformation of ore minerals: a preliminary investigation, 67 p., (1927). (S.M. Thesis at M.I.T. in Course XII, Department of Geology, June 1927.) (See also 4--1928 and 5--1928a.)

1--1927   (and Maury, J. L.) Tin ores of Chocaya, Bolivia. Econ. Geol. 22: 1-13, (1927). (See also preceding T--1925.)

2--1927a   Note on a method of oblique illumination. Am. J. Sci. (5) 13: 262-263, (1927).

3--1927b   Optical notes on some of the variable contact minerals from Edenville, New York. Am. Mineral. 12: 374-378, (1927).

4--1928   The plastic deformation of ore minerals, Part I. Am. Mineral. 13: 1-17, (1928). (See also preceding T_1--1927.)

5--1928a   The plastic deformation of ore minerals, Part II. Am. Mineral. 13: 35-51, il., (1928). (See also preceding T_1--1927.)

6--1928b   The cause of translation striae and translation strain-hardening in crystals. A.I.M.E., Pr. Inst. Met. Div., p. 375-388, (1928).

7--1928c   (with Newhouse, W. H.) Observations on wood tin modules. Econ. Geol. 23: 185-192, (1928).

T_2--1929   Translation-gliding in crystals, 203 p., il., pl., (1929). (Ph.D. Thesis at M.I.T. in Course XII, Department of Geology, June 1929.) (See also 9--1930.)

8--1929   (and Huntsinger, H. A.) A broad source of monochromatic light. Am. Mineral. 14: 329-331, (1929). (See also 11--1930b.)

9--1930   Translation-gliding in crystals. Am. Mineral. 15: 45-64, il., (1930). (See also preceding T_2--1929.)

10--1930a   Translation-gliding in crystals of the NaCl structural type. Am. Mineral. 15: 174-187, 226-238, il., (1930).

11--1930b   (and Harrington, V. F.) A broad source of monochromatic light (Second note; see also 8--1929). Am. Mineral. 15: 579-580, (1930).

12--1931   (with Harrington, V. F.) Immersion liquids of low refraction. Am. Mineral. 16: 45-54, (1931).

13--1931a   The crystal structure of marcasite. Am. Mineral. 16: 361-395, il., (1931).
14--1931b   The chemical identification of solids by crystallography. T.E.N. 12: 154-155, 170, 172, 174, (1931).
15--1932   The crystal structure of löllingite, $FeAs_2$. Zeitschr. Kristal. (A) 82: 165-187, (1932).
16--1932a   The significance of "block structure" in crystals. Am. Mineral. 17: 177-191, il., (1932).
17--1932b   The negative crystal cavities of certain galena and their brine content. Am. Mineral. 17: 228-233, il., (1932).
18--1932c   The cleavage surfaces of galena. Am. Mineral. 17: 391-395, (1932).
19--1933   The optical properties of ideal solution immersion liquids. Am. Mineral. 18: 325-334, il., (1933).
20--1934   The pyrite-marcasite relation. Am. Mineral. 19: 37-61, (1934).
21--1934a   The temperature-structure-composition behavior of certain crystals. Nat. Ac. Sci., Pr. 20: 444-453, (1934).
22--1934b   (with Buerger, N. W.) Crystallographic relations between cubanite segregation plates, chalcopyrite matrix, and secondary chalcopyrite twins. Am. Mineral. 19: 289-303, il., (1934).
23--1934c   Lattice indices and transformations in the gnomonic projection. Am. Mineral. 19: 360-369, (1934).
24--1934d   The Weissenberg reciprocal lattice projection and the technique of interpreting Weissenberg photographs. Zeitschr. Kristal. (A) 88: 356-380, (1934).
25--1934e   The lineage structure of crystals. Zeitschr. Kristal. (A) 89: 195-220, (1934).
26--1934f   The nonexistence of a regular secondary structure in crystals. Zeitschr. Kristal. (A) 89: 242-267, (1934).
27--1934g   Fluid inclusions in pyrite. Am. Mineral. 19: 605, (1934).
28--1935   The unit cell and space group of realgar. Am. Mineral. 20: 36-43, (1935).
29--1935a   Silica framework crystals and their stability fields [abst.]. Am. Mineral. 20: 196-197, (1935); Geol. Soc. Am., Pr. 1934: 420, (1935). (See also the following reference.)
30--1935b   The silica framework crystals and their stability fields. Zeitschr. Kristal. (A) 90: 186-192, (1935). (See also the preceding reference.)
31--1935c   Application of plane groups to the interpretation of Weissenberg photographs [abst.]. Am. Mineral. 20: 212-213, (1935); Geol. Soc. Am., Pr. 1934: 434, (1935). (See also the following reference.)
32--1935d   The application of plane groups to the interpretation of Weissenberg photographs. Zeitschr. Kristal. (A) 91: 255-289, (1935). (See also the preceding reference.)
33--1935e   The cathode assembly of gas x-ray tubes. Rev. Sci. Instr. 6: 385-386, (1935).
34--1935f   A device for drilling oriented holes in spheres required in the construction of crystal structure models. Rev. Sci. Instr. 6: 412-416, (1935).
35--1936   An x-ray powder camera. Am. Mineral. 21: 11-17, il., (1936).

36--1936a   The probable non-existence of arsenoferrite. Am. Mineral. 21: 70-71, (1936).

37--1936b   (and Butler, R. D.) A technique for the construction of models illustrating the arrangement and packing of atoms in crystals. Am. Mineral. 21: 150-172, il., (1936).

38--1936c   The crystal structure of the arsenopyrite group [abst.]. Am. Mineral. 21: 203, (1936).

39--1936d   The crystal structure of cubanite [abst.]. Am. Mineral. 21: 205, (1936). (See also 62--1937k.)

40--1936e   Discussion to remarks by A. Goetz on the article: The non-existence of a regular secondary structure in crystals. Zeitschr. Kristal. (A) 93: 170-173, (1936).

41--1936f   An apparatus for conveniently taking equi-inclination Weissenberg photographs. Zeitschr. Kristal. (A) 94: 87-99, (1936).

42--1936g   Crystallographic data, unit cell and space group for berthierite ($FeSb_2S_4$). Am. Mineral. 21: 442-448, (1936).

43--1936h   Crystals of the realgar type: the symmetry, unit cell, and space group of nitrogen sulfide. Am. Mineral. 21: 575-583, (1936).

44--1936i   A systematic method of investigating superstructures, applied to the arsenopyrite crystal structural type. Zeitschr. Kristal. (A) 94: 425-438, (1936).

45--1936j   (and Lukesh, J. S.) The preparation of oriented polished sections of small single crystals. Am. Mineral. 21: 667-669, il., (1936).

46--1936k   The symmetry and crystal structure of the minerals of the arsenopyrite group. Zeitschr. Kristal. (A) 95: 83-113, (1936).

47--1936ℓ   The symmetry and crystal structure of manganite, Mn(OH)O. Zeitschr. Kristal. (A) 95: 163-174, (1936).

48--1936m   The law of complication. Am. Mineral. 21: 702-714, (1936).

49--1936n   The kinetic basis of crystal polymorphism. Nat. Ac. Sci., Pr. 12: 682-685, (1936).

50--1936o   The general role of composition in polymorphism. Nat. Ac. Sci., Pr. 12: 685-689, (1936).

51--1937   A common orientation and a classification for crystals based upon a marcasite-like packing. Am. Mineral. 22: 48-56, (1937).

52--1937a   The valences of iron in pyrite and marcasite [abst.]. Am. Mineral. 22: 208-209, (1937).

53--1937b   An apparatus for the precision determination of single-crystal constants [abst.]. Am. Mineral. 22: 218, (1937); Geol. Soc. Am., Pr. 1936: 65-66, (1937).

54--1937c   (and Bloom, M. C.) Crystal polymorphism. Zeitschr. Kristal. (A) 96: 182-200, (1937).

55--1937d   (with Bloom, M. C.) On the genesis of polymorphous forms - Sb_2O_3. Zeitschr. Kristal. (A) 96: 365-375, (1937).

56--1937e   The x-ray determination of lattice constants and axial ratios of crystals belonging to the oblique systems. Abst. Am. Mineral. 22: 210-211, (1937); Ibid. 22: 416-435, il., (1937).

57--1937f   (and Lukesh, J. S.) Wallpaper and atoms. Tech. Rev. 39: 338-342, 370, il., (1937).

58--1937g   (with Hendricks, S. B.) Polymorphism of antimony trioxide and the structure of the orthorhombic form. J. Chem. Phys. 5: 600, (1937).

59--1937h   The precision determination of the linear and angular lattice constants of single crystals. Zeitschr. Kristal. (A) 97: 433-468, (1937).

60--1937i   Interatomic distances in marcasite and notes on the bonding in crystals of löllingite, arsenopyrite, and marcasite types. Zeitschr. Kristal. (A) 97: 504-513, (1937).

61--1937j   (and Hendricks, S. B.) The crystal structure of valentinite (orthorhombic Sb_2O_3). Zeitschr. Kristal. (A) 98: 1-30, (1937).

62--1937k   The unit cell and space group of cubanite. Am. Mineral. 22: 1117-1120, (1937).

63--1937ℓ   (and Parrish, W.) The unit cell and space group of tourmaline (an example of the inspective equi-inclination treatment of trigonal crystals). Am. Mineral. 22: 1139-1150, il., (1937); Abst., Ibid. 23: 182, (1938).

64--1937m   Surface reflection areas in Weissenberg photographs [abst.]. Am. Mineral. 22: following p. 1202: [1]-[2], (1937); Ibid. 23: 166-167, (1938).

65--1938   X-ray surface reflection fields and their application to absorption corrections and to background patterns. Zeitschr. Kristal. (A) 99: 189-204, (1938).

66--1938a   (and Butler, R. D.) Data for the construction of models illustrating the arrangement and packing of atoms in crystals (formula types A, AB, and AB_2). Am. Mineral. 23: 471-512, il., (1938).

67--1938b   Spectacular Frobisher Bay. Tech. Rev. 40: 268-270, 279-280, 282, 284, il., (1938). Reprinted in Can. Geogr. J. 17: 2-18, (1938).

68--1939   The crystal structure of gudmundite [abst.]. Am. Mineral. 24: 183-184, (1939).

69--1939a   The crystal structure of gudmundite (FeSbS) and its bearing on the existence field of the arsenopyrite structural type. Zeitschr. Kristal. (A) 101: 290-316, (1939).

70--1939b   The photography of interatomic distance vectors and of crystal patterns. Nat. Ac. Sci., Pr. 25: 383-388, (1939).

71--1939c   The Optical Identification of Crystalline Substances (‡). Cambridge: M.I.T. Letter Shop, 100 p. [not consecutively numbered], il., (1939).

72--1940   Memorial of Waldemar Lindgren. Am. Mineral. 25: 184-188, port., (1940).

73--1940a   The correction of x-ray diffraction intensities for Lorentz and polarization factors. Nat. Ac. Sci., Pr. 26: 637-642, (1940).

74--1941   Optically reciprocal gratings and their application to syntheses of Fourier series. Nat. Ac. Sci., Pr. 27: 117-124, (1941).

75--1941a   Numerical Structure Factor Tables. Geol. Soc. Am., S.P. 33: vii + 119 p., tab., (1941).

76--1942   X-ray Crystallography. An introduction to the investigation of crystals by their diffraction of monochromatic x-radiation. New York: John Wiley & Sons, Inc., xxii + 531 p., 252 fig., 34 tab., (1942). [Translated to Russian.]

77--1942a   The unit cell and space group of claudetite, As_2O_3 [abst.]. Am. Mineral. 27: 216, (1942).

78--1942b (with Lukesh, J. S.) The unit cell and space group of kaliophilite [abst.]. Am. Mineral. 27: 226-227, (1942).

79--1942c The unit cell and space group of orpiment. Am. Mineral. 27: 301-304, (1942).

80--1942d A new Fourier series technique for crystal structure determination. Nat. Ac. Sci., Pr. 28: 281-285, (1942).

81--1942e (and Smith, L. B., De Bretteville, A., Jr., and Ryer, F. V.) The lower hydrates of soap. Nat. Ac. Sci., Pr. 28: 526-529, (1942).

82--1942f The characteristics of soap hemihydrate crystals. Nat. Ac. Sci., Pr. 28: 529-535, (1942).

83--1942g (with Lukesh, J. S.) The tridymite problem. Science 95: 20-21, (1942).

84--1943 (and Buerger, N. W. and Chesley, F. G.) Apparatus for making x-ray powder photographs at controlled, elevated temperatures. Abst., Am. Mineral. 27: 217, (1942); Ibid. 28: 285-302, il., (1943).

85--1944 (and Buerger, N. W.) Low-chalcocite and high-chalcocite. Am. Mineral. 29: 55-65, (1944).

86--1944a The Photography of the Reciprocal Lattice. ASXRED Mon. 1: ix + 37 p., il., (1944). Published by The American Society for X-ray and Electron Diffraction, (1944).

87--1945 Review of Dana's System of Mineralogy. Science 101: 650-652, (1945).

88--1945a (and Klein, G. E.) Correction of x-ray diffraction intensities for Lorentz and polarization factors. J. Appl. Phys. 16: 408-418, (1945).

89--1945b The genesis of twin crystals. Am. Mineral. 30: 469-482, il., (1945).

90--1945c (and Smith, L. B., Ryer, F. V., and Spike, J. E., Jr.) The crystalline phases of soap. Nat. Ac. Sci., Pr. 31: 226-233, (1945).

91--1945d The design of x-ray powder cameras. J. Appl. Phys. 16: 501-510, (1945).

92--1945e (with Gardiner, K. W. and Smith, L. B.) The hydrate nature of soap. J. Phys. Chem. 49: 417-428, (1945).

93--1945f Soap crystals. Am. Mineral. 30: 551-571, (1945).

94--1945g The structure of cubanite, $CuFe_2S_3$, and the coordination of ferromagnetic iron. J. Am. Chem. Soc. 67: 2056, (1945).

95--1945h Artificial metamorphism of minerals [abst.]. Geol. Soc. Am., B. 56 (12/2): 1150, (1945).

96--1945i The cell and symmetry of pyrrhotite. Geol. Soc. Am., B. 56 (12/2): 1150, (1945).

97--1946 Review of Major Instruments of Science and their Applications to Chemistry, by R. E. Burk and Oliver Grummitt. J. Am. Chem. Soc. 68: 157, (1946).

98--1946a (and Klein, G. E.) Correction of diffraction amplitudes for Lorentz and polarization factors. J. Appl. Phys. 17: 285-306, (1946).

99--1946b The interpretation of Harker syntheses. J. Appl. Phys. 17: 579-595, (1946).

100--1946c (and Klein, G. E. and Hamburger, G. E.) The structure of nepheline [abst.]. Geol. Soc. Am., B. 57 (12/2): 1182-1183, (1946); Am. Mineral. 32: 197, (1947).

101--1946d Artificial metamorphism of minerals [abst.]. Am. Mineral. 31: 190, (1946).

102--1947 Derivative crystal structures. J. Chem. Phys. 15: 1-16, (1947).

103--1947a (and Washken, E.) Metamorphism of minerals. Am. Mineral. 32: 296-308, (1947).

104--1947b (and Smith, L. B. and Ryer, F. V.) An investigation of the crystalline phases in the system: sodium myristate-water. J. Am. Oil Chem. Soc. 24: 193-196, (1947).

105--1947c The cell and symmetry of pyrrhotite. Am. Mineral. 32: 411-414, (1947). (Also see 96--1945i.)

106--1947d The crystal structure of cubanite. Am. Mineral. 32: 415-425, il., (1947).

107--1947e The relative importance of the several faces of a crystal. Am. Mineral. 32: 593-606, il., (1947).

108--1947f The genesis of crystal forms and a rational explanation of the "law" of Bravais [abst.]. Am. Mineral. 32: 686, (1947).

109--1947g The role of temperature in mineralogy [abst.]. Geol. Soc. Am., B. 58 (12/2): 1169-1170, (1947). (See also 112--1948b.)

110--1948 Review of *Los Rayos X y la Estructura Fina de los Cristales*, by Julio Garrido and Joaquin Orland. J. Appl. Phys. 19: 112-113, (1948).

111--1948a Review of *X-rays in Practice*, by Wayne T. Sproull. J. Appl. Phys. 19: 218-219, (1948).

112--1948b The role of temperature in mineralogy. Am. Mineral. 33: 101-121, il., (1948). Abst., *Ibid.* 33: 193-194, (1948); Geol. Soc. Am., B. 58 (12/2): 1169-1170, (1947).

113--1948c Phase determination with the aid of implication theory. Phys. Rev. 73: 927-928, (1948).

114--1948d Some relations between the F's and F^2's of x-ray diffraction. Nat. Ac. Sci., Pr. 34: 277-285, (1948).

115--1948e (with Hamburger, G. E.) The structure of tourmaline. Am. Mineral. 33: 532-540, (1948); Abst., *Ibid.* 33: 761-762, (1948).

116--1948f The ambiguity factor in implication theory. Acta Cryst. 1: 259-263, (1948).

117--1948g The structural nature of the mineralizer action of fluorine and hydroxyl. Am. Mineral. 33: 744-747, (1948).

118--1948h Crystals based on the silica structures [abst.]. Am. Mineral. 33: 751-752, (1948).

119--1948i Fourier aspects of implication theory. Indian Ac. Sci., Pr. 28: 324-331, (1948).

120--1948j Algunos aspectos da teoria implicacao. Ac. Brazil. de Ciências, Anais 20: 1-2, (1948).

121--1949 Crystallographic symmetry in reciprocal space. Nat. Ac. Sci., Pr. 35: 198-201, (1949).

122--1949a Fourier summations for symmetrical crystals. Am. Mineral. 34: 771-788, (1949).

123--1949b Disorder in crystals of non-metals. Ac. Brazil de Ciências, Anais 21: 245-266, (1949).

124--1949c General aspects of disorder in minerals [abst.]. Geol. Soc. Am., B. 60 (12/2): 1876-1877, (1949).

125--1949d (with Washken, E.) [The] effect of potassium on the nepheline-carnegieite transformation [abst.]. Geol. Soc. Am., B. 60 (12/2): 1927, (1949); Am. Mineral. 35: 290-291, (1950).

126--1950 Crystallographic symmetry in reciprocal space and in vector space [abst.]. Am. Mineral. 35: 122, (1950).

127--1950a Vector sets. Acta Cryst. 3: 87-97, (1950); a correction, Ibid. 3: 243, (1950).

128--1950b Vector sets, a correction. Acta Cryst. 3: 243, (1950). (See also the preceding reference.)

129--1950c General aspects of disorder in minerals [abst.]. Am. Mineral. 35: 278-279, (1950).

130--1950d The crystallographic symmetries determinable by x-ray diffraction. Nat. Ac. Sci., Pr. 36: 324-329, (1950).

131--1950e The photography of atoms in crystals. Nat. Ac. Sci., Pr. 36: 330-335, (1950).

132--1950f Some new functions of interest in x-ray crystallography. Nat. Ac. Sci., Pr. 36: 376-382, il., (1950).

133--1950g (with Donnay, Gabrielle) The determination of the crystal structure of tourmaline. Acta Cryst. 3: 379-388, (1950).

134--1950h Generalized microscopy and the two-wave-length microscope. J. Appl. Phys. 21: 909-917, (1950).

135--1950i Tables of the characteristics of the vector representations of the 230 space groups. Acta Cryst. 3: 465-471, (1950).

136--1950j Limitation of electron density by the Patterson function. Nat. Ac. Sci., Pr. 36: 738-742, (1950).

137--1950k Photographs of the atoms in the structures of minerals [abst.]. Geol. Soc. Am., B. 61 (12/2): 1446, (1950); Am. Mineral. 36: 311, (1951).

138--1951 Crystallographic aspects of phase transformations, p. 182-211, in Phase Transformation in Solids, by R. Smoluchowski, J. E. Mayer, and W. A. Weyl, eds. New York: John Wiley & Sons, Inc., (1951).

139--1951a A new approach to crystal-structure analysis. Acta Cryst. 4: 531-544, il., (1951).

140--1952 Presentation of Day Medal to Martin J. Buerger: Response by Martin J. Buerger. Geol. Soc. Am., Pr. 1951: 55-56, port., (1952).

141--1952a The application of image theory to crystal-structure analysis, p. 43-56, in Computing Methods and the Phase Problem in X-ray Crystal Analysis, R. Pepinsky, ed. State College: Pennsylvania State College, (1952).

142--1952b Precipitation of segregate phases from solid solution. Pr. Int. Symp. Reactiv. Solids, Gothenburg, Sweden, 1952, p. 225-235, (1952).

143--1953 Image theory of superposed vector sets. Nat. Ac. Sci., Pr. 39: 669-673, (1953).

144--1953a Solution functions for solving superposed Patterson syntheses. Nat. Ac. Sci., Pr. 39: 674-678, (1953).

145--1953b An intersection function and its relations to the minimum function of x-ray crystallography. Nat. Ac. Sci., Pr. 39: 678-680, (1953).

146--1953c (and Niizeki, N.) Crystal structure of livingstonite, $HgSb_4S_7$ [abst.]. Geol. Soc. Am., B. 64 (12/2): 1404, (1953); Am. Mineral. 39: 319-320, (1954).

147--1953d (with Azároff, L. V.) A one-dimensional Fourier Analogue Computer. ONR Project NR032 346, Contract N5ori-07860. M.I.T. Crystallographic Laboratory Technical Report No. 1, 7 p., il., (30 June 1953).

148--1953e Numerical Structure Factor Tables. Geol. Soc. Am., S.P. 33, p. 1-119, 1941; reprinted (1953).

149--1954 The diffraction symmetry of twins. Ac. Brazil de Ciencias, Anais 26: 111-121, il., (1954).

150--1954a Some relations for crystals with substructures. Nat. Ac. Sci., Pr. 40: 125-128, (1954).

151--1954b The stuffed derivatives of the silica structures. Am. Mineral. 39: 600-614, (1954).

152--1954c (and Klein, G. E. and Donnay, Gabrielle) Determination of the crystal structure of nepheline. Am. Mineral. 39: 805-818, il., (1954).

153--1954d Proyecciones de Patterson de cristales simetricos. Anales de la Real Sociedad Española de Fisica y Quimica 50(A): 221-254, (1954).

154--1954e (and Hahn, T. W.) The structure of nepheline, $KNa_3Al_4Si_4O_{16}$ [abst.]. Acta Cryst. 7: 632, (1954).

155--1955 (with Azároff, L.) Refinement of the structure of cubanite, $CuFe_2S_3$. Am. Mineral. 40: 213-225, (1955).

156--1955a (and Hahn, T.) The crystal structure of berthierite, $FeSb_2S_4$. Am. Mineral. 40: 226-238, il., (1955).

157--1955b (and Robinson, D. W.) The crystal structure and twinning of Co_2S_3. Nat. Ac. Sci., Pr. 41: 199-203, il., (1955).

158--1955c (with Hahn, T.) The detailed structure of nepheline, $KNa_3Al_4Si_4O_{16}$. Zeitschr. Kristal. 106: 308-338, (1955).

159--1956 Elementary Crystallography. An introduction to the fundamental geometrical features of crystals. New York: John Wiley & Sons, Inc., xxiii + 528 p., il., (1956).

160--1956a The arrangement of atoms in crystals of the wollastonite group of metasilicates. Nat. Ac. Sci., Pr. 42: 113-116, il., (1956).

161--1956b (with Pearson, A. D.) Confirmation of the crystal structure of pentlandite. Am. Mineral. 41: 804-805, (1956).

162--1956c (and Barney, E. and Hahn, T.) The crystal structure of diglycine hydrobromide. Zeitschr. Kristal. 108: 130-144, (1956).

163--1956d The determination of the crystal structure of pectolite, $Ca_2NaHSi_3O_9$. Zeitschr. Kristal. 108: 248-262, il., (1956).

164--1956e Partial Fourier syntheses and their application to the solution of certain crystal structures. Nat. Ac. Sci., Pr. 42: 776-781, il., (1956).

165--1957 (with Hahn, T.) The crystal structure of diglycine hydrochloride, $2(C_2H_5O_2N)\cdot HCl$. Zeitschr. Kristal. 108: 419-453, (1957).

166--1957a Reduced cells. Zeitschr. Kristal. 109: 42-60, (1957).

167--1957b (with Niizeki, N.) The crystal structure of livingstonite, $HgSb_4S_8$. Zeitschr. Kristal. 109: 129-157, (1957).

168--1957c (with Niizeki, N.) The crystal structure of jamesonite, $FePb_4Sb_6S_{16}$. Zeitschr. Kristal. 109: 161-183, (1957).

169--1958 (and Niizeki, N.) Correction for absorption for rod-shaped single crystals. Am. Mineral. 43: 726-731, (1958).

170--1958a (and Kennedy, G. C.) An improved specimen holder for the focusing type x-ray spectrometer. Am. Mineral. 43: 756-757, (1958).

171--1958b (with Azaroff, L. V.) <u>The Powder Method in X-ray Crystallography</u>. New York: McGraw-Hill Book Co., xv + 342 p., il., (1958). [Translated to Russian.]

172--1958c (and Zoltai, T.) Crystal structure of coesite, the high-density form of silica [abst.]. Geol. Soc. Am., B. 69 (12/2): 1543, (1958).

173--1959 <u>Vector Space</u>, and its application in crystal-structure investigation. New York: John Wiley & Sons, Inc., xiv + 347 p., (1959). [Translated to Russian.]

174--1959a (with Zoltai, T.) The crystal structure of coesite, high-pressure form of silica. Zeitschr. Kristal. 111: 129-141, (1959).

175--1959b [Presentation and] Acceptance of the Roebling Medal of the Mineralogical Society of America. Am. Mineral. 44: 390-395, port., (1959).

176--1959c (and Zoltai, T.) Relative energies of rings of tetrahedra [abst.]. Geol. Soc. Am., B. 70 (12/2): 1706, (1959). (See also 181--1960d.)

177--1960 <u>Crystal-Structure Analysis</u>. New York: John Wiley & Sons, Inc., xvii + 668 p., il., (1960).

178--1960a Note on reduced cells. Zeitschr. Kristal. 113: 52-56, (1960).

179--1960b (Chairman) Symposium on Twinning: Inst. Inv. Geol. "Lucas Malada" Cursillos y Conf. (Madrid), fasc. 7: 3-57, incl. ils., (1960). Includes individual papers that are cited separately.

180--1960c Twinning with special regard to coherence, in Symposium on Twinning: Inst. Inv. Geol. "Lucas Malada" Cursillos y Conf. (Madrid), fasc. 7: 5-7, (1960).

181--1960d (with Zoltai, T.) The relative energies of rings of tetrahedra. Zeitschr. Kristal. 114: 1-8, (1960). (See also 176--1959c.)

182--1960e (with Takeuchi, Y.) The crystal structure of terramycin hydrochloride. Nat. Ac. Sci., Pr. 46: 1366-1370, il., (1960).

183--1960f (with Burnham, C. W.) Refinement of the crystal structure of andalusite [abst.]. Geol. Soc. Am., B. 71 (12/2): 1838, (1960).

184--1960g (with Prewitt, C. T.) Crystal structure of cahnite, $Ca_2BAsO_4(OH)_4$ [abst.]. Geol. Soc. Am., B. 71 (12/2): 1946-1947, (1960). Abst., Acta Cryst. 13: 1007, (1960).

185--1961 Polymorphism and phase transformation. Fort. Mineral. 39: 9-24, (1961).

186--1961a (with Burnham, C. W.) Refinement of the crystal structure of andalusite. Zeitschr. Kristal. 115: 269-290, (1961). (See also 183--1960f.)

187--1961b Review of A Handbook of Lattice Spacings and Structures of Metals and Alloys, by W. B. Pearson. Zeitschr. Kristal. 115: 319-320, (1961).

188--1961c (with Prewitt, C. T.) The crystal structure of cahnite, $Ca_2BAsO_4(OH)_4$. Am. Mineral. 46: 1077-1085, (1961). (See also 184--1960g.)

189--1961d Review of The Theory of Crystal Structure Analysis, by A. I. Kitaigorodskii. Science 134: 1412-1413, (1961).

190--1961e (and Prewitt, C. T.) The crystal structures of wollastonite and pectolite. Nat. Ac. Sci., Pr. 47: 1884-1888, (1961).

191--1961f Image sets [with German abstract]. Zeitschr. Kristal. 116: 430-467, il., (1961).

192--1961g Review of Fourier Transforms and Convolutions for the Experimentalist, by R. C. Jennison. Science 134: 2093-2094, (1961).

193--1962 An algebraic representation of the images of sets of points. Accad. delle Sci. Torino, Atti 96: 175-192, (1962).

194--1962a The algebra and geometry of convolutions. Atti della Accad. Naz. dei Lincei, Mem. VI: 83-95, (1962).

195--1962b Review of Organic Chemical Crystallography, by A. I. Kitaigorodskii. Science 135: 912, (1962).

196--1962c Review of Crystallization, Theory and Practice, by Andrew van Hook. Science 136: 518-519, (1962).

197--1962d (with Prewitt, C. T.) A comparison of the crystal structures of wollastonite and pectolite [abst.]. Am. Mineral. 47: 200, (1962).

198--1962e (with Wuensch, B. J.) The crystal structure of pyrrhotite, Fe_7S_8 [abst.]. Am. Mineral. 47: 209, (1962).

199--1962f (with Wuensch, B. J.) The crystal structure of chalcocite, Cu_2S [abst.]. Am. Mineral. 47: 209, (1962).

200--1962g (with Peacor, D. R.) The determination and refinement of the structure of narsarsukite, $Na_2TiOSi_4O_{10}$. Am. Mineral. 47: 539-556, il., (1962).

201--1962h (and Burnham, C. W. and Peacor, D. R.) Restudy of the study of tourmaline [abst.]. Geol. Soc. Am., S.P. 68: 143, (1962).

202--1962i (and Burnham, C. W. and Peacor, D. R.) Assessment of the several structures proposed for tourmaline. Acta Cryst. 15: 583-590, il., (1962). (See also the preceding reference.)

203--1962j Personal reminiscences, p. 550-555 in Fifty Years of X-ray Diffraction, by P. P. Ewald, ed. Utrecht, Netherlands: Pub. Int. Union of Crystallog. by N.V.A.-Oosthoek's Uitgeversmaatschappij, i-ix + 717 p., (1962).

204--1962k Review of Direct Observation of Imperfections in Crystals, J. B. Newkirk and J. H. Wernick, eds. Science 137: 972-973, (1962).

205--1962l (with Peacor, D. R.) Determination and refinement of the crystal structure of bustamite, $CaMnSi_2O_6$. [with German abstract] Zeitschr. Kristal. 117: 331-343, (1962).

206--1962m Image functions. Zeitschr. Kristal. 117: 358-361, (1962).

207--1962n (with Cid-Dresdner, Hilda) The crystal structure of potassium hexatitanite, $K_2Ti_6O_{13}$. Zeitschr. Kristal. 117: 411-430, (1962).

208--1962o   Review of X-ray Powder Data for Ore Minerals; the Peacock Atlas, by L. G. Berry and R. M. Thompson. Zeitschr. Kristal. 117: 473-474, (1962).

209--1962p   Review of Plasticity of Crystals, R. V. Klassen-Nekyludova, ed. Science 138: 1390-1391, (1962).

210--1963   Review of Molecular Structure and Properties of Liquid Crystals, by G. W. Gray. Science 139: 206-207, (1963).

211--1963a   Review of Cours de Cristallographie, Livre III, by R. Gay (Paris: Gauthier-Villars, 1961). J. Phys. Chem. Solids 24: 337-338, (1963).

212--1963b   Review of Direct Methods in Crystallography, by M. M. Woolfson. Zeitschr. Kristal. 118: 334-335, (1963).

213--1963c   (and Wuensch, B. J.) Distribution of atoms in high chalcocite, Cu_2S. Science 141: 276-277, (1963).

214--1963d   Some properties of image functions, p. 3-14 in "Crystallography and Crystal Perfection", Proc. Int. Symp. Protein Structure and Crystal., Madras [India], Jan. 1963, G. N. Ramachandran, ed. London: Academic Press, (1963).

215--1963e   (with Wuensch, B. J.) The crystal structure of chalcocite, Cu_2S. Mineral Soc. Am., S.P. 1: 164-170, (1963).

216--1963f   (with Prewitt, C. T.) Comparison of crystal structures of wollastonite and pectolite, in Miscellaneous papers - Int. Mineral. Assoc., 3rd Gen. Meeting, Washington, D.C. Mineral. Soc. Am., S.P. 1: 293-302, il., (1963).

217--1963g   Phase transformations [abst.], in Int. Union Crystal., 6th Int. Cong. Symp., Rome, Italy, 1963. Acta Cryst. 16: A180, (1963).

218--1963h   Crystallography, in Encyclopedia Britannica, v. 6: 851-863, (1963). Chicago: Encyclopedia Britannica, Inc.

219--1964   (and Dollase, W. A.) Shape of the recorded area in precession photographs and its application in orienting crystals. Science 145: 264-265, (1964).

220--1964a   The development of methods and instrumentation for crystal-structure analysis. Zeitschr. Kristal. 120: 3-18, (1964).

221--1964b   The Precession Method in X-ray Crystallography. New York: John Wiley & Sons, Inc., xvi + 276 p., il., (1964).

222--1964c   Review of Colored Symmetry, by A. V. Shubnikov, N. V. Belov et al. Science 145: 804-805, (1964).

223--1964d   (with Cid-Dresdner, Hilda) An improved collimator for single-crystal x-ray diffraction work. J. Sci. Instr. 41: 689, (1964).

224--1964e   Image methods in crystal-structure analysis, p. 1-24 in Advanced Methods of Crystallography, G. N. Ramachandran, ed. London: Academic Press, (1964).

224a--1964f Scientists and God. Christianity Today, p. 6[1036]-8[1038], August 28, 1964).

225--1965   (and Dollase, W. A.) The shape of misoriented reciprocal lattice planes as recorded by precession photography [abst.]. Am. Mineral. 50: 281-282, (1965).

226--1965a   Presentation of the Roebling Medal to Clifford Frondel. Am. Mineral. 50: 530-532, (1965).

227--1965b   The geometry of projections [with German abstract]. Tschermaks Mineral. Petrog. Mitt. (3) 10: 595-607, il., (1965).

228--1966   (and Taxer, K.) Rhodizite: structure and composition. Science 152: 500-502, il., (1966).

229--1966a Review of Optical Transforms, Their Preparation and Application to X-ray Diffraction Patterns, by C. A. Taylor and H. Lipson. Acta Cryst. 20: 596-597, (1966).

230--1966b Background for the use of image-seeking functions. Am. Crystal. Assoc., Tr. 2: 1-9, (1966).

231--1967   (with Beger, R. M.) The crystal structure of the mineral pollucite. Nat. Ac. Sci., Pr. 58: 853-854, (1967).

232--1967a Some desirable modifications of the international symbols. Nat. Ac. Sci., Pr. 58: 1768-1773, (1967).

233--1967b (and Dollase, W. A. and Garaycochea-Wittke, Isabel) The structure and composition of the mineral pharmacosiderite. Zeitschr. Kristal. 125: 92-108, il., (1967).

234--1967c (with Taxer, K. J.) The crystal structure of rhodizite [with German abstract]. Zeitschr. Kristal. 125: 423-436, (1967). (See also 225--1968.)

235--1968   (and Dollase, W. A.) The possible bonding between aluminum atoms in rhodizite and other crystals [abst.], in Int. Mineral. Assoc., 5th Gen. Meeting, Cambridge, England, 1966. Papers and Proc.: London, Mineral. Soc., p. 169, (1968).

236--1968a (and Dollase, W. A. and Garaycochea-Wittke, Isabel) Zeolitic nature of pharmacosiderite [abst.]. Geol. Soc. Am., S.P. 101: 29, (1968). (See also 233--1967b.)

237--1968b (with Dollase, W. A.) Crystal structure of some meteoritic tridymites [abst.]. Geol. Soc. Am., S.P. 101: 54-55, (1968).

238--1969   Presentation of the Roebling Medal of the Mineralogical Society of America for 1968 to Tei-ichi Ito. Am. Mineral. 54: 586-589, (1969).

239--1969a (with Redden, Martha J.) Note on the symmetry and cell of calcium orthovanadate. Zeitschr. Kristal. 129: 459-460, (1969).

240--1969b Patterson symmetry, (Part 3 of the Pilot Issue of the International Tables for Crystallography), 39 p., (1969).

241--1969c Diffraction symbols (Contr. 3, p. 27-42, in Physics of the Solid State [Commemoration Volume to Prof. S. Bhagavantam], S. Balakrishna, M. Krishnamurthi, and B. Ramachandra Rao, eds.) London and New York: Academic Press, (1969).

242--1969d Equi-inclination rotating-crystal photographs [with German abstract]. Zeitschr. Kristal. 130: 173-184, il., (1969).

243--1970   Contemporary Crystallography. New York: McGraw-Hill Book Co., Inc., xiv + 364 p., il., (1970). Materials Science and Engineering Series.

244--1970a (with Süsse, P.) The structure of $Ba_3(VO_4)_2$. Zeitschr. Kristal. 131: 161-174, (1970).

245--1971   Crystal-structure aspects of phase transformations. Am. Cryst. Assoc., Tr. 7: 1-20, (1971).

246--1971a Introduction to Crystal Geometry. New York: McGraw-Hill Book Co., Inc., xii + 204 p., il., (1971).

247--1971b Phase transformations. Kristallografia 16: 1084-1096, (1971).

248--1972   Revival of Fedorov's crystallochemical analysis by the use of data from modern methods. Zeitschr. Kristal. 135: 161-171, (1972).

249--1972a Crystals with the warwickite structure. Materials Res. B. 7: 1201-1207, (1972).

250--1972b Thermal effect in opal below room temperature. Nat. Ac. Sci., Pr. 69: 3225-3227, (1972).

251--1972c The crystal structure of $FeCoOBO_3$. Zeitschr. Kristal. 135: 321-338, (1972).

252--1973 Karl Weissenberg and the development of x-ray crystallography, p. 17-39 in "The Karl Weissenberg 80th Birthday Celebration Essays," (John Harris, ed.). Nairobi: East African Literature Bureau, (1973).

253--1974 (and Venkatakrishnan, V.) Serendibite, a new complicated inorganic crystal structure. Nat. Ac. Sci., Pr. 71: 4348-4351, (1974).

254--1975 Review of Crystalline Solids by Duncan McKie and Christine McKie. London: Nelson, x + 628 p., (1974). Acta Crystallog. A31: 397-399, (1975).

255--1975a (with Amorós, J. L. and Amorós, Marisa Canut de) The Laue Method. New York: Academic Press, xii + 376 p., il., (1975).

256--1975b Review of The Structure of the Elements, by Jerry Donohue. New York: Wiley-Interscience, xi + 436 p., il., (1974). Materials Sci. and Eng. 19: 299-300, (1975).

Prof. Buerger adjusts an instrument in his Building 24 laboratory.
(Photo by Jackman, 1952)

620   GEOLOGY AT M.I.T.   1865-1965

1

2

3

4

5

CRYSTAL-MODEL-MAKING APPARATUS AND CRYSTAL MODELS

1. Mark I ball orienter for drilling holes in balls used to represent atoms. (Ref. 34--1935f; Rev. Sci. Instr. 6: 412-416, 1935.)
2. Mark II ball orienter (operated by Newton W. Buerger; not published).
3. Mark III ball orienter.
4. Group of crystal models constructed with the aid of the Mark II ball orienter. The group illustrates the structures of the polymorphous forms of silica.
5. The arrangement of atoms in the complicated rhombohedral structure of tourmaline as seen looking approximately along the c axis of a model made with the aid of the Mark II ball orienter.

(M.I.T. Photos)

(19)
FREDERICK KUHNE MORRIS
(1885-1962)

FREDERICK KUHNE MORRIS

MIT: 1927-1950

Reared in New York City, educated in its public schools, at the American Museum of Natural History, and at Columbia University, trained for teaching in the City's "Hells Kitchen" area and later at Columbia, and finally tested to the full in the American Museum of Natural History's Asiatic Expeditions of the early 1920s, Frederick K. Morris came to M.I.T.'s Department of Geology in 1927 unusually well prepared scientifically to offer work in physical and structural geology, geomorphology, and glacial geology. Besides this training he brought a combination of artistic skill, knowledge of literature, extraordinary memory, and remarkable speaking ability. Although known at first chiefly for his work in the Gobi Desert as a member of Roy Chapman Andrews' famed Asiatic expeditions, Morris quickly established an enviable reputation within M.I.T. as a teacher, and outside as a public lecturer and geological consultant. During his 23 years as a faculty member of the Department of Geology he gave to thousands of students and the non-college public their first introduction to the nature, origin, and history of the earth. He had a fondness for children and for college youth and, because he could quickly gain rapport with them and sense their special problems, it is they who will remember him longest and best. It was with them that he felt he had his greatest success as a teacher and scientist. When war came and he was rejected because of near-sightedness, he volunteered for civilian service and was assigned special work during and after both World Wars. Finally, after retiring from M.I.T. in 1950 at age 65, he spent the next seven years as a Professor and Lecturer at the Maxwell Air Force Base in Montgomery, Alabama. His exciting and eventful life as a geologist was made so by extensive field work and worldwide travel to a score of countries in Europe, Africa, and Asia, as well as research and travel in Canada, Mexico, and the United States. It came to a close on 5 October 1962 at age 78 in Montgomery, Alabama, where he and his devoted wife Florence had endeared themselves to a great host of friends as they had done earlier at M.I.T. The Frederick K. and Florence E. Fund memorializes their many years of devoted service to the Institute (See Volume 2).

BIRTH, ANCESTRY, AND YOUTH

Frederick K.* Morris** was born in Salt Lake City, Utah, on 11 February 1885 and died at his home in Montgomery, Alabama, on 5 October 1962, following a stroke ascribed to "cardio-vascular arteriosclerosis." He was

---

*,** See footnotes on following page.

still physically active and mentally alert to the day of his death and would have been 78 years old on his next birthday.

He was the eldest of three boys and had an older sister. (The four Morris children were: May Josephine, Frederick Kuhne, Harold Clifford, and Alan.) His father, Frederick Kuhne Morris, was of Welsh-German descent and was a fine musician but his son inherited only love and appreciation of music for he played no instrument and was a monotone in singing. His mother, Emma Lang Morris, of Scotch-Irish descent, was the artistic parent and no doubt passed along some of her skills to her first son.

Salt Lake City was a pioneer town when Fred was born and fires had to be fought by a volunteer brigade which probably accounted for the fact that the Morris family home twice burned to the ground, once while the overturned fire-engine lay just around the corner.

In 1892, when Fred was only seven, the family moved to San Francisco, hoping that the change in climate would help the ailing father to regain his health; but the move was to no avail. So the next year, 1893, the family moved again, this time all the way across the country to New York City, where they hoped to find better medical care. Here Fred first entered public school, having been taught previously at home by his parents, both of whom were much interested in their children's education. He graduated from high school in 1900 and spent the next year at the Marine Biological Laboratory in Woods Hole working at the Laboratory and at the nearby Bureau of Fisheries. This opportunity came as a result of winning a scholarship in high school. Here for the first time he could use and improve the drawing skills that would be so helpful in later years and in his drawing could become familiar with a great many of the common organ-

---

* Fred was named after his father, Frederick Kuhne Morris, but the father was so hurt by the anti-German public attitude before and during World War I that he and Fred did not use Kuhne in their names thereafter; merely K.

** Additional information may be found in the following references:
  1. Streeter, D. D. In Memoriam - Frederick Kuhne Morris. Explor. J. 40/4: 53-55, December 1962. New York, N.Y.: The Explorers' Club.
  2. R. G. L. Dr. Frederick Morris. The Montgomery Advertiser, Montgomery, Alabama, November 1962.
  3. Shrock, R. R. and Hooker, Marjorie. Memorial to Frederick Kuhne Morris (1885-1962). Geol. Soc. Am., Pr. 1966: 329-335, port., (1968).

The writer is particularly grateful to Mrs. F. K. Morris for granting full access to her husband's private papers, now preserved in the M.I.T. Institute Archives, and for providing many details of Fred's activities and achievements. She also made many suggestions that greatly improved my final draft, but I alone am responsible for any errors or misstatements.

isms of the ocean. This experience kindled an interest in biology and he kept that interest throughout his life.

COLLEGE YEARS AND EARLY TEACHING

After three years at City College (now called the College of the City of New York, or C.C.N.Y.), Fred was awarded a Bachelor of Science degree in 1904 with honors in Science and English. In his senior year he was awarded medals in English and Biology, was selected to be Commencement Orator, and was later elected to Phi Beta Kappa [election came in 1936 after City College was accredited as a 4-year college.]

Being strongly motivated toward an educational career, Morris would have liked to continue work for a graduate degree, but his father's declining health and the general family situation made it necessary for him to find employment. This he found as a teacher in the New York City Public School System. For ten years, 1904-1914, he taught science to grade school children in the notorious "Hell's Kitchen" area of the city, and years later, when one of the area's inhabitants was electrocuted at Sing Sing Prison, Morris is said to have remarked, "Tony? He was one of my boys in Hell's Kitchen!"

While teaching his young charges their regular lessons, directing the school's athletic program, and helping the students make cameras and other similar things that attract young boys, he managed to do graduate work in evening classes at Columbia and at the American Museum of Natural History. This arrangement was made necessary by conditions at home, which were such that he had to work full-time at teaching during the regular school days, hence he could only register for evening classes. That he did this is early evidence of his determination to continue his education whatever the obstacles he had to overcome. He studied biology at Columbia with T. H. Morgan, then involved in his genetic research with fruit flies, and comparative anatomy at the Museum with W. K. Gregory. One evening during this period, when he read a paper at a meeting of Columbia's Journal Club, Fred attracted the attention of Amadeus W. Grabau (XII S.B. 1896), who remarked:

> "You've the right kind of mind for a geologist! Come to Columbia and work with me."

At this invitation, Morris answered that he had to teach during regular school hours to help out with family finances so that he could not study except after school. Kindhearted Grabau, always generous with students, switched some classes and arranged for Fred to take courses and to assist him in research after hours and in the evenings. Ultimately, in 1910,

Fred was awarded a Master of Science degree in invertebrate paleontology by Columbia largely because of Grabau's encouragement and assistance.

At this juncture, serious illness in the family, a familiar situation for the first third of his life, now made it necessary for him to increase his income substantially. C. P. Berkey found work for him at Columbia, tutoring fees helped, and in the summer of 1912 he was a field assistant to J. E. Hyde of the Geological Survey of Canada who was doing field work in Nova Scotia (Cape Breton Island) at the time.

Through his work with Grabau and Berkey, Morris had become well known in the Columbia Department of Geology and at the American Museum of Natural History. Early in 1914 he was appointed Berkey's laboratory assistant and was made responsible for keeping in order the collections of minerals, rocks, maps, and microscopes that were used in student instruction. After spending the summer of 1914 in Europe and assisting during SY 1914-15, he was advanced to Lecturer (1915-1916), with assignments to lecture and lead field excursions, and then to Instructor (1916-1920). He now had added responsibilities in lecturing, substituting for professors when they were called away for short periods, and opportunities for attending advanced courses and conducting his own field and laboratory investigations. During these Columbia years, 1910-1920, Morris profited greatly from association with C. P. Berkey, A. W. Grabau, D. W. Johnson, and J. F. Kemp. Each introduced him to a broad and exciting division of geology and gave him training that would be used again and again in his later professional career. He often mentioned how greatly these famous teachers had influenced him and how much he appreciated the help they had given him outside the classroom.

WORLD WAR I AND CIVILIAN SERVICE

During World War I Morris tried to enlist several times but was rejected because of severe near-sightedness, a condition that required him to wear corrective glasses and which had prevented him from enjoying any of the contact sports when a youth. Instead, having volunteered his services to the Armed Forces, he taught Topography in the U. S. Air Service Radio School in 1917 and 1918; prepared large confidential maps of parts of the Western and Italian fronts for a group headed by Colonel Edward M. House (1917-1918) and landscape maps for the National Research Council (1917-1918); taught Topography and Military Geology to the SATC (Student Army Training Corps) in 1918; prepared a 50-page chapter on "Map reading and map interpretation" for H. E. Gregory's <u>Military Geology and Topography</u> (1918); and drew topographic maps for D. W. Johnson's <u>Battlefields of the World War</u> (1921).

On 7 February 1920 Morris was appointed a Special Assistant in the U. S. Department of State and was designated "Geographer of the Department of State," a new position created at the time of his appointment. He was assigned to the Department's Division of Political Information and was placed in charge of the Map Section. His task was to build an entirely new service within the Department. He classified and catalogued the maps on hand and indicated those that should be acquired as essential to future work on international boundaries (e.g. delineation of the Italian-Jugoslav boundary); designed a light table for tracing maps and cabinets for storing them; plotted daily changes in military situations on special maps; and prepared brief commentaries on geographic aspects of military actions. The Chief of his Division described his work as follows:

> "He brought a trained perception and a scientific viewpoint to bear upon all the questions submitted to him, which produced results of the highest order."
> [Prentiss B. Gilbert in letter dated July 12, 1920.]

So satisfactory was Fred's work, in fact, that he was promoted to the position of Drafting Officer in the Department of State on 1 July 1920. This appointment was to last only one month, however, because he resigned on 7 April. The double life of spending half of his time at Columbia and the other half in Washington, with the necessary inter-city commuting particularly now that the War was over, had made him want to get back to full-time geology and its possible opportunities for field work.

TEACHING AND EXPLORATION IN CHINA

Word had recently come from his old Columbia friend and mentor, A. W. Grabau, who was then settled in Peking, that there was a position open at Pei Yang University in Tientsin. Would Fred be interested? Emphatically, yes! This looked like the chance of a lifetime. So Morris resigned from his positions at Columbia and at the Department of State in early July, as stated above, packed his bachelor's belongings, and headed for his first great adventure--a professorship of Geology in a Chinese university that didn't even have a Department of Geology! It was his job to build one in what was then China's foremost engineering university. The next two years were going to be busy ones!

In establishing the new Department of Geology Morris had to do about everything imaginable to get a program of instruction going. He had to learn to speak Chinese, although all lectures were in English and American textbooks were used. He taught the practical geology needed by the students in mining and civil engineering, sometimes as many as eight lectures a day; he taught his students, several hundred of them ultimately, about minerals, rocks and ores, using a petrographic microscope in the more advanced work. He had to design and construct a machine for making thin sections and then teach his students how to operate it. He took field trips and collected specimens for classroom instruction, and he and his students mapped the local geology. Alone he traveled with a small donkey to many parts of Northern China to study the regional geology. Outside the University he was able to help numerous industries in Shantung, Hupeh, and Chili with some of their geological problems. In less than two years Pei Yang University had a vigorous Department of Geology with teaching collections, maps, and other necessary teaching materials, and several hundred students, but it was still a one-man Department, and Morris was the one man.*

---

* The quality of the program that Morris established, and something of his own personal attributes, are clearly indicated in the following excerpt from a letter written to me by Donald D. Smythe, who followed Morris at Pei Yang in 1922:

> "Fred was not at Peiyang when I arrived there as he had already left for Peking to prepare for the Gobi expedition. Prof. Barbour had taken over the Geology Dept. as locum tenens (sp?) due to my delayed arrival.
>
> "Peiyang was a Chinese Government Engineering University where Mining, Civil, Mechanical, Electrical, and Chemical Engineering were taught. Teaching was all done in English and American textbooks were used. Some Chinese teachers taught non-engineering and non-scientific courses but technical courses were all given by American professors.
>
> "I was the Geology Dept. as Fred had been before me. All Geology courses normally given to Mining and Civil Engineering students in American Universities were taught at Peiyang. It was a terrific load, particularly at the beginning of a semester when, at times, I lectured as much as 8 hours a day. That of course was not the case after the beginning courses had been organized later on.
>
> "The standard at Peiyang was high, first class American Schools accepting transfers from Peiyang without an examination.
>
> "Some of my Mining students had studied from 1 to 3 years under Fred. They all had an extremely high regard, almost adoration, for Fred as well they might. That was what made it difficult for me, such a high standard to live up to. His ability to illustrate his lectures with blackboard sketches was remarked upon by many of his former students.
>
> "Fred had another gift, the ability to acquire a foreign language rapidly. In the time he was at Peiyang he achieved an almost perfect command of spoken and written Chinese. [According to Mrs. Morris, Fred spoke Chinese fluently but neither read nor wrote the language to any extent.] I have struggled with half a dozen foreign languages in various countries and can truthfully say that Chinese is far and away the most difficult to learn for those speaking any of the European languages."

In early 1922, Roy Chapman Andrews asked C. P. Berkey to recommend an assistant geologist for work on the then "Third Asiatic Expedition" of the American Museum of Natural History, which would carry the Explorer's Club Flag. Berkey, who was to be Chief Geologist, asked for his former student and colleague, Frederick K. Morris. So it came about that Fred resigned his professorship at Pei Yang and left Tientsin in February to join the party of the Third Asiatic Expedition.

Morris worked with the Expedition in the Gobi Desert of Mongolia during the summers of 1922, 1923, and 1925. Between times during field seasons he learned Russian, carried on research, and helped to prepare maps and reports back at the Museum, where he held the appointment of Associate Curator of Geology and Geography. For one year, 1924, he was also a Lecturer at Columbia. In 1927 he accepted a position at M.I.T. but maintained his connection with the Museum for many years as Research Geologist without pay and continued to work on the materials collected by the Expedition. During this interim, 1926-1929, he also taught the Geography of Asia in the Columbia Summer School. After the 1927 Session Fred and Florence led a busload of 31 students up the Hudson River and into New England, culminating in a night on top of Mount Washington, followed by a long and tiresome trip back.

MARRIAGE TO FLORENCE ELISABETH EDDOWES (1922)

At the end of his third year in China, 1922, most of which he had spent as a member of the AMNH's Third Asiatic Expedition, Morris hastened to Shanghai to meet his bride-to-be, Florence Elisabeth Eddowes, a 1911 graduate of Goucher College. She had known Fred at Columbia and was spunky enough to travel halfway around the world, and of all places to Shanghai, China, to marry her chosen mate with whom she was to share forty years of happiness. According to Florence, Fred arrived in Shanghai on 29 December 1922 and they were married the next evening in a Methodist compound with their only guests an enormous sofa full of tiny children who watched the wedding ceremony in wide-eyed wonder. Thereafter, regardless of where Fred and Florence happened to be, young people seemed to be drawn to the Morrises for council and discussion, and on festive occasions when they sat around after a repast and enjoyed the gracious hospitality of a couple who themselves remained childless. As I have written elsewhere ["Memorial to Frederick Kuhne Morris (1885-1962)." See second footnote on second page of this sketch.]:

> "They had no children of their own; so they adopted the children of the world, and their home, whether in Cambridge, in Montgomery, or in a distant land, was regarded by many children and newly married student couples

as a second home. A host of students left M.I.T. with a warm feeling for that institution because of the hospitality shown them by the Morrises."

Fred's never-ending interest in children was shown in various ways. In 1936 at the request of Rollo A. Reynolds, editor of "Our Changing World Library," he wrote a textbook on the Hudson River for use in New York City schools. It was entitled The Making of the Valley. To its writing he devoted a whole summer vacation that he might explore the entire length of the river. This being a public service, his only remuneration was $100 for the illustrative drawings. In contrast were the "dinosaur parties" given when living in Montgomery. For the children who came to these from far and near he laid out a large part of a mososaur skeleton on the floor of the garage.

WORK IN MONGOLIA - THE THIRD ASIATIC EXPEDITION OF THE AMERICAN MUSEUM OF NATURAL HISTORY

In the Gobi Desert Morris made geological studies covering more than 8,000 miles of geologic traverse, sketched some 1,500 miles of route map, and surveyed numerous selected areas by plane-table. His means of travel varied from the running board of automobiles to the back of a camel or a small Mongolian pony to his own feet. Standing 6' 2" tall and weighing only 175 pounds he was a striding speedster and could cover ground at an impressive pace. Many of his students of later years can attest to this as well as to his rock-climbing agility and speed. He was a natural field geologist.

His field notebooks, well illustrated with pen drawings, cross-sections and colored sketches, provided many of the illustrations in Geology of Mongolia (1927), co-authored with C. P. Berkey, and in many Novitates and Bulletins of the American Museum of Natural History (See Bibliography).

While studying the Expedition's collections at the Museum, Fred had to consult the foreign literature on the Gobi, and that meant reading French, German and Russian; there was no important literature on Mongolia in Chinese. From these years of study and research came ultimately more than 25 articles and abstracts describing the geology of Central Asia. Most of these were published jointly with C. P. Berkey between 1924 and 1932 (See Bibliography). More than fifteen years later, writers of The Christian Science Monitor in Boston would still be writing feature stories based on descriptions and sketches of the Gobi provided by Morris,* and today, almost half a century later, the Berkey-Morris articles and the

---

* See footnote on following page.

classic Geology of Mongolia (1927) are still important sources of information about a little known region of Asia.

THE M.I.T. YEARS (1927-1950)

By 1927 Morris had about completed the first and most important part of his work with Berkey on the Gobi and needed to be looking about for a permanent position. An opportunity came much sooner than he could have hoped. One afternoon in the summer of 1927, while he was working in Columbia's Schermerhorn Library, M.I.T.'s Waldemar Lindgren happened to enter in search of a certain reference. Morris noticed that he was seemingly unsuccessful in his search and, knowing that the Librarian was out of the room temporarily, offered his aid to Lindgren and quickly located the desired reference. This chance meeting--or was it contrived?--led to some questions from Lindgren, then some conversation, and finally an invitation to join the M.I.T. faculty as a member of the Department of Geology. After determining that Mrs. Morris would be willing to leave New York City for Cambridge, Fred enthusiastically accepted the invitation. Some months earlier when he had viewed the M.I.T. buildings at sunset from a church across the Charles River, where he was to lecture later in the evening, he had said to Mrs. Morris:

> "I think that if I could teach there, I'd consider it
> the most wonderful thing in life."

So good fortune does sometimes await the unsuspecting!

September 1927 found Morris an Assistant Professor of Geology in Course XII assigned to teach engineering and structural geology and to help out in the introductory subjects. For the next 23 years, until retirement as Professor Emeritus on 30 June 1950, he served the Department of Geology successively as Assistant Professor (1927-1928), Associate Professor (1928-1931), and Professor (1931-1950).

Soon after Morris retired he received the following letter from his old and deeply respected friend and colleague of Gobi days--C. P. Berkey:

> "My dear Fred Morris:
>
> > "Some way I feel a bit sorry to learn from Science
> > of the 31st of March that you are to stop teaching at
> > the Institute. I can't quite see it,--in fact--I can't
> > quite believe it. I don't believe I ever saw you when

---

* "China reported prosperous under apparent adversity" (Anonymous), Monitor, Nov. 12, 1927, p. 1, 4; "Peking Man type of but one of many tribes" (Anonymous), Monitor, Dec. 5, 1931, p. 5; "Geologist tells how Gobi rocks yield bits of extinct monsters" (Anonymous), Monitor, Jan. 15, 1932, p. 5; "Dinosaur's appetite linked to development of Geology" (Smith, Everett M.), Monitor, Feb. 13, 1947, p. 2.

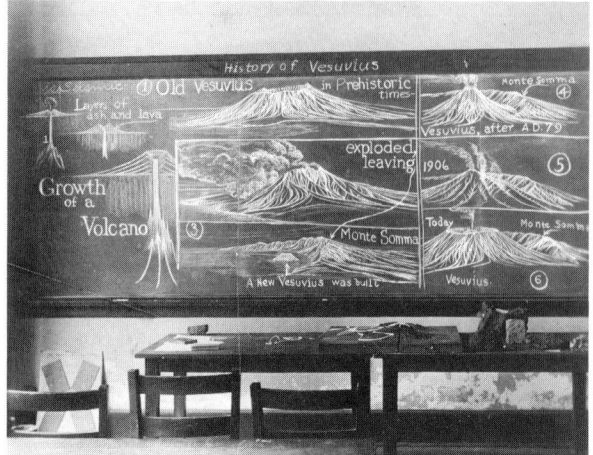

SKETCHES BY F. K. MORRIS

you were not ready but well enough prepared to do a
good job of it day or night or Sunday. I just can't
believe that you are to stop.

"In fact I think I know that you won't. I'll bet
you'll be showing some one what to do or how to do it
better before you get 10 miles from the old campus.
After all these years you can't stop. When you and I
stop, they'll know we're dead.

"What you ought to do (you see I'm at it again) is
to move up into the Catskills and make it rain. I'll
bet you know more about how to do it than any Kats
Killer I ever saw,--and we sure need water.

"Good luck, old man. We've been in the same boat
before,--and stood it. I guess we can pull through.

Sincerely,

[Signed]   Charles P. Berkey"

Morris brought to M.I.T. a combination of skills, knowledge, and personality traits that soon made him one of the most accomplished and sought-after public lecturers at the Institute. He was a skillful artist with chalk, pencil, pen or brush, and his sketches have a characteristic fluidity of line and form that remind one of Tennyson's verse:

> "The hills are shadows, and they flow
> From form to form, and nothing stands;
> They melt like mist, the solid lands,
> Like clouds they shape themselves and go."
>
> [In Memoriam 123]

They also have a certain whimsicality that makes many of them a delight to study, as exemplified by the accompanying Map of M.I.T. drawn in 1941 at the request of Mrs. Karl T. Compton. His blackboard sketches were legend; his whimsical sketches for announcements were always received with a smile; his Christmas cards with their poetic titles were eagerly awaited by his host of admirers; his geological sketches added much to several books mentioned earlier and not infrequently appeared in attractively hand-colored lantern slides; and his colleagues often closed a lecture with the crediting remark--"The slides were prepared by Fred Morris." A few examples of Fred's drawings are shown on several accompanying pages.

There is a poignant story to be told here about Fred and his drawing. He drew so well as a child that his parents took it for granted that he would be an artist and encouraged him to draw. When he was about twelve he viewed an outstanding collection of the child drawings of numerous great artists at the Metropolitan Museum of Art in New York City and when he compared his own efforts with the ones he saw on exhibit he concluded that he could never be a great artist. Having the commendable attribute of being completely honest with himself and trying never to hurt others he kept his deep disappointment from his parents, turned his efforts and interest to geology, and set about preparing himself for a scientific career. Almost half a century later when reminiscing at his retirement dinner in Cambridge's Continental Hotel on 24 May 1950, he candidly remarked:

> "I have always been a creature of half talents--in languages, literature, and the artistic skills, but I have used and enjoyed them to the full."

Many of his friends would feel he surely possessed more than the half talents he mentioned that evening, especially in drawing!

Morris was an exceedingly rapid reader, and he read widely in such diverse fields as science, literature, art, philosophy, religion, history, and anthropology. He had an unusually retentive memory and stored away an amazing amount of knowledge. With almost total recall he could recite poetry by the hundreds of lines, and his taste ranged from Shakespeare through Browning and the Gilbert and Sullivan lyrics to many a choice limerick. It is said that he could tell the stories of a hundred books naming characters and places. Furthermore, he did not confine himself to English for he also knew French, German and Russian, and a little Chinese. There is a limerick in one of the old copies of <u>Dinosaur Dust</u> (the "rag" of Columbia's Department of Geology) that reads:

> "Fred Morris: There was a geologist Morris
> Who could reel you off yards of old Horace
> He went to the Gobi
> Sat down on adobe
> And told it to each dinosaurus."

Marjorie Hooker of the U. S. Geological Survey, who gave me the preceding limerick, also recalled an afternoon at the Oasis of Ghardaia, deep in the Sahara on a field trip during the 19th International Geological Congress in Algeria in 1952. At tea that afternoon, Morris entertained and held the small party spellbound until dinnertime with verse after verse, culminating in a memorable rendition of Gilbert's "Perils of Invisibility." At the celebration marking the end of the field excursion, with some fifty geologists and numerous French military officers assembled at Tamanrasset, a French Army base at the time, Fred was chosen as the geologists' spokesman to thank their leaders and French hosts. As Marjorie has written to me:

> "And then too soon, it was the last day of our excursion. Instead of returning to Algiers as we had come, a second group was to be flown to Tamanrasset to do the excursion in reverse, and we were to return to Algiers by plane. For one night, there would be fifty of us at the hotel and we would have a celebration. Tamanrasset is a French Army base, and the Commandant of the District flew in from Ouargla and with his officers joined us at the banquet. It was a gala affair with champagne and a minimum of speeches. Fred had been chosen as our spokesman to express our thanks to our leaders and their

(19) FREDERICK KUHNE MORRIS   633

Dear Friends—

    Our Mrs. Compton asked us to draw a map of the M.I.T., showing places of interest to the lay visitor, so that copies might be given to the new ladies of the Institute. The time was short, and I sought the help of Professor Brown of Architecture, who enlisted some talented students; and the map was the joint work of all our hands.

    Our plan was not to <u>illustrate</u> an activity, but to offer a whimsical <u>symbol for the</u> activity. Thus our Radiation Laboratory <u>is sending</u> forth beams of unheard-of energy; so we drew Svengali the hypnotist as symbol for power-radiation. The stroboscope stops the humming-bird in one 3000th of a second; the posed bird is the symbol for Dr. Edgerton's laboratory; the abacus is surely a symbol for the brass-brain which solves calculus equations. The wind tunnel is the function of the great God Aeolus, <u>redivivus</u>; and Jove himself brings lightnings to the <u>giant</u> generator of Dr. Van de Graaff. And when two chickens contend for a worm, they are indeed testing materials. Of course we represented all the ladies activities by angels. The Architects are drawing ultra-modern block-houses, and rave over structures such as our generation built at the age of four.

    With these hints we leave you to interpret the pictures or to ask us about them. Florence is in charge of our Red Cross unit this year, and runs it with her whole intemperate soul, keeping all her workers busy, happy and productive, at the expense of uncounted labor of her own. I am still champion crutcher of this region. We send you our love and greetings.

Florence and Frederick Morris
1941-2

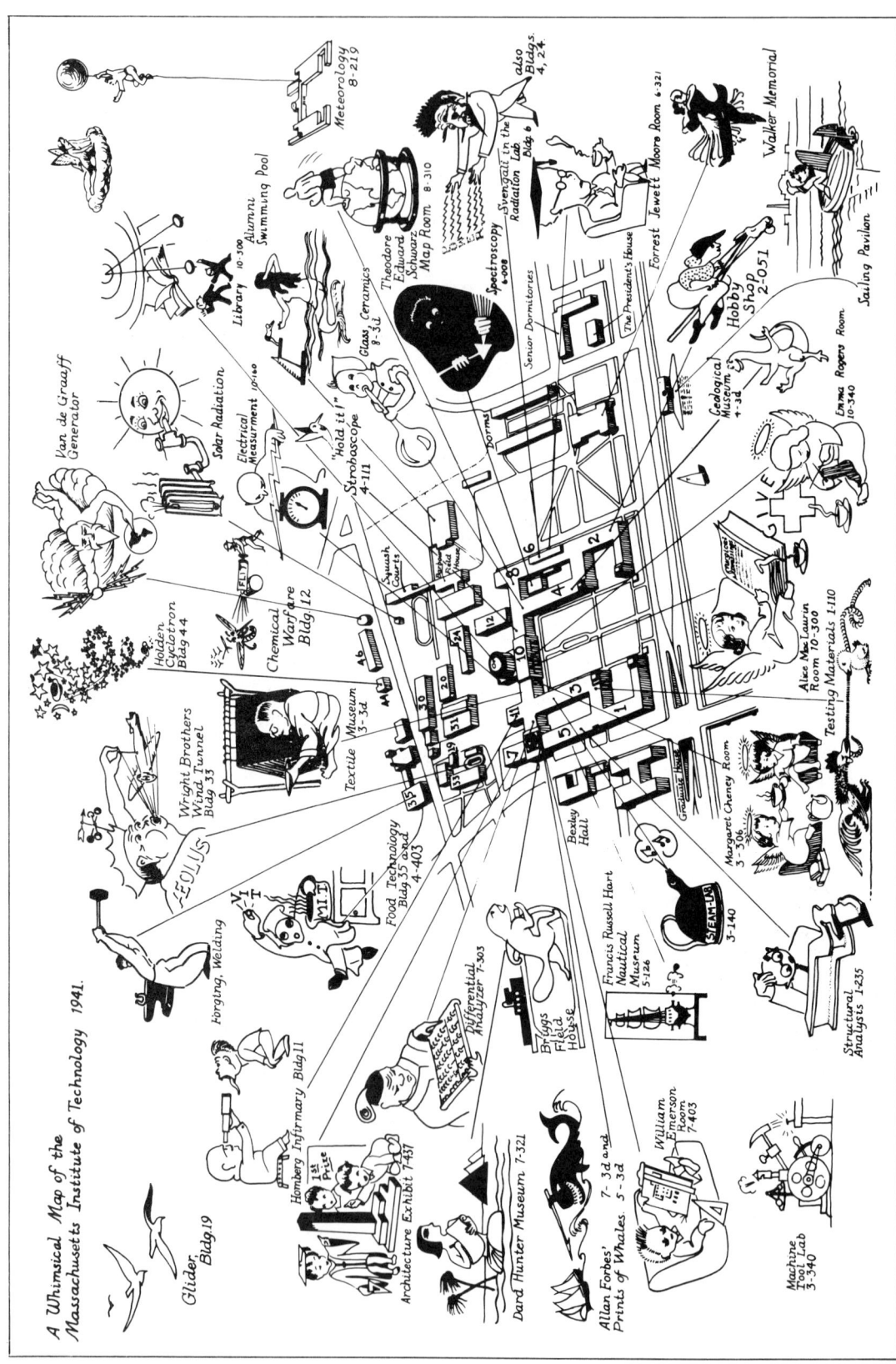

> hosts. In meticulously enunciated (and American accented) French, he spoke simply and from the heart to the rousing cheers of all present. For all of us it will be unforgettable. That excursion came at almost the last possible time that geologists could safely pursue field work in the Sahara." [Letter from Marjorie Hooker to Robert R. Shrock, dated February 12, 1973.]

Further details of that memorable excursion that illustrate additional facets of Fred's character are included in the remainder of Marjorie Hooker's letter which is quoted on a following page. Years after the excursion upon learning of Fred's death Professor Lelubre, one of the trip leaders, wrote to Mrs. Morris as follows:

> "Le décès du Professeur Morris m'a fait beaucoup de peine. J'en garde le souvenir d'un homme de grande culture et de grande intelligence, et d'une amibilité parfaite."

Teacher, Investigator, Consultant

Morris was a member of M.I.T.'s Department of Geology for 23 years, from 1927 to 1950. During this period he taught subjects with the following titles: Physical and Historical Geology, Field Geology, Engineering Geology, Structural Geology, Geomorphology, Glacial Geology, Geology and Evolution, World-Resources and World-Problems, Topography and Strategy in both Great Wars, and Economic Geography (for students in Naval Architecture). He also worked closely with the Department of Civil Engineering in studying soil formation and the field relations of the different types of soil. For some years after joining the M.I.T. Faculty, he retained a close relationship with the American Museum of Natural History and when possible he continued to work on the Gobi material. Ultimately, in 1936, some of this work was presented as a thesis for which Columbia awarded him the Ph.D. degree in geology.

For Fred Morris to teach, to share his knowledge and ideas with others, was the most exciting and satisfying activity he could imagine, and this commonly took the form of lectures. Every lecture was an opportunity to teach and every question was the stimulation for a lecture. It was an evening paper at Columbia's Journal Club early in his career that impressed A. W. Grabau and brought from him the invitation to study paleontology at Columbia. It was his teaching talent that helped him go from C. P. Berkey's laboratory assistant to Lecturer and then Instructor in Columbia's Department of Geology, and it was at M.I.T., from 1927 on, that he came to be known as one of the outstanding teachers and public lecturers at the Institute.

He brought to his teaching and public lecturing every skill and piece of knowledge he had, every bit of drama and ingenuity and imagination he could muster. He passionately believed in the importance of excellent teaching, in which he included lecturing as an indispensable part, urged his students to develop good teaching techniques, and made himself an excellent model. He made his major contribution to Geology as a teacher--as a classroom or platform lecturer, as a laboratory demonstrator, or as a field trip leader. He assembled specimens, prepared attractively colored lantern slides for projection, sketched explanatory diagrams on the blackboard (See a preceding page for an example), and otherwise prepared himself for every lecture as an actor would. He was often called upon for public lectures and invariably captivated his audience. His public lecturing is discussed in a following section.

Standing 6' 2" tall, and seeming even taller because of his spare frame and shock of hair that characteristically stood erect as if electrified, he used his facile hands to sketch on the blackboard, illustrate how Tyrannosaurus rex walked on his hind legs, or trace an invading solution upward through a host rock. He was truly a performing artist and put as much time and effort into preparing and delivering a lecture as would an actor on the stage.

He happily met the challenge of any kind of audience whether children or adults, uninformed or sophisticated, laymen or professionals, scientists or non-scientists. He was always the teacher--the describer, the narrator, the explainer--whether his audience or class was a group of kindergarten children, a troop of Scouts, a service or professional club, a fraternity, an amateur group studying minerals and gemstones, a military group discussing strategy, his regularly scheduled classes, or a group of his peers at a geological convention. It was his compelling urge to explain and to help his hearers to gain a better understanding of minerals, rocks, land forms, evolution, and the like. His classes did not soon forget what he taught them and why. His public audiences always left with a better understanding of what Geology was about and how geologists had built up the knowledge of the earth.

In these days of debate on what marks a good teacher, many contend that students are poor judges of good teaching; that they become competent judges only after many years away from the classroom during which they have had time to reflect on their teachers and evaluate them in the hard cold light of experience and mature judgment. But with Fred Morris and his students things were different! They knew that he was dedicated to teaching, that he relished the opportunity to speak to them, that he loved his audience, every audience, and that he always gave the best there was in him.

His love of knowledge, his sense of the dramatic, his feeling for the whimsical, his incisive vocabulary, his expressive and skillful hands so much a part of his presentation, whether by gesture or through a blackboard sketch, all came into focus when he mounted the platform and began his discourse. His words flowed with ease and grace. Without breaking continuity he would quote from an English poet, shift to an expression in French or German--he liked to quote from the original--insert a side comment of satire or praise regarding a work of art or a piece of music, and then return to the theme of his geological subject as the hound to the trail. No person who ever heard Fred Morris discuss crystals, tektites, dinosaurs, pebbles, granitization, or the importance in military strategy of the northward-facing cuestas of the Paris Basin was likely to forget the experience. They may have forgotten the details but they were not likely to forget the man or what he talked about. The proof of the foregoing statement is to be found repeatedly in the many letters of condolence that came to Mrs. Morris following Fred's death.

In laboratory work, which he regarded as one of the three essential components of complete teaching--class room lecture, laboratory instruction, and field trip--Morris placed great emphasis on map interpretation and study of thin sections by petrographic microscope, both disciplines that he had learned well as a student with D. W. Johnson and J. F. Kemp, respectively. Much of Morris' research involved either maps and map interpretation, as at the Department of State in the 1920s, or microscopic study of granites and other igneous rocks later on. (See Bibliography.)

Field work was as close to Fred's heart as lecturing and laboratory instruction, and he never passed by an opportunity to go into the field. He always kept himself in excellent physical condition and as a consequence he could outstride, outwalk, and outscramble his most athletic students and peer colleagues on a hike or climb. It was one of his greatest joys to gather a small group around himself at an outcrop and interpret the rock for them. It was only natural, therefore, when poliomyelitis struck him down in 1938, for his friends to assume that his days of field work were done. But they underestimated his unconquerable spirit and great will power; he was undaunted. He turned to learning how to use his crutches, forearm aluminum ones that he called his "sticks," and he ultimately was able to get about with surprising agility. Most importantly, he resumed field work with his M.I.T. students and met his greatest challenge when he went to the 19th International Geological Congress in Algiers in 1952 and held his own on the post-Congress field excursion into the Sahara. Not only Fred's agility on his "sticks," but also many of his other abilities, became apparent on that memorable trip which Marjorie Hooker has graphically described in a letter to me dated February 12, 1973:

"It was a typical early September morning in Algiers in 1952 when twenty-two geologists in several varieties of desert garb converged on a designated city street to board the two trans-Saharan buses that were to take us to the Hoggar. This was a post-Congress excursion headed for the central Sahara, and while some of the people were old hands at desert life (Manuel Alía, for instance, had traversed 2500 km. on camel, mapping in the Spanish Sahara), others of us had had no desert experience at all. I was one of these, and I considered it was my exceptional good fortune to be on an excursion that had Doc Schürmann and Fred Morris as the two senior members. Fred, of Gobi Desert fame, knew from personal experience every large desert in the world, and Doc, of course, knew the Egyptian Desert as well as the back of his hand. And there was Dick Bramkamp who knew the Arabian Desert like a well-thumbed book. Under the leadership of three French geologists, Lelubre, Follot, and Bordet, all experienced in the Sahara, we started off on the 1200-mile journey to Tamanrasset in southern Algeria.

"After retiring from his professorship at Massachusetts Institute of Technology, Fred had moved, in 1950, to Montgomery, Alabama, where he was Professor of Geology at the Arctic, Desert, Tropic Research Center of the United States Air Force, located at Maxwell Field. Fred was the official representative of the Air Force at the International Geological Congress in Algiers. Heedless of his age and knowing full well what a Sahara trip in September would be, he had nevertheless signed up for the excursion. While others busily queried the leaders about how such things as laundry and mail were to be handled on the trip, Fred took Lelubre to one side and carefully explained to him that if anything should 'happen' to him in the desert he was to be buried right there - on no account was he to be shipped home. As he was not too sure how well Lelubre really comprehended, he also took two of us into his confidence on the matter. It must have been with some inward alarm, following such instructions, that our leaders viewed the weeks ahead.

"After crossing the Atlas Mountains and spending the first night at Laghouat, it was only a short run to our next overnight stop at the oasis of Chardaia and we were there in time for lunch. South of the Atlas, the Cretaceous strata are gently folded, and where erosion has exposed them, one can follow the lines of the formations for miles. Generally light colored, we had noticed almost continually visible thin dark beds in the sections and were told that these were manganiferous and not just desert varnish. Having the whole afternoon available, Fred suggested that we investigate these dark layers which we could see in a not too distant outcrop. Have you ever had occasion to accompany a geologist who uses 'sticks' because he has had polio, but who averages about eight miles an hour as a casual gait? In no time at all we had reached the outcrop, examined it, and returned to the oasis to have a leisurely tea in the garden. Chardaia is well known as a beauty spot in the desert and the grounds of the hotel are particularly attractive. While we were enjoying our tea in these surroundings, someone recalled and recited a verse, another followed, and each of us contributed something, but it was Fred who entertained and held us spellbound until dinnertime. I can still see him sitting there giving us his rendition of Gilbert's 'Perils of Invisibility' and if you are not acquainted with it, I recommend it.

"From Chardaia southward, our excursion was more and more on its own as we left civilization behind. We traveled at night from El Golea to In Salah to escape the unbearable heat of the day. When we arrived at Arak, cognac was brought out to revive an exhausted troupe and half the party spent the afternoon prostrated in the scant shade of an adobe 'hotel.' What did Fred do? Well, he was busily discussing the origin of granite. A few days later we arrived at Tamanrasset, our base while we made excursions locally to Precambrian granite and gneiss outcrops and went into the Tertiary volcanic area of the Hoggar. There was time, now that we were no longer on a daily mileage schedule, for more leisurely conversations and we learned all manner of things from Fred, one of which I recall was how easy it is to learn to speak Chinese. I am not sure I believe it to this day.

....

"...That excursion came at almost the last possible time that geologists could safely pursue field work in the Sahara. One of our leaders, Follot, was assassinated not long after. Lelubre and Bordet have had to postpone further work there for years. Lelubre, now professor at Toulouse, has written of Fred, 'Le décès du Professeur Morris m'a fait beaucoup de peine. J'en garde le souvenir d'un homme de grande culture et de grande intelligence, et d'une amabilité parfaite.'"

The research that Morris conducted as an M.I.T. professor spread over a broad range of problems and produced much of interest that he could use in his lectures. In contrast he seldom published more than abstracts because he much preferred the lecture audience to the printed page. He listed among his "researches" the following:

1. Metamorphic processes in rocks of central Massachusetts
2. Paragenesis of minerals in the Marlboro series
3. Origin of Fitchburg granite
4. Structural relations of the Dedham granite and the Lynn volcanics, including the origin and history of the inclusions in the Dedham granite
5. Stresses in crystalline rocks of deep-seated origin: studies in the Chelmsford granite, the Medford diabase and other crystalline masses
6. Study of the water supply of Needham
7. Study of sands of the Atlantic and Pacific coasts of the United States
8. Investigation of tektites

With such a varied group of research problems, and with such close contacts with students, some may wonder that in 23 years at M.I.T. Morris supervised only 7 theses--3 bachelors and 4 masters. Quite obviously he made his impact on students less by exciting their research interests than by stimulating their interest in learning, in acquiring knowledge for use in research, not in research itself.

Fred enjoyed acting as a technical consultant in law cases. In this he was quite successful. The lawyers involved always enjoyed working with him, and recognized his great ingenuity in settling cases out of court to the satisfaction of both parties. It was the research and the people he enjoyed. In his consulting Fred would willingly advise on almost any problem that involved application of a geological principle. It mattered little where it was or what the weather; he was always ready and eager to go. Much of his work was done without remuneration. In times of little employment, he commonly turned over jobs to his needy students. To him it was the challenge of the problem that mattered, not the fee, so he looked upon consulting jobs as opportunities for exciting geological work.

Public Lectures

Inasmuch as Morris kept no systematic diary of his public lectures, there is no way of determining the exact number or, for that matter, the subject, group addressed, or size of audience. It has been possible, however, to learn of a number of public lectures that were cited or commented on in a variety of publications, and a few of these may be considered typical.

In 1924-1925, at 5 p.m. on Thursdays, he offered the following lectures in the University Extension program of Columbia University (in 401 Schermerhorn Hall) under the title "Physiographic and Geologic Illustrations and Methods: A series of seven lecture-demonstrations" to show how to make interpretative perspective drawings:

```
1924
    Nov. 13.  Preliminary symbols
    Nov. 20.  Drainage systems and simple surface forms
    Dec.  4.  The simple block-diagram
    Dec. 18.  Unglaciated mountains
1925
    Jan.  8.  Dissected blocks
    Jan. 15.  Field sketching
```

Soon after coming to M.I.T. Morris participated in the Popular Science Lectures of the Society of Arts given to pupils of the secondary schools in and about Boston on Friday and Saturday afternoons. In this series he gave a lecture on "A geologist's travels in the Gobi Desert" [President's Report 1929-1930, M.I.T., B. 66/3: 84-85, (1930)], and in February 1939 he gave a series of lectures to the same group on meteorites and tektites under the title "Rocks from the Sky." These lectures, delivered on Sunday afternoons, were so well received that they were given for three successive years.

At the 93rd meeting of the American Association for the Advancement of Science and Associated Societies, held in Boston on 27 December 1933 to 2 January 1934, Morris presented two papers: "The date of the dinosaur eggs," and "The history of geology and mining in the Colonies." The first seems not to have been published, but the second, with the title "Geology and mining in the Colonies: the geological stage on which America's pageant of colonial progress developed," was published in Tech. Rev. 46/6: 209, 215-217, 236, 238, (March 1934).

During World War II, in the spring of 1945, the Boston Public Library in cooperation with M.I.T. offered a series of illustrated public lectures on "Certain aspects of post-war developments in Science and Engineering" on Sunday afternoons. On 8 April 1945, Morris gave a lecture on "Earth, Moon, Man and Time." [Tech. Rev. 47/6: iv following p. 400, April 1945.]

Fred's lectures often elicited newspaper and other comments, of which the following are typical:

> China reported prosperous under apparent adversity—a news story based on an interview with Prof. Morris published in the Christian Science Monitor for 12 November 1927 (p. 1 of Atlantic Edition).
>
> Peking Man type but one of many tribes—a newspaper interview in anticipation of a visit to the Children's Museum of Boston by Roy Chapman Andrews of Gobi Desert fame, published in the Christian Science Monitor for 5 December 1931 (p. 5 of Atlantic Edition).
>
> Geologist tells how Gobi rocks yield bits of extinct monsters—a news story based on an interview with Prof. Morris and published in the Christian Science Monitor for 15 January 1932 (p. 5 of all editions).
>
> Rocks from the Sky—a series of lectures delivered before the Boston Society of Arts, and reported on in the Christian Science Monitor on 13 February 1939 (p. 9 of the Atlantic Edition) under the heading "Earth proves elusive target for sky rocks."
>
> Gem story found in dinosaur's eggs—a news report on lectures given by Morris to the Eastern Guild of the American Gemological Society, published in the Christian Science Monitor for 13 February 1947 (p. 2 of Atlantic Edition).
>
> Power of wealthiest nations seen key to One World Idea—a report on a lecture to a monthly Club Institute-Forum meeting at the Massachusetts State Federation of Women's Clubs, reported on in the Christian Science Monitor for 4 March 1948 (p. 5 of Atlantic Edition).
>
> Let us think clearly—a Commencement Address at Hiram College on 3 June 1956 (Published in Bulletin of Hiram College, Vol. XLVIII, No. 6, June 1956; copy in the M.I.T. Archives; also published in Vital Speeches of the Day, 22/23: 734-735, 15 September 1956).

At least twice Morris participated in radio broadcasts. In 1935 in a program sponsored by the New England Section of the American Chemical Society, he delivered a lecture on "Into the depths of Time: when did life begin?" The substance of the lecture was later published under the same title in Tech. Rev. 37/4: 121, 139, 152, (Feb. 1935). In January 1939, in a series on "Better Homes for Better Living," over an international broadcast at M.I.T. through the World Wide Broadcasting Foundation, he gave a lecture on "Mother Earth: the foundation and source of supply," which was noticed in Tech. Rev. 41/4: 180, (Feb. 1939).

Early Travels and a Trip around the World

Morris traveled widely and wherever he went he observed, sketched, and remembered his impressions. He met people and adventure with equal eagerness so that his life was considerably more exciting and adventuresome than most. And for more than half of his 77 years, many of his adventures and travel experiences were shared with his wife, whom he affectionately referred to as "My Lady."

His travel started when the family moved across the continent from San Francisco to New York City when he was only eight years old. Field trips in New York and New Jersey while he was a student at C.C.N.Y. and Columbia took him to famous outcrops in that area; a summer in the field with Jesse Hyde in 1912 gave him a chance to study the geology of the Canadian Maritimes. A brief trip to England and France in 1914 was terminated abruptly by the outbreak of World War I. On field trips during annual meetings of The Geological Society of America Morris was a familiar figure; characteristically, he was likely to be the center of a little discussion group. He attended three meetings of the International Geological Congress: the 17th in Russia in 1937, the 19th in Algiers in 1952, and the 20th in Mexico in 1956. At the two last-named he went as delegate of the U. S. Air Force. In addition to his early travels in the Far East, in the 1920s, he and his wife extended their attendance at the 1937 International Geological Congress into a year-long trip around the world. On a memorable trip, during which they met many people they had known earlier at M.I.T., the Morrises had enough experiences to last an ordinary geological couple a lifetime.

Leaving Cambridge in June 1937 they went first to Germany and Austria, where they met Albrecht Penck and Karl and Ruth Terzaghi; then on to the Congress in Moscow. At a great banquet in the Georgian Hall of the Grand Palace of the Kremlin, Florence, not Fred this time, was asked after the dinner started to thank their hosts in the name of the foreign women who had been guests of the Russians at the Congress. (It was said that she was the first woman to speak in the Great Hall of the Kremlin since Catherine the Great; and even Intourist was impressed.)

After the Congress the Morrises took the Siberian field trip across the Urals to Krasnoyarsk and the Volga and back, and then went on to Istanbul where Fred conferred with Professor Charles Whittemore of Harvard who was overseeing restoration of the mosaics in the great Santa Sophia cathedral.* Fred was given a cigar-box full of tessellae to identify the stone ones and to determine how the glass ones had been cut with so little waste. After studying the fracture marks on the glass, he drew a hypothetical instrument that he thought might have cut the glass economically. When Whittemore saw the sketch, he is reported to have exclaimed:

> "By George, Morris, we found one of those. It is now in the Byzantine Museum in Paris, and we could not guess what it had been used for."

Syria, Palestine and Egypt were next on the trip, then India with several more typical incidents. At the Silver Jubilee of the Indian Science Congress in Calcutta Morris gave three of the public lectures. The Maharaja of Bangalore asked him to "diagnose" the spots on Sri Swaraswami, the largest monolithic statue in the world and one of the most sacred statues in Hindu India. Concluding that the disfiguring spots on the statue were due to prolonged weathering, Fred suggested that the faithful be told that Sri had stood on his high hill enduring the wind and rain of the centuries to protect them, and that he would continue to do so indefinitely into the future. Later, while a guest of the Nizam of Hyderabad, he examined the famous carvings in the Ellora and Ajunta rock temples and was able to assure the Nizam that in spite of some obvious vandalism most of the destruction was due to normal weathering of the easily carved rock. Years later, in Cambridge, the Morrises received a letter from the Nizam saying that he planned to visit Boston with his entourage and asking if Fred would help plan his visit. Inasmuch as the Nizam had been especially kind to the Morrises in India, they in turn wished to extend their warmest hospitality, although, as Fred once remarked to me, "It cost us a pretty penny!"

---

* Before their world trip in 1937, Morris had been working with M.I.T.'s Dean of Architecture, William Emerson, in studies of rock samples from the early Christian cathedral, Santa Sophia, in Constantinople. Emerson was chairman of an international committee, based in Paris, established for the purpose of preserving and restoring the early Christian mosaics in Santa Sophia, which had been covered with four inches of plaster by the Mohammedans when they conquered Constantinople in 1453. When he learned that Morris was going to Constantinople he gave him a letter to Professor Charles Whittemore who was in charge of the restorations in the great mosque.

After returning to the United States in 1938, Morris continued as a consultant to Dean Emerson and from 1946 to 1950 served on his Advisory Committee for Santa Sophia, and continued his studies of tessarae until retirement in 1950.

After India came Ceylon, Burma, Cambodia, the Malay States, Java, Bali, Japan, the Philippines and finally home, with their most memorable trip behind them.

World War II

When World War II came Morris again volunteered his services, as he had done in World War I, and for a number of months he taught elementary geography to a group of weather officers of the U. S. Army Air Force at the Institute.* In addition to lectures, he made a series of landscape sketch maps for them, and many of these sketches were carried over "The Hump" during operations in the Far Eastern Theater. Later on, in 1959, when three officers from the Department of Defense visited Fred in the Arctic, Desert, Tropic Information Center in Montgomery, Alabama, in what they feared would be a fruitless search for some rather precise elevations in Mongolia, much to their surprise they found exactly what they sought neatly recorded in his original notebooks and maps from the Gobi expeditions. The officers subsequently told Dr. Paul Nesbitt, who had suggested Fred might be helpful, that they received more information from Fred's material than from anywhere else they had looked. Shortly after his death in 1962, Mrs. Morris presented Fred's maps, notes, and notebooks, including all of his Gobi material, to the American Museum of Natural History, where they are now preserved but available for reference in the Museum's archives. This was a logical as well as a generous thing to do because Fred had been appointed a Research Associate in Geology in the Museum's Department of Central Asiatic Research and Publication, as of January 1942, and had served in this capacity during the following decade.

Organizational Memberships and Outside Activities

Fred Morris was an intellectual activist, although he played an active role in only those few organizations in which he had a special interest. He maintained his membership in more than 20 societies, institutes, academies, etc.,** however, holding the view that scientists should support their professional organizations if they believe in their objectives and programs.

The Children's Museum in Boston was one of his special interests. He served it as Trustee and Vice President and was made a Trustee for Life

---

* "Prof. F. K. Morris has charge of a course in elementary geography, given to more than 200 members of the Meteorology B Group of the U. S. Army Air Forces." [Pres. Rept. 1942-1943, M.I.T. B. 79/1: 119, (1943).]

** See footnote on following page.

when he left the Boston area in 1950. At M.I.T. he was active in The Outing Club because he liked to hike and climb. He served on the New England Council, and as President of the Boston Geological Society and of the Boston Section of the American Institute of Mining and Metallurgical Engineers. He helped to establish the Gemological Institute of America and gave the first lectures on gems to Boston jewelers. In the last decade of his life, while living in Montgomery, Alabama, he joined a group of intellectuals (called simply "The Group") with interests like his own, and over a period of years organized almost a hundred monthly programs on science, history, literature, and world problems.

He was elected member of Phi Beta Kappa (City College of the City of New York), Tau Beta Pi and Phi Beta Epsilon (Honorary Member), Chi Epsilon and Sigma Xi (Past President, M.I.T. Chapter). He was also a Past President of the M.I.T. Faculty Club.

THE CLOSING YEARS (1950-1962)

In 1950, shortly before retirement from M.I.T., Morris gave a series of classified lectures to a military group in the Boston region and so impressed them that he was asked to join the civilian staff at Maxwell Air Force Base in Montgomery, Alabama. He accepted the invitation, moved with Mrs. Morris to Montgomery, and became Chief of the Tropic Section of the Arctic, Desert, Tropic Information Center of the Research Studies Institute. He held the academic title of Professor of Geology and Geography to the U. S. Air Force's Air University Library and Lecturer to the Air War College of the Air University at Maxwell Air Force Base. In 1955 he was a participant in the "Global Strategy Discussions" held 6 to 10 June at the Naval War College in Newport, Rhode Island.

But retirement in 1957 by no means meant the end of activity for Fred Morris. For the remaining five years of his life he interested himself in the affairs of the growing Geology Department at Birmingham-Southern Col-

---

** At one time or another during his more active years Morris held membership in the following organizations: Alabama Academy of Science, American Academy of Arts and Sciences (Member of the Council, 1946-1950), American Association for the Advancement of Science (Fellow, Life Member), American Geological Institute, American Geophysical Union, American Geographical Society, American Institute of Mining and Metallurgical Engineers, Appalachian Mountain Club, Arctic Institute of North America, Boston Geological Society (President, 1931), Byzantine Institute, Children's Museum of Boston (Vice President, 1950, Life Trustee), Ends of the Earth Club, Explorers Club (New York City), Gemological Institute of America (Charter Member), The Geological Society of America (Fellow), Massachusetts Forest and Park Association, Mineralogical Society of America, New York Academy of Science, Sigma Xi, and the Harvard Travellers Club (Cambridge).

lege. He gave the Department his library, petrographic microscope, and geological collection that included a portion of a dinosaur egg from the Gobi, believing, as he put it, "It is better for the young to use than for the old to store." Outside the College he specialized in children as Mrs. Morris characterized his activities. He gave "dinosaur parties," using colored slides of scenes in the Gobi Desert and mososaur bones that a Boy Scout had found near Fort Deposit, Alabama. He helped many girls and boys with their Science Fair exhibits by contributing both advice and specimens from his own collection. To the very end he continued what he had done all his life--giving willingly and generously of what he had and what he knew.

FREDERICK K. MORRIS, THE MAN

Fred Morris was a warm-hearted person whose sympathy and understanding went out quickly to those who needed help. He was particularly fond of children and youth and could involve himself in their problems so completely that he forgot his own. The extraordinary will power that he showed in overcoming the physically disabling effects of poliomyelitis was admired by all who knew him. His philosophy of life, his attitude toward the unfortunate, his tolerance towards those with whom he disagreed, and his widespread generosity endeared him to a wide circle of friends and left a deep impression of a gentle and kind man who wanted above all else to share with others what he owned and knew.

He could also be stubborn as a mule in holding onto or rejecting an idea. He strongly believed that the Pleistocene was of much longer duration than most geologists of his day assigned to it, and glaciologists of today are coming around to his view. He was convinced from geological evidence that man was very much older than most anthropologists believed, and it would please him to know that they now acknowledge this. He was convinced that certain granites resulted from metasomatism rather than from direct crystallization from a magma, a view that was somewhat ahead of its time. He would hear nothing of continental drift, refusing even to discuss it or to have his students discuss it in class, yet I feel sure that if he were alive today he would be proclaiming sea-floor spreading and plate tectonics as vigorously as he once denounced continental drift, for he could and would change his views in the face of new and convincing evidence.

His geological peers will remember him most for his field work in Mongolia with Berkey; his military friends will remember him for his contributions to several books on military geology and for his lectures on military strategy at Maxwell Air Force Base in Alabama; and the many who sought his advice and engaged his services as a geological consultant will remember him as a person who would go to almost any lengths to help them. Most of all, however, will he be remembered by the students in his classes and laboratories and on his field trips, and by those who heard his public lectures. It was in these latter activities that he displayed his greatest skills the combination of which made him so widely recognized as a superb lecturer and teacher. His chief contribution to Geology lay not in his publications, important as were his joint papers with Berkey on the Gobi; nor in the advices to his clients, important as they were in their place and time; nor even in the services he rendered his Government in two great wars. It was, rather, the image of Geology he implanted in the memories of his listeners, children and adults alike.

No biography of Fred Morris would be complete without including some comments about his wife, Florence Elisabeth (Eddowes) Morris,* who was his devoted and loyal helpmate for 39 years. From the day they were married in Shanghai on 30 December 1922 until they were separated by death on 5 October 1962, she was always at his side and ready to do what was needed, whatever the occasion. Together they showed their guests how gentlewomen and gentlemen were supposed to act in a cultured society; how cultured persons should conduct themselves when engaging in spirited discourse or in passionate praise or criticism of some individual or his works. They strongly complemented one another, and the kindness, understanding and sympathy that they extended to all their guests made a visit to their home an unforgettable experience. When Fred's death became known compassionate letters by the score poured into Montgomery and many of these emphasized the great influence that the Morrises had had on the writers.

An award was established in Fred's honor at Birmingham-Southern to recognize an outstanding geology major, and after his death Florence established a loan fund in his memory for geology students there.

---

* Florence Morris was a member of the first Foreign Student Committee established by Mrs. Karl T. Compton, to aid the foreign students with their many adjustment problems, and she and Fred entertained many of them in their apartment in the former Riverside Apartment Hotel, now renamed Burton House. On their world trip in 1937 and 1938, they renewed their friendships with and were entertained by many of the students who had enjoyed their hospitality during their years at M.I.T. Many others mentioned that hospitality in their letters of condolence to Mrs. Morris following Fred's death. An active member of the M.I.T. Matrons and a hard-working knitter in the M.I.T. Red Cross Chapter, she not infrequently stepped out of the conventional female role to run the projector for her husband or even take over a lecture or laboratory section if an emergency arose. Regardless of where they were, or what the occasion, whether in Cambridge, in Montgomery or abroad, the Morrises were as one in many, many ways.

THE FREDERICK K. AND FLORENCE E. MORRIS MEMORIAL FUND

At M.I.T. the present author established a fund to memorialize Frederick K. and Florence E. Morris and invited contributions from their friends. More than 150 individuals have responded thus far and many recalled how much they owed to the Morrises. By the end of fiscal year 1972, i.e. 30 June 1972, the endowment principal had reached a total of $12,000, and the income for the period from 1967 to 1972 had helped numerous students with their field-work expenses. At the beginning of 1973 it was decided that thereafter the annual income would be used to pay for the use of the Department's new Electron Microprobe by students who had no project or other funds to draw on for the fees charged. Thus the Morris Fund will keep alive the memory of a former professor who, if he were living today, would surely be as enthusiastic about the new microprobe as he was about his own special petrographic microscope, one of the best available at the time. (Also see further discussion of the Fund in my Volume 2.)

BIBLIOGRAPHY OF FREDERICK KUHNE MORRIS

(Symbols and abbreviations used in the following references are explained on pages 91-98; in general, abstracts are listed separately, as well as with references to the complete article. This bibliography begins with Morris' first published work in 1918 and ends with a map that was published in 1962, the year of his death.)

1--1918   Map reading and map interpretation, Chap. 7, p. 198-252, il., in Gregory, H. E., ed., Military Geology and Topography, New Haven, Conn.: Yale Univ. Press, 281 p., (1918).

2--1918a   The influence of topography upon the strategy of the European War. Eng. Club of Philadelphia, J. 35/162: 223-234, il., (1918).

3--1918b   The fourth year in France. Eng. Club of Philadelphia [not seen], (1918).

4--1919   (with Kemp, J. F.) Engineering geology, Sec. III, p. 85-145, il., in Blanchard, A. H., ed., American Highway Engineers' Handbook. New York, N.Y.: John Wiley & Sons, 1658 p., il., (1919).

5--1921   Maps and block-diagrams, in Johnson, D. W., ed., Battlefields of the World War. Am. Geogr. Soc., Res. Ser. 3: 648 p., il., (1921).

6--1923   Physiography of Mongolia [abst.]. Geol. Soc. China, B. 2/3-4: 109, (1923); Geol. Soc. Am., B. 35: 87-88, (1924).

7--1924   Notes on the mapping program of the Third Asiatic Expedition in Mongolia. Geogr. Rev. 14/2: 287-292, il., (1924).

8--1924a   (with Berkey, C. P.) Basin structures in Mongolia. Am. Mus. Nat. Hist., B. 51: 103-127, il., (1924; reprinted in Central Asiatic Expeditions, Prelim. Rept. 1918-1925, 1/29: 103-127, il., (1924); [abst.] Geol. Soc. Am., B. 35: 59-60, (1924).

9--1924b   (with Berkey, C. P.) The great bathylith of central Mongolia. Am. Mus. Nat. Hist., No. 119: 11 p., il., (1924); reprinted in Central Asiatic Expeditions, Prelim. Rept. 1918-1925, 1/24, (1924).

10--1924c   (with Berkey, C. P.) Structural elements of the old rock floor of the Gobi region. Am. Mus. Nat. Hist., Nov. 1935: 16 p., il., (1924); reprinted in Central Asiatic Expeditions, Prelim. Rept. 1918-1925, 1/30, (1924).

11--1924d   (with Berkey, C. P.) The peneplanes of Mongolia. Am. Mus. Nat. Hist., Nov. 136: 11 p., il., (1924); reprinted in Central Asiatic Expeditions, Prelim. Rept. 1918-1925, 1/31, (1924).

12--1925   (with Berkey, C. P.) Structural nature and origin of the eastern Altai [abst]. Geol. Soc. Am., B. 36: 133, (1925).

13--1925a   (with Berkey, C. P.) Tectonic history of central Asia [abst.]. Geol. Soc. Am., B. 36: 134, (1925).

14--1925b   (with Berkey, C. P.) Central Asia in Cretaceous time [abst.]. Geol. Soc. Am., B. 36: 158-159, (1925). [Also see item 45--1936a.]

15--1925c   (with Berkey, C. P.) Origin of desert depressions [abst.]. Geol. Soc. Am., B. 36: 169-170, (1925).

16--1926   (with Berkey, C. P.) The geological background of fossil hunting in Mongolia. Nat. Hist. (Am. Mus. Nat. Hist.), J. 26/5: 527-531, il., (1926).

17--1927   (with Berkey, C. P.) Geology of Mongolia, Vol. II of Natural History of Central Asia. New York, N.Y.: Am. Mus. Nat. Hist., 475 p., il., (1927).

18--1927a   (with Berkey, C. P.) Pleistocene deposits in the eastern Altai [abst.]. Geol. Soc. Am., B. 38: 126-127, (1927).

19--1927b   (with Berkey, C. P.) Shabarakh formation [abst.]. Geol. Soc. Am., B. 38: 165-166, (1927).

20--1927c   (with Berkey, C. P.) Climatic pulsations in Mongolia [abst.]. Geol. Soc. Am., B. 38: 211-212, (1927).

21--1928   (with Berkey, C. P.) Review of mountain-making in Asia [abst.]. Geol. Soc. Am., B. 39: 207, (1928).

22--1928a   (with Berkey, C. P. and Granger, W.) New stratigraphic formations in Mongolia [abst.]. Geol. Soc. Am., B. 38: 213-214, (1928).

23--1928b   (with Berkey, C. P.) Time of the last glaciation in central Asia [abst.]. Geol. Soc. Am., B. 39: 221-222, (1929).

24--1928c   Prosperity in Cathay. Tech. Rev. 30/3: 150, (1928).

25--1928d   The biology of racial problems--An interpretation of the Chinese in comparison with other races. Tech. Rev. 30: 226-229, il., (1928).

26--1929   (with Berkey, C. P.) Additional new formations in the later sediments of Mongolia. Am. Mus. Nat. Hist., Nov. 385: 12 p., il., (1929); reprinted in Central Asiatic Expeditions, Prelim. Rept. 1926-1929, 2/95, (1929).

27--1929a   Geology for engineers. Tech. Eng. News 10/7: 362-363, 388, il., (1929).

28--1930   Amygdaloids and cavity fillings [abst.]. Geol. Soc. Am., B. 41: 49, (1930); Pan-Am: Geol. 53: 74, (1930). [Also see next item, 29--1930a.]

29--1930a   Amygdules and pseudoamygdules. Geol. Soc. Am., B. 41: 383-404, il., (1930). [Also see preceding item, 28-1930.]

30--1931   (with Berkey, C. P.) Relations of the Jisu Honguer formation to the general geology of Mongolia, in Grabau, A. W., ed., The Permian of Mongolia, Natural History of Central Asia, Vol. IV, p. 12-32, il., map, (1931).

31--1931a   (with Berkey, C. P. and Grabau, A. W.) Locality list, Jisu Honguer, Mongolia, in Grabau, A. W., ed., The Permian of Mongolia, Natural History of Central Asia, Vol. IV, p. 32-33, (1931).

32--1931b   The Manchurian triangle: its physical background. Tech. Eng. News 12/7: 205-206, il., (1931).

33--1932   (with Berkey, C. P., Grabau, A. W., and Spock, L. E.) Unsolved geologic problems, in Natural History of Central Asia, Vol. I, p. 576-590, (1932).

34--1932a   Some pediments in Arizona [abst.]. Geol. Soc. Am., B. 43: 129-130, (1932); Pan-Am. Geol. 57: 60-61, (1932).

35--1932b   (with Berkey, C. P.) Pediments in the Gobi [abst.]. Geol. Soc. Am., B. 43: 130, (1932); Pan-Am. Geol. 57: 61, (1932).

36--1932c   (with Berkey, C. P.) Paleozoic rocks in Mongolia [abst.]. Geol. Soc. Am., B. 43: 147, (1932); Pan-Am. Geol. 57: 70-71, (1932).

37--1932d   The story of the earth and its creatures: a scientific essay. Part I. From Sun to cell. Tech. Eng. News 13/2: 25-26, 40, il., March 1932. Part II. Whence came the moon? The first seas. Ibid 13/3: 48-49, 59-60, il., April 1932. Part III, The beginning of life. The evolution of the cell. Ibid 13/4: 70-71, 80, il., May 1932.

38--1933   (with Larsen, E. S., Jr.) Origin of the schists and granite of the Wachusett-Coldbrook tunnel, Massachusetts [abst.]. Geo. Soc. Am., B. 44: 92-93, (1933).

39--1934   Date of dinosaur eggs [abst.]. Geol. Soc. Am., Pr. 1933: 452-453, (1934).

40--1934a   Thinking about thinking. Tech. Eng. News 15/3: 52-53, (1934).

41--1934b   Geology and Mining in the Colonies: the geological stage on which America's pageant of colonial progress developed. Tech. Rev. 36/6: 215-217, 236, 238, (1934).

42--1935   Into the depths of time: when did life begin? Tech. Rev. 37/4: 139, 152, (1935. [First presented as a radio broadcast sponsored by the New England Section of the American Chemical Society; also see p. 121, The Tabular View, in Tech. Rev. Ibid.]

43--1935a   Time and our ways of thinking. Am. Scholar 4: 409-418, (1935).

44--1936   Eastern Appalachian geosyncline [abst.], with discussion. 16th Internat. Geol. Cong. 1933, Rept. 2: 996, (1936). The abstract was reprinted in Pan-Am. Geol. 61: 145, (1934).

45--1936a   Central Asia in Cretaceous time. Geol. Soc. Am., B. 47: 1477-1533, il., (1936). [Also see item 14--1925b.]

46--1936b   The Making of the Valley: a billion years along the Hudson. New York, N.Y.: Thomas Nelson & Sons, 75 p., il., (1936).

47--1937   Memorial of Jesse Earl Hyde [1884-1936]. Geol. Soc. Am., Pr. 1936: 163-173, port., (1937).

48--1937a   A scientist's philosophy; glance at philosophy with a scientist. Tech. Eng. News 17/8: 193, 202, (1937).

49--1939   The Pacific Ocean, real and maligned. N. Y. Ac. Sci., Tr. 2/1/8: 121-125, (1939).

50--1939a   Geomorphology of Mount Abu, India [abst.]. Geol. Soc. Am., B. 50: 1924-1925, (1939).

51--1939b   Granite window of Aswan [abst.]. Geol. Soc. Am., B. 50: 1925, (1939).

52--1939c   Pacific Ocean Basin [abst.]. Geol. Soc. Am., B. 50: 1925, (1939).

53--1941   Geologic record and helium time scale [abst.]. Geol. Soc. Am., B. 52: 1925, (1941).

54--1941a   A whimsical map of the Massachusetts Institute of Technology, Cambridge, Mass. 1 sheet, 11" × 17" (1941); the same, 17" × 22" (1944). [Reproduced on page 634 of this biography.]

55--1946   [Sketches in] "This is M.I.T.--Geology" by Mead, W. J. Tech. Eng. News 27/6: 199-203, 220, 228, (1946).

56--1946a   The age of the earth. Physical chemists, astronomers, and geologists unite their knowledge to fix three billion years as a conservative estimate of the Earth's age. Tech. Rev. 48: 223-226, 252, (1946).

57--1949   Cause of an ice age [abst.]. Geol. Soc. Am., B. 60: 1910-1911, (1949).

58--1950   (and Morris, Florence E.) The Rocky Mountain Front. Wilmington, Del.: Hambleton Co., Inc., 15 p., il., (1950).

59--1953   The age of the mountains [abst.]. 19th Internat. Geol. Cong., Algeria, Comp. Rend. 7/7: 41, (1953).

60--1953a   The origin of pediments [abst.], in Capot-Rey, R., ed., Déserts actuals et anciens: 19th Internat. Geol. Cong., Algeria, Comp. Rend. 7/7: 131-133; discussion p. 132-133, (1953).

61--1955   The Ice Age in Sahara and Alabama [abst.]. Ala. Ac. Sci., J. 27: 99, (1955).

62--1956   Let us think clearly: the art and science of thinking. New York, N.Y.: Vital Speeches of the Day 22/23: 734-735, (1956). [Commencement address, Hiram College, Hiram, Ohio, 3 June 1956, and published in Bulletin of Hiram College XLVIII (48)/6: 4 p., (1956).]

63--1961   Memories and reflections of Russia. Summer issue of Lines, Columbia, S.C.: Seibels, Bruce & Co., 6/2: 52-80, (1961).

64--1962   Map of the Gobi basins, following p. 256 in The Desert World, Pond, A. W., ed. New York, N.Y.: Thomas Nelson & Sons, 342 p., (1962).

Geraldine Sullivan in Whitehead's Geochemical Laboratory in Building 24 (Room 24-309).

Irving Breger (XII Ph.D. 1950) in Whitehead's Geochemical Laboratory in Building 24 (Room 24-309).

(Photos by Jackman, 1952)

(20)
WALTER LUCIUS WHITEHEAD*
(1891-1969)

WALTER LUCIUS WHITEHEAD

MIT: 1928-1962

A widely experienced mining geologist, trained in mineralogy and mineral deposits by Lindgren, Warren and Graton, and in petrology by Daly and Lahee, he did field work in South America, Mexico, New Caledonia, Canada and the Caribbean, as well as in many states of the United States. This training and experience he brought to his classes at M.I.T. for 35 years, and to his field charges at the M.I.T. Summer School of Geology in Nova Scotia for a full decade. He supervised the thesis work of 83 individual students and directed two productive research projects--the 10-year API Project 43c, concerning the role of radioactivity in transforming organic materials into petroleum hydrocarbons, and a 7-year project, supported by funds from Nova Scotia, inquiring into the physical and chemical properties of coal. He excelled at suggesting research problems for students to investigate and was at his best with small informal groups. His office door was always open, and many a student and faculty member walked through it to enjoy a conversation whether brief or extended.

Walter Lucius Whitehead was born in Pittsburgh, Pennsylvania, on 4 July 1891, the first son and fourth child of John and Mary (Aitken) Whitehead. In addition to his three sisters--Edith, Helen and Florence--he also had a younger brother, Gilbert, who like himself was a practicing mining geologist before retirement.

Walter's father, John Whitehead, was born in England in 1850, the youngest of twelve children, and came to America in 1863, at age 13, when his parents and siblings emigrated to the United States. John later became a U. S. citizen when his father was naturalized. He died in 1930 at age 80.

---

* Brief memorials have been published in the following journals: Shrock, Robert R. [Memorial to] Walter Lucius Whitehead (1891-1969). Am. Assoc. Petrol. Geol., Bull. 54/2: 363-364, port., (1970).-- Memorial to Walter Lucius Whitehead (1891-1969). Geol. Soc. Am., Pr. 1969: 1-5, port., [preprinted July 1970]; Geol. Soc. Am., Mem. Vol. 1: 105-109, port., (1973).

His mother, Mary (Aitken) Whitehead, was born in the United States in 1848 and died in 1920. She was of Scotch descent and traced her ancestry back through a grandmother Miller to forebears who came to North America before the American Revolution.

Both of Walter's parents were devout members of The Church of The New Jerusalem which is based on the revelations of Emanuel Swedenborg. His father translated many of Swedenborg's writings from Latin into English, served his Church as a minister for many years, and in later years lectured widely on the Swedenborgian faith. Walter himself was confirmed in the faith in 1912 and was a lifelong member of the Cambridge Society of the Church.

It is interesting to note that, while he was reared in a family strongly dominated by the religious beliefs and writings of Swedenborg, Walter followed not the religion of the gifted Swede but the engineering profession that he practiced successfully before turning to religion.

Both Whitehead brothers finished their four undergraduate years as mining engineers, and both started professional careers in the mineral industry. One can not help but wonder if they became interested in science, and especially geology and mining, by hearing their father talk about Swedenborg's scientific work or by looking at the diagrams and woodcuts in the OPERA.*

Walter's primary education was obtained in the schools of Urbana, Ohio, and Detroit, Michigan--cities where the Whitehead family resided for short periods after leaving Pittsburgh. Very early, however, he had attended the New Church School in Waltham, Massachusetts (now the Chapel Hill School for Girls), and when the Whitehead family moved to the Boston area for good in 1906, Walter resumed his education at the school, graduating in 1909. Being an unusually bright student, he soon came to the

---

* The scientific work of Emanuel Swedenborg (1688-1772) is not well known among American geologists, and his name is seldom met in historical sketches of the development of the earth sciences, but he was one of the scientific giants of his age. His 3-volume work, OPERA PHILOSOPHICA ET METALLICA, published in 1734, is comparable in many respects to Agricola's classic DE RE METALLICA, published two centuries earlier in 1556. Detailed discussions of Swedenborg's scientific work can be found in the following articles:

> A. G. Nathorst, "Emanuel Swedenborg as a geologist" in "Emanuel Swedenborg as a scientist--Miscellaneous Contributions," edited by Alfred H. Stroh, Vol. 1, Sec. 1, p. 1-47, (1908), Stockholm.
>
> R. W. Raymond, "Luther, Körner, Humboldt, and Swedenborg," Trans. Amer. Inst. Mining Eng. for New York Meeting, February 1908, p. 1089-1097, (1908).
>
> Herbert Dingle, "The scientific work of Emanuel Swedenborg," Endeavour, July 1958, p. 127-132, (1958).

attention of Headmaster George Beaman who quickly recognized the potential talent of his young charge and urged him to prepare for M.I.T. Walter followed Beaman's advice and was admitted to M.I.T. in the fall of 1909. His undergraduate subjects were largely in mining and geology, and his Bachelor of Science degree, awarded in 1913, was in mining engineering. Continuing his education as a graduate student, he turned away from mining subjects and towards work in mineral deposits as he came under the influence of the great Waldemar Lindgren, who first came to M.I.T. as a Visiting Lecturer on Mineral Deposits during the years 1908 to 1912, and then as Head of the Department of Geology in 1912. Walter took his doctor's thesis work under Lindgren's supervision, receiving a Ph.D. degree in geology in 1918, and was closely associated with him on numerous consulting assignments until his death in 1939.

While a graduate student Walter was an Assistant in geology during the 1913-1914 school year, then was awarded an Austin Fellowship by M.I.T. for the school years 1914-1915 and 1915-1916, in the amount of $250 each year. As part of his graduate program he arranged to do some research in the new laboratory that Prof. L. C. Graton had set up at Harvard for the study of supergene copper enrichment. In a letter to the author dated May 31, 1963, Graton wrote:

> "When material from the field began to pour in for intensive investigation, an effective member of the laboratory staff during 1914 was Walter Whitehead, holder of an Austin Fellowship, who later returned to Technology [M.I.T.] to become a stalwart in its Department of Geology."

Immediately after completing all requirements for the doctor's degree in geology, Walter volunteered for service in the American Expeditionary Force and served as a field artillery and balloon corps officer from 1918 until discharged in 1919. His knowledge of the French language resulted in his being assigned to the artillery school at Besançon, in the Jura, where he became involved in the settling of reparations claims after the war ended.

Soon after returning home from France, he started a decade of consulting work as a mining geologist. His travels took him the length of South America, from Chile and Argentina, through Peru and Bolivia, to Venezuela; thence to Australia; and finally on to New Caledonia in 1922. There he spent several years developing the chromite deposits in the Central Highlands of that island for French interests.*

Whitehead's work on the nitrate deposits in the Chilean desert gave him a superb opportunity to test the validity of W. M. Davis' "desert

---

* See footnote on following page.

cycle" as presented in his classic Geographical Essays (Boston, 1909), and led him to formulate a theory of origin for the extensive deposits of nitre in his well-known article, "The Chilean nitrate deposits" (see Bibliography), in which he concluded that the nitrates had been leached from Mesozoic volcanic rocks under unusual climatic conditions.

While working in New Caledonia, Whitehead met Eugénie Fernande Loupias (b. 10 April 1896), the daughter of the Mayor of Nouméa, the capital city, and they were married there on 3 April 1923 in one of the year's major social functions. The bride's father, Jean-Louis Charles Philippe Marcellin Loupias, besides being a leading politician, was also a bijoutier (jeweler) and an amateur astronomer. He was often called upon to repair and adjust the chronometers on ships that put in at Nouméa, and his astronomic observations, made for time determinations in his backyard observatory, won for him the award of a Chevalier in the Légion d'Honneur and honorary membership in the Société d'astronomique de France. Eugénie Fernande's mother, Eulalie Maria Riviere (Loupias), was a member of one of Nouméa's leading French families.

In 1924 the Whiteheads returned to the United States and took up residence in Malden, a suburb of Boston, where they made their home until Mrs. Whitehead passed away on 13 October 1964. Shortly after returning to America, Whitehead again accepted a foreign consulting assignment, this time with a petroleum company in Venezuela. There, accompanied by three other M.I.T. alumni of Course XII--William F. Jones '09, William B. Millar '26 and Guillermo Zuloaga '27--he made a memorable trip from Caracas to the headwaters of the Apure River at the Colombian border, in the foothills of the Colombian Andes. This trip is interestingly described in "Touring Venezuela in a Model T--An account of a geological expedition just north of the equator" by Jones and Whitehead in the 1928 Technology Review (v. 30, p. 221-225).

With his return to Boston in 1928, Whitehead was appointed Lecturer on Coal and Petroleum in the M.I.T. Department of Geology and henceforth

---

* Whitehead once told the author about an amusing incident that took place while he was working in the field high up on the chromite-capped mountain range. He and his native laborers were sitting around a campfire one night, and the discussion turned to the subject of cannibalism.

    Said Whitehead, to the wrinkled and grizzled old "chief" of the laborers, who was reputed to have indulged in cannibalism in his younger days:
    "Tapoué, did you ever eat white man?"
    Tapoué shook his head and answered:
    "Nah! Him too salty!"

divided his time equally between consulting and teaching for the next 15 years. Consulting assignments took him to Ontario, to investigate nickel and cobalt at Cobalt; to California and Arizona (gold and copper); to Texas (natural gas) and Louisiana (sulphur); in the Boston area (municipal water supply); and to Nova Scotia (petroleum, gypsum and limestone). In some of these assignments he was closely associated with Professor Lindgren.

In 1942 he was appointed Assistant Professor of Geology and planned to devote full time to teaching but World War II changed these plans drastically. July 1943 found him in Haiti in charge of bauxite exploration for the Reynolds Metals Company, an assignment that would ultimately take him and the author to the Dominican Republic, Puerto Rico, Cuba and Mexico. It was not until November 1944 that he again resumed lecturing at M.I.T., and thereafter he devoted essentially full time to teaching and research until final retirement in 1962. He was appointed Associate Professor of Geology in 1947, and Associate Professor Emeritus in 1957 upon reaching age 65, but he continued as Lecturer for another five years until final retirement in 1962 at age 70.

Whitehead's geological activities took three forms--consulting, teaching, and research direction. His consulting record lies hidden in the files of his clients, but the foregoing description of his many and varied assignments should indicate that he was a successful and highly respected mining engineering consultant. As a teacher he made his major impact in seminars and other small groups and in field instruction. Soon after he returned to M.I.T. in 1928 as a Lecturer, he organized courses in field training, based in part on the instruction he had received from his mentor and teacher, Frederic H. Lahee, and in part on what he had learned from his experiences as a consulting mining engineer. He taught field geology from 1928 to 1947, using Cambridge as a base of operations. When, in 1947, it was decided that M.I.T.'s Department of Geology should have its own summer field camp, Whitehead was given the responsibility of establishing the program.

He and the author had done some geological work in Nova Scotia during the summer of 1942 before taking leave to carry on the bauxite exploration program for Reynolds Metals in July of 1943, and had agreed that the coastline north of Antigonish would be an ideal site for a geology summer field camp. It was not until both of us had returned to academic life two years later, however, that our plans could be made and carried out. Discussions were started with appropriate officials in Nova Scotia during the summer of 1947, and by fall all arrangements had been completed, and an official announcement was made from Antigonish and printed on page 11 of <u>The Halifax Herald</u> for Friday, October 17, 1947.

A picture of the founding group and of Crystal Farm Manor House appeared on page 9 of the Herald for Saturday, October 18, 1947. The announcement (in part) was as follows:

> "Antigonish, Oct. 16--The Province of Nova Scotia and the Massachusetts Institute of Technology in a joint statement today announced their agreement to establish an Institute of Geology with headquarters at Crystal Farm on St. George's Bay on Antigonish Harbour.
>
> . . . .
>
> "The statement issued by Premier Angus L. Macdonald and Dr. W. L. Whitehead, professor of geology at M.I.T., said that M.I.T. proposed to give its regular Summer field training in geology and associated sciences at this location."

Whitehead was appointed Director and Professor Roland D. Parks Associate Director. These two started the school the next summer, 1948, and co-directed it for a full decade. The nature of this rather unique international cooperative effort was well described by John J. Rowlands in an article entitled "Acadia--Place of Plenty" that appeared in Technology Review for May 1951 (v. 53, p. 3-8). The present author summarized the work of the first decade in a departmental booklet entitled "Ten Years in Nova Scotia--The Massachusetts Institute of Technology Summer School of Geology, 1948-1957" published by M.I.T. on August 30, 1957. A complete but brief report on the full period during which the school operated, from 1948 to 1961 inclusive, is included in the chapter on Summer Field Camps and Programs in Volume 2.

Of the 350 students who participated in the 14 sessions of the summer school, ten came from as many countries in the Eastern Hemisphere, and the remaining 340 from North America--1 from Mexico, 26 from Canada, and 313 from 34 of the 48 states of the United States. It hardly needs to be emphasized how extensively Whitehead's instruction influenced a decade of budding geologists around the world.

Not long after Robley D. Evans joined the M.I.T. Physics Department in 1934 and started research on radioactivity, one of his graduate students, Clark Goodman, aroused Whitehead's interest in this physical phenomenon. Was there any radioactivity in sedimentary rocks, and could radioactivity play a role in transforming organic materials into petroleum hydrocarbons? Preliminary research directed toward these questions was started by Whitehead and Goodman, and another graduate student, Kenneth G. Bell. By 1940 the three investigators had enough data to publish an article entitled "Radioactivity of sedimentary rocks and associated petroleum," and Whitehead was more than ever desirous of looking further into the question. The opportunity came almost immediately when the Advisory Committee of the American Petroleum Institute invited M.I.T.'s

Department of Geology, through its Head, Warren J. Mead, to submit a proposal for API support along the lines of Whitehead's interest. At Mead's request, Whitehead and Goodman outlined a program of research, Mead submitted it to the API, and the Advisory Committee approved it, along with two others, for financial support.

This research program, designated API Research Project 43c, proposed to study the effects of radioactivity in the transformation of marine organic substances into petroleum hydrocarbons. Research started in 1942 and came to an end a decade later, in June 1952, after the expenditure of some $92,000, the production of more than 40 papers and reports, and the training of half a dozen graduate students. A somewhat detailed history of this important research project is given in Volume 2. Whitehead acted as Geological Director and Goodman as Physical Director throughout the duration of the program.

As the research on API Project 43c was coming to an end, Whitehead and Irving A. Breger, a graduate student at the time, became interested in the thermal analysis of coal and induced the Nova Scotia Research Foundation to fund a research program at M.I.T. Together they started laboratory work in 1948 and continued the program for seven years, until 1955. A total of $16,655 was expended during this period; 15 reports were produced; and half a dozen graduate students gained valuable laboratory experience. A brief history of this project is included in Volume 2. Whitehead directed the project throughout its duration, with much help from I. A. Breger. See illustrations on p. 652.

After retirement in 1957, Whitehead continued for one additional year as Director of M.I.T.'s Summer School of Geology at Crystal Cliffs. He then relinquished that responsibility but continued for several more years to act as a geological consultant to the Nova Scotia Research Foundation and Department of Mines, both organizations with which he had worked closely for almost twenty years. In recognition of his many services to the Province of Nova Scotia, and also to St. Francis Xavier University at Antigonish, that university awarded him an honorary Doctor of Science degree in 1957. Every autumn until 1963, when his health would no longer permit him to drive his indispensable automobile the 900 miles to Antigonish, he returned to St. Francis Xavier University as Visiting Professor and gave a series of lectures on a variety of geological subjects. While there he had an opportunity to renew his long-standing friendship with the late Professor Donald J. MacNeil, Chairman of the Department of Geology, and with a host of friends in the Antigonish community. Whitehead loved Nova Scotia and its friendly Scots and found in the Province's rocky seacoasts many subjects for his brush and crayons. Its flag, with a recumbent X upon a field of light blue, hung on the wall of his M.I.T. office, and after retirement in his Cambridge apartment, and he always thought of the Province as his second home.

Although he was a skilled and successful consultant in mining engineering and mineral exploration, and an imaginative and effective research director, Whitehead was, nevertheless, primarily a teacher.* He had a phenomenal memory, seemingly never having forgotten anything geological he ever read, and he read omnivorously--history, biography, travel and philosophy, as well as science and technology. He thus kept abreast of latest advances in the several earth sciences and passed this knowledge on to his students with appropriate commentary. He spiced his lectures with subtly humorous anecdotes, drawn from his wide experience as an explorationist, and he was a polished raconteur when among those who expected such conversation.

During his 35 years of teaching at M.I.T. he supervised more than 80 theses,** by far the most of any of his contemporary departmental colleagues, and most of these were field theses because he had learned the critical importance of field work for a geologist from Lindgren and Lahee, and proved its importance to himself in his own exploration work, so he tended to encourage students to choose a field problem when they came to him for suggestions. His office door was always open to students and colleagues alike, and he was always willing to discuss their questions and problems. When these same persons were solicited for contributions to a

---

* Often the question is asked--What is a good teacher; what does a good teacher do? I believe that Walter Whitehead was a good teacher, and these are the reasons I think so. He knew his subject thoroughly and could present its many aspects in a clear and unambiguous manner. He liked to speculate, to develop models, but always with definite boundary conditions. He had a keen sense of spatial relationships, both planar and in three dimensions, and because of this could use simple sketches to illustrate complicated structural conditions. His lifelong habit of reading omnivorously, backed by unusual ability to understand and retain what he read, gave him an encyclopedic knowledge on which he drew easily and often in his lectures and more informal discussions. He left no doubt in the minds of his listeners as to his mastery of the subject being discussed. He was an excellent raconteur and conversationalist and could hold the interest of a group without apparent effort. Most important of all, perhaps, was his ability to evaluate the scientific merit of a colleague's work. His criticism, always offered in a sympathetic way, was invariably constructive, for his greatest desire was to help others by sharing with them what he himself thought and knew. All these characteristics, when taken together, are likely to make a good teacher.

** Three of the 83 theses he supervised were done by students in Course III, Mining Engineering, while he was a lecturer in that Department; these were George William Beer '39 ("So called viscosity of drilling muds under high pressure," 35 p., 1940), Francis Lincoln Lee '40 ("Factors affecting the valuation of oil properties," 106 p., 1940), and William Fleming Wingard '39 ("A proposed device for directional drilling," 25 p., 1940).

Prof. W. L. Whitehead and class in Oilfield Reservoirs (12.45), Spring Term 1952. Reading from left to right: Wm. C. Fowler (XII S.B. 1953), John Small, Jr. (XII S.B. 1952), John J. Fritts (XII S.B. 1952), Paul D. Kaminsky (XII S.B. 1952, S.M. 1953), Frank T. Wheby (XII S.B. 1952), Prof. Whitehead, James R. Strawn (XII S.B. 1952) [standing], Wm. B. Farrington (XII Ph.D. 1953), and unidentified student with back to camera.

(Photo by Jackman)

memorial fund to perpetuate his name at M.I.T., more than a hundred responded generously, and he carried knowledge of this evidence of respect and esteem to his grave.*

When one travels the same trail with a fellow geologist, whether on foot or astride an animal, or by jeep, boat or plane, he gets to know him rather well; particularly if it is too hot or too cold, or if the sea is a bit rough or the air a little too bumpy, or if the insects seem to bite

---

* In the spring of 1969, through the efforts of the author (Robert R. Shrock), the WALTER LUCIUS WHITEHEAD MEMORIAL FUND was established in the Department of Geology and Geophysics to perpetuate his name at M.I.T. Income from the Fund is to be used to help students with field and laboratory expenses. A more extended discussion of the Fund is included in the chapter on The Financing of the Geological Sciences at M.I.T. in my Volume 2.

unusually viciously or invade eyes, ears and nose unmercifully or, perhaps worst of all, if the cook performs badly! I travelled many trails with Walter Whitehead, and he was always a genial companion who wore well. Beneath a gentle and sensitive exterior was a physically tough individual who could see a job through when the going became difficult and frustrating. Few of his kind come out of any generation, but, when they appear, they reassure us all that man can still attain that grace that we call human and civilized.

BIBLIOGRAPHY OF WALTER LUCIUS WHITEHEAD

Symbols and abbreviations used in the following references are explained on pages 91 - 98; in general, abstracts are listed separately as well as with the references to the complete article. This bibliography begins with the title of Whitehead's S.B. Thesis (T--1913) followed by his first publication, and includes all known titles through 1969, the year of his death.

T--1913	The geology of Rattlesnake Hill granite, Sharon, Massachusetts, 35 p., (1913). (S.B. Thesis at M.I.T. in Course III, Department of Mining Engineering and Metallurgy, No. 442, 1913.)
1--1914	(with Lindgren, W.) A deposit of jamesonite near Zimapan, Mexico. Econ. Geol. 9: 435-462, (1914).
2--1916	The paragenesis of certain sulphide intergrowths. Econ. Geol. 11: 1-13, (1916).
3--1917	Notes on the techniques of mineragraphy. Econ. Geol. 12: 697-716, (1917).
4--1918	The veins of Chanarcillo, Chile [Ph.D. Thesis at M.I.T. in Course XII, Department of Geology, 87 p., (1918)]; Econ. Geol. 14: 1-45, (1918).
5--1920	The veins of Cobalt, Ontario. Econ. Geol. 15: 103-135, (1920).
6--1920a	The Chilean nitrate deposits. Econ. Geol. 15: 187-224. [This article is not listed in the Bibliography of North American Geology, 1919-1928, because it concerns an area in South America], (1920).
7--1928	(with Jones, W. F.) Touring Venezuela in a Model T--An account of a geological expedition just north of the Equator. Tech. Rev. 30: 221-225, (1928).
8--1929	Oil geology and production. Tech. Eng. News 10/3: 90-91, 118, (1929).
9--1930	Sulphur. Tech. Eng. News 11/7: 258-259, 274, 277, (1930).
10--1939	(with Goodman, Clark and Bell, K. G.) The radioactivity of sedimentary rocks and associated petroleum [abst.]. Econ. Geol. 34: 941, (1939); Am. Mineralogist 24: 7 (at end following Index); 25: 208, (1940).
11--1940	(with Bell, K. G. and Goodman, Clark) Radioactivity of sedimentary rocks and associated petroleum. Am. Assoc. Petroleum Geol., B. 24: 1529-1547, (1940).
12--1942	The Mother Lode System in southern Eldorado and Amador counties, California (p. 178-182), in Ore Deposits as Related to Structural Features, W. H. Newhouse, ed.: Princeton, Princeton Univ. Press, (1942).

13--1942a   The Chanarcillo silver district, Chile (p. 216-220), in Ore Deposits as Related to Structural Features, W. H. Newhouse, ed.: Princeton, Princeton Univ. Press, (1942).

14--1946   (with Sheppard, C. W.) Formation of hydrocarbons from fatty acids by alpha-particle bombardment. Am. Assoc. Petroleum Geol., B. 30: 32-51, (1946); reprinted in Am. Petroleum Inst., Rept. Progress--Fundamental Research on Occurrence and Recovery of Petroleum 1944-1945, p. 115-125, (1946).

15--1949   Valuation of Oil Property: Part III (p. 297-343), in Examination and Valuation of Mineral Property, R. D. Parks; 3rd ed. (1949), 4th ed. (1957): Reading, Mass., Addison-Wesley Pub. Co., (1949).

16--1950   (and Breger, I. A.) Vacuum differential thermal analysis. Science 111: 279-281; reprinted in Am. Petroleum Inst., Rept. Progress--Fundamental Research on Occurrence and Recovery of Petroleum 1950-1951, p. 205-207, (1952).

17--1950a   (and Breger, I. A.) The origin of petroleum--Effects of low temperature pyrolysis on the organic extract of a recent marine sediment. Science 111: 335-337; reprinted in Am. Petroleum Inst., Rept. Progress--Fundamental Research on Occurrence and Recovery of Petroleum 1950-1951, p. 202-204, (1952).

18--1950b   (and Sullivan, Geraldine R.) Potassium content of marine sediments [abst.]. Geol. Soc. Am., B. 61: 1514, (1950).

19--1950c   Biennial Report on Research Project 43c--Studies of the effect of radioactivity in the transformation of marine organic materials into petroleum hydrocarbons, in Am. Petroleum Inst., Rept. Progress--Fundamental Research on Occurrence and Recovery of Petroleum 1948-1949, p. 226-229, (1950).

20--1951 to 25--1951 a-e   The following items by W. L. Whitehead appear in the "[First] Conference on the Origin and Constitution of Coal," sponsored by the Nova Scotia Dept. Mines and Nova Scotia Research Foundation, June 21-23, 1950, Crystal Cliffs, Nova Scotia [Halifax, N. S., 1951].

   20. [Discussion of] The nomenclature and classification of coal petrography, by P. A. Hacquebard, p. 46-47.

   21. [Comments on M.I.T.'s Summer School of Geology at Crystal Cliffs], p. 75.

   22. [Discussion of] Some geological aspects of the Sydney Coal Field with reference to their influence on mining operations, by T. B. Haites, p. 76.

   23. The vacuum differential thermal analysis of coals, p. 100-110.

   24. (with Breger, I. A.) A thermographic study of the role of lignin in coal genesis, p. 120-139. [Discussion by W. L. Whitehead, p. 140]; also Fuel, 30: 247-253, London.

   25. [Discussion of] Pennsylvanian stratigraphy and sedimentation of the northern Appalachian region, by A. T. Cross, p. 154-155.

26--1951f   (with Breger, I. A.) Radioactivity and the origin of petroleum (with discussion and French summary). Proc. 3rd World Petroleum Congress, The Hague, sec. 1, p. 421-427, (1951), Leiden, Netherlands; reprinted in Am. Petroleum Inst., Rept. Progress--Fundamental Research on Occurrence and Recovery of Petroleum 1950-1951, p. 214-220, (1952).

27--1951g (and Goodman, Clark and Breger, I. A.) The decomposition of fatty acids by alpha particles. Jour. Chimie Physique 48: 184-189; reprinted in Am. Petroleum Inst., Rept. Progress--Fundamental Research on Occurrence and Recovery of Petroleum 1950-1951, p. 208-213, (1952).

28--1951h (and King, L. H.) Vacuum differential thermal analysis of coal [abst.]. Geol. Soc. Am., B. 62: 1489, (1951); Econ. Geol. 50: 22-41, (1955).

29--1952 Research Project 43c--Studies of the effect of radioactivity in the transformation of marine organic materials into petroleum hydrocarbons. In Amer. Petroleum Inst., Rept. Progress--Fundamental Research on Occurrence and Recovery of Petroleum 1950-1951, p. 192-201, (1952).

30--1952a (with Breger, I. A.) A thermographic study of the role of lignin in coal genesis (revised): Cong. Av. Études de Stratigraphie et Geologie du Carbonifere, 3e Heerleen, June 25-30, 1951, Comp. Rendu, t. 1, p. 65-71, Maestricht, Netherlands, (1952).

31--1953
32--1953a The following two items by W. L. Whitehead appear in the "Second Conference on the Origin and Constitution of Coal," sponsored by the Nova Scotia Dept. Mines and Nova Scotia Research Foundation, June 18-20, 1952, Crystal Cliffs, Nova Scotia [Halifax, N. S., 1953].

    30. [Discussion of] Some geological aspects of the Inverness county coalfield in comparison with those of the Sydney coalfield, by T. B. Haites, p. 146-147.

    31. [Discussion of] The origin and transportation of bituminous material in coal, by William F. Seyer, p. 336-337.

33--1953b 2 Globe Articles Favorably Impress [A Letter to the Editor, printed on the editorial page of The Morning Globe of Boston, for Feb. 2, 1953].

34--1954 Hydrocarbons formed by the effects of radioactivity and their role in the origin of petroleum. Chapter 7 (p. 195-218), in Nuclear Geology--A Symposium on Nuclear Phenomena in the Earth Sciences, Henry Faul, ed.: New York, John Wiley & Sons, Inc., (1954).

35--1955 Notes on the accomplishments of Project 43c: in K. C. Heald's "Review of Findings of API Research Project $\overline{43c}$," on p. 151-169 of Am. Petroleum Inst., Rept. Progress--Fundamental Research on Occurrence and Recovery of Petroleum 1952-1953, (1955).

36--1955a [Biennial Report on] Research Project 43c--Studies of the effect of radioactivity in the transformation of marine organic materials into petroleum hydrocarbons. In Am. Petroleum Inst., Rept. Progress--Fundamental Research on Occurrence and Recovery of Petroleum 1952-1953, p. 205-207, (1955).

37--1955b (with King, L. H.) Differential thermal analysis of coal [abst.], Geol. Soc. Am., B. 62: 1489, (1951); Econ. Geol. 50: 22-41, (1955).

38--1963 (and Breger, I. A.) Geochemistry of Petroleum - Chap. 7 (p. 248-332) in Organic Geochemistry, Irving A. Breger, ed.: New York, The Macmillan Co. and London, Pergamon Press, (1963).

(21)
LOUIS BYRNE SLICHTER

LOUIS BYRNE SLICHTER

MIT: 1931-1945

Geophysical research in M.I.T.'s Department of Geology actually began in 1931 with the appointment of Louis Byrne Slichter as Associate Professor of Geophysics. Trained in physics and mathematics at the University of Wisconsin, and with practical training in submarine detection with Max Mason's group in the U.S. Navy (1917-1919), as a physicist with the Submarine Signal Corp. (1922-1924), and as a partner in the consulting firm of Mason, Slichter & Gauld (1924-1931), Slichter brought to his M.I.T. professorship a thorough training in classical physics and broad experience in applied geophysics.

During his first ten years as an active faculty member, 1931-1940, he conducted research on both electrical resistivity and seismic-wave propagation, and carried on one of the earliest American investigations of the earth's crust, using quarry blasts for an energy source and specially designed portable seismographs for field receivers. He simultaneously introduced a program of instruction and research in geophysics and, as the program developed and attracted students, he supervised a total of 5 theses - 2 S.B.s, 1 S.M., and 1 doctor's during this period. A feature of the field programs was the large number of cooperating students - Chaim Pekeris, Judson Mead, Edward A. Colson, George E. Best, Donald G. Fink, and more than a dozen others - many of whom are now well-known geophysicists.

Before the fall of France early in 1939, Slichter was drawn again into problems of submarine detection (He had done similar work during World War I). In a letter to Vannevar Bush he outlined the economic advantages of magnetic detection from airplanes. In the fall of 1940 he was a member of a small National Academy of Sciences Committee, consisting of William D. Coolidge, E. A. Colpitts, Vern O. Knudsen and himself, charged with appraising the current effectiveness of the U.S. Navy in antisubmarine attacks. This Committee was able to demonstrate quickly and dramatically by tests at Key West the need for radical improvement. In March 1941, Jack Tate and Slichter were sent to England to report on the antisubmarine work of the British. On his return in May, Slichter was appointed to head the NDRC's airborne magnetic submarine detection laboratory on Quonsett Point, Rhode Island. He remained in a succession of antisubmarine duties during the rest of the war.

He finally resigned from his professorship of geophysics at M.I.T. on 1 October 1945, without resuming academic work at the Institute, and joined the faculty of the University of Wisconsin. After a short stay in Madison he accepted an appointment as Professor of Geophysics and Physics in the University of California, Los Angeles, in 1947, and became the first Director of the newly created Institute of Geophysics centered at U.C.L.A. He is credited with nursing that pioneering research organization - the first of its kind to be established in the United States - through its early critical years and guiding it to the position of high distinction it holds today among America's research organizations. In 1960 he saw "Planetary Physics" added to the title of the Institute, which then became the Institute of Geophysics and Planetary Physics. At the end of 1961 he handed over to his successor, W. F. Libby, a vigorous and distinguished Institute that owed its growth and program in very large part to his patience, persistence, and wisdom.*

The following biographical sketch is essentially limited to a discussion of Slichter's professional activities before and during the years he was a professor of geophysics at M.I.T. because my history is purposely limited to the M.I.T. part of our professors' careers. In the case of Slichter, his war work is described in John Burchard's Q.E.D.: M.I.T. in World War II (p. 36, 47-50, 63, 267, and 268) [The Technology Press (Cambridge, Mass.) and John Wiley & Sons, Inc. (New York), xvi + 354 p., (1948).], and his subsequent career at the Institute of Geophysics (University of California, Los Angeles), is described in the booklet referred to in the accompanying footnote.

In preparing the foregoing précis and following brief biographical sketch I have drawn information from the two foregoing references as well as from the following:

1) Slichter, L. B., The theory of the interpretation of seismic travel-time curves in horizontal structures. Physics 3: 273-295, (1932).

2) Slichter, L. B., An electrical problem in geophysics - A study of the earth to depths of 30 kilometers. Tech. Eng. News 15: 8-9, 17, (March, 1934); See also Physics 4: 307-322, (1933).

3) Slichter, L. B., Seismic studies of crustal structure in New England by means of quarry blasts [abst.]. Geol. Soc. Am., B. 50/12, pt. 2: 1934, (1939).

4) Ingraham, M. H., Charles Sumner Slichter: The Golden Vector [A biography of Louis Slichter's father, the University of Wisconsin's famous graduate dean]. Madison, Wis.: The University of Wisconsin Press, xiii + 316 p., il., (1972).

5) WHO'S WHO IN AMERICA (with World notables), 36th ed., 1970-1971. Chicago: Marquis Who's Who Inc., (1971).

6) THE INTERNATIONAL WHO'S WHO, 35th ed., 1971-72. London: Europa Publications Ltd., (1971).

I am grateful to Dr. Slichter for revising and correcting my first draft and for providing helpful information about his many professional activities. Here I have been able to consider only a small part of his distinguished career; hopefully someone else will prepare a complete biography in the very near future. In one important respect, however, I have attempted complete coverage; namely, his bibliography complete to 1974. This was only possible because of the full cooperation of Slichter and his efficient secretary, Ms. Nila Simpson, to both of whom I am most grateful.

---

* Slichter's role in the development of the Institute of Geophysics and Planetary Physics is discussed at some length by C. E. Palmer in his prefacing remarks on "Louis Byrne Slichter: Builder of the Institute of Geophysics and Planetary Physics" (J. Geophys. Res. 68: 2867-2870, May 15, 1963) that introduce the reader to the "festschrift" booklet, Papers in Geophysics in honor of Louis Byrne Slichter, (J. Geophys. Res. 68: 2867-2983, 3627-3634, (May and June, 1963).

BIRTH, ANCESTRY, AND YOUTH

Louis Byrne Slichter was born in Madison, Wisconsin, on 19 May 1896, the second of four sons of Charles Sumner Slichter and Mary Louise (Byrne) Slichter. His ancestry traces back to Christian Schlichter, a Swiss Mennonite born in Switzerland in 1761, who came to Pennsylvania, probably in the 1780s, married Mary Wanderbach, reared five children, and then moved to Ontario, where he lived the remainder of his life. One of Christian's grandsons, Jacob Bechtel Slichter, with his wife Catherine Huber, moved from Ontario to St. Paul, Minnesota, where they lived for some time before moving on to Illinois in about 1840. All nine of their children were born in the United States. The youngest of the nine, Charles Sumner Slichter - Louis' father - remembered as a small child seeing trains of Indians coming in to St. Paul each spring with their loads of furs trapped during the preceding winter. He grew up in Chicago, received his college education at Northwestern University, and in 1886 accepted an appointment as a teacher of mathematics on the faculty of the University of Wisconsin. There he spent his entire lifetime, advancing up the academic ladder to professor, department head, and finally Dean of the Graduate School, and leaving an imperishable impact on the University. The story of his life is delightfully told in Mark H. Ingraham's biography, <u>Charles Sumner Slichter: The Golden Vector</u> (The University of Wisconsin Press), (1972), and from the story one can almost literally sit within the family circle and sense the conditions and atmosphere under which Louis grew up, got educated, and then motivated to become a practical scientist.

Louis' youthful years in Madison were exciting and challenging, as he held his own with his three brothers, and he was constantly exposed to the many-sided nature of a household in which both mother and father were loving, sympathetic, and enthusiastic supporters of the activities of youth.*

Slichter now asserts that his education in and enthusiasm for science were obtained unconsciously, as an unlisted graduate of that great alma

---

* Recently (17 May 1975) Louis wrote me about his father's attitude toward young activists and cited an example as quoted in the <u>Milwaukee Sentinel</u> of Wednesday, 6 May 1936, in a report on Dean Slichter's talk to a Phi Beta Kappa meeting. Said Slichter:

> "Students today are charged with being radicals and pacifists; the first charge is true, and always has been," said Dean Slichter. "It is an expression of the natural altruism of youth. Youth stands for the betterment of everything, everywhere. We need not less altruism, but a new form of it which will not evaporate at age 30. No, the disease afflicting radical students is not radicalism, but lack of sense of humor, the most devastating of all mental ills."

mater, the Deutsches Museum in Munich, which he visited frequently during SY 1909-10 while the Slichter family were enjoying a year abroad.* Louis' father was able to get a year's leave of absence and arranged to take his family to Europe to visit England first, then to take hiking trips in Switzerland in the summer, and finally to spend the winter in Munich where he could carry on research while the Slichter boys could go to school. One can easily imagine how Louis, just turned 13 years of age and keenly alert to everything around him, became immediately entranced by the exhibits he saw in the famous Museum!

Louis received his pre-college education in the Madison (Wisconsin) schools and then entered the University, where he completed his studies as a mechanical engineer, with fringe benefits in Latin, mathematics and physics. He received a B.A. degree in engineering in 1917, an M.A. in 1920, and a Ph.D. in physics in 1922.

His first job, which was in 1917, was as a student test engineer with the General Electric Company, but he soon left to join Max Mason's anti-submarine work in Lake Mendota, which investigation was transferred in the fall of 1917 to the new Navy Experimental Laboratory in New London, Connecticut. He was commissioned as Ensign and transferred in the summer of 1918 to the Navy's Subchaser Base in Plymouth, England, and was discharged as an Ensign in New London in the summer of 1919.

After completing his doctorate in physics in 1922 he first worked as a physicist with the Submarine Signal Company, 1922-1924, and then became a partner in the consulting firm of Mason, Slichter [Louis] and Gauld from 1924 to 1931. The last year, that is the academic year of 1930-1931, he spent as a Research Associate at Caltech.

THE M.I.T. YEARS (1931-1941-1945)

When Slichter came to M.I.T. in the fall of 1931, as Associate Professor of Geophysics (he was promoted to Professor in 1932), he brought with him sound training in mathematics and physics and extensive experience in applying these two disciplines to a wide range of engineering problems involving physical properties of the earth.**

Instruction in geophysics, if the latter term be used to denote study of the physics of the earth, had been offered in several M.I.T. depart-

---

* Ingraham describes some of the family's activities on pages 261-264 of Charles Sumner Slichter. The Golden Vector: The University of Wisconsin Press, (1972).

** See footnote on facing page.

ments, including Course XII, long before Slichter's arrival, as pointed out in another chapter of this history that deals with the development of that earth science. The devastating Charleston earthquake in 1886, and the Jamaican quake of 1907, had created both interest and anxiety among M.I.T. scientists, and efforts were made in several ways to learn more about earth physics. A small geodetic observatory was established on a hill in the southeastern part of the Middlesex Fells (Tech. Rev. 1: 170-171, il., 1899), and there were proposals to add laboratory instruments to study seismology; the Whitneys and Morsses contributed funds to support seismological research; and Prof. Jaggar was dividing his time between seismology and volcanology in an effort to arouse more interest in these impressive and destructive natural phenomena. Interest died down, however, when Jaggar used the Whitney and Morss funds to establish the Hawaiian Volcano Observatory in 1912, which seems to have been his primary objective all along. Then came World War I, and it was not until cessation of hostilities that the nation came to realize that a renewed search for minerals and petroleum was critical to the country's future security.

So it happened that immediately following the war there was a great surge of interest in exploration geophysics. The use of gravity to locate salt domes and of magnetics to locate certain types of ore bodies was quickly followed by refraction and then reflection seismometry, and a great wave of geophysical exploration was initiated in the early 1920s. At M.I.T. instruction in exploration geophysics was first offered in the Department of Mining and Mining Engineering; at the time, the geologists were much more interested in learning about the origin and nature of known ore deposits than they were in developing methods and instruments to find new ore bodies or oil fields!

The development of an instructional program in geophysical prospecting, as the subject was then called, first took place in the Mining Department (Course III) and is discussed at length in the accompanying chapter on Geophysics at M.I.T. Suffice it to note here that by 1928 W. S. Hutchinson and F. L. Foster were offering two subjects in applied geophysics: a fourth-year subject titled Elements of Geophysical Prospecting (3.13) that was required of all Course III seniors, and a graduate sub-

---

** Under the heading of "Additions to the Faculty" in Tech. Rev. 33: 462, 476, (July 1931), is a brief biographical statement, noting among other things that:

> "With his mathematical training and combination of theoretical and practical interest in geophysics, Dr. Slichter will bring to his work a valuable fund of knowledge in the important field of physics of the earth, in specialized portions of which investigations are now being carried on at the Institute."

ject, <u>Geophysics, Theory and Applications</u> (3.14) [which was dropped when Slichter came in 1931].

Interest in geophysics at M.I.T. in the 1920s was not, however, limited to prospecting for minerals and petroleum. There was also a renewal of interest in seismology, stimulated by the successful use of refraction and reflection seismology in petroleum exploration in the Southwest. Although the Whitney and Morss contributions of the early 1900s had long gone to help establish Jaggar's Hawaiian Volcano Observatory where, to be sure, earthquakes were to be investigated as assiduously as volcanic phenomena, there were still funds available in SY 1925-26 to bring Robert B. Sosman (VIII S.B. 1904; V Ph.D. 1907) of The Geophysical Laboratory of the Carnegie Institution of Washington to give

> "A course of ten lectures on 'Elastic Waves and the Earth's Structure.'" (Rept. Pres. for SY 1925-26, p. 15.)

in April 1926, and in the spring of 1928 Donald C. Barton, son of Professor George H. Barton (III S.B. 1880) gave ten lectures on

> "Geophysical Methods of Determining Underground Structure."

During the same spring term of 1928, V. G. Gabriel gave a seminar of 15 hours on

> "Seismology and its applications to prospecting,"

and he later shared authorship with Slichter on an article on reflected seismic waves (see reference 8--1933 in the Bibliography farther on).

It might also be noted here that as early as 1900 the Institute had a geodetic observatory atop a hill in the southeastern part of Middlesex Fells, as mentioned earlier, and that there was an M.I.T. Eclipse Expedition to Washington, Georgia, to observe the total solar eclipse on 28 May 1900 (<u>Tech. Rev.</u> 2: 194-217, 1900).

Shortly after his arrival at M.I.T., Slichter organized three subjects of instruction in geophysics, initiated field research on electrical resistivity, and started to plan seismological investigations of the earth's crust.

Successive M.I.T. catalogs show that Slichter lost no time in organizing several subjects in geophysics: <u>Elements of Seismology</u> (12.86), <u>Introduction to Geophysical Prospecting</u> (12.87), and <u>Theoretical Geophysics</u> (12.581, 12.582). These subjects appeared first in the catalog for SY 1931-32 and were offered during the entire decade that Slichter taught in the Department, even being listed for SY 1940-41 although he had left for Washington and Key West in December 1940.

Slichter also gave a radio address on <u>Deep Earthquakes</u> over Station WEAF and the Red Network of the National Broadcasting Company on Monday evening, 29 May 1939, at 7:30-7:45 p.m. EDT, under the theme "Frontiers in Geology." (See reference 21--1938b, 1939 in the Bibliography farther on.)

The field research in electrodynamics was initiated to determine the electrical resistivity, and possibly also the dielectric constant, in the upper crust of the earth, to a depth, it was hoped, of about 30 kilometers or more. The power was supplied to ground electrodes at Clinton and West Roxbury, Massachusetts (30 miles = 40 kilometers apart) and, working at night, the resulting voltages were mapped over most of the Commonwealth of Massachusetts, using telephone circuits as lead wires.* According to a note on page 56 of the April 1953 issue of Tech. Eng. News (14: 36, 1933), Slichter used Vannevar Bush's differential analyzer to great advantage in interpreting his field data. Slichter explained his research to the M.I.T. community in a somewhat detailed article in The Tech Engineering News (Vol. 15: 8-9, 17, March 1934) and reported on it to his professional colleagues in the several articles listed in his Bibliography farther on (11--1932; 13--1933a; and 14--1933b).

His seismological program was organized to investigate the structure of the earth's crust in the New England region. In this program he was able to use blasts at Meriden, Conn., New Brandford, Conn., and at Hudson, N.Y. as energy sources and he and his associates** designed and built the twelve portable seismographs that were transported to selected sites throughout New England to receive the shocks from the quarry blasts. The field operation was synchronized by radio with the blast and a time signal from the U.S. Naval Observatory in Arlington, Virginia. This was a pioneering investigation of the deeper part of the earth's crust, surely one of the earliest in eastern North America, and Slichter reported on his

---

* Slichter has informed me that Prof. Dugald C. Jackson of M.I.T.'s Electrical Engineering Department was "extremely influential in persuading the power and telephone companies to turn over their circuits for research, in the wee small hours." (Slichter letter to Shrock, 8 May 1975.)

** In the M.I.T. Catalog for SY 1936-37, opposite page 25, is a photograph of Slichter with one of his portable seismographs, and the accompanying title informs the reader that

> "In the Department of Geology geophysicists designed and built this new and highly sensitive seismograph. It reports earth displacements of less than a millionth of an inch, earth tiltings as slight as a millionth of a degree."

Slichter reported on the seismograph briefly in references 16--1936 and 17--1937 of his Bibliography that follows farther on.

Recently (17 February 1975) I learned from Otto von der Heyde, the instrument maker in the Geology Department in the 1930s, that he made twelve of the portable seismographs for Slichter.

findings in the several references in his Bibliography farther on (20--1938a and 22--1939a), and lectured on them to the Boston Geological Society on Tuesday evening, 23 January 1940, in the Eastman Lecture Room at M.I.T. Basic to the seismic work carried on by Slichter was his 1932 article on "The theory of the interpretation of seismic travel-time curves in horizontal structures" (Physics 3: 273-295, 1932), which Byerly* described (p. 110) as

> "... a classic monograph on travel time curves and velocity-depth relations for the case when the earth's curvature can be neglected."

WORLD WAR II (1940-1945) AND THE CALIFORNIA YEARS (1946-)

When in early 1940 it became apparent that the United States would probably be drawn into World War II, Slichter was asked to join Division 6 of the National Defense Research Committee (NDRC) and assist with research and development on problems involving subsurface warfare.** He was granted a leave of absence for the remainder of the spring term, from 15 April to June 30 1940, and this leave was ultimately extended for the duration of hostilities. On 1 October 1945 he resigned his M.I.T. appointment, without resuming his work at the Institute, and accepted a professorship in geophysics at his alma mater, the University of Wisconsin at Madison. Within a year or so he had moved on to Los Angeles to become Director of the Institute of Geophysics, and Professor of Geophysics and Physics in the University of California, Los Angeles.

---

* Byerly, P., Subcontinental structure in the light of seismological evidence, in Advances in Geophysics (H. E. Landsberg, ed.), Vol. 3: 105-152, (1956); New York: Academic Press.

** As a recent graduate in physics in 1917, Louis assisted Max Mason on the problem of detection and location of enemy submarines in World War I. Ingraham (p. 141 of Charles Sumner Slichter: The Golden Vector, cited earlier) writes:

> "Louis Slichter was Mason's right-hand man in this work, and before the end of the war both he and Mason were abroad, supervising and participating in the installation of instruments on ships and the testing of them in the danger zone."

When the threat of enemy submarines arose again at the advent of World War II, Slichter was asked to join a special group charged with solving the same problem of detecting, locating and tracking down enemy submarines. Because of his distinguished work on this problem he was asked in 1944 to join Division 3 of NDRC and to transfer his group to the Underwater Ballistics Program at Caltech. There he and his research team, including P. M. Hurley (XII Ph.D. 1940), spent more than a year studying the water entry, ricochet, and underwater behavior of bombs, rockets, torpedoes and other underwater projectiles." (See p. 267-268 in Burchard's Q.E.D.: M.I.T. in World War II cited earlier.)

The new Institute of Geophysics was the first institution of its kind in the United States. His success in developing and guiding this unique institution to its present position of prominence among research organizations in America was recognized in 1963 by issuance of a "festschrift" booklet, Papers in Geophysics in honor of Louis Byrne Slichter, published in volume 68 of the Journal of Geophysical Research, and constituting a fitting gesture to the man who built the Institute of Geophysics and Planetary Physics.

RETIREMENT YEARS AND SUMMATION

Looking back to the decade immediately before World War II, it is fair to say that Louis Byrne Slichter initiated true research in earth physics in M.I.T.'s Department of Geology with his classroom instruction in geophysics and his field investigations of electrical resistivity and seismic properties of the earth's crust.

After leaving M.I.T., in 1945, Slichter continued his research activities at the University of Wisconsin and at the University of California, and by the time he reached retirement age in 1963 he was recognized as one of America's most distinguished geophysicists. This reputation was based in part on his original work in earth resistivity, gravity, magnetics and seismology, and in geophysical exploration, as reported in some 50 publications; on his outstanding work on submarine detection in both World Wars; and on the administrative skill, scientific wisdom, and constant patience he displayed in building California's Institute of Geophysics and Planetary Physics.

Although he retired in 1963, Slichter has been recalled to active duty each year since then, to head the U.C.L.A. gravity program, including the unique project at the South Pole, and in 1975 he continues to go to his office and to help his younger colleagues just as he has for the past three decades! Because of his continued interest in geophysical research after retirement, and also because much of this has been reported on jointly with his colleagues, I have seen fit to bring his list of publications up to 1974 in order that his lifetime record of research can be recorded as completely as possible.

Likewise, I have also seen fit to extend this biographical sketch many years beyond Slichter's decade of teaching and research at M.I.T., from 1931 to 1940, because he is such an excellent example of a young scientist who at the very beginning of his academic career at M.I.T. already showed the knowledge, imagination and skills that would lead to his future distinguished career. Never mind the fact that he developed so much of that career after leaving the Institute; he left an important impact on geophysics before he left!

Finally, in the light of the preceding comments, it seems appropriate to mention the more important honors that have been accorded him through the years. He received the Presidential Certificate of Merit in 1946; was appointed a Rockefeller Research Fellow in 1946; was cited "in recognition of eminent professional services" by the University of Wisconsin in 1957; was given the Jackling Award of the AIME, and delivered the Jackling Lecture (see Bibliography), in 1960, and became an honorary member of the Society of Exploration Geophysicists the same year; was honored by a "festschrift" volume of the Journal of Geophysical Research (Vol. 68, Nos. 10, 12) in 1963; was awarded the William Bowie Medal, and had Louis Byrne Slichter Hall at U.C.L.A. named after him, in 1966; was given an Honorary D.Sc. by the University of Wisconsin in June 1967, and an Honorary LL.D. by the University of California, Los Angeles in June 1969; and has been a longtime member of the National Academy of Sciences (1944) and a fellow of the American Academy of Arts and Sciences (1957).

Louis married Martha Mary Buell on 20 October 1926, and they had two children - Susan Merry and Mary Louise (Mrs. Ward Whaling). At this time, June 1975, the Slichters live in Pacific Palisades, California, where Louis continues to visit his office every day, having recently recovered from a bout with an infected foot that kept him at home for a few weeks.

BIBLIOGRAPHY OF LOUIS BYRNE SLICHTER

Symbols and abbreviations used in the following references are explained on p. 91 - 98; in general, abstracts are listed separately as well as with the references to the complete article. Although Slichter left M.I.T. in 1945, and I would normally, therefore, have not included in his Bibliography publications for more than a year or two thereafter, I have made an exception for the reasons given on a preceding page in my summation of his career. Because he started his academic career in M.I.T.'s Department of Geology, and left an important impact on the Course XII curriculum, students and faculty, and also because he ranks today as one of America's most distinguished geophysicists, it has seemed to me altogether proper to include the complete list of publications to date to indicate that his distinguished career was foretold by his first decade in the academic world, which he spent at M.I.T.

1--1927 Geophysical prospecting methods, p. 418-423, in R. Peele's Mining Engineers Handbook, 2nd ed. New York: John Wiley & Sons, Inc., (1927).

2--1928 Certain aspects of magnetic surveying. A.I.M.E., Tech. Pub. 120: 21 p., (1928); See also Tr. 81: 238-260, (1929).

3--1929 Discussion of Max Mason's Geophysical exploration for ores. A.I.M.E., Tr. 81 (Geophysical Prospecting): 9-43, (1929).

4--1929a Certain aspects of magnetic surveying. A.I.M.E., Tr. 81 (Geophysical Prospecting): 238-260 [discussion, 259], il., (1929).

5--1929b   Discussion of N. H. Stearns' A background for the application of geomagnetics to exploration. A.I.M.E., Tr. 81 (Geophysical Prospecting): 315-344, (1929).

6--1930   Observed and theoretical electromagnetic model response of conducting spheres. A.I.M.E. Tech. Pub. 332: 19 p., (1930); Tr. 97 (Geophysical Prospecting): 443-459, (1932).

7--1930a   Discussion of J. J. O'Neill's [Results from the geophysical surveys in] Southern British Columbia. A.I.M.E., Tech. Pub. 369: 12-14; See also Tr. 97 (Geophysical Prospecting): 34, 35, (1932).

8--1932   Discussion of E. Y. Dougherty's [Results from the geophysical surveys at] Gull Lake, North Central Newfoundland. A.I.M.E., Tr. 97 (Geophysical Prospecting): 43, (1932).

9--1932a   Discussion of C. A. Heiland's A new geophone. A.I.M.E., Tr. 97 (Geophysical Prospecting): 237-244, (1932).

10--1932b   The theory of the interpretation of seismic travel-time curves in horizontal structures. Physics 3: 273-295, (1932).

11--1932c   Observed and theoretical electromagnetic model response of conducting spheres, p. 443-459, in Geophysical Prospecting. A.I.M.E., 510 p., il., (1932).

12--1933   (with Gabriel, V. G.) Studies in reflected seismic waves: Pt. I - Some computations of the reflection of seismic waves at solid boundaries; Pt. II - Surface motions due to reflections in a layered crust. Gerland's Beitrage zur Geophysik 38: 228-256, (1933).

13--1933a   The interpretation of the resistivity prospecting method for horizontal structures. Physics 4: 307-322, (1933).

14--1933b   An inverse boundary value problem in electrodynamics. Physics 4: 411-418, (1933).

15--1934   An electrical problem in geophysics - A study of the earth to depths of 30 kilometers. Tech. Eng. News 15: 8-9, 17, il., (1934).

16--1935 1936   Progress report on a three-component seismometer and tiltmeter. Abst., Am. Geophys. Union, Tr. 16th Ann. Mtg. Pt. 1: 78 (‡), N.R.C. (Aug. 1935); Am. Geophys. Union, Tr. 17th Ann. Mtg. Pt. 1: 78 (‡), N.R.C., (July 1936); Earthquake Notes 8: 76 (‡), (June 1936).

17--1937   Exhibition and description of three-component portable seismograph [abst.]. Geol. Soc. Am., Pr. 1936: 102-103, (1957).

18--1937a   Concerning the vibrations of a layered earth [abst.]. Earthquake Notes 9: 1 (‡), (Dec. 1937).

19--1938   (with Lovering, T. S. et al.) Report of the International Committee on Borderland Fields between Geology, Physics, and Chemistry, 1937, 73 p. (‡), N.R.C., Div. Geol. Geog., (1938).

20--1938a   Seismological investigations of the earth's crust, using quarry blasts [abst.]. Geol. Soc. Am., B. 49/12, pt. 2: 1927, (1938).

21--1938b 1939   Deep earthquakes, p. 23-26 in Frontiers of Geology; Ten papers prepared for fifteen-minute radio addresses by Fellows of the Society, 1938-39. The Geol. Soc. Am., 48 p., (1939).

22--1939a   (with Pekeris, C. L.) Problems of ice formation. J. Appl. Phys. 10: 135-137, (1939).

23--1939b   Seismic studies of crustal structure in New England by means of quarry blasts [abst.]. Geol. Soc. Am., B. 50/12, pt. 2: 1934, (1939).

24--1939c   Geological meaning of deep-seated earthquakes. Pan-Am. Geol. 72: 344-348, (1939).

25--1940   Internal heat of the earth [abst.]. Geol. Soc. Am., B. 51/12, pt. 2: 1946, (1940).

26--1941   Cooling of the earth. Geol. Soc. Am., B. 52: 561-600, il., (1941).

27--1941a   Cooling of the earth [abst.]. Am. Geophys. Union, Tr. 22nd Ann. Meeting, Pt. 2: 547, (1941).

28--1942   Magnetic properties of rocks, p. 293-298, in Handbook of Physical Constants (Edited by F. Birch, Chmn., et al.). Geol. Soc. Am., Sp. P. 36: 325 p., il., (1942).

29--1942a   (and Telkes, M.) Electrical properties of rocks and minerals, p. 299-319, in Handbook of Physical Constants (Edited by F. Birch, Chmn., et al.). Geol. Soc. Am., Sp. P. 36: 325 p., il., (1942).

30--1947   Geophysical prospecting for ores. Min. Cong. J. 33: 47-51, (1947).

31--1947a   (and Bullard, E. C.) Frenkel's views on the origin of terrestrial magnetism. Nature 160 (4057): 157, (1947).

32--1950   The Rancho Santa Fé conference on The Evolution of the Earth. Nat. Ac. Sci., Pr. 36: 511-514, (1950).

33--1951   Earth Sciences. Science 113 (2930): 3 (2/23), (1951).

34--1951a   An electromagnetic interpretation problem in geophysics. Geophysics 16: 431-449, (1951).

35--1951b   Crustal structure in the Wisconsin Area. Univ. Calif., Inst. Geophysics, Oct. 31, 1951.

36--1952   An electromagnetic interpretation problem for the sphere. Roy. Soc., Pr. A214 (1118): 356-370, (1952).

37--1953   (with Pettit, J. T. and LaCoste, L.) Earth tides. Am. Geophys. Union, Tr. 34/2: 174-184, il., (1953).

38--1953a   (and Knopoff, L.) An appraisal of electromagnetic prospecting procedures with the aid of scale models [abst.]. A.I.M.E., Min. Geol. Geophys. Div. Ann. Mtg., Feb. 1953, Abst. Tech. Papers, p. 21, (1953).

39--1953b   (with Price, A. T.) Exceptional cases of electromagnetic problems. Roy. Soc., Pr. A216: 434-435, (1953).

40--1954   Seismic interpretation theory for an elastic earth. Roy. Soc., Pr. A224: 43-63, (1954).

41--1955   Geophysics applied to prospecting for ores. Econ. Geol. 50 (Anniversary Vol.)/2: 885-969, il., (1955).

42--1956   Theoretical seismology [abst.]. J. Geophys. Res. 61: 378-379, (1956).

43--1957   Remarks relative to Maxwell's formula for the magnetic susceptibility of disseminated materials. Advances in Physics 6: 333-335 (London), (1957); Phil. Soc. Mag. Suppl. 6: 333-335, (1957).

44--1959   (and Knopoff, L.) Field of an alternating magnetic dipole on the surface of a layered earth. Geophysics 24/1: 77-88, il., (1959).

45--1959a   Mining geophysics. Min. Cong. J. 45/5: 36-39, 44, (1959).

46--1959b   Some aspects, mainly geophysical, of mineral exploration, p. 368-412 (Chap. 15), of Natural Resources (Huberty, M. R. and Flock, W. L., eds.). New York: McGraw-Hill Book Co., Inc., (1959).

47--1960   Earth tides, in McGraw-Hill Encyclopedia of Science and Technology, vol. 4: 346-351. New York: McGraw-Hill Book Co., Inc., (1960). (See also the new edition of 1968.)

48--1960a   The need of a new philosophy of prospecting (The 1960 Jackling Lecture). Min. Eng. 12/6: 570-576, il., (1960); Can. Min. Manual, p. 11-21, il., (1960).

49--1960b   (and Caputo, M.) Deformation of an earth model by surface pressures. J. Geophys. Res. 65/12: 4151-4156, (1960).

50--1961   The fundamental free mode of the earth's inner core. Nat. Ac. Sci., Pr. 47/2: 186-190, il., (1961).

51--1961a   (with Ness, N. F. and Harrison, J. C.) Observations of the free oscillations of the earth. J. Geophys. Res. 66/2: 621-629, (1961).

52--1961b   Concerning a free mode of the earth's inner core possibly observed during the Chilean earthquake [abst.]. Science 133: 1369, (1961).

53--1961c   (and Harrison, J. C. and Ness, N. F.) Does the earth's inner core behave as a Foucault pendulum? [abst.]. Science 134: 1434-1435, (1961).

54--1963   (with Harrison, J. C. et al.) Earth-tide observations made during the International Geophysical Year. J. Geophys. Res. 68/5: 1497-1516, il., (1963).

55--1963a   Secular effects of tidal friction upon the earth's rotation. J. Geophys. Res. 68/14: 4281-4288, (1963).

56--1964   (and MacDonald, G. J. F., Caputo, M., and Hager, C. L.) Report of earth tide results and other gravity observations at U.C.L.A. Observatoire Royal de Belgique Cinquiemen, Communications, 236, (1964).

57--1965   Earth's free modes and a new gravimeter. Geophysics 30/3: 339-347, (1965).

58--1965a   Review of The Earth's Crust and Mantle, by F. A. Vening-Meinesz. Am. Geophys. Union, Tr. 46/3: 582, (1965).

59--1965b   (and Caputo, M. and Hager, C. L.) An experiment concerning gravitational shielding. J. Geophys. Res. 70/6: 1541-1551, (1965).

60--1966   Response, Acceptance, and Address of the Evening. A glimpse at the geophysical scene [Response to award of the William Bowie Medal]. Am. Geophys. Union, Tr. 47/2: 346-355, il., (1966).

61--1966a   Review of Italian Expeditions to the Karakorum (K_2) and Hindu Kush, by A. Marussi. Am. Geophys. Union, Tr. 47/3: 504-505, (1966).

62--1966b   Gravity observations and the dynamics of the earth (Faculty Research Lecture), U.C.L.A., May 1, 1963; revised, U. Calif. Press, (1966).

63--1967   Spherical oscillations of the earth, in Non-elastic processes in the mantle--Int. Upper Mantle Comm. Symposium, Newcastle-upon-Tyne, 1966, Proc. Roy. Astron. Soc. Geophys. J. 14: 171-176, il., (1967).

64--1967a   Earth tides. (U.S. Nat. Rept. 1963-1967, 14th Gen. Assembly, I.U.G.G.). Am. Geophys. Union, Tr. 48/2: 355-358, (1967).

65--1967b   Earth tides gravimeters at the South Pole. Upper Mantle Proj., U.S. Prog. Rept. 7.33: 131 ff., (1967).

66--1967c (and Dixon, W. J. and Meyer, G. H.) Statistics as a guide to prospecting, in Computer short course and symposium on mathematical techniques and computer applications in mining and exploration, 1962, Vol. 1: Tucson, Ariz., Univ. Arizona, Coll. Mines, p. Fl-1--Fl-27, il., (1967).

67--1967d Earth, free oscillations of, in Vol. 1, p. 331-343, of International Dictionary of Geophysics, (K. Runcorn, general ed.). London: Pergamon Press, (1967).

68--1968 Autobiographical data, in Modern Men of Science 2: 500, (1968). See also the 10th and earlier editions of American Men of Science.

69--1968a (and Hager, C. L. and O'Connell, R. V.) Observations of earth tides and free vibrations at South Pole. Antarctic J. of U.S. 3/5: 182-183, (1968).

70--1968b (and Hager, C. L.) Earth tides, in Pan-Am. Symposium on the Upper Mantle, Mexico, D.F., 1968. Geofisica Int. 8: 43-54, il., (1968).

71--1969 (and Hager, C. L. et al.) The long-period earth tide at South Pole. Antarctica J. of U.S. 4/5: 214, il., (1969).

72--1970 (and Hager, C. L. et al.) Gravity tides at South Pole. Antarctica J. of U.S. 5/5: 165-166, (1970).

73--1971 (and Hager, C. L. et al.) Earth tides and earth's free vibrations: observations at the South Pole. Antarctica J. of U.S. 6/5: 227-228, (1971).

74--1971a (and Hager, C. L. et al.) Earth tide observations at South Pole and in Southern California, and Preliminary free mode observations at the South Pole. Upper Mantle Project; U.S. Program--Final Rept., Item 7.22: 166-167, (July 1971).

75--1972 (with Broucke, R. A. and Zürn, W. E.) Lunar tidal acceleration on a rigid earth. Am. Geophys. Union, Geophys. Mon. 16 (The Griggs Volume): 319-324, (1972).

76--1972a Earth tides, p. 285-320, in The Nature of the Solid Earth (E. C. Robertson, ed.). New York: McGraw-Hill Book Co., Inc., (1972).

77--1973 (and Zürn, W. and Ritala, K.) Observations on earth tides and earth's free vibrations at the South Pole. Antarctic J. of U.S. 7: 254-255, (1973).

78--1974 Review of Geomagnetism in Marine Geology, by V. Vacquier. Geoexploration 12: 70, (1974).

79--1974a (with Jackson, B. V.) The residual daily earth tides at the South Pole. J. Geophys. Res. 79: 1711-1715, (1974).

----

(22)
WARREN JUDSON MEAD
(1883-1960)

WARREN JUDSON MEAD

MIT: 1934-1949-1954

Warren J. Mead was a noted structural and engineering geologist, known around the world for his published research and private work as geological consultant and engineer, when he came to M.I.T. in 1934 to succeed Waldemar Lindgren as Head of the Department of Geology. A product of the famed "Wisconsin School of Geology," having been trained by C. R. Van Hise and C. K. Leith, Mead was widely experienced in both practical and experimental geology, in Precambrian, structural and economic geology, and in the teaching of these aspects of the science. He was a worthy successor to the great Waldemar Lindgren, who had made M.I.T.'s Department of Geology a world center for the training of economic geologists and for research on mineral deposits.

During his tenure as Professor and Head of Course XII, from 1934 to 1949, and as Honorary Lecturer, after retirement, for another five years, 1949 to 1954, he brought about many innovations in the Department and taught structural and engineering geology to a generation of students. He liberalized the rigid curriculum of Course XII, recruited promising young faculty members to broaden the Department's offerings, added a corridor museum of mineral and rock exhibits to improve instruction, and saw established the Lindgren Library and the Schwarz Memorial Map Room. He helped equip a machine shop and employed a full-time instrument maker to support experimental work by the staff, secured funds for and had built a high-precision grating spectrograph for study of trace elements in ores, joined strongly in support of an innovative program using radiochemical methods for age-determination of rocks and study of organic substances, and conducted his own research on the mechanics of deformation of soils and rocks. Finally, he expanded and diversified the geology curriculum he found in 1934 to include geochemistry and more geophysics, and got the Department more and better space by having it moved from Building 4 to Building 24. The latter was accomplished after Course XII was almost abolished by the M.I.T. Administration because of its decline during World War II! When he finally retired as Chairman (and Professor) he left a department with an excellent curriculum, a young and vigorous faculty, a growing enrollment of degree candidates, and a considerable array of modern facilities to support an innovative research program of much promise.

In almost fifty years of service to Geology, first at the University of Wisconsin (1906-1934) and then at M.I.T. (1934-1954), Mead served his science well as an outstanding teacher of engineering and structural geology, as a perceptive investigator of iron and aluminum ores, as an imaginative and inventive experimentalist in soil and rock mechanics, as an able consultant on problems involving mine subsidence, slides in open cuts, and geological conditions at dam and tunnel sites, and as an innovative administrator in bringing about many improvements in M.I.T.'s Department of Geology.

BIRTH, ANCESTRY, AND EARLY EDUCATION

Warren Judson Mead* was born in Plymouth, Wisconsin on 5 August 1883 and died in his home in Belmont, Massachusetts, on 16 January 1960 in his 77th year. He was the third of four children born to Major C Mead [not a military title and no period after the C] and Rose Anna (Robinson) Mead. The first child, a boy, died at birth, and Warren's two other siblings were girls, Arlisle Maria (b. 1881) and Jessie Rose (b. 1887). The families of both the Meads and the Robinsons had lived in northeastern Wisconsin for several generations, and both traced their ancestry back to earlier forebears who were mostly practical farmers in New York State.

Mead attended the public schools of his home town and would normally have left after graduating from Plymouth High School in 1899. However, his principal, who must have perceived the latent talent in young Mead, persuaded him to stay on for an additional year in order to participate in a special class in solid geometry and do some tinkering with electricity. This he did and he ascribed to this special instruction his ability at visualizing relations in three dimensions and his lifelong interest in electrical phenomena.

With high school behind him, having graduated in 1900, and no money available for college, he qualified by an examination for a First Grade Teacher's Certificate, which permitted him to teach in any school in his county. He found a position open in a one-room school in the neighboring

---

* The main biographical part of this sketch is based largely on the three following memorials that I have published: "W. J. Mead, Experimental Geologist," Science 132: 1235-1236, port., (1960); "Warren Judson Mead, August 5, 1883-January 16, 1960," Nat. Ac. Sci., Biog. Mem. 35: 252-271, port., (1961); and "Memorial to Warren Judson Mead (1883-1960)," Geol. Soc. Am., Pr. 1960: 125-136, port., (1962). I also published a special short account of one of his consulting jobs, "He found bauxite for Reynolds," in Reynolds Review 20: 10-13, il., (1960), and a brief obituarial note in the Wisconsin Academy Review (Wisconsin Academy of Sciences, Arts and Letters), p. 42, (Winter 1961) entitled "In Memoriam--Warren Judson Mead (1883-1960)." In addition I have drawn freely for personal details on Mead's own privately printed autobiographical letters with the title Dear Family. Brief biographical notes are also included in the following: [Biographical sketch upon being appointed Head of the Geology Department], Tech. Rev. 36: 361, (1933-34); "Prof. Mead's reserved exterior belies his hidden energy as 'Expert Geologist,'" The Tech, (June 6, 1939); and "Warren Judson Mead 1883-1960," Tech. Rev. 62: 6, (March 1960). The several obituarial notes appearing in Boston newspapers contained biographical material like that in the references cited above.

Seeing nothing to be gained by simply recasting the language in my several memorials cited above, I have excerpted many paragraphs verbatim; these are typed with single spacing and are referenced at the end to the appropriate publication included in the aforementioned list.

village of Johnsonville and took the job for SY 1900-01 of 7 months at
$40 per month. After paying his living expenses, consisting largely of
the $8 per month for room and board that he paid to a local German family,
he saved about $200! His pupils, some 20 in number, came mainly from
local farms, and ranged in age from 5 to 18 years. To meet their needs
he had to teach all eight grades. The next year, SY 1901-02, he taught
the "upper" floor in a two-room school at Hingham (Wisconsin), and again
saved a substantial part of his salary. By the end of the year he had
enough money to think seriously of entering the University at Madison in
the fall.

STUDENT YEARS AT THE UNIVERSITY OF WISCONSIN (1902-1908)

I have written at some length about Mead's activities during his high
school and college years as follows in my 1962 Memorial (p. 125 ff.):

> As the Twentieth Century opened, the Age of Electricity was being
> rapidly ushered in, with the introduction of electric bells and
> lights, trolley cars, telephones, and the like. The youthful
> Mead was greatly impressed by all the wonderful changes going on
> around him and became much interested in electricity. When his
> high school teacher--the same one who taught him "visual" geom-
> etry so effectively--took him and a schoolmate to the school
> basement and showed them the wet batteries that made the elec-
> tricity, and the magnets and bells that had been installed,
> Mead's interest was deeply excited. Soon he was volunteering
> to help the electricians who came from Milwaukee to wire the
> local houses, and he and a schoolmate made their own telegraph
> sets, strung up wires between their homes several blocks apart,
> learned the Morse Code, and carried on many lengthy telegraphic
> conversations. It was natural, therefore, that, when he en-
> tered the University of Wisconsin in the fall of 1902, he en-
> rolled as a freshman in electrical engineering. Furthermore,
> Mead's father, a lawyer and politician who had struggled through
> the early years of his own legal career, felt that his son would
> do much better to prepare for a career in electrical engineer-
> ing, so that he could get a good position and start earning a
> reasonable salary immediately after graduation. And so the die
> seemed cast that Mead would become an electrical engineer. How
> then did it happen that he became a geologist instead?
>
> A toy camera, when he was eight years of age, led him to a
> lifelong interest in photography; an imaginative and resource-
> ful teacher of high school geometry helped him to develop an
> unusual ability to visualize relationships in three dimensions;
> a senior student in trouble with his petrographic thesis was
> responsible for creating an immediate interest in geology; an
> able college teacher led him to leave electrical engineering and
> enter geology seriously; another great teacher and research
> geologist, sensing his latent abilities, motivated him to alter
> his course for the last time and to devote his efforts to the
> quantitative aspects of geological problems. Once his course
> was firmly set, he never left it, and during his long and pro-
> ductive life he added something new and important to every prob-
> lem he touched.
>
> The first two years in college were fine so far as mathe-
> matics, physics, and chemistry were concerned, but Mead found the

electrical engineering subjects dry and uninteresting because students were told to "study and remember what is in the books; memorize the formulae; don't try to understand anything, just learn it!" Strongly motivated by the desire to know how and why things worked the way they did, Mead naturally found this kind of instruction dull indeed.

One Sunday morning, near the end of his sophomore year, Mead found his next-door fraternity brother, a senior in geology, trying unsuccessfully to make drawings, for his thesis, of what he saw in a petrographic microscope. "If it is permitted," he asked, "why not photograph what you see?" The senior replied that the Geology Department did not have any equipment for taking photographs through a microscope. But Mead, who had had several cameras as a youngster (the first a toy one obtained at age eight as a prize for getting new subscriptions to a youth's magazine), knew all about them, was an ardent amateur photographer, as he continued to be all of his life, and felt certain that they could photograph anything they could see. So after lunch the two students went to Science Hall, taking the microscope, the thin sections, and Mead's 5×7 view-camera loaded with plate holders and dry plates. Using sunlight, they made a series of excellent photographs, and in doing so Mead became so excited over the problems of the senior's thesis that he decided to take some geology courses the following fall.

Thus it came about that he enrolled in general geology with N. M. Fenneman, then a professor at Wisconsin. He was so stimulated by Fenneman's lectures that he soon changed his course to General Engineering with a major in geology. Incidentally, he never regretted the training in engineering, for many years later, when he sat with planning boards to discuss power dams, he understood the technical language of the electrical engineers as well as that of the geologists.

Fenneman suggested that he investigate the physiography of Sheboygan Marsh, the remnant of a former glacial lake near his home at Plymouth, and write up its history for his thesis. Accordingly, Mead spent the summer of 1905 mapping the marsh and gathering field data. When he returned to Madison in the fall, he had to take a heavy schedule of geology subjects to make up for what he had missed while a student in electrical engineering. Mineralogy was with W. O. Hotchkiss, who was to become his lifelong friend; metamorphic geology and structural geology were with C. K. Leith, head of the Department, with whom he was to be closely associated for the next 30 years.

To his subjects in geology Mead brought an understanding of mechanics and strength of materials and facility with the slide rule, both learned as an engineering student and essentially unknown to geology students at that time. No doubt Leith quickly sensed these capabilities, as well as Mead's desire "to figure things out" and get quantitative results, for he persuaded him to abandon the thesis on the Sheboygan Marsh and attack a challenging problem in metamorphic geology. The problem, in effect, was this--"Assuming that all sedimentary rocks have come from the natural destruction of older rocks, and that the earliest sedimentary rocks came from still older igneous rocks, figure out the relative amounts of the principal types of sedimentary rocks (shale, sandstone, and limestone) that have been produced by the redistribution of the materials of the igneous rocks." Quite a thesis problem for a senior, but Mead tackled it with enthusiasm!

As the end of the spring term of his senior year approached, he had not yet completed his thesis because he couldn't determine

how to relate the chemical data. Then, in the middle of a May
night, the "visual" geometry he had learned with his principal
in that extra year in high school came to him, and he decided to
develop a three-dimensional graph, which would establish limits
within which the shale/sandstone/limestone ratio had to fall.
Within a few days he had constructed a three-dimensional model,
and his thesis was complete. It won him the Science Medal of the
Science Club of the Faculty for "the best Baccalaureate thesis
in Science" for the year 1906.

The thesis, which was the basis for his first published
paper--Redistribution of elements in the formation of sedimentary rocks (1907), soon came to the attention of Charles R. Van
Hise, then President of the University of Wisconsin. Van Hise
had only recently published his monumental A treatise on metamorphism (1904) and had estimated the ratio of shale/sandstone/
limestone to be 65/30/5, whereas the young senior's estimate was
80/11/9, and the latter could demonstrate the validity of his
ratio by an ingenious three-dimensional model that he had thought
up and constructed specially for the problem. Great man that he
was, Van Hise came down the hill from the President's Office and
visited Mead in his student cubby hole in Science Hall, listened
eagerly to the young student explain his model and defend his
conclusion, and then expressed his delight and satisfaction with
the new ratio, which he immediately accepted. This meeting of
student and master made a great impression on Mead, and he often
spoke of it to his students and colleagues as an example of how
a really great scientist reacts when a youngster proves that he
had reached an incorrect conclusion.

Having come under the influence of three of America's great
teachers of geology--Fenneman, Leith,* and Van Hise**--Mead was
now fully convinced that he wanted to follow a geological career.
It was natural that his future work would lead into the quantitative and experimental aspects of geological problems, because of
his engineering training and of his interest in the quantitative
approach.

TEACHING, RESEARCH, AND CONSULTING WORK DURING
WISCONSIN YEARS (1906-1934)

I have written further about Mead's years at Wisconsin, as teacher, investigator and consultant, in my 1962 Memorial (p. 127 ff.) as follows:

Recognizing the unusual abilities and preparation of his young
protégé, Leith appointed Mead a student assistant for the school
years 1906-1907 and 1907-1908 and asked him to devise problems
for the students to solve quantitatively in three new laboratory
courses that he wished to start--one in structural geology, one
in metamorphic geology, and one in ore deposits. These new laboratory courses, with their many interesting and challenging problems and ingenious experimental models, immediately became established parts of the geology curriculum. They are still used in

---

* See Sylvia W. McGrath's Charles Kenneth Leith, Scientific Adviser. Madison: The University of Wisconsin Press, xii + 255 p., il., (1971).

** See Maurice M. Vance's Charles Richard Van Hise, Scientist Progressive. Madison: The State Historical Society of Wisconsin, iv + 246 p., il., (1960).

modified form at Wisconsin and in many other colleges where Wisconsin graduates are teaching. Thus the Mead problems and models have had long and widespread use in geological instruction in North America.

During the decade following graduation, Mead spent the summers doing field work, largely on the Precambrian in the Lake Superior region, and the winters in Madison teaching at the University, developing problems and laboratory aids and carrying on research with Leith that culminated in the joint authorship of Metamorphic geology published in 1915. This textbook, the first of its kind in emphasizing a quantitative treatment of changes that take place in rocks, drew heavily on Van Hise's A treatise on metamorphism (Monograph 47, U. S. Geological Survey, 1904) and was designed to test and check the conclusions of that treatise. It was made possible by the extensive field and laboratory work that Leith and Mead carried on during the period 1907 to 1915. Mead's ability in presenting data and relationships in simple graphical form is nowhere more evident than in this highly original book, now a collector's item, that opened up a whole new field of quantitative geology.

In the summer of 1908, for the third time as an employee of the U. S. Geological Survey, Mead was assigned the task of converting a great mass of data on the physical properties of the Lake Superior iron ores into estimates of cubic-feet-per-ton of ore in place. In typical fashion he accomplished in a few days what at first seemed a full summer's task of laborious calculations by devising a special kind of graph (nomograph) that made possible rapid calculation by graphical means. This novel device soon came to the attention of H. Foster Bain, then Director of the U. S. Bureau of Mines, and W. C. Mendenhall, Director of the U. S. Geological Survey, and both urged that it be made known immediately because of its wide applicability. The nomograph and an accompanying explanation were soon published in ECONOMIC GEOLOGY, in an article entitled The relation of density, porosity, and moisture to the specific volume of ores (1908), and were later included in and credited to Mead in the section on The iron ores of the Lake Superior region (pages 460-572) in Monograph 52 (The geology of the Lake Superior region, C. R. Van Hise and C. K. Leith, 1911).

In early 1910 Mead accompanied Leith to Cuba to investigate certain iron-ore lands held by the Spanish-American Iron Company. The investigation convinced Leith and Mead that the ores had resulted from the weathering of the underlying serpentine, and they published their conclusions in two well-known papers--Origin of the iron ores of central and northeastern Cuba (1911) and Additional data on origin of lateritic ores of eastern Cuba (1915).

In these same years Mead was doing pioneering work with the E. J. Longyear Company, in using diamond drilling in the exploration for mineral deposits, and his assignments took him into many areas of the Lake Superior region that were then virgin territory. In 1912, while thus employed, he was asked to report on some mineral lands near Bauxite, Arkansas. His report, the first in the 100 that he was to make during the next 50 years, soon came to the attention of Arthur V. Davis, then President of the Aluminum Company of America, and resulted in an invitation by Davis to make a complete study of ALCOA's bauxite lands in Arkansas. In carrying out the earlier work and again during his year's exploration program for ALCOA, Mead became convinced that the Arkansas bauxite, like the Cuban iron ore he had investigated earlier with Leith, was a residual product of the surface weathering of the underlying rock (nepheline syenite), rather than the result of deposition as a chemical sediment in a shallow sea, the theory generally held at the time. His conclusions were published in his well-known Occurrence and origin of the bauxite deposits of Arkansas (1915) and have long since become the generally accepted theory of origin of these deposits.

As mentioned earlier, Mead had been asked to prepare problems for a new laboratory course that Leith wanted to start in structural geology. In doing this he became interested in experiments and laboratory models that could be used to illustrate principles involving rock deformation and failure. Out of this early interest in the mechanics of deformation, which later grew to be a major interest, came a series of papers, starting in 1920, which have long since become classic references in structural geology. Included in these are--Notes on the mechanics of geology structures (1920), Determination of attitude of concealed bedded formations by diamond drilling (1921), The geologic role of dilatancy (1925), Some applications of the strain ellipsoid (1930), and Folding, rock flowage, and foliate structures (1940). In the 1923 revision of his Structural geology (Henry Holt and Company), Leith used many of Mead's ideas, illustrated numerous of his laboratory models, and acknowledged his "valuable criticisms and suggestions arising from his own extensive field and experimental work."

It was an outstanding characteristic of Mead that he constantly added to his own skills and knowledge by using every lesson, every problem, as a means of gaining still more and deeper understanding of the field in which he pioneered. Once he developed an idea or built some device, he promptly applied it or used it to get additional information. As one of his former students recently wrote me--"I think of him as my ideal of one who lived accumulatively--no lesson but what it had its application, no device but what it had its use."

Mead's introduction to engineering geology came in late 1915 when Van Hise asked him to assist in a study of the great slides that were occurring in the excavations being made for the Panama Canal. At the request of the National Academy of Sciences, he went to Panama in early 1916 and assisted in making numerous physical measurements of the materials that had been involved in the slides. Unit weight, moisture content, capacity to absorb water, drainage characteristics, and strength--all of these were determined, and their bearing on the slides evaluated. This was before the advent of soil mechanics, but the study was certainly one of the earliest in which attention was specifically directed to the nature and behavior of soils and weak rocks in open cuts. The work done greatly aided the Academy Committee in formulating its report to the Panama Canal Commission. Many years later, in 1936, he returned to the Panama Canal Zone to consult on the problem of leakage from the Madden Dam reservoir, and assisted in developing a better means of detecting the fluorescein being used to trace the escaping water and of estimating its concentration.

When he returned to the University of Wisconsin in the fall of 1916, he was asked to initiate a course in Engineering Geology for civil engineers. A National Committee on Engineering Education had recommended that the curriculum for civil engineers should include a course in engineering geology, and to Dean Turneaure of Wisconsin's School of Engineering Mead was the obvious choice. So he organized such a course, certainly one of the first in the country, started it in the fall of 1916, and continued to teach geology to civil engineers until he retired 38 years later. The last class he lectured to at M.I.T. before retirement in 1954 was a group of civil engineers.

World War I found Mead assigned to develop a concentrated course in "Military Mapping" for the Students Army Training Corps (S.A.T.C.). On short notice he had to organize a program for 1500 students, write an adequate text and get it mimeographed, find surveying instruments, enlist a teaching staff, etc., etc. And on top of all this, he had to give a special course in engineering geology to 40 students who were trying for an officer's commission. In a hectic but well-organized effort he got the program under way, and many students got some excellent training in field mapping before the end of 1918!

Before 1918, Mead's professional activities were largely those connected with teaching, mineral exploration, and mine development, but in 1919 came the first real opportunity to do engineering geology. True, he had started to teach engineering geology in 1916, and had had a few minor consulting assignments involving slides in open pits and cuts, mine subsidence, underground water, etc., but now he was to do his first big dam site.

The site was that of the proposed Calderwood Dam on the Little Tennessee River, and the construction was being done by the Aluminum Company of America. Mead's advice was followed in selecting the ultimate site for the dam, and thereafter he made many dam-site studies for ALCOA in the United States and Canada. As a matter of fact, Mead probably was consulted by ALCOA on every dam site they considered from then until his retirement from active field work about 1950. The last was a study in 1949 of a great project on the Upper Yukon, near Whitehorse, involving a large dam and a 20-mile tunnel.

When Viscount Inouye of Japan visited the United States in 1921 and asked H. Foster Bain, Director of the U. S. Bureau of Mines, to recommend a group of engineers and geologists to advise his company, the South Manchurian Railway Company, on their iron and coal holdings, Mead was invited to join the group. As he described it many times, the trip was one of the highlights of his life, for it not only gave him an unparalleled opportunity to see first hand the great iron-ore and coal deposits of South Manchuria, but also provided an unusually favorable opportunity to see an unfamiliar side of Japan itself.

While serving as an expert witness in some mining litigation in Butte in the summer of 1923, Mead became acquainted with Andrew Lawson, then Professor of Geology in the University of California, and was asked to accept an appointment as Visiting Professor at Berkeley for the school year 1926-1927. He enthusiastically accepted the offer, and he and his family drove from Madison to Berkeley in their air-cooled Franklin--quite a trip in those days when there were no great thruways and few miles of paved highway. It was a wonderful year, but Mead often remarked that he had difficulty in impressing on his students the implications of the ellipsoid of strain and cleavage-bedding relations when they could go to the near-by mountains and see whole anticlines and synclines beautifully exposed. After so many years spent on the folded and metamorphosed rocks of the Lake Superior Precambrian, he never quite adjusted to the highly folded and faulted but little metamorphosed rocks he saw in California. In contrast, he was greatly impressed by the deep and extensive mantle of weathered rock in California, and often pointed out in lectures the great difference between this kind of landscape and that in the Lake Superior country where the rocks lie bare from glaciation or are covered with glacial debris.

Upon the suggestion of David White, who saw the importance of shearing stresses in the origin of petroleum, Mead consented to undertake and direct the first project of the American Petroleum

Institute, which was to investigate "The generation of oil in rocks by shearing pressures." The work started in September 1926 at Berkeley and continued after June 1927, in Madison. Financial assistance was received from an A.P.I. research fund donated by the Universal Oil Products Company. The fund was administered by the A.P.I. with the cooperation of the Central Petroleum Committee of the National Research Council. Mead acted as director, and J. E. Hawley, an American Petroleum Institute Research Fellow at the University of Wisconsin, conducted the investigation, at both Berkeley and Madison, and reported on the work in three separate articles.* In its later phases, W. P. Rand, Junior Research Fellow at the University of Wisconsin, first assisted Hawley in the experimental work and later assumed full responsibility for the project when Hawley went to Queen's University (Canada) as Professor of Mineralogy. Rand, working under Mead's direction, completed the project and reported on it in 1933.**

When he returned to Madison after the year's leave of absence at Berkeley, Mead accepted a consulting arrangement with the Consolidated Mining and Smelting Company of Canada, which continued for 8 years until 1934 when he left Madison to go to the Massachusetts Institute of Technology. During these years he not only mapped and investigated the great lead and zinc sulfide ore body at the Sullivan mine, but also did considerable exploration in the field. These trips took him to such interesting places as the Big Missouri gold mine in the Salmon River district, Great Bear Lake, and Great Slave Lake.

While in British Columbia, in the summer of 1927, Mead was notified that he had been appointed a member of the Colorado River Board, which had been created by President Calvin Coolidge to study the engineering and economic feasibility of, and to select the site for, a great dam, commonly spoken of as Boulder Dam, to be built on the Colorado River. Among the distinguished members of this Board were two of Mead's old friends and professional colleagues, Dr. Charles P. Berkey then Professor of Geology at Columbia University and Dr. Daniel W. Mead (no close relative) then Professor of Hydraulics at the University of Wisconsin. Service on the Board led to other assignments for the U. S. Army Corps of Engineers, and in the following decade Mead examined some 40 dam sites for this agency, including that for the proposed Garrison Dam on the Missouri River in South Dakota.

After almost a lifetime in Wisconsin (1883-1934) including 28 years at the University of Wisconsin, Mead left Madison in 1934 to become Professor of Geology and Chairman of the Department of Geology at the Massachusetts Institute of Technology. Here he was to follow one of America's greatest geologists, Professor Waldemar Lindgren, who had recently retired. (Excerpted from my 1962 Memorial, p. 125 ff.)

---

* Hawley, J. E. Generation of oil in rocks by shearing pressures; I, The problems--methods of determining the soluble organic content of oil shales: American Association of Petroleum Geologists, Bulletin, volume 13, pages 303-328, (1929); II, Effect of shearing pressures on oil shales and oil-bearing rocks: Ibid, volume 13, pages 329-365, (1929); III, Further effects of high shearing pressures on oil shales: Ibid, volume 14, pages 451-481, (1930).

** Rand, W. P. Generation of oil in rocks by shearing pressures. IV-V. Further studies of effects of heat on oil shales: American Association of Petroleum Geologists, Bulletin, volume 17, pages 1229-1250, (1933).

TEACHING, ADMINISTRATION, CONSULTING, AND RESEARCH DURING M.I.T. YEARS (1934-1954-1960)

When Waldemar Lindgren's retirement became imminent, a world-wide search was made for his successor as head of the Department of Geology. Mead was the unanimous choice of the selection committee, and in the summer of 1934 he was appointed Professor of Geology and Chairman of the Department of Geology (Course XII). After retirement from the double appointment in 1949, as Professor Emeritus, he continued another five years as Honorary Lecturer in order to give his well-known course in engineering geology. Finally, in 1954, after 46 years of college teaching he retired to his Belmont home where he kept busy with hobbies until his death in early 1960.

As soon as he got his family temporarily settled in Cambridge (he soon moved out to Belmont), Mead set about acquainting himself with the Geology staff, departmental problems, and the space and facilities available for instruction and research. Lindgren had brought international fame in mineral deposits and economic geology, and the Department was strong in these fields as well as in ore geology and mineralogy (Newhouse and Buerger), paleontology and stratigraphy (Shimer), and geophysics (Slichter). In addition, petroleum geology and field geology were offered by Whitehead, and geomorphology and general geology by Morris. Mead disliked the rigid curriculum that gave the student only a limited choice of elective subjects, so he promptly brought about flexibility in the Course XII curriculum by reducing the number of required subjects and increasing the number of electives.

He himself immediately began lecturing in his own specialties, structural geology and engineering geology, and within a very few years recruited five younger faculty members to broaden the instructional offerings--H. W. Fairbairn (1937) in petrology, R. R. Shrock (1937) in paleontology and sedimentology, R. D. Parks (1940) in mineral industry and mineral economics, P. M. Hurley (1946) in mineral deposits and geochronology, and J. N. Adkins (1946) in geophysics.

In line with his own interests, he encouraged and sponsored faculty research that gave more emphasis to the experimental and quantitative approaches to geological problems. He expanded the small machine shop that Prof. Shimer before him had established, and put the instrument maker on a full-time basis, so that both would be available to Slichter in his geophysical experiments and to Buerger in his work in x-ray crystallography. He raised funds from Boston industrialist Godfrey Cabot for construction of a high-resolution grating spectrograph that Newhouse wanted for study of the variation in minor element content in minerals and ores. Thus began the Department's Cabot Spectrographic Laboratory, which was later to do other kinds of research under the direction of L. H. Ahrens and W. H. Dennen. When R. D. Evans, a recent physics doctorate from Caltech, arrived at M.I.T. in 1934 with a keen interest in investigating whether radioactivity could be used to determine the age of rocks, he found an immediate and enthusiastic supporter in Mead, who not only gave him every assistance possible but did likewise for Evans' associate, C. D. Goodman (VIII Ph.D. 1940), who was later to help start several projects on radiochemistry and age-work with Whitehead and Hurley. When Whitehead became interested in the possible effect of radioactivity on the transformation of marine organic substances into petroleum hydrocarbons,* Mead was willing to sponsor a research project on the problem, American Petroleum Institute Project 43C, just as he had done 20 years earlier at Madison, when he directed the first of A.P.I.'s many significant research projects. (See quoted description on preceding pages.)

Recognizing that one of the original objectives of Rogers' "Technology Plan" was to create a museum, he organized a staff-student team that ultimately put together some twenty exhibits of minerals, rocks, ores, and fossils. These were handsome glass-walled cases with several glass shelves holding the labelled specimens on display. They first lined both sides of the corridor on the third floor of Building 4, then the third and fourth floors of Building 24. A few of them, which have withstood repeated movings, may still be seen in some of the class rooms and elevator lobbies in the Green Building (54).

It was also during Mead's tenure as department head that the Lindgren Library and the adjoining Schwarz Memorial Map Room were established. (See the chapter on "Libraries" in my Volume 2.)

In my 1962 Memorial (p. 131 ff.), I wrote as follows about Mead's consulting activities after he joined the M.I.T. faculty:

> Although devoting much time to teaching and administration at M.I.T., Mead still continued to act as a consulting geologist for ALCOA, the U. S. Army Corps of Engineers, and several other clients. As one of America's leading engineering geologists he was frequently asked to write articles on the subject, and certain of these deserve mention here--Geology of dam sites in hard rock (1937), Geology of dam sites in shale and earth (1937), Engineering geology of dam sites (1938), and the review article on Engineering Geology for the 50th Anniversary Volume of The Geological Society of America in 1941.

. . . . .

One year before Pearl Harbor, R. S. Reynolds, then President of the Reynolds Metals Company, asked Mead to undertake the job

---

* The objective of this research was discussed on page 2 of The Tech for 7 February 1950 (Vol. 70/3: 2, 1950).

of obtaining a supply of bauxite for his company.* So, for the second time, Mead set up an exploration program for bauxite, and ultimately Reynolds Metals Company developed large ore reserves in Arkansas, Jamaica, and Haiti. As evidence of the high respect in which Mead was held by his clients, he continued to do consulting on dam sites for the Aluminum Company of America while looking for bauxite for the Reynolds Metals Company, its principal competitor, and this without any conflict of interest! Later, in 1944-1945, he arranged a year's leave of absence to organize and direct a proposed research center for Reynolds on Long Island, but soon found this kind of activity not to his liking, so he returned to M.I.T. and spent the remainder of the war years on an assortment of consulting assignments. From then until retirement he busied himself with teaching, occasional consulting, and building up the Department of Geology.

Finally, after half a century of academic and applied work in geology, when retirement gave him the leisure he had never enjoyed before, he repaired once more to his basement laboratory and beloved machine shop, this time to design and make a number of devices and instruments for medical research, an area of interest aroused by one of his sons who was engaged in physiological research. One of his first interests was to find and test medical applications of the principle of dilatancy, one of which he described in his last paper, The principles of dilatancy applied to techniques of radiotherapy (1954). Earlier, he had demonstrated to numerous medical people how his famous sand-filled rubber bags could be used without pain to the patient in getting faithful molds of the end of severed members for use in making artificial limbs.

At no time in his long and active life were his imagination, ingenuity, creativity, and skill with machine tools more impressive than when, with eyesight dimming, he was still able to design and build several unique devices for making delicate and difficult measurements in research on the physiology of respiration. His son, Jere, who had published several articles** on research done with these instruments, wrote me recently (11 July 1960) as follows in reply to my inquiry about them:

> "Everything he made worked the first time and still works. Many have been copied. The CO_2 analyzer is now made commercially. The 'interrupter valve' is used in many other labs, and our machinist still gets orders for pressure transducers according to Dad's design."

(Excerpted from my 1962 Memorial, p. 131-132.)

Mead was noted for his generosity to and support of young scientists, both students and staff members. If one had a legitimate request--i.e. one that made sense to Mead, and his judgment was rarely wrong--he was

---

* The author of this memorial has discussed at some length Mead's services to Reynolds in an article entitled He found bauxite for Reynolds published in the March 1960 issue of Reynolds Review, pages 10-13.

** Mead, Jere, and Whittenberger, J. L. Physical properties of human lungs measured during spontaneous respiration: Jour. Applied Physiology, volume 5, pages 779-797, (1953); Evaluation of airway interruption technique as a method for measuring pulmonary air-flow resistance, Ibid, volume 6, pages 408-416, (1954); Mead, Jere. A critical orifice CO_2 analyzer suitable for student use: Science, volume 121, pages 103-104, (1955).

sure to find Mead receptive and enthusiastic. If he could not find the needed funds in his official budgets, he actively sought them from some of his affluent friends. Failing this, he reached into his own pocket. More than one graduate student and young staff member (myself included) received such personal aid under the guise of a gift from some other source.

> When the Meads lived in Madison, Mrs. Mead held open house and served tea every Sunday afternoon, and many graduate students and young staff members enjoyed her gracious hospitality. Later, when the family moved to their home in Belmont, a suburb of Boston, she not only continued the teas but initiated a departmental supper every spring. These famous "bean suppers" brought students and faculty together in a congenial atmosphere, and many M.I.T. graduate students will recall with nostalgia those happy occasions with the movies, the notorious "Scribliography," and other activities full of pleasant banter. As is so often the case, behind the outstanding teacher and successful professional geologist quietly moved a gracious lady who made his home a lovely place to go when invitation came. In a sense it is a pity that too often we are too young to sense at the time the graciousness that in later years of maturity and in retrospect takes on a lasting glow of warmth and appreciation. (Excerpted from my 1962 Memorial, p. 132.)

SUMMATION

Like Lindgren before him, Mead came to M.I.T. already a distinguished geologist, and again like his predecessor he brought international renown to the Department of Geology. It has seemed appropriate, therefore, that part of this biographical sketch should be devoted to his earlier professional achievements while at the University of Wisconsin, in order that the reader may know the kind of scientist selected to be the fifth head of Course XII at M.I.T. Consequently, I shall now comment briefly on his entire career as <u>Teacher of Geology and Administrator</u>, as <u>Geological Consultant and Engineering Geologist</u>, and as <u>Research Scientist and Inventor</u>.

Teacher of Geology and Administrator

Referring again to my 1962 Memorial (p. 132 ff.), I wrote as follows about Mead as a teacher and administrator:

> As a teacher, Mead's activities spanned nearly 50 years. After 2 years in the country schools of Wisconsin and several years as a graduate-student assistant at the University of Wisconsin, he moved up the academic ladder from instructor (1908) to professor (1918). Before leaving Madison to spend a year's leave of absence (1926-1927) at the University of California (Berkeley) as Visiting Professor, he was awarded a Ph.D. degree in geology in 1926 by the University of Wisconsin. In 1934 he was appointed Professor of Geology and Chairman of the Department of Geology at the Massachusetts Institute of Technology. In 1949, on reaching retirement age he relinquished the chairmanship but continued,

as an Honorary Lecturer, to give lectures in engineering geology until June 1954, when he was made Professor Emeritus.

Mention has been made elsewhere of his early contributions to teaching in the form of laboratory problems and demonstration models. He constantly revised his own lecture notes and laboratory problems, developed new instructional aids, and added to his number of case histories and field experiences on which he frequently called for examples to illustrate a certain point. His disarmingly simple presentation of a complex subject not infrequently led to after-class arguments among students who had failed to get the full implications of his remarks or experiments. In class he was an excellent lecturer and always prided himself on his careful preparation. Best of all were those lectures in which he demonstrated one of his many experimental devices and then discussed how it could be used to gain better understanding of some geological phenomenon or problem. He loved to teach and took great pride in those of his students who later became well-known geologists. (Excerpted from my 1962 Memorial, p. 132.)

Geological Consultant and Engineering Geologist

During his long academic career, almost a half century, at Wisconsin (1906-1937) and M.I.T. (1937-1954), Mead frequently did consulting work that took part or all of most of his summers and many shorter periods during the regular school year. While at Wisconsin, he was frequently on leave for part or all of one term, and students carefully planned their schedules so as to be able to take every course he offered while in Madison.

The many professional assignments he accepted as geological consultant or engineering geologist are described in some detail in the lengthy excerpt from my 1962 Memorial quoted earlier; here it will suffice to list a few of the more important of some 25 clients--Aluminum Company of America (ALCOA), 1912-1950, (bauxite exploration; dam sites and tunnels); Panama Canal Commission, 1916, (earth slides); South Manchurian Railway Company (Japan), 1921, (coal; iron ore); Colorado River Board, 1916, (Boulder Dam); U. S. Army Corps of Engineers, 1932-1937, (Garrison Dam; Fort Peck Dam; some 35 other dam sites); Federal Emergency Administration of Public Works, 1935, (flood control and conservation); and Reynolds Metals Company, 1941-1960, (bauxite exploration; research direction).

His professional reports, all carefully bound and numbered, total almost a hundred and are preserved in the M.I.T. Historical Collection where they are sequestered for the time being, together with his autobiography, Dear Family,, privately printed in June 1959, seven months before his death.

Research Scientist and Inventor

Mead had an insatiable curiosity about how things worked and how geological situations and conditions were to be explained. His lifelong interest in such matters let him see quickly how new ideas, methods, techniques, and instruments could be used to solve geological problems. If he himself got an idea and could not find a device or instrument to test or apply it, he went to his machine shop and made the gadget!

Much of his research and most of his inventions grew out of the need for a demonstration model or graph required for laboratory instruction or for some device that would be useful in his professional activities as a geological consultant or engineering geologist. As a consequence, there was a continuous interplay between the needs of the lecture room, the instructional and research laboratory, the company board room, and the actual engineering problems in the field. His scientific contributions and his numerous ingenious devices can rightly be said to have resulted in part from scientific interest and in part from practical necessity.

Marriage and Family

Mead was married to Bertha May Taylor, a member of one of Madison's (Wisconsin) leading families and a teacher of music, on 17 June 1909. Seven months before his death they celebrated their Golden Wedding Anniversary with their 10 grandchildren, 3 daughters-in-law, and 3 sons. Warren, the oldest, is with Station KWWL in Waterloo, Iowa; Judson is Professor of Geophysics at Indiana University (Bloomington); and Jeremiah, the youngest, is Professor of Physiology in Harvard's School of Public Health. Mrs. Mead spent her last years living in the Commander Hotel in Cambridge and died on 18 June 1964.

Honors and Memberships

Mead was highly respected by his professional peers and gratefully esteemed by his students. He was elected a fellow of the Geological Society of America in 1916 and served the Society as a Councilor (1930-1932), as a member and then Chairman of the first Committee on Projects (1933-1935), and as Vice-president (1938). He was President of the Society of Economic Geologists in 1942, and a longtime member of the American Institute of Mining and Metallurgical Engineers and of the American Society of Civil Engineers. He was elected to the American Academy of Arts and Sciences in 1935 and to the National Academy of Sciences in 1938.

The Mead Memorial Funds

Two special departmental funds have been established at M.I.T. to perpetuate the memory of the Warren J. Meads. In early November of 1968, the proceeds from the sale of the residue of Mead's geological library, which had been given to the Department of Geology and Geophysics earlier, were used

> "... to establish a Warren Judson and Bertha Taylor Mead Memorial Fund, the $1500 to be preserved as the principal of an endowment fund and the actual income to be available for purchase of geological items of special interest for the Lindgren Library."

Course XII alumni were invited to contribute to this fund and on 1 July 1974 the principal was $2,090. The income, used as specified above, has been most helpful to the Lindgren Librarian, as discussed more fully in the chapters on "Libraries" and "The Financing of the Geological Sciences at M.I.T." In March 1969 I was able to solicit $10,000 from the Reynolds Metals Company for a Warren J. Mead Memorial Fund with the understanding that the contribution would be preserved as an endowment, and that the income would be used by the Head of Course XII

> "... in support of geological education and research in the Department."

This fund, like the preceding one, is discussed more fully in the chapter on "The Financing of the Geological Sciences at M.I.T." in my Volume 2.

Such then was the life of Warren Judson Mead--inspiring teacher; imaginative and creative scientist; widely respected engineer; skilled craftsman with camera, slide rule and machine tool; canny and far-sighted administrator; and devoted family man.

BIBLIOGRAPHY OF WARREN JUDSON MEAD

Symbols and abbreviations used in the following references are explained on pages 91 - 98. This is the most complete bibliography of W. J. Mead thus far collected and it is believed that it lists every article of any consequence that he published.

1--1907　　Redistribution of elements in the formation of sedimentary rocks. J. Geol. 15: 238-256, (1907).

2--1908　　The relation of density, porosity, and moisture to the specific volume of ores. Econ. Geol. 3: 319-325, il., (1908); Mining Sci. 58: 89-91, (1908).

3--1909   (and Martin, L.) Apparatus for topographic field work on models in the laboratory. J. Geogr. 7: 209-211, (1909).

4--1911   (with Leith, C. K.) Origin of the iron ores of central and northeastern Cuba. A.I.M.E., B. 51: 217-229, il., (1911).

5--1911a   (with Van Hise, C. R. and Leith, C. K.) [Separate sections in] The geology of the Lake Superior region, U. S. Geol. Surv., Mon. 52, (1911), as follows:
   The iron ores of the: Vermilion district, Minnesota, p. 137-143; Michipicoten district, p. 156-158; Mesabi district, p. 179-197; Penokee-Gogebic district, p. 235-250; Marquette district, p. 270-282; Swanzy district, p. 286; Crystal Falls, Iron River and Florence districts, p. 323-326; Felch Mountain and Calumet districts, p. 326-328; Menominee district, p. 346-354; Baraboo district, p. 362-364; and
   The iron ores of the Lake Superior Region, p. 460-572.

6--1912   Some geological short-cuts. Econ. Geol. 7: 136-144, il., (1912).

7--1912a   (with Leith, C. K.) Metamorphic studies. J. Geol. 20: 353-361, (1912).

8--1914   The average igneous rock. J. Geol. 22: 772-781, (1914).

9--1915   Occurrence and origin of the bauxite deposits of Arkansas. Econ. Geol. 10: 28-54, il., (1915).

10--1915a   (with Leith, C. K.) Metamorphic Geology (a text-book). New York: Henry Holt and Co., xxiii + 337 p., il., (1915). [Reviewed by J. P. Iddings, Science n.s. 45: 386-388, (1917); Albert Johannsen, J. Geol. 26: 82-86, (1918).

11--1915b   (with Leith, C. K.) Metamorphic studies; convergence to mineral type in dynamic metamorphism. J. Geol. 23: 600-607, (1915).

12--1915c   (with Leith, C. K.) Additional data on origin of lateritic
    1916   iron ores of eastern Cuba. A.I.M.E., B. 103: 1377-1380, il., (1915); Tr. 53: 75-78, il., (1916).

13--1920   Notes on the mechanics of geologic structures. J. Geol. 28: 505-523, il., (1920).

14--1920a   A simple method for making block diagrams. The Wisconsin Engineer 25: 3-7, il., (1920).

15--1921   Determination of attitude of concealed bedded formations by diamond drilling. Econ. Geol. 16: 37-47, il., (1921).

16--1924   (and Swanson, C. O.) X-ray determination of minerals. Econ. Geol. 19: 486-489, (1924).

17--1925   The geologic rôle of dilatancy. J. Geol. 33: 685-698, il., (1925).

18--1930   Mechanics of gravitational restraint of subterranean fluid pressures [abst.]. Pan-Am. Geol. 53: 75, (1930).

19--1930a   Some applications of the strain ellipsoid [discussion]. Am. Assoc. Petroleum Geol., B. 14: 234-239, (1930).

20--1930b   The role of dilatancy in engineering. B. Assoc. State Eng. Soc., p. 1-5, (July 1930).

21--1937   Geology of dam sites in hard rock. Civil Eng. 7: 331-334, il., (1937).

22--1937a   Geology of dam sites in shale and earth. Civil Eng. 7: 392-395, il., (1937).

23--1938   Engineering geology of dam sites. Second Congress on large dams, Washington, D.C., (1936). [Conference proof published in 1936 by U. S. Government Printing Office, p. 1-22.] Tr. 4: 171-192, Washington, D.C.: U. S. Govt. Printing Office, (1938).

24--1939
     1940   (with Evans, R. D. and Goodman, C.) Critical evaluation of the present state of the Helium Method of age determination of rocks [abst.]. Geol. Soc. Am., B. 50: 1921, (1939). Also in A. C. Lane, Chmn., Rept. Comm. Measurement of Geologic Time, 1939-43: N.R.C., Div. Geol. Geogr. Ann. Rept. 1939-40, App. G., Exhib. 3, p. 74-79, (Sept. 1940). (‡)

25--1940a   Folding, rock flowage, and foliate structures. J. Geol. 48: 1007-1021, (1940).

26--1941   Engineering geology. Geol. Soc. Am., 50th Anniversary Vol.: 571-578, (1941).

27--1944   (with others) Review of API Research Project 43C--Studies of the effect of radioactivity on the transformation of marine organic materials into petroleum hydrocarbons: Fundamental research on occurrence and recovery of petroleum. A.P.I. Ann. Rept. Prog. 1943: 126-140, (1944). Also in A.P.I. Rept. Prog. 1946-47: 192-195, (1949).

28--1946   (with others) Studies of the effect of radioactivity on the transformation of marine organic materials into petroleum hydrocarbons: Fundamental research on occurrence and recovery of petroleum, 1944-45, API Project 43C, p. 105-106, (1946).

29--1946a   This is M.I.T.--Geology (with drawings by F. K. Morris). Tech. Eng. News 27/6: 199-203, 220, 228, (1946).

30--1954   (and Collins, V. P.) The principles of dilatancy applied to technique of radiotherapy. Am. J. Roent., Radium Therapy and Nuclear Medicine 71: 864-866, il., (1954).

Departmental Reports included in the M.I.T. President's
Reports for SYs 1934-35 to 1947-48

As Head of Course XII, Department of Geology, Mead prepared a brief report on Departmental activities following the end of each school year in June. These reports, included in the President's Report for each particular year, were published in October of the same year, and were as follows:

1) Geology. M.I.T. Pres. Rept. SY 1934-35, [M.I.T., B. 71/1: 98-100].
2) Geology. M.I.T. Pres. Rept. SY 1935-36, [M.I.T., B. 72/1: 129-131].
3) Geology. M.I.T. Pres. Rept. SY 1936-37, [M.I.T., B. 73/1: 122-123].
4) Geology. M.I.T. Pres. Rept. SY 1937-38, [M.I.T., B. 74/1: 117-118].
5) Geology. M.I.T. Pres. Rept. SY 1938-39, [M.I.T., B. 75/1: 128-129].
6) Geology. M.I.T. Pres. Rept. SY 1939-40, [M.I.T., B. 76/1: 124-125].
7) Geology. M.I.T. Pres. Rept. SY 1940-41, [M.I.T., B. 77/1: 125].
8) Geology. M.I.T. Pres. Rept. SY 1941-42, [M.I.T., B. 78/1: 108].
9) Geology. M.I.T. Pres. Rept. SY 1942-43, [M.I.T., B. 79/1: 119].
10) Geology. M.I.T. Pres. Rept. SY 1943-44, [M.I.T., B. 80/1: 131].
11) Geology. M.I.T. Pres. Rept. SY 1944-45, [M.I.T., B. 81/1: 134].
12) Geology. M.I.T. Pres. Rept. SY 1945-46, [M.I.T., B. 82/1: 131-132].
13) Geology. M.I.T. Pres. Rept. SY 1946-47, [M.I.T., B. 83/1: 136-137].
14) Geology. M.I.T. Pres. Rept. SY 1947-48, [M.I.T., B. 84/1: 137-138].

----

(23)
HAROLD WILLIAMS FAIRBAIRN*

HAROLD WILLIAMS FAIRBAIRN

MIT: 1937-1972-

Harold W. Fairbairn brought unusually broad and thorough training and experience in petrology to M.I.T.'s Department of Geology in 1937 when he accepted an appointment as Assistant Professor of Geology. Born, reared, and early educated in Ottawa, Canada, he took his first college work in Queen's University (Kingston, Ont.), from which he received a B.Sc. degree (with Honours) in Geology and Mineralogy in 1929. His first practical work was as laboratory assistant with the Metallurgical Division of the Federal Department of Mines, a year before entering Queen's. He later earned money toward his college expenses as a summer field assistant with the Geological Survey of Canada. After a year of graduate study at the University of Wisconsin, SY 1929-30, he entered Harvard in the fall of 1930 and received an A.M. degree in 1931. Using field work done in Quebec as the basis for a doctor's thesis, he carried on petrological work under Larsen and structural work under Billings, and earned a Ph.D. degree in 1932. In his thesis research he attempted a statistical study of grain orientation in some of his rock thin sections, and in the process became acquainted with a Universal Stage and with petrofabrics, an aspect of petrology little known in North America but well known in Europe through Bruno Sander's formidable treatise Gefügekunde der Gesteine (1930). Hearing by chance that the Royal Society of Canada had funds available for post-doctoral study outside the country, he submitted a proposal for petrofabric study at Innsbruck with Sander and got a grant for a year's study in Europe, which was later extended for a second year. The two years of study and travel in Europe, 1932-1934, which consisted of more than a year at Innsbruck University with Sander, and shorter periods with Goldschmidt, Laves and Strock at Göttingen and with Schmidt and Baier at the Technische-Hochschule in Berlin, provided a superb opportunity for Fairbairn to learn about the use of x-rays in fabric analysis and to become more fully acquainted with the most recent European advances in structural petrology.

Back in Kingston in 1934, after his two years abroad, Fairbairn spent the first year, SY 1934-35, giving a series of lectures at Queen's and writing Introduction to Petrofabric Analysis, a brief (146 page) treatment of the subject matter of Sander's Gefügekunde mentioned above. This mimeographed manual would be enlarged later and published as Structural Petrology of Deformed Rocks (Addison-Wesley Press, 1949; see Bibliography later on). He then became an Instructor in Mineralogy on the regular Geology staff at Queen's and taught there for the next two years, SYs 1935-36 and 1936-37.

---

* The introductory précis and the following biographical sketch are based on a brief autobiographical outline kindly prepared for my guidance by Prof. Fairbairn. I am happy to record here my grateful thanks for this most helpful document.

Soon after 1934, when W. J. Mead came from Wisconsin to head M.I.T.'s Department of Geology, he started a search for several young staff members to add to the faculty. One of the staff positions to be filled was in petrology. When he asked one of his former Wisconsin colleagues, J. E. Hawley, to suggest candidates, Hawley then at Queen's strongly recommended Fairbairn for the position, at the same time hoping to keep him at Queen's. A visit to M.I.T. and an interview with Mead resulted in Fairbairn's accepting an appointment as Assistant Professor of Geology starting in the fall of 1937. Thus began Fairbairn's long association with geology at M.I.T. He served successively as assistant professor (1937-1943), associate professor (1943-1955), and professor (1955-1972), and retired as Professor Emeritus on 30 June 1972.* Since then he has been a Senior Lecturer, and as SY 1974-75 begins he will once again offer instruction in petrology to undergraduates in Course XII.

Fairbairn's long teaching career, starting in 1934 and still continuing today, 40 years later, has been devoted primarily to instruction in undergraduate petrology. Whenever needed he also gave a short course on the Universal Stage to graduate students, and participated in special graduate seminars as they came and went. Throughout he has kept abreast of latest advances in petrology and as a consequence has given hundreds of students the best possible instruction in both practical microscopy and interpretative petrology. Many of his teaching assistants and thesis students have achieved distinction in research fields in which the thorough knowledge of mineralogy and petrology insisted upon by Fairbairn has been indispensable. This group of former students includes, among others, Drs. M. L. Keith (1939), +O. F. Tuttle (1948), +H. S. Yoder (1948), W. H. Dennen (1949), +J. B. Thompson, Jr. (1950), +W. F. Brace (1953), the late J. A. Gower (1955), J. A. Wood (1958), W. C. Phinney (1959), D. R. Wones (1960), J. M. Moore, Jr. (1960), and J. L. Powell (1962). Those marked (+) in the foregoing list have been elected members of the National Academy of Sciences.

Fairbairn's research career can be divided into three periods or stages. During the first period, while yet a college student, his primary activity was field mapping, and he spent summers from 1926 to 1942 doing field work for the Geological Survey of Canada and the Ontario and Quebec Departments of Mines. The second stage, starting with his thesis work at Harvard, involved a growing interest in petrofabric analysis, and commanded his chief efforts for some 20 years, culminating in <u>Structural Petrology of Deformed Rocks</u>. During this same period he also devoted some time to a study of packing in ionic minerals, optical crystallography techniques, a survey of precision and accuracy of chemical analyses of rocks, and some exploratory experimental work on deformation of quartz. Most of this research was conducted with only modest financial support from the Geological Survey of Canada, Geological Society of America, and the Office of Naval Research. The third stage commenced in 1953 when he gave up further research in petrofabrics and related work and joined P. M. Hurley in geochronological investigations. Since then to the present (1974), he has collaborated with Hurley, Pinson, and a steady stream of students, supported by massive funding from the United States Atomic Energy Commission for many years, and since 1972 with modest support from the National Science Foundation and M.I.T. This Hurley-Fairbairn-Pinson team, and the tens of graduate assistants, together and individually, had compiled an impressive record by 1972--19 annual reports, a final report in 1972, several hundred abstracts and full-length articles in leading geochemical and geological journals, and some 20 thesis dissertations. Fairbairn himself contributed more than 100 of the published articles and assisted in preparing the annual reports. His particular contributions were concerned with Sudbury and Huronian Age problems in Ontario, Late Precambrian rocks in Massachusetts, Nova Scotia and Newfoundland, Devonian in-

---

* See <u>Tech Talk</u>, May 17, 1972, p. 3, port.

trusives in the Northern Appalachians, and Archean granulites in Uganda. These and his earlier publications are listed in his Bibliography farther on.

In addition to his teaching and research activities, Fairbairn served for many years as Departmental Registration Officer, as Course XII Representative on the important Committee on Graduate School Policy, on several Library committees, and on other less-important but nonetheless necessary Department committees of one kind or another. And he cheerfully assumed the trying job of directing and overseeing the moving of the Department from Building 4 to Building 24 during the summer of 1946. Finally, for more than thirty years he has maintained an x-ray laboratory and optics laboratory open to both staff and student use, and has had charge of the several preparation rooms where samples could be prepared for analysis and where polished and thin sections could be made.

Fairbairn's wife, Sheila May (Sargent) Fairbairn, born in Burma of English-Irish parents, has been an active member of the M.I.T. Family since marriage in early 1939. Besides participation in the affairs of the Technology Matrons and other Institute activities, she and Hal have offered gracious hospitality to a host of M.I.T. students and their friends at afternoon teas and evening dinners. Four children have been born to the Fairbairns--Ann (1940), Patrick William (1941), Elspeth (1944), and Neil Alastair (1948).

BIRTH, ANCESTRY, AND EARLY EDUCATION

Harold Williams Fairbairn, the oldest of the six children of Arthur Edwin Fairbairn, a general merchant, and Maria (Spratt) Fairbairn, stenographer and housewife, was born in Ottawa, Ontario, on 10 July 1906. He traces his ancestry back to Scottish* and Welsh forebears on his father's side and to Northern Irish (Belfast) on his mother's side. His siblings are Eleanor (1908), $^{\times}$Bruce (1910), Bernice (1914), Donald (1916), and $^{\times}$Thomas (1918). ($^{\times}$ Bruce and Thomas are deceased.)

He grew up in Ottawa where he attended the Hopewell Avenue Public School for his elementary education (1911-1919) and the Lisgar Collegiate Institute for his high school work (1919-1924). He gained a First in Science in his final year but did not get a medal, as did the Firsts in all the other fields, because science was then in low esteem!

For a year following graduation he worked for the Federal Department of Mines in their Chemical Metallurgy Laboratory in Ottawa. Most of the

---

* The FAIRBAIRN engraved with a group of other early engineers and scientists on the southwest side of the M.I.T. Great Court refers to Sir William Fairbairn, a noted 19th Century bridge builder, inventor of rivetting, and experimentalist in metallurgy. He was a first cousin of our M.I.T. Harold Fairbairn's great grandfather. The latter emigrated with his family to Montreal in 1827. See "The life of William Fairbairn, partly written by himself, edited and completed by William Pole." London: Longmans, 16 + 507 p., port., (1877).

$60 a month he earned as clean-up boy was carefully put aside for college, and the year's experience gave him time to decide between various scientific fields for a career.

COLLEGE YEARS (1925-1932)

In the fall of 1925, now financially stable, if not affluent, he registered for the program offered by the Faculty of Applied Science at Queen's, and for the first two years, SYs 1925-26 and 1926-27, he took the courses required of all students. At the beginning of the third year he chose Geology and Mineralogy as a major because it offered a combination of outdoor field mapping, and indoor analytical chemical work on minerals, with which he was already familiar from his Mines' Laboratory experience in Ottawa as a clean-up boy. Furthermore, in those days Canadian geologists, at least those in Government service, were getting higher salaries than chemists!

In his senior year he did a "micro"-thesis, as he calls it, on the chemical composition of a granite, and learned the difficulty of getting the amounts of the constituents to add up to 100%. He recalls that his investigation didn't prove anything, but that he enjoyed it, and perhaps therein lies the secret of why even today he can wait patiently in his chemical laboratory on the eleventh floor of the Green Building for solutions to slowly filter down.

The four years spent earning the B.Sc. (with Honours) awarded him in the spring of 1929 cost slightly less than $2,500. He earned about half the amount by summer field work and borrowed the remainder from his family. As he has commented to me--"The tuition was $210/year. It was a happy, uncomplicated time."

Then came the opportunity to spend a year of graduate study at the University of Wisconsin, whose Department of Geology, especially noted for its program in Precambrian, structural, and metamorphic geology, had attracted some of Canada's most promising young geologists. One of these, E. L. Bruce, after spending a year at Madison, had joined the Queen's geology staff and was teaching a course based on one of Wisconsin's noted books, Metamorphic Geology, written by C. K. Leith and W. J. Mead. He urged Fairbairn to go to Madison for a year and helped get him a tuition award of $200 for the fall of 1929. The SY 1929-30 was especially worthwhile for Fairbairn because it gave him the opportunity to become acquainted with and to sit in lectures by some of the leading geologists of the time--Leith, Mead, Twenhofel, Winchell, and Emmons--and to become acquainted with fellow graduate students who would later have distinguished careers in Canada and the United States: e.g. G. Burton, L. Greer, A. E. Jure, A. Leahy, G. McCartney, A. G. Pentland, and J. E. Thomson, Canadians; and Alice Allen, W. F. Anderson, V. E. Barnes, E. Ellsworth, G. O. Raasch, and S. A. Tyler.

Then once again a friendly hand directed Fairbairn's educational program; this time it was Kirtley Mather of Harvard, who had an interest in Queen's University and its graduates, having taught geology there before Fairbairn's time. He helped get Fairbairn a tuition scholarship of $400 at Harvard, and though some of Fairbairn's friends in Madison considered it odd that he should go East for his doctoral degree, go he did, and in the fall of 1930 he registered as a degree candidate in Harvard's Graduate School.

Although he found the same large contingent of able Canadian students in Cambridge as there had been in Madison, he found the geology courses at Harvard totally different from those at Wisconsin. As a consequence, he got the best of two great departments when he added to his educational experience the stimulation of Harvard's distinguished geologists, with some of whom he took course work or did research--Billings, Bryan, Daly, Graton, Larsen, McLaughlin and Palache.

For his doctoral thesis he chose to continue the study of a field area in Quebec that he had commenced for the Geological Survey of Canada during the summer of 1930 under the supervision of T. H. Clark. The area, straddling the extension in Quebec of the Green Mountain geanticline, was an excellent one because of the deformed and metamorphosed condition of the rocks. Clark was glad to turn over the hard-rock geology to Fairbairn, and the latter naturally turned to Larsen for supervision of his petrological work and to Billings for help with structure. The two professors visited the field area in the summer of 1931 and later approved the completed thesis, "The Structure and Metamorphism of Brome County, Quebec," following which Fairbairn received his Ph.D. degree in 1932.

A POST-DOCTORAL INTERIM (1932-1934)

An outgrowth of the petrological work on his thesis-area rocks was an attempt at a statistical study of grain orientation in some of his thin sections. Bruno Sander had just published his formidable Gefügekunde der Gesteine (1930), and Larsen had read enough of the difficult German to conclude that Fairbairn ought to try his hand at applying some of Sander's methods; hence the aforementioned attempt. So Larsen provided Fairbairn with a Universal Stage for his petrographic microscope, and in due course Hal learned how to use this newest device. Further work on the fabric of rocks might well have ended for Fairbairn, after he completed his thesis, but for a chance interview with H. C. Cooke in Ottawa in December 1931. At the time, Cooke, a senior officer on the Geological Survey of Canada, was on an Awards Committee of the Royal Society of Canada which was charged with seeking applicants for some Carnegie funds that had just come to the Society. The money was for a year of post-doctoral study outside Canada. At Cooke's suggestion, Fairbairn submitted a proposal for petrofabric study with Sander at Innsbruck, got the grant, later extended for a year, and went to Austria at the end of the 1932 summer field season.

During his two-year residence in Europe, 1932-1934, Fairbairn spent most of the time at Innsbruck University, working with Sander, learning German, and struggling with the Gefügekunde. He also spent a 3-month period at Göttingen, where he became acquainted with V. M. Goldschmidt, Fritz Laves, and Lester Strock, and another similar period at the Technische Hochschule in Berlin with Walter Schmidt and Ernst Baier, where he was introduced to the x-ray methods which he wished to apply to fabric analysis. Between times he travelled frequently with his Innsbruck colleagues in the Eastern Alps, and by himself in Italy, Germany, Scandinavia, Czechoslovakia and Poland. Altogether, his European experience was exceedingly fruitful because it came at the critical time between the end of his college student days and the beginning of his career as a college professor. He not only learned of the latest developments along the geological frontier in Europe but also became acquainted with many of the leading geologists on the continent. Furthermore, his travels allowed him to expand greatly the first-hand observation so vital for the field and experimental geologist. As he returned to Ontario in the fall of 1934, at the peak of the Depression Years, he was well prepared to start his professional career; the problem was where to find a job.

AS INSTRUCTOR AT QUEEN'S (1934-1937)

The fall of 1934, when Fairbairn returned from Europe, was not a good time to find a geological job in Canada. There just weren't any! Fortunately, however, a geological colleague once again came to Fairbairn's aid. M. B. Baker, then head of the Geology Department at Queen's, was able to raise $1,000 from ten loyal mining alumni, and he made this available to Fairbairn, with no strings attached, for SY 1934-35. During the winter Fairbairn gave a series of lectures on petrofabrics and organized his notes into a mimeographed manual, Introduction to Petrofabric Analysis, a brief 146-page treatment of some of the subject matter of Sander's Gefügekunde, which he printed privately in 1935. Later Fairbairn would revise and expand this first effort into Structural Petrology of Deformed

Rocks (Addison-Wesley Press, 1949). After another summer of field work, this time with the Ontario Bureau of Mines, he returned to Queen's as Instructor in Mineralogy, and for the next two years he taught undergraduate classes there. Then came the episode in 1937 that led to his long academic career at M.I.T.

M.I.T. YEARS (1937-   )

Soon after 1934, when he came to M.I.T. to head the Department of Geology, W. J. Mead began to seek promising young geologists for his growing staff. One post to be filled was in petrology. Among those from whom he requested recommendations was J. E. Hawley, formerly his colleague at Wisconsin, and in 1937 Professor of Mineralogy at Queen's. Hawley strongly recommended Fairbairn, who had just the combination of training, skills, and field experience that Mead wanted in his new professor, but he also made every effort to induce him to remain at Queen's. Fairbairn came down to M.I.T., had an interview with Mead, and when offered an appointment as Assistant Professor of Geology, starting in September 1937, he accepted it. Thus began his long association with geology at M.I.T.*

Starting in 1937, he served successively as assistant professor (1937-1943), associate professor (1943-1955), and professor (1955-1972), and then retired as Professor Emeritus on 30 June 1972.** Since then he has been on a half-time basis as a Senior Lecturer, and as SY 1974-75 begins he will be offering instruction in petrology for the 38th year, surely a remarkable record of continuity!

As Teacher

During his long academic career as an M.I.T. professor, Fairbairn has judiciously divided his time and effort between teaching and research. In this section I shall discuss his teaching; in the next, his research.

Previous to Fairbairn's appointment in 1937, M. J. Buerger had been giving instruction in petrology in three parts--optical crystallography, petrography, and petrology. Upon Fairbairn's arrival, Buerger immediately released petrography and petrology to him, and after a few years, optical crystallography as well. This proved to be an excellent move because it afforded Buerger more time to develop his growing interest in x-ray crystallography, and it gave Fairbairn the opportunity to reorganize the

---

* Fairbairn became a naturalized citizen of the United States in 1949.

** See Tech Talk, May 17, 1972, p. 3, port., (1972).

subject matter into a single course running through two terms. In the new two-term course he had the flexibility needed to modify and modernize the subject matter when such was necessary.

During the 35 or more years since he took over instruction in petrology he has kept abreast of the advances in his special fields of interest and as a result has given his hundreds of students the best possible instruction in both practical microscopy and interpretative petrology.

Holding the view that understanding the composition, associations, origin and history of rocks should be a fundamental part of the training of every geologist, and that use of the petrographic microscope is indispensable in gaining that understanding, Fairbairn has always followed this philosophy in his teaching. It is not surprising, therefore, that many of his teaching assistants and thesis students have achieved distinction in

Prof. Fairbairn (right) with graduate student W. F. Brace (left) (XII Ph.D. 1953) and guest graduate student Hans Eugster, in his Optics Research Laboratory.

(Photo by Jackman, 1952)

fields of research in which the thorough knowledge of mineralogy and petrology insisted upon by him has been of critical importance. He can list with pride the following former students, among others, who have put his thorough training to good use--Drs. M. L. Keith (1939), $^+$O. F. Tuttle (1948), $^+$H. S. Yoder (1948), W. H. Dennen (1949), $^+$J. B. Thompson, Jr. (1950), $^+$W. F. Brace (1953), the late J. A. Gower (1955), J. A. Wood (1958), W. C. Phinney (1959), D. R. Wones (1960), J. M. Moore, Jr. (1960), and J. L. Powell (1962). Those marked (+) in the foregoing list have been elected members of the National Academy of Sciences.

During his 35 years of full-time teaching at M.I.T. Fairbairn supervised 24 theses--2 bachelor's, 3 master's, and 19 doctor's. Through SY 1964-65 he had supervised more doctor's theses than any other professor in the Geology Department. (More is said about thesis supervision in the chapter on "Theses and Degrees" in my Volume 2.

On three occasions Fairbairn has taken on brief assignments outside the Department. In 1942 he prepared diffractograms of unspecified material for unspecified purposes for the Manhattan Project. During the war years, 1943-1945, he was a Visiting Professor in the M.I.T. Department of Physics and taught basic physics to Navy trainees. (He tells me he learned more than they did!) Finally, during SY 1952-53 he was a Visiting Professor of Geology at Harvard, assigned to teach Petrology for his former student, Prof. James B. Thompson, Jr., during the latter's leave-of-absence.

As Research Investigator

Fairbairn's research career began with field mapping, a summer activity that started in 1926 and ended in 1942 with the onset of World War II. The second phase, involving petrofabric analysis, started in 1930 and continued until the early 1950s, when his research interest turned to geochronology. This third phase occupied the remainder of his active academic career until 1972, when he became Professor Emeritus.

Inasmuch as his first professional contacts were with officers of the Geological Survey of Canada, or with members of the Queen's Geology Department doing mining and mapping work, it was natural for him to seek employment as a field geologist, particularly during his college years when the money earned during the summer helped greatly with school expenses. Field work after completing his doctorate not only yielded a salary, but also provided opportunities to publish articles on special problems or areas.

When his interest in petrofabrics was aroused during his doctoral thesis work, and was further advanced during his two years of post-doctoral study abroad, he turned away from general field mapping and concentrated on fabric analysis of rocks for two decades. During this period, 1935-1953, he published successively improved editions of his original Introduction to Petrofabric Analysis (1935), which finally had the title Structural Petrology of Deformed Rocks (1949, 1954). The several editions are listed in the Bibliography farther on as references numbered 7, 14, 27, and 38. During the same period his research in structural petrology produced a dozen articles. As a result of his manual and his separate articles, Fairbairn became one of the acknowledged North American leaders in structural petrology.

Although devoting major effort to petrofabrics, Fairbairn did find time to consider other problems now and then. His bibliography includes articles on packing in ionic minerals, optical crystallographic techniques, a survey (with others) of chemical methods of rock analysis, and some exploratory experimental work on the deformation of quartz. All of this non-petrofabrics research, however, antedated 1953, the year he decided to join P. M. Hurley and commit his full time and energy to the research going on in Hurley's Geochronology Laboratory.

Whereas his research before 1953 had been largely his individual effort, and the financial support only modest, coming partly from the Geological Survey of Canada, and partly from the Geological Society of America (Project No. 466-45) and the Office of Naval Research, the situation changed when he joined Hurley's project.

The Hurley project on "Variations in isotopic abundances of strontium, calcium, argon and related topics" was massively supported by the Atomic Energy Commission, and Fairbairn became a key member of the research team of Hurley-Fairbairn-Pinson. This remarkably productive team, and the numerous research assistants and associates who worked with them, conducted an impressive program of geochronological research which lasted for 20 years (1953-1972), and still continues (1974) but on a much reduced scale and with only modest funding from N.S.F. and M.I.T. The program produced 19 annual reports, a final report in 1972 (See Bibliography farther on), more than 200 abstracts and full-length articles in leading geochemical and geological journals and periodicals, and some 20 or more thesis dissertations.

Fairbairn played a key role in the aforementioned program. He collected rock samples from many localities, lent a hand in preparing samples for analysis, kept close and patient watch on the chemical work, monitored the spectrometers, and operated the x-ray and other supporting facilities.

He also assisted in preparing the annual reports and the manuscripts to be submitted to journals for publication. He shared with Hurley and Pinson the presentation of research results at society meetings, and was always on hand to assist thesis students needing help. He was the first or only author of more than 100 articles, and has advised me that his particular contributions were concerned with Sudbury and Huronian age problems in Ontario; Late Precambrian rocks in Massachusetts, Nova Scotia and Newfoundland; Devonian intrusives in the Northern Appalachians; and Archean granulites in Uganda and elsewhere. These and his earlier publications are listed in his Bibliography farther on.

The spotlight of time has singled out several of his contributions as worthy of special mention. In "Correlation of quartz deformation with its crystal structure" (19--1939a) his prediction of the principal glide line in quartz deformation has been experimentally confirmed. In "Dolomite orientation in deformed rocks" (25--1941b) the inferred glide system for dolomitic rocks was later fully substantiated by experiment. "Packing in ionic minerals" (32--1943c) gave new and semi-quantitative insight into the role of pressure for mineral stability. "A co-operative investigation of precision and accuracy ..." (42--1951) provided the long overdue data showing the importance of careful preparation and meticulous analysis of rock standards, matters which are now routine but at the time were non-existent or chaotic. Two papers, "X-ray petrology of some fine-grained rocks (29--1943) and "The relation of discordant Rb-Sr mineral and whole-rock ages ..." (91--1961), have been selected for reprinting in a symposium series "Benchmark Papers in Geology" (Rhodes Fairbridge, ed.), the first title (29--1943) appearing in a volume entitled "Rock Cleavage" (D. S. Wood, ed.), the second (91--1961) in "Geochronology" (C. T. Harper, ed.).

Other Departmental Activities

As is the case with most M.I.T. professors, Fairbairn served his Department in a variety of ways other than teaching and research. He served for many years as Graduate Admissions Officer, Graduate Registration Officer, and Course XII Representative on the important Committee on Graduate School Policy. He served on several Library committees and on other less important but nonetheless necessary departmental committees of one kind or another. When it came time for the Department to be moved from Building 4 to 24 during the summer of 1946 he cheerfully directed the movers and saw that things were placed where they belonged. For more than 35 years he has maintained x-ray and microscope laboratories always available to staff members and students on short notice, and has had charge of the several preparation rooms where rock samples could be sawed or crushed and prepared for analysis and where polished and thin sections could be made. Finally, he has looked after and made important additions to the several irreplaceable collections of minerals and rocks.

FIELD WORK AND TRAVEL

Field work, geological excursions, and travel have always strongly attracted Fairbairn, and he has participated extensively in all three activities.

In his earlier years he gained valuable field experience as a summertime assistant or party leader with the federal and provincial surveys:

- 1926 – Assistant on Topographic Survey (G.S.C.) around Bay of Chaleur, Quebec.
- 1927 – Assistant to Carl Tolman on granite bodies west of Sudbury. (G.S.C.)
- 1928 – Assistant to Terence Quirke on Grenville gneisses, Parry Sound, Ontario. (G.S.C.)
- 1929 – Assistant to C. H. Stockwell on Precambrian of Great Slave Lake. (G.S.C.)
- 1930 – Thesis study of Sutton Mountains, (Brome Co.), Quebec. T. H.
- 1931   Clark, field supervisor. (G.S.C.)
- 1932 – Assistant to W. H. Collins on Sudbury lopolith, Ontario. (G.S.C.)
- 1933 – Miscellaneous field excursions in the Eastern Alps while a
- 1934   Royal Society of Canada Fellow in Europe.
- 1935 – Assistant to W. D. Harding on general mapping in northwestern Ontario (near Sioux Lookout). (Ont. Dept. Min.)
- 1936 – Party leader on detailed mapping for Ont. Dept. Min.:
- 1939
    - 1936 – Area near Little Long Lake, Ontario
    - 1938-1939 – Sudbury area, Ontario.
- 1940 – Party leader on detailed mapping near Senneterre, Quebec for Que. Dept. Min.
- 1942 – Party leader looking for potential oil structures on Anticosti Island. (G.S.C.)

He has also participated in the following International Congresses:

- 1937 – I.G.C. in Moscow, including field excursions to Ukraine, Crimea, and Urals.
- 1948 – I.G.C. in London, including field excursions to Northwest Highlands of Scotland, northern England, and Wales.
- 1952 – I.G.C. in Algiers, including a field excursion in Morocco and the Atlas Mountains.

A recent trip, in 1967, was a family safari in East Africa, during which he collected granulites in Uganda for age study. His latest field trip, to Nova Scotia and Newfoundland to collect rocks for age work, included a side trip to the Viking village site on the northwestern peninsula of Newfoundland.

MARRIAGE AND FAMILY

Harold married Sheila May Sargent on 18 April 1939. Sheila was born in Burma of English and Irish parentage (her father was an officer in the British Army), lived briefly in Ireland, and was educated in England. Her special interest in Technology Matrons' affairs was the Book Club (now discontinued) of which she was President for a time. She was active for a short period with the English instruction program for wives of foreign students. She has also been active in the League of Women Voters for many years. More recently, with lightened family responsibilities, church affairs and oil painting have been major occupations. She and Hal have offered gracious hospitality to a host of M.I.T. students and their friends at afternoon teas and evening dinners, and many students first became better acquainted with one another while guests in their home.

Four children have been born to the Fairbairns: Ann (1940), S.B. from Simmons College and Master of Social Science from Columbia, now a mother of two and part-time teacher and social worker in Worthington, Ohio, where her husband, David A. Rigney is an associate professor of metallurgy in The Ohio State University (Columbus); Patrick William (1941), A.B. Harvard, M.A. Toronto, Peace Corps alumnus, and soon to be Ph.D. in Land Resources Planning, University of Massachusetts, and recently married to Monica Amiel, who is currently employed as an urban planner for Franklin County, Massachusetts; Elspeth (1944), a busy keyboard technologist (piano and harpsichord); and Neil Alastair, A.B. Harvard, recently teaching basic English in Italy and playing the bassoon with various musical organizations in England.

SUMMATION

Fairbairn has long been one of the most highly respected and scholarly professors in our Department of Geology. His teaching of up-to-date petrology for more than 35 years has prepared many a doctor for future distinction; his papers on field geology, his teaching and writing in the early development of petrofabrics, and his contributions to geochronology that number more than 100 journal articles, all deserve special notice; and his caretaking of Departmental instruments and collections has been greatly appreciated. Finally, to show that he does relax at times, it can be recorded that he is a skillful bird watcher (See the first two articles listed in his following Bibliography), and can do a fair job with classical chamber music on either viola or cello. Once, too, he played a good game of handball with Newhouse, Buerger, and Shrock, but that was long ago!

Fairbairn was elected a fellow of the American Academy of Arts and Sciences, the Geological Society of America, and the Mineralogical Society of America. He is or has been a member of the Geological Association of Canada, the American Geophysical Union, the Geochemical Society, and the Massachusetts Audubon Society.

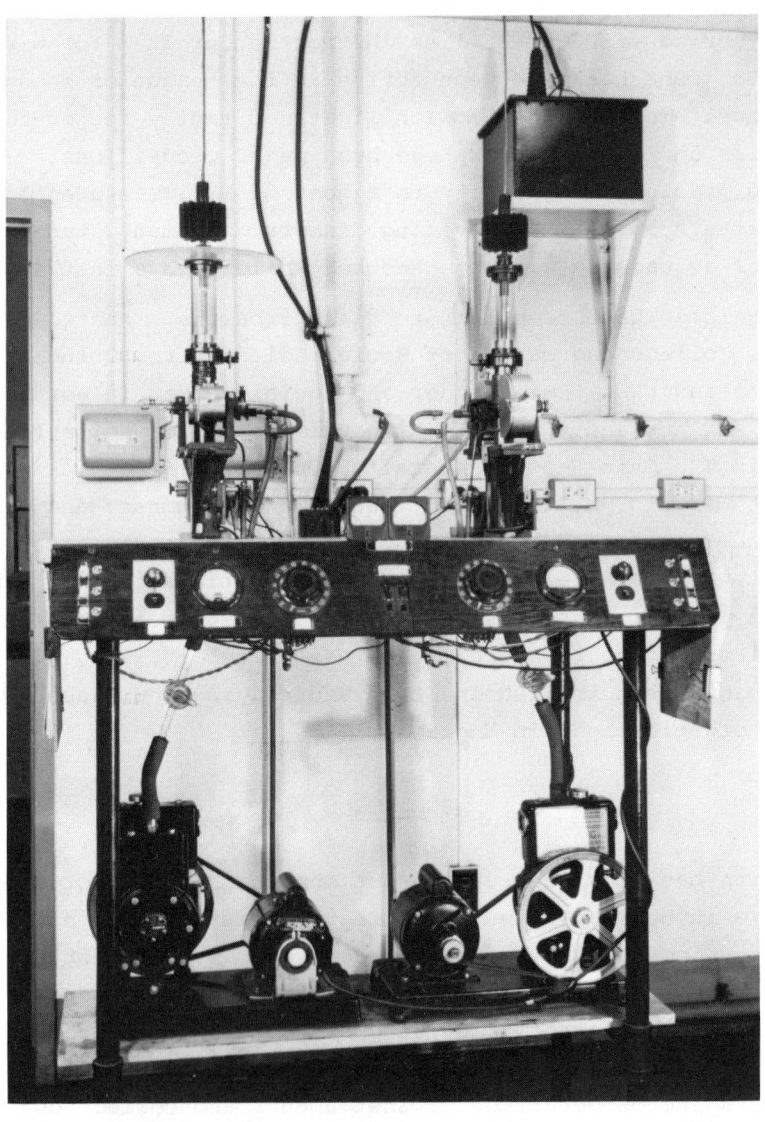

Cold Cathode X-ray Powder Diffraction Units built by the Department of Geology Shop for H. W. Fairbairn in 1946, in collaboration with M. J. Buerger.

(Photo by M.I.T. Graphic Arts)

BIBLIOGRAPHY OF HAROLD WILLIAMS FAIRBAIRN

Symbols and abbreviations used in the following references are explained on pages 91 - 98; in general, abstracts are listed separately, and a few are included with the references to the complete article. This bibliography begins with Fairbairn's first publication, in 1929, and includes all known titles through 1972, the year in which he retired on 1 July.

1--1929   Celestite in central Ontario. Am. Mineral. 14: 286-289, (1929).

2--1930   Bird notes from Parry Sound District [Ontario]. Canadian Field-Nat. 44: 88-91, (1930).

3--1931   Notes on mammals and birds from Great Slave Lake. Canadian Field-Nat. 45: 158-162, (1931).

T--1932   Structure and metamorphism of Brome County, Quebec, A6 + viii + 167 p., il., (1932). (Ph.D. Thesis at Harvard University in Division of Geological Sciences, June 1932.)

4--1932a   Some recent mining developments in southern Quebec. Canada Geol. Surv. Summ. Rept. 1931, Pt. D: 25-27D, (1932).

5--1933   Chemical changes in metabasalt from southern Quebec. J. Geol. 41: 553-555, (1933).

6--1934   Spilite and the average metabasalt. Am. J. Sci. (5) 27: 92-97, (1934).

7--1935   <u>Introduction to petrofabric analysis.</u> Kingston, Ont.: Queen's Univ. Dept. Geol., 142 p. (mimeographed), 47 figs., (1935).

8--1935a   Structural petrology of the Claire River syncline, Tweed, Ontario. Roy. Soc. Canada, Tr. (3) 29: 21-25, (1935).

9--1935b   Petrofabric analysis and some possible applications. Canadian Min. J. 56: 263-267, (1935).

10--1935c   A petrofabric analysis of gypsum. Zeitschr. Kristal. (A) 92: 321-343, (1935).

11--1935d   Notes on the mechanics of rock foliation. J. Geol. 43: 591-608, (1935).

12--1935e   (with Clark, T. H.) The Bolton igneous group of southern Quebec. Roy. Soc. Canada, Tr. (3) 30: 13-18, (1936); Abst., Pr. (3) 29: xcviii, (1935).

13--1936   Elongation in deformed rocks. J. Geol. 44: 670-680, (1936); Abst., Geol. Soc. Am., Pr. 1935: 76, (1936).

14--1937   <u>Structural petrology</u> (revision of "Introduction to petrofabric analysis," 7--1935). Kingston, Ont.: Queen's Univ. Dept. Geol., 150 p. (mimeographed), 54 figs., (1937). [Reviewed by D. T. Griggs <u>in</u> J. Geol. 46: 673-675, (1938).]

15--1937a   Enantimorphous quartz in tectonics [abst.]. Am. Mineral. 22: 211, (1937).

16--1938   Geology of the Northern Long Lake area. Ont. Dept. Min. 46th Ann. Rept., 1939, 46: 1-22, il., (1938).

17--1938a   (with Lovering, T. S. <u>et al</u>.) Report of the Committee on structural petrology, 1937. 103 p. (‡) 17 figs., Nat. Res. Council, Div. Geol. and Geography, (1938).

18--1939   [Review of] Structural petrology by Eleanor Frances Bliss Knopf and Fred Earl Ingerson, 1938. J. Geol. 47: 214-215, (1939).

19--1939a   Correlation of quartz deformation with its crystal structure. Am. Mineral. 24: 351-368, (1939).

20--1939b   Hypotheses of quartz orientation in tectonites. Geol. Soc. Am., B. 50: 1475-1491, (1939); Abst., Roy. Soc. Canada, Pr. (3) 33: 199, (1939).

21--1939c   Geology of the Ashigami Lake area. Ont. Dept. Min. Ann. Rept. 1939, 48: 1-15, il., (1939).

22--1940   (and Hawkes, H. H. [E]) Petrofabric analysis of dolomite [abst.]. Geol. Soc. Am., B. 51: 1926, (1940).

23--1941   Petrofabric relations of nepheline and albite in litchfieldite from Blue Mountain, Ontario. Am. Mineral. 26: 316-320, (1941).

24--1941a   Deformation lamellae in quartz from Ajibik formation, Michigan. Geol. Soc. Am., B. 52: 1265-1277, (1941); Abst., B. 51: 1925-1926, (1940).

25--1941b   (and Hawkes, H. E., Jr.) Dolomite orientation in deformed rocks. Am. J. Sci. 239: 617-632, (1941).

26--1942   Structural petrology applied to ore deposits, in Ore deposits as related to structural features, (W. H. Newhouse, ed.), p. 265-267. Princeton, N. J.: Princeton Univ. Press, (1942).

27--1942a   Structural petrology of deformed rocks. Cambridge, Mass.: Addison-Wesley Press, 143 p. (offset), il., (1942); 2nd ed., ix, 344 p., with supplementary chapters on statistical analysis by Felix A. Chayes, (1949).

28--1942b   (and Robson, G. M.) Breccia at Sudbury, Ontario. J. Geol. 50: 1-33, il., (1942); Ont. Dept. Min. 50th Ann. Rept. 1941, 50: 18-33, (1944); Abst., Roy. Soc. Canada, Pr. (3) 35: 191-192, (1941).

29--1943   X-ray petrology of some fine grained rocks [slate and shales, Vt. and Nova Scotia]. Am. Mineral. 28: 246-256, (1943); Abst., Geol. Soc. Am., B. 53: 1800-1801, (1942).

30--1943a   Gelatin-coated slides for refractive index immersion mounts. Am. Mineral. 28: 396-397, (1943).

31--1943b   Notes on the Felker Di-Met rock saw. Am. Mineral. 28: 398-399, (1943).

32--1943c   Packing in ionic minerals. Geol. Soc. Am., B. 54: 1305-1374, il., (1943).

33--1944   The relations of the Sudbury series to the Bruce series in the vicinity of Sudbury. Ont. Dept. Min. 50th Ann. Rept. 1941, 50: 1-13, il., (1944); Abst., Roy. Soc. Canada, Pr. (3) 34: 162, (1940).

34--1944a   The Bruce series in Falconbridge and Dryden Townships [Ontario]. Ont. Dept. Min. 50th Ann. Rept. 1941, 50: 14-17, il., (1944).

35--1945   (and Sheppard, C. W.) Maximum error in some mineralogic computations. Am. Mineral. 30: 673-703, (1945); Abst., 31: 191-192, (1946); Abst., Geol. Soc. Am., B. 56: 1158, (1945); Abst., Econ. Geol. 40: 590-591, (1945).

36--1946   Wetetnagami River area - Ralleau, Effiat, and Carpiquet Townships, County of Abitibi East [Quebec]. Que. Dept. Min. Geol. Rept. 28, 19 p. (‡), il., (1946).

37--1946a   Six-target cold-cathode X-ray diffraction unit [abst.]. Geol. Soc. Am., B. 57: 1192, (1946); Abst., Am. Mineral. 32: 198-199, (1947).

38--1949   Structural Petrology of Deformed Rocks. Cambridge, Mass.: Addison-Wesley Press, Inc., 344 p., (1949); reprinted in 1954. [With two chapters by F. Chayes.]

39--1950   Synthetic quartzite. Am. Mineral. 35: 735-748, (1950); Abst., Geol. Soc. Am., B. 60: 1886, (1949).

40--1950a   Pressure shadows and relative movements in a shear zone. Am. Geophys. Union, Tr. 31: 914-916, (1950).

41--1950b   A comparative study of chemical and spectrographic methods of determination of the major elements of rocks [abst.]. Roy. Soc. Canada, Pr. (3) 44: 227, (1950).

42--1951   (and others) A cooperative investigation of precision and accuracy in chemical, spectrochemical, and modal analysis of silicate rocks. U.S. Geol. Surv. Bull. 980, vi, 71 p., il., (1951). Contributions to Geochemistry, 1950-51:
  Pt. 1) Preparation and distribution of the samples, p. 1-6.
  Pt. 2) (with Dennen, W. H. and Ahrens, L. H.) Spectrochemical analysis of major constituent elements in rocks and minerals [R.I. and Va.], p. 25-52.
  Pt. 6) Summary of results, p. 69-71.

43--1951a   Preparation and distribution of the samples [R.I. and Va.], in preceding item, 42-1951, p. 1-6, (1951).

44--1951b   (with Dennen, W. H. and Ahrens, L. H.) Spectrochemical analysis of major constituent elements in rocks and minerals [R.I. and Va.], in preceding item, 42--1951, p. 25-52, (1951).

45--1951c   Summary of results, in preceding item, 42--1951, p. 69-71, (1951).

46--1951d   (with Chayes, F.) A test of the precision of thin-section analysis by point counter. Am. Mineral. 36: 704-712, (1951).

47--1951e   Hydrothermal-differential pressure equipment for experimental studies in low-grade rock metamorphism. Geol. Soc. Am., B. 62: 39-43, (1951).

48--1951f   (and Podolsky, T.) Notes on precision and accuracy of optic angle determination with the universal stage. Am. Mineral. 36: 823-832, (1951).

49--1952   Notes on minimum-deviation refractometry. Am. Mineral. 37: 37-47, (1952).

50--1952a   (with Hurley, P. M.) Alpha radiation damage in zircon. J. Appl. Phys. 23: 1408, (1952).

51--1952b   (with Hurley, P. M.) Radiation damage in zircon: a possible age method [abst.]. Geol. Soc. Am., B. 63: 1266, (1952).

52--1952c   (and Schairer, J. F.) A test of the accuracy of chemical analysis of silicate rocks. Am. Mineral. 37: 744-757, (1952).

53--1953   (with Hurley, P. M.) Radiation damage in zircon - a possible age method. Geol. Soc. Am., B. 64: 659-673, (1953).

54--1953a   (and Ahrens, L. H. and Gorfinkle, L. G.) Minor element content of Ontario diabase. Geochim. Cosmochim. Acta 3: 34-46, (1953).

55--1953b   Precision and accuracy of chemical analysis of silicate rocks. Geochim. Cosmochim. Acta 4: 143-156, (1953); Abst., Cong. Géol. International C.R. 19th sess., Sect. XIII, Fasc. XV, Alger 1954: 439, (1954).

56--1954   The stress-sensitivity of quartz in tectonites. Tschermaks Mineral. Petrog. Mitt., F3, Bd 4, H 1-4: 75-80, (1954).

57--1955   Concentration of heavy accessories from large rock samples. Am. Mineral. 40: 458-468, (1955).

58--1955a   (with Hurley, P. M.) Ratio of thorium to uranium in zircon, sphene and apatite [abst.]. Geol. Soc. Am., B. 66: 1578, (1955).

59--1955b (with Handin, J.) Experimental deformation of Hasmark dolomite. Geol. Soc. Am., B. 66: 1257-1273, (1955).

60--1956 (and Hurley, P. M.) Radiation damage in zircon from eastern Massachusetts and Nova Scotia [abst.]. Am. Geophys. Union, Tr. 37: 344, (1956).

61--1956a (with Webber, G. R. and Hurley, P. M.) Relative ages of eastern Massachusetts granites by total lead ratios in zircon. Am. J. Sci. 254: 574-583, (1956).

62--1957 (and Hurley, P. M.) Radiation damage in zircon and its relation to ages of Paleozoic igneous rocks in northern New England and adjacent Canada. Am. Geophys. Union, Tr. 38: 99-107, (1957).

63--1957a (with Hurley, P. M. et al.) Comparison of A^{40}/K^{40} and Sr^{87}/Rb^{87} ages on biotite [abst.]. Am. Geophys. Union, Tr. 38: 396, (1957).

64--1957b (with Hurley, P. M.) Abundance and distribution of uranium and thorium in zircon, sphene, apatite, epidote, and monazite in granitic rocks. Am. Geophys. Union, Tr. 38: 939-944, (1957).

65--1957c (and others) Age of Nova Scotia granites [abst.]. Geol. Soc. Am., B. 68: 1725, (1957).

66--1957d (with Hurley, P. M. et al.) Age study of some crystalline rocks of the Georgia Piedmont [abst.]. Geol. Soc. Am., B. 68: 1781, (1957).

67--1957e (with Powell, R. M. et al.) Test of the half-life of Rb^{87} [abst.]. Geol. Soc. Am., B. 68: 1783, (1957).

68--1958 (with Pinson, W. H., Jr. and Cormier, R. F.) Sr/Rb age measurements on hornblende and feldspars, and the age of syenite at Chicoutimi, Quebec, Canada. Geol. Soc. Am., B. 69: 599-602, (1958).

69--1958a (with Pinson, W. H., Jr. et al.) Sr/Rb age study of tektites. Geochim. Cosmochim. Acta 14: 331-339, (1958).

70--1958b Nova Scotia age program. U.S. Atomic Energy Comm. Rept. NYO-3938: 4-16, il., (1958). (Report prepared for A.E.C. by M.I.T.)

71--1958c Age study from Newfoundland. U.S. Atomic Energy Comm. Rept. NYO-3938: 69, (1958). (Report prepared for A.E.C. by M.I.T.)

72--1958d (with Pinson, W. H., Jr.) Sr/Rb age study of tektites. U.S. Atomic Energy Comm. Rept. NYO-3938: 82-96, il., (1958). (Report prepared for A.E.C. by M.I.T.)

73--1958e Preparation of large biotite sample for interlaboratory analysis. U.S. Atomic Energy Comm. Rept. NYO-3938: 121-122, (1958). (Report prepared for A.E.C. by M.I.T.)

74--1958f (with Bullwinkel, H. J. et al.) Age investigation of syenites from Coldwell, Ontario [abst.]. Geol. Soc. Am., B. 69: 1543-1544, (1958).

75--1958g (with Hurley, P. M. and Pinson, W. H., Jr.) Intrusive and metamorphic rock ages in Maine and surrounding areas [abst.]. Geol. Soc. Am., B. 69: 1591, (1958).

76--1959 (with Hurley, P. M. et al.) Age study of Lower Paleozoic glauconites [abst.]. J. Geophys. Res. 64: 1109, (1959).

77--1959a (and others) Age investigation of syenites from Coldwell, Ontario. Geol. Assoc. Canada, Pr. 11: 141-144, (1959).

78--1959b (with Hurley, P. M. et al.) Minimum age of the lower Devonian slate near Jackman, Maine. Geol. Soc. Am., B. 70: 947-949, (1959).

79--1959c (with Allen, V. T. et al.) Age of Precambrian igneous rocks of Missouri [abst.]. Geol. Soc. Am., B. 70: 1560-1561, (1959).

80--1959d (with Pinson, W. H., Jr. and Hurley, P. M.) Rb/Sr feldspar ages in granitic rocks of Sudbury - Blind River, Ontario, Canada [abst.]. Geol. Soc. Am., B. 70: 1599-1600, (1959).

81--1959e (and others) Authigenic versus detrital illite in sediments [abst.]. Geol. Soc. Am., B. 70: 1622, (1959).

82--1959f (with Pinson, W. H., Jr. et al.) Three ages of rock crystallization in Colombia, South America [abst.]. Geol. Soc. Am., B. 70: 1656, (1959).

83--1960 (and others) A comparison of the ages of coexisting biotite and muscovite in some Paleozoic granite rocks. Geochim. Cosmochim. Acta 19: 7-9, (1960).

84--1960a (and others) Age of the granitic rocks of Nova Scotia. Geol. Soc. Am., B. 71: 399-413, il., (1960).

85--1960b (and Pinson, W. H., Jr. and Hurley, P. M.) Comparison of Rb-Sr mineral and whole-rock ages at Sudbury, Ontario [abst.]. J. Geophys. Res. 65: 2488-2489, (1960).

86--1960c (with Hart, S. R. et al.) Use of amphiboles and pyroxenes for K-Ar dating [abst.]. Geol. Soc. Am., B. 71: 1882, (1960).

87--1960d (with Hurley, P. M. et al.) Reliability of glauconite for age measurement by K-Ar and Rb-Sr methods. Am. Assoc. Petroleum Geol., B. 44: 1793-1808, il., (1960).

88--1960e (and Hurley, P. M. and Pinson, W. H., Jr.) Mineral and rock ages at Sudbury - Blind River, Ontario. Geol. Assoc. Canada, Pr. 12: 41-66, il., (1960).

89--1960f (with Hurley, P. M. et al.) K-Ar and Rb-Sr minimum ages for the Pennsylvanian section in the Narragansett Basin. Geochim. Cosmochim. Acta 18: 247-258, (1960).

90--1960g (and others) A comparison of the ages of coexisting biotite and muscovite in some Paleozoic granite rocks. Geochim. Cosmochim. Acta 19: 7-9, (1960).

91--1961 (and Hurley, P. M. and Pinson, W. H., Jr.) The relation of discordant Rb-Sr mineral and whole-rock ages in an igneous rock to its time of crystallization and to the time of subsequent Sr^{87}/Sr^{86} metamorphism. Geochim. Cosmochim. Acta 23: 135-144, (1961).

92--1961a (with Hurley, P. M. et al.) Geochronology of Proterozoic granites in Northern Territory, Australia. Part I: K-Ar and Rb-Sr age determinations. Geol. Soc. Am., B. 72: 653-662, (1961).

93--1961b (with Herz, N. et al.) Age measurements from a part of the Brazilian Shield. Geol. Soc. Am., B. 72: 1111-1120, (1961).

94--1961c (with Hurley, P. M. et al.) K-Ar age studies of Mississippi and other river sediments. Geol. Soc. Am., B. 72: 1807-1816, (1961).

95--1961d Summary of discussion of age investigations at Sudbury, Ontario, Canada. Geochronology of Rock Systems, N.Y. Acad. Sci., An. 91: 431, (1961).

96--1961e Summary of discussion of geochronology of Proterozoic granites in Northern Territory, Australia. Geochronology of Rock Systems, N.Y. Acad. Sci., An. 91: 521-523, (1961).

97--1962 (with Beall, G. H. et al.) Comparison of K-Ar and whole-rock Rb-Sr dating in New Quebec and Labrador [abst.]. J. Geophys. Res. 67: 3541, (1962). (See also 120--1963g.)

98--1962a   (with Faure, G. et al.) Isotopic compositions of strontium in continental basic intrusives [abst.]. J. Geophys. Res. 67: 3556-3557, (1962).

99--1962b   (with Hurley, P. M. et al.) Radiogenic argon and strontium diffusion parameters in biotite at low temperatures obtained from Alpine Fault Uplift in New Zealand. Geochim. Cosmochim. Acta 26: 67-80, (1962).

100--1962c (with Hurley, P. M. et al.) Radiogenic strontium-87 model of continent formation [abst.]. J. Geophys. Res. 67: 3567-3568, (1962).

101--1962d (with Moorbath, S. et al.) Evidence for the origin of mineralized Tertiary intrusives in Southwestern States from strontium-isotope ratios [abst.]. J. Geophys. Res. 67: 3582, (1962).

102--1962e (with Powell, J. L. et al.) Sr^{87}/Sr^{86} evidence bearing on the genesis of carbonatites [abst.]. J. Geophys. Res. 67: 3588-3589, (1962).

103--1962f (with Hurley, P. M. et al.) Radiogenic model of continent formation. J. Geophys. Res. 67: 5315-5334, (1962).

104--1962g (with Hurley, P. M. et al.) Unmetamorphosed minerals in the Gunflint formation used to test the age of the Animikie. J. Geol. 70: 489-492, (1962).

105--1962h (and others) Evidence of the origin and time of separation of magmas of the Monteregian Hills, Quebec, from development of radiogenic Sr^{87} [abst.]. Geol. Soc. Am., Sp.P. 68: 174, (1962).

106--1962i (with Hower, J. et al.) Effect of mineralogy on K/Ar age as a function of particle size in a shale [abst.]. Geol. Soc. Am., Sp.P. 68: 201-202, (1962).

107--1962j (with Hurley, P. M. et al.) K-Ar age values on the clay fractions in shales ranging in age from Tertiary to Ordovician [abst.]. Geol. Soc. Am., Sp.P. 68: 203-204, (1962).

108--1962k (with Pinson, W. H. and Bottino, M. L.) Rb-Sr ages of Tertiary volcanic rocks [abst.]. Geol. Soc. Am., Sp.P. 68: 246-247, (1962).

109--1962l (with Powell, J. L. and Hurley, P. M.) Isotopic composition of strontium in carbonatites. Nature 196: 1085-1086, (1962).

110--1962m (with Pinson, W. H. et al.) K-Ar and Rb-Sr ages of biotites from Colombia, South America. Geol. Soc. Am., B. 73: 907-910, (1962).

111--1962n (with Bailey, S. W. et al.) K-Ar dating of sedimentary illite polytypes. Geol. Soc. Am., B. 73: 1167-1170, (1962).

112--1962o (with Hurley, P. M. et al.) New approaches to geochronology by strontium isotope variations in whole rocks. Radioactive Dating, International Atomic Energy Agency, Vienna, p. 201-217, (1962).

113--1963   (with Faure, G. et al.) Whole-rock Rb-Sr age of norite and micropegmatite at Sudbury, Ontario [abst.]. Am. Geophys. Union, Tr. 44: 110-111, (1963). (See also 134--1964d.)

114--1963a (with Bottino, M. L. et al.) Whole-rock Rb-Sr ages of some Paleozoic volcanics and related granites in the northern Appalachians [abst.]. Am. Geophys. Union, Tr. 44: 111, (1963).

115--1963b (with Hurley, P. M. and Pinson, W. H., Jr.) Progress Report on analytical accuracy of Sr^{87}/Sr^{86} measurement [abst.]. Am. Geophys. Union, Tr. 44: 111-112, (1963).

116--1963c (with Faure, G. and Hurley, P. M.) An estimate of the isotopic composition of strontium in rocks of the Precambrian Shield of North America. J. Geophys. Res. 68: 2323-2329, (1963).

117--1963d (with Bottino, M. L. et al.) Rb-Sr age study of the Lower Devonian volcanic sequence at Kineo, Maine [abst.]. Geol. Soc. Am., Sp.P. 73: 121, (1963).

118--1963e (with Brookins, D. G. et al.) Whole-rock Rb-Sr investigations of the Collins Hill, Maromas, and Glastonbury formations at Collins Hill, Connecticut [abst.]. Geol. Soc. Am., Sp.P. 73: 123, (1963).

119--1963f (with Faure, G. et al.) Estimate of the isotopic composition of strontium in rocks of the Precambrian basement, Canada [abst.]. Geol. Soc. Am., Sp.P. 73: 150-151, (1963).

120--1963g (with Beall, G. H. et al.) Comparison of K-Ar and whole-rock Rb-Sr dating in New Quebec and Labrador. Am. J. Sci. 261: 571-580, il., (1963). (See also 97--1962.)

121--1963h (with Pinson, W. H. et al.) Evidence on the origin of felsic volcanic rocks from their initial abundance of Sr^{87} [abst.]. Geol. Soc. Am., Sp.P. 73: 216, (1963).

122--1963i (with Hurley, P. M. et al.) K-Ar age values on the clay fractions in dated shales. Geochim. Cosmochim. Acta 27: 279-284, (1963).

123--1963j (with Hurley, P. M. et al.) K-Ar age values in pelagic sediments of the North Atlantic. Geochim. Cosmochim. Acta 27: 393-399, (1963).

124--1963k (with Hower, J. et al.) The dependence of K-Ar age on the mineralogy of various particle size ranges in a shale. Geochim. Cosmochim. Acta 27: 405-410, (1963).

125--1963ℓ (and others) Initial ratio of strontium-87 to strontium-86, whole-rock age, and discordant biotite in the Monteregian igneous province, Quebec. J. Geophys. Res. 68: 6515-6522, (1963).

126--1963m (with Faure, G. et al.) Age of the Great Dyke of Southern Rhodesia. Nature 200: 769-770, (1963).

127--1963n (and Hurley, P. M. and Pinson, W. H., Jr.) Progress Report on initial Sr^{87}/Sr^{86} in igneous rocks [abst.]. Committee on Problems of Geochemistry, I.U.G.G., Berkeley, 9: 32, (1963).

128--1963o (with Hurley, P. M. et al.) Evidence of continuing separation of sial from the mantle from the isotopic composition of common strontium. Nuclear Geophysics, Nuclear Science Series Rept. 30, NAS-NRC Pub. 1075: 83-93, (1963).

129--1963p (with Hurley, P. M. et al.) New approaches to geochronology by strontium isotope variations in whole rocks, in "Radioactive Dating," International Atomic Energy Agency, Proceedings of a Symposium, Athens, p. 201, (1963).

130--1964 (with Hurley, P. M. et al.) Preliminary investigation of Sr^{87}/Rb^{87} relationships in the Sierra Nevada plutonic rocks [abst.]. Geol. Soc. Am., Sp.P. 76: 85, (1964).

131--1964a (and Hurley, P. M. and Pinson, W. H.) Initial Sr^{87}/Sr^{86} and possible sources of granitic rocks in southern British Columbia. J. Geophys. Res. 69: 4889-4893, il., (1964).

132--1964b (with Hurley, P. M. and Pinson, W. H., Jr.) Rb-Sr relationships in serpentinite from Mayagüez, Puerto Rico, and dunite from St. Paul's Rocks--A progress report, in "A study of serpentinite," NAS-NRC Pub. 1188: 149-151, (1964).

133--1964c (and Hurley, P. M. and Pinson, W. H.) Preliminary age study and initial Sr^{87}/Sr^{86} of Nova Scotia granitic rocks by the Rb-Sr whole-rock method. Geol. Soc. Am., B. 75: 253-257, (1964).

134--1964d (with Faure, G. et al.) Whole-rock Rb-Sr age of norite and micropegmatite at Sudbury, Ontario. J. Geol. 72: 848-854, il., (1964). (See also 113--1963.)

135--1965 (and others) Rb/Sr whole-rock isotopic analyses and the Cambrian-Precambrian problem in southeastern Massachusetts [abst.]. Am. Geophys. Union, Tr. 46: 173, (1965).

136--1965a (with Pinson, W. H., Jr. et al.) Rb-Sr age of stony meteorites. Geochim. Cosmochim. Acta 29: 455-466, (1965).

137--1965b (with Hurley, P. M. et al.) Investigation of initial Sr^{87}/Sr^{86} ratios in the Sierra Nevada plutonic province. Geol. Soc. Am., B. 76: 165-174, il., (1965).

138--1965c (with Hurley, P. M. et al.) Radioactive decay of Rb^{87} to Sr^{87} in geological science exclusive of age dating. Primera Conferencia Interamericana de Radioquimica, Montevideo. Pan-American Union, Washington, D.C., p. 175-178, (1965).

139--1965d (and Hurley, P. M. and Pinson, W. H.) Re-examination of Rb-Sr whole-rock ages at Sudbury, Ontario. Geol. Assoc. Canada, Pr. 16: 95-101, (1965).

140--1966 (with Powell, J. L. and Hurley, P. M.) The strontium isotopic composition and origin of carbonatites, in Carbonatites (O. F. Tuttle and J. Gittins, eds.). New York: Interscience Publishers, p. 365-378, (1966).

141--1966a (and others) Whole-rock age and initial $^{87}Sr/^{86}Sr$ of volcanics underlying fossiliferous Lower Cambrian in the Atlantic provinces of Canada. Can. J. Earth Sci. 3: 509-521, (1966). (See also 150--1968.)

142--1966b (with Hurley, P. M. and Pinson, W. H., Jr.) Evidence from Western Ontario of the isotopic composition of strontium in Archean seas [abst.]. Geol. Soc. Am., Sp.P. 87: 84, (1966).

143--1966c (with Hurley, P. M. and Pinson, W. H., Jr.) Rb-Sr isotopic evidence in the origin of potash-rich lavas of western Italy. Earth Planet. Sci. Letters 1: 301-306, (1966).

144--1967 (with Moorbath, S. and Hurley, P. M.) Evidence for the origin and age of some mineralized Laramide intrusives in the southwestern United States from strontium isotope and rubidium-strontium measurements. Econ. Geol. 62: 228-236, (1967).

145--1967a (with Hurley, P. M. et al.) Tracing the history of differentiation of the mantle by Rb-Sr isotopic relationships, in Upper Mantle Project: U.S. Progress Rept.: 103, (1967) (NAS-NRC Pub.). (See also 169--1971d.)

146--1967b (and others) Rb-Sr age of granitic rocks of southeastern Massachusetts and the age of the Lower Cambrian at Hoppin Hill. Earth Planet. Sci. Letters 2: 321-328, il., (1967).

147--1967c (and others) Rb-Sr whole-rock age of the Sudbury lopolith and basin sediments [abst.]. Am. Geophys. Union, Tr. 48: 242, (1967); Can. J. Earth Sci. 5: 707-714, (1968).

148--1967d (with Heath, S. A.) Strontium isotopic evidence bearing on the origin of anorthosite [abst.]. Am. Geophys. Union, Tr. 48: 244, (1967).

149--1967e (with Hurley, P. M. et al.) Test of continental drift by comparison of radiometric ages. Science 157: 495-500, (1967).

150--1968 (and others) Whole-rock age and initial Sr^{87}/Sr^{86} of volcanic rocks underlying fossiliferous Lower Cambrian in the Atlantic Provinces of Canada [abst.]. Geol. Soc. Am., Sp.P. 101: 65, (1968). (See also 141--1966a.)

151--1968a (with Hurley, P. M. et al.) Rb-Sr whole-rock analyses in northern Brazil correlated with ages in West Africa [abst.]. Geol. Soc. Am., Sp.P. 101: 100-101, (1968).

152--1968b (with Brookins, D. G. et al.) Geochronological aspects of the genesis of large granitic pegmatites in non-igneous environments [abst.]. Geol. Soc. Am., Sp.P. 115: 25, (1968).

153--1968c (and others) Radiometric ages of igneous rocks in northeastern Massachusetts [abst.]. Geol. Soc. Am., Sp.P. 115: 260-261, (1968).

154--1968d (with Heath, S. A.) Sr^{87}/Sr^{86} ratios in anorthosites and some associated rocks, in Origin of anorthosites and related rocks (Y. W. Isachsen, ed.). Mem. 18, Univ. State N.Y., State Ed. Dept., Albany, p. 99-110, (1968).

155--1968e (with Hurley, P. M. et al.) Some orogenic episodes in South America by K-Ar and whole-rock Rb-Sr dating. Can. J. Earth Sci. 5: 633-638, (1968).

156--1968f (and others) Rb-Sr whole-rock age of the Sudbury lopolith and basin sediments. Can. J. Earth Sci. 5: 707-714, (1968); Abst., Am. Geophys. Union, Tr. 48: 242, (1967).

157--1969 (with Brookins, D. G. et al.) A Rb-Sr geochronologic study of the pegmatites of the Middletown area, Connecticut. Contr. Mineral. Petrology 22: 157-168, (1969).

158--1969a (and others) Correlation of radiometric ages of Nipissing diabase and Huronian metasediments with Proterozoic orogenic events in Ontario. Can. J. Earth Sci. 6: 489-497, (1969).

159--1969b [Review of] Studies of Appalachian Geology: Northern and Maritime (E-an Zen et al., eds.). New York: Interscience Publishers, 475 p. Am. Geophys. Union EØS 50: 415, (1969).

160--1970 (and Hurley, P. M.) Northern Appalachian geochronology as a model for interpreting ages in older orogens. Eclogae Geol. Helv. 63: 83-90, (1970).

161--1970a (with Spooner, C. M.) Relation of radiometric age of granitic rocks near Calais, Maine, to the time of Acadian orogeny. Geol. Soc. Am., B. 81: 3663-3670, (1970).

162--1970b (with Bottino, M. L. et al.) Blue Hills igneous complex, Massachusetts: Whole-rock Rb-Sr open systems. Geol. Soc. Am., B. 81: 3739-3746, (1970).

163--1970c (with Spooner, C. M.) $Strontium^{87}/strontium^{86}$ initial ratios in pyroxene granulite terranes. J. Geophys. Res. 75: 6706-6713, (1970).

164--1970d (with Spooner, C. M. and Hepworth, J. V.) Whole-rock Rb-Sr isotopic investigation of some East African granulites. Geol. Mag. 107: 511-521, (1970).

165--1971 (and Hurley, P. M.) Evaluation of X-ray fluorescence and mass spectrometric analyses of Rb and Sr in some silicate standards. Geochim. Cosmochim. Acta 35: 149-156, (1971).

166--1971a Radiometric age of mid-Paleozoic intrusives in the Appalachian-Caledonides mobile belt. Am. J. Sci. 270: 203-217, (1971).

167--1971b (with Spooner, C. M. and Berrangé, J. P.) Rb-Sr whole-rock age of the Kanuku Complex, Guyana. Geol. Soc. Am., B. 82: 207-210, (1971).

168--1971c (with Hurley, P. M. et al.) Liberian age province (about 2700 m.y.) and adjacent provinces in Liberia and Sierra Leone. Geol. Soc. Am., B. 82: 3483-3490, (1971).

169--1971d (with Hurley, P. M. and Pinson, W. H., Jr.) Tracing the history of differentiation of the mantle by Rb-Sr isotopic relationships, in Upper Mantle Project: U.S. Program Final Rept.: 214-215, (1971) (NAS-NRC Pub.). (See also 145--1967a.)

The following are items prepared by H. W. Fairbairn, alone or with colleagues, and published only as so-called "gray literature," in Variations in isotopic abundances of strontium, calcium, argon and related topics, the annual reports made by P. M. Hurley to the U.S.A.E.C. under Contract AT(30-1)-1381. These annual reports were prepared in M.I.T.'s Department of Geology and Geophysics, and printed and distributed by the Institute's Graphic Arts. Although, as stated earlier, they constitute "gray literature," they are included in this bibliography because they have been cited in the same way as regularly published articles and also because they indicate a kind of productivity not always known to the scientific community at large.

1) Computation of error in Rb-Sr age calculations. 3rd Ann. Rept. 1955-56: 29-33, (1956).

2) Age data from Newfoundland. 5th Ann. Rept. 1957-58: 69, (1958).

3) New England age program. 6th Ann. Rept. 1958: 28-47, (1958).

4) Preparation and analyses of large biotite sample for interlaboratory analysis. 6th Ann. Rept. 1958: 48-52, (1958).

5) Concentration of Rb and Sr in K-feldspar. 6th Ann. Rept. 1958: 119-120, (1958).

6) Redistribution of radiogenic Sr^{87} between Rb-rich and Rb-poor phases during metamorphism. 8th Ann. Rept. 1960: 225-236, (1960).

7) Rb-Sr age investigation of Massachusetts granites. 9th Ann. Rept. 1961: 249-254, (1961).

8) Geochronology of some granitic plutons, Lake-of-the-Woods region, Ontario. 11th Ann. Rept. 1963: 113-116, (1963).

9) Application of Least-Squares analysis to Rb-Sr whole-rock isochrons. 11th Ann. Rept. 1963: 129-131, (1963).

10) Comparison of Rb/Sr by X-ray spectrographic and mass spectrometric methods. 11th Ann. Rept. 1963: 135-142, (1963).

11) Rb-Sr whole-rock age of Keewatin and Timiskaming volcanics at Kirkland Lake, Ontario. 13th Ann. Rept. 1965: 75-78, (1965).

12) Preliminary whole-rock age of Huronian sediments, Ontario. 14th Ann. Rept. 1966: 127-128, (1966).

13) Age relations of volcanics at Kirkland Lake, Ontario, with the Round Lake pluton. 14th Ann. Rept. 1966: 141-143, (1966).

14) Rb-Sr age investigation of the Brighton volcanic complex. 14th Ann. Rept. 1966: 167-168, (1966).

15) Rb-Sr age of intrusives near Salem, Massachusetts. 14th Ann. Rept. 1966: 173, (1966).

16) Progress report on determination of Rb/Sr ratios by X-ray fluorescence. 14th Ann. Rept. 1966: 187-192, (1966).

17) Rb-Sr age and initial Sr^{87}/Sr^{86} of the Huronian section southwest of Sudbury, Ontario. 15th Ann. Rept. 1967: 53-60, (1967).

18) Rb-Sr isochron ages of metasediments in the Canadian Shield. 15th Ann. Rept. 1967: 61-63, (1967).

19) Preliminary geochronological studies in northeastern Newfoundland. 17th Ann. Rept. 1969: 19-20, (1969).

20) Progress report on rock standards. 17th Ann. Rept. 1969: 99-102, (1969).

21) Preliminary whole-rock Rb-Sr isotopic study of Paleozoic shales and slates in Maine and the Atlantic provinces of Canada. 19th Ann. Rept. 1971: 18-22, (1971).

22) Supplementary Rb-Sr isotopic investigation of the late Precambrian Bull Arm formation, Newfoundland. 19th Ann. Rept. 1971: 34-35, (1971).

Prof. Fairbairn's class in petrology laboratory in Building 24.
Teaching Assistant Paul Cloke (XII Ph.D. 1954) assists Jerry F. Champlin (VII S.B. 1951) at far left; S. Parker Gay, Jr. (XII S.B. 1952) and Joan M. Fleckenstein (XII-A S.B. 1953) study a data strip; and Bruce C. Murray (XII-A S.B. 1953; S.M. 1954; Ph.D. 1955) studies a thin section by himself.

(Photo by Jackman, 1952)

Prof. Fairbairn adjusting equipment in his X-ray Laboratory in Building 24.

(Photo by Jackman, 1952)

(24)
ROBERT RAKES SHROCK*

ROBERT RAKES SHROCK

MIT: 1937-1970-1975

Trained in invertebrate paleontology, stratigraphy and sedimentology at Indiana University, and initiated into the academic world of teaching and research at the University of Wisconsin, Robert R. Shrock came to M.I.T. in 1937 to offer the subjects normally given by Prof. Morris, who was on leave, and to work with Prof. Shimer. He would offer the subjects long given by Shimer when the latter retired in 1942. During World War II Shrock took a leave of absence to work with the War Production Board in Washington (1943) and then with the Reynolds Metals Company in exploration for aluminum ore. Upon returning to M.I.T. in October 1945 he resumed teaching and soon after (1946) became Executive Officer of the Department of Geology when W. J. Mead became ill. After Mead's retirement in 1949, he was appointed Acting Chairman for SY 1949-50, while a search was made for a new head for Course XII, and in 1950 he was appointed Chairman, serving in that position until mid-1965, when he resigned in order to devote his last five years to teaching and writing. Upon reaching retirement age in 1970, he became Professor Emeritus. However, wishing to continue on a part-time basis after retirement, in order to write this history, he was appointed Senior Lecturer, in which capacity he served from 1970 to 1975.

During his 38 years as a member of the Geology Department, successively as Assistant Professor (1937-1943), Associate Professor (1943-1949), and Professor (1949-1970), Shrock saw the Department moved twice; first from Building 4 to 24 in 1946, and then from 24 to 54, the newly constructed 20-story Green Building, in 1964. During his chairmanship (1949-1965) he brought about an extensive revision of the Course XII curriculum that involved strengthening the basic science requirements, adding subjects in geochemistry and geophysics, and establishing a required summer field program in Nova Scotia, which was later replaced by a similar program at the Indiana University Field Station in Montana. In 1956 Shrock and Prof. Houghton, Head of Course XIX, Meteorology, initiated a joint program in oceanography with the Woods Hole Oceanographic Institution, a program which developed into the joint graduate-degree program that was formalized in 1968.

Soon after becoming Head of Course XII, Shrock and Cecil H. Green (VIA S.B. 1923, S.M. 1924), then President of Geophysical Service Inc. of

---

* When I came to the preparation of my own biography, I was presented with a dilemma - should I write in the third person, so that the sketch would conform to all the others in this history, or should I write in the first person? After some thought I decided to use both because I sounded a little too immodest in some statements, whereas he made me feel as though I were writing about a person no longer living. I hope the reader will understand my dilemma and why I have written as I have.

Dallas, Texas, organized a summer training program for geophysics students, The G.S.I. Student Cooperative Plan, which operated for 17 years and gave some 350 students from 80 different U.S. schools an opportunity to get practical field experience in geophysical exploration. The Plan proved to be a most successful coupling of industry and academia in a cooperative educational effort, and ended only when technological advances in field methods and data processing made it desirable to move much of the analytical work to a centralized laboratory. As one result from this program, Green accepted membership on Course XII's Corporation Visiting Committee in 1951, and has served continuously to the present (1975), many of the years as Chairman. By 1962 Green and his wife Ida had given M.I.T. substantial funds for support of Course XII's program, and in that year they funded the towering 20-story Green Building, which was dedicated on 2 October 1964, and now houses the Department of Earth and Planetary Sciences and the Department of Meteorology.

Shrock wrote three textbooks for use in his several subjects, all of which have found wide usage both in North America and abroad: Index Fossils of North America (Shimer and Shrock, 1944) quickly became an indispensable reference work for paleontologic and stratigraphic research; Sequence in Layered Rocks (Shrock, 1948) brought together widely scattered descriptions and illustrations of numerous features of bedded rocks that could be useful to the field geologist involved in structural and stratigraphic work; and Principles of Invertebrate Paleontology (Shrock and Twenhofel, 1953), which treats all the generally recognized phyla of invertebrates roughly in proportion to their representation in the fossil record. The cooperative training program developed with Geophysical Service Inc., referred to in the preceding paragraph, was described in a small booklet titled A Cooperative Plan in Geophysical Education (Shrock, 1966) in which the nature and short-term results of the unique program were described. Most recently Shrock privately published a book on a well-known M.I.T.-trained, father-son team of engineering geologists, The Geologists Crosby of Boston (Shrock, 1972). A copy of this book was sent to all living Course XII alumni in Classes 1891 to 1964; to some hundred or more geology and other libraries in North America and abroad; and to numerous friends. In addition to these five books, Shrock has published more than a hundred articles on geological subjects, which can be categorized as follows: 3 major reports; 37 articles; 20 memorials and biographical sketches; 17 short notes; 20 abstracts; and 14 reviews (in addition to some 30 activities reports of one kind or another).

As a Consulting Editor he got 28 geological books for McGraw-Hill's International Series in the Earth Sciences over a period of 12 years from 1953 through 1965. He also acted in a similar capacity for Paleontology and Paleobotany in McGraw-Hill's Encyclopedia of Science and Technology, which first appeared in 1960 and was updated by annual Yearbooks for a number of years before a revised edition was issued.

During his incumbency as Head of Course XII, 1949-1965, and during the next five years, Shrock gave considerable time and effort to raising funds for the Department. By 1970 he had established 12 Departmental Funds, mostly as memorials for former professors, and at this writing (July 1975) the total endowment principal sums to more than $500,000. Income from these Funds is used to aid faculty members and students in their research and to purchase books, equipment, and instruments. He continues to seek additional funds as he writes on the historical work of which this biographical sketch is a part.

As shown on charts in this biography and in the chapter on "The Financing of the Geological Sciences at M.I.T.," Shrock recruited 25 new faculty members during his chairmanship; 9 faculty members resigned and one retired. During the same period the number of graduate students increased from 27 in SY 1949-50 to 80 in SY 1964-65, and the total annual budget increased from about $150,000 in 1950 to more than $1,000,000 in 1965, as outside funds increased much more than Institute funds.

When Shrock relinquished the chairmanship of Course XII to Prof. Frank Press, at the start of SY 1965-66, the Department of Geology and Geophysics was housed in a new building; was well equipped with facilities and instruments; had an enthusiastic and able faculty; had a growing student body; and was in good financial condition. It was ready for another impressive surge of growth and productivity, and that is what happened during the following decade which ends as this is being written in the fall of 1975.

BIRTH, ANCESTRY, AND EARLY EDUCATION

Robert Rakes Shrock was born in the crossroads village of Wawpecong, Miami County, Indiana, on 27 August 1904. He was the first of two sons and the second of the seven children of Andrew and Stella (Glassburn) Shrock. His father was a self-trained artisan (carpenter, violin-maker, mechanic) of Pennsylvania German ancestry reaching back through several American generations to the Palatinate in Europe.* Andrew's four brothers and four sisters were typical Midwestern farmers and small-town dwellers who were strong believers in the work ethic and in the responsibility of one to look after and care for one's own. Grandfather Levi Shrock and Grandmother Rachel (Bugher) Shrock, like her husband also of German ancestry, had moved their family to north-central Indiana from neighboring northern Ohio before the turn of the century. By 1900, the Shrock children, including Andrew, had settled down in the countryside around Kokomo and Wawpecong, and here Robert grew up among a host of aunts, uncles, and cousins on both sides of his family.

His mother was the oldest of the seven children of Jasper Glassburn, whose ancestry could be traced back through soldiers of the War of 1812 and of the Revolutionary War to German emigrants from the Rhineland,** and Philena (Sims) Glassburn, whose forebears came from England and from the home of the Fraser Clan in southern Scotland. Like the Shrocks, the Glassburns were largely farmers and mechanics, used to hard work and a rather austere style of life, who lived in the same countryside in Miami and Howard counties.

Shrock attended the two-room elementary school in Wawpecong for the usual eight years, then entered Kokomo High School ten miles away because there was not at that time (1918) a high school in his own township. In

---

* See <u>Descendants of Jacob Hochstetler</u>, the Immigrant of 1736, by Harvey Hostetler. Elgin, Ill.: Brethren Publishing House, 1191 p., il., (1912).

** See <u>David Glassburn - Virginia Pioneer</u>, by Oma Glasburn Robinson. Los Angeles: The Ward Ritchie Press, x + 355 p., (1964).

driving a sometimes balky 1914 Model T, with presto lights, coil boxes, and side curtains, on the round trip of twenty miles five days a week, young Shrock quickly learned some important characteristics of that amazing vehicle. Jacking up one hind wheel to make it possible for him to spin the crank, meanwhile being careful not to grasp the crank with his thumb over the handle; removing a punctured tire from the rim, patching the inner tube on the rear fender, then replacing tube in casing and pumping up the tube; backing out of a ditch where the car had gone when it suddenly jumped out of the deep ruts in the mud road; or lighting the presto lights as darkness came on - these were but a few of the experiences 14-year old Robert and his 16-year old sister enjoyed doing during his first year of high school. The Model T Ford was a remarkable machine when everything went as expected, but it could give you some interesting surprises, too!

Graduated With Distinction in 1922, with his lowest grade in General Science, he won a Miami County Scholarship to Indiana University, and in the fall of that year started on his college career. By taking extra subjects during the regular school year and attending summer sessions he completed the requirements for the A.B. degree by the end of the summer of 1925, when he graduated With Distinction, and With Special Honors in Geology as the result of following an experimental program created to give qualified students greater latitude in choosing their subjects of study. He was the first student at the University to be graduated from this program.

After working for the Pure Oil Company during part of the summer of 1925, he registered for graduate work. A year later (1926) he received a Master's degree in Geology, submitting a thesis on the Pennsylvanian West Franklin limestone of southwestern Indiana. The essential substance of the thesis was published later with the title "Structural features of the West Franklin formation of southwestern Indiana," (A.A.P.G., B. 13: 1301-1315, il., 1929; co-authored with C. A. Malott, thesis supervisor).

Next came the decision to commence work toward a doctoral degree. Under the supervision of E. R. Cumings, internationally known for his research on the development of Paleozoic Bryozoa and Brachiopoda, and on Ordovician paleontology and stratigraphy, Shrock undertook investigation of the controversial Silurian "domes" exposed along the Wabash and Mississinewa rivers in northern Indiana. The study led to recognition of a new sequence of Silurian strata and to the conclusion that the "domes" were ancient organic reefs, for which he and Cumings proposed the new term bioherm. Two preliminary papers were published jointly with Cumings, cited as 1--1927 and 6--1928 in the Bibliography farther on,* and Shrock's doctoral thesis was likewise published with Cumings, under the title, The Geology of the Silurian Rocks of Northern Indiana (reference 5--1928Tc). Indiana University awarded Shrock a Ph.D. in Geology in June 1928, and 43 years later recalled him to receive an honorary degree of Doctor of Science in June 1971.

In addition to his doctoral thesis, Shrock published several papers based on field observations made while pursuing his thesis research. These are cited in the Bibliography as: a new Silurian graptolite fauna (2--1928); description of previously unrecognized physiographic features along the upper Wabash valley (3--1928a; 7--1929); and some interesting rock features (11--1930a; 12--1930b).** With his college training behind him, and several publications to his credit, he was now ready to start a career as a college teacher.

EARLY TRAINING FOR TEACHING

My training for teaching and administration started early. During my senior year at Kokomo High School I had charge of a dozen boys who delivered the morning edition of The Kokomo Dispatch, the local Democratic newspaper, long since combined with The Kokomo Tribune. This job provided some valuable experience in financial responsibility and in managing a group of active and imaginative teen-agers. During the spring term of the same year, when one of the older high school teachers of beginning algebra became ill, I was selected to teach several classes in the subject for the remainder of the term.

The next teaching assignment came in my second year at Indiana University when I was appointed a teaching assistant in the Geology Department, an appointment I held during SYs 1923-24 and 1924-25. As fellowships took care of expenses for the graduate years, SYs 1925-26 to 1927-28, I did no more teaching at Indiana. The next assignment came with appointment in the Geology Department at Wisconsin, as related in the next section.

---

* References listed in my Bibliography farther on are cited by number and date in the following autobiographical sketch: e.g. 1--1927.

** During my student days, both Malott and Cumings were publishing regularly, and as my thesis supervisors they insisted that I also publish the results of my thesis work as soon as possible. Their example, and their insistence on publication, greatly influenced me to follow a similar course in my subsequent research.

WISCONSIN YEARS (1928-1937)

The first post-doctoral opportunity to start on a college teaching career came in the spring of 1928 when W. H. Twenhofel, Professor of Geology at the University of Wisconsin, came to Bloomington to install Rho Chapter of Sigma Gamma Epsilon, the national geological fraternity. I had been elected the first president of the newly organized chapter and assigned the responsibility for entertaining Twenhofel. As a result of his visit,* Twenhofel soon after offered me an appointment in the Department of Geology at the University of Wisconsin for SY 1928-29. Thus began a close professional association, and a warm personal relationship, that would continue until Twenhofel's death on 4 January 1957, and would produce two textbooks, several jointly authored articles, and nine years of shared teaching at Madison.

Starting as an assistant in 1928, I advanced to Instructor in 1929, and then to Assistant Professor in 1930, the appointment I held when I resigned in June 1937 to accept a similar appointment at M.I.T.

The Wisconsin years were among the most important of my long professional career. I have long felt that the academicians I came to know and the additional instruction I received from them greatly enhanced my ability to meet the problems that came in later years at the Institute.

Besides sharing teaching duties with Twenhofel, I attended lectures by all the senior members of Wisconsin's famed School of Geology. Sedimentation, Stratigraphy, and Paleontology with Twenhofel led to the joint publications mentioned in a preceding paragraph, particularly the textbook Invertebrate Paleontology (reference 22--1935c in my Bibliography). Structural, Metamorphic, and Precambrian Geology with Leith and Mead stimulated preparation of Sequence in Layered Rocks (reference 57--1948a). Petrology and Petrogeny with A. N. Winchell provided a rigorous training in the use of the petrographic microscope and an early introduction to Bowen's pioneering papers describing what has come to be known as "Bowen's Reaction Series." A. K. Lobeck's instruction in the construction and use of block diagrams proved invaluable again and again in later years. And F. T. Thwaites' work on Pleistocene glacial deposits broadened the excellent earlier training in physiography (now generally called geomorphology) that I had received from C. A. Malott (widely known for his original work on cave development and Chester stratigraphy) in meeting the requirements of a minor subject in my doctoral program at Indiana. Finally, summer em-

---

* Twenhofel was seeking a younger associate at the time and was much interested in my thesis on the Indiana Silurian because he himself had done work on rocks of the same age in Canada (Anticosti Island and Nova Scotia) and in Wisconsin.

ployment with the Wisconsin Geological Survey during four summers gave valuable experience in field work under several different party chiefs and under the overall supervision of State Geologist E. F. Bean, who started so many young geologists on their professional careers in field geology. In sum, the Wisconsin years were really the equivalent of another doctoral program and were especially valuable to me because they gave both breadth and depth to the rather limited program that I had been able to pursue at Indiana.

The year 1937, like 1928, was one of those critical turning points in my career, for I left Madison and came to Cambridge to join the M.I.T. Geology Faculty. At the time of my departure the University of Wisconsin was in a somewhat turbulent state, even for that liberal institution. In fact, though not in action, the University had _five_ presidents! Glenn Frank, the controversial president, had been discharged but refused to move from the presidential residence; Clarence Dykstra, City Manager of Cincinnati, had been elected to the presidency, but would not or could not move to Madison under existing conditions; George Sellery, senior Dean of the College of Letters and Science, had been appointed president _pro tempore_, but he had become seriously ill and had gone to Florida to regain his health; and two other deans, E. B. Fred, Dean of the Graduate School, and H. Glicksman, Junior Dean of Letters and Science, had been asked to serve jointly as acting presidents until the situation could be resolved. It was under these conditions that I received an invitation from W. J. Mead, a colleague who had left Madison some years earlier to head M.I.T.'s Department of Geology, to spend a year at the Institute as a visiting assistant professor. When Bean and Twenhofel assumed that I might be offered a regular appointment at M.I.T., and so told Fred and Glicksman, the leave of absence that had been granted me was cancelled, and special inducements were offered to both me and my wife to remain. As a result of this action, President Compton authorized Mead to offer me a regular appointment as Assistant Professor of Geology. I accepted the offer, resigned from my Wisconsin appointment, and at the end of the summer's work with the Wisconsin Geological Survey moved to Cambridge. The die was cast, and I would spend the next 38 years as a member of the M.I.T. Faculty until final retirement as Professor Emeritus on 30 June 1975.

M.I.T. YEARS (1937-1970-1975)

Introduction

School Year 1937-38 was an exciting and eventful one. With F. K. Morris on leave, to make a trip around the world, and H. W. Shimer within a few years of retirement, I was assigned to teach classes in physical geology and help out otherwise where I could, with the intention that I would assume full responsibility for Shimer's subjects when he retired.

Soon after the term started, I called on J. R. Killian, Jr., then Manager of The Technology Press, to propose that Shimer and I prepare a completely new edition of Grabau and Shimer's North American Index Fossils. This would involve the authors' releasing their royalties to the Press, and M.I.T.'s providing the funds for preparing, publishing, and selling the new book. Happily, Killian and his advisors approved the proposition, made the necessary arrangements for joint publication with John Wiley & Sons, and provided the funds required. Five years later, 1943, I was reading page proof on a completely new book, Index Fossils of North America (reference 46--1944), while riding the buses to and from work at the War Production Board in Washington, having taken leave of absence meanwhile to do war work. Index Fossils was published early in 1944 while I was in Haiti exploring for bauxite for the Reynolds Metals Company, and immediately became a standard reference for invertebrate paleontologists the world over. Thirty years after publication, and in the ninth reprinting, the book is still selling at a steady rate.* Further information on the book is given in another chapter. (See Volume 2.)

Bauxite exploration for the Reynolds Metals Company in the Southern States during the summer of 1941 was followed by Pearl Harbor and an uneasy year of teaching, during which the concern of both faculty and students turned from the Atlantic to the Pacific Theater of warfare. Late in 1942, after some summer field work in Nova Scotia with W. L. Whitehead, I was asked to accept a position with the War Production Board as a Senior Industrial Specialist in the Miscellaneous Minerals Division. This assignment is discussed more fully in the section on "Government Service." Having accomplished the work desired by the WPB, I resigned and immediately joined the Reynolds Metals Company to aid in their search for aluminum ore. Soon after 1 July 1943 I joined W. L. Whitehead and Forbes Robertson, a Course XII graduate student, and flew to Haiti to start exploration in the Caribbean area.

After eleven months of exploration in Hispaniola, Puerto Rico, Cuba and Mexico, Whitehead and I returned to the United States on 31 May 1944. From Little Rock, Arkansas, Whitehead returned to Cambridge and terminated his employment with Reynolds, whereas I went to Washington whence I continued working with the Company, as discussed farther on, until August 1945 when I returned to M.I.T. to resume full-time teaching.

---

* Shrock, R. R., A 27-year report on sales of "Index Fossils of North America," J. Paleont. 46: 453-455, (1972).

In my Headquarters Office, Building 24-302, in 1952, getting ready to dictate galley proof corrections to Ms. Pauline Richmond, the Department's Senior Secretary from 1947 to 1965. (Photo by Jackman)

During the summer and fall of 1946 a series of events ensued that directly affected my future at the Institute. The Department of Geology was asked to move from Building 4, where it had been since 1916 when the Institute moved from Boston to Cambridge, to Building 24, which had been constructed quickly to house war research. Mead had developed a serious case of asthma and by August was in the hospital under special care. There being no person in the Department with authority to make decisions, I was asked to serve as Executive Officer, starting in October 1946, while Mead was hospitalized. Later on, when Mead reached retirement on 30 June 1949, I was asked to serve as Acting Chairman during SY 1949-50 while the search for a new Head of Course XII was conducted. In the end, the post of Chairman was offered to me, and I accepted it. For the next 15 years, from 1 July 1950 to 1 September 1965, I served as Head of Course XII. I asked to be relieved of the chairmanship on the latter date, but continued as Professor of Geology until retirement on 30 June 1970 as Professor Emeritus. For the next five years I served as Senior Lecturer on a half-time basis, after which I retired permanently as Professor Emeritus on 30 June 1975.

As Department Head (1949-1965)

When I assumed the chairmanship of the Department of Geology, I was faced with a number of critical problems and needs: 1) how to get the Lindgren Library and Schwarz Memorial Map Room reestablished; 2) space needed for new laboratories and for expanding crowded existing laboratories; 3) space needed for expanding the machine shop that was rapidly becoming an indispensable research facility; 4) additional office space for new faculty members and for an increasing number of graduate students; 5) a careful but ruthless upgrading of collections of minerals, rocks, ore specimens, and fossils; 6) the need for disciplined field training in geology and geophysics during the summer for Course XII undergraduates; 7) the need to revise, improve, and diversify the Course XII program; and 8) the need to recruit promising young staff members to help achieve the Department's major academic objectives. In addition to all the preceding was a critical need for special funds under the direct control of the Department that could be used for unexpected needs that could not be anticipated or provided for in the regular departmental budget. And finally, as research project funds began to flow into the Department, from industrial sources and from federal agencies and bureaus, new pressures arose for more space, more facilities, and more personnel.

From 1950 on, Course XII grew steadily away from a department oriented primarily toward classical geology to one that expanded and diversified successively as geophysics, geochemistry, and oceanography were added, and instruction in planetary science anticipated. By the end of M.I.T.'s first century, 1965, and at the end of Course XII's 75th year of existence, the Department of Geology had been renamed the Department of Geology and Geophysics, and in another year became the Department of Earth and Planetary Sciences, designations that indicated the changing emphasis of the Course XII program of instruction and research.

Let us now review briefly what happened during my 16 years as head of Course XII (1949-1965). The single greatest event was when Cecil H. Green (VIA S.B. 1923, S.M. 1924) visited his alma mater in 1949, for the first time in 25 years, in search of graduates in electrical engineering and geophysics for his young but growing company, Geophysical Service Inc. of Dallas, Texas. Having no success in his old Department, he called at Geology Headquarters, met me and other members of the geology faculty, learned that a dozen or more students in the geophysics option were looking for summer employment, and forthwith established a cordial relationship with the personnel of Course XII that has broadened and deepened steadily through the years.

As a result of the first two or three meetings, Green and I organized the G.S.I. Cooperative Plan (first called the MIT-GSI Cooperative Plan) for the purpose of providing disciplined summer field training for undergraduates interested in geophysical exploration.* This program, financed by Geophysical Service Incorporated, of which Green was President, started in the summer of 1951, operated for 17 consecutive summers, produced some

Cecil H. Green, Geophysical Service Inc. honorary board chairman; R. C. Dunlap, Jr., GSI president; and Dr. Robert R. Shrock of Massachusetts Institute of Technology with plaques presented to Green and Shrock at the 16th annual Earth Sciences Orientation Conference of the GSI Student Cooperative Plan, June 14-17, 1966, in Dallas. Green and Shrock are co-founders of the cooperative program. Plaque bears signatures of speakers at Co-op orientation sessions held since the plan was established in 1951.
(Photo and caption by Texas Instruments Inc.)

---

* Cecil and I, as co-founders of the MIT-GSI Cooperative Plan, were presented with plaques (shown in the photograph above), at the 16th annual orientation session in Dallas in June 1966. A similar photograph, with an accompanying news story, appeared in the May 1967 issue of Technology Review (p. 59-60).

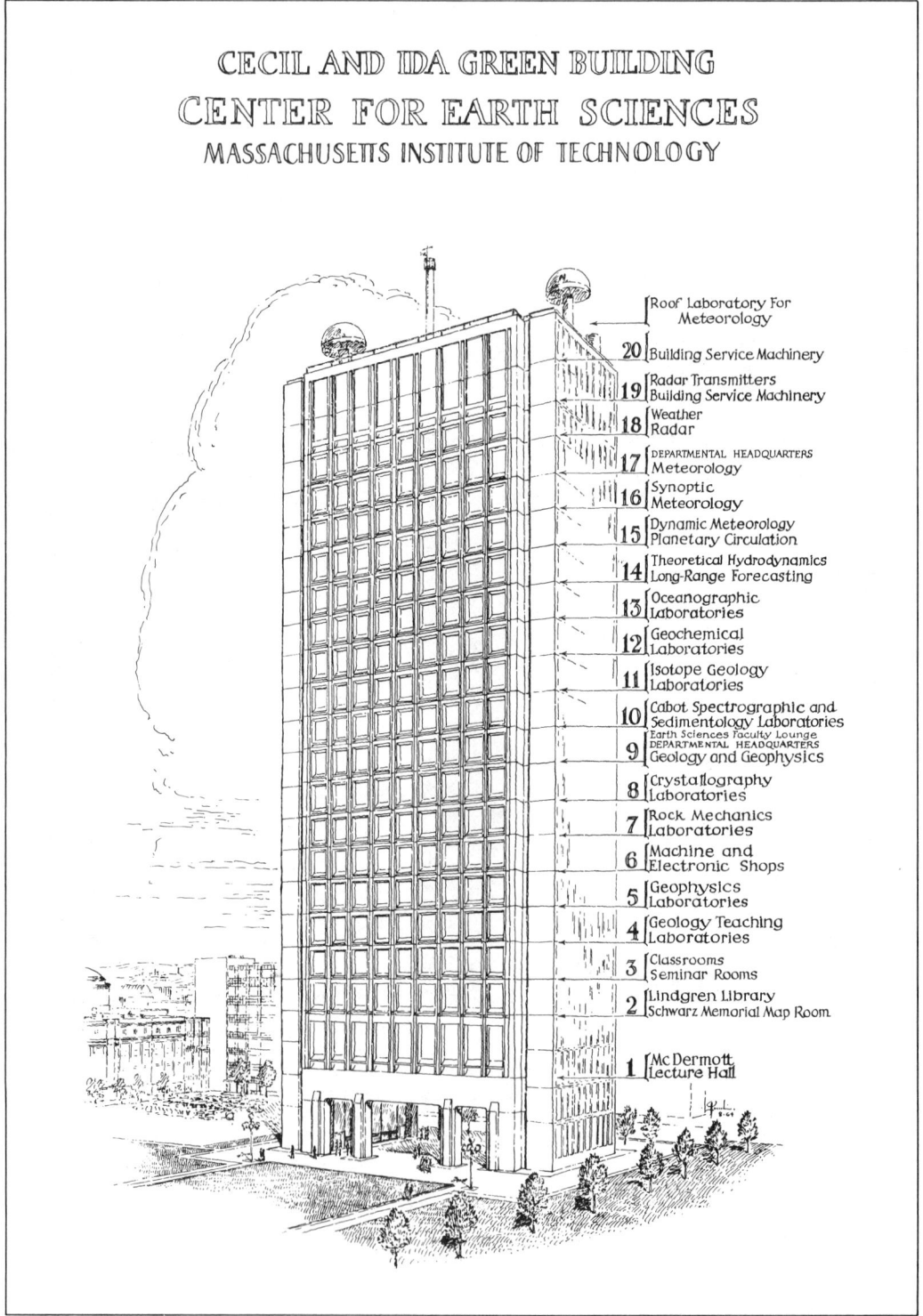

Percy Lund's sketch of the Green Building showing the space assignments as planned when construction was under way. After occupancy, some of the assignments were changed in response to new space requirements.

350 "alumni," and ceased only when advances in the art of geophysical prospecting dictated a change from field to laboratory training for student geophysicists. The details of this unique and innovative training program have been described in my booklet, A Cooperative Plan in Geophysical Education (see reference 105--1966b).

Not long after we became acquainted, I induced Green to serve on the Corporation Visiting Committee for Course XII, and since 1951 he has served continuously on the Committee to the present (1975), many years as Chairman. When he heard year after year of the Department's past accomplishments and of its plans and needs for the future, as itemized in a preceding paragraph, and witnessed the immediate success of the Cooperative Plan he had patterned on his own educational training in Course VIA, he began to think seriously about how he might aid Course XII in a substantial way. When he was given the assurance of President J. A. Stratton (his Course VI classmate of 1923) that the M.I.T. Administration was prepared to support the strengthening of Course XII, he and his wife Ida made two successive gifts to be used for the construction and maintenance of a building for the earth sciences at M.I.T. The electrifying announcement of their munificent gift by President Stratton* on 5 April 1959 set in motion a series of activities which culminated with the dedication, on 2 October 1964, of the impressive 20-story Cecil and Ida Green Center for the Earth Sciences. Thus in one fell swoop of magnificent proportion the Greens solved many of the problems cited at the beginning of this section: reestablishment of the Lindgren Library and the Schwarz Memorial Map Room; well equipped air-conditioned space for both existing and future laboratories, classrooms and shops; adequate storage space for the Department's valuable teaching and research collections; special preparation rooms; an excellent lecture hall seating some 294 persons; space and facilities for the Department of Meteorology; and a Lounge where students and staff members could mingle during the afternoon coffee-tea hour.

Having adequate space has made it possible for both Geology and Meteorology to expand and diversify existing research and to start new

---

* Prophetically, President Stratton announced:
> "The gift of Mr. and Mrs. Green will enable M.I.T. to build a multi-story Center which will house the laboratories on its campus that are now actively exploring the physical environment. Geologists, chemists, physicists, meteorologists, and oceanographers will now be able to work side by side in a basic and applied scientific program which will have, I am certain, the greatest impact on our economy and society as a whole." (Page 2 of an announcement from M.I.T.'s Office of Public Relations, for release in morning papers of Sunday, April 5, 1959.)

investigations. Air-conditioning has greatly improved both the comfort and the efficiency of personnel, and has been directly responsible for increased productivity because both students and staff members work longer hours than they would if the environment were less comfortable. Restoration of the Lindgren Library and the Schwarz Memorial Map Room has given a much-needed study facility for all M.I.T.'s earth scientists. McDermott Hall has become a heavily used lecture room for both regularly scheduled classes and special conferences and symposia. The roof of the building provides an excellent outside laboratory and launching platform for the atmospheric scientists. The ninth-floor lounge, recently named the <u>Ida Green Room</u> to honor one of the donors, has proved unusually valuable in bringing students and staff members together in an informal way that has greatly improved intercommunication and clearly enhanced the morale of the students, who have come to think of it as their common meeting place. In sum, the Green Building has definitely fulfilled the hopes of the donors as it has become in fact a Center for the Earth Sciences!

Three different programs of summer field training for Course XII undergraduates were initiated during my chairmanship. The first of these, primarily designed for all Course XII majors after they had completed their second year of study, was the M.I.T. Summer School of Geology, organized in 1948 when I was acting as Executive Officer while Mead was ill. This program, which was conducted from a base camp near Antigonish, Nova Scotia, for 14 consecutive summers, 1948 to 1961 inclusive, is discussed briefly in the chapter on "Summer Field Trips, Field Camps, and Field Programs," and has been described at some length in two published reports - "Acadia - Place of Plenty," by J. J. Rowlands, (<u>Tech. Rev.</u> 53/7: 1-8, il., May 1951); and "Ten Years in Nova Scotia: The Massachusetts Institute of Technology Summer School of Geology, 1948-1957," by R. R. Shrock (Published by the Dept. Geol. & Geophys., M.I.T., Cambridge, Mass., viii + 96 p., il., Aug. 1957). The second program, The G.S.I. Student Cooperative Plan, is discussed in foregoing paragraphs. The third program, like the first, primarily arranged for students who had completed the second year of the Course XII program, is actually a continuation of the first program, but in a different area and under different direction. In this program the students spend six to eight weeks at the Indiana University Field Station near Caldwell, Montana. Here they participate in a disciplined program under the direction of an M.I.T. alumnus, Prof. Judson Mead (XII S.B. 1940; Ph.D. 1949). This program started during the summer of 1962, following termination of the School in Nova Scotia at the end of the preceding summer, and continued this summer (1975) for the 14th time. It also is discussed briefly in the chapter on "Summer Field Trips, Field Camps, and Field Programs." (See my Volume 2.)

A fourth training program, which is now nationwide, and has involved students from many U.S. colleges and universities, was first presented to the Director of the U.S. Geological Survey in the early 1960s by several of us who had hoped to interest the Survey in providing supervised in-the-field training for a limited number of outstanding undergraduate geology majors to be selected by the National Association of Geology Teachers. The project, which has now been in operation for ten years, was discussed early in its history by John H. Moss in an article entitled "The NAGT-USGS Cooperative Project, a new focus on education in field geology," Jour. Geol. Education 16/3: 99-100, (June 1968).

Soon after I became Chairman, I started a weekly luncheon meeting at which the Course XII faculty considered how to improve the departmental program. After many vigorous discussions, the group agreed on a curriculum with two options, one in geology and one in geophysics, and when this was officially approved the name of Course XII became the Department of Geology and Geophysics in July 1952.*

As shown on the chart on the next page, the Course XII faculty roster grew from 9 professors in SY 1949-50 to 23 in SY 1964 as geochemists, geophysicists, and oceanographers were recruited. The personnel flux was actually greater than the increase from 9 to 23 implies, however, because we actually appointed 25 new professors. While doing this, however, two faculty members reached retirement, and 9 resigned to take posts elsewhere, so the Department's total roster in July 1965 was 23 (9 + 25 = 34 - 11 = 23).

The research program in Course XII increased impressively during my 16-year period as Chairman, 1949-1965, and even more so after 1965 under Prof. Press, as shown on the chart on following page 739 and in the chapter on "The Financing of the Geological Sciences at M.I.T." in Volume 2. This impressive increase in funds was due primarily to the financial inflow from federal agencies and bureaus in support of faculty members other than myself, who were conducting important research. There were two projects, however, in which I did play a minor role - "The Geophysical Analysis Group project (GAG)," described in detail in another chapter; and the initial ONR-supported oceanographic program which was quite productive and which led to the joint M.I.T.-W.H.O.I. graduate program officially established shortly after Press became Chairman. This early program is described at some length in the chapter on "Oceanography in Course XII at M.I.T." (See my Volume 2 for further discussion of the subjects above.)

---

* The results of the many luncheon discussions were used as a basis for a special full-page story in <u>The Tech</u> for Friday, Feb. 17, 1956 (p. 5), with the boldface headline, "Liberalization of Course XII Contemplated; New Plan Emphasizes Wide Choice of Elective Subjects."

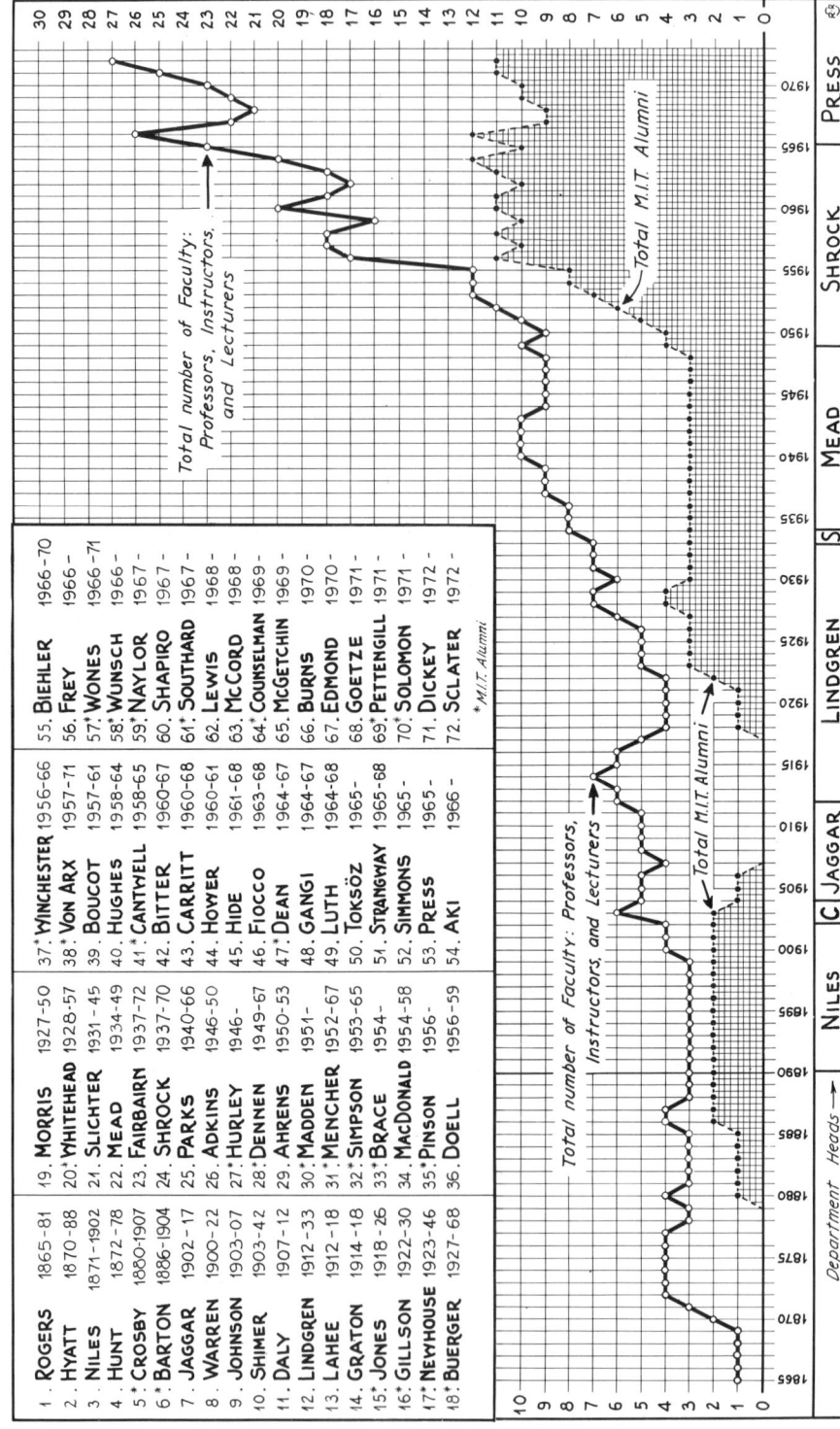

Chart Showing Order of Appointment of Course XII Professors and Number of M.I.T. Alumni Serving on the Faculty during 1865-1972

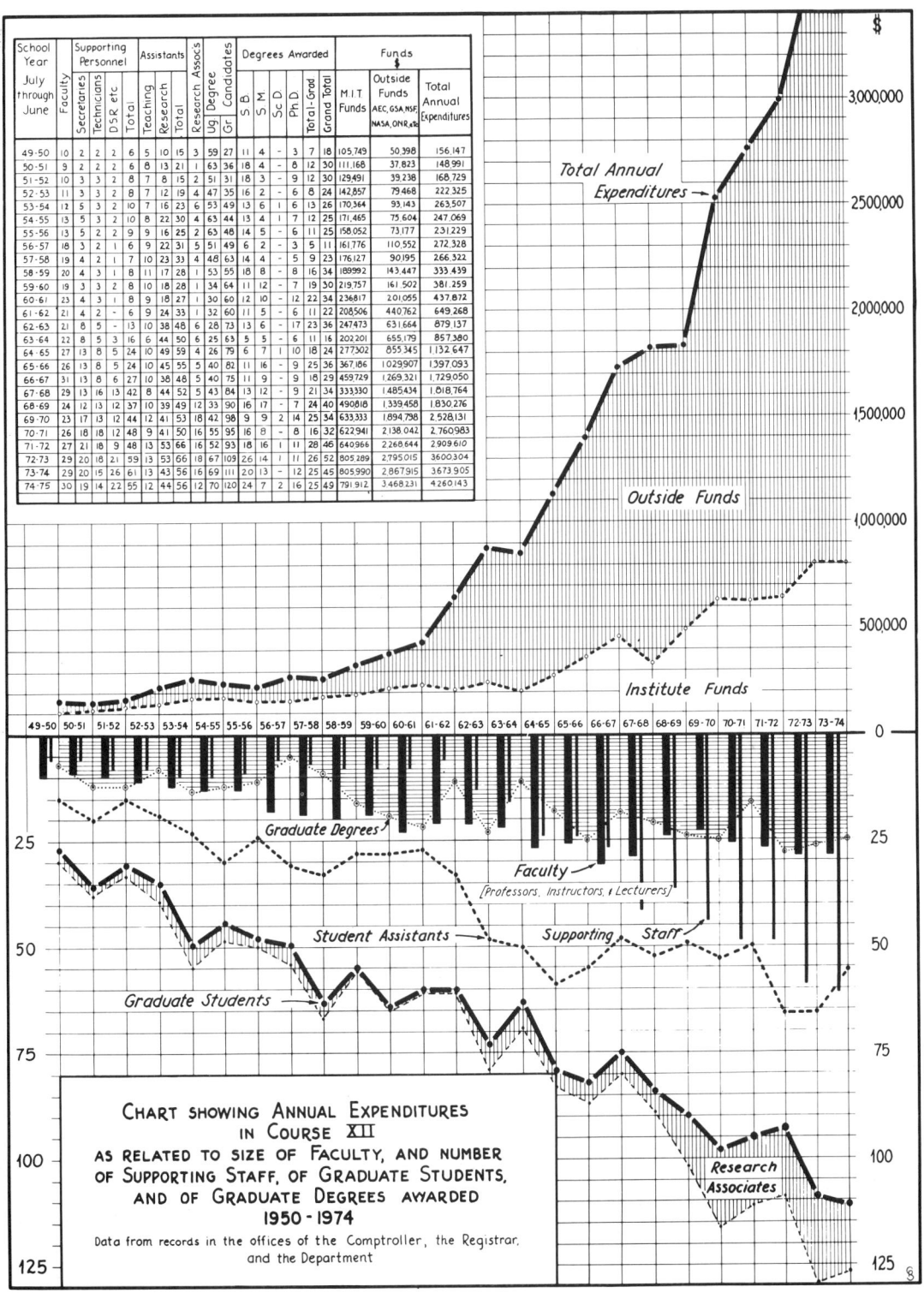

CHART SHOWING ANNUAL EXPENDITURES IN COURSE XII AS RELATED TO SIZE OF FACULTY, AND NUMBER OF SUPPORTING STAFF, OF GRADUATE STUDENTS, AND OF GRADUATE DEGREES AWARDED 1950-1974

Data from records in the offices of the Comptroller, the Registrar, and the Department

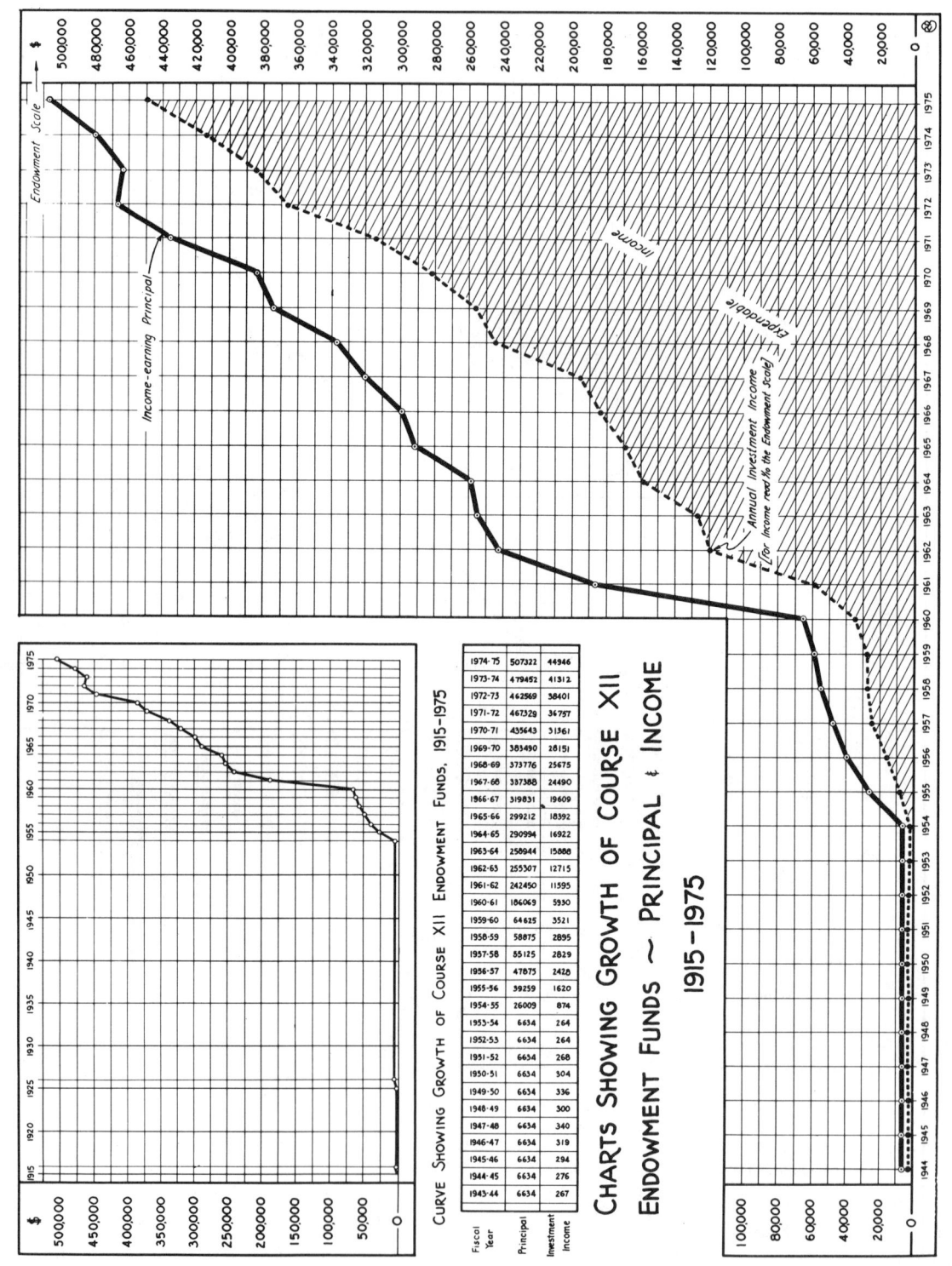

The need for special funds under direct control of the Department was of special concern to me during my chairmanship, and still remains so. The Department is now the richer by more than $500,000 of endowed funds, yielding an annual expendable income of almost $45,000, as shown on the accompanying chart. These funds came largely from loyal Course XII and Course III alumni who responded generously to repeated solicitations. A much fuller account of these special Departmental Funds is included in the chapter on "The Financing of the Geological Sciences at M.I.T." (Vol. 2.)

As with many faculty members and all department heads, I was called upon for a wide range of services outside Course XII; some within the Institute, others outside. Suffice it here to mention the more important of these to give some idea of the nature of these services. I served for a decade or more on the Committee on Graduate School Policy and on its Sub-Committee on Financial Aid. I was an interested member of the Institute's Library Committee for several years. When M.I.T. acquired Endicott House in Dedham on Christmas Eve 1954 (the formal opening was held in June 1955), a Board of Governors was appointed to oversee the House's activities. I served on this Board from its beginning until my retirement in 1970.

Soon after we decided to add instruction and research in oceanography to the Course XII program in 1956, I was invited to become a Trustee and a Corporation Member of the Woods Hole Oceanographic Institution. Subsequently I have served W.H.O.I. in the following ways and for the periods indicated:

Member of the Corporation	1957-1975
Member of the Board of Trustees	1957-1964
Member of the Executive Committee	1959-1962
Member of the Nominating Committee	1960-1962
Member of the Sub-Committee on the Bigelow Award	1961
Member of the Sub-Committee on Housing	1963
Honorary Trustee	1975-
Honorary Member of the Corporation	1975-

I was for some years a member of the Massachusetts Association for the Marine Sciences, and have served from its beginning on Edgerton's Science Advisory Committee of the New England Aquarium. At the American Academy of Arts and Sciences, I served on committees having to do with membership and grants for research. I was called in as an advisor when Boston College was considering whether to establish a Department of Geology (I recommended in favor, and a Department was established). Harvard asked me to serve on several ad hoc committees charged with evaluating candidates being considered for appointment or for tenure. In May 1967 I was a member of a Visiting Committee for the Institute of Earth Sciences at the University of Toronto, and more recently I have made several visits to Lubbock, Texas, to consult with the President of Texas Tech University and with members of Tech's Department of Geological Sciences.

So it has gone through the years! Clearly a typical M.I.T. professor does a good deal more than teach and conduct research, and a department head is certain to find himself involved in a wide range of activities in addition to teaching and administration in his own department. And it should be noted that I have said nothing about official activities in professional societies and service on federal agencies and bureaus. These are discussed in following sections.

I taught paleontology to Course XII students for more than thirty years (1937-1970). Here I am discussing a specimen with Donna G. Burt (technical assistant) and Claude P. T. Hill (XII S.M. 1952).

(1952 photo by Jackman)

As Teacher (1937-1970)

During my first six years at M.I.T., 1937-1943, I shared with Shimer the teaching of Paleontology, Sedimentology, and Historical Geology, and also took over General Geology and Economic Geography for Morris during SY 1937-38 while he was on leave. Then came World War II, and I left the Institute on 1 January 1943 to do war-related work until the fall of 1945. At that time I returned to the Institute and resumed a full-time teaching load. Shimer had retired during my leave, so I now took full responsibility for Paleontology, Sedimentation, and Historical Geology, and also organized several seminars that were attended by quite a few graduate students from Harvard. Later on, when the work as Department Head demanded an increasing amount of time, much of my teaching was turned over to Ely Mencher, but I continued to teach Paleontology until retirement in 1970.

Starting as Assistant Professor of Geology in 1937, I was promoted to Associate Professor in 1943, and to Professor in 1949 upon being appointed Acting Chairman. I became Professor Emeritus on 1 July 1970, but continued on a part-time basis as Senior Lecturer for the next five years following retirement.

During my 33 years of teaching at M.I.T., I took only one leave of absence, 1943-1945, but I was a Visiting Lecturer at Harvard during SY 1948-49, and was also an Associate in Invertebrate Paleontology in Harvard's Museum of Comparative Zoology for 20 years, from 1950 to retirement in 1970. Like most other members of the Geology Department, I was frequently invited to give lectures at other schools, and during my long teaching career I gave such lectures at Indiana University, University of Kentucky, University of Eastern Kentucky, Texas Tech University, Miami University, University of Michigan, Wesleyan University, Middlebury College, Bowdoin College, University of Maine, and St. Francis Xavier University in Nova Scotia.

As Professional Geologist

The first field work I did, while still a graduate student at Indiana University, appealed to me and led me to accept many consulting assignments during my academic career. These assignments involved a wide variety of geological problems and natural materials: oil shale, and high alumina shale for bricks and rock wool; sand for concrete work and road material; limestone for lime, building stone, and road material; salt deposits; lignite and coal; bauxite; fluorite; structural and stratigraphic work for petroleum companies; and cave exploration. I numbered more than

50 companies among my clients, of which the following are typical examples: Pacific Western Oil Co., Pure Oil Co., and Sun Oil Co.; Angas Corp.; Glen Alden Coal Co., Lehigh Valley Coal Co., and The Hudson Coal Co.; M. A. Hanna Co.; Apex Aluminum Co., Kaiser Aluminum and Chemical Corp., and Reynolds Metals Co.; Halifax Stone Co., National Gypsum Co., Union Fibre Co., and Universal Atlas Cement Co.; and C. and C. Super Corp. I also conducted investigations of one sort or another for the American Geographical Society (lake sediments); cities of Black River, Wisconsin, and Winchester, Massachusetts (water supply); Milwaukee Public Museum (fossil collection); U.S. Navy Yard, Boston (drydock foundation); Indiana Geological Survey (limestone, shale, and coal); Wisconsin Geological Survey and Wisconsin Highway Commission (sand, gravel, lime, limestone, and road materials); Nova Scotia Department of Mines (numerous assignments in connection with the M.I.T. Summer School of Geology at Antigonish, N.S.); and National Fire Insurance (building stone).

I have always regarded consulting assignments as excellent ways of learning about practical applications of geology as well as serving clients and the general public. Experience gained from consulting, as well as many excellent rock specimens, were made available to my students in their class work.

As Author and Editor

My complete Bibliography appearing farther on includes some 125 items which can be categorized as follows: 6 books, 3 major reports; 37 articles; 20 memorials and biographical sketches; 17 short notes; 20 abstracts; 14 reviews; and 8 miscellanea. In addition are listed some 30 reports of one sort or another that were prepared, and in some cases published, because of being required as part of the services I was rendering as department head or society officer.

Jointly or alone I wrote six books: four textbooks, a report on an educational experiment, and a double biography. The details are given in the aforementioned Bibliography; here it will suffice merely to give the titles: Invertebrate Paleontology (with W. H. Twenhofel, 1935); Index Fossils of North America (with H. W. Shimer et al., 1944); Sequence in Layered Rocks (1948); Principles of Invertebrate Paleontology (and W. H. Twenhofel, 1952); A Cooperative Plan in Geophysical Education (1966); and The Geologists Crosby of Boston (1972).

I regard the following as most important among my publications, excluding the books and memorials mentioned in the preceding paragraph (citations are to numbered references in my Bibliography; e.g. 1--1927):

1) Silurian stratigraphy and coral reefs of Indiana and Wisconsin (1--1927; 5--1928Tc; 6--1928d; 32--1939a).

2) West Franklin formation (T--1926; 9--1929b).

3) A new graptolite fauna (2--1928).

4) Physiography of the upper Wabash Valley (3--1928a; 7--1929).

5) The disturbed area of Ordovician (and younger) strata near Kentland, Indiana (14--1933a; 24--1936; 26--1937; 27--1937a).

6) Great Basin lake sediments (20--1935c).

7) Newfoundland Silurian (28--1937b; 33--1939b).

8) Geology of Washington Island [Wis.] (43--1941).

9) Classification of sedimentary rocks (56--1948).

10) Oceanography (64--1950; 111--1971).

11) Presidential addresses (79--1957b; 96--1961d).

Older men, who are fortunate enough to have outlived their still older teachers and colleagues, are commonly asked to prepare suitable memorials, necrologies, or biographical sketches of the departed ones. I have written a dozen or more of these for the following individuals for the reasons given (citations are to references in my Bibliography):

1) E. R. Cumings (109--1970) - my oldest professor who taught me geology and paleontology; drove me around for my field work; supervised my doctoral thesis; and helped me get my first academic appointment, at Wisconsin.

2) C. A. Malott (66--1951) - he taught me geomorphology; how to do field work and make a good map; and supervised my Master's thesis.

3) W. H. Twenhofel (75--1957a; 76--1957b; 78--1957d) - he brought me to Wisconsin; taught me much about sedimentation; and gave me the advantage of his prestige by allowing me to share authorship on articles and books (Invertebrate Paleontology, 1935, and Principles of Invertebrate Paleontology, 1953).

4) W. J. Mead (86--1960; 88--1960b; 94--1961b; 98--1962) - he taught me structural geology at Wisconsin and later brought me to M.I.T., and, when he reached retirement, recommended me as his successor. Together we found bauxite for the Reynolds Metals Company during World War II.

5) H. W. Shimer (104--1966; 108--1968a) - my colleague in stratigraphy and paleontology at M.I.T.; and coauthor of Index Fossils of North America.

6) W. L. Whitehead (110--1970a; 113--1971b) - my colleague at M.I.T., who showed me the geology of New England and Nova Scotia, and arranged for and directed the M.I.T. Summer School of Geology near Antigonish, Nova Scotia.

7) F. K. Morris (108--1968) - my colleague at M.I.T., and one of the Institute's best-known teachers.

8) I. B. Crosby (89--1960c) - a Course XII graduate who became a well-known engineering geologist, like his father, William O. Crosby (VII S.B. 1876). See The Geologists Crosby of Boston (120--1972c).

9) C. O'D. Iselin (114--1971c) - he was one of those from the Woods Hole Oceanographic Institution who helped Course XII greatly in starting and then developing the program in oceanography.

As an editor, I have not only evaluated numerous manuscripts (for proposed books) for different publishing companies, as have many of my geological colleagues, but I also served the McGraw-Hill Book Co., Inc. as Consulting Editor for some 15 years. In this capacity I worked closely with the Company in starting a special series of books on geology. This series was first announced in June 1953 as the McGraw-Hill Series in the Geological Sciences. By 1959, 9 books had been published in the Series. Then to broaden the subject matter, the name was changed to the McGraw-Hill Series in the Earth Sciences. By 1965, 15 additional titles had been added, and it was decided to change the name again, this time to the McGraw-Hill International Series in Earth and Planetary Sciences, in order to include works by foreign authors and on subjects involving planetary science. Shortly after I resigned as Chairman of Course XII in August 1965, and after four new titles had been committed to the Series, I also resigned as Consulting Editor. Fortunately, Frank Press, who succeeded me as Chairman of Course XII in 1965, agreed also to assume the consulting editorship of the Series, which by this time listed 28 titles on earth sciences and was well regarded by the geological community.

In 1958 I also started as a Consulting Editor in Paleontology and Paleobotany for the McGraw-Hill Encyclopedia of Science and Technology, which first appeared in 1960, and continued in this capacity as annual Yearbooks were published, until 1966, at which time I resigned after eight years of service. As Consulting Editor in Paleontology and Paleobotany I solicited desired articles from my colleagues and contributed a few myself. The latter are listed in my Bibliography (87--1960a).

As an Officer in Professional Societies

Like many of my academic colleagues, I have served numerous professional societies in one official capacity or another.

While a member of the geology faculty at the University of Wisconsin I was elected Secretary-Treasurer of the Wisconsin Academy of Sciences, Arts and Letters and served during SYs 1935-36 and 1936-37. Reports on these years are listed after my Bibliography.

Soon after joining the M.I.T. faculty, I was elected Treasurer of the Paleontological Society. I served from 1 January 1938 through December 1942, whereupon C. O. Dunbar was appointed Treasurer pro tempore for the next two years, 1943 and 1944, while I was away from Cambridge on war work. Five Treasurer's Reports were prepared, presented at annual meetings, and published in the Proceedings of the Society. These are listed after my Bibliography.

I became a member of the Society of Economic Paleontologists and Sedimentologists in 1948. I was elected Vice President for 1955-1956, President for 1956-1957 and Honorary Member in 1969, and have been designated the Twenhofel Medalist for 1976. My Presidential Address, "New Geological Horizons," was published in the A.A.P.G. Bulletin for July 1957 (see reference 79--1957e in my Bibliography).

I was elected a Vice President of the American Association for the Advancement of Science, and Chairman of Section E, Geology and Geography, for 1958. I presided at the Indianapolis meeting of the Section in late December 1957, and delivered my Vice-Presidential Address at the annual meeting in Washington, D.C. on 27 December 1958. The address, "Primary structures in sedimentary deposits (A review and a report of progress with some remarks)," was later included in the Mahadevan [Commemmorative] Volume, published in Hyderabad, India, in 1961 (see reference 96--1961d in my Bibliography).

I was elected President of the National Association of Geology Teachers for 1959, and in a meeting at the Ohio State University, on 5 December 1959, I delivered my Presidential Address, never published, which emphasized the importance of the smaller colleges in producing graduates who then went on to the larger and more prestigious universities for their advanced degrees.

Government Service

Soon after Pearl Harbor, I was asked to join the Miscellaneous Minerals Division of the War Production Board. I served that Division as a Senior Industrial Specialist for the first half of 1943, 1 January to 30 June, in charge of the quartz section. This involved following the flow of quartz crystals from Brazil, through testing at the National Bureau of Standards, to distribution to numerous companies where the crystals were cut into radio oscillators of various types.

In 1946 I became a member of the Panel on Oceanography of the Research and Development Board, and during the next six years attended meetings throughout the country from coast to coast. One result of this service was the jointly prepared paper, "Education and training for oceanographers," (V. O. Knudsen, A. C. Redfield, R. Revelle, and R. R. Shrock), published in Science 111: 700-703, (1950). A second result was generation of interest in oceanography that led me to initiate a program of instruction and research in that earth science in Course XII in SY 1956-57. This beginning led a decade later to establishment of the joint graduate program between M.I.T. and the Woods Hole Oceanographic Institution.

For three years, July 1958 - June 1961, I served as a member of the Advisory Panel for the Earth Sciences of the National Science Foundation. This panel evaluated and advised on applications for research funds. I also served short terms on ad hoc committees of the Department of Health, Education and Welfare charged with site visiting and advising regarding institutional applications for building and facilities grants. My last assignment was membership on an ad hoc committee charged with evaluating applications for block grants to college departments of science and engineering.

MARRIAGE, FAMILY, AND SOCIAL ACTIVITIES

I was married to Theodora Antoinette Weidman of Lakewood, Ohio, on 2 February 1933. She is the oldest of the three children, all girls, of William Roe Weidman (1879-1944) and Ruth Victoria Stubbings (1887-1965). Theodora's father was of Pennsylvania German and English ancestry, and some of his English forebears fought on the American side in the Revolutionary War. He received a B.S. degree in Mechanical Engineering from the University of Michigan in 1899 and practiced as a bridge and foundation engineer. Her mother, of English and Swedish ancestry, was trained in home economics at the former Lewis Institute (now the Illinois Institute of Technology), and took graduate work at the University of Chicago. Theodora's sisters were Phyllis Adelaide (Mrs. J. R. Glasgow), for many years office manager and tax expert with a petroleum company in Tulsa, and William Ruth (deceased).

Theodora holds a B.A. degree in Comparative Literature and an M.A. degree in English from the University of Wisconsin, and an A.M. degree in English from Radcliffe College. She stopped preparation of her doctoral thesis at Harvard in 1942 in order to work at the National Security Agency in Washington for the duration of World War II.

For more than 20 years, and especially during the 16 years (1949-1965) when I was Head of Course XII, she was the hostess at numerous departmental social affairs at our home in Lexington. At the Institute she participated in the usual social affairs at which wives of department chairmen were expected to be present. For 17 successive years, 1950-1966, we held a Geology Department picnic in the fall in Lexington. These were especially helpful in providing an opportunity for the members of the M.I.T. Geology Family to become better acquainted with one another.

Two children were born to the Shrocks: Wendolyn Theodora (b. 1947) and Robert Ellsworth (b. 1950). Wendolyn earned an A.B. degree (summa cum laude) in history from Indiana University in 1967 and an A.M. degree in the same subject from Harvard in 1968. She has been teaching history at the Northfield-Mt. Hermon School (Massachusetts) for the past six years. Robert E. earned an A.B. degree (summa cum laude) in physics from Harvard in 1971 and a Ph.D. in physics from Princeton in 1975. After a two-year post-doctoral appointment at the Fermi National Accelerator Laboratory at Batavia, Illinois, he will join Princeton's Department of Physics this fall (1977) as an assistant professor. All four Shrocks are members of Phi Beta Kappa.

MEMBERSHIPS, FELLOWSHIPS, HONORS AND AWARDS

At one time or another I have been a member of quite a few professional organizations, all of which had a more or less direct connection with geology.

I have held membership in the following: American Association of Petroleum Geologists, American Geological Institute, American Geophysical Union, Boston Geological Society, Indiana Academy of Science, National Association of Geology Teachers (Pres. 1959), Society of Economic Paleontologists and Mineralogists (Pres. 1957; Honorary Member 1969; Twenhofel Medalist 1976), and Wisconsin Academy of Sciences, Arts and Letters (Secy.-Treas. 1936-1937). I was elected a fellow of the American Academy of Arts and Sciences (1954), American Association for the Advancement of Science (1930; Vice Pres., Sect. E, 1958), Geological Society of America (1931), and the Paleontological Society (1931; Treas. 1938-1942).

In 1961 I was made an Honorary Member of the M.I.T. Alumni Association (Class of 1926) and in 1970 I received the Bronze Beaver Award of the Association. Upon retirement from M.I.T. in 1970 I was given a Certificate of Recognition for 33 years of service to the Institute, and in June of the next year (1971) I was awarded an Honorary Doctor of Science degree from Indiana University, my alma mater.

Three unusual honors have come to me; three fossils, a research vessel, and an endowed professorship carry my name. The three fossils were identified and named by three of America's outstanding paleontologists:

    Kentlandoceras schrocki [sic] A. F. Foerste 1933
     (an Ordovician cephalopod)
    Atrypella shrocki G. A. Cooper 1944
     (a Silurian brachiopod)
    Dictyonema shrocki R. Ruedemann 1947
     (a Silurian graptolite)

In addition to the aforementioned new species described by others, I myself, alone or with co-authors, have described 26 new species or varieties of fossils: 1 alga, 1 foraminifer, 4 corals, 1 bryozoan, 5 brachiopods, 1 gastropod, 3 trilobites, and 10 graptolites.

On 10 December 1965, a 50-foot vessel built by the U.S. Navy in 1955, as an underwater ordnance research vessel, was christened the R/V. R. R. SHROCK, and became M.I.T.'s first oceanographic research vessel.* It was so named in recognition of my role in first starting and then developing a program of instruction and research in oceanography at M.I.T. Since its christening, the vessel has been in almost constant use in oceanographic research by M.I.T. professors and their students. (See also the chapter on "Oceanography at M.I.T." in my Volume 2.)

Early in 1970 Dr. and Mrs. Cecil H. Green of Dallas, Texas, gave M.I.T. $1,200,000 for the endowment of two distinguished chairs: one to honor the donor, the Cecil H. Green Professorship of Electrical Engineering; the other to be the Robert R. Shrock Professorship of Earth and Planetary Sciences. The latter recognizes the leadership I provided for Course XII, during my 16-year chairmanship, 1949-1965, while Green served as Chairman of the Course XII Corporation Visiting Committee. This cooperation between Green and myself is discussed at some length in several other chapters of this history.

The 50-foot R.R.SHROCK at Lewis' Wharf in Boston.
(Photo by E. L. Mollo-Christensen)

---

* "'R.R.Shrock' becomes first boat of MIT Navy." The Tech 85/22: 1, Nov. 3, 1965; Tech Talk 18/3: 3, July 25, 1973.

RETIREMENT AND SUMMATION

I passed my 65th birthday on 27 August 1969 and became Professor Emeritus on 1 July 1970. I have continued beyond this date, on a year by year basis, as Senior Lecturer, looking after the Department's endowed funds and preparing this history. This post-retirement appointment came to an end after the maximum permissible period of five years, with permanent retirement as Professor Emeritus on 30 June 1975. On that date I had served the Institute a total of 38 years, from September 1937 through June 1975.

Let me now summarize what I regard as the major accomplishments of the Geology Department in which I played some role during those 38 years. Here I shall shift from the pronoun I to the pronoun we, because the accomplishments came about as a result of magnificent cooperation by everyone involved: The M.I.T. Administration; the Course XII faculty, staff, students, and alumni; and our loyal friends inside and outside the Institute.

Shortly before I became Executive Officer of the Department, in 1946, Fairbairn directed the Department's move from Building 4 to 24, and expeditiously upgraded our collections by appropriate elimination of useless and poor quality specimens. With the move we lost the Lindgren Library and the Schwarz Memorial Map Room, and did not get them back until some 18 years later when they were reestablished in the new Green Building in the summer of 1964. With a minimum of impatience and a remarkable attitude of cooperation, the faculty members developed laboratories in the inadequate spaces assigned them in Building 24 and got on with their teaching and research. It was not until the Green Building was constructed that these and other laboratories could be expanded and diversified to meet the demands of the steadily increasing enrollment of graduate students. The departmental machine shop was doubled in size and quickly became an indispensable component of our research facilities.

When the Green Building was ready for occupancy in the spring and summer of 1964, Geology and Meteorology moved into the new space quickly. By the time the building was dedicated on 2 October 1964 both departments were ready to start on what would prove to be probably the most exciting and productive decade for the earth sciences at M.I.T.

During my own chairmanship, from 1949 to 1965, we recruited 23 new faculty members, but lost most of them (20) later as they resigned to take positions elsewhere. They were succeeded by another 22 new professors by 1974, 4 of whom have resigned in the interim since their appointment. In spite of this impressive flux of personnel, the faculty has not only maintained its high quality, with more than a doubling in numbers from 10 in 1949 to 23 in 1965 and 28 in 1974, but has also become diversified with development of programs in geochemistry, geophysics, oceanography, and planetary science. It should be emphasized that it has been Frank Press, who succeeded me as Head of Course XII in October 1965, who has recruited the 22 new professors since that date, accepted 14 resignations, helped to establish a joint graduate program in oceanography with the Woods Hole Oceanographic Institution, established an international center for seismic research at M.I.T. and added planetary science to Course XII's program, all of which necessitated the change of name to Department of Earth and Planetary Sciences. Thus Course XII changed dramatically in numerous aspects during my chairmanship from 1949 to 1965; changed equally dramatically, if not more so, during Press' first decade as Chairman, from 1965 to 1975; and is entering its 85th year, and the Institute's 110th, as one of the world's leading centers for instruction in the earth and planetary sciences, much as our great benefactors, Cecil and Ida Green, envisioned when they made their first gifts twenty-five years ago.

While occupied with the responsibilities of the chairmanship, as well as before and after, I managed to teach several subjects and conduct several seminars, and to write three major textbooks, two smaller books, and a fair number of articles, all dealing with paleontology and sedimentology. With the help of a great many Course XII and Course III alumni and other friends, we established a dozen departmental funds which had at the end of FY 1974-75 a total endowment principal of more than $500,000 and annual income of almost $45,000 (see chart on a preceding page). We also acquired a vessel for oceanographic research, the R.V. R. R. SHROCK. Finally, many other activities and accomplishments too numerous to mention here are discussed throughout the second volume of this history.

It is appropriate to end this summation by mentioning the celebration of the tenth anniversary of the dedication of the Green Building, and the beginning of Frank Press's tenth year as Chairman of Course XII, on 3-4 December 1974. On that auspicious occasion eleven distinguished earth scientists delivered lectures addressed to the theme of the symposium, "The New Wave of Exploration in the Earth Sciences"; these lectures were published later by the M.I.T. Alumni Association, first in the Technology Review (Vol. 77, No. 5, p. 3, 14-49; and No. 6, 29-37; 1975), then later in a booklet titled The New Wave in the Earth Sciences (Cambridge: M.I.T. Alum. Assoc., 77 p., il., Nov. 1975). At the Tuesday evening dinner, on 3 December 1974, Mr. Green was given a special volume titled The First Decade in the Green Building, 1964-1974, which contained reprints of articles judged by their authors, some 35 professors of Course XII (Earth and Planetary Sciences) and Course XIX (Meteorology), as the most important of the past decade (1964-1974). At the close of the symposium on the following Wednesday afternoon, the lounge on the 9th floor of the Cecil and Ida Green Building was dedicated as THE IDA GREEN ROOM (reported in the Technology Review, Vol. 77/4: 70,80; 1975). At the ceremony, attention was called to the bronze dedicatory plaque on the wall, and to the special portrait of Mrs. Green that she had had made for the room. We also distributed a 19-page booklet entitled The Ida Green Room, that I had prepared and had printed by M.I.T.'s Graphic Arts Service. The booklet includes a brief account of Mrs. Green's life as the wife of exploration geophysicist Cecil H. Green (S.B. VI-A 1923, S.M. VI-A 1924).

BIBLIOGRAPHY OF ROBERT RAKES SHROCK

Symbols and abbreviations used in the following references are explained on pages 91 - 98. If only an abstract was published, it is cited as a separate item; if a complete article with essentially the same title was also published, the abstract is generally cited along with the complete article. This is the most complete list of my scientific publications yet compiled.

T--1926   The stratigraphy and structure of the West Franklin limestone of Indiana, iii + 76 p., il., (1926). (A.M. Thesis at Indiana University in the Department of Geology, June 1926.) (See also 9--1929b.)

1--1927   (with Cumings, E. R.) The Silurian coral reefs of northern Indiana and their associated strata. Ind. Ac. Sci., Pr. 36: 71-85, il., (1927).

2--1928   A new graptolite fauna from the Niagaran of northern Indiana. Am. J. Sci. (5) 16: 1-38, il., (1928).

3--1928a Some interesting physiographic features of the Upper Wabash drainage basin in Indiana (Waterman Foundation, Indiana University, Contr. 32). Ind. Ac. Sci. 37: 125-139, il., (1928).

4--1928b (with Cumings, E. R.) Niagaran coral reefs of Indiana and adjacent states and their stratigraphic relationships [abst.]. Geol. Soc. Am., B. 39: 211-212, (1928). (See also 6--1928d.)

5--1928Tc (with Cumings, E. R.) The Geology of the Silurian Rocks of Northern Indiana (Ph.D. thesis in the Department of Geology, Indiana University, 293 p., il., June 1928). Dept. Conserv., State of Indiana, Div. Geol., Pub. 75: 226 p., il., (1928).

6--1928d (with Cumings, E. R.) Niagaran coral reefs of Indiana and adjacent states and their stratigraphic relations. Geol. Soc. Am., B. 39: 579-620, il., (1928). (See also 4--1928b.)

7--1929   The klintar of the upper Wabash valley in northern Indiana (Waterman Foundation, Indiana University, Contr. 34). J. Geol. 37: 17-29, il., (1929).

8--1929a (with Malott, C. A.) Features of the Wabash sluiceway of northern Indiana [abst.]. Geol. Soc. Am., B. 40: 101-102, (1929).

9--1929b (and Malott, C. A.) Structural features of the West Franklin formation of southwestern Indiana. Am. Assoc. Petroleum Geol., B. 13: 1301-1315, il., (1929). (See also T--1926.)

10--1930   (with Malott, C. A.) Origin and development of Natural Bridge, Virginia (Waterman Foundation, Indiana University, Contr. 47). Am. J. Sci. (5) 19: 257-273, il., (1930). Abst., Geol. Soc. Am., B. 41: 106-107, (1930).

11--1930a Polyhedral pisolites. Am. J. Sci. (5) 19: 368-372, il., (1930).

12--1930b (and Malott, C. A.) Notes on some northwestern Indiana rock exposures (Waterman Foundation, Indiana University, Contr. 45). Ind. Ac. Sci., Pr. 39: 221-227, (1930).

13--1933   (with Malott, C. A.) Mud stalagmites (Waterman Foundation, Indiana University, Contr. 58). Am. J. Sci. (5) 25: 55-60, il., (1933).

14--1933a (and Malott, C. A.) The Kentland area of disturbed Ordovician rocks in northwestern Indiana. J. Geol. 41: 337-370, il., (1933).

15--1933b Geology of the "Niagara" dolomite (p. 32-37) in George Barker's "Wisconsin magnesium lime mortars," Univ. Wisconsin Bull., Eng. Exp. Sta. Ser. 75: 39 p., il., (1933).

16--1934   (and Malott, C. A.) West Franklin formation of southwestern Indiana [abst.]. Geol. Soc. Am., Pr. 1933: 363-364, (1934). (See also T--1926 and 9--1929b.)

17--1935   Silurian geology of Wisconsin [abst.]. Geol. Soc. Am., Pr. 1934: 107-108, (1935).

18--1935a (and Raasch, G. O.) Correlation of Ordovician sequence at Kentland, Indiana [abst.]. Geol. Soc. Am., Pr. 1934: 355-356, (1935).

19--1935b (and Havard, J.) Washington Island - its geology and natural history [abst.]. Wis. Ac. Sci. Arts Lett., Program of 61st Ann. Meeting at Beloit, Wis., p. 4, (1935). (‡)

20--1935c (and Hunzicker, A. A.) A study of some Great Basin lake sediments of California. J. Sed. Pet. 5: 9-30, tab., il., (1935).

21--1935d Probable worm castings ("coprolites") in the Salem limestone of Indiana. Ind. Ac. Sci., Pr. 44: 174-175, il., (1935).

22--1935e (with Twenhofel, W. H.) *Invertebrate Paleontology*. New York: McGraw-Hill Book Co., Inc., xvi + 507 p., 175 fig., (1935).

23--1935f Insoluble residues from Wisconsin sedimentary rocks [p. 257-271]:
  Part I. Insoluble residues as an aid in the study of sedimentary rocks (R. R. Shrock);
  Part II. Studies of Wisconsin sedimentary rocks -
    1. Insoluble residues from Wisconsin Silurian dolomites (G. R. Burpee; abstracted by R. R. Shrock);
    2. The insoluble residues of the Oneota dolomite of western Wisconsin (J. J. Drindak; abstracted by R. R. Shrock);
    3. A sedimentational study of a part of the Trempealeau formation in southern Wisconsin (B. O. Hougen; abstracted by R. R. Shrock);
    4. Insoluble residues of the Mendota (St. Lawrence) dolomite (R. E. Wilcox; abstracted by R. R. Shrock).
  Wis. Ac. Sci. Arts Lett. 29: 257-271 (257-260, 260-262, 262-266, 266-268, and 268-271, respectively), il., (1935).

24--1936   Kentland structure, Indiana (p. 1072-1074) in W. H. Bucher's "Cryptovolcanic structures in the United States," in Rept. XVI Int. Geol. Cong., Washington, [D.C.], 1933, vol. II: 1055-1084, il., (1936).

25--1936a   (with Twenhofel, W. H.) Silurian strata of Notre Dame Bay, northern Newfoundland [abst.]. Geol. Soc. Am., Pr. 1935: 112, 379, (1936). (See also 28--1937b and 33--1939b.)

26--1937   Stratigraphy and structure of the area of disturbed Ordovician rocks near Kentland, Indiana. Am. Midland Nat. 18: 471-531, il., (1937).

27--1937a   (and Raasch, G. O.) Paleontology of the disturbed Ordovician rocks near Kentland, Indiana. Am. Midland Nat. 18: 532-607, 11 pl., (1937).

28--1937b   (with Twenhofel, W. H.) Silurian strata of Notre Dame Bay and Exploits Valley, Newfoundland. Geol. Soc. Am., B. 48: 1743-1772, il., (1937).

29--1938   Fossil algae from the Salem limestone of Indiana. Science 87: 438-439, (1938).

30--1938a   Wisconsin Silurian bioherms (organic reefs) [abst.]. Geol. Soc. Am., B. 49: 1922, (1938). (See also 32--1939a.)

31--1939   (with Ruedemann, R.) A new Wisconsin Upper Cambrian foraminifer. Am. J. Sci. 237: 66-71, il., (1939).

32--1939a   Wisconsin Silurian bioherms (organic reefs). Geol. Soc. Am., B. 50: 529-562, il., (1939). (See also 30--1938a.)

33--1939b   (and Twenhofel, W. H.) Silurian fossils from northern Newfoundland. J. Pal. 13: 241-266, il., (1939).

34--1939c   Geological aspects of Washington Island, Wisconsin [abst.]. A.A.A.Sci. Program 104th Meeting in Milwaukee, June 1939 (p. 39), jointly with the Geol. Soc. Am., B. 50: 2009-2010, (1939).

35--1939d   Niagaran bioherms of the Milwaukee region [abst.]. A.A.A.S., Program 104th Meeting in Milwaukee, June 1939 (p. 24), jointly with the Geol. Soc. Am., B. 50: 2010, (1939).

36--1939e   Life through the ages (A review of P. E. Raymond's Prehistoric Life [1939]). New England Nat. 5: 30-31, (1939).

37--1940   "Lucite" as an aid in studying hard parts of living and fossil animals. J. Pal. 14: 86-88, (1940).

38--1940a   Note on "Paleontology of the disturbed Ordovician rocks near Kentland, Indiana." Am. Midland Nat. 23: 493, (1940).

39--1940b   (with Shimer, H. W.) Revision of North American Index Fossils. J. Pal. 14: 286, (1940).

40--1940c   Publication dates of some of Hyatt's cephalopod genera (Discussions). Am. J. Sci. 238: 676-678, (1940).

41--1940d   Weathering of ferruginous beds in the Pennsylvanian of Greene County, Indiana. Ind. Ac. Sci. 49: 163-168, il., (1940).

42--1940e   (and Vonnegut, B. and Herpers, H. F.) Calcareous incrustations formed on cascades at the Indiana State Soldiers and Sailors Monument, Indianapolis. Ind. Ac. Sci. 49: 169-174, il., (1940).

43--1941   (assisted in the field by J. H. R. Havard) Geology of Washington Island and its neighbors, Door County, Wisconsin. Wis. Ac. Sci. Arts Lett., Tr. 32: 199-227, il., (1941).

44--1941a   Rectangular mudcracks. Wis. Ac. Sci. Arts Lett., Tr. 32: 229-232, il., (1941).

45--1942   Correlation of the Silurian formations of North America. (C. K. Swartz, Chairman, et al., including R. R. Shrock.) Geol. Soc. Am., B. 53: 533-538, 1 large correlation table, (1942).

46--1944   (with Shimer, H. W., assisted by numerous paleontologists) Index Fossils of North America. M.I.T.: The Technology Press and New York: John Wiley and Sons, Inc., ix + 837 p., including 303 pl., (1944).

47--1946   Calcitic pisolites forming in travertine cascade deposits [Haiti]. Ind. Ac. Sci., Pr. 55: 102-106, il., (1946).

48--1946a   Surficial breccias produced from chemical weathering of Eocene limestone in Haiti, West Indies. Ind. Ac. Sci., Pr. 55: 107-110, il., (1946).

49--1946b   Karst features in Mayan region of Yucatan Peninsula, Mexico. Ind. Ac. Sci., Pr. 55: 111-116, il., (1946).

50--1946c   Sedimentation and wind action around Volcan Parícutin, Mexico. Ind. Ac. Sci., Pr. 55: 117-120, il., (1946).

51--1946d   Classification of sedimentary rocks [abst.]. Geol. Soc. Am., B. 57: 1231, (1946).

52--1947   William Henry Twenhofel - Honorary Member. Am. Assoc. Petroleum Geol., B. 31: 835-840, (1947).

53--1947a   Corrigenda for Index Fossils of North America. J. Pal. 21: 494-497, (1947).

54--1947b   Fossils in laterite and bauxite [abst.]. Geol. Soc. Am., B. 58: 1227-1228, (1947).

55--1947c   Loiponic deposits [abst.]. Geol. Soc. Am., B. 58: 1228, (1947).

56--1948   A classification of sedimentary rocks. J. Geol. 56: 118-129, il., (1948).

57--1948a   Sequence in Layered Rocks. New York: McGraw-Hill Book Co., Inc., xiii + 507 p., 397 fig., (1948). (Translated into Russian in 1950.)

58--1948b   Brèches superficielles du Calcaire éocène dues à l'altération chimique en Haiti, Antilles. Rev. Soc. Haitienne d'Hist.et de Geographie 19/70: 41-47, il., (1948). (Translated by C. Pressoir.)

59--1948c   Pisolites calcitiques en formation dans des dépôts de travertin de cascade. Rev. Soc. Haitienne d'Hist. et de Geographie 19/70: 48-54, il., (1948). (Translated by C. Pressoir.)

60--1948d   Review of "Reptile and Amphibian trackways from the Lower Triassic Moenkopi formation of Arizona and Utah," by F. E. Peabody; Berkeley and Los Angeles: U. Cal. Press, 1948. U.S. Quart. Book List 4/4: 502, (1948).

61--1949   Review of Genetics, Paleontology, and Evolution, by G. L. Jepsen, E. Mays, and G. G. Simpson; Princeton: Princeton Univ. Press, 1949. U.S. Quart. Book List 5/3: 390-391, (1949).

62--1949a   Review of Historical Geology, by C. O. Dunbar; New York: John Wiley and Sons, 1949. Am. J. Sci. 248: 75-77, (1949).

63--1950   Some physical aspects of ancient reef complexes [abst.]. Program A.A.P.G. Meeting at Chicago, 25 April 1950, p. 13; Oil and Gas Jour. 48/51: 118, (1950).

64--1950a   (with Knudsen, V. O., Redfield, A. C., and Revelle, R.) Education and training for oceanographers. Science 111: 700-703, (1950).

65--1950b   Review of "Stratigraphy and Paleontology of the Brownsport formation (Silurian) of Western Tennessee," by T. W. Amsden; New Haven: Yale Univ. Press, 1949. U.S. Quart. Book Rev. 6/2: 238, (1950).

66--1951   Memorial to Clyde Arnett Malott [1887-1950]. Geol. Soc. Am., Pr. 1950: 105-110, port., (1951).

67--1951a   Review of "The morphology of ostracod molt stages," by R. V. Kesling; Urbana: Univ. Illinois Press, 1951. U.S. Quart. Book Rev. 7/4: 420, (1951).

68--1953   (and Twenhofel, W. H.) Principles of Invertebrate Paleontology. New York: McGraw-Hill Book Co., Inc., xx + 816 p., il., (1953).

69--1953a   Review of Life of the Past, an introduction to Paleontology, by G. G. Simpson; New Haven: Yale Univ. Press, 1953. U.S. Quart. Book Rev. 9/3: 363-364, (1953).

70--1954   Review of "Silicified Middle Ordovician trilobites," by H. B. Whittington and W. R. Evitt II; Geol. Soc. Am., Mem. 59, 1953. U.S. Quart. Book Rev. 10/2: 260-261, (1954).

71--1956   Review of "Handbook of Ostracod Taxonomy," by H. V. Howe; Baton Rouge: La. St. Univ. Press, 1955. U.S. Quart. Book Rev. 12/1: 106-107, (1956).

72--1956a   Review of "The Mascall fauna from the Miocene of Oregon," by T. Downs; Berkeley: U. Cal. Press, 1956. U.S. Quart. Book Rev. 12/244-245, (1956).

73--1956b   Review of "Classification of Clypeasteroid echinoids," by J. W. Durham; Berkeley: U. Cal. Press, 1955. U.S. Quart. Book Rev. 12/2: 245, (1956).

74--1957   Notice concerning Acta Palaeontologica Polonica. J. Pal. 31: 1029-1030, (1957).

75--1957a   William Henry Twenhofel (1875-1957). Am. Assoc. Petroleum Geol., B. 41: 978-980, port., (1957).

76--1957b   William Henry Twenhofel. Tulsa Geol. Soc. Digest 25: 32-33, (1957).

77--1957c   (Robert R. Shock [Shrock]) Geological education. Tulsa Geol. Soc. Digest 25: 116-120, (1957).

78--1957d   Memorial to William Henry Twenhofel (1875-1957). J. Sed. Pet. 27: 203, port., (1957).

79--1957e   New geological horizons [S.E.P.M. Presidential Address]. Am. Assoc. Petroleum Geol., B. 41: 1403-1408, (1957).

80--1957f   Ten Years in Nova Scotia. The Massachusetts Institute of Technology Summer School of Geology, 1948-1957. M.I.T.: Dept. Geol. Geophysics, 96 p., il., (1957).

81--1958   A Co-operative Training Program for exploration geophysicists. Tech. Rev. 61/2: 83-88, il., (1958).

82--1958a   Some expanded palaeontological horizons. J. Pal. Soc. India, Birbal Sahni Memorial Number, Vol. 5: 119-122, (1958). (An invited paper.)

83--1959   Earth Science Research Center proposal for M.I.T. [Prepared by R. R. Shrock]. Geotimes 3/8: 15, (1959).

84--1959a   Educating future earth scientists [abst.]. Am. Assoc. Petroleum Geol., B. 43: 1771-1772, (1959).

85--1959b   Earth Science Center at Massachusetts Institute of Technology. Am. Geophys. Union, Tr. 40: 120-122, (1959).

86--1960   He found bauxite for Reynolds [A short biography of W. J. Mead]. Reynolds Review 20/3: 10-13, il., (1960).

87--1960a   Earth sciences (p. 345-346) in McGraw-Hill Encyclopedia of Science and Technology, Vol. 4, (1960). New York: McGraw-Hill Book Co., Inc., (1960); Ibid., 4: 345-346, (1966); Ibid., 4: 388-389, (1971).

88--1960b   W. J. Mead, Experimental Geologist. Science 132: 1235-1236, port., (1960).

89--1960c   Memorial to Irving Ballard Crosby (1891-1959). Geol. Soc. Am., Pr. 1959: 117-120, port., (1960).

90--1960d   Review of Giants of Geology, by C. L. and M. A. Fenton; New York: Doubleday & Co., Inc., 1952. Tech. Rev. 63/2: 42, (1960).

91--1960e   [Letter (p. 155-156) to Senator Warren G. Magnuson, re S2692 (Marine Science).] Hearings before the Committee on Interstate and Foreign Commerce, U.S. Senate, 86th Congress, 2nd Sess. on S2692, April 20-22, 1960, U.S. Government Printing Office, p. 151-165, (1960).

92--1961   In Memoriam - Warren Judson Mead (1883-1960). Wis. Ac. Rev. (Wis. Ac. Sci. Arts Lett.), p. 42, (winter 1961).

93--1961a   Citation of Arthur Carleton Trowbridge for the Neil Miner Award, 1960. J. Geol. Educ. 9/1: 51-52, (1961).

94--1961b   Warren Judson Mead, August 5, 1883-January 16, 1960. Nat. Ac. Sci., Biog. Mem. 35: 252-271, port., (1961).

95--1961c   M.I.T. to offer Earth Science Degree [Prepared by R. R. Shrock]. Geotimes 6/1: 34, (1961).

96--1961d   Primary structures in sedimentary deposits (A review and a report of progress with some remarks). Mahadevan Volume - A collection of geological papers in commemoration of the sixty-first birthday of Prof. C. Mahadevan; Dr. M. S. Krishnan, ed. Published by the Publications Committee, Hyderabad, India, 6th May 1961. (R. R. Shrock's article, which was invited, comprises pages 1-12 in the Volume.)

97--1961e   William Otis Crosby Lectureship [Prepared by R. R. Shrock]. Geotimes 6/2: 33, (1961).

98--1962   Memorial to Warren Judson Mead (1883-1960). Geol. Soc. Am., Pr. 1960: 125-136, port., (1962).

99--1963   Oceanographic Research at M.I.T., 15 June 1961 to 31 December 1962 (Robert R. Shrock, Principal Investigator, et al.). NR083-157 Prog. Rept. 1, O.N.R. Contract Nonr 1841(74), 68 p., il., (Jan. 1963), (‡), (Printed at M.I.T.).

100--1964   Oceanographic Research at M.I.T. in Department of Geology and Geophysics (Robert R. Shrock, Principal Investigator, et al.). NR083-157 Prog. Rept. 2, O.N.R. Contract Nonr 1841(74), 38 p., il., (Feb. 1964), (‡), (Printed at M.I.T.).

101--1964a Sales record of Index Fossils of North America. J. Pal. 38/3: 613-616, il., (1964).

102--1964b Doctor's programs in the earth sciences [abst.]. Program, Nat. Assoc. Geol. Teachers Meeting in Miami Beach, Fla., 19-21 November 1964, p. 184-185, (1964). (An invited paper.)

103--1965   Oceanographic Research at M.I.T., in the Department of Geology and Geophysics, 1964 (Robert R. Shrock, Principal Investigator, et al.). NR083-157 Prog. Rept. 3, O.N.R. Contract Nonr 1841 (74), 42 p., il., (Feb. 1965), (‡), (Printed at M.I.T.).

104--1966   Memorial to Hervey Woodburn Shimer (1872-1965). Geol. Soc. Am., B. 77/5: 73-80, port., (1966); Ibid., Pr. 1966: 379-385, port., (1968).

105--1966a The GSI Student Cooperative Plan. A summer training program in geophysical education pays off [Prepared by R. R. Shrock]. Geotimes 11/5: 15-20, (1966).

106--1966b   A Cooperative Plan in Geophysical Education, The G.S.I. Student Cooperative Plan; The first fifteen summer programs, 1951-1965. Dallas, Tex.: Geophysical Service Inc., 143 p., il., (1966).

107--1968   Martin J. Buerger, '24; Institute Professor; Professor of Mineralogy and Crystallography. Tech. Rev. 70/9: 81, (1968).

108--1968a   (and Hooker, Marjorie) Memorial to Frederick Kuhne Morris (1885-1962). Geol. Soc. Am., Pr. 1966: 329-335, port.,(1968).

109--1970   Memorial to Edgar Roscoe Cumings (1874-1967). Geol. Soc. Am., Pr. 1967: 177-185, port., (1970).

110--1970a   [Memorial to] Walter Lucius Whitehead (1891-1969). Am. Assoc. Petroleum Geol., B. 54/2: 363-364, port., (1970).

111--1971   Final and Summary Report on Oceanographic Research at M.I.T. in the Department of Earth and Planetary Sciences and the Department of Meteorology, 15 June 1961 - 31 December 1970. Sponsored by the U.S. Office of Naval Research Nonr 1841(74), Contract Authority NR083-157. M.I.T. Rept. 71-5, 43 p., il., (April 1971), (‡), (Printed at M.I.T.).

112--1971a   Citation of Enders A. Robinson [XII Ph.D. 1954] and Sven Treitel [XII Ph.D. 1958] for the Society of Exploration Geophysicists' Award. Geophysics 34: 1047-1048, (1969). (Printed in 1971.)

113--1971b   Memorial to Walter Lucius Whitehead (1891-1969). Geol. Soc. Am., Pr. 1969 [never published]; preprinted in June 1971 (5 p.). Published in Geol. Soc. Am., Memorials 1: 105-109, port., (1973).

114--1971c   Go forth and be like Columbus [A tribute to Columbus O'Donnell Iselin]. Oceanus 16/2: 34, (1971).

115--1971d   [Review of] The Elements of Palaeontology, by Rhona M. Black. London: Cambridge Univ. Press, viii + 339 p., il., (1970). Am. Geophys. Union, Tr. 52/6: 470, (1971).

116--1971e   [Review of] Evolution of the Earth, by R. H. Dott, Jr. and R. L. Batten. New York: McGraw-Hill Book Co., xiv + 649 p., il., (1971). J. Geol. Educ. 19/5: 240-242, (1971).

117--1972   A way to help The Paleontological Society meet publication costs. J. Pal. 46/3: 451-453, il., (1972).

118--1972a   A 27-year report on sales of Index Fossils of North America. J. Pal. 46/3: 453-455, il., (1972).

119--1972b   Corrigenda II for Index Fossils of North America. J. Pal. 46/3: 455-459, (1972).

120--1972c   The Geologists Crosby of Boston. Cambridge: M.I.T. Graphic Arts Service, xii + 96 + A1-A79, il., (1972).

121--1973   [Obituary of] Roland D. Parks, in the Lexington Minute-man for 18 January 1973, p. 6, (1973).

122--1973a   Memorial to Walter Lucius Whitehead. Geol. Soc. Am., Memorials 1: 105-109, port., (1973). Preprinted in June 1971, as noted in reference 113--1971b.

123--1973b   Bones for Stratton and Green, p. 91, and [Comments re Cecil H.
124--1973c   Green '23], p. 190-191, A Great History of the Great Class of 1923 (by Arthur W. and Phyllis Davenport). Cambridge: Massachusetts Institute of Technology, 352 p., il., (1973).

125--1974   [Citation of] Cecil H. Green: Human Needs Award [of the American Association of Petroleum Geologists. A.A.P.G., B. 58/9: 1881-1882, (1974).

126--1974a Preface and Front Matter in The First Decade in the Green Building, 1964-1974. (Selected papers by M.I.T. authors in honor of Cecil H. and Ida M. Green.) Dept. Earth and Planetary Sciences and Dept. Meteorology. Cambridge, Mass.: M.I.T. Graphic Arts Service, 36 separate articles with illustrations, (1974).

127--1974b The Ida Green Room, in the Cecil and Ida Green Building, Center for the Earth and Planetary Sciences, Massachusetts Institute of Technology. Cambridge, Mass.: M.I.T. Graphic Arts Service, 20 p., il., (1974).

128--1975 Twenhofel, William Henry, in Dictionary of Scientific Biography XII, New York: Scribners. (In press.)

ADDITIONAL MISCELLANEOUS PUBLICATIONS OF ROBERT RAKES SHROCK

1. Reports as an Officer of a Scientific Society (1938-1960)

Most scientific and professional societies require that their officers - President, Secretary, Treasurer, etc. - publish annual reports, and some also require that the President deliver and publish an address upon leaving office. Following are such items:

(1937) The Wisconsin Academy of Sciences, Arts and Letters (R. R. Shrock, Secretary-Treasurer). Science 85: 455, (1937).

(1938) Proceedings of the Academy [Wisconsin Academy of Science, Arts and Letters], 67th Annual Meeting (R. R. Shrock, Secretary-Treasurer). Wis. Ac. Sci. Arts Lett., Tr. 31: 567-570, (1938).

(1939) Treasurer's Report. Proc. Pal. Soc. in Geol. Soc. Am., Pr. 1938: 220-222, (May 1939).

(1940) Treasurer's Report. Proc. Pal. Soc. in Geol. Soc. Am., Pr. 1939: 264-265, (June 1940).

(1941) Treasurer's Report. Proc. Pal. Soc. in Geol. Soc. Am., Pr. 1940: 262-264, (June 1941).

(1942) Treasurer's Report. Proc. Pal. Soc. in Geol. Soc. Am., Pr. 1941: 222-224, (March 1942).

(1943) Treasurer's Report. Proc. Pal. Soc. in Geol. Soc. Am., Pr. 1942: 260-261, (April 1943).

(1957) Special Resolution on death of W. H. Twenhofel adopted at S.E.P.M. Council Meeting, 31 March 1957. J. Pal. 31: 1179, (1957).

(1957) Comment on election of Philip Henry Kuenen as Correspondent, incorporated in the Reports and Minutes of the 31st Ann. Meeting of S.E.P.M. in St. Louis, Mo., 1-4 April 1957. J. Sed. Pet. 27: 205-208, port., (1957); J. Pal. 31: 816-819, (1957).

(1957) S.E.P.M. President's Report (Robert R. Shrock, President). J. Sed. Pet. 27: 219-221, (1957); J. Pal. 31: 830-832, (1957).

(1957) Report on the American Geological Institute, included in the Society Records and Activities of the S.E.P.M. for 1958. J. Sed. Pet. 28: 251-252, (1958); J. Pal. 32: 784-785, (1958).

(1960) Report of the President [National Association of Geology Teachers, by Robert R. Shrock, President]. J. Geol. Educ. 8/1: 37-38 under Transactions, (1960).

2. REGISTERS of Course XII Alumni (1955-1970)

In order to keep our Course XII alumni informed of what was going on at the Institute and about their fellow graduates, my secretary, Pauline Richmond, and I prepared and sent to all Course XII graduates, Registers in 1955, 1960, 1965 and 1970, as listed below.

(1955) A REGISTER of the Department of Geology and Geophysics in the Massachusetts Institute of Technology at Cambridge, Massachusetts 1865-1955, 85 p., il., (March 1955), (‡), (Printed at M.I.T.).

(1960) A REGISTER of the Department of Geology and Geophysics in the Massachusetts Institute of Technology at Cambridge, Massachusetts 1865-1960, 90 p., il.,(June 1960), (‡), (Printed at M.I.T.).

(1965) A REGISTER of the Department of Geology and Geophysics in the Massachusetts Institute of Technology at Cambridge, Massachusetts 1865-1964, 38 p., il., (February 1965), (‡), (Printed at M.I.T.).

(1970) A REGISTER of the Department of Earth and Planetary Sciences in the Massachusetts Institute of Technology at Cambridge, Massachusetts 1865-1970, 74 p., il., (June 1970), (‡), (Printed at M.I.T.).

3. Departmental Reports in the annual M.I.T. President's Reports (1949-1964)

Each summer, while I served as Head of Course XII (1949-1965), department heads were asked to submit to their respective deans a brief report on departmental activities during the previous school year. These annual reports were later included in the President's Report Issue, that appeared regularly in the following fall or winter, except for SYs 1959-60, 1964-65, and 1965-66, when much abbreviated President's Reports omitted them. My reports are listed below:

GEOLOGY [A school-year Report on Departmental activities].

    in President's Report Issue 1948-1949
      M.I.T. Bulletin, vol. 85, no. 1, p. 154-156. October, 1949.

    in President's Report Issue 1949-1950
      M.I.T. Bulletin, vol. 86, no. 1, p. 184-187. October, 1950.

    in President's Report Issue 1950-1951
      M.I.T. Bulletin, vol. 87, no. 1, p. 186-188. October, 1951.

GEOLOGY AND GEOPHYSICS

    in President's Report Issue 1951-1952
      M.I.T. Bulletin, vol. 88, no. 1, p. 59-60. October, 1952.

SCHOOL OF SCIENCE [Geol. & Geophys. mentioned repeatedly].

    in President's Report Issue 1952-1953
      M.I.T. Bulletin, vol. 89, no. 2, p. 63-72. October, 1953.

    in President's Report Issue 1953-1954
      M.I.T. Bulletin, vol. 90, no. 2, p. 100-112. October, 1954.

in The Reports of the President and of the Deans of the Schools for the year ending Oct. 1, 1955, p. 153-173. October, 1954.
[M.I.T. Bulletin, vol. 91, no. 1, p. 153-173.]

DEPARTMENT OF GEOLOGY AND GEOPHYSICS

in President's Report Issue, for the year ending October 1, 1956, in School of Science Report.
M.I.T. Bulletin, vol. 92, no. 3, p. 157-161. November, 1956.

in President's Report Issue for the year ending October 1, 1957, in School of Science Report.
M.I.T. Bulletin, vol. 93, no. 2, p. 147-151. November, 1957.

in President's Report Issue for the academic year ending July 1, 1958.
M.I.T. Bulletin, vol. 94, no. 2, p. 166-171. November, 1958.

in President's Report Issue, 1959.
M.I.T. Bulletin, vol. 95, no. 2, p. 191-194. November, 1959.

in The President's Report, 1960. (No publication date indicated.)
[Spec. M.I.T. Pub.], p. 24-26.

in President's Report Issue, 1961
M.I.T. Bulletin, vol. 97, no. 2, p. 192-195. November, 1961.

in President's Report Issue, 1962
M.I.T. Bulletin, vol. 98, no. 2, p. 256-261. November, 1962.

in President's Report Issue, 1963
M.I.T. Bulletin, vol. 99, no. 2, p. 305-312. November, 1963.

in President's Report Issue, for the year ending July 1, 1964.
M.I.T. Bulletin, vol. 100, no. 2, p. 320-327. November, 1964.

4. International Conference on the Earth Sciences on the occasion of the dedication of the Cecil H. and Ida F. Green Building, Center for Earth Sciences, on 30 September - 2 October 1964

Numerous items had to be prepared and distributed in connection with the dedication of the new Green Building and holding of the International Conference on the Earth Sciences held on 30 September and 1 and 2 October, 1964. Among the more important of these, in which I took an active part in preparing, were the following:

1) [Program of the] International Conference on the Earth Sciences on the occasion of the dedication of the Green Building, Massachusetts Institute of Technology, Cambridge, [Massachusetts], September 30 to October 2, 1964. A pamphlet of 8 pages, with several illustrations, printed at M.I.T., (September 1964).

2) [Program of the] Dedication Exercises, The Cecil and Ida Green Building, Center for Earth Sciences, Massachusetts Institute of Technology; the South Plaza of the Green Building, Cambridge, Massachusetts, Friday, October 2, 1964, 3:00 P.M. A pamphlet of 12 pages, with several illustrations, printed at M.I.T., (September 1964).

3) The Cecil and Ida Green Building Dedication, Massachusetts Institute of Technology, Cambridge, Massachusetts, October 2, 1964. A 48-page booklet, including remarks and addresses and a number of pictures of the participants. Printed at M.I.T. and distributed shortly after the dedication. ("Remarks by Robert R. Shrock" are included on pages 21-23 of the booklet.)

4) See also references 126--1974 and 127--1974a in the preceding Bibliography.

(25)
ROLAND DANE PARKS
(1900-1972)

ROLAND DANE PARKS

MIT: 1940-1966

Born and educated in the Keweenaw Peninsula, son of Upper Michigan's copper country, Parks early became familiar with mines, miners, and the hidden wealth that was to be found in the famous copper-bearing lavas of the region. What he saw above and below ground left an indelible impression on him. The romance of exploration, especially with new techniques; the challenges of production, especially in the economic applications of the suitable methods--these two aspects of the mineral industry were uppermost in his personal approach and ever-present in the teaching of his students. In addition, throughout his academic career he argued for regulations and laws that would help to protect the miners in their hazardous activities at the working face and to assure them a fair wage for their labor.

Well-trained in the fundamentals of minerals and ore deposits, and in the importance of these commodities in international affairs, by virtue of college training and faculty service at the Michigan College of Mines and Technology, and of graduate work at the University of Wisconsin, Parks brought to M.I.T. in 1940 a much needed aspect of the overall mineral industry. For nearly 25 years he taught courses in Mining Methods and Practices, Mineral Economics, Mine Valuation, and Mineral Resources. Today all these subjects would fall neatly into the present national concern with our environmental and ecological problems.

World War II found him in Washington as one of the chief administrators of the Miscellaneous Minerals Branch of the War Production Board. From this post he moved to the Minerals Bureau and then to the Office for Metals and Minerals. After five years in Washington (1941-1946) he returned to M.I.T. to resume teaching. Twice later he took annual leaves of absence to assist two foreign universities in their mineral engineering problems. In SY 1955-56 he was guest professor at the Indian School of Mines and Applied Geology at Dhanbad, Bihar, India, and in SY 1961-62 he served as a Fulbright lecturer at the University of Assiut in Egypt.

In July 1964 Parks took early leave and accepted a position with the Mineral Resources Department of the Internal Revenue Service in Washington, continuing in his work until 1970. Meanwhile, he retired formally from M.I.T. as of 30 June 1966 when he became Associate Professor Emeritus.

Parks contributed importantly to departmental programs and Institute affairs, and to national and international activities. In the Department he brought breadth and the economic importance of minerals to the overall program of economic geology; he served as Assistant Director of the M.I.T. Summer School of Geology at Crystal Cliffs, near Antigonish, Nova Scotia, for more than a decade (1948-1959); he supervised 14 theses; he served on numerous departmental and Institute committees; and he kept up to date his widely used textbook, unique in its field, <u>Examination and Valuation of Mineral Property</u>. In national and international activities he not only served our Government during World War II, in fields of his special knowledge, but also carried his expertise to academic institutions in India and Egypt, as mentioned earlier, and traveled to the Far East to evaluate mineral properties for the U. S. Treasury Department. Active to the day of his death as a consultant on evaluation of mineral properties, Parks served many private clients during a period of more than thirty years. At their home in Lexington, and at Crystal Cliffs in Nova Scotia, the Parks were gracious hosts to a generation of students, who remember afternoons and evenings with good food and stimulating conversation.

Roland Dane Parks* was born on 26 September 1900, in Lake Linden, Michigan, a small town halfway up that finger of land which projects northeastward into Lake Superior like a bent finger, and known to geologists as the copper-rich Keweenaw Peninsula. Lake Linden was in the heart of this copper country, and the youthful Parks early became acquainted with the names of the great copper-producing companies of the Upper Peninsula-- Calumet and Hecla, Quincy, Copper Range, and Isle Royale.

He was the only child of William John Parks and Anna Elizabeth (Cool) Parks, and his was a genealogical mix of forebears from the British Isles and Western Europe. His father's people were Irish, Scotch-Irish and French, and the name Parks was thought to stem from the French Huguenot Parc, which changed in spelling as the name-bearers moved via Ireland and Canada to the Lake Linden region. Grandfather Parks, by name Thomas Parks, came to Michigan's Upper Peninsula from lower Canada as a "land-looker," one of the early timber cruisers. He was a Paul Bunyan of a man, a true woodsman and lumberman, and his son, Roland's father, William John Parks, followed in his ways and also found time to indulge in the exciting avocation of racing fine harness horses. Although Roland did not follow the woodsman tradition of his father and grandfather, nor the racing interests of his father, he did have a love of automobiles, like his father's love of fine horses, and always kept up to date on the special features of the latest model. He was a superb driver, as the present author can attest, having been a front-seat occupant when only the quickest action on Parks' part landed us front-end buried in a soft snowdrift rather than draped over the front part of an onrushing truck.

---

* In preparing this biography I drew freely on an autobiographical sketch that Parks prepared at my request in April of 1972. The original nine-page typescript, which bears the title "Biographical sketch: ROLAND DANE PARKS," together with the letter sent with the typescript and dated April 11, 1972, have been presented to the M.I.T. Institute Archives. A copy is included in Parks' file in Course XII Departmental Headquarters. I am particularly grateful to Ruth Martin Parks for reading the first draft, answering numerous questions, and suggesting important changes that greatly improved the narrative.

Grandmother Parks, born Mary Ann Quinlan, fled the potato famine in Ireland and ultimately arrived in the Upper Peninsula by way of Montreal, where she disembarked, and then Vermont, where she worked as a domestic.

On his mother's side, Parks' ancestry traced back to families from England named Dane, Dean, Butler and Trull, which settled in New Boston, New Hampshire, and around Ipswich, Massachusetts. Grandmother Mary Elizabeth Trull of New Boston met and married Isaac Newton Cool of Decatur, Illinois. The name Cool is thought to have come from the Germanic Kuhl, but the reason why the name was changed is unknown.

One of Parks' more distant forebears on his mother's side, always spoken of with great respect in family discussions, was Nathan Dane (1752-1835), a distinguished lawyer and statesman from Beverly, Massachusetts, who had much to do with helping the young Harvard Law School survive during its years of distress in the late 1820s.*

In the autobiographical sketch mentioned in a preceding footnote, Parks wryly added the following to the mention of Dane's name--"Another in this genealogy not quite so often mentioned is reputed to have been a juror at the Salem trials. It has not been recorded just how he voted."

Parks received his early education in the schools of his hometown and graduated from the Lake Linden High School in 1917, no doubt having been inspired to good effort by his mother, an 1896 graduate from the University of Michigan, who taught Latin in the same high school. When it came time to go to college, it was thought best that Roland enter the University of Wisconsin because his uncle, Charles Dean Cool, was a Professor in the Spanish Department there and could perhaps keep a distant eye on him. This arrangement lasted only a few weeks, however, because young Parks

---

* Dane was a member of the Continental Congress, draftsman of the 1787 "Ordinance for the Government of the Territory Northwest of the Ohio," and author of Dane's "Abridgement," 40 years in the making, which has been described as follows:

> "... General Abridgement and Digest of American Law, with Occasional Notes and Comments, the first really great systematic treatise on the general law of the United States." [p. 94 in A. E. Sutherland's "The Law at Harvard: A History of Ideas and Men, 1810-1967." Cambridge, Mass., The Belknap Press of Harvard University Press, xv + 408 p., (1967).]

On 3 June 1829, when Harvard's recently established Law School (1817) was having a hard time to survive, Dane gave the Harvard Corporation $10,000, representing profits from his "Abridgement," for founding a professorship of American Law, and nominated famous Judge Joseph Story (1779-1845) for appointment as the first Dane Professor. [p. 10 in "The Centennial History of the Harvard Law School, 1817-1917." Cambridge, Mass. (Prepared by the Harvard Law School Faculty and) Published by the Harvard University Harvard Law School Association, x + 412 p., (1917).]

became homesick--Madison, even in 1917, was "big city" and the University a large and energetic place compared to the quiet sylvan environment of Lake Linden--so he went home and enrolled at the Michigan College of Mines at Houghton some fifteen miles to the south. Here he would be a happier undergraduate because of the skating, hockey and other winter sports that he had known from his earliest youth, and he would not be far from home!

At M.C.M., as the College was called in those days, Parks fell under the influence of Frederick W. Sperr, then Head of the Civil and Mining Engineering Department, whom he found to be a stimulating and inspiring teacher. Sperr's influence, no doubt strengthened by great admiration for his uncle Charles at Madison, led Parks to the conclusion that he would like to be a college professor and, naturally, to teach subjects that would be concerned with minerals, ore deposits, and mining. Given the environment of his youth, in the heart of the copper-mining country, and the traditions and lore of men who burrow underground for precious metals, and aware of the dangers and rewards of mining, it is not at all surprising that a sensitive and observant young college student should react as Parks did to Professor Sperr's influence. And later on, he would receive further stimulation from two other outstanding professors, C. K. Leith and W. J. Mead, at the University of Wisconsin.

The four undergraduate years at M.C.M. were demanding because the curriculum was designed to train students for practical engineering, and the College had the reputation of producing well-trained graduates who were competent to enter professional work in engineering immediately after Commencement. However difficult and demanding were his academic activities, Parks nevertheless found time to take apart and repair his first automobile, the redoubtable Model T Ford from which so many of his generation, including the author, learned their first practical lessons in auto mechanics. And in sports there was hockey, four years of it, on the varsity team that was good enough to play such teams as Michigan, Minnesota, and Notre Dame. In his own words--

> "I gave my all for the M.C.M. team (later the Michigan Tech Huskies) and, if I failed to score, as I often did, I was indefatigable in feeding the puck to our Canadian center who was a more proficient stickhandler."

Staying on at M.C.M. after earning a Bachelor of Science and Mining Engineering degree in 1921, Parks first served as Instructor in Civil and Mining Engineering (1921-1927), and then successively as Assistant Professor (1927-1931) and Associate Professor of Mining Engineering (1931-1940). During these years he supplemented classroom instruction with visits to mines in neighboring states (Michigan, Wisconsin, and Minnesota) as well as farther away (Colorado, Montana, and Utah). These visits, which he continued later at M.I.T., gave the students excellent opportunities to study mining methods and practices in active mines.

While a member of the M.C.M. faculty, he took leave to pursue graduate work at the University of Wisconsin, from which he received a Master of Science degree in 1925. Here he took geology courses with two well-known professors who would as mentioned before greatly influence his future career--C. K. Leith (mineral economics) and W. J. Mead (structural geology).

Resuming his teaching duties at Houghton, the school by then renamed Michigan College of Mining and Technology (MCMT), and presided over by Dr. William Otis Hotchkiss, himself a graduate of the University of Wisconsin, Parks soon came in close association with the new Head of the Mining Department, Professor Charles H. Baxter, a mining engineer and mine operator. (Professor Sperr had died in the interim.) From this association came a series of lectures on mine valuation which culminated in a unique and widely used textbook that has gone through four editions (1933, 1939, 1949, and 1957--See Bibliography). Today, the present author would agree with Parks' own judgment that this book, now 40 years old and worldwide in its usage, is his most significant and important publication.

In the fall of 1940 Parks joined M.I.T.'s Department of Geology as Assistant Professor of Mineral Industry, having been recruited by one of his former Wisconsin professors, W. J. Mead, now Geology Head at M.I.T., upon recommendation of the other of his Wisconsin professors, C. K. Leith, whereupon he took responsibility for the Mineral Industry option of Course XII. His first task was to organize lecture subjects on Mining Methods and Practice, Mineral Economics, Mine Valuation, and Mineral Resources. These subjects were designed to broaden the training of geology students who planned to follow professional careers in practical geology. Today, such courses, which are no longer offered, would no doubt be deemed quite relevant to current concern about environmental and ecological problems.

Hardly had his academic program got underway when, in the summer of 1941, he was recruited by Leith for the Office of Production Management (OPM), which after Pearl Harbor became the War Production Board (WPB).*
His first assignment was in Washington as Deputy Director of the Miscella-

---

* "... in January 1941 [President] Roosevelt established the Office of Production Management to stimulate and control defense production [of mineral commodities]. One year later, in January 1942, Roosevelt replaced that agency with the War Production Board. Leith and the ever-enlarging group of mineral specialists [of which R. D. Parks was one, serving as Deputy Director of Miscellaneous Minerals Division] that he organized continued as advisers to the Office of Production Management and the War Production Board." Quoted from page 201 of Sylvia W. McGrath's Charles Kenneth Leith - Scientific Adviser, The University of Wisconsin Press, Madison, 1971.

768   GEOLOGY AT M.I.T.   1865-1965

In the classroom (1960)

At his desk (1960)

On a mine visit
with a class
November 11, 1957

M.I.T. Student Group
Bethlehem-Cornwall
#4 Mine

Professor Parks is
seated at the lower
righthand corner of
the car

neous Minerals Division from 1941 to 1943. Next he moved on to the Minerals Bureau as Assistant Director of the Mineral Resources Coordinating Division, (1943-1944) and then to the Office of the Vice Chairman for Metals and Minerals as Assistant Deputy (1944-1946), where he was Alternate Chairman of the Quota Committee for copper, lead and zinc, and also of the Mineral Resources Advisory Committee and the Mineral Resources Operating Committee. In these assignments he was involved in production problems, import programs, and stockpiling for almost the entire mineral program, with the result that he was able to broaden immensely his knowledge of the whole diverse field of mineral industry on a worldwide scale.

After five years in Washington, Parks returned to M.I.T. in March 1946 and resumed his academic work along the lines laid out before Pearl Harbor. Now, however, he could bring greater breadth and depth to his subjects because of his Washington experience. For the next 19 years, with the exception of two leaves for missions abroad, he devoted the major part of his time to his M.I.T. responsibilities.

Every summer from 1948 to 1959, excepting 1955, he conducted the surveying part of the Department's Summer School of Geology at Crystal Cliffs, near Antigonish, Nova Scotia, and simultaneously aided Professor Walter L. Whitehead as Assistant Director. As a result of his instruction, more than 300 geology undergraduates from the United States and Canada got their first taste of field work, regardless of the weather.

Soon after returning to M.I.T. from Washington, Parks became an associate in the New York consulting firm of Behre Dolbear & Co., and for many years rendered expert advice on problems involving the mineral industry.

Twice during his post-war years at M.I.T. he took leaves to accept appointments abroad. From 1 February 1955 to 30 June 1956 he joined the University of Wisconsin program under the Technical Cooperation Mission of the U. S. Government, and served as Guest Professor of Mining and Civil Engineering for the Indian School of Mines and Applied Geology at Dhanbad, Bihar, India. His second assignment came during SY 1961-62 when he held a Fulbright Lectureship at the University of Assiut in Egypt. This assignment not only provided an opportunity to work with advanced undergraduates in mining engineering at a new university, but also to participate in the promotion of an academic exchange program between a consortium of eight Mid-Western universities and the University of Assiut.

As he neared retirement, an opportunity came to do mineral evaluation work for the U. S. Treasury Department, so he took early leave, as of 1 July 1964, and joined the Mineral Resources Department of the Internal Revenue Service in Washington, D.C. Among his duties was to act as an expert witness for the Government in tax cases going to litigation, and to travel abroad to make on-site evaluations of mineral properties. One of these latter assignments took him to Malaysia in 1964. Retirement from Government service came in 1971, five years after retirement from M.I.T. on 30 June 1966, but this only resulted in change of activity. He again became an Associate Consultant with Behre Dolbear & Co. in New York--the consulting firm he had been associated with from 1948 to 1964, before going into Government service--and late in 1971 he gave a paper on "Valuation of Mineral Property" at the International Association of Assessing Officers Symposium in Phoenix, Arizona.

Death came unexpectedly and suddenly from cerebral hemorrhage on 18 December 1972 as he was looking forward to reunion with his family at Christmas time.*

While a graduate student at the University of Wisconsin in the early 1920s, Parks met Ruth Dickey Martin, a student in the School of Journalism, and they were married in Chicago on 19 February 1927. Ruth's family were Iowans, born and bred, and farther back, homesteaders from Pennsylvania, both the Martins and the Dickeys on her mother's side. All were Scotch-Irish Presbyterians, rugged individualists, inheritors of Middle West farm traditions, owners of rich Iowa soil, and classic followers of the work ethic. In the genealogy were several judges and a link to General George B. McClellan of Civil War repute and disrepute, but for the most part the Martins and Dickeys were typical children of the Middle West. In 1930 an only child was born to the Parks, Nancy Dane Parks, now married to Richard P. Valelly, and the mother of three children.

As a member of the M.I.T. faculty, Parks offered a broad range of subjects to his students, subjects aimed at acquainting them with the importance of mineral resources on both national and international scales. He directed 14 theses--7 S.B.s, 5 S.M.s, and 2 Ph.D.s. Several of his thesis advisees now hold positions of high responsibility in the mineral industry. As Assistant Director of the M.I.T. Summer School of Geology in Nova Scotia (1948-1959) he taught more than 300 students the surveying they needed to pursue a career as practicing geologists. Outside the Department he served as a Freshman Advisor (1952-1954), on the Student Aid Committee (1952-1955), and on the Faculty Fulbright Committee (1963-1966).

In the Boston area he served the Boston Section of the A.I.M.E. as Secretary-Treasurer (1947-1950), Vice Chairman (1951), and Chairman (1952). He also served the national office of A.I.M.E. on the following committees: Mineral Economics, Mineral Industry Education, and Mining Methods.

---

* Obituaries appeared on page C16 of The Washington Post for December 20, 1972; on page 8 of Tech Talk for January 17, 1973; and on page 6 of the Lexington Minute-Man for January 18, 1973.

As activities outside the Institute, he gave many lectures on aspects of the mineral industry, served on several special academic committees concerned with the education and accrediting of engineers, and visited some 250 mines and affiliated plants in the United States, Canada, Venezuela, France, Yugoslavia, Egypt, India and Malaysia.

He was a member of Tau Beta Pi and Sigma Xi, and of the American Association for the Advancement of Science, American Institute of Mining, Metallurgical, and Petroleum Engineers (AIME), Mining, Geological, and Metallurgical Institute of India (Life Member), Pan American Institute of Mining Engineering and Geology, Society of Economic Geologists, Geochemical Society, and Mining Club of New York.

Parks was a conscientious and devoted teacher who insisted on good work by his students and supplemented class instruction with field work in the belief that geologists should come in contact with the real world. His students knew him as a helpful counselor who could always be counted on as an understanding and thoughtful advisor. His professional peers knew him best for his classic handbook, Examination and Evaluation of Mineral Property, and his Washington colleagues respected him for his broad knowledge and sound judgment on matters concerning mineral properties and mineral resources policy. And finally, both his students and colleagues knew him and his wife as friendly and gracious hosts who entertained with a special style both at their Lexington home and at Dawson cottage at Crystal Cliffs in Nova Scotia.

BIBLIOGRAPHY OF ROLAND DANE PARKS

(Symbols and abbreviations used in the following references are explained on p. 91-98; in general, abstracts are listed separately as well as with references to the complete articles. This bibliography begins with Parks' first publication in 1928 and includes one posthumous paper in 1971.)

1--1928 Recent developments in methods of mining in the Michigan iron mines. Mich. Coll. Min. Tech., B. 1928-29: 1-38, (1928); Lake Superior Min. Inst., Pr. 26: 115-152, (1928).

2--1928a Timbering practice in the Michigan iron mines. Mich. Coll. Min. Tech., B. 1928-29: 39-58, (1928); Lake Superior Min. Inst., Pr. 26: 153-172, (1928).

3--1929 Drilling and blasting in the Michigan iron mines. Explosives Eng. 7: 12-14, 64-68, (Jan., Feb., 1929).

4--1933 Yieldable metal props for underground supports. A.I.M.E. Tech. Paper 44: 8 p., (May 1933). [U. S. Patents No. 1677796 and No. 2029789 issued July 17, 1928 and February 4, 1936, respectively.]

5--1933 (with Baxter, C. H.) Mine Examination and Evaluation, with interest and annuity tables. Appendix by F. G. Pardee. Houghton, Mich.; published by C. H. Baxter and R. D. Parks; Mich. Coll. Min. Tech., xii + 316 p., il., (1933); 2nd ed., xiv + 331 p., il., (1939).

6--1936   (with Hotchkiss, W. O.) Total profits vs. present value in mining. A.I.M.E., Tech. Pub. 708: 9 p., il., (1936).

7--1937   (with Sperr, F. W.) Mine Surveying Notes (by the late F. W. Sperr, with Revisions and Additions by Roland D. Parks), 6th ed., 48 p., (1937). Published by Department of Civil and Mining Engineering, Michigan College of Mining and Technology, Houghton, Michigan.

8--1939   Improved equipment more noticeable than changes in mining methods, p. 3-5, in Mining progress in Annual Review, A.I.M.E., Min. Metallurgy for January 1939, 3 p., (1939).

9--1939a   Second edition of Mine Examination and Evaluation, See preceding item 5--1933.

10--1945   Corundum - a vital wartime abrasive. A.I.M.E. Tech. Pub. 1883, Min. Tech. 9, 8 p., (May 1945).

11--1947   [Consultant on mining terms], in American College Dictionary. New York, N.Y.: Random House, (1947).

12--1949   Source materials for nuclear power [Chap. 1, p. 1-18], in The Science and Engineering of Nuclear Power, Goodman, Clark, Ed., et al., Vol. 2. Cambridge, Mass.: Addison-Wesley Press, Inc., (1949).

13--1949a   Mineral Policy. Tech. Eng. News 30: 138-139, 148, (1949).

14--1949b   Examination and Valuation of Mineral Property [by] Baxter and Parks. 3rd ed. by R. D. Parks. Oil property valuation [by] W. L. Whitehead. The Michigan mine appraisal system [by] F. G. Pardee. Cambridge, Mass.: Addison-Wesley, xv + 504 p., il., (1949); 4th ed., Reading, Mass.: Addison-Wesley Pub. Co., 507 p., il., (1957).

15--1950   Chromium [p. 244-253], in Minerals Yearbook, Bur. Min., U. S. Dept. Int. 1948, (1950).

16--1950a   Iron ore [p. 627-661], in Minerals Yearbook, Bur. Min., U. S. Dept. Int. 1948, (1950).

17--1952   Chromium - ranks among the most strategic of metals. A.I.M.E., Min. Eng. 4: 469-476,(1952).

18--1953   Materials Survey, Manganese, 1950, compiled for Materials Office, National Security Resources Board, by the U. S. Department of the Interior, Bureau of Mines, with the cooperation of the Geological Survey. Washington, D.C., 536 p., Oct. 1952 (issued March 1953).

19--1957   Fourth edition of Examination and Valuation of Mineral Property. (See preceding item 14--1949b).

20--1960   Field sampling and ore estimation [p. 497-498], in Vol. 8 of Encyclopedia of Science and Technology, New York: McGraw-Hill Book Co., Inc., (1960).

21--1960a   Mine valuation [p. 498-499], in Vol. 8 of Encyclopedia of Science and Technology, New York: McGraw-Hill Book Co., Inc., (1960).

22--1962   (and Galbraith, J. N.) Computer programming in evaluation of mineral property. Univ. Arizona Symposium on Computer Applications in Mineral Industry, Vol. 2, sec. F2: 1-11, (1962).

23--1971   Valuation of mineral property [paper delivered], at International Assoc. Assessing Officers, Symposium, Phoenix, Arizona, Dec. 1971. (Published in 1972 by IAAO, 1313 E. 60th St., Chicago, Illinois.)

(26)
JOHN NATHANIEL ADKINS

JOHN NATHANIEL ADKINS

MIT: 1946-1948-1950

John Nathaniel Adkins was appointed an Assistant Professor of Geophysics in 1946 and hardly got settled down before a request for his services came to President Karl T. Compton from the Office of Naval Research. Adkins requested a leave of absence from 1 March 1948 to 1 February 1949, in order to work in Washington, and it was granted. Again at the Navy's request, the leave was extended until July 1949, and was further extended an additional year to give Adkins an opportunity to complete important tasks for the Navy. At the end of this last leave he decided to resign in March 1950 so as to continue as Director of the Earth Sciences Division of the Office of Naval Research. He had earlier served as Head of ONR's Geophysics Branch, from 1948 to 1950, before becoming Director of its Earth Sciences Division.

John Nathaniel Adkins, the son of John Dauford and Martha Ellen (Cabbage) Adkins, was born in Spokane, Washington, on 23 July 1911. After attending Stanford (1927-1928) and Sacramento Junior College (1930-1932), he entered the University of California at Berkeley, from which he received a B.A. degree in physics in 1936 and a Ph.D. degree in seismology in 1939, having been a University Fellow during the years of graduate work, 1936-1939. He came to M.I.T. as a National Research Council Fellow for two years, 1939-1941, then left, as World War II came on, to join the staff of the Division of War Research at Columbia University, where he was employed until 1945, at which time he started a one-year assignment as Supervisor of the Antenna Section of Airborne Instruments Laboratory, Inc. of New York.

Known favorably by Prof. W. J. Mead, then Head of Course XII, as a result of the research he conducted earlier as a NRC Fellow, from 1939 to 1941, Adkins was offered and accepted an appointment as Assistant Professor of Geophysics, starting on 1 July 1946. For the next two years he divided his time between academic duties at M.I.T. and consulting work with American Smelting and Refining Company in Boston.

Hardly had he got his academic work organized when a letter came to President Karl T. Compton from Rear Admiral P. F. Lee, Chief of Naval Research (dated 26 November 1947), asking if Adkins could be granted a year's leave of absence to succeed Commander Roger Revelle as Head of ONR's Geophysics Branch. Admiral Lee supported his request with the following statement that indicates the Navy's high opinion of Adkins:

> "If Professor Adkins could come to the Office of Naval Research it would be most beneficial, not only to the Navy's research program, but to the science of geophysics and to the whole development of federal support of free fundamental research. His wide knowledge of the various fields of geophysics, his modesty and idealism, and his outstanding analytical ability are precisely the characteristics needed for the responsible position we hope he will assume."

The request for leave was granted and Adkins joined ONR on 1 March 1948. By July it was evident that Adkins would be needed beyond the period of his leave, and a request was granted to extend it until July 1949. Appointment as Director of the Earth Sciences Division of ONR brought a request for still another year of leave until the end of March 1950, at which time Adkins resigned his assistant professorship at M.I.T. and thus ended his membership on our Geology faculty.

On 18 May 1941 Adkins married Katherine Owens Jackson, and our latest information (November 1973) is that John is now Assistant Chief Scientist of ONR and that he and Katherine live at 103 East Melrose Avenue, Baltimore, Maryland 21212.

BIBLIOGRAPHY OF JOHN NATHANIEL ADKINS

Symbols and abbreviations used in the following references are explained on p. 91-98. This bibliography includes only those articles published before or during the time Adkins was a member of the M.I.T. Geology faculty.

1--1938 (with Byerly, P.) Earthquakes in northern California and the registration of earthquakes at Berkeley, Mount Hamilton, Palo Alto, San Francisco, Ferndale, Fresno, from January 1, 1937 to March 31, 1937. Calif. Univ. Seismogr. Sta., B. 7/1: 1-46, (1938).

2--1938a (with Byerly, P.) Ibid. B. 7/2: 47-97, (1938).

3--1938b (with Byerly, P.) Northern California earthquakes, January 1 to December 31, 1937. Seismol. Soc. Am., B. 28/4: 263-268, (1938).

4--1939 (with Byerly, P.) Earthquakes in northern California and the registration of earthquakes at Berkeley, Mount Hamilton, Palo Alto, San Francisco, Ferndale, Fresno, from October 1, 1937 to December 31, 1937. Calif. Univ. Seismogr. Sta., B. 7/4: 151-216, (1939).

5--1940 The Alaskan earthquake of July 22, 1937. Seismol. Soc. Am., B. 30/4: 353-376, il., (1940).

6--1946 Training the geologist for geophysical work, in Mather, K. F., Chmn. Proc. 5th Conference on training in geology. Geol. Soc. Interim Proc., pt. 1: 77-78, (1947); Abst., Geol. Soc. Am., B. 57/12-2: 1277, (1946)

(27)
PATRICK MASON HURLEY

PATRICK MASON HURLEY

MIT: 1937-1942, 1946-

British born, Canadian educated Patrick Mason Hurley came to M.I.T. in the fall of 1937 as a doctoral candidate in geology with B.A. and B.A.Sc. degrees from the University of British Columbia. With excellent training in both the arts and applied sciences (mining engineering), and with considerable experience as a mining engineer, he easily completed all requirements for the Ph.D. degree in economic geology in three years, and received that degree in 1940. His thesis on the helium method of age determination of minerals and rocks set the pattern for his future research which has brought him worldwide distinction as a global geochemist who has contributed significantly to geochronology, continental structure, plate tectonics and crustal evolution.

He started his academic career as a teaching fellow in M.I.T.'s Department of Geology, and his research career as a Research Associate, following his Ph.D. in 1940. He had scarcely launched his research on the radioactivity of minerals and rocks, however, when he was asked to join the NDRC and conduct investigations in undersea warfare - detecting, locating and tracking enemy submarines, work which he carried on in the Atlantic Ocean. Later on he worked with Louis Slichter's Underwater Ballistics Program group at Caltech and there investigated the entry, ricochet, and underwater behavior of bombs, torpedoes, and other similar missiles.

At War's end, in 1945, he accompanied Slichter to the University of Wisconsin and resumed his academic career. A year later, on 1 October 1946, he accepted an appointment at M.I.T. as an Assistant Professor of Geology, advancing to Associate Professor in July 1951 and to Professor in July 1953. He served as Executive Officer of Course XII from 1950 to 1961, and as Chairman of the M.I.T. Faculty during SYs 1960-61 and 1961-62.

In his 29 years as a geology professor, Hurley has been an outstanding teacher, an imaginative and productive research scientist, an able administrator and laboratory director, and an active member of the M.I.T. faculty. His teaching has involved mineral deposits, isotope geology, earth chemistry, and global tectonics. His research has been mainly devoted to the determination of mineral and rock ages, using radiometric measurements, and the application of these ages to the problems of crustal and mantle history, origin of mineral deposits, and plate tectonics. He founded his Geochronology Laboratory in 1947, with initial funding from G.S.A., and during the 25 years since, with funding from G.S.A., O.N.R., A.E.C., and N.S.F., he and his professorial colleagues (H.W. Fairbairn and W. H. Pinson) and research students have produced more than 200 articles and some 20 annual reports. At least 25 doctors, 10 masters, and 32 seniors did their thesis work in his Laboratory.

In addition to his own research on radioactive measurements for age determination, he foresaw the potential use of neutron activation analysis and was instrumental in getting J. W. Winchester to initiate this method to study trace elements in earth materials. He sensed the possibility of using certain electrical methods, developed during World War II, to detect certain metallic minerals and helped T. R. Madden initiate research that revealed the nature of what is now known as the Induced Polarization (IP) method of exploration. Finally, with Prof. George P. Wadsworth and Joseph G. Bryan he initiated the Geophysical Analysis Group, whose student members ultimately produced a series of papers on the signal-to-noise problem in seismic records that have revolutionized geophysical exploration for petroleum.

Known around the world for his brief but rigorous paperback, How Old is the Earth? (Anchor Books), which has been translated into a dozen foreign languages, Hurley has also become one of the leading proponents of plate tectonics. He has achieved this distinction by virtue of analyzing rocks from every part of the earth. Many of his samples came from cooperating colleagues, but he himself has travelled to many remote places to collect desired specimens. As a result of his distinguished work he has given many public lectures, at which he excels, and has participated in most of the more important national and international colloquia, conferences, and symposia on the broad subject of earth dynamics, crustal and mantle evolution, and plate tectonics. Thus he represents our Department of Earth and Planetary Sciences in an impressive manner both nationally and internationally, and his students are already gaining their own distinctions as they follow in his footsteps.

BIRTH, ANCESTRY, AND EARLY EDUCATION

Patrick Mason Hurley was born a British subject in Hong Kong, China, on 12 January 1912, the second of three children of Frederick Charles Mason Hurley and Anne (Peacock) Hurley, both of whom were engaged in business pursuits when they met in Hong Kong.

Frederick C. M. Hurley was born an Englishman, and like his youthful companions of the time yearned for adventure on the high seas and beyond, so at 15 he misrepresented his age a bit, got himself accepted into the British Army, and was soon on his way to the Boer War. After participating in all the major engagements, and being decorated by Queen Victoria, he next escorted a group of prisoners to India, and then proceeded to Hong Kong, where he took employment in a British trading firm (Hughes and Hough). In due course he worked up in the firm to be one of the partners, meanwhile indulging his interests in yachting, painting, and collecting Chinese objects of art - some of the last named objects now adorn the Patrick Hurley home in Lexington, Massachusetts.

Anne Peacock, Patrick's mother, was no ordinary English woman, for after training to be a commercial secretary she struck out on her own and went to Johannesburg, South Africa, to work for a British gold mining company. While there she became interested in a young mining engineer and, though nothing came of the romance, she became so much interested in mining engineering that years later, when young Patrick asked what he should study in college, she enthusiastically urged him to prepare for a career in that profession, and he did. After some years with the gold mining company, Miss Peacock took employment with a major American petroleum company, and in due course moved to Hong Kong, where she met and married Frederick C. M. Hurley. To this union were born the three children mentioned earlier - Denis Mason-Hurley, Patrick Mason Hurley, and Isabel Mason Hurley (now Mrs. F. W. Reuter). Obviously the name Mason was important in the Hurley genealogy, and Denis used the hyphenated familial form to emphasize ancestral pride in the name.

The Hurley family remained in Hong Kong until 1921, when they moved to Canada, and Pat remembers enough of his first nine years to recall how he played with his two siblings in a large rambling house full of a dozen or more servants, who occupied the lowest floors in the house. Each Hurley child had an amah, or nursemaid, and they all enjoyed going "downstairs" to play and eat with the servants.

Then came the move to Canada in 1921 when Pat was nine years old. The family had actually intended to return home to England, where Frederick could retire and indulge his numerous interests, but he had to postpone his departure from Hong Kong for several years in order to close his business affairs. Meanwhile Anne Peacock Hurley and their three children stopped off in British Columbia and settled down on Vancouver Island at Shawnigan Lake. When the father finally joined his family, and had some time to relish retirement and the attractions of Vancouver Island, he decided to stay in Canada and rear his family there. Whereupon he built the Shawnigan Beach Hotel and operated it until he died.

UNDERGRADUATE WORK IN BRITISH COLUMBIA (1927-1934)

Young Patrick grew up on Vancouver Island, where he attended several boys preparatory schools, including the Oak Bay High School in Victoria. He then started on a six-year double degree program consisting of two years at Victoria College (B.C.) (1927-1929), an affiliate of McGill University, followed by four years at the University of British Columbia (1929-1931, 1932-1934). This program led to a B.A. degree in the arts and a B.A.Sc. in applied science, with four years in mining engineering - a profession that his mother had recommended, as mentioned earlier on. Hurley received his B.A. and B.A.Sc. degrees in 1934, with First Class Honors. He also received the Dunsmuir Scholarship for highest rank in the U.B.C. Mining Department and the Engineering Institute of Canada Award for highest rank in the U.B.C. Engineering School. His superb academic record, which made him a first-class applicant for admission to M.I.T., gave clear evidence that he could be expected to achieve a distinguished career in the earth sciences.

But Pat was not a "grind" as an undergraduate. He indulged in tennis, golf, swimming, football, skiing, rowing, and squash; and today, at almost 64, still plays a vigorous game of tennis and rides the ski slopes with considerable skill. As an undergraduate he was on the 1st Team of English Rugby, and served as Treasurer of the Swimming Club. He also found time to serve as President of his social fraternity (Beta Theta Pi).

Hurley was one of the best trained and most experienced Canadian graduate students to be admitted to M.I.T.'s Department of Geology in the Institute's first century (1865-1965), and his subsequent career in Course XII has demonstrated repeatedly the great importance of that excellent preparation. Reference has already been made to his academic training; let us now review briefly his practical experience.

A PERIOD OF PRACTICAL EXPERIENCE (1931-1937)

Hurley started his practical experience during SY 1931-32 by working in a smelter of the Consolidated Mining and Smelting Company in British Columbia, where he was transferred through various testing jobs in the lead and zinc plants, mostly on smoke and sulphur. He spent the summer of 1933 reporting on gold prospects in various parts of British Columbia for the Canada Lode Gold Company. Instead of attending his Commencement in 1934, and receiving in person the several awards mentioned in a foregoing paragraph, he went into the field as soon as possible - this was the depth of the Great Depression and he had to earn some money to pursue graduate work later. From 1934 to 1937 he worked as a mining engineer in charge of the properties of the Vancouver Island Gold Mines, Ltd., and the Golden Zone Mines, Ltd. For the first-named company he supervised the opening of a mineral property to the extent of several thousand feet of underground workings and the completion of a mill. For a few months during this same three-year period he was also mining engineer and geologist at the Relief Arlington Mines, a subsidiary of American Smelting and Refining Company. In each case the operation was small enough that he had to do the geology and also supervise the mapping.

During the summers of 1938 and 1939, after he became a doctoral candidate at M.I.T., he continued field work. In 1938, as a geologist on the Canadian Geological Survey, he mapped in detail some 50 square miles of the Cadillac belt in northern Quebec. The summer of 1939 found him employed by the British Columbia Department of Mines as instructor in a training project. In this assignment he was in charge of a party of twelve and mapped the geology of 40 square miles of mountainous terrain at 600-ft. intervals.

Since 1939 Hurley has largely confined his geological research to the measurement of geological age by radioactivity methods and the application of radiometric measurements to geological problems. More will be said about this doctoral and post-doctoral research a little farther on.

GRADUATE WORK AT M.I.T. (1937-1940)

Hurley was admitted to M.I.T. in September 1937 as a candidate for the Ph.D. degree in Economic Geology, and was appointed a Teaching Fellow for the second term of SY 1937-38, with a tuition scholarship of $250. SY 1938-39 brought him a second appointment as a Teaching Fellow, with a tuition scholarship award of $500. During the next two years, SYs 1939-40 and 1940-41, he held a $1500 Royal Society of Canada Fellowship. He was awarded a Ph.D. degree in Economic Geology in October 1940, with a minor in Mining Engineering. His doctoral thesis, supervised by Prof. Robley D. Evans, of the M.I.T. Department of Physics, and Prof. W. J. Mead, was entitled, "Investigations on the helium method of age determination." (See reference T--1940 in the Bibliography farther on.)

The research that Hurley did for his doctoral thesis launched him on the career in geochronology that he continues to follow today, in mid-1975, thirty five years and more than 200 articles later! As has been typical since his student days, Hurley promptly reported on his thesis research at the May 1940 meeting of the Royal Society of Canada in Ontario, at the Conference on Applied Nuclear Physics at M.I.T. in October 1940, and at the Geological Society of America meetings in Texas in 1940. Abstracts were published in each case, and a complete paper was published in the Bulletin of the Geological Society of America for 1941. (See the Bibliography that follows farther on.)

THE BEGINNING OF AN ACADEMIC CAREER (1940-1942)

As a graduate student, Hurley wanted to find a thesis project that would give him an opportunity to apply the basic sciences he had learned at U.B.C. to some important and challenging geological problem. Accordingly, he talked the matter over with Mead, who was then Head of Geology, and it was suggested that he visit Prof. Robley D. Evans in the Department of Physics, because Evans was involved in determining the amount of radium in sediments and rocks. At the time, Evans, and Clark Goodman (VIII Ph.D. 1940) were boiling helium out of rocks and measuring uranium and thorium by alpha particle counts on their radon and thoron emanations. This research excited Hurley's interest, and he wondered if all the minerals in a rock would show the same age, or put another way, would the age obtained by measuring the radioactivity of a given rock as a whole be the same as for its component minerals. Soon Hurley was working alongside Goodman and N. B. Keevil as another of Evans' doctoral thesis students, and in due course he completed his thesis and reported his first research results as mentioned in a preceding paragraph.

In January 1940 Hurley was appointed a Research Assistant and in May 1941 a Research Associate. The latter appointment ended in June 1942 when he left M.I.T. to join Louis Slichter's research group in the National Defense Research Committee (NDRC). During these years immediately preceding the entrance of the United States into World War II, i.e. from early 1940 to June 1942, Hurley devoted full time to research on helium age measurements, sharing with Goodman half a dozen brief reports on the work done (see Bibliography).

Hurley's radioactivity research ended abruptly in June 1942, and did not resume again until he took a position as Research Associate at the University of Wisconsin after the end of World War II. Here he built a rock-dating laboratory on WARF (Wisconsin Alumni Research Foundation) funds. In 1946 he was offered an assistant professorship at M.I.T., which he accepted, and joined the geology faculty in the fall of that year.

WAR WORK AND A YEAR AT THE UNIVERSITY OF WISCONSIN (1942-1946)

As mentioned above, Hurley was asked to join Slichter's research group in Division 6 of the National Defense Research Committee (NDRC) in June 1942 and to participate in research and development on problems involving subsurface warfare - detecting, locating, and tracking down enemy submarines. In 1944 Slichter was asked to join Division 3 of NDRC and transfer his research group to the Underwater Ballistics Program at Caltech. There, he and his team, which included Hurley, spent more than a year studying the entry, ricochet, and underwater behavior of bombs, rockets, torpedoes, and other underwater projectiles.*

In 1945, at War's end, Hurley went with Slichter to the University of Wisconsin's Department of Geology, but when Slichter moved on to U.C.L.A. in 1946, M.I.T. was able to induce Hurley to return as an Assistant Professor of Geology, and he has been a member of the Geology faculty ever since.

---

* See a more detailed discussion of this research on pages 267-268 in John Burchard's Q.E.D.: M.I.T. in World War II. Cambridge, Mass.: The Technology Press and New York: John Wiley & Sons, Inc., (1948).

AS A MEMBER OF THE M.I.T. FACULTY (1946-present)

Immediately on rejoining his old Department, in October 1946, Hurley set about resuming research and initiating instruction in mineral deposits and in isotope geology. Both instruction and research would focus on the application of nuclear physics to geological problems - determination of mineral and rock ages, earth heat produced by radioactivity, etc., using mass spectrometry.

During these same years, 1946 to the present (1975), Hurley first established and instrumented a mass-spectrometer laboratory and then carried on a vigorous and imaginative program of research on the variations in isotopic abundances of strontium, calcium, and argon in earth materials. A more extended account of Hurley's Geochronology Laboratory, now known around the world for its sustained productivity for more than 25 years, is included in another chapter in this history. Suffice it to record here that at least 200 articles have resulted from work done in the laboratory by Hurley and his two professorial colleagues, Profs. Harold W. Fairbairn and William H. Pinson; and by graduate students working under the supervision of one or more of the professors just named.

Worth further notice is the fact that in the past 25 years, 1950-1975, at least 10 masters and 25 doctoral thesis investigations in Course XII have had the use of Hurley's several geochronology laboratories in Buildings 24 and 54. The nature and breadth of Hurley's research interests may be ascertained by a perusal of his Bibliography that appears farther on.

As Professor of Geology (1946-)

Hurley was appointed Assistant Professor of Geology on 1 October 1946, Associate Professor on 1 July 1951 and Professor on 1 July 1953. He served as Executive Officer of Course XII from 1 July 1950 until 30 September 1961, contributing importantly to departmental activities and policies, and he served as Chairman of the M.I.T. Faculty during SYs 1960-61 and 1961-62.

Immediately following his first professorial appointment in 1946, Hurley assumed responsibility for the several subjects dealing with <u>Mineral Deposits</u> (12.343, 12.40, 12.41, and 12.432--12.434) that had been taught previously by Prof. W. H. Newhouse, and continued to offer them until the mid-1960s when the curriculum of Course XII was revised with the coming of Prof. Frank Press as the new head. During the years when we were trying to get a program in geophysics underway without a professor in the subject, <u>i.e.</u> in 1952-1957, Hurley offered a beginning subject

Hurley making an adjustment on the gas train for argon analysis (early 1950s).
(Photo by Jackman, 1952)

The Hurley-Fairbairn Isotope Geology and Geochronology Laboratory (54-1117), equipped and supported by grants from the U.S. Atomic Energy Commission (1953-1971).
(Photo by M.I.T. Graphic Arts)

Introduction to Geophysics (12.87) for interested undergraduates. Early on (1954) he also organized a lecture subject in Isotope Geology (12.89) and he continues to offer both lectures and laboratory research (12.418) in this important discipline, in which his students learn the fundamental knowledge needed to conduct research in his Geochronology Laboratory. From 1954 to 1967 he shared with H. W. Fairbairn a two-term Mineral Deposits and Petrology Seminar (12.161, 12.162), and for a short period, 1960-1963, he offered the subject Geological Evolution of the Earth (12.841, 12.842). Since 1965 he has also offered lectures on the Chemistry of the Earth (12.352) that couple closely with the subjects in petrology given by others in the Department.

Hurley brought to his instruction on mineral deposits a thorough training in physical chemistry, mineralogy, and petrology, and an extensive practical experience gained from field work on a variety of mineral deposits. In these respects he followed ably in the footsteps of his mentors, Waldemar Lindgren and Walter H. Newhouse. He only gave up formal instruction in this aspect of geology when the emphasis in Course XII shifted to other earth science disciplines. Moving with the times, he replaced his subjects dealing with mineral deposits by ones dealing with broader geochemical and tectonic aspects of the earth (e.g. Geological Evolution of the Earth (12.841, 12.842) and Chemistry of the Earth (12.352, 12.40)), and today is regarded as one of the leading proponents of plate tectonics, as will be mentioned again farther on in discussing his research work.

As a Research Scientist and Director of Research

Hurley's research for his doctoral thesis, and his prompt publication of his results, foretold the career that he would follow and in which he would win worldwide distinction - the application of radioactivity to fundamental geological problems. Perusal of his extensive Bibliography farther on will show the variety of problems that he has investigated and discussed. He can well be called a global geochemist who has used the principles of nuclear physics and physical chemistry to explain the behavior of the earth's crust and mantle through its multi-billion year history.

Hurley's earliest articles dealt with the numerous aspects of the helium age method for determining the age of rocks; later, during the 1950s, he turned his attention to variations in isotopic abundances of strontium, calcium, and argon, and for the next two decades this general line of research was conducted in his Geochronology Laboratory, first in the basement of Building 24, then on the 11th floor of the Green Build-

784   GEOLOGY AT M.I.T.   1865-1965

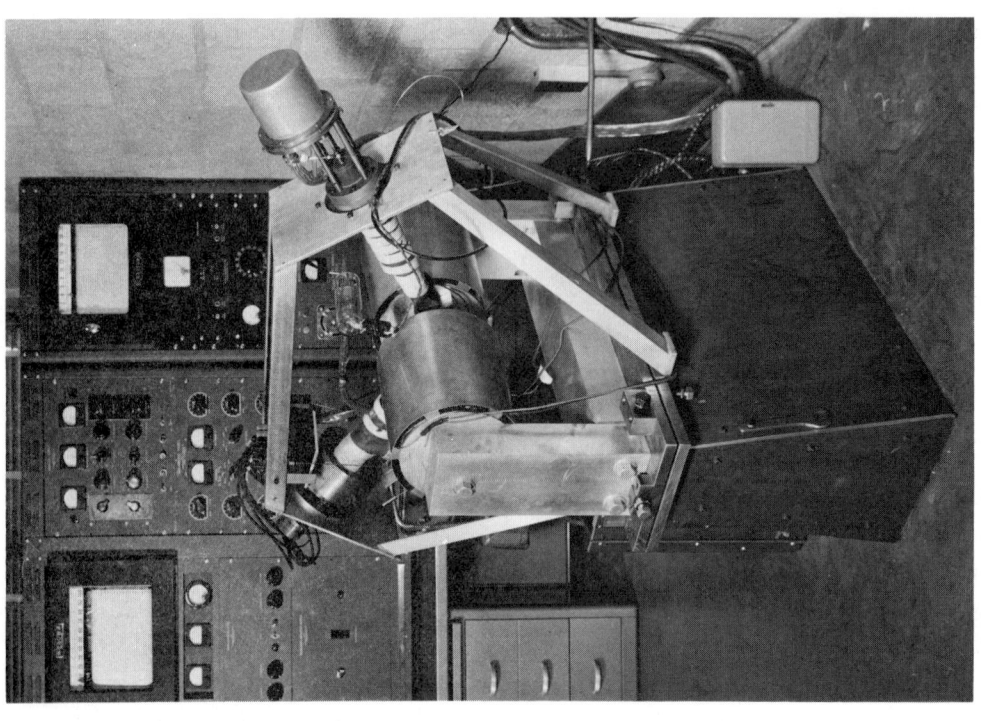

Two views of the first solid-source mass spectrometer built in Hurley's laboratory for Rb-Sr radiometric dating. The spectrometer was designed by graduate student L. F. Herzog II (Ph.D. XII 1952). (Photos from M.I.T. Historical Collections)

ing (54). The main purpose of the research involved the measurement of radioactive-radiogenic pairs of isotopes, and was conducted by the use of mass spectrometers. In the earlier days these were "home-built," and indeed one graduate student (L. F. Herzog II, XII Ph.D. 1952) from the laboratory started the successful Nuclide Corporation (State College, Pa.) which has specialized in the commercial construction of these instruments. The research was directed toward measuring the geologic age of minerals and rocks, and the use of the changing ratio of Sr^{87}/Sr^{86} in the earth in a study of the origin of rock types, and of the overall chemical differentiation history of the earth. The program was at first limited to the study of materials from the United States and Canada, but as it progressed samples were gathered from all the continents, through the kind cooperation of foreign scientists. Hurley, himself, as well as his professorial colleagues, Fairbairn and Pinson, also went abroad to collect rocks for study.

Inasmuch as the age determinations made in the M.I.T. Geochronology Laboratory were of critical importance in testing the new concepts of sea-floor spreading and plate tectonics, Hurley and his co-workers were soon deeply involved in the ever-increasing revolution that has fundamentally changed the idea of how the earth's crust and mantle have behaved through geologic time.

The wide interest and great importance in knowing the ages of non-fossiliferous rocks is demonstrated by the fact that Hurley's little paperback book entitled How Old is the Earth? (Garden City, N.Y.: Doubleday & Co., Inc. - Anchor Books [Science Study Series S5, 160 p., il., 1959]) has been translated into more than a dozen foreign languages.

Hurley's first laboratory for geologic age determination was made possible by an $8,000 grant from the Geological Society of America in 1947. Once the laboratory was operational further funds were provided by the G.S.A. and the O.N.R. until 1954, when major funding began to be provided by the A.E.C. For the next 17 years, 1954-1971, the laboratory was funded by U.S.A.E.C. Contract AT(30-1)-1381, and the 19 annual reports and the final report are listed directly following the Bibliography farther on. All the more important findings discussed in the annual reports were later published in more detailed form in the usual journals and are included in the bibliographies of Hurley, Fairbairn, and Pinson. Throughout the period of this AEC-supported program Hurley directed the laboratory and had the major administrative responsibility for the research program.

Since 1971 financial support for the laboratory has come from the National Science Foundation.

Prof. Hurley holds in his right hand rock specimens from Africa representing two sides of a boundary that separates provinces of 2 billion and 500 million years in age, respectively. In his left hand the same boundary is demonstrated by a pair of rocks found in northeastern Brazil at the expected location. These specimens lead to the conclusion that Africa and South America once fitted together as part of a single supercontinent but have drifted apart since the younger rocks were formed 500 million years ago. (See H. J. Sanders, Chemistry and the Solid Earth, Sp. Repts. Chem. & Eng. News, Am. Chem. Soc., 50A P., 1967.)

(Photo by Robert Lyon)

As Geological and Educational Consultant

Ever since his undergraduate days, Hurley has accepted limited assignments as a geological consultant, and these have involved a wide variety of natural resource problems throughout North America. Typical have been supervising an airborne magnetic survey of Maine (see bibliographic items 16--1949d and 19--1950b) to locate magnetite-bearing asbestos deposits; evaluating fluorite deposits in Colorado and titaniferous coastal sands in Georgia; reporting on gold prospects in British Columbia and salt deposits in Mexico; and investigating manganese deposits in Cuba while dodging a then unknown guerrilla leader named Fidel Castro.

In reply to a recent query from me, Hurley listed what consulting and exploration assignments he could remember; the list shows that he has worked in more than 20 different countries, on every continent except Antarctica, in seven of the nine Canadian provinces, and in 16 of the 50 states of the United States.

As a result of his experience as Chairman of the M.I.T. Faculty and as a member of numerous Institute committees, and as an officer or committee member in a number of professional societies, Hurley has been called on from time to time to aid educational institutions and their departments with such problems as science curricula (see reference 171--1968c in Bibliography), departmental structure, new degree programs, and research financing. A recent assignment took him to England on a National Environmental Research Council committee reviewing research grants in isotope geology for all of the British universities.

As A Public Lecturer

Hurley, an accomplished speaker, has long been in great demand as a public lecturer; earlier in his career as a discussant of the application of radiometric age dating to geological problems, and more recently as one of the leading American proponents of plate tectonics.

In addition to presenting many papers at regularly scheduled meetings of different earth science societies, Hurley has given numerous invited lectures at special conferences, symposia and other similar meetings, and has often reported on his research to college audiences throughout North America.

As A Participant at National and International Conferences and Symposia on Geochronology and Plate Tectonics

As one of the world's leaders in geochronology and plate tectonics, Hurley has participated actively in some 35 national and international conferences, symposia, and other special meetings devoted to these broad areas of earth history and dynamics. A few examples will suffice to show how he has brought the research work of his laboratory, and of the scientists working there, to his colleagues around the world.

1) International Conference on Applied Nuclear Physics. M.I.T., 1940. (The first conference of its kind ever held was organized by Prof. Robley D. Evans of M.I.T.'s Department of Physics.)

2) Conference on Nuclear Processes in Geologic Settings, cosponsored by N.R.C. and N.A.S. and the University of Chicago, in Williams Bay, Wisconsin, 21-23 Sept. 1953. (Hurley was a member of the Subcommittee on Nuclear Geophysics, but did not give a paper.)

3) Conference on Nuclear Processes in Geologic Settings, held at Pennsylvania State University on 8-10 Sept. 1955. (Hurley was a member of the NAS-NRC Committee on Nuclear Science and chairman of the Subcommittee on Nuclear Geophysics; he did not read a paper.)

4) Informal Conference on Nuclear Physics, held at M.I.T. on 13-15 June 1957 by Subcommittee of Nuclear Physics, N.A.S.-N.R.C. (See NAS-NRC Pub. 572, 1958.)

5) Conference on Geochronology of Rock Systems, held by the New York Academy of Science on 3-5 March 1960. (See references 89--1961b, 90--1961c, and 91--1961d.)

6) Symposium on Radioactive Dating, held by the International Atomic Energy Agency in cooperation with the joint Committee on Applied Radioactivity (ICSU) in Athens, 19-23 Nov. 1962. (See reference 126--1963m in Bibliography.)

7) Primera Conferencia Interamericana de Radioquimica, held at Montevideo, Uruguay in 1963, under the auspices of the Pan-American Union. (See reference 151--1965j in Bibliography.)

8) Colloquium on the Geochronology of Phanerozoic Orogenic Belts, held in Bern, Switzerland, 1969. (See reference 189--1970 in Bibliography.)

MARRIAGE AND FAMILY

Hurley married Margaret McCurda on 9 August 1941. He first met her while working in Robley Evans' radioactivity laboratory where she was Evans' secretary. The Hurleys have three children - David Mason, now married and living in New Hampshire; Peter Mason, a graduate of M.I.T.'s School of Business Management (XV S.B. 1968), now married and employed by Digital Equipment Corporation (Maynard, Mass.), and Pamela Mason (B.S. in Chemistry, Univ. New Hampshire, 1971), a recent doctorate in toxicology in M.I.T.'s Department of Nutrition and Food Science.

Pat and Peg Hurley, as they are fondly known to their host of friends everywhere, have been especially hospitable to both students and staff of the Geology Department through the years, and their social affairs have added a special grace to departmental activities. They have been particularly solicitous of foreign students and visiting scientists, by entertaining them in their Lexington home, and have done much to help them adapt to living in the New England area. At M.I.T. Mrs. Hurley has served on numerous committees and has participated actively in the social events at the Institute.

SUMMATION

Patrick M. Hurley has been one of the most imaginative, innovative, and productive members of M.I.T.'s Geology Department. In addition to work in his own special fields, he helped greatly in starting instruction and research in geophysics, and later on in geochemistry. Outside the Department he has served on or chaired half a dozen Institute committees, and for two school years (1960-61 and 1961-62) he was Chairman of the Faculty. Nationally and internationally, he has participated actively in the affairs of numerous professional societies, and has reported regularly on his research work in public lectures.

He early grasped the potential use of stable isotopes in dating minerals and rocks, and in gaining a deeper insight into many complicated geological processes. As a result he developed his world-famous Geochronology Laboratory which has made a major impact on some of geology's most challenging and perplexing problems.

In a carpool conversation with mathematics professor George P. Wadsworth, he sensed that Wiener's time-series analysis, that had been used in weather prediction during World War II, might help the exploration geophysicist retrieve desired information from seismic records that seemed to be too noisy to be useful. Together with Wadsworth and Bryan he organized the M.I.T. Geophysical Analysis Group (GAG), whose research later revolutionized seismic exploration for petroleum reservoirs. (See the discussion of the Geophysical Analysis Group in Volume 2.)

When at the turn of the 1950s he heard rumors that the Newmont Mining Corporation had developed an electrical method for delineating orebodies, he suggested to several of his students that they try to determine the nature of the secret technique (the now-familiar "IP method"), suggested how they might go about experimentation, and then found them the money to fund their laboratory and field experiments. (See biography of Theodore Richard Madden.) In the early 1950s, when we were trying to get started a program of instruction and research in geophysics, Hurley organized an elementary lecture course to help undergraduates get started in that field.

Turning from mere age-dating of minerals and rocks, in which he became a world leader by addressing himself effectively to many of the criticisms that questioned the validity of current techniques and measurements, he took up serious consideration of some of the major questions about how the earth, as a radioactively heated, chemically differentiating system, has evolved through time.

Using the principles of physics, chemistry and geology that he mastered early as a student at the University of British Columbia and M.I.T., and adding to them constantly by continuing to be a student as well as a teacher and investigator; leaning on his extensive and diversified experience as a practicing field geologist; and depending on the impressive expertise he has accumulated from the development and management of the array of mass spectrometers and other measuring devices in his Geochronology Laboratory; Hurley brings an unusually impressive and effective combination of skills, experience, knowledge, and imagination to bear on such current forefront problems as plate tectonics, seafloor spreading, continental displacements, and the thermodynamic history of the earth.

The marks of an outstanding teacher were quite apparent even in his student years, when he started his academic career as a Teaching Fellow - clear explanations delivered in a pleasant and well-modulated voice that showed the evidence of excellent language training; imaginative and well-conceived laboratory exercises; good blackboard work in the form of relevant diagrams; and excellent leadership in field work. As a consequence, Hurley is regarded as one of Course XII's best teachers, and outside M.I.T. he has gained an enviable reputation as an outstanding public lecturer. It is not an overstatement to say that his public lectures, delivered in many places in the United States and Canada, must now total several scores in number.

One aspect of Hurley's stature as a teacher is indicated by the number of thesis investigations he had supervised through 1965 - 32 Bachelor's, 4 Master's, and 18 Doctor's, for a total of 54, the third largest after Whitehead (109) and Lindgren (64), and with Newhouse (18) second to Fairbairn (19) with 18 Doctor's.

A second measure of his rapport with Course XII students is to be found in the impressive number of publications of which he has shared authorship with one or more of his students - 50 in number!

In research Hurley displays three distinctive qualities not always possessed by the productive scientific investigator. First, he is an imaginative and innovative scientist who freely shares his ideas with others; second, he quickly establishes rapport with an interested peer or student and by his enthusiasm and logic gets them to start an inquiry either with him jointly or on their own, always, however, being willing to supervise and otherwise assist if need be; and finally, he is an effective and productive investigator on his own. His skill at managing research projects is manifest by the 17-year support he has had from the Atomic Energy Commission, and his productivity is demonstrated by the many reports and articles he has published alone and with others, as listed in his Bibliography to be found on following pages.

Within the Department, Hurley has served in many important ways in addition to his teaching and research activities. He was Executive Officer from 1950 to 1961; he served on numerous committees involved with departmental problems - admissions, curriculum, examinations and the like; he edited the papers contributed to the International Conference on the Earth Sciences held at M.I.T. on 30 September - 2 October 1964 on the occasion of the dedication of the new Cecil and Ida Green Building; and he has guided the Geochronology Laboratory since its inception in 1947.

Outside the Department, but within the Institute, he has contributed more than a fair share of service to special committees and as Chairman of the Faculty during school years 1960-61 and 1961-62. Earlier he was Chairman of the Committee on the First Two Years that was set up by the Lewis Committee. This Committee was active for two years or more and led to the formation of the Committee on Undergraduate Policy, mentioned a little farther on.

While Chairman of the Faculty he was an ex officio member of the Committee on Graduate School Policy (SYs 1960-61 and 1961-62), and the Committee on Undergraduate Policy (SY 1960-61). When the latter changed its name to the Committee on Educational Policy, he acted as Chairman during SY 1961-62. Later he served as Chairman of the 1970 Search Committee for a new M.I.T. President, and similarly for the ad hoc Committee to evaluate the proposal for a doctoral program in Philosophy. He was Chairman of The Student Activities Development Board during SY 1968-69, and was a member of the Student Aid Board for a three-year period (1970-1973). He was President of the M.I.T. Faculty Club during SY 1958-59.

Beyond the boundaries of Boston and Cambridge Hurley has given scores of public lectures, served on advisory committees on academic matters, and held office in several professional organizations: e.g. President of the Tectonophysics Section of the American Geophysical Union (1959,'60 and '61), and of the Geochemical Society of America (1966,'67). He also served on the editorial and advisory boards of Precambrian Research, Earth and Planetary Science Letters, and McGraw-Hill's Earth Sciences Paper Back Series.

From his academic training and practical experience came a number of closely related skills that Hurley has developed in his own work, as professor and research scientist, and has insisted that his advanced students develop - drawing neat and informative maps, charts, and other diagrams; constructing relevant tables and curves to emphasize the significance of data; preparing abstracts, briefs, and longer articles that report tersely and clearly the results of research; and finally learning how to present those results before scientific audiences in the most convincing and understandable manner within a fixed time limit. As a consequence of his insistence on the preceding, Hurley's students are well known for the good quality of their presentations at scientific meetings and for the high quality of their published articles.

Hurley holds or has held membership in the American Academy of Arts and Sciences, American Association for the Advancement of Science (fellow), American Geophysical Union (fellow), American Institute of Mining and Metallurgical Engineers, Geochemical Society, Geological Society of America (fellow), and Society of Economic Geology.

He became a naturalized citizen in 1945, and served the United States as a civilian scientist in World War II. More recently he had a NASA contract for work on certain lunar samples.

Prof. Hurley adjusts equipment which he uses to determine the age of minerals and rocks by measuring their radioactivity.
(Photo from M.I.T. Historical Collections)

BIBLIOGRAPHY OF PATRICK MASON HURLEY

Symbols and abbreviations used in the following references are explained on pages 91 - 98; in general, abstracts are listed separately, and a few are included with the references to the complete article. This bibliography begins with the title of Hurley's Ph.D. thesis in 1940, and includes all known titles through 1970.

T--1940 Investigations on the helium method of age determination, 6 + 128 p., il., (1940). (Ph.D. Thesis at M.I.T. in Course XII, December 1940.)

1--1940a (and Goodman, C. and Evans, R. D.) Further investigations of the helium method of age determination [abst.]. Roy. Soc. Canada, Pr. (3) 34: 162, (1940).

2--1940b (and Goodman, C.) Investigation of helium in common rock minerals for the purpose of age determination [abst.]. Geol. Soc. Am., B. 51: 1909, (1940).

3--1940c (and Goodman, C.) Helium retention in common rock minerals [abst.]. Geol. Soc. Am., B. 51: 1913, (1940); B. 52: 545-559, (1941).

4--1941 (and Goodman, C.) Proposed helium time scale [abst.]. Geol. Soc. Am., B. 52: 1909, (1941).

5--1941a (and Goodman, C.) Helium age measurements; 1. Preliminary magnetite index [abst.]. Roy. Soc. Canada, Pr. (3) 35: 191, (1941); Geol. Soc. Am., B. 54: 305-323, (1943).

6--1941b Helium retention in common rock minerals. J. Appl. Phys. 12: 300, (1941).

7--1941c (and Goodman, C.) Helium retention in common rock minerals. Geol. Soc. Am., B. 52: 545-559, (1941). (See also 3--1940c.)

8--1941d (and Goodman, C. and Evans, R. D.) Helium age measurements [abst.]. Phys. Rev. 59: 920, (1941).

9--1943 (and Goodman, C.) Helium age measurements; 1. Preliminary magnetite index. Geol. Soc. Am., B. 54: 305-323, (1943). (See also 5--1941a.)

10--1947 (with Nogami, H. H.) Experimental test of predicted absorption of alpha-rays in minerals [abst.]. Geol. Soc. Am., B. 58: 1214, (1947).

11--1948 (with Nogami, H. H.) The absorption factor in counting alpha rays from thick mineral sources. Am. Geophys. Union, Tr. 29: 335-340, (1948).

12--1949 Age of Canada's principal gold-producing belt. Science 110: 49-50, (1949).

13--1949a [prepared by Hurley, P. M.] Geology - piecing continents together. Time 90/7: 36, (18 Aug. 1967).

14--1949b Radioactivity and time. Scientific American 181/2: 48-51, (1949).

15--1949c Some problems of age measurements on eastern North American magnetites, in Marble, J. P., chmn., Report of the Committee of Measurement of Geologic Time, 1943-49: NRC-Div. Geol. Geogr. [Ann. Rept.] 1948-49, Exhibit E, p. 79-82 (‡), (1949).

16--1949d Airborne magnetic survey in Maine. Eng. M. J. 150: 52-55, il., (1949); summary with map, in World Oil 130: 66-67, (1950).

17--1950 Distribution of radioactivity in granites and possible relation to helium age measurements. Geol. Soc. Am., B. 61: 1-8, (1950).

18--1950a   Appendix A - Progress report on age measurements. (Rept. Special Committee on Geophysical and Geological Study of the Continents.) Am. Geophys. Union, Tr. 31: 142-144, (1950).

19--1950b   (and Thompson, J. B., Jr.) Airborne magnetometer and geological reconnaissance survey in northwestern Maine. Geol. Soc. Am., B. 61: 835-841, il., (1950).

20--1950c   Progress report on geologic time measurement [abst. in Nat. Acad. Sci. meeting]. Science 112: 453, (1950).

21--1950d   Progress report to the Committee on Measurement of Geologic Time, in Marble, J. P., chmn., Rept. Comm. Measurement of Geologic Time, 1949-1950. NRC-Div. Geol. Geogr. [Ann. Rept.], Exhibit A, p. 25-28 (‡), (1950).

22--1951   Formation of continents. M.I.T. Repts. on Research 2/3: 3-4, (Jan. 1951).

23--1951a   On the origin of continents. Tech. Eng. News 32/5: 8, 22, (1951).

24--1951b   Radioactivity and the origin of continents. 5 p. (‡), il., (1951). [Privately printed and circulated.]

25--1951c   Alpha ionization damage as a cause of low helium ratios - technical report. [13] p. (‡), il., (1951). Cambridge: Dept. Geol., M.I.T.

26--1951d   Heat flow and chemical segregation of the mantle [abst.]. Am. Geophys. Union, Tr. 32: 328-329, (1951).

27--1952   Alpha ionization damage as a cause of low helium ratios. Am. Geophys. Union, Tr. 33: 174-183, (1952).

28--1952a   (and Shorey, R. R.) Discrimination of thoron alpha activity in presence of radon. Am. Geophys. Union, Tr. 33: 722-724, (1952).

29--1952b   (and Fairbairn, H. W.) Alpha-radiation damage in zircon. J. Appl. Phys. 23: 1408, (1952).

30--1952c   Heat production in basalts and their origin [abst.]. Geol. Soc. Am., B. 63: 1265-1266, (1952); Am. Mineral. 38: 345, (1953).

31--1952d   (and Fairbairn, H. W.) Radiation damage in zircon: a possible age method [abst.]. Geol. Soc. Am., B. 63: 1266, (1952); B. 64: 659-673, il., (1953).

32--1953   (with Backus, M. M. and Stetson, H. C.) Relationships of radioactive elements in three cores from the Gulf of Mexico [abst.]. Geol. Soc. Am., B. 64: 1391, (1953).

33--1953a   (with Wadsworth, G. P. et al.) Detection of reflections on seismic records by linear operators. Geophysics 18: 539-586, (1953).

34--1954   Geologic time measurements. M.I.T. Repts. on Res. 5: 2, (1954).

35--1954a   Geologic time measurements. Tech. Rev. 57: 392, (1954).

36--1954b   Relationships of radioactive elements in three cores from the Gulf of Mexico [abst.]. Geophysics 19: 356, (1954).

37--1954c   (with Herzog, L. F. and Pinson, W. H.) Preliminary survey, Pt. I of Isotopic variations in strontium [abst.]. Am. Geophys. Union, Tr. 35: 380, (1954).

38--1954d   (with Herzog, L. F. et al.) Variations in isotopic abundances of strontium, calcium, and argon and related topics--Annual progress report for 1953-54, Pt. 2, Annual research progress report: U.S. Atomic Energy Comm. Rept. NYO-3934 (pt. 2), [163] p., il., (1954). (Report prepared for the A.E.C. by M.I.T.)

39--1954e   Use of the gamma-ray scintillation spectrometer in the separate measurement of the uranium and thorium series in geological materials [abst.]. Econ. Geol. 49: 802, (1954); Geol. Soc. Am., B. 65: 1265-1266, (1954).

40--1954f   The helium age method and the distribution and migration of helium in rocks, in Nuclear Geology - a symposium on nuclear phenomena in the earth sciences, (H. Faul, ed., et al.), New York: John Wiley & Sons, p. 301-329, (1954).

41--1954g   (and Larsen, E. S., Jr. and Gottfried, D.) Comparison of radiogenic helium and lead in zircon. Trace Elements Investigations Rept. 475, U.S. Geol. Surv., 12 p., (1954).

42--1955   Direct radiometric measurement of uranium and thorium series in equilibrium by gamma-ray scintillation spectrometer. Trace Elements Investigations Rept. 499, U.S. Geol. Surv., 30 p., (1955).

43--1955a   Scintillation spectrometer II: Simultaneous measurement of uranium, thorium, and potassium in common rocks [abst.]. Am. Geophys. Union, Program of Ann. Meeting 1955: 37, (1955).

44--1955b   Discussion: [New geologic time scale.] Geol. Assoc. Canada, Pr. 7: 125-126, (1955).

45--1955c   [Review of] Isotope Geology (K. Rankama, author), New York: McGraw-Hill Book Co., Inc., 535 p., (1955). Nucleonics 13/6: 115-117, (June 1955).

46--1955d   (and Fairbairn, H. W.) Ratio of thorium to uranium in zircon, sphene and apatite [abst.]. Geol. Soc. Am., B. 66: 1578, (1955).

47--1955e   (with Webber, G. R. and Fairbairn, H. W.) Relative ages of eastern Massachusetts granites by total lead ratios in zircon [abst.]. Geol. Soc. Am., B. 66: 1632, (1955). (See also 54--1956d.)

48--1955f   Gamma-ray spectrometry: U.S. Geol. Surv. Rept. T.E.I. 590: 314-315, (1955). (Report prepared for A.E.C. by U.S.G.S.)

49--1955g   (as collaborator) The Earth is Born, in The World We Live In, Part 1: 1-18. New York: Time Inc., (1955).

50--1956   (and Larsen, E. S., Jr. and Gottfried, D.) Comparison of radiogenic helium and lead in zircon. Geochim. Cosmochim. Acta 9: 98-102, (1956).

51--1956a   Direct radiometric measurement by gamma-ray scintillation spectrometer - Pt. 1, Uranium and thorium in equilibrium; Pt. 2, Uranium, thorium, and potassium in common rocks. Geol. Soc. Am., B. 67: 395-411, (1956).

52--1956b   (with Fairbairn, H. W.) Radiation damage in zircon from eastern Massachusetts and Nova Scotia [abst.]. Am. Geophys. Union, Tr. 37: 344, (1956).

53--1956c   (and Pinson, W. H., Jr.) Variations in radioactive elements between mafic rock provinces [abst.]. Am. Geophys. Union, Tr. 37: 350, (1956).

54--1956d   (with Webber, G. R. and Fairbairn, H. W.) Relative ages of eastern Massachusetts granites by total lead ratios in zircon. Am. J. Sci. 254: 574-583, (1956).

55--1956e   (with Cormier, R. F. et al.) Rubidium-strontium age determinations on the mineral glauconite [abst.]. Geol. Soc. Am., B. 67: 1681-1682, (1956).

56--1957   (with Fairbairn, H. W.) Radiation damage in zircon and its relation to ages of Paleozoic igneous rocks in northern New England and adjacent Canada. Am. Geophys. Union, Tr. 38: 99-107, (1957).

57--1957a   Test on the possible chondritic composition of the Earth's mantle and its abundance of uranium, thorium and potassium. Geol. Soc. Am., B. 68: 379-382, (1957).

58--1957b   (and others) Comparison of A^{40}/K^{40} and Sr^{87}/Sr^{86} ages on biotite [abst.]. Am. Geophys. Union, Tr. 38: 396, (1957).

59--1957c   (and Fairbairn, H. W.) Abundance and distribution of uranium and thorium in zircon, sphene, apatite, epidote, and monazite in granitic rocks. Am. Geophys. Union, Tr. 38: 939-944, (1957).

60--1957d   (with Fairbairn, H. W. et al.) Age of Nova Scotia granites [abst.]. Geol. Soc. Am., B. 68: 1725, (1957).

61--1957e   (with Pinson, W. H., Jr. et al.) Age study of some crystalline rocks of the Georgia Piedmont [abst.]. Geol. Soc. Am., B. 68: 1781, (1957).

62--1958   Progress report on argon analysis: U.S. Atomic Energy Comm. Rept. NYO-3938: 17-25, il., (1958). (Report prepared for A.E.C. by M.I.T.)

63--1958a   Collection of glauconite from known stratigraphic horizons for the dating of the geologic time scale: U.S. Atomic Energy Comm. Rept. NYO-3938: 132-166, (1958). (Report prepared for A.E.C. by M.I.T.)

64--1958b   (with Bullwinkel, H. J. et al.) Age investigation of syenites from Coldwell, Ontario [abst.]. Geol. Soc. Am., B. 69: 1543-1544, (1958).

65--1958c   (and Fairbairn, H. W. and Pinson, W. H., Jr.) Intrusive and metamorphic rock ages in Maine and surrounding areas [abst.]. Geol. Soc. Am., B. 69: 1591, (1958).

66--1959   (and others) Minimum age of the Lower Devonian slate near Jackman, Maine. Geol. Soc. Am., B. 70: 947-949, (1959).

67--1959a   How old is the earth? [1st ed.]. Garden City, N.Y.: Anchor Books, 160 p., (1959). [Translated into 13 different foreign languages.]

68--1959b   (and others) Age study of Lower Paleozoic glauconites [abst.]. J. Geophys. Res. 64: 1109, (1959).

69--1959c   (with Allen, V. T. et al.) Age of Precambrian igneous rocks of Missouri [abst.]. Geol. Soc. Am., B. 70: 1560-1561, (1959).

70--1959d   (with Fairbairn, H. W. and Pinson, W. H., Jr.) Rb-Sr feldspar ages in granitic rocks of Sudbury-Blind River, Ontario, Canada [abst.]. Geol. Soc. Am., B. 70: 1599-1600, (1959).

71--1959e   (and others) Authigenic versus detrital illite in sediments [abst.]. Geol. Soc. Am., B. 70: 1622, (1959).

72--1959f   (with Moore, J. M., Jr. et al.) Potassium-argon ages in northern Manitoba, Canada [abst.]. Geol. Soc. Am., B. 70: 1647-1648, (1959).

73--1959g   (with Pinson, W. H., Jr. et al.) Three ages of rock crystallization in Colombia [abst.]. Geol. Soc. Am., B. 70: 1656, (1959).

74--1959h   (with Fairbairn, H. W. et al.) Age investigation of syenites from Coldwell, Ontario. Geol. Assoc. Can., Pr. 11: 141-144, (1959).

75--1959i   (with Tupper, W. M. and Jensen, M. L.) The genesis of the sulfide deposits of northern New Brunswick - an interpretation based on sulfur isotopic studies [abst.]. Min. Eng. 11: 1230-1231, (1959).

76--1959j [Review of] Geology of the manganese deposits of Cuba (Simonds, F. S. and Straczek, J. A.). Econ. Geol. 54: 753-757, (1959).

77--1960 (with Moore, T. M., Jr. et al.) Potassium-argon ages in northern Manitoba. Geol. Soc. Am., B. 71: 225, (1960).

78--1960a (with Herzog, L. F., 2d. and Pinson, W. H., Jr.) Rb-Sr analyses and age determinations of certain lepidolites, including an international interlaboratory comparison suite. Am. J. Sci. 258: 191-208, il., (1960).

79--1960b (with Fairbairn, H. W. et al.) Age of the granitic rocks of Nova Scotia. Geol. Soc. Am., B. 71: 399-413, il., (1960).

80--1960c (with Fairbairn, H. W. and Pinson, W. H., Jr.) Comparison of Rb-Sr mineral and whole-rock ages at Sudbury, Ontario [abst.]. J. Geophys. Res. 65: 2488-2489, (1960).

81--1960d (and others) Reliability of glauconite for age measurement by K-Ar and Rb-Sr methods. Am. Assoc. Petroleum Geol., B. 44: 1793-1808, (1960).

82--1960e (with Fairbairn, H. W. and Pinson, W. H., Jr.) Mineral and rock ages at Sudbury-Blind River, Ontario. Geol. Assoc. Can., Pr. 12: 41-66, (1960).

83--1960f (with Hart, S. R. et al.) Use of amphiboles and pyroxenes for K-Ar dating [abst.]. Geol. Soc. Am., B. 71: 1882, (1960).

84--1960g (and others) K-Ar and Rb-Sr minimum ages for the Pennsylvanian section in the Narragansett Basin. Geochim. Cosmochim. Acta 18: 247-258, (1960).

85--1960h (with Fairbairn, H. W. et al.) A comparison of the ages of coexisting biotite and muscovite in some Paleozoic granite rocks. Geochim. Cosmochim. Acta 19: 7-9, (1960).

86--1960i (with Schürmann, H. M. E. et al.) Fourth preliminary note on age determinations of magmatic rocks by means of radioactivity. Geol. en Mijnbouw 39: 93-104, (1960).

87--1961 (with Herz, N. et al.) Age measurements from a part of the Brazilian Shield. Geol. Soc. Am., B. 72: 1111-1120, (1961).

88--1961a (with Fairbairn, H. W. and Pinson, W. H., Jr.) The relation of discordant Rb-Sr mineral and whole-rock ages in igneous rock to its time of crystallization and to the time of subsequent Sr^{87}/Sr^{86} metamorphism. Geochim. Cosmochim. Acta 23: 135-144, (1961).

89--1961b Glauconite as a possible means of measuring the age of sediments, in Geochronology of rock systems. N.Y. Acad. Sci., An. 91: 294-297, (1961).

90--1961c The northern Appalachians, in Geochronology of rock systems. N.Y. Acad. Sci., An. 91: 397-399, (1961).

91--1961d The basement of Central and South America, or, how not to date a continent, in Geochronology of rock systems. N.Y. Acad. Sci., An. 91: 571-575, (1961).

92--1961e (with Faure, G.) The ratio Sr^{87}/Sr^{86} in oceanic and continental basalts [abst.]. J. Geophys. Res. 66: 2527, (1961).

93--1961f The five-billion year clock. The Saturday Evening Post 234/11: 99-100, (March 18, 1961).

94--1961g (and others) Geochronology of Proterozoic granites in Northern Territory, Australia, Part I: K-Ar and Sb-Sr age determinations. Geol. Soc. Am., B. 72: 653-662, (1961).

95--1961h (and others) K-Ar age studies of Mississippi and other river sediments. Geol. Soc. Am., B. 72: 1807-1810, (1961).

96--1961i   (and Hughes, H. and Pinson, W. H., Jr.) Argon diffusion coefficients in micas at low temperatures obtained from Alpine Fault Uplift in New Zealand [abst.]. J. Geophys. Res. 66: 2538, (1961).

97--1962   (with Fairbairn, H. W. et al.) Evidence of the origin and time of separation of magmas of the Monteregian Hills, Quebec, from development of radiogenic Sr^{87} [abst.]. Geol. Soc. Am., Sp.P. 68: 174, (1962).

98--1962a   (with Hower, J. et al.) Effect of mineralogy on K/Ar age as a function of particle size in a shale [abst.]. Geol. Soc. Am., Sp.P. 68: 201-202, (1962).

99--1962b   (and others) K-Ar age values on the clay fractions in shales ranging in age from Tertiary to Ordovician [abst.]. Geol. Soc. Am., Sp.P. 68: 203-204, (1962).

100--1962c (and others) Unmetamorphosed minerals in the Gunflint formation used to test the age of the Animikie. J. Geol. 70: 489-492, (1962).

101--1962d (with Beall, G. H. et al.) Comparison of K-Ar and whole-rock Rb-Sr dating in New Quebec and Labrador [abst.]. J. Geophys. Res. 67: 3541, (1962).

102--1962e (with Faure, G. et al.) Isotopic composition of strontium in continental basic intrusives [abst.]. J. Geophys. Res. 67: 3556-3557, (1962).

103--1962f (and others) Radiogenic strontium 87 model of continent formation [abst.]. J. Geophys. Res. 67: 3567-3568, (1962).

104--1962g (and Hughes, H. et al.) Radiogenic Sr^{87} model of continent formation. J. Geophys. Res. 67: 5315-5334, (1962).

105--1962h (with Moorbath, S. et al.) Evidence for the origin of mineralized Tertiary intrusives in Southwestern States from strontium isotope ratios [abst.]. J. Geophys. Res. 67: 3582, (1962).

106--1962i (with Powell, J. L. et al.) Sr^{87}/Sr^{86} evidence bearing on the genesis of carbonatite [abst.]. J. Geophys. Res. 67: 3588-3589, (1962).

107--1962j (with Powell, J. L. and Fairbairn, H. W.) Isotopic composition of strontium in carbonatites. Nature 196: 1085-1086, (1962).

108--1962k (and others) Radiogenic argon and strontium diffusion parameters in biotite at low temperatures obtained from Alpine Fault Uplift in New Zealand. Geochim. Cosmochim. Acta 26: 67-80, (1962).

109--1962ℓ (as consultant to Hyde, R. W. et al.) Iron ore resources of the world. Eng. M. J. 163: 84-88, (1962).

110--1962m (with Pinson, W. H.,Jr. et al.) K-Ar and Rb-Sr ages of biotites from Colombia, South America. Geol. Soc. Am., B. 73: 907-910, (1962).

111--1962n (with Bailey, S. W. et al.) K-Ar dating of sedimentary illite polytypes. Geol. Soc. Am., B. 73: 1167-1170, (1962).

112--1962o The radioactive earth, in Study of the Earth, Readings in Geological Sciences, (J. F. White, ed.). Englewood Cliffs, N.J.: Prentice-Hall, p. 73-78, (1962).

113--1963   (with Bottino, M. L. et al.) Rb-Sr age study of the Lower Devonian volcanic sequence at Kineo, Maine [abst.]. Geol. Soc. Am., Sp.P. 73: 121, (1963).

114--1963a (with Brookins, D. G. et al.) Whole-rock Rb-Sr investigations of the Collins Hill, Maromas, and Glastonbury formations at Collins Hill, Connecticut [abst.]. Geol. Soc. Am., Sp.P. 73: 123, (1963).

115--1963b (with Faure, G. et al.) Estimate of the isotopic composition of strontium in rocks of the Precambrian basement, Canada [abst.]. Geol. Soc. Am., Sp.P. 73: 150-151, (1963).

116--1963c (with Pinson, W. H. et al.) Evidence on the origin of felsic volcanic rocks from their initial abundance of Sr^{87} [abst.]. Geol. Soc. Am., Sp.P. 73: 216, (1963).

117--1963d (with Faure, G.) The isotopic composition of strontium in oceanic and continental basalts - Application to the origin of igneous rocks. J. Petrol. 4: 31-50, (1963).

118--1963e (with Faure, G. et al.) Whole-rock Rb-Sr age of norite and micropegmatite at Sudbury, Ontario [abst.]. Am. Geophys. Union, Tr. 44: 110-111, (1963). (See also 139-1964h.)

119--1963f (with Bottino, M. L. et al.) Whole-rock Rb-Sr ages of some Paleozoic volcanics and related granites in the northern Appalachians [abst.]. Am. Geophys. Union, Tr. 44: 111, (1963).

120--1963g (and Pinson, W. H., Jr. and Fairbairn, H. W.) Progress report on analytical accuracy of Sr^{87}/Sr^{86} measurement [abst.]. Am. Geophys. Union, Tr. 44: 111-112, (1963).

121--1963h (and others) K-Ar age values on the clay fractions in dated shales. Geochim. Cosmochim. Acta 27: 279-284, (1963).

122--1963i (and others) K-Ar age values in pelagic sediments of the North Atlantic. Geochim. Cosmochim. Acta 27: 393-399, (1963).

123--1963j (with Faure, G. et al.) Age of the Great Dyke of Southern Rhodesia. Nature 200: 769-770, (1963).

124--1963k The evolution of continents. Tech. Rev. 66: 28-31, (1963).

125--1963ℓ (and others) Evidence of continuing separation of sial from the mantle from the isotopic composition of common strontium. Nuclear Geophysics, Nuclear Science Series Rept. 38, NAS-NRC Pub. 1075: 83-92, (1963).

126--1963m (and others) New approaches to geochronology by strontium isotope variations in whole rocks, in Radioactive Dating, International Atomic Energy Agency in cooperation with the joint Committee on Applied Radioactivity (ICSU), Proc. Symposium held in Athens, 19-23 Nov. 1962; published by Intl. A.E.A. in Vienna, 1963, p. 201-217, (1963).

127--1963n (with Hower, J. et al.) The dependence of K-Ar age on the mineralogy of various particle size ranges in a shale. Geochim. Cosmochim. Acta 27: 405-410, (1963).

128--1963o (with Faure, G. and Fairbairn, H. W.) An estimate of the isotopic composition of strontium in rocks of the Precambrian shield of North America. J. Geophys. Res. 68: 2323-2329, (1963).

129--1963p (with Fairbairn, H. W. et al.) Initial ratio of strontium 87 to strontium 86, whole-rock age, and discordant biotite in the Monteregian igneous province, Quebec. J. Geophys. Res. 68: 6515-6522, il., (1963).

130--1963q (with Beall, G. H. et al.) Comparison of K-Ar and whole-rock Rb-Sr dating in New Quebec and Labrador. Am. J. Sci. 261: 571-580, il., (1963).

131--1964 (with Backus, M. M. et al.) Calcium isotope ratios in the Homestead and Pasamonte meteorites and Devonian limestone. Geochim. Cosmochim. Acta 28: 735-742, (1964).

132--1964a (and others) Preliminary investigation of Sr^{87}/Rb^{87} relationships in the Sierra Nevada plutonic rocks [abst.]. Geol. Soc. Am., Sp.P. 76: 85, (1964).

133--1964b   (with Powell, J. L.) Sr-87/Sr-86 ratios in carbonate rocks of possible igneous origin [abst.]. Geol. Soc. Am., Sp.P. 76: 132, (1964).

134--1964c   (with Fairbairn, H. W. and Pinson, W. H.) Preliminary age study and initial Sr^{87}/Sr^{86} of Nova Scotia granitic rocks by the Rb-Sr whole-rock method. Geol. Soc. Am., B. 75: 253-257, (1964).

135--1964d   (with Faure, G. and Powell, J. L.) The isotopic composition of strontium in surface water from the North Atlantic Ocean [abst.]. Am. Geophys. Union, Tr. 45: 113-114, (1964).

136--1964e   (with Powell, J. L. and Faure, G.) Strontium 87 in a suite of Hawaiian volcanic rocks [abst.]. Am. Geophys. Union, Tr. 45: 114, (1964).

137--1964f   (with Fairbairn, H. W. and Pinson, W. H.) Initial Sr^{87}/Sr^{86} and possible sources of granitic rocks in southern British Columbia. J. Geophys. Res. 69: 4889-4893, (1964).

138--1964g   (and Fairbairn, H. W. and Pinson, W. H., Jr.) Rb-Sr relationships in serpentinite from Mayagüez, Puerto Rico, and dunite from St. Paul's Rocks -- A progress report, in A study of serpentinite. NAS-NRC Pub. 1188: 149-151, (1964).

139--1964h   (with Faure, G. et al.) Whole-rock Rb-Sr age of norite and micropegmatite at Sudbury, Ontario. J. Geol. 72: 848-854, (1964). (See also 118--1963e.)

140--1964i   (with Whitney, P. R.) The problem of inherited radiogenic strontium in sedimentary age determinations. Geochim. Cosmochim. Acta 28: 425-436, (1964).

141--1965   (with Brookins, D. G.) Rb-Sr geochronological investigations in the middle Haddam and Glastonbury quadrangles, eastern Connecticut. Am. J. Sci. 263: 1-16, (1965).

142--1965a   (with Fairbairn, H. W. and Pinson, W. H.) Re-examination of Rb-Sr whole-rock ages at Sudbury, Ontario. Geol. Assoc. Can., Pr. 16: 95-101, (1965).

143--1965b   (with Crocket, J. H. and Faure, G.) Some aspects of the marine and fresh water isotopic geochemistry of strontium [abst.]. Am. Geophys. Union, Tr. 46: 168, (1965).

144--1965c   (with Fairbairn, H. W. et al.) Rb/Sr whole-rock isotopic analyses and the Cambrian-Precambrian problem in southeastern Massachusetts [abst.]. Am. Geophys. Union, Tr. 46: 173, (1965).

145--1965d   (with Roe, G. D. and Pinson, W. H., Jr.) Rb-Sr evidence for the origin of peridotites [abst.]. Am. Geophys. Union, Tr. 46: 186, (1965).

146--1965e   (with Shields, R. M. and Pinson, W. H.) The Rb^{87}/Sr^{87} age of stony meteorites [abst.]. Am. Geophys. Union, Tr. 46: 124, (1965).

147--1965f   (with Faure, G. and Powell, J. L.) The isotopic composition of strontium in surface water from the North Atlantic Ocean. Geochim. Cosmochim. Acta 29: 209-220, (1965).

148--1965g   (with Pinson, W. H., Jr. et al.) Rb-Sr age of stony meteorites. Geochim. Cosmochim. Acta 29: 455-466, (1965).

149--1965h   (and others) Investigation of initial Sr^{87}/Sr^{86} ratios in the Sierra Nevada plutonic province. Geol. Soc. Am., B. 76: 165-174, (1965).

150--1965i   (with Powell, J. L. and Faure, G.) Strontium 87 abundance in a suite of Hawaiian volcanic rocks of varying silica content. J. Geophys. Res. 70: 1509-1513, (1965).

151--1965j (and others) Radioactive decay of Rb^{87} to Sr^{86} in geological science exclusive of age dating. Primera Conferencia Interamericana de Radioquimica, Montevideo. Pan-Am. Union, Washington, p. 175-178, (1965).

152--1966 (and Fairbairn, H. W. and Pinson, W. H., Jr.) Evidence from western Ontario of the isotopic composition of strontium in Archean seas [abst.]. Geol. Soc. Am., Sp.P. 87: 84, (1966).

153--1966a K-Ar dating of sediments, in Potassium Argon Dating (O. A. Schaeffer and J. Zahringer, eds.). New York: Springer-Verlag, p. 134-151, (1966).

154--1966b Advances in Earth Science - Contributions to the International Conference on the Earth Sciences, Massachusetts Institute of Technology, (P. M. Hurley, ed.), (1964). Cambridge, Mass: M.I.T. Press, 502 p., il., (1966).

155--1966c (with Powell, J. L. and Fairbairn, H. W.) The strontium isotopic composition and origin of carbonatites, in Carbonatites (O. F. Tuttle and J. Gittins, eds.). New York: Interscience Publishers, p. 365-378, (1966).

156--1966d (with Fairbairn, H. W. et al.) Whole-rock age and initial $^{87}Sr/^{86}Sr$ of volcanics underlying fossiliferous Lower Cambrian in the Atlantic provinces of Canada. Can. J. Earth Sci. 3: 509-521, (1966); Abst., Geol. Soc. Am., Sp.P. 101: 65, (1968).

157--1966e (and Fairbairn, H. W. and Pinson, W. H., Jr.) Rb-Sr isotopic evidence in the origin of potash-rich lavas of western Italy. Earth Planet. Sci. Lett. 1: 301-306, (1966).

158--1966f (with Schnetzler, C. C. and Pinson, W. H., Jr.) Rubidium-strontium age of the Bosumtwi Crater area, Ghana, compared with the age of the Ivory Coast tektites. Science 151: 817-819, (1966).

159--1966g (with Shields, R. M. and Pinson, W. H., Jr.) Rubidium-strontium analyses of the Bjurböle chondrite. J. Geophys. Res. 71: 2163-2167, (1966).

160--1967 (with Fairbairn, H. W. et al.) Rb-Sr age of granitic rocks of southeastern Massachusetts and the age of the Lower Cambrian at Hoppin Hill. Earth Planet. Sci. Lett. 2: 321-328, (1967).

161--1967a (with Fairbairn, H. W. et al.) Rb-Sr whole-rock age of the Sudbury lopolith and basin sediments [abst.]. Am. Geophys. Union, Tr. 48: 242, (1967); See also Can. J. Earth Sci. 5: 707-714, (1968).

162--1967b (with Bence, A. E.) Rubidium-strontium isotopic relationships in oceanic basalts. Am. Geophys. Union, Tr. 44: 242, (1967).

163--1967c Rb^{87}-Sr^{87} relationships in the differentiation of the mantle, in Ultramafic and Related Rocks (J. P. Wylie, ed.). New York and London: John Wiley & Sons, p. 372-375, (1967).

164--1967d (and others) Test of continental drift by comparison of radiometric ages. Science 157: 495-500, (1967).

165--1967e (with Faure, G. and Crocket, J. H.) Some aspects of the geochemistry of strontium and calcium in the Hudson Bay and the Great Lakes. Geochim. Cosmochim. Acta 31: 451-461, (1967).

166--1967f (and others) Tracing the history of differentiation of the mantle by Rb-Sr isotopic relationships, in Upper Mantle Project: U.S. Progress Rept.: 103, (1967); See also U.S. Program, Final Report [NAS-NRC Pub.]: 214-215, (1971).

167--1967g (with Moorbath, S. and Fairbairn, H. W.) Evidence for the origin and age of some mineralized Laramide intrusives in the southwestern United States from strontium isotope and Rb-Sr measurements. Econ. Geol. 62: 228-236, (1967).

168--1968 (with Fairbairn, H. W. et al.) Whole-rock age and initial Sr^{87}/Sr^{86} of volcanic rocks underlying fossiliferous Lower Cambrian in the Atlantic Provinces of Canada [abst.]. Geol. Soc. Am., Sp.P. 101: 65, (1968).

169--1968a (and others) Rb-Sr whole-rock analyses in northern Brazil correlated with ages in West Africa [abst.]. Geol. Soc. Am., Sp.P. 101: 100-101, (1968).

170--1968b (with Brookins, D. G. et al.) Geochronological aspects of the genesis of large granitic pegmatites in nonigneous environments [abst.]. Geol. Soc. Am., Sp.P. 115: 25, (1968).

171--1968c (with Dehlinger, P. et al.) Guidelines for the establishment of a university program of excellence in the geophysical sciences. Am. Geophys. Union, Tr. 49: 465-468, (1968).

172--1968d (with Fairbairn, H. W. et al.) Rb-Sr whole-rock age of the Sudbury lopolith and basin sediments. Can. J. Earth Sci. 5: 707-714; Abst., Am. Geophys. Union, Tr. 48: 242, (1967).

173--1968e The confirmation of continental drift. Scientific American 218: 52-64, (April 1968).

174--1968f Absolute abundance and distribution of Rb, K and Sr in the earth. Geochim. Cosmochim. Acta 32: 273-283, (1968).

175--1968g Correction to: Absolute abundance and distribution of Rb, K and Sr in the earth. Geochim. Cosmochim. Acta 32: 1025, (1968).

176--1968h (and others) Some orogenic episodes in South America by K-Ar and whole-rock Rb-Sr dating. Can. J. Earth Sci. 5: 633-638, (1968).

177--1968i (and Rand, J. R.) Review of age data in West Africa and South America relative to a test of continental drift, in The History of the Earth's Crust (R. A. Phinney, ed.). Princeton, N.J.: Princeton Univ. Press, p. 153-160, (1968).

178--1968j (with Krogh, T. M.) Strontium isotope variation and whole-rock isochron studies, Grenville Province of Ontario. J. Geophys. Res. 73: 7107-7125, (1968).

179--1969 (and Rand, J. R.) Radiometric age data on two-thirds of the continental areas of the Earth. Geol. Soc. Am., Sp.P. 121: 145-146, (1969).

180--1969a (with Fairbairn, H. W. et al.) Correlation of radiometric ages of Nipissing diabase and Huronian metasediments with Proterozoic orogenic events in Ontario. Can. J. Earth Sci. 6: 489-497, (1969).

181--1969b (with Brookins, D. G. et al.) A Rb-Sr geochronologic study of the pegmatites of the Middletown area, Connecticut. Contr. Mineral. Petrology 22: 157-168, (1969).

182--1969c (and Rand, J. R.) Pre-drift continental nuclei. Science 164: 1229-1242, il., (1969).

183--1969d The first continents. M.I.T. Repts. on Research, p. 1-2, (June 1969).

184--1969e [Review of] Radiometric dating for Geologists (E. I. Hamilton and R. M. Farquhar, eds.). New York: Interscience Pubs., (1969). Geotimes 14/2: 32-34, (Feb. 1969).

185--1969f   Some observations on the geological history of Laurasia. IAVCEI Symposium, Oxford, England, (1969). (See also 190--1970a.)

186--1969g   (and Rand, J. R.) Evidence against dispersal of continental nuclei prior to the last great drift [abst.]. Am. Geophys. Union, Tr. 50: 334, (1969).

187--1969h   (with MacDonald, W. D.) Precambrian gneisses from northern Colombia, South America. Geol. Soc. Am., B. 80: 1867-1871, (1969).

188--1969i   The Primitive Earth, in The Age of the Earth's Crust [A symposium report (‡)]. Oxford, Ohio: Miami Univ., Dept. Geol., 4 p., il., (1969); Hurley's Summarization: Sum I^3 of 3 p. comes at end of report, (1969).

189--1970   (with Fairbairn, H. W.) Northern Appalachian geochronology as a model for interpreting ages in older orogens. [Colloquium on the Geochronology of Phanerozoic Orogenic Belts in Bern, Switzerland, 1969.] Eclogae Geol. Helv. 63: 83-90, (1970).

190--1970a   Distribution of age provinces in Laurasia. Earth Planet. Sci. Lett. 8: 189-196, (1970).

191--1970b   (and Pinson, W. H., Jr.) Rubidium-strontium relations in Tranquillity Base samples. Science 167: 473-474, (1970).

192--1970c   (and Pinson, W. H., Jr.) Whole-rock Rb-Sr isotopic age relationships in Apollo 11 lunar samples, in Apollo Lunar Sci. Conf., Pr. 2: 1311-1315, New York, Pergamon Press, (1970).

193--1970d   (with Rand, J. R.) Bibliography of Precambrian radiometric age dates, through 1969. 20 p. (‡), (Privately published, M.I.T. Graphic Arts), (1970).

194--1970e   (with Zartman, R. E. et al.) A Permian disturbance of K-Ar radiometric ages in New England: Its occurrence and cause. Geol. Soc. Am., B. 81: 3359-3373, (1970).

195--1970f   (with Bottino, M. L. et al.) Blue Hills igneous complex, Massachusetts: Whole-rock Rb-Sr open systems. Geol. Soc. Am., B. 81: 3739-3746, (1970).

The following are items prepared by P. M. Hurley, alone or with colleagues, and published only as so-called "gray literature" in Variations in isotopic abundances of strontium, calcium, argon and related topics, the annual reports made by him to the U.S.A.E.C. under Contract AT(30-1)-1381. These annual reports were prepared in M.I.T.'s Department of Geology and Geophysics, and printed and distributed by the Institute's Graphic Arts. Although, as stated earlier, they constitute "gray literature," they are included in this bibliography because they have been cited in the same way as regularly published articles and also because they indicate a kind of productivity not always known to the scientific community at large.

1) Progress report on argon analysis. 5th Ann. Rept.: 17-25, (1958).

2) Mica samples from the Billiton, Lausitz, and Dara granites. 6th Ann. Rept.: 96-101, (1958).

3) The Paleozoic time scale problem. 6th Ann. Rept.: 107-109, (1958).

4) Effects of hydrogen in argon analysis. 7th Ann. Rept.: 224-228, (1959).

5) Ar^{38} spike release system. 7th Ann. Rept.: 229-231, (1959).

6) Redistribution of radiogenic Sr^{87} between Rb-rich and Rb-poor phases during metamorphism. 8th Ann. Rept.: 225-236, (1960).

7) Observed migration of Sr^{87} in metamorphic rocks in Vermont. 9th Ann. Rept.: 187-192, (1961).

8) New approaches to geochronology by strontium isotope variations in whole rocks. 10th Ann. Rept.: 109-114, (1962).

In addition to the preceding items, Hurley, as supervisor of U.S. A.E.C. Contract AT(30-1)-1381, had the major responsibility for directing and administering the research program, and for preparing the 19 annual progress reports and the final report, all of which are listed below.

First Annual Progress Report for 1954-55	March 25, 1955
Second Annual Progress Report for 1954-55	30 June, 1955
Third Annual Progress Report for 1955-56	March 1, 1956
Fourth Annual Progress Report for 1956-57	March 1, 1957
Fifth Annual Progress Report for 1957-58	March 1, 1958
Sixth Annual Progress Report for 1958	December 1, 1958
Seventh Annual Progress Report for 1959	December 1, 1959
Eighth Annual Progress Report for 1960	December 1, 1960
Ninth Annual Progress Report for 1961	December 1, 1961
Tenth Annual Progress Report for 1962	December 1, 1962
Eleventh Annual Progress Report for 1963	December 1, 1963
Twelfth Annual Progress Report for 1964	December 1, 1964
Thirteenth Annual Progress Report for 1965	December 1, 1965
Fourteenth Annual Progress Report for 1966	December 1, 1966
Fifteenth Annual Progress Report for 1967	December 1, 1967
Sixteenth Annual Progress Report for 1968	December 1, 1968
Seventeenth Annual Progress Report for 1969	December 1, 1969
Eighteenth Annual Progress Report for 1970	December 1, 1970
Nineteenth Annual Progress Report for 1971	1 December 1971
Final Report	March 31, 1972

----

(28)
WILLIAM HENRY DENNEN

WILLIAM HENRY DENNEN

MIT: 1946-1967

Outstanding teacher of mineralogy and physical geology; highly respected senior counselor and registration officer; imaginative thesis advisor and supervisor; innovative director of the Cabot Spectrographic Laboratory and of the M.I.T. Summer School of Geology in Nova Scotia; long-time member of the Institute's Committee on Commencement, as well as member of numerous other Institute and Departmental committees, to all of which he brought unusual organizational skills; efficient Executive Officer of the Department of Geology and Geophysics; author of a well-received introductory textbook of mineralogy and of more than 30 articles on geological subjects; consultant to numerous industrial firms; and retired officer of the U. S. Marine Corps; Professor W. H. Dennen ably served his Department, the Institute, his Country, and the business world from 1946 to 1967 while a member of M.I.T.'s Department of Geology and Geophysics. He resigned his M.I.T. appointment as Associate Professor in 1967 to accept a position as Professor and Head of the Department of Geology at the University of Kentucky. Less than three years were required there for his organizational and administrative skills to be recognized, with the result that he was asked to assume the duties of Acting Dean of the Graduate School in late 1970.

William Henry Dennen was born in Gloucester, Massachusetts, on 8 April 1920, the first of four children of William Llewellyn Dennen and Ruth Louise (Lufkin) Dennen. After William H. came Richard Llewellyn (b. 1922; West Point, 1944), Nathalie (who died at one year of age), and David Warren (b. 1932; VII S.B. 1954).

Both the Dennens and the Lufkins were of Yankee stock, tracing their ancestry back to English forebears who came to the fishing community of Gloucester in the mid-1600s. They were not, however, of the sea-going type; rather, they established themselves in the West Parish of Gloucester as farmers, cobblers, coppersmiths, boat makers, chandlers and sail makers, thus serving the local community in a variety of useful trades. Here Dennen's father drove a delivery wagon for his own father, Henry Parker Dennen, who was an expert on meat, and operated a highly successful butchershop and grocery in Gloucester, and here young William H. was born. His mother, Ruth L. Lufkin, grew up in the same community. After being trained as a teacher at Salem State College, she taught elementary school at nearby Rockport and Rowley, Massachusetts, before being married.

William L. Dennen, the father of our subject, received his S.B. in Mining Engineering (Course III_3, Geology Option) in 1917, before joining the American Expeditionary Force. Upon returning from service he started graduate work and in 1920 earned an S.M. degree in Geology in Course XII. He found early employment (1921) in Mexico (Parral) and Guatemala, doing mine property evaluation, but soon returned (1922) to the United States to take employment in the coal industry. He spent most of his subsequent professional career as a mining engineer with the Hudson Coal Company in Pennsylvania, moving his family to Scranton in 1924 and to Clarks Summit in 1925.

At Clarks Summit, Pennsylvania, young William attended grade school from 1926 to 1934, and high school from 1934 to graduation in 1938. During these youthful years he commonly spent Sunday forenoons accompanying his father on inspection tours of some fifty different coal mines. As a result of seeing his father performing his professional work, young Dennen acquired an excellent knowledge of geology firsthand, and was well equipped for summer employment during his high school years. Because he was a skinny lad he always got the call to go into the tight places when something had to be repaired, and this kind of experience proved valuable later on when he worked underground at the Star Pointer Shaft of the Nevada Consolidated Copper Corporation at Ruth, Nevada. With such a background of experience, and with his father an M.I.T. alumnus and a mining engineer with geological training, it was not surprising that he applied for admission to M.I.T. and, when accepted, chose to work for a degree in geology.

While an undergraduate, Dennen frequently did drafting work for faculty members, with pay at the going rate of 35 cents an hour, and his handiwork is to be seen in Buerger's X-ray Crystallography (1942) and Newhouse's Ore Deposits as Related to Structural Features (1942).

Just as his father had enlisted in the U. S. Army in 1917, after completing requirements for an S.B. degree, so William followed the same course in World War II, being commissioned a 2nd Lieutenant in the U. S. Army in May 1942 upon graduation with an S.B. degree in geology. Transferring from the Army to the U. S. Marine Corps in July 1942, he served in that branch of the Service in the Pacific Theatre, then remained in the Active Reserve as Captain (April 1944-Aug. 1950), Major (Aug. 1950-

Dec. 1954), and Lt. Col. (Dec. 1954-Sept. 1964) until retirement in 1964. His duties in the Marine Corps included the following:

> Instructor, AA Artillery School, Camp Le Jeune, N.C., 1943-1944
>
> C.O., C Battery, 18th AA Bn, FMF Pac, Tinian
>
> Asst. G-3, Mariannas AA Artillery Group, Guam
>
> Asst. G-3, Fleet Marine Force, Pacific, Pearl Harbor

After discharge at the end of World War II, Dennen started graduate work in M.I.T.'s Department of Geology in the fall of 1946. With an outstanding undergraduate record--he had held the AIME Women's Auxiliary Scholarship in his senior year--and with teaching experience in the Marine Corps, he was a logical candidate for a Teaching Fellowship when he started graduate work in September 1946. Thus he combined teaching duties with study and research for two years, SYs 1946-1947 and 1947-1948, working under F. K. Morris, W. J. Mead, and H. W. Fairbairn, after which he served as a Research Assistant for SY 1948-1949, receiving his Ph.D. in geology in June 1949.

Having done his Bachelor's thesis on the Dracut [Massachusetts] norite intrusive under W. L. Whitehead, Dennen followed his interest in igneous geochemistry by choosing to investigate variations in chemical composition across igneous contacts for his Ph.D. thesis (see Bibliography). This research, carried out in the Cabot Spectrographic Laboratory under the supervision of Prof. H. W. Fairbairn and Research Associate L. H. Ahrens, got Dennen well acquainted with spectrographic analysis and prepared him for later direction of the same Laboratory.

Starting his teaching career in 1946, while still a graduate student, he progressed through the usual successive stages of Teaching Fellow (1946-1948), Research Assistant (1948-1949), Instructor (1949-1952), and Assistant Professor (1952-1957), to Associate Professor (1957-1967), the rank he held when he resigned to accept an appointment as Professor and Head of the Department of Geology at the University of Kentucky.

Dennen's services to the Department of Geology and Geophysics were many and varied; he was always ready and willing when asked to take on a new assignment. He served as Executive Officer of the Department from the fall of 1961 to the end of 1966, during which he assumed a wide variety of departmental duties including undergraduate thesis advisor, guidance counselor and advisor to Course XII seniors, departmental representative in the Boston Mineral Club and on the Sigma Xi selection committee.

He assisted in directing the M.I.T. Summer School of Geology in Nova Scotia during the summer of 1955, when Prof. R. D. Parks was on leave and

Prof. W. L. Whitehead became ill, and he assumed full responsibility for the 1961 session, the last to be held in Nova Scotia.

A noteworthy innovation during the 1961 session was student investigation of the tidal estuary in front of the Crystal Cliffs property--the first oceanographic research to be done in this part of the Province. Dennen and his students had to start from scratch so far as oceanographic research equipment was concerned. They managed to scrounge a leaky rowboat, borrow an alidade and stadia rod from the Department of Mines, and improvise several measuring devices from materials at hand. A hollow metal broomstick with plunger served as a bottom sampler. A wooden broomstick, properly weighted with a rock-filled metal can nailed to one end, and surmounted by a graduated pole with an attached flag, became a current meter. A graduated pole, equipped with a crude floating device which both damped the waves and responded to the tidal fluctuations, was driven into the soft bottom to serve as a tidal gauge. To equip the gauge the two ends of an ordinary tin can were cut out and the resulting cylinder then made the central element of a rectangular wooden frame. This crude float, with the cylinder impaled on the graduated pole, slid up and down the pole as the tide ebbed and flowed, meanwhile damping the waves and allowing a reasonably good determination of tide level. Although the actual research accomplished was not spectacular, the students who had to improvise their own measuring devices will probably not soon forget that summer.

Dennen was a superb teacher when measured by the characteristics generally cited as required for good teaching. He presented the subject matter in a clear and well-organized fashion, in an enthusiastic and pleasantly modulated voice, and with blackboard work that quickly demonstrated his drawing skill. His students came away from his lectures with a well-organized, logically developed body of theoretical and practical knowledge, and many in later years were able to recall especially interesting items from his courses.

Inasmuch as the beginning courses in physical geology and mineralogy are among the most important that any geology undergraduate student takes, because they provide a kind of frame of reference or a set of boundary conditions for all subsequent study and research in the geosciences, the wise Department Head will assign these important subjects to his best teacher. It was for this reason that Dennen was assigned the introductory courses in Mineralogy and Physical Geology. From time to time he also conducted lecture-laboratory courses in Spectrochemical Analysis, Photogeology, Geochemistry, Geological Surveying, and Exploration Geology.

Good as he was in formal classes, though, Dennen was always at his best in his laboratory, where, with a few students around, and a pot of coffee nearby, he carried on that informal and stimulating kind of teacher-student discussion that so often results in arousing a genuine interest in learning and in research. It was no accident, therefore, that during the 13 years when he was Director of the Cabot Spectrographic Laboratory, that departmental facility was one of the most productive laboratories in the Department, both in respect to training students as spectrographic technicians and providing a facility with which they could do their thesis work. A more detailed and extended account of Dennen's activities in the Cabot Laboratory is included in the chapter on the Godfrey Lowell Cabot Spectrographic Laboratory in Volume 2.

As guidance counselor and thesis advisor for Course XII seniors, Dennen played an important role in the closing phases of many geology students' college education by helping them find and then supervising an investigation that developed into their thesis. During his 18 years on the M.I.T. Faculty, from 1949 to 1967, he supervised 29 Bachelor's, 7 Master's, and 6 Doctor's theses, most of which were done in the Cabot Spectrographic Laboratory while he served as Director. Following are the names of the graduates; the titles of their theses are given in the chapter on Theses and on The Cabot Spectrographic Laboratory. (See my Volume 2.)

S.B.				S.M.	
Adey, W. H.	(1955)	Mielke, J. E.	(1962)	Anderson, P. J.	(1961)
Aitken, A. R.	(1956)	Murray, B. C.	(1953)	Jokela, A. W.	(1965)
Bauschatz, P. C.	(1957)	Phillips, J. J.	(1958)	Nalbandian, M.R.	(1966)
Bruneau, R. E.	(1967)	Reed, D. E.	(1964)	Quesada, A. L.	(1965)
Chng, K. M.	(1965)	Rose, E. R.	(1959)	Ritter, C. J.	(1962)
Essene, E. J.	(1961)	Sax, R. L.	(1952)	Rose, E. R.	(1959)
Fowler, W. C.	(1953)	Small, J., Jr.	(1952)	Zidle, T.	(1967)
Fritts, J. J.	(1952)	Trask, N.J., Jr.	(1953)		
Kimball, P. P.	(1956)	Webber, D. A.	(1960)	Sc.D.	
Koster, H.	(1959)	Wones, D. R.	(1954)	James, A. H.	(1954)
Landy, R. A.	(1953)	Worsh, F. M.	(1956)		
Linder, H. W.	(1958)	Yut, L.	(1954)	Ph.D.	
McCarl, H. N.	(1962)			Hagen, J. C.	(1954)
Bisson, W. J.	(XXI-B, 1960)			Perry, E. C.	(1963)
Byrd, W. E., Jr.	(XXI-A, 1963)			Short, N. M.	(1958)
Moody, M. M.	(VIII, 1961)			Webber, G. R.	(1955)
Rocchio, R. A.	(VIII, 1961)			Blackburn, W.H.	(1967)

Dennen's services to the Institute were varied and numerous; following is a list of the more important of these:

1953-55	Member, Undergraduate Course Committee
1953-56	Cooperating Staff Member of Admissions Office, Secondary School Visiting Program
1955	Member, Faculty Club Membership Committee
1955-57	Local Representative, Alumni Club of Northeastern Pennsylvania
1956-61	Member, Committee on Commencement; Chairman 1958-61
1960	Member, Centennial Committee
1962-65	Member, Committee on Alumni Seminars
1962-65	Member, Committee on Discipline
1963-64	Member, Committee for Dedication of the Green Building
1966	Member, Committee for Inauguration of Howard W. Johnson, President
1966	Treasurer of M.I.T. Faculty Club

He took an interest in local affairs, serving as Assistant District Commissioner for the Lexington [Mass.] Area of the Boy Scouts of America (1959-1962), President of the Nahant [Mass.] P.T.A. (1963) and member of the Nahant [Mass.] Conservation Commission (1964-1967). In 1949 he was Instructor at Tufts College; in 1961 he served on a Visiting Committee for Geology at Carleton University (Ottawa, Canada); and for the first $4\frac{1}{2}$ months of 1967, was a Visiting Professor of Geology at the Universidad Central of Caracas, Venezuela.

Field work and consulting assignments gave him a great range of geological experience and took him to many widely separated areas. Following is a list of the more important of these activities.

1947	Areal geology for Quebec Geological Survey as an Assistant Party Chief
1949-50	Engineering geology studies in Alaska and Yukon Territory for Aluminum Company of America as Party Chief; clay investigations in Tennessee and South Carolina
1950	Geophysical work on overburden and gypsum thickness for Victoria Gypsum Co. on Cape Breton Island, Nova Scotia
1951	Evaluation of Colorado fluorite deposits for Reynolds Aluminum Co.
1952	Evaluation of natural gas potential on lands of 12 major coal producers in northeastern Pennsylvania for ANGAS Corp. as Party Chief
1953	Review of limestone resources of New England for Federal Reserve Bank of Boston
1954	Geochemical exploration in New Brunswick and the Gaspé Peninsula for Selection Trust Ltd.

1956-60	Geochemical and geophysical exploration and follow-up drilling to evaluate the mineral potential of lands in Maine for Scott Paper Co., as Exploration Manager
1962	Study of potash deposits for Nova Scotia Research Foundation
1962-65	Geochemical reconnaissance of large areas in Maine for U. S. Geological Survey, as Project Chief (WAE)
1965	Evaluation of mercury properties in Spain, coal lands in Tennessee, gold deposits in Venezuela, and tantalum deposits in North America, as a consultant for Geoscience Inc.
1945-67	Spectrochemical and geochemical consultant to numerous individuals and organizations; mostly free but a few for a fee

After leaving M.I.T. in August of 1967, Dennen continued his active and successful professional career at the University of Kentucky as Professor of Geology, Head of the Department of Geology, and Acting Dean of the Graduate School.

On 20 December 1942 Dennen married Charlotte Davidson, whose ancestry goes back to Scotch forebears in Nova Scotia. Their children are William Sheldon (b. 11 February 1950), Peter Davidson (b. 16 Oct. 1951), and Susan Olding (b. 12 July 1957). All are living in Lexington, Kentucky, as this is being written (April 1971).

BIBLIOGRAPHY OF WILLIAM HENRY DENNEN

Symbols and abbreviations used in the following references are explained on p. 91 - 98; in general abstracts are listed separately as well as with the references to the complete article. This bibliography begins with 2 thesis titles (T--1942, etc.), continues with his first publication in 1943, and includes all known titles through 1967, the year he left M.I.T.

T--1942	The Dracut norite intrusive, 41 p., (1942). (S.B. Thesis at M.I.T. in Course XII, August 1942.) (Also see item 1--1943.)
T--1949	Spectrographic investigation of major element variations across igneous contacts, 106 + 13 p., (June 1949). (Ph.D. Thesis at M.I.T. in Course XII, June 1949.) (Also see item 3--1951a.)
1--1943	A nickel deposit near Dracut, Massachusetts. Econ. Geol. 38: 25-55, il., (1943). [Based on S.B. thesis; see preceding T--1942.]
2--1951	(and Ahrens, L. H. and Fairbairn, H. W.) Spectrochemical analysis of major constituent elements in rocks and minerals [R.I. and Va.], in Fairbairn, H. W., A cooperative investigation of precision and accuracy in chemical, spectrochemical, and modal analysis of silicate rocks. U. S. Geol. Surv., B. 980: 25-52, il., (1951).

3--1951a   Variations in chemical composition across igneous contacts. Geol. Soc. Am., B. 62: 547-557, il., (1951). [Based on Ph.D. thesis; see preceding T--1949.]

4--1954   Laboratory mapping. J. Geol. Educ. 2: 75-76, il., (1954).

5--1955   (and Fowler, W. C.) Spectrographic analysis by use of mutual standard method. Geol. Soc. Am., B. 66: 655-662, il., (1955).

6--1956   (and Shields, R.) Yttria in zircon. Am. Mineral. 41: 655-657, il., (1956).

7--1957   Spectrographic determination of carbon in sedimentary rocks, using direct-current arc excitation. Spectrochim. Acta 9: 89-97, il., (1957).

8--1957a   Illustration of periodic properties. J. Geol. Educ. 5: 22-23, il., (1957).

9--1958   The direct current arc in spectrochemical analysis of sands, soils, ores. Florida Seminars on Spectroscopy 1955-57, B. 100: 53-63, (1958).

10--1959   Principles of Mineralogy. New York: Ronald Press Co., 453 p., il., (1959). Revised printing with determinative tables, (1960).

11--1959a (and Linder, H.) Relationship of carbon in soils to graphitic zones [abst.]. Min. Eng. 11: 1232, (1959).

12--1961   Mineralogy: 2000 word entry in the Encyclopedic Dictionary of Physics, J. Thewlis, Editor-in-Chief. London: The Pergamon Press, (1961).

13--1962   (and Linder, H.) Relationship of graphite in soils to graphitic zones. A.I.M.E., Tr. 223: 252-254, (1962).

14--1962a [Book Review of] The External Characters of Minerals by A. G. Warner [translated by A. V. Carozzi]. Science 137: 419, (1962).

15--1962b (with Bisson, W. J.) Newton and spectral lines. Science 135: 921-922, (1962).

16--1962c (and Anderson, P. J.) Chemical changes in incipient rock weathering. Geol. Soc. Am., B. 73: 375-384, il., (1962).

17--1962d (with James, A. H.) Trace ferrides in the magnetic ores of the Mount Hope mine and the New Jersey Highlands. Econ. Geol. 57: 439-449, il., (1962).

18--1963   (and Sankaran, A. V.) A method for matrix correction in the spectrographic determination of uranium. Appl. Spectroscopy 17: 44-47, (1963).

19--1963a Movement of the Morristown Horst, Cape George, Nova Scotia. Geol. Assoc. Canada, Pr. 15 [pt. 1]: 87-90, il., (1963).

20--1963b (with Post, E. V. et al.) Mineral investigations field studies map MF-278. U. S. Geol. Surv., (1963).

21--1964   (with Post, E. V.) Geochemical mapping in Maine [abst.]. Min. Eng. 16: 67, (1964).

22--1964a (with Canney, F. C. and Post, E. V.) Some observations on the evaluation of stream sediment geochemical data [abst.]. Min. Eng. 16: 92, (1964).

23--1964b Impurities in quartz. Geol. Soc. Am., B. 75: 241-245, il., (1964).

24--1964c (with Ritter, C. J.) Color center zonation in quartz. Geol. Soc. Am., B. 75: 915-916, il., (1964).

25--1964d (with Van Sickle, G. and Post, E. V.) Heavy metals in stream sediments, southeastern Maine. U. S. Geol. Surv., Open File Report (MA), (1964).

26--1964e A brief pre-history of Nahant [Mass.]. Rept. Conserv. Comm., Nahant Ann. Rept., (1964).

27--1964f (with Brookins, D. G.) Trace element variation across some igneous contacts. Kan. Ac. Sci. 67: 70-91, (1964).

28--1966 (with Ritter, C. J.) Blackening of natural quartz by [gamma] y-irradiation. Am. Mineral. 51: 220-227, il., (1966).

29--1966a Stoichiometric substitution in natural quartz. Geochim. Cosmochim. Acta 30: 1235-1241, (1966).

30--1967 Trace elements in quartz as indicators of provenance. Geol. Soc. Am., B. 78: 125-130, il., (1967).

31--1967a (with Quesada, A.) Spectrochemical determination of water in minerals and rocks. Appl. Spectroscopy 21: 155-156, il., (1967).

32--1967b (with Post, E. V. et al.) A map of southeastern Maine showing heavy metals in stream sediments. U.S.G.S. Min. Inv. Field Studies Map MF-301, scale 1:250,000, text, (1967).

33--1967c Mineralogy. A 2000 word article in the World Book Encyclopedia, (1967).

NOTE:

Dennen's research since leaving M.I.T. has continued to yield publications; these can be found in the usual abstracts and bibliographies of North American Geology.

Prof. Dennen lecturing to a mineralogy class in Building 24 (Room 24-405).

(Photo by Jackman, 1952)

Prof. Dennen speaking at the 1963 M.I.T. Alumni Seminar.
(Photo from M.I.T. Historical Collections)

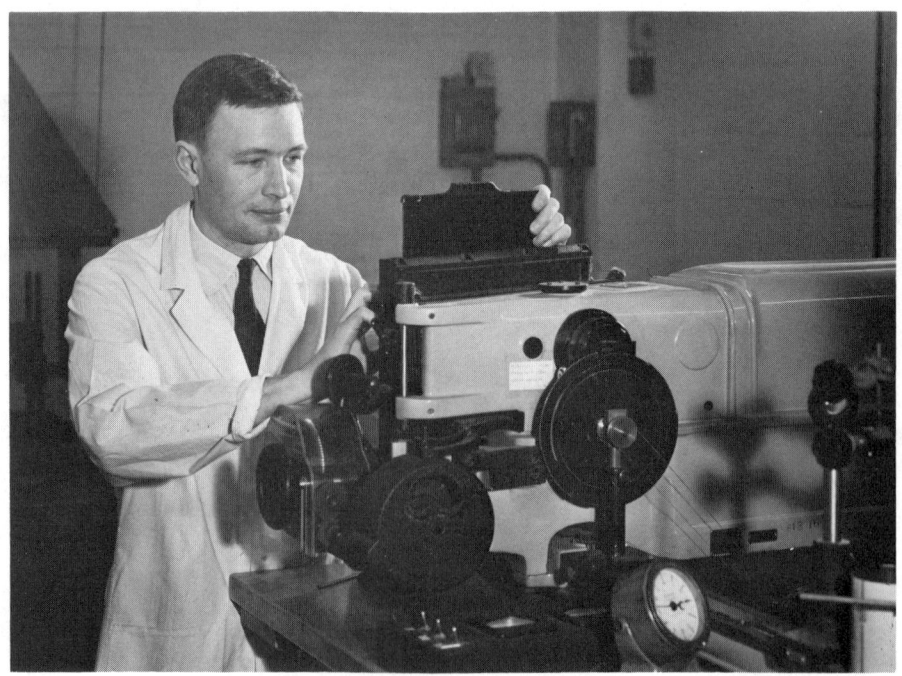

Prof. Ahrens preparing to use the Hilger spectrometer in the Godfrey Cabot Laboratory in Building 24 (Room 24-014).
(Photo by Jackman, 1952)

(29)
LOUIS HERMAN AHRENS

LOUIS HERMAN AHRENS

MIT: 1948-1953; 1962-1963

In September 1946 Louis H. Ahrens, a brilliant young South African analytical chemist from the University of Pretoria, came to M.I.T. to carry on spectrochemical research in our Cabot Spectrographic Laboratory. For two years, 1946-1948, he was a Post-Doctoral Fellow of the South Africa Council for Scientific and Industrial Research, then a Research Associate in the Department of Geology and Geophysics for another two-year period, 1948-1950. In July 1950 he was appointed an Assistant Professor of Geochemistry and served in that capacity for the next three and one-half years until he resigned as of 31 December 1953 to accept a Readership in Mineralogy at England's Oxford University. Ten years later he returned to M.I.T., from the University of Cape Town where he had gone from Oxford in 1956, as Visiting Professor of Geochemistry for most of SY 1962-63. During the time of his two appointments he lectured on geochemistry and conducted a vigorous program of spectrochemical analysis in the Cabot Spectrographic Laboratory which resulted in a dozen student theses, more than 15 published articles, and three books. Since leaving M.I.T. in 1953 Ahrens has achieved international renown for his contributions to theoretical and experimental geochemistry as applied to geological problems and today holds first rank among the world's geochemists.* A recently updated list of his many publications through 1964 constitutes the Bibliography at the end of this biographical sketch.

Louis Herman Ahrens was born in Pietermaritzburg, Natal, Union of South Africa on 24 April 1918. His father was a District Magistrate who had to travel widely over Southwest Africa and the Union of South Africa, so young Louis had to get his pre-college education pretty much on the run. In the light of Ahrens' distinguished career in geochemistry, it is interesting to read in the Sunday Tribune (Durban, South Africa), of 19 April 1953 that

> "According to his father, Mr. F. W. Ahrens, who lives in Maritzburg, Louis never shone at school and

---

* Ahrens was cited as "Spectroscopist-of-the-Month" in Arcs & Sparks (p. 4-5, vol. 12, December 1967), published by the Ultra Carbon Corporation of Bay City, Michigan. The citation is in the form of a brief statement of his professional training and achievements.

> 'only just' matriculated. In fact Mr. Ahrens (senior),
> whose two elder sons had both gained scholarships to
> university, was so sure that young Louis had no aca-
> demic future that he put him into an office job. But
> Louis begged to be allowed to go on with his studies."

Go on he did, and in 1939 he received a B.Sc. degree in Chemistry and Geology from the University of Natal. Soon after (1940) he was awarded the Captain Scott Memorial Medal for the best pass in the final year (B.Sc.) Geology examination with respect to all constituent Colleges of the then University of South Africa.

His professional career began in 1940 as an analytical chemist at the Government Metallurgical Laboratory (now the National Institute of Metallurgy) in Johannesburg. He also served as officer in charge of the Analytical Section and from 1945 to 1946 he was Senior Analyst. During these same years he carried on graduate work at the University of Pretoria, and in 1944 received a D.Sc. degree in Chemistry from that institution.

Having become interested in spectrochemical analysis during his graduate work (his doctoral degree was actually in this special field, i.e. spectrochemical analysis), he sought and was awarded the first Post-Doctoral Research Fellowship by the Council for Scientific and Industrial Research of South Africa to do research abroad. Leaving home in 1946 he went first to the United Kingdom and during a two-month stay visited several laboratories (Cambridge, Oxford and Durham Universities and Macaulay Institute, Aberdeen), then came to Cambridge, Massachusetts and took up research in M.I.T.'s Cabot Spectrographic Laboratory. When his South African Fellowship expired in 1948 he continued on as an M.I.T. Research Associate for two years until 1950 when he was appointed Assistant Professor of Geochemistry in the Department of Geology and Geophysics.

As a faculty member he not only taught undergraduate and graduate subjects in geochemistry and spectrochemistry, but also was made Director of the Cabot Spectrographic Laboratory and assigned major responsibility for directing an ongoing research project supported by federal funds. Although Ahrens remained on our faculty for only three and a half years, from 1 July 1950 through 31 December 1953, he and the students he attracted brought international recognition to the Cabot Spectrographic Laboratory by their thesis research and publications. The activities of the Laboratory during Ahrens' years are described more fully in the Chapter on The Godfrey Lowell Cabot Spectrographic Laboratory. Back home, in South Africa, Ahrens was not forgotten, as he received the Jubilee Gold Medal of the Geological Society of South Africa in 1948, in recognition of his research on the strontium method for age determination, and the Commonwealth Degree F.R.I.C. (Fellow of the Royal Institute of Chemistry), for achievement in analytical chemistry. In 1953, before leaving the United States, he received the Mineralogical Society of America Award* (See following page).

While at M.I.T. Ahrens' interests ranged over a broad spectrum of geochemical and geological problems that were suitable for study by precision techniques available with the spectrographs in the Cabot Laboratory. Included in this list of problems were the following:

1) Development of precision techniques by which the spectrograph could be used more widely and effectively on geological problems.
2) Precise measurements of not only minor or trace elements but also of common or major elements.
3) Determination of the age of rocks by the Sr/Rb method.
4) Determination of the relative abundance and distribution of elements in the earth's crust, in soils, and in deep-sea sediments.
5) Trace elements in meteorites.
6) Geochemistry of rarer elements--gallium, fluorine, etc.
7) Decay rate of radioactive elements.

As evidence of the activity in the Cabot Laboratory, and the intensity of his own research, during his rather brief stay in our Department, Ahrens supervised 10 student theses--1 Bachelor's, 1 Master's and 8 Doctor's, published a dozen articles, and produced three books--<u>Spectrochemical Analysis</u> (1950); <u>Wavelength Tables of Sensitive Lines</u> (1951); and <u>Quantitative Spectrochemical Analysis of Silicates</u> (1954). One of the most important of his articles of this period was "The lognormal distribution of the elements (1) (A fundamental law of geochemistry and its subsidiary)," (Geochim. Cosmochim. Acta 5: 49-73, 1954), published shortly after he left M.I.T., but written earlier, which presents the evidence for what is now referred to as "Ahrens' Law." Equally important, however, were two earlier papers: "Measuring geological time by the strontium method," (Geol. Soc. Am., B. 60: 217-266, 1949) and "The use of ionization potentials, Part 2, Anion affinity and geochemistry," (Geochim. Cosmochim. Acta 3: 1-29, 1953).

Ahrens left M.I.T. early in 1954 to accept an appointment as Reader in Mineralogy at famed Oxford University in England. He was not yet fully settled when an offer came from his homeland that he could not reject-- Professor of Inorganic Chemistry at the University of Cape Town--and so he returned to South Africa in 1956, after but two years at Oxford, and in

---

* The citation for this Award is so informative regarding Ahrens' achievements that it is considered worth reproduction here, with permission of the <u>American Mineralogist</u> (Vol. 30: 300-303, port., 1954).

PRESENTATION OF THE MINERALOGICAL SOCIETY
OF AMERICA AWARD TO LOUIS H. AHRENS*

Esper S. Larsen, Jr., *U. S. Geological Survey, Washington, D. C.*

Mr. President, Ladies, and Gentlemen:

It is a pleasure to a teacher to watch young men and women grow in their science and become the leaders of their group and the promise of the coming generation. Though I have never been a teacher to Dr. Ahrens, I have had the pleasure of seeing him grow from a promising young geochemist to a leader in his field. Dr. Ahrens was trained in South Africa and he received his Doctor of Science degree in chemistry. He early concentrated in geochemistry and especially on the application of spectroscopy to the solution of problems in geochemistry.

He has studied the distribution of some of the rare elements in rocks and minerals, such as the association of thallium and rubidium. He has contributed much to the problem of determining the age of rocks, using chiefly the ratios of strontium to rubidium, and this method of age determination was largely developed by him.

His two papers on ionization potential will prove useful to all geochemists, and his two books on spectrochemical analysis and wavelength tables of sensitive lines are much used by all spectroscopists, especially those working with rocks and minerals.

Much of Dr. Ahren's work has been carried on in South Africa and at the Massachusetts Institute of Technology. He is soon moving to Oxford, England, as reader in Mineralogy. We are sorry to see him leave America, but wish him a happy and successful time in England and shall follow his work with pleasure and profit. We still claim some part of his success!

The Mineralogical Society of America Award is made to Dr. Ahrens because of his contributions to the measurement of geological time and for the particular paper "Measurements of geologic time by the strontium method" (*Geol. Soc. Am. Bull.*, **60**, 217–266, 1949). When he began his work on the strontium method many geologists thought that the wide distribution of strontium in rocks and minerals would make the method impractical but Dr. Ahrens has shown that lepidolite and some other minerals can be used for age determinations by the strontium method. At first he was obliged to make his determinations without correction for primary strontium but later he has been able to secure isotopic data on the strontium and so to correct for primary strontium. He has lately developed the method so that he can use biotite for age determinations. This makes the method applicable to many more rocks.

In the past eight years Dr. Ahrens has contributed a long list of excellent papers, chiefly on the subjects of geochemistry and spectroscopy. He is a steady worker, intensely interested in his problems, and he finishes his work in published papers. We can look with confidence to the future and expect many contributions from him on the determinations of geologic time and the distribution of the elements in rocks and minerals and other problems in geochemistry.

Mr. President, I have the honor and the pleasure of presenting Dr. Louis H. Ahrens for the Third Mineralogical Society of America Award.

* Ahrens, Dr. Louis Herman, Dept. of Geology and Geophysics, Mass. Inst. Tech. (from Jan. 1, 1954; Dept. of Geology and Mineralogy, Oxford University). Born in Pietermaritzburg, Natal, Union of South Africa, April 24, 1918. B. Sc. (Geology and Chemistry) Univ. Natal, 1939; D. Sc. (Chemistry) Univ. Pretoria, 1944. Analytical Chemist, Gov't Met. Lab., Johannesburg, 1940–1945; Senior Chemist 1946. Post Doc. Fellowship, S. Afr. Council Sci. Ind. Res., 1947–48. Res. Assoc., M. I. T. 1949–1950; Assis't. Prof. 1950–1953. Fellow Roy. Soc. Chem. Gt. Brit. and Ireland, Geol. Soc. Am., Mineral. Soc. Am., Geol. Soc. S. Afr. mem., S. Afr. Chem. Inst. and Fin. Geol. Soc. Jubilee Gold Medal (1948), Geol. Soc. S. Afr. Spectrochemical analysis, geological age, geochemistry and cosmochemistry.

1960 became Professor of Geochemistry, and Head of the Department of Geochemistry. Although he continues to hold those positions today (1974), he returned to M.I.T. as a Visiting Professor of Geochemistry for SY 1962-63 (October 1962 - May 1963), and was Guest Professor at Göttingen University during the summer of 1961.

While serving as Visiting Professor, Ahrens presented a series of lectures on geochemistry each term, started and completed seven papers (which are listed in the Bibliography later on), and initiated an investigation on the distribution of lithium in 45 specimens of diabase with K. F. Thompson; research, unfortunately, which was never reported on in print.

Inasmuch as this history is primarily an account of persons and events at M.I.T., my story of Louis Ahrens does not include an account of his activities and achievements since he left in 1954, except as they directly involve M.I.T. Nevertheless, I would emphasize that he is another of those brilliant and promising young scientists who came to M.I.T. early in their professional careers, demonstrated their great abilities and skills, left an impressive mark, and then departed to achieve renown elsewhere. With four books, more than 175 published papers, and numerous awards from professional societies, Louis Herman Ahrens stands today among the leading geochemists of the world.

Louis was married to Evelyn Millicent McCulloch of Mid-Illovo (Union of South Africa) in 1941, and they have three children--Yolande Margaret (b. 1944), Wendy Mae (b. 1949), and Ian Louis (b. 1955). Our latest information is that the Ahrens reside in Claremont, a suburb of Cape Town.

BIBLIOGRAPHY OF LOUIS HERMAN AHRENS

Symbols and abbreviations used in the following references are explained on pages 91 - 98; in general, abstracts are listed separately as well as with the references to the complete article. This bibliography begins with Ahrens' first article, published in 1942 four years before he came to M.I.T. from South Africa as a Post-Doctoral Fellow, and includes all publications through calendar year 1964. Normally, the list of titles in this bibliography would include only those articles and books written while Ahrens was at M.I.T., i.e. 1948-1953 and 1962-1963; however, because he lives in another land and has published many articles that would not expectedly be seen by American readers, I have requested and received the following updated list from Ahrens and herewith extend sincerest thanks for his cooperation.

1--1942   The spectrochemical analysis of fluorine in phosphate rock. S. African Chem. Inst., J. 25: 18-32, (1942).

2--1943   The use of a strontium subfluoride band in the spectrochemical analysis of fluorine. S. African J. Sci. 39: 98, (1943).

3--1943a   Spectrochemical determination of wolfram in siliceous materials. S. African Chem. Inst., J. 26: 27-32, (1943).

4--1944   The spectrochemical determination of iron in glass sands.
            S. African Chem. Inst., J. 27: 28-38, (1944).

5--1945   Trace elements in clays. S. African J. Sci. 41: 152-160,
            (1945).

6--1945a   Quantitative spectrochemical examination of the minor constituents in pollucite [Norway, Maine]. Am. Mineral. 30: 616-622,
            (1945).

7--1945b   Isomorphic relationship between rubidium and thallium in igneous minerals. Nature 155: 610, (1945).

8--1945c   (and Liebenberg, W. R.) Geochemical studies on some of the
            rarer elements in South African minerals and rocks. I. Lithium
            in mica and feldspar. Geol. Soc. S. Africa, Tr. 48: 75-82,
            (1945).

9--1946   Geochemical studies on some of the rarer elements in South
            African minerals and rocks. II. The geochemical relationship
            between thallium and rubidium in minerals of igneous origin.
            Geol. Soc. S. Africa, Tr. 48: 207-231, (1946).

10--1946a   Determination of the age of minerals by means of the radioactivity of rubidium. Nature 157: 269, (1946).

11--1946b   (and Liebenberg, W. R.) Qualitative spectrochemical analysis
            of minerals and rocks. Geol. Soc. S. Africa, Tr. 49: 133-154,
            il., (1946).

12--1947   Analyses of the minor constituents in pollucite. Am. Mineral.
            32: 44-51, (1947).

13--1947a   The abundance of thallium in the earth's crust. Science 106:
            268, (1947).

14--1947b   Geological age: The extreme antiquity of pegmatites from Manitoba. Nature 160: 874-875, (1947).

15--1947c   The determination of geological age by means of the natural
            radioactivity of rubidium: A report of preliminary investigations. Geol. Soc. S. Africa, Tr. 50: 24-54, (1947).

16--1947d   The unique association of the thallium and rubidium in minerals [abst.]. Geol. Soc. Am., B. 58: 1161, (1947); Am. Mineral. 33: 191, (1948). See also 22--1948e.

17--1948   A summary of the use of the Rb/Sr method for the determination of geologic age, in Marble, J. P., Chmn., Rept. Comm. Measurement of Geological Time, 1943-49, NRC, Div. Geol. Geogr. [Ann.
            Rept.] 1946-47, Ex. C.: 47-54 (‡), (1948).

18--1948a   The geochemistry of radiogenic strontium. Mineral. Mag. [London] 28: 277-295, (1948).

19--1948b   Molecular spectroscopic evidence of the existence of strontium
            isotopes Sr^{88}, Sr^{87} and Sr^{86}. Phys. Rev. 74: 74-77, (1948).

20--1948c   (with Evans, R. D.) The radioactive decay constants of K^{40}
            as determined from the accumulation of Ca^{40} in ancient minerals. Phys. Rev. 74: 279-286, (1948).

21--1948d   Evidence of geological age against decay of Tin-115 to Indium-115 by electron capture. Nature 162: 413-414, (1948).

22--1948e   The unique association of thallium and rubidium in minerals.
            J. Geol. 56: 578-590, (1948); correction 57: 99, (1949).
            See also Abst., Geol. Soc. Am., B. 58: 1161, (1947); Am. Mineral. 33: 191, (1948).

23--1949   Measuring geologic time by the strontium method. Geol. Soc.
            Am., B. 60: 217-266, (1949).

24--1949a   Report to Committee on Measurement of Geologic Time, in Marble, J. P., chmn., Rept. Comm. Measurement of Geologic Time 1948-49; NRC, Div. Geol. Geogr. [Ann. Rept.] 1943-46, Ex.D.: 75-78, (1949).

25--1950   Spectrochemical Analysis. Cambridge, Mass.: Addison-Wesley Press, xxiv + 269 p., il., (1950).

26--1950a   What to expect from a standard spectrochemical analysis of common silicate rock types. Am. J. Sci. 248: 142-145, (1950).

27--1950b   (and Liebenberg, W. R.) Tin and indium in mica, as determined spectrochemically. Am. Mineral. 35: 571-578, (1950).

28--1950c   (and Gorfinkle, Lorraine G.) The abundance of several relatively rare elements in igneous rocks of North America. Science 112: 565, (1950).

29--1950d   (with Shaw, D. M. and Joensuu, O. I.) A double-arc method for spectrochemical analysis of geological materials. Spectrochim. Acta 4: 233-236, (1950).

30--1950e   (and Gorfinkle, Lorraine G.) Age of extremely ancient pegmatites from southeastern Manitoba. Nature 166: 149, (1950).

31--1950f   Report to Committee on Measurement of Geologic Time, in Marble, J. P., chmn., Rept. Comm. Measurement of Geologic Time 1949-50; NRC, Div. Geol. Geogr. [Ann. Rept.] 1949-50, Ex.B. 29-36, (1950).

32--1950g   (and Whiting, F.) Use of biotite for geological age measurements by the strontium method [abst.]. Geol. Soc. Am., B. 61: 1439, (1950).

33--1951   (and Gorfinkle, Lorraine G.) Quantitative spectrochemical analysis of rubidium in lepidolite (for geological age measurements by the strontium method). Am. J. Sci. 249: 451-456, (1951).

34--1951a   Quantitative spectrochemical analysis of silicates (minerals, rocks, soils, meteorites and allied materials)--preliminary report on a scheme of analysis. 7 p., tabs., ($\ddagger$). Cambridge, Mass.: M.I.T., (1951); Spectrochim. Acta 4: 302-306, (1951)..

35--1951b   Some newer developments of the strontium method applied to lepidolite and biotite, and the feasibility of a calcium method [abst.]. Am. Geophys. Union, Tr. 32: 314, (1951).

36--1951c   (and Macgregor, A. M.) Probable extreme age of pegmatites from Southern Rhodesia. Science 114: 64-65, (1951).

37--1951d   The feasibility of a calcium method for the determination of geological age. Geochim. Cosmochim. Acta 1: 312-316, (1951).

38--1951e   Wavelength Tables of Sensitive Lines. Cambridge, Mass.: Addison-Wesley Press, 86 p., tab., (1951).

39--1951f   (with Dennen, W. H. and Fairbairn, H. W.) Spectrochemical analysis of major constituent elements in rocks and minerals [R.I. and Va.], in Fairbairn, H. W., A cooperative investigation of precision and accuracy in chemical, spectrochemical, and modal analysis of silicate rocks. U.S. Geol. Surv., B. 980: 25-52, il., (1951).

40--1951g   Spectrochemical analysis of some of the rarer elements in the granite and diabase samples [R.I. and Va.], in Fairbairn, H. W., A cooperative investigation of precision and accuracy in chemical, spectrochemical, and modal analysis of silicate rocks. U.S. Geol. Surv., B. 980: 53-57, tab., (1951).

41--1952   The use of ionization potentials, Pt. 1, Ionic radii of the elements. Geochim. Cosmochim. Acta 2: 155-169, tab., (1952).

42--1952a   Ionic radii of the elements [abst.]. Am. Mineral. 37: 283, (1952).

43--1952b   Of what is the earth made? M.I.T. Repts. on Research 3/6: 1, (1952).

44--1952c   Anion affinity and polarizing power of cations. Nature 169: 463, (1952).

45--1952d   The oldest rocks, in Marble, J. P., Symposium on the measurement of geologic time. Am. Geophys. Union, Tr. 33: 193-195, (1952).

46--1952e   (and Pinson, W. H., Jr. and Kearns, Margaret M.) Association of rubidium and potassium and their abundance in common igneous rocks and meteorites. Geochim. Cosmochim. Acta 2: 229-242, (1952).

47--1952f   Spectrochemical analysis as a tool in geochemistry. Appl. Spectroscopy 6/5: [3 p.], (1952).

48--1953   The use of ionization potentials, Pt. 2, Anion affinity and geochemistry. Geochim. Cosmochim. Acta 3: 1-29, (1953).

49--1953a   (with Fairbairn, H. W. and Gorfinkle, Lorraine G.) Minor element content of Ontario diabase. Geochim. Cosmochim. Acta 3: 34-46, (1953).

50--1953b   Variations in isotopic abundances of strontium. Phys. Rev. 89: 631-632, (1953).

51--1953c   Elements in the earth's crust. M.I.T. Repts. on Research 5/2: 3-4, (1953).

52--1953d   (with Herzog, L. F. 2nd et al.) Radiogenic Sr^{87} in biotite, feldspar, and celestite, Pt. 2 of Variations in strontium abundance in minerals. Am. Geophys. Union, Tr. 34: 461-470, (1953).

53--1953e   (with Holyk, W.) Potassium in ultramafic rocks. Geochim. Cosmochim. Acta 4: 241-250, (1953).

54--1953f   (with Pinson, W. H. and Franck, Mona L.) The abundances of Li, Sc, Sr, Ba and Zr in chondrites and some ultramafic rocks. Geochim. Cosmochim. Acta 4: 251-260, (1953).

55--1953g   Uniquely old Precambrian age determinations and some geological implications [abst.]. Geol. Soc. Am., B. 64: 1390, (1953).

56--1953h   A fundamental law of geochemistry. Nature 172: 1148, (1953).

57--1954   The lognormal distribution of the elements (1) A fundamental law of geochemistry and its subsidiary--[Pt. 1]: Geochim. Cosmochim. Acta 5: 49-73, il., (1954); discussion by F. Chayes, Ibid. 6: 119-120, (1954); [Pt. 2]: Ibid. 6: 121-131, il., with reply by Ahrens, Sept. 1954; summary, Nature 172: 1148, (1954); discussion by K. V. Aubrey, Nature 174: 141-142, (1954); Geochim. Cosmochim. Acta 9: 83-89, il., (1956); [Pt. 3]: Ibid. 11: 205-212, il., (1957).

58--1954a   Discussion of "Statistical Methods Applied to Geochemistry" by Shaw, D. M. and Bankier, J. D., authors. Geochim. Cosmochim. Acta 6: 121-123, (1954).

59--1954b   Elements in the earth's crust. Tech. Rev. 56: 340, (1954).

60--1954c   Shielding efficiency of cations. Nature 174: 644-645, (1954).

61--1954d   Distribution of minor elements in rocks. Nature 174: 811-813, (1954).

62--1954e   The abundance of potassium (Chap. 3, p. 128-132), in Nuclear Geology--A symposium on nuclear phenomena in the earth sciences (H. Faul, ed.). New York: John Wiley & Sons, Inc., xvii + 414 p., (1954).

63--1954f   The strontium method for determining geological age (p. 331-341), in Nuclear Geology--A symposium on nuclear phenomena in the earth sciences (H. Faul, ed.). New York: John Wiley & Sons, Inc. xvii + 414 p., (1954).

64--1954g   [Presentation of the Mineralogical Society of America Award to Louis H. Ahrens, by Esper S. Larsen, Jr.] Acceptance of the Mineralogical Society of America Award. Am. Mineral. 39: 302-303, port., (1954).

65--1954h   A note on the relationship between the precision of classical methods of rock analysis and the concentration of each constituent. Mineral. Mag. [London] 30: 467-470, (1954).

66--1954i   Quantitative Spectrochemical Analysis of Silicates. A scheme of quantitative DC arc analysis of the silicate minerals, rocks, soils and meteorites. London: Pergamon Press, 122 p., il., (1954). See also 107--1961g.

67--1955   Uniquely old pre-Cambrian age determinations and some geological implications (a summary). Geol. Assoc. Canada 7/2: 25, (1955).

68--1955a   The convergent lead ages of the oldest monazites and uraninites (Rhodesia, Manitoba, Madagascar, and Transvaal). Geochim. Cosmochim. Acta 7: 294-300, (1955).

69--1955b   Implications of the Rhodesia age pattern. Geochim. Cosmochim. Acta 8: 1-15, (1955).

70--1955c   Analytical error as a possible cause of the $t(206/238) > t(207/235) > t(207/206)$ age distribution. Geochim. Cosmochim. Acta 8: 299, (1955).

71--1955d   Oldest rocks exposed (p. 155-168), in Crust of the Earth--A symposium (A. Poldervaart, ed.). Geol. Soc. Am., Sp. P. 62: viii + 762 p., il., (1955).

72--1956   The Sc abundance in chondrites and the neutron excess of principal isotopes. Geochim. Cosmochim. Acta 9: 273-278, (1956).

73--1956a   Some ionization potential variations and relationships. J. Inorg. Nucl. Chem. 2: 290-314, (1956).

74--1956b   Ionization potentials and the chemical binding and structure of simple inorganic crystalline compounds--I. Chemical binding. J. Inorg. Nucl. Chem. 3: 263-269, (1956).

75--1956c   (and Morris, D. F. C.) Ionization potentials and the chemical binding and structure of simple inorganic crystalline compounds--II. Crystal structure. J. Inorg. Nucl. Chem. 3: 270-280, (1956).

76--1956d   (with McKerrow, W. S., Taylor, S. R., and Blackburn, A. L.) Occurrence of caesium in fossils. Nature 178: 204, (1956).

77--1956e   (with McKerrow, W. S., Taylor, S. R., and Blackburn, A. L.) Rare alkali elements in trilobites. Geol. Mag. 93: 504-516, (1956).

78--1956f   Radioactive methods of determining geological age. The Physical Society, Repts. Prog. Physics 19: 80-106, (1956).

79--1956g   Radioactive methods for determining geological age (Chap. 3, p. 44-67), in Physics and Chemistry of the Earth, Vol. 1. (L. H. Ahrens, K. Rankama, and S. K. Runcorn, eds.). London: Pergamon Press, viii + 317 p., (1956).

80--1957   Studies on the relative abundances of isotopes. Geochim. Cosmochim. Acta 11: 1-27, (1957).

81--1957a   The spectrograph in geochemistry and cosmochemistry. Soil Science 83: 33-41, (1957).

82--1957b   Lognormal-type distributions--III. Geochim. Cosmochim. Acta 11: 205-212, (1957).

83--1957c   (with Russell, R. D.) Additional regularities among discordant lead-uranium ages. Geochim. Cosmochim. Acta 11: 213-218, (1957).

84--1957d   Some features of the relationship between mass defect and mass number. Roy. Soc. S. Africa, Tr. 35/2: 109-113, (1957).

85--1957e   Some ionization potential variations and relationships--II. J. Inorg. Nucl. Chem. 4: 264-272, (1957).

86--1957f   A survey of the quality of the principal abundance data of geochemistry (p. 30-45), in Physics and Chemistry of the Earth, Vol. 2 (L. H. Ahrens, F. Press, K. Rankama, and S. K. Runcorn, eds.). London: Pergamon Press, viii + 259 p., (1957).

87--1957g   (with Deleon, G.) The distribution of Li, Rb, Cs and Pb in some Yugoslav granites. Geochim. Cosmochim. Acta 12: 94-96, (1957).

88--1958   Variation of refractive index with ionization potential in some isostructural crystals. Mineral. Mag. [London] 31: 929-936, (1958).

89--1958a   (and Cherry, R. D.) Evidence for $^{6}_{2}$He clusters in nuclei. Nature 182: 1434-1435, (1958).

90--1959   (and Edge, R. A.) Observations on the use of LaO bands for d-c arc spectrochemical analysis. Spectrochim. Acta 13: 304-307, (1959).

91--1959a   (with Edge, R. A., Brooks, R. R., and Amdurer, S.) Some reconnaissance observations on the combined use of ion-exchange enrichment and spectrochemical analysis for the determination of trace constituents in silicate rocks. Geochim. Cosmochim. Acta 15: 337-341, (1959).

92--1959b   (with Taylor, S. R.) The significance of K/Rb ratios for theories of tektite origin. Geochim. Cosmochim. Acta 15: 370-372, (1959).

93--1959c   (with Besaire, H. and Burger, A. J.) Measurement of age of monazites from Madagascar. Extr. Comptes des Séances de l'Academie de Sciences, Meeting held 1 June 1959.

94--1959d   The possible significance of the rare alkali metals for an understanding of the origin of eruptive rocks (p. 59-63), in The geochemistry of the rare elements in relation to the problem of petrogenesis. Moscow: U.S.S.R. Academy of Sciences, (1959).

95--1960   (with Brooks, R. R. and Taylor, S. R.) The determination of trace elements in silicate rocks by a combined spectrochemical-anion exchange technique. Geochim. Cosmochim. Acta 18: 162-175, (1960).

96--1960a   (with Taylor, S. R.) Spectrochemical analysis (Chap. 4, p. 81-110), in Methods in Geochemistry (A. A. Smales and L. R. Wager, eds.). New York: Interscience Pub. Inc., vii + 464 p., il., (1960).

97--1960b   (and Fleischer, M.) Trace element content of the standard granite (G-1) and standard diabase (W-1), in Second report on a cooperative investigation of the composition of two silicate rocks. U. S. Geol. Surv., B. 1113: 83-111, tab., (1960).

98--1960c   (with Brooks, R. R.) The application of anion-exchange techniques to the spectrochemical detection of the noble metals in silicate rocks. Spectrochim. Acta 16: 783-788, (1960).

99--1960d	(and Edge, R. A. and Taylor, S. R.) The uniformity of concentration of lithophile elements in chondrites--with particular reference to Cs. Geochim. Cosmochim. Acta 20: 260-272, (1960).
100--1961	Regularities involving ^6_2He and ^4_2He and their possible applications for nuclear structure. J. Inorg. Nucl. Chem. 16: 368-369, (1961).
101--1961a	(with Nicolaysen, L. O., Burger, A. J., and Tatsumi, T.) Age measurements on pegmatites and a basic charnockite lens occurring near Lütsow-Holm Bay, Antarctica. Geochim. Cosmochim. Acta 22: 94-98, (1961).
102--1961b	(with Brooks, R. R.) Some observations on the distribution of thallium, cadmium and bismuth in silicate rocks and the significance of covalency on their degree of association with other elements. Geochim. Cosmochim. Acta 23: 100-115, (1961).
103--1961c	(with Brooks, R. R.) The determination of indium and thallium in G-1, W-1 and other silicate rocks by a new technique. Geochim. Cosmochim. Acta 23: 145-147, (1961).
104--1961d	(and Edge, R. A. and Brooks, R. R.) The combination of ion exchange and spectrographic techniques for the estimation of trace elements in geological material. S. Africa Indust. Chemist 15: 102-105, (1961).
105--1961e	(and Edge, R. A.) The K/Cs ratio in some basic rocks. Geochim. Cosmochim. Acta 25: 91-94, (1961).
106--1961f	Regularities involving ^6_2He and ^4_2He and their possible implications for nuclear structure. Roy. Soc. S. Africa, Tr. 36: 163-178, (1961).
107--1961g	(and Taylor, S. R.) <u>Spectrochemical Analysis</u>--A treatise on the d-c arc analysis of geological and related materials, 2d edition. Reading, Mass.: Addison-Wesley Pub. Co., 454 p., il., (1961). See also 66--1954i.
108--1962	(with Edge, R. A.) The determination of Sc, Y, Nd, Ce and La in silicate rocks by a combined cation exchange-spectrochemical method. Anal. Chim. Acta 26: 355-362, (1962).
109--1962a	(with Willis, J. P.) Some investigations on the composition of manganese nodules, with particular reference to certain trace elements. Geochim. Cosmochim. Acta 26: 751-764, (1962).
110--1962b	(with Edge, R. A.) Studies on the trace element content of some South African rocks. Geol. Soc. S. Africa, Tr. 65: 113-124, (1962).
111--1962c	Possible Zr-Hf fraction between earth and meteorites. Geochim. Cosmochim. Acta 26: 1077-1079, (1962).
112--1962d	(with Edge, R. A. and Dunn, J. D.) The determination of molybdenum in granitic and related rocks by a combined solvent extraction-spectrochemical technique. Anal. Chim. Acta 27: 551-558, (1962).
113--1963	Lognormal-type distributions in igneous rocks--IV. Geochim. Cosmochim. Acta 27: 333-343, il., (1963).
114--1963a	Neutron numbers 98, 108 and 116. <u>Nature</u> 197: 993-994, (1963).
115--1963b	Negatively skewed distributions of silica and potassium in igneous rocks. <u>Nature</u> 198: 373-374, (1963).
116--1963c	The significance of the chemical bond for controlling the geochemical distribution of the elements. Phys. Chem. Earth (V): 1-54, (1963).

117--1963d (and Edge, R. A. and Brooks, R. R.) Investigations on the development of a scheme of silicate analysis based principally on spectrographic and ion exchange techniques. Anal. Chim. Acta 28: 551-573, (1963).

118--1963e Lognormal-type distributions in igneous rocks--V. Geochim. Cosmochim. Acta 27: 877-890, il., (1963).

119--1963f Element distributions in igneous rocks--VI. Negative skewness of SiO_2 and K. Geochim. Cosmochim. Acta 27: 929-938, (1963).

120--1964 Element distributions in igneous rocks--VII. A reconnaissance survey of the distribution of SiO_2 in granitic and basaltic rocks. Geochim. Cosmochim. Acta 28: 271-290, il., (1964).

121--1964a Aspects of the Geochemistry of the Rare Earths (Chap. 1, p. 1-29), in Prog. Sci. Tech. Rare Earths, Vol. 1. New York: The Macmillan Co., (1964).

122--1964b Si-Mg fractionation in chondrites. Geochim. Cosmochim. Acta 28: 411-423, il., (1964).

123--1964c Negatively skewed distributions of silica and potassium in igneous rocks. Nature 201: 1313, (1964).

124--1964d Element distribution in igneous rocks. Adv. Sci., (May 1964).

125--1964e (with Willis, J. P. and Kaye, M.) The spectrochemical estimation of thallium in granites and in Mn nodules. Appl. Spectroscopy 18/3: 84-87, (1964).

126--1964f Notes on Na-K fractionation in chondrites. Geochim. Cosmochim. Acta 28: 1869-1870, (1964).

127--1964g The significance of the chemical bond for controlling the geochemical distribution of the elements, Pt. 1 (p. 1-54, il.), in Physics and Chemistry of the Earth (L. H. Ahrens, F. Press, and S. K. Runcorn, eds.). New York: The Macmillan Co., v + 398 p., (1964).

Publications as a Co-Editor

For fifteen years (1956-1971) Ahrens has been serving as a co-editor of Physics and Chemistry of the Earth, a series of volumes intended to cover the more important developments in the chemistry and physics of the earth. To 1971, eight volumes have appeared, as follows:

Vol. 1. Ahrens, L. H., Rankama, K., and Runcorn, S. K., eds. London: Pergamon Press, viii + 317 p., (1956).

Vol. 2. Ahrens, L. H., Press, F., Rankama, K., and Runcorn, S. K., eds. London: Pergamon Press, viii + 259 p., (1957).

Vol. 3. Ahrens, L. H., Press, F., Rankama, K., and Runcorn, S. K., eds. London: Pergamon Press, viii + 464 p., (1958).

Vol. 4. Ahrens, L. H., Press, F., Rankama, K., and Runcorn, S. K., eds. New York: Pergamon Press, v + 317 p., (1961).

Vol. 5. Ahrens, L. H., Press, F., and Runcorn, S. K., eds. New York: The Macmillan Co., v + 398 p., (1964).

Vol. 6. Ahrens, L. H., Press, F., Runcorn, S. K., and Urey, H. C., eds. Oxford (England): Pergamon Press, v + 510 p., (1965).

Vol. 7. Ahrens, L. H., Press, F., Runcorn, S. K., and Urey, H. C., eds. Oxford (England): Pergamon Press, v + 337 p., (1966).

Vol. 8. Ahrens, L. H., Press, F., Runcorn, S. K., and Urey, H. C., eds. Oxford (England): Pergamon Press, v + 337 p., (1971).

(30)
THEODORE RICHARD MADDEN

THEODORE RICHARD MADDEN

MIT: 1950–

Soon after the end of World War II, in the early 1950s, when it was decided to develop instruction and research in geophysics as an integral part of Course XII's program, we sought help from one of M.I.T.'s graduates in physics, Theodore R. Madden (VIII S.B. 1949), who as an undergraduate had become interested in geophysics as a result of several summer cruises with Maurice Ewing of Columbia University. From this introduction he became interested in geophysical exploration for mineral deposits and turned to summer work, first with Phelps Dodge and then with Bear Creek Mining.

Starting at M.I.T. as a Teaching Fellow in Geophysics in SY 1950-51, he served successively as Instructor (1951-1954), Lecturer (1954-1957), Assistant Professor (1957-1962), Associate Professor (1962-1967), and finally Professor (1967). In the earlier years of this period of service he carried much of the load of both lecturing and supervision of student research in geophysics because there was no one else in the Department of Geology to do it. Students found him an imaginative and stimulating teacher and research leader and were attracted to his laboratory where some geophysical experimentation seemed always to be going on. As a consequence, many students sought his advice and assistance in finding a thesis problem and then asked him to be their supervisor. Thus it happened that in the fourteen years from 1951 to 1965 he supervised 26 theses--11 bachelors, 7 masters, and 8 doctors--an impressive record, particularly since he carried on his own research and teaching in the meantime. His earliest research involved investigation of the induced polarization (IP) method of exploring for mineral deposits. Later he turned to studies of deep earth conductivity by means of measurements of magnetotelluric and magnetic variations, and to investigations of the ionosphere and magnetosphere by means of low frequency electromagnetic and magnetohydrodynamic wave studies. His most recent research (1973-1974) has involved investigation of atmospheric gravity-wave phenomena.

Important as his research results have been, however, and continue to be, Madden's greatest contribution to the earth sciences at M.I.T. has been as teacher and friend of a generation of students. Many of his students first became interested in geophysics from his lectures and laboratory and field investigations, then went on to thesis work under his supervision, and quite a few completed graduate work before entering a professional career based on the foundation he gave them as students.

Theodore Richard Madden was born in Boston, Massachusetts, on 14 March 1925, the son of Thomas Richard and Mercedes (Johnson) Madden. He is one of four children--two older sisters, "Baba" (Mercedes) and "Sissy" (Margarita [deceased]); himself; and a younger brother, Manuel. His forebears on his father's side were Irish, his paternal grandfather having come to America from Ireland; his mother was born in Cuba of Scottish and Spanish ancestry.

He was graduated from Milton Academy (Massachusetts) in 1942 and was admitted to M.I.T. in the fall of the same year to start undergraduate studies. At the end of SY 1942-43, however, World War II intervened to interrupt his academic career for the next three years, 1943-1946, during which he served in the U. S. Marine Corps with the 4th Marine Air Wing in the Pacific Theater (Western Carolines). After being discharged as Staff Sergeant he resumed his studies at M.I.T. in the fall of 1946 and received an S.B. degree in physics in June 1949. His senior thesis, written jointly with Edward T. Miller, was on neutron well logging and has the title

> "Neutron Logging: Experiments on gamma intensity caused by a fast neutron source in a cased well through hydrogenous media."

Having become interested in geophysics while still an undergraduate, he spent the summers of 1948, 1949, 1950 and 1952 as a geophysicist on cruises from the Lamont Geological Observatory which were led by Columbia's famed geophysicist, W. Maurice Ewing. The summer of 1951 found him involved in geophysical exploration for mineral deposits with the Phelps Dodge Corporation, and this experience was followed by field work in Nova Scotia in the summers of 1953 and 1954 for the purpose of testing certain electrical methods for ore finding that had been suggested by Prof. Hurley (These are discussed farther on). Still later, in 1955, 1956 and 1957 came summer field work with the Bear Creek Mining Company (Kennecott).

During these same years, 1949-1957, Madden was busy at M.I.T. during the regular school year as both graduate student (i.e. degree candidate) and instructor in geophysics. His academic appointments during this period, and for the next five years until 1961, when he finally found time to write his doctoral thesis and receive a Ph.D. degree, make an unusual record. He was simply so much occupied with his teaching and research, and with helping his students get their thesis investigations and other degree requirements completed, that he didn't get around to finishing his own doctoral thesis* until 1961. As the following record shows, his appointments changed rather surprisingly to meet his changing status as degree candidate and student instructor:

---

* The thesis Madden presented for his Ph.D. degree awarded in September 1961 bears the following title: "Electrode polarization and its influence on the electrical properties of mineralized rocks." For further information see reference T--1961 in the following Bibliography.

```
Teaching Fellow in Geology                        Oct.  1950 to June  1951
Instructor in Geology                             Sept. 1951 to March 1952
Instructor in Geology and Geophysics              March 1952 to June  1953
Research Assistant in Geology and Geophysics      June  1953 to Sept. 1953
Instructor in Geology and Geophysics              Oct.  1953 to Sept. 1955
Lecturer in Geology and Geophysics                Sept. 1955 to June  1958
Assistant Professor of Geophysics                 July  1958 to June  1962
Associate Professor of Geophysics                 July  1962 to June  1967
Professor of Geophysics                           July  1967 to date
```

Madden was on leave as Visiting Associate Professor at the University of California at San Diego during SY 1963-64 and as Visiting Professor at Stanford during the fall term of 1973.

As teacher, Madden has been involved to some extent in most of the subjects and seminars in geophysics; has supervised more than twenty-five theses; and has been a sympathetic and helpful advisor to a great many students who regard him as one of the outstanding teaching scientists in the Geology Department.

As a research scientist, Madden is widely known for his investigations into and applications of the induced polarization method* of geophysical exploration for ore bodies, for his investigation of the electrical conductivity structure of the crust and upper mantle by means of magnetotelluric measurements, and most recently for his studies of propagation of low frequency magnetohydrodynamic waves in the earth-ionosphere-magnetosphere environment, and for studies of atmospheric gravity waves.

It may be of some interest at this point to mention briefly how Madden got involved in research on the so-called IP method of geophysical exploration because the story is an excellent example of how the secrecy of discoveries in an industrial

Madden as Teacher.
(Photo by Jackman, 1952)

---

* Worth special notice is the research on induced polarization conducted for the U. S. Atomic Energy Commission by Madden and Marshall during 1957 to 1959. The results were published in four reports, as indicated in items 2--1957, 3--1958, 4--1958a, and 5--1959 in Madden's Bibliography on later pages. Also see discussion of induced polarization in the biography of Prof. Patrick Mason Hurley.

Madden as Investigator, discussing a record with his colleague, Dr. T. Cantwell.

laboratory contrasts with the openness of research results in academic laboratories.

In the interim between summer employment with Phelps Dodge in 1951 and Bear Creek Mining in 1954, Madden learned from Prof. P. M. Hurley that rumors were being circulated to the effect that the Newmont Mining Corporation was using a new method of geophysical exploration that could delineate metallic ore bodies by means of observing their electrical polarization effects. Inasmuch as there was much secrecy and speculation about the new method, Hurley suggested to Madden and two of his fellow graduate students that an investigation of this polarization phenomenon might be a useful research area. Whereupon, Madden, aided by P. G. Hallof and K. Vozoff, made some laboratory measurements on mineralized rock samples and became convinced, as a result of these measurements, that it might be worth while to conduct certain tests in the field. The trio decided to make their test in Nova Scotia because the Department was then running its Summer School of Geology in that Province.

Accordingly, in the summer of 1953 the trio visited a mine on Cape Breton Island and made variable frequency resistivity measurements in and around the sulphide ore zone. Their measurements indicated that the sulphide mineral polarization effects produced measurable frequency variations in the rock resistivity, but that very low frequency measurements were needed to prevent electromagnetic coupling effects from masking the polarization effect. In order to make such low frequency measurements, Madden and Hallof returned to Cape Breton the following summer (1954), and found that the desired measurements gave excellent results and showed the "induced polarization" or IP method to be a very sensitive indicator of the presence of metallic or sulphide minerals. A little later Prof. R. D. Parks arranged for Madden and his student associates to make appropriate measurements at certain lead mines in New York and Missouri, and again they came away with convincing results.

When the team of Madden, Hallof and Vozoff reported on their research at the Annual Meeting of the A.I.M.E. in New York in February 1956 (See item 1--1956 in Madden's Bibliography later on), scientists at the U. S. Atomic Energy Commission became interested and suggested that the A.E.C. would provide funds for Madden if he wished to continue research on induced polarization. Whereupon he engaged the interest of one of his fellow students, D. J. Marshall, and for the next three years, 1956-1959, the two of them conducted research reported on in the four A.E.C. Annual Reports listed in Madden's Bibliography later on--items 2--1957, 3--1958, 4--1959 and 5--1959a.

Such then is the story of how a rumor, heard by Hurley and handed on to Madden to ponder, resulted in experimentation that led to the understanding of what the IP method of geophysical exploration was all about and to publications on the method that Newmont had tried hard to keep secret.

It should be emphasized that neither Hurley nor Madden knew about the phenomenon mentioned in the preceding discussion at the time the rumors were being spread; they had to figure out what the Newmont scientists were doing and why, and thereby hangs the story of Madden's scientific detective work.

In addition to teaching and research, Madden has done his fair share of work in both Institute and Departmental committees, particularly serving the Department well in recent years as its representative on the Institute's Committee on Graduate School Policy.

Ted and Sheila Anne Murphy were married in 1959, and their daughter Jennifer is now (1974) in the eighth grade. Sheila, one of a family of four children, two girls and two boys, received an A.B. degree in history from Newton College of the Sacred Heart in 1956, and in recent years has been working at Harvard in the Department of History. At their home in Weston the Maddens have been gracious hosts at numerous delightful social gatherings.

BIBLIOGRAPHY OF THEODORE RICHARD MADDEN

Symbols and abbreviations used in the following references are explained on p. 91 - 98; in general, abstracts are listed separately as well as with the references to the complete article. This bibliography begins with the title of Madden's S.B. thesis in Course VIII, and includes all known publications through 1970. T = Thesis.

T--1949 (and Miller, E. T.) Neutron Logging: Experiments on gamma intensity caused by a fast neutron source in a cased well through hydrogenous media, iii + 50 p., il., (1949). (S.B. Thesis at M.I.T. in Course VIII, Physics, June 1949).

1--1956 (with Hallof, P. G. and Vozoff, K.) The interfacial polarization of metallic minerals [abst.]. A.I.M.E., Min. Geol. Geophys. Div. Ann. Mtg., Feb. 1956, Min. Br. Abst., p. 41, (1956).

2--1957 (and others) Background effects in the induced polarization method of geophysical exploration--Ann. Prog. Rept. 1956-1957, U. S. Atomic Energy Comm. Rept. RME-3150, 80 p., il., (1957). (Rept. prepared for A.E.C. by M.I.T.)

3--1958 (and Marshall, D. J.) A laboratory investigation of induced polarization--An interim report for 1957-1958, U. S. Atomic Energy Comm. Rept. RME-3156, 37 p., il., (1958). (Rept. prepared for A.E.C. by M.I.T.)

4--1959 (and Marshall, D. J.) Electrode and membrane polarization--Interim report for 1958, U. S. Atomic Energy Comm. Rept. RME-3157, 115 p., il., (1959). (Rept. prepared for A.E.C. by M.I.T.)

5--1959a (and Marshall, D. J.) Induced polarization--A study of its causes and magnitudes in geological materials--Final Rept., U. S. Atomic Energy Comm. Rept. RME-3160, 80 p., il., (1959). (Rept. prepared for A.E.C. by M.I.T.)

6--1959b (and Marshall, D. J.) Induced polarization, A study of its causes. Geophysics 24: 790-816, (1959).

7--1960 (and Cantwell, T. and Hauck, A. M.) Electrical conductivity structure of the crust under Massachusetts determined from resistivity and magnetotelluric measurements [abst.]. J. Geophys. Res. 65: 2509, (1960).

8--1960a (with Cantwell, T.) Preliminary report on crustal magnetotelluric measurements. J. Geophys. Res. 65: 4204-4205, il., (1960).

T--1961 Electrode polarization and its influence on the electrical properties of mineralized rocks, 4 + 11 + 158 + 3 p., (1961). (Ph.D. Thesis at M.I.T. in Course XII, September 1961; M.I.T. Abst. Theses 1961-1962, p. 203-204, 1963.)

9--1963 (and Cantwell, T.) Practical consideration in applying induced polarization [abst.]. Min. Eng. 15/1: 60, (1963).

10--1963a (with Eckhardt, D. and Larner, K.) Long-period magnetic fluctuations and mantle electrical conductivity estimates [abst.]. Am. Geophys. Union, Tr. 44: 39, (1963).

11--1963b (with Eckhardt, D. and Larner, K.) Long-period magnetic fluctuations and mantle electrical conductivity estimates. J. Geophys. Res. 68: 6279-6286, il., (1963).

12--1964 Spectral, cross-spectral, and bispectral analysis of low frequency electromagnetic data (p. 429-450) in <u>Natural Electromagnetic Phenomena</u>, New York, Plenum Press, (1964).

13--1965   (with Orange, A. S. and Brace, W. F.) A preliminary study of the effect of pressure on electrical resistivity of saturated rock [abst.]. Am. Geophys. Union, Tr. 46/1: 70, (1965).

14--1965a   (with Brace, W. F. and Orange, A. S.) The effect of pressure on the electrical resistivity of water-saturated crystalline rocks. J. Geophys. Res. 70: 5669-5698, il., (1965).

15--1965b   (and Thompson, W. B.) Low-frequency electromagnetic oscillations of the earth-ionosphere cavity. Rev. Geophys. 3: 211-254, (1965).

[15a--1966] (and Nourbehecht, B.) Irreversible thermodynamics in inhomogeneous media and geoelectric applications, a chapter in Mass Transport and Membrane Phenomena in Geology [Prepared in 1966 but never published].

16--1967   (and Cantwell, T.) Induced polarization--A review, in Min. Geophys. Vol. 2, Theory: Tulsa, Okla., Soc. Explor. Geophysicists, p. 373-400, il., (1967).

17--1967a   (with Swift, C. M., Jr.) A magnetotelluric investigation of the electrical conductivity anomaly in the upper mantle in the southwestern United States [abst.]. Am. Geophys. Union, Tr. 48: 210, (1967).

18--1968   (with Nelson, P.) A resonance mode tracker for the electromagnetic earth-ionosphere cavity. IEEE, Tr. on Geosci. Electronics VGE-6/2: 70-77, (1968).

19--1968a   (and Claerbout, J.) Jet stream gravity waves and implications concerning jet stream stability (p. 121-134) in Acoustic-Gravity Waves in the Atmosphere, Pr. ESSA/ARPA Symposium, T. M. Georges (ed.), Washington, D.C., Gov't. Printing Office, (1968).

20--1968b   (with Claerbout, J.) Electromagnetic effects of atmospheric gravity waves (p. 135-155), in Acoustic-Gravity Waves in the Atmosphere, Pr. ESSA/ARPA Symposium, T. M. Georges, (ed.), Washington, D.C., Gov't. Printing Office, (1968).

21--1969   (and Swift, C. M., Jr.) Magnetotelluric studies of the electrical conductivity structure of the crust and upper mantle (p. 469-479), in The Earth's Crust and Upper Mantle (P. J. Hart, ed.), A.G.U. Geophys. Union Mon. 13 (NAS-NRC, Pub. 1708), Washington, D.C., Am. Geophys. Union, (1969).

22--1970   Geoelectric upper mantle anomalies in the United States. J. Geomag. Geoelect. 22: 91-95, (1970).

Prof. Mencher arranging his notes for a lecture in historical geology, in Room 420 of the Green Building (54).

(Photo by John Dingler in May 1967)

(31)
ELY MENCHER

ELY MENCHER

MIT: 1952-1967

Soon after World War II ended, when it became obvious that the petroleum industry would have to expand its exploration program, it seemed appropriate that M.I.T.'s Department of Geology and Geophysics add to its staff an experienced petroleum geologist who could organize and conduct instruction and research in the broad field of petroleum geology. In 1952, after a wide search at home and abroad, and at a time when experienced petroleum geologists were in short supply, the Department was fortunate to engage the services of Ely Mencher (XII Ph.D. 1938), then employed as Research Geologist with the Socony-Vacuum Oil Company of Venezuela. Mencher had gone directly to Venezuela in 1938, after receiving his doctorate in stratigraphy under Professor H. W. Shimer, and had taken an active part in organizing the Escuela de Geologia of the Universidad Central in Caracas. He served the School as Professor of Geology from 1938 to 1943 and was Technical Director in 1941 and 1942. He left in 1943 to accept a position with Socony-Vacuum. With such a background, he brought both academic and industrial experience to his new position at M.I.T. He immediately revised existing subjects and organized new ones to create a program suitable for training students interested in a career in the petroleum industry. He soon accepted other duties in the Department, as well as in the Institute as a whole, taking on work in historical geology, conducting numerous field trips, and serving on numerous committees. In 1961 he took charge of the summer field program in Maine, that Professor Boucot had started earlier, when the latter left for Caltech, and with his students carried on field work there for the next five years, 1962-1967. As the oceanography program got underway in Course XII, Mencher organized a graduate lecture subject in Marine Geology and Sedimentation and a research program on the bottom sediments in Boston Harbor. The Maine project and the oceanographic work in Boston Harbor provided excellent thesis problems for Mencher's students, and during his fifteen years at M.I.T. he supervised 17 theses: 14 S.B.'s, 2 S.M.'s, and 1 Ph.D. On 31 August 1967 Mencher resigned his associate professorship at M.I.T. to accept a professorship at his alma mater, the City College of the City of New York (CCNY). A year later he was appointed Head of that Institution's Department of Geology, and served in that capacity for the next four years, 1968-1972, before resigning to assume a full-time teaching schedule.

Ely Mencher was born in New York City on 14 December 1913, the second of three children, all sons, of Morris Mencher, a dentist, and Rachel (April) Mencher, a housewife. His father was Austrian by birth, but ultimately became a naturalized citizen of the United States; his mother was born in the U.S.A.

After attending Public School No. 40 in New York City, from 1919 to 1926, young Mencher went on to the George Washington High School, from which he graduated in 1930. The next four years brought him a Bachelor of Science degree from City College of the City of New York in 1934, with the Ward Medal for outstanding achievement in geology. He entered M.I.T. in September 1934 and for the next four years worked under the supervision of Professor Hervey W. Shimer, whom he helped as a Teaching Assistant for three successive years, 1935-1938. During this same period he was an instructor in geology at CCNY in the summer sessions of 1936 and 1937.

His thesis problem for the Ph.D. degree in geology, which was awarded in December 1938, concerned the stratigraphy and sedimentology of the Devonian Catskill sequence of New York. His work was later published under the title "Catskill Facies of New York State" (See Bibliography), and was excellent preparation for the field investigations he carried on some twenty years later involving the Devonian and associated strata of northeastern Maine.

Immediately after completing requirements for the doctorate, Mencher went to Venezuela. There he joined two alumni of M.I.T.--Victor M. Lopez (XII S.M. 1936 and Ph.D. 1937), a former classmate, and Guillermo Zuloaga (XII Ph.D. 1930)--in Caracas, and helped them to organize the Escuela de Geologia in the Universidad Central, a government institution. He was Professor of Geology in the School from 1938 to 1943 and Technical Director during 1941 and 1942. When the School became a branch of the Engineering Faculty in 1943, Mencher left to take the position of Senior Field Geologist with the Socony-Mobil Company of Venezuela. Five years later, in 1948, he was advanced to Research Geologist, the position he held until 1952, when he left Venezuela to become Associate Professor of Geology at M.I.T.

While in Venezuela, Mencher carried on field and laboratory work in that country, as well as in Colombia and Puerto Rico, and became well-known as one of the leading authorities on the geology of the Colombia-Venezuela region. Five of his more important papers deal with this region; the latest, a 1963 summary article entitled "Tectonic History of Venezuela" (See Bibliography).

He served as a member of the Organizing Committee of the Venezuelan Petroleum Congress that met in Caracas in 1951, and was chairman and editor of the geological part of the Proceedings that came out of this Congress. These Proceedings, in both English and Spanish, were subsequently published by the Venezuelan Government. He was a founding member of the Venezuelan Society of Geology, Mining and Petroleum, and for five years a director, as well as a founding member, of the Venezuelan Association for the Advancement of Science, serving that organization as co-ordinator of the Geological Section during the 1952 session. (Extracted from The Technology Review for January 1953, p. 159). He was also Honorary Curator of Paleontología y Conquiliologia, Muséo de Historía Naturel in Caracas from 1941 to 1944.

As stated earlier Mencher was brought to M.I.T. to meet Course XII's need for an experienced petroleum geologist who could organize and conduct instruction and research in the broad field of petroleum geology. Immediately after his arrival, in the fall of 1952, he set about revising several existing subjects and organized new ones suitable for training students who were interested in a career in the petroleum industry. Particularly noteworthy were a new subject in practical laboratory methods (12.45 Petroleum Geology Laboratory) and a two-term sequence of lectures on petroleum geology (12.441 and 12.442, Petroleum Geology I and II, respectively). In the laboratory, students were taught how to use the latest techniques and methods in common use in the petroleum industry--logging methods of various sorts; microscopic examination of cuttings, insoluble residues, heavy accessory minerals, microfossils, etc.; and interpretation of the field and laboratory data generally made available to the petroleum geologist. The lectures, supplemented with field trips, were primarily aimed at presenting the latest ideas in sedimentology and stratigraphy as these applied to petroleum geology. Mencher also assumed responsibility for historical geology soon after Professor Shrock had to give up teaching the subject because of his increased work load as Head of Course XII.

In 1961, when Professor Boucot resigned to accept a position at Caltech, he interested Mencher in continuing and expanding a program of field work that he had started on the Paleozoic geology of northern Maine. As a result, Mencher and his students took up the work left by Boucot, and during the next five years, from 1962 to 1967, spent the summers doing field work on the Ordovician, Silurian, and Devonian strata of northeastern Maine, chiefly in Aroostook County. This work, supported the first year by a small grant from the Maine Geological Survey, and in subsequent years by grants from the National Science Foundation, resulted in a number of student theses at M.I.T.

As the oceanography program got underway in Course XII in the years following 1962, Mencher organized and offered a graduate lecture subject in Marine Geology and Sedimentology (12.361), and a graduate seminar in

Advanced Sedimentology (12.582). He also organized and directed a research program in marine sedimentation aimed primarily at investigating the bottom sediments in Boston Harbor. This latter program, like the field project in Maine, was supported by grants from the National Science Foundation and provided several students with thesis problems.

During his fifteen years as a member of the M.I.T. Faculty, 1952-1967, Mencher was an able and dedicated teacher of geology and a loyal and dependable staff member. He led his classes on field trips to the Catskills and the Connecticut Valley for ten successive years and did much to restore the importance of student field work as an integral part of the introductory subjects in geology. He was Graduate Registration Officer for the Department of Geology and Geophysics for many years and during the same period represented Course XII on the Committee for Graduate School Policy. At different times he was also called upon to serve on the Institute's Committee for Selection of Fulbright Scholars, on the Library Committee, as a United Fund solicitor, and as Treasurer of M.I.T.'s chapter of Sigma Xi. He could always be counted on to do more than his fair share of such extra-departmental chores.

At the invitation of Harvard College, Mencher served that Institution as Visiting Lecturer for both terms of SY 1955-56.

In 1967, when an opportunity came to return to his alma mater as Professor of Geology, Mencher resigned his tenured associate professorship at M.I.T. on 31 August 1967 and accepted the position at City College of the City of New York. A year later he assumed the chairmanship of the Department of Geology and served in that capacity until the beginning of the summer of 1972 at which time he resigned as Chairman and returned to full-time teaching together with resumption of summer field work in Maine.

Mencher was married to Miriam Beatrice Pollak, a New York City native, on 10 May 1951, a little more than a year before he returned to M.I.T. They have one child, Frederick Marshall Mencher, who at this writing, October 1973, has started his senior year at Harvard as a biology major.

BIBLIOGRAPHY OF ELY MENCHER

Symbols and abbreviations used in the following references are explained on p. 91-98; in general, abstracts are listed separately as well as with the reference to the complete article. This bibliography begins with Mencher's first paper in 1938 and includes all known titles through 1967, the year he left M.I.T.

T--1938   Sedimentary study of the Catskill facies in New York State [Ph.D. thesis at M.I.T.] 3+156+5 p., maps, pl., diagr., (1938). (Also see item 2 below.)

1--1938a   The salinity of the ocean in relation to water vapor in the atmosphere and the level of the sea. J. Geol. 46: 106-108, (1938).

2--1939   Catskill facies of New York State. Geol. Soc. Am., B. 50: 1761-1793, il., maps, (1939). (Also see preceding item T--1938.)

3--1942   (with López, V. M., and Brineman, J. H., Jr.) Geology of southeastern Venezuela. Geol. Soc. Am., B. 53: 849-872, il., (1942).

4--1950   Sucesos Cretácicos - Eocénicos en al norte de Venezuela. Assoc. Venezuela Geol. Min. Petrol., B. 2: 91-99, (1950).

5--1951   (Editor and senior author with Fichter, H. J., Renz, H. H., Wallis, W. E., Patterson, J. M., Renz, H. H., and Robie, R. H.). Geological Review, in symposium volume on the National Petroleum Convention, Ministerio de Minas e Hidrocarburos de Venezuela, p. 1-75, (1951). [Published in both English and Spanish.]

6--1953   (Senior author with same co-authors as in the preceding article - 5--1951). Geology of Venezuela and its oilfields. Am. Assoc. Petroleum Geol., B. 37: 670-777, il., (1953).

7--1962   (with Pinson, W. H. et al.) K-Ar and Rb-Sr ages of biotites from Colombia, South America. Geol. Soc. Am., B. 73: 907-910, (1962).

8--1963   Tectonic history of Venezuela, in Backbone of the Americas Symposium. Am. Assoc. Petroleum Geol., Mem. 2: 73-87, (1963).

9--1964   (with Pavlides, L. et al.) Outline of the stratigraphic and tectonic features of northeastern Maine. U.S. Geol. Surv., P.P. 501-C: C28-C38, map, (1964).

10--1966   (with Schopf, J. M. et al.) Erect plants in the early Silurian of Maine, in Geological Survey Research 1966: U.S. Geol. Surv., P.P. 550-D: D69-D75, il., (1966).

Articles published after 1966 are listed in the standard "Bibliographies" of the U.S. Geological Survey.

Dr. Stephen M. Simpson, Jr. (XII Ph.D. 1953) discussing a function of two variables.

(Photo by Studio One, Plymouth, Mass.)

(32)
STEPHEN MILTON SIMPSON, JR.

STEPHEN MILTON SIMPSON, JR.

MIT: 1952-1964

Shortly after entering M.I.T. as a doctoral candidate, Simpson became a marked student because of his expertise in mathematical analysis with high-speed digital computers. This skill, supplemented by unusual insight in applying mathematical analysis to physical problems, made him an obvious choice to assume directorship of the M.I.T. Geophysical Analysis Group (GAG) in 1954 when Enders Robinson resigned. The GAG Program, conceived in 1952 and initiated in early 1953, was supported by 20 major petroleum and geophysical companies, with the aim of developing techniques, based on modern advances in time series methods, for interpreting otherwise worthless seismic records taken in geophysical exploration. As Director he was responsible for controlling the direction and extent of research, training personnel (chiefly graduate students), supervising individual research projects, writing reports, organizing annual meetings of the M.I.T. GAG personnel and members of the Industry Advisory Committee, preparing budgets, and for seeing that the research program achieved its major objectives. Actually, Simpson was a member of the GAG throughout its entire existence from 1952 to 1957, and by the end of the project was considered such an expert in the field that he was selected as Director for VELA UNIFORM sub-project Contract AF 19 (604) 7378 at M.I.T. from July 1960 to June 1965. In this position he directed a research group in computer and modern time series methods for the determination of source polarity (and other methods of discriminating) for underground nuclear blasts and earthquakes from seismic recording. Eleven scientific reports were prepared on this project; Simpson himself wrote five, was first author of two, and supervised the other four reports.

After serving as a Research Assistant in Geophysics for a year, 1952-1953, he became successively: Instructor (1953-1954), Assistant Professor (1954-1962), Associate Professor (1962-1964), and finally Lecturer (1964-1965). During these years he taught subjects in seismology, elementary geophysics and statistical problems in geophysics, and conducted graduate seminars on applications of generalized harmonic analysis to geophysics, particularly to seismology. In 1959, at the invitation of the Mexican Government, he gave a series of lectures on computers to scientists in Mexico City. He acted as supervisor for 3 S.B. theses, 10 S.M.s, and 9 Doctor's, and was a consultant to several projects at M.I.T. and to a number of outside industrial firms.

More than any other person in Course XII, Simpson early saw the great potential future use of the high-speed digital computer, when programmed to analyze geophysical problems, and he was the principal programmer for the GAG Project, for the VELA UNIFORM Contract AF 19 (604) 7378, and for crystallographic research in M. J. Buerger's laboratory. Also in class as well as in laboratory he taught a generation of students how to program the modern computer to analyze their data. In short, his most important contribution in the Department of Geology and Geophysics, and his greatest impact on the earth sciences, lay in his leading role in bringing the potential of the programmed computer to the attention of staff members and students alike. Second only was his important role in helping to reorganize and restructure the curriculum and research program in geophysics.

Stephen Milton Simpson, Jr. was born in New York City on 29 January 1929. He was one of four children of Stephen Milton Simpson, a tax lawyer trained in the Georgetown University Law School, and Marjorie (Ewen) Simpson, a stenographer who had college training at Bryn Mawr. He attended Landon School for Boys in Bethesda, Maryland, from 1942 to 1946, compiling an outstanding scholastic record, majoring in science, and serving as both President of his Senior Class and as Valedictorian. Four years at Yale, 1946-1950, with a major program in physics, brought him a Bachelor of Science degree, election to Phi Beta Kappa and Sigma Xi, and award of the freshman Barge Prize for Mathematical Excellence.

In the autumn of 1950 he entered M.I.T. and started graduate work in Course XII in geophysics. It is interesting to note that Simpson's earliest interest at Yale was mathematics, but he abandoned that subject as a major field because of poorly taught mathematics courses plus the death of the professor who had inspired his freshman mathematics work.

Finding problems in geophysics that attracted his strong mathematical interest, and appearing on the scene as academically oriented computations were being accelerated in the M.I.T. Computation Center with its WHIRLWIND digital computer, he quickly turned back to his earlier interest in mathematics and directed this interest toward problems of wave propagation.

Professor George P. Wadsworth of M.I.T.'s Department of Mathematics and Professor Patrick M. Hurley of our Department of Geology had just started to look into the signal/noise problem that was bothering exploration geophysicists when Simpson started graduate work. Upon becoming acquainted with the problem, and with J. G. Bryan and E. A. Robinson who were experimenting with some low quality seismic records, he immediately got involved in the research and in it found the kind of mathematical challenge that excited his interest.

In Simpson's development from a graduate student to a professor, and from a partner in a small company to his independent consulting practice, we have an example of one of the great strengths of M.I.T.; namely, the training of professionals who leave in due course to found and develop entirely new enterprises. A brief statement of Simpson's stages of development is considered worth recording here because his experience illustrates how many graduate students at M.I.T. in the 1950s and 1960s had most unusual opportunities to participate in the so-called "digital revolution" that ensued as the mathematics invented by Norbert Wiener, simplified by Norman Levinson and applied to weather forecasting by George Wadsworth was used, with the WHIRLWIND computer, to analyze seismic records for the first time. The problems, the mathematics, the high-speed digital computer, and the competent and interested people all existed, as World War II ended and veterans returned to resume their interrupted education. It only remained to bring the four together in a working combination, and that is exactly what happened when Simpson first became acquainted with Wadsworth, Hurley, Bryan, and Robinson.

Early in 1950 Simpson applied for admission to M.I.T. to start graduate work in physics in September. After being admitted, however, he requested transfer to the Department of Geology and advice on what course of study he should follow to meet the requirements for a Master of Science degree in geophysics. In conformity with his wishes he was permitted to transfer to Course XII, but as a doctoral candidate, and was advised to expend major effort preparing himself in geology inasmuch as he had had no work in this field as a physics major at Yale. Thus it was that he spent his first two years of graduate work at M.I.T., SYs 1950-51 and 1951-52, taking first introductory and then advanced subjects in geology, and at the same time further strengthening his earlier preparation in mathematics and physics. During this period he learned of the experiments previously mentioned that Wadsworth and Bryan were performing on some low quality seismic records and soon became involved to the extent that he was appointed a Research Assistant for the SY 1952-53. Soon he was programming for the WHIRLWIND computer in order to carry out the computations needed by the Geophysical Analysis Group which was then being organized. This programming involved polynomial procedures, spectral analysis, auto- and cross-correlations, linear operators, spectral factorization, frequency distributions, iteration techniques, matrix manipulations, and various print and scope display routines.

By the time he had completed his doctoral thesis on "Statistical approaches to certain problems in geophysics," and received his Ph.D. degree (1953), he had become familiar with the general methods of mineral exploration with special emphasis on seismic techniques. His physics background had been slanted toward problems of elastic wave propagation, and his formal training in mathematics had been supplemented by independent studies of differential and integral equations, linear operator theory, computer techniques, numerical methods, and time series statistics. This program of study and research, involving the use of the rapidly developing computers in analyzing seismic data, gave him exactly the experience he needed to direct and extend the GAG program.

Accordingly, when Enders Robinson relinquished direction of the program in the summer of 1954, Simpson was the ideal candidate to succeed him, which he did in August 1954. For the next three years, until June 1957, Simpson acted as the GAG Project Director, having overall responsibility for the evolving research program that was supported by 20 major petroleum and geophysical companies. His unsuspected administrative ability became obvious as he reorganized and redirected the program, trained graduate students how to prepare programs for and use the computation facilities of M.I.T.'s Computation Center, prepared budgets for the overall project, organized periodic meetings at which the GAG participants presented their latest research results to representatives of the participating companies, wrote the annual reports, and supervised several graduate student theses. At the termination of the project he wrote the summary report and saw to it that all financial and other obligations were met. The success achieved by the GAG Project, particularly in its later stages, was as much due to the outstanding administrative ability of Simpson as to his fine training in geophysics, applied mathematics, and computer use. A more detailed history of the GAG Project is included elsewhere in this history in Volume 2. Suffice it to list below the eight GAG Reports (out of a total of eleven) that Simpson had a direct hand in preparing, either as author, co-author, or project supervisor:

3. Case study of Henderson County [Texas] seismic record, Part I, by GAG; E. A. Robinson, Director. MIT GAG Rept. 3: 5 p., il., 8 July 1953. [Personnel: Simpson, Smith, Walsh, Calnan, and Halpern.]

4. Linear operator study of a Texas Company seismic profile, Part I, by GAG; E. A. Robinson, Director. MIT GAG Rept. 4: numerous sections containing unnumbered pages, il., 21 July 1953. [Personnel: Simpson, Smith, Calnan, and Halpern.]

5. On the theory and practice of linear operators in seismic analysis, by GAG; E. A. Robinson, Director. MIT GAG Rept. 5: 8 sections individually paged, approx. 108 p., il., 4 August 1953. [Personnel: Simpson, Smith, Calnan, and Halpern.]

6. Further research on linear operators in seismic analysis, by GAG; E. A. Robinson, Director; M. K. Smith and S. M. Simpson, Assistant Directors. MIT GAG Rept. 6: 9 sections individually paged, approx. 200 p., il., 10 March 1954. [Personnel: Smith, Simpson, Briscoe, Gilbert, Bowman, Bowker, Treitel, Turyn, Walsh, Calnan, and Halpern.]

8. A multiple trace criterion for linear operator selection, by S. M. Simpson, Jr. MIT GAG Rept. 8: 27 p., 9 August 1954. [Published in its entirety in Geophysics 20/2: 254-269, (1955).]

9. Linear operators and seismic noise, by GAG; S. M. Simpson, Jr., Director. MIT GAG Rept. 9: 11 sect., approx. 300 p., il., 15 March 1955. [Personnel: Simpson, Bowman, Gilbert, Treitel, Tooley, Grine, Calnan, and Halpern.]

10. Properties, origin, and treatment of certain types of seismic noise, by GAG; S. M. Robinson, Jr., Director. MIT GAG Rept. 10a-b: 4 sect. approx. 200 p., il., 1 April 1956. [Personnel: Simpson, Bowman, Bryan, Tooley, Treitel, Wylie, Posen, Grine, Lopez-Linares, Fink, Calnan, Halpern, and Savage.] (This Report includes Robert Bowman's Ph.D. thesis, "Scattering of seismic waves by small inhomogeneities," in MIT GAG Rept. 10b, sect. 1, p. 1-0 to 1-76, il., September 1955.)

11. The interrelation of the deterministic and probabilistic approaches to seismic problems, by GAG; S. M. Simpson, Jr., Director. MIT GAG Rept. 11 [terminal]: 150 p., il., 30 June 1957. [Personnel: Simpson, Gilbert, Grine, Sax, Treitel, and Wylie.]

In the mid-1950s, when Professor M. J. Buerger and his graduate students were laboriously determining crystal structures by desk calculators, the labor being Fourier summations of trigonometric functions, Simpson asked Buerger if the summations might be accelerated by using the facilities of the M.I.T. Computation Center. At Buerger's request Simpson prepared the necessary program for the WHIRLWIND computer, and thereafter the research crystallographers turned completely to that computer and Simpson's program for the Fourier summations. This was certainly the first use of a recently developed high-speed digital computer in crystallographic research at M.I.T. and surely one of the earliest anywhere. Previously the much slower analog computer had been in wide use in England and Europe as well as in America.

By the time the GAG Project was completed Simpson had gained considerable expertise in the use of M.I.T.'s high-speed computers and had also demonstrated unusual administrative skill in organizing and directing research and supervising personnel. Recognition of this performance resulted in his being asked to direct a subproject of the VELA UNIFORM PROJECT for the U. S. Air Force, Contract AF 19 (604) 7378, with an annual budget of $175,000 and a staff of twelve persons. This research program, which ran from July 1960 through June 1965 produced the following eleven reports:*

1961 (VELA UNIFORM Rept. 1) Initial studies on underground nuclear detection with seismic data prepared by a novel digitization system, by S. M. Simpson, Jr., AFCRL--863, 450 p., June 1961.

1961a (VELA UNIFORM Rept. 2) Time series techniques applied to underground nuclear detection and further digitized seismic data, by S. M. Simpson, Jr., AFCRL--62-262, 500 p., Dec. 1961.

---

* Requests for copies of these reports should be directed to the Clearinghouse for Federal Scientific and Technical Information (CFSTI), Sills Building, 5285 Port Royal Road, Springfield, Virginia 22151.

1962 (VELA UNIFORM Rept. 3) Continued numerical studies on underground nuclear detection and further digitized seismic data, by S. M. Simpson, Jr., AFCRL--62-879, 363 p., June 1962.

1962a (VELA UNIFORM Rept. 4) Magnetic tape copies of M.I.T. Geophysics Program Set I (Time Series Programs for the IBM 709, 7090), by S. M. Simpson, Jr., AFCRL--63-282, 47 p., Dec. 1962.

1963 (VELA UNIFORM Rept. 5) Digital filters and applications to seismic detection and discrimination, by J. F. Claerbout [S.M. thesis at M.I.T.], AFCRL--63-604, 89 p., Feb. 1963.

1963a (VELA UNIFORM Rept. 6) Computer studies of microseism statistics with application to prediction and detection, by J. N. Galbraith, Jr. [Ph.D. thesis at M.I.T.], AFCRL--63-673, 283 p., May 1963.

1963b (VELA UNIFORM Rept. 7) Studies in optimum filtering of single and multiple stochastic processes, by S. M. Simpson, Jr., E. A. Robinson, R. A. Wiggins, and C. I. Wunsch, AFCRL--64-241, 140 p., June 1963.

1964 (VELA UNIFORM Rept. 8) Seismic arrays for the detection of nuclear explosions, by E. A. Robinson, AFCRL--64-855, 107 p., June 1964.

1965 (VELA UNIFORM Rept. 9) On factoring the correlations of discrete multivariable stochastic processes, by R. A. Wiggins [Ph.D. thesis at M.I.T.], AFCRL--65-207, 196 p., Feb. 1965.

1965a (VELA UNIFORM Rept. 10) Magnetic tape copies of M.I.T. Geophysics Program Set II (Time Series Programs for the IBM 709, 7090, 7094), by S. M. Simpson, Jr., AFCRL--65-306, 79 p., March 1965.

1965b (VELA UNIFORM Rept. 11) Sampling events from U. S. C. & G. S. [U. S. Coast & Geodetic Survey] earthquake cards, by S. M. Simpson, Jr., R. A. Wiggins, and C. Pan, AFCRL--65-463, 70 p., June 1965.

In May 1961, in response to communications from the Advanced Research Projects Agency [ARPA] in Washington, President J. A. Stratton requested our Department of Geology and Geophysics to work with the Departments of Physics, of Mathematics, and of Electrical Engineering in preparing <u>A Preliminary Proposal for the Establishment of a Center for VELA UNIFORM Research and Evaluation</u> to be submitted to ARPA by M.I.T. Simpson, aided by T. Cantwell and T. R. Madden of our Department, played the major role in the preparation of a 61-page Proposal that was submitted to ARPA by President Stratton on 29 June 1961. Throughout the document, a copy of which is preserved in the M.I.T. Institute Archives, is evidence of the innovative and imaginative skill and the impressive organizational ability of Simpson in computer technology. Unfortunately ARPA dropped the idea of such a center, and nothing ever came of the proposal. A year earlier, however, Simpson had been provided with ARPA funds to study the problem of nuclear detection on Contract AF 19 (604) 7378, as mentioned in preceding paragraphs.

Soon after completion of the ARPA contract Simpson resigned his appointment as Lecturer, on 30 June 1965, and proceeded to establish a private consulting practice.

Let us next review his teaching career at M.I.T. before going on to discussion of his more recent consulting practice. Serving first as a Research Assistant in Geophysics, from October 1952 to June 1953, Simpson programmed the WHIRLWIND computer in order to handle the GAG computations mentioned in earlier paragraphs. Then with his doctorate in hand he was appointed Instructor in Geophysics for SY 1953-54. During this year, and for several following, he taught courses in seismology, elementary geophysics, statistical problems in geophysics, and graduate seminars. He also took a leading role in reorganizing the geophysics curriculum, as Chairman of the Department's Committee on Graduate Students in Geophysics, and in seeking research funds for graduate programs. He continued these activities as Assistant Professor of Geophysics from 1954 to 1962, taking a leave of absence from June 1957 to September 1958 to be a Senior Project Member of RCA's Airborne Systems Laboratory at Waltham, Massachusetts, where he developed a group of eight or nine persons for computer programming and analysis for the Long Range Interceptor aircraft. When he returned to M.I.T. to resume teaching in September 1958 he continued as a consultant to RCA, in Burlington, Massachusetts, for the next four years, concerning himself with mathematical and statistical analysis, programming, and digital computer system design.

He was promoted to Associate Professor of Geophysics in July 1962 and continued in that rank until July 1964. When in 1963, one of his schoolmates, Thomas Cantwell (XII Ph.D. 1960), took a year's leave of absence from the Department to develop his own company, Geoscience Incorporated, Simpson took a similar leave, but for only half-time, for SY 1963-64 to help get the company started, serving as Vice-President and working on applied geophysics and time series analysis in private and government problems.

His interest in the work at Geoscience Incorporated was so challenging that at the end of the school year he requested that his rank be changed from Associate Professor (half-time) to Lecturer (half-time) so that he could devote more time to company activities. After one year as Lecturer he resigned, on 30 June 1965, and established his own private consulting practice, which he later named SIMPSON PROGRAMS, in Boston (now located in South Duxbury, Massachusetts).

Since establishing his own practice Simpson has been fully occupied providing clients with time-series analysis computer programs and consulting on data processing and mathematical analysis. Among his numerous assignments were two at M.I.T.--1) teaching a subject "Mathematical aspects of visual design" in the School of Architecture in the spring term of SY 1965-66 and 2) consulting in vision problems in the design of a robot for the Artificial Intelligence Group of Project MAC. In the early 1970s he was spending most of his time working as a consultant to Western Geophysical Company of Houston, Texas.

In 1962 Simpson married Jacqueline Clark, an M.I.T. mathematics student who helps him with mathematical problems when not engaged in developing a fine singing voice. At the time of this writing they are living in and restoring a fine old house built in 1755 in South Duxbury, Massachusetts, whence Steve carries on his consulting work. In a recent letter to Professor Whitehead he summed up his activities as follows:

> "I work (partially in Houston, mostly at home)... helping to guide the computers in their search through veritable torrents of seismograms for indications of oil bearing structures. There is much satisfaction in following through to massive industrial implementation...those concepts which began 17 years ago with a small group in building 24." [Letter to W. L. Whitehead dated May 10, 1969.]

Stephen M. Simpson, Jr., mathematical geophysicist, is a splendid example of those brilliant young students who, having received their undergraduate training elsewhere, find an area of interest at M.I.T., quickly develop their latent skills, and become leaders in one way or another. By his imaginative and creative use of the modern high-speed computer, in applying the statistical methods set forth by Wiener, Levinson and employed by Wadsworth, Simpson had a leading role in starting and then guiding the so-called "digital revolution" in geophysical exploration that resulted from the experiments of the M.I.T. Geophysical Analysis Group (GAG). Not only did he bring to the project mathematical insight and imagination of unusual depth and perception; he also quickly demonstrated unanticipated administrative skill that was so important to the success of the MIT GAG Project and later on of the VELA UNIFORM subproject. He it was who first brought the modern digital computer to M.I.T.'s geophysicists, geologists and geochemists, who have continued to use it ever since and on an increasingly broader and more effective scale.

BIBLIOGRAPHY OF STEPHEN MILTON SIMPSON, JR.

Symbols and abbreviations used in the following references are explained on p. 91-98; in general, abstracts are listed separately as well as with the references to the complete articles. This bibliography begins with the title of Simpson's doctoral thesis (T), submitted in 1953, and includes all publications through 1970. Not included are the numerous mimeographed reports on the MIT GAG Project and on Contract AF 19 (604) 7378; these are listed on preceding pages of the foregoing narrative. The bulk of Simpson's writings are unpublished military and industrial documents and computer programs (around 500 in number).

T--1953   Statistical approaches to certain problems in geophysics, 142 p., (1953). (Ph.D. Thesis at M.I.T. in Course XII, September 1953.)

1--1954   Least squares polynomial fitting to gravitational data and density plotting by digital computers. Geophysics 19/2: 255-269, (1954).

2--1955   Similarity of output traces as a seismic operator criterion. Geophysics 20/2: 254-269, (1955).

3--1966   Time Series Computations in FORTRAN and FAP: Vol. 1--A Program Library (7022). Reading, Mass.: Addison-Wesley Publishing Co., Inc., xi + 1120 p., (1966).

4--1967   Traveling signal-to-noise ratio and signal power estimates. Geophysics 32/3: 485-493, (1967). (First produced as Sect. 2 of MIT GAG Rept. 9, 15 March 1955.)

5--1967a   (and Fink, D. and Treitel, S.) Moveout averaging experiments. Geophysics 32/3: 494-498, (1967). (First produced as Sect. 1 of MIT GAG Rept. 10a, 1 April 1956.)

6--1967b   Fifteen years growth of Geophysics at M.I.T., a talk given... September 7, 1967, at a Western Geophysical Company Seminar in Houston, Texas...11 p., (1967). (Copy in M.I.T. Institute Archives.)

NOTE:

Simpson's unpublished lecture, "Fifteen Years Growth of Geophysics at MIT," mentioned in the foregoing bibliography is included in Volume 2, in which the history of geophysics at M.I.T. is discussed.

Graduate Student Kate Hadley (XII S.B. 1971; Ph.D. 1975) conducting an experiment on the high pressure-temperature equipment in Prof. Brace's laboratory in the Green Building (Room 54-720).

(Photo by C. Goetze, 1973)

(33)
WILLIAM FRANCIS BRACE

WILLIAM FRANCIS BRACE

MIT: 1954-

Following undergraduate and graduate training at M.I.T. (XIII S.B. 1946; I S.B. 1949; XII Ph.D. 1953), and a year with Bruno Sander in Innsbruck, Austria, W. F. Brace joined the Department of Geology in 1954 as Assistant Professor, advanced to Associate Professor in 1962, and became Professor of Geology in 1966. He took leave during SYs 1960-61 and 1961-62 to conduct experimental work on rock mechanics at Harvard, and is currently on leave (SY 1974-75) at Stanford, where he is conducting research on the role of rock mechanics in earthquakes.

From his student days as a trainee in engineering, he has been interested in the behavior of materials, from surface soils to minerals and rocks under high confining pressures. He has pursued this interest for 20 years (1954-1974) as an M.I.T. faculty member and has extended it to field work in both North America and Europe. In pursuing this interest he has brought to his teaching and research thorough training in geology, physics and the appropriate mathematics and a fine combination of imagination, engineering ingenuity and experimental skill. These attributes and skills he has used to gain deeper insight into how minerals and rocks in the earth's crust have behaved under high pressure and temperature during the geologic past. The students he trained have themselves made important contributions to a better understanding of rock mechanics. Finally, the articles that he has published, alone and with his students and research associates, have brought him recognition as one of the world leaders in research on rock mechanics, and election to the American Academy of Arts and Sciences and to the National Academy of Sciences.

BIRTH AND EARLY EDUCATION

William Francis Brace, one of Course XII's most distinguished alumni, was born in Littleton, New Hampshire, on 26 August 1926, to Frank Charles and Frances (Badger) Brace. When Bill was six years old his parents separated; his father lived elsewhere and held a number of business posts in Boston and New York City.

His mother, working full-time as a professional tax accountant and private secretary to a Boston businessman, could not take care of her son at home, so she sent him off to a private boarding school for his primary and secondary school education. After seven years at St. Clement's in

Canton (Mass.), Bill entered St. John's Preparatory School at Danvers
(Mass.), from which he was graduated in the spring of 1943. The instruction, particularly at the high school level, was good, and he acquired a
firm foundation in mathematics and developed a keen interest in the sciences. As a consequence, when he considered going on to college, he decided upon M.I.T.

STUDENT YEARS AT M.I.T. (1943-1953)

Entering M.I.T. as a freshman in the fall of 1943, Brace prepared for
an engineering career and ultimately received S.B. degrees in Naval Architecture and Marine Engineering (February 1946) and Civil Engineering (June
1949). During his undergraduate years he interrupted his academic training to serve in the U.S. Navy from 1944 to 1946, being separated as Ensign. He resumed his academic work in the fall of 1949, changing his
course of study to geology. As an undergraduate student in Prof. F. K.
Morris' introductory geology, he had become interested in the subject because of the dramatic way in which Morris lectured, and being much interested in outdoor activities, he decided to work towards a doctorate in
geology. Classroom discussions and field trips with Morris, lectures under W. J. Mead, and both lectures and laboratory work with H. W. Fairbairn,
led him to choose structural geology as his special field. Combining
field work, which took him to the out-of-doors which he liked so much,
with microscopic work in the laboratory, he did his doctoral thesis under
Fairbairn's supervision on "Rock deformation in the Rutland, Vermont,
area," and received his Ph.D. degree in June 1953.

Brace had taken engineering geology with Mead when working on his
S.B. degree in Civil Engineering so he was an obvious candidate for a
teaching assistantship in that subject when he became a graduate student
in Course XII. He assisted Mead from February 1950 to June 1952, while
completing his doctoral requirements, and he recalls that this teaching
experience convinced him he did not want to follow a career in the kind
of applied work that an engineering geologist was called upon to do. The
reason for the decision was his discovery of the exciting research possibilities in the Sander* type of structural petrology which he had learned
about in Fairbairn's advanced courses (Fairbairn had studied with Sander
earlier and had written an abridged version of Sander's work as a laboratory manual for students - Structural Petrology of Deformed Rocks (1949);
see Fairbairn's Bibliography in a preceding sketch). Furthermore, he

---

* Sander, B., Einführung in die Gefügekunde der geologischen Körper.
Vienna and Innsbruck: Springer-Verlag OHG, (1948).

found his doctoral thesis work much to his liking because it required extensive field work and application of Sander's principles, and this led to a desire to learn more about Sander's work. How better to do this than go directly to the master in Innsbruck, Austria. Accordingly, he applied for and won a Fulbright Fellowship for study abroad and spent SY 1953-54 at Innsbruck studying with Sander.

AS A FACULTY MEMBER AT M.I.T.

With such a promising young structural geologist in front of us, and particularly since we had not replaced Mead, who had finally to retire on 30 June 1954, we offered Brace an appointment as Assistant Professor of Geology, to start on 1 September 1954 on his return from Europe, and he accepted. Being Head of Course XII at the time, and being well aware of Brace's scientific interests and future promise, I suggested that he extend and expand his knowledge of mechanics and the mathematics needed for the kind of research he wished to do in rock deformation. Even though assigned a very heavy teaching load, he followed the suggestion and for the next six years, 1954-1960, he combined his teaching duties with classwork in physics and mathematics, and with field work, meanwhile preparing himself for the laboratory work in rock mechanics that he saw ahead. To carry on the latter, he took leave of absence for SY 1960-61 (later extended for SY 1961-62) to return to Europe for further study and then to work under Francis Birch's direction at Harvard, where Bridgman and Birch had earlier carried on notable work on the deformation of solids.

The investigations he carried on at Harvard, in a laboratory made available for his special work, produced half a dozen articles in 1960 and 1961 (See Bibliography farther on) that immediately drew favorable comment from leading structural geologists throughout America.

Upon his return to M.I.T. in the fall of 1962 and resumption of his professorial duties, meanwhile having been advanced to Associate Professor (1 July 1962), Brace did a number of things that would aid him in his future work.

With financial support from federal agencies he engaged Joseph B. Walsh (II Sc.D. 1958) as a Research Associate, to help him with many mathematical problems that arise in the study of rock mechanics. Walsh continues to this day (1975) as an indispensable partner in Brace's research, having been advanced meanwhile to Senior Research Scientist and having published by 1970 some 20 articles of which 8 were co-authored with Brace.

Brace immediately started planning a special laboratory, and a supporting machine shop, with which he could investigate the deformation of minerals and rocks under high confining pressure. These facilities were installed on the seventh floor of the new Green Building upon its completion in 1964, and since then have been in constant and productive use.

He immediately attracted a number of well-trained and promising graduate students to his laboratory work, and by 1970 seven of them had done their thesis work under his supervision.

Finally, since moving into the Green Building in 1964, ten years ago, he has pursued his own investigations of the deformation of solids under high confining pressure, and has published some 30 articles in that period. His research achievements brought him election in 1971 to the American Academy of Arts and Sciences and to the National Academy of Sciences. His most recent work on the important role of dilatancy in the rock mechanics associated with earthquakes has made his name a familiar one among seismologists, and some of his students are rapidly gaining similar recognition.

MARRIAGE AND FAMILY

Brace married Margaret Grant, from Saint John, N.B., a Radcliffe alumna (A.B. June 1955), on 3 September 1955, and they have three children: Colin William, Nathaniel Charles, and Sarah Tremaine. At the present time (March 1975) they reside at 136 Lakeview Avenue, Cambridge, where their backyard has been the scene of delightful fall picnics replete with good food, drink, and companionship.

SUMMATION

By 1970 W. F. Brace had established himself as one of America's leading investigators of rock mechanics by virtue of fundamental laboratory experimentation coupled with sound theoretical analysis shared with his M.I.T. research partner, Senior Research Associate Joseph B. Walsh (II S.B. 1952; S.M. 1954; M.E. 1956; Sc.D. 1958). The 70 (by 1974) or more articles, written alone or co-authored with his thesis students or with Walsh, have brought worldwide recognition of his experimental work and to the laboratory in which the investigations were conducted. In addition to the deeper insight into rock mechanics that he has produced by his own research work, he has trained a dozen students who are continuing and extending his work in important new directions. I have no hesitation in designating him one of Course XII's most distinguished alumni, and in predicting that he will continue to contribute importantly to better understanding of mineral and rock behavior under varying natural conditions.

BIBLIOGRAPHY OF WILLIAM FRANCIS BRACE

Symbols and abbreviations used in the following references are explained on pages 91 - 98; in general, abstracts are listed separately as well as with the reference to the complete article. This bibliography begins with Brace's first paper in 1953 and includes all known titles through the year 1970.

T--1946 (and Humphreys, W. Y., III) Cavitation pressure measurements, 4 + 48 p., il., (1946). (S.B. Thesis at M.I.T. in Course XIII, Naval Architecture and Marine Engineering, June 1946.)

T--1953 Rock deformation in the Rutland, Vermont area, 164 + 20 p., il., (1953). (Ph.D. Thesis at M.I.T. in Course XII, June 1953.)

1--1953a The geology of the Rutland area, Vermont. Vt. Geol. Surv., B. 6: 1-120, il., map, (1953).

2--1954 Die Einregelung von (100) der Hellglimmer .... Oest. Akad. Miss. 9: 79-85, (1954).

3--1955 Quartzite pebble deformation in central Vermont. Am. J. Sci. 253: 129-145, (1955).

4--1958 Interaction of basement and mantle during folding near Rutland, Vermont. Am. J. Sci. 256: 241-256, (1958).

5--1958a Indentation and plastic properties of certain geologic materials [abst.]. Am. Geophys. Union, Tr. 39: 507, (1958).

6--1958b (with Boucot, A. J. and Demar, R. E.) Distribution of brachiopod and pelecypod shells by currents. J. Sed. Petrology 28: 321-332, (1958).

7--1958c Plastic deformation of quartz during indentation [abst.]. Geol. Soc. Am., B. 69: 1539, (1958).

8--1958d Use of Mohr circles in the analysis of large geologic strain [abst.]. Geol. Soc. Am., B. 70: 1573, (1958). (See also 16--1961a.)

9--1960 [Discussion of] Orientation of anisotropic minerals in a stress field. Geol. Soc. Am., Mem. 79: 9-20, (1960).

10--1960a Behavior of rock salt, limestone, and anhydrite during indentation. J. Geophys. Res. 65: 1773-1788, (1960).

11--1960b An extension of the Griffith theory of fracture to rocks. J. Geophys. Res. 65: 3477-3480, il., (1960).

12--1960c Analysis of large two-dimensional strain in deformed rocks. Internat. Geol. Cong., 21st, Copenhagen, 1960, Rept., pt. 18: 261-269, il., (1960).

13--1960d Orientation of anisotropic minerals in a stress field--Discussion, Chap. 2 in Griggs, D. T., ed., Rock deformation - A symposium. Geol. Soc. Am., Mem. 79: 9-20, il., (1960).

14--1960e (and Dulaney, E. N.) The velocity behavior of a growing crack. J. Appl. Phys. 31: 2233-2236, (1960).

15--1961 Dependence of fracture strength of rocks on grain size, in Mining Eng. Ser. - Pr. 4th symposium on rock mechanics, 1961. Penn. State Univ. Mineral Ind. Exper. Sta., B. 76: 99-103, il., tabs., (1961).

16--1961a Mohr construction in the analysis of large geologic strain. Geol. Soc. Am., B. 72: 1059-1080, (1961). (See also 8--1958d.)

17--1961b   Experimental study of the indentation of rocks and minerals. Cambridge, Mass.: M.I.T., 81 p. (‡), (Oct. 1961).

18--1962   (and Walsh, J. B.) Some direct measurements of the surface energy of quartz and orthoclase. Am. Mineral. 47: 1111-1122, (1962).

19--1962a   (and Dulaney, E. N.) Velocity behavior of a growing crack: comments on a discussion by J. P. Berry. J. Appl. Phys. 33: 227, (1962).

20--1963   Behavior of quartz during indentation. J. Geol. 71: 581-595, (1963).

21--1963a   (and Bombolakis, E. G.) A note on brittle crack growth in compression. J. Geophys. Res. 68: 3709-3713, (1963).

22--1964   Effect of pressure on electric-resistance strain gages. Exper. Mechanics 4: 212-216, (1964).

23--1964a   Indentation hardness of minerals and rocks. Neues Jahrb. Mineral. Monatsh. 1964: 257-269, (1964).

24--1964b   (with Walsh, J. B.) A fracture criterion for brittle anisotropic rock. J. Geophys. Res. 69: 3449-3456, (1964).

25--1964c   Brittle fracture of rocks [with French and German abstracts], in State of stress in the earth's crust--International Conference, Santa Monica, Calif. 1963, Proceedings (W. R. Judd, ed.) New York: American Elsevier Pub. Co., p. 110-178, il., tabs., discussion, (1964).

26--1965   (with Orange, A. S. and Madden, T. R.) A preliminary study of the effect of pressure on electrical resistivity of saturated rock [abst.]. Am. Geophys. Union, Tr. 46: 70, (1965). (See also 33--1965g.)

27--1965a   (with Paulding, B. W., Jr.) Volume changes associated with fracture of rocks [abst.]. Am. Geophys. Union, Tr. 46: 162, (1965). (See also 34--1966.)

28--1965b   (with Walsh, J. B.) The effects of cracks on the elastic properties of rocks [abst.]. Am. Geophys. Union, Tr. 46: 162-163, (1965).

29--1965c   Some new measurements of linear compressibility of rocks. J. Geophys. Rev. 70: 391-398, (1965).

30--1965d   (with Walsh, J. B. and England, A. W.) Effect of porosity on compressibility of glass. J. Am. Ceramic Soc. 48: 605-608, (1965).

31--1965e   (with Simmons, G.) Comparison of static and dynamic measurements of compressibility of glass. J. Geophys. Res. 70: 5649-5656, (1965).

32--1965f   Relation of elastic properties of rocks to fabric. J. Geophys. Res. 70: 5657-5667, il., tabs., (1965).

33--1965g   (and Orange, A. S. and Madden, T. R.) The effect of pressure on the electrical resistivity of water-saturated crystalline rocks. J. Geophys. Res. 70: 5669-5678, il., tabs., (1965). (See also 26--1965.)

34--1966   (and Paulding, B. W., Jr. and Scholz, C.) Dilatancy in the fracture of crystalline rocks. J. Geophys. Res. 71: 3939-3953, il., tabs., (1966). (See also 27--1965.)

35--1966a   (with Press, F.) Earthquake prediction. Science 152: 1575-1584, (1966).

36--1966b   (and Byerlee, J. D.) Stick-slip as a mechanism for earthquakes. Science 153: 990-992, (1966).

37--1966c (and Orange, A. S.) Electrical resistivity: changes in saturated rock due to stress. Science 153: 1525-1526, (1966).

38--1966d (with Walsh, J. B.) Cracks and pores in rocks [with French and German abstracts], in Internat. Soc. Rock Mechanics Cong., 1st, Lisbon, 1966, Tr 1: Lisbon, Lab. Nac. Engenharia Civil: 643-646, (1966).

39--1966e (with Walsh, J. B.) Elasticity of rock--A review of some recent theoretical studies [with French and German abstracts]. Felsmechanik Ingenieurgeol. 4: 283-297, (1966).

40--1966f Laboratory studies of frictional sliding and of the effect of stress on the electrical resistivity of saturated crystalline rocks. Proc. 2nd U.S.-Japan Conference on research related to earthquake prediction, June 1966: 64-65, (1966).

41--1967 (and Byerlee, J. D.) Recent experimental studies of brittle fracture of rocks, Chap. 2 in Failure and breakage of rock--Symposium on rock mechanics, 8th, Univ. Minnesota, 1966, Pr.: New York, A.I.M.M.P.E., p. 58-81, il., (1967).

42--1968 (and Orange, A. S.) Electrical resistivity changes in saturated rocks during fracture and frictional sliding. J. Geophys. Res. 73: 1433-1445, (1968).

43--1968a Current laboratory studies pertaining to earthquake prediction. Tectonophysics 6: 75-87, (1968).

44--1968b (and Walsh, J. B. and Frangos, W. T.) Permeability of granite under high pressure. J. Geophys. Res. 73: 2225-2236, (1968).

45--1968c (and Martin, R. J., III) A test of the law of effective stress for crystalline rocks of low porosity. Int. J. Rock Mech. Min. Sci. 5: 415-426, (1968).

46--1968d (and Orange, A. S.) Further studies of the effect of pressure on electrical resistivity of rocks. J. Geophys. Res. 73: 5407-5420, il., (1968).

47--1968e (with Byerlee, J. D.) Stick slip, stable sliding and earthquakes--Effect of rock type, pressure, strain rate, and stiffness. J. Geophys. Res. 73: 6031-6037, (1968).

48--1968f (and Walsh, J. B.) Mechanical properties of rock of importance in decoupling analysis. Final Rept. Contract No. DASA-67-C-0090, Dept. Geol. Geophys., M.I.T., (1 Dec. 1968) (‡).

49--1968g (and Luth, W. C. and Unger, J. D.) Melting of granite under an effective confining pressure [abst.]. Geol. Soc. Am., S.P. 115: 21, (1968).

50--1968h (with Martin, R. J., III) Test of law of effective stress for crystalline rocks of low porosity [abst.]. Geol. Soc. Am., S.P. 115: 140, (1968).

51--1968i (with Unger, J. D. and Luth, W. C.) Melting of granite under an effective confining pressure [abst.]. Geol. Soc. Am., S.P. 115: 225-226, (1968).

52--1968j (and Scholz, C. H. and Weidner, D. J.) Compressibility of synthetic mineral aggregates [abst.]. Am. Geophys. Union, Tr. 49: 322, (1968).

53--1969 The mechanical effects of pore pressure on fracturing of rocks. Proc. Conference on Research in Tectonics, Ottawa, March 1968. Geol. Surv. Canada, C.G.S.P. 68-52: 113-123, il., (1969).

54--1969a (and Scholz, C. H. and La Mori, P. N.) Isothermal compressibility of kyanite, andalusite, and sillimanite from synthetic mineral aggregates. J. Geophys. Res. 74: 2089-2098, il., (1969).

55--1969b (with Greenberg, R. J.) Archie's Law for rocks modeled by simple networks. J. Geophys. Res. 74: 2099-2102, il., (1968).

56--1969c (with Byerlee, J. D.) High-pressure mechanical instability in rocks. Science 164: 713-715, (1969).

57--1969d Laboratory studies pertaining to earthquakes. N.Y. Ac. Sci., Tr. (2) 31: 892-906, il., (1969).

58--1969e (and Byerlee, J. D.) Mechanical instabilities observed in laboratory experiments on rocks [abst.]. EOS (Am. Geophys. Union), Tr. 50: 400-401, (1969).

59--1969f (with Wittels, Razel) Tectonic overpressure in Franciscan rocks [abst.]. EOS (Am. Geophys. Union), Tr. 50: 325, (1969). (See also 61--1970.)

60--1969g (with Byerlee, J. D.) Embrittlement of rocks at high confining pressure [abst.]. Geol. Soc. Am., S.P. 121: 44-45, (1969).

61--1970 (and Ernst, W. G. and Kallberg, Razel Wittels) An experimental study of tectonic overpressure in Franciscan rocks. Geol. Soc. Am., B. 81: 1325-1328, (1960). (See also 59--1969f.)

62--1970a (and Byerlee, J. D.) California earthquakes: why only shallow focus? Science 168: 1573-1575, (1970).

63--1970b (and Walsh, J. B. and Wawersik, W. R.) Attenuations of stress waves in Cedar City quartz diorite. Tech. Rept. No. AFWL-TR-70-8 (Air Force Weapons Laboratory, Kirkland AFB, N. Mex.), 76 p., (‡), (July 1970).

----

(34)
GORDON JAMES FRASER MACDONALD

GORDON J. F. MACDONALD

MIT: 1954-1958

After a brilliant student career at Harvard - A.B. summa cum laude, 1950; A.M., 1952; Junior Fellow in the Society of Fellows, 1952-1954; and Ph.D., 1954 - Gordon James Fraser MacDonald was appointed Assistant Professor of Geology at M.I.T. on 1 July 1954. The next year he was promoted to Associate Professor and continued as such until 1 July 1958 when he resigned to accept a professorship in geophysics at the University of California, Los Angeles.

During his four years in our Department of Geology and Geophysics he organized and taught courses in Geodynamics and Theoretical Geology; carried on quantitative and theoretical research on several of the more complex and challenging problems in mineralogy and petrology; and brought to both teaching and research effective application of mathematics, physics and chemistry to the attack on geological problems. He was a strong influence in upgrading the Course XII curriculum and in formulating a cooperative program in the earth and planetary sciences with the departments of physics and meteorology, a plan that was followed after he left. He impressed his M.I.T. colleagues by his insight into complex field problems and his imaginative attack on them. Our better prepared students recognized him as an exciting and stimulating teacher, and he quickly became a thesis supervisor for several of them. When he left us, we lost a brilliant young scientist of prodigious industry and superior talents, whose future seemed exceedingly promising.

That he lived up to those expectations has been proved dramatically by his meteoric rise among the younger scientists of the United States, as he has gone from one position of great responsibility to another, in both educational institutions and federal bureaus. Although these achievements are not a part of his M.I.T. services, I have added a brief review of them in order to demonstrate that his M.I.T. colleagues were fully aware of his remarkable talents and that he has more than fulfilled their expectations.

BIRTH AND EARLY EDUCATION

Gordon James Fraser MacDonald was born a British subject in Mexico City, D.F. on 30 July 1929, the son of Gordon and Josephine (Bennett) MacDonald. His British father was a mining administrator with American Smelting and Refining Company (ASARCO) in Mexico, and his mother was born in Texas. He has a younger sister.

He received his high school training at San Marcos Academy in San Marcos, Texas, then entered Harvard in the fall of 1946 as a Mexican citizen on an Immigrant Visa (later on he became a naturalized U.S. citizen).

While still a high school student he started working in the summer for the American Smelting and Refining Company in San Luis Potosí, Mexico, and for five summers, 1945-1949, he did quantitative chemical analyses of silicates and sulphides and research on various smelting problems. He alternated this inside laboratory work with visits to neighboring mines and work with company geologists in both surface and underground mapping.

THE HARVARD YEARS (1946-1954)

As a Harvard undergraduate, 1946-1950, he majored in geology, but, being interested in a broad range of science, he also took courses in the other physical sciences. Upon receiving an A.B. degree summa cum laude in 1950, he immediately entered Harvard's Graduate School, from which he received an A.M. degree in geology in 1952. During his two years of graduate study he concentrated his course work in mathematics and physics, having become intensely interested in application of thermodynamics and the quantitative approach to geological problems. Simultaneously he gained his first teaching experience by conducting the laboratory sections for optical crystallography and elementary petrology. In both instruction and research he demonstrated unusual ability in applying the principles of physics and chemistry to a variety of geological problems, and again used the summers to get more field experience. He spent the summer of 1950 working under George C. Kennedy (with whom he would later be a colleague at U.C.L.A.), mapping sedimentary and volcanic rocks in southern Montana, and the two subsequent summers, 1951 and 1952, mapping the Brattleboro area of southern Vermont in an attempt to correlate the stratigraphy of Vermont with that of adjoining New Hampshire.

By the time he had completed requirements for the A.M. degree in 1952 he had demonstrated such outstanding ability in the quantitative application of physics and chemistry to geological problems, and such promise for future achievement in geological research, that he was elected a Junior Fellow in Harvard's prestigious Society of Fellows in June 1952.

There next followed a two year period, 1952-1954, as a Junior Fellow, during which MacDonald had the time and opportunity to attack a variety of geological problems that excited his imagination and ingenuity. It took only a few moments of attentive listening to one of his enthusiastic discussions to conclude that here was a brilliant young scientist ready and eager to attack some of the most difficult problems in geology, particularly in mineralogy and petrology. His plans for the future stand forth clearly in a communication he sent me in May 1953, in which he wrote:

> "My general plan of attack is to use the methods of physics and chemistry to help solve problems in geology. Thermodynamics is not the only technique applicable. Many problems of mineral stability can be handled using statistical mechanics. Methods of irreversible thermodynamics and chemical kinetics will no doubt help answer various questions in metamorphic and igneous geology. Classical physics will be of great aid in interpreting problems in fields other than those commonly assigned to geophysics, in particular, structural geology. I believe that analysis of geologic problems in light of these disciplines will aid in evaluating accumulated data and guide future experimentation and field work.*
>
> "I am particularly interested in guiding students into using these methods in attacking geologic problems. I have found teacher-student relationship extremely valuable, in that questions of students have often led me to better understanding of the problems under discussion."

THE M.I.T. YEARS (1954-1958)

Little wonder that the young scientist who wrote the foregoing statement, when not yet 24 and without a doctorate, which however came from Harvard the next year, was elected a Junior Fellow so early. Little wonder, too, that when our Department sought a candidate for the vacancy left by L. H. Ahrens' departure for Oxford at the end of 1953, we unanimously agreed on MacDonald.

Thus it happened that he joined our Department of Geology in July 1954 as Assistant Professor of Geology, and within a year was promoted to Associate Professor, the appointment he held until he resigned on 1 July 1958 to accept a similar post at the University of California, Los Angeles.

MacDonald's impact on our Department was immediate, impressive, and highly constructive in several different ways. He quickly attracted the best of our graduate students to his several lecture subjects, and was soon supervising the theses of several of them. During his four years on our faculty he supervised a total of 5 theses: 1 S.B. and 4 doctorates.

---

* This educational philosophy appealed to me strongly, and when MacDonald joined our Geology faculty I initiated a series of weekly luncheon meetings during which our faculty hammered out a revised curriculum that called for more preparation in mathematics and the physical sciences. At the same time continued efforts were made to attract physicists and chemists into our graduate program. I gladly give MacDonald much credit for stimulating our Course XII faculty, myself included, to follow the broad outline that he set down in his "general plan of attack."

He took an active part in discussions of department planning and curriculum revision, and was instrumental in initiating changes that substantially improved the substance and quality of a number of Course XII subjects. His service on two ad hoc committees charged with formulating a plan for bringing the earth sciences into closer relations with physics, was especially helpful. Finally, he initiated instruction and research in Theoretical Geology (12.801-12.804) and Geodynamics (12,841, 12.842), supervised the five theses mentioned earlier, and himself published half a dozen articles on research in these fields (see Bibliography farther on).

M.I.T. lost a brilliant young scientist with great future promise and the Department of Geology and Geophysics lost one of its truly outstanding younger staff members when MacDonald decided to go to U.C.L.A. where he concluded that conditions and opportunities were more favorable for his future development.

THE YEARS AFTER M.I.T. (1958--)

Normally my biographical sketch would end here, but MacDonald's meteoric rise in both academic and public service has been so impressive that I feel a few additional brief comments are justified so that the reader may know how completely the young scientist who came to M.I.T. from Harvard's Society of Fellows at 24 years of age has lived up to everyone's expectations.

His ten years at U.C.L.A., 1958-1968, were productive of a series of outstanding articles on fundamental geophysics and won him election to the National Academy of Sciences at the almost unheard of age of 32. He served as Associate Director of the Institute of Geophysics and Planetary Physics from 1960 to 1968, and as Director of the Institute's Atmospheric Research Laboratory from 1960 to 1966. In 1968 he left Los Angeles to become Vice Chancellor for Research and Graduate Affairs at the Santa Barbara campus of the University of California, a post he held until 1970 when he was appointed by the President of the United States to serve as a member of the Council on Environmental Quality. In 1972 he resigned this office to accept appointment as the Henry R. Luce Third Century Professor of Environmental Studies and Policy and Director of the Environmental Studies Program at Dartmouth College.

Outside the academy during the 1950s and 1960s he was much on the go as an advisory scientist. He was a staff associate at the Geophysical Laboratory of the Carnegie Institution of Washington from 1955 to 1963; consultant to the U.S. Geological Survey from 1955-1960; member of PSAC (President's Science Advisory Committee) from 1965 to 1969; and he served IDA (Institute for Defense Analysis) successively as Vice President of

Research, 1966-1967, Executive Vice President, 1967-1968, and Trustee, 1966-1970. Other services included membership on the Lunar and Planetary Missions Board, 1967; member of the Defense Science Board of the Department of Defense, 1967-1970; consultant to the Department of State, 1967-1970; and member of the National Academy of Science's Space Science Board, 1962-1970, and Environmental Studies Board, 1969-1970, 1972-1973, serving as Chairman in 1970 and again from 1972-1973. MacDonald currently is serving as Chairman of the Commission on Natural Resources of the National Academy of Sciences and National Research Council.

SUMMATION

With the foregoing brief review of MacDonald's activities within and outside the academy, it is understandable that a recent notice in Science (vol. 176: 892, 26 May 1972) included the following:

> "For a man of 42, MacDonald has had an unusually varied and peripatetic career. By any measure, he is one of the fastest rising and most ubiquitous figures in science policy circles."

But with all his varied activities, MacDonald still maintained a keen interest in geophysics, and during the 16 years since leaving M.I.T. he has continued to publish important articles on some of geology's most complex and challenging problems.

Gordon J. F. MacDonald has occupied with distinction many positions of great responsibility in both educational institutions and governmental bureaus, and I see him going on to even greater achievements in the years ahead as he brings his remarkable talents to bear on one of the more critical problems of our time - the preservation and improvement of our environment.

BIBLIOGRAPHY OF GORDON JAMES FRASER MACDONALD

Symbols and abbreviations used in the following references are explained on p. 91 - 98; in general, abstracts are listed separately as well as with the complete article. This bibliography begins with MacDonald's earliest publication and includes all known articles through 1960, a year and a half after he left M.I.T. I have extended the cutoff time eighteen months beyond his actual resignation on the assumption that he could get published in this period any research left uncompleted or in press when he left.

1--1953   (with Boucot, A. J., Harner, R. S., and Hon, D. M.) Age of the Bernardston formation [Massachusetts] [abst.]. Geol. Soc. Am., B. 64/12; pt. 2: 1397-1398, (1953).

2--1953a   Anhydrite-gypsum equilibrium relations. Am. J. Sci. 251: 884-898, (1953).

T--1954   A critical review of geologically important thermochemical data, 248 p., (1954). (Ph.D. Thesis at Harvard University in the Division of Geological Sciences, June 1954.)

3--1955   Gibbs' free energy of water at elevated temperatures and pressures with applications to the brucite-periclase equilibrium. J. Geol. 63: 244-252, (1955).

4--1956   Equations of state of solids in the earth [summary]. J. Geophys. Res. 61: 387-391, tab., (1956).

5--1956a   Experimental determinations of calcite-aragonite equilibrium relations at elevated temperatures and pressures. Am. Mineral. 41: 744-756, il., (1956).

6--1956b   Quartz-coesite stability relations at high temperatures and pressures. Am. J. Sci. 254: 713-721, il., (1956).

7--1957   (with Robertson, E. C. and Birch, F.) Experimental determination of jadeite stability relations to 25,000 bars. Am. J. Sci. 255: 115-137, il., (1957).

8--1957a   Thermodynamics of solids under non-hydrostatic stress with geologic applications. Am. J. Sci. 255: 266-281, il., (1957).

9--1957b   Orientation of anisotropic minerals in a stress field [abst.]. Geol. Soc. Am., B. 68/12, pt. 2: 1762, (1957). (See also 20--1960a.)

10--1958   (with Knopoff, L.) The magnetic field and the central core of the earth [abst.]. Am. Geophys. Union, Tr. 39: 522, (1958). (See also 14--1958d.)

11--1958a (and Knopoff, L.) The chemical composition of the outer core [abst.]. Am. Geophys. Union, Tr. 39: 524, (1958). (See also 15--1958e.)

12--1958b (with Boucot, A. J. et al.) Metamorphosed middle Paleozoic fossils from central Massachusetts, eastern Vermont, and western New Hampshire [New England]. Geol. Soc. Am., B. 69: 855-864, il., (1958).

13--1958c (with Knopoff, L.) Attenuation of small amplitude stress waves in solids. Rev. Mod. Phys. 30: 1178-1192, (1958).

14--1958d (with Knopoff, L.) The magnetic field and the central core of the earth. Roy. Astron. Soc., Geophys. J. 1: 216-223, (1958). (See also 10--1958.)

15--1958e (with Knopoff, L.) On the chemical composition of the outer core. Roy. Astron. Soc., Geophys. J. 1: 284-297, (1958). (See also 11--1958a.)

16--1959   Chondrites and the chemical composition of the earth, p. 476-494 in Abelson, P. H., ed., Researches in Geochemistry. New York: John Wiley & Sons, Inc., x + 511 p., (1959).

17--1959a (with Knopoff, L.) An Equation of state for the earth's core [abst.]. J. Geophys. Res. 64: 1112, (1959). (See also 23--1960d.)

18--1959b Calculations on the thermal history of the earth. J. Geophys. Res. 64: 1967-2000, il., (1959).

19--1960   (with Knopoff, L.) An equation of state for the core of the earth. Roy. Astron. Soc., Geophys. J. 3: 68-77, il., (1960). (See also 23--1960d.)

20--1960a The orientation of anisotropic minerals in a stress field, Chap. 1 in Griggs, D. T., ed., Rock Deformation--A symposium. Geol. Soc. Am., Mem. 79: 1-8, (1960).

21--1960b   (with Munk, W. H.) Rotation of the Earth; a geophysical discussion. (Cambridge monographs on mechanics and applied mathematics.) Cambridge, England: Cambridge Univ. Press, 323 p., il., (1960). [Awarded the A.Ac.A.S. Monograph Prize in Physical and Biological Sciences, 1959.]

22--1960c   (with Gilbert, F.) Free oscillations of the earth: 1. Toroidal oscillations. J. Geophys. Res. 65: 675-693, (1960).

23--1960d   (with Knopoff, L.) An equation of state for the core of the earth. Roy. Astron. Soc., Geophys. J. 3: 68-77, (1960). (See also 17-1959a and 19--1960.)

24--1960e   Stress history of the moon. J. Planet. Space Sci. 2: 249-255, (1960).

25--1960f   (with Munk, W. H.) Continentality and the gravitational field of the earth. J. Geophys. Res. 65: 2169-2172, (1960).

26--1960g   (and Ness, N. F.) Stability of phase transitions within the earth. J. Geophys. Res. 65: 2173-2190, (1960).

27--1960h   (with Knopoff, L.) Models for acoustic loss in solids. J. Geophys. Res. 65: 2191-2197, (1960).

28--1960i   Tectonic theories. Am. Geophys. Union, Tr. 41: 168-169, (1960).

29--1960j   NASA Lunar Research Conference. Astronautics: 71-75, (May 1960).

30--1960k   Seismic activity on the moon. Proc. 8th Lunar Planet. Expl. Colloq. 2: 45-47, (1960).

31--1960l (with Jastrow, R.) Highlights of the planetary sciences program. Am. Geophys. Union, Tr. 41: 430-434, (1960).

NOTE:

References to articles published after 1960 can be found in the Bibliographies of North American Geology and other similar standard bibliographies.

----

(35)
WILLIAM HAMET PINSON

WILLIAM HAMET PINSON

MIT: 1956-

When the Department of Geology and Geophysics was asked to offer a freshman elective in Astronomy in 1956, William H. Pinson (XII Ph.D. 1952) was appointed to the geology faculty to give the subject. As a result of the initial success of this elective, he was asked later to develop a similar one in earth science. Today, 18 years later, as an associate professor, he offers several introductory subjects in astronomy and on occasion may teach a specialized course in earth science. Excellent lectures coordinated with work in local observatories and with field trips to areas of special geological interest have helped to maintain the popularity of Pinson's courses in the face of steadily growing numbers of elective subjects available to freshmen. In addition, he developed a productive program of research on meteorites and tektites, and supervised a dozen theses dealing with these interesting objects.

Pinson first came to M.I.T. in 1949 as a doctoral candidate in geochemistry. With essentially four undergraduate majors--geology, mathematics, physics and chemistry--and with A.B. and S.M. degrees in geology from Emory University, he was well prepared to start his doctoral program on some aspect of earth science. Having a longtime interest in astronomy, he took advantage of the opportunity to study the subject at Harvard, and there became acquainted with famed Harlow Shapley, who would greatly influence his future career. At the same time he pursued a regular program of doctoral work at M.I.T. He soon learned of the exciting research in geochemistry being conducted by L. H. Ahrens, then Director of the Department's Cabot Spectrographic Laboratory, and became much interested. Ahrens introduced him to some of the more challenging problems in geochemistry and cosmochemistry and emphasized how further knowledge of meteorites could be of great importance in learning more about the history of the earth and of the solar system, and about geochemical properties in general. As a consequence, Pinson decided to investigate meteorites for his doctoral thesis under Ahrens' supervision. In due course he submitted a thesis on "Trace element studies in meteorites and rocks and the origin of meteorites" for the Ph.D. degree, which he received in January 1952.

While still working on his doctoral thesis at M.I.T. as a degree candidate, Pinson was simultaneously a research fellow in geology and an instructor in astronomy at Harvard (1951-1952), and he continued at Harvard for an additional year (SY 1952-53) before returning to M.I.T. as a research associate in October 1953. At that time he joined P. M. Hurley, who was rapidly developing a geochronological laboratory for determining mineral and rock ages by radiochemical methods, and thus began an association with Hurley and his research that would continue until the late 1960s. Almost immediately he became closely associated with L. F. Herzog, 2nd (XII Ph.D. 1952), a fellow doctoral candidate of previous years, who was helping Hurley build the spectrometers and supporting equipment needed for the Geochronology Laboratory in Building 24. Pinson gives much credit to Herzog for making him aware of the rigorous approach and precise laboratory methods he would have to bring to the study of meteorites if he wished to extend the research he had started in his doctoral work with Ahrens.

There followed a brief period, 1953-1956, in which several changes in staff and in Pinson's activities set the course for his future career at M.I.T. Ahrens left at the end of 1953, just two months after Pinson's return, and Herzog left in 1956 as Pinson started his teaching career. Both Ahrens and Herzog had greatly influenced Pinson, and he credits them with guiding him to his subsequent research on meteorites and tektites. In 1953 H. W. Fairbairn also joined Hurley's research group, and it was not long until the triumvirate, Hurley-Fairbairn-Pinson, became well known around the world for their AEC-supported research on "Variations in isotopic abundances of strontium, calcium, argon, and other topics." Pinson's own contributions to the impressive record made by the triumvirate consist of some 120 articles, published alone or jointly with others, and supervision of a dozen theses. Included in his contributions is the program of instruction and research on meteorites and tektites that he initiated and then developed into an internationally recognized effort. Several of the doctorates trained in the Pinson program are now national leaders in meteoritic research.

In addition to teaching and conducting research Pinson has delivered numerous public lectures (generally on meteorites and tektites) and participated as an invited lecturer in a number of conferences and symposia on meteorites and other astronomic and earth science subjects. He worked one summer (1953) for the Geological Survey of Canada as an assistant to D. G. Kelley (XII Ph.D. 1959), doing field work on Cape Breton Island, Nova Scotia; served as one of the instructors at the Indiana University Summer Field Camp in Montana (1963), directed by J. Mead (XII Ph.D. 1949), where our Course XII juniors were in attendance; and spent SY 1968-69 on leave as a Visiting Scientist at the NASA Goddard Space Flight Center at Greenbelt, Maryland. He has served on various Department and Institute committees; assisted the M.I.T. Admissions Office in evaluating applications; and most recently has been active in the IAP (Independent Activities Period) and UROP (Undergraduate Research Opportunities Program).

He can still be enticed to lead a group of Scouts (girls or boys) on a mineral-collecting trip, or a group of older people on a geological field excursion, and he is not likely to turn down a request to take a group on an all-night observing session to one of the local observatories (if the weather is right). And in some of his spare moments he has returned to two of his interests that go back to his boyhood--making telescopes and experimenting with violins, activities that he enthusiastically shares with interested students. He is a "student's professor" in every sense of that highly complimentary phrase.

BIRTH, ANCESTRY, AND EARLY YOUTH

William Hamet Pinson, Jr.* was born in Atlanta, Georgia, on 6 September 1919. He was the only son and the fourth of five children of William Hamet Pinson, a broker in cotton seed products, and Mary Rhett (Jones) Pinson, an enterprising dealer in antiques and other items as well as a housewife and mother. The first three children were girls, two of

---

* See footnote on following page.

whom died early, one in infancy and the other at about age five; then came William in 1919 and Rhett in 1922. The older sister, born in 1912, is named Virginia.

Pinson's ancestry can be traced back through both parents to English forebears who came first to Virginia in early colonial days, then moved on to other states in the South. A family tradition has it that the Pinsons were direct descendants of Vincent and Alonzo Pinzon, commanders of the Nina and Pinta, sister ships in the fleet of Christopher Columbus. However, whether ancestors of our William H. Pinson, Jr. did in fact command those famous ships remains to be proved.

Bill's father, the senior William Hamet Pinson, attended Alabama Polytech, now Auburn University, for two years and then left school to work as a travelling salesman dealing in fertilizer, cottonseed oil, meal and hulls for several years before organizing his own company. He was in business for a while with his brother John Pinson.

As a young man he engaged in football and baseball in college, and followed cockfighting and other sporting activities pursued by some young men of his time. He had no interest in science, music, or the arts, but he did appreciate English literature and read widely among the works of Tennyson, Scott, Dickens, and Thackeray. He did not, however, especially encourage his own son to read his books and even went so far as to ridicule such arts as music, painting, sculpture, and ballet. On the other hand, he did instill in his son an intensive interest in nature, especially bird life, and took him along on many hunting and fishing trips. Later in his own life, Bill gave up hunting in 1946, first for philosophical reasons and second because he became intensively interested in so many aspects of nature. It took him another ten years to give up fishing. In 1957 he gave up eating meat altogether, except for fish, which remains an acceptable food, but for dietary rather than philosophical reasons.

The senior William Hamet Pinson devoted most of his time to his business responsibilities--Bill remembers him as a hard worker and a good provider for his family--and to the amusement he found in the sporting activities mentioned earlier. When he was born in 1877 the Civil War and Re-

---

* Much of what is written in the biographical part of this sketch is based upon several hours of conversation with Prof. Pinson in which he told me in much detail of his childhood and early youth. It is with his permission and approval that I comment on some of the attitudes and activities of his parents, who, as he emphasizes, were good people but typical of the persons he knew as a boy in Georgia. He feels, and I think he is right, that his compassion for the oppressed and his concern about human rights can be traced back to what he saw and experienced as he grew up in the city of Atlanta in the decades immediately following World War I.

construction that followed were still fresh in the minds of his parents, and he grew up in a generation that had to make great adjustments to the hard times of the post-bellum period. As a speculator in cotton and cotton futures he had his successes and failures, at times reaching some affluence and on one occasion going through bankruptcy. Bill junior was aware of his father's business activities and of the kind of life he led, and as a result had no desire to follow him into business. Nor did he develop any interest in medicine, the profession practiced by two paternal relatives, his grandfather and his uncle, Chris Pinson. Likewise, there was nothing of interest to him in the urban development activities and politics followed by his father's older brother, John Pinson, who served in the Alabama Legislature for more than twenty years, and during the same period raised funds for the Tuskegee Institute. It is probably fair to say that in his youth Bill had a greater desire to read books about all sorts of subjects than to think seriously about a future profession.

Pinson's mother, some ten years younger than his father, was one of a family of 13 girls, and lived to an age of 81 years, having been born in 1888. The Jones family was "land-poor," *i.e.* they had inherited large acreages but little money, and the members had to struggle for their economic existence, particularly after the father of the family died when only a little more than 40 years old.

As a youngster, Bill once asked one of his maternal aunts, whom he remembers as a rather strong-willed lady, and who was greatly interested in their family tree, and in the purported ancestral linkage to Alonzo and Vincent Pinzon mentioned earlier,

> "Aunt Katie, I have heard a lot about my Grandfather Jones but I have never known what he did for a living. What did he do for a living?"

Aunt Katie replied vehemently,

> "Do? Why he didn't do anything! He was a gentleman!"

and in her statement, every word of which she meant, she clearly revealed the attitude of the Southern land-owner of ante-bellum days. Actually, he farmed and taught school for a living.

In spite of the fact that the Jones family was not affluent, all thirteen Jones girls received some education beyond the grades, generally in a secretarial school, and all ultimately married men who achieved success in business.

Of the thirteen Jones sisters, Pinson's mother Charlotte Rhett, who was about midway in the sibling sequence, probably had the hardest life of any of them, due to her husband's business difficulties and socially embarrassing sporting activities such as cockfighting. Nevertheless, she tried to keep her children properly dressed, enrolled them in the best possible schools, restricted them to proper play areas, and allowed them to enter only what she considered the better social circles [Even at a very early age, however, Pinson's ideas did not coincide with his mother's as to which "social circles" to move in!]. She arranged that they had music and dancing lessons, and tried to expose them to the arts in contrast to the attitude of her husband, who was wont to poke fun at the arts. To help support the family she sold corsets and bought and sold antiques, and even raised Persian cats for sale. Between times she also worked in her husband's office. From about 1938 until her death she rented out rooms in their home. Quite obviously, Charlotte Rhett (Jones) Pinson was a lady of strong character and with a determination to see that her children were accepted into "decent society" and prepared to make their own way by gaining skill in some profession. With such a social background it does not seem too surprising that Bill's mother was intolerant politically, racially, and socially. She literally dictated what her children could and could not do. Bill still remembers how, after a certain age (4 or 5 years), he was no longer permitted to play with a little black boy, and how years later when he was offered an opportunity to become an apprentice machinist, at twice the salary he was earning as a white-collar clerk, his mother adamantly refused to let him seize the opportunity, because doing so would be socially demeaning!

Both parents were pleased when their son decided to go to college, but their pleasure came from the assumption that he would be training for making a livelihood; they may have respected scholarship, but survival of the family by having its members able to earn their livelihood was the dominant ambition of the parents.

Today, Bill ascribes his longtime interest in astronomy to the fact that when he was quite young his mother bought him a book on the subject, Simon Newcomb's Astronomy for Everybody, which immediately aroused his interest, and a telescope that he quickly put to use. Ever since, he has been interested in making telescopes and in teaching astronomy. His current interest in making and playing violins, an interest he has been sharing with M.I.T. students participating in the recently introduced IAP (Independent Activites Period) and UROP (Undergraduate Research Opportunities Program), goes back to the violin lessons he was given when a boy.

EARLY EDUCATION

Pinson received his primary education in the Spring Street School in a middle-class neighborhood in the north-central part of Atlanta (1925-1930), after which he attended the O'Keefe Junior High School in the same area for three years (1930-1933). He then entered the Boys High School in Atlanta, where he prepared for college and graduated in 1937. After one year at the Georgia Institute of Technology, 1937-1938, during which he did not do well for lack of motivation, he left school and tried several kinds of employment (attendant in a filling station; clerk in two gas companies in Atlanta) in order to earn some money and to discover what he wanted to do with his life. World War II brought an end to this uncertain two-year period of his early education.

WORLD WAR II SERVICE (1941-1946)

He enlisted in the U.S. Army in January 1941 but soon applied for Officer's Training School. In the following December he was transferred to the U.S. Air Force to start training as an Aviation Cadet. At that time he wanted a piece of the action! He was commissioned a 2nd Lieutenant Bomber Pilot in July 1942, after training in Arkansas and Mississippi and in the Pacific Northwest, and in early 1943 was sent to the European Theater with the 8th Air Force. His plane was shot down over Wilhelmshaven, Germany, 11 June 1943, after completing its bombing run on the submarine docks, and he jumped safely into the North Sea where he was picked up by a German patrol boat. He was first imprisoned in Stalag Luft 3, until January 1945, and was then marched across Germany to Mooseburg, near Munich. The Russian army would soon have liberated his camp! He was liberated, instead, by American troops on 29 April 1945, returned to the United States soon after, and discharged as Captain in April 1946.

COLLEGE EDUCATION (1946-1949) - (EMORY)

While a prisoner of war Pinson studied from books made available by the Y.M.C.A. and also attended classes organized among themselves by some of the prisoners. With much time on his hands, and a renewed interest in further education, he studied diligently and as a result decided that as soon as possible after discharge he would resume the college training he had interrupted when he left Georgia Tech in 1938. Accordingly, in September 1946, having been released and discharged meanwhile, and with the help of the G.I. Bill of Rights, he entered Emory University. He was awarded an A.B. degree in Geology, while at the same time completing major programs in mathematics and chemistry. Continuing his studies, he received an M.S. degree in geology in 1949, meanwhile having further broadened his scientific education with four courses in physics and his humanities with courses in philosophy and literature. This latter interest in the humanities is even more manifest today in Pinson's wide reading in the social and political sciences and in the ever growing shelves of books that he has collected.

As an Emory student Pinson was greatly stimulated and influenced by James Lester, his professor of geology. When Lester became aware of Pinson's intense interest in the more fundamental and quantitative, physical and chemical, aspects of the earth, he urged him to stay a year beyond his bachelor's degree, so as to strengthen his knowledge of physics, and then apply for admission to M.I.T. for the doctorate. Following Lester's perceptive advice, he applied to M.I.T. in early 1949 and was accepted as a doctoral candidate in Course XII, Geology and Geophysics.

GRADUATE STUDENT YEARS AT M.I.T. AND HARVARD (1949-1952)

Pinson entered the Institute in September 1949 and during the next three years concentrated his graduate work in geochemistry, radiogeology, and spectrometry. During this same period, and also later, during SYs 1951-52 and 1952-53, he followed his boyhood interest in telescopes and astronomy by first taking several formal Harvard courses in astrophysics, and then later by auditing several other courses. His education in astrophysics, therefore, was fairly extensive during his post-doctoral period at Harvard. This extra-M.I.T. work brought him into close contact with famed Harvard astronomer Harlow Shapley, whose scientific prowess and liberal social views deeply influenced Pinson's own future attitude. Later on his great admiration for Shapley led him to name his son Harlow after the famous astronomer. Later, also, Shapley would write two of the strongest letters of support for Pinson's appointment as assistant professor at M.I.T. and for his promotion later to associate professor.

Soon after starting graduate work at M.I.T., Pinson became aware of the geochemical research interests of L. H. Ahrens, then Director of the Department's Cabot Spectrographic Laboratory, and was attracted by the work going on in the Laboratory. Ahrens informed him of some of the more important and exciting problems in geochemistry and cosmochemistry and particularly emphasized that the study of meteorites was of fundamental importance in gaining a better understanding of the history of the earth and of the solar system. It should be remembered, incidentally, that this was many years before the first lunar landing! With Ahrens' encouragement and promise of supervisory assistance, Pinson decided to investigate the trace elements in meteorites for his doctoral thesis. For the next three years he carried on his research, having a research assistantship during SY 1950-51, and in January 1952 he submitted his doctoral thesis, "Trace element studies in meteorites and rocks and the origin of meteorites" and received his Ph.D. degree in geochemistry.

Meanwhile he had been serving as a Research Fellow in Geology and as an Instructor in Astronomy at Harvard during SY 1951-52, and after receiving his doctorate he continued to hold these two appointments for SY 1952-53. During these two years at Harvard he gained the teaching and observatory experience that made him an ideal choice later on for teaching astronomy in our Department.

TEACHING AND RESEARCH AT M.I.T. (1953 TO PRESENT)

In October 1953 Pinson returned to M.I.T. as a research associate, joining his former graduate schoolmate, L. F. Herzog, 2nd (XII Ph.D. 1952), in Hurley's developing geochronological laboratory. Herzog had chosen as his doctoral thesis the construction of a mass spectrometer to investigate the natural variations in strontium isotope abundances in minerals as a possible age method. After constructing the instrument and receiving his doctorate in June 1952, he had been appointed a research associate in geology and was working full-time in Hurley's laboratory in Building 24 when Pinson returned. For the next three years (1953-1956), both Herzog and Pinson worked with Hurley in operating the newly developed spectrometric facilities. Pinson has remarked to me that Herzog showed him the great importance of rigorous thinking and rigorous instrumental work in his research, and constantly emphasized the need for careful work in the laboratory.

Pinson worked as a D.I.C. staff member from July 1955 to February 1956, and as Lecturer in Geology until 1 July 1956 at which time he was appointed an assistant professor. From then until now (1974) he has been a full-time professor except for SY 1967-68, during which he was on leave as a Visiting Scientist to work at the NASA Goddard Space Flight Center in Greenbelt, Maryland. On 1 July 1960 he was advanced to associate professor with tenure, the appointment he holds as this is being written (July 1974).

In 1956, when the Dean of Science, George R. Harrison, asked if Course XII could offer a freshman elective subject in astronomy, and we asked Prof. Harlow Shapley of Harvard if he knew a young scientist competent to teach such a subject, he strongly recommended Pinson. As a consequence of his recommendation, Pinson was appointed to the Course XII faculty, as discussed in the preceding paragraph. He first offered the subject in the spring term of SY 1955-56, and drew 120 students. Because such an unexpectedly large group could not be adequately handled in night observations on the roof of Building 24, where Pinson set up his portable telescopes, or at the Harvard College Observatory and the Smithsonian Astrophysical Observatory (Harvard), where his students were permitted to use the permanently mounted instruments, it was decided to offer the subject in both terms of SY 1956-57. Thus began instruction in astronomy in Course XII in 1956.

During the following decade (1956-1966) Pinson offered additional subjects in astronomy and initiated both instruction and research in "meteoritics," a new science which required the very combination of training and skills he possessed.

When he joined the Geology faculty in 1956, after having gained valuable experience under Herzog's tutelage in operating Hurley's mass spectrometer, Pinson immediately began looking for geochemical research that would be related to his teaching in astronomy. By this time Hurley's geochronology laboratory was in full operation, financial support was being given by the A.E.C., and Fairbairn was now an active participant in the research project. Pinson continued to work with this laboratory group and became a productive member of the now well-known geochemical triumvirate, Hurley-Fairbairn-Pinson, which has been responsible for 20 annual reports, several hundred publications, and supervision of a hundred or more theses in geochemistry and geology through 1970. Pinson's own participation in the aforementioned program has resulted in some 120 articles, published alone or jointly with others, and has involved supervision of a dozen theses. Included in his record is the program of instruction and research in meteoritics and on tektites that he initiated in 1950 and subsequently developed into an internationally recognized effort. As mentioned earlier, several of the doctorates trained in this particular program are now national leaders in research on meteorites and tektites.

MARRIAGES AND FAMILY

In 1942 Pinson was married to Mary Eugenia Latta, and two daughters were born to them: Merri Eugenia (1943) and Naomi Ruth (1947). After divorce he married Gertrude Herz in 1963, and son Harlow Earnest Williams, mentioned earlier, was born to this union. Today Pinson is the proud grandfather of three girls born to his daughter Naomi.

SUMMATION

During the 15 or more years that Pinson has been a professor at M.I.T. he has contributed importantly in three academic activities: teaching, research, and miscellaneous professional services. First, he has been an outstanding teacher of introductory astronomy, earth science, and meteoritics. His devotion to teaching is widely known and respected throughout the Greater Boston Community. He has spent many nights with his astronomy students in the observatories at Harvard and Wellesley, and in M.I.T.'s Wallace Observatory, and many weekends with his geology students on field trips. With both groups his own van, driven by himself, and frequently at his own expense, has generally been one, if not the only, means of transportation. And many students have had their first experience with a sleeping bag under an open sky or in a tent on a Pinson field trip. Outside M.I.T. he has taught several times at the Boston Center for Adult Education—Survey of science in 1954; astronomy and mineralogy during SY 1958-59—and has lectured on astronomy at the Cambridge Center for Adult Education (1953). On numerous occasions he has responded to requests for lectures on astronomy or mineralogy to youth groups, and he has enthusiastically accompanied Scout troops on mineral-hunting hikes. During 1972, 1973 and 1974 he has conducted an "Earth Science Club" for a 4-H group. Its present membership is 16, consisting of both girls and boys, and his son Harlow is one of the group. Clearly, Pinson is a natural teacher and lover of nature who gets deep satisfaction from informing young and old alike about the world around them, with no thought of his own efforts and expenditure of time. He is rightly described as a "student's professor."

His second activity has been research. Possessed of a deep curiosity about the physical world, and of a desire to learn more about it by direct study, he has carried on geochemical research in Hurley's Geochronology Laboratory on minerals, rocks, meteorites, and lunar samples, employing the latest spectrometric instruments and radiochemical methods. As stated earlier, he has produced an impressive number of publications on age determinations of minerals and rocks by study of variations in isotopic abundances of strontium, calcium, and argon, both alone and jointly with Hurley and Fairbairn. Inspection of his Bibliography will indicate the wide variety and impressive volume of his research contributions.

His third academic activity within the Institute has consisted of services of several kinds outside the Department. His longtime interest in books has been of help to him as a member of several Library Committees. For a number of years he visited high schools for the Institute, evaluated applications for the Admissions Office, and served on the Institute's Admissions Committee. More recently he has been participating in the IAP and UROP activities as mentioned heretofore.

Away from the Institute he has delivered many invited lectures, chiefly on astronomical and geochemical subjects, and has participated in a number of symposia and conferences on the same subjects: e.g. "Meteorites," at the Harvard College Observatory (1951, 1955); "Trace elements in meteorites," at Yerkes Observatory (1952); Star Island Conference on Science and Religion (1956); and an NSF-sponsored Institute on Astronomy at Eau Claire, Wisconsin, in June and July, 1956.

Pinson has been a champion of the oppressed and of minority groups since childhood; he has never forgotten how his mother forbade him to continue playing with his little 4-year old black chum, when both boys were too young to understand the reasoning behind the prohibition; or again when she adamantly refused to let him accept a much better paying job as a machinist's apprentice than his white-collar job as a clerk, because the former would in her eyes be socially demeaning. As a youth in Atlanta, he saw the plight of black people and poor working-class whites, and recoiled in anger from what he observed. As a prisoner of war in Germany he witnessed indescribable cruelty and oppression in even greater degree. In the turbulent 1960s his sympathies lay with those who were protesting by civil disobedience against the denial of human rights, and he often marched with them in their demonstrations. It is no secret that he has openly and actively opposed certain M.I.T. administrators and officers on a number of political issues. However, he has had the courage to stand strongly for his beliefs, even though his actions have brought harsh criticism and hostile attitudes. A less sensitive and determined man might have lost faith in his beliefs, but not Pinson!

Prof. Pinson, right of center with upraised hands, demonstrating the use of observational instruments to his astronomy class in front of the Green Building at M.I.T.

(Photo by James W. A. Fullmer, 1976)

BIBLIOGRAPHY OF WILLIAM HAMET PINSON

Symbols and abbreviations used in the following references are explained on pages 91 - 98; in general, abstracts are listed separately, and a few are included with the reference to the complete article. This bibliography begins with Pinson's first publication, in 1950, and includes all known titles through 1970.

1--1950   A criticism of "Gondwana land bridges." Ga. Geol. Surv., B. 56: 134-140, (1950).

T--1952   Trace element studies in meteorites and rocks and the origin of meteorites, 161 + 13 p., il., (1952). (Ph.D. Thesis at M.I.T. in Course XII, January 1952.)

2--1952a (with Ahrens, L. H. and Kearns, Margaret M.) Association of rubidium and potassium and their abundance in common igneous rocks and meteorites. Geochim. Cosmochim. Acta 2: 229-242, (1952).

3--1953   (and Ahrens, L. H. and Franck, Mona L.) The abundances of Li, Sc, Sr, Ba and Zr in chondrites and some ultramafic rocks. Geochim. Cosmochim. Acta 4: 251-260, (1953).

4--1953a Trace element studies in meteorites and rocks and the origin of meteorites [abst.]. M.I.T. Abstracts of Theses 1951-52: p. 100-101, (1953).

5--1954   Trace-element composition of a carbonaceous chondrite [abst.]. Am. Geophys. Union, Tr. 35: 380, (1954).

6--1954a (with Herzog, L. F., 2d and Hurley, P. M.) Preliminary survey, Pt. 1 of Isotopic variations in strontium [abst.]. Am. Geophys. Union, Tr. 35: 380, (1954).

7--1954b (with Herzog, L. F., 2d et al.) Variations in isotopic abundances of strontium, calcium, and argon and related topics - Ann. Prog. Rept. 1953-54, Pt. 2, Ann. Res. Prog. Rept., U.S. A.E.C. Rept. NYO-3934 (pt. 2), [163] p., (1954). (Report prepared for A.E.C. by M.I.T.)

8--1954c (with Herzog, L. F., 2d) Isotopic composition of strontium in Homestead meteorite [abst.]. Geol. Soc. Am., B. 65: 1262, (1954); Am. Mineral. 40: 320, (1955).

9--1955   (with Herzog, L. F., 2d) The Rb and Sr content and geologic age of certain biotites, and a correction to the "cosmic" and "crustal" abundances of Rb [abst.]. Am. Geophys. Union, Tr. 36: 513, (1955).

10--1955a (and Backus, M. M. and Herzog, L. F., 2d) The Rb and Sr content and geologic age of certain lepidolites and their radiogenic Ca^{40} content [abst.]. Am. Geophys. Union, Tr. 36: 523, (1955).

11--1955b (with Aldrich, L. T. et al.) Radiogenic Sr^{87} in micas from granites. Am. Geophys. Union, Tr. 36: 875-876, (1955).

12--1955c (with Herzog, L. F., 2d) The Sr and Rb contents of the granite G-1 [R.I.] and the diabase W-1 [Va.]. Geochim. Cosmochim. Acta 8: 295-298, (1955).

13--1955d (with Herzog, L. F.) Rb and Sr content and a minimum age for the Homestead meteorite [abst.]. Astronomical Jour. 60: 163, (1955).

14--1955e [Article on] Astronomy, in The Lincoln Library of Essential Information, Buffalo, N.Y.: The Frontier Press Co., 22nd ed., p. 911-924, (1955).

15--1955f [Review of] Earth as a Planet (G. Kuiper, ed.). Chicago: The University of Chicago Press, 1953. Sky and Telescope 15: 29-31, (1955).

16--1956 (with Hurley, P. M.) Variations in radioactive elements between mafic rock provinces [abst.]. Am. Geophys. Union, Tr. 37: 350, (1956).

17--1956a [Review of] Nuclear Geology (H. Faul, ed.). New York: John Wiley and Sons, Inc., 1944. Am. J. Sci. 254: 516-518, (1956).

18--1956b (with Herzog, L. F. 2d) Rb/Sr age, elemental and isotopic abundance studies of stony meteorites. Am. J. Sci. 254: 555-556, (1956).

19--1956c (with Cormier, R. F. et al.) Rubidium-strontium age determinations on the mineral glauconite [abst.]. Geol. Soc. Am., B. 67: 1681-1682, (1956).

20--1956d (and Herzog, L. F., 2d and Cormier, R. F.) Age study of a tektite [abst.]. Geol. Soc. Am., B. 67: 1725-1726, (1956).

21--1957 (with Hurley, P. M. et al.) Comparison of A^{40}/K^{40} and Sr^{87}/Rb^{87} ages on biotite [abst.]. Am. Geophys. Union, Tr. 38: 396, (1957).

22--1957a (with Fairbairn, H. W. et al.) Age of Nova Scotia granites [abst.]. Geol. Soc. Am., B. 68: 1725, (1957).

23--1957b (and others) Age study of some crystalline rocks of the Georgia Piedmont [abst.]. Geol. Soc. Am., B. 68: 1781, (1957).

24--1957c (and others) Rb, Sr, Ca, and K contents and the isotopic relative abundances of Ca and Sr in a sea-water sample [abst.]. Geol. Soc. Am., B. 68: 1781-1782, (1957).

25--1957d (with Powell, R. M. et al.) Test of the half-life of Rb^{87} [abst.]. Geol. Soc. Am., B. 68: 1782-1783, (1957).

26--1958 Flame photometric analysis for potassium in micas. U.S.A.E.C. Comm. Rept. NYO-3938, p. 26-45, (1958). (Report prepared for A.E.C. by M.I.T.)

27--1958a (and others) Sr/Rb age study of tektites. U.S.A.E.C. Rept. NYO-3938, p. 82-96, (1958). (Report prepared for A.E.C. by M.I.T.)

28--1958b (with Herzog, L. F. 2d and Cormier, R. F.) Sediment age determination by Rb/Sr analysis of glauconite. Am. Assoc. Petroleum Geol., B. 42: 717-733, (1958).

29--1958c (and Fairbairn, H. W. and Cormier, R. F.) Sr/Rb age measurements on hornblende and feldspar, and the age of syenite at Chicoutimi, Quebec, Canada. Geol. Soc. Am., B. 69: 599-601, (1958).

30--1958d (and others) Sr/Rb age study of tektites. Geochim. Cosmochim. Acta 14: 331-339, (1958).

31--1958e (with Bullwinkel, H. J. et al.) Age investigation of syenites from Coldwell, Ontario [abst.]. Geol. Soc. Am., B. 69: 1543-1544, (1958).

32--1958f (with Hurley, P. M. and Fairbairn, H. W.) Intrusive and metamorphic rock ages in Maine and surrounding areas [abst.]. Geol. Soc. Am., B. 69: 1591, (1958).

33--1959 (with Fairbairn, H. W. et al.) Age investigation of syenites from Coldwell, Ontario. Geol. Assoc. Can., Pr. 11: 141-144, (1959).

34--1959a (with Hurley, P. M. et al.) Age study of Lower Paleozoic glauconites [abst.]. J. Geophys. Res. 64: 1109, (1959).

35--1959b (with Hurley, P. M. et al.) Minimum age of the Lower Devonian slate near Jackman, Maine. Geol. Soc. Am., B. 70: 947-949, (1959).

36--1959c (with Allen, V. T. et al.) Age of Precambrian igneous rocks of Missouri [abst.]. Geol. Soc. Am., B. 70: 1560-1561, (1959).

37--1959d (with Fairbairn, H. W. and Hurley, P. M.) Rb-Sr feldspar ages in granitic rocks of Sudbury-Blind River, Ontario, Canada [abst.]. Geol. Soc. Am., B. 70: 1599-1600, (1959).

38--1959e (with Hurley, P. M. et al.) Authigenic versus detrital illite in sediments [abst.]. Geol. Soc. Am., B. 70: 1622, (1959).

39--1959f (and Schnetzler, C. C.) Chemical and physical studies of tektites [abst.]. Geol. Soc. Am., B. 70: 1656, (1959).

40--1959g (and others) Three ages of rock crystallization in Colombia, South America [abst.]. Geol. Soc. Am., B. 70: 1656, (1959).

41--1960 (with Hurley, P. M. et al.) Reliability of glauconite for age measurement by K-Ar and Rb-Sr methods. Am. Assoc. Petroleum Geol., B. 44: 1793-1808, (1960).

42--1960a (with Herzog, L. F., 2d and Hurley, P. M.) Rb-Sr analyses and age determinations of certain lepidolites, including an international interlaboratory comparison suite. Am. J. Sci. 258: 191-208, (1960).

43--1960b (with Fairbairn, H. W. et al.) Age of granitic rocks of Nova Scotia. Geol. Soc. Am., B. 71: 399-413, (1960).

44--1960c (with Fairbairn, H. W. and Hurley, P. M.) Comparison of Rb-Sr mineral and whole-rock ages at Sudbury, Ontario [abst.]. J. Geophys. Res. 65: 2488-2489, (1960).

45--1960d (with Fairbairn, H. W. and Hurley, P. M.) Mineral and rock ages at Sudbury-Blind River, Ontario. Geol. Assoc. Can., Pr. 12: 41-66, (1960).

46--1960e (with Hart, S. R. et al.) Use of amphiboles and pyroxenes for K-Ar dating [abst.]. Geol. Soc. Am., B. 71: 1882, (1960).

47--1960f (with Hurley, P. M. et al.) K-Ar and Rb-Sr minimum ages for the Pennsylvanian section in the Narragansett Basin. Geochim. Cosmochim. Acta 18: 247-258, (1960).

48--1960g (with Fairbairn, H. W. et al.) A comparison of the ages of coexisting biotite and muscovite in some Paleozoic granite rocks. Geochim. Cosmochim. Acta 19: 7-9, (1960).

49--1961 The potassium-argon method - The problem of potassium analysis, in Geochronology of rock systems. N.Y. Acad. Sci., An. 91: 221-224, (1961).

50--1961a Some points on the geological time scale from Nova Scotia and New England, in Geochronology of rock systems. N.Y. Acad. Sci., An. 91: 372-377, (1961).

51--1961b (with Fairbairn, H. W. and Hurley, P. M.) The relation of discordant Rb-Sr mineral and whole-rock ages in an igneous rock to its time of crystallization and to the time of subsequent Sr^{87}/Sr^{86} metamorphism. Geochim. Cosmochim. Acta 23: 135-144, (1961).

52--1961c (with Hurley, P. M. and Hughes, H.) Argon diffusion coefficients in micas at low temperatures obtained from Alpine Fault Uplift in New Zealand [abst.]. J. Geophys. Res. 66: 25-38, (1961).

53--1961d (and Schnetzler, C. C.) Rb-Sr correlation studies of tektites [abst.]. J. Geophys. Res. 66: 2553, (1961).

54--1961e (with Hurley, P. M. et al.) Geochronology of Proterozoic granites in Northern Territory, Australia, Part I: K-Ar and Rb-Sr age determinations. Geol. Soc. Am., B. 72: 653-662, (1961).

55--1961f (with Hurley, P. M. et al.) K-Ar age studies of Mississippi and other river sediments. Geol. Soc. Am., B. 72: 1807-1816, (1961).

56--1961g (with Herz, N. et al.) Age measurements from a part of the Brazilian Shield. Geol. Soc. Am., B. 72: 1111-1120, (1961).

57--1962 (with Beall, G. H. et al.) Comparison of K-Ar and whole-rock Rb-Sr dating in New Quebec and Labrador [abst.]. J. Geophys. Res. 67: 3541, (1962).

58--1962a (with Faure, G. et al.) Isotopic compositions of strontium in continental basic intrusives [abst.]. J. Geophys. Res. 67: 3556-3557, (1962).

59--1962b (with Hurley, P. M. et al.) Radiogenic strontium 87 model of continent formation [abst.]. J. Geophys. Res. 67: 3567-3568, (1962).

60--1962c (with Moorbath, S. et al.) Evidence for the origin of mineralized Tertiary intrusives in Southwestern States from strontium-isotope ratios [abst.]. J. Geophys. Res. 67: 3582, (1962).

61--1962d (with Winchester, J. W.) Preparation of carrier-free 65-day strontium-85 radiotracer. Anal. Chim. Acta 27: 93-94, (1962).

62--1962e (and Schnetzler, C. C.) Rubidium-strontium correlation of three tektites and their supposed sedimentary matrices. Nature 193: 233-234, (1962).

63--1962f (with Hurley, P. M. et al.) Radiogenic argon and strontium diffusion parameters in biotite at low temperatures obtained from Alpine Fault Uplift in New Zealand. Geochim. Cosmochim. Acta 26: 67-80, (1962).

64--1962g (with Hurley, P. M. et al.) Unmetamorphosed minerals in the Gunflint Formation used to test the age of the Animikie. J. Geol. 70: 489-492, (1962).

65--1962h (and others) K-Ar and Rb-Sr ages of biotites from Colombia, South America. Geol. Soc. Am., B. 73: 907-910, (1962).

66--1962i (with Bailey, S. W. et al.) K-Ar dating of sedimentary illite polytypes. Geol. Soc. Am., B. 73: 1167-1170, (1962).

67--1962j (with Fairbairn, H. W. et al.) Evidence of the origin and time of separation of magmas of the Monteregian Hills, Quebec, from development of radiogenic Sr^{87} [abst.]. Geol. Soc. Am., Sp.P. 68: 174, (1962).

68--1962k (with Hower, J. et al.) Effect of mineralogy on K/Ar age as a function of particle size in a shale [abst.]. Geol. Soc. Am., Sp.P. 68: 201-202, (1962).

69--1962l (with Hurley, P. M. et al.) K-Ar age values on the clay fractions in shales ranging in age from Tertiary to Ordovician [abst.]. Geol. Soc. Am., Sp.P. 68: 203-204, (1962).

70--1962m (and Bottino, M. L. and Fairbairn, H. W.) Rb-Sr ages of Tertiary volcanic rocks [abst.]. Geol. Soc. Am., Sp.P. 68: 246-247, (1962).

71--1963 (with Bottino, M. L. et al.) Rb-Sr age study of the Lower Devonian volcanic sequence at Kineo, Maine [abst.]. Geol. Soc. Am., Sp.P. 73: 121, (1963).

72--1963a (with Brookins, D. G. et al.) Whole-rock Rb-Sr investigations of the Collins Hill, Maromas, and Glastonbury formations at Collins Hill, Connecticut [abst.]. Geol. Soc. Am., Sp.P. 73: 123, (1963).

73--1963b (with Faure, G. et al.) Estimate of the isotopic composition of strontium in rocks of the Precambrian basement, Canada [abst.]. Geol. Soc. Am., Sp.P. 73: 150-151, (1963).

74--1963c (and others) Evidence on the origin of felsic volcanic rocks from their initial abundance of Sr^{87} [abst.]. Geol. Soc. Am., Sp.P. 73: 216, (1963).

75--1963d (with Faure, G. et al.) Whole-rock Rb-Sr age of norite and micropegmatite at Sudbury, Ontario [abst.]. Am. Geophys. Union, Tr. 44: 110-111, (1963).

76--1963e (with Bottino, M. L. et al.) Whole-rock Rb-Sr ages of some Paleozoic volcanics and related granites in the Northern Appalachians [abst.]. Am. Geophys. Union, Tr. 44: 111, (1963).

77--1963f (with Hurley, P. M. and Fairbairn, H. W.) Progress report on analytical accuracy of Sr^{87}/Sr^{86} measurement [abst.]. Am. Geophys. Union, Tr. 44: 111-112, (1963).

78--1963g (with Beall, G. H. et al.) Comparison of K-Ar and whole-rock Rb-Sr dating in New Quebec and Labrador. Am. J. Sci. 261: 571-580, il., (1963).

79--1963h (with Hurley, P. M. et al.) K-Ar age values on the clay fractions in dated shales. Geochim. Cosmochim. Acta 27: 279-284, (1963).

80--1963i (with Hurley, P. M. et al.) K-Ar age values in pelagic sediments of the North Atlantic. Geochim. Cosmochim. Acta 27: 393-399, (1963).

81--1963j (with Hower, J. et al.) The dependence of K-Ar age on the mineralogy of various particle size ranges in a shale. Geochim. Cosmochim. Acta 27: 405-410, (1963).

82--1963k (with Fairbairn, H. W. et al.) Initial ratio of strontium 87 to strontium 86, whole-rock age, and discordant biotite in the Monteregian igneous province, Quebec. J. Geophys. Res. 68: 6515-6522, (1963).

83--1963l (with Faure, G. et al.) Age of the Great Dyke of Southern Rhodesia. Nature 200: 769-770, (1963).

84--1963m (with Hurley, P. M. et al.) Evidence of continuing separation of sial from the mantle from the isotopic composition of common strontium. Nuclear Geophysics, Nuclear Science Series, Rept. 38, NAS-NRC Pub. 1075: 83-92, (1963).

85--1963n (with Hurley, P. M. et al.) New approaches to geochronology by strontium isotope variations in whole rocks, in Radioactive Dating, International Atomic Energy Agency, Vienna, p. 201-217, (1963).

86--1963o (with Schnetzler, C. C.) The chemical composition of tektites, Chapter 4 in Tektites (J. A.O'Keefe, ed.). Chicago and London: Univ. Chicago Press, p. 95-129, (1963).

87--1964 (with Hurley, P. M. et al.) Preliminary investigation of Sr^{87}/Rb^{87} relationships in the Sierra Nevada plutonic rocks [abst.]. Geol. Soc. Am., Sp.P. 76: 85, (1964).

88--1964a (with Beiser, Erna) Rb-Sr age of the Murray carbonaceous chondrite [abst.]. Am. Geophys. Union, Tr. 45: 91, (1964).

89--1964b (with Fairbairn, H. W. and Hurley, P. M.) Preliminary age study and initial Sr^{87}/Sr^{86} of Nova Scotia granitic rocks by the Rb-Sr whole-rock method. Geol. Soc. Am., B. 75: 253-257, (1964).

90--1964c (with Backus, M. M. et al.) Calcium isotope ratios in the Homestead and Pasamonte meteorites and devonian limestone. Geochim. Cosmochim. Acta 28: 735-742, (1964).

91--1964d (with Schnetzler, C. C.) A report on some recent major element analyses of tektites. Geochim. Cosmochim. Acta 28: 793-806, (1964).

92--1964e (with Schnetzler, C. C.) Variation of strontium isotopes in tektites. Geochim. Cosmochim. Acta 28: 953-969, (1964).

93--1964f (with Faure, G. et al.) Whole-rock Rb-Sr age of norite and micropegmatite at Sudbury, Ontario. J. Geol. 72: 848-854, (1964).

94--1964g (with Fairbairn, H. W. and Hurley, P. M.) Initial Sr^{87}/Sr^{86} and possible sources of granitic rocks in southern British Columbia. J. Geophys. Res. 69: 4889-4893, (1964).

95--1964h (with Hurley, P. M. and Fairbairn, H. W.) Rb-Sr relationships in serpentinite from Mayagüez, Puerto Rico, and dunite from St. Paul's Rocks - A progress report, in A study of serpentinite. NAS-NRC Pub. 1188: 149-151, (1964).

96--1965 (with Hurley, P. M. et al.) Investigation of initial Sr^{87}/Sr^{86} ratios in the Sierra Nevada plutonic province. Geol. Soc. Am., B. 76: 165-174, (1965).

97--1965a (with Fairbairn, H. W. and Hurley, P. M.) Re-examination of Rb-Sr whole-rock ages at Sudbury, Ontario. Geol. Assoc. Can., Pr. 16: 95-101, (1965).

98--1965b (with Shields, R. M. and Hurley, P. M.) The Rb^{87}/Sr^{87} age of stony meteorites [abst.]. Am. Geophys. Union, Tr. 46: 124, (1965).

99--1965c (with Fairbairn, H. W. et al.) Rb-Sr whole-rock isotopic analyses and the Cambrian-Precambrian problem in southeastern Massachusetts [abst.]. Am. Geophys. Union, Tr. 46: 173, (1965).

100--1965d (with Roe, G. D. and Hurley, P. M.) Rb-Sr evidence for the origin of peridotites [abst.]. Am. Geophys. Union, Tr. 46: 186, (1965).

101--1965e (and others) Rb-Sr age of stony meteorites. Geochim. Cosmochim. Acta 29: 455-456, (1965).

102--1965f (and Philpotts, J. A. and Schnetzler, C. C.) K/Rb ratios in tektites. J. Geophys. Res. 70: 2889-2894, (1965).

103--1965g (with Hurley, P. M. and Fairbairn, H. W.) Radioactive decay of Rb^{87} to Sr^{87} in geological science exclusive of age dating. Primera Conferencia Interamericana de Radioquimica, Montevideo, Pan-American Union, Washington, p. 175-178, (1965).

104--1966 (with Hurley, P. M. and Fairbairn, H. W.) Evidence from western Ontario of the isotopic composition of strontium in Archean seas [abst.]. Geol. Soc. Am., Sp.P. 87: 84, (1966).

105--1966a (with Philpotts, J. A.) New data on the chemical composition and origin of moldavites. Geochim. Cosmochim. Acta 30: 253-266, (1966).

106--1966b (with Fairbairn, H. W. et al.) Whole-rock and initial $^{87}Sr/^{86}Sr$ of volcanics underlying fossiliferous Lower Cambrian in the Atlantic provinces of Canada. Can. J. Earth Sci. 3: 509-521, (1966).

107--1966c (with Hurley, P. M. and Fairbairn, H. W.) Rb-Sr isotopic evidence in the origin of potash-rich lavas of western Italy. Earth Planet. Sci. Lett. 1: 301-306, (1966).

108--1966d (with Schnetzler, C. C. and Hurley, P. M.) Rubidium-strontium age of the Bosumtwi Crater area, Ghana, compared with the age of the Ivory Coast tektites. Science 151: 817-819, (1966).

109--1966e (with Shields, R. M. and Hurley, P. M.) Rubidium-strontium analyses of the Bjurböle chondrite. J. Geophys. Res. 71: 2163-2167, (1966).

110--1967 (with Fairbairn, H. W. et al.) Rb-Sr age of granitic rocks of southeastern Massachusetts and the age of the Lower Cambrian at Hoppin Hill. Earth Planet. Sci. Lett. 2: 321-328, (1967).

111--1967a (with Hurley, P. M. et al.) Test of continental drift by comparison of radiometric ages. Science 157: 494-500, (1967).

112--1967b (with Fairbairn, H. W. et al.) Rb-Sr whole-rock age of the Sudbury lopolith and basin sediments [abst.]. Am. Geophys. Union, Tr. 48: 242, (1967); See also Can. J. Earth Sci. 5: 707-714, (1968).

113--1967c (with Hurley, P. M. et al.) Tracing the history of differentiation of the mantle by Rb-Sr relationships, in Upper Mantle Project, U.S. Progress Rept.: 103, (1967); See also U.S. Program, Final Rept. [NAS-NRC Pub.]: 214-215, (1971).

114--1967d (with Kolbe, P. et al.) Rb-Sr study on country rocks of the Bosumtwi Crater, Ghana. Geochim. Cosmochim. Acta 31: 869-875, (1967). (See also 117--1968b.)

115--1968 (with Fairbairn, H. W. et al.) Whole-rock age and initial Sr^{87}/Sr^{86} of volcanic rocks underlying fossiliferous Lower Cambrian in the Atlantic Provinces of Canada [abst.]. Geol. Soc. Am., Sp.P. 101: 65, (1968).

116--1968a (with Hurley, P. M. et al.) Rb-Sr whole-rock analyses in northern Brazil correlated with ages in West Africa [abst.]. Geol. Soc. Am., Sp.P. 101: 100-101, (1968).

117--1968b (with Kolbe, P.) Rb-Sr correlation of Bosumtwi Crater rocks with Ivory Coast tektites [abst.]. Geol. Soc. Am., Sp.P. 101: 113, (1968). (See also 114--1967d.)

118--1968c (with Brookins, D. G. et al.) Geochronological aspects of the genesis of large granitic pegmatites in non-igneous environments [abst.]. Geol. Soc. Am., Sp.P. 115: 25, (1968).

119--1968d (with Hurley, P. M. et al.) Radiometric ages of igneous rocks in northeastern Massachusetts [abst.]. Geol. Soc. Am., Sp.P. 115: 260-261, (1968).

120--1968e (with Hurley, P. M. et al.) Some orogenic episodes in South America by K-Ar and whole-rock Rb-Sr dating. Can. J. Earth Sci. 5: 633-638, (1968).

121--1968f (with Fairbairn, H. W. et al.) Rb-Sr whole-rock age of the Sudbury lopolith and basin sediments. Can. J. Earth Sci. 5: 707-714, (1968); Abst., Am. Geophys. Union, Tr. 48: 242, (1967).

122--1969 (with Brookins, D. G. et al.) A Rb-Sr geochronologic study of the pegmatites of the Middletown area, Connecticut. Contr. Mineral. Petrology 22: 157-168, (1969).

123--1969a (and Griswold, T. B.) The relationship of nickel and chromium in tektites: New data on the Ivory Coast tektites. J. Geophys. Res. 74: 6811-6815, (1969).

124--1970   (with Hurley, P. M.) Rubidium-strontium relations in Tranquillity Base samples. Science 167: 473-474, (1970).

125--1970a (with Hurley, P. M.) Whole-rock Rb-Sr isotopic age relationships in Apollo 11 lunar samples, in Apollo 11 Lunar Sci. Conf., Pr. 2: 1311-1315, New York: Pergamon Press, (1970).

126--1970b (with Bottino, M. L. et al.) Blue Hills igneous complex, Massachusetts: Whole-rock Rb-Sr open systems. Geol. Soc. Am., B. 81: 3739-3746, (1970).

The following are items prepared by W. H. Pinson, alone or with colleagues, and published only as "gray literature" in Variations in isotopic abundances of strontium, calcium, argon, and related topics, the annual reports made by P. M. Hurley to the U.S.A.E.C. under Contract AT(30-1)-1381. These annual reports were prepared in M.I.T.'s Department of Geology and Geophysics, and printed and distributed by the Institute's Graphic Arts. Although, as stated earlier, they constitute "gray literature," they are included in this bibliography because they have been cited in the same way as regularly published articles and also because they indicate a kind of productivity not always known to the scientific community at large.

1) Sr and Rb contents and ages of certain lepidolites. 1st Ann. Rept.: 31-34, (1955).

2) Relative abundance of Rb and Sr in vitrain ashes from Nova Scotia. 5th Ann. Rept.: 123-126, (1958).

3) Flame photometric analysis for potassium in minerals and rocks. 6th Ann. Rept.: 4-27, (1958).

4) Standard biotite B 3203. 7th Ann. Rept.: 209-211, (1959).

5) Sources of error in the preparation of spike and shelf solutions for geochronometric work. 8th Ann. Rept.: 237-244, (1960).

6) Age measurements of the Grenville metamorphism near Dolbeau, Quebec. 8th Ann. Rept.: 277-282, (1960).

7) Age of sylvite from Palangana salt dome, Duval County, Texas. 8th Ann. Rept.: 287, (1960).

8) Determination of the isotopic composition of a strontium shelf solution using three mass spectrometers. 9th Ann. Rept.: 263-266, (1961).

9) Rb-Sr study of the Westerly, Rhode Island, granite (G-1) and new Rb-Sr values for G-1 and W-1. 10th Ann. Rept.: 71-74, (1962).

10) A review of the preparation and calibration of shelf and spike solutions currently in use in the geochronology laboratory. 10th Ann. Rept.: 91-96, (1962).

(36)
RICHARD RAYMAN DOELL

RICHARD RAYMAN DOELL

MIT: 1956-1959

In the early 1950s, when it was decided to broaden and diversify Course XII's offerings in geophysics, Richard R. Doell, a recent doctor in geophysics from the University of California at Berkeley, was invited to join our faculty as an Assistant Professor of Geophysics. With excellent training in both mathematics and physics, and with degrees in geophysics (A.B. 1952; Ph.D. 1955), Doell not only brought strength to our academic program in general geophysics, but also brought practical experience in geophysical exploration with several petroleum companies and a special research interest in paleomagnetism. He quickly established himself in the Department as an able teacher and imaginative research scientist, and soon gathered around himself a small but enthusiastic group of students. Even though he was in our Department for only three years, 1956-1959, he supervised 5 theses in that short period--4 S.B.s and 1 doctorate. Illness forced him to request a leave of absence early in 1959 and he resigned in August 1959 when his future health became uncertain. Returning to California, he recovered his health, and within the year joined the staff of the Branch of Geophysics of the U.S. Geological Survey at Menlo Park. Since 1960 he has continued his research on the earth's magnetic field, and has become one of the leaders in developing a new geologic time scale based on the remanent magnetism in rocks.

Richard Rayman Doell was born in Oakland, California, on 28 June 1923, the younger of the two sons of Raymond A. and Mabel (Frost) Doell. His father, now retired, was in the real estate business, and his mother, now deceased, was a housewife.

Doell grew up in California and received all his earlier education in the schools of that State. He first attended the Carpinteria Union Grammar School from 1928 to 1936, and then the Santa Barbara High School from 1936 to 1940. Entering the University of California at Los Angeles in September 1940, he spent the next three years working toward a bachelor's degree as a mathematics major, then went to the University of Oklahoma to participate in the Army Specialized Training Program.

His service in the Army of the United States, which started in February 1943, and included the aforementioned year of training at the

University of Oklahoma, ended in November 1945 when he was discharged as a Staff Sergeant. From October 1944 to July 1945 he served in the 411th Infantry Regiment in the European Theater of Operations.

Shortly after returning home from Europe he spent 20 months (Feb. 1946 - Oct. 1947) as a physics major at Berkeley, then left the academic halls to take employment with the United Geophysical Company of Pasadena. He worked as a seismologist on both land and off-shore seismic operations in California, the Persian Gulf, and Canada, until September 1950, at which time he again returned to Berkeley and continued his studies until June 1952 when he was awarded an A.B. degree in geophysics and elected to Phi Beta Kappa.

Having spent the previous summer (1951) as a seismologist with his former employer, United Geophysical Company, he decided to diversify his field experience by accepting summer employment in 1952 with the U.S. Geological Survey, acting as a geological field assistant in the Bear Paw Mountains of Montana. September 1952 found him back at Berkeley as a teaching assistant and doctoral candidate in the Department of Geological Sciences. After a second year at Berkeley, SY 1953-54, this time as a research assistant, he again took summer employment in the petroleum industry, working as a junior geophysicist on gravity and vertical magnetic intensity operations. He returned to Berkeley for SY 1954-55, having been awarded the Standard Oil Company of California Scholarship in Geological Sciences, and proceeded to complete the requirements for a Ph.D. degree in geophysics by the end of the summer (September 1955), at which time he successfully defended his doctoral dissertation on "Remanent Magnetism in Sediments," a special field of geophysics in which he would gain international renown later on.

Now came a crucial time in his career. Would he stay in the academic halls, seek a research position somewhere, or return to the kind of practical work that he had sampled in previous summers with petroleum companies? Academia won out, at least for the time being, when he accepted appointment as a Lecturer in Geophysics at the University of Toronto for SY 1955-56. This appointment lasted for only one year, however, because in our search for a promising young geophysicist, Doell was so highly recommended for our position that we lost no time in offering him an assistant professorship even before he moved to Toronto. Happily for M.I.T. he accepted our offer with the result that he joined our Department in July 1956 as Assistant Professor of Geophysics. In typical fashion he quickly organized several courses in geophysics--12.831, 12.832 Elements of Geophysics I and II; 12.831 Geophysical Lab.; and 12.86 Elements of Seismology. He lost no time in seeking research funds from the National Science Foundation, and soon had laboratory and field investigations of remanent magnetism in rocks under way. During his first year in the Department, SY 1956-57, he participated in symposia on rock magnetism in England (London) and California (Berkeley); gave a lecture in June to the 1957 group of students attending the G.S.I. Cooperative Plan in Dallas; and in late July travelled to Crystal Cliffs, Nova Scotia, to give a short course on "Field Geophysics" to our Course XII undergraduates who were participating in our M.I.T. Summer School near Antigonish. Hurrying across Canada from Nova Scotia, he led the 1957 University of Toronto Expedition to the Salmon Glacier in northwestern British Columbia during the month of August. On his way back to Cambridge, he attended the 11th General Assembly of the IUGG at Toronto, 3-7 September, as an official U.S. delegate.

SY 1957-58 was a busy one for Doell as he gathered a group of able students about himself and got several thesis projects started. Just when everything seemed to be interlocking into a coordinated program, Doell's health became impaired with melanoma, and he had to undergo surgery early in September. With his future health uncertain, he requested a leave of absence, starting on 1 March 1959, and later resigned from the Department as of 31 August 1959.

Happily, he recovered from his illness and before the end of the year (1959) he was able to accept an appointment as Geophysicist in the Geophysics Branch of the U.S. Geological Survey in Menlo Park, California. Since then he has devoted most of his time to the study of remanent magnetism in rocks of various kinds. In the past decade, 1960 to 1970, he worked closely with Allan Cox, and between them they produced more than 70 publications which established them as North American leaders in research on geomagnetic reversals and, with the collaboration of G. Brent Dalrymple on Potassium-Argon dating, on the geochronology based on that physical phenomenon. The interested reader will find these publications listed in the annual volumes of <u>Bibliography of North American Geology</u>, published as Bulletins of the United States Geological Survey.

In recognition of the aforementioned work Doell was elected a member of the National Academy of Sciences in 1970, and in 1971 he was a co-recipient of the prestigious Vetlesen Prize. In late 1970, he ceased research on paleomagnetism and began studies in the field of environmental geology, especially as related to fuel resources development.

Although active in our Department for only three years, Doell supervised four theses in that interval--4 S.B.s and 1 doctors--and got several other students interested enough in rock magnetism to do theses in that field under the supervision of his successor. He also demonstrated the potential importance of remanent magnetism in rocks, with the result that an immediate successor was sought and ultimately added to the staff in order that instruction and research in this aspect of geophysics be continued.

In 1950, while an undergraduate at the University of California at Berkeley, Doell married Ruth Gertrude Jones, an outstanding fellow student in biochemistry. Ruth, who was born in Vancouver, British Columbia, on 24 March 1926, had come to Berkeley from Canada because of her deep interest in biochemistry. By the time she and Richard had decided to come to the Boston area, in 1956, she had received a Ph.D. degree in Comparative Biochemistry from the University of California (Berkeley). While in the Greater Boston area, during Richard's three years at M.I.T. (1956-1959), she continued biochemical research, and when the Doells returned to California in March 1959, she was again able to resume research after getting her family settled down in the San Francisco area. At this writing (September 1974) Ruth pursues her academic career as Professor of Biology at San Francisco State University. The Doells have two daughters, both now students in the University of California; one at Santa Cruz and the other at San Diego.

BIBLIOGRAPHY OF RICHARD RAYMAN DOELL

Symbols and abbreviations used in the following references are explained on pages 91 - 98; in general, abstracts are listed separately as well as with the references to the complete article. This bibliography includes only items written by Doell before and during his brief appointment of three years, 1956-1959, at M.I.T. Subsequent publications are listed in the annual volumes of Bibliography of North American Geology, published as Bulletins of the U.S. Geological Survey.

T--1955   Remanent magnetism in sediments--Dissertation for the Ph.D. degree in geophysics at the University of California, Berkeley, September 1955.

1--1955   Paleomagnetic study of rocks from the Grand Canyon of the Colorado River [Ariz.]. Nature 176: 1167, (1955).

2--1956   Remanent magnetization of the upper-Miocene 'blue' sandstones of California. Am. Geophys. Union, Tr. 37: 156-167, il., (1956); condensed with title, Crystallization magnetization, Advances Physics 6: 327-332, il., [London], (1957).

3--1958   Paleomagnetic interpretations [abst.]. Am. Geophys. Union, Tr. 39: 513, (1958).

4--1959   (and Cox, A. V.). Analysis of paleomagnetic data [abst.]. Geol. Soc. Am., B. 70/12/2: 1590-1591, (1959).

5--1960   (with Cox, A. V.). Review of paleomagnetism. Geol. Soc. Am., B. 71: 645-768, il., (1960).

6--1960a (and Cox, A. V.). Paleomagnetism, polar wandering, and continental drift: Art. 193 in U.S. Geol. Surv., P.P. 400-B: B426-B427, (1960).

7--1960b (and Altenhofen, R. E.). Preparation of an accurate equal-area projection: Art. 194 in U.S. Geol. Surv., P.P. 400-B: B427-B429, il., (1960).

(37)
JOHN WIDMER WINCHESTER

JOHN WIDMER WINCHESTER

MIT: 1956-1966

With A.B. (1950) and S.M. (1952) degrees, and five summers of experience as a chemical technician, John W. Winchester entered M.I.T. in September 1952 as an NSF Predoctoral Fellow, received a doctorate in physical chemistry in June 1955, spent SY 1955-56 as a Fulbright Postdoctoral Fellow in the Netherlands, and then returned to M.I.T. in September 1956 as Assistant Professor of Geochemistry. For the next decade, 1956-1966, he offered lecture and laboratory work in geochemistry, worked closely with colleagues in the M.I.T. Reactor, and carried on research alone and with thesis students in which neutron activation analysis was the principal research method. He supervised a total of 24 theses--8 S.B.s, 6 S.M.s, and 10 doctors--and published more than 100 articles. Thrice during his 14 years at M.I.T., as graduate student (1952-1956), assistant professor (1956-1963), and associate professor (1963-1966), respectively, he took leaves of absence: first as a Fulbright Grantee to the Netherlands during SY 1955-56 as mentioned above; next as a Fulbright Professor to Taiwan during SY 1962-63; and finally with a similar appointment to Argentina during the fall term of SY 1966-67. He resigned on 31 December 1966 to accept an appointment as Associate Professor of Oceanography in the Department of Meteorology and Oceanography at the University of Michigan at Ann Arbor. Four years later, in 1970, he moved on to Florida State University as Professor of Oceanography and Chairman of that institution's Department of Oceanography.

Winchester is a typical example of the brilliant student who comes to M.I.T. to complete his graduate education, stays on for a few years as a fledgling professor, meanwhile developing into a mature scientist, then leaves to achieve a distinguished career elsewhere.

John Widmer Winchester was born in Chicago, Illinois on 8 October 1929, the son of Harold J. and L. Darlene (Eastes) Winchester. He has an older sister, Mrs. Barbara (Winchester) Swords, and a younger brother, James T. Winchester. On 1 February 1958 he married Ms. Ellen M. Sullivan, who had a daughter, Kathleen Jo Sullivan.

Jack, as he is known by his friends, spent his first eight grades in the Public Schools of Western Springs, Illinois; then entered the Lyons Township High School in La Grange, Illinois, and graduated in 1947. He early demonstrated his quick learning abilities, and was elected to the

National Honor Society in 1946. In 1947 he won the Bausch and Lomb Science Award while a senior at both the Lyons Township High School and Junior College. The next five years, 1947-1952, found him enrolled at the University of Chicago, from which he received an A.B. degree in chemistry in 1950 and an S.M. degree in the same science in 1952. In the fall of 1952 he entered M.I.T. as a graduate student and in June 1955 received a Ph.D. in the Department of Chemistry, with a doctoral thesis in physical chemistry (nuclear chemistry), titled "A study of fission products in the region of mass numbers 103 to 131."

During his college years at Chicago he was a member of the Student Government for two years (1950-1952), was elected to both Phi Beta Kappa and Sigma Xi as a senior (1952), and also won the American Institute of Chemists Award (1952). During the same period he worked summers as a chemical technician at the Union Oil Products Co., Inc. in Riverside, Illinois; as a technician in Chicago's Institute for the Study of Metals during SY 1950-51; and as a Teaching Assistant in the Department of Chemistry during SY 1951-52.

At M.I.T. he was an NSF Predoctoral Fellow during SYs 1952-53 and 1954-55, and a Teaching Assistant in the Department of Chemistry during the year between, *i.e.* SY 1953-54. After completing his doctoral program, and serving as a Research Assistant in Chemistry for a few months during the summer of 1955, he went to the Municipal University of Amsterdam (The Netherlands) for SY 1955-56 as a Fulbright Postdoctoral Fellow in Chemistry. There he conducted research at the famed Instituut voor Kernphysisch Onderzoek.

Upon his return to the United States, and to M.I.T., he was appointed an Assistant Professor of Geochemistry, in the Department of Geology and Geophysics, commencing in the fall of 1956, and was advanced to the rank of Associate Professor on 1 July 1963. He resigned from the Department on 31 December 1966 to accept an associate professorship of oceanography in the Department of Meteorology and Oceanography at the University of Michigan. Four years later in the fall of 1970 he moved to Florida State University as Professor and Chairman of that institution's Department of Oceanography.

During his ten years on the M.I.T. faculty Winchester took two leaves of absence, the first to Taiwan and the second to Argentina. During SY 1962-63 he was a Fulbright Professor at the National Tsinghua University and at the National Taiwan University, where he taught nuclear chemistry and geochemistry. Four years later, during the fall term of SY 1966-67, and again on a Fulbright grant, he gave lectures and conducted research in geochemistry at the Universidad Nacional de La Plata, in La Plata, Argentina.

While at M.I.T., Winchester initiated a research program that utilized the Institute's Reactor for neutron activation analysis. He supervised 24 theses--8 Bachelor's, 6 Master's, and 10 Doctor's, and also acted as Research Supervisor for five projects: DSR 7971 (1958-1961)--Development of methods for trace element analysis by neutron activation; DSR 8054 (1958-1961)--Analysis of metamorphic minerals by neutron activation and related methods; DSR 8383 (1959-1961)--Radioactivity of rain water from individual storms (with Prof. R. E. Newell); DSR 8390 (1959-1960)--Activation analysis research; and DSR 8841 (1961-1966)--Oceanographic research (chemical part of program).

In the classroom he lectured on geochemistry and in the laboratory he and his students, as mentioned in the preceding paragraph, conducted a diversified research program based largely on neutron activation analysis. During the summers of 1958, 1959 and 1961 he was a University Research Participant at the Oak Ridge Institute of Nuclear Studies and Oak Ridge National Laboratory. He was a Visiting Scientist at the Institute of Marine Science, University of Alaska, in the summer of 1964.

Within the Institute, both inside and outside the Department of Geology and Geophysics, Winchester served at one time or another on numerous committees: Radiation Protection; Student Aid; N.R.C. Research Committee on Nuclear Science; Earth Sciences Library; Course XII's Geochemistry Committee; Secretary of the M.I.T. Chapter of Sigma Xi; and Freshman Advisory Council. He also served as Abstractor for Chemical Abstracts (1960-1965), as Consulting Editor for Earth Sciences to Addison-Wesley Publishing Company (1961-1966), and for one year as Managing Editor of Geochemistry - A Translation of Geokhimiya (1960).

All of the activities cited in the preceding paragraphs indicate that Jack Winchester is an exceedingly active individual. This was obvious even in his youth when he was Head Carrier for the Chicago Daily News in Western Springs, Illinois, in 1943 as a high school freshman; as Junior Assistant House Head on the Residence Halls Staff of the University of Chicago from 1950 to 1952; and as Head Usher at Chicago's Rockefeller Memorial Chapel. Soon after becoming a resident of Concord, Massachusetts, he became Chairman of that Town's Fair Housing Practices Committee in 1960, and also served as Chairman of the Finances Committee for the Fair Housing Federation of Greater Boston from 1961 to 1966. His concern about his less fortunate fellowmen and his desire to help them by education prompted him to spend a year in Taiwan and a term in Argentina.

One would think that Winchester's time at M.I.T. was so fully taken up with the varied interests and activities mentioned above that he would have had little left for individual research and publication. Quite the opposite proves to be the case, for his Bibliography appearing later on lists more than 100 publications: scientific papers, articles in books, letters and reviews, abstracts, progress reports and translations written before he left the Institute at the end of 1966.

Winchester was recognized as an imaginative and stimulating teacher and research scientist who excited the interest of both his students and his faculty colleagues, and attracted them to a working relationship with him. He opened the way to the M.I.T. Reactor for his students, and initiated original research on certain chemical aspects of the atmosphere, of the world ocean, and of the minerals and rocks of the solid earth. Himself a skillful analyst, he trained many of his students in the art of accurate chemical analysis and as a result sent them forth with an extra skill to offer a potential employer. In sum, he was the brilliant and promising young scientist every Department Head hopes to recruit, and we predict for him an outstanding career in whatever he chooses to do.*

While at M.I.T. he was a member of Phi Beta Kappa, Sigma Xi, American Chemical Society, American Geological Institute, American Geophysical Union, American Institute of Physics, American Physical Society, Geochemical Society, and Geological Society of America.

---

* During four years at the University of Michigan, first as Associate Professor of Oceanography (1967-1969) and then as Professor of Oceanography (1969-1970), Winchester published a dozen articles and became an active participant in the national program in oceanography. When he left Michigan to chair the Department of Oceanography at Florida State University, [in his own words] he "embarked on still a higher level of national involvement as well as on a more difficult 'administrative task than at Michigan."

Today (1974), four years later, we find him directing a well-organized and productive Department of Oceanography, and busy as ever in teaching and research. Currently he is involved in atmospheric and marine geochemistry, using a nuclear particle accelerator for analysis of aerosols for trace metals. And as predicted, significant contributions in the broad field of geochemistry continue to come from his research work.

Finally, as would be predicted from earlier activities, he serves on many national and international committees, commissions, councils and panels, all of which are concerned to some extent with chemical aspects of our environment.

BIBLIOGRAPHY OF JOHN WIDMER WINCHESTER

Symbols and abbreviations used in the following references are explained on pages 91 - 98. In general, abstracts are listed separately as well as with the complete article. This bibliography includes all items involving work done through 31 December 1966, at which time Winchester left M.I.T. Because of delay in publication, however, numerous of his articles did not appear in print until later; these, therefore, bear publication dates later than 1966. T = Thesis.

T--1955   A study of fission products in the region of mass numbers 103 to 131, 140 p., (1955). (Ph.D. Thesis at M.I.T. in Course V, Chemistry, June 1955.)

1--1956   (as translator) Determination of the climatic conditions of some regions of the USSR during the Upper Cretaceous period by the method of isotopic paleothermometry, by Naidin, D. P., Teiss, R. V., and Tchupakhin, in Geokhimiya 8: 23-25, (1956).

2--1956a (as translator) Determination of the age of a pegmatite vein of the Koita-Tundra by cyrtolite, orthite, and uraninite, by Zykov, S. I. and Stupnikova, N. I., in Geokhimiya 8: 35-38, (1956).

3--1957   (with Schindewolf, U. L. and Coryell, C. D.) Decay properties of 74-second Ag-111m. Phys. Rev. 105: 1963-1965, (1957).

4--1957a (and Aten, A. H. W., Jr.) The content of tin in iron meteorites. Geochim. Cosmochim. Acta 12: 57-60, (1957).

5--1957b B-decay-energy systematics near A = 40 and a possible long-lived K^{38}, p. 12-13, in M.I.T. Lab. Nuclear Sci., Prog. Rept., Feb. 28, 1957.

6--1957c Radiochemical Studies: (1) On the possible existence of long-lived K-38 in nature, p. 82; (2) The use of radioactive Sr tracer in ion exchange separations, p. 83; (3) Determination of sodium and potassium in micas by neutron activation, p. 83-84. NYO-3937, 4th Ann. Prog. Rept. for 1956-57, USAEC, Dept. Geol. Geophys., M.I.T., March 1, 1957. (‡)

7--1958   Radiochemical Studies: (1) Deuteron reactions with Sr and Rb isotopes (with Crocket, J. and Gowen, K.), p. 70-71; (2) Determination of K in common rocks and minerals by neutron activation (with Goldstein, M. I. and Anderson, D.), p. 72-74; (3) Determination of Nb in silicate materials by alpha particle activation (with Buyrn, A.), p. 75-76. NYO-3938, 5th Ann. Prog. Rept. for 1957-58, USAEC, Dept. Geol. Geophys., M.I.T., March 1, 1958. (‡)

8--1958a Microanalysis of geological materials. [M.I.T.] Rept. Res. 9/6: 3, (1958).

9--1958b Trace elements analysis, p. 298-299, in Encyclopedia of Chemistry (Supplement), Clark, G. L., Hawley, G. G., and Hamor, W. A., eds. New York: Reinhold Pub. Corp., (1958).

10--1958c Sodium and potassium determination by neutron activation [abst.]. Am. Geophys. Union, Tr. 39: 536, (1958).

11--1958d Determination of K in biotites by neutron activation (and Anderson, D. H., Cohen, L. H., Hurley, P. M., Faure, G., Green, E. J., and Hilar, E. P.), p. 70-71. NYO-3939, 6th Ann. Prog. Rept. for 1957-58, USAEC, Dept. Geol. Geophys., M.I.T., Dec. 1, 1958. (‡)

12--1958e A preliminary investigation of a chromatographic column separation of rare earths using Di(2-ethylhexyl) phosphoric acid. USAEC Oak Ridge Nat. Lab., ORNL-CF-58-12-43: 11 p., (Dec. 30, 1958).

13--1959   Determination of potassium in biotite by neutron activation [abst.]. J. Geophys. Res. 64: 1130, (1959).

14--1959a   Trace element analysis in geochemistry by neutron activation-- Discussion of sensitivities and errors [abst.]. J. Geophys. Res. 64: 1131, (1959).

15--1959b   (and Bate, L. C. and Leddicotte, G. W.) Determination of Li-6 in aqueous solution by neutron activation analysis, USAEC Oak Ridge Nat. Lab., ORNL-CF-59-7-127: 7 p., (July 10, 1959).

16--1959c   (and Meyer, R. E., Bate, L. C., and Leddicotte, G. W.) Determination of oxygen in oxide films by neutron activation analysis. USAEC Oak Ridge Nat. Lab., ORNL-CF-59-7-128: 8 p., (July 15, 1959).

17--1960   Radioactivation analysis in inorganic geochemistry. Prog. Inorganic Chem. 2: 1-32, (1960). New York, Interscience Pub., (1960).

18--1960a   Importance of Chinese for scientific communication. Science 131: 1561-1562, (1960).

19--1960b   The growing importance of Chinese as a scientific language. Tech. Rev. 62/9: 3, 58, (1960).

20--1960c   (and Bottino, M. L.) Cyclotron charged particle activation analysis for determining oxygen, carbon, and silicon [abst.]. J. Geophys. Res. 65: 2533, (1960). See also 25--1961a.

21--1960d   (and Crocket, J. H.) Study of the dilute solid solution $ZnCO_3$ in $CaCO_3$ by radiotracer techniques [abst.]. J. Geophys. Res. 65: 2533, (1960). See also 28--1961d.

22--1960e   Rare earth chromatography using Bis(2-ethylhexyl) orthophosphoric acid. USAEC Oak Ridge Nat. Lab., ORNL-CF-60-3-158: 10 p., (March 14, 1960).

23--1960f   A progress report on the use of 15 M.E.V. deuterons for the determination of carbon, oxygen, and silicon in solid materials by radioactivation analysis. Dept. Geol. Geophys., M.I.T., 26 October, 1960. (‡)

24--1961   [Review of] Soviet oceanography. Science 134: 277-278, (1961).

25--1961a   (and Bottino, M. L.) Determination of carbon, oxygen, and silicon in solids by activation analysis with 15 M.E.V. deuterons. Anal. Chem. 33: 472-473, (1961). See also 20--1960c.

26--1961b   Preparation of high specific activity Sn-125 radiotracer. Anal. Chim. Acta 24: 388, (1961).

27--1961c   Determination of potassium in silicate minerals and rocks by neutron activation analysis. Anal. Chem. 33: 1007-1012, (1961).

28--1961d   (with Crocket, J. H.) Study of the dilute solid solution of $ZnCO_3$ in $CaCO_3$ and its relevance to the marine geochemistry of Zn [abst.]. J. Geophys. Res. 66: 2523, (1961). See also 21--1960d.

29--1962   (with Schroeder, G. L.) Determination of sodium in silicate minerals and rocks by neutron activation analysis. Anal. Chem. 34: 96-99, (1962).

30--1962a   Undergraduate preparation for the student of geochemistry. J. Geol. Educ. 10/1: 13-17, (1962).

31--1962b   (and Pinson, W. H., Jr.) Preparation of carrier-free 65-day strontium-85 radiotracer. Anal. Chim. Acta 27: 93-94, (1962).

32--1962c   (with Chase, J. W. and Coryell, C. D.) Lanthanum, europium, and dysprosium distributions in igneous rocks and minerals [abst.]. J. Geophys. Res. 67: 3549, (1962). See also 38--1963a.

33--1962d (with Corless, J. T. and Rahn, K. A.) Determination of the relative isotopic abundance of Ca^{48} in natural materials by neutron activation analysis [abst.]. J. Geophys. Res. 67: 3551, (1962).

34--1962e (and Burns, F. and Duce, R.) The determination of iodine, bromine, and chlorine in sea water and rain water by neutron activation analysis [abst.]. J. Geophys. Res. 67: 3610, (1962).

35--1962f (with Coryell, C. D. and Chase, J. W.) A procedure of geochemical interpretation of rare-earth abundances in rocks and minerals--Abst. 141st Meeting Am. Chem. Soc., Washington, D.C., 20-29 March, 1962. Cf. Meteorites give clues to geochemistry, Aid studies of mineral-producing processes. Chem. & Eng. News 40/14: 56-57, (1962).

36--1962g The ocean as a chemical buffer system--A report for use by the National Science Foundation Institute for Secondary School Teachers of Chemistry. Dept. Geol. Geophys., M.I.T., 30 p., 25 May 1961; 34 p. (revised), 20 May 1962.

37--1963 (with Coryell, C. D. and Chase, J. W.) A procedure for geochemical interpretation of terrestrial rare-earth abundance patterns. J. Geophys. Res. 68: 559-566, (1963).

38--1963a (with Chase, J. W. and Coryell, C. D.) Lanthanum, europium, and dysprosium in igneous rocks and minerals. J. Geophys. Res. 68: 567-575, (1963). See also 32--1962c.

39--1963b (with Chase, J. W., Schnetzler, C. C. and Czamanske, G. K.) The lanthanum, europium, and dysprosium contents of two tektites. J. Geophys. Res. 68: 577-579, (1963).

40--1963c (with Corless, J. T. and Rahn, K.) Variations in the ratio Ca^{48}/(total calcium) in natural materials [abst.]. Am. Geophys. Un., Tr. 44/1: 69, (1963). See also 54--1964c.

41--1963d (and Youh, C. C.) A review of neutron activation analysis as a technique for the study of economic raw materials [abst.]. Ann. Meeting, Geol. Soc. China, Taipei, Taiwan, 27 April 1963.

42--1963e (and Youh, C. C.) A review of chemical problems in the industrial utilization of pyrite [abst.]. Ann. Meeting, Geol. Soc. China, Taipei, Taiwan, 27 April 1963.

43--1963f The nuclear research reactor at Tsing Hua University, [a talk presented to the Chinese Assoc. Adv. Nat. Sci., Taipei, Taiwan, 25 May 1963], (1963).

44--1963g Rare-earth chromatography using Bis(2-ethylhexyl) orthophosphoric acid. J. Chromatog. 10: 502-506, (1963).

45--1963h (with Duce, R. A., Wasson, J. T., and Burns, F.) Atmospheric iodine, bromine, and chlorine. J. Geophys. Res. 68: 3943-3947, (1963).

46--1963i (with Cheng, H-s, Ke, C-h, and Lin, C-y) Characteristics of Bis(2-ethylhexyl) orthophosphoric acid chromatographic columns for rare-earth separation. J. Chinese Soc., (2)10: 80-85, (1963).

47--1963j (and Corless, J. T.) High precision neutron activation analysis for calcium isotope abundance determination [abst.]. Program, 2nd Nat. Meeting, Soc. Appl. Spectroscopy, San Diego, (14-18 October 1963).

48--1963k (with Towell, D. G. and Volfovsky, Regina) Rare-earth distributions in some rocks and associated minerals of the batholith of Southern California [abst.]. Am. Geophys. Un., Tr. 44/4: 891, (1963). See also 61--1965f.

49--1963ℓ (with Duce, R. A. and Van Nahl, T.) Iodine, bromine, and chlorine contents of some Hawaiian rains [abst.]. Am. Geophys. Un., Tr. 44/4: 892, (1963).

50--1963m (and Winchester, Ellen M.) Taiwan--How can we hold scientists when U.S. policies conflict? Int. Sci. Tech., 88-94, (December 1963).

51--1964 Rare earth analysis by neutron activation [abst.]. The Nucleus of the Northeastern Section of the Am. Chem. Soc. 41: 146, (1964). See also Ibid. 41: 103, (1964).

52--1964a (with Duce, R. A. and Van Nahl, T.) Iodine, bromine, and chlorine in the aerosol and gaseous phase of marine air from Hawaii [abst.]. Am. Geophys. Un. 45: 67, (1964). See also 58--1965c.

53--1964b (and Youh, C. C.) Determination of selenium and tellurium in the Chinkuashih pyrite ores by neutron activation analysis. Geol. Soc. China, Pr. 7: 82-84, (1964).

54--1964c (with Corless, J. T.) Variations in the ratio $^{48}Ca/$(total Ca) in the natural environment. Pure Appl. Chem. 8: 317-323, (1964). See also 40--1963c.

55--1965 [Review of] The Natural Radiation Environment, Adams, J. A. S. and Lowder, W. M., eds. Chicago: Univ. Chicago Press, (1964). Nuclear Sci. Eng. 21: 409, (1965).

56--1965a (and Duce, R. A.) Geochemistry of iodine, bromine, and chlorine in the air-sea-sediment system. Symp. Marine Chem. Pr., Univ. Rhode Island, (1965).

57--1965b (and Duce, R. A.) Determination of iodine, bromine, and chlorine in atmospheric samples by neutron activation. Radiochim. Acta 4: 100-104, (1965).

58--1965c (with Duce, R. A. and Van Nahl, T.) Iodine, bromine, and chlorine in the Hawaiian marine atmosphere. J. Geophys. Res. 70: 1775-1799, (1965). See also 52--1964a.

59--1965d (with Towell, D. G. and Spirn, Regina V.) Rare earth abundances in the Standard Granite G-1 and Standard Diabase W-1. Geochim. Acta 29: 569-572, (1965).

60--1965e (and Duce, R. A.) Halogen geochemistry. Tech. Eng. News 47/3: 27-31, (1965).

61--1965f (with Towell, D. G. and Spirn, Regina V.) Rare earth distributions in some rocks and associated minerals of the batholith of Southern California. J. Geophys. Res. 70: 3485-3496, (1965). See also 48--1963k.

62--1965g [Review of] University Chemistry, by Mahan, B. H., Reading, Mass.: Addison-Wesley Pub. Co., (1965). Geotimes 10/1: 22, (1965).

63--1965h (and Hashimoto, Y.) Packaging liquid samples for reactor neutron irradiation. Radiochim. Acta 4: 108-109, (1965).

64--1965i (with Duce, R. A. and Van Nahl, T.) Iodine, bromine, and chlorine in winter aerosols and snow from Barrow, Alaska [abst.]. Program, CARC Symp., Atmospheric Chemistry, Circulation and Aerosols, Visby, Sweden, (18-25 August 1965). See also 79--1966ℓ.

65--1965j (and Duce, R. A.) Coherence of iodine and bromine in the atmospheres of Hawaii, Northern Alaska, and Massachusetts [abst.]. Program, CARC Symp., Atmospheric Chemistry, Circulation and Aerosols, Visby, Sweden, (18-25 August 1965). See also 80--1966m.

66--1965k   Neutron activation analysis of halogens in the atmosphere [abst.]. Am. Nuclear Soc., Tr. 8: 326-327, (1965).

67--1966   (with Schilling, J.-G.) Rare earth distributions in Hawaiian volcanic rocks [abst.]. Am. Geophys. Un., Tr. 47: 201, (1966). See also 83--1966p.

68--1966a   (with MacIntyre, F.) Phosphate enrichment in drops from breaking bubbles [abst.]. Am. Geophys. Un., Tr. 47: 201, (1966). See also 97--1969a.

69--1966b   (and Duce, R. A.) The global distribution of iodine, bromine, and chlorine in marine aerosols [abst.]. Am. Geophys. Un., Tr. 47: 201, (1966). See also 85--1967a.

70--1966c   (with Duce, R. A., Woodcock, A. H., and Zoller, W. H.) Variation of ion ratios with age and location in tradewind rain showers [abst.]. Am. Geophys. Un., Tr. 47: 202, (1966).

71--1966d   (with Spirn, Regina V.) Rare earths in marine sedimentary materials [abst.]. Am. Geophys. Un., Tr. 47: 202, (1966).

72--1966e   (with Hashimoto, Y.) Determination of selenium in precipitation and atmospheric air [abst.]. Chem. Soc. Japan, Pr. April 1966.

73--1966f   Terrestrial heat flow, radioactivity, and the chemical composition of the earth's interior. J. Geol. Educ. 14: 200-204, (1966).

74--1966g   Earth Science Bibliography, in Source Material for Radiochemistry, revision 2, NAS-NRC Nuclear Science Series Rept., (1966).

75--1966h   Trace elements analysis (p. 1057-1058), in Encyclopedia of Chemistry, 2nd ed., Clark, G. L., and Hawley, G. G., eds., New York: Reinhold Pub. Corp., (1966).

76--1966i   (with Lininger, R. L., Duce, R. A., and Matson, W. R.) Chlorine, bromine, iodine, and lead in aerosols from Cambridge, Massachusetts. J. Geophys. Res. 71: 2457-2463, (1966).

77--1966j   (and Zoller, W. H. and Duce, R. A.) Lead and halogens in pollution aerosols and snow from Fairbanks, Alaska [abst.]. Western National Meeting, Am. Geophys. Un., Tr. 47: 473, (1966). See also 89--1967e.

78--1966k   (with Schilling, J.-G.) Rare earth evidence for crystallization processes in the earth's crust. Western National Meeting, Am. Geophys. Un., Tr. 47: 495, (1966).

79--1966ℓ   (with Duce, R. A. and Van Nahl, T.) Iodine, bromine, and chlorine in winter aerosols and snow from Barrow, Alaska. Tellus 18: 238-248, (1966). See also 64--1965i.

80--1966m   (and Duce, R. A.) Coherence of iodine and bromine in the atmosphere of Hawaii, Northern Alaska, and Massachusetts. Tellus 18: 287-291, (1966). See also 65--1965j.

81--1966n   (with Schilling, J.-G.) Rare earth evidence for fractional crystallization in Hawaiian basalts [abst.]. Program, Ann. Meeting Geol. Soc. Am., San Francisco, 14-16 November, 1966.

82--1966o   (with Crocket, J. H.) Coprecipitation of zinc with calcium carbonate. Geochim. Cosmochim. Acta 30: 1093-1109, (1966).

83--1966p   (with Schilling, J.-G.) Rare earths in Hawaiian basalts. Science 153: 867-869, (1966). See also 67--1966.

84--1967   Chemical processes in the sea air boundary region [abst.]. Am. Meteorol. Soc., B. 48: 26, (1967).

85--1967a   (and Duce, R. A.) The global distribution of iodine, bromine, and chlorine in marine aerosols. Die Naturwiss. 54/5: 110-113, (1967). See also 69--1966b.

86--1967b   (with Ehrlich, A. M.) Rare earths in manganese nodules [abst.]. Am. Geophys. Un., Tr. 48/1: 239, (1967).

87--1967c   (with Schilling, J.-G.) Can rare earth fractionation distinguish magma types and differentiation processes? [abst.]. Am. Geophys. Un., Tr. 48/1: 257, (1967).

88--1967d   (with Schilling, J.-G.) Rare earth fractionation and magnetic processes (p. 267-283), in Mantles of the Earth and Terrestrial Planets, Runcorn, S. K., ed. New York: Interscience Pub., (1967).

89--1967e   (and Zoller, W. H., Duce, R. A., and Benson, C. S.) Lead and halogens in pollution aerosols and snow from Fairbanks, Alaska. Atmosph. Environ. 1: 105-119, (1967). See also 77--1966j.

90--1967f   (with Hashimoto, Y.) Selenium in the atmosphere. Environ. Sci. Tech. 1: 338-340, (1967).

91--1967g   (with Hashimoto, Y.) Preparation of carrier-free radioactive tracers of selenium and tellurium. Radiochim. Acta 7: 217, (1967).

92--1967h   (and Duce, R. A.) Neutron activation analysis of lead halide pollution aerosols. Nuclear Activation Techniques in the Life Sciences, Symp. Pr. 631-643, Int. Atomic Energy Agency, (1967).

93--1968   (with Towell, D. G., Ehrlich, A. M., Schilling, J.-G., and Spirn, Regina V.) Rare earth analysis of rocks and minerals by neutron activation analysis: A method and a comparison with other procedures [abst.]. Am. Geophys. Un., Tr. 49/1: 338, (1968).

94--1968a   (with Walters, L. J., Jr.) Bound halogens in sediments [abst.]. Am. Geophys. Un., Tr. 49/1: 366, (1968).

95--1968b   [Review of] Geochemistry and Mineralogy of Rare Elements and Genetic Types of Their Deposits: Vol. 1, Geochemistry of Rare Elements; Vol. 2, Mineralogy of Rare Elements; Vlasov, K. A., ed. Moscow, (1966). Am. Scientist 56: 333A, (1968).

96--1969   (and Catoggio, J. A.) The application of neutron activation analysis to geochemical studies of mineral resources, Nuclear Techniques and Mineral Resources, Symp. Pr. 435-449, Int. Atomic Energy Agency, (1969).

97--1969a   (with MacIntyre, F.) Phosphate ion enrichment in drops from breaking bubbles. J. Phys. Chem. 73: 2163-2169, (1969). See also 68--1966a.

98--1969b   (with Towell, D. G. and Spirn, Regina V.) Europium anomalies and the genesis of basalt: A discussion. Chem. Geol. 4: 461-464, (1969).

99--1969c   (with Schilling, J.-G.) Rare earth contribution to the origin of Hawaiian lavas. Contr. Mineral. Petrol. 23: 27-37, (1969).

100--1971   (with Walters, L. J., Jr.) Neutron activation analysis of sediments for halogens using Szilard-Chalmers reactions. Anal. Chem. 43: 1020-1025, (1971).

101--1972   Buffer system (p. 95-98), in The Encyclopedia of Geochemistry and Environmental Sciences, Fairbridge, R. W., ed. New York: Van Nostrand Reinhold Co., (1972).

102--1972a   Geochemistry (p. 402-410), in The Encyclopedia of Geochemistry and Environmental Sciences, Fairbridge, R. W., ed. New York: Van Nostrand Reinhold Co., (1972).

(38)
WILLIAM STELLING VON ARX

WILLIAM STELLING VON ARX

MIT: 1956-1970

William Stelling von Arx, trained in physics and geology at Brown and Yale, first came to M.I.T. in 1953 as a doctoral candidate in meteorology. After receiving his Sc.D. degree in 1955 he returned to his earlier position as physical oceanographer at the Woods Hole Oceanographic Institution. However, one year later, in September 1956, he was appointed a Visiting Lecturer in order to help our Department of Geology and Geophysics start a program of instruction in Oceanography. For the next 15 years he served M.I.T. successively as Visiting Lecturer, 1956-1957; Associate Professor (part-time), 1957-1959; Professor (part-time), 1959-1963; Professor (full-time), 1963-1967; and Professor (part-time), 1968-1970 inclusive. During this 15-year period he lectured, conducted his own research, assisted students with their research, supervised theses, and in other ways proved a tower of strength in the joint effort of the Geology and Meteorology departments to establish at M.I.T. and with W.H.O.I. a joint academic program in Oceanography. Soon after the M.I.T.-W.H.O.I. Joint Program in Oceanography was formally established on 8 May 1968, von Arx gave up his full-time appointment as Professor of Physical Oceanography, but continued on a part-time basis for several years before finally resigning as of 31 December 1970, and returning to W.H.O.I. as a full-time senior scientist.

Although the sketch that follows is primarily concerned with von Arx's activities during the 15 years he was a faculty member in our Department of Geology and Geophysics, I have judged it appropriate that some reference, however brief and incomplete, should be made to the distinguished career in physical oceanography that he has developed through the years as a senior scientist at W.H.O.I. Consequently, I make brief reference to that career, but leave to others the full evaluation of von Arx's impressive contributions to Oceanography as a major earth science. These contributions, as defined by von Arx himself in Who's Who and elsewhere, fall into the following categories: methods and instruments for measurement of ocean currents; laboratory studies of the ocean circulation; investigation of the short-period variations in the structure of the Gulf Stream; ultra-wide field optics; stabilized optical systems for geophysical measurements at sea; heat balance of the earth; geodesy of ocean areas; and the problems of human survival.*

While at M.I.T. von Arx made three major contributions to our initial program in Oceanography. He aroused much interest among our students by telling them what study of the oceans entails, and stimulated a fair number of them to pursue the science further. He demonstrated by his own imagination and ingenuity that instruments can be invented and then used

---

* His most recent paper deals with "Energy: natural limits and abundances," Am. Geophys. Union, E⊕S 55/9: 828-832, (Sept. 1974), based on a similar article in the July 1974 issue of Oceanus.

to measure important properties of the ocean, and encouraged some of his students to design and build their own devices. Finally, he left on everyone who came to know him the impact of his own brilliant mind and energetic attitude as he applied them to study of the world ocean. In sum, whether the sea was smooth or rough, whether the new device worked or failed the first time, and whether the problem of the moment was solved or not, for him it was fun and a worthy intellectual challenge, and he has always wanted his associates to share with him that attitude toward scientific research.

BIRTH, ANCESTRY, AND EARLY EDUCATION

William Stelling von Arx* was born in Highland Mills, New York, on 27 September 1916. He traces his paternal ancestry back to great-grandfather Urs von Arx, who came to America from Canton Solothurn, Switzerland, where the von Arx name is well known and is recorded in the archives and church bibles of Solothurn going back many centuries. Urs von Arx joined his skills with those of John Hart to produce the wood engravings for the early publications of the U.S. Geological Survey. Bill's father, Arthur William von Arx, was a business man who first worked as General Eastern Agent of the Chicago Great Western Railroad and later as an officer in the Isbrandtsen-Moller Steamship Company.

The von Arx family had courtesy passes on railroads, so one day young Bill, at age 12, used one to take the train from New York to Boston, where he visited the Harvard College Observatory, to satisfy his interest in telescopes, and got his first view of the Great Court (Killian Court) at M.I.T. He recalls that he got spanked when he returned home, but that later he overheard his father telling a friend about his son's adventure with a certain pride in his voice. Even at twelve the young von Arx was showing the insatiable appetite for observational knowledge and the independent spirit for adventure that would become the driving forces of his future career as a distinguished scientist.

Bill's mother, Minnie Helen (Stelling) von Arx, was the daughter of Claus Herman and Minnie Stelling, grocers and tenement owners, in Brooklyn, where they looked after immigrant families and in many cases got them started in small businesses. Grandfather Claus Herman Stelling had immi-

---

* I am indebted to von Arx himself for much of the information in this biographical sketch; information and impressions that have come from many discussions with him, from a curriculum vitae that he provided, and from the list of publications he helped me assemble. An excellent informal and highly personal sketch, titled "William S. von Arx, Ocean Scientist," was published in <u>Woods Hole Notes</u> 2/2: 1-3, (April 1970), and brief biographies are included in the latest editions of <u>Who's Who</u> and <u>American Men of Science</u>.

grated to the United States from Hanover, Germany, when he was 13, and was self-taught. Grandmother Stelling helped her husband establish a chain of fine grocery stores that became the start of Thomas Rolston and Son, and also bore him 13 children, of whom 11 lived to a ripe old age. Minnie and Claus Herman ("Oma" and "Opa" as they were affectionately called by members of their family) were sufficiently well off by age 45 that they retired and thereafter lived "up the Hudson" in Central Valley, New York, reaching ages well into the eighties.

Bill's mother, next to the youngest of the 13 Stelling children, is remembered as a veritable book worm, a wonderful story-teller, and a queenly person who loved people, literature, and nature, and was interested in the art and poetry of the time. These same interests she quite obviously stimulated and encouraged in her own daughter and son.

Bill's father, who had "the head for business" in the von Arx family, is remembered as a person who enjoyed a circle of close professional friends, doctors and lawyers, with whom he spent many hours of conversation which commonly ended with food and drink at midnight. He also found time to do woodworking and gardening, played the violin fairly well, and loved trout fishing (with son Bill doing most of the rowing). Although he found Bill's interest in natural science "too much for himself" he nevertheless encouraged that interest in his son by giving him a small microscope and a one-inch telescope for his birthday and Christmas of his tenth year. He also saw that his son did his fair share of work around the house, such as bringing up coal for the fires, emptying ice water, mowing the lawn, or shoveling snow in winter. But for young Bill there was always time left over for reading, and "Oma" and "Opa," as well as his mother, saw to it that appropriate books to whet a youth's intellectual appetite were always visibly at hand.

With such parents and grandparents, and with a home environment that provided every stimulation and encouragement that any intellectually motivated and inquiring boy could want, even to an adventuresome train ride on a courtesy pass, it was natural for Bill to exploit every opportunity for learning, particularly because of his insatiable desire for knowledge and understanding. To understand how something worked, and to create something new that worked, were youthful desires that grew into powerful driving forces as he went through college and then developed his career as an imaginative and creative scientist, engineer, teacher and artist.

His early years were spent in wintertime study and other school activities in Brooklyn Public School 129, from 1922 to 1930, and in the Brooklyn Friends School, from 1930 to 1934, and in summertime vacations with his maternal grandparents, "Opa" and "Oma" Stelling, across the

Hudson River in Central Valley, New York. Grandfather Stelling, though self-taught, was quite active in the affairs of the Brooklyn Academy of Arts and Sciences, and young Bill, still in his early "'teens" and with an intense interest in instruments that allowed him to see the smallest and the largest objects around him, was a frequent visitor to the Academy. As early as 1926, at age 10, he made his own telescope following a short note in <u>Science Newsletter</u> on "An effective astronomical telescope made with ordinary spectacle lenses, pipe fittings and cardboard tubing," and in 1933 he wrote papers for the Astronomy Section of the Academy on "An inexpensive 'blink microscope' for amateurs" and "Electronic amplification of the images of faint stars."

During the depression years of the early 1930s, (1934-1938), after he had graduated from high school, and had completed the fall term of 1934 at the University of Rochester, he returned home and worked in order to earn and save money for additional college study. However, one has only to examine his Bibliography, listed later on following this sketch, to see that he found time after work to pursue his growing interest in science and to write articles about what he was doing.

THE COLLEGE YEARS - BROWN, YALE, AND M.I.T. (1938-1945)

When he entered Brown University in 1938 as a freshman, he had already published half a dozen articles on astronomy, and was eager to get on with his formal education. He reminisced as follows about those depression years (<u>Woods Hole Notes</u> 2/2, April 1970):

> "I was older than most of my classmates [he was 22], and by the time I finally reached college, I was suffering from intellectual starvation. I reveled for four years in intellectual gluttony."

Soon he was in the thick of things, both study and research. Physics came naturally because of his interest in astronomy and astronomical instruments, and geology proved exciting when he came under the influence of scholarly Alonzo W. Quinn, head of Brown's Department of Geology. But other sciences and their research activities also excited his interest, and soon after matriculation he was engaged in research in the Department of Psychology involving construction of an apparatus for measuring the nerve impulses from the <u>Limulus</u> eye at low light levels (with David L. Griggs), and of a frequency generator for small auditory tone differences at various sound levels; and research (with several psychology professors) on auditory discrimination of the human ear over a background of random noise (the so-called "cocktail party" experiment).

After four years at Brown, during which he was a Francis Wayland Scholar, he completed majors in both physics and geology and was awarded an A.B. degree magna cum laude, and with high honors in geology, in June 1942. For his double major he submitted a qualifying thesis in physics titled "On the determination of the distances of stars from one photographic observation employing a crossed diffraction grating" in 1940* and an A.B. thesis in the same science titled "The X-ray image method of chemical analysis of rocks, ores, and alloys" in 1942.* Simultaneously he submitted an honors thesis in geology on "A spectrochemical study of certain igneous rocks of central New Hampshire" in 1942,* meanwhile getting his first teaching experience as an Assistant in Geology.

After Brown, von Arx entered Yale in the fall of 1942 as a Dana Fellow in Geology, and received an Sc.M. degree in June 1943. He stayed on at Yale to gain additional teaching experience as an Instructor in Physics, 1943-1945, during which time he also conducted research on ice with Max Desmoret in the Department of Geology. From Yale he joined the Columbia University Radiation Laboratory for a part of 1945, working as a physicist on a classified project involving methods for continuous annealment hobbing and lost wax casting of magnetron anodes.

AS PHYSICAL OCEANOGRAPHER AT W.H.O.I. (1945-1963; 1968-)

Near the end of 1945 von Arx was appointed a Research Associate at the Woods Hole Oceanographic Institution, and in 1947 he became a physical oceanographer, a post he has held until the present (1975) except for two relatively short interruptions - a 2-year period from September 1953 to June 1955, during which he fulfilled requirements for an Sc.D. degree in Oceanography awarded by M.I.T. in 1955; and a 4-year period, 1963-1967, during which he was Professor of Physical Oceanography at M.I.T.

During these 30 years he has brought to focus on a group of diverse but closely related problems a unique combination of knowledge of the fundamentals of classical physics and geology, of creative and innovative skill in instrument design and construction, and truly outstanding imagination and the ability to define a problem needing investigation. Only a cursory inspection of his impressive Bibliography, which appears farther on, is necessary to show the diversity and difficulty of the many problems he has investigated and reported on. In 1970 his peers in oceanography accorded him their highest accolade, the Albatross Award of the Miscellaneous Society,** and the editor of the Woods Hole Notes (vol. 2, no. 2)

---

* See references T--1940, T--1942b, and T--1942c in the Bibliography farther on.

** For the history of AMSOC's Albatross Award see R.G.'s "Do oceanographers have more fun?" in Science 181: 926, (7 Sept. 1973).

for April 1970 has him speaking his eloquent and logical philosophy as to what future research in oceanography should be and what rewards it offers to the inquiring scientist who will observe and measure the many aspects of the World Ocean. As he puts it:

> "But most important is the prospect that if and when a satellite can provide either two maps each day of the world ocean or a continuous map of each hemisphere showing the sea surface molding and remolding itself in response to all the forces that act upon it, we will not only have achieved the synoptic dream but will almost certainly see some features of the ocean change in ways that we now know nothing about." (p. 2 of Woods Hole Notes 2/2, April 1970.)

AS PROFESSOR OF OCEANOGRAPHY AT M.I.T. (1956-1970)

When, in 1953, as recently appointed head of M.I.T.'s Department of Geology, I decided to develop a degree program in Oceanography as an integral part of the Course XII curriculum in the Earth Sciences, I naturally sought the cooperation of Prof. Henry G. Houghton, who headed Course XIX, Department of Meteorology. As related in a separate chapter on Oceanography at M.I.T., we first gained approval for developing a cooperative oceanographic program between our two departments. We next sought the advice and guidance of certain of the senior staff members at the Woods Hole Oceanographic Institution, inasmuch as some of our Course XII and Course XIX students had found thesis problems at Woods Hole as early as 1953. We were aware that Columbus O'D. Iselin had been giving an introductory course in Oceanography in the spring term at Harvard for many years, and that von Arx had been a Lecturer in Oceanography there during SY 1947-48. More recently, some of our geophysics students were working with J. B. Hersey at Woods Hole and at sea. Why not arrange to have these three internationally known oceanographers come to M.I.T. on a part-time basis and lecture to our students on their special fields in Oceanography? Perhaps the experiment could be conducted on an informal basis for a few years to determine if the arrangement produced satisfactory results.

Accordingly, von Arx was selected to start the experiment during the fall term of SY 1956-57. As a Visiting Lecturer in the Department of Geology and Geophysics he offered a series of regularly scheduled lectures on physical oceanography. These were so favorably received that they were repeated the next fall (1957), and von Arx was appointed Associate Professor (part-time) in 1957. He was followed by Hersey and Iselin during SY 1958-59, when it became obvious that our cooperative experiment was succeeding.

On the basis of the enthusiastic student response to von Arx's lectures, Houghton and I, in the fall of 1958, sought approval of our joint academic program from the M.I.T. Administration, through Dean of Science George R. Harrison, and from Director Paul M. Fye of the Woods Hole Oceanographic Institution. After several meetings in late 1958, Houghton and I were authorized to arrange part-time appointments with von Arx, Iselin, and Hersey in geology and geophysics and with Henry Stommel and Willem Malkus in meteorology. This arrangement, which started in 1958, proved so successful that it led to the M.I.T.-W.H.O.I. Joint Program in Oceanography that was formalized at Woods Hole on 8 May 1968 as discussed in another chapter titled Oceanography at M.I.T. (See my Volume 2.)

As a result of the earlier 1958 arrangement, the following five senior scientists at W.H.O.I. were given part-time appointments at M.I.T.:

W. S. von Arx	Associate Professor (1957-1958); Professor (1958-1963)
J. B. Hersey	Associate Professor (1959-1963); Professor (1963-1966)
C. O'D. Iselin	Visiting Lecturer (1959); Professor (1959-1966)
H. Stommel	Professor (1959-1961)
W. Malkus	Professor (1959-1961)

Other W.H.O.I. scientists also participated in the program as discussed in the chapter mentioned in the preceding paragraph.

It needs to be emphasized here that von Arx, Hersey, and Iselin, in that order as participants in our experimental program, were the ones who successfully brought Oceanography to the students of Course XII and XIX, and whose efforts were of primary importance in the establishment of the M.I.T.-W.H.O.I. Joint Program in 1968.

In every appointment von Arx held in our Department during those critical years, 1956-1968, when our oceanographic program was slowly taking form, and then later, 1968-1970, when he continued as an active participant in the M.I.T.-W.H.O.I. Joint Program in Oceanography, he was a tower of strength as an enthusiastic and exciting teacher and an imaginative and creative experimentalist, who always kept his students' needs and interests foremost in his mind. Perhaps the best way to describe von Arx's philosophy of teaching is to quote his own thoughts on the subject, as he wrote them out for me for this particular section of his biography (27 March 1975):

> "After some very happy years at the Oceanographic Institution in Woods Hole, I found I was suddenly forty years old! I looked about me and saw that most of my fellow workers were at least my age or older. It seemed necessary for someone to gather up younger people - just for the good of the place - fresh ideas and strong backs. One fall, as luck would have it, Columbus Iselin, then on the Harvard faculty, could not meet his teaching schedule and asked me to fill in

for him. I was pleased to try, and went the distance to Cambridge three times each week [This was during SY 1947-48]. The students were wonderful. We sweated out the course of study together and I learned more than they did.

"But more impressive than that happy circumstance was the fact that the lecture room assigned to me was the very one I had entered years before, when a student at Brown, to learn at the feet of Esper Larsen, Henry Stetson, M. P. Billings, Alfred Romer, and, once or twice, R. A. Daly! With that heritage to live up to, I did my utmost. But even more important was the visit I always paid to Henry Bigelow [Founder and first Director of W.H.O.I.] before my lectures. He would look up when I entered his lab and say,

> 'Well, young fellah, watcha gonna teach 'em today?'

I would explain my plan and he would invariably respond,

> 'That sounds reasonable, but what about this, and this, and this?'

As a result I gave a completely different lecture; one better than I had dreamed.

"A good many of the oceanography students at Harvard were from M.I.T. After two years at Harvard (in the Agassiz Museum [M.C.Z.] to be exact) and later on after a year's sabbatical at MIT to learn something about meteorology and geophysical fluid dynamics, I signed on at Professor Shrock's invitation as a part-time teacher of oceanography at MIT along with C. O'D. Iselin, J. B. Hersey and J. Kanwisher. Half the class turned out to be Harvard students, so it didn't really matter where oceanography was offered, the usual mix showed up. I did find though that the Harvard breed liked philosophical matters best while the 'MITs' favored numerical problems. When I asked them about it, the 'Harvards' said,

> 'If the route to an answer is that direct, why ask the question?,'

and the 'MITs' response to philosophic queries was,

> 'Well, if there is no answer, why ask the question?'

By Christmas recess the difference in attitudes had been pretty well washed away, and both groups seemed about the same. But I still marvel at the admission processes at the two institutions; each functioning so bravely in support of that imaginary and false schism between mental attitudes.

"Although not primarily an educator, I have taught a goodly number of students - it must be many thousands by now - in physics, astronomy, geology, geodesy, meteorology and, of course oceanography. I always lecture with the door open; just for fresh air. After one of the early fall lectures on oceanography at MIT, a little fellow peeked around the door frame and said,

> 'Sir, what you are talking about in here is much more interesting than what I am

doing across the hall [He was grinding up some rocks for spectrometric analysis]. Could I take your course?'

It was soon agreed that he could. He became my first Ph.D. and is now my boss - Ferris Webster, Associate Director of the Woods Hole Oceanographic Institution - which shows how careful you must be about letting in strays!

"Some of the best young men we have recruited into marine science have been won by either deliberate or just plain fortunate proselytization. I think it may have been a mistake to have formalized oceanographic education in the MIT-WHOI program. Now it is only those who already know they are interested in the marine world who are the candidates we deal with, while some of the best and most innovative in earlier years 'didn't even know the ocean world could be such an exciting object of study.'

"Now that the WHOI-MIT program is flourishing and students come to Woods Hole by bus twice each week, I no longer make the tiresome journey to Cambridge. I like teaching at Woods Hole because there are <u>no bells</u>; you start a class <u>when and where</u> the day <u>suggests</u> (on the dock, under a <u>tree, or when</u> the coffee has ceased perking) and work along until either the subject or the participants are exhausted. I don't teach crowds anymore. Having lectured on elementary physics at Yale to 250 students each hour for four consecutive hours, six days per week during the World War II years, and on the earth sciences to 150 or so at a time in rooms 10-250 or 54-100 at MIT, I don't appreciate crowds. I like the more intimate Socratic approach. A little group of three, or at most five, carefully chosen students is a 'knowable' class; you can understand each person and show each one how to learn by self-instruction. After a bit they find that every question they ask will be met by a question from me, so that little by little, question by question, they reorganize <u>the knowledge they already have</u> and turn it into a fresh insight or perhaps understanding. I love to see the light in their eyes when that moment finally dawns. It is also fun to see how much thinking they learn to do before asking a question the next time around.

"Of course, there are times when none of us has enough information at hand to deal with an issue - but that's what libraries are for; or friends. It is a gratifying experience, which students often fail to recognize early on, that an honest question reinforces friendship and, more often than not, results in a deeper knowledge of the subject than either had at the start. In the library it is a pretty dull student who comes away with only the answer he sought.

"A library is like the dictionary where the search for the meaning of one word is often side-tracked by the discovery of some other word, totally unrelated to the first and often much more interesting. All libraries should be open-stacked and a bit puzzling in their organization so that browsing is encouraged. I like to think of libraries as filled with people - behind each binding lies the richness of a man's mind however long ago he lived. (Compendia are abominations because they bind crowds and are uneven in thought and content.) What a pleasure it is to pull down a copy of some great man's work and realize as you read that the very same typographic impressions that greet your eyes were once freshly met by his; and then wonder at his thoughts when he read them; the ink still wet. Often it is footnotes that say most about that.

"The only thing that bothers me about libraries is that they are not designed to help people with limited capacities of visual accommodation. Hypermetropes [far-sighted persons] with bifocals develop an awful crick in the neck from trying to read titles on high shelves - enough to severely discourage thorough scholarship even when you are still physically agile enough to reach that high.

"But among students today I find an awesome calm; a maturity beyond their years. The G.I. Bill students were mature but really eager. Today some young students are not only very bright but almost 'judicial' in their approach to learning. It may be the sobering effect of our time that does it, but I still think learning should be just plain fun."

A summary of von Arx's appointments and activities during the 15 years he was a member of the faculty of our Department of Geology and Geophysics reads as follows:

Visiting Lecturer: 1956-1957. Gave lectures in fall term.

Associate Professor of Oceanography, part-time: 1957-1959.
Gave lectures in fall of 1957, and supervised thesis work during entire period.

Professor of Oceanography, part-time: 1959-1963.
Gave lectures in fall of 1959 and 1962 and supervised thesis work during entire period.

Professor of Physical Oceanography, full time: 1963-1967.
Lectured and conducted and directed research on a full-time basis at M.I.T.

Professor of Physical Oceanography, part-time: 1 January 1968 through 31 December 1970. Lectured and supervised student research on a part-time basis.

During his 15 years of service as a Course XII faculty member, all but 4 of which were on a part-time, and therefore limited, basis, he supervised three doctoral theses, helped substantially with many others, and excited an interest in oceanography in a host of students, some of whom decided to follow the science as a career. His energetic and imaginative approach to a problem, his ability to invent and construct needed instruments, and the high standards of performance he demanded of both himself and of others, are all admired and respected by those who know him. More than once I have heard a former student speak glowingly of one or another of his outstanding talents and traits. Although M.I.T. lost his services as an active faculty member when he decided to resume a full-time appointment as a Senior Scientist at W.H.O.I. on 1 January 1971, his educational

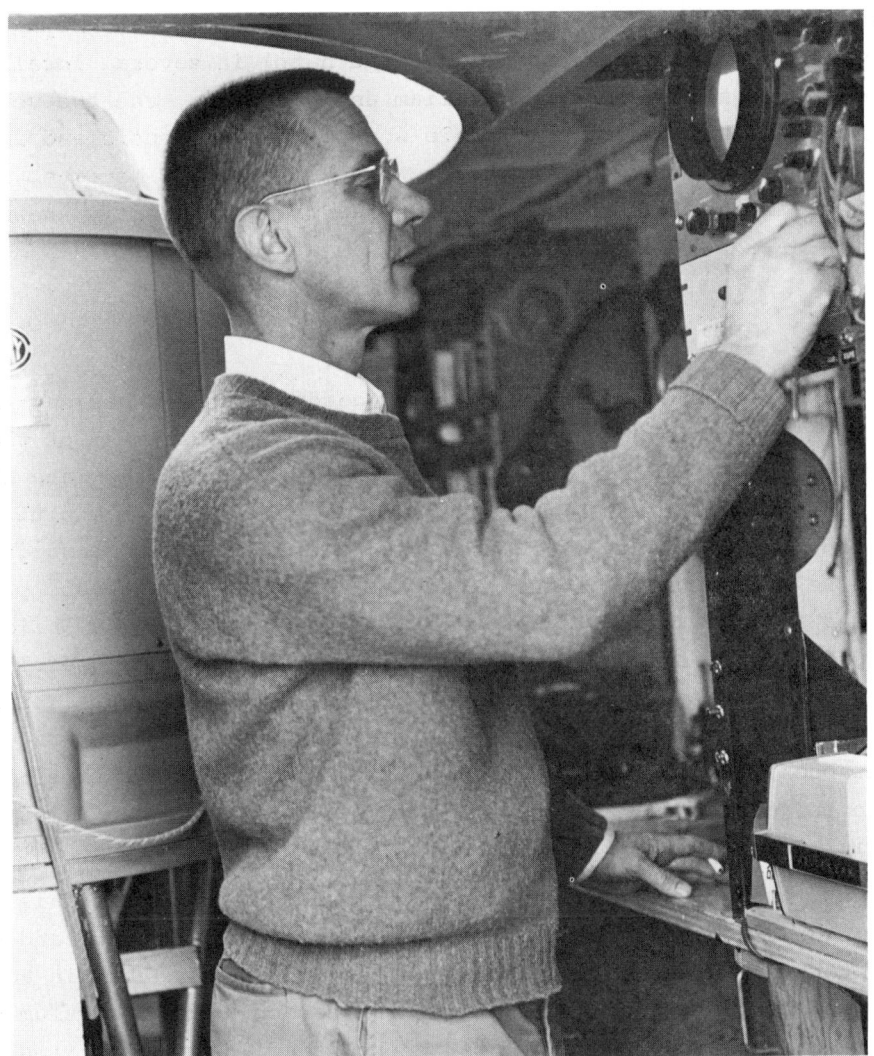

Von Arx - adjusting an instrument aboard ship.
(Photo by Jan Hahn)

philosophy left an important impact on the thinking of his M.I.T. students and faculty associates, and he remains as interested as ever in enticing future scientists into having a serious look at the unsolved problems of the World Ocean.

MARRIAGE AND FAMILY

William Stelling von Arx married Ruth Marie Lineback in 1944, while they were both students at Yale. After she was graduated from Agnes Scott College and awarded a master's degree in biology by Yale, she accompanied him to Woods Hole in 1945. Here she quickly made a place for herself in the community both as a professional biologist at the Oceanographic Institution and as a musician (viola, piano, and harp) in several local musical ensembles. Here also, Frederick William and Katherine, the two von Arx children, grew up in a home atmosphere where classical music and intellectual discussion were the common rule. When Ruth died in November 1971 after a long illness, a touching concert featuring nationally known harpist, Sylvia Meyer, was specially arranged as a dedication to her memory.*

SUMMATION

In this brief sketch I have tried to describe and emphasize the great importance of the role that von Arx played in the development of the oceanographic part of M.I.T.'s earth sciences program in the late 1950s and early 1960s. He first stirred up student interest with exciting lectures, then motivated some of our best students to take up oceanography as a career; he helped them to discover problems suitable as thesis projects, then guided and supervised them to their degrees; and all the while he conducted his own research, designed and built a variety of useful instruments for measuring characteristics of the ocean, steadily publishing articles describing his achievements; and he even found time to write an excellent book, Introduction to Physical Oceanography. All this and much more with the energy and zest that only a true scientist can bring to his work, which in the doing becomes fun.

I must leave for others the description and evaluation of his long and distinguished career as one of W.H.O.I.'s most imaginative and creative investigators of the sea. I can only mention in passing that in addition to a hundred articles (see Bibliography farther on) his curriculum vitae lists some 50 technical reports (most unpublished), 25 movie films (some of exquisite beauty as the viewer sees the changing surface pattern of a sector of moving water coupled with a background of classical music), and several patents on instruments. And his cello has given him many hours of pleasure as a member of the Little Harbor String Quartet, and the Cape Cod Symphony Orchestra, ensembles in which his wife Ruth joined in playing one of her favored instruments.

Thirty years of serious scientific work and relaxing but rewarding play have fallen lightly on von Arx's shoulders, and he can look back on an impressive array of achievements as he ponders his next invention or bounds up the gangplank for his next cruise.

---

*   See the Falmouth Enterprise, p. 2, for Tuesday, November 16, 1971 (Ruth's obituary), and p. 1 for Friday, January 7, 1972 (the concert dedicated to Ruth's memory).

BIBLIOGRAPHY OF WILLIAM STELLING VON ARX

Symbols and abbreviations in the following references are explained on p. 91 - 98. In general, abstracts are listed separately as well as with the reference to the complete article. I have included all references through 31 December 1971, one year after the date when von Arx resigned his part-time appointment at M.I.T. to return to full-time activities as a Senior Scientist at Woods Hole.

1--1932 Light and the universe. Popular Astron. 50: 438-439, (1932).

2--1935 A comparative study of calligraphy and classical type faces used in contemporary hand-set, monotype and linotype fonts. Am. Artist, (1935).

3--1937 Stellar photography, p. 556-572, in Amateur Telescope Making - Advanced, A. G. Ingalls, ed. [First ed., (1937).] Scientific American, Inc., vi + 650 p., (1970).

4--1937a A photographic atlas of cloud forms (privately published).

5--1934-1938 The cliche verre process, and other contributions to Art Instruction and American Artist, (1934-1938).

6--1938a Amateurs turn cameras skyward. The Sky 2: 7, 14, 28, (1938).

7--1938b A camera sees where eyes cannot. The Sky 2: 10-11, 25-26, (1938).

8--1939 Proposed mounting and sketch of 20-inch photographic and $12\frac{1}{2}$-inch visual telescope for Amateur Astronomers Association Observatory. The Sky 3: 24, (1939).

T--1940 On the determination of the distance of stars from one photographic observation employing a crossed diffraction grating. Qualifying Thesis in Physics Department, Brown University, Providence, Rhode Island, (1940).

9--1941 An atlas of the iron spark spectrum from 2300 to 6200 A.U. in grating dispersion. Dept. Geol., Brown University, Rhode Island, (1941).

10--1942 A grating spectrograph for qualitative analysis. J. Chem. Ed. 19: 407-410, (1942).

11--1942a The crystal structure of pond ice observed with polaroid spectacles. J. Chem. Ed. 19: 478-479, (1942).

T--1942b A spectrochemical study of certain igneous rocks of central New Hampshire, 55 p., (1942). Honors Thesis A.B., Department of Geology, Brown University, Providence, Rhode Island, (June 1942).

T--1942c The X-ray image method of chemical analysis of rocks, ores, and alloys. Undergraduate Thesis, Department of Physics, Brown University, Providence, Rhode Island, (1942). [no degree]

12--1944 The technical spectrograph, p. 140-141, in Contributions to Dept. for the Amateur Telescope Maker, A. G. Ingalls, ed. Sci. American 170, (1944).

13--1945 A molecular motion projector. J. Chem. Ed. 22: 57-63, (1945).

14--1947 A salinometer for use in brackish water. J. Marine Res. 6: 139-145, (1947).

15--1948 The circulation systems of Bikini and Rongelap Lagoons. Am. Geophys. Union, Tr. 29: 861-870, (1948).

16--1950 The properties of hydraulic models as aids to civilian harbor defense. Colloquium on Tidal Flushing, M.I.T. Hydrodynamics Lab., p. 24-42, (1950).

17--1950a   Some current meters designed for suspension from an anchored ship. J. Marine Res. 9: 93-99, (1950).

18--1950b   Synoptic models of the circulation in estuaries, p. 24-42, in Proc. Colloquium on the Flushing of Estuaries, M.I.T. Hydrodynamics Lab., (Sept. 1950), Henry Stommel, ed. Published as W.H.O.I. Ref. 50-37: 206 p., il., (1950).

19--1950c   An electromagnetic method for measuring the velocities of ocean currents from a ship underway. Pap. Phys. Oceanogr. Meteor. 11: 1-62, (1950).

20--1951   Dead reckoning by surface current observation. J. Inst. Navigation (London) 4: 117-125, (1951).

21--1952   Notes of the surface velocity profile and horizontal shear across the width of the Gulf Stream. Tellus 4: 211-214, (1952).

22--1952a   A laboratory study of the wind-driven ocean circulation. Tellus 4: 311-318, (1952).

23--1952b   Measurement of the oceanic circulation in temperate and tropical latitudes. Appendix: Some suggestions for future current measuring techniques. Symposium of Oceanographic Instrumentation, Rancho Santa Fé, 21-23 July 1952, Chap. 2.

24--1953   (and Stommel, H., Parson, D., and Richardson, W. S.) Rapid aerial survey of the Gulf Stream with camera and radiation thermometer. Science 117: 639-640, (1953).

25--1953a   Cartographic principles applied to wide-field photography. Photo. Eng. 4: 60-73, (1953).

26--1953b   Some techniques for laboratory study of the primary ocean circulation, in Fluid Models in Geophysics; Proc. 1st Symp. on the use of Models in Geophys. Fluid Dynamics, Johns Hopkins University, Sept. 1953.

27--1954   A laboratory model of the wind-driven ocean circulation. Weather 9: 170-176, (1954).

28--1954a   The circulation systems of Bikini and Rongelap Lagoons. Bikini and nearby atolls, Part 2. Oceanography (Physical). U.S. Geol. Surv., P.P. 260-B-I: 265-273, (1954).

29--1954b   Measurement of the oceanic circulation in temperate and tropical latitudes, in Oceanographic Instrumentation, N.R.C. 309: 13-15, (1954).

T--1955   An experimental study of the dependence of the primary ocean circulation on the mean zonal wind field, 120 p., il., (1955). (Sc.D. Thesis at M.I.T. in Course XIX, Department of Meteorology, June 1955.)

30--1955a   (and Bumpus, D. F. and Richardson, W. S.) On the fine-structure of the Gulf Stream front. Deep-Sea Res. 3: 46-65, (1955).

31--1955b   [Secretary's] Report of the Committee on Current Measurements. A.I.O.P. Proc.-Verb., I.U.G.G., Rome, 1954, No. 6: 107-108, (1955).

32--1955c   Observational aspects of the current measurements by the electromagnetic method. A.I.O.P. Proc.-Verb., I.U.G.G., Rome, 1954, No. 6: 221-222, (1955).

33--1956   [Editor] Proc. Symposium on Aspects of Deep-Sea Research, NAS-NRC Pub. 473. Washington, D.C., (1956).

34--1956a   On the promise and limitations of ocean model experiments, p. 45-49 in Proc. Symposium on Aspects of Deep-Sea Research (W. S. von Arx, ed.), NAS-NRC Pub. 473. Washington, D.C., (1956).

35--1956b   A small world. Oceanus 4/4: 5-13, (1956).

36--1957   An experimental approach to problems in physical oceanography, p. 1-29, in Physics and Chemistry of the Earth, vol. 2, L. H. Ahrens, et al., eds., viii + 259 p., il., (1957).

37--1958   Dister: A displacement sequence stereoscope. J. Meteor. 15: 230-231, (1958).

38--1958a   Synoptic photography. Weather 13: 179-196, (1958).

39--1958b   [Editor] Historic Paintings of Woods Hole by Franklin Lewis Gifford. Woods Hole Public Library (1958); 12 p. + 17 paintings, private printing, (1962).

40--1958c   (and Malone, T. F. and Braham, R. R.) Preliminary plans for a National Institute for Atmospheric Research, prepared for the National Science Foundation by the University Committee for Atmospheric Research, (1958).

41--1959   (and Faller, A. J.) The modelling of fluid flow on a planetary scale, p. 53-72 in Proc. 7th Hydraulic Conference, Iowa Inst. Hydraulic Res., Iowa State University, Iowa City, Iowa, (1959).

42--1960   Recommendations for research, p. 154-155 in Proc. 3rd Conference on Great Lakes Research Present Status and Future Needs. Great Lakes Res. Inst., Univ. Michigan, (1960).

43--1960a   Effects of abnormal wind torque on the circulation of a barotropic model of the North Pacific Ocean, p. 103-104 in Symposium of the Changing Ocean, 1957-1958, California Cooperative Oceanic Fisheries Investigations Reports 7, (1960).

44--1960b   The line of zero set. Deep-Sea Res. 7: 219-220, (1960).

45--1961   Stars and gravity. Oceanus 8/1: 12-13, (1961).

46--1961a   The levels of sea level at sea. Oceanus 8/2: 2-8, (1961).

47--1962   Navigation and leveling at sea. Naval Res. Rev., p. 1-6, (Jan. 1962).

48--1962a   Citation for the Bigelow Medal. Oceanus 9/1: 17, (1962).

49--1962b   Introduction to Physical Oceanography. Reading, Mass.: Addison-Wesley Pub. Co., Inc., 422 p., il., (1962).

50--1962c   GEON: A new navigation system. Navigation, J. Inst. Nav. 9: 224-230, (1962).

51--1963   Discussion of F. D. Braddon's Ship Inertial Navigation. J. Soc. Nav. Arch. & Mar. Eng. 70: 273-274, (1963).

52--1963a   Applications of the gyropendulum, p. 325-345 in The Sea 2/16, (1963). New York: Interscience Pub.

53--1963b   Measurement of sub-surface currents by submarine. Deep-Sea Res. 10: 189-194, (1963).

54--1963c   Un nouveau système de navigation. Navigation 11: 43, (1963).

55--1963d   Current measurements, in Vol. 7, Encyclopaedic Dictionary of Physics. London: Pergamon Press, (1963).

56--1964   The origins of the ocean. Tech. Rev. 66/3: 15-17, (1964).

57--1964a   Given: The Earth. Tech. Rev. 67/2: 13-16, (1964).

58--1965   Practical astronomy from shipboard. Sky and Telescope 29/6: 340-349, (1965).

59--1965a   Absolute dynamic topography. Limnol. & Oceanogr. 10: (Redfield Volume): R265-R273, (1965).

60--1965b   (with Frey, E. J. and Harrington, J. V.) A study of satellite altimetry for geophysical and oceanographic measurement, p. 53-72 in Proc. 16th Cong. Int. Astronaut. Fed., (1965).

61--1966   Level-surface profiles across the Puerto Rico Trench. Science 154: 1651-1654, il., (1966).

62--1966a   (with Spilhaus, A. F., Jr.) Measurements of the forward scattering of a laser beam in sea water. Deep-Sea Res. 13: 755-759, (1966).

63--1967   Relationship between marine geodesy and oceanographic measurements, p. 37-42 in Proc. 1st Marine Geod. Symp., Columbus, Ohio, 28-30 Sept. 1967.

64--1967a   The Ocean, p. 88-96 in The Earth in Space (H. Odishaw, ed.). New York: Basic Books, Inc., ix + 340 p., (1967).

65--1969   A technique for finding gravity vertical at sea. Deep-Sea Res. 16 (Suppl.): 325-330, (1969).

66--1969a   The Comser Report - NOAA and EPA, from the NAS/NAE Committee Advisory to the Environmental Science Services Administration, Memorandum No. 3, (1969). (N.A.S. publication.)

67--1970   An oceanographic satellite? Abst., Proc.-Verb. 11, in Symp. on Remote Sensing, 15th Int. Cong. IAPSO, Tokyo, Japan, (1970).

68--1970a   [Editor and Contributor] Ocean physics, p. 3-1--3-18, in The Terrestrial Environment, Solid-Earth and Ocean Physics, Application of Space and Astronomic Techniques. NASA, CR-1579, Report of a study at Williamstown, Mass. to NASA, (Aug. 1970).

69--1971   An oceanographic satellite? Abst., Proc.-Verb. 12, in Symp. on Marine Geodesy, I.U.G.G., Moscow, 1971.

70--1971a   Marine Geodesy, p. 597-635, in The Sea (A. E. Maxwell, ed.). New York: John Wiley & Sons (Wiley-Interscience), vol. 4, pt. 1, xiii + 791 p., il., (1971).

71--1971b   The water in seawater. Oceanus 16/1: 12-13, (1971).

NOTE:

References to articles published after 1971 can be found in the standard bibliographies of oceanographic literature.

(39)
ARTHUR JAMES BOUCOT

ARTHUR JAMES BOUCOT

MIT: 1957-1961

Trained at Harvard (A.B. 1948, A.M. 1949, Ph.D. 1953) and with the U.S. Geological Survey, (1949-1951 [w.a.e.] and 1951-1956), Arthur James Boucot joined M.I.T.'s Department of Geology and Geophysics as an Assistant Professor in July 1957. For four years he offered courses in historical geology and paleontology, mounted an extensive program of research on Silurian and Devonian stratigraphy and paleontology, and directed the Department's M.I.T. Summer School of Geology near Antigonish, Nova Scotia during the summers of 1958 and 1959. Although in our Department for only four years, Boucot was outstandingly successful in getting students interested in research and as a result supervised 23 theses - 15 S.B.s, 6 S.M.s and 2 doctorates. He himself published some 20 papers in the same period of time, and laid the foundation for a major research program on the Silurian and Devonian of the world, a program he later carried to completion in the early 1970s, meanwhile establishing himself as one of the leading authorities on the history of those two geological periods.

Boucot left M.I.T. in 1961 to accept an appointment at Caltech. After seven years there he moved back east to a professorship at the University of Pennsylvania in 1968. After a year at Penn he once again moved across the continent, this time to Oregon State University, where, since 1969, he has established, with his paleontological associate, J. G. Johnson, a productive center of research on the geology of the Silurian and Devonian rocks and fossils of the world.

He represents yet another outstanding geologist who started an academic career at M.I.T., soon after completing his formal college training, then left after only a few years to develop a distinguished career elsewhere. Although the resignation of such promising younger professors is always a loss, perhaps the Department should nevertheless find some satisfaction in the fact that it provided the opportunity for a young scientist to make the start on his chosen career.

BIRTH AND EARLY EDUCATION

Arthur James Boucot was born in Philadelphia, Pennsylvania, on 26 May 1924, the older of two children (he has a sister Nancy George) of Joseph Ronald and Katherine (Rosenbaum) Boucot. His mother, an M.D., has had a distinguished career as a physician, medical administrator, and professor of medicine.*

---

* See WHO'S WHO IN AMERICA, vol. 36, 1970-1971, p. 231. Chicago: Marquis Who's Who, xxxii + 2585 p., (1970-1971).

Following public school education in the Philadelphia area, Arthur entered the Elkins Park High School in 1937 and graduated in 1941 as World War II was in progress. After one year (SY 1941-42) as a freshman in chemical engineering at the University of Pennsylvania, he took a job as a crystal finisher at the Camden, New Jersey plant of Radio Corporation of America, where his work was to finish quartz oscillator plates for use in radios that were being made for the Armed Forces. This job lasted from July 1942 through January 1943, at which time he was drafted into the U.S. Army.

MILITARY SERVICE IN WORLD WAR II

In the Army he was assigned to the Infantry, but before he had completed his Infantry Basic Training he volunteered and was accepted as an Air Crew trainee. On completion of the Air Crew training program, he joined the Eighth Air Force in Europe as a Navigator, and flew as Lead Navigator on forty-six missions (about one and one-half tours of duty) by war's end. All his missions were flown in B-24 bombers. For his services in World War II he was awarded an Air Medal with 6 oak leaf clusters and a Distinguished Flying Cross. He was discharged shortly after the end of the war, in November 1945, but remained as a Captain in the U.S. Air Force Reserve until 1962.

THE INFLUENCE OF HONESS, SWARTZ, GORDON, HOWELL AND CLOUD

As a child growing up in Philadelphia, Boucot early developed a keen interest in mineralogy. When his mother was completing her pre-medical training at the Pennsylvania State College in 1934-1935, he came under the influence of Arthur Honess (mineralogist) and Frank Swartz (paleontologist), professors at State College. Both of these men encouraged and helped him to identify minerals and fossils and to collect specimens locally. At the same time he became acquainted with Samuel G. Gordon, mineralogist at the Academy of Natural Sciences in Philadelphia from 1935 to 1941. Gordon, no doubt sensing he had an unusual young man on his hands, permitted Boucot to assist in various "chores" suitable for a youngster "hanging about" the Museum, while at the same time slowly getting him involved with minerals. During these same years, 1935-1939, Boucot attended the weekly evening lectures on paleontology given by Benjamin F. Howell, Princeton Professor, in connection with the geology course of the Wagner Free Institute of Science, and successfully passed the examinations terminating this series of lectures. By the time he joined the U.S. Air Force, Boucot had developed a keen interest in minerals, so immediately after his discharge from the Air Corps in November 1945 Gordon directed him to Harvard, where it was assumed he would prepare for a career in mineralogy. This did not happen, however, because Boucot came under the stimulating influence of Preston E. Cloud, Jr., then a professor of paleontology at Harvard, and changed his primary geological interest from mineralogy to paleontology. Under Cloud's guidance and encouragement he started the paleontological career that he continues today with outstanding distinction, as will be discussed a little farther on.

HARVARD, THE U.S.G.S., AND M.I.T.

Following Gordon's direction, Boucot entered Harvard late in the fall of 1945 as an undergraduate in geology and qualified for an A.B. degree, magna cum laude, in February 1948, having held a Veteran's National Scholarship from September 1946 to February 1948. During this undergraduate period he attended a six-week field course at the Louisiana State University Field Camp near Colorado Springs, Colorado, in the summer of 1946.

Immediately after receiving his A.B. degree from Harvard in February 1948, he started graduate work there, continuing to hold a Veteran's National Scholarship until June 1949, at which time he received an A.M. degree in geology. He spent the summer of 1948 taking a six-week course in invertebrate zoology at the Marine Biological Laboratory in Woods Hole, Massachusetts.

When the Veteran's scholarship ran out in June 1949, Boucot was appointed a Teaching Fellow in Petrography, meanwhile having earned his S.M. degree, and for the next two years, SYs 1949-50 and 1950-51, he served in this capacity. Obviously he had not lost his touch with minerals and the petrographic microscope, even though paleontology had by this time become his primary geological interest.

During these years of graduate work as a doctoral candidate, 1949-1951, he became associated with the U.S. Geological Survey on a w.a.e. basis through the assistance of Cloud, and then served as a full-time Geologist for the next five years, 1951-1956, before again returning to a w.a.e. basis during the years 1956 to 1967.

During the winter of 1951-1952 he studied Paleozoic gastropods with J. Brookes Knight at the U.S. National Museum, and thereafter for the next four years, 1952-1956, he learned about brachiopods from the master of them all, G. Arthur Cooper, who quietly but firmly set for Boucot the high standards of research that he would follow in his own work later on the brachiopods and stratigraphy of the Silurian and Devonian around the world. An aspiring young paleontologist could not have found better teachers than Cloud, Cooper, and Knight!

Having completed his doctoral thesis on "The Lower Devonian rocks of west central Maine," and having received his Ph.D. degree in paleontology in June 1953, Boucot was employed as a Geologist with the U.S. Geological Survey until June 1956. Then followed a year as a Research Associate at the Smithsonian Institution, during which he collected fossils from the Silurian and Devonian rocks of Western Europe on a John Simon Guggenheim Fellowship. (By this time the mid-Paleozoic brachiopods had him firmly in their grasp!)

At this juncture in his developing paleontological career, Boucot was invited to enter the academic world and try his hand at enticing students to take up the study of fossils and their enclosing rocks. Accepting our offer, he came to M.I.T. in July 1957, as an Assistant Professor of Paleontology, in the Department of Geology and Geophysics, and moved up to Associate Professor in July 1960.

Immediately upon his arrival, accompanied by boxes and boxes and boxes of fossilferous rocks, he set about in his characteristically aggressive manner to retrieve the eagerly sought fossils, chiefly brachiopods, that he knew the rocks in those boxes contained. In no time he had the Department's machinist design and make a greatly improved rock splitter, and the many wooden-drawered cases began to fill with the fossils that resulted from his rock-breaking activities. Not only did he throw his own 6-foot 190-pound body into the fray, on a 10 to 15-hour a day basis; he kept his wife Barbara busy when she was not typing manuscript, and when her parents came for a visit they were promptly drafted for whatever work they could do! And woe to the innocent and curious undergraduate who looked in on the noisy and dusty operation in temporary Building 20, where the foot-square beams were beginning to sag under the weight of the boxes of rocks. Before he knew it, he too was drawn into the attack on those rocks, and having been somehow inoculated with the same enthusiasm and excitement that burst forth from Boucot, he quickly found that senior thesis he was seeking. Little wonder, then, that during Boucot's all too brief sojourn of only four years in the Department, he supervised an impressive number of theses, 23 in all - 15 Bachelor's, 6 Masters, and 2 Doctors. And what to me is even more impressive, he insisted on his students publishing their findings, and shared authorship with them to assure that they did (See his Bibliography farther on).

Out of this amazing rock-breaking operation came not only the training of two dozen students, a fair number of whom have gone on through graduate work to important positions in the academic world, but also the retrieval of new fossils that provided material for a number of Boucot's twenty papers he published while at M.I.T.

In the classroom, Boucot was the same enthusiastic and driving force for geology that he was in the laboratory. He wanted to wrap his arms around all the knowledge of Silurian and Devonian geology there was in the world, and he tried to get his students to develop in the same attitude, but this proved a bit frustrating, even for the typical M.I.T. undergraduate. Art just didn't seem to sense his own or his students' physical limits at this stage in his career.

Finally, at the M.I.T. Summer School of Geology in Nova Scotia, Boucot insisted that his student charges follow the high standards of the field geologists of the U.S. Geological Survey by whom he himself had been trained, and many a student is the better geologist today because of the training he received at Crystal Cliffs.

Even as early as his M.I.T. years, Boucot began to visualize a long-term comprehensive study of the Silurian and Devonian stratigraphy and paleontology of the entire world. He felt certain that brachiopods could not only help in correlation but could also provide basic information on facies changes, paleo-ecological variations, and other aspects of sedimentation and organic evolution during middle Paleozoic times.

In 1960 he became a member of the Subcommittee on Silurian-Devonian and Lower Devonian-Middle Devonian boundaries of the Stratigraphic Commission of the International Geological Commission. This membership gave him the incentive and opportunity to start his ambitious worldwide study of the Silurian and Devonian, the prosecution of which has won him international acclaim and recognition as one of the world's leading stratigraphic paleontologists.

CALTECH, UNIVERSITY OF PENNSYLVANIA, AND OREGON STATE

In July 1961, he decided to resign and go to the California Institute of Technology where he felt the academic atmosphere was more favorable for carrying on his ambitious project. So in the summer of 1961 Boucot and his rocks travelled across the continent to Pasadena, and shortly thereafter he was again characteristically at work breaking out and dissolving the brachiopods he sought. For seven years he flourished in the encouraging atmosphere of Caltech's Division of Geological Sciences, and contributed more than 80 articles bearing mainly on his worldwide study of the Silurian and Devonian.

Then came what seemed to be an ideal combination of opportunities, and the summer of 1968 found the Boucot collection, now grown almost to box-car size, again crossing the continent, this time from Pasadena to Philadelphia, where Boucot would be Professor of Paleontology on the faculty of the University of Pennsylvania, and a colleague of Henry Faul, Department of Geology Chairman, and George de Vries Klein, already attracting favorable notice for excellent research on sedimentation. And of special significance was the arrangement that made it possible for J. G. Johnson, an amazingly productive research assistant, to continue to work with Boucot. As a result of a bout with polio, Johnson depends on a power-driven respirator to breathe and stay alive, and Boucot had assumed responsibility for his health and maintenance, knowing that high-quality research could be expected from Johnson's efforts as long as he could breathe.

Unfortunately, the absence of research facilities made it impossible for Boucot and Johnson to continue their research at Penn, and so in the summer of 1969, they and those cases and cases of fossils and rocks moved back across the continent, this time from East to Northwest, and came to rest in Corvallis, Oregon, where Boucot became Professor of Paleontology and Johnson Assistant Professor of Paleontology in the Department of Geology at Oregon State University. Here the Boucot collection is now preserved as a nucleus for an even more extensive and representative assemblage of the world's Silurian and Devonian rocks and fossils. Since then, as would be expected, important contributions continue to appear, making Corvallis a focal center for research on the world's Silurian and Devonian history.

SUMMATION

By the time Boucot came to M.I.T. in 1957 he had already become known as a well-trained, indefatigable, ambitious and aggressive stratigraphic paleontologist who was prepared to dedicate his professional life to learning everything possible about the fossils, especially the brachiopods, and the stratigraphy of the Silurian and Devonian. At M.I.T. he started his academic career in auspicious manner, leaving an important impact in the form of 23 student theses supervised, 20 articles published, and a worldwide research project started. Thereafter, at Caltech, University of Pennsylvania, and Oregon State University, he has continued to be probably the world's most enthusiastic and devoted student of Silurian and Devonian stratigraphy and paleontology. In all, he has now (1975) authored more than 200 papers and has described many new species; and he continues in 1975 to be as active, enthusiastic, and productive as ever. What energy! The renown he has won is clearly well deserved.

However, it would be unfair if I ended this brief account of Arthur Boucot's activities and accomplishments without mentioning two individuals who have quietly worked for him in the background - his devoted wife, Barbara, and his accomplished and productive research partner, J. G. Johnson, mentioned in a foregoing paragraph.

Arthur married Barbara Pierce of Wilton, New Hampshire, on 12 June 1948, shortly after he graduated from Harvard. Barbara has a B.A. degree from Radcliffe in American History and Literature. She immediately became his typist, draftswoman, preparator of fossils, and general assistant, all of which she continues to be on occasion, and she has also borne him four fine children - Hannah Gray, Katherine Marsh, Samuel Gordon, and Peter Morris - and kept the family on the road to education while Arthur pursued fossil brachiopods in North America, Europe, Asia, Australia, Malaya, New Zealand and Antarctica. So a tip of the hat to Barbara!

In the laboratory, working from his wheel-chair and at all times depending for life-sustaining air on an electrically-powered pump, Johnson has produced an amazing amount of high-quality work on fossil brachiopods, as attested by more than 40 publications that he has co-authored with Boucot and others. A truly remarkable record!

Since 1970 Boucot has progressively become concerned with the biologic significance, particularly the evolutionary consequences, of the fossil record. His recent volume, Evolution and Extinction Controls (Elsevier, 1975), provides the documentation for concluding that the first order control exercised over rate of evolution is worldwide interbreeding population size as estimated in large part from his familiarity with communities and biogeographic units based on Silurian and Devonian brachiopods. Lately (1975) he has turned more and more to the study of Phanerozoic animal community and biogeographic history in an attempt to learn more about the evolutionary process in theory, and to develop better methods of using fossils in stratigraphic studies.

BIBLIOGRAPHY OF ARTHUR JAMES BOUCOT

Symbols and abbreviations used in the following references are explained on p. 91 - 98; in general, abstracts are listed separately as well as with the complete article. This bibliography includes all known publications by Boucot through 1962, the year after he left M.I.T. Included are 18 articles he published before coming to M.I.T. in 1957 and 20 articles he published while at the Institute. Subsequently, from 1962 to 1975 inclusive, he added another 175 titles to his list of publications; these can be found in the annual Bibliographies of North American Geology, etc.

1--1947   Triplite from El Paso County, Colorado. Rocks and Minerals 22: 517, (1947).

2--1949   Allanite from Godhaab, South Greenland. Rocks and Minerals 24: 35, 61, (1949).

3--1949a   Notes on the Ecton Mine, Montgomery County, Pennsylvania. Rocks and Minerals 24: 492-495, (1949).

4--1951   (with Menard, H. W.) Experiments on the movement of shells by water. Am. J. Sci. 249: 131-151, il., (1951).

5--1951a   A new zeolite locality on Disko Island, Greenland. Rocks and Minerals 26: 526-527, (1951).

6--1953   Life and death assemblages among fossils. Am. J. Sci. 251:
             25-40; correction, p. 248, (1953).

T--1953   The Lower Devonian rocks of west central Maine, viii + 135 p.,
             pl., il., (1953). (Ph.D. Thesis at Harvard University in the
             Division of Geological Sciences, May 1953.)

7--1953a   Fossils in metamorphic rocks. Geol. Soc. Am., B. 64: 997-998,
             (1953), being a discussion of "Fossils in metamorphic rocks,"
             by W. H. Bucher, in Geol. Soc. Am., B. 64: 275-300, (1953).

8--1953b   (and Cumming, L. M.) Contributions to the age of the Gaspé
             sandstone [Quebec] [abst.]. Geol. Soc. Am., Pr. 1953: 31,
             (1953); B. 64: 1397, (1953).

9--1953c   (and Harner, R. S., MacDonald, G., and Milton, C.) Age of the
             Bernardston formation, Mass.[achusetts] [abst.]. Geol. Soc.
             Am., Pr. 1953: 31, (1953); B. 64: 1397-1398, (1953).

10--1953d   (with Switzer, G. W.) Mineralogy of some microfossils [abst.].
             Geol. Soc. Am., Pr. 1953: 115, (1953).

11--1953e   Problems in New England Paleozoic stratigraphy [abst.].
             A.A.A.S., Proc. 1953, Sec. E: 5, (1953); Geol. Soc. Am., B.
             64: 1559, (1953).

12--1954   Age of the Katahdin granite [Maine]. Am. J. Sci. 252: 144-
             148, (1954).

13--1955   (with Switzer, G.) The mineral composition of some microfos-
             sils. J. Paleont. 29: 525-533, (1955).

14--1956   Gyrospira, a new genus of bellerophontid (Gastropoda) from
             Bolivia. J. Wash. Acad. Sci. 46: 41-47, (1956).

15--1956a   (and Gill, E. D.) Austrocoelia, a new Lower Devonian brachio-
             pod from South Africa, South America, and Australia. J. Pale-
             ont. 30: 1173-1178, (1956).

16--1957   A Devonian brachiopod, Cyrtinopsis, redescribed. Senckenb.
             Leth., B. 38: 37-48, (1957).

17--1957a   Position of North Atlantic Silurian-Devonian boundary [abst.].
             Geol. Soc. Am., Pr. 1957: 129, (1957); B. 68: 1702, (1957).

18--1958   Revision of some Silurian and early Devonian spiriferid genera
             and erection of Kozlowskiellinae, new subfamily. Senckenb.
             Leth., B. 38: 311-334, il., (1957); correction, with title
             Kozlowskiellina, a new name for Kozlowskiella Boucot 1957.
             J. Paleont. 32: 1031, (1958).

19--1958a   (and Amdsen, T. W.) New genera of brachiopods, Pt. 4 of
             Stratigraphy and paleontology of the Hunton group in the Ar-
             buckle Mountain region. Okla. Geol. Surv., B. 78: 159-170,
             il., (1958).

20--1958b   (and others) Metamorphosed middle Paleozoic fossils from cen-
             tral Massachusetts, eastern Vermont, and western New Hampshire
             [New England]. Geol. Soc. Am., B. 69: 855-864, il., (1958).

21--1958c   (and Thompson, J. B., Jr.) Late lower Silurian fossils from
             sillimanite zone near Claremont, New Hampshire. Science 128:
             362-363, (1958).

22--1958d   Age of the Bainbridge limestone [Missouri]. J. Paleont. 32:
             1029-1030, (1958); discussion of Conodonten aus dem oberen
             Gotlandium Deutschlands und der Karnischen Alpen, by O. H.
             Walliser, Hesse Landesamt Bodenforsch, Notizblatt, B. 85: 28-
             52, il., (1957).

23--1958e   (and Brace, W. F. and DeMar, R. E.) Distribution of brachio-
             pod and pelecypod shells by currents. J. Sed. Petrology 28:
             321-332, il., (1958).

24--1958f   Beginning of the Acadian orogeny in the Northern Appalachians [abst.]. Geol. Soc. Am., Pr. 1958: 33, (1958); B. 69: 1537, (1958).

25--1959   Early Devonian Ambocoeliinae (Brachiopoda). J. Paleont. 33: 16-24, il., (1959).

26--1959a   A new family and genus of Silurian orthotetacid brachiopods. J. Paleont. 33: 25-28, il., (1959).

27--1959b   (and Harper, C. and Rhea, R.) Geology of the Beck Pond area, Township 3, Range 5, Somerset County, Maine. Me. Geol. Surv., Spec. Geol. Studies Ser. 1: 33 p., il., (1959).

28--1959c   (and others) The geology of a six-mile section along Spencer Stream, Somerset County, Maine. Me. Geol. Surv., Spec. Geol. Studies Ser. 2: 28 p., il., (1959).

29--1959d   (with Hurley, P. M. et al.) Minimum age of the Lower Devonian slate near Jackman, Maine. Geol. Soc. Am., B. 70: 947-950, (1959).

30--1959e   Brachiopods of the Lower Devonian rocks at Highland Mills, New York. J. Paleont. 33: 727-769, il., (1959).

31--1959f   (and Fletcher, R. and Griffin, J.) Middle or Upper Ordovician in Nova Scotia [abst.]. Geol. Soc. Am., Pr. 1959: 14A, (1959); B. 70: 1572, (1959).

32--1960   A new Lower Devonian stropheodontid brachiopod [Nova Scotia]. J. Paleont. 34: 483-485, il., (1960).

33--1960a   Implications of Rhenish Lower Devonian brachiopods from Nova Scotia. Int. Geol. Cong., 21st, Copenhagen, 1960, Rept., pt. 12: 129-137, (1960).

34--1960b   (and Arndt, R.) Fossils of the Littleton formation (Lower Devonian) of New Hampshire. U.S. Geol. Surv., P.P. 334-B: 41-51, il., (1960).

35--1960c   (and others) A late Silurian fauna from the Sutherland River formation, Devon Island, Canadian Arctic Archipelago [Northwest Territories]. Can. Geol. Surv., B. 65: 51 p., il., (1960). [Contains 6 papers which are not cited separately.]

36--1960d   Lower Gedinnian brachiopods of Belgium. Inst. Geol. Univ. Louvain, T. XXI: 283-324, il., (1960).

37--1961   Stratigraphy of the Moose River synclinorium, Maine. U.S. Geol. Surv., B. 1111-E: 153-188, il., (1961).

38--1961a   (and Caster, K.) Relationships of a new Lower Devonian terebratuloid from Antarctica [abst.]. [Cincinnati Meeting, Nov. 1961, p. 139, (1961).] Geol. Soc. Am., Sp. P. 68: 139, (1962).

39--1962   [Review of] H. M. Muir-Woods' On the morphology and classification of the brachiopod Suborder Chonetoidea. Science 137: 744-745, (1962).

40--1962a   (and Pankiwskyj, K.) Llandoverian to Gedinnian stratigraphy of Podolia and adjacent Moldavia: Symposiums Band 2, Int. Arbeitstagung über die Silur/Devon-Grenze und die Stratigraphie von Silur und Devon, Stuttgart, p. 1-11, (1962).

41--1962b   (with Pankiwskyj, K.) Llandoverian to Gedinnian fossil localities of Podolia and adjacent Moldavia. Pasadena, Calif., 128 p., (1962).

42--1962c   (with Berry, W. B. N.) Great Basin Silurian correlations [abst.]. [Los Angeles Meeting, April 1962, p. 24-25, (1962).] Geol. Soc. Am., Sp.P. 73: 24-25, (1963).

43--1962d   Hunton Group (Silurian and Devonian) and related strata in Oklahoma [discussion of paper by J. P. Shannon, Jr., 1962]. Am. Assoc. Petroleum Geol., B. 46: 1528-1530, (1962).

44--1962e   (and Siehl, A.) Zdimir Barrande (Brachiopoda) redefined [with German abstract]. Hesse Landesamt Bodenforschung Notizbl., Bd. 90: 117-131, (1962).

45--1962f   Appalachian Siluro-Devonian, in Some aspects of the Variscan Fold Belt - 10 lectures delivered to the 9th Inter-University Geological Congress: Manchester, England: Manchester Univ. Press, p. 155-163, (1962).

46--1962g   Observations regarding some Silurian and Devonian spiriferoid genera. Senckenb. Leth., B. 43: 411-432, il., (1962).

47--1962h   (with Berry, W. B. N.) Distribution of Silurian rocks in North America [abst.]. [Houston Meeting, Nov. 1962, p. 116, (1962).] Geol. Soc. Am., Sp. P. 73: 116, (1963).

NOTE:

References to articles published after 1962 can be found in the Bibliographies of North American Geology and other similar standard bibliographies.

----

(40)
HARRY HUGHES

HARRY HUGHES

MIT: 1957-1964

Trained as a physicist and geophysicist at Manchester and Cambridge Universities, and with several years of subsequent work as a research physicist, English-born Harry Hughes was appointed an Assistant Professor of Geophysics in September 1958. He brought to Course XII academic training and practical laboratory experience that we badly needed in bringing geology and geophysics much closer together. After a year as Visiting Lecturer (SY 1957-58), he was appointed Assistant Professor of Geophysics and given responsibility for two major courses involving the application of physics to problems in geology. He offered a two-term subject on Theoretical Geology (12.801 and 12.802) and two one-term subjects, The Earth's Interior (12.803) and The Earth's Crust (12.804). In addition he was most helpful in advising thesis students in geophysics and serving as a member of examining committees, meanwhile pursuing his own research on applying thermodynamics and statistical mechanics to geological problems and to estimation of physical constants at depth in the earth. After seven years in the Department he resigned in 1964 to return to England, where he accepted a research position with the United Kingdom Atomic Energy Authority (UKAEA) at Salwick near Preston, Lancashire.

Harry Hughes was born at Leyland, Lancashire, England on 1 October 1929. After six years at the Leyland Methodist School (1934-1940), he was awarded a Lancashire County Junior Scholarship to Balshaw's Grammar School in Leyland, which he attended from 1940 to 1947. On graduation he won the A. and E. Hamer Open Scholarship to Manchester University and entered that institution as a scholar at Hulme Hall of Residence in 1947. Three years later, in 1950, he was awarded a First Class Honors B.Sc. degree in physics, and was elected to the Shell Studentship in Geophysics at Cambridge University. While studying at Manchester he attended Shell's course in practical seismology at De Bilt and Putten, Holland, in July and August of 1949, with special attention to the weathered layer and its effects in geophysical exploration; and in the summer of 1950 he did magnetic surveys around coal mines. While at Cambridge he worked on a study of drowned river estuaries at Easter 1952. He spent three years at Cambridge

University, 1950-1953, working in the Department of Geodesy and Geophysics with B. C. Browne, and prepared a doctoral thesis on "The electrical properties of the Earth's interior," the essentials of which were presented at several international meetings. He received a Ph.D. degree in geophysics in 1953 from Cambridge, having been a member of Gonville and Caius College while a graduate student.

Immediately upon completion of his doctoral work, Hughes came to the United States to work with Prof. Francis Birch at Harvard as a Research Fellow. The results of his research during SY 1953-54, published as "The pressure effect on the electrical conductivity of peridot," (see Bibliography), enables an estimate of the temperature of the mantle to be made.

Not only did Hughes complete his research project in Birch's laboratory; he also won the hand of his professor's daughter, Anne Campaspe Birch, whom he married on 19 June 1954.

Returning to England soon after the wedding, he accepted a 3-month appointment as a physicist in the Wembley Research Laboratories of the General Electric Company to do classified work under contract to the Admiralty on semiconductors and vacuum physics. This assignment was followed in February 1955 by appointment as Scientific Officer to the U. K. Atomic Energy Authority at Preston, where the fuel elements for all the British nuclear reactors are made.

In early 1957 Hughes was invited to come to M.I.T. as a Visiting Lecturer for SY 1957-58, during which he would pursue research and lecture on the electrical conductivity of minerals at high temperatures and pressures and how these relate to the electrical properties of the Earth's interior as inferred from analysis of the geomagnetic field. Before the year was up he was offered and accepted a three-year appointment as Assistant Professor of Geophysics.

Immediately upon appointment as Assistant Professor, Hughes organized a series of lectures which he described as follows in the Catalog for SY 1958-59:

> "12.801, 12.802   Theoretical Geology I and II
> Application of the physics of solids to geological problems, including estimation of changes of elastic constants with depth, magnetism in rocks, phase transitions, optical absorption and radiative heat transfer, and excitation of solids giving rise to diffusion and electrical conduction in the earth. (Alternate years; 12.802 not offered in 1958-59.) [3-0-9.]"
>
> "12.803   The Earth's Interior (A)
> Interpretation of the seismic velocity depth curve and geomagnetic data to infer the constitution and temperature of the Earth's interior. The degree of

uncertainty of the conventional earth model will be specially considered so that the plausibility of current hypotheses in this field can be assessed. (Alternates with 12.804. Not offered 1958-59.) [3-0-9.]"

"12.804   The Earth's Crust (A)
Discussion of the physics of the crust and its deformation. Poldervaart, The Crust of the Earth. (Alternates with 12.803.) [3-0-9.]"

From the foregoing descriptions it is evident that Hughes intended to offer a series of searching lectures on the physical nature of the Earth, and that is just what he did during his six years as a professor in the Department. His approach to the whole subject of geology was quantitative and theoretical and did much to stimulate both colleagues and students to follow his lead. That he attracted some of the best graduate students in Course XII is attested by the fact that he supervised six doctoral theses during the period he was a regular member of the Department; i.e. 1958-1964. These were the theses of J. P. Downs (1960), David Greenewalt (1960), D. E. Bowker (1960), D. C. Uhri (1961), Wm. A. Schneider (1961), and G. H. Beall (1962). The Department lost one of its most promising younger members when Hughes decided to return to England in 1964 to accept a position as Senior Scientific Officer in the Reactor Fuel Element Laboratories (RFEL) of the United Kingdom Atomic Energy Authority (UKAEA) at Salwick, Preston.

As stated earlier, Hughes married Anne Campaspe Birch, of Cambridge, Massachusetts, on 19 June 1954. The Hughes now have three daughters, Helen (a first year student in Queen Mary College, University of London, 1973), Barbara, and Diana. Anne is pursuing her academic interests at the University of Lancashire, where she expects to complete the requirements for a Bachelor's degree in Russian language and literature, a field of study that she had entered earlier as an undergraduate at Radcliffe College.

BIBLIOGRAPHY OF HARRY HUGHES

Symbols and abbreviations used in the following references are explained on p. 91-98. This bibliography starts with Hughes' doctoral thesis (T--1953) and includes only those articles known to have been published while he was at M.I.T., i.e. 1957-1964.

T--1953   The electrical conductivity of the earth's interior. Ph.D. Thesis, University of Cambridge, Cambridge, England, (1953).

1--1953a   The nature of the electrical conductivity in the earth (abst.). Science 118: 572, (1953).

2--1955   The pressure effect on the electrical conductivity of peridot. J. Geophys. Res. 60: 187-191, (1955).

3--1959   The conductivity mechanism in the earth's mantle (abst.). J. Geophys. Res. 64: 1108, (1959).

(41)
THOMAS CANTWELL, JR.

THOMAS CANTWELL, JR.

MIT: 1960-1965

With a rare combination of interests and skills in both applied science and the mineral industry, Cantwell was able to present to the Department's student geophysicists a broad range of subject matter and practical experience. His interest in business administration developed early in his college career, and led later to the founding of his own multi-national group of companies to carry on geophysical exploration and mineral development. As a consequence, he taught his students the practical side of geophysics, encouraged them to consider professional careers in applied science, and gave quite a few of them their first opportunity to do practical field work. He gave five years of valuable service to the Department of Geology and Geophysics, from 1960 to 1965, lecturing and directing research in exploration geophysics, elastic wave propagation, and potential theory. He resigned in 1965 in order to devote full time to development of his newly founded company, Geoscience Incorporated. As could be predicted from his activities and interests while at M.I.T., he later enjoyed a successful professional career in exploration geophysics and eventually he helped to found a group that is now heavily involved in international exploration and development of fuel and mineral resources.

Thomas Cantwell, Jr. was born in Buffalo, New York on 25 June 1927, the first of two boys of Thomas and Helen (Robinson) Cantwell. Both parents, of Irish and English ancestry, were bankers and no doubt influenced their young son to consider a career in business, a path that he followed with success after college by organizing and directing several companies involved in geophysical exploration.

Cantwell graduated with honors from Kenmore High School in Kenmore, New York, in the spring of 1944. In September of the same year he entered M.I.T. and finished the first two years of work before entering the United States Army, in 1946. After his tour of duty in Japan as a Private he resumed his academic program at M.I.T. in 1947 and in 1949 received both an S.B. and an S.M. in Chemical Engineering. Being only 22 years of age and without management training, he decided to enter Harvard University from which he received a Master's degree in Business Administration in 1951.

Wishing to put his training to practical use, he took a position with Ionics Incorporated of Cambridge and worked as a design and development engineer on equipment for demineralizing sea water. A year of this kind of work was enough, and in 1952 he returned to M.I.T. as a staff member of the Industrial Liaison Office. Two years later came the opportunity to follow both his engineering and systems management interests, and in March 1955 he joined the group who were to develop the M.I.T. Reactor. First as a Research Associate in Chemical Engineering, from March 1955 to June 1958, and later in Nuclear Engineering, he acted as Project Engineer and Business Manager for the M.I.T. Reactor Project. During this period he also made economic studies of power reactors and chemical separation plants. Then came his transfer to the Department of Geology and Geophysics in December 1958 as a Research Associate, and candidate for a doctor's degree.

After completing all requirements for the degree during the fall term of 1959, he was granted a Ph.D. in February 1960, having been previously appointed an Assistant Professor of Geophysics on 1 January 1960. He served in this capacity until July 1963 when he took a year's leave of absence to develop his own company, Geoscience Incorporated.

During the 5-year period from January 1960 to June 1965 Cantwell taught courses in seismology, gravity, magnetics and exploration geophysics; conducted laboratory and field research in exploration geophysics; and in the meantime, through Geoscience Incorporated, he was becoming involved in mineral exploration for hard rock minerals and petroleum.

For example, during September 1962 he was invited to Australia by a state governmental body for the purpose of giving a series of lectures at the Bureau of Mines of South Australia. These lectures concerned geophysical exploration for mineral deposits and emphasized the induced polarization and resistivity methods of exploration. This visit not only gave M.I.T.'s Department of Geology and Geophysics excellent publicity, but also gave Cantwell opportunities which led to the establishment of the first overseas affiliate of his geophysical exploration business at a time when Australian mineral exploration techniques were not well developed.

During his last year in the Department, 1964-1965, when his appointment was changed from Assistant Professor to Lecturer in order that he have most of his time available for developing his Geoscience Incorporated as part of a multinational geophysical exploration and research group, he offered a special course in exploration geophysics. This course included the fields of interest in which Cantwell had special competence and experience--seismology, geomagnetism, gravimetrics and magneto-tellurics, and the improvement of old and the development of new methods and instrumentation for geophysical exploration.

During his 5-year period in the Department of Geology and Geophysics, a critical period for geophysics at M.I.T., when we lacked equipment and manpower, Cantwell was a tower of strength in both instruction and research, and his resignation left a gap in exploration geophysics that has not been filled to this day.

For the record, his appointments at M.I.T. were as follows:

Assistant, Dept. Chemical Engineering	1/1/49- 9/30/49
Staff Member, Industrial Liaison Office	9/1/52- 5/31/54
Research Associate, Dept. Chemical Engineering	3/1/55- 6/30/58
Research Associate and Project Engineer, Dept. Nuclear Engineering	7/1/58-11/30/58
Research Associate, Dept. Geology and Geophysics	12/1/58-12/31/59
Assistant Professor of Geophysics, Dept. Geology and Geophysics (partial leave during 7/1/63-6/30/64)	1/1/60- 6/30/64
Lecturer, Dept. Geology and Geophysics	7/1/64- 6/30/65

Resigned on 6/30/65

To review the business side of Cantwell's career, Geoscience Incorporated, the first company he organized and then directed as President, was started as a consulting firm in 1961; soon it was carrying out government research, performing geophysical exploration for mining companies on a contract basis, conducting geophysical research for both oil and mining companies, and making instruments for mining geophysics. Annual sales went from an initial $50,000 to about $1,000,000 in 1966 when the Company was sold to Ampex Corporation. After the sale Cantwell continued as President and Treasurer for a short time before becoming Vice President, for Data Processing and Research, of Mandrel Industries of Houston, Texas, a subsidiary of Ampex. Later, in 1969, he moved up to Executive Vice President and then to President and Chief Operating Officer of Mandrel during years when annual sales approached $50 million.

Under Cantwell's direction Mandrel established computer centers in Houston, London, and Libya, and centralized profit planning and cash control. Seismic field operations were carried out in the United States, Canada, Australia, Argentina, Libya, Algeria, Saudi Arabia, Oman, Abu Dhabi, Madagascar, and offshore Africa and Southwest Asia.

In April 1970 after the 3-year period as an executive of Mandrel, during which time he went from Vice President to President, he resigned to become President and Chief Operating Officer of a new corporation that merged a number of U.S. and overseas groups including exploration interests and mineral companies. This corporation, called Pexamin (for Petroleum Exploration and Minerals) is involved in multinational exploration, management, financing, and development. It is active in the United States, Alaska, Australia, Indonesia, and Europe.

Thus goes the story of an M.I.T. graduate whose early interest in engineering and business, supplemented by instruction and research in geophysics, led to a business career in worldwide exploration for petroleum and mineral resources. In this career, Cantwell followed a typical path for M.I.T. graduates--first, assisting in teaching and research as a graduate student; next, gaining professional experience as a faculty member and outside consultant; then leaving the Institute to found new companies built on his academic and early consulting experience backed by his ability to build professional technical teams.

On 10 May 1970 Cantwell married Ann Japhet of Houston, Texas. He has three children by a former marriage; Elizabeth Raye Cantwell (b. 1955), Thomas Cantwell III (b. 1957), and Douglas (b. 1961).

BIBLIOGRAPHY OF THOMAS CANTWELL, JR.

Symbols and abbreviations used in the following references are explained on p. 91 - 98. In general, abstracts are listed separately as well as with the complete article. The bibliography is complete only through 1965, when Cantwell left M.I.T. T = thesis.

1--1954 Power plants with thermal reactors. M.I.T. Group Report to A.E.C., (1954).

2--1955 Processing of spent power reactor fuel. M.I.T. Group Report to A.E.C., (1955).

3--1955a (with Thompson, T. J.) Hazard Summary Report for the M.I.T. Nuclear Reactor. M.I.T. Group Report to A.E.C., (1955).

4--1957 The M.I.T. Nuclear Reactor. Nucleonics 15: 38-40, (1957).

5--1959 Nuclear detector for beryllium minerals. A.I.M.E., Tr. 214: 938-940, (1959).

6--1960 (and Madden, T. R.) Preliminary Report on crustal magnetotelluric measurements. J. Geophys. Res. 65: 4202-4205, (1960).

T--1960a Detection and analysis of low frequency magnetotelluric signals, 170 p., il., (1960). (Ph.D. Thesis at M.I.T. in Course XII--Dept. Geology and Geophysics, February 1960.)

7--1962 Operations analysis of a limited class of exploration programs (Read at the New York meeting of the A.I.M.E. on 22 February 1962). Typescript of 11 p. and il., (1962).

8--1962a Progress Report on geomagnetic studies and electrical conductivity in the Earth's Crust and Upper Mantle (1962), M.I.T. Report to ONR, (1962).

9--1962b   (and Dulaney, E. N.) Deep resistivity investigation in Nebraska. Cambridge, Mass.: Geoscience Inc., (1962).

10--1963   On-site resistivity and self-potential measurements (1963), Report to ARPA by M.I.T. and Allied Research Associates, (1963).

11--1963a   Resistivity investigation on Cape Cod (1963), AFCRL, 63-370, (1963).

12--1963b   (with Galbraith, J. N., Jr.) Resistivity calculations. Cambridge, Mass.: Geoscience Inc., (1963).

13--1964   (with Galbraith, J. N., Jr., and Nelson, P.) Deep resistivity results from New York and Virginia. J. Geophys. Res. 69/20: 4367-4376, (1964).

14--1964a   (with Orange, A. S., Komack, R. L., and Bostick, F. X.) Simultaneous measurements and spectral analysis of micropulsation activity. Nature 201/4918: 460-462, (1964).

15--1964b   (with Komack, R. L., Orange, A. S., and Bostick, F. X.) Simultaneous measurements and spectral analysis of micropulsation activity. Nature 204/4958: 534-537, (1964).

16--1965   (with Nelson, P., Webb, J., and Orange, A. [S.]) Deep resistivity measurements in the Pacific Northwest. J. Geophys. Res. 70/8: 1931-1938, (1965).

17--1965a   (with Orange, A. [S.]) Further deep resistivity measurements in the Pacific Northwest. J. Geophys. Res. 70/16: 4068-4072, (1965).

18--1966   (with Orange, A. [S.] et al.) Ground current in HVDC transmission - I, Preliminary Report, I.E.E.E., Tr. on Power Apparatus and Systems, P.A.S. - 85/3, (1966).

19--1967   (with Madden, T. R.) Induced polarization - A review, in Mining Geophysics, Vol. 2, Theory: Tulsa, Soc. Explor. Geophys., p. 373-400, il., tabs., (1967).

20--1968   (with Keller, R. F. and Bradshaw, J.) Digital "Revolution" enters period of refinement. Oil and Gas Jour., 19 Feb. 1968: 68-70, (1968).

21--1969   (with Vozoff, K. and Mebane, W. M.) Magnetotellurics provides a useful new tool for petroleum hunters. Oil and Gas Jour., 15 Sept. 1969: 66-80, (1969).

(42)
FRANCIS BITTER
(1902-1967)

MIT: 1934-1960-1967

FRANCIS BITTER

Francis Bitter, a physicist renowned worldwide for his special work in the field of magnetism, joined our Department of Geology and Geophysics as Professor of Geophysics in July 1960. He had requested transfer from the Department of Physics to our Department because he wished to broaden his research program to include earth and space science as well as electronics and atomic physics and to have adequate time to assist in directing the building of a facility that would produce a continuous magnetic field of unprecedented strength. This facility was constructed under his supervision, on a $9.5 million contract with the U. S. Air Force, and was completed in 1963. A year later a Bitter magnet at the new laboratory achieved a record-breaking magnetic field of 250,000 gauss, and plans were underway for an even more powerful magnet, to produce a field of 325,000 gauss, when Bitter died on 26 July 1967. Three months later, the facility he had envisioned, helped to design, and supervised during construction, was named the Francis Bitter National Magnet Laboratory, in recognition of his preeminence in the field of high intensity magnetism.

Francis Bitter was an unusual man--even among the many such at M.I.T. He was a creative genius, a seeker of the subtlest truths of his physical environment, a scholar in the finest sense of that word, a humanitarian much concerned about the hopes and problems of students, and most of all a gentleman at a time when being one seemed to be going out of fashion in much of our Society.

Normally, I would present a brief biography here, then follow with an evaluation of Bitter's work as it concerned geology and geophysics, and end with a list of his publications. Fortunately for us all, however, Bitter has discussed his life in his delightful and informative autobiography, <u>Magnets: The Education of a Physicist</u> (Doubleday Anchor Books, Doubleday & Co., Inc., Garden City, N.Y., 155 p., 1959), and his colleagues and peers have evaluated his scientific achievements in <u>Francis Bitter: Selected Papers and Commentaries</u> (The M.I.T. Press, Cambridge, Mass., 551 p., port., 1969), edited by T. Erber and C. M. Fowler. The interested reader will find satisfying accounts of Bitter's activities, interests, attitudes, and achievements in these two references. A third reference that I shall mention again a little farther on, is the Resolution prepared shortly after his death by four of his M.I.T. colleagues (George R. Harrison, Victor F. Weisskopf, Jerrold R. Zacharias, and Benjamin Lax) and read into the minutes of the Faculty Meeting of 20 September 1967. This Resolution is quoted on an accompanying page.

Course XII graduate students were immediately attracted to Bitter's fields of interest, and there would surely have been more had he lived and been able to help them. As it was, he did find time in his busy schedule involving the design and construction of the Magnet Laboratory to supervise four graduate theses--those of Timothy Fohl (Ph.D. 1963), Roger G. Little (S.M. 1964), Ward D. Halverson (Sc.D. 1965), and Jamie C. Chapman (Ph.D. 1966)--and to help many other graduate students interested in magnetism.

It is sad to recall that World War II diverted Bitter from a longtime interest in geomagnetism, and poor health afterwards kept him from pursuing it. As T. Erber writes in "Francis Bitter: A biographical sketch," the introductory chapter in <u>Francis Bitter: Selected Papers and Commentaries</u>, alluded to in a preceding paragraph,

> "His own interests [Bitter's] began to turn to another novel direction, the connection between magnetism and geology. This rested on the fact that the magnetization of clays and rocks preserved information about the ambient magnetic environment at the time they had been formed. Studies of samples from many scattered locations could be used to reconstruct something of the history of the earth's magnetic field. Unfortunately, before Bitter could really get started on this project, war [World War II] broke out in earnest, and the Navy called on him to cope with magnetic weapons launched by the enemy." (p. 14-15.)

Bitter hoped to return to his longtime interest in geomagnetism and geology when he joined our Department in 1960 and he discussed this possibility with much enthusiasm at the time we arranged his transfer from the Department of Physics, but the new magnet facility and associated responsibilities again thwarted his research interests, as they had been thwarted by World War II, and by the time the Laboratory was truly operational his health had declined to such an extent that he could no longer think of starting the research program mentioned in Erber's foregoing comment.

Bitter's long and brilliant career was best summarized in the Resolution referred to earlier that was read to the M.I.T. Faculty on its regular meeting on 20 September 1967, and because it relates so fully his achievements as an experimental physicist and his attributes as a man, it is reproduced on an accompanying page.

The National Magnet Laboratory at M.I.T. came into existence officially on 1 July 1960 under a contract between M.I.T. and the Air Force through its Office of Scientific Research. At the time the $9.5 million contract was announced, Bitter, who became Professor of Geophysics on the same day, was made responsible for the design and construction of the new laboratory and was appointed head of its scientific advisory board. The project was widely publicized locally as well as nationally. The Technology Review described it at some length on pages 29 and 30 of its November 1960 number.

In the next three years the laboratory got built and in 1964 a Bitter magnet at the laboratory achieved a record-breaking field of 250,000 gauss.

Bitter continued to participate actively in the research program of the Laboratory almost to the day of his death on 26 July 1967. Soon after, the Institute held a Symposium on Physics and Magnetic Fields in his honor, and more than 200 distinguished scientists and educators gathered to pay him tribute for his pioneering work in magnetism and also for his part in establishing the National Magnet Laboratory, which in ceremonies following the symposium was renamed the Francis Bitter National Magnet Laboratory, in his honor. At the side of his photograph, on the plaque that hangs in the vestibule of the Laboratory, is the following dedicatory statement:

FRANCIS BITTER
1902-1967
PIONEER IN THE PRODUCTION OF INTENSE MAGNETIC FIELDS
CREATIVE EXPERIMENTAL PHYSICIST IN THE STUDY OF MAGNETISM
TEACHER, COUNSELOR AND FRIEND OF M.I.T. STUDENTS FOR THIRTY-THREE YEARS
INSTRUMENTAL IN FOUNDING THIS LABORATORY
NAMED IN HIS HONOR

21 NOVEMBER 1967

No statement about Francis Bitter's life would be complete, however brief, without mentioning the great influence on his last years of Mrs. Katherine Welchman, a singer and pianist active in musical affairs, whom he married in 1959. She matched and accentuated the artistic strain that characterized Bitter's family and background, shared his deep concern for student affairs, and willingly moved with him to the M.I.T. Graduate House (Ashdown House), where he served as Resident Master from 1962 to 1965, and she was the gracious Resident Mistress.

Ordinarily this biography would end with the biographee's complete bibliography, but inasmuch as Bitter's publications were all in the field of physics, and written for the most part before he became Professor of Geophysics, the reader who wishes to refer to any of Bitter's publications, of which there are some 90 articles and 6 books, is advised to consult his Bibliography recorded on pages 20 to 28 in Francis Bitter: Selected Papers and Commentaries (M.I.T. Press, Cambridge, Mass., 1969). Below are listed only those publications that appeared after he was appointed Professor of Geophysics on 1 July 1960.

"RESOLUTION ON THE DEATH OF FRANCIS BITTER, PROFESSOR OF GEOPHYSICS
(to be presented at the Faculty Meeting, 20 September 1967)

"The faculty records with deep regret the death, on 26 July 1967, of Francis Bitter, Professor of Geophysics.

"Francis Bitter was born in Weehawken, New Jersey, on 22 July 1902, and spent his childhood in New York City. He obtained his Bachelor's Degree from Columbia University in 1924. A year of study in Berlin, listening to lectures by Max Planck, Einstein, and Von Laue and interacting with Leo Szilard and others, had a profound influence on his scientific career. His career in magnetism began with his thesis on the magnetic susceptibility of gases at Columbia where he received his doctorate in 1928. He continued this work as a National Research Postdoctoral Fellow at the California Institute of Technology under Robert Millikan. After he joined the Westinghouse Research Laboratory in 1930, he discovered the ingenious Bitter powder pattern for visually observing magnetic domains. Then, as a Guggenheim Fellow at the Cavendish Laboratory in Cambridge during 1933-34, his interest in high field magnets was aroused by the work of the great Russian physicist, Peter Kapitza. Consequently, when Francis Bitter came to M.I.T. as an Associate Professor of Metallurgy in 1934, he developed the remarkable design of the water-cooled solenoids, known as the Bitter magnet, which is still the prototype for the present high field magnets throughout the world. Soon after, with the encouragement and support of Vannevar Bush, he built the first magnet laboratory at M.I.T. The 100 kilogauss magnet was then put to use by Bitter and George Harrison on the optical Zeeman effect of atoms and molecules.

"World War II interrupted this work, and Professor Bitter joined the Naval Ordnance Laboratory to devote his talents to degaussing the fleet and to other important magnetic problems related to submarine warfare. He became a Navy Commander in 1943 and retained his commission until 1951.

"In 1945 he returned to M.I.T. as a member of the Physics Department. At that time the work of Jerrold Zacharias and Ed Purcell inspired him to begin experiments on nuclear resonance. Together with his student, Jean Brossel, he embarked on his pioneering work of double resonance with optics and microwaves. When Professor Kastler received his 1966 Nobel Prize, he acknowledged the important contributions of these early experiments by Bitter and Brossel to his prize-winning work on optical pumping.

"In 1951 Francis Bitter became a full professor, and from 1956 to 1960 he served as Associate Dean of Science at M.I.T. It was during this period that the group at Lincoln Laboratory reawakened his interest in high magnetic fields. Their work on resonance experiments with pulse fields led to the decision to construct a modern high field facility with Bitter magnets. Assisting this group, Francis Bitter, as the Chairman of the Planning and Design Group, played a prominent role in the creation of the National Magnet Laboratory. Together with Bruce Montgomery he conceived the design of the quarter million gauss solenoid that stands as the record today. It is therefore very appropriate that the National Magnet Laboratory will soon be named after him.

"In 1960 Francis Bitter joined the Department of Geology and Geophysics. From 1962 to 1965 he also served as Master of Ashdown House. He became involved with the study of the earth's magnetism and related plasma problems. With Professor Fiocco he recently developed electrical techniques for generating megagauss fields.

"Francis Bitter was a member of the American Physical Society, the Association of American Physics Teachers, the American Academy of Arts and Sciences, Phi Beta Kappa, Sigma Xi, and Phi Gamma Delta.

"The qualities that made Francis Bitter the extraordinary experimental physicist were his enthusiasm, creative imagination, and dedication. His scientific and intellectual vigor was transmitted to his students and colleagues. His influence as a teacher and researcher transcended the classroom and the laboratory. He was a sensitive human being who was aware of the value of science in today's cultural life. He had a deep understanding and appreciation of art and literature and all the cultural trends of this age. He always regarded science as an important part of our cultural heritage. M.I.T. lost in him not only one of its best scientists, but also one of those rare personalities who have given science and technology a meaning beyond its utilitarian aspects.

"We, his friends and colleagues of the faculty at M.I.T., express our deep sorrow and sympathy to his wife, Katherine Bitter, who in the last two years of his illness helped him to carry on his work with remarkable courage and effectiveness.

    G. R. Harrison
    V. F. Weisskopf
    J. R. Zacharias
    B. Lax, Chairman"

[From Records of the M.I.T. Faculty, 1967-1968.]

BIBLIOGRAPHY (1960-1967) OF FRANCIS BITTER

In this partial bibliography I have used the method of citation followed in the Bibliography on pages 20 to 28 in the reference cited in the preceding paragraph, since this is the one familiar to physicists. Symbols and abbreviations are explained on pages 91-98.

1--1961 Magnetic Resonance: Chapter II-D of Dielectric Materials and Applications, A. R. Von Hippel, ed., M.I.T. Press, Cambridge, Mass., 1961.

2--1961a New developments in high magnetic field research. Physics Today, 14, 9, 22; Bull. Phil. Soc. Washington, 16, May 1961.

3--1962 Magnetic fields. Saturday Review, p. 40, February 1962.

4--1962a Magnetic resonance in radiating and absorbing atoms. Applied Optics 1, 1, 1962 [M.I.T. Research Lab. Electronics Rept. 394, Feb. 1, 1962].

5--1962b High field magnets (Strong magnets). Int. Sci. and Technology, p. 58, April 1962.

6--1962c High Magnetic Fields (Proc. Int. Conf. High Magnetic Fields), edited by H. Kolm, B. Lax, F. Bitter, and R. Mills. M.I.T. Press, Cambridge, Mass., and John Wiley & Sons Inc., New York, 1962.

7--1962d Water-cooled magnets. Review of Scientific Instruments, 33, 3, 342, 1962. [Proc. Int. Conf. High Magnetic Fields], edited by H. Kolm, B. Lax, F. Bitter, and R. Mills, M.I.T. Press, p. 85-99, 1962.

8--1963 Flows in a steady plasma, in Lectures in Materials Science, W. A. Benjamin, New York, 1963.

9--1963a Mathematical Aspects of Physics. Doubleday & Co., Inc., Garden City, N.Y., 1963.

10--1963b The National Magnet Laboratory at the Massachusetts Institute of Technology. British Jour. Appl. Physics, 14, 759, November 1963.

11--1965 Ultrastrong magnetic fields. Scientific American, 213, 1, 65, July 1965.

12--1965a Megagauss magnets using capacitor discharges. Proc. Int. Symposium on Magnet Technology, Stanford Linear Accelerator Center, Sept. 1965 [AEC document edited by H. Brechna and H. S. Gordon], p. 680-682, September 1965.

13--1967 Histoire des champs magnétiques intenses, et leurs contributions à la physique [History of intense magnetic fields and their contribution to physics]. Les Champs Magnétiques Intenses, CNRS, Paris, p. 19-29, 1967.

(43)
DAYTON ERNEST CARRITT

DAYTON ERNEST CARRITT

MIT: 1960-1968

Formal and organized instruction and research in marine chemistry (chemical oceanography) started at M.I.T. in the fall of 1960 when Harvard-trained chemist Dr. Dayton E. Carritt joined the Department of Geology and Geophysics as Associate Professor of Chemical Oceanography. He came from The Johns Hopkins University where he had been teaching chemical oceanography and marine ecology in their Department of Oceanography and serving as Assistant Director of the Chesapeake Bay Institute.

During his first year, SY 1960-61, Carritt divided his time between the Institute, where he started his teaching program, and the Woods Hole Oceanographic Institution, where he carried on research as a member of the staff. In 1962 he moved his research project to M.I.T., organized special chemical laboratories, accepted thesis students, and entered on a full-time program of instruction and research on the chemistry of the world ocean. During the next six years, 1962-1968, until he resigned to accept a combined teaching-administrative position in Nova University in Ft. Lauderdale, Florida, he developed a vigorous and diversified program in marine chemistry that produced five theses under his supervision-- 2 S.B.s, 1 S.M., and 2 Ph.D.s--and helped greatly in getting the Department's overall oceanographic program established.

Dayton Ernest Carritt was born in Boston, Massachusetts, on 12 March 1915, the only child of Ernest Henry Carritt, a minister of the gospel, and Laura Agnes (Cave) Carritt, a professional singer and school teacher. Both parents traced their ancestry to English forebears. Being a minister's son, young Dayton moved about extensively during his earliest years with the result that his elementary education was divided among schools in Boston, Milton and Peabody, Massachusetts; Barrington and Providence, Rhode Island; Middleport, New York; and Oxford and Lewiston, Maine. After attending the Providence Technical High School from 1929 to 1933, he was graduated in June 1933. In the fall of the same year he entered Rhode Island State College and in 1937 he was given a B.S. degree in chemistry. He next took off a year (1937-1938) to work at the Woods Hole Oceanographic Institution as a chemical technician, then returned to Rhode Island State to start graduate work. As a graduate assistant in physical chemistry he added to the teaching experience he had gained earlier (SY 1936-37) as a senior laboratory assistant in general chemistry. In 1940 he entered the Harvard Graduate School to start on a doctoral program in chemistry. During the summer of 1940 he assisted Prof. G. S. Forbes in qualitative analysis, following which he was appointed a Teaching Fellow for SY 1940-41 to assist Prof. G. P. Baxter in undergraduate quantitative analysis and advanced work in gas analysis. The following summer, 1941, he returned to Woods Hole as a research chemist, then later became an Instructor in chemistry at his alma mater, Rhode Island State College, where he taught general chemistry in both the regular college program and in the Army Specialized Training Program (ASTP) in 1942.

His next assignment was to the Manhattan District Project in Los Alamos, New Mexico, as a scientist, and there he worked full-time during the war years, 1943-1946.

Returning to the academic halls after the end of the war, he served as Instructor (1948) and then Assistant Professor (1949) at Scripps Institution of Oceanography, meanwhile managing to qualify for a Ph.D. degree in Chemistry at Harvard in 1948. He then came back across the continent to Baltimore, where he spent a decade at The Johns Hopkins University. Along with Donald W. Pritchard and Wayne V. Burt, he helped to organize Hopkins' Department of Oceanography and to found its Chesapeake Bay Institute, which would soon become a leader in estuarine research. Starting as Assistant Professor (1949-1955), he advanced to the rank of Associate Professor and Assistant Director of the Institute, posts held until 1960 when he resigned to accept a joint appointment at M.I.T. and Woods Hole.

During the decade at Hopkins, when there was great national concern about oceanography, he played a major role in the instructional program. He taught a course in chemical oceanography every year (1950-1960), and one in marine ecology for the second five years (1955-1960), in the Department of Oceanography. He also taught the chemistry part of a course in chemistry and bacteriology of water supplies with Prof. C. E. Renn in the Department of Sanitary Engineering for several years (1950-1952).

Carritt joined the M.I.T. faculty at a time when the Department of Geology and Geophysics, as Course XII was then called, was starting a program of instruction and research in oceanography. Inasmuch as no facilities for his kind of chemical work existed in our Department at the time of his appointment, it was arranged that he share his time equally with the Institute and the Woods Hole Oceanographic Institution in order that he could conduct his laboratory research at the latter. As a consequence he spent his first two M.I.T. years as a half-time Associate Professor of Oceanography, giving lectures and directing student research at M.I.T.

while conducting his own research at Woods Hole meanwhile as a half-time staff member there.

When the new Green Building was completed in the summer of 1964, and laboratory space and facilities for marine chemistry became available, Carritt moved his research from Woods Hole to M.I.T. and organized an ambitious program of instruction and research in marine chemistry that immediately became an integral part of Course XII's overall program in oceanography. This program was well described in The Tech for 20 March 1963 (p. 2) under the heading

> "Professor Carritt Investigates Oceans: Dissolved Ions, Sea-to-Air Transfers."

At this time, the chemistry of the world ocean was essentially unknown because few samples had been analyzed and analytical methods and facilities were inadequate for determining the percentages of even the more abundant ions--sodium, potassium, magnesium--in the presence of one another, let alone the less abundant and trace ions.

Although it was known that six ions--chlorine, sodium, magnesium, sulphate, calcium, and potassium--make up 99% of the matter dissolved in sea water, and that the ratio of any one of these six to any other, or to all the rest of them, is remarkably constant, little was known about either the abundance or ratio of the many other elements present in trace amounts in the world ocean.

There was a need to collect many water samples from all the seas and oceans of the world, as well as from water bodies in adjoining land areas, and to develop new techniques and instruments to analyze these samples. Carritt and his faculty colleague Winchester, together with their graduate students, working with colleagues at Woods Hole, organized themselves into a research group to learn more about the chemistry of the oceans. With financial support from the Office of Naval Research and the National Science Foundation, the group started a research program that was described as follows in the news story mentioned earlier:

> "The project is aimed at providing information on the routes and rates of passage of the ocean's solid constituents from the land, through the sea to the ocean floor."

Another important aspect of the overall program was to investigate air-sea interactions and how these affect exchange of material between oceanic waters and the atmosphere.

While carrying on his regular academic activities at Johns Hopkins, from 1950 to 1960, and at M.I.T., from 1960 to 1968, Carritt also devoted considerable time and effort to furthering oceanography by serving on both national and international action and advisory groups, among which were included the following:

> NAS-NRC; Chairman of a working group of the Committee on Oceanography to study the Radioactive Waste Disposal into Atlantic and Gulf Coastal Waters. 1955-1957.
>
> NAS-NRC; Member of the Committee on the Biological Effects of Atomic Radiation on Oceanography and Fisheries. 1955-1957.
>
> Amer. Chem. Society (ACS); Chairman, Symposium on the Chemistry of Sea Water, held at Miami, Fla. Feb. 1957.
>
> Amer. Soc. Limnology and Oceanography; Member, Editorial Board of Limnology and Oceanography. 1958-1963.
>
> A.A.A.S.; Convener of the session on chemical oceanography of the International Oceanographic Congress held in New York in September 1959.
>
> Joint Committee of Special Committee on Oceanic Research (SCOR) of the International Council of Scientific Unions and the International Association of Physical Oceanographers. Member, 1959-1961.
>
> International Association of Physical Oceanographers: Secretary of Committee on Chemical Oceanography, 1958-1960.
>
> The President's Science Advisory Committee (PSAC); Member of an ad hoc advisory panel to consider oceanographic programs in the agencies of the Federal Government, 1959, 1960, and 1961.
>
> Sears Foundation--*Journal of Marine Research*; Member, Advisory Committee, 1960-1966.
>
> *Deep Sea Research*; Member, Editorial Advisory Board, 1960-1974.
>
> NAS-NRC, Office of Saline Water; Chairman of Summer Study Conference at Woods Hole, Mass., 19 June to 14 July, 1961.
>
> NAS--Committee on Oceanography; Chairman of Panel on Intercomparison of Chemical Methods, 1960-1962.
>
> NAS--Committee on Pollution; Member, 1965.
>
> AGU--Visiting Scientist Program, 1965-1969.
>
> Massachusetts Association for Marine Sciences (MAMS); Member, 1966-1968, Co-Chairman, 1967-1968.
>
> Oceanography 2000--A symposium at the U. S. Naval Academy; Convener and Participant, Annapolis, Md., December 1969.
>
> Gordon Research Conference; Leader of Section on Chemical Oceanography at Meriden, N. H., Summer 1969.
>
> Gordon Research Conference; Leader of Section on General Review and Future Plans, Chemical Oceanography, at Santa Barbara, Calif., Jan. 1971.

In addition to the preceding activities, Carritt also served as consultant to numerous clients on a variety of problems, including the following:

Du Pont in Virginia--disposal of zinc wastes, (1953).

Bay Food Products in Maryland and Morgan Bros., Virginia--oyster quality standards, (1957-1960).

USAEC--radioactive waste disposal problem, expert witness at hearings in Hartford, Conn. and Houston, Tex., (1958-1959).

Franklin Systems, Inc., West Palm Beach, Fla.--instrumentation using radioactivity, (1958-1959).

Holt, Rhinehart and Winston, Inc., Publisher, New York--consultant on Modern Earth Science, (1960).

Boston Edison Co., Boston, Mass.--environmental characteristics of a proposed nuclear power plant, (1967-1968).

American Dynamics International, Inc., Fort Lauderdale, Fla.--development of visuant and sealant systems, (1968-1969).

Publications produced by Carritt during his pre-M.I.T. period (i.e. 1939-1960) as well as during his M.I.T. years (1960-1968) are included in his Bibliography at the end of this biographical sketch. This list of publications clearly indicates that Carritt contributed important research results to the growing science of marine chemistry, or chemical oceanography, during the decades of 1950-1960 and 1960-1970, while rendering important services as an academician at Johns Hopkins and M.I.T., and as an expert consultant to numerous clients who needed advice on oceanographic problems.

Carritt went from M.I.T. to Nova University in Fort Lauderdale, Florida, where he served as Gold Key Professor of Oceanography, as well as Provost-Vice President for Academic Affairs for two years, 1969-1970. But he did not find Florida to his liking, and 1971 found him back in Massachusetts as Professor of Marine Sciences, and Director of the Institute for Man and His Environment, at the Amherst campus of the University of Massachusetts.

Carritt married Jeanne Roberta Brooks on 5 August 1939, and they have one child, Jan Brooks Carritt, born in 1948. Jeanne received an S.B. degree in biology from Oberlin College in 1936 and an A.M. in the same field from Smith College in 1938. After teaching at the St. Timothy and the Park high schools in Baltimore, she taught freshman biology at Simmons College (Boston) for a little while, then joined the staff of the Education Development Center in Boston. In this privately funded corporation she worked with others in developing science curricula for elementary school children. Since going with her husband to Amherst (Mass.) she has become much interested in educational counseling and human relations and in how to help adults return to school for more training. As a result of this interest she received a Master of Education degree from the University of Massachusetts in 1973.

BIBLIOGRAPHY OF DAYTON ERNEST CARRITT

Symbols and abbreviations used in the following references are explained on p. 91 - 98; in general, abstracts are listed separately as well as with the references to the complete article. This bibliography lists all of Carritt's publications through 1968, the year he left M.I.T.

1--1939   (with Parks, W. G.) An improved stopcock substitute. Rev. Sci. Instruments 10: 148, (1939).

2--1946   (with Ferry, J. D.) Action of antifouling paints. Solubility and rate of solution of cuprous oxide in sea water. Indust. Eng. Chem. 38: 612-617, (1946).

3--1948   (with Rakestraw, N. W.) Some seasonal chemical changes in the open ocean. J. Marine Res. 7: 362-369, (1948).

4--1951   (with Wooster, W. S. and Isaacs, J. D.) An automatic reagent dispenser for shipboard use. J. Marine Res. 10: 194-196, (1951).

5--1952   Chemical measurements (Chap. X, p. 166-193) in Office of Naval Research Symposium on Oceanographic Instrumentation (Rancho Santa Fé, Calif., June 21-23, 1952). NRC, Pub. 309, 233 p., (1952).

6--1953   Separation and concentration of trace metals from natural waters--Partition chromatographic technique. Anal. Chem. 25: 1927-1928, (1953).

7--1954   (with Goodgal, S. and Gloyna, E.) Reduction of radioactivity in water. J. Am. Water Wks. Assoc. 46: 66-78, (1954).

8--1954a (and Goodgal, S.) Sorption reactions and some ecological implications. Deep Sea Res. 1: 224-243, (1954).

9--1954b Atmospheric pressure changes and gas solubility. Deep Sea Res. 2: 59-62, (1954).

10--1954c (with Snodgrass, J. and Wooster, W. S.) Automatic servo-operated filter photometer. Anal. Chem. 26: 249-250, (1954).

11--1956 Recent developments in the chemistry and hydrography of estuaries (p. 420-436), in Tr. 21st North American Wildlife Conf., Philadelphia, March 5-7, 1956.

12--1957 (and Harley, J. H.) Precipitation of fission product elements on the ocean bottom by physical, chemical, and biological processes (p. 60-68), in The Effects of Atomic Radiation on Oceanography and Fisheries. NAS-NRC, Pub. 551, 137 p. (1957).

13--1958 Analytical chemistry in oceanography. J. Chem. Ed. 35: 119-122, (1958).

14--1959 (and Kanwisher, J. W.) An electrode system for measuring dissolved oxygen. Anal. Chem. 31: 5-9, (1959). [U. S. Patent No. 3,000,805 held by D. E. Carritt and J. W. Kanwisher.]

15--1959a (and Carpenter, J. H.) The composition of sea water and the salinity-chlorinity-density problems. NAS-NRC, Pub. 600: 67-86, (1959).

16--1959b (and others) Radioactive waste disposal into Atlantic and Gulf Coastal waters. NAS-NRC, Pub. 655, 38 p., (1959).

17--1960 Analysis of water samples, under "Sea Water," McGraw-Hill Encyclopedia of Science and Technology, v. 12, p. 116-117, (1960). New York: McGraw-Hill Book Co., Inc., 1960 and 1969 (revised).

18--1960a   Oceanographic research needed for the safe disposal of radioactive wastes into the ocean. Disposal of Radioactive Wastes Conf. Pr., Monaco, 16-21 Nov. 1959, vol. 2, I.A.E.A. (Int. Atomic Energy Agency), Vienna, (1960).

19--1962   Use of anion-exchange resins in the analysis of sea water [abst.]. J. Geophys. Res. 67: 3548, (1962).

20--1963   Chemical instrumentation (Chap. 5, p. 109-127), in <u>The Sea, Ideas and Observations on Progress in the Study of the Seas</u>, Hill, M. N., ed. New York and London: Interscience Pub., Vol. II, 554 p., (1963).

21--1965   (with Matson, W. and Roe, D. K.) Composite graphite-mercury electrode for anodic stripping voltammetry. Anal. Chem. 37: 1594-1595, (1965).

22--1966   (and Carpenter, J. H.) Comparison and evaluation of currently employed modifications of the Winkler method of determining dissolved oxygen in sea water. A NASCO Rept. J. Marine Res. 24: 286-318, (1966).

23--1966a   (with Green, E. J.) An improved iodine determination flask for whole bottle titration. Analyst 91: 207-208, (1966).

24--1967   (with Green, E. J.) New tables for oxygen saturation of seawater. J. Marine Res. 25: 140-147, (1967).

25--1967a   (with Green, E. J.) Oxygen solubility in seawater. Thermodynamic influence of sea salt. <u>Science</u> 157: 191-193, (1967).

Eel Pond Bridge, Woods Hole, Massachusetts.
(Sketch by Joan T. Kanwisher)

(44)
JOHN HOWER, JR.

JOHN HOWER, JR.

MIT: 1960-1961

When the need developed for a clay mineralogist to participate in certain research then being planned by Hurley's geochronology group, Dr. John Hower, Jr., an assistant professor at Montana State University, was invited to join the M.I.T. Department of Geology and Geophysics as assistant professor of geochemistry, starting in June 1960. Because of his college training in physics and clay mineralogy, his industrial experience in the research laboratories of Pan-American Oil Company, and his more recent academic experience as an assistant professor at Montana State, he joined the Department well-equipped to develop his own special research interest, which centered on geochemistry and clay mineralogy. He quickly and effectively became involved in the instructional and research program of the Department. He offered instruction in clay mineralogy and built up an x-ray laboratory for research on clay minerals, but, finding the Boston and Cambridge environment not to his liking, he resigned at the end of the first year and returned to the position he formerly held at Montana State. Thus the Department lost the services of a promising young scientist who subsequently moved from Montana State to Case Western Reserve University where he has continued active research on clay minerals.

John Hower, Jr. was born in Englewood, New Jersey, on 2 December 1927. After receiving his secondary education at the Canal Zone Junior College in Balboa, Canal Zone, he entered Syracuse University to study physics, and was awarded an S.B. degree in 1952. He then registered for graduate work in geophysics at Washington University in St. Louis and while working for an A.M. degree there also attended classes in geophysics at neighboring St. Louis University. Upon receiving an S.M. degree in geophysics in 1954, he changed his field of interest to clay mineralogy, then a highly-regarded field of study at Washington University, and in 1955 was awarded a Ph.D. degree in geochemistry and clay mineralogy. His doctoral thesis, as listed in his Bibliography (T--1955), involved "The fixation of heavy metal cations by some clay minerals."

Immediately after completing the requirements for the doctoral degree, Hower accepted a position as Research Engineer at the Pan-American Research

Center in Tulsa. From 1955 to 1957 he carried on research involving clay minerals and their use in the petroleum industry, but the academic environment proved more attractive, and the fall of 1957 found him an assistant professor of geology at Montana State University.

In our search for a promising young clay mineralogist, with fundamental training in the physical sciences and with both industrial and academic experience, Hower seemed an excellent candidate and was invited to join the staff of the Department in 1960 as an Assistant Professor of Geochemistry. He would become involved in the study of glauconite and illite, clay minerals then being investigated by Hurley and his geochronological colleagues. Soon Hower had an x-ray laboratory in operation and research under way, and seemed to have made the first step on a long career at M.I.T.

As the year wore on, however, he became dissatisfied with the necessity of commuting from a distant suburb, where he had settled his family, and found the exceedingly competitive spirit and intense intellectual and academic environment of the Cambridge-Boston area not to his liking. He longed for the open country of the West and for the more relaxed way of life there. By June he had decided to resign from his appointment, and he returned in August to the position he had formerly held at Montana State. The Department had lost a promising young scientist, and research on clay mineralogy had to be postponed for some future date.

In Hower's case there is an example of how a faculty goes about trying to extend and improve its membership and its program. The promising young men who are recruited are given an opportunity to develop their interests and to prove their competence in a friendly but competitive atmosphere. Some accept the challenge and become distinguished members of the M.I.T. faculty. Others move on to other institutions and make their mark there. So an active and progressive faculty must always try to gain strength from those young scientists who wish to test themselves and their ideas in the intensely competitive atmosphere that pervades M.I.T. and is partly responsible for its greatness.

As stated earlier, Hower moved his family back to Missoula at the end of the summer of 1961--he was married to Joann T. Hower and they had two children at the time: Mark, age 6, and Brigitt, age 4. He remained at Montana State for several years, then joined the Department of Geology at Case Western Reserve University in Cleveland, where he is now Professor of Geochemistry (July 1972).

BIBLIOGRAPHY OF JOHN HOWER, JR.

(Symbols and abbreviations used in the following references are explained on p. 91-98. This bibliography begins with Hower's first publication in 1955 and includes all known titles through 1961, when he left M.I.T., and two of later date.)

T--1955   The fixation of heavy metal cations by some clay minerals [abst.]. Dissertation Absts. 15/9: 1597 (1955). [Ph.D. thesis at Washington Univ., St. Louis, Mo.]

1--1957   (and Fancher, T. W.) Analysis of standard granite and standard diabase for trace elements. Science 125: 498, (1957).

2--1957a (with Burnham, H. D. and Jones, L. C.) Generalized X-ray emission spectrographic calibration applicable to varying compositions and sample forms. Anal. Chem. 29: 1827-1834, (1957).

3--1959   Matrix corrections in the X-ray spectrographic trace element analysis of rocks and minerals. Am. Mineral. 44: 19-32, (1959).

4--1959a (with Toler, L. G.) Determination of mixed layers in glauconites by index of refraction. Am. Mineral. 44: 1314-1318, (1959).

5--1960   (with Pilkey, O. H.) The effect of environment on concentration of skeletal magnesium and strontium in Dendraster. J. Geol. 68: 203-216, (1960).

6--1960a (with Hurley, P. M. et al.) Reliability of glauconite for age measurement by K-Ar and Rb-Sr methods. Am. Assoc. Petroleum Geol., B. 44: 1793-1808, (1960).

7--1961   Some factors concerning the nature and origin of glauconite. Am. Mineral. 46: 313-334, (1961).

8--1961a (with Stehli, F. G.) Mineralogy and early diagenesis of carbonate sediments. J. Sed. Petrol. 31: 358-371, (1961).

9--1962   (and Hurley, P. M. et al.) Effect of mineralogy on K/Ar age as a function of particle size in a shale [abst.]. Geol. Soc. Am., Sp.P. 68: 201-202, (1962). (Also see 10--1963.)

10--1963 (and Hurley, P. M. et al.) The dependence of K-Ar age on the mineralogy of various particle size ranges in a shale. Geochim. Cosmochim. Acta 27: 405-410, (1963). (Also see 9--1962.)

NOTE:

Articles on work done after he left M.I.T. in 1961 are listed in the standard "Bibliographies" of the U. S. Geological Survey.

----

(45)
RAYMOND HIDE

RAYMOND HIDE

MIT: 1960, 1961-1967

Experimental work in fluid dynamics in our Department of Geology and Geophysics started in 1961 when English-born, Manchester- and Cambridge-trained Raymond Hide was appointed Professor of Geophysics and Physics. Immediately on arrival, Hide started setting up a laboratory and machine shop in Building 20, where he and his research associates and graduate students could investigate the behavior of rotating fluids. By the spring of 1964 he had assembled an array of machine tools, built several pieces of special apparatus, and developed a lively research program. During the summer of 1964 the equipment was moved to new laboratory space on the fifth floor of the newly constructed Green Building, and additional apparatus was developed in the space for continued research on the dynamics of rotating fluids. This change of location brought about a closer coupling of his experimental work with the theoretical work being carried on in the Department of Meteorology, on the higher floors of the Green Building, and with the hydrodynamicists from the Woods Hole Oceanographic Institution who as part-time Visiting Professors were coming to M.I.T. on a regular schedule to help us develop a program of instruction and research in oceanography. The results of the investigations carried on in the two laboratories are to be found in the numerous articles listed in the Bibliography farther on.

In the classroom Hide taught undergraduate physics and both undergraduate and graduate geophysics; in the laboratory he conducted his own research as well as supervising that of a number of doctoral candidates; and in broader departmental activities he organized and chaired the Committee for Geophysics and Planetary Physics (1962-1966), chaired COMPASS, Committee for Planetary and Space Science (1962-1966), chaired the Committee for Oceanography (1966-1967), and directed the Department's Geophysical Fluid Dynamics Laboratory.

He left M.I.T. in July 1967 on sabbatical and resigned soon after in order to accept the position of Director of the Geophysical Fluid Dynamics Laboratory, Meteorological Office, Bracknell, Berkshire, England. As this is being written (June 1974) Dr. and Mrs. Hide with their three children are living in nearby Wokingham, Berkshire. When he left M.I.T. for his homeland, the Institute lost another of its unusually promising younger scientists and his distinguished career at the Meteorological Office since then is a measure of that loss.

Raymond Hide was born on 17 May 1929 in Doncaster, Yorkshire, England, the eldest of four sons of Stephen and Rose Edna (Cartlidge) Hide. He received his early education at local elementary and grammar schools (Percy

Jackson Grammar School, Woodlands, near Doncaster) and went on to university education at Manchester University (1947-1950), from which he received a first-class honours Bachelor of Science degree in physics in 1950. He next entered Gonville and Caius College, Cambridge University, in 1950 and was awarded a Doctor of Philosophy degree in geophysics in 1953.

After completing his doctorate at Cambridge he came to the United States as a Research Associate in Astrophysics at Yerkes Observatory (University of Chicago) for SY 1953-54. During April and May of 1954 he was a Visiting Scientist at the Institute for Advanced Study in Princeton, New Jersey.

He then returned to England as Senior Research Fellow in the General Physics Division of the Atomic Energy Research Establishment at Harwell, Berkshire, and held that post until 1957. He next accepted an appointment as Lecturer in Physics at King's College, University of Durham, Newcastle upon Tyne, where he remained until the fall of 1960. September of that year found him back in the United States, this time on a five-month appointment as a Visiting Lecturer in Geology and Geophysics in our Department.

At the end of the fall term of SY 1960-61, he spent February and March of 1961 as Research Meteorologist at the University of California in Los Angeles. Next came an offer of a permanent position at M.I.T., and in October 1961 he took up his duties as Professor of Geophysics in our Department of Geology and Geophysics. The next year his title was changed to Professor of Geophysics and Physics, to indicate that he also participated in the physics program, but he remained in our Department until he resigned and returned to England in June 1967.

During this second period at M.I.T., from October 1961 to June 1967, Hide participated in teaching undergraduate physics to freshmen and both undergraduate and graduate geophysics to Course XII students. While conducting his own research on the dynamics of rotating fluids, on planetary atmospheres, and on magnetohydrodynamics, he supervised the thesis investigations of nine doctorates, two of whom received their degree from Durham University (now the University of Newcastle upon Tyne) in England; these two were William W. Fowlis and Alan Ibbetson.

Instruction and research in fluid dynamics in the Department started with Hide's appointment as Professor of Geophysics in 1961. He was primarily interested in the behavior of rotating fluids under both stable and changing temperature conditions. . Accordingly, upon his arrival he proceeded to create a laboratory in which he and his students could experiment with rotating fluids under carefully controlled conditions.

The laboratory and its adjacent machine shop first occupied four bays of space in Building 20; in 1964, however, it was moved to the fifth floor of the newly constructed Green Building. Rotating tanks and annuli were built in the laboratory's own machine shop, by machinists John Burke, Jack Duplin and Tony Cieri, and these facilities were used to study a wide variety of hydrodynamical processes with particular emphasis on rotating fluids. During the existence of this Hydrodynamics of Rotating Fluids Laboratory, as it was called, its research program received support from the National Science Foundation and resulted in a series of important publications by Hide and his students. These are listed in the Bibliography farther on for the years 1961 through 1968, and constitute a substantial body of fundamental information on the physical behavior of rotating fluids.

In addition to lecturing, conducting his own research and supervising that of his graduate students and associates, and directing the Department's Hydrodynamics of Rotating Fluids Laboratory, Hide contributed much time and effort to committee work. He helped to organize and then chaired the Institute's Committee for Geophysics and Planetary Physics (1962-1966) and its Committee for Planetary and Space Science ["COMPASS"] (1962-1966), both of which played important roles in developing what ultimately became Course XII's program in the Planetary Sciences. The history of COMPASS is discussed in the chapter on "Geophysics at M.I.T." in Volume 2. He also chaired the interdepartmental Committee on Oceanography its first year, SY 1966-67, and helped greatly in getting successfully launched the joint M.I.T.-W.H.O.I. Graduate Program in Oceanography initiated in the fall of 1966. The interested reader is referred to the chapter on "Oceanography at M.I.T." for a discussion of the joint program with Woods Hole. (Volume 2.)

Soon after being granted a sabbatical leave for SY 1967-68, Hide decided to resign from his professorship in order to accept the appointment as Director of the Geophysical Fluid Dynamics Laboratory of the Meteorological Office at Bracknell, Berkshire, England. June 1967 found the Hide family on their way back to England, and our latest information (June 1974) is that Ray is as involved and interested as ever in experimental fluid dynamics and the origin of the earth's magnetism, as he carries on his work in the Meteorological Office. Ann, to whom he was married in 1958, has been teaching French in a private school near their home in Wokingham (Berkshire) until recently, and is now kept busy looking after three bright and vigorous youngsters as they carry on their varied school activities--Julia Ann (b. 1959), Stephen James (b. 1961) and Kathryn Margaret Rosanne (b. 1962).

As teacher, investigator, research supervisor and director, and committee chairman, Hide contributed greatly to Course XII's overall program of instruction and research, and provided exceptional strength in the specialized area of experimental geophysical fluid dynamics. The successful careers of the students he trained and the publications by him and them attest his outstanding performance as an M.I.T. professor, and were prophetic of the distinguished career he would make for himself in his homeland after leaving M.I.T.

BIBLIOGRAPHY OF RAYMOND HIDE

Symbols and abbreviations used in the following references are explained on pages 91 - 98. Abstracts are generally listed separately and may also be listed again in the reference to the complete article. The bibliography includes all of Hide's publications through 1968, a year and a half after he left M.I.T.

1--1952 A model experiment on thermal convection in the Earth's core. An. Geophysique 8: 17-18, (1952).

2--1952a (with Runcorn, S. K.) Towards a theory of the earth's magnetism [Summary of a paper read at the Royal Astronomical Society summer meeting in Leeds, 1952]. The Observatory 72, No. 870, (1952).

T--1953 Some experiments on thermal convection in a rotating liquid. Ph.D. Dissertation at Cambridge University, 1953. (See also the following reference.)

3--1953a Some experiments on thermal convection in a rotating liquid. Roy. Meteorol. Soc., Q.J. 79: 161, (1953). (See also the directly preceding and following references.)

4--1953b Geomagnetism, fluid motions in the Earth's core, and some experiments on thermal convection in a rotating fluid, p. 101-116, in Proc. First Symposium on the Use of Models in Geophysical Fluid Dynamics, Baltimore, Sept. 1953, (R. R. Long, ed.), (1953). (See also p. 101-116 in Fluid Models in Geophysics, Washington, D.C.: U.S. Govt. Printing Office, 1956, and the preceding reference.)

5--1955 The character of the equilibrium of an incompressible heavy viscous fluid of variable density: an approximate theory. Cambridge Phil. Soc., Pr. 51: 179-201, (1955).

6--1955a Waves in a heavy viscous incompressible electrically-conducting fluid of variable density in the presence of a magnetic field. Roy. Soc. London, Pr. A233: 376-396, (1955).

7--1956 The character of the equilibrium of a heavy viscous incompressible rotating fluid of variable density. I. General Theory, p. 22-34; II. Two special cases, p. 35-50--Mechanics and Applied Mathematics, Q.J. 60: 22-34, 35-50, (1956).

8--1956a Hydrodynamics of the Earth's core, p. 94-137, in Physics and Chemistry of the Earth, Vol. 1 (L. H. Ahrens, K. Rankama, and S. K. Runcorn, eds.). London: Pergamon Press, viii + 317 p., (1956).

9--1957 Experiments on convection in rotating liquids [Summary of paper read at the Royal Astronomical Society meeting on Planetary Atmospheres and Convection in Rotating Fluids, May, 1957]. The Observatory 77: 176-177, (1957).

10--1958   An experimental study of thermal convection in a rotating liquid. Roy. Soc. London, Phil. Tr. A250: 441-478, (1958).

11--1958a   (with Dolder, K.) An experiment on the interaction between a plane shock and a magnetic field. Nature 181: 1116-1118, (1958).

12--1960   (with Dolder, K.) Experiments on the passage of a shock wave through a magnetic field. Rev. Mod. Phys. 32/4: 770-779, (1960).

13--1960a   (and Roberts, P. H.) Hydromagnetic flow due to an oscillating plane. Rev. Mod. Phys. 32/4: 799-806, (1960).

14--1960b   Fluid motion in the Earth's core and the geomagnetic secular variation over the polar caps, in Proc. 12th General Assembly I.U.G.G., Helsinki. Comm. 4 on Geomagnetism and Aeronomy, (1960).

15--1961   (and Roberts, P. H.) The origin of the main magnetic field, p. 27-98 in Physics and Chemistry of the Earth, Vol. 4 (L. H. Ahrens, F. Press, K. Rankama, and S. K. Runcorn, eds.). New York: Pergamon Press, v + 317 p., (1961).

16--1961a   Hydrodynamical flow in the Earth's core, p. 1083-1088 in Dictionary of Physics. London: Pergamon Press, (1961).

17--1961b   Origin of Jupiter's Great Red Spot. Nature 190: 895-896, (1961).

18--1962   (with Roberts, P. H.) Some elementary problems in magnetohydrodynamics. Adv. Appl. Mech. 7: 215-316, (1962).

19--1962a   On the hydrodynamics of Jupiter's atmosphere, in La Physique des Planetes, Mém. Soc. Roy. Sci. Liége (5) 7: 481-505, (1962).

20--1964   The viscous boundary layer at the free surface of a rotating baroclinic fluid. Tellus 16: 523-529, (1964).

21--1965   The viscous boundary layer at the free surface of a rotating baroclinic fluid: effects due to the temperature dependence of surface tension. Tellus 17: 440-442, (1965).

22--1965a   (with Fowlis, W. W.) Thermal convection in a rotating annulus of liquid: effect of viscosity on the transition between axisymmetric and non-axisymmetric flow regimes. J. Atmos. Sci. 22: 541-558, (1965).

23--1965b
   (1967)   On the theory of the geomagnetic secular variation (p. 141-147) in Magnetism and the Cosmos (W. R. Hindmarsh et al., eds.). Edinburgh and London: Oliver and Boyd Ltd., xiv + 436 p., (1965, 1967).

24--1965c   (and Ibbetson, A.) Taylor columns (p. 343-347), in Magnetism and the Cosmos (W. R. Hindmarsh et al., eds.). Edinburgh and London: Oliver and Boyd Ltd., xiv + 436 p., (1965, 1967).

25--1965d
   (1967)   On the dynamics of Jupiter's interior and the origin of his magnetic field (p. 378-395), in Magnetism and the Cosmos (W. R. Hindmarsh et al., eds.). Edinburgh and London: Oliver and Boyd Ltd., xiv + 436 p., (1965, 1967).

26--1966   (and Ibbetson, A.) An experimental study of "Taylor Columns." Icarus 5: 279-290, (1966).

27--1966a   Free hydromagnetic oscillations of the Earth's core and the theory of the geomagnetic secular variation. Roy. Soc. London, Phil. Tr. A259: 615-647, (1966).

28--1966b   Planetary magnetic fields. Planet. Space Sci. 14: 579-586, (1966).

29--1966c   On the circulation of the atmospheres of Jupiter and Saturn. Planet. Space Sci. 14: 669-675, (1966).

30--1966d   Planetary magnetic fields (p. 224-226), in Science Year Book. New York: McGraw-Hill Pub. Co., (1966).

31--1966e   On the dynamics of rotating fluids and related topics in geophysical fluid dynamics. Am. Meteor. Soc., B. 47: 873-885, (1966).

32--1967   On the vertical stability of a rotating fluid subject to a horizontal temperature gradient. J. Atmos. Sci. 24: 6-9, (1967).

33--1967a   Theory of axisymmetric thermal convection in a rotating fluid annulus. Phys. Fluids 10: 56-68, (1967).

34--1967b   Motions of the Earth's core and mantle, and variations of the main geomagnetic field. Science 157: 55-56, (1967).

35--1967c   (with Titman, C. W.) Detached shear layers in a rotating fluid. J. Fluid Mech. 29: 39-60, (1967).

36--1967d   Jupiter and Saturn (p. 189-198), in The Earth in Space (H. Odishaw, ed.). New York: Basic Books Inc., (1967).

37--1967e   Electrical conductivity of the Earth's liquid core (p. 358) in International Dictionary of Geophysics (S. K. Runcorn, ed.). London: Pergamon Press, vol. 1, (1967).

38--1967f   Magnetohydrodynamic waves within the Earth (p. 897) in International Dictionary of Geophysics (S. K. Runcorn, ed.). London: Pergamon Press, vol. 2, (1967).

39--1967g   Inertial effects in shear layers in rotating fluids. Phys. Fluids 10,S: 306, (1967).

40--1968   Jupiter's Great Red Spot. Sci. Am. 218: 75-82, (1968).

41--1968a   (with Ibbetson, A. and Lighthill, M. J.) On slow transverse flow past obstacles in a rapidly rotating fluid. J. Fluid Mech. 32: 251-272, (1968).

42--1968b   On source-sink flows in a rotating fluid. J. Fluid Mech. 32: 737-764, (1968).

43--1968c   (with Horai, K. I.) On the topography of the core-mantle interface. Phys. Earth Planet. Int. 1: 305-308, (1968).

44--1968d   Planetary Circulation - Film and teachers' notes on global circulation of the Earth's atmosphere, commissioned by the American Meteorological Society, (1968).

45--1969   On the dynamics of the Earth's deep interior, in The Application of Modern Physics to the Earth and Planetary Interiors, (S. K. Runcorn, ed.), p. 651-652. London: John Wiley and Sons, Ltd., (1969).

46--1969a   Interaction between the Earth's liquid core and solid mantle. Nature 222: 1055-1056, (1969).

47--1969b   Dynamics of the atmospheres of the major planets. J. Atmos. Sci. 26: 841-847, (1969).

48--1969c   The viscous boundary layer at the rigid bounding surface of an electrically-conducting rotating fluid in the presence of a magnetic field. J. Atmos. Sci. 26: 847-853, (1969).

49--1969d   On hydromagnetic waves in a stratified rotating incompressible fluid. J. Fluid Mech. 39: 283-287, (1969).

50--1969e   Some laboratory experiments on free thermal convection in a rotating fluid subject to a horizontal temperature gradient and their relation to the theory of the global atmospheric circulation, in <u>The Global Circulation of the Atmosphere</u> (G. A. Corby, ed.). London: The Royal Meteorological Society, p. 196-221, (1969).

51--1970 Equatorial jets in planetary atmospheres. <u>Nature</u> 225: 254-255, (1970).

52--1970a (and Malin, S. R. C.) Novel correlations between global features of the Earth's gravitational and magnetic fields. <u>Nature</u> 225: 605-609, (1970).

53--1970b Planetary magnetic fields, in <u>Surfaces and Interiors of the Planets and Satellites</u> (A. Dollfus, ed.). London and New York: Academic Press, Inc., p. 511-534, (1970).

54--1970c Geophysical data and long-wave heterogeneities of the Earth's mantle [comment on a paper by Toksöz, M. N. <u>et al</u>.], J. Geophys. Res. 75: 2141, (1970).

55--1970d On the Earth's core-mantle interface (Symons Memorial Lecture). Q. J. Roy. Meteorol. Soc. 96: 579-590, (1970).

56--1970e (and Mason, P. J.) Baroclinic waves in a rotating fluid subject to internal heating. Phil. Trans. Roy. Soc. London A268: 201-232, (1970).

57--1971 Magnetohydrodynamic oscillations of neutron stars. <u>Nature Phys. Sci.</u> 229: 114-115, (1971).

58--1971a (and Malin, S. C. R.) Novel correlations between global features of the Earth's gravitational and magnetic fields: further statistical considerations. <u>Nature Phys. Sci.</u> 230: 63, (1971).

59--1971b Magnetohydrodynamics and the 1970 Nobel Prize for Physics. Weather 26: 128, (1971).

60--1971c On planetary atmospheres and interiors. Lect. Appl. Math. 13: 229-353, (American Mathematical Society), (1971).

61--1971d On geostrophic motion of a non-homogeneous fluid. J. Fluid Mech. 49: 745-751, (1971).

62--1971e Motions in planetary atmospheres: a review. Met. Mag. 100: 268-276, (1971).

63--1971f Viscosity of the Earth's core. <u>Nature Phys. Sci.</u> 233: 100-101, (1971).

64--1971g Magnetohydrodynamics of rotating fluids: a summary of some recent work. Q. J. Roy. Astron. Soc. 12: 380-383, (1971).

65--1971h (and Malin, S. R. C.) Correlations between the Earth's gravitational and magnetic fields: effect of rotation in latitude. <u>Nature Phys. Sci.</u> 232: 31-33, (1971).

66--1971i (with Malin, S. R. C.) Bumps on the core-mantle boundary. Comments on Earth Sciences: Geophysics 2/1: 1-13, (1971).

<u>Miscellaneous Writings, chiefly Research Reports, of limited circulation not submitted for publication.</u>

A.E.R.E. = Atomic Energy Research Establishment
H.R.F.P. = Hydrodynamics of Rotating Fluids Project

In the course of his employment with different agencies and educational institutions, Hide wrote numerous reports of one kind or another that were never submitted for publication. These were characteristically "in-house" research reports that were of chief interest to the organization for which the work was done. The titles in the list on the next page serve to indicate the nature and extent of these unpublished writings.

1) The experimental shock wave programme at Harwell. A.E.R.E., Res. Rept. G17/P5, (1955).

2) (with Millar, W.) A preliminary investigation of shocks in a curved channel. A.E.R.E., Res. Rept. GP/R.1918, (1956).

3) (with Allen, J. E. and Reynolds, P.) The production of controlled thermonuclear energy. A.E.R.E., Res. Rept. GP/R.2073, (1956).

4) Note on converging shocks. A.E.R.E., Res. Rept. GP/M.191, (1956).

5) Note on the production of strong shocks. A.E.R.E., Res. Rept. GP/M.202, (1956).

6) Shock heating of a wriggling discharge. A.E.R.E., Res. Rept. GP/R.2328, (1957).

7) (with Dolder, K.) Bibliography on shock waves, shock tubes, and allied topics. A.E.R.E., Res. Rept. GP/R.2055, (1957).

8) (in collaboration with Bowden, M., Eden, E. F., Fowlis, W. W., Ibbetson, A., and Titman, C. W.) Research on thermal convection in rotating fluids on rotating barotropic fluids. Technical Summary Report to the U.S.A.F., Contract AF 61(052)216, AFCRL, OAR (European Office), from Physics Dept., Univ. Durham, Newcastle upon Tyne, England, (1961).

9) (with Price, N. H. and Shatford, P. A.) The design, construction and performance of the A.E.R.E. Rept. Z/R.2798, (1962).

10) Research on thermal convection in rotating fluids and on rotating barotropic fluids (in collaboration with Bowden, M., Eden, H. F., Fowlis, W. W., Ibbetson, A., and Titman, C. W.). Final Report to U.S.A.F., Contract AF 61(052)216, AFCRL, OAR (European Office), from Physics Dept., Univ. Durham, Newcastle upon Tyne, England, (1962).

11) Some thoughts on rotating fluids. Sci. Rept. 2, H.R.F.P., Dept. Geol. Geophys., M.I.T., (Nov. 1962).

12) On the transition between the upper symmetrical regime and the steady waves regime of thermal convection in a rotating fluid annulus. Sci. Rept. 3, H.R.F.P., Dept. Geol. Geophys., M.I.T., (July 1963).

13) Thermal convection in a rotating fluid annulus: notation and dimensionless parameters. Sci. Note 1, H.R.F.P., Dept. Geol. Geophys., M.I.T., (July 1963).

14) Salt convection due to radial cooling. Sci. Note 4, H.R.F.P., Dept. Geol. Geophys., M.I.T., (July 1964).

15) Topics in geophysical fluid dynamics. A.E.R.E., Memorandum CLM-16 [Outline of two Research Group lectures], (1965).

16) (with Fowlis, W. W.) Thermal convection in a rotating annulus of liquid: effect of viscosity on the transition between axisymmetric and non-axisymmetric flow regimes: Appendices. Sci. Rept. 8, H.R.F.P., Dept. Geol. Geophys., M.I.T., (1965).

17) Free and forced reversals of the Earth's magnetic field. Sci. Note 5, H.R.F.P., Dept. Geol. Geophys., M.I.T., (Sept. 1966).

(46)
GIORGIO FIOCCO

GIORGIO FIOCCO

MIT: 1963-1968

Trained in electrical engineering at the University of Rome, Fiocco came to M.I.T. in 1961 by way of the Aeronautical Laboratory at Cornell, where he conducted research on weather radars and electromagnetic wave propagation in plasmas. Being trained and interested in radar, he joined L. D. Smullin's research group in the Research Laboratory of Electronics, and assisted in the experiment in which for the first time a laser light beam was bounced off the moon. The success of this project led him to recognize the possibility of using the same technique to observe Thomson scattering of light by electrons, and of using a simpler laser system to explore the upper atmosphere for scattering layers--dust layers, noctilucent clouds, etc. According to Smullin (letter of May 14, 1963, addressed to F. Bitter), Fiocco and E. Thompson carried out two original experiments-- 1) detection of laser light scattered from an electron beam, and 2) from a plasma. These experiments were clear "firsts" in a field of research in which many groups around the world had been actively trying to do the same thing.

With such a background of successful experimentation, and with a keen interest in probing the atmosphere with optical radar and conducting other geophysical research, Fiocco was an excellent candidate for Bitter's research group in the National Magnet Laboratory. Having strong recommendations from Bitter and Smullin, Fiocco was appointed Assistant Professor of Geophysics in 1963, and immediately became an important member of Bitter's teaching and research team. From 1963 to 1968 he carried on a vigorous and productive research program, as indicated by the 40 titles in his bibliography--a program that can be described as "probing the upper atmosphere with optical radar." In this period he took an active role in assisting eight Course XII graduate students with their thesis work. A year before his second 3-year appointment as assistant professor was to end, he decided to return to Italy. There, in 1968, he accepted the position of Senior Scientist with the European Space Research Organization (ESRO) at the European Space Research Institute (ESRIN), in Frascati, Roma, Italy. He also took charge of a research group at the Centro Nazionale per la Fisica dell'Atmosfera e Meteorologia of the Consiglio Nazionale delle Ricerche, and became a part-time lecturer at the University of Rome. Thus M.I.T. and Course XII lost a promising young geophysicist who rendered valuable service to the Department of Geology and Geophysics in the 1960s when strenuous efforts were being made to build up a strong program in planetary science.

Giorgio Fiocco was born in Rome, Italy, on 13 June 1931, the first and only child of Camillo Fiocco, a civil engineer, and Maria (Paglia) Fiocco, a housewife. After elementary and intermediate schooling, he entered the Liceo scientifico C. Cavour in Rome in 1944, and graduated in 1949 in the highest hundredth of the course. He immediately entered the Università di Roma as an engineering student and six years later, in December 1955, won a Doctor of Engineering degree, ranking in the highest tenth of the class. His thesis, done in radio engineering, concerned a "Broad-band panoramic analyzer."

After serving a year's tour of duty in the Italian Navy, 1956-1957, as a civilian scientist, Fiocco left his homeland to accept a position with the Baddow Research Laboratory of Marconi's Wireless Telegraph Company in Great Baddow, Essex, about 20 miles from London. During his three years at this laboratory in England he wrote nine research reports concerning such subjects as Doppler radar, navigational aids, and a new FM radar system, and had issued to him three British patents relating to radar systems.

In 1960 he emigrated to the United States to work in the Aeronautical Laboratory of Cornell University. There his research, all unpublished, dealt with analyses of weather radars and electromagnetic wave propagation in plasmas. After a year at Cornell he came to M.I.T. as a D.S.R. project engineer in L. D. Smullin's research group in the Research Laboratory of Electronics. His accomplishments during the next two years, from September 1961 to July 1963, when he was appointed Assistant Professor of Geophysics in Course XII, are best described by the following excerpt from a letter dated May 14, 1963, from Smullin to Bitter:

> "Dear Francis:
>
> I am writing to endorse the appointment of Dr. G. Fiocco as a member of the faculty in the Department of Geology and Geophysics. Giorgio came to work in my lab on September 1961 as a DSR engineer. In the ensuing period, I have developed a very great respect for him both as a person, and as a scientist-engineer.
>
> "He worked with me on the project of bouncing a laser light beam off the moon, and his efforts had a great deal to do with the success of that venture. With the completion of that project he recognized the possibility of using the same technique to observe Thomson scattering of light by electrons, and of using a simpler laser system to explore the upper atmosphere for scattering layers, etc. He and E. Thompson immediately carried out the two experiments of detecting laser light scattered from an electron beam, and from a plasma. These experiments were clear 'firsts' in a field, where many groups around the world had been hard at work for some time, trying to do the same thing. At present, his apparatus for studying atmospheric scattering is virtually complete, and first tests should be made within a week or so. If they are successful, we will have an important new geophysical tool. Thus, during Fiocco's short stay here ($\sim$ 20 months) he has had a major part in two highly

> successful experiments, and is on the threshold of a
> third. In addition, he has helped Prof. H. A. Haus
> to carry out some important studies of noise in lasers, and has helped the graduate students in my laboratory with their various difficulties. Finally, he
> has assisted me, this term, in running our required
> Junior Electromagnetic Laboratory 6.72."

The research achievements referred to in the foregoing letter are recorded in the first dozen titles in his Bibliography. Upon joining the faculty of the Department of Geology and Geophysics, Fiocco immediately started several research projects with Bitter in the National Magnet Laboratory, where he was assigned office and laboratory space. He assisted four of Bitter's graduate students in their thesis work and joined his discussion group on low-density magnetically contained plasmas.

During the five years following his appointment as an assistant professor, 1963-1968, Fiocco conducted his research as a member of the geophysics team that Bitter brought together, and used the space and facilities assigned to him in the National Magnet Laboratory and the Research Laboratory of Electronics. The nature of his research and the scientists with whom he collaborated can be determined by noting the items in the Bibliography for the period. Especially to be noted are the numerous articles concerned with aerosol layers, dust, and noctilucent clouds in the upper atmosphere. He investigated scattering layers in the 60-140 km region with Smullin (1964) and the aerosol layer at 20 km with Grams (1964-1967). He next turned his attention to dust layers and noctilucent clouds, in a research program that attracted several Course XII graduate students, and travelled as far as Alaska, Norway, and Sweden to observe and measure these phenomena.

Because Fiocco conducted his research outside the Green Building, most of the geology faculty saw little of him and did not learn much about what he was doing except as they could read about it in The Tech, Tech Talk, Reports on Research, and similar M.I.T. publications. In contrast, Course XII graduate students quickly discovered him, and during the five years before he left in 1968, he assisted in supervising four masters and four doctoral candidates: R. G. Little, R. J. Breeding, J. B. DeWolf, and D. F. Kitrosser (S.M.s); W. D. Halverson and S. J. Bless (Sc.D.s); and H. C. Koons and J. B. DeWolf (Ph.D.s).

When an opportunity came to return to a post in his homeland within a year of the time his second three-year appointment as assistant professor would end, he resigned his professorship in early 1968, and accepted the position of Senior Scientist with the European Space Research Organization (ESRO) at the European Space Research Institute (ESRIN) in Frascati, Roma, Italy. At about the same time he also took charge of a research group at the Centro Nazionale per la Fisica dell'Atmosfera e Meteorologia of the Consiglio Nazionale delle Ricerche. He continued his academic connections, however, as a part-time lecturer at the University of Rome and the University of L'Aquila.

Giorgio married Gabriella Sanna-Solinas on 6 August 1956, and daughter Sylvia was born a year later on 12 November 1957. The Department lost a fine family and a productive young geophysicist when the Fioccos returned to their homeland. Giorgio had contributed importantly to the planning and development of the program in planetary physics that later grew to such size as to cause the designation of Course XII to be changed to the Department of Earth and Planetary Sciences.

BIBLIOGRAPHY OF GIORGIO FIOCCO

Symbols and abbreviations used in the following references are mostly explained on p. 91 - 98; several items, however, need special comment inasmuch as they are not used in most other bibliographies in this work. The numerous citations of M.I.T., R.L.E., QPR refer to Massachusetts Institute of Technology, Research Laboratory of Electronics, Quarterly Progress Report No. ___, etc. The following list includes only those papers published while Fiocco was at M.I.T.

1--1962   High energy cyclotron resonance of electrons in a plasma. M.I.T., R.L.E., QPR 64: 103-104, (1962).

2--1962a Electron cyclotron heating of a plasma. M.I.T., R.L.E., QPR 65: 95-97, (1962).

3--1962b (with Smullin, L. D.) Optical echoes from the Moon. Nature 194: 1267, (1962).

4--1962c (with Smullin, L. D.) Project Luna See. Inst. Radio Engineers [IRE], Pr. 50: 1703-1704, (1962); also M.I.T., R.L.E., QPR 66: 60-66, (1962).

5--1962d Some applications of optical radar to astronomical and geophysical research. Northeast Electronics Research and Engineering Meeting [NEREM], Boston, 5-7 Nov. 1962, Record, p. 53, (1962).

6--1962e Scintillating probe for plasmas. M.I.T., R.L.E., QPR 67: 89, (1962).

7--1962f (with Thompson, E.) Production of ion beams. M.I.T., R.L.E., QPR 67: 89-90, (1962).

8--1962g (and Thompson, E.) Scattering of light from electrons. M.I.T., R.L.E., QPR 67: 90-91, (1962).

9--1963   Applicazioni dei "Maser" ottici alle ricerche spaziali. Missili 2: 73-78, (1963).

10--1963a (and Thompson, E.) Thomson scattering of light from electrons. M.I.T., R.L.E., QPR 68: 68, (1963).

11--1963b (with Rose, D. J. and Smullin, L. D.) Electron cyclotron resonant discharge. M.I.T., R.L.E., QPR 68: 68-69, (1963).

12--1963c (and Thompson, E.) Scattering of light from electrons II. M.I.T., R.L.E., QPR 68: 74-77, (1963).

13--1963d   Optical radar to study the earth's atmosphere. M.I.T., R.L.E., QPR 69: 28-29, (1963).

14--1963e   (and Haus, H. A.) Photon statistics of optical maser output. M.I.T., R.L.E., QPR 69: 31-33, (1963).

15--1963f   (and Smullin, L. D.) Electron-cyclotron plasma heating. M.I.T., R.L.E., QPR 69: 63-65, (1963).

16--1963g   (with Thompson, E.) Scattering of light from plasma electrons III. M.I.T., R.L.E., QPR 69: 74-80, (1963).

17--1963h   (with Thompson, E.) Errata: Scattering of light from (plasma) electrons. M.I.T., R.L.E., QPR 70: 152, (1963).

18--1963i   (and Thompson, E.) Thomson scattering of optical radiation from an electron beam. Phys. Rev. Letters 10: 89-91, (1963).

19--1963j   (and Thompson, E.) Techniques for observing Thomson scattering of optical radiation from electrons [abst.]. Am. Phys. Soc., B ser 11, 8: 372, (1963).

20--1963k   (with Thompson, E.) Thomson scattering of optical radiation from a thermal plasma [abst.]. Am. Phys. Soc., B ser 11, 8: 372, (1963).

21--1963l   Electron-beam and plasma diagnostics by scattering of optical radiation, Symposium of the Division of Plasma Physics, APS, Buffalo, June 24, 1963. Am. Phys. Soc., B ser 11, 8: 434, (1963).

22--1963m   (with Thompson, E.) Measurements of large-angle scattering of laser radiation from D.C. plasmas [abst.]. Proc. VI International Conference on Ionization Phenomena in Gases, Paris, (1963).

23--1963n   (and Smullin, L. D.) Detection of scattering layers in the upper atmosphere (60-140 km). M.I.T., R.L.E., QPR 71: 76-80, (1963); Nature 199: 1275-1276, (1963).

24--1963o   (with Colombo, G.) Optical radar results and meteoric fragmentation. Smithsonian Astrophysical Observatory, Sp. Rept. 139: 1-25, (1963).

25--1964   (and Colombo, G.) Optical radar results and meteoric fragmentation. J. Geophys. Res. 69: 1795-1803, (1964).

26--1964a   Scattering of light in the earth's atmosphere. M.I.T., R.L.E., QPR 72: 45, (1964).

27--1964b   An interpretation of some optical radar results. M.I.T., R.L.E., QPR 72: 48-52, (1964).

28--1964c   (and Grams, G. W.) Observations of the aerosol layer at 20 km. M.I.T., R.L.E., QPR 73: 20-21, (1964).

29--1964d   (and Grams, G.) Observations of the aerosol layer at 20 km by optical radar. J. Atmos. Sci. 21: 323-324, (1964).

30--1965   A summary of research with optical radar. Instrument Soc. Am., Proc. An. Meeting, Los Angeles, 4-11 Oct., (1965).

31--1965a   (with Colombo, G.) Reply [to letter by D. Diermendjian]. J. Geophys. Res. 70: 746, (1965).

32--1965b   (with Bitter, F.) Van Allen radiation belts. M.I.T., R.L.E., QPR 76: 41, (1965).

33--1965c   Optical investigation of the upper atmosphere. M.I.T., R.L.E., QPR 76: 41-42, (1965).

34--1965d   Optical radar results and ionospheric sporadic E. M.I.T., R.L.E., QPR 76: 42, (1965).

35--1965e (and Grams, G., Urbankek, K., and Breeding, R. J.) Observations of the upper atmosphere by optical radar in Alaska and Sweden (Pt. I). M.I.T., R.L.E., QPR 76: 43-46, (1965); (Pt. II), QPR 77: 53-58, (1965).

36--1965f (with Koons, H. C.) Proton flow into the magnetosphere. M.I.T., R.L.E., QPR 78: 67-70, (1965).

37--1965g Optical radar results and ionospheric sporadic E. J. Geophys. Res. 70: 2213-2215, (1965).

38--1965h (and Koons, H. C. and Meeks, L. M.) A note on the search for Comet Ikeya-Seki [1966b]. Internat. Astron. Union, Circular, (November 1965).

39--1966 (with Bitter, F.) Research objectives. M.I.T., R.L.E., QPR 80: 17, (1966).

40--1966a (with Chiang, R.) Electromagnetic back-scattering from spheres. M.I.T., R.L.E., QPR 80: 17-18, (1966).

41--1966b (and Koons, H. C. and Meeks, L. M.) Search for Comet Ikeya-Seki [1965h] at 8 Gc/sec and 15 Gc/sec. M.I.T., R.L.E., QPR 80: 18-19, (1966).

42--1966c A note on the influx of extraterrestrial dust as an energy source in the E region. Radio Sci. 1: 252, (1966).

43--1966d (and Grams, G.) Observations of the upper atmosphere by optical radar in Alaska and Sweden during the summer of 1964. Tellus 18: 34-38, (1966).

44--1966e (with Grams, G.) Studies of stratospheric aerosols and their correlation with ozone. M.I.T., R.L.E., QPR 81: 36-37, (1966).

45--1967 (and Grams, G.) Optical radar and airglow observations during noctilucent cloud displays [abst.]. Am. Geophys. Union, Tr. 48: 188-189, (1967).

46--1967a Application of laser radars to the study of the atmosphere, in Aerospace Measurement Techniques (G. Manelli, ed.), Symposium held at M.I.T., 7-8 July 1967, NASA SP-132, National Aeronautics and Space Administration, Washington, D.C., and Electronics Research Center, Cambridge, Mass., p. 123-131, (1967).

47--1967b (with Bitter, F.) Research objectives. M.I.T., R.L.E., QPR 84: 63, (1967).

48--1967c Research objectives and summary of research. M.I.T., R.L.E., QPR 84: 64-66, (1967).

49--1967d (and Grams, G.) Possible relation between dust and rainfall. M.I.T., R.L.E., QPR 84: 66-67, (1967).

50--1967e (and Grams, G.) Optical radar and airglow observations in Norway during the presence of noctilucent clouds. M.I.T., R.L.E., QPR 85: 49-50, (1967).

51--1967f (with Bitter, F. and Bless, S. J.) Preliminary experiments with a fast Z-pinch capacitor discharge: the generation of high pressures in solids. M.I.T., R.L.E., QPR 86: 53-57, (1967).

52--1967g (and Grams, G.) Optical radar observations of mesospheric aerosols in Norway during the summer of 1966. M.I.T., R.L.E., QPR 86: 59-66, (1967).

53--1967h Possibility of continuous measurement by optical radar of the influx on Earth of extraterrestrial dust, Chap. 20 in The Zodiacal Light and the Interplanetary Medium (J. L. Weinberg, ed.), Proc. Symposium 30 Jan.-2 Feb. 1967 in Honolulu, Hawaii, NASA SP-150: 115-117, (1967).

54--1967i   On the production of ionization by micrometeorites. J. Geophys. Res. 72: 3497-3501, (1967).

55--1967j   (with Grams, G.) Stratospheric aerosol layer in 1964 and 1965. J. Geophys. Res. 72: 3523-3542, (1967).

56--1967k   (with DeWolf, J. B.) Experiments to determine the aerosol content of air from spectral analysis of laser echoes. M.I.T., R.L.E., QPR 87: 37-40, (1967).

57--1967ℓ   (with Koons, H. C.) Measurement of the density and temperature of the electrons in a low-density reflex discharge by scattering of continuous-wave A^+ laser light. M.I.T., R.L.E., QPR 87: 40-43, (1967).

58--1968   Research objectives. M.I.T., R.L.E., QPR 88: 59, (1968).

59--1968a   (and DeWolf, J. B.) Frequency spectrum of laser echoes from atmospheric constituents and determination of the aerosol content of air. J. Atmos. Sci. 25: 488-496, (1968).

60--1968b   (with Koons, H. C.) Anisotropy of the electron velocity distribution in a reflex discharge measured by continuous-wave laser scattering. M.I.T., R.L.E., QPR 89: 43-45, (1968); Phys. Letters 26 A: 614-615, (1968).

61--1968c   (with Koons, H. C.) Measurements of the density and temperature of electrons in a reflex discharge by scattering of cw Ar^+ laser light. J. Appl. Phys. 39: 3389-3392, (1968).

62--1968d   Optical radar observations of mesospheric aerosols in Norway during the summer 1966. M.I.T., R.L.E., QPR 90: 47-49, (1968).

63--1969   (and Grams, G.) Optical radar observations of mesospheric aerosols in Norway during the summer 1966. J. Geophys. Res. 74: 2453-2458, (1969).

----

(47)
LEE WALLACE DEAN III

LEE WALLACE DEAN III

MIT: 1960-1967

In the mid-1960s, when every effort was being made to build up a strong group of younger faculty in geophysics, Dean was recommended as a well-trained physicist who could bring expertise in sound propagation to our program, particularly in experimental research. With some seven years of experience as a member of a research team in M.I.T.'s Research Laboratory of Electronics, he had the kind of experimental experience needed to complement other aspects of our geophysics program. During his three-year appointment as Assistant Professor of Geophysics, from 1964 to 1967, he contributed importantly to instruction by giving several lecture subjects and aiding a number of graduate students in their thesis work. In July 1967 he left M.I.T. to accept a position as a research scientist with United Aircraft (Pratt and Whitney) in Hartford, Connecticut.

Lee Wallace Dean III was born in St. Louis, Missouri, on 15 August 1932. He was the first son of Dr. Lee Wallace Dean II, a nose and throat physician, and Sarah (McGarvey) Dean, whose father was born in Ireland. He has a younger brother, David (b. 1934) and a younger sister, Linda (b. 1937). Lee married Fleda Asbury of Colonia, New Jersey, in 1955, and they have three children: Michael (b. 1957), David (b. 1960), and Carol (b. 1963).

After graduation from John Burrough High School in St. Louis in June, 1950, Dean entered Amherst in the fall and four years later was graduated cum laude with a Bachelor of Arts in physics. Immediately after graduation he registered as a degree candidate in M.I.T.'s Department of Electrical Engineering. He enrolled in the Cooperative Program (Course VI-A) and did the off-campus part of the program at the Air Force Cambridge Research Center, where he worked on electromagnetic radiation (infrared and radar). When he completed the program in 1957 he received a Bachelor of Science degree in Electrical Engineering and a Master of Science without specification. His S.M. thesis involved "Excitation of acoustic resonators by wind" and was supervised by Prof. Uno Ingard, with whom he would do much more work later.

Not wishing to interrupt his academic program, he held a temporary staff position with the Division of Sponsored Research during the summer, then registered in the Department of Physics as a doctoral candidate in September 1957. Within the year he got started on the investigation of the scattering of sound by sound, working with Ingard's group in R.L.E., and held a research assistantship in that Laboratory from September 1957 to February 1960 when he received the Ph.D. degree in physics with his thesis on "Scattering of Sound by Sound."

After a brief appointment as a D.S.R. staff member in R.L.E. during the spring and summer of 1960, he became an Instructor in the Department of Physics, and for the next four years taught undergraduate physics and carried on acoustical experimentation in R.L.E. as a member of Ingard's research group. As is evident from his bibliography, most of his research was done in the Research Laboratory of Electronics, and the results recorded in the Quarterly Progress Reports of that Laboratory.

At about this same time in the earlier 1960s, with student interest in oceanography and geophysics growing rapidly, it became necessary for the Department of Geology and Geophysics to seek a full-time staff member who could offer instruction and supervise research in sound propagation, particularly hydroacoustics. At the time, the Department had to depend entirely on Dr. J. B. Hersey of the Woods Hole Oceanographic Institution, who, because of his extensive research program there, could give only a limited amount of effort and attention to our on-campus program in geophysics while serving M.I.T. as a part-time Professor.

Accordingly, Dean Wiesner was requested to approve transfer of Lee from Physics to Geology, and thus it came about that on 1 July 1964, Dean was appointed an Assistant Professor of Geophysics in the Department of Geology and Geophysics and immediately became involved in the program of geophysical instruction and research being developed in Course XII. In addition to participating in the introductory subjects in geophysics, and helping graduate students with their thesis work, he organized a lecture subject in Physical Acoustics (12.621), assisted in Wave Propagation in Fluids (12.721), and continued his own research on acoustical waves. At a time when the Department was struggling to get a program of instruction and research in seismology underway Dean contributed importantly by his teaching and research supervision in the field of sound propagation.

At the end of his three-year appointment, in June 1967, he accepted a position as research scientist with the Pratt & Whitney division of United Aircraft at Hartford, Connecticut. At the time this is being written (August 1972), he lives in Hartford with his wife, the former Fleda Asbury, whom he married in 1955, and their three children.

BIBLIOGRAPHY OF LEE WALLACE DEAN III

In the following references, the sequence M.I.T., R.L.E., QPR etc. is to be read: Massachusetts Institute of Technology, Research Laboratory of Electronics; Quarterly Progress Report Number ___: pages ___, (year).

1--1958   (and Ingard, Uno) Excitation of resonators by flow. [Report on] WADC Contract AF33(616)-3307, Task No. 71705. M.I.T., Acoustics Lab., March 1958, 21 p., (‡) (1958).

2--1958a   (with Ingard, Uno) Excitation of acoustic resonators. 2nd Symposium on Naval Hydrodynamics--Hydrodynamic noise cavity flow, sponsored by the Office of Naval Research and the National Academy of Science-National Research Council, August 25-29, 1958, Washington, D.C., p. 137-150, (1958). ACR-38, ONR-Dept. Navy, Washington, D.C., (1958).

3--1958b   Measurement of the attenuation of sound in metal rods. M.I.T., R.L.E., QPR 51: 105-106, (1958).

4--1959   Scattering of sound by sound. M.I.T., R.L.E., QPR 53: 170, (1959). (Also see 7--1959c.)

5--1959a   (with Scop, M. M.) Acoustic attenuation in the vicinity of an order-disorder transition in a copper-gold alloy (Cu_3Au). M.I.T., R.L.E., QPR 54: 155-157, (1959).

6--1959b   (with Schaefer, J. B.) Attenuation of sound in aluminum. M.I.T., R.L.E., QPR 54: 157-158, (1959).

7--1959c   Scattering of sound by sound. M.I.T., R.L.E., QPR 54: 158-159; QPR 55: 140, (1959). (Also see 4--1959.)

T--1960   Scattering of sound by sound. Ph.D. Thesis at M.I.T. in Course VIII, Department of Physics, 68 p., (January 1960).

8--1960a   Scattering of sound by sound. M.I.T., R.L.E., QPR 56: 167-171; QPR 57: 117-119, (1960); QPR 60: 175, (1961).

9--1960b   (with Ingard, U. and Lehman, A. B.) Sound reception by a moving cavity, Part II of Final Report on Contract Nonr 1841 (03), iv + 21 p. (‡), M.I.T., Acoustics Lab. and R.L.E., (1960).

10--1961   (with Ingard, U.) Properties of a moving acoustic resonator. M.I.T., R.L.E., QPR 60: 171-175, (1961).

11--1961a   (with Ball, N. A.) A low resonant frequency barium titanite transducer. M.I.T., R.L.E., QPR 63: 131-132, (1961).

12--1962   (and Ball, N. A.) A low resonant-frequency barium-titanite transducer. J. Acoustical Soc. Am. 34: 347, (1962).

13--1962a   (and Matchett, G. A.) Sound propagation in Cu_3Au. M.I.T., R.L.E., QPR 66: 69, (1962).

14--1962b   Interactions between sound waves. J. Acoustical Soc. Am. 34: 1039-1044, (1962).

15--1964   (and Friedlander, S. B.) Interaction between concentric cylindrical sound waves. M.I.T., R.L.E., QPR 75: 39-42, (1964).

(48)
ANTHONY FRANK GANGI

ANTHONY FRANK GANGI

MIT: 1964-1967

During the early 1960s, when the staff in geophysics was being enlarged and diversified, a need developed for several young scientists who could offer instruction in and conduct research on seismic wave propagation and detection. In the search for candidates, our attention was called to A. F. Gangi, a recent doctorate from the University of California at Los Angeles, who was reported to have outstanding talents in electronics, electronic instrumentation, array design and performance, and radar. He was appointed associate professor of geophysics on 1 July 1964 for a three year term ending 30 June 1967. During this period he taught Theoretical Seismology (12.723), supervised both senior and graduate thesis research in seismology and concentrated his personal research in the same area. At the termination of his three-year appointment on 30 June 1967 he accepted a position as Associate Professor in the Department of Geophysics at Texas Agricultural and Mechanical University [Texas A & M] in College Station, Texas. During his short but productive stay at M.I.T., Gangi was particularly helpful in developing a cooperative program of seismological research with the Lincoln Laboratory.

Anthony Frank Gangi was born in Newark, New Jersey, on 19 February 1929. He was one of five children of Francesco Paolo and Bestianna Lucia Cosenza diGangi. He is second among the children, having an older brother, Charles Robert (b. 1927), two younger twin sisters, Marie Rose (b. 1931) and Rose Marie (b. 1931), and a younger brother, Salvadore Frank (b. 1933).

On 26 January 1961 he married Enrichetta (de Gange) Grove who had three children by her first husband: Robert Francis Grove (b. 1955) and his twin sister Theresa Ann Grove (b. 1955), and a younger son, Stephen John.

Gangi attended the West Side High School in Newark, N.J., and was graduated in June 1946. He served in the Army of the United States from August 1946 to February 1948. Immediately on discharge he enrolled in Glendale City College, Glendale, California, and studied engineering until September 1949, when he entered the University of California at

Los Angeles, where he continued his studies in electrical engineering. In 1950 he changed his major interest to physics and mathematics, and was awarded a Bachelor of Science degree in Applied Physics in February 1953. Continuing as a graduate student in physics, he was awarded an M.S. degree in applied physics in 1954 and a Ph.D. in physics in 1960. His doctor's thesis bears the title "Elastic waves in wedges," (an investigation concerned with the solutions of boundary value problems dealing with the diffraction of elastic waves by wedges and partly with an experimental investigation of the phenomena of elastic wave diffraction by wedges).

While pursuing graduate work at U.C.L.A., from 1953 to 1956, Gangi assisted in the Department of Physics, first as a Teaching Assistant and later as a Senior Electronics Technician. He was a Shell Foundation Fellow in U.C.L.A.'s Institute of Geophysics from September 1956 to September 1957.

From September 1959 to June 1961, Gangi was a member of the Technical Staff of Space Electronics Corporation, Glendale, California. His work there involved study of very low frequency (VLF) electromagnetic wave propagation from underground antennas, and analysis of an electronic, self-focusing antenna array. From this position he went to Space-General Corporation (Glendale and later El Monte, California), as Manager of the Antenna Techniques Department. There he supervised senior members of the technical staff, had responsibility for performance of contractual requirements, generated new ideas and techniques to be investigated, and kept track of equipment research and design.

It was with such a varied background of academic and industrial experience that Gangi joined M.I.T.'s Department of Geology and Geophysics as Associate Professor in June 1964. He immediately began to participate in the Department's instructional program, lecturing in Theoretical Seismology (12.723) and leading discussions in the geophysics seminar, and simultaneously organized a diversified program of geophysical research for himself and his thesis students consisting of the following projects:

1. Use of Dyadic Green's Theorems in Static and Dynamic Elasticity
2. Attenuation of elastic waves
3. Array processing of seismic data to enhance signal-to-noise ratio and determine particular characteristics of the data
4. Determination of the causes of station corrections in Large Aperture Seismic Array
5. Seismic modeling techniques for research and instruction
6. Investigation of the inverse problems of seismology

In addition to the foregoing research program, Gangi quickly established rapport with Lincoln Laboratory, and helped to initiate a cooperative program of seismological research. By December 1964 signals from the Tonto Forest Observatory were being received in Room 54-611 of the Green Building. Subsequently Gangi was instrumental in making further cooperative arrangements with staff members of Lincoln Laboratory, arrangements which made the unique facilities of that laboratory available to our advanced students.

BIBLIOGRAPHY OF ANTHONY FRANK GANGI

Symbols and abbreviations used in the following references are explained on p. 91 - 98. This bibliography begins with his first publication in 1957 and includes all known titles through June 1967, the time at which he left M.I.T.

1--1957   (with Knopoff, L., Fredericks, R. W. and Porter, L. D.) Surface amplitudes of reflected body waves. Geophysics 22: 842-847, (1957).

2--1959   (with Knopoff, L.) Seismic reciprocity. Geophysics 24: 681-691, (1959).

T--1960   Elastic waves in wedges. Ph.D. thesis in physics at U. Cal. Los Angeles, (1960).

3--1960   (with Knopoff, L.) Transmission and reflection of Rayleigh waves by wedges. Geophysics 25: 1203-1214, (1960).

4--1963   The Active Adaptive Antenna Array System. [Inst. Electrical and Electronic Engs.] On Antennas and Propagation, Tr. AP-11: 405-414, (1963).

5--1963a   (with Brown, G. L.) Electromagnetic modeling studies of lithospheric propagation. IEEE, PTGGE, Tr. GE-1: 17-23, (1963).

6--1965   (with Sensiper, S. and Dunn, G. R.) The characteristics of electrically short, umbrella top-loaded antennas. IEEE, PTGAP, Tr. AP-13: 864-871, (1965).

7--1966   Doppler tracking loops for the Active Adaptive Antenna System. IEEE, PTGAP, Tr. AP-14: 502-503, (1966).

----

(49)
WILLIAM CLAIR LUTH

WILLIAM CLAIR LUTH

MIT: 1964-1968

Trained as a geology undergraduate at the State University of Iowa (B.A. 1958; M.S. 1960), and at Penn State for the doctorate (Ph.D. 1963), Luth was appointed Assistant Professor of Geochemistry in our Department of Geology and Geophysics on 1 July 1964. During the first year of his appointment (SY 1964-65), however, he took a leave of absence to continue his appointment as Research Associate at Penn State in order to complete several research projects before coming to M.I.T. At Penn State he had worked closely with one of our distinguished alumni, O. F. Tuttle (XII Ph.D. 1948), in laboratory investigations of the pressure-temperature stability limits of several oxide systems bearing on the origin of granitic-type rocks, and it was the intent that he would establish an appropriate laboratory and continue such research at M.I.T. Coming to the Institute just as the Department was moving into the new Green Building, and being held up by the delays that typically accompany occupancy of new space, he barely got his laboratory established before he received and accepted an offer of an appointment as Associate Professor in Stanford University's Department of Geology. Once again an able and exceedingly promising young scientist started his academic career in our Department of Geology and Geophysics, then left after a short time to start on a distinguished career at a sister institution.

William Clair Luth* was born in Winterset, Iowa, on 28 June 1934, the son of William Henry and Ora (Klingaman) Luth of German ancestry. William was the first of two children; his sister Barbara joined the family three years later. His father was an independent painter and paper-hanger in his small-town birthplace, Winterset.

After attending the public schools there, and graduating from Winterset High School in 1952, he enrolled in the undergraduate program in photo-

---

* As a response to a number of my questions, Luth prepared a brief autobiographical sketch which I received in September 1974. At his suggestion I have drawn heavily on the sketch for details of his biography, either paraphrasing without complete quotation or quoting directly. I am happy to acknowledge his most helpful sketch and to follow his request that I not delete his complimentary remarks about the M.I.T. Department of Geology and Geophysics.

journalism at the State University of Iowa in the fall of 1952. After completing one semester, he enlisted in the U.S. Army, and during his period of service, which he describes as "two years, eleven months and one day," he spent most of his time as a topographic surveyor in Alaska and California. In February 1956 he resumed his studies at the State University of Iowa, intending to pursue a pre-law program; however, hearing a student rumor that the introductory course in Geology was the easiest science requirement, he registered in that course during the summer session of 1956, and quickly decided to change his major as the result of the interest aroused by one of his teachers. This teacher, the late Professor A. K. Miller, a distinguished paleontologist and international authority on fossil cephalopods, had the enthusiasm and ability to make geology a live and exciting subject for his students, and many like Luth decided to change their majors to some aspect of geology.

Having decided to pursue a geological career, Luth took the usual sequence of courses required for a bachelor's degree, and as he progressed he became particularly interested in geochemistry, mineralogy and petrology. He completed degree requirements by the end of the 1958 summer session, meanwhile having come under the influence of Prof. R. A. Hoppin, and the fall found him a graduate student working on a field and petrographic thesis under Hoppin's direction. During the later stages of his thesis research, which led to an M.S. degree in June 1960, he became interested in experimental studies, even though facilities for such research were not available at Iowa. Undeterred, however, and while finishing his thesis work, he wrote about his newly developed interest to O. F. Tuttle (XII Ph.D. 1948), then at Penn State and a leader in experimental petrology. The correspondence led to an offer of a research assistantship at Penn State with Tuttle, and after a summer of structural mapping in southwestern Colorado with the Shell Oil Company, Luth entered Penn State in the fall of 1960 to start what would become an enthusiastic and productive association with Tuttle and would lead to a doctorate in petrology in 1963.

There is an interesting story to tell here, of master teacher and motivated student, and I ask the reader's indulgence for a little academic detour before proceeding with Luth's career, because he himself is an important character in the story.

> When Luth was accepted as a doctoral candidate at Penn State, and granted a research assistantship to work under O. F. Tuttle's supervision, he could hardly have imagined how he would one day become another link in a chain of distinguished experimental petrologists, all of whom were products of M.I.T.
>
> Tuttle himself was a Penn State alumnus, as well as an M.I.T. doctorate, and had made an enviable name for

himself as a result of brilliant laboratory work at the Carnegie Institution's Geophysical Laboratory in Washington, D.C. during the decade from 1942 to 1952, at the end of which he returned to administrative and research duties at his alma mater, Penn State.

Tuttle had been introduced to optical mineralogy by Prof. A. P. Honess, a much admired lecturer at Penn State, and after completing the course in the middle of his junior year had decided he wanted to be a geologist and mineralogist. After completing his master's degree he came to M.I.T. in 1940 as an assistant to Prof. M. J. Buerger, and again came under the influence of a stimulating mineralogist and crystallographer. But Pearl Harbor interrupted activities in M.I.T.'s Geology Department, and Tuttle was assigned to work on a crystal-growing project with Prof. D. C. Stockbarger, who was trying to grow single large crystals of fluorite for aerial camera lenses. While working on this project, Tuttle learned of the potential usefulness of single crystals of willemite ($ZnSiO_4$), particularly long needles of the mineral, whereupon he borrowed a pressure vessel from Prof. W. H. Newhouse and proceeded to synthesize some willemite needles hydrothermally. When word of Tuttle's successful experiment got to George W. Morey and F. Earl Ingerson, who were interested in the same and related crystal-growing experiments at the Geophysical Laboratory in Washington, he was invited to join the two at that famous laboratory. Thus it happened that Tuttle arrived at the Geophysical Laboratory in November 1942, just a year after Pearl Harbor. While working with Morey and Ingerson, Tuttle became acquainted with N. L. Bowen (XII Ph.D. 1912), who was working on another project. This acquaintance quickly developed into a common interest in "the granite problem" and was a turning point in Tuttle's career, for by 1 January 1947, both Bowen and Tuttle were back at the Geophysical Laboratory, after short appointments elsewhere (Bowen at the University of Chicago and Tuttle at the Naval Research Laboratory).

Thus began that fruitful collaboration between master and apprentice that produced a series of fundamental papers that revolutionized the thinking on how granites are formed. When Bowen retired in 1952 and moved to Florida, Tuttle accepted a position at his alma mater, The Pennsylvania State University, meanwhile having received his M.I.T. doctorate in 1948, and there initiated a new program of laboratory research that won him a place among the world leaders in experimental petrology. Recognition came when he received the first Mineralogical Society of America Award in 1951, the Day Medal of the Geological Society of America in 1967, and election to the National Academy of Sciences in 1968.

Now comes the point of the story just told and how it fits into an interesting chain of circumstances. Bowen had been sent to the Geophysical Laboratory, by his M.I.T. Profs. Noyes (chemistry), Warren, and Lindgren, when as a doctoral candidate he declared his interest in making a phase-equilibrium study of the silicate system nepheline-anorthite. This study, on which his Ph.D. thesis at M.I.T. was based, ["The binary system: $Na_2Al_2Si_2O_8$ (Nephelite-Carnegieite)-$Ca_2Al_2Si_2O_8$ (Anorthite)"], was supervised at the Laboratory by E. S. Shepherd and F. E. Wright, and led to Bowen's joining the Laboratory staff immediately after receiving his doctorate in 1912.

Although Tuttle's doctoral thesis at M.I.T. was on the "Structural petrology of planes of liquid inclusions," for which he received his Ph.D. degree in 1948, he had become deeply interested in Bowen's research on "the granite problem" a year earlier and proceeded to design the needed equipment to operate in the pressure range up to 4000 bars and at temperatures up to 900°C.

As mentioned earlier, by the time he arrived at Penn State in 1952, after a decade of intermittent employment at the Geophysical Laboratory, Tuttle was recognized as a leader among the younger experimental petrologists. As he continued his research on granite genesis and related problems at Penn State, his work attracted Luth, the Iowa graduate student looking for a doctoral thesis problem. So once again came into existence a collaborative effort between master and apprentice, and soon (1963) the first jointly authored papers were published. It was the outstanding ability at research shown by Luth that brought Tuttle's strong recommendation that we offer him an appointment at M.I.T., which we did. Thus ends the story which started with Noyes, Warren, and Lindgren at M.I.T., continued with Bowen and then Tuttle at the Geophysical Laboratory, changed to Tuttle and Luth at Penn State, and then jumped to Stanford where Luth again became a colleague of Tuttle and Jahns with whom he coauthored several of his earlier papers.

Now (1974), eight years since he left M.I.T., Luth has more than lived up to the early expectations of his teachers and peers, and there is every indication that he is already stimulating younger geologists to become the fourth generation link in the Bowen-Tuttle-Luth chain.

Let us now return to discussion of Luth's education and preparation for the career that lay ahead of him when he resumed his studies after his military service.

While still in the Army, Luth married Betty Lou Heubrock on 23 August 1953, and like so many young married couples in service, they had a rather hectic time of it before they could settle down, at least for a year or two back at Iowa City. Betty recalls that during the first two and a half years of their married life they were together a total of six months and lived in <u>seven</u> different apartments in San Francisco. Their first daughter, Linda Diane, was born on 19 December 1955 in San Francisco. Back in Iowa, Betty supplemented their meager income from the "G.I. Bill" by working as a secretary while Bill worked outside school hours part-time as a meat-cutter, and for a short period as a door-to-door salesman of pots and pans. Son William Robert was born in July 1957, and Betty suspended her secretarial work for a few months, but thereafter returned to her regular work and continued it until they left Iowa for Pennsylvania in the fall of 1960.

Luth first met Tuttle in the fall of 1960, when he arrived at Penn State to begin his duties as a research assistant, and before the day was over enthusiastic professor and eager student were in the laboratory starting an experiment. Luth's course work brought him into direct contact with a group of outstanding teachers--Jahns, Wyllie, Roy, Barnes, Burnham and Gibbs, and with an equally promising group of fellow graduate students and post-doctoral researchers. While pursuing his work toward the doctorate, Luth attended the annual meeting of the Geological Society of America in Cincinnati in November 1961 and while there was introduced to me by Tuttle. We were planning to recruit a younger staff member to work in the general area of experimental petrology, and I had suggested to Tuttle that I should like to meet any of his outstanding graduate students who might be candidates. That is how it came about that Luth and I met, and it might be of interest to mention how he felt about this meeting (paraphrasing his recent letter to me):

> Professor Tuttle introduced me to Prof. Shrock, then Chairman of the Department at M.I.T., who to my surprise exhibited considerable interest in me as a potential future faculty member at M.I.T. Since M.I.T. represented the mecca of the scientific world and I was a mere untested graduate student, my euphoria was understandable, though quite immature.
>
> At any rate, I continued both the experimental research and academic work at Penn State, which led to a doctoral degree in June 1963. Meanwhile, Tuttle and I had submitted a research proposal to the National Science Foundation to investigate the two systems: $Na_2O-Al_2O_3-SiO_2-H_2O$, and $K_2O-Al_2O_3-SiO_2-H_2O$, as soon as I completed my doctoral thesis. With favorable action by the NSF, and completion of all requirements for my doctorate, awarded in June 1963, I started the proposed research as a research associate in July 1963. [Luth had been a National Science Foundation Fellow during SY 1962-63.]
>
> Of course this was very much a temporary position, and I began to think seriously about an academic position. To my great surprise, Prof. Shrock again exhibited interest in me and my work and invited me to present a lecture at M.I.T. As a result of the visit to Cambridge, I became very enthusiastic about both the faculty and the students in the Geology Department, and particularly noticed the strong element of cooperation between M.I.T. and Harvard. Shortly thereafter I accepted an appointment as Assistant Professor of Geochemistry at M.I.T., but was granted a one-year leave of absence to complete my post-doctoral research with Tuttle.

Again departing from discussion of Luth's academic progress for a moment, let us mention his family. Betty had continued working as a secretary during Bill's years at Penn State, until their third child, Sharon Jean, was born in January 1965, after which she worked only part-time until the Luth family moved to the Boston area in the fall of 1965.

Luth's arrival at M.I.T. came at a critical time in the history of the Geology Department. Great changes had taken place and more were to come. The 20-story Green Building had been dedicated on 2 October 1964; Professors Dean, Gangi, Simmons, Strangway, and Toksöz had been added to the staff; I had asked to be relieved of the Departmental Chairmanship after sixteen years in the position; and Press had come to head the Department in September 1965. Space in the new building had been reserved for Luth, and a group of eager and interested students were awaiting his first lectures.

His first year at M.I.T., SY 1965-66, was a busy and stimulating one. He spent much time preparing for publication the research he had just finished at Penn State. He set up an experimental laboratory with the aid of a grant from the National Science Foundation. And he quickly became acquainted with and respected by his colleagues at M.I.T. and Harvard and the graduate students at both institutions.

Full-scale teaching responsibilities began in the second year, SY 1966-67, and he was granted an Alfred P. Sloan Fellowship in recognition of his published research and future promise. His first regularly scheduled contact with the students brought him to the realization that he was going to be tested vigorously as a young professor. In his own words, as he recently wrote me:

> "I quickly became aware of the outstanding quality of the M.I.T. undergraduates; they provided an extremely critical and knowledgeable group which continued to force my development. It was quite clear to me that the M.I.T.-Harvard axis provided the highest degree of intellectual stimulation to which I have been exposed, and probably unique in the geological fraternity."

During his short stay at M.I.T., Luth maintained close contacts with two of his former Penn State professors, Jahns and Tuttle, who had moved to Stanford in the fall of 1965, and actually spent the summer of 1966 conducting research with them at Stanford. He also spent a part of the spring term of 1967 engaged in experimental studies at the Geophysical Laboratory in Washington. There he worked with Boyd, Bell and Lindslye in a pressure regime unavailable in his own laboratory at M.I.T. As indicated in his Bibliography a little farther on, numerous important articles resulted from these different research projects.

As SY 1967-68 got underway, Luth's future at M.I.T. seemed bright and full of promise but, unfortunately for the Institute and our Department of Geology, it would be his last year, for Jahns and Tuttle would attract him away to Stanford. It happened this way, as Luth wrote me in September 1974. In 1968, Tuttle's physical condition had deteriorated to such a degree, due to Parkinsonism, that he was unable to continue at the level of activity he wished in the academic community, and he requested medical leave. In doing so he recommended Luth as the person to continue his work at Stanford, and after due consideration Luth was offered and accepted an appointment as Associate Professor in the Department of Geology. He accepted the offer, not because of any dissatisfaction with M.I.T., but rather because, as he wrote me:

> "Frank Tuttle was my scientific godfather, so to speak, and if he had thought it desirable for me to take a position at East Podunk University I probably would have done so."

Luth went on to write in his letter to me that

> "My brief stay at M.I.T. was one of the best things that ever happened to me; the chance to become acquainted with a truly great group of undergraduate and graduate students was unique. Interaction with an outstanding faculty where I was able to work closely with people such as Brace, Wones, Frey, Hurley, Buerger, Fairbairn and many other outstanding individuals resulted in a significant change toward the quantitative approach. This was supplemented by interaction with Jim Thompson and Dave Waldbaum at Harvard. I suspect that my professional growth during the stay at M.I.T. was probably even greater than during my graduate and postgraduate studies at Penn State."

Immediately after resigning from the M.I.T. faculty, effective on 30 June 1968, Luth went to Stanford and spent the summer doing experimental research in the laboratories that had been established there by Jahns and Tuttle. He also became immediately responsible for supervision of the thesis research of four doctoral candidates. During the four years since 1968 he had supervised the doctoral theses of four Course XII alumni--Richard D. Warner (XII S.B. 1966), James A. Whitney (XII S.B. 1968), Phillip M. Fenn (XII S.B. 1968), and John C. Eichelberger (XII S.B. 1970). He writes that two recent doctorates from Course XII, Jon Claerbout (XII Ph.D. 1967) and Amos Nur (XII Ph.D. 1969) are now faculty members in Stanford's Department of Geophysics; that Frank Press is on the Advisory Board for the School of Earth Science; and that Theodore Madden and William Brace, both Course XII doctorates, have been recent Visiting Professors in the Department of Geophysics.

Quite clearly, Luth gained much from his brief period of service in our Department of Geology and Geophysics, and just as clearly he is living up to the expectations of his peers and older colleagues as he steadily develops a distinguished career in experimental petrology at Stanford. He is another of those well-trained, well-disciplined and highly devoted young scientists who start on their academic career at the Institute, and then go elsewhere, after a few years, to achieve distinction in their special field. Although M.I.T. suffers a loss when such young scientists leave to seek a career elsewhere, at the same time it fulfills one of the primary educational functions—*i.e.* the training of young scientists for a successful career, wherever they ultimately find it.

BIBLIOGRAPHY OF WILLIAM CLAIR LUTH

Symbols and abbreviations used in the following references are explained on p. 91 - 98; in general, abstracts are listed separately as well as with the complete article. This bibliography includes all of Luth's publications through 1969, the year after he left M.I.T.

T--1960   Mafic and ultramafic rocks of the Trailside Area, Bighorn Mountains, Wyoming. M.S. Thesis in Department of Geology, The State University of Iowa, Iowa City, Iowa, (1960).

T_1--1963   The system $KAlSiO_4$-Mg_2SiO_4-SiO_2-H_2O from 500 to 3000 bars and 800° to 1200°C and its petrologic significance. Ph.D. thesis in Department of Geochemistry and Mineralogy, The Pennsylvania State University, University Park, Pennsylvania, (1963). (See also 7--1964d.)

1--1963a   (and Jahns, R. H. and Tuttle, O. F.). The 'Granite System' to 10,000 bars P_{water} [abst.]. Am. Geophys Union, Tr. 44/1: 118, (1963).

2--1963b   (and Tuttle, O. F.). Externally heated cold-seal pressure vessels for use to 10,000 bars and 750°C. Am. Mineral. 48: 1401-1403, il., (1963).

3--1964   System $KAlSiO_4$-Mg_2SiO_4-SiO_2-H_2O Pressure-temperature stability limits [abst.]. Geol. Soc. Am., Sp. P. 76: 106, (1964).

4--1964a   (with Tuttle, O. F. and Jahns, R. H.). The hypersolvus granite--granophyre--rhyolite association [abst.]. Am. Geophys. Union, Tr. 45: 124, (1964).

5--1964b   Invariant and univariant equilibria in the system $KAlSiO_4$-Mg_2SiO_4-SiO_2-H_2O [abst.]. Am. Geophys. Union, Tr. 45: 125-126, (1964).

6--1964c   (and Jahns, R. H. and Tuttle, O. F.). The granite system at pressures of 4 to 10 kilobars. J. Geophys. Res. 69: 759-773, (1964).

7--1964d   The system $KAlSiO_4$-Mg_2SiO_4-SiO_2-H_2O from 500 to 3000 bars and 800° to 1200°C and its petrologic significance [abst.]. Dissertation Abst. 24/8: 3353, (1964). [Ph.D. Thesis at The Pennsylvania State University, 1964.]

8--1965   (and Tuttle, O. F.). The effects of excess Al_2O_3 and excess alkali silicate on the alkali feldspar solvus [abst.]. Am. Geophys. Union, Tr. 46: 179, (1965).

9--1965a   (with Peters, Tj. and Tuttle, O. F.). The melting of analcite solid solutions in the system $NaAlSiO_4$-$NaAlSi_3O_8$-H_2O [abst.]. Am. Geophys. Union, Tr. 46: 179, (1965).

10--1965b   (and Ingamells, C. O.). Gel preparation of starting materials for hydrothermal experimentation. Am. Mineral. 50: 255-258, tab., (1965).

11--1966   (with Scarfe, C. M. and Tuttle, O. F.). An experimental study bearing on the absence of leucite in plutonic rocks. Am. Mineral. 51: 726-735, il., tab., (1966).

12--1966a   (with Peters, Tj. and Tuttle, O. F.). The melting of analcite solid solutions in the system $NaAlSiO_4$-$NaAlSi_3O_8$-H_2O. Am. Mineral. 51: 736-753, il., tab., (1966).

13--1966b   (and Tuttle, O. F.). The alkali feldspar solvus in the system $Na_2O-K_2O-Al_2O_3-SiO_2-H_2O$. Am. Mineral. 51: 1359-1373, il., tab., (1966).

14--1967   Studies in the system $KAlSiO_4-Mg_2SiO_4-SiO_2-H_2O$ --[Pt.] 1, Inferred phase relations and petrologic applications. J. Petrology 8: 372-416, il., tab., (1967).

15--1967a   The system $KAlSiO_4-Mg_2SiO_4-KAlSi_2O_6$. Am. Ceramic Soc. J. 50: 174-176, il., (1967).

16--1967b   (and Tuttle, O. F.). The hydrous phase in equilibrium with granite and granite magmas [abst.]. Am. Geophys. Union, Tr. 48: 245, (1967).

17--1968 P_{H_2O} and P_{total} in water undersaturated granitic liquids [abst.]. Am. Geophys. Union, Tr. 49: 331, (1968).

18--1968a   The influence of pressure on the composition of eutectic liquids in the binary systems sanadine--silica and albite--silica. Ann. Rept. Director Geophys. Lab., Carnegie Institution of Washington, Yearbook 66: 480-484, (1968).

19--1969   (with Querol-Sune, F.). Composition-unit cell parameter relations for the alkali feldspars [abst.]. Am. Geophys. Union, EOS 50: 350, (1969).

20--1969a   (with Fenn, P. M.). Three-phase perthites of the Mt. Doherty, Montana, stock [abst.]. Am. Geophys. Union, EOS 50: 351, (1969).

21--1969b   The systems $NaAlSi_3O_8-SiO_2$ and $KAlSi_3O_8-SiO_2$ to 20 kb and the relationships between H_2O content, P_{water}, and P_{total} in granitic magmas. Am. J. Sci. 267-A (Schairer Vol.): 325-341, il., tab., (1969).

22--1969c   (with Walters, L. J., Jr.). Unit-cell dimensions, optical properties, and halogen concentrations in several natural apatites. Am. Mineral. 54: 156-162, il., tab., (1969).

23--1969d   (and Tuttle, O. F.). The hydrous vapor phase in equilibrium with granite and granite magmas, in Igneous and Metamorphic Geology--A volume in honor of Arie Poldervaart. Geol. Soc. Am., Mem. 115: 513-548, il., tab., (1969).

24--1969e   (and Simmons, G.). Melting relations in natural anorthosite, in Origin of anorthosite and related rocks. N.Y. State Mus. Sci. Service Mem. 18: 31-37, il., tab., (1968) [1969].

----

(50)
DAVID WILLIAM STRANGWAY

DAVID WILLIAM STRANGWAY

MIT: 1965-1968

Experienced in mining geophysics in particular and exploration geophysics more generally, Strangway brought to the Department special expertise in geomagnetism and the magnetic properties of rocks. Besides his regular academic duties he acted as a consultant to Bear Creek Mining Co., a subsidiary of Kennecott Copper Corp., thus making a useful coupling of academic work with practical industrial problems for both himself and his students. Although a member of our Department for only three and a half years, he quickly established a productive research program and organized much-needed lectures on rock magnetism (1966-1968), regional geophysics (1966-1968) and geoelectricity and geomagnetism (1967-1968). After serving two two-year appointments at M.I.T. as Assistant Professor of Geophysics (1 February 1965 through 31 August 1968), Strangway resigned to accept an appointment as Associate Professor of physics at his alma mater, the University of Toronto.

David William Strangway was born on 7 June 1934 in Simcoe, Ontario, the second of three sons of Walter Earl Strangway and Alice Kathleen (Skinner) Strangway, both Canadians by birth. His older brother, Paul, born in 1931, died at the age of two; his younger brother, Donald Walter, was born in 1936. Both parents are college educated; his father is a medical doctor and his mother has a Master's degree in food chemistry.

After one year of primary education in the Simcoe schools, young David accompanied his parents to Angola, Portuguese West Africa, where his father accepted a post as a medical missionary. He completed grades 2 through 8 in a special school for the children of missionaries in Angola, then returned to Canada in 1948 to enter high school in Toronto. After a year and a half he returned to his parents' home in Africa, and for the next two and one-half years attended the Milton School in Bulawayo, Southern Rhodesia, graduating in 1952. After completing high school there, he once again returned to his homeland and entered the University of Toronto in September 1952. Four years later he earned a B.A. degree in physics and geology, then entered graduate work in the Department of Physics. He was awarded an M.A. degree in geophysics in 1958, with a thesis entitled:

"Theoretical design and calibration of an apparatus to measure the variation of the susceptibility of rocks with temperature,"

and a Ph.D. in geophysics in 1960, with a thesis on

"Magnetic properties of some Canadian diabase dikes."

While a graduate student he held the Garnet McKee-Lachlan Gilchrist Scholarship for two years, and was the recipient of a National Research Council of Canada summer fellowship for the summer of 1959.

Once started on his college education, Strangway used the summers to get field experience in geology and geophysics. During the summers of 1953 to 1958 he worked in succession for the Ontario Department of Highways (surveying--1953), Holannah Mines Ltd. (magnetometer work in New Quebec and Labrador--1954), Shell Oil Co. (seismic work in Alberta--1955), Dominion Gulf Co. (mining geophysics in Ontario and Quebec), and Ventures Ltd. (in charge of mining geophysics in 1956-1957, and airborne electromagnetic work in 1958, assignments which took him to several areas in eastern Canada as well as to Greenland).

Immediately after receiving his Ph.D. degree he joined the Kennecott Copper Corp. as a research geophysicist, working full-time for about a year, 1960-1961, and as a consultant since then to Kennecott's subsidiary, Bear Creek Mining Co. Work for Bear Creek involved electromagnetic scale modelling, remote sensing, and electrical techniques, particularly as these are applied in mining geophysics.

In 1961 he joined the faculty of the University of Colorado, as an Assistant Professor of Geology, and for the next four years, from 1961 through 1964, taught physical sciences and geophysics. On 1 February 1965 he joined M.I.T.'s Department of Geology and Geophysics as an Assistant Professor of Geophysics, and on 31 August 1968 resigned to accept an appointment as Associate Professor of Physics at his alma mater, the University of Toronto.

On 22 September 1957, David married Alice Norine Gow, who was born in Fergus, Ontario and graduated in history at the University of Toronto. Son Richard Paul came along on 5 November 1959, daughter Susan Kathleen followed on 21 September 1961, and daughter Patricia Ruth on 26 August 1967.

Strangway brought to M.I.T.'s Department of Geology and Geophysics a broad and varied experience in geophysical exploration, especially involving the geophysics that is applicable to mining, and a special research interest in the magnetic properties of rocks. He quickly attracted a number of students to his lecture subjects and laboratory research, and had an active and productive research program on remanent magnetism underway when he left M.I.T.

While still at the University of Colorado, and later at M.I.T., Strangway joined geophysicists from the United States and Japan in a research program involving rock magnetism, archeomagnetism, paleomagnetism, and geomagnetism of the Pacific area. At a meeting of the research group, held 27-29 October 1966 in Kyoto, Japan, Strangway's M.I.T. group, in discussing "the fundamental mechanisms by which rocks become stably magnetized", demonstrated

> "... that the extreme stability of the thermoremanent magnetization in volcanic rocks may be due to the formation in large titano-magnetite grains of single domains of magnetite separated by ilmenite lamellas; ..."

Allan Cox and Naoto Kawai, in their report on the Kyoto Conference ("Paleomagnetism: United States--Japan Committee on Scientific Cooperation", Science 155: 724 [1969]), from which the preceding statement was quoted, go on to state that

> "One of the main objectives of the program was to determine a radiometric time scale for reversals of the geomagnetic field. This has now been done for the interval back to 4 million years ago. During this time there have been four broad epochs of alternating polarity, as well as four much briefer polarity fluctuations termed events. The fourth event was identified on the basis of new data presented at Kyoto by the U.S. Geological Survey group and the Tokyo-M.I.T.-Colorado group. The two sets of results were complementary, each group having identified one of the boundaries of the same short polarity event occurring about 3.8 million years ago."

Strangway early demonstrated his research ability, publishing his first paper on Magnetic properties of diabase dikes, the subject of his Ph.D. thesis, in 1961, scarcely a year after gaining his degree, and by 1965, when he joined the M.I.T. faculty, he had five substantial papers and two abstracts to his credit. Five years later it could be written in Geophysics 35/4: 736 (1970) that

> "He is the author or coauthor of 40 papers and a monograph on the history of the earth's magnetic field. He is a principal investigator on returned lunar samples and has acted as a consultant to NASA [National Aeronautics and Space Agency] and the UN [United Nations]."

As commonly happens at M.I.T., a young scientist of much promise joined our faculty, quickly made a place for himself as a fine teacher and a reputation for high quality research; then went on to what he regarded as an even more challenging appointment. M.I.T.'s loss was Toronto's gain, but all who knew David Strangway at the Institute wish him every success in his new position.

BIBLIOGRAPHY OF DAVID WILLIAM STRANGWAY

Symbols and abbreviations used in the following references are explained on p. 91-98; in general, abstracts are listed separately as well as with the references to the complete article. This bibliography begins with Strangway's first publication in 1961, based on his doctoral thesis, and includes all known titles through 1969, some sixteen months after he left M.I.T.

T--1961   Magnetic properties of diabase dikes. J. Geophys. Res. 66/9: 3021-3032, il., tab., (1961). [Ph.D. thesis, Univ. Toronto, 1960.]

1--1961   (with Gross, W. H.) Remanent magnetism and the origin of hard hematites in Precambrian banded iron formation. Econ. Geol. 56/8: 1345-1362, (1961).

2--1962   Rock magnetism and dike classification [abst.]. J. Geophys. Res. 67/9: 3601, (1962).

3--1964   Rock magnetism and dike classification. J. Geol. 72/5: 648-663, il., (1964).

4--1965   (and Holmer, R. C.) Infrared geology, in Symposium on remote sensing of environment, 3rd, 1964, Pr--U.S. Office Naval Research and Air Force Cambridge Research Labs., Ann Arbor, Mich., Univ. Michigan, Willow Run Labs: 293-319, il., tab., (1965).

5--1965a   The interpretation of the magnetic anomalies over some Precambrian dikes. Geophysics 30/5: 783-796, il., (1965).

6--1965b   (and Larson, E. E.) A paleomagnetic study of some late Cenozoic basalts from Oregon [abst.]. Am. Geophys. Union, Tr. 46/1: 66-67, (1965).

7--1966   (and Holmer, R. C.) The search for ore deposits using thermal radiation. Geophysics 31: 225-242, (1966).

8--1966a   Electromagnetic scale modeling, in Methods and Techniques in Geophysics 2: 1-31, (1968), edited by S. K. Runcorn. New York, N.Y.: John Wiley and Sons, Inc., (1966).

9--1966b   (with Larson, E. E.) Magnetic polarity and igneous petrology. Nature 212: 750-757, (1966).

10--1966c Rock magnetism and geologic correlation. Min. Geophys. 1: 54-66, (1966).

11--1966d Electromagnetic parameters of some sulphide ore bodies. Min. Geophys. 1: 227-242, (1966).

12--1966e (with Gross, W. H.) Remanent magnetism and the origin of hard hematites in Precambrian banded iron formation. Min. Geophys. 2: 366-378, (1966). [Reprinted from Econ. Geol. 56/8: 1345-1362, (1961). See item 2 above.]

13--1967   (with Cox, A. and Kawai, N.) Paleomagnetism: United States--Japan Committee on Scientific Cooperation. Science 155: 724, (1967).

14--1967a (and McMahon, B. E. and Honea, R. M.) Stable magnetic remanence in antiferromagnetic goethite. Science 155: 785-787, (1967).

15--1967b Mineral magnetism. Min. Geophys. 2: 437-445, (1967).

16--1967c Magnetic characteristics of rocks. Min. Geophys. 2: 454-473, (1967).

17--1967d Field tests for stability, in Methods in paleomagnetism, edited by Collinson, D. W., Creer, K. M., and Runcorn, S. K.--Developments in Solid Earth Geophysics 3: 209-216. Amsterdam: Elsevier, (1967).

18--1967e (with McMahon, B. E.) Kiaman magnetic interval in the western United States. Science 155: 1012-1013, (1967).

19--1967f (with Ozima, M., Kono, M., Kaneoka, I., Kinoshita, H., Kobayashi, K., Nagata, I., and Larson, E. E.) Paleomagnetism and potassium-argon ages of some volcanic rocks from the Rio Grande Gorge, New Mexico. J. Geophys. Res. 72: 2615-2621, (1967).

20--1967g (with Kono, M., Kobayashi, K., Ozima, M., Kinoshita, H., Nagata, T., and Larson, E. E.) Paleomagnetism of Pliocene basalts from the southwestern U.S.A. J. Geomag. Geoelect. 19: 357-375, (1967).

21--1967h (and McMahon, B. E., Honea, R., and Larson, E. E.) Superparamagnetism in hematite. Earth and Planetary Science Letters 2: 367-371, (1967).

22--1968 (with McMahon, B. E.) Investigation of the Kiaman magnetic division in Colorado redbeds. Geophys. J. 15: 265-285, (1968).

23--1968a (with McMahon, B. E.) Stratigraphic implications of paleomagnetic data from Upper Paleozoic-Lower Triassic redbeds of Colorado. Geol. Soc. Am., B. 79: 417-428, (1968).

24--1968b (with England, A. W., and Simmons, G.) Electrical conductivity of the moon. J. Geophys. Res. 73: 3219-3226, (1968).

25--1968c (and Larson, E. E., and Goldstein, M.) A possible cause of high magnetic stability in volcanic rocks. J. Geophys. Res. 73: 3787-3795, (1968).

26--1968d The geophysicist in the mining industry. Geoexploration 6: 65-67, (1968).

27--1968e (and Honea, R. M., McMahon, B. E., and Larson, E. E.) The magnetic properties of naturally occurring goethite. Geophys. J. 15: 345-359, (1968).

28--1968f (with Larson, E. E.) Discussion of "Correlation of petrology and natural polarity in Columbia Plateau basalts" by L. R. Wilson and N. D. Watkins. Geophys. J. 15: 437-441, (1968).

29--1968g (and McMahon, B. E., and Larson, E. E.) Magnetic paleointensity studies on a recent basalt from Flagstaff, Arizona. J. Geophys. Res. 73: 7031-7037, (1968).

30--1969 (with Larson, E. E.) Magnetization of the Spanish Peaks dike swarm, Colorado, and Shiprock Dike, New Mexico. J. Geophys. Res. 74: 1505-1514, (1969).

31--1969a (with Larson, E. E., Ozima, M., and Nagata, T.) Stability of remanent magnetization of igneous rocks. Geophys. J. 17: 263-292, (1969).

32--1969b Moon: Electrical properties of the uppermost layers. Science 165: 1012-1013, (1969).

33--1969c (with Goldstein, M. A., and Larson, E. E.) Paleomagnetism of a Miocene transition zone in southeastern Oregon. Earth and Planetary Science Letters 7: 231-239, (1969).

34--1969d (and McMahon, B. E., and Bischoff, J. L.) Magnetic properties of minerals from the Red Sea thermal brines, in Hot Brines and Heavy Metal Deposits in the Red Sea, edited by Degans and Ross. Springer-Verlag, p. 460-473, (1969).

(51)
M. NAFI TOKSÖZ

M. NAFI TOKSÖZ

MIT: 1965-

As M.I.T. reached the end of its first century, and geology, as a course of study, its 75th year at the Institute, the program of instruction and research in geophysics was in a transition period. The year 1964 was marked by dedication of the 20-story Green Building in which it was hoped there would develop a great center for study of the earth sciences. Geophysics was to be one of the more important of these. But in 1964 three of the promising young geophysicists resigned--Cantwell, Hughes, and Simpson--and their replacements had to be recruited. Word came that a recent doctorate at Caltech was highly regarded by those familiar with him and his work, and on the basis of this information, M. Nafi Toksöz was offered a position as Assistant Professor of Geophysics, starting in January 1965. He accepted the offer and brought much needed strength in seismology. Soon he was deeply involved in teaching the introductory aspects of that subject, and participating in others, as the new chairman, Dr. Frank Press, began to build what would soon become one of the leading groups of geophysicists to be found anywhere. Toksöz quickly proved to be an excellent teacher, a highly productive investigator in seismology and solid earth geophysics, and a professor who attracted some of the best graduate students in the Department. As a result he was promoted to Associate Professor in July 1967 and to Professor in 1971. With forty publications to his credit by the end of 1970, and seventy-five as this is being written (May 1974), he has more than lived up to our earliest expectations. In addition, he and his wife are currently serving as Senior Faculty Residents at Baker House where their presence and counsel are much appreciated by the students living there.

M. Nafi Toksöz was born in Antakya, Turkey on 18 April 1934. He was the first of eight children born to Mahmut and Nazime Toksöz, farmer and farmer's wife, who had a hard time winning enough from the soil to maintain even a subsistence level for their active and growing children--four boys and four girls. It was clear to both parents and children that their life of poverty could only be improved by education, which hopefully would open the doors to employment opportunities that would break their bondage to the land. Today all of the children either have college degrees or are in college and are demonstrating what education can mean when there is recognition of its value and the will and persistence to meet its demands.

In the primary and secondary public schools of Antakya, young Nafi excelled in the mathematics and sciences that were required, and did so well in his school work that he was singled out for special study abroad. He was awarded a Turkish Government Fellowship to prepare himself for a career in exploration geophysics, with the expectation, of course, that he would return home when his training was completed. The fall of 1954 found him registered at the Colorado School of Mines and starting on the program of study that would lead to an S.B. degree in geophysics in May 1958. Desiring to extend his training and to change the direction of his studies somewhat, and still enjoying the support of the Turkish Government Fellowship, he entered the California Institute of Technology in the fall of 1958 for graduate study. During the next two years he came to realize that his real interests were much broader than exploration geophysics, which called for the practical application of his knowledge and training and the development of the art of exploration. Instead, he became deeply interested in the more fundamental aspects of seismic waves, especially as they were related to earthquakes. By the time he received his M.S. degree in June 1960, he had definitely decided on his future course, no doubt influenced by the mathematical and physical challenges of more theoretical work. He would turn away from the practical and field-oriented geophysics and take up what he considered the more interesting and challenging problems involving seismology, and seismic wave propagation through the earth. Accordingly, he relinquished the Government fellowship he had held for six years, from 1954 to 1960, accepted a research assistantship, and registered for more graduate work at Caltech that led to a Ph.D. degree in geophysics in June 1963. His doctoral thesis involved "Velocities of long-period waves and microseisms and their use in structural studies." Then following the common practice in the 1960s, he became a Postdoctoral Research Fellow (in geophysics) at Caltech, and for the next two years, 1963 and 1964, he carried on the research that led to some of his first publications (see Bibliography).

Now came an offer of a position back home that would require him to conduct geophysical exploration for the Government. He declined this offer and decided to follow an academic career in the United States. It was at this time, in 1964, that we were looking for promising recent doctors in geophysics to fill the vacancies left by the resignation of Profs. Cantwell, Hughes, and Simpson. Happily for M.I.T. and our Department of Geology and Geophysics, as it was then called, Toksöz accepted the invitation to join our faculty as Assistant Professor of Geophysics on 16 January 1965. He quickly established himself in our Department, and when Frank Press became Head of Course XII in September 1965, Nafi became the first member of the group of brilliant young geophysicists who would later bring worldwide distinction to our Department under Press' able leadership. He was advanced to Associate Professor in July 1967 and to Professor in 1971. His teaching and other activities in the Department, his services to the Institute in other ways, and his impressive list of publications show that he has more than met advanced expectations.

In 1969 Toksöz was married to Helena Terzian, a chemistry graduate from the University of Massachusetts at Amherst, who was then teaching in the Avon (Massachusetts) school system. As this is being written (1974) the Toksözs are living in Baker House, where they are serving as Senior Faculty Residents. Meanwhile, Helena is back in school, but taking courses instead of teaching.

BIBLIOGRAPHY OF M. NAFI TOKSÖZ

Symbols and abbreviations used in the following references are explained on pages 91 - 98. This bibliography begins with Toksöz's first publication in 1961 and includes all his papers through 1970. Titles after 1970 are listed in his curriculum vitae filed in Course XII Headquarters, Room 54-912 M.I.T.

1--1961 (with Press, F. and Ben-Menahem, A.) Experimental determination of earthquake fault length and rupture velocity. J. Geophys. Res. 66: 3471-3485, (1961).

2--1962 (with Ben-Menahem, A.) Source-mechanism from spectra of long-period seismic surface-waves: The Mongolian earthquake of December 4, 1957. J. Geophys. Res. 67: 1943-1955, (1962).

T--1963 Velocities of long-period surface waves and microseisms and their use in structural studies, 173 p., (1963). (Ph.D. thesis at California Institute of Technology in Geophysics, June 1963.)

3--1963a (and Anderson, D. L.) Generalized two-dimensional model seismology with application to anisotropic earth models. J. Geophys. Res. 68: 1121-1130, (1963).

4--1963b (and Ben-Menahem, A.) Velocities of mantle Love and Rayleigh waves over multiple paths. Seism. Soc. Am., B. 53: 741-764, (1963).

5--1963c (with Anderson, D. L.) Surface waves on a spherical earth: 1. Upper mantle structure from Love waves. J. Geophys. Res. 68: 3483-3500, (1963).

6--1963d (with Ben-Menahem, A.) Source mechanism from spectrums of long-period surface waves: 2. The Kamchatka earthquake of November 4, 1952. J. Geophys. Res. 68: 5207-5222, (1963).

7--1963e (with Ben-Menahem, A.) Source-mechanism from spectra of long-period seismic surface waves: 3. The Alaska earthquake of July 10, 1958. Seism. Soc. Am., B. 53: 905-919, (1963).

8--1963f (with Anderson, D. L. and Kovach, R. L.) Upper mantle structure from long period surface waves. Int. Union Geol. Geophys., XIII General Assembly, Berkeley, (1963).

9--1964 Microseisms and an attempted application to exploration. Geophysics 29: 154-177, (1964).

10--1964a   (and Schwab, F.) Bonding of two-dimensional seismic modeling. Geophysics 29: 405-413, (1964).

11--1964b   (and Ben-Menahem, A.) Excitation of seismic surface waves by atmospheric nuclear explosions. J. Geophys. Res. 69: 1639-1648, (1964).

12--1964c   (and Ben-Menahem, A. and Harkrider, D. G.) Determination of source parameters of explosions and earthquakes by amplitude equalization of seismic surface waves: 1. Underground nuclear explosions. J. Geophys. Res. 69: 4355-4366, (1964).

13--1965   (and Harkrider, D. G. and Ben-Menahem, A.) Determination of source parameters by amplitude equalization of seismic surface waves: 2. Release of tectonic strain by underground nuclear explosions and mechanisms of earthquakes. J. Geophys. Res. 70: 907-922, (1965).

14--1965a   (with Dewart, G.) Crustal structure in East Antarctica from surface wave dispersion. Geophys. J. Roy. Astron. Soc. 10: 127-139, (1965).

15--1966   (and Anderson, D. L.) Phase velocities of long period surface waves and structure of the upper mantle: 1. Great-circle Love and Rayleigh wave data. J. Geophys. Res. 71: 1649-1658, (1966).

16--1967   (with Chinnery, M. A.) P-wave velocities in the mantle below 700 km. Seism. Soc. Am., B. 57: 199-226, (1967).

17--1967a   (and Chinnery, M. A. and Anderson, D. L.) Inhomogeneities in the earth's mantle. Geophys. J. Roy. Astron. Soc. 13: 31-59, (1967).

18--1967b   Radiation of seismic waves from underground explosions. Pr. VESIAC Conference on the Current Status and Future Prognosis of Shallow Seismic Events. VESIAC report, Willow Run Laboratories, University of Michigan, p. 65-84, (1967).

19--1967c   (and Clermont, K.) Radiation of seismic waves from the Bilby explosions. Teledyne, Seismic Data Laboratory Report No. 183, 32 p., (1967).

20--1967d   (and Arkani-Hamed, J.) Seismic delay times: Correlation with other data. Science 158: 783-785, (1967).

21--1968   (with Thomson, K. C. and Ahrens, T. J.) A near field study of optic techniques of the generation and propagation of seismic waves from explosions in prestressed models. Seismic Coupling, VESIAC, Willow Run Laboratories, University of Michigan, p. 211-243, (January, 1968).

22--1968a   Seismic waves from atmospheric explosions and air-coupled Rayleigh waves, in International Dictionary of Geophysics. Elmsford, N.Y.: Pergamon Press, (1968).

23--1968b   (and Lacoss, R. T.) Microseisms: Mode structure and sources. Science 159: 872-873, (1968).

24--1968c   (with Solomon, S. C.) On the density distribution in the Moon. Phys. Earth Planet. Interiors 1: 475-484, (1968).

25--1968d   (with Arkani-Hamed, J.) Analysis and correlation of geophysical data. Supplemento al Nuovo Cimento, Ser. I/6: 22-66, (1968).

26--1968e   Review of Mantles of the Earth and Terrestrial Planets, S. K. Runcorn, ed. New York: Interscience Pub., ix + 584 p., (1967). Am. Geophys. Un., Tr. 49: 613, (1968).

27--1969   (with Lacoss, R. T. and Kelly, E. J.) Estimation of seismic noise structure using arrays. Geophysics 34: 21-38, (1969).

28--1969a   (and Boore, D. M.) Rayleigh wave particle motion and crustal structure. Seism. Soc. Am., B. 59: 331-346, (1969).

29--1969b   (and Arkani-Hamed, J. and Knight, C. A.) Geophysical data and long wave heterogeneities of the Earth's mantle. J. Geophys. Res. 74: 3751-3770, (1969).

30--1969c   (with Thomson, K. C. and Ahrens, T. J.) Dynamic photoelastic studies of P and S wave propagation in prestressed media. Geophysics 34: 696-712, (1969).

31--1969d   (and Wiggins, R. A.) Seismic arrays and the structure of the Earth's interior. NEREM Record, p. 170-171, (1969).

32--1970   (with Latham, G. et al.) Apollo 11 passive seismic experiment. Apollo 11 Lunar Sci. Conf., Pr. 3: 2309-2320, (1970).

33--1970a   (with Latham, G. et al.) Passive seismic experiment [Apollo 11]. Science 167: 455-457, (1970).

34--1970b   The Earth's Mantle, in Encyclopedia of Earth Sciences, Vol. 2, Geophysics of the Solid Earth. New York: Van Nostrand Reinhold Co., (1970).

35--1970c   (with Minear, J. W.) Thermal regime of a downgoing slab and new global tectonics. J. Geophys. Res. 75: 1397-1419, (1970).

36--1970d   (with Minear, J. W.) Thermal regime of a downgoing slab. Tectonophysics 10: 367-390, (1970).

37--1970e   (with Schlien, S.) Frequency-magnitude statistics of earthquake occurrences. Earthquake Notes 41: 5-18, (1970).

38--1970f   (with Solomon, S. C.) Lateral variation of attenuation of P and S waves beneath the United States. Seism. Soc. Am., B. 60: 819-838, (1970).

39--1970g   (with Latham, G. et al.) Seismic data from man-made impacts on the Moon. Science 170: 620-626, (1970).

40--1970h   (with Schlien, S.) A clustering model for earthquake occurrences. Seism. Soc. Am., B. 60: 1765-1787, (1970).

41--1970i   (with Solomon, S. C. and Ward, R. W.) Earthquake magnitudes: The effect of lateral variation of seismic attenuation. Pr. Woods Hole Conf. Seismic Discrimination (Working Paper), Vol. I, (July 20-23, 1970).

42--1970j   Crustal effects on long period chirp filters. Pr. Woods Hole Conf. Seismic Discrimination (Working Paper), Vol. I, (July 20-23, 1970).

----

(52)
GENE SIMMONS

GENE SIMMONS

MIT: 1965-

Gene Simmons, trained at Texas A & M, Southern Methodist, and Harvard, joined M.I.T.'s geology faculty in 1965 as the Institute's first century of existence ended and the second began. He was the 52nd and last professor of geology to be appointed during M.I.T.'s first century. Already a proven earth scientist in the field, in the laboratory and in the class room, when he joined the M.I.T. faculty he quickened the pace of his professional activities dramatically by organizing and directing a varied program of research by himself and his associates that attracted some $8,000,000 during his first decade at the Institute, 1965-1975. Especially noteworthy was his two-year period of service with NASA, 1969-1971, during which he played a critical role as Chief Scientist in Apollo flights 15, 16, and 17, for which he and his associates prepared outstanding Guidebooks for the general public. His research program during his first decade at M.I.T. produced a hundred publications and a dozen doctorates, and his services on many departmental committees added greatly to the accomplishments of those unavoidably time-consuming but necessary activities.

Gene Simmons, as he prefers to be called, was born Marvin Gene Simmons, in Dallas, Texas, on 15 May 1929, the only child of Burt H [no name] and Mabel (Marshall) Simmons. He received his pre-college education in Dallas, and then entered Texas Agricultural and Mechanical College, graduating with a B.S. degree in electrical engineering in 1949. For the next two years, 1949-1951, he worked as a petroleum engineer with Humble Oil & Refining Company in Houston, Texas, after which he joined the U.S. Air Force and served a two-year tour of duty as Communications Officer, being separated in March 1953 as a First Lieutenant. After a five-year period of self employment in the gravel business, 1953-1958, during which he completed requirements for and received an M.S. degree in geology (with a minor in physics) from Southern Methodist University in June 1958, he decided to apply for graduate work at Harvard. September 1958 found him in Cambridge engaged in a full-time doctoral program in geophysics. By October 1961 he had completed academic requirements for a Ph.D. degree in geophysics, which he received in February 1962, meanwhile having won an

N.S.F. Postdoctoral Fellowship for research work in Harvard's Dunbar Laboratory for 1961-1962 under the supervision of famed geophysicist, Dr. Francis Birch. His doctoral research involved investigation of regional gravity in northern New York, and his thesis is titled, "Gravity Survey in Northern New York" [v + 70 p., + numerous illustrations, Harvard University, Ph.D., October 1961]. Following completion of postdoctoral research at Harvard, he accepted a three-year appointment as Assistant Professor in the Department of Geology and Geophysics at Southern Methodist University. In July 1965 he joined M.I.T.'s Department of Geology and Geophysics as Professor under exciting and challenging conditions that were to bring great changes in the immediate future.

In the spring of 1964, as I approached my 60th birthday and we started to move into the recently completed 20-story Green Building, I informed the new Dean of Science, Dr. Jerome B. Wiesner, that I wished to relinquish my duties as Head of Course XII (Department of Geology and Geophysics) as soon thereafter as possible and not later than 30 June 1965. I had been Head of Course XII since 1949, and as I would be reaching retirement age on 30 June 1970, I wished to have the last five years free to return to full-time teaching and research. A new department head was in the offing; our new building awaited new programs in every one of the earth sciences; the program in oceanography was off to a good start; and there were several vacancies to be filled, especially in geology and geophysics. We had been making every effort possible to build up geophysics, and wished particularly to find a young scientist who was competent to bridge the gap between geology and geophysics.

Accordingly, when Simmons was recommended to us in the highest terms by colleagues who knew him and our needs, it was not difficult to get Dean Wiesner's approval to make him an offer. When I submitted the usual documents to support my recommendation for his appointment as an associate professor, Wiesner said essentially, "If he is that competent, why not offer him a full professorship at the start rather than to propose him for promotion a year or two hence, with all the attendant letters, paper work, etc.?" I, of course, was delighted at Wiesner's sagacious question, made Simmons the offer, he accepted, and that is how an assistant professor at another institution surprisingly jumped over an associate-professorship to become a Professor of Geology at M.I.T. And if I may be allowed a retroactive judgment, I think we were right, for Simmons has more than lived up to our expectations.

As the 52nd professor of geology at M.I.T. Simmons accepted our offer as the first century of the Institute was ending and took up his duties in the fall of 1965 as the second century began and as I relinquished the chairmanship of the Department to my successor, Dr. Frank Press. He was the last of the 25 professors that I recruited during my 16 years as department chairman from 1949 to September 1965; thereafter additional faculty members would be recruited by Press.

Inasmuch as Simmons' activities at M.I.T. did not start until September 1965, which was already several months into the Institute's second century, his career at M.I.T. belongs totally to this century. Nevertheless, a few comments seem in order to support my foregoing statement that he has more than lived up to expectations.

In the decade since he joined the M.I.T. faculty in 1965, Simmons has taught subjects on properties of rocks and minerals, introductory geophysics, marine geophysics, properties of solids, the energy crisis, physics of the earth and solar system, solid state geophysics, geophysical field techniques, and the moon. He has also served on numerous departmental committees and has supervised his fair share of theses. After serving NASA in an advisory capacity one day a week during SY 1969-70, he took a leave of absence from July 1970 to September 1971 to serve full-time as Chief Scientist in NASA's Manned Spacecraft Center in Houston, Texas. In this position he took a leading role in planning experiments and overseeing the scientific programs for the Apollo missions. For this service he received the NASA Medal for Exceptional Scientific Achievement in 1971.

In addition to all the foregoing activities he has thus far obtained more than $8,000,000 from half a dozen federal agencies to support a widely diverse program of research on earth and moon problems and materials. From this research program, during the decade 1965-1975, have come a total of some 100 articles, many shared with his students and associates; a 415-page Report on Seismic Coupling for the VELA program; a 370-page Handbook on Single Crystal Elastic Constants and Calculated Aggregate Properties (M.I.T. Press); and three NASA Guidebooks - On the Moon with Apollo 15 (June 1971); On the Moon with Apollo 16 (April 1972); and On the Moon with Apollo 17 (December 1972).

From the foregoing it hardly seems necessary to state that Simmons has added immeasurably to the strength of Course XII's program of instruction and research in the earth sciences, and has brought added distinction to it by his publications and public service. In addition to his NASA Award in 1971, he was given a Distinguished Alumnus Award by Southern Methodist University in 1973.

BIBLIOGRAPHY OF GENE SIMMONS

Symbols and abbreviations used in the following references are explained on p. 91 - 98; in general, abstracts are listed separately as well as with the references to the complete article. This bibliography begins with Simmons' first publication in 1959, and includes all titles through 1967, two years after his arrival at M.I.T. Normally I would not include references after 1965, but inasmuch as that is the year he joined our staff, and also because he is the last professor added to our staff in the first century of the Institute's existence, I have chosen to make him an exception in order to indicate his research productivity during his first two years in our Department. Publications after 1967 can be found in the standard geological and geophysical bibliographies.

1--1959   The photo-extinction method for the measurement of silt-sized particles. J. Sed. Petrol. 29: 233-245, (1959).

T--1961   Gravity Survey in New York, v + 70 p., (October 1961). (Ph.D. Thesis at Harvard University, February 1962.)

2--1961a   Anisotropic thermal conductivity. J. Geophys. Res. 66: 2269-2270, (1961).

3--1962   Implications for the anorthosite problem of a gravity survey in northern New York [abst.]. Geol. Soc. Am., S.P. 68: 272, (1962).

4--1962a   On Darcy's Law. J. Geophys. Res. 67: 4516, (1962).

5--1963   Velocity of shear waves in rocks [abst.]. Am. Geophys. Union, Tr. 44: 94, (1963).

6--1963a   (and Bell, P.) Calcite-aragonite equilibrium. Science 139: 1197-1198, il., (1963).

7--1963b   (and Birch, F.) Elastic constants of pyrite, FeS_2. J. Appl. Phys. 34: 2736-2738, (1963).

8--1963c   Gravity data collected in New York State during June 1963. U.S. Geol. Surv., Open File, 12 p., (1963).

9--1964   Gravity survey and geological interpretation, northern New York. Geol. Soc. Am., B. 75: 81-98, il., (1964).

10--1964a   Velocity of compressional waves in various minerals at pressures to 10 kilobars. J. Geophys. Res. 69: 1117-1121, il., (1964).

11--1964b   Velocity of shear waves in rocks to 10 kilobars, Pt. 1. J. Geophys. Res. 69: 1123-1130, il., (1964).

12--1965   Single crystal elastic constants and calculated aggregate properties. J. Grad. Res. Center 34: 1-269, (1965).

13--1965a   Effect of thermal conductivity contrasts on measured heat flow [abst.]. Am. Geophys. Union, Tr. 46: 176, (1965).

14--1965b   (with Bell, P. M. and England, J. L.) Experimental observations of a fast reaction with $CaCO_3$ [abst.]. Geol. Soc. Am., S.P. 82: 9, (1965).

15--1965c   Ultrasonics in geology. IEEE, Pr. 53: 1337-1345, il., (1965).

16--1965d   Continuous temperature-logging equipment. J. Geophys. Res. 70: 1349-1352, (1965).

17--1965e   (with Bell, P. M. and Hays, J. F.) Shearing squeezer experiments with quartz and coesite. Carnegie Inst. Washington Year Book 64, 1964-65: 141-144, il., (1965).

18--1965f   (and Brace, W. F.) Comparison of static and dynamic measurements of compressibility of rocks. J. Geophys. Res. 70: 5649-5656, (1965).

19--1965g   Crustal drilling in the United States, in Drilling for scientific purposes - International Upper Mantle Symposium in Ottawa, September 1965. Rept. Can. Geol. Surv., Paper 66-13: 42-51, il., (1966).

20--1966   The interpretation of heat flow anomalies due to contrasts in heat production [abst.]. Am. Geophys. Union, Tr. 47: 183, (1966).

21--1966a   Temperature logging and heat flow, in Soc. Prof. Well Log Analysts (SPWLA), Logging Symposium, 7th Ann., Tulsa, May 3-11, 1966. Tr. Houston, Tex., SPWLA: 11-19, il., (1966).

22--1966b   (as Chmn.) Deep drilling on land for scientific purposes. Am. Geophys. Union, Tr. 47: 373-378, (1966).

23--1966c   Energy relationships in the Earth [abst.]. Geol. Soc. Am., S.P. 87: 156, (1966).

24--1966d   Heat flow in the earth. J. Geol. Ed. 14: 105-110, (1966).

25--1967   Interpretation of heat flow anomalies - [Pt.] 1: Contrasts in heat production. Rev. Geophys. 5: 43-52, (1967).

26--1967a   Interpretation of heat flow anomalies - [Pt.] 2: Flux due to initial temperature of intrusives. Rev. Geophys. 5: 109-120, il., (1967).

27--1967b   Hashin bounds for aggregates of cubic crystals. J. Grad. Res. Center 36: 1-87, (1967).

28--1967c   Which graduate school? Geotimes 12: 14-15, (1967).

29--1967d   (with Horai, Ki-iti) Measurement of thermal conductivity of monomineralic aggregates [abst.]. Am. Geophys. Union, Tr. 48: 210-211, (1967).

30--1967e   (with Gretener, P. E.) Short-time temperature variations in the waters of the Gulf of Mexico. J. Geophys. Res. 72: 2263-2266, (1967).

Since 1967 Simmons has published some 90 or more articles, abstracts, and books, and has prepared numerous unpublished reports of one kind or another. Noteworthy among his books are: **Seismic Coupling**, VELA Information Analysis Center for ARPA, Willow Run Labs., Univ. Michigan, 345 p., (1968); (and Wang, H.), **Single Crystal Elastic Constants and Calculated Aggregate Properties**: A Handbook, 2nd ed. Cambridge, Mass.: M.I.T. Press, 370 p., (1971); and the three following Guidebooks for NASA, all published in Washington, D.C. by the U.S. Government Printing Office - **On the Moon with Apollo 15**, A Guidebook to the Hadley Rille and the Apennine Mountains, 46 p., il., (June 1971), **On the Moon with Apollo 16**, A Guidebook to the Descartes Region, 90 p., il., (April 1972), and **On the Moon with Apollo 17**, A Guidebook to Taurus-Littrow, 111 p., il., (December 1972).

POSTSCRIPT
THE SECOND CENTURY BEGINS: 1965 -

When the fall term of SY 1965-66 began on 15 September 1965, M.I.T. started its second century, and Course XII its 75th year. I had resigned as Chairman of our Department of Geology and Geophysics, effective as soon as possible after 30 June 1965, after having served in that capacity for 15 years, 1949-1965, desiring to devote my last five years to teaching and starting research for this history.

Dean of Science Jerome B. Wiesner, later to become M.I.T.'s 13th President, had meanwhile induced Dr. Frank Press, then Professor of Geophysics at the California Institute of Technology and Director of that institution's famed Seismological Laboratory, to accept an M.I.T. appointment as Professor of Geophysics and Head of Course XII, then called the Department of Geology and Geophysics. The appointment met with the unanimous and enthusiastic approval of the Course XII Faculty, and Press took up his duties in September 1965 as the new school year began.

I had concluded some time before that 15 years was about as long as I wanted to serve as department head, particularly if I could achieve my major goals by the end of that period of time, i.e. by June 1965. The goals were:

1) a revised curriculum, with more preparation in mathematics and physics, and more electives in the earth sciences;

2) an expanded and more diversified faculty, with professors added in geochemistry, geophysics, and oceanography;

3) new and expanded space for classrooms and research laboratories, and reestablishment of the Lindgren Library and the Schwarz Memorial Map Room within our Department;

4) increased enrollment, particularly of well-qualified graduate students;

5) a substantial increase in research funds for faculty, staff members, and student assistants;

6) a summer program for undergraduate training in practical field work; and

7) increased endowment funds under direct control of the Department, which would yield annual income that could be used for a variety of unexpected and recurrent needs that could not be met from the regular departmental budget.

Good progress had been made towards all of the above mentioned goals by 1965. The new $5,000,000, 20-story Green Building had been occupied and dedicated by Geology and Meteorology the year before (2 October 1964), with reestablishment of the Lindgren Library and Schwarz Memorial Map Room, a 300-seat auditorium, excellent classrooms, and numerous well-equipped research laboratories. The total departmental staff had grown in size and become more diversified; the total departmental budget had grown an order of magnitude, from some $120,000 to about $1,200,000; the number and quality of graduate students had increased substantially; grant and contract funds from federal bureaus had increased greatly, allowing much more research; and endowment funds for the Department were increased from less than $3,000 to more than $350,000 by 1965. An informal arrangement with the Woods Hole Oceanographic Institution made it possible for our graduate students to carry on thesis research at Woods Hole and on the Institution's ships at sea, and this arrangement was formalized later into a joint graduate-degree program.

Accordingly, the year 1965 seemed just the right time for a new department head, a change in emphasis and direction, and all the other changes that should come from time to time if a Department is to remain viable.

The leadership that Press has provided during the decade since 1965 has been both dramatic and highly beneficial, with the result that our current Department of Earth and Planetary Sciences, still Course XII, ranks equally with the best such departments anywhere.

With the end of M.I.T.'s first century and the beginning of the second, and with the Geosciences doing so well under Press' leadership, I deem it appropriate to include a brief biography of him, which follows, even though he belongs to the beginning decade of the Institute's second century.

The Cecil and Ida Green Building for the Earth Sciences, with McDermott Court and Calder's stabile, "The Big Sail," in the foreground, as they appeared in 1965, at the beginning of M.I.T.'s second century.

(Photo from the M.I.T. Historical Collections)

(53)
FRANK PRESS

Photo by Ivan Massar
FRANK PRESS

MIT: 1965-

As M.I.T. reached the end of its first century, in 1965, and Course XII (Geology) its 75th anniversary, famed geophysicist Frank Press came from Caltech to become Professor of Geophysics and Head of Course XII, then called the Department of Geology and Geophysics, and to lead the Department into the Institute's second century.

Educated in the superb primary and secondary school system of Greater New York, trained for a professional career at Columbia's unique Lamont Geological Observatory under renowned geophysicist and oceanographer, W. Maurice Ewing, and experienced in the academic world of the earth sciences as Professor of Geophysics and Director of Caltech's great Seismological Laboratory, Press was uniquely qualified by training and experience to take charge of Course XII, which he soon had renamed the Department of Earth and Planetary Sciences. At this critical time, in 1965, the Department was just settling down in the year-old 20-story tower, the Green Building, built with a 5-million dollar gift from Cecil and Ida Green of Dallas, Texas. Forty professors of earth science (geology, geochemistry, geophysics, oceanography, and meteorology) and their 150 graduate students awaited Press' arrival in September 1965.

During the decade since 1965 the Department has undergone important and impressive changes under Press' outstanding leadership, and today, in 1975, ranks with the best in the world. Early in the decade instruction in and professors of space science were added to Course XII, and a joint graduate program in oceanography was formalized with the Woods Hole Oceanographic Institution. The Department's name was appropriately changed to Earth and Planetary Sciences. Federally funded research grew dramatically in volume and diversity, graduate student enrollment increased impressively, and publications by faculty and students increased concomitantly. Private gifts from Cecil and Ida Green made possible the establishment of three distinguished chairs in the Department, the first of which is deservedly occupied by Press himself, and gifts from George R. Wallace, Jr. made possible construction and equipment of an Astrophysical Observatory and a Geophysical Observatory, both on M.I.T. property in Westford, Massachusetts. The U.S. Navy provided the Department with a much needed 50-foot research vessel, which not only served well the early needs of the Department's developing program in oceanography, but happily continues in active service in M.I.T.'s Ocean Engineering Program. Finally, the high quality of Course XII undergraduate and graduate students has been maintained in the face of substantial increase in both categories, and the overall budget of the Department has quadrupled during the decade.

The complete story of the Department's impressive growth in size, diversity, and quality during Press' first decade of leadership must be left to others; hopefully at some future date someone else will include it in a complete record of his outstanding accomplishments at M.I.T., after he has handed his responsibilities on to his successor. In the following brief biographical sketch I review his education, training and experience before coming to M.I.T.; mention a few of the more significant happenings during his first decade as Head of Course XII; and discuss briefly his more important accomplishments as student, teacher, scientific investigator, author, administrator, public servant, and scientific statesman. A brief section on "Honors and Awards" makes evident his distinguished achievements in all of the aforementioned activities, and the "Bibliography" records his outstanding contributions to the scientific literature.

BIRTH, ANCESTRY, AND EARLY EDUCATION (1924-1944)*

Frank Press, the youngest of the three children, all sons, of Russian emigrants Solomon Press and Dora (Sternholz) Press, was born in Brooklyn, New York, on 4 December 1924. His father was educated through high school in Russia, which was unusual for the time, and then came to America in 1916, settling in the Greater New York area and becoming a wholesale grocer and auctioneer. His mother, although educated only through grade school, was, like his father, a strong influence in encouraging Frank to be proficient and to excel in whatever he did. Motivation toward education also came from his two older brothers - Samuel (b. 1916), now an attorney, and Paul (b. 1918), a magazine producer. Both brought home the excitement of the academic world, as they trained for their respective careers, and Frank remembers that he caught some of that excitement when hearing them talk about what they were learning.

Frank started his own education in Brooklyn, first attending grammar school there and then going on to Samuel Tilden High School. He remembers that at that time the New York City high schools had an extraordinary group of teachers, all of whom were well educated and highly motivated, and that he not only received an excellent education but was also strongly encouraged to go on to college.

Even as early as high school Frank already knew that he wanted a life career in some intellectual area with a scientific bias. He was one of a small group that ran the usual student affairs and was editor of the school paper. He found mathematics and physics especially interesting and

---

* I am most grateful to Dr. Press for his willingness to let me tape an hour-long interview with him, and to both him and his wife Billie for helping me improve my original draft. While much of my sketch is based on the tape, which will be preserved in the M.I.T. Historical Collections, I have also found information in numerous news stories, journal items, citations, and the like. These sources are specifically cited at the appropriate places in the text.

won several medals in science. He was also an active participant in several discussion groups, so that his overall interests went well beyond those in science. Clearly the die for his future career in science and in involvement in important public affairs was cast during his high school years.

Having graduated from high school in 1941, with an excellent academic record and a keen interest in science, he chose to enter C.C.N.Y. (College of the City of New York) because it had free tuition and was near enough to his home that he could commute to classes. Again, as had been the case in his high school, he found the level of teaching and the quality of faculty extraordinarily high at the College - as a matter of fact, C.C.N.Y. was then famous because of having more of its graduates go on to doctoral degrees than any other American college. Little wonder then that, with his excellent high school preparation, Press was not only able to complete all requirements for a Bachelor of Science degree in two and one-half years, being graduated magna cum laude in 1944, but he also found he was at least a year ahead of most of his classmates who entered Columbia with him in the fall of 1944.

Having enjoyed the strong home influence of his parents and brothers, the excellent program and environment of Samuel Tilden High School, and the superb curriculum and faculty at City College, Press was extraordinarily well motivated and prepared for an academic career years in advance of his actual age and of his peers. At this stage there was no question about his continuing to a doctoral degree; it was simply a matter of where. Columbia seemed a natural choice. It was nearby, it was strong in science, and it had one of the leading graduate schools in North America.

THE COLUMBIA YEARS (1944-1955)

Although he had majored in physics at C.C.N.Y., Press had been captivated by several courses in geology. When he learned that physics and geology could be combined in a career in geophysics, he decided that that was the career he wanted to follow, and Columbia was the place he wanted to attend to get his academic training.

Good fortune now smiled on him because when he applied for admission to Columbia's Graduate School he learned that Dr. Maurice Ewing, a well-known geophysicist, was soon to join the Columbia faculty. Although he could not have known it then, his decision to prepare for a career in geophysics would inevitably bring him into a close personal and professional relationship with Ewing, who would become one of the world's foremost geophysicists during the next decade. The Ewing-Press relationship is now history and is intensely interesting, but that history is much too long to be discussed adequately in this brief biographical sketch. Suffice it to assert here that Ewing's influence on Press was exceedingly powerful during the latter's formative years - first when he was a graduate student and research associate, and then when he became an academic colleague - and went far in launching Press on his own distinguished career. The interested reader can sense some of the warmth and depth of this teacher-student relationship by reading the eloquent eulogy that Press delivered at the memorial service for Ewing in Houston in 1974 (see bibliographic reference 226--1974b). Here I confine my discussion of their relationship mostly to their cooperation in research projects and joint publications.

Entering Columbia in 1944, Press immediately combined formal classwork with research and within two years completed all requirements for an M.A. degree in Geophysics, which he received in 1946. His master's thesis on "Magnetic anomalies over oceanic structures" was ultimately published in 1952 with Ewing as the second author (ref. 36--1952d).

For the next three years, 1946-1949, Press was a doctoral candidate and was awarded a Ph.D. degree in Geophysics in 1949. Two of his publications during the three year period (ref. 3--1948 and 4--1948a; also ref. T--1949) were accepted as his doctoral thesis.

While a doctoral candidate he was a member of the Research Staff of Columbia University's Division of Government Aided Research: first as a research assistant (1946-1948), then as a University Fellow (1948) and a Research Associate in Geology (1948-1949).

In 1949, with his doctorate in hand, he was appointed Instructor in Geology, in Columbia University, and served as such until 1951, when he was advanced to Assistant Professor of Geology for SY 1951-52. Advancement to Associate Professor followed in 1952, and he continued in this rank until 1955 when he left for a professorship at Caltech.

As repeatedly emphasized throughout this biographical sketch, Press worked closely with Maurice Ewing from the day the two started joint research until he left for Caltech. The extent and diversity of their cooperative research is evident in the titles of the 46 articles and 15 abstracts they published during the period 1946-1956. It should be noted, however, that although Ewing was playing a dominant role as joint author, Press was the first author of 16 of the joint articles and 6 of the joint abstracts, and the sole author of 4 articles. Press' outstanding publication record for his Columbia duodecade, 1944-1956, is discussed more fully farther on in the section on "Research and Publications."

Although Ewing's influence on Press was probably strongest in their jointly conducted research and writing, it was also substantial in other ways. How better than to hear directly from Press as I taped his comments on Ewing and their relationship:

> "Ewing was an extraordinary human being and mentor. The strength of his personality was such that he influenced the lives of anyone who worked with him. From my point of view the most important thing he did for me was essentially to teach me a style of doing science. He gave me the attitude that there were always enormously important problems to be solved; that I could solve them as well as anyone else; and that if experiments were required for their solution or if theoretical advances were required, I shouldn't hesitate to prepare myself to do both and to try both.
>
> "He had enormous curiosity, which translates into breadth of interest, and which I also inherited. He wouldn't hesitate to jump into fields that he knew nothing about and within a short time he would be competing with the leaders of the field, and I did that with him in a number of different areas.
>
> "We had different cultural and social backgrounds which we both understood, but at no time did these differences interfere with our collaboration, which was long and extremely fruitful.
>
> "So he perhaps was the most important individual in fashioning my life as a scientist."*

Their relationship actually started in the summer of 1945, after Press had completed a year of graduate work but before Ewing had arrived at Columbia. At the time Ewing had a small group of geophysicists working out of Woods Hole on the outer Continental Shelf, and he invited Press to join the summer cruise and participate in the work at sea. Subsequently, Press went on similar cruises with Ewing in the summers of 1946, 1947, and 1948, and on several of his own cruises during the same period. After receiving his doctoral degree in 1949, and being appointed Instructor in Geology, he turned away from geophysical work at sea and began to study the different aspects of earthquakes and their attendant seismic waves, and how the latter could be used to determine the crustal and mantle structure of the earth.

Only a cursory inspection of his bibliography to 1955 is necessary to see the breadth and fundamental nature of his research interests and

---

* Press recalls that he was also strongly influenced by Columbia's Prof. Walter H. Bucher, a structural geologist in the broadest sense, who challenged his students to explain ocean swells, rifts in the earth's crust, large scale faulting, orogenic belts, and other similar structural features. Press also remembers that Bucher gave him a strong bias against continental drift, which at the time, in the 1940s, was not an unusual attitude of many physical geologists. Were Bucher living today, I would guess that he would have abandoned that bias and would be supporting the concept of sea-floor spreading and plate tectonics.

accomplishments, and to see why he was offered a professorship in geophysics at Caltech. The master had done his work well, and his protégé was superbly prepared for the greater opportunities in Pasadena.

THE CALTECH YEARS (1955-1965)

When in 1955 Press was invited to join the faculty of the California Institute of Technology, he accepted appointment as Professor of Geophysics. Two years later he was made Director of the Institute's famed Seismological Laboratory, and thereafter, until 1965, he performed his academic duties at the Institute and his administrative duties and research investigations at the Laboratory.

On the campus he quickly attracted a group of unusually talented students who found his lectures interesting and informative and his research ideas exciting and challenging. At the Laboratory he became well acquainted with Caltech's renowned geophysical trio - Beno Gutenberg, Hugo Benioff, and Charles F. Richter. It was no small honor for Press, only 33 years of age and almost 25 years younger than the youngest of the trio, to be put in charge of the famed Laboratory. But any doubts there may have been about his competence as a teacher, or ability as an administrator and director of research, were quickly dispelled as he involved himself in his own and the research of his students and other associates. Important and significant research results were published at regular intervals, and by the end of his Pasadena decade, he had more than equalled his publication record of the preceding decade at Columbia - 89 titles in all, consisting of 71 articles and 18 abstracts, of which he was the first author of 43 articles and 14 abstracts, as shown in a table farther on in the section on "Research and Publications."

The recognition accorded him because of his publications and other accomplishments brought requests for his assistance and counsel from professional societies, state and federal agencies and bureaus, and international organizations. Among his more important such services was membership on President Kennedy's Science Advisory Committee (1961-1964) and membership on the U.S. Delegation to Nuclear Test Ban Conferences in Geneva (1959, 1960, and 1961) and in Moscow (1963). Other such assignments are discussed farther on in the section on "Public Service."

With such an enviable and impressive record of accomplishments to his credit, Press was the obvious choice of his physics colleague and close friend, Prof. Jerome B. Wiesner, then Dean of M.I.T.'s School of Science, to head the Institute's Department of Geology and Geophysics, in 1965, when I asked to be relieved of the chairmanship of that Department. Fortunately for both the Department and the Institute, Press accepted Wiesner's offer.

THE M.I.T. YEARS (1965-   )

As pointed out in the preceding "Postscript - the Second Century Begins: 1965- ," and in the précis to this biographical sketch of Press, the Department of Geology and Geophysics was ready for a change of leadership and direction in 1965, as M.I.T.'s first century was ending and its second century beginning. Press arrived on the scene at this critical and propitious time, as Professor of Geophysics and Head of Course XII, and soon set in motion some of the actions that would greatly change and improve the Department.

In a transition period in which a Department changes leadership and direction, it is typical to see change of personnel, of curriculum, of research emphasis, and of funding. All of these happened during Press' first decade as Head of Course XII, and definitely raised our Department to the foremost position it enjoys today. Although it is not my purpose here to discuss at length the history of geology at M.I.T. during the first decade of the Institute's second century, I do feel, nevertheless, that I should briefly review the more important things that greatly affected Course XII during the decade 1965-1975 because they are a part of Press' scientific career at M.I.T. and should be recorded.

Changes in Course XII faculty personnel are briefly discussed earlier on, in Chapters I and II, and are indicated on the faculty charts in those chapters. Here I briefly review the flux of faculty members in order to show how dramatically the Course XII faculty roster changed during the transition period, 1965-1970. (See accompanying Charts.)

On 1 January 1965, the Faculty of Course XII consisted of 23 professors; on 1 July 1970 the number was the same, 23, but the changes during the 5-year transition period were dramatic! One professor died (Bitter), three retired (Parks, Buerger and Shrock), and 13 resigned to accept positions elsewhere; in short 17 of the 23 professors on the 1965 roster were gone or retired by 1 July 1970. In contrast, Press recruited 14 new professors during the period 1966-1970, and these, added to the two I had recruited in 1965 (Toksöz and Simmons), and counting Press and the 6 professors remaining from the 1965 roster, gave a total of 23 in mid-1970, the same as five years earlier, but what a dramatic turnover of personnel during the 5-year transition period!

When the M.I.T.-W.H.O.I. joint graduate degree program in oceanography was formalized on 8 May 1968, and the first regularly scheduled subjects were added to the Course XII curriculum, all this after subjects and professors in space science had been added, the name of the Department was appropriately changed to Earth and Planetary Sciences.

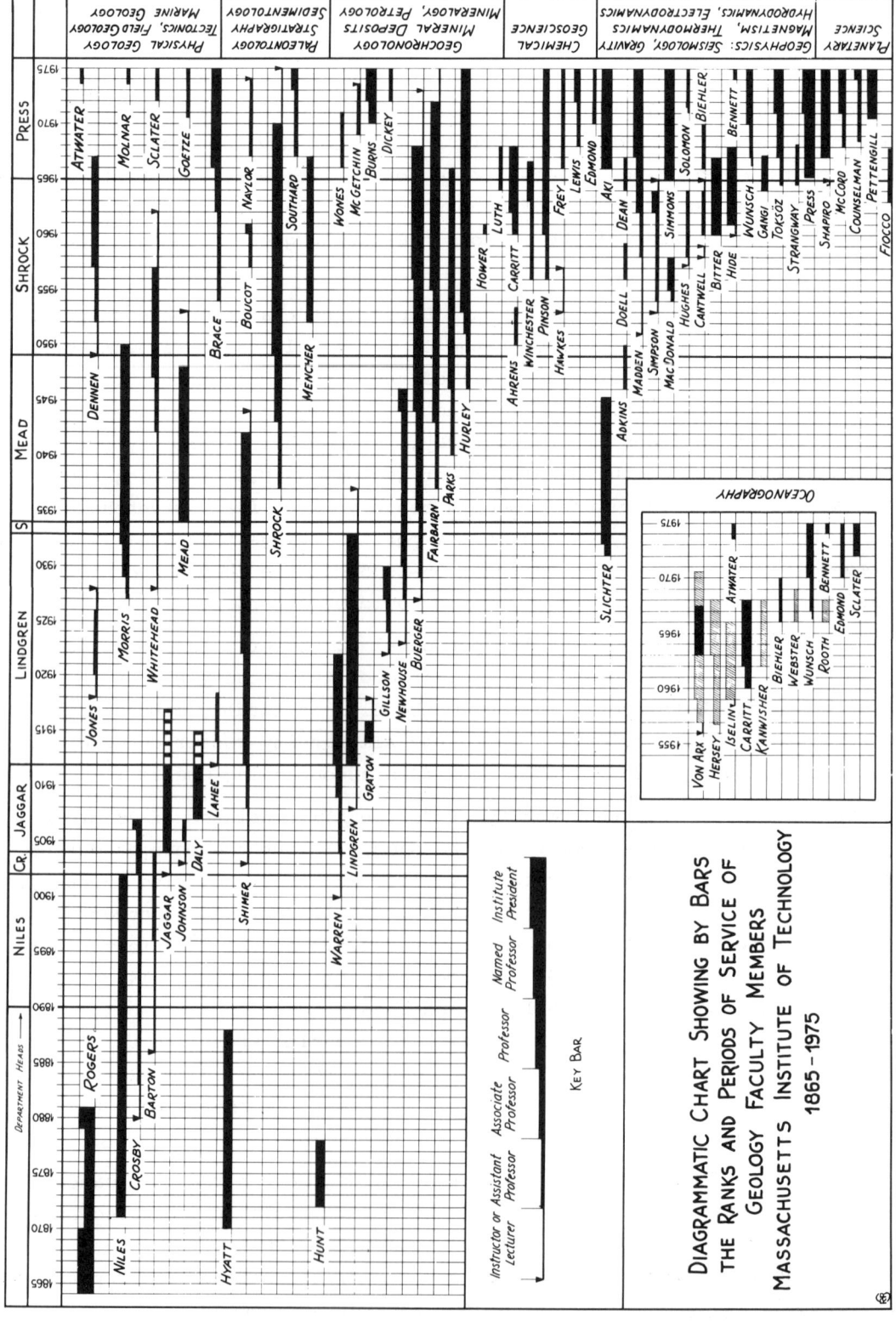

1004   GEOLOGY AT M.I.T.   1865–1965

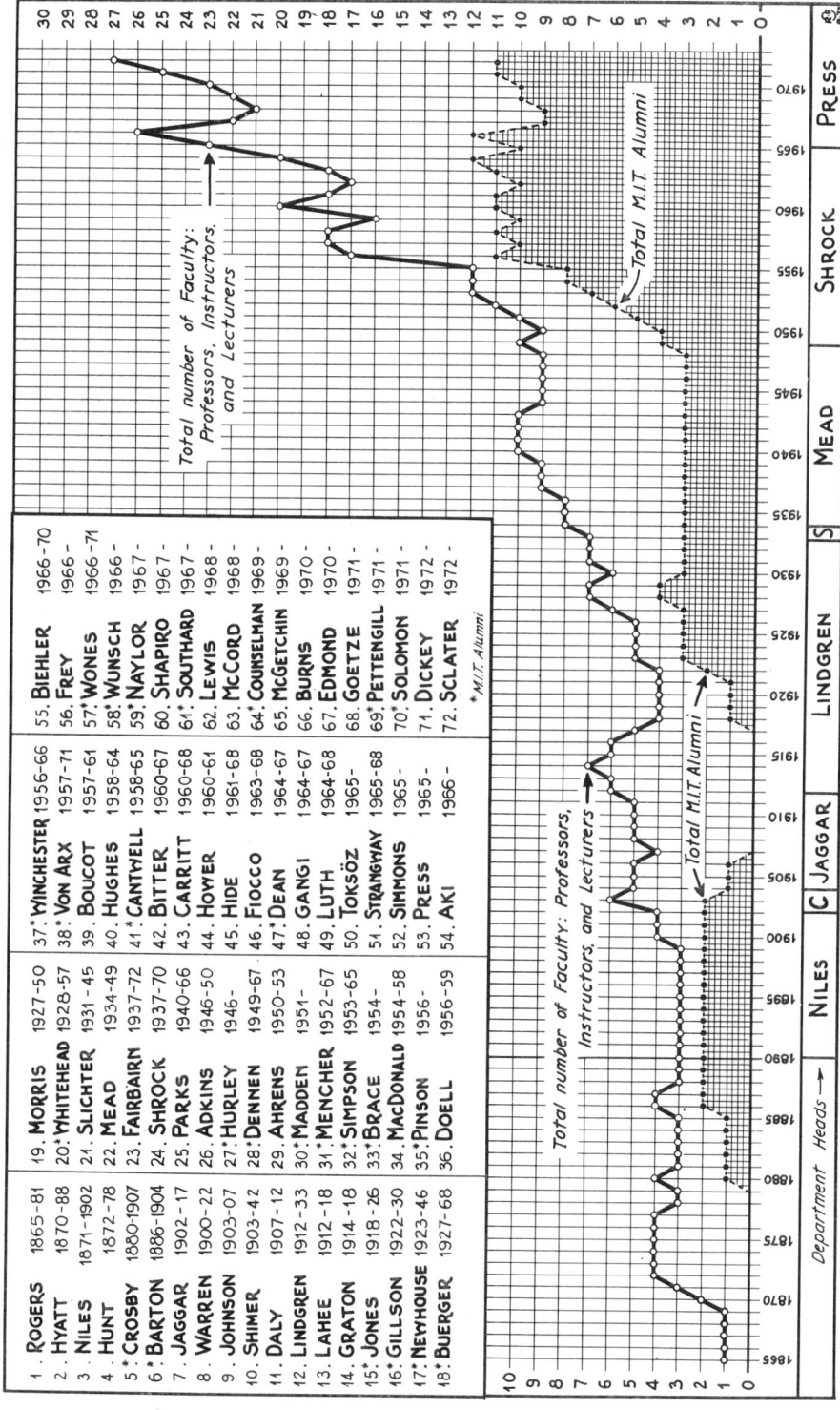

Chart showing order of appointment of Course XII Professors and number of M.I.T. Alumni serving on the faculty during 1865-1972

Acquisition of a 50-foot oceanographic research vessel, christened the RV R.R. SHROCK on 10 December 1965, made it possible for students of Course XII and XIX to do geoscientific and meteorological research at sea, and added a much-needed facility to those that were available to our students in the joint oceanographic program mentioned in the preceding paragraph.

The George R. Wallace, Jr. Astrophysical Observatory at Westford, Massachusetts, was dedicated on 14 October 1971. Its two telescopes (a 16-inch and a 24-inch housed in separate domes, with a computer facility designed to control the 24-inch and process data on-line) and a support building provide the most advanced facilities for space research. Nearby, on the same M.I.T. property, the George R. Wallace, Jr. Geophysical Observatory, one of the best equipped geophysical laboratories in the world, was dedicated on 20 May 1975 (see Tech Talk 19/45: 1, May 21, 1975). The underground facility is capable of detecting earthquakes anywhere in the world; is being used to evaluate earthquake risk in New England; and serves as an excellent test bed for the development of instruments to be used at other locations.

Three of the greatest departmental changes brought about by Press were formalization of the joint oceanographic program with the Woods Hole Oceanographic Institution; recruitment of an outstanding group of space scientists to go with the new Wallace Astrophysical Observatory; and recruitment of an equally outstanding group of seismologists to go with the new Wallace Geophysical Observatory. Only a leader of Press' achievements and renown in geophysics could have recruited the personnel, acquired the new observational facilities, and attracted the financial support that have become such important components in our current Department of Earth and Planetary Sciences.

In this all too brief review of Press' accomplishments during his first decade at M.I.T., 1965-1975 (see also the section "Summation" farther on), it should be noted further that he continued his research productivity, with 62 publications (54 regular articles and 8 abstracts); that the total departmental budget quadrupled from $1,200,000 in FY 1965 to some $4,250,000 in FY 1975; that student enrollment increased from 26 undergraduate majors in 1965 to 70 in 1975 and graduate students from 79 to 110 for the same 10-year period. Finally, along with all of the preceding activities and achievements, he has continued to give generously of his time and thought to public service. And the end is not yet, as clearly indicated by what happened during SY 1975-76 and what is in prospect for SY 1976-77!

RESEARCH AND PUBLICATIONS (1946-1975)

Press' principal research activities during the past thirty years, *i.e.* from 1946 to 1975, have been concerned with problems in the following aspects of geophysics: 1) theory of elastic wave propagation; 2) exploration geophysics, with special reference to submarine geology; 3) earthquake phenomena and their use in deducing crustal and mantle structure; 4) earthquake prediction; 5) planetary physics and earth models; 6) lunar structure and dynamics; and 7) pattern analysis and recognition. In each of these aspects of geophysics he has made important contributions. These have brought him recognition, both nationally and internationally, and have given him highest standing among the leading seismologists of the world, as indicated by his numerous services on important commissions and committees, and by the many awards and honors he has received - both aspects of his accomplishments that are discussed elsewhere in this sketch.

Examination of his Bibliography, which follows on later pages, shows a total of 208 articles through 1970, *i.e.* for 1944-1970, a period for which I feel we have an essentially complete list of articles (including abstracts and notes), and of some 232 to the present (through 1975). The totals include 43 abstracts and 2 textbooks: in the latter, Press was the third of three authors of Elastic Waves in Layered Media (McGraw-Hill Book Co., Inc., 1957; see reference 93--1957c in his Bibliography) and the first of the two authors of Earth (W. H. Freeman and Co., 1974; see reference 224--1974 in his Bibliography).

As one of the leaders in his particular field of geophysics, Press has been called upon repeatedly to serve as editor or co-editor for some special series or individual publication, or to contribute reviews, summaries, or special articles to symposia, colloquia, congresses, conferences and the like; to encyclopedias; and to volumes honoring some distinguished scientist. Examples of all these kinds of publications are included in his Bibliography.

While he served as Consulting Editor of the McGraw-Hill International Series in the Earth and Planetary Sciences, during 1965-1973, the Company added 10 books to their well-known series of 24 titles. The first of these was Krauskopf's Introduction to Geochemistry (1967) and the most recent, Hyndman's Petrology of Igneous and Metamorphic Rocks (1972).

He served as a co-editor, with L. H. Ahrens and others, of Vols. 2-8 of Physics and Chemistry of the Earth (see references 99--1957i and 163--1964h); with H. Benioff and others, of Contributions in Geophysics (reference 100--1958); and recently has been serving on the Editorial Board of Science.

AN ANALYSIS OF THE BIBLIOGRAPHY OF FRANK PRESS, 1944-1970

(165 articles, 43 abstracts)
Total publications: 208

Authorship (60 different authors)	COLUMBIA 1944-1955		CALTECH 1955-1965		M.I.T. 1965-1970		TOTAL 1944-1970	
	Articles	Abstracts	Articles	Abstracts	Articles	Abstracts	Articles	Abstracts
As sole author	4		28	7	17	4	49	11
As 1st author and M. Ewing only	13	3					13	3
As 1st author and M. Ewing et al.	3	3					3	3
As 1st author and others excluding M. Ewing	4	1	15	7	2		21	8
As 2nd author with M. Ewing only	17	5					17	5
As 2nd or 3rd author with M. Ewing et al.	13	4					13	4
As 2nd or 3rd author with others excluding M. Ewing	10	1	28	4	11	4	49	9
TOTALS	64	17	71	18	30	8	165	43

Typical of the many and varied requests he has received for special articles are the following examples. He contributed articles on geophysical subjects to the Encyclopedia of Physics (ref. 66--1955b, 67--1955c, and 68--1955d) and to the Encyclopaedic Dictionary of Physics (ref. 160--1964e); a chapter on "Geophysics" to Listen to Leaders in Science (ref. 166--1965b); and a section on "Seismic velocities" to Handbook of Physical Constants (S. P. Clark, Jr., ed.) (ref. 174--1966c). He contributed to a monograph - Contributions in Geophysics - in honor of Beno Gutenberg (ref. 112--1958ℓ); to Perspectives in Modern Physics - Essays in honor of Hans A. Bethe (ref. 173--1966b); to The Earth Beneath the Continents - A volume of geophysical studies in honor of Merle A. Tuve (ref. 177--1966f); and to The Nature of the Solid Earth - Geophysical essays in honor of Francis Birch (ref. 216--1972). Finally, as examples of papers presented at international meetings may be cited his review of advances in seismology at the M.I.T. International Conference on the Earth Sciences in 1964, held on the occasion of the dedication of our Cecil and Ida Green Building (ref. 171--1966); and his paper on "Earth models" presented at the 1969 Symposium on "Phase Transformations and the Earth's Interior," held in Canberra, Australia (ref. 205--1970f).

Analysis of the authorship of the articles in Press' Bibliography clearly shows several significant aspects of his research and publishing activities. The 208 references listed for the period 1944 through 1970 consist of 43 abstracts and 165 regular articles. He was the sole author of 11 of the abstracts; the first of two or more authors of 14; and a second or third author of 18, as tabulated on the facing page. He was the sole author of 49 of the regular articles; the first in joint authorship with Maurice Ewing of 16; and the first in joint authorship with authors other than Ewing of 21. In addition, he was the second or third author with Ewing alone of 17, with Ewing et al. of 13, and with authors other than Ewing of 49. These data are displayed in the table for easy reference.

It is evident from the data shown in the table, and from scrutiny of the Bibliography on following pages, that - 1) Maurice Ewing played a dominant role in Press' early publishing activities; 2) Press was the sole author of only a third of his regular articles (49/165) and a fourth of his abstracts (11/43); and 3) he involved himself in the research of his students and colleagues to such an extraordinary degree that he shared authorship with more than 60 different individuals! It should be further noted that although by far the largest number of Press' publications have concerned seismology, he has not ignored other important aspects of geophysics such as density, gravity, crustal and mantle structure, earth magnetism, and the dynamics of the solid earth, the world ocean, and the atmosphere.

Again referring to the accompanying table and to Maurice Ewing's dominant role mentioned above, it should be pointed out that Press spent the first decade of his professional career, 1946-1955, at Columbia in a remarkably close relationship with Ewing. Sixty-one of his first 81 publications (46 articles and 15 abstracts) were co-authored with Ewing, who was the first author of 39 of the 81 publications, and the second or third author after Press of 22 additional ones. Even after Press left Columbia for Caltech in 1955, he continued to publish jointly with Ewing until the latter's untimely death in 1974, and the final record shows that they shared in the authorship of 45 regular articles, one textbook, and 15 abstracts.

In the exemplary relationship described above, the master teacher saw to it that his student published his first papers alone; then gave to the brilliant young scientist the prestige of his own name in joint authorship as they reported their research results. By the end of the decade of their cooperative work as master and apprentice, with 81 titles to his credit, Press was superbly prepared to start the second decade of his career, 1955-1965, which would be spent at Caltech, and would give him the opportunity to develop independently that potential for greatness that Ewing had astutely recognized and constantly encouraged in his young protégé.

At Caltech Press became teacher, research supervisor and administrative leader, and soon attracted a group of outstanding graduate students and post-doctorates. Following the pattern that Ewing, as preceptor, had developed so effectively at Columbia's unique Lamont Geological Observatory, Press quickly became involved in the research work of his associates. The result was that during his decade at Caltech, 1955-1965, he not only gave regularly scheduled lectures, supervised student theses, and directed the Institute's Seismological Laboratory, but also managed to conduct an active research program of his own, which ultimately produced a total of 89 publications (71 regular articles and 18 abstracts), only a few of which, surprisingly, dealt with California earthquakes.

Inspection of his publications during his Pasadena decade shows a broadening of interest in geophysics, particularly in elastic waves and what they could reveal about the structure and physical characteristics of the earth's crust, mantle and core. He also became involved in the plans for the lunar missions and in the problems associated with the detection of nuclear tests. Well before the end of the decade he had become a world leader in geophysics, and his publications for the period clearly show the broadening and diversification of his scientific interests and research.

Far from Columbia, and from the close association with Ewing that he had enjoyed for a decade, he himself now became the teacher and research leader of a group of outstanding young scientists. However, as Ewing had done, he carried on his own investigations while supervising the research of his students and associates: the results were much the same, as displayed on the accompanying table - 28 articles and 7 abstracts of which he was the sole author; 15 articles and 7 abstracts jointly published with others, but as first author; and 28 articles and 4 abstracts with others of which he was a second or third author. Clearly he was completely on his own and a distinguished geophysicist by the end of his Pasadena decade, with an impressive publication record fully comparable with his earlier one at Columbia, and even greater opportunities lay ahead at M.I.T.

September of 1965 found him Professor of Geophysics and Head of M.I.T.'s Department of Geology and Geophysics; which soon became the Department of Earth and Planetary Sciences as programs in oceanography and planetary sciences were added to those in the geosciences. In a preceding section I have briefly discussed how Press has coordinated faculty, students, supporting staff, curriculum, space, facilities, and funds into one of the foremost departments of earth and planetary sciences anywhere. In this section I wish to emphasize that while accomplishing the preceding, he has continued unabated his intense desire to gain an ever deeper insight into the physical nature, internal structure, and dynamic behavior of the bodies of our solar system. And while pursuing this quest, and becoming one of the world's leading geophysicists, he has become even more the master of his scientific discipline as is evident from the publication data in the table. Little wonder, then, that so many honors and awards have come to the scientist who has led Course XII so dramatically through the first decade of the Institute's second century.

PUBLIC SERVICE

One of the prices and responsibilities of greatness is public service, a kind of <u>savoir oblige</u>, which can have many forms - 1) public lectures; 2) participation in the activities of professional societies, associations, etc.; 3) services of various kinds on committees, panels, councils, work groups, etc.; and 4) advisory and consulting services to state, national, and international organizations of various kinds. Press has been called upon repeatedly during the past twenty years for the kinds of public services just noted, and has responded most generously to a variety of requests. The following list, not altogether complete, but containing the more important assignments (in rough chronological order), indicates the nature and extent of these services.

PUBLIC LECTURES

As might be expected from his renown as a geophysicist, and more specifically as a seismologist, Press has delivered many lectures on earthquakes and related subjects to public audiences both at home and abroad. Inasmuch as he has added substantially to his repertoire of subjects as a result of his recent trip to China, to confer with Chinese seismologists, his several visits to the Soviet Union, and his recent research on pattern recognition, invitations for future lectures seem sure to increase.

MEMBERSHIP IN PROFESSIONAL SOCIETIES

In addition to membership in the several honorary organizations mentioned farther on in the section on "Honors and Awards," Press has held membership in the American Association for the Advancement of Science (A.A.A.S.), American Association of University Professors (A.A.U.P.), American Geophysical Union (A.G.U.; President, 1974-1976), Geological Society of America (Councillor, 1965-1967), Royal Astronomical Society (Britain), Seismological Society of America (Secretary, 1957 and 1958; Vice President, 1959-1961; and President, 1962), and Society of Exploration Geophysicists (Honorary Member, 1972).

EDITORIAL SERVICES

A professional service often requested of scholarly leaders, more often than not as a "labor of love," is advice and assistance of an editorial nature on manuscripts submitted to the publishers of journals, periodicals, proceedings, symposia, etc. As with so many other professional services, Press has given generously of his time as an editor, co-editor, or editorial consultant. He assisted with the Arden House Microseism Symposium Volume of the National Research Council (1952); he was a Co-Editor, with L. H. Ahrens et al. of volumes 2 to 8 of Physics and Chemistry of the Earth (see ref. 99--1957i and 163--1964h), from 1956 to 1971; he was an Associate Editor of the Transactions of the American Geophysical Union in 1956 and 1957; Co-Editor of the Journal of Geophysical Research from 1962 to 1964; Consulting Editor of McGraw-Hill's International Series in the Earth and Planetary Sciences, 1965-1973, as discussed in the section on "Research and Publications"; Co-Editor of Physics of the Earth and Planetary Interiors, 1967 to present; and has been serving on the Editorial Board of Science since 1973.

SERVICES ON STATE, NATIONAL, AND INTERNATIONAL ORGANIZATIONS OF DIFFERENT KINDS FOR A VARIETY OF PURPOSES

During the past twenty years or more, 1953 to 1975, Press has served as an expert in geophysics on more than twenty different advisory boards, commissions, committees, councils, corporations, delegations, panels, and visiting committees, which together have required considerable travel and time away from his regular professorial and administrative duties at Caltech and M.I.T.

Suffice it here merely to mention the more important of these assignments, in chronological order, so as to indicate the changing pattern and growing importance of Press' services to the general public.

He was a member of the UNESCO Technical Assistance Mission of 1953, and of the group that conducted the worldwide program of the International Geophysical Year (IGY) and the Upper Mantle Project. In the latter program he was a member of the Glaciology and Seismology Panel, 1955-1959; of the Committee on Polar Research, 1957-1959; of the Interdisciplinary Research Panel, 1958-1959; and of the Seismology Working Group of the Upper Mantle Project, 1964-1970. He was a member of the Governor's Advisory Council on Atomic Activities for the State of California in 1959; the International Geophysics Committee of the International Council of Scientific Unions, 1959-1962; the U.S. Delegation, Nuclear Test Ban Conferences in Geneva in 1959, 1960 and 1961, and in Moscow in 1963 (see ref. 128--1960h); the Geophysical Research Board Panel on Solid Earth Problems, of the National Academy of Sciences, in 1961; the President's [Kennedy's] Science Advisory Committee (PSAC), 1961-1964; the U.S. Delegation to the United Nations Conference on Science and Technology for Underdeveloped Nations, in 1963; as Chairman, Earthquake Prediction Panel, Office of Science and Technology, in 1965, and of Board of Advisors, National Center for Earthquake Research of the U.S. Geological Survey, since 1966; as member of N.A.S.A.'s Lunar and Planetary Missions Board, 1966-1969, and the Planetology Subcommittee, in 1966; participant in SIPRI (Swedish International Peace Research Institute), later renamed International Institute for Peace and Conflict Research, as reported in Science 162: 1465-1466, (27 Dec. 1968); member of the National Science Board, 1970-1976; member of the Corporation of the Woods Hole Oceanographic Institution, since 1973; member IUGG Committee on Mathematical Geophysics, since 1966; of the Commission on Natural Resources, National Academy of Sciences, 1973-1975; member, U.S./U.S.S.R. Working Group in Earthquake Prediction, since 1973, and U.S./U.S.S.R. Working Group in Marine Geology, since 1975; Chairman, Committee on Scholarly Communication with the People's Republic of China from 1975; and member, Executive Council, National Academy of Sciences, since 1975.

CONSULTANT TO GOVERNMENT BUREAUS

Among the services rendered to the nation were those as consultant to the following bureaus - U.S. Navy, 1956-1957; U.S. Geological Survey, 1957-1959; U.S. Department of State, 1958-1962; U.S. Department of Defense, 1958-1962; President's [Kennedy's] Assistant for Science and Technology, 1956-1960 and 1964; Agency for International Development, 1962-1963; Arms Control and Disarmament Agency, 1962-1963; and National Aeronautics and Space Administration (NASA), 1960-1962 and 1965 to present.

HONORS AND AWARDS

Recognition of excellence in academic work and unusual promise in scientific research came early in Press' professional career. Election to Phi Beta Kappa and graduation magna cum laude from C.C.N.Y. in 1944 indicated outstanding academic performance. Election to Sigma Xi; appointment to the research staff of Columbia's Division of Government Aided Research during the period 1946-1949; election as University Fellow in 1948; and appointment as Research Associate in Geology at Columbia for SY 1948-49, at the end of which he was awarded a Ph.D. degree in Geology, all recognized Press' unusual abilities in scientific research.

Starting his teaching career in 1949, immediately after receiving his doctorate, he advanced unusually rapidly from Instructor (1949-1951), to Assistant Professor (1951-1952), to Associate Professor (1952-1955), and then left Columbia to become Professor of Geophysics at Caltech in 1955. Such rapid academic advancement is almost always indicative of unusual promise of some sort; in Press' case it no doubt came from the impressive publication record that he established during his duodecade at Columbia from 1944 to 1955 inclusive - 64 articles (not counting a dozen or more abstracts) - a most impressive record for a young scientist by the time he was only 31 years old and but 6 years beyond his doctorate! Clearly he had established himself early as an unusually productive investigator.

A year after he joined the Caltech faculty he was honored by being appointed an Alfred P. Sloan Fellow (1956-1959), and the next year he was appointed Director of Caltech's Seismological Laboratory, a post he held from 1957 to 1965, the year he came to M.I.T. Election to the National Academy of Sciences came in 1958; the Columbia University Medal of Science for Excellence was awarded in 1959; he received the California Scientist of the Year Award in 1960; Mount Press, located in West Antarctica, Latitude 78°05' South, Longitude 86°05' West, was named on 6 February 1961; the Towsend Harris Medal of the College of the City of New York, his alma mater, in 1962; and was included in LIFE magazine's list of the 100 most important young people in the United States, also in 1962.

Honors and awards continued to come after he joined the M.I.T. faculty in 1965. He was elected a fellow of the American Academy of Arts and Sciences in 1966; in December 1970 he was chosen to occupy the Robert R. Shrock Chair in Earth and Planetary Sciences, endowed by the Department's generous benefactors, Cecil and Ida Green of Dallas (see Tech Talk, p. 1, Dec. 17, 1970), the appointment to begin on 1 July 1971; and in the same year he was given the Distinguished Service Award of the U.S. Department of the Interior,* and the Gold Medal of Britain's Royal Astronomical Society (see Tech Talk, p. 4, March 18, 1971). The year 1972 brought an Honorary LL.D. Degree from his alma mater (C.C.N.Y.) and the prestigious Arthur L. Day Medal of the Geological Society of America. Next came an Honorary Sc.D. Degree from the University of Notre Dame; N.A.S.A.'s Distinguished Public Service Medal;** and election to the American Philosophical Society - all in 1973. His latest honor was M.I.T.'s coveted James R. Killian Faculty Achievement Award for 1975 (see Tech Talk 20/7: 1, Sept. 17, 1975). Such then is the distinction of the Chairman who has led Course XII into its fourth quarter century, starting in 1965.

THE FRANK PRESS FAMILY

While in high school, Press first became acquainted with one of his classmates, Billie Kallick, and they were married in 1946 when he was a second-year graduate student at Columbia. Billie was born in St. Louis, but her family moved back East and settled in New York five years after her birth. Both her parents were born in Philadelphia and came from

---

* According to a news item on p. 40B of The Boston Herald Traveler, Tuesday, June 8, 1971, Press "was named recipient of one of four new public service awards by interior Secretary Rogers Morton for contributions to conservation of human and natural resources. ... He was cited for his contributions in earthquake research and for his assistance to the U.S. geological survey in aiding it to take a leading role in such research."

** Press received one of two 1973 Distinguished Public Service Medals of N.A.S.A., being cited for major contributions
> "to the success of Apollo by providing distinguished leadership in the early identification of pertinent lunar geophysical studies, perceptive guidance to overall exploration planning, mature professional expertise in establishing objectives for and interpreting results from the lunar seismic network, and leadership of a department excelling in lunar geophysics, remote sensing, and sample investigations."
> (Tech Talk 18/17: 1, Oct. 31, 1973.)

Polish ancestors. Her father, William Kallick, was a long-time newspaperman with The New York Times, and her mother, Mary (Enders) Kallick, occupied herself with taking care of Billie and her younger sister, Evelyn.

After graduation from Brooklyn's Samuel Tilden High School, Billie entered New York University and received a B.A. degree in English in 1946. Having acquired an interest in child development by this time, she next attended Bank Street College in New York City, famous for child development studies, and received an M.S. degree in Child Development in 1953. While Frank pursued his scientific studies and research work at Columbia, Billie first taught English in a New York City high school, then became involved in curriculum development and the education of gifted children in the Pasadena school system - this during their ten-year stay in California. When Frank came to Cambridge in the fall of 1965, she continued to follow her educational interests, earning a doctorate (Ed.D.) from Boston University in 1967. She then served as educational coordinator of Boston's first full year Head Start program (1967-1968) and then as Associate Professor of Early Childhood Education at Salem State College from 1968 to 1970. Since then she has worked for the Cambridge School Department where she is in charge of Early Childhood and Parent Education Programs.

The two Press children, Bill and Paula, have had quite varied educational experiences, and like their parents are also following teaching careers.

William Henry Press ("Bill," b. 1948) graduated from Pasadena (Calif.) High School; received an A.B. degree (magna cum laude) in Physics from Harvard in 1969, with election to Phi Beta Kappa; and Master's (1970) and Doctor's (1972) degrees, in Physics, from Caltech, having one of the prestigious Hertz Fellowships during his graduate student years (1969-1972). After a year as a post-doctoral fellow at Caltech, he stayed on an additional year as an Assistant Professor, then returned East to accept a similar appointment at Princeton (1974-1976). On 1 July 1976 he became Professor of Physics and Astrophysics at Harvard. Meanwhile he had married Margaret Ann Lauristen, a doctorate in linguistics, who is now teaching at Boston State College. They have one child, Sara, born in 1974.

Paula Evelyn Press, born in 1950, attended junior high school in Pasadena, then finished her secondary education at the Buckingham School in Cambridge (Mass.). She then entered Boston University and received a B.A. degree in History of Art in 1972. Following a year of graduate work in education at Southern Connecticut State College, she took up teaching in a Kindergarten in Chapel Hill (N. Carolina), where her husband, Harvey Checkoway, whom she married in 1973, is now working on a doctoral degree in Epidemiology at the University of North Carolina.

SUMMATION

Frank Press is a distinguished scientist who stands high among the world's leaders in geophysics for the following reasons: 1) because of the important impact his scientific work, particularly in seismology, has had on the development of modern geophysics; 2) because of his personal involvement as teacher, and co-investigator and co-author, with a great many of the leading American geophysicists of the past quarter century; and 3) because of the influence of his leadership in the planning and administration of scientific research at a national level and in cooperative programs designed to share and advance technological knowledge on an international scale.

His brilliant career is a fascinating example of how an extraordinarily promising young scientist, already marked for greatness as early as high school, developed into one of the world's most distinguished geophysicists. First, as the protégé of a peerless teacher and researcher, renowned Maurice Ewing of Columbia's Lamont Geological Observatory, he learned the distinctive style and philosophy of a master of research. Next, on his own, as professor and administrator at Caltech, he quickly matured into one of America's leading seismologists, with calls for his advice and counsel from many quarters, even from President John F. Kennedy himself. Then well on his way to world renown, he came to M.I.T. in 1965 to start a decade of activity that has brought the Institute's Department of Earth and Planetary Sciences first rank among the world's leaders and has brought him both the honors and awards, and the responsibilities, that come from being a world leader in earthquake research.

Inspection of his Bibliography, which follows, shows that he has involved himself in the scientific work of more than sixty geophysicists, and that his research has been consistently directed toward gaining deeper insight into the nature, structure, and physical behavior of the earth and moon.

He was a member of Ewing's early group which first showed the great thickness of sediments on the Continental Shelf off the east coast of North America - a discovery that has brought great pressure for off-shore drilling for oil. He was a member of that extremely exciting group that established the thinness of the crust under the oceans in contrast to that of the continents; i.e. the shallow depth of the "M" (Mohorovičić) discontinuity under the oceans, which is so important in understanding modern plate structure, was established by a single experiment at sea in which he participated with Ewing and Worzel. He participated in developing the Press-Ewing seismograph, which opened up the whole area of long waves and free oscillations of the earth, and was a member of the first team to discover those free oscillations, which led to high resolution methods of determining the internal constitution of the earth. At Caltech he started a program in regional geophysics which led to the study of thickness of sediments and subsurface faulting in the Imperial Valley and in Owens Valley, two areas now being investigated for geothermal energy. He participated actively in the research connected with nuclear test detection, a magnificent story in American geophysical history, namely the successful solution to the problem of distinguishing explosions from earthquakes. He became interested in the lithosphere and the asthenosphere, i.e. the 100-kilometer thick rigid crustal layer underlain by a partially molten region, and was the first to suggest that the partly molten asthenosphere was the region over which the rigid lithospheric plates have slid. He next became interested in using free oscillations and long waves to determine the details of the earth's interior: e.g. the precise radius of the core, and the density distribution in the mantle and the lithosphere; and he first introduced the use of Monte Carlo methods in connection with free oscillations of the earth to show the range of possible solutions as a means of overcoming the lack of uniqueness question.

In recent years he has become interested in the way the lithosphere changes its properties as it ages. In the concept of plate tectonics, discrete lithospheric plates are created, then changed physically through time, and finally destroyed. Press and his students at M.I.T. have worked out methods for studying just how the plates change their properties with time, from the place of their origin along a mid-ocean ridge up to the point where they become old and start sinking back into the mantle.

He was one of the first scientists to point out the need, and to develop techniques and instruments, for seismic studies on the moon and the planets, and he helped to guide experiments that have revealed that the moon is like the earth in having a crust and mantle, but unlike it in being much less active seismically.

A big interest in recent years has been the problem of predicting earthquakes. He has been a prime mover in seeking support for research directed toward this worldwide problem (see "Quake Funds Urged" on page 1 of Tech Talk 19/45:1, May 21, 1975; and ref. 228--1975), and has led international cooperative exchange of information and research with Japan, the U.S.S.R., and the People's Republic of China, with the hope of developing more accurate methods of earthquake prediction. His latest research has involved "pattern recognition," a new technique in geophysics, which serves to sift through huge amounts of geophysical data in search of common patterns that might be significant in predicting earthquakes and other natural phenomena. A recent guest editorial in Science (ref. 231--1975c) suggests a provocative change in science instruction and research in American universities, and deserves serious consideration.

During his first decade as Head of Course XII, Department of Earth and Planetary Sciences, Press has expanded and diversified the curriculum and the faculty; saw formalized a joint graduate-degree program in oceanography with the Woods Hole Oceanographic Institution; and developed a strong faculty and program in planetary sciences. A 50-foot research vessel was obtained from the U.S. Navy for oceanographic research within a day's cruise of Boston; and two modern facilities for off-campus research were made possible by contributions from George R. Wallace, Jr. - the Wallace Astrophysical Observatory and the Wallace Geophysical Observatory. Existing facilities of all sorts were upgraded where needed; the ninth floor lounge was named the Ida Green Room to honor our generous benefactress; and new space was acquired in Building 24 for Prof. McCord and his research group. Both undergraduate and graduate groups increased to optimum size; the Lindgren Library tripled its holdings; the Department's own endowed funds continued to grow steadily, now totalling in excess of $500,000; and the Department's overall annual budget had quadrupled to some $4,000,000 by FY 1974-75. During the decade the Department was again dramatically recognized by Cecil and Ida Green in the form of three endowed professorships and a magnificently endowed Ida Green Fellowship Fund for women graduate students.

Awards and honors have come to Press in recognition of his scientific leadership and accomplishments, and more are certain to come, for he is now at the very height of his career. Inasmuch as his past leadership has brought the M.I.T. Department of Earth and Planetary Sciences to a foremost position among the world's best, his continued guidance augurs well for the future of all the earth and planetary sciences at M.I.T.

BIBLIOGRAPHY OF FRANK PRESS

The abbreviations, acronyms, and symbols used in the following references are explained on pages 91-98. In general, abstracts are listed as separate references but may also be included in the reference to the complete article. With Dr. Press' kind assistance, I hope we have included in the following list every one of his publications of any importance through 1970. In view of the fact that he is recognized as one of the world's leading geophysicists, with special distinction in seismology, I have considered it appropriate to include his up-to-date bibliography together with the preceding biographical sketch.

1--1944 The Geiger-Muller counter and its geological applications. Geol. Rev., Geol. Soc. City College, New York City, 4: 22-23, (‡), (June 1944).

2--1946 (with Ewing, M. et al.) Geophysical investigations in the emerged and submerged Atlantic Coastal Plain: Pt. V. [abst.]; Cape May [N.J.], New York, and Woods Hole [Mass.] sections [abst.]. Geol. Soc. Am., B. 57: 1192, (Dec. 1946).

3--1948 (and Ewing, M.) A theory of microseisms with geologic applications. Am. Geophys. Union, Tr. 29: 163-174, (‡), (Apr. 1948).

4--1948a (and Ewing, M.) Low-speed layer in water-covered areas. Geophysics 13: 404-420, (July 1948).

T--1949 (and Ewing, M.) Two applications of normal mode sound propagation in the ocean. Under this title, Press presented the two directly preceding articles, references 3--1948 and 4--1948a, as a Ph.D. thesis, to the Department of Geology in Columbia University in June 1949.

5--1949a (with Hersey, J. B.) Seismic studies of geologic structure of the ocean floor [abst.], in Symposium on the earth's crust. Am. Geophys. Union, Tr. 30: 171, (‡), (Apr. 1949).

6--1949b (with Ewing, M.) Notes on surface waves. N.Y. Ac. Sci., An. 51: 453-462, (May 1949).

7--1949c (with Ewing, M. et al.) Short notes: [Seismic refraction measurements in the Atlantic Ocean Basin, Pt. I.] Geol. Soc. Am., B. 60: 1303, (Aug. 1949).

8--1949d (and Ewing, M. and Tolstoy, I.) Airy phase of shallow-focus submarine earthquakes [abst.]. Geol. Soc. Am., B. 60: 1956, (Dec. 1949). (See also 11--1950a.)

9--1949e (with Tolstoy, I. and Ewing, M.) "T" phase of shallow-focus submarine earthquakes [abst.]. Geol. Soc. Am., B. 60: 1957, (Dec. 1949).

10--1950 (with Ewing, M. and Tolstoy, I.) Proposed use of the T phase in tsunami warning systems. Seism. Soc. Am., B. 40: 53-58, (Jan. 1950); discussion by Leet, D., Ibid. 41: 165-167, (Apr. 1951).

11--1950a (and Ewing, M. and Tolstoy, I.) The Airy phase of shallow-focus submarine earthquakes. Seism. Soc. Am., B. 40: 111-148, (Apr. 1950). (See also 8--1949d.)

12--1950b (and Ewing, M.) Propagation of explosive sound in a liquid layer overlying a semi-infinite elastic solid. Geophysics 15: 426-446, (July 1950).

13--1950c (with Ewing, M. et al.) Seismic refraction measurements in the Atlantic Ocean Basin (Part I). Seism. Soc. Am., B. 40: 233-242, (July 1950).

14--1950d (with Ewing, M. et al.) Woods Hole, New York, and Cape May sections, Part V of Geophysical investigations in the emerged and submerged Atlantic Coastal Plain. Geol. Soc. Am., B. 61: 877-892, (Sept. 1950).

15--1950e (with Ewing, M.) Crustal structure and surface-wave dispersion. Seism. Soc. Am., B. 40: 271-280, (Oct. 1950).

16--1950f (with Owen, H.) [Earthquake Notes in] Seism. Bull. - Lamont Geol. Observ., Sept. 1 - Dec. 31, 1950. Lamont Geol. Observ. Contr. 28: 16 p., (Dec. 1950).

17--1950g (with Katz, S. and Edwards, R. S.) Crustal structure beneath the Gulf of Maine. Columbia Univ., Lamont Geol. Observ. Tech. Rept. Seismology 9: 14 p., (Dec. 1950).

18--1951 (with Owen, H.) [Earthquake Notes in] Seism. Bull. - Lamont Geol. Observ., Jan. 1 - Apr. 30, 1951. Lamont Geol. Observ. Contr. 31: 9 p., (1951).

19--1951a (with Luskin, B.) Conversion of Bosch-Omori mechanical seismographs to electromagnetic seismographs. Earthquake Notes 22: 3-4, (Mar. 1951).

20--1951b (and Crary, A. P., Oliver, J., and Katz, S.) Air-coupled flexural waves in floating ice. Am. Geophys. Union, Tr. 32: 166-172, (Apr. 1951). (See also Geophys. Res. Paper 6A, Geophys. Res. Dir., AFCRL.)

21--1951c (and Ewing, M.) Ground roll coupling to atmospheric compressional waves. Geophysics 16: 416-430, (July 1951).

22--1951d (and Ewing, M.) Theory of air-coupled flexural waves. J. Appl. Phys. 22: 892-899, (July 1951).

23--1951e (and McGinnis, Lois) [Earthquake Notes in] Seism. Bull. - Lamont Geol. Observ., May 1 - Aug. 31, 1951. Lamont Geol. Observ. Contr. 41: 11-25, (1951).

24--1951f (and Ewing, M.) Two slow surface waves across North America. Columbia Univ., Lamont Geol. Observ. Tech. Rept. Seismology 15: 17 p., (Aug. 1951); Seism. Soc. Am., B. 42: 219-228, (July 1952).

25--1951g (with Ewing, M.) Solomon Islands earthquake of 29 July 1950, Part II of Crustal structure and surface-wave dispersion. Columbia Univ., Lamont Geol. Observ. Tech. Rept. Seismology 16: 11 p., (Aug. 1951); slightly revised, Seism. Soc. Am., B. 42: 315-325, (Oct. 1952).

26--1951h (with Benioff, H. and Ewing, M.) Sound waves in the atmosphere generated by a small earthquake. Nat. Ac. Sci., Pr. 37: 600-603, (Sept. 1951).

27--1951i (with Burg, K. et al.) A seismic wave guide phenomenon. Geophysics 16: 594-611, (Oct. 1951).

28--1951j (and Ewing, M.) Propagation of elastic waves in a flowing ice sheet. Am. Geophys. Union, Tr. 32: 673-678, (Oct. 1951).

29--1851k (and Ewing, M.) Surface waves as aids in epicenter location. Earthquake Notes 22: 33, (Dec. 1951).

30--1951ℓ (with Ewing, M.) Surface waves at Honolulu [abst.]. Geol. Soc. Am., B. 62: 1526, (Dec. 1951).

31--1951m (with Ewing, M.) Surface waves from Atlantic earthquakes [abst.]. Geol. Soc. Am., B. 62: 1526, (Dec. 1951).

32--1952 (with Ewing, M.) Propagation of earthquake waves along oceanic paths. Int. Geod. Geophys. Union, Assoc. Seismology, Sér. A., Travaux Scientifiques, Bur. Ent. Seism. Int. 18: 41-46, Toulouse, France, (1952).

33--1952a (with Ewing, M. and Worzel, J. L.) Further study of the T phase. Seism. Soc. Am., B. 42: 37-51, (Jan. 1952); discussion in French by P. Mollard, An. Géophys., t. 8: 335-336, Paris, (July-Sept. 1952); reply in English by authors, Ibid. t. 9: 248-249, (July-Sept. 1953); full paper reprinted in Columbia Univ., Lamont Geol. Observ. Contr. 51; (Jan. 1952); reply as Ibid. 83, (July-Sept. 1953).

34--1952b (with Jardetsky, W. S.) Theoretical dispersion curves for suboceanic Rayleigh waves, Part III of Crustal structure and surface-wave dispersion. Columbia Univ., Lamont Geol. Observ. Tech. Rept. Seismology 18: 10 p., (Jan. 1952); slightly revised, Seism. Soc. Am., B. 43: 137-144, (Apr. 1953). (See also 49--1953d.)

35--1952c (with Jardetsky, W. S.) Rayleigh-wave coupling to atmospheric compression waves. Seism. Soc. Am., B. 42: 135-144, (Apr. 1952).

36--1952d (and Ewing, M.) Magnetic anomalies over oceanic structures. Am. Geophys. Union, Tr. 33: 349-355, (June 1952). [This article is based on Press' M.A. thesis in June 1946.]

37--1952e [Discussion of] Geophysical measurements, by R. W. Raitt, in Isaacs and Iselin, eds., Symposium on Oceanographic Instrumentation, June 1952, p. 70-84, (1952).

38--1952f (and Ewing, M.) Two slow surface waves across North America. Seism. Soc. Am., B. 42: 219-228, (July 1952).

39--1952g (with Ewing, M.) Recent results from earthquake surface-wave investigations [abst.]. Geol. Soc. Am., B. 63: 1248, (Dec. 1952); Am. Mineralogist 38: 337, (Mar.-Apr. 1953).

40--1952h (with Ewing, M.) Shear-wave propagation and continental structure [abst.]. Geol. Soc. Am., B. 63: 1352-1353, (Dec. 1952).

41--1952i (with Jardetsky, W. S.) Theoretical dispersion curves for suboceanic Rayleigh waves, Part III of Crustal structure and surface-wave dispersion [abst.]. Geol. Soc. Am., B. 63: 1354, (Dec. 1952).

42--1952j (and Ewing, M.) Note on refracted waves in a layer [abst.]. Geol. Soc. Am., B. 63: 1356, (Dec. 1952); Geophysics 18: 737, (July 1953).

43--1952k (and Ewing, M.) Surface waves and mantle structure [abst.]. Geol. Soc. Am., B. 63: 1356, (Dec. 1952).

44--1952ℓ (with Ewing, M.) Propagation of earthquake waves along oceanic paths. Internat. Geod. Geophys. Union, Assoc. Seismology, Bur. Cent. Séism. Int., Sér. A, Travaux Scientifiques, Fasc. 18: 41-46, [Mém. presentés à l'Assemblée de Bruxelles, 1951], (1952).

45--1953 An ultra-short period seismograph. Earthquake News 24: 4-5, (1953).

46--1953a (with Oliver, J. and Ewing, M.) The Atlantic and Pacific Ocean Basins, Part IV of Crustal structure and surface-wave dispersion. Columbia Univ., Lamont Geol. Observ. Tech. Rept. Seismology 26: 39 p., (Jan. 1953); Geol. Soc. Am., B. 66: 913-946, (July 1955).

47--1953b (with Katz, S. and Edwards, R. S.) Seismic-refraction profile across the Gulf of Maine. Geol. Soc. Am., B. 64: 249-252, (Feb. 1953).

48--1953c (with Ewing, M.) Further study of atmospheric pressure fluctuations recorded on seismographs. Am. Geophys. Union, Tr. 34: 95-100, (Feb. 1953).

49--1953d (with Jardetsky, W. S.) Theoretical dispersion curves for suboceanic Rayleigh waves, Part III of Crustal structure and surface-wave dispersion. Seism. Soc. Am., B. 43: 137-144, (Apr. 1953). (See also 34--1952b.)

50--1953e (with Ewing, M.) The oceans as an acoustic system: p. 109-111, in Symposium on Microseisms, Nat. Res. Council, Washington, D.C., (1953).

51--1953f (with Ewing, M.) Mechanisms of T wave propagation. An. Géophys. 9: 248-249, (July-Sept. 1953).

52--1954 (and Beckmann, W. C.) Grand Banks and adjacent shelves, Part VIII of Geophysical investigations in the emerged and submerged Atlantic Coastal Plain. Geol. Soc. Am., B. 65: 299-313, (Mar. 1954).

53--1954a (with Oliver, J. and Ewing, M.) Two-dimensional model seismology. Geophysics 19: 202-219, (Apr. 1954).

54--1954b (with Ewing, M.) An investigation of mantle Rayleigh waves. Seism. Soc. Am., B. 44: 127-147, (Apr. 1954).

55--1954c (and Oliver, J. E.) Model studies of seismic wave propagation, in Symposium on geophysical models [abst.]. Am. Geophys. Union, Tr. 35: 363, (Apr. 1954).

56--1954d (with Ewing, M.) Mantle Rayleigh waves from the Kamchatka earthquake of November 4, 1952. Seism. Soc. Am., B. 44: 471-479, (July 1954).

57--1954e (and Oliver, J. and Ewing, M.) Seismic model study of refractions from a layer of finite thickness. Geophysics 19: 388-401, (July 1954).

58--1954f (with Donn, W., Rommer, R., and Ewing, M.) Atmospheric oscillations and related synoptic patterns. Am. Meteorol. Soc., B. 35: 301-309, (Sept. 1954).

59--1954g (with Donn, W. L. and Ewing, M.) Performance of resonant seismometers. Geophysics 19: 802-819, (Oct. 1954).

60--1954h Geophysical measurements discussion: Oceanographic Instrumentation. NAS-NRC Pub. 309: 79-80, (1954).

61--1954i (with Ewing, M.) Propagation of elastic waves in the ocean with reference to microseisms. La Semaine d'Étude sur les Problèmes des Microséismes. Pont. Academiae Scientiarum Scripta Varia 12, (1954).

62--1954j (with Ewing, M. and Donn, W. L.) An explanation of the Lake Michigan wave of 26 June 1954. Science 120: 684-686, (29 Oct. 1954).

63--1954k (and Oliver, J. E. and Ewing, M.) Model seismology studies [abst.]. Science 120: 786, (Nov. 1954).

64--1955 (and Ewing, M.) Waves with P_n and S_n velocity at great distances. Nat. Ac. Sci., Pr. 41/1: 24-27, (Jan. 1955).

65--1955a (and Oliver, J.) Model study of air-coupled surface waves. Acoust. Soc. Am., J. 27/1: 43-46, (Jan. 1955).

66--1955b (with Ewing, M.) Surface waves and guided waves, p. 118-139, in Vol. 47 of Encyclopedia of Physics. Heidelberg: Springer Verlag, (1955).

67--1955c (with Ewing, M.) Seismic prospecting, p. 153-168, in Vol. 47 of Encyclopedia of Physics. Heidelberg: Springer Verlag, (1955).

68--1955d (with Ewing, M.) Structure of the earth's crust, p. 246-257 in Vol. 47 of Encyclopedia of Physics. Heidelberg: Springer Verlag, (1955).

69--1955e   (and Ewing, M.) Tide-gage disturbances from the great eruption of Krakatoa. Am. Geophys. Union, Tr. 36: 53-60, (Feb. 1955).

70--1955f   (with Oliver, J. E. and Ewing, M.) Crustal structure and surface-wave dispersion, Part IV of Atlantic and Pacific Ocean Basins. Geol. Soc. Am., B. 66: 913-946, (July 1955).

71--1955g   (with Ewing, M.) Geophysical contrasts between continents and ocean basins, p. 1-6 in The Earth's Crust, Geol. Soc. Am. Sp.P. 62, (July 1955).

72--1955h   (and Ewing, M.) Earthquake surface waves and crustal structure, p. 51-60 in The Earth's Crust, Geol. Soc. Am., Sp.P. 62, (July 1955).

73--1955i   (with Oliver, J. and Ewing, M.) Crustal structure of the Arctic regions from the Lg phase. Geol. Soc. Am., B. 66: 1063-1074, (Sept. 1955).

74--1955j   (with Ewing, M.) Seismic measurements in ocean basins. W.H.O.I. Symposium. J. Marine Res. 14: 417-422, (Dec. 1955).

75--1955k   (with Ewing, M. and Oliver, J. E.) Dispersion of Love waves crossing the United States [abst.]. Geol. Soc. Am., B. 60: 1649, (Dec. 1955).

76--1955ℓ   (with Oliver, J. and Ewing, M.) Crustal structure of the Indian Ocean basin from Rayleigh-wave dispersion [abst.]. Geol. Soc. Am., B. 60: 1658, (Dec. 1955).

77--1955m   (and Ewing, M. and Oliver, J.) Dispersion of Rayleigh waves crossing Africa [abst.]. Geol. Soc. Am., B. 60: 1660, (Dec. 1955).

78--1956   (and Dobrin, M. B.) Seismic wave studies over a high-speed surface layer. Geophysics 21: 285-298, (Apr. 1956).

79--1956a   (with Ewing, M.) Rayleigh wave dispersion in the period range 10 to 500 seconds. Am. Geophys. Union, Tr. 37: 213-215, (Apr. 1956).

80--1956b   (with Benioff, H., Gutenberg, B., and Richter, C. F.) Progress Report, Seism. Lab. Cal. Inst. Tech., 1955. Am. Geophys. Union, Tr. 37: 232-238, (Apr. 1956).

81--1956c   (and Ewing, M. and Oliver, J.) Crustal structure and surface-wave dispersion in Africa. Seism. Soc. Am., B. 46: 97-103, (Apr. 1956).

82--1956d   (with Ewing, M.) The long-period nature of S waves [abst.]. Am. Geophys. Union, Tr. 37: 343, (June 1956).

83--1956e   (and Ewing, M.) A mechanism for G-wave propagation [abst.]. Am. Geophys. Union, Tr. 37: 355-356, (June 1956).

84--1956f   Some new trends in seismology. J. Geophys. Res. 61: 377-378, (June 1956).

85--1956g   Velocity of Lg waves in California. Am. Geophys. Union, Tr. 37: 615-618, (Oct. 1956).

86--1956h   Volcanoes, ice and destructive waves. Eng. Sci. Month. (Caltech) 19: 26-30, (Nov. 1956).

87--1956i   Rigidity of the Earth's core. Science 124: 1204, (Dec. 1956).

88--1956j   Southern California, Part I of Determination of crustal structure from phase velocity of Rayleigh waves. Geol. Soc. Am., B. 67: 1647-1658, (Dec. 1956).

89--1956k   (and Gutenberg, B.) Channel P waves IIg in the Earth's crust. Am. Geophys. Union, Tr. 37: 754-756, (Dec. 1956).

90--1957   (with Benioff, H., Gutenberg, B., and Richter, C. F.) Progress Report, Seism. Lab. Cal. Inst. Tech., 1956. Am. Geophys. Union, Tr. 38/2: 248-254, (Apr. 1957).

91--1957a San Francisco Bay region [California], Part II of Determination of crustal structure from phase velocity of Rayleigh waves. Seism. Soc. Am., B. 47/2: 87-88, (Apr. 1957).

92--1957b A seismic model study of the phase velocity method of exploration. Geophysics 22/2: 275-285, (Apr. 1957).

93--1957c (with Ewing, W. M. and Jardetsky, W. S.) Elastic Waves in Layered Media. New York: McGraw-Hill Book Co., Inc., 380 p., (1957).

94--1957d Antarctic seismology. Eng. Sci. Month. (Caltech) 20: 10-11, (June 1957).

95--1957e Internal friction in the mantle and rigidity of the earth's core [abst.]. Am. Geophys. Union, Tr. 38: 402-403, (June 1957).

96--1957f (and Healy, J.) Absorption of Rayleigh waves in low-loss media. J. Appl. Phys. 28/11: 1323-1325, (Nov. 1957).

97--1957g (with Ewing, M.) Regional measurements of crustal thickness [abst.]. Geol. Soc. Am., B. 68: 1816, (Dec. 1957).

98--1957h (and Benioff, V. H.) New results from long-period seismographs [abst.]. Geol. Soc. Am., B. 68: 1841-1842, (Dec. 1957).

99--1957i (with Ahrens, L. H. et al., Eds) Physics and Chemistry of the Earth.
Vol. 2: 259 p., (1957).   London & New York: Pergamon Press.
Vol. 3: 464 p., (1959).   London & New York: Pergamon Press.
Vol. 4: 317 p., (1961).   London & New York: Pergamon Press.
Vol. 5: 398 p., (1964).   London & New York: Pergamon Press, New York: The Macmillan Co.
Vol. 6: 510 p., (1965).   London & New York: Pergamon Press.
Vol. 7: 333 p., (1966).   London & New York: Pergamon Press.
Vol. 8: 337 p., (1971).   London & New York: Pergamon Press.

100--1958 (Co-editor with Benioff, H., Ewing, M., and Howell, B. F., Jr.) Contributions in Geophysics. London: Pergamon Press, 244 p., (1958).

101--1958a (and Ewing, M. and Lehner, F.) A long-period seismograph system. Am. Geophys. Union, Tr. 39: 106-108, (Feb. 1958).

102--1958b Velocity distribution in the crust [abst.]. Am. Geophys. Union, Tr. 39: 528, (June 1958).

103--1958c (with Gutenberg, B., Benioff, H., and Richter, C. F.) Progress Report, Seism. Lab. Cal. Inst. Tech., 1957. Am. Geophys. Union, Tr. 39: 721-725, (Aug. 1958).

104--1958d Geological aspects of the interior of the moon. Lunar and Planetary Exploration Colloquium, Pr. 1: 15-18, (1958).

105--1958e Seismic investigation of the earth's crust (in Russian). Priroda 8: 33-37, (1958).

106--1958f (with Benioff, H.) Progress report on long period seismographs. Roy. Astronom. Soc., Geophys. Jour. 1: 208-215, (1958).

107--1958g Remarks on refraction arrivals from a layer of finite thickness. J. Geophys. Res. 63: 631-634, (1958).

108--1958h Elastic wave radiation from faults in ultrasonic models. Publ. Dom. Obs. (Canada) 20: 271-277, (1958). [Reprinted from A Symposium on the Mechanics of Faulting, with Special Reference to the Fault-Plane Work; Toronto, Canada, 1957.]

109--1958i (with Healy, J. H.) Further model study of radiation pattern from faults [abst.]. Geol. Soc. Am., B. 69: 1687-1688, (Dec. 1958).

110--1958j (with Pakiser, L. C., Jr. et al.) Geophysical investigation of Mono Basin, California [abst.]. Geol. Soc. Am., B. 69: 1699-1700, (Dec. 1958).

111--1958k Continental crust [abst.]. Geol. Soc. Am., B. 69: 1700, (Dec. 1958).

112--1958l (with Benioff, H. V. et al., eds.) Contributions in Geophysics - in honor of Beno Gutenberg. Internat. Ser. Mon. Earth Sci. 1: viii + 244 p., (1958).

113--1959 (and Dewart, G.) Extent of the Antarctic continent. Science 129/3347: 462-463, (Feb. 1959).

114--1959a (with Ewing, M.) The United States, Part III of Determination of crustal structure from phase velocity of Rayleigh waves. Geol. Soc. Am., B. 70: 229-244, (Mar. 1959).

115--1959b (with Healy, J.) Further model study of the radiation of elastic waves from a dipole source. Seism. Soc. Am., B. 49: 193-198, (Apr. 1959).

116--1959c Some implications on mantle and crustal structure from G waves and Love waves. J. Geophys. Res. 64: 565-568, (May 1959).

117--1959d The need for fundamental research in seismology - Preface. Am. Geophys. Union, Tr. 40: 212-213, (Sept. 1959).

118--1959e (and Oliver, J. and Romney, C.) The need for fundamental research in seismology - Summary Report. Am. Geophys. Union, Tr. 40: 213-221, (Sept. 1959).

119--1959f (with Takeuchi, H. and Kobayashi, N.) Rayleigh-wave evidence for the low-velocity zone in the mantle. Seism. Soc. Am., B. 49: 355-364, (Oct. 1959).

120--1960 (and Takeuchi, H.) Note on the variational and homogeneous layer approximations for the computation of Rayleigh wave dispersion. Seism. Soc. Am., B. 50: 81-85, (Jan. 1960).

121--1960a (with White, J. E.) Geophysical research and progress in exploration. Geophysics 25: 168-180, (Feb. 1960).

122--1960b Crustal structure in California-Nevada region. J. Geophys. Res. 65: 1039-1051, (Mar. 1960).

123--1960c (with Pakiser, L. C. and Kane, M. F.) Geophysical investigation of the Mono Basin, California. Geol. Soc. Am., B. 71: 415-448, (Apr. 1960).

124--1960d Seismic wave propagation. Am. Geophys. Union, Tr. 41: 150-151, (June 1960).

125--1960e (with Griggs, D. T.) Probing the earth with nuclear explosions. California Univ., Livermore Radiation Lab. Rept. UCRL-6013, 40 p., (July 1960).

126--1960f (with Healy, J. H.) Two-dimensional seismic models with continuously variable velocity depth and density functions. Geophysics 25: 987-997, (Oct. 1960).

127--1960g (and Buwalda, Phyllis and Neugebauer, Marcia) A lunar seismic experiment. J. Geophys. Res. 65: 3097-3105, (Oct. 1960).

128--1960h Scientific aspects of the Nuclear Test Ban. Eng. Sci. Mag. (Caltech), p. 26, 28, 30, 32, 36, (Dec. 1960).

129--1961 (with Griggs, D. T.) Probing the earth with nuclear explosions. J. Geophys. Res. 66: 237-258, (Jan. 1961).

130--1961a [Tsunamis] The seismic source [abst.]. Pacific Sci. Cong., 10th, Honolulu, 1961. Abst. Symposium Papers, p. 351, (1961); Internat. Geod. Geophys. Mon. 24:7, (1963).

131--1961b Crustal and upper mantle structure of the eastern Pacific Ocean and adjacent continent [abst.]. Pacific Sci. Cong., 10th, Honolulu, 1961. Abst. Symposium Papers, p. 363, (1961).

132--1961c (and Benioff, V. H.) Report on the free oscillations of the earth [abst.]. Science 133: 1368, (1961). (See also 133--1961d.)

133--1961d (with Benioff, H. and Smith, S.) Excitation of the free oscillations of the earth by earthquakes. J. Geophys. Res. 66: 605-619, (Feb. 1961). (See also 132--1961c.)

134--1961e The earth's crust and upper mantle. Science 133: 1455-1463, (May 1961).

135--1961f (with Kovach, R. L.) Rayleigh wave dispersion and crustal structure in the eastern Pacific and Indian Oceans. Roy. Astronom. Soc., Geophys. Jour. 4: 202-216, (1961).

136--1961g (with Kovach, R. L.) A note on ocean sediment thickness from surface wave dispersion. J. Geophys. Res. 66: 3073-3074, (Sept. 1961).

137--1961h (with Aki, K.) Upper mantle structure under oceans and continents from Rayleigh waves. Roy. Astronom. Soc. Geophys. Jour. 5: 292-305, (1961).

138--1961i (and Ben-Menahem, A. and Toksöz, M. N.) Experimental determination of earthquake fault length and rupture velocity. J. Geophys. Res. 66: 3471-3485, (Oct. 1961).

139--1961j (and Harkrider, D. and Seafeldt, C. A.) A fast, convenient program for computation of surface-wave dispersion curves in multilayered media. Seism. Soc. Am., B. 51: 495-502, (Oct. 1961).

140--1961k (with Kovach, R. L.) Surface wave dispersion and crustal structure in Antarctica and the surrounding oceans. Annali di Geofisica 14: 211-224, (1961).

141--1961l (and Archambeau, C.) Release of tectonic strain by underground nuclear explosions. Abst., Science 134: 1433, (1961); J. Geophys. Res. 67: 337-343, (Jan. 1962).

142--1962 (and Ben-Menahem, A. and Toksöz, N.) Deriving fault length and rupture velocity from seismograms - Experiment [abst.]. Geol. Soc. Am., Sp.P. 68: 49, (1962).

143--1962a (and Harkrider, D. and Seafeldt, C. A.) Computation of dispersion of surface waves on digital computers [abst.]. Geol. Soc. Am., Sp.P. 68: 49-50, (1962).

144--1962b (with Knopoff, L.) A proposal for the study of the crustal and upper mantle structure of the western Mediterranean by surface wave dispersion. Ext. Rapports et Proces-verbaux des Reunions de la C.I.E.S.M.M. 16/3: 711-715, (1962).

145--1962c (with Kovach, R. L. and Allen, C. R.) Geophysical investigations in the Colorado delta region. J. Geophys. Res. 67: 2845-2871, (July 1962).

146--1962d (with Kovach, R. L.) Lunar seismology. Tech. Rept. No. 32-328, Contract No. NAS 7-100, Aug. 10, 1962. [Jet Propulsion Laboratory, Caltech.]

147--1962e Introduction - The technical basis for a research program in detection, 16 p., figs., (‡), (1962). [In Press' Reprint Collection - Vol. 3, No. 32.]

148--1962f (and Harkrider, D.) Propagation of acoustic-gravity waves in the atmosphere. J. Geophys. Res. 67: 3889-3908, (Sept. 1962).

149--1963 (with Harkrider, D. G. and Hales, A.) On detecting soft layers in the mantle with Rayleigh waves. Seism. Soc. Am., B. 53: 539-548, (Apr. 1963).

150--1963a (and Dewart, G. and Gilman, R.) A study of diagnostic techniques for identifying earthquakes. J. Geophys. Res. 68: 2909-2928, (May 1963).

151--1963b The dimensions of the seismic source [abst.]. Am. Geophys. Union, Tr. 44: 98, (1963).

152--1963c (with Kovach, R. L. and Lehner, F.) Seismic exploration of the moon. Am. Inst. Aeronaut. Astronaut. Summer Meeting, Los Angeles, Calif., June 17-20, 1963, 8 p., ( ), 1963.

153--1963d (with Biehler, S.) P wave anomalies in California as an indication of crustal structure [abst.]. Geol. Soc. Am., Sp.P. 73: 25, (1963).

154--1963e Diagnostic aids for distinguishing explosions and earthquakes [abst.]. Geol. Soc. Am., Sp.P. 73: 58-59, (1963).

155--1964 Long-period waves and free oscillations of the Earth, in Research in Geophysics - Vol. 2, Solid earth and interface phenomena. Cambridge, Mass.: M.I.T. Press, p. 1-26, (1964).

156--1964a (and Biehler, S.) Velocity reversals in California batholiths [abst.]. Am. Geophys. Union, Tr. 45: 94, (1964).

157--1964b (with Healy, J. H.) Geophysical studies of basin structures along the eastern front of the Sierra Nevada, California. Geophysics 29: 337-359, (June 1964).

158--1964c (and Biehler, S.) Inferences on crustal velocities and densities from P wave delays and gravity anomalies. J. Geophys. Res. 69: 2979-2995, (July 1964).

159--1964d Seismic wave attenuation in the crust [letter to the editor]. J. Geophys. Res. 69: 4417-4419, (Oct. 1964).

160--1964e Seismology, Model in Encyclopaedic Dictionary of Physics. New York: Pergamon Press, 1964.

161--1964f The world of the scientist. New Scientist 23: 388-389, (Aug. 1964).

162--1964g (and Smith, S.) Artificially induced thermoelastic strains in the earth [abst.]. Geol. Soc. Am., Sp.P. 76: 219, (1964).

163--1964h (with Ahrens, L. H. and Runcorn, S. K., eds.) Physics and Chemistry of the Earth, Vol. 5. New York: Macmillan Co., 398 p., (1964). (See also 99--1957i.)

164--1965 (and Jackson, D.) Alaskan earthquake, 27 March 1964: Vertical extent of faulting and elastic strain energy release. Science 147: 867-868, (Feb. 1965).

165--1965a Displacements, strains, and tilts at teleseismic distances. J. Geophys. Res. 70: 2395-2412, (May 1965).

166--1965b [Chapter] 7 - Geophysics, p. 87-95 in Listen to Leaders in Science. (Albert Love and James Saxon Childers, Editors). New York: David McKay Co., (1965); Atlanta, Ga.: Tupper and Love, (1965).

167--1965c Twenty-seventh Award of the William Bowie Medal [to Hugo Benioff], April 20, 1965 - Citation. Am. Geophys. Union, Tr. 46: 361-362, (June 1965).

168--1965d   Resonant vibrations of the earth. Sci. American 213: 28-37, (1965).

169--1965e   Research leading to earthquake prediction in Earthquake and Geologic Hazards Conference, 1964: San Francisco, Calif. Resources Agency, p. 44-46, (1965).

170--1965f   (and Biehler, S.) Thermal argument for a velocity reversal in the Sierra Nevada crust [abst.]. Geol. Soc. Am., Sp.P. 82: 270-271, (1965).

171--1966   Seismological information and advances, p. 247-286 in Advances in Earth Science - Internat. Conf. M.I.T., 1964. (P. M. Hurley, ed.). Cambridge: M.I.T. Press, (1966).

172--1966a   (and Brace, W. F.) Earthquake prediction. Science 152: 1575-1584, (June 1966).

173--1966b   A plastic decoupling zone in the upper mantle, p. 657-664, in Perspectives in Modern Physics (Essays in honor of Hans A. Bethe on his 60th birthday); edited by R. E. Marshak and J. W. Blaker. New York: John Wiley & Sons, Inc., (1966).

174--1966c   Seismic velocities (Sect. 9, p. 195-218) in Handbook of Physical Constants (edited by S. P. Clark, Jr.). Geol. Soc. Am., Mem. 97, (1966).

175--1966d   (with Lehner, F. L.) A mobile seismograph array. Seism. Soc. Am., B. 56: 889-897, (Aug. 1966).

176--1966e   (and Harkrider, D.) Air-sea waves from the explosion of Krakatoa. Science 154: 1325-1327, (Dec. 1966).

177--1966f   Free oscillations, aftershocks, and Q, p. 498-501 in The Earth Beneath the Continents - A volume of geophysical studies in honor of Merle A. Tuve (edited by J. S. Steinhart and T. J. Smith). A.G.U. Mon. 10 (NAS-NRC Pub. 1467), (1966).

178--1966g   The Scientist Today: Who is he? What is he? [An address given on the occasion of the dedication of the Loutit Hall of Science at Grand Valley State College [Michigan], May 13, 1966. Pamphlet of 12 pages, (1966).

179--1966h   Strain energy, strain release, and vertical extent of faulting for the Alaskan earthquake of March 28, 1964 [abst.]. Geol. Soc. Am., Sp.P. 87: 325, (1966).

180--1967   Geophysics in man's expanding domain. [Keynote Address.] Geophysics 32: 8-11, (1967).

181--1967a   A strategy for lunar scientific exploration in the post-Apollo period. Astronautics and Aeronautics, May 1967, p. 26-31, (May 1967).

182--1967b   (with Harkrider, D.) The Krakatoa air-sea waves: an example of pulse propagation in coupled systems. Roy. Astronom. Soc. Geophys. Jour. 13: 149-159, (1967).

183--1967c   Spectra of free oscillations from an aftershock sequence, in Internat. Symp. Geophys. Theory and Computers, 3rd, Cambridge, England, 1966, Proc.: Roy. Astronom. Soc., Geophys. Jour. 13: 219-222, (1967).

184--1967d   Dimensions of the source region for small shallow earthquakes. Proc. of the VESIAC Conf. on the Current Status and Future Prognosis for Understanding the Source Mechanism of Shallow Seismic Events in the 3-5 Magnitude Range, VESIAC [VELA Seismic Information Analysis Center, U. Michigan, Lansing], 1967.

185--1968   A strategy for an earthquake prediction research program, in Earthquake prediction - Proceedings of a symposium, Zurich, Switzerland, 1967. Tectonophysics 6: 11-15, (1968).

186--1968a The best earth model after 1,000,000 random trials. Supplements al Nuovo Cimento 6: 152-153, (1968).

187--1968b (with Laster, S.) A new estimate of lunar seismicity due to meteorite impact. Physics Earth & Planet. Interiors 1: 151-154, (1968).

188--1968c Density distribution in the Earth. Science 160: 1218-1221, (June 1968).

189--1968d Earth models obtained by Monte Carlo inversion. J. Geophys. Res. 73: 5223-5234, (Aug. 1968).

190--1968e (with Brune, J. N. and Allen, R. C.) Microearthquakes survey of the southern San Andreas fault [abst.]. Geol. Soc. Am., S.P. 101: 292, (Dec. 1968).

191--1969 Models of the earth's interior [abst.]. Am. Geophys. Union, Tr. EθS 50/4: 244, (July 1969).

192--1969a The sub-oceanic mantle. Science 165: 174-176, (July 1969).

193--1969b (with Latham, G., Ewing, M., and Sutton, G.) The Apollo passive seismic experiment. Science 165: 241-250, (July 1969).

194--1969c Search for seismic signals at pulsar frequencies. J. Geophys. Res. 74: 5351-5352, (Oct. 1969).

195--1969d (with Latham, G. V., Ewing, M., et al.) The Apollo lunar seismic experiment [abst.]. Am. Geophys. Union, Tr. EθS 50: 679, (Oct. 1969).

196--1969e A mechanism for creation of the suboceanic lithosphere [abst.]. Geol. Soc. Am., Abst. with Program 1969, pt. 7: 182, (Dec. 1969).

197--1969f Prospects for earthquake prediction and control [abst.]. Geol. Soc. Am., Abst. with Program 1969, pt. 7: 182-183, (Dec. 1969). Also Geophysics 35: 1145-1146, (May 1970); and in Searching the Seventies from Moho to Mars, Soc. Explor. Geophys., Program, p. 53-54, New Orleans, (Dec. 1970).

198--1969g Zero frequency seismology, p. 171-173, in The Earth's Crust and Upper Mantle (P. J. Hart, ed.). Am. Geophys. Union, Geophys. Mon. 13 (NAS-NRC Pub. 1708), (1969).

199--1970 (with Latham, G. V. and 6 other authors) Passivnyy seysmicheskiy eksperiment [The passive seismic experiment]. Akad. Nauk SSSR Izv. Ser. Geol. 8: 51-55, (1970).

200--1970a (with Latham, G. V. et al.) Passive seismic experiment. Science 167: 455-457, (Jan. 1970).

201--1970b (with Latham, G. V. et al.) Results from the Apollo passive seismic experiment [abst.]. Am. Geophys. Union, Tr. EθS 51: 363, (May 1970).

202--1970c (with Latham, G. V. et al.) Passive seismic experiment, p. 39-55 in Apollo 12 Preliminary Science Report, NASA, SP-235, (1970); also Apollo Lunar Science Conference Pr. 3: 2309-2320. New York - Oxford: Pergamon Press, (Aug. 1970).

203--1970d (with Benioff, H.) Earthquake, p. 853-861, in Encyclopaedia Britannica, (1970).

204--1970e (with Kanamori, H.) How thick is the lithosphere? [abst.]. Nature 226: 330-331, (Aug. 1970).

205--1970f Earth models consistent with geophysical data, p. 3-22, in Phase Transformations and the Earth's Interior: Symposium, Canberra, Australia, 1969, Proc. Physics of Earth and Planetary Interiors 3 (K. E. Bullen, F. Press, S. K. Runcorn, and D. W. Collinson, eds.), 518 p., (1970).

206--1970g (with Latham, G. V. et al.) Apollo 11 passive seismic experiment, in Apollo 11 Lunar Scientific Conference, Houston, Texas, 1970, Proc. 3, Physical Properties. New York - Oxford: Pergamon Press (Geochim. Cosmochim. Acta Suppl. 1), p. 2309-2320, (1970).

207--1970h (with Latham, G. V. et al.) Seismic data from man-made impacts on the Moon. Science 170: 620-626, (Nov. 1970).

208--1970i Regionalized earth models. J. Geophys. Res. 75: 6575-6581, (Nov. 1970).

NOTE:

I believe that Press' bibliography through 1970 as given above is essentially complete; for the years since then, however, the following list should not be regarded as complete, because I have not attempted to make it so.

209--1971 The Chandler Wobble [Book review of Earthquake Displacement Fields and the Rotation of the Earth]. A NATO Advanced Study Institute, London, Ontario, June 1969, (L. Mansinha, D. E. Smylie, and A. E. Beck, eds.). New York: Reidel, Dordrecht, and Springer - Verlag, xii + 310 p., (1970). Science 172: 693-694, (May 1971).

210--1971a An introduction to earth structure and seismotectonics, p. 209-241, in Mantle and Core in Planetary Physics. Proc. Enrico Fermi International School of Physics. New York: Academic Press, Inc., (1971).

211--1971b The Earth and the Moon. Roy. Astronom. Soc., Q.J. 12: 232-243, (1971).

212--1971c The earth and the moon (summary). The Observatory 91: 135-138, (1971).

213--1971d (with Ewing, M. et al.) Seismology of the Moon and implications on internal structure, origin, and evolution, p. 155-172, in Reports on Astronomy (De Jager, ed.). Int. Astronom. Union, Tr. 14A: 566 p. Boston: D. Reidel Pub. Co., (1971).

214--1971e (with Latham, G. et al.) Moonquakes. Science 174: 687-692, (Nov. 1971).

215--1971f (with Forsyth, D. W.) Geophysical tests of petrological models of the spreading lithosphere. J. Geophys. Res. 76: 7963-7979, (Nov. 1971).

216--1972 The earth's interior as inferred from a family of models, Art. 7, p. 147-171, in The Nature of the Solid Earth (E. C. Robertson, ed.). New York: McGraw-Hill Book Co., Inc., (1972).

217--1972a (with Latham, G. et al.) Moonquakes and lunar tectonism. The Moon 4: 373-382, (1972).

218--1972b (with Toksöz, M. N. et al.) Velocity structure and properties of the lunar crust. The Moon 4: 490-504, (1972).

219--1973 (with Hart, R. S.) S_n velocities and the composition of the lithosphere in the regionalized Atlantic. J. Geophys. Res. 78: 407-411, (Jan. 1973).

220--1973a (with Nakamura, Y. et al.) New seismic data on the state of the deep lunar interior. Science 181: 49-51, (July 1973).

221--1973b (with Latham, G. et al.) Lunar structure and dynamics - Results from the Apollo Passive Seismic Experiment. The Moon 7: 396-420, (1973).

222--1973c The gravitational instability of the lithosphere, p. 7-16, in Gravity and Tectonics (K. A. DeJong and R. Scholten, eds.). New York: John Wiley & Sons, Inc., xxx + 502 p., (1973).

223--1973d Victor Hugo Benioff (September 14, 1899 - February 29, 1968). Nat. Ac. Sci., Biog. Mem. 43: 27-40, port., (1973).

224--1974   (and Siever, R.) Earth. San Francisco: W. H. Freeman and Co., xiv + 945 p., (1974).

225--1974a Natural hazards reduction, p. 71-73 in Earth Science in the Public Service, U.S. Geol. Surv. P.P. 921, (1974).

226--1974b [Eulogy of William Maurice Ewing (1906-1974)], in Dedication. Focus on the Gulf, The University of Texas Marine Science Institute, (May 1974).

227--1974c William Maurice Ewing (1906-1974), p. 165-169, in Year Book of the American Philosophical Society, (1974).

228--1975   Earthquake prediction. Sci. American 232: 5, 14-24, (1975).

229--1975a On predicting earthquakes. The New York Times, February 11, 1975.

230--1975b (and Briggs, P.) Chandler Wobble, earthquakes, rotation, and geomagnetic changes. Nature 256: 270-273, (1975).

231--1975c New Arrangements for Science in the Universities. [Editorial Page] Science 189: 177, (18 July 1975).

232--1975d (as Chairman, et al.) Earthquake Research in China. EθS 56/11: 838-881, (Nov. 1975).

233--1976   (with Gelfand, I. et al.) Pattern Recognition applied to earthquake epicenters in California, p. 227-283 in Physics of the Earth and Planetary Interiors 11, (1976). Amsterdam: Elsevier Sci. Pub. Co.

SPECIAL NOTE:

Some six months after I had completed the preceding biography of Press, President Carter offered him the appointment of science adviser during a White House conference on 9 February 1977. Although there was no official statement about the offer at the time, Press was soon mentioned in the media as a "long-shot" candidate for the post (Science 195: 763, 765-766, 25 February 1977). Official announcement of his appointment as science adviser and director of the newly created Office of Science and Technology Policy (OSTP) was made on 18 March 1977, and soon thereafter was discussed at some length in The New York Times for 19 March 1977, in Tech Talk (30/21: 1, 7) for 23 March 1977, and in Time ("The President's Scientist," p. 73) for 11 April 1977.

Inasmuch as Press will be on leave from M.I.T. for the time being, I have not indicated a terminal date for his professorial appointment in the preceding biography, or in the faculty lists in Chapter II (p. 22, 24, and 37) and in Chapter IV (p. 88 and 89).

RAYMOND H. FOGLER LIBRARY
DATE DUE

BOOKS ARE SUBJECT TO
RECALL AFTER TWO WEEKS